Quantum Information Processing, Quantum Computing, and Quantum Error Correction

Quantum Information Processing, Quantum Computing, and Quantum Error Correction

An Engineering Approach

Second Edition

Ivan B. Djordjevic

ELSEVIER

ACADEMIC PRESS
An imprint of Elsevier

Academic Press is an imprint of Elsevier
125 London Wall, London EC2Y 5AS, United Kingdom
525 B Street, Suite 1650, San Diego, CA 92101, United States
50 Hampshire Street, 5th Floor, Cambridge, MA 02139, United States
The Boulevard, Langford Lane, Kidlington, Oxford OX5 1GB, United Kingdom

Library of Congress Cataloging-in-Publication Data
A catalog record for this book is available from the Library of Congress

British Library Cataloguing-in-Publication Data
A catalogue record for this book is available from the British Library

ISBN: 978-0-12-821982-9

For information on all Academic Press publications visit our website at https://www.elsevier.com/books-and-journals

Publisher: Mara Conner
Acquisitions Editor: Tim Pitts
Editorial Project Manager: Rachel Pomery
Production Project Manager: Prem Kumar Kaliamoorthi
Cover Designer: Greg Harris

Typeset by TNQ Technologies

To Milena

Contents

Preface

Quantum information is related to the use of quantum mechanics concepts to perform information processing and transmission of information. Quantum information processing (QIP) is an exciting research area with numerous applications, including quantum key distribution (QKD), quantum teleportation, quantum computing, quantum lithography, and quantum memories. This area currently experiences rapid development, which can be judged from the number of published books on QIP and the number of conferences devoted solely to QIP concepts. Given the novelty of underlying QIP concepts, it is expected that this topic will be a subject of interest to a broad range of scientists, not just of those involved in QIP research. Moreover, based on Moore's law, which claims that the number of chips that can be etched on a single chip doubles every 18 months, leading to the doubling of both memory and computational speed, the ever-increasing demands in miniaturization of electronics will eventually lead us to the point where quantum effects become important. Given this fact, it seems that a much broader range of scientists will be forced to turn to the study of QIP much sooner than expected. Note that, on the other hand, as multicore architecture is becoming a prevailing high-performance chip design approach, improvement in computational speed can be achieved even without reducing the feature size through parallelization. Therefore thanks to multicore processor architectures, the need for QIP might be prolonged for a certain amount of time. Another point might be that, despite intensive development of quantum algorithms, the number of available quantum algorithms is still small compared to that of classical algorithms. Furthermore, current quantum gates can operate only on several tens of quantum bits (also known as qubits), which is too low for any meaningful quantum computation operation. Until recently, it was widely believed that quantum computation would never become a reality. However, recent advances in various quantum circuits implementations, as well as the proof of accuracy threshold theorem, have given rise to the optimism that quantum computers might soon become a reality.

Because of the interdisciplinary nature of the fields of QIP, quantum computing, and quantum error correction, this book aims to provide the right balance between quantum mechanics, quantum error correction, quantum computing, and quantum communication. This book has the following objectives:

1. It describes the trends in QIP, quantum error correction, and quantum computing.
2. It represents a self-contained introduction to QIP, quantum computing, and quantum error correction.
3. It targets a very wide range of readers: electrical engineers, optical engineers, applied mathematicians, computer scientists, and physicists.
4. It does not require prior knowledge of quantum mechanics. The basic concepts of quantum mechanics are provided in Chapter 2.

5. It does not require any background knowledge, except for understanding the basic concepts of vector algebra at an undergraduate level. An appendix is provided with basic descriptions of abstract algebra at a level sufficient to follow the book easily.
6. It offers in-depth exposition on the design and realization of QIP and quantum error correction circuits.
7. For readers not familiar with the concepts of information theory and channel coding, Chapter 6 provides basic concepts and definitions from coding theory. Only the concepts from coding theory of importance in quantum error correction are covered in this chapter.
8. Readers not interested in quantum error correction can approach the chapters related to QIP and quantum computing.
9. Readers interested only in quantum error correction do not need to read the chapters on quantum computing, except Chapters 3 and 4. An effort has been made to create self-independent chapters, while ensuring a proper flow between the chapters.
10. Various concepts are gradually introduced starting from intuitive and basic concepts, through medium difficulty topics, to highly mathematically involved topics.
11. At the end of each chapter, a set of problems is provided so that readers can have a deeper understanding of underlying concepts and be able to perform independent research in quantum computing and/or quantum error correction.
12. Several different courses can be offered by using this book: (1) quantum error correction, (2) quantum information processing, (3) quantum computing, and (4) integrating the concepts of quantum information processing, quantum computing, and quantum error correction.

This book is a self-contained introduction to quantum information, quantum computation, and quantum error correction. After completion of the book readers will be ready for further study in this area and be prepared to perform independent research. They will also be able to design QIP circuits, stabilizer codes, Calderbank–Shor–Steane (CSS) codes, subsystem codes, topological codes, surface codes, and entanglement-assisted quantum error correction codes, as well as propose corresponding physical implementation, and be proficient in fault-tolerant design.

The book starts with basic principles of quantum mechanics, including state vectors, operators, density operators, measurements, and dynamics of a quantum system. It continues with fundamental principles of quantum computation, quantum gates, quantum algorithms, quantum teleportation, and quantum information theory fundamentals. Significant space is given to quantum error correction codes, in particular stabilizer codes, CSS codes, quantum low-density parity-check codes, subsystem codes (also known as operator-quantum error correction codes), topological codes, surface codes, and entanglement-assisted quantum error correction codes. Other topics covered are fault-tolerant quantum error correction coding, fault-tolerant quantum computing, and the cluster states concept, which is applied to one-way quantum computation. Relevant space is devoted in investigating physical realizations of quantum computers, encoders, and decoders, including nuclear magnetic resonance, ion traps, photonic quantum realization, cavity quantum electrodynamics, and quantum dots. Focus is also given to quantum machine learning algorithms. The final chapter in the book is devoted to QKD.

The following are some unique opportunities available to readers of the book:

1. They will be prepared for further study in these areas and will be qualified to perform independent research.
2. They will be able design the information processing circuits, stabilizer codes, CSS codes, topological codes, surface codes, subsystem codes, quantum LDPC codes, and entanglement-assisted quantum error correction codes, and will be proficient in fault-tolerant design.
3. They will be able to propose the physical implementation of QIP, quantum computing, and quantum error correction circuits.
4. They will be proficient in quantum machine learning.
5. They will be proficient in QKD too.
6. They will have access to extra material to support the book on an accompanying website.

The author would like to acknowledge the support of the National Science Foundation.

Finally, special thanks are extended to Rachel Pomery and Tim Pitts of Elsevier for their tremendous effort in organizing the logistics of the book, including editing and promotion, which has been indispensible in making this book a reality.

About the Author

Ivan B. Djordjevic is a professor of electrical and computer engineering and optical sciences at the University of Arizona; director of the Optical Communications Systems Laboratory and Quantum Communications Lab; and codirector of the Signal Processing and Coding Lab. He is both an IEEE Fellow and OSA Fellow. He received his PhD degree from the University of Nis, Yugoslavia, in 1999.

Professor Djordjevic has authored or coauthored eight books, more than 540 journal and conference publications, and 54 US patents. He presently serves as an Area Editor/Associate Editor/Member of Editorial Board for the following journals: *OSA/IEEE Journal of Optical Communications and Networking, IEEE Communications Letters, Optical and Quantum Electronics, and Frequenz.* He served as an editorial board member/associate editor for *IOP Journal of Optics* and *Elsevier Physical Communication Journal* from 2016 to 2021.

Prior to joining the University of Arizona, Dr. Djordjevic held appointments at the University of Bristol and University of the West of England in the United Kingdom, Tyco Telecommunications in the United States, National Technical University of Athens in Greece, and State Telecommunication Company in Yugoslavia.

CHAPTER

Introduction

1

CHAPTER OUTLINE

This chapter is devoted to the introduction to quantum information processing (QIP), quantum computing, and quantum error correction coding (QECC) [1−12]. *Quantum information* is related to the use of quantum mechanics concepts to perform information processing and transmission of information. QIP is an exciting research area with numerous applications, including quantum key distribution (QKD), quantum teleportation, quantum computing, quantum lithography, and quantum memories. This area currently experiences intensive development, and given the novelty of underlying concepts, it might become of interest to a broad range of scientists, not just those involved in research. Moreover, it is based on Moore's law, which claims that the number of chips that can be etched on a single chip doubles every 18 months, leading to the doubling of both memory and computational speed. Therefore the ever-increasing demands in miniaturization of electronics will eventually lead to the point where quantum effects become important. Given this fact, it seems that a much broader range of scientists will be forced to study QIP much sooner than it appears right now. Note that because multicore architecture is becoming a prevailing high-performance chip design approach, improving the computational speed can be achieved even without reducing the feature size through parallelization. Therefore thanks to multicore processor architectures, the need for QIP might be prolonged for a certain amount of time. Another point might be that, despite intensive development of quantum algorithms, the number of available quantum algorithms is still small compared to that of classical algorithms. Until

Quantum Information Processing, Quantum Computing, and Quantum Error Correction
https://doi.org/10.1016/B978-0-12-821982-9.00016-2

recently, it was widely believed that quantum computation would never become a reality. However, recent advances in various quantum gate implementations, as well as proof of the accuracy threshold theorem, have given rise to the optimism that quantum computers will soon become a reality.

Fundamental features of QIP are different from that of classical computing and can be summarized as: (1) linear superposition, (2) entanglement, and (3) quantum parallelism. The following are basic details of these features:

1. *Linear superposition.* Contrary to the classical bit, a quantum bit or *qubit* can take not only two discrete values 0 and 1 but also *all* possible *linear combinations* of them. This is a consequence of a fundamental property of quantum states: it is possible to construct a *linear superposition* of quantum state $|0\rangle$ and quantum state $|1\rangle$.
2. *Entanglement.* At a quantum level it appears that two quantum objects can form a single entity, even when they are well separated from each other. Any attempt to consider this entity as a combination of two independent quantum objects, given by a tensor product of quantum states, fails unless the possibility of signal propagation at superluminal speed is allowed. These quantum objects, which cannot be decomposed into a tensor product of individual independent quantum objects, are called *entangled* quantum objects. Given the fact that arbitrary quantum states cannot be copied, which is the consequence of no-cloning theorem, communication at superluminal speed is not possible, and as a consequence the entangled quantum states cannot be written as the tensor product of independent quantum states. Moreover, it can be shown that the amount of information contained in an entangled state of N qubits grows exponentially instead of linearly, which is the case for classical bits.
3. *Quantum parallelism. Quantum parallelism* is the possibility of performing a large number of operations in parallel, which represents the key difference from classical computing. Namely, in classical computing it is possible to know what is the internal status of a computer. On the other hand, because of no-cloning theorem, it is not possible to know the current state of a quantum computer. This property has led to the development of Shor's factorization algorithm, which can be used to crack the Rivest–Shamir–Adleman (RSA) encryption protocol. Some other important quantum algorithms include Grover's search algorithm, which is used to perform a search for an entry in an unstructured database; quantum Fourier transform, which is a basis for a number of different algorithms; and Simon's algorithm. These algorithms are the subject of Chapter 5. The quantum computer is able to encode all input strings of length N simultaneously into a single computation step. In other words, the quantum computer is able simultaneously to pursue 2^N classical paths, indicating that the quantum computer is significantly more powerful than the classical one.

Although QIP has opened up some fascinating perspectives, as indicated earlier, there are certain limitations that should be overcome before QIP becomes a commercial reality. The first is related to existing quantum algorithms, whose number is significantly lower than that of classical algorithms. The second problem is related to physical implementation issues. There are many potential technologies such as nuclear magnetic resonance (NMR), ion traps, cavity quantum electrodynamics, photonics, quantum dots, and superconducting technologies, just to mention a few. Nevertheless, it is not clear which technology will prevail. Regarding quantum teleportation, most probably the photonic implementation will prevail. On the other hand, for quantum computing applications there are many potential technologies that compete with each other. Moreover, presently, the number of qubits that can be efficiently manipulated is in the order of tens, which is well below that needed for meaningful quantum computation, which is in the order of thousands. Another problem, which can be considered as problem number one, is related to *decoherence*. Decoherence is related to the interaction of qubits with the environment that blurs the fragile superposition states. Decoherence also introduces errors, indicating that the quantum register should be sufficiently isolated from the environment so that only a few random errors occur occasionally, which can be corrected by QECC techniques. One of the most powerful applications of quantum error correction is the protection of quantum information as it dynamically undergoes quantum computation. Imperfect quantum gates affect the quantum computation by introducing errors in computed data. Moreover, the imperfect control gates introduce the errors in a processed sequence since wrong operations can be applied. The QECC scheme now needs to deal not only with errors introduced by the quantum channel by also with errors introduced by imperfect quantum gates during the encoding/decoding process. Because of this, the reliability of data processed by a quantum computer is not a priori guaranteed by QECC. The reason is threefold: (1) the gates used for encoders and decoders are composed of imperfect gates, including controlled imperfect gates; (2) syndrome extraction applies unitary operators to entangle ancilla qubits with code blocks; and (3) error recovery action requires the use of a controlled operation to correct for the errors. Nevertheless, it can be shown that arbitrary good quantum error protection can be achieved even with imperfect gates, providing that the error probability per gate is below a certain *threshold*; this claim is known as accuracy threshold theorem and will be discussed later in Chapter 11, together with various fault-tolerant concepts.

This introductory chapter is organized as follows. In Section 1.1, photon polarization is described as it represents a simple and most natural connection to QIP. In the same section, we introduce some basic concepts of quantum mechanics, such as the concept of state. We also introduce Dirac notation, which will be used throughout the book. In Section 1.2, we formally introduce the concept of qubit and provide its geometric interpretation. In Section 1.3, another interesting example of the representation of qubits, the spin-1/2 system, is provided. Section 1.4 is devoted to basic quantum gates and QIP fundamentals. The basic concepts of quantum teleportation are introduced in Section 1.5. Section 1.6 is devoted to basic QECC concepts. Section 1.7 is devoted to QKD, also known as quantum cryptography. The organization of the book is described in Section 1.8.

1.1 PHOTON POLARIZATION

The electric/magnetic field of plane linearly polarized waves is described as follows [13]:

$$A(r,t) = pA_0 \exp[j(\omega t - k \cdot r)], \quad A \in \{E, H\} \tag{1.1}$$

where $E(H)$ denotes the electric (magnetic) field, p denotes the polarization orientation, $r = xe_x + ye_y + ze_z$ is the position vector, and $k = k_x e_x + k_y e_y + k_z e_z$ denotes the wave propagation vector whose magnitude is $k = 2\pi/\lambda$ (λ is the operating wavelength). For the x-polarization waves ($p = e_x$, $k = ke_z$), Eq. (1.1) becomes:

$$E_x(z,t) = e_x E_{0x} \cos(\omega t - kz), \tag{1.2}$$

while for y-polarization ($p = e_y$, $k = ke_z$) it becomes:

$$E_y(z,t) = e_y E_{0y} \cos(\omega t - kz + \delta), \tag{1.3}$$

where δ is the relative phase difference between the two orthogonal waves. The resultant wave can be obtained by combining Eqs. (1.2) and (1.3) as follows:

$$E(z,t) = E_x(z,t) + E_y(z,t) = e_x E_{0x} \cos(\omega t - kz) + e_y E_{0y} \cos(\omega t - kz + \delta). \tag{1.4}$$

The *linearly polarized* wave is obtained by setting the phase difference to an integer multiple of 2π, namely $\delta = m \cdot 2\pi$:

$$E(z,t) = (e_x E \cos\theta + e_y E \sin\theta)\cos(\omega t - kz); \quad E = \sqrt{E_{0x}^2 + E_{0y}^2}, \quad \theta = \tan^{-1}\frac{E_{0y}}{E_{0x}}. \tag{1.5}$$

By ignoring the time-dependent term, we can represent the linear polarization as shown in Fig. 1.1A. On the other hand, if $\delta \neq m \cdot 2\pi$, the *elliptical polarization* is obtained. From Eqs. (1.2) and (1.3), by eliminating the time-dependent term we obtain the following equation of ellipse:

$$\left(\frac{E_x}{E_{0x}}\right)^2 + \left(\frac{E_y}{E_{0y}}\right)^2 - 2\frac{E_x}{E_{0x}}\frac{E_y}{E_{0y}}\cos\delta = \sin^2\delta, \tan 2\phi = \frac{2E_{0x}E_{0y}\cos\delta}{E_{0x}^2 - E_{0y}^2}, \tag{1.6}$$

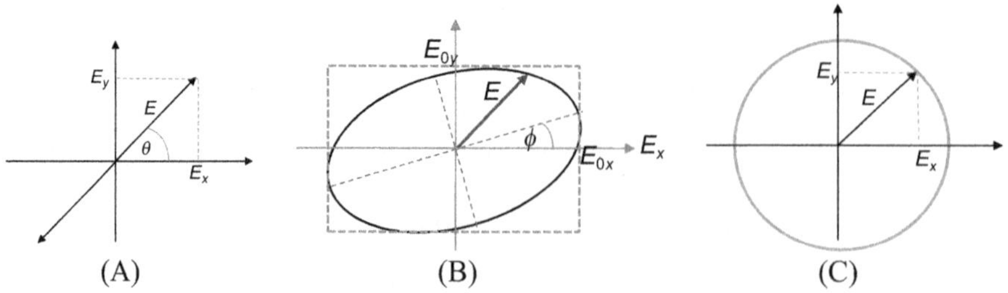

FIGURE 1.1

Various forms of polarizations: (A) linear polarization, (B) elliptic polarization, and (C) circular polarization.

which is shown in Fig. 1.1B. By setting $\delta = \pm\pi/2, \pm 3\pi/2, \ldots$ the equation of ellipse becomes:

$$\left(\frac{E_x}{E_{0x}}\right)^2 + \left(\frac{E_y}{E_{0y}}\right)^2 = 1. \tag{1.7}$$

By setting further $E_{0x} = E_{0y} = E_0$; $\delta = \pm\pi/2, \pm 3\pi/2, \ldots$ the equation of ellipse becomes the circle:

$$E_x^2 + E_y^2 = 1, \tag{1.8}$$

and corresponding polarization is known as *circular polarization* (Fig. 1.1C). *The right circularly polarized* wave is obtained for $\delta = \pi/2 + 2m\pi$:

$$\boldsymbol{E} = E_0[\boldsymbol{e}_x \cos(\omega t - kz) - \boldsymbol{e}_y \sin(\omega t - kz)]. \tag{1.9}$$

Otherwise, for $\delta = -\pi/2 + 2m\pi$, the polarization is known as left circularly polarized.

Very often, the *Jones vector representation* of the polarization wave is used:

$$\boldsymbol{E}(t) = \begin{bmatrix} E_x(t) \\ E_y(t) \end{bmatrix} = E\begin{bmatrix} \sqrt{1-\kappa} \\ \sqrt{\kappa}e^{j\delta} \end{bmatrix} e^{j(\omega t - kz)}, \tag{1.10}$$

where κ is the power-splitting ratio between states of polarizations (SOPs), with the complex phasor term being typically omitted in practice.

Another interesting representation is *Stokes vector representation*:

$$\boldsymbol{S}(t) = \begin{bmatrix} S_0(t) \\ S_1(t) \\ S_2(t) \\ S_3(t) \end{bmatrix}, \tag{1.11}$$

where the parameter S_0 is related to the optical intensity by:

$$S_0(t) = |E_x(t)|^2 + |E_y(t)|^2. \tag{1.12a}$$

The parameter $S_1 > 0$ is related to the preference for horizontal polarization and is defined by:

$$S_1(t) = |E_x(t)|^2 - |E_y(t)|^2. \tag{1.12b}$$

The parameter $S_2 > 0$ is related to the preference for $\pi/4$ SOP:

$$S_2(t) = E_x(t)E_y^*(t) + E_x^*(t)E_y(t). \tag{1.12c}$$

Finally, the parameter $S_3 > 0$ is related to the preference for right circular polarization and is defined by:

$$S_3(t) = j\left[E_x(t)E_y^*(t) - E_x^*(t)E_y(t)\right]. \tag{1.12d}$$

The parameter S_0 is related to other Stokes parameters by:

$$S_0^2(t) = S_1^2(t) + S_2^2(t) + S_3^2(t). \tag{1.13}$$

The *degree of polarization* is defined by:

$$p = \frac{\left[S_1^2 + S_2^2 + S_3^2\right]^{1/2}}{S_0}, \quad 0 \le p \le 1 \tag{1.14}$$

For $p = 1$ the polarization does not change in time. The Stokes vector can be represented in terms of Jones vector parameters as:

$$s_1 = 1 - 2\kappa, \quad s_2 = 2\sqrt{\kappa(1-\kappa)}\cos\delta, \quad s_3 = 2\sqrt{\kappa(1-\kappa)}\sin\delta. \tag{1.15}$$

After *normalization* with respect to S_0, the normalized Stokes parameters are given by:

$$s_i = \frac{S_i}{S_0}; \quad i = 0, 1, 2, 3. \tag{1.16}$$

If the normalized Stokes parameters are used, the polarization state can be represented as a point on a *Poincaré sphere*, as shown in Fig. 1.2. The points located at the opposite sides of the line crossing the center represent the orthogonal polarizations.

The polarization ellipse is very often represented in terms of ellipticity and azimuth, which are illustrated in Fig. 1.3. The ellipticity is defined by the ratio of half-axes lengths. The corresponding angle is called the ellipticity angle and denoted by ε. The small ellipticity means that the polarization ellipse is highly elongated, while for zero ellipticity the polarization is linear. For $\varepsilon = \pm\pi/4$, the polarization is circular. For $\varepsilon > 0$, the polarization is right elliptical. On the other hand, the azimuth angle η defines the orientation of the main axis of ellipse with respect to E_x.

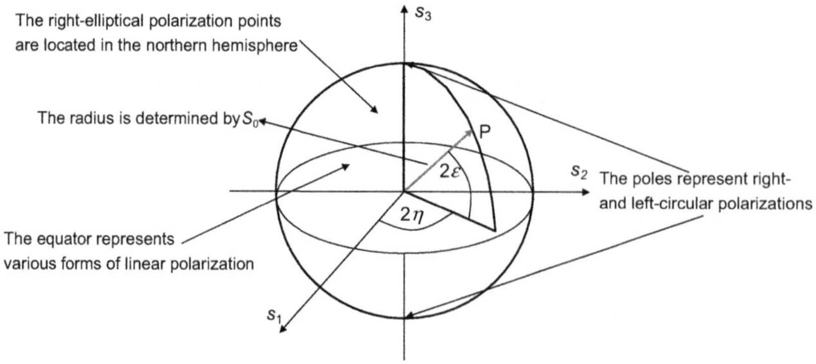

FIGURE 1.2

Representation of polarization state as a point on a Poincaré sphere.

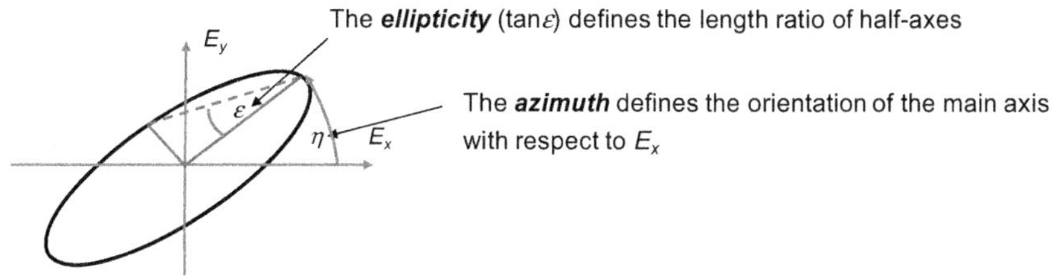

FIGURE 1.3

The ellipticity and azimuth of the polarization ellipse.

The parameters of the polarization ellipse can be related to the Jones vector parameters by:

$$\sin 2\,\varepsilon = 2\sqrt{\kappa(1-\kappa)}\sin\delta,\quad \tan 2\,\eta = \frac{2\sqrt{\kappa(1-\kappa)}\sin\delta}{1-2\kappa}. \tag{1.17}$$

Finally, the parameters of the polarization ellipse can be related to the Stokes vector parameters by:

$$s_1 = \cos 2\eta \cos 2\varepsilon,\;\; s_2 = \sin 2\eta \cos 2\varepsilon,\;\; s_3 = \sin 2\varepsilon, \tag{1.18}$$

and corresponding geometrical interpretation is provided in Fig. 1.2.

Let us now observe the *polarizer–analyzer ensemble*, shown in Fig. 1.4. When an electromagnetic wave passes through the polarizer, it can be represented as a vector in the x–y plane, transversal to the propagation direction, as given by Eq. (1.5), where the angle θ depends on the filter orientation. By introducing the unit vector $\widehat{p} = (\cos\theta, \sin\theta)$, Eq. (1.5) can be rewritten as:

$$\boldsymbol{E} = E_0\widehat{p}\cos(\omega t - kz). \tag{1.19}$$

If $\theta = 0$ rad, the light is polarized along the x-axis, while for $\theta = \pi/2$ rad it is polarized along the y-axis. The *natural light* is *unpolarized* as it represents

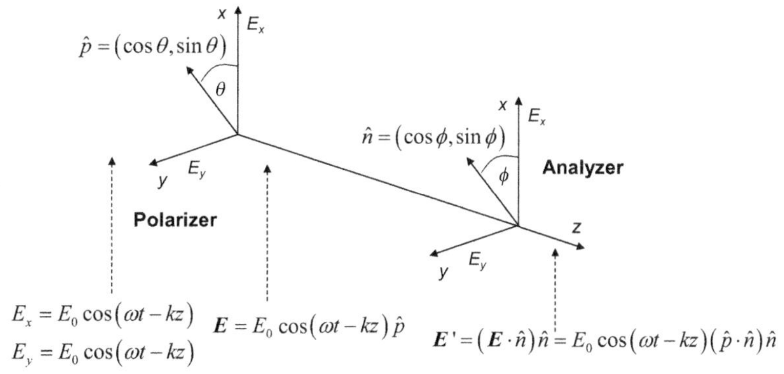

FIGURE 1.4

The polarizer–analyzer ensemble for the study of photon polarization.

incoherent superposition of 50% of light polarized along the x-axis and 50% of light polarized along the y-axis. After the analyzer, whose axis makes an angle ϕ with respect to the x-axis, which can be represented by the unit vector $\hat{n} = (\cos\phi, \sin\phi)$, the output electric field is given by:

$$\begin{aligned} \boldsymbol{E}' = (\boldsymbol{E}\cdot\hat{n})\hat{n} &= E_0\cos(\omega t - kz)(\hat{p}\cdot\hat{n})\hat{n} \\ &= E_0\cos(\omega t - kz)[(\cos\theta, \sin\theta)\cdot(\cos\phi, \sin\phi)]\hat{n} \\ &= E_0\cos(\omega t - kz)[\cos\theta\cos\phi + \sin\theta\sin\phi]\hat{n} \\ &= E_0\cos(\omega t - kz)\cos(\theta - \phi)\hat{n}. \end{aligned} \tag{1.20}$$

The intensity of the analyzer output field is given by:

$$I' = |\boldsymbol{E}'|^2 = I\cos^2(\theta - \phi), \tag{1.21}$$

which is commonly referred to as *Malus' law.*

The polarization decomposition by a birefringent plate is now studied (Fig. 1.5). Experiment shows that photodetectors PD_x and PD_y are never triggered simultaneously, which indicates that an entire photon reaches either PD_x or PD_y (a photon never splits). Therefore the corresponding probabilities that a photon is detected by photodetector PD_x and PD_x can be determined by:

$$p_x = \Pr(\mathrm{PD}_x) = \cos^2\theta, \quad p_y = \Pr(\mathrm{PD}_y) = \sin^2\theta. \tag{1.22}$$

If the total number of photons is N, the number of detected photons in x-polarization will be $N_x \cong N\cos^2\theta$, and the number of detected photons in y-polarization will be $N_y \cong N\sin^2\theta$. In the limit, as $N \to \infty$ we would expect Malus' law to be obtained.

Let us now study polarization decomposition and recombination by means of birefringent plates, as illustrated in Fig. 1.6. Classical physics prediction of total probability of a photon passing the polarizer–analyzer ensemble is given by:

$$p_{\mathrm{tot}} = \cos^2\theta\cos^2\phi + \sin^2\theta\sin^2\phi \neq \cos^2(\theta - \phi), \tag{1.23}$$

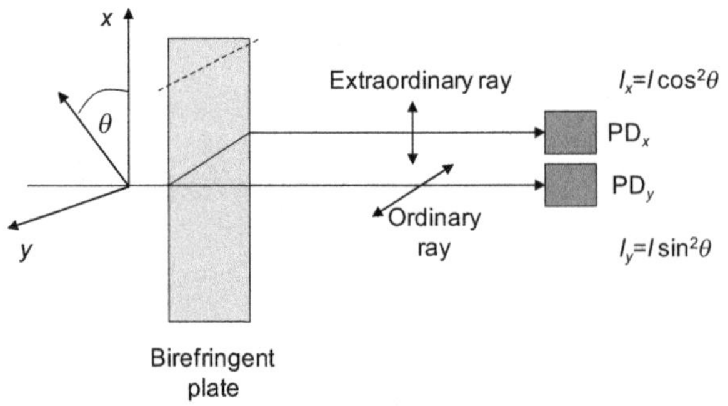

FIGURE 1.5

Polarization decomposition by a birefringent plate. *PD*, Photodetector.

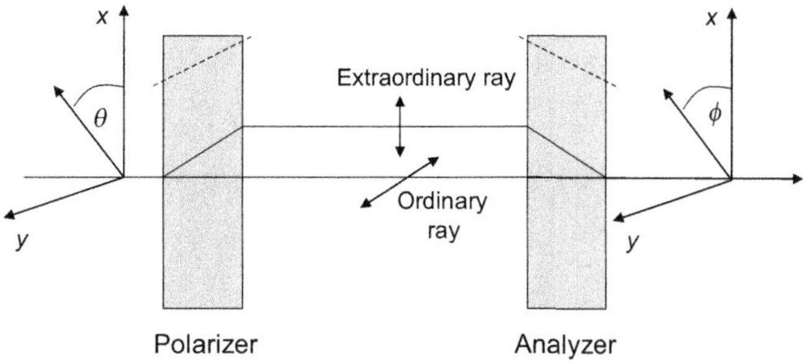

FIGURE 1.6

Polarization decomposition and recombination by a birefringent plate.

which is inconsistent with Malus' law, given by Eq. (1.21). To reconstruct the results from wave optics, it is necessary to introduce into quantum mechanics a concept of probability amplitude that α is detected as β, which is denoted as $a(\alpha \rightarrow \beta)$, and is a complex number. The probability is obtained as the squared magnitude of probability amplitude:

$$p(\alpha \rightarrow \beta) = |a(\alpha \rightarrow \beta)|^2. \tag{1.24}$$

The relevant probability amplitudes related to Fig. 1.6 are:

$$\begin{aligned} a(\theta \rightarrow x) &= \cos\theta \quad a(x \rightarrow \phi) = \cos\phi \\ a(\theta \rightarrow y) &= \sin\theta \quad a(x \rightarrow \phi) = \sin\phi \end{aligned} \tag{1.25}$$

The basic principle of quantum mechanics is to sum up the probability amplitudes for indistinguishable paths:

$$a_{\text{tot}} = \cos\theta \cos\phi + \sin\theta \sin\phi = \cos(\theta - \phi). \tag{1.26}$$

The corresponding total probability is:

$$p_{\text{tot}} = |a_{\text{tot}}|^2 = \cos^2(\theta - \phi), \tag{1.27}$$

and this result is consistent with Malus' law!

Based on the previous discussion, the *state vector* of photon polarization is given by:

$$|\psi\rangle = \begin{pmatrix} \psi_x \\ \psi_y \end{pmatrix}, \tag{1.28}$$

where ψ_x is related to x-polarization and ψ_y is related to y-polarization, with the normalization condition as follows:

$$|\psi_x|^2 + |\psi_y|^2 = 1. \tag{1.29}$$

In this representation, the x- and y-polarization photons can be represented by:

$$|x\rangle = \begin{pmatrix} 1 \\ 0 \end{pmatrix}, |y\rangle = \begin{pmatrix} 0 \\ 1 \end{pmatrix}, \tag{1.30}$$

and the right and left circular polarization photons are represented by:

$$|R\rangle = \frac{1}{\sqrt{2}} \begin{pmatrix} 1 \\ j \end{pmatrix}, |L\rangle = \frac{1}{\sqrt{2}} \begin{pmatrix} 1 \\ -j \end{pmatrix}. \tag{1.31}$$

In Eqs. (1.28)–(1.31) we used Dirac notation to denote the column vectors (kets). In Dirac notation, with each column vector ("ket") $|\psi\rangle$, we associate a row vector ("bra") $\langle\psi|$ as follows:

$$\langle\psi| = \begin{pmatrix} \psi_x^* & \psi_y^* \end{pmatrix}. \tag{1.32}$$

The *scalar (dot) product* of ket $|\phi\rangle$ bra $\langle\psi|$ is defined by "brackets" as follows:

$$\langle\phi|\psi\rangle = \phi_x^*\psi_x + \phi_y^*\psi_y = \langle\psi|\phi\rangle^*. \tag{1.33}$$

The normalization condition can be expressed in terms of scalar product by:

$$\langle\psi|\psi\rangle = 1. \tag{1.34}$$

Based on Eqs. (1.30) and (1.31) it is evident that:

$$\langle x|x\rangle = \langle y|y\rangle = 1, \quad \langle R|R\rangle = \langle L|L\rangle = 1.$$

Because the vectors $|x\rangle$ and $|y\rangle$ are *orthogonal*, their dot product is zero:

$$\langle x|y\rangle = 0, \tag{1.35}$$

and they form the *basis*. Any state vector $|\psi\rangle$ can be written as a *linear superposition* of basis kets as follows:

$$|\psi\rangle = \begin{pmatrix} \psi_x \\ \psi_y \end{pmatrix} = \psi_x|x\rangle + \psi_y|y\rangle. \tag{1.36}$$

We can now use Eqs. (1.34) and (1.36) to derive an important relation in quantum mechanics, known as *completeness relation*. The projections of state vector $|\psi\rangle$ along basis vectors $|x\rangle$ and $|y\rangle$ are given by:

$$\langle x|\psi\rangle = \psi_x \underbrace{\langle x|x\rangle}_{1} + \psi_y \underbrace{\langle x|y\rangle}_{0} = \psi_x, \quad \langle y|\psi\rangle = \psi_x\langle y|x\rangle + \psi_y\langle y|y\rangle = \psi_y. \tag{1.37}$$

By substituting Eq. (1.37) into Eq. (1.36) we obtain:

$$|\psi\rangle = |x\rangle\langle x|\psi\rangle + |y\rangle\langle y|\psi\rangle = \underbrace{(|x\rangle\langle x| + |y\rangle\langle y|)}_{I}|\psi\rangle, \tag{1.38}$$

and from the right side of Eq. (1.38) we derive the completeness relation:

$$|x\rangle\langle x| + |y\rangle\langle y| = I. \tag{1.39}$$

The probability that the photon in state $|\psi\rangle$ will pass the x-polaroid is given by:

$$p_x = \frac{|\psi_x|^2}{|\psi_x|^2 + |\psi_y|^2} = |\psi_x|^2 = |\langle x|\psi\rangle|^2, \tag{1.40}$$

the probability amplitude of the photon in state $|\psi\rangle$ to pass the x-polaroid is:

$$a_x = \langle x|\psi\rangle. \tag{1.41}$$

Let $|\phi\rangle$ and $|\psi\rangle$ be two physical states. The probability amplitude of finding ϕ in ψ, denoted as $a(\phi \rightarrow \psi)$, is given by:

$$a(\phi \rightarrow \psi) = \langle \phi|\psi\rangle, \tag{1.42}$$

and the probability for ϕ to pass the ψ test is given by:

$$p(\phi \rightarrow \psi) = |\langle \phi|\psi\rangle|^2. \tag{1.43}$$

1.2 THE CONCEPT OF QUBIT

Based on the previous section, it can be concluded that the quantum bit, also known as qubit, lies in a 2D Hilbert space H, isomorphic to C^2-space, where C is the complex number space, and can be represented as:

$$|\psi\rangle = \alpha|0\rangle + \beta|1\rangle = \begin{pmatrix} \alpha \\ \beta \end{pmatrix}; \quad \alpha, \beta \in C; \quad |\alpha|^2 + |\beta|^2 = 1, \tag{1.44}$$

where $|0\rangle$ and $|1\rangle$ states are computational basis (CB) states, and $|\psi\rangle$ is a superposition state. If we perform the measurement of a qubit, we will get $|0\rangle$ with probability $|\alpha|^2$ and $|1\rangle$ with probability $|\beta|^2$. Measurement changes the state of a qubit from a superposition of $|0\rangle$ and $|1\rangle$ to the specific state consistent with the measurement result. If we parametrize the probability amplitudes α and β as follows:

$$\alpha = \cos\left(\frac{\theta}{2}\right), \quad \beta = e^{j\phi} \sin\left(\frac{\theta}{2}\right), \tag{1.45}$$

where θ is a polar angle and ϕ is an azimuthal angle, we can geometrically represent the qubit by the Bloch sphere (or the Poincaré sphere for the photon) as illustrated in Fig. 1.7. (Note that the Bloch sphere from Fig. 1.7 is a little bit different from that of the Poincaré sphere from Fig. 1.2.) Bloch vector coordinates are given by ($\cos\phi \sin\theta$, $\sin\phi \sin\theta$, $\cos\theta$). This Bloch vector representation is related to CB by:

$$|\psi(\theta,\phi)\rangle = \cos(\theta/2)|0\rangle + e^{j\phi}\sin(\theta/2)|1\rangle \doteq \begin{pmatrix} \cos(\theta/2) \\ e^{j\phi}\sin(\theta/2) \end{pmatrix}, \tag{1.46}$$

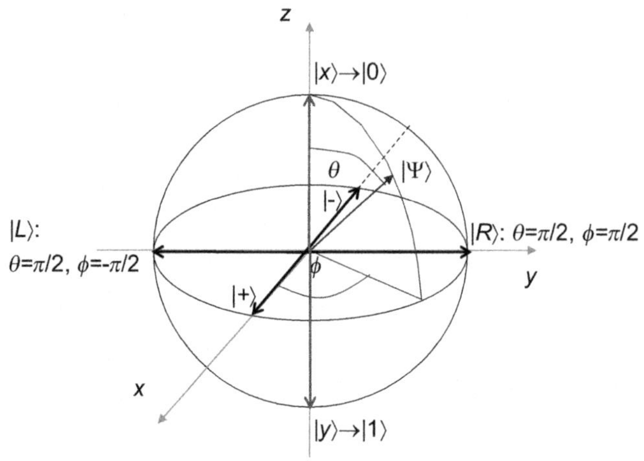

FIGURE 1.7

Bloch (Poincaré) sphere representation of a single qubit.

where $0 \le \theta \le \pi$ and $0 \le \phi < 2\pi$. The north and south poles correspond to computational $|0\rangle$ ($|x\rangle$-polarization) and $|1\rangle$ ($|y\rangle$-polarization) basis kets, respectively. Other important bases are the *diagonal basis* $\{|+\rangle, |-\rangle\}$, very often denoted as $\{|\nearrow\rangle, |\searrow\rangle\}$, related to CB by:

$$|+\rangle = |\nearrow\rangle = \frac{1}{\sqrt{2}}(|0\rangle + |1\rangle), \quad |-\rangle = |\searrow\rangle = \frac{1}{\sqrt{2}}(|0\rangle - |1\rangle), \tag{1.47}$$

and the *circular basis* $\{|R\rangle, |L\rangle\}$, related to the CB as follows:

$$|R\rangle = \frac{1}{\sqrt{2}}(|0\rangle + j|1\rangle), \quad |L\rangle = \frac{1}{\sqrt{2}}(|0\rangle - j|1\rangle). \tag{1.48}$$

1.3 THE SPIN-1/2 SYSTEMS

In addition to the photon, an important realization of the qubit is a spin-1/2 system. NMR is based on the fact that the proton possesses a magnetic moment $\boldsymbol{\mu}$ that can take only two values along the direction of the magnetic field. Namely, the projection along the $\hat{n}$-axis obtained by $\boldsymbol{\mu} \cdot \hat{n}$ takes only two values, and this property characterizes a spin-1/2 particle. Experimentally, this was confirmed by *Stern–Gerlach experiment*, shown in Fig. 1.8. A beam of protons is sent into a nonhomogenic magnetic field in a direction $\hat{n}$ orthogonal to the beam direction. It can be observed that the beam splits into two beams: one deflected to the direction $+\hat{n}$ and the other in the opposite direction $-\hat{n}$. Therefore the proton never splits. The basis in spin-1/2 systems can be written as $\{|+\rangle, |-\rangle\}$, where the corresponding

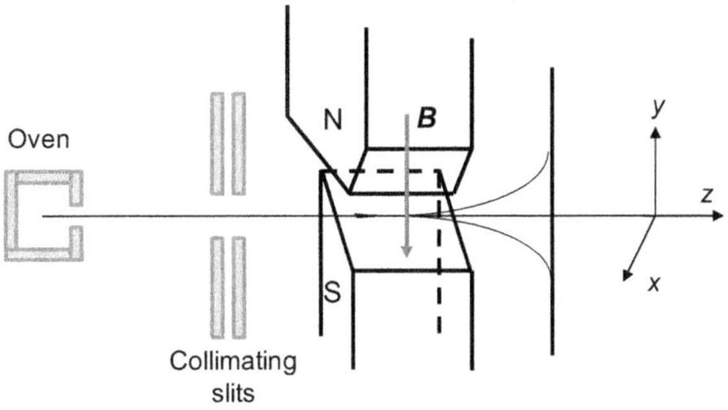

FIGURE 1.8

The Stern–Gerlach experiment.

basis kets represent the spin-up and spin-down states. The superposition state can be represented in terms of these bases as follows:

$$|\psi\rangle = \begin{pmatrix} \psi_+ \\ \psi_- \end{pmatrix} = \psi_+|+\rangle + \psi_-|-\rangle, \quad |\psi_-|^2 + |\psi_+|^2 = 1, \tag{1.49}$$

where $|\psi_+|^2$ ($|\psi_-|^2$) denotes the probability of finding the system in the spin-up (spin-down) state. By using the trigonometric identity $\sin^2(\theta/2) + \cos^2(\theta/2) = 1$, the expansion coefficients ψ_+ and ψ_+ can be expressed as follows:

$$\psi_+ = e^{-j\phi/2}\cos(\theta/2) \quad \psi_- = e^{j\phi/2}\sin(\theta/2), \tag{1.50}$$

so that:

$$|\psi\rangle = e^{-j\phi/2}\cos(\theta/2)|+\rangle + e^{j\phi/2}\sin(\theta/2)|-\rangle. \tag{1.51}$$

Therefore the single superposition state of the spin-1/2 system can also be visualized as the point (θ,φ) on a unit sphere (Bloch sphere).

1.4 QUANTUM GATES AND QIP

In quantum mechanics, the primitive undefined concepts are *physical system*, *observable*, and *state*. The concept of state has been introduced in previous sections. An observable, such as momentum and spin, can be represented by an *operator*, denoted by A, in the vector space of question. An operator, or gate, acts on a ket from the left: $(A) \cdot |\alpha\rangle = A\ |\alpha\rangle$, and results in another ket. A linear operator (gate) B can be expressed in terms of eigenkets $\{|a^{(n)}\rangle\}$ of an Hermitian operator A. (An operator A is said to be *Hermitian* if $A^\dagger = A, A^\dagger = (A^{\mathrm{T}})^*$.) The *operator* X is associated with a *square matrix* (albeit infinite in extent), whose elements are:

$$X_{mn} = \left\langle a^{(m)} \middle| X \middle| a^{(n)} \right\rangle, \tag{1.52}$$

and can be explicitly written as:

$$X \doteq \begin{pmatrix} \langle a^{(1)}|X|a^{(1)}\rangle & \langle a^{(1)}|X|a^{(2)}\rangle & \cdots \\ \langle a^{(2)}|X|a^{(1)}\rangle & \langle a^{(2)}|X|a^{(2)}\rangle & \cdots \\ \vdots & \vdots & \ddots \end{pmatrix}, \tag{1.53}$$

where we use the notation $\doteq$ to denote that operator X is represented by the foregoing matrix.

Very important single-qubit gates are: Hadamard gate H, phase shift gate S, $\pi/8$ (or T) gate, controlled-NOT (or CNOT) gate, and Pauli operators X, Y, Z. The Hadamard gate H, phase shift gate, T gate, and CNOT gate have the following matrix representation in CB $\{|0\rangle, |1\rangle\}$:

$$H \doteq \frac{1}{\sqrt{2}}\begin{bmatrix} 1 & 1 \\ 1 & -1 \end{bmatrix}, \quad S \doteq \begin{bmatrix} 1 & 0 \\ 0 & j \end{bmatrix}, \quad T \doteq \begin{bmatrix} 1 & 0 \\ 0 & e^{j\pi/4} \end{bmatrix}, \quad \text{CNOT} \doteq \begin{bmatrix} 1 & 0 & 0 & 0 \\ 0 & 1 & 0 & 0 \\ 0 & 0 & 0 & 1 \\ 0 & 0 & 1 & 0 \end{bmatrix}. \tag{1.54}$$

The Pauli operators, on the other hand, have the following matrix representation in CB:

$$X \doteq \begin{bmatrix} 0 & 1 \\ 1 & 0 \end{bmatrix}, \quad Y \doteq \begin{bmatrix} 0 & -j \\ j & 0 \end{bmatrix}, \quad Z \doteq \begin{bmatrix} 1 & 0 \\ 0 & -1 \end{bmatrix}. \tag{1.55}$$

The action of Pauli gates on an arbitrary qubit $|\psi\rangle = a|0\rangle + b|1\rangle$ is given as follows:

$$\begin{aligned} X(a|0\rangle + b|1\rangle) = a|1\rangle + b|0\rangle, \quad & Y(a|0\rangle + b|1\rangle) \\ = j(a|1\rangle - b|0\rangle), \quad & Z(a|0\rangle + b|1\rangle) = a|0\rangle - b|1\rangle. \end{aligned} \tag{1.56}$$

So, the action of the X gate is to introduce the bit flip, the action of the Z gate is to introduce the phase flip, and the action of the Y gate is to simultaneously introduce the bit and phase flips.

Several important single-, double-, and triple-qubit gates are shown in Fig. 1.9. The action of a single-qubit gate is to apply the operator U on qubit $|\psi\rangle$, which results in another qubit. Controlled-U gate conditionally applies the operator U on target qubit $|\psi\rangle$, when the control qubit $|c\rangle$ is in the $|1\rangle$ state. One particularly important controlled-U gate is the CNOT gate. This gate flips the content of the target qubit $|t\rangle$ when the control qubit $|c\rangle$ is in the $|1\rangle$ state. The purpose of SWAP gate is to interchange the positions of two qubits, and can be implemented by using three CNOT gates as shown in Fig. 1.9D. Finally, the Toffoli gate represents the

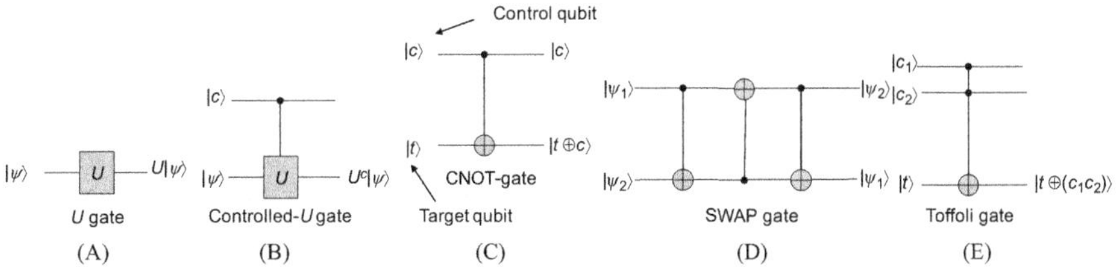

FIGURE 1.9

Important quantum gates and their actions: (A) single-qubit gate, (B) controlled-*U* gate, (C) CNOT gate, (D) SWAP gate, and (E) Toffoli gate.

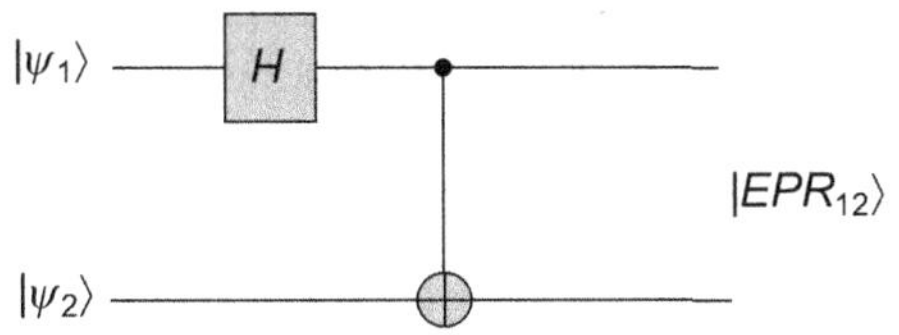

FIGURE 1.10

Bell states [Einstein-Podolsky-Rosen (EPR) pairs] preparation circuit.

generalization of the CNOT gate, where two control qubits are used. The minimum set of gates that can be used to perform an arbitrary quantum computation algorithm is known as *universal set of gates*. The most popular sets of universal quantum gates are: {H, S, CNOT, Toffoli} gates, {H, S, $\pi/8$ (T), CNOT} gates, Barenco gate [14], and Deutsch gate [15]. By using these universal quantum gates, more complicated operations can be performed. As an illustration, Fig. 1.10 shows the Bell states (Einstein–Podolsky–Rosen [EPR] pairs) preparation circuit, which is of high importance in quantum teleportation and QKD applications.

So far, single-, double-, and triple-qubit quantum gates have been considered. An arbitrary quantum state of K qubits has the form $\sum_s \alpha_s |s\rangle$, where s runs over all binary strings of length K. Therefore there are 2^K complex coefficients, all independent except for the normalization constraint:

$$\sum_{s=00...00}^{11..11} |\alpha_s|^2 = 1. \tag{1.57}$$

For example, the state $\alpha_{00}|00\rangle + \alpha_{01}|01\rangle + \alpha_{10}|10\rangle + \alpha_{11}|11\rangle$ (with $|\alpha_{00}|^2 + |\alpha_{01}|^2 + |\alpha_{10}|^2 + |\alpha_{11}|^2 = 1$) is the general two-qubit state (we use $|00\rangle$ to denote the tensor product $|0\rangle \otimes |0\rangle$). The multiple qubits can be *entangled* so that they cannot be decomposed into two separate states. For example, the Bell state or EPR pair $(|00\rangle + |11\rangle)/\sqrt{2}$ cannot be written in terms of tensor product $|\psi_1\rangle\,|\psi_2\rangle = (\alpha_1|0\rangle + \beta_1|1\rangle) \otimes (\alpha_2|0\rangle + \beta_2|1\rangle) = \alpha_1\alpha_2|00\rangle + \alpha_1\beta_2|01\rangle + \beta_1\alpha_2|10\rangle + \beta_1\beta_2|11\rangle$, because $\alpha_1\alpha_2 = \beta_1\beta_2 = 1/\sqrt{2}$, while $\alpha_1\beta_2 = \beta_1\alpha_2 = 0$, which a priori has no reason to be valid. This state can be obtained by using the circuit shown in

Fig. 1.10 for two-qubit input state $|00\rangle$. For more details on quantum gates and algorithms, interested readers are referred to Chapters 3 and 5, respectively.

1.5 QUANTUM TELEPORTATION

Quantum teleportation [16] is a technique to transfer quantum information from source to destination by employing the entangled states. Namely, in quantum teleportation, the entanglement in the Bell state (EPR pair) is used to transport arbitrary quantum state $|\psi\rangle$ between two distant observers A and B (often called Alice and Bob), as illustrated in Fig. 1.11. The quantum teleportation system employs three qubits: qubit 1 is an arbitrary state to be teleported, while qubits 2 and 3 are in the Bell state $|B_{00}\rangle=(|00\rangle+|11\rangle)/\sqrt{2}$. Let the state to be teleported be denoted by $|\psi\rangle = a|0\rangle+b|1\rangle$. The input to the circuit shown in Fig. 1.11 is therefore $|\psi\rangle|B_{00}\rangle$, and can be rewritten as:

$$\begin{aligned}|\psi\rangle|B_{00}\rangle &= (a|0\rangle + b|1\rangle)(|00\rangle + |11\rangle)/\sqrt{2} \\ &= (a|000\rangle + a|011\rangle + b|100\rangle + b|111\rangle)/\sqrt{2}.\end{aligned} \tag{1.58}$$

The CNOT gate is then applied with the first qubit serving as control and the second qubit as target, which transforms Eq. (1.58) into:

$$\begin{aligned}\mathrm{CNOT}^{(12)}|\psi\rangle|B_{00}\rangle &= \mathrm{CNOT}^{(12)}(a|000\rangle + a|011\rangle + b|100\rangle + b|111\rangle)/\sqrt{2} \\ &= (a|000\rangle + a|011\rangle + b|110\rangle + b|101\rangle)/\sqrt{2}.\end{aligned} \tag{1.59}$$

In the next stage, the Hadamard gate is applied to the first qubit, which maps $|0\rangle$ to $(|0\rangle+|1\rangle)/\sqrt{2}$ and $|1\rangle$ to $(|0\rangle-|1\rangle)/\sqrt{2}$, so that the overall transformation of Eq. (1.59) is as follows:

$$\begin{aligned}H^{(1)}\mathrm{CNOT}^{(12)}|\psi\rangle|B_{00}\rangle = &\frac{1}{2}(a|000\rangle + a|100\rangle + a|011\rangle + a|111\rangle \\ &+ b|010\rangle - b|110\rangle + b|001\rangle - b|101\rangle).\end{aligned} \tag{1.60}$$

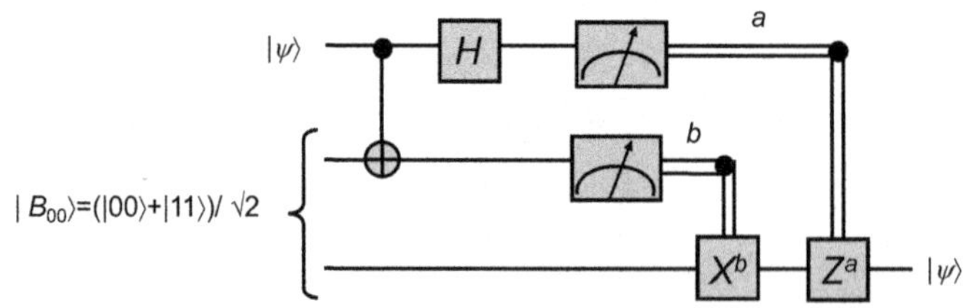

FIGURE 1.11

Illustration of quantum teleportation principle.

The measurements are performed on qubits 1 and 2, and based on the results of measurements, denoted respectively as a and b, the controlled-X (CNOT) and controlled-Z gates are applied conditionally to lead to the following content on qubit 3:

$$\frac{1}{2}(2a|0\rangle + 2b|1\rangle) = a|0\rangle + b|1\rangle = |\psi\rangle, \tag{1.61}$$

indicating that the arbitrary state $|\psi\rangle$ is teleported to the remote destination and can be found at qubit 3 position.

1.6 QUANTUM ERROR CORRECTION CONCEPTS

QIP relies on delicate superposition states, which are sensitive to interactions with the environment, resulting in decoherence. Moreover, the quantum gates are imperfect and the use of QECC is necessary to enable fault-tolerant computing and to deal with quantum errors [17–22]. QECC is also essential in quantum communication and quantum teleportation applications. The elements of quantum error correction codes are shown in Fig. 1.12A. The (N,K) QECC code performs encoding of the quantum state of K qubits, specified by 2^K complex coefficients α_s, into a quantum state of N qubits, in such a way that errors can be detected and corrected, and all 2^K complex coefficients can be perfectly restored up to the global phase shift. Namely, from quantum mechanics (see Chapter 2) we know that two states $|\psi\rangle$ and $e^{j\theta}|\psi\rangle$ are equal up to a *global phase shift* as the results of measurement on both states are the same. A quantum error correction consists of four major steps: encoding, error detection, error recovery, and decoding. The sender (Alice) encodes quantum information in state $|\psi\rangle$ with the help of local ancilla qubits $|0\rangle$, and then sends the encoded qubits over a noisy quantum channel (say free-space optical channel or

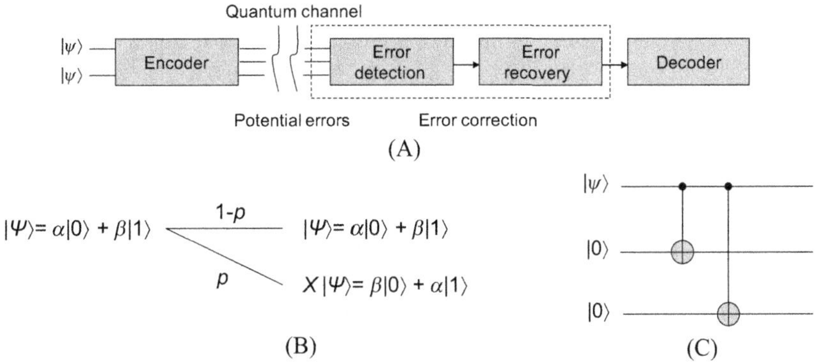

FIGURE 1.12

(A) A quantum error correction principle. (B) Bit-flipping channel model. (C) Three-qubit flip-code encoder.

optical fiber). The receiver (Bob) performs multiqubit measurement on all qubits to diagnose the channel error and performs a recovery unitary operation R to reverse the action of the channel. The quantum error correction is essentially more complicated than classical error correction. Difficulties for quantum error correction can be summarized as follows: (1) the no-cloning theorem indicates that it is impossible to make a copy of an arbitrary quantum state, (2) quantum errors are continuous and a qubit can be in any superposition of the two bases states, and (3) the measurements destroy the quantum information. The quantum error correction principles will be more evident after the following simple example.

Assume we want to send a single qubit $|\Psi\rangle = \alpha|0\rangle + \beta|1\rangle$ through the quantum channel in which during transmission the transmitted qubit can be flipped to $X|\Psi\rangle = \beta|0\rangle + \alpha|1\rangle$ with probability p. Such a quantum channel is called a *bit-flip channel* and it can be described as shown in Fig. 1.12B. Three-qubit flip code sends the same qubit three times, and therefore represents the *repetition code* equivalent. The corresponding codewords in this code are $|\bar{0}\rangle = |000\rangle$, $|\bar{1}\rangle = |111\rangle$. The three-qubit flip-code encoder is shown in Fig. 1.12C. One input qubit and two ancillas are used at the input of encoder, which can be represented by $|\psi_{123}\rangle = \alpha|000\rangle + \beta|100\rangle$. The first ancilla qubit (the second qubit at the encoder input) is controlled by the information qubit (the first qubit at encoder input) so that its output can be represented by $\text{CNOT}_{12}(\alpha|000\rangle + \beta|100\rangle) = \alpha|000\rangle + \beta|110\rangle$ (if the control qubit is $|1\rangle$ the target qubit is flipped, otherwise it stays unchanged). The output of the first CNOT gate is used as input to the second CNOT gate in which the second ancilla qubit (the third qubit) is controlled by the information qubit (the first qubit) so that the corresponding encoder output is obtained as $\text{CNOT}_{13}(\alpha|000\rangle + \beta|110\rangle) = \alpha|000\rangle + \beta|111\rangle$, which indicates that basis codewords are indeed $|\bar{0}\rangle$ and $|\bar{1}\rangle$. With this code, we are capable of correcting a single qubit flip error, which occurs with probability $(1-p)^3 + 3p(1-p)^2 = 1 - 3p^2 + 2p^3$. Therefore the probability of an error remaining uncorrected or wrongly corrected with this code is $3p^2 - 2p^3$. It is clear from Fig. 1.12C that a three-qubit bit-flip encoder is a *systematic encoder* in which the information qubit is unchanged, and the ancilla qubits are used to impose the encoding operation and create the parity qubits (the output qubits 2 and 3).

Let us assume that a qubit flip occurred on the first qubit leading to the received quantum word $|\Psi_r\rangle = \alpha|100\rangle + \beta|011\rangle$. To identify the error measurements need to be taken of the following observables Z_1Z_2 and Z_2Z_3, where the subscript denotes the index of qubit on which a given Pauli gate is applied. The result of measurement is the eigenvalue ± 1, and corresponding eigenvectors are two valid codewords, namely $|000\rangle$ and $|111\rangle$. The observables can be represented as follows:

$$\begin{aligned} Z_1Z_2 &= (|00\rangle\langle 11| + |11\rangle\langle 11|) \otimes I - (|01\rangle\langle 01| + |10\rangle\langle 10|) \otimes I \\ Z_2Z_3 &= I \otimes (|00\rangle\langle 11| + |11\rangle\langle 11|) - I \otimes (|01\rangle\langle 01| + |10\rangle\langle 10|) \end{aligned} \tag{1.62}$$

It can be shown that $\langle\psi_r|Z_1Z_2|\psi_r\rangle = -1$, $\langle\psi_r|Z_2Z_3|\psi_r\rangle = +1$, indicating that an error occurred on either the first or second qubit, but not on the second or third

Table 1.1 The three-qubit flip-code look-up table.

Z_1Z_2	Z_3Z_4	Error
+1	+1	I
+1	−1	X_3
−1	+1	X_1
−1	−1	X_2

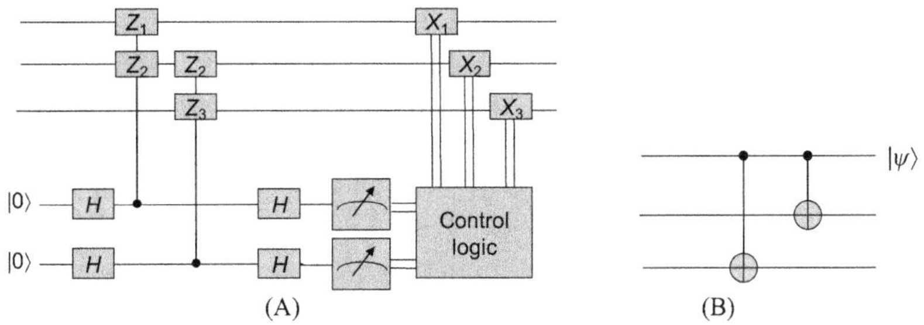

FIGURE 1.13

(A) Three-qubit flip-code error detection and error correction circuit. (B) Decoder circuit configuration.

qubit. The intersection reveals that the first qubit was in error. By using this approach we can create the three-qubit look-up table (LUT), given as Table 1.1.

Three-qubit flip-code error detection and error correction circuits are shown in Fig. 1.13. The results of measurements on ancillas (Fig. 1.13A) will determine the error syndrome $[\pm 1 \pm 1]$, and based on the LUT given by Table 1.1, we identify the error event and apply a corresponding X_i gate on the i th qubit being in error, and the error is corrected since $X^2 = I$. The control logic operation is described in Table 1.1. For example, if both outputs at the measurement circuits are -1, the operator X_2 is activated. The last step is to perform decoding as shown in Fig. 1.13B by simply reversing the order of elements in the corresponding encoder.

1.7 QUANTUM KEY DISTRIBUTION

QKD exploits the principles of quantum mechanics to enable provably secure distribution of a private key between remote destinations. Private key cryptography is much older than the public key cryptosystems commonly used today. In a private key cryptosystem, Alice (sender) must have an *encoding key*, while Bob (receiver) must have a matching *decoding key* to decrypt the encoding message. The simplest private key cryptosystem is the *Vernam cipher* (*one time pad*), which operates as follows [2]: (1) Alice and Bob share n-bit key strings, (2) Alice encodes her n-bit message by adding the message and the key together, and (3) Bob decodes the information by subtracting the key from the received message. There are several *drawbacks* to this scheme: (1) secure distribution of the key, and the length of the key must be at least as long as the message length, (2) the key bits cannot be reused, and (3) the keys must be delivered in advance, securely stored until used, and destroyed after use.

On the other hand, if the classical information, such as key, is transmitted over the quantum channel, thanks to the no-cloning theorem, which states that a device cannot be constructed to produce an exact copy of an arbitrary quantum state, the eavesdropper (Eve) cannot get an exact copy of the key in a QKD system. Namely, in an attempt to distinguish between two nonorthogonal states, information gain is only possible at the expense of introducing the disturbance to the signal. This observation can be proved as follows: let $|\psi_1\rangle$ and $|\psi_2\rangle$ be two nonorthogonal states Eve is interested to learn about and $|\alpha\rangle$ be the standard state prepared by the Eve. The Eve tries to interact with these states without disturbing them, and as a result of this interaction the following transformation is performed [2]:

$$|\psi_1\rangle \otimes |\alpha\rangle \rightarrow |\psi_1\rangle \otimes |\alpha\rangle, \quad |\psi_2\rangle \otimes |\alpha\rangle \rightarrow |\psi_2\rangle \otimes |\beta\rangle. \tag{1.63}$$

The Eve hopes that the states $|\alpha\rangle$ and $|\beta\rangle$ would be different in an attempt to learn something about the states. However, any unitary transformation must preserve the dot product so that from Eq. (1.63) we obtain:

$$\langle\alpha|\beta\rangle\langle\psi_1|\psi_2\rangle = \langle\alpha|\alpha\rangle\langle\psi_1|\psi_2\rangle, \tag{1.64}$$

which indicates that $\langle\alpha|\alpha\rangle = \langle\alpha|\beta\rangle = 1$, and consequently $|\alpha\rangle = |\beta\rangle$. Therefore in an attempt to distinguish between $|\psi_1\rangle$ and $|\psi_2\rangle$, the Eve will disturb them. The key idea of QKD is therefore to transmit the nonorthogonal qubit states between Alice and Bob, and by checking for disturbance in the transmitted state, they can establish an upper bound on noise/eavesdropping level in their communication channel. Once the raw key transmission is completed, the transmitter A and receiver B perform a series of classical steps, which are illustrated in Fig. 1.14. As the transmission distance increases and for higher key distribution speeds, the error correction becomes increasingly important. By performing information reconciliation by low-density parity-check (LDPC) codes, the higher input bit error rates can be tolerated compared to other coding schemes. *Privacy amplification* is further performed

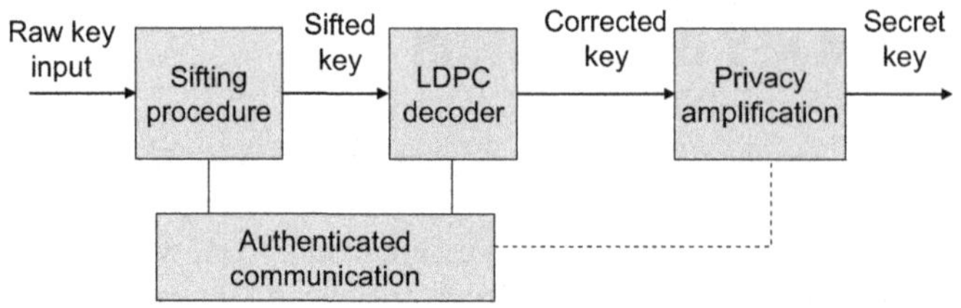

FIGURE 1.14

Illustration of the classical postprocessing steps. *LDPC*, Low-density parity-check.

to eliminate any information obtained by an Eve. Privacy amplification must be performed between Alice and Bob to distill from the generated key a smaller set of bits whose correlation with the Eve's string is below the desired threshold. One way to accomplish privacy amplification is through the use of universal hash functions. The simplest family of hash functions is based on a multiplication with a random element of the Galois field $GF(2^m)$ $(m > n)$, where n denotes the number of bits remaining after information reconciliation is completed. The threshold for maximum tolerable error rate is dictated by the efficiency of the implemented information reconciliation and privacy amplification protocols. The BB84 protocol [23,24] is described next as an illustration of the QKD procedure.

The BB84 protocol consists of the following steps [2]:

1. Alice chooses $(4 + \delta)n$ random data bits $\boldsymbol{a}$. Alice also chooses at random a $(4 + \delta)n$-long bit string $\boldsymbol{b}$ $(\delta > 0)$ and encodes each data bit (in $\boldsymbol{a}$) as $\{|0\rangle,|1\rangle\}$ if the corresponding bit in $\boldsymbol{b}$ is 0, or $\{|+\rangle, |-\rangle\}$ if the corresponding bit in $\boldsymbol{b}$ is 1. Alice sends the resulting quantum state to Bob.
2. Bob receives the $(4 + \delta)n$ qubits, and measures each qubit in either $\{|0\rangle, |1\rangle\}$ or $\{|+\rangle, |-\rangle\}$ basis at random.
3. Alice announces the vector $\boldsymbol{b}$ over a public channel.
4. Alice and Bob discard any bits where Bob is measured on a different basis from the Alice prepared basis. With high probability there are at least $2n$ bits left (otherwise abort protocol), and they keep $2n$ bits.
5. Alice then selects a subset of n bits to be used against the Eve's interference, and provides Bob the information about which ones are selected.
6. Alice and Bob compare the values of n check bits. If they disagree in the number of locations exceeding the error correction capability of the error correction scheme, they abort the protocol.
7. Otherwise, Alice and Bob perform information reconciliation and privacy amplification on the remaining n bits to obtain m $(m < n)$ shared key bits.

Readers interested in learning more about QKD are referred to Chapter 15.

1.8 ORGANIZATION OF THE BOOK

This book is a self-contained introduction to quantum information, quantum computation, and quantum error correction. Readers of this book will be ready for further study in these areas, and will be prepared to perform independent research. After completion of the book, readers will be able design information processing circuits, stabilizer codes, Calderbank–Shor–Steane (CSS) codes, subsystem codes, topological codes, surface codes, and entanglement-assisted quantum error correction codes, and propose corresponding physical implementation. They will also be proficient in fault-tolerant design. The book starts with the basic principles of quantum mechanics, including state vectors, operators, density operators, measurements, and dynamics of a quantum system. It continues with fundamental principles of quantum computation, quantum gates, quantum algorithms, quantum teleportation, and fault-tolerant quantum computing. The concepts from classical information theory and channel coding are introduced at a level to better understand the corresponding concepts from QIP. The book continues with quantum information theory. Significant space has been allocated to quantum error correction codes, in particular on stabilizer codes, CSS codes, LDPC codes, subsystem codes (also known as operator-quantum error correction codes), topological codes, surface codes, and entanglement-assisted quantum error correction codes. The next topic in the book is devoted to fault-tolerant QECC and fault-tolerant quantum computing. We then discuss cluster state-based computing. The next part of the book is spent investigating physical realizations of quantum computers, encoders, and decoders, including NMR, ion traps, photonic quantum realization, cavity quantum electrodynamics, and quantum dots. Furthermore, the fundamental concepts from quantum machine learning are introduced. The final chapter in the book is devoted to QKD. For better understanding of the material in each chapter, a set of problems is provided at the end of every chapter. The chapters of the book are organized as follows.

Chapter 2 is devoted to quantum mechanics fundamentals. The following topics from quantum mechanics are covered: state vectors, operators, density operators, measurements, and dynamics of a quantum system. In addition, we describe harmonic oscillators, orbital angular momentum, spin-1/2 systems, and hydrogen-like systems. These systems are of high importance in various QIP systems and in quantum communications.

Chapter 3 is devoted to quantum gates and modules. We first describe basic single-qubit gates and provide Bloch sphere representation of the single qubit. Next, we describe binary and ternary quantum gates. We then provide the generalization to N-qubit gates and QIP fundamentals. Quantum measurement circuits are provided next, and the principles of deferred measurement and implicit measurement are formulated. We also formulate Gottesman−Knill and Solovay−Kitaev theorems. Furthermore, various sets of universal quantum gates are introduced. We then describe the Bell state preparation circuit and quantum relay. We also describe basics concepts of quantum teleportation. Finally, we briefly discuss the computation-in-place concept.

Chapter 4 is devoted to the basic concepts of QIP. After describing elementary QIP features, the chapter introduces the superposition principle and quantum parallelism concept as well as QIP basics. It continues with formulation and proof of no-cloning theorem. We further formulate and prove an important theorem, which claims that it is impossible unambiguously to distinguish nonorthogonal quantum states. The next section is devoted to the various aspects of entanglement, including the Schmidt decomposition theorem, purification, superdense coding, and entanglement swapping. How entanglement can be used in detecting and correcting quantum errors is described as well. Various measures of entanglement are introduced, including the entanglement of formation, concurrence, and entanglement fidelity. Furthermore, we discuss the operator-sum representation of a quantum operation and apply this concept to describe the quantum errors and decoherence. The bit-flip, phase-flip, phase-damping, depolarization, and amplitude-damping channels are described as well. Then, we describe different quantum computing libraries, including *Qiskit*, *Cirq*, *Forest*, and *Quantum development kit*.

Chapter 5 is devoted to various quantum algorithms and methods. The chapter starts by revisiting the quantum parallelism concept and describing its power in calculating a global property of a certain function by performing only one evaluation of that function, namely Deutsch and Deutsch–Jozsa algorithms. We also describe the Bernstein–Vazirani algorithm that is able to determine a string encoded in a function in only one computational step. In addition, the Grover search algorithm to perform a search for an entry in an unstructured database is described. Next, quantum Fourier transform is described, which is a basic algorithm used in many other quantum algorithms. Furthermore, an algorithm to evaluate the period of a function is provided. How to crack the RSA encryption protocol is also described. Then, Shor's factorization algorithm is provided. Furthermore, Simon's algorithm is described. Next, the quantum phase estimation algorithm is described. Quantum computational complexity and Turing machine representation are discussed as well. The quantum circuits are imperfect, which prevents us from running well-known quantum algorithms using the gates-based quantum computing approach. To solve this problem, adiabatic quantum computing and variational (quantum) circuits are introduced.

In Chapter 6, we provide basic concepts of classical information theory and coding theory. These topics are described to the level needed to more easily understand in later chapters various topics such as quantum information theory, quantum error correction, fault-tolerant error correction, fault-tolerant computing, and QKD. The chapter starts with definitions of entropy, joint entropy, conditional entropy, relative entropy, mutual information, and channel capacity, followed by information capacity theorem. We also briefly describe the source coding and data compaction concepts. We then discuss the channel capacity of discrete memoryless channels, continuous channels, and some optical channels. After that, the fundamentals of block codes are introduced, such as linear block codes (LBCs), as well as a definition of generator and parity-check matrices, syndrome decoding, distance properties of LBCs, and some important coding bounds. Additionally, cyclic codes are introduced. BCH codes are described next. RS, concatenated, and product codes are then described.

Chapter 7 is devoted to quantum information theory. The chapter starts with classical and von Neumann entropies definitions and properties, followed by quantum representation of classical information. Next, we introduce accessible information as the maximum mutual information over all possible measurement schemes. We further introduce an important bound, Holevo's bound, and prove it. Next, we introduce typical sequence, typical state, typical subspace, and projector on typical subspace. Then, we prove the typical subspace theorem and formulate and prove Schumacher's source coding theorem, the quantum equivalent of Shannon's source coding theorem. We further introduce the concept of quantum compression and decompression and introduce the concept of reliable compression. We then describe various quantum channel models, including bit-flip, phase-flip, depolarizing, and amplitude-damping channel models. We then formulate and prove the Holevo–Schumacher–Westmoreland theorem, the quantum equivalent of Shannon's channel capacity theorem.

Chapter 8 is devoted to quantum error correction. The concept of this chapter is to gradually introduce readers to QECC principles, from an intuitive description to a rigorous mathematical description. The chapter starts with Pauli operators, basic definitions, and representation of quantum errors. Although the Pauli operators were introduced in Chapter 3, they are put here in the context of quantum errors and quantum error correction. Next, basic quantum codes, such as three-qubit flip code, three-qubit phase-flip code, and Shor's nine-qubit code, are presented. The projection measurements are used to determine the error syndrome and perform corresponding error correction action. Furthermore, stabilizer formalism and stabilizer group are introduced. The basic stabilizer codes are described as well. The whole of the next chapter is devoted to the stabilizer codes; here, only the basic concepts are presented. An important class of codes, the class of CSS codes, is described next. The connection between classical and quantum codes is then established, and two classes of CSS codes, dual-containing and quantum codes derived from classical codes over GF(4), are described. The concept of quantum error correction is then formally introduced, followed by the necessary and sufficient conditions for quantum code to correct a given set of errors. Then, the minimum distance of a quantum code is defined and used to relate it to the error correction capability of a quantum code. The CSS codes are then revisited by using this mathematical framework. The next section is devoted to important quantum coding bounds, such as the Hamming quantum bound, the quantum Gilbert–Varshamov bound, and the quantum Singleton bound (also known as the Knill–Laflamme bound). Next, the concept of operator-sum representation is introduced and used to provide the physical interpretation and describe the measurement of the environment. Finally, several important quantum channel models are introduced, such as the depolarizing channel, amplitude-damping channel, and generalized amplitude-damping channel.

In Chapter 9, a well-studied class of quantum codes is described, namely stabilizer codes and related quantum codes. The basic concept of stabilizer codes was already introduced in Chapter 8. Here the stabilizer codes are described in a more rigorous way. After rigorous introduction of stabilizer codes (Section 9.1), their properties and encoded operations are introduced in Section 9.2. Furthermore, the stabilizer codes are described by using the finite geometry interpretation in Section 9.3. The standard form of stabilizer codes is introduced in Section 9.4, which is further used to provide the efficient encoder and decoder implementations in Section 9.5. Section 9.6 is devoted to nonbinary stabilizer codes. Next, the subsystem codes are described in Section 9.7, followed by the topological codes in Section 9.8. Section 9.9 is related to surface codes. Entanglement-assisted codes are described in Section 9.10.

Chapter 10 is devoted to quantum LDPC codes, which have many advantages compared to other classes of quantum codes thanks to the sparseness of their quantum-check matrices. Both semirandom and structured quantum LDPC codes are described. Key advantages of structured quantum LDPC codes compared to other codes include: (1) regular structure in corresponding parity-check (***H***-) matrices leads to low complexity encoders/decoders, and (2) their sparse ***H***-matrices require a small number of interactions per qubit to determine the error location. The chapter begins with the introduction of classical LDPC codes in Section 10.1, including their design and decoding algorithms. Furthermore, dual-containing quantum LDPC codes are described in Section 10.2. The next section (Section 10.3) is devoted to the entanglement-assisted quantum LDPC codes. Additionally, Section 10.4 introduces the probabilistic sum-product algorithm based on the quantum-check matrix instead of the classical parity-check matrix. Notice that encoders for dual-containing and entanglement-assisted quantum LDPC codes can be implemented based on either a standard form method or conjugation method (described in Chapter 9). Since there is no difference between encoder implementation of quantum LDPC codes and other classes of quantum block codes described in Chapter 9, we omit discussion of encoder implementation and concentrate instead on design and decoding algorithms for quantum LDPC codes. The quantum spatially coupled LDPC codes are briefly introduced in Section 10.5.

Chapter 11 is devoted to one of the most important applications of quantum error correction, namely the protection of quantum information as it dynamically undergoes computation through so-called fault-tolerant quantum computing. The chapter starts with introduction of fault-tolerance basics and traversal operations (Section 11.1). It continues with fault-tolerant quantum computation concepts and procedures (Section 11.2). In Section 11.2, the universal set of fault-tolerant quantum gates is introduced, followed by fault-tolerant measurement, fault-tolerant state preparation, and fault-tolerant encoded state preparation by using Steane's code as an illustrative example. Section 11.3 is devoted to a rigorous description of fault-tolerant quantum error correction. This section starts with a short review of some basic concepts from stabilizer codes. In Subsection 11.3.1, fault-tolerant syndrome extraction is described, which is followed by a description of fault-tolerant encoded operations in Subsection 11.3.2. Subsection 11.3.3 is concerned with the application of quantum gates on a quantum register by means of a measurement protocol. This method, when applied transversally, is used in Subsection 11.3.4 to enable fault-tolerant error correction based on arbitrary quantum stabilizer code. The fault-tolerant stabilizer codes are described in Subsection 11.3.4. In Subsection 11.3.5, the [5,1,3] fault-tolerant stabilizer code is described as an illustrative example. Section 11.4 is devoted to fault-tolerant computing, in particular the fault-tolerant implementation of the Tofolli gate is discussed. Furthermore, in Section 11.5, the accuracy threshold theorem is formulated and proved. Finally, in Section 11.6, surface code-based large-scale quantum computing is discussed.

Chapter 12 is devoted to cluster state-based quantum computing. The chapter starts with the definition of cluster states, followed by a description of their relationship with graph states and stabilizer formalism. We also discuss the relationship between cluster states and qubit teleportation circuits. We further formulate the conditions to be met for a corresponding set of unitaries to be able to generate the cluster state. We then relate the problem of generating the cluster state to the Ising model. The focus is then moved to the universality of cluster state-based quantum computing. We prove that a 1D cluster state is sufficient to implement an arbitrary single-qubit gate, while a 2D cluster state is needed for an arbitrary two-qubit gate by applying the proper sequence of measurements. We demonstrate that a five-qubit linear chain cluster is enough to implement arbitrary single-qubit rotation by employing the Euler decomposition theorem. We also demonstrate that T-shape and H-shape cluster states can be used to implement the CNOT gate. We then discuss cluster state processing by first describing the roles of *X*-, *Z*-, and *Y*-measurements. We further provide a generic 2D cluster state to be used in one-way quantum computation (1WQC) and provide details of the corresponding quantum computational model. During 1WQC, the random Pauli by-product operator naturally arises, and we show that the performance of 1WQC is not affected by it. After that we discuss various physical implementations suitable for 1WQC with special focus being devoted to photonic 1WQC implementations, including resource-efficient linear optics implementation. We describe how the Bell states can be used to form linear chain, T-shape, and H-shape, and arbitrary 2D cluster states through type I and type II fusion processes. We describe how the implementation of cluster states and 1WQC can be experimentally demonstrated. Furthermore, we briefly describe the basic concepts of continuous variable cluster state-based quantum computing.

Chapter 13 is devoted to several promising physical implementations of QIP. The chapter starts with a description of physical implementation basics, *di Vincenzo criteria*, and an overview of physical implementation concepts (Section 13.1). The next section (Section 13.2) is related to NMR implementation, whose basic concepts are used in various implementations. In Section 13.3, the use of ion traps in quantum computing is described. Next, the various photonic implementations (Section 13.4) are described, followed by quantum relay implementation (Section 13.5) and the implementation of quantum encoders and decoders (Section 13.6). Implementation based on optical cavity electrodynamics is further described in Section 13.7. Finally, the use of quantum dots in quantum computing is discussed in Section 13.8.

Chapter 14 is related to quantum machine learning. The chapter starts by reviewing relevant classical machine learning algorithms, categorized as supervised, unsupervised, and reinforcement learning algorithms. The following topics from classical machine learning are described: principal component analysis (PCA), support vector machines (SVMs), clustering, boosting, regression analysis, and neural networks. In the PCA section, we describe how to determine the principal components from the correlation matrix, followed by a description of the singular value decomposition-based PCA and scoring phase. We describe SVMs from both geometric and Lagrangian method-based points of view and introduce both hard and soft margins as well as the kernel method. In the clustering section, we describe

k-means, expectation maximization, and k-nearest neighbor algorithms. We also describe how to evaluate clustering quality. In the boosting section, we describe how to build strong learners from weak learners, with special attention devoted to the AdaBoost algorithm. In the regression analysis section, we describe least squares estimation, the pseudoinverse approach, and the ridge regression method. In the neural networks' section, we describe in detail the perceptron, activation functions, and feedforward networks. The focus then moves to quantum machine learning (QML) algorithms. We first describe the Ising model and relate it to the quadratic unconstrained binary optimization (QUBO) problem. We then study how to solve the QUBO problem by adiabatic quantum computing and quantum annealing. To perform QML using imperfect and noisy quantum circuits we describe the variational quantum eigensolver and quantum approximate optimization algorithm (QAOA). To illustrate the impact of QAOA we describe how to use it in combinatorial optimization problems and how to solve the MAX-CUT problem. Next, quantum boosting is discussed, which is related to the QUBO problem. Quantum random access memory is described next, allowing us to address the superposition of memory cells with the help of a quantum register. Quantum matrix inversion, also known as the Harrow—Hassidim—Lloyd algorithm, is then described, which is used as a basic ingredient for other QML algorithms such as quantum PCA, which is described as well. In the quantum optimization-based clustering section, we describe how the MAX-CUT problem can be related to clustering, and thus be solved by adiabatic computing, quantum annealing, and QAOA. Grover algorithm-based quantum optimization is discussed next. In the quantum k-means section we describe how to calculate the dot product and quantum distance, followed by the Grover search-based k-means algorithm. In the quantum SVM section, we formulate the SVM problem using least squares and describe how to solve it using quantum matrix inversion. In the quantum neural networks (QNNs) section, we describe feedforward QNNs, quantum perceptron, and quantum convolutional networks.

Chapter 15 is devoted to the fundamentals of QKD. The chapter starts with a description of key differences between conventional cryptography and QKD. In the section on QKD basics, after a historical overview, we review different QKD types. We then describe two fundamental theorems on which QKD relies, namely the no-cloning theorem and the theorem of inability to unambiguously distinguish nonorthogonal quantum states. In the section on discrete variable QKD systems, we describe BB84, B92, Ekert (E91), EPR, and time-phase encoding protocols. In the section on QKD security, the secret key rate (SKR) is represented as the product of raw key rate and fractional rate. Moreover, the generic expression for the fractional rate is provided, followed by a description of different eavesdropping strategies, including individual (independent or incoherent) attacks, collective attacks, coherent attacks, and quantum hacking/side-channel attacks. For individual and coherent attacks, corresponding secrete fraction expressions are provided. After that the decoy-state protocol is described together with corresponding SKR calculation. Next, the key concepts for measurement device-independent (MDI)-QKD protocols are introduced, including polarization-based and time-phase encoding-based MDI-QKD protocols as well as secrecy fraction calculation. The twin-field QKD protocols are further described, whose performance is evaluated against decoy-state and MDI-QKD protocols. The focus is then moved to the information reconciliation and privacy amplification steps. To facilitate the description of continuous variable (CV)-QKD protocols, the fundamentals of quantum optics and Gaussian information theory are introduced first. The topics include quadrature operators, Gaussian and squeezed states, Gaussian transformation, generation of quantum states, thermal decomposition of Gaussian states, two-mode Gaussian states, and the measurements on Gaussian states. In the section on CV-QKD protocols, homodyne and heterodyne detection schemes are described first, followed by a brief description of squeezed state-based protocols. Given that the coherent states are much easier to generate and manipulate, the coherent state-based protocols are described in detail. For a lossy transmission channel the corresponding covariance matrices are derived for both homodyne and coherent detection schemes, followed by the SKR derivation for prepare-and-measure Gaussian modulation (GM)-based CV-QKD. Some illustrative SKR results are provided for GM-based CV-QKD schemes.

References

[1] I.B. Djordjevic, Quantum Information Processing and Quantum Error Correction: An Engineering Approach, Elsevier/Academic Press, 2012.

[2] M.A. Neilsen, I.L. Chuang, Quantum Computation and Quantum Information, Cambridge University Press, Cambridge, 2000.

[3] M. Le Bellac, An Introduction to Quantum Information and Quantum Computation, Cambridge University Press, 2006.

[4] F. Gaitan, Quantum Error Correction and Fault Tolerant Quantum Computing, CRC Press, 2008.

[5] G. Jaeger, Quantum Information: An Overview, Springer, 2007.

[6] D. Petz, Quantum Information Theory and Quantum Statistics, Theoretical and Mathematical Physics, Springer, Berlin, 2008.

[7] P. Lambropoulos, D. Petrosyan, Fundamentals of Quantum Optics and Quantum Information, Springer-Verlag, Berlin, 2007.

[8] G. Johnson, A Shortcut Thought Time: The Path to the Quantum Computer, Knopf, N. York, 2003.

[9] J. Preskill, Quantum Computing, 1999. Available at: http://www.theory.caltech.edu/~preskill/.

[10] J. Stolze, D. Suter, Quantum Computing, Wiley, N. York, 2004.

[11] R. Landauer, Information is physical, Phys. Today 44 (5) (1991) 23–29.

[12] R. Landauer, The physical nature of information, Phys. Lett. A 217 (1991) 188–193.

[13] G. Keiser, Optical Fiber Communications, McGraw Hill, 2000.

[14] A. Barenco, A universal two-bit quantum computation, Proc. Roy. Soc. Lond. A 449 (1937) (1995) 679–683.

[15] D. Deutsch, Quantum computational networks, Proc. Roy. Soc. Lond. A 425 (1868) (1989) 73–90.
[16] C.H. Bennett, G. Brassard, C. Crépeau, R. Jozsa, A. Peres, W.K. Wootters, Teleporting an unknown quantum state via dual classical and Einstein-Podolsky-Rosen channels, Phys. Rev. Lett. 70 (13) (1993) 1895–1899.
[17] D.J.C. MacKay, G. Mitchison, P.L. McFadden, Sparse-graph codes for quantum error correction, IEEE Trans. Inf. Theor. 50 (2004) 2315–2330.
[18] I.B. Djordjevic, Photonic implementation of quantum relay and encoders/decoders for sparse-graph quantum codes based on optical hybrid, IEEE Photon. Technol. Lett. 22 (19) (2010) 1449–1451.
[19] I.B. Djordjevic, Photonic entanglement-assisted quantum low-density parity-check encoders and decoders, Opt. Lett. 35 (9) (2010) 1464–1466.
[20] I.B. Djordjevic, Quantum LDPC codes from balanced incomplete block designs, IEEE Commun. Lett. 12 (2008) 389–391.
[21] I. B. Djordjevic, Cavity quantum electrodynamics (CQED) based quantum LDPC encoders and decoders, IEEE Photon. J., (accepted for publication).
[22] I.B. Djordjevic, Photonic quantum dual-containing LDPC encoders and decoders, IEEE Photon. Technol. Lett. 21 (13) (July 1, 2009) 842–844.
[23] C.H. Bennet, G. Brassard, Quantum Cryptography: public key distribution and coin tossing, in: Proc. IEEE International Conference on Computers, Systems, and Signal Processing, Bangalore, India, 1984, pp. 175–179.
[24] C.H. Bennett, Quantum cryptography: uncertainty in the service of privacy, Science 257 (1992) 752–753.

Further reading

[1] D.P. DiVincenzo, Two-bit gates are universal for quantum computation, Phys. Rev. A 51 (2) (1995) 1015–1022.
[2] A. Barenco, C.H. Bennett, R. Cleve, D.P. DiVincenzo, N. Margolus, P. Shor, T. Sleator, J.A. Smolin, H. Weinfurter, Elementary gates for quantum computation, Phys. Rev. A 52 (5) (1995) 3457–3467.

CHAPTER

Quantum Mechanics Fundamentals

2

CHAPTER OUTLINE

Quantum Information Processing, Quantum Computing, and Quantum Error Correction
https://doi.org/10.1016/B978-0-12-821982-9.00014-9

This chapter is devoted to quantum mechanics fundamentals. The following topics from quantum mechanics are covered: state vectors, operators, density operators, measurements, and dynamics of a quantum system. In addition, we describe harmonic oscillators, orbital angular momentum, spin-1/2 systems and hydrogen-like systems. These systems are of high importance in various quantum information processing systems and in quantum communications.

The chapter is organized as follows. In Section 2.1 we introduce the basic concepts in quantum mechanics. In Section 2.2 we describe how eigenkets can be used as basis kets. The matrix representation of operators, kets, and bras is described in Section 2.3. Section 2.4 is devoted to Pauli operators, while Section 2.5 is related to density operators. Quantum measurements are described in Section 2.6. The uncertainty principle is introduced in Section 2.7. Section 2.8 deals with change of basis and diagonalization. Section 2.9 is devoted to the dynamics of quantum systems and the Schrödinger equation. Harmonic oscillator theory is used in many quantum computing applications, therefore Section 2.10 is devoted to the harmonic oscillator. Orbital angular momentum is described in Section 2.11, while spin-1/2 systems are described in Section 2.12. In Section 2.13 we describe hydrogen-like atoms. Finally, Section 2.14 concludes the chapter. Some problems for self-study are provided at the end of the chapter (see Section 2.15).

2.1 INTRODUCTION

In quantum mechanics, the primitive undefined concepts are *physical system*, *observable*, and *state* [1–24]. A physical system is any sufficiently isolated quantum object, say an electron, a photon, or a molecule. An observable will be associated with a measurable property of a physical system, say energy or z-component of the spin. The state of a physical system is a trickier concept in quantum mechanics compared to classical mechanics. The problem arises when considering composite physical systems. In particular, states exist, known as *entangled states*, for a bipartite physical system in which neither of the subsystems is in a definite state. Even in cases where physical systems can be described as being in a state, two classes of states are possible: pure and mixed. The condition of a quantum-mechanical system is completely specified by its *state vector* $|\psi\rangle$ in a Hilbert space H (a vector space on which a positive-definite scalar product is defined) over the field of complex numbers. Any state vector $|\alpha\rangle$, also known as a *ket*, can be expressed in terms of basis vectors $|\phi_n\rangle$ by:

$$|\alpha\rangle = \sum_{n=1}^{\infty} a_n |\phi_n\rangle. \tag{2.1}$$

An *observable*, such as momentum and spin, can be represented by an *operator*, such as A, in the vector space in question. Quite generally, an operator acts on a ket from the left: $(A) \cdot |\alpha\rangle = A|\alpha\rangle$, which results in another ket. An operator A is said to be *Hermitian* if:

$$A^\dagger = A, \quad A^\dagger = \left(A^{\mathrm{T}}\right)^*. \tag{2.2}$$

Suppose that the Hermitian operator A has a discrete set of eigenvalues $a^{(1)}, \ldots, a^{(n)}, \ldots$. The associated eigenvectors (eigenkets) $|a^{(1)}\rangle, \ldots, |a^{(n)}\rangle, \ldots$ can be obtained from:

$$A\left|a^{(n)}\right\rangle = a^{(n)}\left|a^{(n)}\right\rangle. \tag{2.3}$$

The Hermitian conjugate of a ket $|\alpha\rangle$ is denoted by $\langle\alpha|$ and called the "bra." The space dual to ket space is known as *bra* space. There exists is a one-to-one correspondence, dual correspondence (D.C.), between a ket space and a bra space:

$$\begin{aligned}
&|\alpha\rangle \overset{\text{D.C.}}{\leftrightarrow} \langle\alpha| \\
&|a^{(1)}\rangle, |a^{(2)}\rangle, \ldots \overset{\text{D.C.}}{\leftrightarrow} \langle a^{(1)}|, \langle a^{(2)}|, \ldots \\
&|\alpha\rangle + |\beta\rangle \overset{\text{D.C.}}{\leftrightarrow} \langle\alpha| + \langle\beta| \\
&c_\alpha|\alpha\rangle + c_b|\beta\rangle \overset{\text{D.C.}}{\leftrightarrow} c_\alpha^*\langle\alpha| + c_\beta^*\langle\beta|.
\end{aligned} \tag{2.4}$$

The *scalar (inner) product* of two state vectors $|\phi\rangle$ and $|\psi\rangle$ is defined by:

$$\langle\beta|\alpha\rangle = \sum_{n=1}^{\infty} a_n b_n^*. \tag{2.5}$$

Let $|\alpha\rangle$, $|\beta\rangle$, and $|\gamma\rangle$ be the state kets. The following *properties of inner product* are valid:

1. $\langle\beta|\alpha\rangle = \langle\alpha|\beta\rangle^*$
2. $\langle\beta|(a|\alpha\rangle + b|\gamma\rangle) = a\langle\beta|\alpha\rangle + b\langle\beta|\gamma\rangle$
3. $(\langle a\alpha| + \langle b\beta|)|\gamma\rangle = a^*\langle\alpha|\gamma\rangle + b^*\langle\beta|\gamma\rangle$
4. $\langle\alpha|\alpha\rangle \geq 0$
5. $\sqrt{\langle\alpha+\beta|\alpha+\beta\rangle} \leq \sqrt{\langle\alpha|\alpha\rangle} + \sqrt{\langle\beta|\beta\rangle}$ (Triangle inequality)
6. $|\langle\beta|\alpha\rangle|^2 \leq \langle\alpha|\alpha\rangle\langle\beta|\beta\rangle$ (Cauchy-Schwartz inequality)

(2.6)

Property 5 is known as the triangle inequality, and property 6 as *Cauchy–Schwartz inequality*, which is important in proving the Heisenberg uncertainty relationship, and as such will be proved here. The starting point in the proof is to observe the linear combination of two kets $|\alpha\rangle$ and $|\beta\rangle$, that is, $|\alpha\rangle + \lambda|\beta\rangle$, and apply property 4 to obtain:

$$(\langle\alpha| + \lambda^*\langle\beta|)(|\alpha\rangle + \lambda|\beta\rangle) \geq 0 \Leftrightarrow \langle\alpha|\alpha\rangle + \lambda\langle\alpha|\beta\rangle + \lambda^*\left\langle\beta|\alpha\right\rangle + |\lambda|^2\left\langle\beta|\beta\right\rangle. \quad (2.7)$$

The inequality (2.7) is satisfied for any complex λ, including $\lambda = -\langle\beta|\alpha\rangle/\langle\beta|\beta\rangle$:

$$\langle\alpha|\alpha\rangle - \frac{\langle\beta|\alpha\rangle}{\langle\beta|\beta\rangle}\langle\alpha|\beta\rangle - \frac{\langle\alpha|\beta\rangle}{\langle\beta|\beta\rangle}\langle\beta|\alpha\rangle + \frac{|\langle\beta|\alpha\rangle|^2}{|\langle\beta|\beta\rangle|^2}\langle\alpha|\alpha\rangle \geq 0. \quad (2.8)$$

By multiplying Eq. (2.8) with $\langle\beta|\beta\rangle$ we obtain:

$$|\langle\beta|\alpha\rangle|^2 \leq \langle\alpha|\alpha\rangle\langle\beta|\beta\rangle, \quad (2.9)$$

proving therefore the Cauchy–Schwartz inequality.

The norm of ket $|\alpha\rangle$ is defined by $\sqrt{\langle\alpha|\alpha\rangle}$, and the normalized ket by $|\widetilde{\alpha}\rangle = |\alpha\rangle/\sqrt{\langle\alpha|\alpha\rangle}$. Clearly, $\langle\widetilde{\alpha}|\widetilde{\alpha}\rangle = 1$. Two kets are said to be *orthogonal* if their dot product is zero, that is, $\langle\alpha|\beta\rangle = 0$. Two operators X and Y are equal $X = Y$ if their action on a ket is the same: $X|\alpha\rangle = Y|\alpha\rangle$. The identity operator I is defined by $I|\psi = |\psi$. Operators can be *added*; addition operations are commutative and associative. Operators can be *multiplied*; multiplication operations are in general *noncommutative*, but associative. The associative property in quantum mechanics, known as the *associative axiom*, is postulated to hold quite generally as long as we are dealing with "legal" multiplications among kets, bras, and operators as follows:

$$\begin{aligned} &(|\beta\rangle\langle\alpha|)\cdot|\gamma\rangle = |\beta\rangle\cdot(\langle\alpha|\gamma\rangle) && (\langle\beta|)\cdot X|\alpha\rangle = (\langle\beta|X)\cdot|\alpha\rangle = \left\{\langle\alpha|X^\dagger|\beta\rangle\right\}^* \\ &X = |\beta\rangle\langle\alpha| \Rightarrow X^\dagger = |\alpha\rangle\langle\beta| && \langle\beta|X|\alpha\rangle = \langle\alpha|X^\dagger|\beta\rangle^* \\ & && X^\dagger = X:\ \langle\beta|X|\alpha\rangle = \langle\alpha|X|\beta\rangle^*. \end{aligned} \quad (2.10)$$

2.2 EIGENKETS AS BASE KETS

The eigenkets $\{|\xi^{(n)}\rangle\}$ of operator Ξ form the basis so that arbitrary ket $|\psi\rangle$ can be expressed in terms of eigenkets by:

$$|\psi\rangle = \sum_{n=1}^{\infty} c_n\left|\xi^{(n)}\right\rangle. \quad (2.11)$$

By multiplying Eq. (2.11) with $\langle\xi^{(n)}|$ from the left we obtain:

$$\left\langle\xi^{(n)}\middle|\psi\right\rangle = \sum_{j=1}^{\infty} c_j\left\langle\xi^{(n)}\middle|\xi^{(j)}\right\rangle = c_n\left\langle\xi^{(n)}\middle|\xi^{(n)}\right\rangle + \sum_{j=1, j\neq n}^{\infty} c_j\left\langle\xi^{(n)}\middle|\xi^{(j)}\right\rangle. \quad (2.12)$$

Since the eigenkets $\{|\xi^{(n)}\rangle\}$ form the basis, the principle of orthonormality is satisfied $\left\langle\xi^{(n)}|\xi^{(j)}\right\rangle = \delta_{nj}$, $\delta_{nj} = \begin{cases} 1, & n = j \\ 0, & n \neq j \end{cases}$ so that Eq. (2.12) becomes:

$$c_n = \left\langle \xi^{(n)} \middle| \psi \right\rangle. \tag{2.13}$$

By substituting Eq. (2.13) into Eq. (2.11) we obtain:

$$|\psi\rangle = \sum_{n=1}^{\infty} \left\langle \xi^{(n)} \middle| \psi \right\rangle \left| \xi^{(n)} \right\rangle = \sum_{n=1}^{\infty} \left| \xi^{(n)} \right\rangle \left\langle \xi^{(n)} \middle| \psi \right\rangle. \tag{2.14}$$

Because $|\psi\rangle = I|\psi\rangle$, from Eq. (2.14) it is clear that:

$$\sum_{n=1}^{\infty} \left| \xi^{(n)} \right\rangle \left\langle \xi^{(n)} \right| = I, \tag{2.15}$$

and the foregoing relation is known as the *completeness relation*. The operators under summation in Eq. (2.15) are known as *projection* operators P_n:

$$P_n = \left| \xi^{(n)} \right\rangle \left\langle \xi^{(n)} \right|, \tag{2.16}$$

which satisfy the relationship $\sum_{n=1}^{\infty} P_n = I$. It is easy to show that the ket (2.11) with c_n determined with Eq. (2.13) is of unit length:

$$\langle \psi | \psi \rangle = \sum_{n=1}^{\infty} \left\langle \psi \middle| \xi^{(n)} \right\rangle \left\langle \xi^{(n)} \middle| \psi \right\rangle = \sum_{n=1}^{\infty} \left| \left\langle \psi \middle| \xi^{(n)} \right\rangle \right|^2 = 1. \tag{2.17}$$

The following theorem is an important theorem that will be used a lot throughout the chapter.

Theorem. The eigenvalues of a Hermitian operator A are real, while the eigenkets are orthogonal:

$$\left\langle a^{(m)} \middle| a^{(n)} \right\rangle = \delta_{nm}.$$

This theorem is quite straightforward to prove by starting from the eigenvalue equation:

$$A \left| a^{(i)} \right\rangle = a^{(i)} \left| a^{(i)} \right\rangle, \tag{2.18a}$$

and its dual conjugate:

$$\left\langle a^{(j)} \right| A^{\dagger} = a^{(j)^*} \left\langle a^{(j)} \right|. \tag{2.18b}$$

By multiplying Eq. (2.18a) by $\left\langle a^{(j)} \right|$ from the left and Eq. (2.18b) by $\left| a^{(i)} \right\rangle$ from the right and subtracting them we obtain:

$$\left(a^{(i)} - a^{(j)^*} \right) \left\langle a^{(j)} \middle| a^{(i)} \right\rangle = 0. \tag{2.19}$$

For $i = j$, it is clear from Eq. (2.19) that $a^{(i)} = a^{(j)^*}$ proving that eigenvalues are real. For $i \neq j$ from Eq. (2.19) we conclude that $\left\langle a^{(j)} \middle| a^{(i)} \right\rangle = 0$ proving that the eigenkets are orthogonal. We can represent now the operator A in terms of its eigenkets as follows:

$$A=\sum_i a^{(i)}\left|a^{(i)}\right\rangle\left\langle a^{(i)}\right|, \tag{2.20}$$

which is known as *spectral decomposition.*

2.3 MATRIX REPRESENTATIONS

A linear operator X can be expressed in terms of eigenkets $\{|a^{(n)}\rangle\}$ by applying the completeness relation twice as follows:

$$\begin{aligned} X=IXI &= \sum_{m=1}^{\infty}\left|a^{(m)}\right\rangle\left\langle a^{(m)}\right|X\sum_{n=1}^{\infty}\left|a^{(n)}\right\rangle\left\langle a^{(n)}\right| \\ &= \sum_{n=1}^{\infty}\sum_{m=1}^{\infty}\left|a^{(m)}\right\rangle\left\langle a^{(m)}\right|X\left|a^{(n)}\right\rangle\left\langle a^{(n)}\right|. \end{aligned} \tag{2.21}$$

Therefore the *operator* X is associated with a *square matrix* (albeit infinite in extent), whose elements are:

$$X_{mn}=\left\langle a^{(m)}\right|X\left|a^{(n)}\right\rangle, \tag{2.22a}$$

and can explicitly be written as:

$$X\doteq\begin{pmatrix} \langle a^{(1)}|X|a^{(1)}\rangle & \langle a^{(1)}|X|a^{(2)}\rangle & \cdots \\ \langle a^{(2)}|X|a^{(1)}\rangle & \langle a^{(2)}|X|a^{(2)}\rangle & \cdots \\ \vdots & \vdots & \ddots \end{pmatrix}, \tag{2.22b}$$

where we use the notation $\doteq$ to denote that operator X is represented by the foregoing matrix.

Operator multiplication $Z=XY$ representation in matrix form can be obtained by employing Eq. (2.22b) and the completeness relation to obtain:

$$Z_{mn}=\left\langle a^{(m)}\right|Z\left|a^{(n)}\right\rangle=\left\langle a^{(m)}\right|XY\left|a^{(n)}\right\rangle=\sum_k\left\langle a^{(m)}\right|X\left|a^{(k)}\right\rangle\left\langle a^{(k)}\right|Y\left|a^{(n)}\right\rangle. \tag{2.23}$$

The *kets* are represented as *column vectors* and corresponding representation can be obtained by multiplying the ket $|\gamma\rangle=X|\alpha\rangle$ from the left by $\langle a^{(i)}|$ and by applying the completeness relation as follows:

$$\left\langle a^{(i)}|\gamma\right\rangle=\left\langle a^{(i)}\right|X|\alpha\rangle=\sum_j\underbrace{\left\langle a^{(i)}\right|X\left|a^{(j)}\right\rangle}_{\text{matrix}}\ \underbrace{\left\langle a^{(j)}|\alpha\right\rangle}_{\text{column vector}}. \tag{2.24}$$

Therefore the ket $|\alpha\rangle$ can be a column vector based on Eq. (2.24) by:

$$|\alpha\rangle = \begin{pmatrix} \langle a^{(1)}|\alpha\rangle \\ \langle a^{(2)}|\alpha\rangle \\ \vdots \end{pmatrix}. \tag{2.25}$$

The *bras* are represented as *row vectors* and corresponding representation can be obtained by multiplying the bra $\langle\gamma| = \langle\alpha|X$ from the right by $|a^{(i)}\rangle$ and by applying the completeness relation as follows:

$$\langle\gamma|a^{(i)}\rangle = \langle\alpha|X|a^{(i)}\rangle = \sum_j \underbrace{\langle\alpha|a^{(j)}\rangle}_{\langle a^{(j)}|\alpha\rangle^*,\text{row vector}} \underbrace{\langle a^{(j)}|X|a^{(i)}\rangle}_{\text{matrix}}. \tag{2.26}$$

From Eq. (2.26) is clear that the bra $\langle\alpha|$ can be represented as a row vector by:

$$\langle\alpha| \doteq \left(\langle a^{(1)}|\alpha\rangle^* \langle a^{(2)}|\alpha\rangle^* \cdots\right). \tag{2.27}$$

The *inner product* of the ket $|\beta\rangle$ and bra $\langle\alpha|$ can be represented by applying the completeness representation as a conventional dot product as follows:

$$\begin{aligned} \langle\alpha|\beta\rangle &= \sum_i \langle\alpha|a^{(i)}\rangle\langle a^{(i)}|\beta\rangle = \sum_i \langle a^{(i)}|\alpha\rangle^* \langle a^{(i)}|\beta\rangle \\ &= \left(\langle a^{(1)}|\alpha\rangle^* \langle a^{(2)}|\alpha\rangle^* \cdots\right) \begin{pmatrix} \langle a^{(1)}|\beta\rangle \\ \langle a^{(2)}|\beta\rangle \\ \vdots \end{pmatrix}. \end{aligned} \tag{2.28}$$

The *outer product* $|\beta\rangle\langle\alpha|$ represented in the matrix by using the completeness relation twice is as follows:

$$\begin{aligned} |\beta\rangle\langle\alpha| &= \sum_i\sum_j |a^{(i)}\rangle\langle a^{(i)}|\beta\rangle\langle\alpha|a^{(j)}\rangle\langle a^{(j)}| = \sum_i\sum_j |a^{(i)}\rangle\langle a^{(j)}| \langle a^{(i)}|\beta\rangle\langle a^{(j)}|\alpha\rangle^* \\ &= \begin{pmatrix} \langle a^{(1)}|\beta\rangle\langle a^{(1)}|\alpha\rangle^* & \langle a^{(1)}|\beta\rangle\langle a^{(2)}|\alpha\rangle^* & \cdots \\ \langle a^{(2)}|\beta\rangle\langle a^{(1)}|\alpha\rangle^* & \langle a^{(2)}|\beta\rangle\langle a^{(2)}|\alpha\rangle^* & \cdots \\ \vdots & \vdots & \ddots \end{pmatrix}. \end{aligned} \tag{2.29}$$

Finally, the *operator* can be represented in terms of *outer products* by:

$$\begin{aligned} X &= \sum_{n=1}^{\infty}\sum_{m=1}^{\infty} |a^{(m)}\rangle\langle a^{(m)}|X|a^{(n)}\rangle\langle a^{(n)}| = \sum_{n=1}^{\infty}\sum_{m=1}^{\infty} X_{mn}|a^{(m)}\rangle\langle a^{(n)}|, \\ X_{mn} &= \langle a^{(m)}|X|a^{(n)}\rangle. \end{aligned} \tag{2.30}$$

To illustrate these various representations we study photons and spin-1/2 systems as described in the rest of this section.

2.3.1 Photons

The x- and y-polarizations can be represented by:

$$|E_x\rangle = \begin{pmatrix} 1 \\ 0 \end{pmatrix} \quad |E_y\rangle = \begin{pmatrix} 0 \\ 1 \end{pmatrix}.$$

On the other hand, the right- and left-circular polarizations can be represented by:

$$|E_R\rangle = \frac{1}{\sqrt{2}}\begin{pmatrix} 1 \\ j \end{pmatrix} \quad |E_L\rangle = \frac{1}{\sqrt{2}}\begin{pmatrix} 1 \\ -j \end{pmatrix}.$$

The 45-degree polarization ket can be represented as follows:

$$|E_{45^\circ}\rangle = \cos\left(\frac{\pi}{4}\right)|E_x\rangle + \sin\left(\frac{\pi}{4}\right)|E_y\rangle = \frac{1}{\sqrt{2}}(|E_x\rangle + |E_y\rangle) = \frac{1}{\sqrt{2}}\begin{pmatrix} 1 \\ 1 \end{pmatrix}.$$

The bras corresponding to the left and right polarization can be written by:

$$\langle E_R| = \frac{1}{\sqrt{2}}(1 \ -j) \quad \langle E_L| = \frac{1}{\sqrt{2}}(1 \ \ j).$$

It can be easily verified that the left and right states are orthogonal and that the right polarization state is of unit length:

$$\langle E_R|E_L\rangle = \frac{1}{2}(1 \ -j)\begin{pmatrix} 1 \\ -j \end{pmatrix} = 0 \quad \langle E_R|E_R\rangle = \frac{1}{2}(1 \ -j)\begin{pmatrix} 1 \\ j \end{pmatrix} = 1.$$

The completeness relation is clearly satisfied because:

$$|E_x\rangle\langle E_x| + |E_y\rangle\langle E_y| = \begin{pmatrix} 1 \\ 0 \end{pmatrix}(1 \quad 0) + \begin{pmatrix} 0 \\ 1 \end{pmatrix}(0 \quad 1) = \begin{pmatrix} 1 & 0 \\ 0 & 1 \end{pmatrix} = I_2.$$

An arbitrary polarization state can be represented by:

$$|E\rangle = |E_R\rangle\langle E_R|E\rangle + |E_L\rangle\langle E_L|E\rangle.$$

For example, for $E = E_x$ we obtain:

$$|E_x\rangle = |E_R\rangle\langle E_R|E_x\rangle + |E_L\rangle\langle E_L|E_x\rangle.$$

For the photon spin operator S matrix representation we have to solve the following eigenvalue equation:

$$S|\psi\rangle = \lambda|\psi\rangle.$$

The photon spin operator satisfies $S^2 = I$ so that we can write:

$$|\psi\rangle = S^2|\psi\rangle = S(S|\psi\rangle) = S(\lambda|\psi\rangle) = \lambda S|\psi\rangle = \lambda^2|\psi\rangle.$$

It is clear from the previous equation that $\lambda^2 = 1$ so that the corresponding eigenvalues are $\lambda = \pm 1$. By substituting the eigenvalues into an eigenvalue equation we obtain that corresponding eigenkets are the left and right polarization states:

$$S|E_R\rangle = |E_R\rangle \quad S|E_L\rangle = -|E_L\rangle.$$

The photon spin representation in $\{|E_x\rangle, |E_y\rangle\}$ basis can be obtained by:

$$S \doteq \begin{pmatrix} S_{xx} & S_{xy} \\ S_{yx} & S_{yy} \end{pmatrix} = \begin{pmatrix} \langle E_x|S|E_x\rangle & \langle E_x|S|E_y\rangle \\ \langle E_y|S|E_x\rangle & \langle E_y|S|E_y\rangle \end{pmatrix} = \begin{pmatrix} 0 & -j \\ j & 0 \end{pmatrix}.$$

2.3.2 Spin-1/2 Systems

The S_z basis in spin-1/2 systems can be written as $\{|E_z;+\rangle, |E_z;-\rangle\}$, where the corresponding basis kets represent the spin-up and spin-down states. The eigenvalues are $\{\hbar/2, -\hbar/2\}$, and the corresponding eigenket/eigenvalue relation is:

$S_z|S_z;\pm\rangle = \pm\frac{\hbar}{2}|S_z;\pm\rangle$, where S_z is the spin operator that can be represented in the foregoing basis as follows:

$$S_z = \sum_{i=+,-}\sum_{j=+,-} |i\rangle\langle j| \underbrace{S_z|i\rangle}_{i\frac{\hbar}{2}|i\rangle} \langle j| = \sum_{i=+,-} i\frac{\hbar}{2}|i\rangle\langle i| = \frac{\hbar}{2}(|+\rangle\langle +| - |-\rangle\langle -|).$$

The matrix representation of spin-1/2 systems is obtained by:

$$|S_z;+\rangle = \begin{pmatrix} \langle S_z;+|S_z;+\rangle \\ \langle S_z;-|S_z;+\rangle \end{pmatrix} \doteq \begin{pmatrix} 1 \\ 0 \end{pmatrix} \quad |S_z;-\rangle \doteq \begin{pmatrix} 0 \\ 1 \end{pmatrix}$$

$$S_z \doteq \begin{pmatrix} \langle S_z;+|S_z|S_z;+\rangle & \langle S_z;+|S_z|S_z;-\rangle \\ \langle S_z;-|S_z|S_z;+\rangle & \langle S_z;-|S_z|S_z;-\rangle \end{pmatrix} = \frac{\hbar}{2}\begin{pmatrix} 1 & 0 \\ 0 & -1 \end{pmatrix}.$$

2.4 PAULI OPERATORS AND HADAMARD GATE

A basic unit of information in a quantum computer is known as a *quantum bit* or *qubit*. The corresponding state space is 2D with the following basis $\{|0\rangle = [1\ 0]^T, |1\rangle = [0\ 1]^T\}$. The arbitrary state ket can be written in terms of basis kets as follows:

$$|\psi\rangle = \alpha|0\rangle + \beta|1\rangle, \tag{2.31}$$

with α and β satisfying the following normalization condition: $|\alpha|^2 + |\beta|^2 = 1$. The probability that a qubit upon measurement is in a state $|0\rangle$ is determined by $|\langle 0|\psi\rangle|^2 = |\alpha|^2$.

2.4.1 Pauli Operators

The *Pauli operators* X, Y, Z (very often denoted as σ_x, σ_y, and σ_z or σ_1, σ_2, and σ_3) correspond to the measurement of the spin along x-, y-, and z-axes, respectively. Their actions on basis states are given by:

$$\begin{aligned} X|0\rangle &= |1\rangle, \quad X|1\rangle = |0\rangle \\ Y|0\rangle &= -j|1\rangle, \quad Y|1\rangle = j|0\rangle \\ Z|0\rangle &= |0\rangle, \quad Z|1\rangle = -|1\rangle. \end{aligned} \tag{2.32}$$

It is clear that basis states are eigenkets of Z. We have shown earlier that an operator Ξ can be represented in matrix form in $\{|a^{(k)}\rangle\}$ basis with matrix elements given by $\Xi_{ij} = \langle a^{(i)}|\Xi|a^{(j)}\rangle$. Based on the action of the Pauli X-operator in Eq. (2.32), it can be represented in matrix form by:

$$X \doteq \begin{pmatrix} \langle 0|X|0\rangle & \langle 0|X|1\rangle \\ \langle 1|X|0\rangle & \langle 1|X|1\rangle \end{pmatrix} = \begin{pmatrix} \langle 0|1\rangle & \langle 0|0\rangle \\ \langle 1|1\rangle & \langle 1|0\rangle \end{pmatrix} = \begin{pmatrix} 0 & 1 \\ 1 & 0 \end{pmatrix}. \tag{2.33}$$

In similar fashion, Pauli X- and Y-operators can be represented by:

$$Y \doteq \begin{pmatrix} 0 & -j \\ j & 0 \end{pmatrix}, \quad Z \doteq \begin{pmatrix} 1 & 0 \\ 0 & -1 \end{pmatrix}. \tag{2.34}$$

Since any operator Ξ can be written in terms of the outer product as follows: $\Xi = \sum_{n=1}^{\infty}\sum_{m=1}^{\infty} \Xi_{mn}|a^{(m)}\rangle\langle a^{(n)}|$, the Pauli X-operator can be represented as:

$$X = \begin{pmatrix} 0 & X_{01} \\ X_{10} & 0 \end{pmatrix} = \begin{pmatrix} 0 & 1 \\ 1 & 0 \end{pmatrix} = X_{01}|0\rangle\langle 1| + X_{10}|1\rangle\langle 0| = |0\rangle\langle 1| + |1\rangle\langle 0|. \tag{2.35}$$

We have shown that if Ξ has eigenvalues $\xi^{(k)}$ and eigenkets $|\xi^{(k)}\rangle$ determined from $\Xi|\xi^{(k)}\rangle = \xi^{(k)}|\xi^{(k)}\rangle$, the spectral decomposition Ξ is given by $\Xi = \sum_k \xi^{(k)}|\xi^{(k)}\rangle\langle\xi^{(k)}|$, so that the spectral decomposition of the Pauli Z-operator will be:

$$Z = |0\rangle\langle 0| - |1\rangle\langle 1|. \tag{2.36}$$

The projection operators corresponding to measurements of 1 and -1 can be defined by:

$$P_0 = |0\rangle\langle 0|, \quad P_1 = |1\rangle\langle 1|. \tag{2.37}$$

The projector P_0 (P_1) performs projection of the arbitrary state to the $|0\rangle$ ($|1\rangle$) state:

$$\begin{aligned} &|\psi\rangle = \alpha|0\rangle + \beta|1\rangle \\ &P_0|\psi\rangle = (|0\rangle\langle 0|)|\psi\rangle = \alpha|0\rangle, \quad P_1|\psi\rangle = (|1\rangle\langle 1|)|\psi\rangle = \beta|1\rangle. \end{aligned} \tag{2.38}$$

The sum of projection operators clearly results in the identity operator as follows:

$$P_0 + P_1 = |0\rangle\langle 0| + |1\rangle\langle 1| = \begin{pmatrix} 1 \\ 0 \end{pmatrix}\begin{pmatrix} 1 & 0 \end{pmatrix} + \begin{pmatrix} 0 \\ 1 \end{pmatrix}\begin{pmatrix} 0 & 1 \end{pmatrix}$$
$$= \begin{pmatrix} 1 & 0 \\ 0 & 0 \end{pmatrix} + \begin{pmatrix} 0 & 0 \\ 0 & 1 \end{pmatrix} = \begin{pmatrix} 1 & 0 \\ 0 & 1 \end{pmatrix} = I. \tag{2.39}$$

We have shown earlier that we can express the state ket $|\psi\rangle$ in terms of operator Ξ as follows: $|\psi\rangle = \sum_k \alpha_k \left|\xi^{(k)}\right\rangle$. The probability of obtaining the measurement result $\xi^{(k)}$ can be expressed in terms of projection operator $P_k = |\xi^{(k)}\rangle\langle\xi^{(k)}|$ by:

$$\begin{aligned} |\alpha_k|^2 &= \left|\left\langle \xi^{(k)} | \psi \right\rangle\right|^2 = \left\langle \xi^{(k)} | \psi \right\rangle \left\langle \xi^{(k)} | \psi \right\rangle^* = \left\langle \xi^{(k)} | \psi \right\rangle \left\langle \psi \middle| \xi^{(k)} \right\rangle \\ &= \left\langle \psi \middle| \xi^{(k)} \right\rangle \left\langle \xi^{(k)} | \psi \right\rangle = \langle \psi | P_k | \psi \rangle. \end{aligned} \tag{2.40}$$

The final state of the system of the measurement will be $P_k|\psi\rangle \Big/ \sqrt{\langle\psi|P_k|\psi\rangle}$. For a 2D system, if the measurement result $+1$ is obtained, the result after the measurement will be:

$$|\psi'\rangle = \frac{1}{\sqrt{\langle\psi|P_0|\psi\rangle}} P_0|\psi\rangle \overset{|\psi\rangle=\alpha|0\rangle+\beta|1\rangle}{=} \frac{\alpha}{|\alpha|}|0\rangle, \tag{2.41}$$

and if the measurement result -1 is obtained, the corresponding state will be:

$$|\psi'\rangle = \frac{1}{\sqrt{\langle\psi|P_1|\psi\rangle}} P_1|\psi\rangle = \frac{\beta}{|\beta|}|1\rangle. \tag{2.42}$$

2.4.2 Hadamard Gate

Matrix representation of the Hadamard operator (gate) is given by:

$$H = \frac{1}{\sqrt{2}}\begin{bmatrix} 1 & 1 \\ 1 & -1 \end{bmatrix}. \tag{2.43}$$

It can easily be shown that the Hadamard gate is Hermitian and unitary as follows:

$$H^\dagger = \frac{1}{\sqrt{2}}\begin{bmatrix} 1 & 1 \\ 1 & -1 \end{bmatrix} = H$$

$$H^\dagger H = \frac{1}{\sqrt{2}}\begin{bmatrix} 1 & 1 \\ 1 & -1 \end{bmatrix} \frac{1}{\sqrt{2}}\begin{bmatrix} 1 & 1 \\ 1 & -1 \end{bmatrix} = \begin{bmatrix} 1 & 0 \\ 0 & 1 \end{bmatrix} = I.$$

The eigenvalues for the Hadamard gate can be obtained from $\det(H - \lambda I) = 0$ to be $\lambda_{1,2} = \pm 1$. By substituting the eigenvalues into an eigen-value equation, namely $H|\Psi_{1,2}\rangle = \pm|\Psi_{1,2}\rangle$, the corresponding eigenkets are obtained as follows:

$$|\Psi_1\rangle = \begin{bmatrix} \dfrac{1}{\sqrt{4-2\sqrt{2}}} \\ \dfrac{1}{\sqrt{2\sqrt{2}}} \end{bmatrix} \quad |\Psi_2\rangle = \begin{bmatrix} \dfrac{1}{\sqrt{4+2\sqrt{2}}} \\ -\dfrac{1}{\sqrt{2\sqrt{2}}} \end{bmatrix}. \tag{2.44}$$

2.5 DENSITY OPERATORS

Let the large number of quantum systems of the same kind be prepared, each in one of a set of orthonormal states $|\phi_n\rangle$, and let the fraction of the system being in state $|\phi_n\rangle$ be denoted by probability P_n ($n = 1,2,\ldots$):

$$\langle\phi_m|\phi_n\rangle = \delta_{mn}, \quad \sum_n P_n = 1. \tag{2.45}$$

Therefore this ensemble of quantum states represents a classical *statistical mixture* of kets. The probability of obtaining ξ_k from the measurement of Ξ will be:

$$\Pr(\xi_k) = \sum_{n=1}^{\infty} P_n |\langle\xi_k|\phi_n\rangle|^2 = \sum_{n=1}^{\infty} P_n \langle\xi_k|\phi_n\rangle\langle\phi_n|\xi_k\rangle = \langle\xi_k|\rho|\xi_k\rangle, \tag{2.46}$$

where the operator ρ is known as a *density operator* and is defined by:

$$\rho = \sum_{n=1}^{\infty} P_n |\phi_n\rangle\langle\phi_n|. \tag{2.47}$$

The expected value of operator Ξ is given by:

$$\langle\Xi\rangle = \sum_{k=1}^{\infty} \xi_k \Pr(\xi_k) = \sum_{k=1}^{\infty} \xi_k \langle\xi_k|\rho|\xi_k\rangle = \sum_{k=1}^{\infty} \langle\xi_k|\rho\Xi|\xi_k\rangle = \mathrm{Tr}(\rho\Xi). \tag{2.48}$$

The density operator *properties* can be summarized as follows:

1. The density operator is Hermitian ($\rho^+ = \rho$), with the set of orthonormal eigenkets $|\phi_n\rangle$ corresponding to the nonnegative eigenvalues P_n and $\mathrm{Tr}(\rho) = 1$.
2. Any Hermitian operator with nonnegative eigenvalues and trace 1 may be considered as a density operator.
3. The density operator is positive definite: $\langle\psi|\rho|\psi\rangle \geq 0$ for all $|\psi\rangle$.
4. The density operator has the property $\mathrm{Tr}(\rho^2) \leq 1$, with equality iff one of the prior probabilities is 1, and all the rest 0: $\rho = |\phi_n\rangle\langle\phi_n|$, and the density operator is then a projection operator.
5. The eigenvalues of a density operator satisfy: $0 \leq \lambda_i \leq 1$.

To prove these properties is quite straightforward, and the proof is left as a homework problem. When ρ is the projection operator we say that it represents the system in a *pure state*; otherwise with $\mathrm{Tr}(\rho^2) < 1$ it represents a *mixed state*. A mixed state in which all eigenkets occur with the same probability is known as a *completely mixed state* and can be represented by:

$$\rho = \sum_{k=1}^{\infty} \frac{1}{n} |\phi_n\rangle\langle\phi_n| = \frac{1}{n} I \Rightarrow \text{Tr}(\rho^2) = \frac{1}{n} \Rightarrow \frac{1}{n} \leq \text{Tr}(\rho^2) \leq 1. \tag{2.49}$$

If the density matrix has off-diagonal elements different from zero we say that it exhibits *quantum interference*, which means that state terms can interfere with each other. Let us observe the following pure state:

$$\begin{aligned} |\psi\rangle = \sum_{i=1}^{n} \alpha_i |\xi_i\rangle \Rightarrow \rho = |\psi\rangle\langle\psi| &= \sum_{i=1}^{n} |\alpha_i|^2 |\xi_i\rangle\langle\xi_i| + \sum_{i=1}^{n} \sum_{j=1, j\neq i}^{n} \alpha_i \alpha_j^* |\xi_i\rangle\langle\xi_j| \\ &= \sum_{i=1}^{n} \langle\xi_i|\rho|\xi_i\rangle |\xi_i\rangle\langle\xi_i| + \sum_{i=1}^{n} \sum_{j=1, j\neq i}^{n} \langle\xi_i|\rho|\xi_j\rangle |\xi_i\rangle\langle\xi_j|. \end{aligned} \tag{2.50}$$

The first term in Eq. (2.50) is related to the probability of the system being in state $|\xi_i\rangle$, and the second term is related to quantum interference. It appears that the off-diagonal elements of a mixed state will be zero, while those of the pure state will be nonzero. Notice that the existence of off-diagonal elements is base dependent, therefore to check for purity it is a good idea to compute $\text{Tr}(\rho^2)$ instead.

In quantum information theory, the density matrix can be used to determine the amount of information conveyed by the quantum state, i.e., to compute the *von Neumann entropy*:

$$S = -\text{Tr}(\rho \log \rho) = -\sum_i \lambda_i \log_2 \lambda_i, \tag{2.51}$$

where λ_i are the eigenvalues of the density matrix. The corresponding Shannon entropy can be calculated by:

$$H = -\sum_i p_i \log_2 p_i. \tag{2.52}$$

Let us further consider the observable Ξ with the set of eigenvalues $\{\xi_k; d\}$ where d is the *degeneracy* index. For simplicity of derivation without losing the generality, let us assume that $\rho = |\psi\rangle\langle\psi|$ (the pure state). The probability that the measurement of Ξ will lead to eigenvalue ξ_k is given by:

$$\begin{aligned} \Pr(\xi_k) &= \sum_d |\langle\xi_k; d|\psi\rangle|^2 = \sum_d \langle\xi_k; d|\psi\rangle\langle\psi|\xi_k; d\rangle \\ &= \sum_d \langle\xi_k; d| \left[\sum_n |\phi_n\rangle\langle\phi_n| \right] |\psi\rangle\langle\psi|\xi_k; d\rangle = \sum_n \langle\phi_n|\psi\rangle \sum_d \langle\psi|\xi_k; d\rangle\langle\xi_k; d|\phi_n\rangle \\ &= \sum_n \langle\phi_n|\psi\rangle\langle\psi| \left[\sum_d |\xi_k; d\rangle\langle\xi_k; d| \right] |\phi_n\rangle = \sum_n \langle\phi_n|\rho P_k|\phi_n\rangle, \end{aligned}$$

$$\rho = |\psi\rangle\langle\psi|, P_k = \sum_d |\xi_k; d\rangle\langle\xi_k; d|. \tag{2.53}$$

From the third line of the previous expression, it is clear that:

$$\Pr(\xi_k) = \mathrm{Tr}(\rho P_k). \tag{2.54}$$

For a *two-level system*, the density operator can be written as:

$$\rho = \frac{1}{2}(I + \vec{r} \cdot \vec{\sigma}), \quad \vec{\sigma} = [X \quad Y \quad Z], \tag{2.55}$$

where X, Y, Z are the Pauli operators, and $\vec{r} = [r_x \ r_y \ r_z]$ is the *Bloch vector*, whose components can be determined by:

$$r_x = \mathrm{Tr}(\rho X), \quad r_y = \mathrm{Tr}(\rho Y), \quad r_z = \mathrm{Tr}(\rho Z). \tag{2.56}$$

The magnitude of the Bloch vector is $\|\vec{r}\| \leq 1$, with equality corresponding to the pure state.

Suppose now that S is a *bipartite composite system* with component subsystems A and B. For example, subsystem A can represent the quantum register Q and subsystem B the environment E. The composite system can be represented by $AB = A \otimes B$, where $\otimes$ stands for the tensor product. If the dimensionality of Hilbert space H_A is m and the dimensionality Hilbert space H_B is n, then the dimensionality of Hilbert space H_{AB} will be mn. Let $|\alpha\rangle \in A$ and $|\beta\rangle \in B$, then $|\alpha\rangle|\beta\rangle = |\alpha\rangle \otimes |\beta\rangle \in AB$. If the operator A acts on kets from H_A and the operator B acts on kets from H_B, then the action of AB on $|\alpha\rangle|\beta\rangle$ can be described as follows:

$$(AB)|\alpha\rangle|\beta\rangle = (A|\alpha\rangle)(B|\beta\rangle). \tag{2.57}$$

The norm of state $|\psi\rangle = |\alpha\rangle|\beta\rangle \in AB$ is determined by:

$$\langle\psi|\psi\rangle = \langle\alpha|\alpha\rangle\langle\beta|\beta\rangle. \tag{2.58}$$

Let $\{|\alpha_i\rangle\}$ ($\{|\beta_i\rangle\}$) be a basis for the Hilbert space H_A (H_B) and let E be an ensemble of physical systems S described by the density operator ρ. The *reduced density operator* ρ_A for subsystem A is defined to be the partial trace of ρ over B:

$$\rho_A = \mathrm{Tr}_B(\rho) = \sum_j \langle\beta_j|\rho|\beta_j\rangle. \tag{2.59}$$

Similarly, the *reduced density operator* ρ_B for subsystem B is defined to be the partial trace of ρ over A:

$$\rho_B = \mathrm{Tr}_A(\rho) = \sum_i \langle\alpha_j|\rho|\alpha_j\rangle. \tag{2.60}$$

2.6 QUANTUM MEASUREMENTS

This section is devoted to quantum measurements, including projective measurements, generalized measurements, and the positive operator-valued measure (POVM).

2.6.1 Projective Measurements

Each measurable physical quantity—observable (such as position, momentum, or angular momentum) —is associated with a Hermitian operator that has a complete set of eigenkets. According to P. A. Dirac "A measurement always causes the system to jump into an eigenstate of the dynamical variable that is being measured [3]." Dirac's statement can formulated as the following *postulate*: an exact measurement of an observable with operator A always yields as a result one of the eigenvalues $a^{(n)}$ of A. Thus the measurement changes the state when the measurement system is "thrown into" one of its eigenstates, which can be represented by $|\alpha\rangle \xrightarrow{A \text{ measurement}} |a^{(j)}\rangle$. If before measurement the system was in state $|\alpha\rangle$, the probability that the result of a measurement will be the eigenvalue $a^{(i)}$ is given by:

$$\Pr\left(a^{(i)}\right) = \left|\left\langle a^{(i)} \middle| \alpha\right\rangle\right|^2, \tag{2.61}$$

and this rule is known as the *Born rule*. Since at least one of the eigenvalues must occur as the result of the measurements, these probabilities must sum to 1:

$$\sum_i \Pr\left(a^{(i)}\right) = \sum_i \left|\left\langle a^{(i)} \middle| \alpha\right\rangle\right|^2 = 1. \tag{2.62}$$

The expected value of the outcome of the measurement of A is given by:

$$\langle A\rangle = \sum_i a^{(i)} \Pr\left(a^{(i)}\right) = \sum_i a^{(i)} \left|\left\langle a^{(i)} \middle| \alpha\right\rangle\right|^2 = \sum_i a^{(i)} \left\langle \alpha \middle| a^{(i)}\right\rangle \left\langle a^{(i)} \middle| \alpha\right\rangle. \tag{2.63}$$

By applying the eigenvalue equation $a^{(i)}|a^{(i)}\rangle = A|a^{(i)}\rangle$, Eq. (2.63) becomes:

$$\langle A\rangle = \sum_i \langle \alpha|A|a^{(i)}\rangle \left\langle a^{(i)} \middle| \alpha\right\rangle. \tag{2.64}$$

By using further the completeness relation $\sum_i |a^{(i)}\rangle\langle a^{(i)}| = I$ we obtain the expected value of the measurement of A to be simply:

$$\langle A\rangle = \langle \alpha|A|\alpha\rangle. \tag{2.65}$$

In various situations, like initial state preparations for quantum information processing applications, we need to select one particular outcome of the measurement. This procedure is known as *selective measurement* (or filtration) and can be conducted as shown in Fig. 2.1.

The result of selective measurement can be interpreted as applying the *projection operator* $P_{a'}$ to $|\alpha\rangle$ to obtain:

$$P_{a'}|\alpha\rangle = |a'\rangle\langle a'|\alpha\rangle. \tag{2.66}$$

The probability that the outcome of the measurement of observable Ξ with eigenvalues $\xi^{(n)}$ lies between (a,b) is given by:

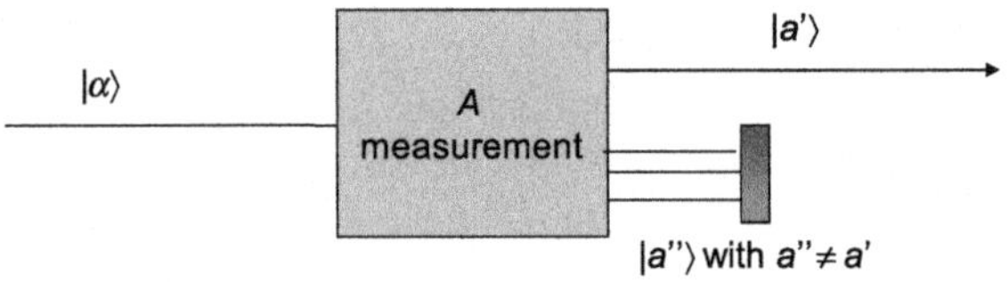

FIGURE 2.1

Concept of a selective measurement (filtration).

$$\Pr(\xi \in R(a,b)) = \sum_{\xi^{(n)} \in R(a,b)} \left|\left\langle \xi^{(n)} \middle| \alpha \right\rangle\right|^2 = \sum_{\xi^{(n)} \in R(a,b)} \left\langle \alpha \middle| \xi^{(n)} \right\rangle \left\langle \xi^{(n)} \middle| \alpha \right\rangle \\ = \langle \alpha | P_{ab} | \alpha \rangle = \langle P_{ab} \rangle, \tag{2.67}$$

where with P_{ab} we denote the following projection operator:

$$P_{ab} = \sum_{\xi^{(n)} \in R(a,b)} \left|\xi^{(n)}\right\rangle \left\langle \xi^{(n)}\right|. \tag{2.68}$$

It can simply be shown that the projection operator P_{ab} satisfies:

$$P_{ab}^2 = P_{ab} \Leftrightarrow P_{ab}(P_{ab} - I) = \mathbf{0}. \tag{2.69}$$

Therefore the eigenvalues of projection operator P_{ab} are either 0 (corresponding to the "false proposition") or 1 (corresponding to the "true proposition"), and it is of high importance in *quantum detection theory* [5].

In terms of projection operators, the state of the system after the measurement is given by:

$$|\alpha\rangle \xrightarrow{A \text{ measurement}} \frac{1}{\sqrt{\langle \alpha | P_j | \alpha \rangle}} P_j |\alpha\rangle, \quad P_j = \left|a^{(j)}\right\rangle \left\langle a^{(j)}\right|. \tag{2.70}$$

In case operator A has the same eigenvalue a_i for the following eigenkets $\left\{\left|a_i^{(j)}\right\rangle\right\}_{j=1}^{d_i}$, with the corresponding characteristic equation:

$$A\left|a_i^{(j)}\right\rangle = a_i\left|a_i^{(j)}\right\rangle; \; j = 1, \ldots, d_i \tag{2.71}$$

we say that eigenvalue a_i is *degenerate* of order d_i. The corresponding probability of obtaining the measurement result a_i can be found by:

$$\Pr(a_i) = \sum_{j=1}^{d_i} \left|\left\langle a_i^{(j)} \middle| \alpha \right\rangle\right|^2. \tag{2.72}$$

2.6.2 Generalized Measurements

Projective measurements can be generalized as follows. Let the set of measurement operators be given by $\{M_m\}$, where index m stands for possible measurement result, satisfying the property:

$$\sum_m M_m^\dagger M_m = I. \tag{2.73}$$

The probability of finding the measurement result m of the observed state $|\psi\rangle$ is given by:

$$p_m = \Pr(m) = \langle\psi|M_m^\dagger M_m|\psi\rangle. \tag{2.74}$$

After measurement, the system will be left in the following state:

$$|\psi_f\rangle = \frac{M_m|\psi\rangle}{\sqrt{\langle\psi|M_m^\dagger M_m|\psi\rangle}}. \tag{2.75}$$

When the system is described in terms of density operator ρ, the probability of getting the measurement result m is:

$$p_m = \Pr(m) = \mathrm{Tr}\left(M_m^\dagger M_m \rho\right), \tag{2.76}$$

and the system will be left in the following density state:

$$\rho_f = \frac{M_m \rho M_m^\dagger}{\mathrm{Tr}\left(M_m \rho M_m^\dagger\right)}. \tag{2.77}$$

For projective measurements, clearly, $M_m = P_m = |a^{(m)}\rangle\langle a^{(m)}|$, and from the foregoing property we obtain:

$$\sum_m M_m^\dagger M_m = \sum_m |a^{(m)}\rangle \underbrace{\langle a^{(m)}|a^{(m)}\rangle}_{1} \langle a^{(m)}| = \sum_m |a^{(m)}\rangle\langle a^{(m)}| = \sum_m P_m = I, \tag{2.78}$$

which is the completeness relationship. Clearly, projective measurement is a special case of generalized measurement. The probability of obtaining the mth result of measurement will then be:

$$\Pr(m) = \mathrm{Tr}\left(P_m^\dagger P_m \rho\right) = \mathrm{Tr}(P_m \rho) = \mathrm{Tr}\left(|a^{(m)}\rangle\langle a^{(m)}|\rho\right) = \langle a^{(m)}|\rho|a^{(m)}\rangle, \tag{2.79}$$

and the system will be left at the following density operator:

$$\rho_f = \frac{P_m \rho P_m^\dagger}{\mathrm{Tr}\left(P_m \rho P_m^\dagger\right)} \overset{P_m = |a^{(m)}\rangle\langle a^{(m)}|}{=} \frac{|a^{(m)}\rangle\langle a^{(m)}|\rho|a^{(m)}\rangle\langle a^{(m)}|}{\mathrm{Tr}\left(|a^{(m)}\rangle\langle a^{(m)}|\rho|a^{(m)}\rangle\langle a^{(m)}|\right)}$$
$$= \frac{|a^{(m)}\rangle\langle a^{(m)}|\rho|a^{(m)}\rangle\langle a^{(m)}|}{\langle a^{(m)}|\rho|a^{(m)}\rangle}. \tag{2.80}$$

2.6.3 Positive Operator-Valued Measure

Another important type of measurement is known as a POVM. A POVM consists of the set of operators $\{E_m\}$, where each operator E_m is positive semidefinite, i.e., $\langle\psi|E_m|\psi\rangle \geq 0$, satisfying the relationship:

$$\sum_m E_m = I. \tag{2.81}$$

The probability of getting the mth measurement result will be:

$$p_m = \Pr(m) = \langle\psi|E_m|\psi\rangle. \tag{2.82}$$

The POVM can be constructed from generalized measurement operators $\{M_m\}$ by setting $E_m = M_m^\dagger M_m$. The probability of obtaining the mth result of measurements is given by $\mathrm{Tr}(E_m\rho)$. The POVM concept is in particular suitable to situations when the measurements are not repeatable. For instance, by performing the measurement on a photon, it can be destroyed so that the repeated measurements are not possible.

The POVM can be used to obtain information about the state without forcing the collapse of a wavefunction, a concept known as *weak measurement*. Let us observe the single qubit $|\psi\rangle = a|0\rangle + b|1\rangle$, and construct the following POVM $\{E_0,E_1\}$ with corresponding operators being:

$$E_0 = |0\rangle\langle 0| + (1-\zeta)|1\rangle\langle 1|,\ \ E_1 = |1\rangle\langle 1|,\ \ 0 < \zeta << 1. \tag{2.83}$$

It can be easily be shown that $E_0 + E_1 = I$, and that eigenvalues for operators E_0 and E_1 are $\{1,1 - \zeta\}$ and $\{0,\zeta\}$, respectively. The probability of obtaining the measurement result corresponding to E_0 is:

$$p_0 = \langle\psi|E_0|\psi\rangle = |a|^2 + (1-\zeta)|b|^2, \tag{2.84}$$

and the state after the measurement will be:

$$\frac{E_0|\psi\rangle}{\sqrt{\langle\psi|E_0|\psi\rangle}} = \frac{a}{\sqrt{p_0}}|0\rangle + \frac{b(1-\zeta)}{\sqrt{p_0}}, \tag{2.85}$$

which is still the superposition state. Clearly, although the initial state is perturbed, the E_0-operator has not caused the collapse of the wavefunction. On the other hand, the probability of getting the measurement result corresponding to E_1 is:

$$p_1 = \langle\psi|E_1|\psi\rangle = \zeta|b|^2, \tag{2.86}$$

which is very low given that $\zeta << 1$. However, the postmeasurement state is $|1\rangle$, and this operator caused the wavefunction to collapse.

2.7 UNCERTAINTY PRINCIPLE

Before we derive the uncertainty principle we briefly describe basic commutator properties, which will be needed later in the section.

2.7.1 Commutators

Let A and B be two operators, which in general do not commute, i.e., $AB \neq BA$. The quantity $[A,B] = AB - BA$ is called the *commutator* of A and B, while the quantity $\{A, B\} = AB + BA$ is called the *anticommutator*. Two observables A and B are said to be *compatible* when their corresponding operators commute: $[A,B] = 0$. Two observables A and B are said to be *incompatible* when $[A,B] \neq 0$. If in the set of operators $\{A,B,C,\ldots\}$ all operators commute in pairs, namely $[A,B] = [A,C] = [B,C] = \ldots = 0$, we say the set represents a *complete set of commuting observables.*

Let A, B, and C be the operators, then the following *properties* of the commutator are valid:

$$\begin{aligned}
&1.\ [A,B] = -[B,A]\\
&2.\ [A+B,C] = [A,C] + [B,C]\\
&3.\ [A,BC] = B[A,C] + [A,B]C\\
&4.\ [X,P] = j\hbar, [X,X] = [P,P] = 0,
\end{aligned} \tag{2.87}$$

where X and P are the position and the momentum operators. Properties 1–3 are quite straightforward to prove from the definition of the commutator, and as such will be left for homework. Here we provide the proof of property 4. At this point it is convenient to establish the *connection* between *wave quantum mechanics* and *matrix quantum mechanics*. In wave mechanics, the information about the state of a particle is described by the corresponding *wavefunction* $\psi(x,t) = \langle x|\psi\rangle$. The wavefunction gives the information about the location of the particle, namely the magnitude squared of the wavefunctions $|\psi(x,t)|^2$ is related to the *probability density*. The probability of finding the particle within the interval x and $x + dx$ is given by:

$$dP(x,t) = |\psi(x,t)|^2 dx. \tag{2.88}$$

In wave quantum mechanics, the position X and the momentum P operators are defined by:

$$X\psi(x,t) = x\psi(x,t) \quad P\psi(x,t) = -j\hbar\frac{\partial}{\partial x}\psi(x,t). \tag{2.89}$$

We apply the commutator to the test wavefunction $\psi(x,t)$ and obtain:

$$\begin{aligned}
[X,P]\psi(x,t) &= (XP - PX)\psi(x,t)\\
&= XP\psi(x,t) - PX\psi(x,t)\\
&= -j\hbar x\frac{\partial}{\partial x}\psi(x,t) + j\hbar\frac{\partial}{\partial x}(X\psi(x,t))\\
&= -j\hbar x\frac{\partial}{\partial x}\psi(x,t) + j\hbar\frac{\partial}{\partial x}(x\psi(x,t))\\
&= -j\hbar x\frac{\partial}{\partial x}\psi(x,t) + j\hbar\left(\psi(x,t) + x\frac{\partial}{\partial x}\psi(x,t)\right) = j\hbar\psi(x,t),
\end{aligned} \tag{2.90}$$

proving therefore that $[X,P] = j\hbar$.

2.7.2 Uncertainty Principle Derivation

If two observables, say A and B, are to be measured simultaneously and exactly on the same system, the system after the measurement must be left in the state $|a^{(n)}; b^{(n)}\rangle$, that is, an eigenstate of both observables:

$$\begin{aligned} A\left|a^{(n)}; b^{(n)}\right\rangle &= a^{(n)}\left|a^{(n)}; b^{(n)}\right\rangle \\ B\left|a^{(n)}; b^{(n)}\right\rangle &= b^{(n)}\left|a^{(n)}; b^{(n)}\right\rangle. \end{aligned} \tag{2.91}$$

This will be true only if $AB = BA$ or equivalently the commutator $[A,B] = AB - BA = 0$, that is, when two operators *commute* as follows:

$$\begin{aligned} AB\left|a^{(n)}; b^{(n)}\right\rangle &= A\left(B\left|a^{(n)}; b^{(n)}\right\rangle\right) = Ab^{(n)}\left|a^{(n)}; b^{(n)}\right\rangle = b^{(n)} \cdot A\left|a^{(n)}; b^{(n)}\right\rangle \\ &= a^{(n)}b^{(n)}\left|a^{(n)}; b^{(n)}\right\rangle \\ BA\left|a^{(n)}; b^{(n)}\right\rangle &= a^{(n)}b^{(n)}\left|a^{(n)}; b^{(n)}\right\rangle \Rightarrow AB = BA. \end{aligned} \tag{2.92}$$

When two operators do not commute, they cannot be simultaneously measured with complete precision. Given an observable A we define the operator $\Delta A = A - \langle A\rangle$, and the corresponding expectation value of $(\Delta A)^2$, which is known as the *variance* (*dispersion*) of A:

$$\left\langle(\Delta A)^2\right\rangle = \left\langle A^2 - 2A\langle A\rangle + \langle A\rangle^2\right\rangle = \langle A^2\rangle - \langle A\rangle^2. \tag{2.93}$$

Then for any state, the following inequality is valid:

$$\left\langle(\Delta A)^2\right\rangle\left\langle(\Delta B)^2\right\rangle \geq \frac{1}{4}|\langle[\Delta A, \Delta B]\rangle|^2, \tag{2.94}$$

which is known as the *Heisenberg uncertainty principle*.

Example. The commutation relation for coordinate X and momentum P observables is $[X, P] = j\hbar$, as shown earlier. By substituting this commutation relation into Eq. (2.94) we obtain:

$$\langle X^2\rangle\langle P^2\rangle \geq \frac{\hbar^2}{4}.$$

If we observe a large ensemble of N independent systems, all of them will be in the state $|\psi\rangle$. On some systems, X is measured, and on some systems, P is measured. The uncertainty principle asserts that for none state the product of dispersions (variances) cannot be less than $\hbar^2/4$.

For the uncertainty principle derivation, the following two lemmas are important.

Lemma 1 (L1). The expectation operator of a Hermitian operator B is purely real.

This lemma can easily be proved by using the definition of Hermitian operator $B^\dagger = B$ and the definition of the expectation of operator B: $\langle B\rangle = \langle\alpha|B|\alpha\rangle$. By using

Eq. (2.10) we obtain $\langle\alpha|B|\alpha\rangle^* = \langle\alpha|B^\dagger|\alpha\rangle^* = \langle\alpha|B|\alpha\rangle$, which proves that the expectation of a Hermitian operator is indeed real.

Lemma 2 (L2). The expectation operator of an anti-Hermitian operator, defined by $C^\dagger = -C$, is purely imaginary.

This lemma can also be proved by using the definition of anti-Hermitian operator $C^\dagger = -C$ and the definition of the expectation of operator C: $\langle C\rangle = \langle\alpha|C|\alpha\rangle$. By using again Eq. (2.10) we obtain $\langle\alpha|C|\alpha\rangle^* = \langle\alpha| - C^\dagger|\alpha\rangle^* = -\langle\alpha|C^\dagger|\alpha\rangle^* = -\langle\alpha|C|\alpha\rangle$, which proves that the expectation of an anti-Hermitian operator is indeed imaginary.

By assuming that A and B are Hermitian operators, the uncertainty principle can be now derived by starting from the *Cauchy–Schwartz inequality* (see Eq. 2.6) as follows. Let $|\alpha\rangle = \Delta A|\psi\rangle$ and $|\beta\rangle = \Delta B|\psi\rangle$. From the Cauchy–Schwartz inequality we know that:

$$\langle\alpha|\alpha\rangle\langle\beta|\beta\rangle = \left\langle(\Delta A)^2\right\rangle\left\langle(\Delta B)^2\right\rangle \geq |\langle\Delta A\Delta B\rangle|^2. \quad (2.95)$$

It is easy to show that $\langle\Delta A\Delta B\rangle = (\langle[\Delta A, \Delta B]\rangle + \langle\{\Delta A, \Delta B\}\rangle)/2$. Since $[A, B]^\dagger = (AB - BA)^\dagger = BA - AB = -[A, B]$, from (L2) it is clear that expectation of $[A,B]$ is purely imaginary. In similar fashion we show that $\{A, B\}^\dagger = (AB + BA)^\dagger = BA + AB = \{A, B\}$, indicating that the expectation of $\{A, B\}$ is purely real (based on (L1)). Therefore the first term in expectation of $\Delta A\Delta B$ is purely imaginary and the second term is purely real as follows:

$$\langle\Delta A\Delta B\rangle = \underbrace{\frac{1}{2}\langle[\Delta A, \Delta B]\rangle}_{\text{purely imaginery}} + \underbrace{\frac{1}{2}\langle\{\Delta A, \Delta B\}\rangle}_{\text{purely real}}. \quad (2.96)$$

From complex numbers theory we know that the magnitude squared of a complex number $z = x + jy$ (j is the imaginary unit) is given by $|z|^2 = x^2 + y^2$, so that the magnitude squared of $|\langle\Delta A\Delta B\rangle|^2$ is given by:

$$|\langle\Delta A\Delta B\rangle|^2 = \frac{1}{4}|\langle[\Delta A, \Delta B]\rangle|^2 + \frac{1}{4}|\langle\{\Delta A, \Delta B\}\rangle|^2. \quad (2.97)$$

If we omit the anticommutator term in Eq. (2.97) we obtain the uncertainty relation given by Eq. (2.94).

2.8 CHANGE OF BASIS

Suppose that we have two incompatible observables A and B. The ket space can be viewed as being spanned by $|a^{(i)}\rangle$ or $|b^{(i)}\rangle$. The two different sets of base kets span the same ket. We are interested in finding out how these two descriptions can be related. The following theorem is going to provide the answer.

Theorem. Given two sets of base kets, $\{|a^{(i)}\rangle\}$ and $\{|b^{(i)}\rangle\}$, both satisfying orthonormality and completeness, there exists a unitary operator U (the unitary operator is defined by $U^\dagger U = I$) such that:

$$\left|b^{(i)}\right\rangle = U\left|a^{(i)}\right\rangle. \tag{2.98}$$

This theorem can be proved by construction. Let the unitary operator be chosen as follows:

$$U = \sum_k \left|b^{(k)}\right\rangle\left\langle a^{(k)}\right|. \tag{2.99}$$

Since $U|a^{(l)}\rangle = |b^{(l)}\rangle$ we can show that such chosen operator U is unitary as follows:

$$U^\dagger U = \sum_k\sum_l \left|a^{(l)}\right\rangle\left\langle b^{(l)}\middle|b^{(k)}\right\rangle\left\langle a^{(k)}\right| \overset{\langle b^{(l)}|b^{(k)}\rangle=\delta_{lk}}{=} \sum_k \left|a^{(k)}\right\rangle\left\langle a^{(k)}\right| = I. \tag{2.100}$$

The matrix representation of the U-operator in the initial basis can be obtained, based on Eq. (2.99), as follows:

$$\left\langle a^{(k)}\middle|U\middle|a^{(l)}\right\rangle = \left\langle a^{(k)}\middle|b^{(l)}\right\rangle. \tag{2.101}$$

The ket $|\alpha\rangle$ in the old basis can be represented by:

$$|\alpha\rangle = \sum_i \left|a^{(i)}\right\rangle\left\langle a^{(i)}\middle|\alpha\right\rangle. \tag{2.102}$$

By multiplying Eq. (2.102) by $\langle b^{(k)}|$ from the left we obtain:

$$\left\langle b^{(k)}\middle|\alpha\right\rangle = \sum_i \left\langle b^{(k)}\middle|a^{(i)}\right\rangle\left\langle a^{(i)}\middle|\alpha\right\rangle \overset{\langle b^{(k)}|=\langle a^{(k)}|U^\dagger}{=} \sum_i \left\langle a^{(k)}\middle|U^\dagger\middle|a^{(i)}\right\rangle\left\langle a^{(i)}\middle|\alpha\right\rangle. \tag{2.103}$$

Therefore the column- vector for $|\alpha\rangle$ in the new basis can be obtained by applying U^+ to the column vector in the old basis:

$$(\text{New}) = \left(U^\dagger\right)(\text{Old}).$$

The relationship between the old matrix elements and the new matrix elements can be established by:

$$\begin{aligned}\left\langle b^{(k)}\middle|X\middle|b^{(l)}\right\rangle &= \sum_m\sum_n \left\langle b^{(k)}\middle|a^{(m)}\right\rangle\left\langle a^{(m)}\middle|X\middle|a^{(n)}\right\rangle\left\langle a^{(n)}\middle|b^{(l)}\right\rangle \\ &\underset{\langle b^{(k)}|=\langle a^{(k)}|U^\dagger}{\overset{|b^{(k)}\rangle=U|a^{(k)}\rangle}{=}} \sum_m\sum_n \left\langle a^{(k)}\middle|U^\dagger\middle|a^{(m)}\right\rangle\left\langle a^{(m)}\middle|X\middle|a^{(n)}\right\rangle\left\langle a^{(n)}\middle|U\middle|a^{(l)}\right\rangle.\end{aligned} \tag{2.104}$$

Therefore the matrix representation X' in new bases can be obtained from the matrix representation in old basis X as follows:

$$X' = U^{\dagger} X U, \tag{2.105}$$

where the transformation $U^{\dagger}XU$ is known as the *similarity transformation*.

The trace of an operator X, Tr(X), is independent of the basis as follows:

$$\begin{aligned}\mathrm{Tr}(X) &= \sum_i \left\langle a^{(i)} \middle| X \middle| a^{(i)} \right\rangle = \sum_i \sum_l \sum_k \left\langle a^{(i)} \middle| b^{(l)} \right\rangle \left\langle b^{(l)} \middle| X \middle| b^{(k)} \right\rangle \left\langle b^{(k)} \middle| a^{(i)} \right\rangle \\ &= \sum_i \sum_l \sum_k \left\langle b^{(k)} \middle| a^{(i)} \right\rangle \left\langle a^{(i)} \middle| b^{(l)} \right\rangle \left\langle b^{(l)} \middle| X \middle| b^{(k)} \right\rangle = \sum_k \left\langle b^{(k)} \middle| X \middle| b^{(k)} \right\rangle.\end{aligned} \tag{2.106}$$

The properties of the trace can be summarized as follows:

$$\begin{aligned}&\mathrm{Tr}(XY) = \mathrm{Tr}(YX) \qquad \mathrm{Tr}(XYZ) = \mathrm{Tr}(ZXY) = \mathrm{Tr}(YZX) \quad \text{(The trace is cyclic)} \\ &\mathrm{Tr}\left(U^{\dagger} X U\right) = \mathrm{Tr}(X) \\ &\mathrm{Tr}\left(\left|a^{(i)}\right\rangle\left\langle a^{(j)}\right|\right) = \delta_{ij} \qquad \mathrm{Tr}\left(\left|b^{(i)}\right\rangle\left\langle a^{(j)}\right|\right) = \left\langle a^{(j)} \middle| b^{(i)} \right\rangle\end{aligned} \tag{2.107}$$

2.8.1 Photon Polarization (Revisited)

In Fig. 2.2 we illustrate the transformation of bases using photon polarization. In Fig. 2.2A, the coordinate system is rotated for θ counterclockwise, while in Fig. 2.2B, the ket $|\psi\rangle$ is rotated for θ in a clockwise direction.

The ket $|\psi\rangle$ can be represented in the original basis by:

$$|\psi\rangle = |x\rangle\langle x|\psi\rangle + |y\rangle\langle y|\psi\rangle, \tag{2.108}$$

where $\langle x|\psi\rangle$ ($\langle y|\psi\rangle$) denotes the projection along the x-polarization (y-polarization). By multiplying Eq. (2.108) by $\langle x'|$ and $\langle y'|$ from the left side we obtain:

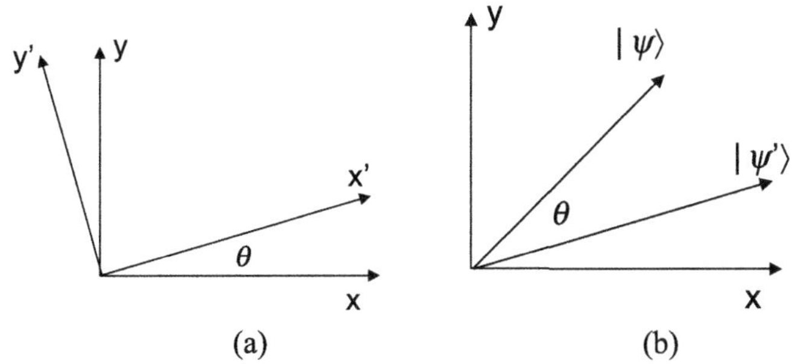

FIGURE 2.2

Transformation basis using photon polarization: (a) the rotation of a coordinate system and (b) the rotation of the ket.

$$\langle x'|\psi\rangle = \langle x'|x\rangle\langle x|\psi\rangle + \langle x'|y\rangle\langle y|\psi\rangle \quad \langle y'|\psi\rangle = \langle y'|x\rangle\langle x|\psi\rangle + \langle y'|y\rangle\langle y|\psi\rangle, \tag{2.109}$$

or by expressing Eq. (2.109) in matrix form we have that:

$$\begin{pmatrix} \langle x'|\psi\rangle \\ \langle y'|\psi\rangle \end{pmatrix} = \begin{pmatrix} \langle x'|x\rangle & \langle x'|y\rangle \\ \langle y'|x\rangle & \langle y'|y\rangle \end{pmatrix} \begin{pmatrix} \langle x|\psi\rangle \\ \langle y|\psi\rangle \end{pmatrix}, \tag{2.110}$$

which is consistent with Eq. (2.103). From Fig. 2.2A is clear that the old basis $\{|x\rangle,|y\rangle\}$ is related to the new basis $\{|x'\rangle,|y'\rangle\}$ by:

$$|x\rangle = \cos\theta|x'\rangle - \sin\theta|y'\rangle \quad |y\rangle = \sin\theta|x'\rangle + \cos\theta|y'\rangle. \tag{2.111}$$

The corresponding projections are:

$$\langle x'|x\rangle = \cos\theta \quad \langle x'|y\rangle = \sin\theta \quad \langle y'|x\rangle = -\sin\theta \quad \langle y'|y\rangle = \cos\theta. \tag{2.112}$$

From Fig. 2.2, it is clear that rotation of the basis counterclockwise for θ is equivalent to the rotation of the ket $|\psi\rangle$ for the same angle but clockwise, so that the ket in the new basis can be expressed as follows:

$$|\psi'\rangle = R(\theta)|\psi\rangle, \tag{2.113}$$

where $R(\theta)$ is the rotation operator, which can be represented in matrix form, based on Eqs. (2.110) and (2.113), by:

$$R(\theta) = \begin{pmatrix} \cos\theta & \sin\theta \\ -\sin\theta & \cos\theta \end{pmatrix}. \tag{2.114}$$

In other words, Eq. (2.110) can be written as:

$$\begin{pmatrix} \langle x'|\psi\rangle \\ \langle y'|\psi\rangle \end{pmatrix} = \begin{pmatrix} \cos\theta & \sin\theta \\ -\sin\theta & \cos\theta \end{pmatrix} \begin{pmatrix} \langle x|\psi\rangle \\ \langle y|\psi\rangle \end{pmatrix} \tag{2.115}$$

The rotation operator $R(\theta)$ can be represented in terms of the photon spin operator S, introduced in Section 2.3, as follows:

$$R(\theta) = \cos\theta I + j\sin\theta \begin{pmatrix} 0 & -j \\ j & 0 \end{pmatrix} = \cos\theta I + jS\sin\theta, \quad S = \begin{pmatrix} 0 & -j \\ j & 0 \end{pmatrix}. \tag{2.116}$$

We have shown earlier that the left-circular $|E_L\rangle$ and the right-circular $|E_R\rangle$ polarizations are eigenvectors of the spin operator, namely $S|E_L\rangle = -|E_L\rangle$ and $S|E_R\rangle = |E_R\rangle$. The action of the rotation operator on eigenkets of S is then:

$$\begin{aligned} R(\theta)|E_R\rangle &= (\cos\theta I + j\sin\theta S)|E_R\rangle = (\cos\theta I + j\sin\theta)|E_R\rangle = e^{j\theta}|E_R\rangle \\ R(\theta)|E_L\rangle &= (\cos\theta I + j\sin\theta S)|E_L\rangle = (\cos\theta I - j\sin\theta)|E_L\rangle = e^{-j\theta}|E_L\rangle. \end{aligned} \tag{2.117}$$

2.8.2 Diagonalization and Unitary Equivalent Observables

We are often concerned in quantum mechanics with the problem of determination the eigenvalues and eigenkets of an operator B, whose matrix elements in the old basis $|a^{(j)}\rangle$ are known. Let the eigenvalue equation of B be given as:

$$B\left|b^{(i)}\right\rangle = b^{(i)}\left|b^{(i)}\right\rangle. \tag{2.118}$$

By multiplying Eq. (2.118) with $\langle a^{(j)}|$ from the left we obtain:

$$\left\langle a^{(j)}\middle|B\middle|b^{(i)}\right\rangle = b^{(i)}\left\langle a^{(j)}\middle|b^{(i)}\right\rangle. \tag{2.119}$$

By applying further the completeness relation we obtain:

$$\sum_k \left\langle a^{(j)}\middle|B\middle|a^{(k)}\right\rangle\left\langle a^{(k)}\middle|b^{(i)}\right\rangle = b^{(i)}\left\langle a^{(j)}\middle|b^{(i)}\right\rangle, \tag{2.120}$$

which can be expressed in matrix form as:

$$\begin{pmatrix} \langle a^{(1)}|B|a^{(1)}\rangle & \langle a^{(1)}|B|a^{(2)}\rangle & \cdots \\ \langle a^{(2)}|B|a^{(1)}\rangle & \langle a^{(2)}|B|a^{(2)}\rangle & \cdots \\ \vdots & \vdots & \ddots \end{pmatrix}\begin{pmatrix} \langle a^{(1)}|b^{(i)}\rangle \\ \langle a^{(2)}|b^{(i)}\rangle \\ \vdots \end{pmatrix} = b^{(i)}\begin{pmatrix} \langle a^{(1)}|b^{(i)}\rangle \\ \langle a^{(2)}|b^{(i)}\rangle \\ \vdots \end{pmatrix}. \tag{2.121}$$

We can perform *diagonalization* by the following matrix, in which eigenkets are used as columns:

$$U = \begin{pmatrix} \langle a^{(1)}|b^{(1)}\rangle & \langle a^{(1)}|b^{(2)}\rangle & \cdots \\ \langle a^{(2)}|b^{(1)}\rangle & \langle a^{(2)}|b^{(2)}\rangle & \cdots \\ \vdots & \vdots & \ddots \end{pmatrix}. \tag{2.122}$$

Nontrivial solutions are possible if the characteristic equation:

$$\det(B - \lambda I) = 0 \tag{2.123}$$

is solvable. Namely, from matrix theory we know that if $n \times n$ matrix A has n linearly independent eigenkets, and eigenkets are used as columns of U, then $U^{-1}AU$ is diagonal. If A is Hermitian:

$$\left(U^{-1}AU\right)^{\dagger} = U^{\dagger}A^{\dagger}\left(U^{-1}\right)^{\dagger} = U^{\dagger}A\left(U^{-1}\right)^{\dagger},$$

then matrix U is a unitary matrix since $U^{\dagger} = U^{-1}$.

Example. Consider a 2×2 rotation matrix $R = \begin{bmatrix} \cos\theta & \sin\theta \\ -\sin\theta & \cos\theta \end{bmatrix}$. We can determine the unitary transformation that diagonalizes R as follows. We first find that the roots of the characteristic polynomial $\det(R - \lambda I) = 0$ are $\lambda_{1,2} = \cos\theta \pm j\sin\theta$.

The eigenvectors can be found from eigenvalue relation $R|\psi_{1,2}\rangle = \lambda_{1,2}|\psi_{1,2}\rangle$ as follows:

$$|\psi_1\rangle = \frac{1}{\sqrt{2}}\begin{bmatrix}1\\ j\end{bmatrix} \quad |\psi_2\rangle = \frac{1}{\sqrt{2}}\begin{bmatrix}1\\ -j\end{bmatrix}.$$

The unitary matrix that diagonalizes R is then given by:

$$U = [\psi_1 \quad \psi_2] = \frac{1}{\sqrt{2}}\begin{bmatrix}1 & 1\\ j & -j\end{bmatrix}.$$

We finally demonstrate that the unitary matrix indeed diagonalizes R by:

$$U^{\dagger}RU = \frac{1}{\sqrt{2}}\begin{bmatrix}1 & -j\\ 1 & j\end{bmatrix}\begin{bmatrix}\cos\theta & \sin\theta\\ -\sin\theta & \cos\theta\end{bmatrix}\frac{1}{\sqrt{2}}\begin{bmatrix}1 & 1\\ j & -j\end{bmatrix} = \begin{bmatrix}e^{j\theta} & 0\\ 0 & e^{-j\theta}\end{bmatrix}.$$

Consider two sets of orthonormal bases $\{|a'\rangle\}$ and $\{|b'\rangle\}$ connected by a U-operator. Knowing U we can construct a unitary transform of A, UAU^{-1}; then A and UAU^{-1} are said to be *unitary equivalent observables*. For the unitary equivalent observables the following theorem is important.

Theorem. The unitary equivalent observables, A and UAU^{-1}, have identical spectra.

This theorem can be proved by starting from eigenvalue relation $A|a^{(l)}\rangle = a^{(l)}|a^{(l)}\rangle$ and base transformation $|b^{(i)}\rangle = U|a^{(i)}\rangle$. By applying the U-operator from the left on eigenvalue relation and by inserting an identity operator $I = UU^{\dagger}$ we obtain:

$$UAU^{-1}\underbrace{U|a^{(l)}\rangle}_{b^{(l)}} = a^{(l)}U|a^{(l)}\rangle \Leftrightarrow (UAU^{-1})|b^{(l)}\rangle = a^{(l)}|b^{(l)}\rangle, \tag{2.124}$$

which confirms that the unitary equivalent objects have the same spectrum $\{a^{(l)}\}$.

2.9 TIME EVOLUTION—SCHRÖDINGER EQUATION

Time-evolution operator $U(t,t_0)$ transforms the initial ket at time instance t_0, $|\alpha,t_0\rangle$ into the final ket at time instance t by:

$$|\alpha,\ t_0;\ t\rangle = U(t,\ t_0)|\alpha,\ t_0\rangle. \tag{2.125}$$

This time-evolution operator must satisfy the following two properties:

1. *Unitary property:* $U^{+}(t,t_0)\ U(t,t_0) = I$
2. *Composition property:* $U(t_2,t_0) = U(t_2,t_1)\ U(t_1,t_0)$, $t_2 > t_1 > t_0$.

Following Eq. (2.125), the action of infinitesimal time-evolution operator $U(t_0 + dt,t_0)$ can be described by:

$$|\alpha,\ t_0;\ t_0 + dt\rangle = U(t_0 + dt,\ t_0)|\alpha,\ t_0\rangle. \tag{2.126}$$

The following operator satisfies all the foregoing propositions when $dt \to 0$:

$$U(t_0 + dt,\ t_0) = I - j\Omega dt, \quad \Omega^\dagger = \Omega, \tag{2.127}$$

where the operator Ω is related to the Hamiltonian H by $H = \hbar\Omega$, and the Hamiltonian eigenvalues correspond to the energy $E = \hbar\omega$. For the infinitesimal time-evolution operator $U(t_0 + dt, t_0)$ we can derive the time-evolution equation as follows. The starting point in derivation is the composition property:

$$U(t + dt,\ t_0) = U(t + dt,\ t)U(t, t_0) = \left(1 - \frac{j}{\hbar}Hdt\right)U(t,\ t_0). \tag{2.128}$$

Eq. (2.128) can be rewritten in the following form:

$$\lim_{dt\to 0}\frac{U(t + dt,\ t_0) - U(t, t_0)}{dt} = -\frac{j}{\hbar}HU(t,\ t_0), \tag{2.129}$$

which by taking the partial derivative definition into account, Eq. (2.129) becomes:

$$j\hbar\frac{\partial}{\partial t}U(t,\ t_0) = HU(t,\ t_0), \tag{2.130}$$

and this equation is known as the *Schrödinger equation for the time-evolution operator.*

The Schrödinger equation for a state ket can be obtained by applying the time-evolution operator on the initial ket:

$$j\hbar\frac{\partial}{\partial t}U(t,\ t_0)|\alpha,\ t_0\rangle = HU(t,\ t_0)|\alpha,\ t_0\rangle, \tag{2.131}$$

which based on Eq. (2.125) can be rewritten as:

$$j\hbar\frac{\partial}{\partial t}|\alpha,\ t_0;\ t\rangle = H|\alpha,\ t_0;\ t\rangle. \tag{2.132}$$

For *conservative systems*, for which the Hamiltonian is time invariant, we can easily solve Eq. (2.130) to obtain:

$$U(t, t_0) = e^{-\frac{j}{\hbar}H(t-t_0)}. \tag{2.133}$$

The time evolution of kets in conservative systems can therefore be described by applying Eq. (2.133) in Eq. (2.125), which yields to:

$$|\alpha(t)\rangle = e^{-\frac{j}{\hbar}H(t-t_0)}|\alpha(t_0)\rangle. \tag{2.134}$$

Therefore the operators do not explicitly depend on time and this concept is known as the *Schrödinger picture*.

In the *Heisenberg picture*, the state vector is independent of time, but operators depend on time:

$$A(t) = e^{\frac{j}{\hbar}H(t-t_0)}Ae^{-\frac{j}{\hbar}H(t-t_0)}. \tag{2.135}$$

The time-evolution equation in the Heisenberg picture is given by:

$$j\hbar\frac{dA(t)}{dt}=[A(t),\ H]+j\hbar\frac{\partial A(t)}{\partial t}. \tag{2.136}$$

The density operator ρ, representing the statistical mixture of states, is independent of time in the Heisenberg picture. The expectation value of a measurement of an observable $\Xi(t)$ at time instance t is given by:

$$\begin{aligned} E_t[\Xi] &= \mathrm{Tr}[\rho\Xi(t)] \\ E_t[\Xi] &= \mathrm{Tr}[\rho(t)\Xi],\ \rho(t)=e^{\frac{j}{\hbar}H(t-t_0)}\rho e^{-\frac{j}{\hbar}H(t-t_0)}. \end{aligned} \tag{2.137}$$

For a *time-variant Hamiltonian H*(*t*), the integration leads to a time-ordered exponential:

$$\begin{aligned} U(t,\ t_0) &= T\left\{\exp\left[-\frac{j}{\hbar}\int_{t_0}^{t}H(t')dt'\right]\right\} \\ &= \lim_{N\to\infty}\prod_{n=0}^{N-1}\exp\left[-\frac{j}{\hbar}H(t_0+n\Delta_N)\Delta_N\right],\quad \Delta_N=(t-t_0)/N. \end{aligned} \tag{2.138}$$

Example. The Hamiltonian for a two-state system is given by:

$$H=\begin{bmatrix}\omega_1 & \omega_2\\ \omega_2 & \omega_1\end{bmatrix}.$$

The basis for this system is given by: $\{|0\rangle=[1\ 0]^{\mathrm{T}},\ |1\rangle=[0\ 1]^{\mathrm{T}}\}$.

(a) Determine the eigenvalues and eigenkets of H, and express the eigenkets in terms of basis.

(b) Determine the time evolution of the system described by the Schrödinger equation:

$$j\hbar\frac{\partial}{\partial t}|\psi\rangle=H|\psi\rangle,\ |\psi(0)\rangle=|0\rangle.$$

To determine the eigenkets of H we start from the characteristic equation $\det(H-\lambda I)=0$ and find that eigenvalues are $\lambda_{1,\,2}=\omega_1\pm\omega_2$. The corresponding eigenvectors are:

$$|\lambda_1\rangle=\frac{1}{\sqrt{2}}\begin{bmatrix}1\\1\end{bmatrix}=\frac{1}{\sqrt{2}}(|0\rangle+|1\rangle)\quad \left|\lambda_1\right\rangle=\frac{1}{\sqrt{2}}\begin{bmatrix}1\\-1\end{bmatrix}=\frac{1}{\sqrt{2}}(|0\rangle-|1\rangle).$$

We now have to determine the time evolution of the arbitrary ket $|\psi(t)\rangle=[\alpha(t)\beta(t)]^{\mathrm{T}}$. The starting point is the Schrödinger equation:

$$j\hbar\frac{\partial}{\partial t}|\psi\rangle = j\hbar\begin{bmatrix}\dot{\alpha}(t)\\ \dot{\beta}(t)\end{bmatrix},\quad H|\psi\rangle = \begin{bmatrix}\omega_1 & \omega_2\\ \omega_2 & \omega_1\end{bmatrix}\begin{bmatrix}\alpha(t)\\ \beta(t)\end{bmatrix}$$
$$= \begin{bmatrix}\omega_1\alpha(t)+\omega_2\beta(t)\\ \omega_2\alpha(t)+\omega_1\beta(t)\end{bmatrix} \Rightarrow j\hbar\begin{bmatrix}\dot{\alpha}(t)\\ \dot{\beta}(t)\end{bmatrix} = \begin{bmatrix}\omega_1\alpha(t)+\omega_2\beta(t)\\ \omega_2\alpha(t)+\omega_1\beta(t)\end{bmatrix}.$$

By substitution $\alpha(t)+\beta(t)=\gamma(t)$ and $\alpha(t)-\beta(t)=\delta(t)$ we obtain the ordinary set of differential equations:

$$j\hbar\frac{d\gamma(t)}{dt} = (\omega_1+\omega_2)\gamma(t)\quad j\hbar\frac{d\delta(t)}{dt} = (\omega_1-\omega_2)\delta(t),$$

whose solution is $\gamma(t) = C\exp\left(\frac{\omega_1+\omega_2}{j\hbar}\right)$ and $\delta(t) = D\exp\left(\frac{\omega_1-\omega_2}{j\hbar}t\right)$. From initial state $|\psi(0)\rangle = |0\rangle = [1\quad 0]^{\mathrm{T}}$ we obtain the unknown constants $C=D=1$ so that state time evolution is given by:

$$|\psi(t)\rangle = \exp\left(-\frac{j}{\hbar}\omega_1 t\right)\begin{bmatrix}\cos\left(\frac{\omega_2 t}{\hbar}\right)\\ -j\sin\left(\frac{\omega_2 t}{\hbar}\right)\end{bmatrix}.$$

Before concluding this section, it is interesting to see how the base kets evolve in time. This is particularly simple to determine if the base kets are chosen to be eigenkets of A that commute with Hamiltonian H, namely $[A,H]=0$. Then, both A and H have simultaneous eigenkets, called the *energy eigenkets*, with eigenvalues denoted by $E_{a'}$ that satisfy the eigenvalue equation:

$$H|a'\rangle = E_{a'}|a'\rangle. \tag{2.139}$$

The time-evolution operator can now be expanded in terms of projection operators as follows:

$$e^{-\frac{j}{\hbar}Ht} = \sum_{a'}\sum_{a''}|a''\rangle\langle a''|e^{-\frac{j}{\hbar}Ht}|a'\rangle\langle a'| = \sum_{a''}|a'\rangle e^{-\frac{j}{\hbar}E_{a'}t}\langle a'|. \tag{2.140}$$

By applying this time-evolution operator on initial state:

$$|\alpha,\ t_0=0\rangle = \sum_{a\psi}|a'\rangle\langle a'|\alpha,\ t_0=0\rangle = \sum_{a'}c_{a'}|a'\rangle \tag{2.141}$$

we obtain:

$$|\alpha,\ t_0=0;\ t\rangle = e^{-\frac{j}{\hbar}Ht}\left|\alpha,\ t_0=0\right\rangle = \sum_{a'}|a'\rangle\langle a'|\alpha\rangle e^{-\frac{j}{\hbar}E_{a'}t}. \tag{2.142}$$

The evolution of expansion coefficient is therefore as follows:

$$c_{a'}(t=0)\to c_{a'}(t) = c_{a'}(t=0)e^{-\frac{j}{\hbar}E_{a'}t}. \tag{2.143}$$

If the initial state was a base state, then from Eq. (2.142) it is clear that:

$$|\alpha,\, t_0=0\rangle = |a'\rangle \rightarrow |\alpha,\, t_0=0;\, t\rangle = |a'\rangle e^{-\frac{j}{\hbar}E_{a'}t}. \tag{2.144}$$

Therefore the system initially being in a simultaneous eigenstate of A and H remains so all the time!

With this section, we conclude the introduction to quantum mechanics. In the following several sections we study different quantum systems that are of high importance in various quantum information processing systems and in quantum communications. These examples can also be used for a deeper understanding of the underlying concepts of quantum mechanics presented earlier.

2.10 HARMONIC OSCILLATOR

The Hamiltonian of a particle in a 1D parabolic potential well $V(x) = m\omega^2x^2/2$ is given by:

$$H = \frac{p^2}{2m} + \frac{m\omega^2 x^2}{2}, \tag{2.145}$$

where p is the momentum operator and x is the position operator. The *annihilation* a and *creation* $a^\dagger$ operators, often used in quantum mechanics, are related to the momentum and position operators as follows:

$$a = \sqrt{\frac{m\omega}{2\hbar}}\left(x + \frac{jp}{m\omega}\right) \quad a^\dagger = \sqrt{\frac{m\omega}{2\hbar}}\left(x - \frac{jp}{m\omega}\right). \tag{2.146}$$

It can be shown that these two operators satisfy the following commutation relation:

$$\left[a,\, a^\dagger\right] = \frac{1}{2\hbar}\left(-j[x,\, p] + j[p,\, x]\right) \overset{[x_i,\, p_j]=j\hbar\delta_{ij}I}{=} I. \tag{2.147}$$

Another important operator is the *number* operator defined by:

$$N = a^\dagger a. \tag{2.148}$$

From Eq. (2.146), it is clear that:

$$a^\dagger a = \left(\frac{m\omega}{2\hbar}\right)\left(x^2 + \frac{p^2}{m^2\omega^2}\right) + \frac{j}{2\hbar}[x,\, p] = \frac{H}{\hbar\omega} - \frac{1}{2}I, \tag{2.149}$$

so that we can relate the Hamiltonian and the number operators as follows:

$$H = \hbar\omega\left(N + \frac{1}{2}I\right). \tag{2.150}$$

By denoting the energy eigenket of N by its eigenvalue n, namely $N|n\rangle = n|n\rangle$, the energy eigenvalues E_n can be determined from:

$$H|n\rangle = \hbar\omega\left(N+\frac{1}{2}I\right)|n\rangle = \hbar\omega\left(n+\frac{1}{2}I\right)|n\rangle = E_n|n\rangle, \tag{2.151}$$

by:

$$E_n = \left(n+\frac{1}{2}\right)\hbar\omega. \tag{2.152}$$

To come up with the physical significance of a, a^+, and N we first prove the following commutation relation:

$$[N, a] = \left[aa^\dagger, a\right] = a^\dagger[a, a] + \left[a^\dagger, a\right]a = -a. \tag{2.153}$$

In a similar fashion we can prove the following commutation relation $\left[N, a^\dagger\right] = a^\dagger$. We now observe the action of operators $Na^\dagger$ and Na on $|n\rangle$ and by using the commutation relations just derived we obtain:

$$\begin{aligned} Na^\dagger|n\rangle &= \left(\left[N, a^\dagger\right] + a^\dagger N\right)|n\rangle = (n+1)a^\dagger|n\rangle \\ Na|n\rangle &= ([N, a] + aN)|n\rangle = (n-1)a|n\rangle. \end{aligned} \tag{2.154}$$

Therefore the creation (annihilation) operator increases (decreases) the quantum energy for one unit. By using Eq. (2.152) and commutation relations we have just derived, it is easy to show that the following commutation relation is valid: $\left[H, a^\dagger\right] = \hbar\omega a^\dagger$, which can be used to determine the action of operators $Ha^\dagger$ and Ha on a ket $|\psi\rangle$ as follows:

$$\begin{aligned} Ha^\dagger|\psi\rangle &= \left(\left[H, a^\dagger\right] + a^\dagger H\right)|\psi\rangle = (E+\hbar\omega)a^\dagger|\psi\rangle \\ Ha|\psi\rangle &= (E-\hbar\omega)a|\psi\rangle. \end{aligned} \tag{2.155}$$

This is the reason for the creation and annihilation operators to be called also as *rising* and *lowering* operators, respectively.

What remains is to determine the simultaneous eigenkets of N and H. From Eq. (2.154) $a|n\rangle$ and $|n-1\rangle$ differ only in a multiplicative constant as follows:

$$a|n\rangle = c|n-1\rangle. \tag{2.156}$$

From $N|n\rangle = n|n\rangle$, it is clear that $\langle n|N|n\rangle = n$, while from Eq. (2.156) is clear that:

$$\left\langle n\left|a^\dagger a\right|n\right\rangle = |c|^2 \geq 0, \tag{2.157}$$

and by combining these two equations we obtain that $c = \sqrt{n}$ and that n is never negative, which indicates that the ground state of the harmonic oscillator has the energy $E_0 = \hbar\omega/2$. Therefore the action of annihilation and creation operators on $|n\rangle$ can be described as follows:

$$a|n\rangle = \sqrt{n}|n-1\rangle \quad a^\dagger|n\rangle = \sqrt{n+1}|n+1\rangle. \tag{2.158a}$$

If we continue applying the annihilation operator we obtain the following sequence:

$$\begin{aligned} a^2|n\rangle &= \sqrt{n(n-1)}|n-2\rangle \\ a^3|n\rangle &= \sqrt{n(n-1)(n-2)}|n-3\rangle \\ &\vdots \\ a^k|n\rangle &= \sqrt{n(n-1)\cdots(n-k+1)}|n-k\rangle. \end{aligned} \tag{2.158b}$$

In a similar fashion, if we keep applying the creation operator on the ground state we obtain:

$$\begin{aligned} a^\dagger|0\rangle &= |1\rangle \\ \left(a^\dagger\right)^2|0\rangle &= \sqrt{1\cdot 2}|2\rangle \\ \left(a^\dagger\right)^3|0\rangle &= \sqrt{1\cdot 2\cdot 3}|3\rangle \\ &\vdots \\ \left(a^\dagger\right)^n|0\rangle &= \sqrt{n!}|n\rangle. \end{aligned} \tag{2.159}$$

In other words we can express the state $|n\rangle$ in terms of ground state as:

$$|n\rangle = \frac{\left(a^\dagger\right)^n}{\sqrt{n!}}|0\rangle. \tag{2.160}$$

The matrix elements of annihilation and creation operators can be obtained by using Eq. (2.158a) and employing the orthogonality principle for basis $\{|n\rangle\}$ as follows:

$$\langle m|a|n\rangle = \sqrt{n}\delta_{m,\,n-1} \quad \langle m|a^\dagger|n\rangle = \sqrt{n+1}\delta_{m,\,n+1}. \tag{2.161}$$

The time evolution of eigenkets, based on the previous section, can be described by:

$$|\psi(t)\rangle = e^{-\frac{j}{\hbar}H}\Big|\psi(0)\Big\rangle. \tag{2.162}$$

We can express the initial state $|\psi(0)\rangle$ in basis $\{|n\rangle\}$ as follows:

$$|\psi(0)\rangle = \sum_n c_n(0)|n\rangle, \tag{2.163}$$

where $c_n(0)$ is the expansion coefficient. Based on Eq. (2.144), the evolution of expansion coefficient can be described as:

$$c_n(t) = c_n(0)e^{-\frac{j}{\hbar}(E_n-E_0)t} \overset{E_n=(n+1/2)\hbar\omega}{=} c_n(0)e^{-jn\omega t}, \tag{2.164}$$

so that Eq. (2.162) becomes:

$$|\psi(t)\rangle = \sum_n c_n(0)e^{-jn\omega t}|n\rangle. \tag{2.165}$$

By using Eq. (2.146) we can express position x and momentum p operators in terms of annihilation and creation operators as follows:

$$x = \sqrt{\frac{\hbar}{2m\omega}}(a^\dagger + a) \quad p = j\sqrt{\frac{m\hbar\omega}{2}}(a^\dagger - a). \tag{2.166}$$

Based on Eq. (2.166) we can determine the matrix elements of x- and p-operators by:

$$\langle m|x|n\rangle = \sqrt{\frac{\hbar}{2m\omega}}\left(\sqrt{n+1}\,\delta_{m,\,n+1} + \sqrt{n}\,\delta_{m,\,n-1}\right)$$

$$\langle m|p|n\rangle = j\sqrt{\frac{m\hbar\omega}{2}}\left(\sqrt{n+1}\,\delta_{m,\,n+1} - \sqrt{n}\,\delta_{m,\,n-1}\right). \tag{2.167}$$

Clearly, neither x nor p are diagonal in number operator representation, which is consistent with the fact that neither annihilation nor creation operators commute with N. Therefore the matrix representations of x and p, based on Eq. (2.167), are as follows:

$$x \doteq \sqrt{\frac{\hbar}{2m\omega}}\begin{bmatrix} 0 & 1 & 0 & 0 & \cdots \\ 1 & 0 & \sqrt{2} & 0 & \cdots \\ 0 & \sqrt{2} & 0 & \sqrt{3} & \cdots \\ 0 & 0 & \sqrt{3} & 0 & \ddots \\ \vdots & \vdots & & \ddots & \ddots \end{bmatrix} \quad p \doteq j\sqrt{\frac{m\hbar\omega}{2}}\begin{bmatrix} 0 & -1 & 0 & 0 & \cdots \\ 1 & 0 & -\sqrt{2} & 0 & \cdots \\ 0 & \sqrt{2} & 0 & -\sqrt{3} & \cdots \\ 0 & 0 & \sqrt{3} & 0 & \ddots \\ \vdots & \vdots & & \ddots & \ddots \end{bmatrix}.$$

Before concluding this section we determine the uncertainty product for x and p. From Eq. (2.161), it is clear that $\langle x\rangle = \langle p\rangle = 0$ for the ground state. On the other hand, from Eq. (2.166) we obtain $x^2 = \left((a^\dagger)^2 + a^\dagger a + aa^\dagger + a^2\right)\hbar\big/(2m\omega)$ so that $\langle x^2\rangle = \hbar/(2m\omega)$. In similar fashion we obtain for expectation of p^2 for the ground state to be $\langle p^2\rangle = \hbar m\omega/2$ so that the uncertainty product is:

$$\left\langle(\Delta x)^2\right\rangle\left\langle(\Delta p)^2\right\rangle = \left(\frac{\hbar}{2}\right)^2. \tag{2.168}$$

Following a similar procedure we can determine the uncertainty product of the nth excited state as follows:

$$\left\langle(\Delta x)^2\right\rangle\left\langle(\Delta p)^2\right\rangle = \left(n + \frac{1}{2}\right)^2\left(\frac{\hbar}{2}\right)^2. \tag{2.169}$$

2.11 ANGULAR MOMENTUM

The angular momentum in classical physics is defined by:

$$\boldsymbol{L} = \boldsymbol{r} \times \boldsymbol{p} = \begin{bmatrix} \widehat{x} & \widehat{y} & \widehat{z} \\ x & y & z \\ p_x & p_y & p_z \end{bmatrix} = (yp_z - zp_y)\widehat{x} + (-xp_z + zp_x)\widehat{y} + (xp_y - yp_x)\widehat{z}$$

$$= L_x\widehat{x} + L_y\widehat{y} + L_z\widehat{z}; \qquad L_x = yp_z - zp_y,\ L_y = zp_x - xp_z,\ L_z = xp_y - yp_x. \tag{2.170}$$

The position operator x can be defined as follows:

$$x|x'\rangle = x'|x'\rangle, \tag{2.171}$$

where x' is the eigenvalue. We assume that eigenkets form the complete set, and we can represent an arbitrary continuous ket $|\alpha\rangle$ as follows:

$$|\alpha\rangle = \int_{-\infty}^{\infty} \langle x'|\alpha\rangle |x'\rangle dx'. \tag{2.172}$$

The corresponding 3D generalization leads to:

$$|\alpha\rangle = \int_{-\infty}^{\infty} \langle \boldsymbol{r}'|\alpha\rangle |\boldsymbol{r}'\rangle dx'dy'dz',\ \boldsymbol{r}' = [x', y', z'], \tag{2.173}$$

where $|\boldsymbol{r}'\rangle$ is the simultaneous eigenket of observables x, y, and z, namely:

$$x|\boldsymbol{r}'\rangle = x'|\boldsymbol{r}'\rangle, \quad y|\boldsymbol{r}'\rangle = y'|\boldsymbol{r}'\rangle, \quad z|\boldsymbol{r}'\rangle = z'|\boldsymbol{r}'\rangle. \tag{2.174}$$

In Eq. (2.174) we implicitly assumed that all three coordinates can be measured simultaneously with arbitrary small accuracy, indicating that:

$$[x_m, x_n] = 0;\ x_m, x_n = x, y, z. \tag{2.175}$$

We now introduce the *infinitesimal translation operator* $T(\mathbf{d}\boldsymbol{r}')$ as follows:

$$T(d\boldsymbol{r}')|\boldsymbol{r}'\rangle = |\boldsymbol{r}' + d\boldsymbol{r}'\rangle, \tag{2.176}$$

which changes the position from $\boldsymbol{r}'$ to $\boldsymbol{r}' + d\boldsymbol{r}'$. This operator must satisfy the following *properties*:

1. *Unitarity property*, namely $T(d\boldsymbol{r}')T^{\dagger}(d\boldsymbol{r}') = I$,
2. *Composition property*: $T(d\boldsymbol{r}')T(d\boldsymbol{r}'') = T(d\boldsymbol{r}' + d\boldsymbol{r}'')$,
3. *Opposite direction translation property*: $T^{-1}(d\boldsymbol{r}') = T(-d\boldsymbol{r}')$, and
4. $\lim_{d\boldsymbol{r}' \to 0} T(d\boldsymbol{r}') = I$.

The following translation operator:

$$T(d\boldsymbol{r}') = 1 - \frac{j}{\hbar}\boldsymbol{p}\cdot d\boldsymbol{r}',\ \boldsymbol{p} = [p_x, p_y, p_z], \tag{2.177}$$

satisfies all the foregoing properties. In Eq. (2.177), p_j denotes the momentum operator along the jth axis ($j = x,y,z$). Therefore we can establish the following relationship between classical and quantum momentum by the following mapping:

$$p_i \rightarrow -j\hbar\frac{\partial}{\partial i}; \quad i = x,\ y,\ z, \tag{2.178}$$

which can be used in Eq. (2.170) to determine the quantum angular momentum operator. For example, the quantum angular momentum along the z-axis is given by:

$$L_z = xp_y - yp_x = x\left(-j\hbar\frac{\partial}{\partial y}\right) - y\left(-j\hbar\frac{\partial}{\partial x}\right) = -j\hbar\left(x\frac{\partial}{\partial y} - y\frac{\partial}{\partial x}\right). \tag{2.179}$$

2.11.1 Commutation Relations

Given that angular momentum operators can be expressed in terms of position x and linear momentum operators p_i ($i = x,y,z$), and that x and p_i obey the following commutation relation:

$$[x_m,\ p_n] = j\hbar\delta_{mn}; \quad x_m,\ x_n = x,\ y,\ z, \tag{2.180}$$

it can be shown that components of angular momentum satisfy the following commutation relations:

$$[L_x,\ L_y] = j\hbar L_z \quad [L_z,\ L_x] = j\hbar L_y \quad [L_y,\ L_z] = j\hbar L_x. \tag{2.181}$$

The first commutation relation can be proved as follows. Since $L_y = zp_x - xp_z$ we obtain:

$$[L_x,\ L_y] = [L_x,\ zp_x - xp_z] = [L_x,\ zp_x] - [L_x,\ xp_z]. \tag{2.182}$$

By applying property (3) of Eq. (2.87), $[A,\ BC] = [A,\ B]C + B[A,\ C]$, $[L_x,\ zp_x]$ can be written as:

$$[L_x,\ zp_x] = [L_x,\ z]p_x + z[L_x,\ p_x]. \tag{2.183}$$

By using now $L_x = yp_z - zp_y$ we obtain:

$$\begin{aligned}[L_x,\ z] &= [yp_z - zp_y,\ z] = [yp_z,\ z] - [zp_y,\ z] = -[z,\ yp_z] + [z,\ zp_y] \\ &= -[z,\ y]p_z - y[z,\ p_z] + [z,\ z]p_y + z[z,\ p_y] = -j\hbar y.\end{aligned} \tag{2.184}$$

Similarly we show that:

$$[L_x,\ p_x] = [yp_z - zp_y,\ p_x] = 0. \tag{2.185}$$

By substituting Eqs. (2.184) and (2.185) into Eq. (2.183) we obtain:

$$[L_x,\ zp_x] = -j\hbar y. \tag{2.186}$$

Applying again property (3) of Eq. (2.87) on the second term in Eq. (2.182) we obtain:

$$[L_x, xp_z] = [L_x, x]p_z + x[L_x, p_z]. \tag{2.187}$$

Since $[L_x, x] = [yp_z - zp_y, x] = 0$, it follows that:

$$[L_x, p_z] = [yp_z - zp_y, p_z] = -j\hbar p_y, \tag{2.188}$$

which upon substitution in Eq. (2.187) yields to:

$$[L_x, xp_z] = -j\hbar xp_y. \tag{2.189}$$

Now by substituting Eqs. (2.189) and (2.186) into Eq. (2.182) we obtain:

$$\begin{aligned}[L_x, L_y] &= [L_x, z]p_x + z[L_x, p_x] - ([L_x, x]p_z + x[L_x, p_z]) = -j\hbar yp_x + j\hbar xp_y \\ &= j\hbar(xp_y - yp_x) = j\hbar L_z,\end{aligned} \tag{2.190}$$

which proves the first commutation relation in Eq. (2.181). The other commutation relations can be proved in a similar fashion. Because the components of angular momentum do not commute we can specify only one component at a time. It can simply be shown that every component of angular momentum commutes with $L^2 = L_x^2 + L_y^2 + L_z^2$. Therefore to specify a state of angular momentum, it is sufficient to use L^2 and L_z. In quantum mechanics, in addition to *orbital angular momentum* (OAM) $\boldsymbol{L}$, *spin angular momentum* (SAM) $\boldsymbol{S}$ is also present. We can therefore define the *total angular momentum* (TAM) as follows $\boldsymbol{J} = \boldsymbol{L} + \boldsymbol{S}$. The components of TAM satisfy the same set of commutation relations as OAM:

$$[J_x, J_y] = i\hbar J_z \quad [J_z, J_x] = i\hbar J_y \quad [J_y, J_z] = i\hbar J_x \tag{2.191}$$

(the imaginary unit is denoted by i to avoid any confusion with corresponding quantum number j to be introduced soon).

The states of OAM are specified by two quantum numbers: (1) the *orbital quantum number l* and (2) the *azimuthal (magnetic) quantum number m*. Similarly, the states of TAM are specified by two quantum numbers j and m. The eigenvalues of J^2 are labeled by j, while the eigenvalues of J_z are labeled by m. The corresponding eigenvalue equations are given by:

$$\begin{aligned}J^2|j, m\rangle = \hbar^2 j(j+1)|j, m\rangle \quad J_z|j, m\rangle = m\hbar|j, m\rangle \\ m = -j, -j+1, \ldots, -1, 0, 1, \ldots j-1, j.\end{aligned} \tag{2.192}$$

In the case of OAM, both quantum numbers l and m are integers. For example, for $l = 3$, the possible values of m are $\{-3, -2, -1, 0, 1, 2, 3\}$, and the corresponding "angular momentum" is $\hbar\sqrt{l(l+1)} = 2\hbar\sqrt{3}$. The spin, on the other hand, can have half-odd-integral values. Since $j = l + s$, the same is true for TAM. For example, the electron has spin $s = 1/2$ and for $l = 1$ we obtain $j = 3/2$, so that possible values for j are $\{-3/2, -1/2, 1/2, 3/2\}$.

To move up and down in states of TAM we introduce the *ladder* operators J_+ and J_-, also known as *raising* and *lowering* operators, respectively. The ladder operators are defined as follows:

$$J_+ = J_x + iJ_y \quad J_- = J_x - iJ_y. \tag{2.193}$$

The ladder operators satisfy the following commuting relations:

$$[J_z, J_+] = \hbar J_+ \quad [J_z, J_-] = -\hbar J_- \quad [J_+, J_-] = 2\hbar J_z. \tag{2.194}$$

It is often very useful to express the J^2, J_x, and J_y operators in terms of ladder operators as follows:

$$\begin{aligned} J^2 &= J_-J_+ + J_z^2 + \hbar J_z = J_z^2 + (J_+J_- + J_-J_+)/2 \\ J_x &= (J_+ + J_-)/2 \quad J_y = (J_+ - J_-)/2i. \end{aligned} \tag{2.195}$$

The ladder operators act on TAM states as follows:

$$\begin{aligned} J_+|j,\ m\rangle &= \hbar\sqrt{j(j+1) - m(m+1)}\,|j,\ m+1\rangle \\ J_-|j,\ m\rangle &= \hbar\sqrt{j(j+1) - m(m-1)}\,|j,\ m-1\rangle. \end{aligned} \tag{2.196}$$

Because $m = -j,\ldots,0,\ldots,+j$, meaning that the maximum possible value for m is j and minimum possible value for m is $-j$, we require:

$$J_+|j,\ m=j\rangle = 0 \quad J_-|j,\ m=-j\rangle = 0. \tag{2.197}$$

The TAM states $|j,m\rangle$ satisfy the following orthonormality relation:

$$\langle j_1,\ m_1|j_2,\ m_2\rangle = \delta_{j_1,\ j_2}\delta_{m_1,\ m_2}, \tag{2.198}$$

and the completeness relation as follows:

$$\sum_{j=0}^{\infty}\sum_{m=-j}^{j}|j,\ m\rangle\langle j,\ m| = I. \tag{2.199}$$

2.11.2 Matrix Representation of Angular Momentum

We have shown in Section 2.3 that an operator A is associated with a square matrix, and this representation is dependent on eigenkets of choice. For TAM, it is natural to use standard basis so that the A_{mn} element satisfies:

$$A_{mn} = \langle j,\ m|A|j,\ n\rangle. \tag{2.200}$$

For every constant j we can associate a $(2j+1)\times(2j+1)$ matrix for J^2 and J_z as follows:

$$\begin{aligned} (J^2)_{mn} &= \langle j,\ m|J^2|j,\ n\rangle = \langle j,\ m|\hbar^2 j(j+1)|j,\ n\rangle = \hbar^2 j(j+1)\delta_{mn} \\ (J_z)_{mn} &= \langle j,\ m|J_z|j,\ n\rangle = \langle j,\ m|n\hbar|j,\ n\rangle = n\hbar\delta_{mn}. \end{aligned} \tag{2.201}$$

For example, for $l = 1$, the matrix representations for operators L^2 and L_z can be written as follows:

$$L^2 = \begin{bmatrix} \langle 1,\ 1|L^2|1,\ 1\rangle & \langle 1,\ 1|L^2|1,\ 0\rangle & \langle 1,\ 1|L^2|1,\ -1\rangle \\ \langle 1,\ 0|L^2|1,\ 1\rangle & \langle 1,\ 0|L^2|1,\ 0\rangle & \langle 1,\ 0|L^2|1,\ -1\rangle \\ \langle 1,\ -1|L^2|1,\ 1\rangle & \langle 1,\ -1|L^2|1,\ 0\rangle & \langle 1,\ -1|L^2|1,\ -1\rangle \end{bmatrix} = 2\hbar^2 \begin{bmatrix} 1 & 0 & 0 \\ 0 & 1 & 0 \\ 0 & 0 & 1 \end{bmatrix}$$

$$L_z = \begin{bmatrix} \langle 1,\ 1|L_z|1,\ 1\rangle & \langle 1,1|L_z|1,0\rangle & \langle 1,\ 1|L_z|1,\ -1\rangle \\ \langle 1,\ 0|L_z|1,\ 1\rangle & \langle 1,0|L_z|1,0\rangle & \langle 1,\ 0|L_z|1,\ -1\rangle \\ \langle 1,\ -1|L_z|1,\ 1\rangle & \langle 1,-1|L_z|1,0\rangle & \langle 1,\ -1|L_z|1,\ -1\rangle \end{bmatrix} = \hbar \begin{bmatrix} 1 & 0 & 0 \\ 0 & 0 & 0 \\ 0 & 0 & -1 \end{bmatrix}. \tag{2.202}$$

Example. Let us consider the quantum system with angular momentum $l = 1$. Determine the matrix representation of the basis of simultaneous eigenkets of L_z and L^2.

The eigenkets of the system with $l = 1$ are:

$$|1\rangle = \begin{pmatrix} 1 \\ 0 \\ 0 \end{pmatrix} \equiv |l = 1,\ m = 1\rangle,\quad |0\rangle = \begin{pmatrix} 0 \\ 1 \\ 0 \end{pmatrix} \equiv |l = 1,\ m = 0\rangle,$$

$$|-1\rangle = \begin{pmatrix} 0 \\ 0 \\ 1 \end{pmatrix} \equiv |l = 1,\ m = -1\rangle.$$

Based on Eqs. (2.192) and (2.193), the following set of relations can be used to determine the matrix representation of L_x:

$$L^2|l,\ m\rangle = \hbar^2 l(l+1)|l,\ m\rangle \quad L_z|l,\ m\rangle = m\hbar|l,\ m\rangle$$

$$L_+|l,\ m\rangle = \hbar\sqrt{l(l+1) - m(m+1)}|l,\ m+1\rangle$$

$$L_-|j,\ m\rangle = \hbar\sqrt{l(l+1) - m(m-1)}|l,\ m-1\rangle;$$

$$L_x = \frac{1}{2}(L_+ + L_-)$$

To determine the matrix representation of L_x, let us observe the action of L_x on base kets:

$$L_x|1\rangle = \frac{1}{2}(L_+ + L_-)|1\rangle = \frac{1}{2}L_-|1\rangle = \frac{1}{2}\hbar|0\rangle,\ L_x|-1\rangle = \frac{1}{2}(L_+ + L_-)|-1\rangle$$
$$= \frac{1}{2}L_+|-1\rangle = \frac{1}{2}\hbar\sqrt{2}|0\rangle;$$

$$L_x|0\rangle = \frac{1}{2}(L_+ + L_-)|0\rangle = \frac{1}{2}L_+|0\rangle + \frac{1}{2}L_-|0\rangle = \frac{1}{2}\hbar\sqrt{2}|1\rangle + \hbar\sqrt{2}|-1\rangle$$
$$= \frac{\hbar}{\sqrt{2}}(|1\rangle + |-1\rangle).$$

The matrix representation of L_x, based on previous equations, is as follows:

$$L_x \doteq \begin{bmatrix} \langle 1|L_x|1\rangle & \langle 1|L_x|0\rangle & \langle 1|L_x|-1\rangle \\ \langle 0|L_x|1\rangle & \langle 0|L_x|0\rangle & \langle 0|L_x|-1\rangle \\ \langle -1|L_x|1\rangle & \langle -1|L_x|0\rangle & \langle -1|L_x|-1\rangle \end{bmatrix} = \frac{\hbar}{\sqrt{2}} \begin{bmatrix} 0 & 1 & 0 \\ 1 & 0 & 1 \\ 0 & 1 & 0 \end{bmatrix}.$$

Example. Let the system with angular momentum $l = 1$ be initially in the state $[1\ 2\ 3]^{\mathrm{T}}$. Determine the probability that a measurement of L_x will be equal to zero.

First of all, the initial state is not normalized; upon normalization the initial state is $[1\ 2\ 3]^{\mathrm{T}}/\sqrt{14}$. Based on the matrix representation of L_x from the previous example we determine the eigenvalues from:

$$\frac{\hbar}{\sqrt{2}} \begin{bmatrix} -\lambda & 1 & 0 \\ 1 & -\lambda & 1 \\ 0 & 1 & -\lambda \end{bmatrix} = 0$$

to be $\lambda_{1,\,2,\,3} = \hbar,\ -\hbar,\ 0$. The corresponding eigenkets can be found from $L_x|\lambda_{1,\,2,\,3}\rangle = \lambda_{1,\,2,\,3}|\lambda_{1,\,2,\,3}\rangle$ as follows:

$$|1\rangle_x = \frac{1}{2}\begin{bmatrix} 1 \\ \sqrt{2} \\ 1 \end{bmatrix}, \quad |-1\rangle_x = \frac{1}{2}\begin{bmatrix} 1 \\ -\sqrt{2} \\ 1 \end{bmatrix}, \quad |0\rangle_x = \frac{1}{2}\begin{bmatrix} 1 \\ 0 \\ -1 \end{bmatrix}.$$

The initial state in L_x basis is now:

$$|\psi\rangle_x = {}_x\langle 1|\psi\rangle|1\rangle_x + {}_x\langle 0|\psi\rangle|0\rangle_x + {}_x\langle 1|\psi\rangle|-1\rangle_x, \quad |\psi\rangle = \frac{1}{\sqrt{14}}\begin{bmatrix} 1 \\ 2 \\ 3 \end{bmatrix}.$$

Since:

$${}_x\langle 1|\psi\rangle = \left(2+\sqrt{2}\right)/\sqrt{14}, \quad {}_x\langle 0|\psi\rangle = -1/\sqrt{7}, \quad {}_x\langle -1|\psi\rangle = \left(2-\sqrt{2}\right)/\sqrt{14},$$

we obtain that:

$$\Pr(|0\rangle_x) = |{}_x\langle 0|\psi\rangle|^2 = 1/7.$$

2.11.3 Coordinate Representation of OAM

We are concerned here with coordinate representation of OAM. Because of spherical symmetry, it is convenient to work with spherical coordinates, which are related to Cartesian coordinates as follows:

$$x = r\sin\theta\cos\phi \quad y = r\sin\theta\sin\phi \quad z = r\cos\theta. \tag{2.203}$$

With these substitutions, the components of OAM can be represented in spherical coordinates by:

$$L_x = j\hbar\left(\sin\phi\frac{\partial}{\partial\theta} + \frac{\cos\phi}{\tan\theta}\frac{\partial}{\partial\phi}\right) \quad L_y = j\hbar\left(-\cos\phi\frac{\partial}{\partial\theta} + \frac{\sin\phi}{\tan\theta}\frac{\partial}{\partial\phi}\right) \quad L_z = -j\hbar\frac{\partial}{\partial\phi}. \tag{2.204}$$

The corresponding L^2 and leader operators can be expressed in spherical coordinates as follows:

$$\begin{aligned} L^2 &= -\hbar^2\left(\frac{\partial^2}{\partial\theta^2} + \frac{1}{\tan\theta}\frac{\partial}{\partial\theta} + \frac{1}{\sin^2\theta}\frac{\partial^2}{\partial\phi^2}\right) \\ L_+ &= \hbar e^{j\phi}\left(\frac{\partial}{\partial\theta} + j\frac{1}{\tan\theta}\frac{\partial}{\partial\phi}\right) \quad L_- = -\hbar e^{-j\phi}\left(\frac{\partial}{\partial\theta} - j\frac{1}{\tan\theta}\frac{\partial}{\partial\phi}\right). \end{aligned} \tag{2.205}$$

Because the foregoing various operators are only functions of angular variables, the corresponding eigenfunctions of OAM will also be functions of θ and ϕ only:

$$\langle\theta,\ \phi|l,\ m\rangle = Y_l^m(\theta,\ \phi), \tag{2.206}$$

where $Y_l^m(\theta,\ \phi)$ are spherical harmonics defined by:

$$Y_l^m(\theta,\ \phi) = \begin{cases} (-1)^m\sqrt{\dfrac{(2l+1)(l-m)!}{4\pi(l+m)!}}P_l^m(\cos\theta)e^{jm\phi}, & m>0 \\ \\ (-1)^{|m|}\sqrt{\dfrac{(2l+1)(l-|m|)!}{4\pi(l+m)!}}P_l^{|m|}(\cos\theta)e^{jm\phi}, & m<0 \end{cases}. \tag{2.207}$$

With $P_l^m(x)$ we denote the associated Legendre polynomials:

$$P_l^m(x) = \left(1-x^2\right)^{m/2}\frac{d^m}{dx^m}P_l(x), \tag{2.208}$$

where $P_l(x)$ are the Legendre polynomials defined by:

$$P_l(x) = \frac{1}{2^l l!}\frac{d^l}{dx^l}\left[\left(x^2-1\right)^l\right]. \tag{2.209}$$

The Legendre polynomial can be determined recursively as follows:

$$(l+1)P_{l+1}(x) = (2l+1)xP_l(x) - lP_{l-1}(x); \quad P_0(x) = 1, \quad P_1(x) = x. \tag{2.210}$$

Example. A symmetrical top with moments of inertia $I_x = I_y$ and I_z in the body of the axis frame is described by the following Hamiltonian:

$$H = \frac{1}{2I_x}\left(L_x^2 + L_y^2\right) + \frac{1}{2I_z}L_z^2.$$

Determine the eigenvalues and eigenkets of the Hamiltonian. Determine the expected measurement of $L_x + L_y + L_z$ for any state. Finally, if initially the top was in state $|l = 3, m = 0\rangle$, determine the probability that the result of a measurement of L_x at time instance $t = 4\pi I_x/\hbar$ is $\hbar$.

We first express the Hamiltonian in terms of L^2 and L_z:

$$H = \frac{1}{2I_x}\left(L_x^2 + L_y^2\right) + \frac{1}{2I_z}L_z^2 = \frac{1}{2I_x}\left(L_x^2 + L_y^2 + L_z^2 - L_z^2\right) + \frac{1}{2I_z}L_z^2$$
$$= \frac{1}{2I_x}L^2 + \left(\frac{1}{2I_x} - \frac{1}{2I_x}\right)L_z^2.$$

We already know that when A is an operator with eigenvalues λ_i, the eigenvalues for the function of A, $f(A)$, are $f(\lambda_i)$. Therefore based on Eq. (2.192) we obtain the following energy eigenvalues:

$$E_{lm} = \frac{1}{2I_x}\hbar^2 l(l+1) + \left(\frac{1}{2I_x} - \frac{1}{2I_x}\right)m^2\hbar^2.$$

Furthermore, since L^2 and L_z have simultaneous eigenkets, the eigenkets of H will be $|Y_l^m(\theta, \phi)\rangle$ with eigenenergies being E_{lm}. The measurement of $L_x + L_y + L_z$ yields to:

$$\langle Y_l^m(\theta, \phi)|L_x + L_y + L_z|Y_l^m(\theta, \phi)\rangle = \left\langle Y_l^m(\theta, \phi)\left|\frac{L_+ + L_-}{2} + \frac{L_+ - L_-}{2j} + L_z\right|Y_l^m(\theta, \phi)\right\rangle$$
$$= \langle Y_l^m(\theta, \phi)|L_z|Y_l^m(\theta, \phi)\rangle = m\hbar.$$

Since the state at $t = 0$, $|Y_0^3(\theta, \phi)\rangle$, is an eigenket of H, it will remain so all the time. The measurement of L_z for $m = 0$ is zero so that the probability of getting $\hbar$ is zero.

2.11.4 Angular Momentum and Rotations

The angular momentum operator L is often called the generator of rotations. The state ket $|\psi\rangle$ in one coordinate system O can be rotated to a new coordinate system O' by means of rotation operator U_R as follows:

$$|\psi'\rangle = U_R|\psi\rangle. \tag{2.211}$$

The infinitesimal rotation operator can be defined by:

$$U_R(d\theta, \hat{n}) = I - \frac{j}{\hbar}d\theta \mathbf{L}\cdot\hat{n}. \tag{2.212}$$

This operator can be used to describe the rotation of the state ket $|\psi\rangle$ in coordinate system O' around $\hat{n}$ for an angle θ (relative to O):

$$U_R(\theta, \hat{n}) = \lim_{N\to\infty} U_R(d\theta = \theta/N, \hat{n})$$
$$= \lim_{N\to\infty}\left(I - \frac{j}{\hbar}\frac{\theta}{N}\mathbf{L}\cdot\hat{n}\right)^N \overset{\lim_{N\to\infty}(1+a/N)^N = e^a}{=} \exp\left(-\frac{j}{\hbar}\theta\mathbf{L}\cdot\hat{n}\right). \tag{2.213}$$

For example, the rotation around the z-axis by an angle φ can be described by:

$$U_R(\varphi,\ z)=\exp\left(-\frac{j}{\hbar}\varphi L_z\right). \tag{2.214}$$

The rotation operator must satisfy:

$$U_R(0,\ \widehat{n})=U_R(2\pi,\ \widehat{n})=I. \tag{2.215}$$

An observable A in coordinate system O is transformed into A' in coordinate system O' by:

$$A'=U_R A U_R^{\dagger}. \tag{2.216}$$

Example. Determine the rotation operator around $\widehat{n}=\widehat{y}$ for $l=1$. By using this operator, determine the representation of L_x in standard L_z basis.

The rotation operator (2.213) for $\widehat{n}=\widehat{y}$ is as follows:

$$U_R(\theta,\ \widehat{y})=\exp\left(-\frac{j\theta}{\hbar}L_y\right)=\sum_{n=0}^{\infty}\frac{(-j\theta)^n}{n!}\left(\frac{L_y}{\hbar}\right)^n.$$

The matrix representation of L_y in L_z basis is given by:

$$L_y\doteq\begin{bmatrix}\langle 1|L_y|1\rangle & \langle 1|L_y|0\rangle & \langle 1|L_y|-1\rangle\\ \langle 0|L_y|1\rangle & \langle 0|L_y|0\rangle & \langle 0|L_y|-1\rangle\\ \langle -1|L_y|1\rangle & \langle -1|L_y|0\rangle & \langle -1|L_y|-1\rangle\end{bmatrix}=\frac{\hbar}{\sqrt{2}}\begin{bmatrix}0 & -j & 0\\ j & 0 & -j\\ 0 & j & 0\end{bmatrix}$$

$$=\frac{j\hbar}{\sqrt{2}}\begin{bmatrix}0 & -1 & 0\\ 1 & 0 & -1\\ 0 & 1 & 0\end{bmatrix}.$$

The first three powers of $L_y/\hbar$ are therefore:

$$\frac{L_y}{\hbar}=\frac{j}{\sqrt{2}}\begin{bmatrix}0 & -1 & 0\\ 1 & 0 & -1\\ 0 & 1 & 0\end{bmatrix},\quad \left(\frac{L_y}{\hbar}\right)^2=-\frac{1}{2}\begin{bmatrix}-1 & 0 & 1\\ 0 & -2 & 0\\ 1 & 0 & -1\end{bmatrix},$$

$$\left(\frac{L_y}{\hbar}\right)^3=\frac{j}{\sqrt{2}}\begin{bmatrix}0 & -1 & 0\\ 1 & 0 & -1\\ 0 & 1 & 0\end{bmatrix}.$$

It is clear that the third power $L_y/\hbar$ is the same as the first power. The rotation operator can therefore be written as:

$$U_R(\theta,\ \widehat{y})=\sum_{n=0}^{\infty}\frac{(-j\theta)^n}{n!}\left(\frac{L_y}{\hbar}\right)^n=I+\sum_{n=0}^{\infty}\frac{(-j\theta)^{2n+1}}{(2n+1)!}\left(\frac{L_y}{\hbar}\right)+\sum_{n=0}^{\infty}\frac{(-j\theta)^{2n}}{(2n)!}\left(\frac{L_y}{\hbar}\right)^2$$

$$=I-j\sin\theta\frac{L_y}{\hbar}-(1-\cos\theta)\left(\frac{L_y}{\hbar}\right)^2.$$

The corresponding representation in matrix form is as follows:

$$U_R(\theta,\hat{y})=\begin{bmatrix}1&0&0\\0&1&0\\0&0&1\end{bmatrix}+\frac{\sin\theta}{\sqrt{2}}\begin{bmatrix}0&-1&0\\1&0&-1\\0&1&0\end{bmatrix}+\frac{1-\cos\theta}{2}\begin{bmatrix}-1&0&1\\0&-2&0\\1&0&-1\end{bmatrix}$$

$$=\begin{bmatrix}\frac{1+\cos\theta}{2}&-\frac{\sin\theta}{\sqrt{2}}&\frac{1-\cos\theta}{2}\\\frac{\sin\theta}{\sqrt{2}}&\cos\theta&-\frac{\sin\theta}{\sqrt{2}}\\\frac{1-\cos\theta}{2}&\frac{\sin\theta}{\sqrt{2}}&\frac{1+\cos\theta}{2}\end{bmatrix}.$$

To obtain the eigenvectors of L_x by using the eigenkets of L_z we have to rotate the eigenkets of L_z by $\theta=\pi/2$, that is, to apply the rotation operator:

$$U_R(\pi/2,\hat{y})=\begin{bmatrix}\frac{1}{2}&-\frac{1}{\sqrt{2}}&\frac{1}{2}\\\frac{1}{\sqrt{2}}&0&-\frac{1}{\sqrt{2}}\\\frac{1}{2}&\frac{1}{\sqrt{2}}&\frac{1}{2}\end{bmatrix}$$

on base kets to obtain:

$$|1\rangle_x=U_R(\pi/2,\hat{y})|1\rangle=\begin{bmatrix}1/2\\1/\sqrt{2}\\1/2\end{bmatrix},\quad |0\rangle_x=U_R(\pi/2,\hat{y})|0\rangle=\begin{bmatrix}-1/\sqrt{2}\\0\\1/\sqrt{2}\end{bmatrix},$$

$$|-1\rangle_x=U_R(\pi/2,\hat{y})|-1\rangle=\begin{bmatrix}1/2\\-1/\sqrt{2}\\1/2\end{bmatrix}.$$

2.12 SPIN-1/2 SYSTEMS

Spin is an intrinsic property of particles, which was deduced from the Stern–Gerlach experiment. Similarly to OAM, to completely characterize the state of the spin, it is sufficient to observe operators S^2 and S_z, whose simultaneous eigenkets can be denoted by $|s,m\rangle$. The corresponding eigenvalue equations are given as follows:

$$S^2|s,\ m\rangle = \hbar^2 s(s+1)|s,\ m\rangle \quad S_z|s,\ m\rangle = m\hbar|s,\ m\rangle$$
$$m = -s,\ -s+1,\ \ldots,\ -1,\ 0,\ 1,\ \ldots s-1,\ s. \tag{2.217}$$

Unlike OAM, spin is not a function of spatial coordinates. To move up and down in states of SAM we introduce the *ladder* operators S_+ and S_-, also known as *raising* and *lowering* operators, by:

$$S_+ = S_x + iS_y \quad S_- = S_x - iS_y. \tag{2.218}$$

The ladder operators act on SAM states as follows:

$$S_\pm|s,\ m\rangle = \hbar\sqrt{s(s+1) - m(m\pm 1)}\left|s,\ m\pm 1\right\rangle \quad S_+|s,\ s\rangle = 0 \quad S_-|s,\ -s\rangle = 0. \tag{2.219}$$

The components of the spin operator satisfy commutation relations similar to those of OAM:

$$[S_x,\ S_y] = j\hbar S_z \quad [S_z,\ S_x] = j\hbar S_y \quad [S_y,\ S_z] = j\hbar S_x. \tag{2.220}$$

Unlike OAM, the spin for a given particle is fixed. For example, the force-carrying boson has spin $s = 1$, the graviton has spin $s = 2$, while electrons and quarks have spin $s = 1/2$. The eigenvalue equation for spin-1/2 particles is:

$$S^2|1/2,\ m\rangle = \frac{3}{4}\hbar^2\left|1/2,\ m\right\rangle \quad S_z|1/2,\ m\rangle = m\hbar|1/2,\ m\rangle \quad m = -1/2,\ 1/2. \tag{2.221}$$

Therefore the measurement of the spin of any spin-1/2 system can have only two possible results: spin-up and spin-down. The basis states for spin-1/2 systems can be labeled by:

$$|+\rangle = |1/2,\ +1/2\rangle = \begin{bmatrix}1\\0\end{bmatrix} \quad |-\rangle = |1/2,\ -1/2\rangle = \begin{bmatrix}0\\1\end{bmatrix}. \tag{2.222}$$

These states are orthonormal, since:

$$\langle +|+\rangle = \langle -|-\rangle = 1 \quad \langle +|-\rangle = \langle -|+\rangle = 0. \tag{2.223}$$

The superposition state $|\psi\rangle$ can be written in terms of bases states by:

$$|\psi\rangle = \alpha|+\rangle + \beta|-\rangle = \begin{bmatrix}\alpha\\ \beta\end{bmatrix}, \quad |\alpha|^2 + |\beta|^2 = 1, \tag{2.224}$$

where $|\alpha|^2$ ($|\beta|^2$) denotes the probability of finding the system in the spin-up (spin-down) state. By using the trigonometric identity $\sin^2(\theta/2) + \cos^2(\theta/2) = 1$, the expansion coefficients α and β can be expressed as follows:

$$\alpha = e^{-j\phi/2}\cos(\theta/2) \quad \beta = e^{j\phi/2}\sin(\theta/2), \tag{2.225}$$

so that:

$$|\psi\rangle = e^{-j\phi/2}\cos(\theta/2)|+\rangle + e^{j\phi/2}\sin(\theta/2)|-\rangle. \tag{2.226}$$

Therefore the single superposition state, also known as quantum bit (qubit), can be visualized as the point (θ,φ) on a unit sphere (Bloch or Poincaré sphere).

The arbitrary operator A can be represented in this basis by:

$$A \doteq \begin{bmatrix} \langle +|A|+\rangle & \langle +|A|-\rangle \\ \langle -|A|+\rangle & \langle -|A|-\rangle \end{bmatrix}. \tag{2.227}$$

For example, the matrix representation of the S_z-operator is given by:

$$S_z \doteq \begin{bmatrix} \langle +|S_z|+\rangle & \langle +|S_z|-\rangle \\ \langle -|S_z|+\rangle & \langle -|S_z|-\rangle \end{bmatrix} = \begin{bmatrix} \langle +|\frac{\hbar}{2}|+\rangle & \langle +|-\frac{\hbar}{2}|-\rangle \\ \langle -|\frac{\hbar}{2}|+\rangle & \langle -|-\frac{\hbar}{2}|-\rangle \end{bmatrix} = \frac{\hbar}{2}\begin{bmatrix} 1 & 0 \\ 0 & -1 \end{bmatrix}. \tag{2.228}$$

The action of ladder operators for spin-1/2 systems can be described as follows:

$$S_+|-\rangle = \hbar|+\rangle \quad S_-|+\rangle = \hbar|-\rangle, \tag{2.229}$$

which can be used for matrix representation in $\{|+\rangle,|-\rangle\}$ basis:

$$\begin{aligned} S_+ &\doteq \begin{bmatrix} \langle +|S_+|+\rangle & \langle +|S_+|-\rangle \\ \langle -|S_+|+\rangle & \langle -|S_+|-\rangle \end{bmatrix} = \begin{bmatrix} 0 & \langle +|\hbar|+\rangle \\ 0 & \langle -|\hbar|+\rangle \end{bmatrix} = \hbar\begin{bmatrix} 0 & 1 \\ 0 & 0 \end{bmatrix} \\ S_- &\doteq \begin{bmatrix} \langle +|S_-|+\rangle & \langle +|S_-|-\rangle \\ \langle -|S_-|+\rangle & \langle -|S_-|-\rangle \end{bmatrix} = \begin{bmatrix} \langle +|\hbar|-\rangle & 0 \\ \langle -|\hbar|-\rangle & 0 \end{bmatrix} = \hbar\begin{bmatrix} 0 & 0 \\ 1 & 0 \end{bmatrix}. \end{aligned} \tag{2.230}$$

The S_x- and S_y-operators can be represented in the same basis by:

$$S_x = (S_+ + S_-)/2 = \frac{\hbar}{2}\begin{bmatrix} 0 & 1 \\ 1 & 0 \end{bmatrix} \quad S_y = (S_+ - S_-)/2j = \frac{\hbar}{2}\begin{bmatrix} 0 & -j \\ j & 0 \end{bmatrix}. \tag{2.231}$$

The action of the S_x-operator on basis kets is given by:

$$\begin{aligned} S_x|+\rangle &= \frac{S_+ + S_-}{2}|+\rangle = \frac{1}{2}S_+|+\rangle + \frac{1}{2}S_-|+\rangle = \frac{\hbar}{2}|-\rangle \\ S_x|-\rangle &= \frac{S_+ + S_-}{2}|-\rangle = \frac{1}{2}S_+|-\rangle + \frac{1}{2}S_-|-\rangle = \frac{\hbar}{2}|+\rangle. \end{aligned} \tag{2.232}$$

The unitary matrix that can be used to transform the S_z basis to the S_x basis is given by:

$$U = \begin{bmatrix} \langle +_x|+\rangle & \langle +_x|-\rangle \\ \langle -_x|+\rangle & \langle -_x|-\rangle \end{bmatrix}, \tag{2.233}$$

where:

$$|+_x\rangle = \frac{1}{\sqrt{2}}(|+\rangle + |-\rangle) \quad \left|-_x\right\rangle = \frac{1}{\sqrt{2}}(|+\rangle - |-\rangle). \tag{2.234}$$

Since $\langle \pm_x | \pm \rangle = \pm 1/\sqrt{2}$ and $\langle \pm_x | \mp \rangle = 1/\sqrt{2}$ we obtain:

$$U = \frac{1}{\sqrt{2}} \begin{bmatrix} 1 & 1 \\ 1 & -1 \end{bmatrix}. \tag{2.235}$$

The application of U to an arbitrary state gives:

$$U|\psi\rangle = U\begin{bmatrix} \alpha \\ \beta \end{bmatrix} = \frac{1}{\sqrt{2}} \begin{bmatrix} \alpha + \beta \\ \alpha - \beta \end{bmatrix}. \tag{2.236}$$

The operator U, also known as the Hadamard gate in quantum computing, is clearly Hermitian since $UU^\dagger = I$, and can be used to diagonalize S_x as follows:

$$US_xU^\dagger = \frac{1}{\sqrt{2}} \begin{bmatrix} 1 & 1 \\ 1 & -1 \end{bmatrix} \frac{\hbar}{2} \begin{bmatrix} 0 & 1 \\ 1 & 0 \end{bmatrix} \frac{1}{\sqrt{2}} \begin{bmatrix} 1 & 1 \\ 1 & -1 \end{bmatrix} = \frac{\hbar}{2} \begin{bmatrix} 1 & 0 \\ 0 & -1 \end{bmatrix}.$$

From the matrix representation of the spin operators S_x, S_y, S_z and the matrix representation of an operator A we can represent the spin operators in outer product representation as follows:

$$\begin{aligned} S_x &= \frac{\hbar}{2} \begin{bmatrix} 0 & 1 \\ 1 & 0 \end{bmatrix} = \frac{\hbar}{2}(\langle + | - \rangle + \langle - | + \rangle), \\ S_y &= \frac{\hbar}{2} \begin{bmatrix} 0 & -j \\ j & 0 \end{bmatrix} = \frac{j\hbar}{2}(-\langle + | - \rangle + \langle - | + \rangle), \\ S_z &= \frac{\hbar}{2} \begin{bmatrix} 1 & 0 \\ 0 & -1 \end{bmatrix} = \frac{\hbar}{2}(\langle + | + \rangle - \langle - | - \rangle). \end{aligned} \tag{2.237}$$

The projection operators $P_+ = |+\rangle\langle+|$ and $P_- = |-\rangle\langle-|$ can be represented in matrix form as follows:

$$\begin{aligned} P_+ &\doteq \begin{bmatrix} \langle + |(| + \rangle\langle + |)| + \rangle & \langle + |(| + \rangle\langle + |)| - \rangle \\ \langle - |(| + \rangle\langle + |)| + \rangle & \langle - |(| + \rangle\langle + |)| - \rangle \end{bmatrix} = \begin{bmatrix} 1 & 0 \\ 0 & 0 \end{bmatrix} \\ P_+ &\doteq \begin{bmatrix} \langle + |(| - \rangle\langle - |)| + \rangle & \langle + |(| - \rangle\langle - |)| - \rangle \\ \langle - |(| - \rangle\langle - |)| + \rangle & \langle - |(| - \rangle\langle - |)| - \rangle \end{bmatrix} = \begin{bmatrix} 0 & 0 \\ 0 & 1 \end{bmatrix}. \end{aligned} \tag{2.238}$$

Clearly, the projection operators satisfy the completeness relation:

$$P_+ + P_- = I.$$

2.12.1 Pauli Operators (Revisited)

By using the standard basis of the S_z-operator we have shown that spin operators can be represented as follows:

$$S_x = \frac{\hbar}{2}\begin{bmatrix} 0 & 1 \\ 1 & 0 \end{bmatrix} = \frac{\hbar}{2}X, \quad S_y = \frac{\hbar}{2}\begin{bmatrix} 0 & -j \\ j & 0 \end{bmatrix} = \frac{\hbar}{2}Y, \quad S_z = \frac{\hbar}{2}\begin{bmatrix} 1 & 0 \\ 0 & -1 \end{bmatrix} = \frac{\hbar}{2}Z; \quad X$$
$$= \begin{bmatrix} 0 & 1 \\ 1 & 0 \end{bmatrix}, \quad Y = \begin{bmatrix} 0 & -j \\ j & 0 \end{bmatrix}, \quad Z = \begin{bmatrix} 1 & 0 \\ 0 & -1 \end{bmatrix}, \tag{2.239}$$

where X, Y, and Z denote the Pauli matrices, which are often denoted in the literature by $\sigma_{x\text{-}}$, σ_y, and σ_z, respectively. It can be simply shown that the Pauli matrices satisfy the following commutation and anticommutation relations:

$$[\sigma_k, \sigma_l] = 2j\varepsilon_{klm}\sigma_m, \ \varepsilon_{klm} = \begin{cases} 1, \text{ for cyclic permutations of } k, \ l \text{ and } m \\ -1, \text{ for anticyclic permutations of } k, \ l \text{ and } m. \\ 0, \text{ otherwise} \end{cases}$$
$$\{\sigma_k, \sigma_l\} = 2\delta_{kl} \tag{2.240}$$

By simple inspection of Pauli matrices we conclude that $\mathrm{Tr}(\sigma_k) = 0$. Furthermore, we show that $\mathrm{Tr}(\sigma_k\sigma_l) = 2\delta_{kl}$. This property can be proved by summing the commutation and anticommutation relations to obtain:

$$\begin{aligned}[\sigma_k, \sigma_l] + \{\sigma_k, \sigma_l\} &= \sigma_k\sigma_l + \sigma_l\sigma_k + \sigma_k\sigma_l - \sigma_l\sigma_k = 2\sigma_k\sigma_l \\ &= 2j\varepsilon_{klm}\sigma_m + 2\delta_{kl} \Rightarrow \sigma_k\sigma_l = j\varepsilon_{klm}\sigma_m + \delta_{kl}.\end{aligned} \tag{2.241}$$

By applying the Tr-operator on the last equality in Eq. (2.241) we obtain:

$$\mathrm{Tr}(\sigma_k\sigma_l) = \mathrm{Tr}(j\varepsilon_{klm}\sigma_m + \delta_{kl}) = j\varepsilon_{klm}\mathrm{Tr}(\sigma_m) + \mathrm{Tr}(\delta_{kl}I) = \begin{cases} 2, \ k = l \\ 0, \ k \neq l \end{cases} = 2\delta_{kl}. \tag{2.242}$$

Because the eigenvalues of spin operators S_k ($k = x,y,z$) are $\pm\hbar/2$ and $S_k = \hbar\sigma_k/2$, it is clear that eigenvalues of Pauli matrices are ± 1, while the eigenkets are the same as those of S_k, namely $|+\rangle$ and $|-\rangle$:

$$Z|+\rangle = |+\rangle \quad Z|-\rangle = -|-\rangle. \tag{2.243}$$

From the definition of Pauli matrices, it is clear that $X^2 = Y^2 = Z^2 = I$. By defining the ladder operator as $\sigma_\pm = (\sigma_x \pm \sigma_y)/2$, it is obvious that the action of ladder operators on basis states is as follows:

$$\sigma_+|+\rangle = 0 \quad \sigma_+|-\rangle = |+\rangle \quad \sigma_-|+\rangle = |-\rangle \quad \sigma_-|-\rangle = 0. \tag{2.244}$$

Matrix representation of ladder operators can be easily obtained from definition formulas and matrix representation of Pauli operators:

$$\sigma_+ \doteq \begin{bmatrix} 0 & 1 \\ 0 & 0 \end{bmatrix} \quad \sigma_- \doteq \begin{bmatrix} 0 & 0 \\ 1 & 0 \end{bmatrix}. \tag{2.245}$$

An arbitrary operator A from a 2D Hilbert space can be represented in terms of Pauli operators as follows:

$$A = \frac{1}{2}(a_0 I + \boldsymbol{a} \cdot \boldsymbol{\sigma}), \quad a_0 = \mathrm{Tr}(A), \quad \boldsymbol{a} = \mathrm{Tr}(A\boldsymbol{\sigma}), \quad \boldsymbol{\sigma} = (X, Y, Z). \tag{2.246}$$

This property can be proved by starting from matrix representation of operator A:

$$A = \begin{bmatrix} a & b \\ c & d \end{bmatrix}.$$

By calculating the dot product of $\boldsymbol{a}$ and $\boldsymbol{\sigma}$ we obtain:

$$\begin{aligned}
\boldsymbol{a} \cdot \boldsymbol{\sigma} = \; & \mathrm{Tr}(AX)X + \mathrm{Tr}(AY)Y + \mathrm{Tr}(AZ)Z = \mathrm{Tr}\left(\begin{bmatrix} a & b \\ c & d \end{bmatrix}\begin{bmatrix} 0 & 1 \\ 1 & 0 \end{bmatrix}\right)\sigma_x \\
& + \mathrm{Tr}\left(\begin{bmatrix} a & b \\ c & d \end{bmatrix}\begin{bmatrix} 0 & -j \\ j & 0 \end{bmatrix}\right)\sigma_y + \mathrm{Tr}\left(\begin{bmatrix} a & b \\ c & d \end{bmatrix}\begin{bmatrix} 1 & 0 \\ 0 & -1 \end{bmatrix}\right)\sigma_z \\
& = \mathrm{Tr}\left(\begin{bmatrix} b & a \\ d & c \end{bmatrix}\right)\sigma_x + \mathrm{Tr}\left(\begin{bmatrix} jb & -ja \\ jd & -jc \end{bmatrix}\right)\sigma_y + \mathrm{Tr}\left(\begin{bmatrix} a & -b \\ c & -d \end{bmatrix}\right)\sigma_z \\
& = (b + c)\sigma_x + (jb - jc)\sigma_y + (a - d)\sigma_z.
\end{aligned} \tag{2.247}$$

By using Eq. (2.247), it follows that:

$$\begin{aligned}
\frac{1}{2}(a_0 I + \boldsymbol{a} \cdot \boldsymbol{\sigma}) &= \frac{1}{2}\left(\begin{bmatrix} a+d & 0 \\ 0 & a+d \end{bmatrix} + \begin{bmatrix} a-d & b+c+b-c \\ b+c-b+c & -a+d \end{bmatrix}\right) \\
&= \begin{bmatrix} a & b \\ c & d \end{bmatrix} = A,
\end{aligned} \tag{2.248}$$

which proves Eq. (2.246).

The projection operators can be represented in this form as follows:

$$P_+ = \frac{1}{2}(I + Z) \quad P_- = \frac{1}{2}(I - Z). \tag{2.249}$$

Another interesting result that will be used later in the quantum circuits chapter is:

$$e^{j\theta X} = I\cos\theta + jX\sin\theta. \tag{2.250}$$

This identity can be proved by starting from the following expansions:

$$e^x = 1 + x + \frac{x^2}{2!} + \frac{x^3}{3!} + \cdots \quad \sin x = x - \frac{x^3}{3!} + \frac{x^2}{5!} + \cdots$$
$$\cos x = 1 - \frac{x^2}{2!} + \frac{x^4}{4!} - \cdots. \tag{2.251}$$

By applying Eq. (2.251) we obtain:

$$\begin{aligned} e^{j\theta X} &= I + j\theta X + \frac{(j\theta X)^2}{2!} + \frac{(j\theta X)^3}{3!} + \frac{(j\theta X)^4}{4!} + \frac{(j\theta X)^5}{5!} + \cdots \\ &= I\left(1 - \frac{\theta^2}{2!} + \frac{\theta^4}{4!} - \cdots\right) + jX\left(\theta - \frac{\theta^3}{3!} + \frac{\theta^2}{5!} + \cdots\right) \\ &= I\cos\theta + jX\sin\theta, \end{aligned} \tag{2.252}$$

where we have used the identity $X^{2n} = (X^2)^n = I$.

2.12.2 Density Operator for Spin-1/2 Systems

Let us observe the pure state given by density operator $\rho_1 = |+_y\rangle\langle +_y|$ and a completely mixed state given by density operator $\rho_2 = 0.5|+\rangle\langle+| + 0.5|-\rangle\langle-|$. The corresponding matrix representations are given by:

$$\begin{aligned} \rho_1 &= |+_y\rangle\langle +_y| = \frac{1}{\sqrt{2}}(|+\rangle + j|-\rangle)\frac{1}{\sqrt{2}}(\langle+| - j\langle-|) \\ &= \frac{1}{2}(|+\rangle\langle+| - j|+\rangle\langle-| + j|-\rangle\langle+| + |-\rangle\langle-|) = \begin{bmatrix} 1/2 & -j/2 \\ j/2 & 1/2 \end{bmatrix} \\ \rho_2 &= \frac{1}{2}|+\rangle\langle+| + \frac{1}{2}|-\rangle\langle-| = \begin{bmatrix} 1/2 & 0 \\ 0 & 1/2 \end{bmatrix}. \end{aligned} \tag{2.253}$$

It can easily be verified that both density operators are the Hermitian and have unit trace. The corresponding traces for ρ^2 can be found by:

$$\begin{aligned} \mathrm{Tr}(\rho_1^2) &= \mathrm{Tr}\left(\begin{bmatrix} 1/2 & -j/2 \\ j/2 & 1/2 \end{bmatrix}\begin{bmatrix} 1/2 & -j/2 \\ j/2 & 1/2 \end{bmatrix}\right) = 1 \\ \mathrm{Tr}(\rho_2^2) &= \mathrm{Tr}\left(\begin{bmatrix} 1/2 & 0 \\ 0 & 1/2 \end{bmatrix}\begin{bmatrix} 1/2 & 0 \\ 0 & 1/2 \end{bmatrix}\right) = \frac{1}{2} < 1. \end{aligned} \tag{2.254}$$

If we measure S_z, the probability of finding the result $\hbar/2$ is:

$$\mathrm{Tr}(\rho_1|+\rangle\langle+|) = \langle+|\rho_1|+\rangle = \begin{bmatrix}1 & 0\end{bmatrix}\begin{bmatrix}1/2 & -j/2 \\ j/2 & 1/2\end{bmatrix}\begin{bmatrix}1 \\ 0\end{bmatrix} = \frac{1}{2}$$
$$\mathrm{Tr}(\rho_2|+\rangle\langle+|) = \langle+|\rho_1|+\rangle = \begin{bmatrix}1 & 0\end{bmatrix}\begin{bmatrix}1/2 & 0 \\ 0 & 1/2\end{bmatrix}\begin{bmatrix}1 \\ 0\end{bmatrix} = \frac{1}{2}. \tag{2.255}$$

If, on the other hand, we measure S_y, the probability of finding the result $\hbar/2$ is:

$$\mathrm{Tr}(\rho_1|+_y\rangle\langle+_y|) = \langle+_y|\rho_1|+_y\rangle = \langle+_y|(|+_y\rangle\langle+_y|)|+_y\rangle = 1$$
$$\mathrm{Tr}(\rho_2|+_y\rangle\langle+_y|) = \langle+_y|\rho_1|+_y\rangle = \frac{1}{\sqrt{2}}\begin{bmatrix}1 & -j\end{bmatrix}\begin{bmatrix}1/2 & 0 \\ 0 & 1/2\end{bmatrix}\frac{1}{\sqrt{2}}\begin{bmatrix}1 \\ j\end{bmatrix} = \frac{1}{2}. \tag{2.256}$$

2.12.3 Time Evolution of Spin-1/2 Systems

The Hamiltonian of a particle of spin $\boldsymbol{S}$ in a magnetic field $\boldsymbol{B}$ is given by:

$$H = -\gamma \boldsymbol{S}\cdot\boldsymbol{B}, \quad \gamma = 2e/mc. \tag{2.257}$$

If a constant magnetic field is applied in the z-direction, the corresponding Hamiltonian is given by:

$$H = -\gamma S_z B_z. \tag{2.258}$$

Because B_z and gyromagnetic ratio γ are constant, the eigenkets of Hamiltonian H are the same as the eigenkets of S_z:

$$H|+\rangle = -\gamma B_z S_z|+\rangle = -\gamma B_z\frac{\hbar}{2}|+\rangle = E_+|+\rangle,\ E_+ = -\gamma B_z\frac{\hbar}{2}$$
$$H|-\rangle = -\gamma B_z S_z|-\rangle = \gamma B_z\frac{\hbar}{2}|-\rangle = E_-|+\rangle,\ E_- = \gamma B_z\frac{\hbar}{2}. \tag{2.259}$$

The time evolution of spin kets is governed by the Schrödinger equation:

$$j\hbar\frac{d}{dt}|\psi\rangle = H|\psi\rangle, \quad |\psi\rangle = \begin{bmatrix}\alpha \\ \beta\end{bmatrix}, \tag{2.260}$$

and by solving it we obtain:

$$|\psi(t)\rangle = \begin{bmatrix}\alpha e^{-\frac{j}{\hbar}E_+ t} \\ \beta e^{-\frac{j}{\hbar}E_- t}\end{bmatrix}. \tag{2.261}$$

For example, if the initial state was $|+_y\rangle$, it would interesting to determine the probability of finding $\pm\hbar/2$ if we measure S_x or S_z at time instance t. For initial state $|\psi\rangle = |+_y\rangle = \begin{bmatrix} 1/\sqrt{2} & 1/\sqrt{2} \end{bmatrix}^{\mathrm{T}}$, by applying Eq. (2.261) we obtain:

$$|\psi(t)\rangle = \frac{1}{\sqrt{2}}\begin{bmatrix} e^{-\frac{j}{\hbar}E_+ t} \\ e^{-\frac{j}{\hbar}E_- t} \end{bmatrix} = \frac{1}{\sqrt{2}}\begin{bmatrix} e^{j\gamma B_z t/2} \\ e^{-j\gamma B_z t/2} \end{bmatrix}, \tag{2.262}$$

so that the probability of finding $\pm\hbar/2$ is:

$$\begin{aligned} \Pr_y(\hbar/2) &= |\langle +_y|\psi(t)\rangle|^2 = \cos^2\left(\frac{\gamma B_z t}{2}\right), \\ \Pr_y(-\hbar/2) &= |\langle -_y|\psi(t)\rangle|^2 = \sin^2\left(\frac{\gamma B_z t}{2}\right). \end{aligned} \tag{2.263}$$

In similar fashion:

$$\langle +|\psi(t)\rangle = \frac{1}{\sqrt{2}} e^{j\gamma B_z/t}, \tag{2.264}$$

so that:

$$\Pr_z(\hbar/2) = |\langle +|\psi(t)\rangle|^2 = \frac{1}{2}, \quad \Pr_z(-\hbar/2) = |\langle -|\psi(t)\rangle|^2 = \frac{1}{2}. \tag{2.265}$$

For the state $|+_y\rangle$ we can determine $\langle S_x\rangle$ and $\langle S_x\rangle$ by:

$$\begin{aligned} \langle S_x\rangle &= \sum_i p_i\lambda_i = \cos^2\left(\frac{\gamma B_z t}{2}\right)\cdot\frac{\hbar}{2} + \sin^2\left(\frac{\gamma B_z t}{2}\right)\cdot\left(-\frac{\hbar}{2}\right) = \frac{\hbar}{2}\cos(\gamma B_z t) \\ \langle S_z\rangle &= \sum_i p_i\lambda_i = \frac{1}{2}\cdot\frac{\hbar}{2} + \frac{1}{2}\cdot\left(-\frac{\hbar}{2}\right) = 0. \end{aligned} \tag{2.266}$$

In the rest of this section we provide several examples that will be of interest for various physical implementations of quantum computers.

Example. A particle is under the influence of a periodic magnetic field $\boldsymbol{B} = B_0 \sin(\omega t)\hat{z}$. Determine the Hamiltonian and the state of the particle at time instance t. If the initial state was $|+_y\rangle$, determine the probability that the result of measurement S_y at t is $\hbar/2$.

To determine the Hamiltonian we use Eq. (2.257) to obtain:

$$H = -\gamma \boldsymbol{S}\cdot\boldsymbol{B} = -\gamma B_0 \sin(\omega t) S_z. \tag{2.267}$$

We now solve the Schrödinger equation:

$$j\hbar\frac{d}{dt}|\psi\rangle = H|\psi\rangle,$$

for arbitrary state $|\psi\rangle = \begin{bmatrix} \alpha & \beta \end{bmatrix}^{\mathrm{T}}$ as follows:

$$j\hbar\begin{bmatrix}\dot{\alpha}(t)\\ \dot{\beta}(t)\end{bmatrix} = -\gamma B_0 \sin(\omega t)\frac{\hbar}{2}\begin{bmatrix}\alpha(t)\\ -\beta(t)\end{bmatrix}. \tag{2.268}$$

The first differential equation:

$$\frac{d\alpha}{dt} = j\frac{\gamma B_0}{2}\sin(\omega t)$$

can simply be solved by the method of separation variables:

$$\frac{d\alpha}{\alpha} = j\frac{\gamma B_0}{2}\sin(\omega t)dt,$$

as follows:

$$\alpha(t) = \alpha(0)\exp\left(-j\frac{\gamma B_0}{2\omega}\cos(\omega t)\right). \tag{2.269}$$

In similar fashion, by solving the second differential equation we obtain:

$$\beta(t) = \beta(0)\exp\left(j\frac{\gamma B_0}{2\omega}\cos(\omega t)\right). \tag{2.270}$$

For initial state $|\psi(0)\rangle = |+_y\rangle = [1 \ \ j]^{\mathrm{T}}$, meaning that $\alpha(0) = 1/\sqrt{2}$, $\beta(0) = j/\sqrt{2}$, we obtain from Eqs. (2.269) and (2.270):

$$|\psi(t)\rangle = \frac{1}{\sqrt{2}}\begin{bmatrix}\exp\left(-j\frac{\gamma B_0}{2\omega}\cos(\omega t)\right)\\ -j\exp\left(j\frac{\gamma B_0}{2\omega}\cos(\omega t)\right)\end{bmatrix}$$

$$= \frac{1}{\sqrt{2}}\begin{bmatrix}\cos\left(\frac{\gamma B_0}{2\omega}\cos(\omega t)\right) - j\sin\left(\frac{\gamma B_0}{2\omega}\cos(\omega t)\right)\\ \sin\left(\frac{\gamma B_0}{2\omega}\cos(\omega t)\right) - j\cos\left(j\frac{\gamma B_0}{2\omega}\cos(\omega t)\right)\end{bmatrix}$$

$$= \cos\left(\frac{\gamma B_0}{2\omega}\cos(\omega t)\right)\frac{1}{\sqrt{2}}\begin{bmatrix}1\\ j\end{bmatrix} - j\sin\left(\frac{\gamma B_0}{2\omega}\cos(\omega t)\right)\frac{1}{\sqrt{2}}\begin{bmatrix}1\\ -j\end{bmatrix}$$

$$= \cos\left(\frac{\gamma B_0}{2\omega}\cos(\omega t)\right)|+_y\rangle - j\sin\left(\frac{\gamma B_0}{2\omega}\cos(\omega t)\right)|-_y\rangle,$$

and the corresponding probability is:

$$|\langle +_y|\psi(t)\rangle|^2 = \cos^2\left(\frac{\gamma B_0}{2\omega}\cos(\omega t)\right).$$

Example. A particle is under the influence of a periodic magnetic field $\boldsymbol{B} = B_0\cos(\omega t)\widehat{x} + B_0\sin(\omega t)\widehat{y}$. Determine the Hamiltonian of the system. Determine the transformation that will allow us to write a time-invariant Hamiltonian. If the initial state was $|+\rangle$, determine the probability that at time instance t, the system is in state $|-\rangle$. Finally, determine the first time instance $t > 0$ for which the system is with certainty in state $|-\rangle$.

To determine the Hamiltonian we again use Eq. (2.257) to obtain:

$$\begin{aligned} H &= -\gamma \boldsymbol{S}\cdot\boldsymbol{B} = -\gamma B_0 \cos(\omega t) S_x - \gamma B_0 \sin(\omega t) S_y \\ &= \omega_0 (\cos(\omega t) S_x + \sin(\omega t) S_y), \quad \omega_0 = -\gamma B_0. \end{aligned} \tag{2.271}$$

Based on matrix representation of S_x and S_y, the Hamiltonian can be represented by:

$$\begin{aligned} H &= \frac{\hbar\omega_0}{2}\left(\begin{bmatrix} 0 & \cos(\omega t) \\ \cos(\omega t) & 0 \end{bmatrix} + \begin{bmatrix} 0 & -j\sin(\omega t) \\ j\sin(\omega t) & 0 \end{bmatrix} \right) \\ &= \frac{\hbar\omega_0}{2} \begin{bmatrix} 0 & \exp(-j\omega t) \\ \exp(j\omega t) & 0 \end{bmatrix}. \end{aligned} \tag{2.272}$$

We now need to solve the Schrödinger equation:

$$j\hbar \frac{d}{dt}|\psi\rangle = H|\psi\rangle,$$

which for arbitrary state $|\psi\rangle = [\alpha \quad \beta]^{\mathrm{T}}$ can be written as follows:

$$j\hbar \begin{bmatrix} \dot{\alpha}(t) \\ \dot{\beta}(t) \end{bmatrix} = \frac{\hbar\omega_0}{2} \begin{bmatrix} 0 & \exp(-j\omega t) \\ \exp(j\omega t) & 0 \end{bmatrix} \begin{bmatrix} \alpha(t) \\ -\beta(t) \end{bmatrix}, \tag{2.273}$$

and the corresponding differential equations resulting from Eq. (2.273) are:

$$\frac{d\alpha}{dt} = \frac{-j\omega_0}{2}\exp(-j\omega t)\beta \quad \frac{d\beta}{dt} = \frac{-j\omega_0}{2}\exp(j\omega t)\alpha, \tag{2.274}$$

which are mutually dependent. By substituting $\alpha = \exp(-j\omega t/2)\alpha'$, $\beta = \exp(j\omega t/2)\beta'$, from Eq. (2.274) we obtain:

$$\begin{aligned} \dot{\alpha} &= -\frac{j\omega}{2}\exp(-j\omega t/2)\alpha' + \exp(-j\omega t/2)\dot{\alpha}', \\ \dot{\beta} &= \frac{j\omega}{2}\exp(j\omega t/2)\beta' + \exp(j\omega t/2)\dot{\beta}', \end{aligned} \tag{2.275}$$

which is equivalent to:

$$\frac{j\omega_0}{2}\exp(-j\omega t/2)\beta = -\frac{j\omega}{2}\alpha' + \dot{\alpha}', \quad \frac{j\omega_0}{2}\exp(j\omega t/2)\alpha = \frac{j\omega}{2}\beta' + \dot{\beta}', \tag{2.276}$$

or expressed in more compact form:

$$j\dot{\alpha}' = -\frac{\omega}{2}\alpha' + \frac{\omega_0}{2}\beta' \quad \frac{\omega_0}{2}\alpha' + \frac{\omega}{2}\beta' = j\dot{\beta}'. \tag{2.277}$$

The Hamiltonian corresponding to Eq. (2.276) is:

$$H = \begin{bmatrix} -\dfrac{\omega}{2} & \dfrac{\omega_0}{2} \\ \dfrac{\omega_0}{2} & \dfrac{\omega}{2} \end{bmatrix}. \tag{2.278}$$

To solve the Schrödinger equation in transform coordinates we differentiate the corresponding component equations to obtain:

$$\ddot{\alpha}' = \frac{\omega}{2}\left(-\frac{\omega}{2}\alpha' + \frac{\omega_0}{2}\beta'\right) - \frac{\omega_0}{2}\left(\frac{\omega_0}{2}\alpha' + \frac{\omega}{2}\beta'\right),$$
$$\ddot{\beta}' = -\frac{\omega_0}{2}\left(-\frac{\omega}{2}\alpha' + \frac{\omega_0}{2}\beta'\right) - \frac{\omega}{2}\left(\frac{\omega_0}{2}\alpha' + \frac{\omega}{2}\beta'\right),$$

which is equivalent to:

$$\ddot{\alpha}' = -\frac{\omega^2 + \omega_0^2}{4}\alpha' \quad \ddot{\beta}' = -\frac{\omega^2 + \omega_0^2}{4}\beta'. \tag{2.279}$$

The corresponding solutions for Eq. (2.279) are:

$$\alpha'(t) = A_1 \cos(\Gamma t) + A_2 \sin(\Gamma t), \quad \Gamma = \sqrt{\frac{\omega^2 + \omega_0^2}{4}}. \tag{2.280}$$

Since $\alpha'(0) = \alpha(t)\exp(j\omega t/2)|_{t=0} = 1$ we obtain $A_1 = 1$, $A_2 = 0$ so that:

$$\alpha'(t) = \cos(\Gamma t) \Rightarrow \alpha(t) = \exp(-j\omega t/2)\alpha'(t) = \exp(-j\omega t/2)\cos(\Gamma t). \tag{2.281}$$

Because $|\alpha(t)|^2 + |\beta(t)|^2 = 1$, it follows that $\beta(t) = \exp(j\omega t/2)\sin(\Gamma t)$ and the state evolution is described by:

$$|\psi(t)\rangle = \begin{bmatrix} \exp(-j\omega t/2)\cos(\Gamma t) \\ \exp(j\omega t/2)\sin(\Gamma t) \end{bmatrix}, \tag{2.282}$$

and the corresponding probability is:

$$|\langle -|\psi\rangle|^2 = \sin^2(\Gamma t). \tag{2.283}$$

Since $|\langle -|\psi\rangle|^2 = 1$, it follows that $\Gamma t = \pi/2$ and the first time instance for which the system is in state $|-\rangle$ is:

$$t = \frac{4\pi}{\sqrt{\omega^2 + \omega_0^2}}.$$

2.13 HYDROGEN-LIKE ATOMS AND BEYOND

A hydrogen atom is a bound system, consisting of a proton and a neutron, with potential given by:

$$V(r) = -\frac{1}{4\pi\varepsilon_0}\frac{e^2}{r}, \tag{2.284}$$

where e is an electron charge. Therefore potential is the only function of the radial coordinate system, and because of spherical symmetry it is convenient to use the spherical coordinate system in which the Laplacian is defined by:

$$\nabla^2 = \frac{1}{r^2}\frac{\partial}{\partial r}\left(r^2\frac{\partial}{\partial r}\right) + \frac{1}{r^2 \sin\theta}\frac{\partial}{\partial \theta}\left(\sin\theta\frac{\partial}{\partial \theta}\right) + \frac{1}{r^2 \sin^2\theta}\frac{\partial^2}{\partial \phi^2}. \tag{2.285}$$

The angular momentum operator L^2 in spherical coordinates is given by:

$$L^2 = -\hbar^2\left[\frac{1}{\sin\theta}\frac{\partial}{\partial \theta}\left(\sin\theta\frac{\partial}{\partial \theta}\right) + \frac{1}{\sin^2\theta}\frac{\partial^2}{\partial \phi^2}\right]. \tag{2.286}$$

The Hamiltonian can be written by:

$$H = -\frac{\hbar^2}{2m}\frac{1}{r^2}\frac{\partial}{\partial r}\left(r^2\frac{\partial}{\partial r}\right) + \frac{1}{2mr^2}L^2 + V(r). \tag{2.287}$$

Because the operators L^2 and L_z have common eigenkets, the Hamiltonian leads to the following three equations:

$$\begin{aligned} H\Psi(r,\theta,\phi) &= E\Psi(r,\theta,\phi) \\ L^2\Psi(r,\theta,\phi) &= \hbar^2 l(l+1)\Psi(r,\theta,\phi) \\ L_z\Psi(r,\theta,\phi) &= m\hbar\Psi(r,\theta,\phi). \end{aligned} \tag{2.288}$$

With this problem we can associate three quantum numbers: (1) the *principal quantum number*, n, corresponding to the energy (originating from H); (2) the *azimuthal quantum number*, l, representing the angular momentum (originating from L^2), and (3) the *magnetic quantum number*, m, originating from L_z. To solve Eq. (2.288) we can use the method of separation of variables: $\Psi(r,\theta,\phi) = R(r)\Theta(\theta)\Phi(\phi)$. Since the $\Theta(\theta)\Phi(\phi)$ term has already been found in the section on OAM to be related to spherical harmonics $\Theta(\theta)\Phi(\phi) = Y_l^m(\theta,\phi)$ we are left with the radial equation to solve:

$$-\frac{\hbar^2}{2mr}\frac{d^2}{dr^2}[rR_{nl}(r)] + \left[\frac{l(l+1)\hbar^2}{2mr^2} + V(r)\right]R_{nl}(r) = ER_{nl}(r). \tag{2.289}$$

By substituting the Coulomb potential into the radial equation we obtain:

$$-\frac{\hbar^2}{2mr}\frac{d^2}{dr^2}[rR_{nl}(r)] + \left[\frac{l(l+1)\hbar^2}{2mr^2} - \frac{1}{4\pi\varepsilon_0}\frac{e^2}{r}\right]R_{nl}(r) = ER_{nl}(r). \tag{2.290}$$

The solution of the radial equation can be written as:

$$R_{nl}(r) = \sqrt{\left(\frac{2}{na_H}\right)\frac{(n-l-1)!}{2n[(n-1)!]^2}}e^{-r/2a_H}\left(\frac{r}{a_H}\right)^l L_{n+1}^{2l+1}\left(\frac{r}{a_H}\right), \tag{2.291}$$

where a_H is the first Bohr radius (the lowest energy orbit radius) ($a_H = 0.0529$ nm) and $L_{n+1}^{2l+1}(r/a_H)$ are corresponding associated Lageurre polynomials defined by:

$$L_n^{\alpha}(x) = \frac{x^{-\alpha}e^x}{n!}\frac{d^n}{dx^n}\left(e^{-x}x^{n+\alpha}\right). \tag{2.292}$$

The radial portion of the wavefunction is typically normalized as follows:

$$\int_0^{\infty} r^2|R(r)|^2 dr = 1, \tag{2.293}$$

while the expectation value is defined by:

$$\langle r^k \rangle = \int_0^{\infty} r^{2+k}|R(r)|^2 dr. \tag{2.294}$$

The following expectation values of the hydrogen atom can be determined in analytic form:

$$\langle r \rangle = \frac{a_H}{2}\left[3n^2 - l(l+1)\right] \quad \langle r^2 \rangle = \frac{a_H^2 n^2}{2}\left[5n^2 + 1 - 3l(l+1)\right] \quad \langle r^{-1} \rangle = \frac{1}{a_0 n^2}. \tag{2.295}$$

The complete wavefunction is the product of radial wavefunction and spherical harmonics:

$$\Psi(r, \theta, \phi) = \sqrt{\left(\frac{2}{na_H}\right)\frac{(n-l-1)!}{2n[(n-1)!]^2}}e^{-r/2a_H}\left(\frac{r}{a_H}\right)^l L_{n+1}^{2l+1}\left(\frac{r}{a_H}\right)Y_l^m(\theta, \phi). \tag{2.296}$$

By substituting Eq. (2.207) into Eq. (2.296) we obtain:

$$\Psi(\rho, \theta, \phi) = C_{n,l,m}e^{-\rho/2}(\rho)^l L_{n+1}^{2l+1}(\rho)P_l^{|m|}(\cos\theta)e^{jm\phi}, \tag{2.297}$$

where $\rho = r/a_H$, and $C_{n,m,l}$ is the normalization constant obtained as a product of normalization constants in Eqs. (2.207) and (2.296). The energy levels in the hydrogen atom are only functions of n and are given by:

$$E_n = \frac{E_1}{n^2}, \quad E_1 = -\frac{m^2 e^4}{32\pi^2\varepsilon_0^2\hbar^2} = -13.6\text{ eV}. \tag{2.298}$$

Because the values that l can take are $\{0,1,\ldots,n-1\}$, while the values that m can take are $\{-l, -l+1,\ldots,l-1, l\}$, and since the radial component is not a function of m, the number of states with the same energy (the total *degeneracy* of the energy level E_n) is:

$$2\sum_{l=0}^{n-1}(2l+1) = 2n^2. \tag{2.299}$$

Table 2.1 Hydrogen atom eigenfunctions.

State	n	l	m	Eigenfunction
1s	1	0	0	$e^{-\rho/2}$
2s	2	0	0	$e^{-\rho/2}(1-\rho)$
2p	2	1	−1 0 1	$e^{-\rho/2}\rho\begin{cases}\sin\theta e^{-j\phi}\\ \cos\theta\\ \sin\theta e^{j\phi}\end{cases}$
3s	3	0	0	$e^{-\rho/2}(\rho^2-4\rho+2)$
3p	3	1	−1 0 1	$e^{-\rho/2}(\rho^2-2\rho)\begin{cases}\sin\theta e^{-j\phi}\\ \cos\theta\\ \sin\theta e^{j\phi}\end{cases}$
3d	3	2	−2 −1 0 1 2	$e^{-\rho/2}\rho^2\begin{cases}\sin^2\theta e^{-j2\phi}\\ \sin\theta\cos\theta e^{-j\phi}\\ 1-3\cos^2\theta\\ \sin\theta\cos\theta e^{j\phi}\\ \sin^2\theta e^{j2\phi}\end{cases}$

The states in which $l = 0,1,2,3,4$ are traditionally called s, p, d, f, and g, respectively. The simpler eigenfunctions of the hydrogen atom, by ignoring the normalization constant, are given in Table 2.1, which is obtained based on Eq. (2.297).

Example. The electron in a hydrogen atom is in the state $\psi_{nlm} = R_{32}\left(\frac{1}{\sqrt{6}}Y_2^1 + \frac{1}{\sqrt{2}}Y_2^0 + \frac{1}{\sqrt{3}}Y_2^{-1}\right)$. Determine the energy of the electron. If L^2 is measured, determine the possible results of measurement. If, on the other hand, L_z is measured, determine the results of measurements and corresponding probabilities. Finally, determine the expectation of L_z.

The energy level is only a function of n, so that for $n = 3$ we obtain:

$$E_3 = \frac{E_1}{3^2} = -13.6/9\ \text{eV} = -1.51\ \text{eV}.$$

The measurements of L^2 will give the result $\hbar^2 l(l+1)|_{l=2} = 6\hbar^2$. The measurements results for L_z will only be a function of the angular portion of wavefunction $\frac{1}{\sqrt{6}}Y_2^1 + \frac{1}{\sqrt{2}}Y_2^0 + \frac{1}{\sqrt{3}}Y_2^{-1}$. Because L_z satisfies the following eigenvalue equation:

$L_z Y_l^m = m\hbar Y_l^m$, it follows that possible results of measurements will be $\hbar$, 0, and $-\hbar$, which occur with probabilities 1/6, 1/2, and 1/3, respectively. Therefore the expectation value of L_z is found to be $\langle L_z\rangle = \frac{1}{6}\hbar + \frac{1}{2}0 + \frac{1}{3}(-\hbar) = -\frac{\hbar}{6}$.

These results are applicable to many two-particle systems with attraction energy being inversely proportional to the distance between them, providing that parameters are properly chosen. For instance, if the charge of the nucleus is Z, then in the foregoing calculations we need to substitute e^2 by Ze^2. Examples include deuterium, tritium, ions that contain only one electron, positronium, and muonic atoms.

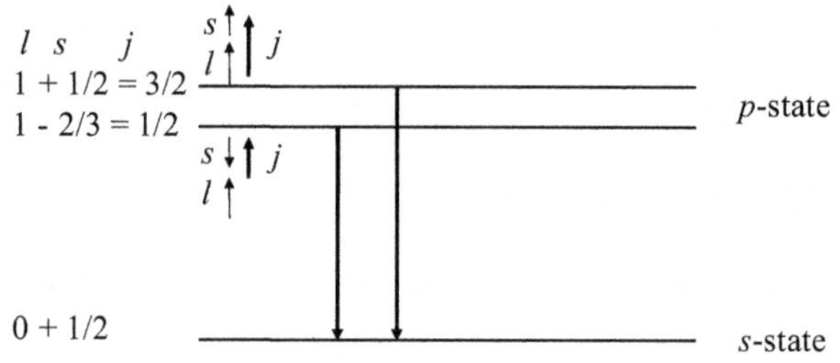

FIGURE 2.3

Spin–orbit interaction.

The total angular momentum of an electron $\boldsymbol{j}$ in an atom can be found as the vector sum of orbital angular momentum $\boldsymbol{l}$ and spin $\boldsymbol{s}$ by:

$$\boldsymbol{j} = \boldsymbol{l} + \boldsymbol{s}. \tag{2.300}$$

For a given value of azimuthal quantum number l, there exist two values of total angular momentum quantum number of an electron: $j = l + 1/2$ and $j = l - 1/2$. Namely, as the electron undergoes orbital motion around the nucleus, it experiences the magnetic field, and this interaction is known as spin–orbit interaction. The result of this interaction is two states $j = l + s$ and $j = l - s$ with slightly different energies as shown in Fig. 2.3.

For atoms containing more than one electron, the total angular momentum $\boldsymbol{J}$ is given by the sum of individual orbital momenta $\boldsymbol{L} = l_1 + l_2 + \ldots$ and spins $\boldsymbol{S} = s_1 + s_2 + \ldots$, so that we can write:

$$\boldsymbol{J} = \boldsymbol{L} + \boldsymbol{S}; \quad \boldsymbol{L} = l_1 + l_2 + \cdots, \quad \boldsymbol{S} = s_1 + s_2 + \cdots. \tag{2.301}$$

This type of coupling is known as *LS* coupling. Another type of coupling, namely *JJ* coupling, occurs when individual ***j***s add together to produce the resulting $\boldsymbol{J}$. *LS* coupling typically occurs in the lighter elements, while *JJ* coupling typically occurs in heavy elements. In *LS* coupling, the magnitudes of **L**, **S**, and $\boldsymbol{J}$ are given by:

$$|\mathbf{L}| = \hbar\sqrt{L(L+1)}, \quad |\mathbf{S}| = \hbar\sqrt{S(S+1)}, \quad |\boldsymbol{J}| = \hbar\sqrt{J(J+1)}, \tag{2.302}$$

where L, S, and J are quantum numbers satisfying the following properties: (1) L is always a nonnegative integer, (2) the spin quantum number S is either integral or half-integral depending on whether the number of electrons is even or odd, and (3) the total angular momentum quantum number J is either integral or half-integral depending on whether the number of electrons is even or odd, respectively. The *spectroscopic notation* of a state characterized by the quantum numbers L, S, and J is as follows:

$$^{2S+1}L_J, \tag{2.303}$$

where the quantity $2S + 1$ is known as *multiplicity*, and determines the numbers of different Js for a given value of L. If $L << S$, different values of J are $L + S$, $L + S -$

1,…,$L-S$, meaning that there are $2S+1$ possible values for J. If, on the other hand, $L<S$, then the possible values of J are $L+S$, $L+S-1$, …,$|L-S|$, meaning only $2L+1$ different values for J exist. The states in which $L=0,1,2,3,4,5,6,7,8,\ldots$ are traditionally called S, P, D, F, G, H, I, K, M,…, respectively. The states with multiplicity $2S+1=0,1,2,34,5$, and 6 are typically called singlet, doublet, triplet, quartet, quintet, and sextet states, respectively.

Example. Let us observe the hydrogen-like atom whose nucleus charge is Ze and which contains only one electron. The wavefunction of the electron in this atom is given by $\psi(r)=Ce^{-r/a}$, $a=a_H/Z$, $a_H=0.05$ nm. If the nucleus number is $A=173$ and $Z=70$, determine the probability that the electron is in the nucleus. By assuming that the radius of the nucleus is $R=1.2\times A^{1/3}$ fm, determine the probability that the electron is in the region x, y, $z>0$.

We first determine the normalization constant from:

$$\iiint \psi^*\psi d^3r = C^2 \underbrace{\int_0^\infty r^2 e^{-2r/a}dr}_{(a/2)^3\Gamma(3)} \int_0^{2\pi} d\phi \int_0^\pi \sin\theta d\theta = 1,$$

as follows: $C=\left(\pi a^3\right)^{-1/2}$. The probability that the electron is found in the nucleolus of radius R is:

$$P=C^2\int_0^R r^2e^{-2r/a}dr\int_0^{2\pi}d\phi\int_0^\pi \sin\theta d\theta = 4\pi C^2\int_0^R r^2e^{-2r/a}dr \approx \frac{4}{a^3}\int_0^R r^2dr$$
$$=\frac{4}{3}\left(\frac{R}{a}\right)^3=\frac{4}{3}\left(\frac{Zr_0}{a_H}\right)^3 A = 1.1\times10^{-6}.$$

Since the wavefunction is independent of θ and ϕ, the probability of the electron being in the region x, y, $z>0$ is 1/8.

2.14 SUMMARY

This chapter was devoted to quantum mechanics fundamentals. The basic topics from quantum mechanics were covered, including state vectors (Sections 2.1 and 2.2), operators (Sections 2.3 and 2.4), and density operators (Section 2.5). We further described quantum measurements (Section 2.6), the uncertainty principle (Section 2.7), change of basis (Section 2.8), and dynamics of a quantum system (Section 2.9). In addition we described the harmonic oscillator (Section 2.10), orbital angular momentum (Section 2.11), spin-1/2 systems (Section 2.12), and hydrogen-like atoms (Section 2.13). For a deeper understanding of this material we provide a set of problems to be used for self-study in Section 2.15. Students not interested in physical implementation, but only in either quantum computing or quantum error correction aspects, can skip Sections 2.10–2.13.

2.15 PROBLEMS

1. Show that states $|\mathrm{R}\rangle$ and $|\mathrm{L}\rangle$, representing right- and left-handed polarized photons, can be represented in terms of $|x\rangle$ and $|y\rangle$ state vectors (representing x- and y-polarized photons) as follows:

$$|R\rangle = \frac{1}{\sqrt{2}}(|x\rangle + j|y\rangle) \quad |L\rangle = \frac{1}{\sqrt{2}}(|x\rangle - j|y\rangle).$$

2. Let the states $|\theta\rangle$ and $|\theta_\perp\rangle$, representing the photons linearly polarized along directions making an angle θ and $\theta_\perp$ with Ox and Oy, respectively, be defined as:

$$|\theta\rangle = \cos\theta|x\rangle + \sin\theta|y\rangle \quad |\theta_\perp\rangle = -\sin\theta|x\rangle + \cos\theta|y\rangle.$$

How are $|R'\rangle = (|\theta\rangle + j|\theta_\perp\rangle)/\sqrt{2}$ and $|L'\rangle = (|\theta\rangle - j|\theta_\perp\rangle)/\sqrt{2}$ related to $|\mathrm{R}\rangle$ and $|\mathrm{L}\rangle$ states?

3. Let the Hermitian operator $\sum$ be defined as:

$$\Sigma = P_{\mathrm{R}} - P_{\mathrm{L}},$$

where P_{R} and P_{L} are projection operators with respect to right- and left-handed polarized photons, respectively. What is the action of $\sum$ on $|\mathrm{R}\rangle$ and $|\mathrm{L}\rangle$? Determine the action of $\exp(-\mathrm{j}\theta\sum)$ on these kets.

4. Using the matrix representation of $\sum$ from Problem 3 in the basis $\{|x\rangle, |y\rangle\}$, show that $\sum^2 = I$ and express $\exp(-\mathrm{j}\theta\sum)$ in terms of $\sum$. What is the action of $\exp(-\mathrm{j}\theta\sum)$ on $|x\rangle$ and $|y\rangle$?

5.
   - **(a)** Write down the basis corresponding to 45- and 135-degree polarizations.
   - **(b)** Write down the basis that is neither plane nor circularly polarized.

6.
   - **(a)** Show that matrix $|\phi\rangle\langle\phi|$ is Hermitian.
   - **(b)** Show that the photon spin operator S is Hermitian.

7. Show that the transformation matrix from one base to another is unitary.

8.
   - **(a)** Determine the transformation matrix from the x, y basis to the R, L basis.
   - **(b)** Determine the transformation matrix from the R, L basis to the basis devised in Problem 5(b).
   - **(c)** Determine the transformation matrix from the x, y basis to the basis devised in Problem 5(b) and show that it can be written as a product of the matrix determined in Problem 8(b) and 8(a).
   - **(d)** Show that the product of the transformation matrix from a basis 1 to a basis 2 with the transformation matrix from base 2 to base 3 is the same as the transformation matrix from base 1 to base 3.

9. The probability that a photon in state $|\psi\rangle$ passes through an x-polaroid is the average value of a physical observable called "x-polarizedness." Write down

the operator P_x corresponding to this observable. What are its eigenstates? Write down its expression in terms of its eigenvalues and eigenstates. Verify the probability that a photon in state $|\psi\rangle$ passing through the x-polaroid is $\langle\psi|P_x|\psi\rangle$.

10. Consider a system of two photons both traveling along the z-axis. The photons are entangled, and their state vector is:

$$|\psi\rangle = \frac{1}{\sqrt{2}}(|xx\rangle + |yy\rangle), \tag{2.P10}$$

where $|xx\rangle$ denotes the state in which both photons are polarized along the x-axis, $|xy\rangle$ denotes the state in which the first photon is polarized along x and the second along y, etc.

(a) Show that $|xx\rangle$, $|xy\rangle$, $|yx\rangle$, $|yy\rangle$ form an orthonormal basis.

(b) Let the axes x' and y' be obtained by rotating x and y by an angle θ. Show that the state $|\psi\rangle$ is rotationally invariant by showing that it has the same form as Eq. (2.P10) in the $|x'x'\rangle$, $|x'y'\rangle$, $|y'x'\rangle$, $|y'y'\rangle$ basis.

(c) Write $|\psi\rangle$ in the circularly polarized basis: $|RR\rangle$, $|RL\rangle$, $|LR\rangle$, $|LL\rangle$. How does this transform under the rotation of part (b)?

(d) The two photons are absorbed by a piece of matter. What are all the possible values of the total angular momentum, L_z, transferred to the matter?

11. This problem is related to the properties of a density operator, defined as:

$$\rho = \sum_i p_i |i\rangle\langle i|, \sum_i p_i = 1.$$

Prove that the most general density operator must satisfy the following properties:

(a) It must be Hermitian: $\rho^\dagger = \rho$.

(b) It must have unit trace: $Tr\rho = 1$.

(c) It must be positive: $\langle\varphi|\rho|\varphi\rangle \geq 0 \ \forall |\varphi\rangle$.

Show that the expectation value of operator M is given by $\langle M\rangle = Tr(\rho M)$. Show that if $\rho^2 = \rho$, then all p_i are zero except one that is equal to unity. Prove that condition $\rho^2 = \rho$ is the necessary and sufficient condition for a state to be pure. Also show that $Tr\rho^2 = 1$ is a necessary and sufficient condition for the density operator to describe a pure state.

12. We would like to determine the most general form of density operator ρ for a qubit.

(a) Show that the most general Hermitian matrix of unit trace has the form:

$$\rho = \begin{pmatrix} a & c \\ c^* & 1-a \end{pmatrix},$$

where a is a real number and c is a complex number. Show that the positivity of eigenvalues of ρ introduces the following constraint on matrix elements:

$$0 \le a(1-a) - |c|^2 \le 1/4.$$

Show that the necessary and sufficient condition for the quantum state described by ρ to be represented by a vector in Hilbert space H is $a(1-a) = |c|^2$. Determine a and c for the matrix ρ describing the normalized state vector $|\psi\rangle = \lambda|0\rangle + \mu|1\rangle$, $|\lambda|^2 + |\mu|^2 = 1$, and show that in this case $a(1-a) = |c|^2$.

(b) Show that ρ can be written as a function of Bloch vector $\vec{b}$ as follows:

$$\rho = \frac{1}{2}\left(I + \vec{b} \bullet \vec{\sigma}\right) = \frac{1}{2}\begin{pmatrix} 1+b_z & b_x - jb_y \\ b_x + jb_y & 1 - b_z \end{pmatrix}, \quad \left|\vec{b}\right| \le 1.$$

Show that a quantum state represented by a vector in Hilbert space H corresponds to the case $\left|\vec{b}\right|^2 = 1$. To interpret the vector $\vec{b}$ physically we calculate the expectation of $\vec{\sigma}$: $\langle\sigma_i\rangle = \mathrm{Tr}(\rho\sigma_i)$. Show that the expectation value of $\vec{\sigma}$ is $\vec{b}$.

13. Let the composite system S be composed of two subsystems A and B, and let A and B be themselves composite systems with corresponding Hilbert spaces H_A and H_B, respectively. The orthonormal basis for H_A and H_B are given as $\{|e_i$: $i = 1,\ldots,N\rangle\}$ and $\{|f_j$: $j = 1,\ldots,M\rangle\}$, respectively.

(a) Let $|\psi\rangle = |\phi > \otimes\, |\chi\rangle$, where $|\phi\rangle$ and $|\chi\rangle$ are normalized states in H_A and H_B, respectively. Show that $E(|\psi\rangle) = 0$.

(b) Let $|ij\rangle$ be the computational basis for $H_A \otimes H_B$ and let:

$$|\psi\rangle = \frac{1}{\sqrt{N}}\sum_i |ii\rangle.$$

Show that $E(|\psi\rangle) = \log(N)$.

14. Show that operator:

$$\frac{1}{2}\left(I + \overrightarrow{\sigma_A} \cdot \overrightarrow{\sigma_B}\right)$$

permutes the qubits A and B ($\overrightarrow{\sigma_A} \cdot \overrightarrow{\sigma_B}$ stands for both scalar and tensor product). This operator is also known as the SWAP operator. Its matrix representation in the basis $\{|00\rangle, |01\rangle, |10\rangle, |11\rangle\}$ is given by:

$$U_{\mathrm{SWAP}} = \begin{pmatrix} 1 & 0 & 0 & 0 \\ 0 & 0 & 1 & 0 \\ 0 & 1 & 0 & 0 \\ 0 & 0 & 0 & 1 \end{pmatrix}.$$

15. Show that for a two-level system, the density operator can be written as:

$$\rho = \frac{1}{2}(I + \vec{r} \cdot \vec{\sigma}), \quad \vec{\sigma} = [X \quad Y \quad Z],$$

where X, Y, Z are the Pauli operators, and $\vec{r} = [r_x \quad r_y \quad r_z]$ is the Bloch vector, whose components can be determined by:

$$r_x = \mathrm{Tr}(\rho X), \quad r_y = \mathrm{Tr}(\rho Y), \quad r_z = \mathrm{Tr}(\rho Z).$$

16. Using the Pauli matrices, prove:

(a) $(\boldsymbol{\sigma}\cdot\boldsymbol{A})(\boldsymbol{\sigma}\cdot\boldsymbol{B}) = (\boldsymbol{A}\cdot\boldsymbol{B})I + j\boldsymbol{\sigma}\cdot(\boldsymbol{A}\times\boldsymbol{B})$; $\boldsymbol{\sigma} = (X, Y, Z)$,
$\boldsymbol{A} = (A_x, A_y, A_z)$, $\boldsymbol{B} = (B_x, B_y, B_z)$.

(b) $\exp\left(-\frac{j\theta}{2}\boldsymbol{n}\cdot\boldsymbol{\sigma}\right) = \cos(\theta/2)I - j\boldsymbol{n}\cdot\boldsymbol{\sigma}\sin(\theta/2)$, $\boldsymbol{n} = (n_x, n_y, n_z)$.

(c) $(\boldsymbol{n}\cdot\boldsymbol{\sigma})^2 = I$.

17. Consider the rotation operator $U_R(\theta, \hat{u}) = \exp\left(\frac{j\theta}{2}\hat{u}\cdot\boldsymbol{\sigma}\right)$, $\boldsymbol{\sigma} = (X, Y, Z)$. By rotating the eigenkets of S_z, determine the eigenkets of S_x and S_y in the standard basis.

18. Determine the normalized momentum distribution for a hydrogen atom electron in states 1*s*, 2*s*, and 2*p*.

19. The parity operator is obtained by replacing $\boldsymbol{r}$ with $-\boldsymbol{r}$. How does this operator affect the wavefunction of an electron in a hydrogen atom?

20. Let the hydrogen atom be in a state specified by quantum numbers n and l. Determine the dispersion for the distance of the electron for the nucleus defined as $\sqrt{\langle r^2\rangle - \langle r\rangle^2}$.

21. For a 2D hydrogen-like atom, the Schrödinger equation in atomic units is given by $(-\nabla^2 - 2Z/r)\psi = E\psi$. Using the cylindrical coordinate system and the method of separation of variables, determine the differential equations for $R(r)$ and $\Phi(\phi)$.

22. Consider the Hamiltonian of a 3D isotropic harmonic oscillator:

$$H = \frac{1}{2m}\left(p_x^2 + p_y^2 + p_z^2\right) + \frac{m\omega^2}{2}\left(x^2 + y^2 + z^2\right).$$

Represent the Hamiltonian in spherical coordinates and determine the eigenfunctions and energy eigenvalues.

23. In a quantum system with an angular momentum $l = 1$, the eigenkets of L_z are given by $|1\rangle$, $|0\rangle$, $|-1\rangle$ and the action of L_z on base kets is given by $L_z|1\rangle = \hbar$, $L_z|0\rangle = 0$, $L_z|-1\rangle = -\hbar|-1\rangle$. The Hamiltonian is given by $H = \omega_0\left(L_x^2 - L_y^2\right)/\hbar$, where ω_0 is a constant. Determine the matrix representation of H and the corresponding eigenvalues and eigenkets in the foregoing bases.

24. Compute $\langle lm|L_x^2|lm\rangle$ and $\langle lm|L_xL_y|lm\rangle$ in the standard angular momentum basis.

25. Consider a particle in a central potential. Given that $|lm\rangle$ is an eigenket of L_z and L^2, determine the sum $\Delta L_x^2 + \Delta L_y^2$. For which values of l and m does this sum vanish?

26. The classical equation of motion for a particle with a mass m and charge q in the presence of an electric field $\boldsymbol{E}$ and magnetic field $\boldsymbol{B}$ is given by:

$$m\boldsymbol{a} = q\boldsymbol{E} + \frac{q}{c}\boldsymbol{v} \times \boldsymbol{B}, \quad \boldsymbol{v} = \frac{d\boldsymbol{r}}{dt} = \dot{\boldsymbol{r}}, \quad \boldsymbol{a} = \frac{d\boldsymbol{v}}{dt} = \ddot{\boldsymbol{r}}.$$

The $\boldsymbol{E}$ and $\boldsymbol{B}$ must satisfy Maxwell's equation so that it is possible to find the vector potential $\boldsymbol{A}(\boldsymbol{r},t)$ and scalar potential $\phi(\boldsymbol{r},t)$ such that:

$$\boldsymbol{E} = -\nabla\phi - \frac{1}{c}\frac{\partial \boldsymbol{A}}{\partial t} \quad \boldsymbol{B} = \nabla \times \boldsymbol{A}.$$

By using the Hamilton equations $\dot{\boldsymbol{r}} = \partial H/\partial \boldsymbol{p}$ and $\boldsymbol{p} = -\partial H/\partial \boldsymbol{r}$, show that the following Hamiltonian leads to the equation of motion:

$$H = \frac{1}{2m}\left(\boldsymbol{p} - \frac{q}{c}\boldsymbol{A}\right)\cdot\left(\boldsymbol{p} - \frac{q}{c}\boldsymbol{A}\right) + q\phi.$$

References

[1] J.J. Sakurai, Modern Quantum Mechanics, Addison-Wisley, 1994.
[2] G. Baym, Lectures on Quantum Mechanics, Westview Press, 1990.
[3] P.A.M. Dirac, Quantum Mechanics, fourth ed., Oxford Univ. Press, London, 1958.
[4] F. Gaitan, Quantum Error Correction and Fault Tolerant Quantum Computing, CRC Press, 2008.
[5] C.W. Helstrom, Quantum Detection and Estimation Theory, Academic, New York, 1976.
[6] Y. Peleg, R. Pnini, E. Zaarur, Schaum's Outline of Theory and Problems of Quantum Mechanics, McGraw-Hill, New York, 1998.
[7] D. McMahon, Quantum Mechanics Demystified, McGraw-Hill, New York, 2006.
[8] D. McMahon, Quantum Field Theory Demystified, McGraw-Hill, New York, 2008.
[9] C.W. Helstrom, J.W.S. Liu, J.P. Gordon, Quantum mechanical communication theory, Proc. IEEE 58 (Oct. 1970) 1578–1598.
[10] M.A. Neilsen, I.L. Chuang, Quantum Computation and Quantum Information, Cambridge University Press, Cambridge, 2000.
[11] S. Fleming, PHYS 570: Quantum Mechanics (Lecture Notes), University of Arizona, 2008.
[12] D. Griffiths, Introduction to Quantum Mechanics, Prentice-Hall, Englewood Cliffs, 1995.
[13] N. Zettilli, Quantum Mechanics, Concepts and Applications, John Willey, New York, 2001.
[14] J. von Neumann, Mathematical Foundations of Quantum Mechanics, Princeton University Press, Princeton, 1955.
[15] J.M. Jauch, Foundations of Quantum Mechanics, Addison-Wesley, Reading, 1968.
[16] A. Peres, Quantum Theory: Concepts and Methods, Kluwer, Boston, 1995.
[17] A. Goswami, Quantum Mechanics, McGraw-Hill, New York, 1996.
[18] M. Jammer, Philosophy of Quantum Mechanics, John Willey, New York, 1974.
[19] R. McWeeny, Quantum Mechanics, Principles and Formalism, Dover Publications, Inc., Mineola, N.Y., 2003.

[20] H. Weyl, Theory of Groups and Quantum Mechanics, Dover Publications, Inc., New York, 1950.
[21] R.L. Liboff, Introductory Quantum Mechanics, Addison-Wesley, Reading, 1997.
[22] G.R. Fowles, Introduction to Modern Optics, Dover Publications, Inc., New York, 1989.
[23] M.O. Scully, M.S. Zubairy, Quantum Optics, Cambridge University Press, Cambridge, 1997.
[24] M. Le Bellac, A Short Introduction to Quantum Information and Quantum Computation, Cambridge University Press, Cambridge, 2006.

CHAPTER

Quantum Circuits and Modules

3

CHAPTER OUTLINE

This chapter is devoted to the quantum circuits and quantum modules. We first describe basic single-qubit gates and provide Bloch sphere representation of the single-qubit. Next, we describe binary and ternary quantum gates. We then provide the generalization to N-qubit gates and quantum information processing fundamentals. The quantum measurements circuits are provided then, and the principle of deferred measurement and the principle of implicit measurement are formulated. We also formulate Gottesman-Knill and Solovay-Kitaev theorems. Further, various sets of universal quantum gates are introduced. We then describe Bell state preparation circuit and quantum relay. We also describe basics concepts of quantum teleportation. Finally, we briefly discuss the computation-in-place concept.

3.1 SINGLE-QUBIT OPERATIONS

A single qubit is a state ket (vector) [1–18]:

$$|\psi\rangle = a|0\rangle + b|1\rangle, \tag{3.1}$$

Quantum Information Processing, Quantum Computing, and Quantum Error Correction
https://doi.org/10.1016/B978-0-12-821982-9.00003-4

and is parameterized by two complex numbers a and b satisfying the normalization condition $|a|^2 + |b|^2 = 1$. The parameters a and b are known as (quantum) *probability amplitudes* as their magnitudes squared $|a|^2$ and $|b|^2$ represent probabilities of finding the state $|\psi\rangle$ in basis states $|0\rangle$and $|1\rangle$, respectively. The computational basis (CB) is represented in matrix form as follows:

$$|0\rangle \doteq \begin{pmatrix} 1 \\ 0 \end{pmatrix}, \quad |1\rangle \doteq \begin{pmatrix} 0 \\ 1 \end{pmatrix}. \tag{3.2}$$

Another important basis is a *diagonal basis* $\{|+\rangle, |-\rangle\}$, very often denoted as $\{|\nearrow\rangle, |\searrow\rangle\}$, which is related to CB by:

$$|+\rangle = |\nearrow\rangle = \frac{1}{\sqrt{2}}(|0\rangle + |1\rangle), \ |-\rangle = |\searrow\rangle = \frac{1}{\sqrt{2}}(|0\rangle - |1\rangle). \tag{3.3}$$

The circular basis, $\{|R\rangle, |L\rangle\}$, is related to CB as follows:

$$|R\rangle = \frac{1}{\sqrt{2}}(|0\rangle + j|1\rangle), \quad |L\rangle = \frac{1}{\sqrt{2}}(|0\rangle - j|1\rangle). \tag{3.4}$$

The *most important single-qubit quantum gates* are Pauli operators *X*, *Y*, *Z*, the Hadamard gate *H*, the phase-shift gate *S* (sometimes denoted by *P*), and the $\pi/8$ (or *T*) gate. The Pauli operators can be represented in matrix form in CB by:

$$\begin{aligned} X = \sigma_x = \sigma_1 \doteq \begin{bmatrix} 0 & 1 \\ 1 & 0 \end{bmatrix}, \quad Y = \sigma_y = \sigma_2 \doteq \begin{bmatrix} 0 & -j \\ j & 0 \end{bmatrix}, \\ Z = \sigma_z = \sigma_3 \doteq \begin{bmatrix} 1 & 0 \\ 0 & -1 \end{bmatrix}, \quad I = \sigma_0 \doteq \begin{bmatrix} 1 & 0 \\ 0 & 1 \end{bmatrix}. \end{aligned} \tag{3.5}$$

Based on their representation (Eq. 3.5), the following properties of Pauli operators (gates) can easily be proved:

$$\begin{aligned} & X(a|0\rangle + b|1\rangle) = a|1\rangle + b|0\rangle, \quad Y(a|0\rangle + b|1\rangle) = j(a|1\rangle - b|0\rangle), \\ & \qquad Z(a|0\rangle + b|1\rangle) = a|0\rangle - b|1\rangle \\ & X^2 = I, \quad Y^2 = I, \quad Z^2 = I \\ & XY = jZ \quad YX = -jZ \end{aligned} \tag{3.6}$$

The Hadamard gate *H*, phase shift gate *S*, and *T* gate in CB are represented by:

$$H \doteq \frac{1}{\sqrt{2}} \begin{bmatrix} 1 & 1 \\ 1 & -1 \end{bmatrix}, \ S \doteq \begin{bmatrix} 1 & 0 \\ 0 & j \end{bmatrix}, \ T \doteq \begin{bmatrix} 1 & 0 \\ 0 & e^{j\pi/4} \end{bmatrix} = e^{j\pi/8} \begin{bmatrix} e^{-j\pi/8} & 0 \\ 0 & e^{j\pi/8} \end{bmatrix}, \tag{3.7}$$

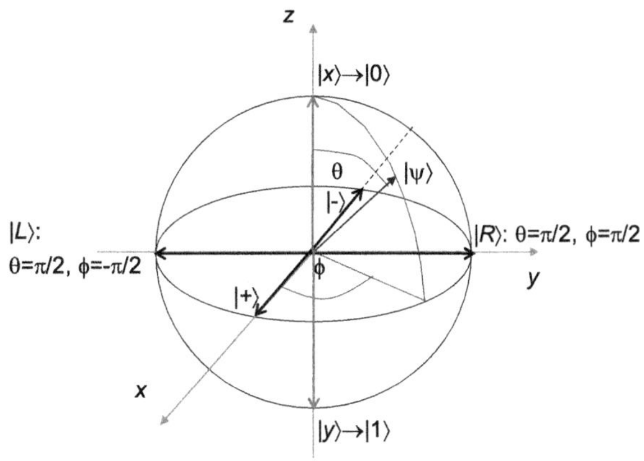

FIGURE 3.1

Bloch (Poincaré) sphere representation of a single qubit.

respectively. The action of the Hadamard gate on single-qubit $|\psi\rangle$ is to perform the following transformation:

$$\begin{aligned} H(a|0\rangle + b|1\rangle) &= a\frac{1}{\sqrt{2}}(|0\rangle + |1\rangle) + b\frac{1}{\sqrt{2}}(|0\rangle - |1\rangle) \\ &= \frac{1}{\sqrt{2}}[(a+b)|0\rangle + (a-b)|1\rangle]. \end{aligned} \tag{3.8}$$

The Hadamard gate therefore interchanges the CB and diagonal basis: $|0\rangle \leftrightarrow |+\rangle$ and $|1\rangle \leftrightarrow |-\rangle$.

A single qubit in the state $a|0\rangle + b|1\rangle$ can be visualized as the point (θ,φ) on a unit sphere (Bloch or Poincaré sphere), where $a = \cos(\theta/2)$ and $b = e^{j\varphi}\sin(\theta/2)$, and corresponding Bloch vector coordinates are $(\cos\varphi \sin\theta, \sin\varphi \sin\theta, \cos\theta)$. Bloch sphere representation of a single qubit (or the Poincaré sphere for the photon) is shown in Fig. 3.1. This spinor representation is related to CB by:

$$|\psi(\theta,\phi)\rangle = \cos(\theta/2)|0\rangle + e^{j\phi}\sin(\theta/2)|1\rangle \doteq \begin{pmatrix} \cos(\theta/2) \\ e^{j\phi}\sin(\theta/2) \end{pmatrix}, \tag{3.9}$$

where $0 \le \theta \le \pi$ and $0 \le \phi < 2\pi$. The north and south poles correspond to computational $|0\rangle$ ($|X\rangle$-polarization) and $|1\rangle$ ($|Y\rangle$-polarization) basis kets, respectively. The pure qubit states lie on the Bloch sphere, while the mixed qubit states lie in the interior of the Bloch sphere. The maximally mixed state $(I/2)$ lies in the center of the Bloch sphere. The orthogonal states are antipodal. From Fig. 3.1 we see that CB, diagonal basis, and circular bases are 90 degrees apart from each other, and we often say that these three bases are mutually *conjugate bases*. These bases are used as three pairs of signal states for the six-state quantum key distribution (QKD) protocol. Another important basis used in QKD and for eavesdropping is the *Breidbart basis* given by $\{\cos(\pi/8)|0\rangle + \sin(\pi/8)|1\rangle, -\sin(\pi/8)|0\rangle + \cos(\pi/8)|1\rangle\}$.

The common single-qubit gates and their matrix representations are summarized in Fig. 3.2.

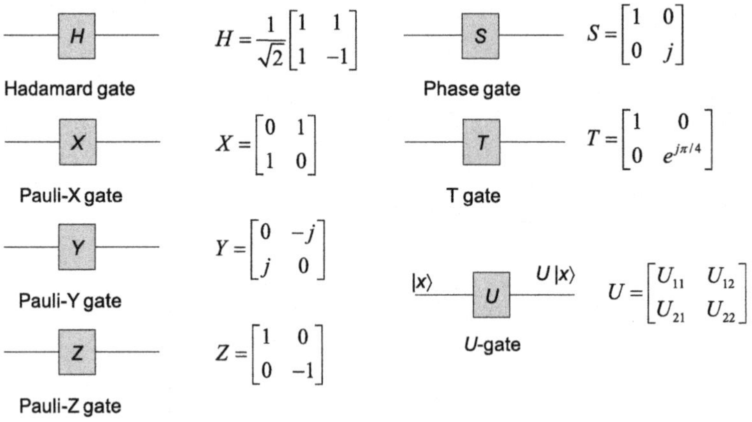

FIGURE 3.2

Common single-qubit gates and their matrix representations.

The *rotation operators* for θ about the x-, y-, and z-axes, denoted as $R_x(\theta)$, $R_y(\theta)$, and $R_z(\theta)$, respectively, can be represented in terms of Pauli operators as follows:

$$R_x(\theta) = e^{-j\theta X/2} = \cos\frac{\theta}{2}I - j\sin\frac{\theta}{2}X = \begin{bmatrix}\cos\frac{\theta}{2} & -j\sin\frac{\theta}{2}\\ -j\sin\frac{\theta}{2} & \cos\frac{\theta}{2}\end{bmatrix},$$

$$R_y(\theta) = e^{-j\theta Y/2} = \cos\frac{\theta}{2}I - j\sin\frac{\theta}{2}Y = \begin{bmatrix}\cos\frac{\theta}{2} & -\sin\frac{\theta}{2}\\ \sin\frac{\theta}{2} & \cos\frac{\theta}{2}\end{bmatrix},$$

$$R_z(\theta) = e^{-j\theta Z/2} = \cos\frac{\theta}{2}I - j\sin\frac{\theta}{2}Z = \begin{bmatrix}e^{-j\frac{\theta}{2}} & 0\\ 0 & e^{j\frac{\theta}{2}}\end{bmatrix}. \tag{3.10}$$

Eq. (3.10) can be proved by Taylor series expansion. Namely, the Taylor expansion of $\exp(j\alpha A)$, for $A^2 = I$, where $\alpha = -\theta/2$ and $A \in \{X,Y,Z\}$, is given by:

$$e^{j\alpha A} = \sum_{n=0}^{\infty}\frac{(j\alpha A)^n}{n!} = \sum_{n-\text{even}}\frac{(j\alpha A)^n}{n!} + \sum_{n-\text{odd}}\frac{(j\alpha A)^n}{n!} = 1 - \frac{(\alpha A)^2}{2!} + \frac{(\alpha A)^4}{4!} - \cdots$$

$$+j\left[(\alpha A) - \frac{(\alpha A)^3}{3!} + \frac{(\alpha A)^5}{5!} - \cdots\right]$$

$$\overset{A^2=I}{=} \underbrace{\left[1 - \frac{\alpha^2}{2!} + \frac{\alpha^4}{4!} - \cdots\right]}_{\cos(\alpha)} I + j\underbrace{\left[\alpha - \frac{\alpha^3}{3!} + \frac{\alpha^5}{5!} - \cdots\right]}_{\sin(\alpha)} A = \cos(\alpha)I + j\sin(\alpha)A. \tag{3.11}$$

By substituting $\alpha = \theta/2$ and $A = X,Y,Z$ we obtain Eq. (3.10).

The rotation by θ around unit vector $\hat{\boldsymbol{n}} = (n_x, n_y, n_z)$ can be described by:

$$R_{\hat{n}}(\theta) = \exp(-j\theta\hat{\boldsymbol{n}}\cdot\boldsymbol{\sigma}/2) = \cos\left(\frac{\theta}{2}\right)I - j\sin\left(\frac{\theta}{2}\right)(n_xX + n_yY + n_zZ). \tag{3.12}$$

$$\hat{\boldsymbol{n}} = (n_x, n_y, n_z),\ \boldsymbol{\sigma} = (X, Y, Z)$$

Notice that Eq. (3.12) is just a special case of Eq. (3.11), obtained by setting $\alpha = -\theta/2$, $A = \hat{\boldsymbol{n}}\cdot\boldsymbol{\sigma}$. Using this substitution, since $A^2 = I$, clearly the following property holds as well:

$$(\hat{\boldsymbol{n}}\cdot\boldsymbol{\sigma})^2 = I. \tag{3.13}$$

By performing the following matrix multiplication, $R_z(\beta)R_y(\gamma)R_z(\delta)$, we obtain:

$$\begin{aligned} R_z(\beta)R_y(\gamma)R_z(\delta) &= \begin{bmatrix} e^{-j\frac{\beta}{2}} & 0 \\ 0 & e^{j\frac{\beta}{2}} \end{bmatrix}\begin{bmatrix} \cos\frac{\gamma}{2} & -\sin\frac{\gamma}{2} \\ \sin\frac{\gamma}{2} & \cos\frac{\gamma}{2} \end{bmatrix}\begin{bmatrix} e^{-j\frac{\delta}{2}} & 0 \\ 0 & e^{j\frac{\delta}{2}} \end{bmatrix} \\ &= \begin{bmatrix} e^{j(-\beta/2-\delta/2)}\cos\frac{\gamma}{2} & -e^{j(-\beta/2+\delta/2)}\sin\frac{\gamma}{2} \\ e^{j(+\beta/2)-\delta/2}\sin\frac{\gamma}{2} & e^{j(+\beta/2+\delta/2)}\cos\frac{\gamma}{2} \end{bmatrix}, \end{aligned} \tag{3.14}$$

which is clearly a unitary matrix. The unitarity will not be changed if we insert a global phase factor $\exp(j\alpha)$. Therefore an arbitrary single-qubit operation can be decomposed as:

$$U = \begin{bmatrix} e^{j(\alpha-\beta/2-\delta/2)}\cos\frac{\gamma}{2} & -e^{j(\alpha-\beta/2+\delta/2)}\sin\frac{\gamma}{2} \\ e^{j(\alpha+\beta/2-\delta/2)}\sin\frac{\gamma}{2} & e^{j(\alpha+\beta/2+\delta/2)}\cos\frac{\gamma}{2} \end{bmatrix} = e^{j\alpha}R_z(\beta)R_y(\gamma)R_z(\delta), \tag{3.15}$$

and this decomposition is sometimes called the single-qubit *Z-Y decomposition theorem* [1]. By simple matrix multiplication we can prove the following important *circuit identities*:

$$HXH = Z,\ HYH = -Y,\ HZH = X,\ HTH = R_x(\pi/4), \tag{3.16}$$

where with $R_x(\theta)$ we denoted the rotation operator for θ about the x-axis.

The *quantum state purity* is specified by the density operator ρ (in some books called the state operator [3]), which was defined in Chapter 2, as follows:

$$\mathcal{P}(\rho) = \mathrm{Tr}\rho^2,\ 1/D \le \mathcal{P}(\rho) \le 1, \tag{3.17}$$

where D is the dimensionality of the corresponding Hilbert space H. If $\mathcal{P}(\rho) = 1$ we say that the state is *pure*; on the other hand, if $\mathcal{P}(\rho) < 1$ we say that the quantum state is *mixed*. Alternatively, instead of purity one can use *mixedness* defined as $\mathcal{M}(\rho) = 1 - \mathcal{P}(\rho)$, which represents the complement of purity. What is interesting is that the purity does not change under *conjugation mapping*: $\rho \to U\rho U^\dagger$, where U is a unitary operator. This is also true for the time-evolution operator $U = \exp[-(j/\hbar)H(t-t_0)]$, where H is the Hamiltonian operator. We have shown in the previous chapter that for pure states $\rho^2 = \rho$, indicating that the density operators serve as projection operators for pure states. The projector operator is defined by:

$$P|\psi_i\rangle \equiv |\psi_i\rangle\langle\psi_i|, \tag{3.18}$$

where $|\psi_i\rangle$ is a pure state. Let a finite state of projectors $\{P|\psi_i\rangle\}$ be given. Any state ρ' can be written in terms of projectors as follows:

$$\rho' = \sum_i p_i P|\psi_i\rangle;\ 0 < p_i < 1,\ \sum_i p_i = 1. \tag{3.19}$$

A quantum state is said to be in a (partially) *coherent superposition* of states $|a_i\rangle$, where $|a_i\rangle$ are eigenkets of Hermitian operator A, if the corresponding density matrix is not diagonal. In the discussion related to Fig. 3.1 we said that mixed qubit states lie inside of Bloch sphere and can be represented as weighted convex combinations of pure states. We can use *Stokes parameters* to represent the density operator:

$$\rho = \frac{1}{2}\sum_{i=0}^{3} S_i\sigma_i = \frac{1}{2}(S_0 I + S_1\sigma_1 + S_2\sigma_2 + S_3\sigma_3) = \frac{1}{2}\begin{pmatrix} S_0 + S_3 & S_1 - jS_2 \\ S_1 - jS_2 & S_0 - S_3 \end{pmatrix}, \tag{3.20}$$

where S_i $(i = 0,1,2,3)$ are Stokes parameters that can be obtained as:

$$S_i = \mathrm{Tr}(\rho\sigma_i);\ i = 0, 1, 2, 3 \tag{3.21}$$

as we have shown in the previous chapter. The Euclidean distance with respect to the origin is given by:

$$r = \sqrt{S_1^2 + S_2^2 + S_3^2}, \tag{3.22}$$

and the corresponding three-component vector $\boldsymbol{S} = (S_1,S_2,S_3)$ is known as the Stokes (Bloch) vector. In optical applications, the vector $\boldsymbol{S} = (S_1,S_2,S_3)$ describes a polarization state of a photon. The ratio r/S_0 is known as the *degree of polarization*. When the state is normalized, $S_0 = 1$ represents the total quantum probability. Going back to Fig. 3.1, we conclude that Stokes coordinates are given by $S_0 = 1$, $S_1 = \sin\theta\,\cos\phi$, $S_2 = \sin\theta\,\sin\phi$, and $S_3 = \cos\theta$.

3.2 TWO-QUBIT OPERATIONS

The mathematical foundation for two-qubit states relies on the tensor product concept. The first qubit lives in the Hilbert space H_A with orthonormal basis $\{|0_A\rangle, |1_A\rangle\}$, and the second qubit lives in the Hilbert space H_B with orthonormal basis $\{|0_A\rangle, |1_A\rangle\}$. The bipartite state representing the fact that subsystem A is in the state $|m_A\rangle$ $(m = 0, 1)$, while at the same time subsystem B is in the state $|n_A\rangle$ $(n = 0, 1)$, can be denoted as $|m_A m_B\rangle = |m_A\rangle \otimes |m_B\rangle$, where $\otimes$ stands for tensor product, which is often omitted in writing. The qubits from subsystem A and B can be represented as:

$$\begin{aligned} |\psi_A\rangle &= \alpha_A|0_A\rangle + \beta_A|1_A\rangle,\ |\alpha_A|^2 + |\beta_A|^2 = 1; \\ |\psi_B\rangle &= \alpha_B|0_B\rangle + \beta_B|1_B\rangle,\ |\alpha_B|^2 + |\beta_B|^2 = 1; \end{aligned} \tag{3.23}$$

respectively. The bipartite state will then be given by:

$$\begin{aligned}|\psi_A\psi_B\rangle &= |\psi_A\rangle \otimes |\psi_B\rangle = (\alpha_A|0_A\rangle + \beta_A|1_A\rangle) \otimes (\alpha_B|0_B\rangle + \beta_B|1_B\rangle) \\ &= \alpha_A\alpha_B|0_A0_B\rangle + \alpha_A\beta_B|0_A1_B\rangle + \beta_A\alpha_B|1_A0_B\rangle + \beta_A\beta_B|1_A1_B\rangle.\end{aligned} \tag{3.24}$$

With Eq. (3.24) we defined the space $H_A \otimes H_B$ as the tensor product of subspaces H_A and H_B. Unfortunately, state (3.24) is not the most general state of $H_A \otimes H_B$. Interestingly enough, the state of the form (3.24) does not even create the subspace of $H_A \otimes H_B$. The most general state in space $H_A \otimes H_B$ has the form [3]:

$$|\psi\rangle = \alpha_{00}|0_A0_B\rangle + \alpha_{01}|0_A1_B\rangle + \alpha_{10}|1_A0_B\rangle + \alpha_{11}|1_A1_B\rangle. \tag{3.25}$$

The state given by Eq. (3.25) has the form given by Eq. (3.24) only when the following condition, which is necessary and sufficient, is satisfied:

$$\alpha_{00}\alpha_{11} = \alpha_{01}\alpha_{10}, \tag{3.26}$$

which a priori has no reason to be satisfied. The two-qubit state $|\psi\rangle$, which cannot be written in the form $|\psi_A\rangle \otimes |\psi_B\rangle$, is called an *entangled state*. Important examples of two-qubit states are *Bell states*, also known as Einstein–Podolsky–Rosen (EPR) states (pairs):

$$\begin{aligned}|B_{00}\rangle &= \frac{1}{\sqrt{2}}(|00\rangle + |11\rangle),\ |B_{01}\rangle = \frac{1}{\sqrt{2}}(|01\rangle + |10\rangle), \\ |B_{10}\rangle &= \frac{1}{\sqrt{2}}(|00\rangle - |11\rangle),\ |B_{11}\rangle = \frac{1}{\sqrt{2}}(|01\rangle - |10\rangle).\end{aligned} \tag{3.27}$$

Bell states are *maximally entangled states* because if we disregard the information on one qubit and perform the measurement on the second qubit the result of the measurement will be purely random. Notice that in Eq. (3.27) we omitted the subscripts, which is commonly done in quantum information processing. The fundamental property of an entangled state is that the subsystem A (B) cannot be in a definite quantum state $|\psi_A\rangle$ ($|\psi_B\rangle$). Let us observe the Bell state $|B_{01}\rangle$, and let M be the physical property of subsystem A, which in space H_AH_B is denoted by $M \otimes I_B$. The expected value of M is given by [3]:

$$\begin{aligned}\langle M\rangle = \langle B_{01}|M|B_{01}\rangle &= \frac{1}{2}(\langle 01| + \langle 10|)M(|01\rangle + |10\rangle) \\ &= \frac{1}{2}(\langle 01| + \langle 10|)[(M|0\rangle)|1\rangle + (M|1\rangle)|0\rangle] \\ &= \frac{1}{2}[\langle 0_A|M|0_A\rangle + \langle 1_A|M|1_A\rangle],\end{aligned} \tag{3.28}$$

where we used $\langle 0_B|0_B\rangle = \langle 1_B|1_B\rangle = 1$ and $\langle 0_B|1_B\rangle = \langle 1_B|0_B\rangle = 0$. Now we have to prove that there is no state $|\psi_A\rangle$ such that:

$$\langle\psi_A|M|\psi_A\rangle = \langle B_{01}|M|B_{01}\rangle. \tag{3.29}$$

Let us now evaluate the expected value of M with respect to state $|\psi_A\rangle$ [3]:

$$\begin{aligned}\langle\psi_A|M|\psi_A\rangle &= (\alpha_A^*\langle 0_A| + \beta_A^*\langle 1_A|)M(\alpha_A|0_A\rangle + \beta_A|1_A\rangle) \\ &= |\alpha_A|^2\langle 0_A|M|0_A\rangle + \alpha_A^*\beta_A\langle 0_A|M|1_A\rangle \\ &\quad + \alpha_A\beta_A^*\langle 1_A|M|0_A\rangle + |\beta_A|^2\langle 1_A|M|1_A\rangle.\end{aligned} \tag{3.30}$$

For Eq. (3.30) to be in the same form as Eq. (3.28), it is required that $|\alpha_A|^2 = |\beta_A|^2 = 1/2$, but the terms $\alpha_A^*\beta_A$ and $\alpha_A\beta_A^*$ will not vanish. So, it appears that Eq. (3.28) is an incoherent mixture of 50% of state $|0_A\rangle$ and 50% of state $|1_A\rangle$, not a linear superposition. In conclusion, in general, it is not possible to represent the subsystem of an entangled state by a state ket. An example of an incoherent mixture is unpolarized (natural) light that is a mixture of 50% of light along the x-axis and 50% of light along the y-axis.

An n-qubit register has 2^n mutually orthogonal states, which in CB can be represented by $|x_1x_2\ldots x_n\rangle$ $(x_i = 0{,}1)$. Any state of the quantum register can be specified by 2^n complex amplitudes $c_x(x \equiv x_1x_2\ldots x_n)$ by:

$$|\Psi\rangle = \sum\nolimits_x c_x|x\rangle,\ \sum\nolimits_x |c_x|^2 = 1. \tag{3.31}$$

Even for moderate n (close to 1000) the number of complex amplitudes c_x specifying the state of quantum register is enormous. The classical computer will require enormous resources to store and manipulate such large amounts of complex coefficients. The n-qubit ($n > 2$) analogs of Bell states will now be briefly reviewed. One popular family of entangled multiqubit states is Greenberger–Horne–Zeilinger (GHZ) states:

$$|\mathrm{GHZ}\rangle = \frac{1}{\sqrt{2}}(|00\cdots 0\rangle \pm |11\cdots 1\rangle). \tag{3.32}$$

Another popular family of multiqubit entangled states is known as W-states:

$$|\mathrm{W}\rangle = \frac{1}{\sqrt{N}}(|00\cdots 01\rangle + |00\cdots 10\rangle + \cdots + |01\cdots 00\rangle + |10\cdots 00\rangle). \tag{3.33}$$

The W-state of n qubits represents a superposition of single-weighted CB states, each occurring with a probability amplitude of $N^{-1/2}$.

In the rest of this section, we describe several important two-qubit gates. We start our description with a controlled-NOT (CNOT) gate. The CNOT quantum gate, with circuit representation and operation principle shown in Fig. 3.3, has two input qubits, known as *control qubit* $|c\rangle$ and *target qubit* $|t\rangle$.

The action of a CNOT gate can be described by $|c,t\rangle \to |c, t\oplus c\rangle$. Therefore if the control qubit is set to one, then the target qubit is flipped. The matrix representation of a CNOT gate is given by:

$$\mathrm{CNOT} = \begin{bmatrix} 1 & 0 & 0 & 0 \\ 0 & 1 & 0 & 0 \\ 0 & 0 & 0 & 1 \\ 0 & 0 & 1 & 0 \end{bmatrix} = \begin{bmatrix} I & 0 \\ 0 & X \end{bmatrix}. \tag{3.34}$$

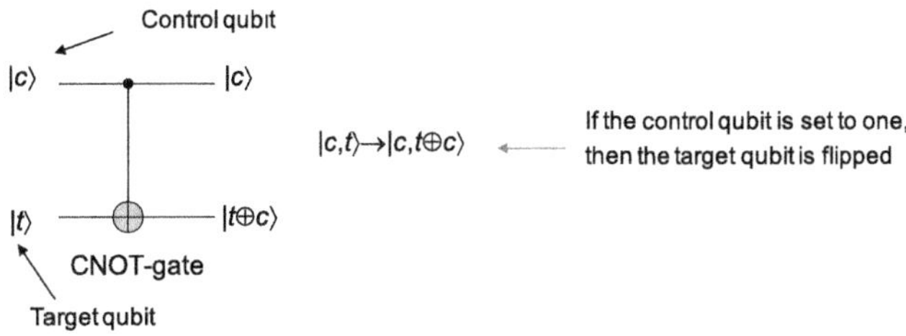

FIGURE 3.3

Circuit representation and operation principle of a controlled-NOT gate.

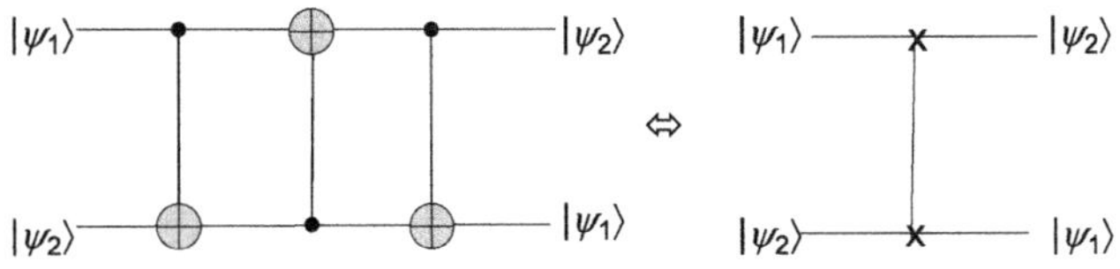

FIGURE 3.4

Operation principle and circuit representation of a SWAP-gate.

Another important two-qubit gate is the quantum swapping circuit (SWAP gate), whose circuit representation and operation principle are provided in Fig. 3.4. The action of a SWAP gate is to interchange the states of two input qubits:

$$U_{\text{SWAP}}^{(12)}|\psi_1\rangle|\psi_2\rangle = |\psi_2\rangle|\psi_1\rangle. \tag{3.35}$$

In certain technologies, the controlled-Z gate, $C(Z)$, is easier to implement than the CNOT gate. Its operation principle and circuit representation are shown in Fig. 3.5. The corresponding matrix representation is given by:

$$C(Z) = \begin{bmatrix} 1 & 0 & 0 & 0 \\ 0 & 1 & 0 & 0 \\ 0 & 0 & 1 & 0 \\ 0 & 0 & 0 & -1 \end{bmatrix} = \begin{bmatrix} I & 0 \\ 0 & Z \end{bmatrix}. \tag{3.36}$$

We can establish the connection between CNOT gate and $C(Z)$ as shown in Fig. 3.6. This equivalence can easily be proved as follows:

$$\begin{aligned} \text{CNOT}|c,t\rangle &= |c\rangle \otimes X^c|t\rangle = |c\rangle \otimes (HZH)^c|t\rangle = |c\rangle \otimes HZ^cH|t\rangle \\ &= I \otimes H \cdot C(Z) \cdot I \otimes H|c,t\rangle. \end{aligned} \tag{3.37}$$

In general, an arbitrary controlled-U gate can be implemented as shown in Fig. 3.7. The action of a U gate can be described by $|c\rangle|t\rangle \rightarrow |c\rangle U^c|t\rangle$. Therefore if the control qubit is set to one, then the U is applied to the target qubit, otherwise it is left unchanged. The corresponding matrix representation of a controlled-U gate is given by:

$$C(Z) = |c\rangle \otimes Z^c |t\rangle$$

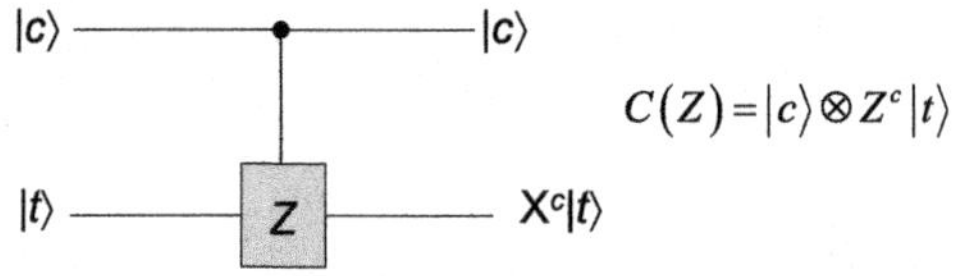

FIGURE 3.5

Operation principle and circuit representation of a controlled-Z gate.

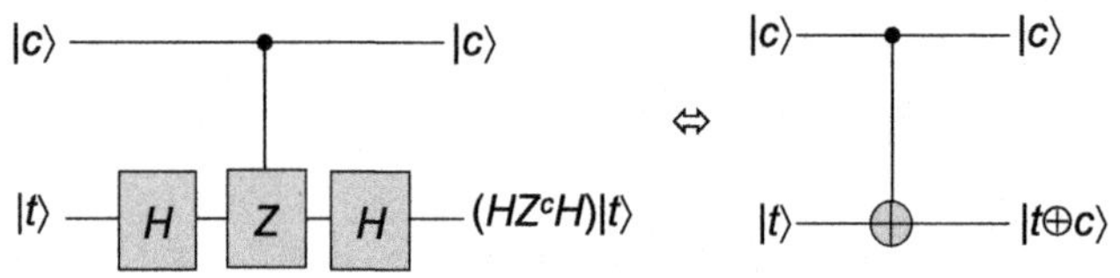

FIGURE 3.6

Implementation of a controlled-NOT gate with the help of a controlled-Z gate.

$|c\rangle|t\rangle \rightarrow |c\rangle U^c |t\rangle$

$\uparrow$

If the control qubit is set to one then the U is applied to target qubit, otherwise it is left as it is.

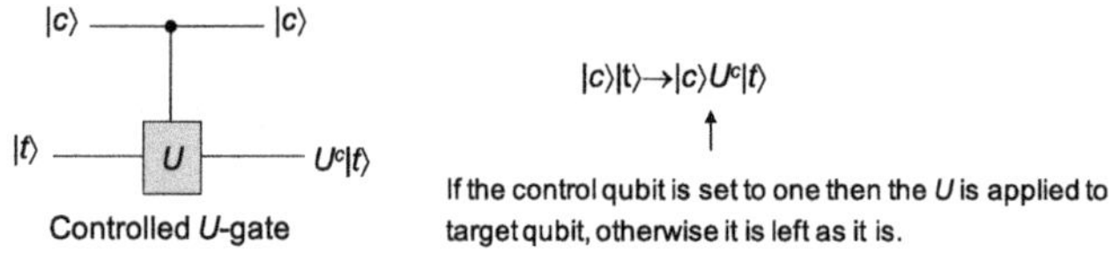

FIGURE 3.7

Controlled-U gate implementation and operation principle.

$$C(U) = \begin{bmatrix} 1 & 0 & 0 & 0 \\ 0 & 1 & 0 & 0 \\ 0 & 0 & U_{00} & U_{01} \\ 0 & 0 & U_{10} & U_{11} \end{bmatrix} = \begin{bmatrix} I & 0 \\ 0 & U \end{bmatrix}, \quad U = \begin{bmatrix} U_{00} & U_{01} \\ U_{10} & U_{11} \end{bmatrix}. \qquad (3.38)$$

By using the gates we have described so far, we can implement more complicated quantum circuits. For example, the entangled EPR pairs (Bell states) preparation circuit is shown in Fig. 3.8. In the same figure we provide the corresponding implementation in the optical domain. The operation principle of this circuit can be described as follows:

$$\mathrm{CNOT}(H \otimes I)|\psi_1\rangle|\psi_2\rangle = \begin{bmatrix} 1 & 0 & 0 & 0 \\ 0 & 1 & 0 & 0 \\ 0 & 0 & 0 & 1 \\ 0 & 0 & 1 & 0 \end{bmatrix} \otimes \left(\frac{1}{\sqrt{2}} \begin{bmatrix} 1 & 1 \\ 1 & -1 \end{bmatrix} \otimes \begin{bmatrix} 1 & 0 \\ 0 & 1 \end{bmatrix} \right) \left(\begin{bmatrix} a_1 \\ b_1 \end{bmatrix} \otimes \begin{bmatrix} a_2 \\ b_2 \end{bmatrix} \right)$$

$$= \frac{1}{\sqrt{2}} \begin{bmatrix} 1 & 0 & 0 & 0 \\ 0 & 1 & 0 & 0 \\ 0 & 0 & 0 & 1 \\ 0 & 0 & 1 & 0 \end{bmatrix} \begin{bmatrix} 1 & 0 & 1 & 0 \\ 0 & 1 & 0 & 1 \\ 1 & 0 & -1 & 0 \\ 0 & 1 & 0 & -1 \end{bmatrix} \begin{bmatrix} a_1 a_2 \\ a_1 b_2 \\ b_1 a_2 \\ b_1 b_2 \end{bmatrix} = \frac{1}{\sqrt{2}} \begin{bmatrix} a_1 a_2 + b_1 b_2 \\ a_1 b_2 + b_1 b_2 \\ b_1 a_2 - b_1 b_2 \\ a_1 a_2 - b_1 a_2 \end{bmatrix}. \qquad (3.39)$$

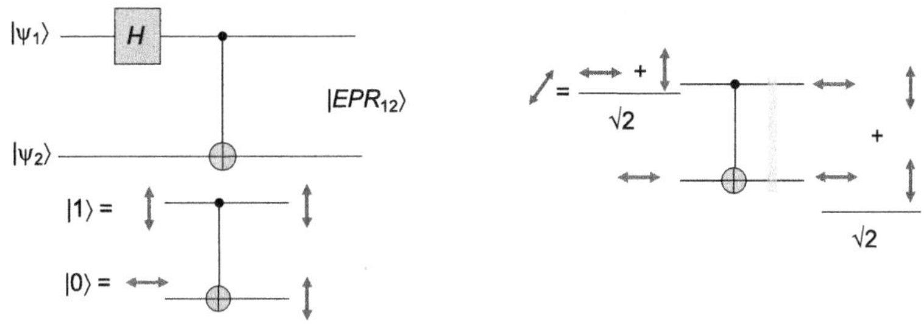

FIGURE 3.8

Entangled Einstein–Podolsky–Rosen pairs (Bell states) preparation circuit.

For example, by setting $a_1 = a_2 = 1$ and $b_1 = b_2 = 0$ we obtain:

$$\mathrm{EPR}_{12}|0\rangle|0\rangle = \mathrm{CNOT}(H \otimes I)\begin{bmatrix}1\\0\\0\\0\end{bmatrix} = \frac{|00\rangle + |11\rangle}{\sqrt{2}} = |B_{00}\rangle.$$

Therefore by setting $|\psi_1\rangle = |a\rangle$ and $|\psi_2\rangle = |b\rangle$, where $a, b \in \{0,1\}$, the output of the circuit shown in Fig. 3.8 will be the Bell state $|B_{ab}\rangle$. Regarding the optical implementation, the vertical photon corresponds to state $|1\rangle$, while the horizontal photon corresponds to state $|0\rangle$. If the control input to the CNOT gate is the vertical photon and the target input is the horizontal photon, both outputs of the CNOT gate will be the vertical photons. If both inputs of the EPR circuit are the horizontal photons, then the control photon will be transformed by the Hadamard gate to the $|45°\rangle$ photon, and after the CNOT gate at the output we will obtain the Bell $|B_{00}\rangle$ state.

By using one CNOT gate and four Hadamard gates we can change the roles of control and target qubits in the CNOT gate, which is illustrated in Fig. 3.9. We learned in the previous section that the Hadamard gate interchanges the CB and diagonal basis, $|0\rangle \to |+\rangle$ and $|1\rangle \to |-\rangle$. After this basis transformation, the action of the CNOT gate is to perform the following mappings:

$$\begin{array}{ll} |+\rangle|+\rangle \to |+\rangle|+\rangle & |-\rangle|+\rangle \to |-\rangle|+\rangle \\ |+\rangle|-\rangle \to |-\rangle|-\rangle & |-\rangle|-\rangle \to |+\rangle|-\rangle \end{array}. \tag{3.40}$$

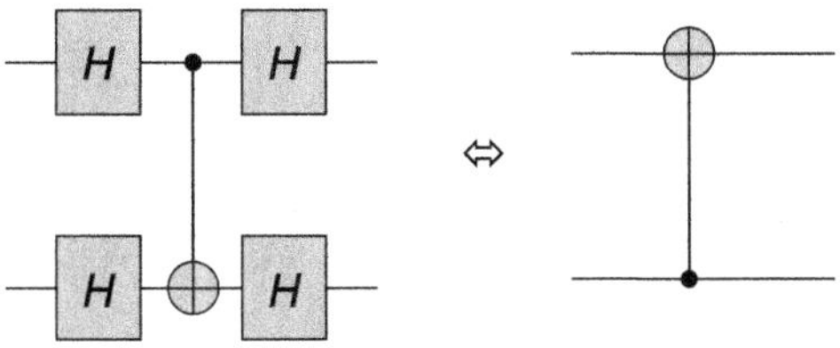

FIGURE 3.9

Quantum circuit to swap the roles of control and target qubits in the controlled-NOT gate.

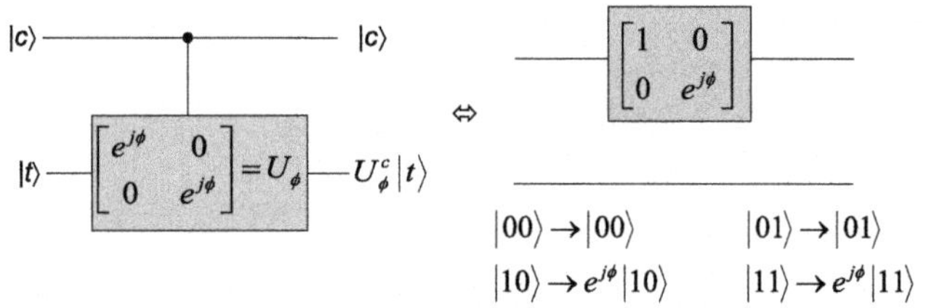

FIGURE 3.10

Control phase-shift gate and its representation.

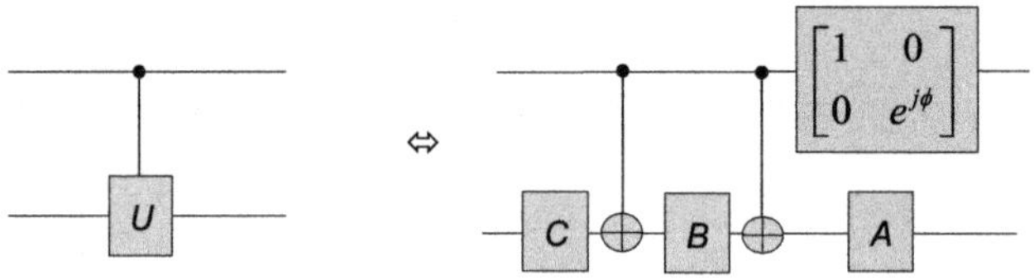

FIGURE 3.11

Quantum circuit to implement the controlled-U gate by means of single-qubit gates A, B, C satisfying the property $ABC = I$ and the following equality $U = e^{j\phi}AXBXC$.

Therefore the second qubit now behaves as control qubit and the first as target qubit. The second pair of Hadamard gates maps diagonal basis back to CB, and the roles of input control and target qubits are interchanged.

As another illustrative example, in Fig. 3.10 we provide the control phase-shift gate and its representation. If the control qubit is one, we introduce the phase shift $\exp(j\phi)$, otherwise we leave the target qubit unchanged. On the other hand, the circuit on the right side of Fig. 3.10 performs the following mapping:

$$\begin{matrix} |00\rangle \rightarrow |00\rangle & |01\rangle \rightarrow |01\rangle \\ |10\rangle \rightarrow e^{j\phi}|10\rangle & |11\rangle \rightarrow e^{j\phi}|11\rangle \end{matrix}, \tag{3.41}$$

which is the same as that of the circuit on the left side of Fig. 3.10.

Before concluding this section, we provide the *corollary of Z-Y single-qubit decomposition theorem*, which is useful in the physical implementation of various controlled-U circuits. The proof of this corollary is left as homework problems (see Problem 1). Let U be a unitary gate on a single qubit. Then there exist unitary operators A, B, C on a single qubit such that $ABC = I$ and $U = e^{j\phi}AXBXC$, where ϕ is overall phase factor. This equivalence is illustrated in Fig. 3.11.

3.3 GENERALIZATION TO *N*-QUBIT GATES AND QUANTUM COMPUTATION FUNDAMENTALS

A generic N-qubit gate is shown in Fig. 3.12. This quantum gate performs the following computation:

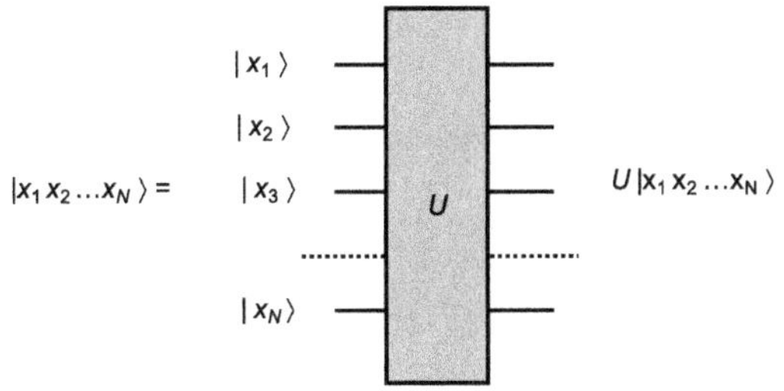

FIGURE 3.12

Representation of the generic N-qubit gate.

$$\begin{pmatrix} O_1(x) \\ \vdots \\ O_N(x) \end{pmatrix} = U \begin{pmatrix} x_1 \\ \vdots \\ x_N \end{pmatrix}; \ (x)_{10} = (x_1 \cdots x_N)_2 \tag{3.42}$$

This circuit maps the input string $x_1 \ldots x_N$ to the output string $O_1(x), \ldots, O_N(x)$, where $x \equiv x_1 \ldots x_N$.

Quantum computation can be implemented through a unitary transformation U on a $2N$-qubit quantum register:

$$U|x_1 \cdots x_N\rangle \otimes |0 \cdots 0\rangle = |x_1 \cdots x_N\rangle \otimes |O_1(x) \cdots O_N(x)\rangle, \tag{3.43}$$

where we added additional N ancilla qubits in state $|0\rangle$ to preserve the unitarity of operator U at the output of the quantum circuit. Namely, when computation is performed according to Eq. (3.42), with N qubits, when $x \neq y$ if:

$$O_1(x) \cdots O_N(x) = O_1(y) \cdots O_N(y), \tag{3.44}$$

then:

$$\langle O_1(x) \cdots O_N(x) | O_1(y) \cdots O_N(y) \rangle = 1, \tag{3.45}$$

indicating that unitarity is not preserved upon quantum computation. On the other hand, when computation is performed based on Eq. (3.43), in situations when $O_1(x) \cdots O_N(x) = O_1(y) \cdots O_N(y)$ $(x \neq y)$, we obtain:

$$\begin{aligned} &\langle x_1 \cdots x_N O_1(x) \cdots O_N(x) | y_1 \cdots y_N O_1(y) \cdots O_N(y) \rangle \\ &= \langle x_1 \cdots x_N | y_1 \cdots y_N \rangle \langle O_1(x) \cdots O_N(x) | O_1(y) \cdots O_N(y) \rangle = 0 \bullet 1 = 0, \end{aligned} \tag{3.46}$$

and the unitarity of the output is preserved. The linear superposition allows us to create the following $2N$-qubit state:

$$|\psi_{in}\rangle = \left[\frac{1}{\sqrt{2^N}} \sum_x |x_1 \cdots x_N\rangle \right] \otimes |0 \cdots 0\rangle, \tag{3.47}$$

and upon the application of quantum operation U, the output can be represented by:

$$|\psi_{\text{out}}\rangle = U(C)|\psi_{\text{in}}\rangle = \frac{1}{\sqrt{2^N}}\sum_x |x_1 \cdots x_N\rangle \otimes |O_1(x) \cdots O_N(x)\rangle. \tag{3.48}$$

Therefore we were able to encode all input strings generated by U into $|\psi_{out}\rangle$; in other words, we have simultaneously pursued 2^N classical paths. This ability of a quantum computer to perform multiple function evaluations into a single quantum computational step is known as *quantum parallelism*. More details about quantum parallelism and its applications will be provided in Chapter 5.

The quantum circuit to perform the N-qubit control operation, $C^N(U)$, is shown in Fig. 3.13. If N control qubits are set to one, then the U gate is applied to the quantum state $|y\rangle$. Action of the U gate on multiqubit state $|y\rangle$ has already been discussed.

For example, the special case for $N = 2$, the $C^2(U)$ gate, is shown in Fig. 3.14, together with its equivalent representation.

When $U = X$, then the $C^2(X)$ gate is called a *Toffoli gate*, and its representation is shown in Fig. 3.15. A Toffoli gate can be implemented based on Fig. 3.14 for $V = (1 - j)(I + jX/2)$. Its implementation is shown in Fig. 3.15, on the right.

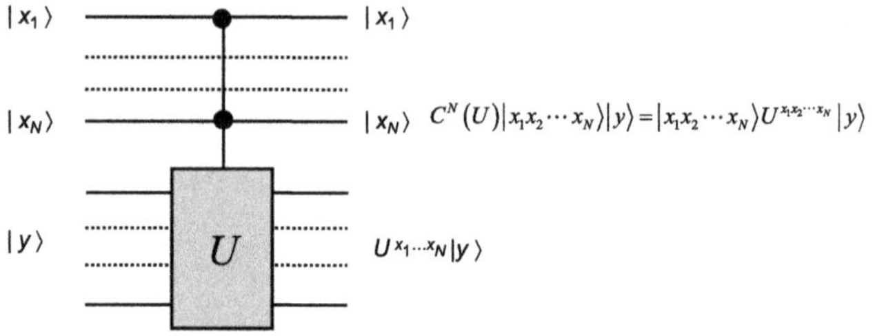

FIGURE 3.13

Quantum circuit to perform N-qubit control operation, $C^N(U)$.

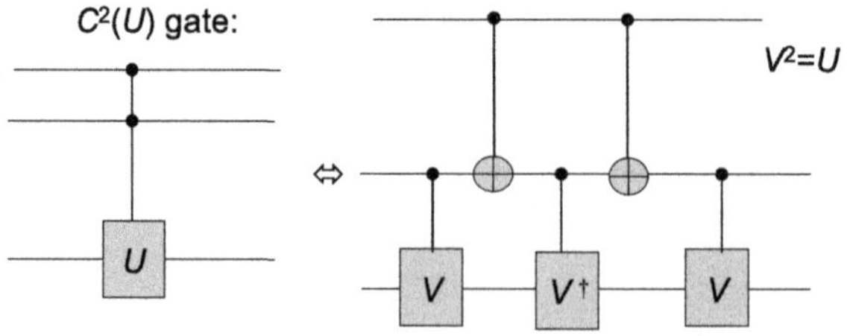

FIGURE 3.14

Quantum circuit to perform two-qubit controlled-U operation, $C^2(U)$, and its equivalent representation.

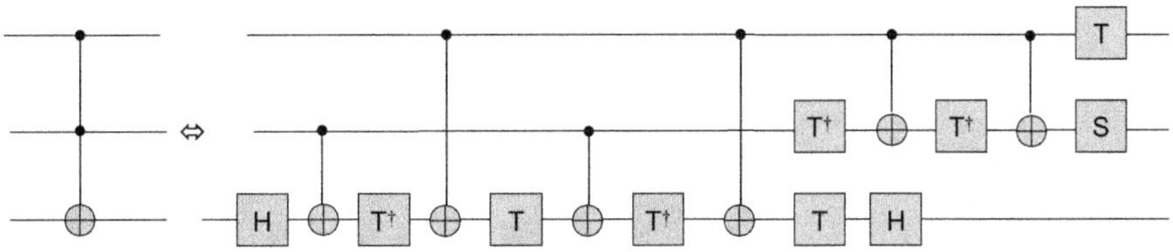

FIGURE 3.15

Toffoli gate representation and its implementation.

3.4 QUBIT MEASUREMENT (REVISITED)

When measuring a single qubit in an arbitrary state $|\psi\rangle = a|0\rangle + b|1\rangle$, the probability of outcome 0 is $|a|^2$ and the probability of outcome 1 is $|b|^2$. Such a measurement is done by Hermitian projection on the basis kets $|0\rangle$ and $|1\rangle$. The axioms of quantum mechanics, described in Chapter 2, indicate that after the measurement an arbitrary qubit collapses to the measured bases state, so that a qubit will be destroyed with this measurement. The measurement is therefore an *irreversible* operation. For example, if during the measurement someone finds that the result of the measurement is 0, then the postmeasurement state will be $|\psi_{out}\rangle = P_0|\psi\rangle / \sqrt{\langle\psi|P_0|\psi\rangle}$. On the other hand, if the result of the measurement is 1, then the postmeasurement state will be $|\psi_{out}\rangle = P_1|\psi\rangle / \sqrt{\langle\psi|P_1|\psi\rangle}$. With P_i $(i = 0,1)$ we denoted the projection operators, namely $P_i = |i\rangle\langle i|$ $(i = 0,1)$. The density operator upon measurement will be $\rho_{out} = P_0\rho P_0 + P_1\rho P_1$, where with ρ we denoted the density operator of the input state. This problem cannot be solved by copying because of the *no-cloning theorem*. The no-cloning theorem claims that a quantum device cannot be constructed to generate an exact copy of an arbitrary quantum state $|\psi\rangle$, namely to perform the following mapping $|\psi\rangle\otimes|0\rangle \rightarrow |\psi\rangle\otimes|\psi\rangle$. This theorem can easily be proved by contradiction (see Problem 3). Notice that this theorem does not rule out the possibility of building a device that can copy a particular set of orthonormal quantum states. Important consequences of no-cloning theorem can be summarized as follows [2]: (1) it does not allow quantum computing to perform copying an arbitrary number of times as a way of backing data in case of errors, (2) it rules out the possibility of using the Bell states (EPR pairs) to transmit the signal faster than the speed of light, and (3) it prevents an eavesdropper obtaining an exact copy of the key in a QKD system.

The symbol for projective measurement is shown in Fig. 3.16. Without loss of generality, any undetermined quantum wires at the end of the quantum circuit are assumed to be measured. This claim is sometimes called the *principle of implicit measurement*. It is also clear that classically conditioned quantum operations can be replaced by quantum conditioned ones. This claim is known as the *principle of*

FIGURE 3.16

Symbol for projective measurement (the *double line* denotes the classical bit).

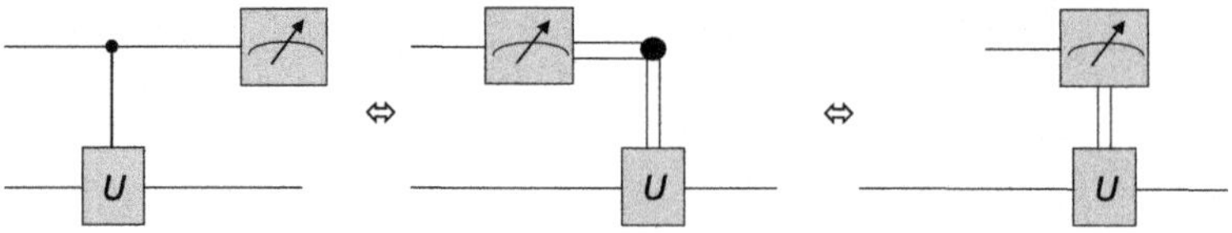

FIGURE 3.17

Measurement and control commute when the measurement is performed on a control qubit.

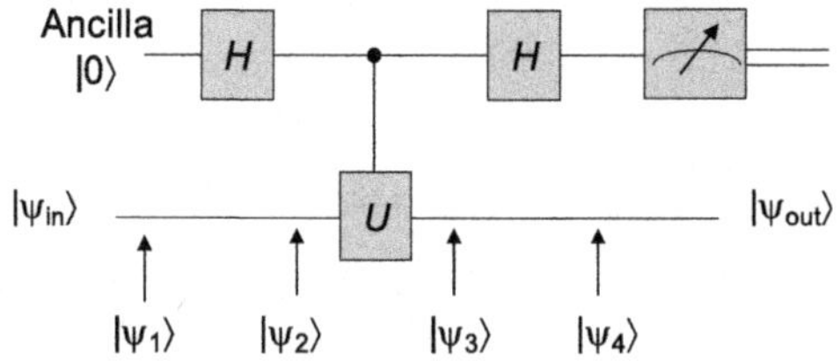

FIGURE 3.18

Measurement of an observable of the operator *U*.

differed measurement, which can also be formulated as: measurements can always be moved from an intermediate stage of the quantum circuit to the end of the quantum circuit. It can also be shown that the *measurement commutes with control operation* when the measurement is performed on a control qubit, which is illustrated in Fig. 3.17 (see also Problem 4).

Suppose now that we are concerned with the measurement of an observable associated with an operator U, which is a Hermitian and a unitary with eigenvalues ± 1, acting on the quantum state $|\psi_{in}\rangle$. The quantum circuit shown in Fig. 3.18 can be used for measuring its eigenvalues without completely destroying the measured qubit. (After measurement, the postmeasurement qubit will be in the corresponding eigenket.)

As an illustration, let us observe the measurement of an observable of operator U with eigenvalues ± 1 and corresponding eigenkets $|\psi_{\pm}\rangle$. The sequence of states at different points of quantum circuits from Fig. 3.18 is given as follows:

$$
\begin{aligned}
|\psi_1\rangle &= |0\rangle|\psi_{\text{in}}\rangle = a|0\rangle|0\rangle + b|0\rangle|1\rangle \\
|\psi_2\rangle &= \frac{1}{\sqrt{2}}(|0\rangle + |1\rangle)|\psi_{\text{in}}\rangle = \frac{1}{\sqrt{2}}(a|0\rangle|0\rangle + a|1\rangle|0\rangle + b|0\rangle|1\rangle + b|1\rangle|1\rangle) \\
|\psi_3\rangle &= \frac{1}{\sqrt{2}}(a|0\rangle|0\rangle + a|1\rangle|0\rangle + b|0\rangle|1\rangle - b|1\rangle|1\rangle) \\
|\psi_4\rangle &= \frac{1}{2}(a(|0\rangle + |1\rangle)|0\rangle + a(|0\rangle - |1\rangle)|0\rangle + b(|0\rangle + |1\rangle)|1\rangle - b(|0\rangle - |1\rangle)|1\rangle) \\
&= \frac{1}{2}(|0\rangle(a|0\rangle + a|0\rangle + b|1\rangle - b|1\rangle) + |1\rangle(a|0\rangle - a|0\rangle + b|1\rangle + b|1\rangle)) \\
&= a|0\rangle|0\rangle + b|1\rangle|1\rangle
\end{aligned}
\tag{3.49}
$$

By performing the measurement of the first qubit the result of the measurement corresponds to eigenvalues of U, while the corresponding postmeasurement state of the second qubit is the eigenket of U:

$$|\psi_{out}\rangle = |\psi_{\pm}\rangle. \tag{3.50}$$

Notice that we did not explicitly say anything about the operator U, whether it is applied to single qubits or many qubits. Therefore this measurement circuit can be applied to observables U acting on any number of qubits with two eigenvalues ± 1.

3.5 GOTTESMAN–KNILL AND SOLOVAY–KITAEV THEOREMS

The *Gottesman–Knill theorem* is related to the *Clifford group* [19,21]. Before we define the Clifford group, we need to introduce the concept of a Pauli operator on N qubits. A *Pauli operator on N qubits* has the following form $cO_1O_2 \ldots O_N$, wherein $O_i \in \{I,X,Y,Z\}$, with operator O_i acting on the ith qubit, and $c = j^l$ ($l = 1,2,3,4$). The corresponding group related to the Pauli operators is called the Pauli group on N qubits, denoted by P_N. The *Clifford group* C_N is composed of unitaries for which the elements of a Pauli group on N qubits are invariant under conjugation mapping, that is we can write $cP_Nc^{\dagger} = P_N \; \forall \; c \in C_N$. In other words, an operator from the Clifford group commutes with all operators from the Pauli group. This is why the operators from the Clifford group can be used as encoding operators for quantum error correction (see Chapter 8). The Clifford group on N qubits can be generated by the following gates: H, S, and CNOT acting on any combination of the qubits [20].

The Gottesman–Knill theorem tells us that a quantum computation that solely uses gates from the Clifford group $\{H, S, \text{CNOT}\}$ and the measurements of observables that belong to the Pauli group P_N can be efficiently simulated in polynomial time on a probabilistic classical computer. This theorem can be formally formulated as [20]: the quantum circuit employing only the following elements can be efficiently simulated on a quantum computer:

- State preparation in the CB;
- Quantum gates from the Clifford group; and
- Measurements in the CB.

Unfortunately, this theorem does not tell why quantum computing can provide exponential speed-up over classical computers. However, it tells us that quantum algorithms employing the entanglement created with the help of H and CNOT gates do not give us any advantage with respect to classical computers.

Another relevant theorem is the Solovay–Kitaev theorem [22]. Before we can formulate it, we need the following definition: we say that a subset S of a topological space X is *dense* if every point $x \in X$ either belongs to S or is a limit point of S. We also need to define the *special unitary* group [23], denoted as SU(d), as the set of all $d \times d$ unitary matrices with positive unit determinant. The generators of SU(2) are

Pauli matrices X, Y, and Z. Solovay–Kitaev theorem can be now formulated as follows [22]: if the single-qubit quantum gates set generates a dense subset of SU(2), then this set is guaranteed to fill SU(2) quickly. In other words, by using the generating set a reasonably good approximation to any desired gate can be made. Let us first introduce the concept of instructional set. We say that the set of gates in d-dimensional space represents the *instructional set* G if it satisfies the following conditions:

- All gates belong to SU(d);
- For each $g \in$ G the corresponding inverse $g^{\dagger}$ is also from G; and
- G is a universal set for SU(d).

The Solovay–Kitaev theorem can now be formulated more rigorously [22]. Let G be an instruction set for SU(d) and let the desired accuracy be $\varepsilon > 0$. Then there is a constant c such that for every $U \in$ SU(d), we can specify the finite sequence of gates S of length $O(\log^{c}(1/\varepsilon))$ such that:

$$\|U - S\| \doteq \sup_{\|\psi\|=1} \|(U - S)\psi\| < \varepsilon.$$

The constant c is found to be ~ 3.97 [22]. Clearly, this theorem tells us how many gates will be needed to approximate a given operation for prespecified accuracy ε.

3.6 UNIVERSAL QUANTUM GATES

To perform an arbitrary quantum computation, a minimum number of gates known as *universal quantum gates* [12–15] are needed. A set of quantum gates is said to be universal if it contains a minimum finite number of gates so that arbitrary unitary operation can be performed with arbitrary small error probability by using only gates from that set. The most popular sets of universal quantum gates are: {Hadamard (H), phase (S), CNOT, Toffoli (U_T)} gates, {H, S, $\pi/8$ (T), CNOT} gates, Barenco gate [12], and Deutsch gate [13]. Typically, for quantum error correction, the use of CNOT, S, and H gates is enough. Unfortunately, for arbitrary quantum computation the Clifford group gates are not enough. However, if the set N_U is extended by either Toffoli or $\pi/8$ gates, we obtain a universal set of quantum gates. Barenco [12] has shown that the following two-qubit gate (with matrix elements relative to the computational basis $\{|ij\rangle\text{: } i,j = 0,1\}$) is universal:

$$A(\phi,\alpha,\theta) = \begin{pmatrix} 1 & 0 & 0 & 0 \\ 0 & 1 & 0 & 0 \\ 0 & 0 & e^{j\alpha}\cos(\theta) & -je^{j(\alpha+\phi)}\sin(\theta) \\ 0 & 0 & -je^{j(\alpha-\phi)}\sin(\theta) & e^{j\alpha}\cos(\theta) \end{pmatrix}. \tag{3.51}$$

For example, if we impose the following phase shifts: $\theta = \pi/2$, $\phi = 0$ rad, and $\alpha = \pi/2$, the Barenco gate operates as a CNOT gate:

$$\mathrm{CNOT} = \begin{pmatrix} 1 & 0 & 0 & 0 \\ 0 & 1 & 0 & 0 \\ 0 & 0 & 0 & 1 \\ 0 & 0 & 1 & 0 \end{pmatrix}.$$

The controlled-Y gate, $C(Y)$, can be obtained from the Barenco gate by setting $\theta = \pi/2$, $\phi = 3\pi/2$, and $\alpha = \pi/2$:

$$C(Y) = \begin{pmatrix} 1 & 0 & 0 & 0 \\ 0 & 1 & 0 & 0 \\ 0 & 0 & 0 & -j \\ 0 & 0 & j & 0 \end{pmatrix}.$$

Various single-qubit gates can be obtained from the Barenco gate by omitting the control qubit and by appropriately setting the parameters α, θ, ϕ.

The three-qubit Deutsch gate [13], denoted by $D(\theta)$, represents the generalization of the Barenco gate, and its matrix representation is given by:

$$D(\theta) = \begin{pmatrix} 1 & 0 & 0 & 0 & 0 & 0 & 0 & 0 \\ 0 & 1 & 0 & 0 & 0 & 0 & 0 & 0 \\ 0 & 0 & 1 & 0 & 0 & 0 & 0 & 0 \\ 0 & 0 & 0 & 1 & 0 & 0 & 0 & 0 \\ 0 & 0 & 0 & 0 & 1 & 0 & 0 & 0 \\ 0 & 0 & 0 & 0 & 0 & 1 & 0 & 0 \\ 0 & 0 & 0 & 0 & 0 & 0 & j\cos(\theta) & \sin(\theta) \\ 0 & 0 & 0 & 0 & 0 & 0 & \sin(\theta) & j\cos(\theta) \end{pmatrix}. \tag{3.52}$$

By using the Z-Y decomposition theorem:

$$U = \begin{pmatrix} e^{j(\alpha-\beta/2-\delta/2)}\cos\left(\frac{\gamma}{2}\right) & -e^{j(\alpha-\beta/2+\delta/2)}\sin\left(\frac{\gamma}{2}\right) \\ e^{j(\alpha+\beta/2-\delta/2)}\sin\left(\frac{\gamma}{2}\right) & \cos\left(\frac{\gamma}{2}\right)e^{j(\alpha+\beta/2+\delta/2)} \end{pmatrix}, \tag{3.53}$$

where α, β, γ, and δ are phase shifts, we can implement an arbitrary single-qubit gate. By setting $\gamma = \delta = 0$ rad, $\alpha = \pi/4$, and $\beta = \pi/2$ rad the foregoing U gate operates as a phase gate:

$$S = \begin{pmatrix} 1 & 0 \\ 0 & j \end{pmatrix}.$$

By setting $\gamma = \delta = 0$ rad, $\alpha = \pi/8$, and $\beta = \pi/4$ rad the U gate operates as a $\pi/8$ gate:

$$T = \begin{pmatrix} 1 & 0 \\ 0 & e^{j\pi/4} \end{pmatrix}.$$

By setting $\gamma = \pi/2$, $\alpha = \pi/2$ rad and $\beta = 0$, $\delta = \pi$, the U gate given by Eq. (3.53) operates as a Hadamard gate:

$$H = \frac{1}{\sqrt{2}}\begin{pmatrix} 1 & 1 \\ 1 & -1 \end{pmatrix}.$$

The Pauli gates X, Y, and Z can be obtained from the U gate given by Eq. (3.53) by appropriately setting the phase shifts α, β, γ, and δ. The X gate is obtained by setting $\gamma = \pi$, $\beta = \delta = 0$ rad, and $\alpha = \pi/2$:

$$X = \begin{pmatrix} 0 & 1 \\ 1 & 0 \end{pmatrix}.$$

The Z gate is obtained by setting $\gamma = \delta = 0$ rad, $\alpha = \pi/2$, and $\beta = \pi$:

$$Z = \begin{pmatrix} 1 & 0 \\ 0 & -1 \end{pmatrix}.$$

The Y gate is obtained by setting $\gamma = \pi$, $\delta = 0$ rad, $\alpha = \pi/4$, and $\beta = -\pi/2$:

$$Y = \begin{pmatrix} 0 & -j \\ j & 0 \end{pmatrix}.$$

3.7 QUANTUM TELEPORTATION

Quantum teleportation [24,25] is a technique to transfer quantum information from source to destination by employing the entangled states. Namely, in quantum teleportation, entanglement in the Bell state (EPR pair) is used to transport arbitrary quantum state $|\psi\rangle$ between two distant observers A and B (often called Alice and Bob). The quantum teleportation system employs three qubits: qubit 1 is an arbitrary state to be teleported, while qubits 2 and 3 are in the Bell state $|B_{00}\rangle = (|00\rangle|11\rangle)/\sqrt{2}$. The operation principle is shown in Fig. 3.19 for an input state $|x\rangle$, where $x = 0{,}1$. Evaluation of the three-qubit state in five characteristic points of the quantum circuit is shown as well. The input three-qubit state at point (1) can be written as:

$$|\psi_1\rangle = \frac{|x00\rangle + |x11\rangle}{\sqrt{2}}. \tag{3.54}$$

The first qubit is then used as control qubit and the second qubit as target qubit for the CNOT gate, so that the state at point (2) becomes:

$$|\psi_2\rangle = \frac{|xx0\rangle + |x(1-x)1\rangle}{\sqrt{2}}. \tag{3.55}$$

The Hadamard gate performs the following mapping on the first qubit: $H|x\rangle = (|0\rangle + (-1)^x|1\rangle)/\sqrt{2}$, so that the state in point (3) can be represented by:

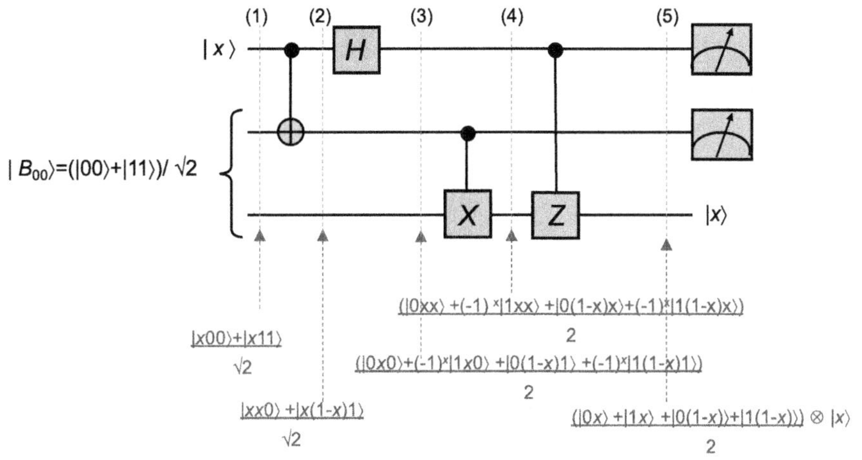

FIGURE 3.19

The quantum teleportation principle.

$$|\psi_3\rangle = \frac{|0x0\rangle + (-1)^x|1x0\rangle + |0(1-x)1\rangle + (-1)^x|1(1-x)1\rangle}{2}. \tag{3.56}$$

The second qubit is then used as control qubit and the third qubit as target qubit in a controlled-X gate (or CNOT gate) so that the quantum state at point (4) can be written as:

$$|\psi_4\rangle = \frac{|0xx\rangle + (-1)^x|1xx\rangle + |0(1-x)x\rangle + (-1)^x|1(1-x)x\rangle}{2}. \tag{3.57}$$

Furthermore, the first qubit is used as control qubit and the third qubit as target qubit for the controlled-Z gate so that the state at point (5) is given by:

$$\begin{aligned} |\psi_5\rangle &= \frac{|0xx\rangle + |1xx\rangle + |0(1-x)x\rangle + |1(1-x)x\rangle}{2} \\ &= \frac{|0x\rangle + |1x\rangle + |0(1-x)\rangle + |1(1-x)\rangle}{2} \otimes |x\rangle. \end{aligned} \tag{3.58}$$

Finally, we perform the measurements on the first two qubits that are destroyed, and the third qubit is delivered to the destination. By comparing the destination and source qubits we can conclude that the correct quantum state is teleported. Notice that in this analysis we assume there is no error introduced by the quantum channel. Another interesting point is that the teleported state is not a superposition state. Following the same procedure we can show that the arbitrary quantum state $|\psi\rangle$ can be teleported by using the scheme from Fig. 3.19 (see Problem 6).

Another alternative quantum teleportation scheme is shown in Fig. 3.20. In this version, we perform the measurement in the middle of the circuit and based on results of the measurement we conditionally execute X- and Z-operators. We can interpret this circuit as the application of principle of deferred measurement, but now in the opposite direction.

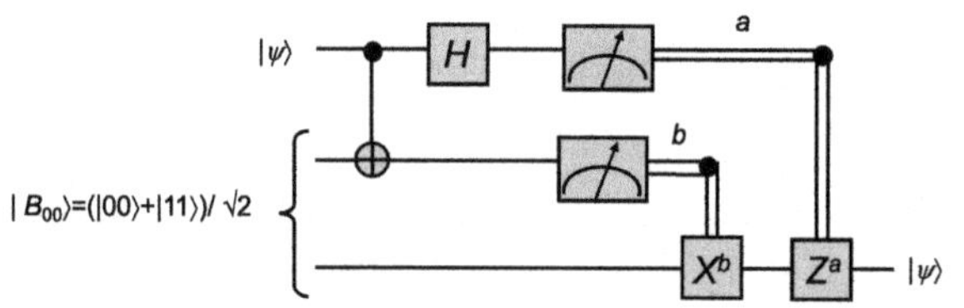

FIGURE 3.20

Quantum teleportation scheme performing the measurement in the middle of the circuit.

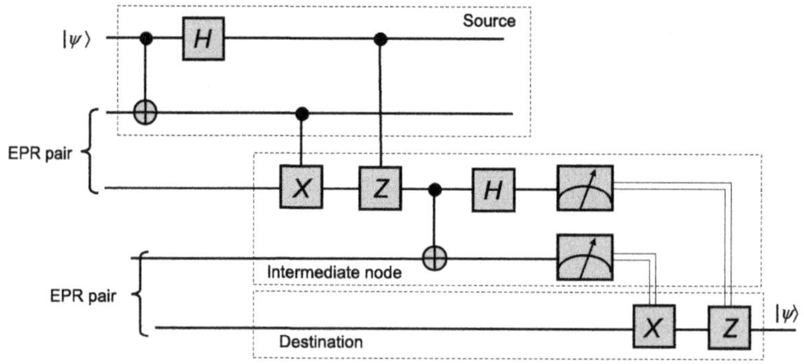

FIGURE 3.21

Operation principle of a quantum relay. *EPR*, Einstein–Podolsky–Rosen.

To extend the total teleportation distance, *quantum repeaters* [25], entanglement purification [26], and *quantum relay* [27–29] can be used. The key obstacle quantum repeater and relay for implementation is the no-cloning theorem. However, we can use the quantum teleportation principle described earlier to implement the quantum relay. One such quantum relay, based on the quantum teleportation circuit from Fig. 3.19, is shown in Fig. 3.21. Its operation principle is the subject of Problem 8.

Quantum teleportation can also be used in universal quantum computing [30].

3.8 COMPUTATION-IN-PLACE

One of the key differences of quantum computing with respect to classical computing is computation-in-place. The basic classical computer with a CPU has an input register, output register, and memory unit in addition to the CPU [31]. The CPU, on the other hand, is composed of processor and control unit, while the processor contains combinational logic and registers. The input data are stored in the corresponding register, and the control unit indicates when the data can be processed by the combinational logic, based on instructions written in the memory.

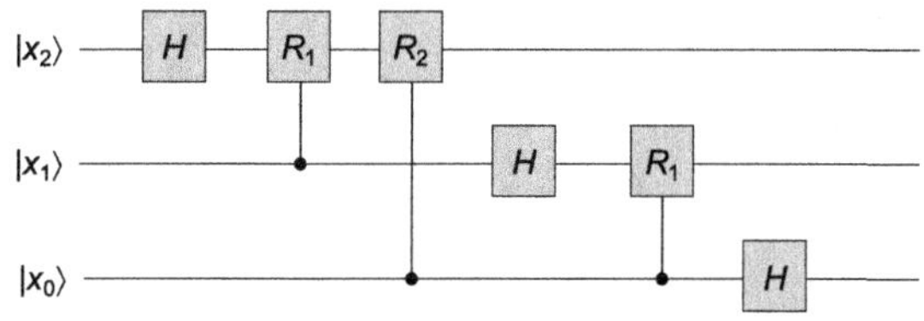

$$R_d = \begin{pmatrix} 1 & 0 \\ 0 & e^{j\pi/2^d} \end{pmatrix}$$

FIGURE 3.22

Quantum circuit to perform the Fourier transform for $n = 3$.

Once the computation is done, the result is stored in the output register. In most quantum gates-based quantum computing, the information is encoded into the quantum states of qubits and these states evolve based on the action of unitary operators. Once the measurements are done, their results are classical and can be further processed by a classical computer.

Quantum measurements are based on modulus squared of probability amplitude for each qubit state, and the key question arises: Which of the qubits will manifest itself upon the measurements? To solve this problem, we can rely on the quantum Fourier transform (QFT) gate, which will be described in Chapter 5. By applying the QFT across all qubits before the measurement takes place we can obtain the phase information at the end. The QFT gate is an efficient quantum module operating on n qubits, which requires n Hadamard gates and $\sim n^2/2$ controlled-phase gates as illustrated in Fig. 3.22 for $n = 3$. The radix-2 classical algorithm to perform the same functionality will require $O(2^{n-1}\log 2^n)$ complex multiplications.

3.9 SUMMARY

This chapter was devoted to quantum gates and quantum information processing fundamentals. We described basic single-qubit (Section 3.1), two-qubit (Section 3.2), and many-qubit gates (Section 3.3). The fundamentals of quantum information processing (Section 3.3) were discussed as well. Quantum measurements and the principles of deferred implicit measurements were discussed in Section 3.4. The Bloch sphere representation of a single qubit was provided in Section 3.1. Gottesman—Knill and Solovay—Kitaev theorems were formulated in Section 3.5. Various sets of universal quantum gates were discussed in Section 3.6. The Bell state preparation circuit and quantum relay were described in Sections 3.2 and 3.6, respectively. The basic concepts of quantum teleportation were described in Section 3.6. In addition to the foregoing topics, we introduced the concepts of quantum parallelism and entanglement (see Section 3.2). We also introduced the no-cloning theorem in Section 3.4, and discussed its consequences. In Section 3.8 we discussed computation-in-place. For a deeper understanding of underlying concepts, Section 3.10 provides a series of problems for self-study.

3.10 PROBLEMS

1. Prove the corollary of *Z-Y* single-qubit decomposition theorem, which can be formulated as follows. Let U be a unitary gate on a single qubit. Then there exist unitary operators A, B, C on a single qubit such that $ABC = I$ and $U = e^{j\phi}AXBXC$, where ϕ is the overall phase factor.
2. Verify that the quantum circuit shown in Fig. 3.P2 implements the Toffoli gate.

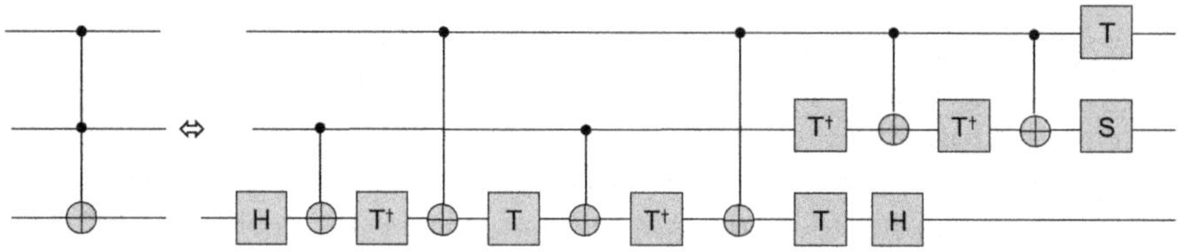

FIGURE 3.P2

3. This problem is related to the *no-cloning theorem*, which states that the quantum device cannot be constructed with the action of creating an exact copy of an arbitrary quantum state. Prove this claim. *Hint*: Use the contradiction approach.
4. A consequence of the principle of deferred measurements is that measurements commute with quantum gates when the measurement is performed on a control qubit, which is shown in Fig. 3.P4. Prove this claim.

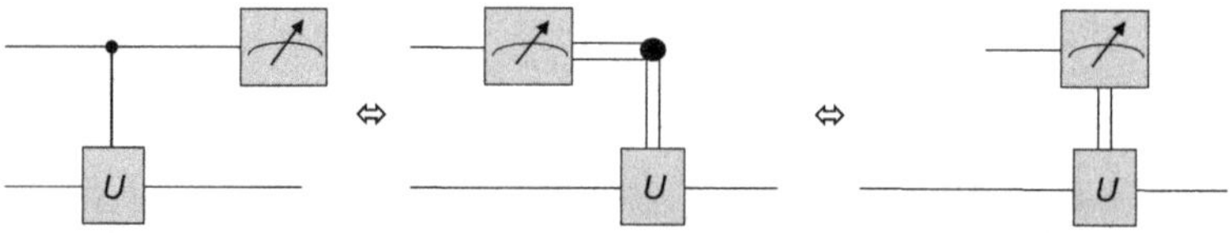

FIGURE 3.P4

5. This problem is related to the *Gottesman–Knill theorem* stating that a quantum computation that solely uses gates from the Clifford group $N_U(G_N) = \{H, S, \text{CNOT}\}$ and the measurements of observables that belong to the Pauli group G_N (*N*-qubit tensor product of Pauli operators) can be simulated efficiently on a classical computer. Prove the theorem.
6. A quantum teleportation circuit is shown in Fig. 3.P6. By using the similar procedure described in Section 3.6, show that this circuit can be used to teleport an arbitrary quantum state to a remote destination.

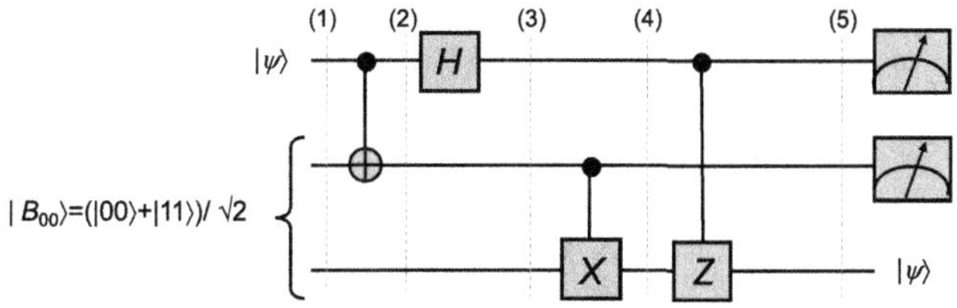

FIGURE 3.P6

7. Prove that the quantum circuit shown in Fig. 3.P7 can be used as a quantum teleportation circuit. Are the circuits from Figs. 3.P6 and 3.P7 equivalent?

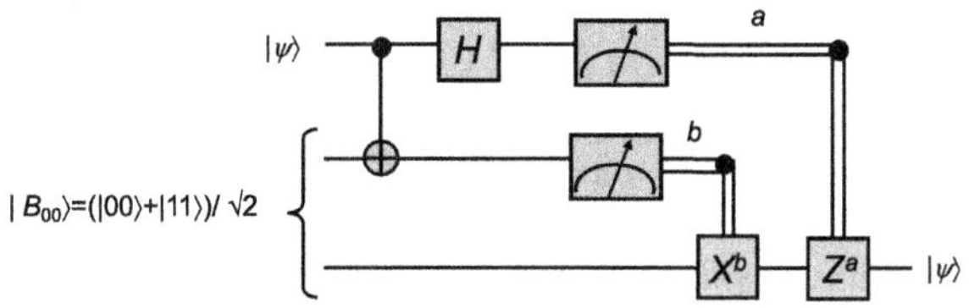

FIGURE 3.P7

8. By using a similar approach to that from the previous two problems, describe the operation principle of *quantum relay* shown in Fig. 3.P8. Prove that indeed an arbitrary quantum state can be teleported to a remote destination using this circuit. Can this circuit be implemented without any measurement; if yes, how?

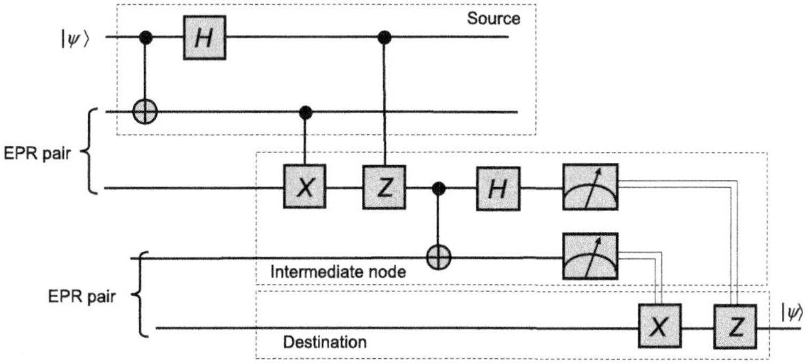

FIGURE 3.P8

9. Show that the following operator:

$$\frac{1}{2}\left(I+\overrightarrow{\sigma_A}\cdot\overrightarrow{\sigma_B}\right),$$

where $\overrightarrow{\sigma_A}\cdot\overrightarrow{\sigma_B}$ stands for simultaneous scalar and tensor product, swaps the qubits A and B. This operator is known as a SWAP operator. Determine its matrix representation in the basis $\{|00\rangle, |01\rangle, |10\rangle, |11\rangle\}$. Determine also $U_{SWAP}^{1/2}$. Show further how the controlled-X gate can be constructed from the SWAP gate.

10. Consider a 2D Hilbert space and define the state as follows:

$$|\psi\rangle = \cos(\theta/2)|0\rangle + \sin(\theta/2)|1\rangle,$$

and let the density matrix be given by:

$$\rho = p|0\rangle\langle 0| + (1-p)|1\rangle\langle 1|,\ 0 < p < 1.$$

Determine the Shannon H_{Sh} and von Neumann H_{vN} entropies and show that $H_{\text{Sh}} \geq H_{\text{vN}}$. The Shannon entropy is defined as $H_{\text{Sh}} = -\sum_i p_i \log p_i$, while the von Neumann entropy is defined by $H_{\text{vN}} = -\text{Tr}\rho\log\rho$.

11. Mutual information is defined as $I(X,Y) = H_{sh}(X) - H_{sh}(X|Y)$, where X is the channel input and Y is the channel output. Let us evaluate the amount of information that Eve can get from Alice, denoted as $I(A,E)$, for the following two situations:

- Let i represent the bit sent in CB or in diagonal basis. Therefore bit i can take different values with probability $p(i) = 1/4$. Let Eve use the CB to measure result r, which takes two different values. Create a table of conditional probabilities $p(r|i)$ and determine $p(i|r)$. What is Eve's mutual information in this case?
- Let Eve now use the Breidbart basis instead. What is Eve's mutual information in this case? Which one is higher?

12. Show that if the perfect quantum copying machine exists, then two distant parties (say Alice and Bob), sharing the pair of entangled qubits in state $|B_{11}\rangle$, could communicate at superluminal velocity.

13. This problem is related to *superdense coding* [26]. Two distant parties (say Alice and Bob) share a pair of entangled qubits in state $|B_{00}\rangle$. Alice intends to send to Bob two classical bits i,j $(i,j = 0,1)$, while using a single qubit. To do so she applies the operator A_{ij} acting on her half of $|B_{00}\rangle$ as follows:
$A_{ij} = X^i Z^j$, where i and j are exponents.
She then sends such transformed qubit to Bob, who actually receives $A_{ij}|B_{00}\rangle$.

(a) Provide explicit expressions for $A_{ij}|B_{00}\rangle$ $(i,j = 0,1)$ in terms of states $|ij\rangle$ $(i,j = 0,1)$.

(b) Let us assume that Bob has the circuit shown in Fig. 3.P13 available. Show that CNOT gates transform $A_{ij}|B_{00}\rangle$ into a tensor product, and that the measurement on the second qubit reveals the transmitted bit i. Finally, show that the second measurement on the first qubit reveals the value of bit j.

Therefore the procedure indeed allows two classical bits to be transmitted over one entangled state.

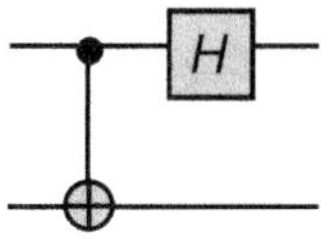

FIGURE 3.P13

References

[1] M.A. Neilsen, I.L. Chuang, Quantum Computation and Quantum Information, Cambridge University Press, Cambridge, 2000.

[2] F. Gaitan, Quantum Error Correction and Fault Tolerant Quantum Computing, CRC Press, 2008.

[3] M. Le Bellac, An Introduction to Quantum Information and Quantum Computation, Cambridge University Press, 2006.

[4] G. Jaeger, Quantum Information: an Overview, Springer, 2007.

[5] D. Petz, Quantum Information Theory and Quantum Statistics, Theoretical and mathematical physics, Springer, Berlin, 2008.

[6] P. Lambropoulos, D. Petrosyan, Fundamentals of Quantum Optics and Quantum Information, Springer-Verlag, Berlin, 2007.

[7] G. Johnson, A Shortcut Thought Time: the Path to the Quantum Computer, Knopf, N. York, 2003.

[8] J. Preskill, Quantum Computing, 1999. Available at: http://www.theory.caltech.edu/~preskill/.

[9] J. Stolze, D. Suter, Quantum Computing, Wiley, N. York, 2004.

[10] R. Landauer, Information is physical, Phys. Today 44 (5) (May 1991) 23–29.

[11] R. Landauer, The physical nature of information, Phys. Lett. 217 (1991) 188–193.

[12] A. Barenco, A universal two-bit quantum computation, Proc. Roy. Soc. Lond. A 449 (1937) (1995) 679–683.

[13] D. Deutsch, Quantum computational networks, Proc. Roy. Soc. Lond. A 425 (1868) (1989) 73–90.

[14] D.P. DiVincenzo, Two-bit gates are universal for quantum computation, Phys. Rev. A 51 (2) (1995) 1015–1022.

[15] A. Barenco, C.H. Bennett, R. Cleve, D.P. DiVincenzo, N. Margolus, P. Shor, T. Sleator, J.A. Smolin, H. Weinfurter, Elementary gates for quantum computation, Phys. Rev. A 52 (5) (1995) 3457–3467.

[16] I.B. Djordjevic, Quantum Information Processing and Quantum Error Correction: an Engineering Approach, Elsevier/Academic Press, 2012.

[17] J.A. Jones, D. Jaksch, Quantum Information, Computation, and Communication, Cambridge University Press, Cambridge-New York, 2012.

[18] I.B. Djordjevic, Physical-Layer Security and Quantum Key Distribution, Springer International Publishing, Switzerland, 2019.

[19] D. Gottesman, The Heisenberg representation of quantum computers, in: S.P. Corney, R. Delbourgo, P.D. Jarvis (Eds.), Proceedings of the XXII International Colloquium on Group Theoretical Methods in Physics, International Press, Cambridge, MA, 1999, pp. 32–43.

[20] D. Gottesman, Stabilizer Codes and Quantum Error Correction, PhD dissertation, California Institute of Technology, Pasadena, CA, 1997.

[21] D. Gottesman, Theory of fault-tolerant quantum computation, Phys. Rev. A 57 (1) (January 1998) 127–137.

[22] C.M. Dawson, M.A. Nielsen, The Solovay-Kitaev algorithm, Quant. Inf. Comput. 6 (1) (January 2006) 81–95.

[23] D. McMahon, Quantum Field Theory Demystified, McGraw Hill, New York, 2008.

[24] C.H. Bennett, G. Brassard, C. Crépeau, R. Jozsa, A. Peres, W.K. Wootters, Teleporting an unknown quantum state via dual classical and Einstein-Podolsky-Rosen channels, Phys. Rev. Lett. 70 (13) (1993) 1895–1899.

[25] H.-J. Briegel, W. Dür, J.I. Cirac, P. Zoller, Quantum repeaters: the role of imperfect local operations in quantum communication, Phys. Rev. Lett. 81 (26) (1998) 5932–5935.

[26] J.-W. Pan, C. Simon, Č. Brunker, A. Zeilinger, Entanglement purification for quantum communication, Nature 410 (April 26, 2001) 1067–1070.

[27] I.B. Djordjevic, Photonic implementation of quantum relay and encoders/decoders for sparse-graph quantum codes based on optical hybrid, IEEE Photon. Technol. Lett. 22 (19) (October 1, 2010) 1449–1451.

[28] I.B. Djordjevic, On the photonic implementation of universal quantum gates, Bell states preparation circuit and quantum LDPC encoders and decoders based on directional couplers and HNLF, Optic Express 18 (8) (April 1, 2010) 8115–8122.

[29] I.B. Djordjevic, On the photonic implementation of universal quantum gates, Bell states preparation circuit, quantum relay and quantum LDPC encoders and decoders, IEEE Photon. J. 2 (1) (February, 2010) 81–91.

[30] D. Gottesman, I.L. Chuang, Demonstrating the viability of universal quantum computation using teleportation and single-qubit operations, Nature 402 (November 25, 1999) 390–393.

[31] A. Clements, The Principles of Computer Hardware, fourth ed., Oxford University Press, Oxford-New York, 2006.

CHAPTER

Quantum Information Processing Fundamentals

4

CHAPTER OUTLINE

This chapter is devoted to the basic concepts of quantum information processing (QIP). After describing the elementary QIP features, the chapter introduces the superposition principle and quantum parallelism concept as well as QIP basics. It continues with formulation and proof of no-cloning theorem. We further formulate and prove an important theorem, which claims that it is impossible unambiguously to distinguish nonorthogonal quantum states. The next section is devoted to the various aspects of entanglement, including the Schmidt decomposition theorem, purification, superdense coding, and entanglement swapping. How entanglement can be used in detecting and correcting quantum errors is described as well. Various measures of entanglement are introduced, including the entanglement of formation, concurrence, and entanglement fidelity. Furthermore, we discuss the operator-sum representation of a quantum operation and apply this concept to describe the quantum errors and decoherence. The bit-flip, phase-flip, phase-damping, depolarizing, and amplitude-damping channels are described as well. Then we describe different quantum computing libraries, including Qiskit, Cirq, Forest, and Quantum Development Kit. After summarizing the chapter, a set of problems is provided for better understanding of QIP concepts.

Quantum Information Processing, Quantum Computing, and Quantum Error Correction
https://doi.org/10.1016/B978-0-12-821982-9.00008-3

4.1 QUANTUM INFORMATION PROCESSING FEATURES

The *fundamental features* of quantum information processing (QIP) are different from that of classical computing and can be summarized into the following three [1–4]: (1) linear superposition, (2) quantum parallelism, and (3) entanglement. Next, we provide some basic details of these features.

1. Linear superposition. Contrary to the classical bit, a quantum bit or *qubit* can take not only two discrete values 0 and 1 but also *all* possible *linear combinations* of them. This is a consequence of a fundamental property of quantum states: it is possible to construct a *linear superposition* of quantum state $|0\rangle$ and quantum state $|1\rangle$.
2. Quantum parallelism. *Quantum parallelism* is the possibility of performing a large number of operations in parallel, which represents the key difference from classical computing. Namely, in classical computing it is possible to know what the internal status of the computer is. On the other hand, because of no-cloning theorem, it is not possible to know the current state of a quantum computer. This property has led to the development of Shor's factorization algorithm, which can be used to crack the Rivest–Shamir–Adleman encryption protocol. Some other important quantum algorithms include Grover's search algorithm, which is used to perform a search for an entry in an unstructured database; the quantum Fourier transform, which is the basis for a number of different algorithms; and Simon's algorithm. The quantum computer is able to encode all input strings of length N simultaneously into a single computation step. In other words, the quantum computer is able simultaneously to pursue 2^N classical paths, indicating that the quantum computer is significantly more powerful than the classical one.
3. Entanglement. At a quantum level it appears that two quantum objects can form a single entity, even when they are well separated from each other. Any attempt to consider this entity as a combination of two independent quantum objects, given by the tensor product of quantum states, fails unless the possibility of signal propagation at superluminal speed is allowed. These quantum objects that cannot be decomposed into a tensor product of individual independent quantum objects are called *entangled* quantum objects. Given the fact that arbitrary quantum states cannot be copied, which is the consequence of no-cloning theorem, communication at superluminal speed is not possible, and as a consequence the entangled quantum states cannot be written as a tensor product of independent quantum states. Moreover, it can be shown that the amount of information contained in an entangled state of N qubits grows exponentially instead of linearly, which is the case for classical bits.

In the following sections we will describe these fundamental features in more detail. Before that, let us briefly review the roots of QIP, as outlined in [2]. The roots of QIP lie with information theory [5–9], computer science [10–12], and quantum mechanics [13–22]. Shannon's development of information theory [5,6] allowed us to formally define "information," while Turing's seminal work [10–12], creating the foundation of computer science, provided us with a computational mechanism to process information. Together they have led to breakthroughs in efficient and secure data communication over a communication channel such as a memory storage device and a wire (or wireless) communication link. These breakthroughs, however, are based on the classical unit of information, a binary digit or a bit that can be either 0 or 1. In this classical setup, to increase the computational speed of a classical computing machine we need to proportionally increase the processing power. For example, given classical computers that can perform a certain number of computations per second, the number of computations can be approximately doubled by adding a similar computer. This picture dramatically changes with the introduction of the computational consequences of quantum mechanics [23]. Now, instead of working with bits, we work with quantum bits or qubits that represent quantum states (such as photon polarization or electron spin) in a 2D Hilbert space.

4.2 SUPERPOSITION PRINCIPLE, QUANTUM PARALLELISM, AND QIP BASICS

We say that the allowable states $|\mu\rangle$ and $|\nu\rangle$ of the quantum system satisfy the *superposition principle* if their linear combination $\alpha\,|\mu\rangle + \beta\,|\nu\rangle$, where α and β are the complex numbers ($\alpha, \beta \in C$), is also an allowable quantum state. Without loss of generality we typically observe the computational basis (CB) composed of the orthogonal canonical states $|0\rangle = \begin{bmatrix} 1 \\ 0 \end{bmatrix}$, $|1\rangle = \begin{bmatrix} 0 \\ 1 \end{bmatrix}$, so that a quantum bit, also known as a *qubit*, lies in a 2D Hilbert space H, isomorphic to C^2 space, and can be represented as:

$$|\psi\rangle = \alpha|0\rangle + \beta|1\rangle = \begin{pmatrix} \alpha \\ \beta \end{pmatrix};\ \alpha,\ \beta \in C;\ |\alpha|^2 + |\beta|^2 = 1. \tag{4.1}$$

If we perform the measurement of a qubit, we will obtain $|0\rangle$ with probability $|\alpha|^2$ and $|1\rangle$ with a probability of $|\beta|^2$. Measurement changes the state of a qubit from a superposition of $|0\rangle$ and $|1\rangle$ to the specific state consistent with the measurement result. If we parametrize the probability amplitudes α and β as follows:

$$\alpha = \cos\left(\frac{\theta}{2}\right), \quad \beta = e^{j\phi} \sin\left(\frac{\theta}{2}\right), \tag{4.2}$$

where θ is a polar angle and ϕ is an azimuthal angle, we can geometrically represent the qubit by the Bloch sphere (or the Poincaré sphere for the photon) as illustrated in Fig. 4.1. Bloch vector coordinates are given by ($\cos\phi \sin\theta$, $\sin\phi \sin\theta$, $\cos\theta$). This Bloch vector representation is related to CB by:

$$|\psi(\theta,\ \phi)\rangle = \cos(\theta/2)|0\rangle + e^{j\phi}\sin(\theta/2)|1\rangle \doteq \begin{pmatrix} \cos(\theta/2) \\ e^{j\phi}\sin(\theta/2) \end{pmatrix}, \tag{4.3}$$

where $0 \le \theta \le \pi$ and $0 \le \phi < 2\pi$. The north and south poles correspond to computational $|0\rangle$ ($|x\rangle$-polarization) and $|1\rangle$ ($|y\rangle$-polarization) basis kets, respectively.

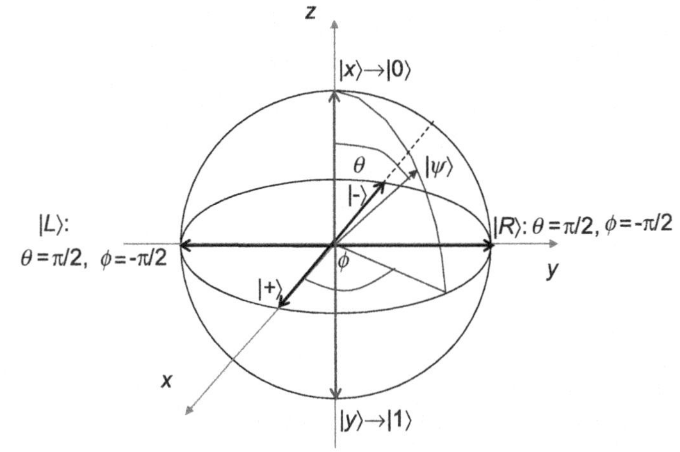

FIGURE 4.1

Bloch (Poincaré) sphere representation of the single qubit.

Other important bases are the *diagonal basis* $\{|+\rangle, |-\rangle\}$, very often denoted as $\{|\nearrow\rangle, |\searrow\rangle\}$, related to CB by:

$$|+\rangle=|\nearrow\rangle=\frac{1}{\sqrt{2}}(|0\rangle+|1\rangle),\ |-\rangle=|\searrow\rangle=\frac{1}{\sqrt{2}}(|0\rangle-|1\rangle). \tag{4.4}$$

and the *circular basis* $\{|R\rangle,|L\rangle\}$, related to CB as follows:

$$|R\rangle=\frac{1}{\sqrt{2}}(|0\rangle+j|1\rangle),\ |L\rangle=\frac{1}{\sqrt{2}}(|0\rangle-j|1\rangle). \tag{4.5}$$

The pure qubit states lie on the Bloch sphere, while the mixed qubit states lie in the interior of the Bloch sphere. The maximally mixed state $I/2$ (I denotes the identity operator) lies in the center of the Bloch sphere. The orthogonal states are antipodal. From Fig. 4.1 we see that CB, diagonal basis, and circular bases are 90 degrees apart from each other, and we often say that these three bases are mutually *conjugate bases*. These bases are used as three pairs of signal states for the six-state quantum key distribution (QKD) protocol. Another important basis used in QKD and for eavesdropping is the *Breidbart basis* given by: $\{\cos(\pi/8)|0\rangle+\sin(\pi/8)|1\rangle,\ -\sin(\pi/8)|0\rangle+\cos(\pi/8)|1\rangle\}$

The superposition principle is the key property that makes quantum parallelism possible. To see this, let us juxtapose n qubits lying in n distinct 2D Hilbert spaces $H_0,H_1,\ldots,H_{n-1}$ that are isomorphic to each other. In practice this means the qubits have been prepared separately, without any interaction, which can be mathematically described by the tensor product:

$$|\psi\rangle=|\psi_0\rangle\otimes|\psi_1\rangle\otimes\cdots\otimes|\psi_{n-1}\rangle\in H_0\otimes H_1\otimes\cdots\otimes H_{n-1}. \tag{4.6}$$

Any arbitrary basis can be selected as the CB for H_i, $i=0,1,\ldots,n-1$. However, to facilitate explanation, we assume the computational basis to be $|0_i\rangle$ and $|1_i\rangle$.

Consequently, we can represent the ith qubit as $|\psi_i\rangle = \alpha_i|0_i\rangle + \beta_i|1_i\rangle$. Introducing a further assumption, $\alpha_i = \beta_i = 2^{-1/2}$, without loss of generality, we now have:

$$|\psi\rangle = \prod_{i=0}^{n-1} \frac{1}{\sqrt{2}}(|0_i\rangle + |1_i\rangle) = 2^{-n/2}\sum_x |\boldsymbol{x}\rangle, \quad \boldsymbol{x} = x_0x_1\cdots x_{n-1},\ x_j \in \{0,1\}. \tag{4.7}$$

This composite quantum system is called the *n-qubit register* and as can be seen from Eq. (4.7), it represents a superposition of 2^n quantum states that exist simultaneously! This is an example of quantum parallelism. In the classical realm a linear increase in size corresponds roughly to a linear increase in processing power. In the quantum world, due to the power of quantum parallelism, a linear increase in size corresponds to an exponential increase in processing power. The downside, however, is accessibility to this parallelism. Remember that superposition collapses the moment we attempt to measure it. The quantum circuit to create the foregoing superposition state, in other words the Walsh–Hadamard transform, is shown in Fig. 4.2. Therefore the Walsh–Hadamard transform on n ancilla qubits in state $|00\ldots0\rangle$ can be implemented by applying the Hadamard operators (gates) H, whose action is described in Fig. 4.2, on ancillary qubits.

More generally, a linear operator (gate) B can be expressed in terms of eigenkets $\{|a^{(n)}\rangle\}$ of a Hermitian operator A. The *operator B* is associated with a *square matrix* (albeit infinite in extent), whose elements are:

$$B_{mn} = \left\langle a^{(m)} \middle| B \middle| a^{(n)} \right\rangle, \tag{4.8}$$

and can explicitly be written as:

$$B \doteq \begin{pmatrix} \left\langle a^{(1)} \middle| B \middle| a^{(1)} \right\rangle & \left\langle a^{(1)} \middle| B \middle| a^{(2)} \right\rangle & \cdots \\ \left\langle a^{(2)} \middle| B \middle| a^{(1)} \right\rangle & \left\langle a^{(2)} \middle| B \middle| a^{(2)} \right\rangle & \cdots \\ \vdots & \vdots & \ddots \end{pmatrix}, \tag{4.9}$$

where we use the notation $\doteq B$ to denote that operator B is represented by the foregoing matrix. Very important single-qubit gates are: Hadamard gate H, the phase shift gate S, the $\pi/8$ (or T) gate, controlled-NOT (or CNOT) gate, and Pauli operators

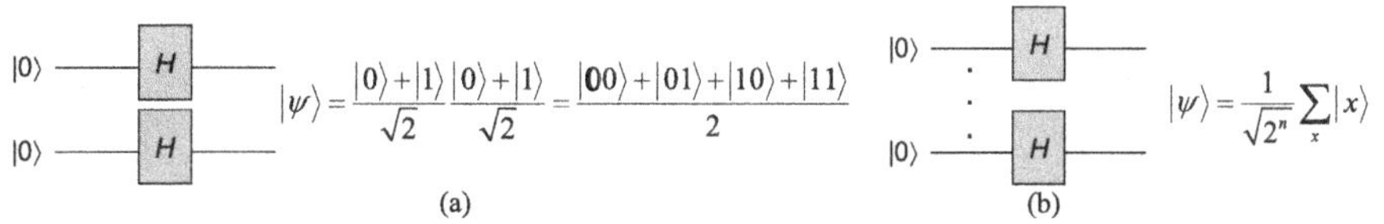

FIGURE 4.2

The Walsh–Hadamard transform: (A) on two qubits and (B) on n qubits. The action of the Hadamard gate H on computational basis kets is given by: $H|0\rangle = 2^{-1/2}(|0\rangle + |1\rangle)$ and $H|1\rangle = 2^{-1/2}(|0\rangle - |1\rangle)$.

X, Y, Z. The Hadamard gate H, phase shift gate, T gate, and CNOT gate have the following matrix representation in CB $\{|0\rangle, |1\rangle\}$:

$$H \doteq \frac{1}{\sqrt{2}}\begin{bmatrix} 1 & 1 \\ 1 & -1 \end{bmatrix}, \; S \doteq \begin{bmatrix} 1 & 0 \\ 0 & j \end{bmatrix}, \; T \doteq \begin{bmatrix} 1 & 0 \\ 0 & e^{j\pi/4} \end{bmatrix}, \; \text{CNOT} \doteq \begin{bmatrix} 1 & 0 & 0 & 0 \\ 0 & 1 & 0 & 0 \\ 0 & 0 & 0 & 1 \\ 0 & 0 & 1 & 0 \end{bmatrix}. \tag{4.10}$$

The Pauli operators, on the other hand, have the following matrix representation in CB:

$$X \doteq \begin{bmatrix} 0 & 1 \\ 1 & 0 \end{bmatrix}, \; Y \doteq \begin{bmatrix} 0 & -j \\ j & 0 \end{bmatrix}, \; Z \doteq \begin{bmatrix} 1 & 0 \\ 0 & -1 \end{bmatrix}. \tag{4.11}$$

The action of Pauli gates on an arbitrary qubit $|\psi\rangle = \alpha|0\rangle + \beta|1\rangle$ is given as follows:

$$\begin{aligned} &X(\alpha|0\rangle + \beta|1\rangle) = \alpha|1\rangle + \beta|0\rangle, \; Y(\alpha|0\rangle + \beta|1\rangle) = j(\alpha|1\rangle - \beta|0\rangle), \\ &Z(\alpha|0\rangle + \beta|1\rangle) = a|0\rangle - b|1\rangle. \end{aligned} \tag{4.12}$$

So the action of the X gate is to introduce the bit flip, the action of the Z gate is to introduce the phase flip, and the action of the Y gate is to simultaneously introduce the bit and phase flips.

Several important single-, two-, and three-qubit gates are shown in Fig. 4.3. The action of the single-qubit gate is to apply the operator U on qubit $|\psi\rangle$, which results in another qubit. The controlled-U gate conditionally applies the operator U on the target qubit $|\psi\rangle$, when the control qubit $|c\rangle$ is in the $|1\rangle$ state. One particularly important controlled-U gate is the controlled-NOT (CNOT) gate. This gate flips the content of the target qubit $|t\rangle$ when the control qubit $|c\rangle$ is in the $|1\rangle$ state. The purpose of the SWAP gate is to interchange the positions of two qubits, and can be implemented by using three CNOT gates as shown in Fig. 4.3D. Finally, the Toffoli gate represents the generalization of the CNOT gate, where two control qubits are used.

The minimum set of gates that can be used to perform an arbitrary quantum computation algorithm is known as the *universal set of gates*. The most popular sets of universal quantum gates are: {H, S, CNOT, Toffoli} gates, {H, S, $\pi/8$ (T),

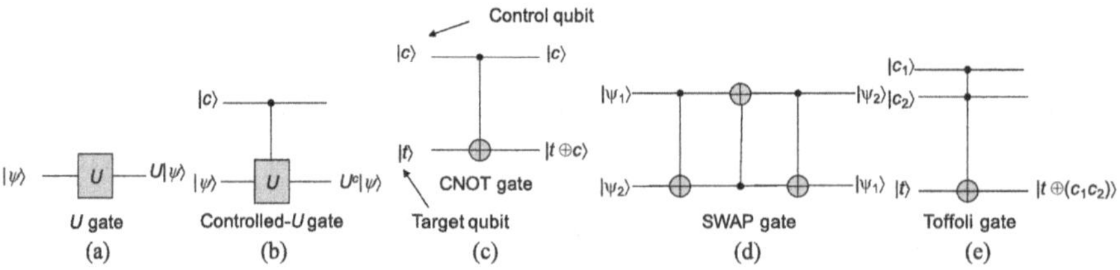

FIGURE 4.3

Important quantum gates and their action: (A) single-qubit gate, (B) controlled-U gate, (C) CNOT gate, (D) SWAP gate, and (E) Toffoli gate.

CNOT} gates, Barenco gate, and Deutsch gate. By using these universal quantum gates, more complicated operations can be performed. As an illustration, Fig. 4.4 shows the *Bell states* (Einstein–Podolsky–Rosen [(EPR] pairs) preparation circuit, which is of high importance in quantum teleportation and QKD applications.

So far single-, two-, and three-qubit quantum gates have been considered. An arbitrary quantum state of K qubits has the form $\sum_s \alpha_s |s\rangle$, where s runs over all binary strings of length K. Therefore there are 2^K complex coefficients, all independent except for the normalization constraint:

$$\sum_{s=00...00}^{11..11} |\alpha_s|^2 = 1. \tag{4.13}$$

For example, the state $\alpha_{00}|00\rangle+ \alpha_{01}|01\rangle+ \alpha_{10}|10\rangle+ \alpha_{11}|11\rangle$ (with $|\alpha_{00}|_2 + |\alpha_{01}|^2 + |\alpha_{10}|^2 + |\alpha_{11}|^2 = 1$) is the general two-qubit state (we use $|00\rangle$ to denote the tensor product $|0\rangle \otimes |0\rangle$). The multiple qubits can be *entangled* so that they cannot be decomposed into two separate states. For example, the Bell state or EPR pair $(|00\rangle + |11\rangle)/\sqrt{2}$ cannot be written in terms of tensor product $|\psi1\rangle\ |\psi2\rangle = (\alpha_1|0\rangle + \beta_1|1\rangle) \otimes (\alpha_2|0\rangle + \beta_2|1\rangle) = \alpha_1\alpha_2|00\rangle + \alpha_1\beta_2|01\rangle + \beta_1\alpha_2|10\rangle + \beta_1\beta_2|11\rangle$, because it would require $\alpha_1\alpha_2 = \beta_1\beta_2 = 1/\sqrt{2}$, while $\alpha_1\beta_2 = \beta_1\alpha_2 = 0$, which a priori has no reason to be valid. This state can be obtained by using the circuit shown in Fig. 4.4 for a two-qubit input state $|00\rangle$. We will return to the concept of entanglement in Section 4.5.

Quantum parallelism can now be introduced more formally as follows. QIP implemented on a quantum register maps the input string $i_1 \ldots i_N$ to the output string $O_1(i),\ldots,O_N(i)$:

$$\begin{pmatrix} O_1(i) \\ \vdots \\ O_N(i) \end{pmatrix} = U(QIP)\begin{pmatrix} i_1 \\ \vdots \\ i_N \end{pmatrix};\ (i)_{10} = (i_1 \cdots i_N)_2 \tag{4.14}$$

The CB states are denoted by:

$$|i_1 \cdots i_N\rangle = |i_1\rangle \otimes \cdots \otimes |i_N\rangle;\ i_1,\ \cdots,\ i_N \in \{0,1\}. \tag{4.15}$$

Linear superposition allows us to form the following 2^N-qubit state:

$$|\psi_{in}\rangle = \left[\frac{1}{\sqrt{2^N}}\sum_i |i_1 \cdots i_N\rangle\right] \otimes |0 \cdots 0\rangle, \tag{4.16}$$

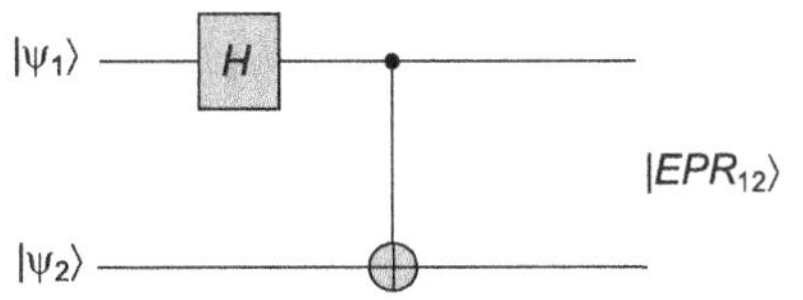

FIGURE 4.4

Bell states (Einstein–Podolsky–Rosen pairs) preparation circuit.

and upon the application of quantum operation $U(QIP)$, the output can be represented by:

$$|\psi_{out}\rangle = U(QIP)|\psi_{in}\rangle = \frac{1}{\sqrt{2^N}}\sum_i |i_1 \cdots i_N\rangle \otimes |O_1(i) \cdots O_N(i)\rangle. \quad (4.17)$$

The *QIP* circuit (or a quantum computer) has been able to encode all input strings generated by *QIP* into $|\psi_{\text{out}}\rangle$; in other words, it has simultaneously pursued 2^N classical paths. This ability of a *QIP* circuit to encode multiple computational results into a quantum state in a single quantum computational step is known as *quantum parallelism*, and this concept was introduced in the previous chapter.

In the classical realm a linear increase in size corresponds roughly to a linear increase in processing power. In the quantum world, due to the power of quantum parallelism, a linear increase in size corresponds to an exponential increase in processing power. The downside, however, is accessibility to this parallelism. Remember that the superposition state collapses the same moment we attempt to measure it. This raises the question whether quantum parallelism has practical uses. The first insight into this question was provided by the seminal works of Deutch and Josza [24]. In 1985 Deutch proposed the first quantum algorithm to determine whether or not a Boolean function is constant by exploiting quantum parallelism [25]. In 1992 Deutch and Josza generalized Deutsch's algorithm to n qubits [26]. Both these algorithms only showed the feasibility of quantum computing without any practical applications. Shor's groundbreaking work [27] changed that when he showed how quantum parallelism could be exploited to achieve a polynomial-time quantum algorithm for prime factorization that achieved the promise of exponential speed-up over classical factorization algorithms. Furthermore, Grover then proposed a quantum algorithm [28] that achieved a quadratic speed-up over classical algorithms to perform unstructured database searches. Various quantum algorithms are the subject of investigation in Chapter 5.

4.3 NO-CLONING THEOREM

Just like in quantum parallelism, quantum superposition is also the key concept behind our inability to clone arbitrary quantum states. To see this let us think of a quantum copier that takes as input an arbitrary quantum state and outputs two copies of that state, resulting in a clone of the original state. For example, if the input state is $|\psi\rangle$, then the output of the copier is $|\psi\rangle|\psi\rangle$. For an arbitrary quantum state, such a copier raises a fundamental contradiction. Consider two arbitrary states $|\psi\rangle$ and $|\chi\rangle$ that are input to the copier. When they are inputted individually we expect to get $|\psi\rangle|\psi\rangle$ and $|\chi\rangle|\chi\rangle$. Now consider a superposition of these two states given by:

$$|\varphi\rangle = \alpha|\psi\rangle + \beta|\chi\rangle. \quad (4.18)$$

Based on the foregoing description that the quantum copier clones the original state we expect the output:

$$
\begin{aligned}
|\varphi\rangle|\varphi\rangle &= (\alpha|\psi\rangle + \beta|\chi\rangle)(\alpha|\psi\rangle + \beta|\chi\rangle) \\
&= \alpha^2|\psi\rangle|\psi\rangle + \alpha\beta|\psi\rangle|\chi\rangle + \alpha\beta|\chi\rangle|\psi\rangle + \beta^2|\chi\rangle|\chi\rangle. \qquad (4.19)
\end{aligned}
$$

On the other hand, linearity of quantum mechanics, as evidenced by the Schrödinger wave equation, tells us that the quantum copier can be represented by a unitary operator that performs the cloning. If such a unitary operator were to act on the superposition state $|\varphi\rangle$, the output would be a superposition of $|\psi\rangle|\psi\rangle$ and $|\chi\rangle|\chi\rangle$, that is:

$$|\varphi'\rangle = \alpha|\psi\rangle|\psi\rangle + \beta|\chi\rangle|\chi\rangle. \qquad (4.20)$$

As is clearly evident the difference between the previous two equations leads to the foregoing contradiction. As a consequence, there is no unitary operator that can clone $|\varphi\rangle$. We therefore formulate the no-cloning theorem as follows:

No-cloning theorem. No quantum copier exists that can clone an arbitrary quantum state.

This result raises a related question: Do there exist some specific states for which cloning is possible? The answer to this question is (surprisingly) yes. Remember, a key result of quantum mechanics is that unitary operators preserve probabilities. This implies that inner (dot) products $\langle\varphi\,|\varphi\rangle$ and $\langle\varphi'\,|\varphi'\rangle$ should be identical. The inner products $\langle\varphi\,|\varphi\rangle$ and $\langle\varphi'\,|\varphi'\rangle$ are, respectively, given by:

$$
\begin{aligned}
\langle\varphi|\varphi\rangle &= (\langle\psi|\alpha^* + \langle\chi|\beta^*)(\alpha|\psi\rangle + \beta|\chi\rangle) \\
&= |\alpha|^2\langle\psi|\psi\rangle + |\beta|^2\langle\chi|\chi\rangle + \alpha^*\beta\langle\psi|\chi\rangle + \alpha\beta^*\langle\chi|\psi\rangle \\
\langle\varphi'|\varphi'\rangle &= (\langle\psi|\langle\psi|\alpha^* + \langle\chi|\langle\chi|\beta^*)(\alpha|\psi\rangle|\psi\rangle + \beta|\chi\rangle|\chi\rangle) \\
&= |\alpha|^2|\langle\psi|\psi\rangle|^2 + |\beta|^2|\langle\chi|\chi\rangle|^2 + \alpha^*\beta|\langle\psi|\chi\rangle|^2 + \alpha\beta^*|\langle\chi|\psi\rangle|^2.
\end{aligned}
\qquad (4.21)
$$

We know that $\langle\psi|\psi\rangle = \langle\chi|\chi\rangle = 1$. Therefore the discrepancy lies in the cross terms. Specifically, to avoid the contradiction that resulted in the no-cloning theorem we require that $|\langle\psi|\chi\rangle|^2 = \langle\psi|\chi\rangle$. This condition can only be satisfied when the states are orthogonal. Thus cloning is possible only for mutually orthogonal states. It is, however, important to remember a subtle point here. Even if we have a mutually orthogonal set of states we need a quantum copier (or unitary operator) specifically for those states. If the unitary operator is specific to a different set of mutually orthogonal states, cloning would fail. It would seem that the no-cloning theorem would prevent us from exploiting the richness of quantum mechanics. It turns out that this is not the case. A key example is the QKD that with very high probability guarantees secure communication.

4.4 DISTINGUISHING THE QUANTUM STATES

Nonorthogonal quantum states cannot be cloned; they also cannot be reliably distinguished. There is no measurement device we can create that can reliably distinguish nonorthogonal states. This fundamental result plays an important role in quantum cryptography. Its proof is based on contradiction. Let us assume that the measurement operator M is the Hermitian operator (with corresponding eigenvalues m_i and corresponding projection operators P_i) of an observable M, which allows us unambiguously to distinguish between two nonorthogonal states $|\psi_1\rangle$ and $|\psi_2\rangle$. The eigenvalue m_1 (m_2) unambiguously identifies the state $|\psi_1\rangle$ ($|\psi_2\rangle$) as the premeasurement state. We know that for projection operators the following properties are valid:

$$\begin{aligned}\langle\psi_1|P_1|\psi_1\rangle &= 1 \quad \langle\psi_2|P_2|\psi_2\rangle = 1\\ \langle\psi_1|P_2|\psi_1\rangle &= 0 \quad \langle\psi_2|P_1|\psi_2\rangle = 0.\end{aligned} \tag{4.22}$$

Since $|\psi_1\rangle$ and $|\psi_2\rangle$ are nonorthogonal states, then $\langle\psi_1|\psi_2\rangle \neq 0$, and $|\psi_2\rangle$ can be represented in terms of $|\psi_1\rangle$ and another state $|\chi\rangle$ that is orthogonal to $|\psi_1\rangle$ ($\langle\psi_1|\chi\rangle = 0$) as follows:

$$|\psi_2\rangle = \alpha|\psi_1\rangle + \beta|\chi\rangle. \tag{4.23}$$

From the projection operator's properties, listed earlier, we can conclude the following:

$$\begin{aligned}0 &= \langle\psi_1|P_2|\psi_1\rangle \overset{P_2^2=P_2}{=} \langle\psi_1|P_2P_2|\psi_1\rangle = \||P_2|\psi_1\rangle\|^2 \Rightarrow P_2|\psi_1\rangle = 0\\ 1 &= \langle\psi_2|P_2|\psi_2\rangle = \langle\psi_2|P_2P_2|\psi_2\rangle = (\alpha^*\langle\psi_1| + \beta^*\langle\chi|)(\alpha P_2|\psi_1\rangle + \beta P_2|\chi\rangle)\\ &= |\beta|^2\langle\chi|P_2|\chi\rangle.\end{aligned} \tag{4.24}$$

Now we use the completeness relationship:

$$1 = \langle\chi|\chi\rangle = \langle\chi|\sum_i^I P_i|\chi\rangle = \sum_i \langle\chi|P_i|\chi\rangle \overset{\langle\chi|P_i|\chi\rangle\geq 0}{\geq} \langle\chi|P_2|\chi\rangle. \tag{4.25}$$

By combining the previous two equations we obtain:

$$1 = \langle\psi_2|P_2|\psi_2\rangle \leq |\beta|^2 \Rightarrow |\beta|^2 = 1, \tag{4.26}$$

indicating that the probability of finding $|\psi_2\rangle$ in $|\chi\rangle$ is 1. Therefore we conclude that $|\psi_2\rangle = |\chi\rangle$, which is a contradiction. Therefore, indeed, *it is impossible to unambiguously distinguish nonorthogonal quantum states.*

4.5 QUANTUM ENTANGLEMENT

Let $|\psi_0\rangle,\ldots,|\psi_{n-1}\rangle$ be n qubits lying in the Hilbert spaces $H_0,\ldots,H_{n-1}$, respectively, and let the state of the joint quantum system lying in $H_0 \otimes \ldots \otimes H_{n-1}$ be denoted by $|\psi\rangle$. The qubit $|\psi\rangle$ is then said to be entangled if it cannot be written in the product state form:

$$|\psi\rangle = |\psi_0\rangle \otimes |\psi_1\rangle \otimes \cdots \otimes |\psi_{n-1}\rangle. \tag{4.27}$$

Important examples of two-qubit states are Bell states, also known as EPR states (pairs):

$$\begin{aligned}|B_{00}\rangle &= \frac{1}{\sqrt{2}}(|00\rangle + |11\rangle),\ |B_{01}\rangle = \frac{1}{\sqrt{2}}(|01\rangle + |10\rangle),\\ |B_{10}\rangle &= \frac{1}{\sqrt{2}}(|00\rangle - |11\rangle),\ |B_{11}\rangle = \frac{1}{\sqrt{2}}(|01\rangle - |10\rangle).\end{aligned} \tag{4.28}$$

The entangled states just introduced are famously called EPR pairs, after Einstein, Podolsky, and Rosen who raised doubts about quantum mechanics itself [29]. Their fundamental objection regarded entanglement of qubits and their apparent nonlocal interaction that intrigued the quantum physics community for quite some time. The objection raised by them is referred to as the hidden variable theory. To put the debate to rest, Bell developed a set of inequalities that, if violated, would validate quantum mechanics [30,31]. Bell's theoretical predictions were experimentally validated [32–35] leading to the establishment of quantum mechanics as we know it today, representing the best physical description of the nature.

The n-qubit ($n > 2$) analogs of Bell states will now be briefly reviewed. One popular family of entangled multiqubit states is Greenberger–Horne–Zeilinger (GHZ) states:

$$|\mathrm{GHZ}\rangle = \frac{1}{\sqrt{2}}(|00\cdots0\rangle \pm |11\cdots1\rangle). \tag{4.29}$$

Another popular family of multiqubit entangled states is known as W states:

$$|\mathrm{W}\rangle = \frac{1}{\sqrt{N}}(|00\cdots01\rangle + |00\cdots10\rangle + \cdots + |01\cdots00\rangle + |10\cdots00\rangle). \tag{4.30}$$

The W state of n qubits represents a superposition of single-weighted CB states, each occurring with a probability amplitude of $N^{-1/2}$.

The two-qubit quantum system density operator can be expanded in terms of Bell states as follows:

$$\rho = \sum_{i,j} c_{ij} |B_{ij}\rangle\langle B_{ij}|, \tag{4.31}$$

where the corresponding outer products are given by $|B_{00}\rangle\langle B_{00}| = 0.25(II + XX - YY + ZZ)$, $|B_{01}\rangle\langle B_{01}| = 0.25(II + XX + YY - ZZ)$, $|B_{10}\rangle\langle B_{10}| = 0.25(II - XX + YY + ZZ)$, and $|B_{11}\rangle\langle B_{11}| = 0.25(II - XX - YY - ZZ)$. The state represented by density operator ρ is separable if and only if $c_{00} < 0.5$.

For a bipartite system, we can elegantly verify whether or not the qubit $|\psi\rangle$ is a product state or an entangled one by the Schmidt decomposition. The *Schmidt decomposition theorem* states that a pure state $|\psi\rangle$ of the composite system $H_A \otimes H_B$ can be represented as:

$$|\psi\rangle = \sum_i c_i |i_A\rangle |i_B\rangle, \tag{4.32}$$

where $|i_A\rangle$ and $|i_B\rangle$ are orthonormal bases of the subsystems H_A and H_B, respectively, and $c_i \in R^+$ (R^+ is the set of nonnegative real numbers) are *Schmidt coefficients* that satisfy the following condition: $\sum c_i^2 = 1$. The Schmidt coefficients can be calculated from the partial density matrix $\mathrm{Tr}_B(|\psi\rangle\langle\psi|)$, whose eigenvalues are c_i^2. The number of nonzero eigenvalues is known as the *Schmidt rank* (number). A corollary of the Schmidt decomposition theorem is that a pure state in a composite system is a product state if and only if the Schmidt rank is 1. On the other hand, it is an entangled state if and only if the Schmidt rank is greater than 1.

As an illustration let us verify if the Bell state $|B_{11}\rangle$ is an entangled one. We first determine the density matrix:

$$\begin{aligned}\rho = |B_{11}\rangle\langle B_{11}| &= \frac{1}{\sqrt{2}}(|01\rangle - |10\rangle)\frac{1}{\sqrt{2}}(\langle 01| - \langle 10|) \\ &= \frac{1}{2}(|01\rangle\langle 01| - |01\rangle\langle 10| - |10\rangle\langle 01| + |10\rangle\langle 10|)\end{aligned}$$

By tracing out the subsystem B we obtain:

$$\begin{aligned}\rho_A &= \mathrm{Tr}_B(|B_{11}\rangle\langle B_{11}|) = \langle 0_B|B_{11}\rangle\langle B_{11}|0_B\rangle + \langle 1_B|B_{11}\rangle\langle B_{11}|1_B\rangle \\ &= \frac{1}{2}(|1\rangle\langle 1| + |0\rangle\langle 0|) = \frac{1}{2}I\end{aligned}$$

The eigenvalues of ρ are $c_1^2 = c_2^2 = 1/2$, and the Schmidt rank is 2 indicating that the Bell state $|B_{11}\rangle$ is an entangled state.

The *proof of Schmidt decomposition theorem* is straightforward. Let $|j\rangle$ and $|k\rangle$ be fixed orthonormal bases for subsystems H_A and H_B, then $|\psi\rangle$ can be represented as:

$$|\psi\rangle = \sum_{j,\,k} a_{jk}|j\rangle|k\rangle, \tag{4.33}$$

where a_{jk} complex coefficient can be interpreted as the matrix $\boldsymbol{A} = (a_{jk})$, which can be decomposed using the singular value decomposition as:

$$\boldsymbol{A} = \boldsymbol{U\Sigma V}^{\dagger}, \tag{4.34}$$

where the $m \times n$ diagonal $\boldsymbol{\Sigma}$ matrix corresponds to the scaling (stretching) operation and is given by $\boldsymbol{\Sigma} = \mathrm{diag}(\sigma_1, \sigma_2, \ldots)$, with $\sigma_1 \geq \sigma_2 \geq \ldots \geq 0$ being the *singular values* of $\boldsymbol{A}$. The singular values are in fact the square roots of eigenvalues of $\boldsymbol{AA}^{\dagger}$. The $m \times m$ matrix $\boldsymbol{U} = (u_{ji})$ and $n \times n$ matrix $\boldsymbol{V} = (v_{jk})$ are unitary matrices, which describe the rotation operations. By substituting Eq. (4.34) into Eq. (4.33) we obtain:

$$|\psi\rangle = \sum_{i,\,j,\,k} u_{ji}\sigma_i v_{jk}|j\rangle|k\rangle. \tag{4.35}$$

Now by substituting $\sum_j u_{ji}|j\rangle = |i_A\rangle$, $\sum_j v_{jk}|k\rangle = |i_B\rangle$, and $\sigma_i = c_i$, we obtain Eq. (4.32). The set $\{|i_A\rangle\}$ ($\{|i_B\rangle\}$) forms an orthonormal set due to orthonormality of $|j\rangle$ ($|k\rangle$) and unitarity of $\boldsymbol{U}(\boldsymbol{V})$.

An important concept in QIP is the *purification concept*, in which we create the reference system H_B such that given the original system H_A, the state $|\psi\rangle \in H_A \otimes H_B$ is a pure state. Similarly to the Schmidt decomposition theorem, let $|i_A\rangle$ and $|i_B\rangle$ be orthonormal bases of the systems H_A and H_B, respectively. The state $|\psi\rangle$ is a purification of ρ_A if:

$$\rho_A = \mathrm{Tr}_B(|\psi\rangle\langle\psi|), \tag{4.36}$$

where we use $\mathrm{Tr}_B(\bullet)$ to denote the partial trace over B, introduced in Chapter 2. If ρ_A is the mixed state, it can be represented by:

$$\rho_A = \sum_i p_i|i_A\rangle\langle i_A|. \tag{4.37}$$

The purification is then given by:

$$|\psi\rangle = \sum_i \sqrt{p_i}|i_A\rangle|i_B\rangle. \tag{4.38}$$

Direct application of the definition of trace gives:

$$\begin{aligned}\mathrm{Tr}_B(|\psi\rangle\langle\psi|) &= \mathrm{Tr}_B\left(\sum_i \sqrt{p_i}|i_A\rangle|i_B\rangle\sum_j \sqrt{p_j}\langle j_A|\langle j_B|\right) = \sum_{i,j}\sqrt{p_ip_j}\underbrace{\langle j_B|i_A\rangle}_{\delta_{i_A,j_B}}(|i_B\rangle\langle j_A|)\\ &= \sum_i p_i|i_A\rangle\langle i_A| = \rho_A,\end{aligned} \tag{4.39}$$

thus proving the claim of Eq. (4.36).

The Schmidt rank is a relevant approach to determine whether a certain two-qubit state is entangled. For multiqubit states, we need to use other measures of entanglement such as concurrence, entanglement of formation, and negativity [36–39]. The amount of information conveyed by the quantum state can be computed by using the *von Neumann entropy*:

$$S(\rho) = -\mathrm{Tr}(\rho \log \rho) = -\sum_i \lambda_i \log_2\lambda_i, \tag{4.40}$$

where λ_i are the eigenvalues of the density matrix. To take the amount of resources needed to form a given entanglement state, the *entanglement of formation* is used, which is for a given mixed density operator ρ_{AB} of a bipartite system composed of two subsystems H_A and H_B defined as:

$$E(\rho_{AB}) = \min\sum_i p_iE(|\psi_i\rangle), \tag{4.41}$$

where $E(|\psi_k\rangle)$ denotes the von Neumann's entropy of either of two subsystems A and B:

$$E(|\psi_i\rangle) = -\mathrm{Tr}(\rho_{i,A}\log\rho_{i,A}) = -\mathrm{Tr}(\rho_{i,B}\log\rho_{i,B}), \tag{4.42}$$

for reduced density operators $\rho_{i,A} = \mathrm{Tr}_B(|\psi_i\rangle\langle\psi_i|)$ and $\rho_{i,B} = \mathrm{Tr}_B(|\psi_i\rangle\langle\psi_i|)$, and the p_i denotes the probability of occurrence of state $|\psi_i\rangle$. The minimization in Eq. (4.41) is performed over all possible pure-state decompositions:

$$\rho_{AB} = \sum_i p_i|\psi_i\rangle\langle\psi_i|. \tag{4.43}$$

For the two-qubit case, we can come up with the closed-form expression for the entanglement of formation [39]:

$$E(\rho_{AB}) = h_2\left(\frac{1+\sqrt{1-C^2(\rho_{AB})}}{2}\right),\ h_2(x) = -p\log p - (1-p)\log(1-p), \tag{4.44}$$

where $h_2(x)$ is the binary entropy function, introduced in Chapter 6. We use $C(\rho_{AB})$ to denote the *concurrence*, the measure of entanglement, determined by:

$$C(\rho_{AB}) = \max(0,\ \lambda_1 - \lambda_2 - \lambda_3 - \lambda_4), \tag{4.45}$$

where λ_i are eigenvalues of the Hermitian matrix:

$$R = \left(\sqrt{\rho_{AB}}\,\widetilde{\rho}_{AB}\sqrt{\rho_{AB}}\right)^{1/2},\ \widetilde{\rho}_{AB} = XY\rho_{AB}^{*}XY, \tag{4.46}$$

arranged in decreasing order ($\lambda_1 \geq \lambda_2 \geq \lambda_3 \geq \lambda_4$). We used $\widetilde{\rho}_{AB}$ to denote the density operator after the nonunitary *spin-flip transformation*, which maps each component qubit in its orthogonal state. In other words, the spin-flipped version of $|\psi\rangle$ is defined by:

$$|\widetilde{\psi}\rangle = YY|\psi^{*}\rangle. \tag{4.47}$$

The concurrence can also be defined as:

$$C(|\psi\rangle) = |\langle\psi|\widetilde{\psi}\rangle|, \tag{4.48}$$

and thus represents the amount of overlap between observed state and the spin-flipped version of it, and clearly $0 \leq C \leq 1$.

As an illustration, let us again observe Bell state $|B_{11}\rangle$, and determine both the concurrence and entanglement of formation. By using the definition expression (4.48) we find that the $C(|B_{11}\rangle) = 1$. Now by using Eq. (4.44) we determine that the entanglement of formation is $E(|B_{11}\rangle) = h_2(1/2) = 1$.

Entanglement can be used to *detect and correct the errors* arising during a quantum computation [1]. At the beginning, the quantum computer Q, being in initial state $|q\rangle$, and its environment E, being in initial state $|e\rangle$, are not entangled, but due to coupling interaction they become entangled:

$$|e\rangle|q\rangle \rightarrow \sum_{s}|e_s\rangle\{E_s|q\rangle\}, \tag{4.49}$$

where $\{E_s\}$ are the errors introduced by the environment. The states of the environment $\{|e_s\rangle\}$ are not necessarily orthonormal. To deal with errors we introduce the ancilla qubits $|a\rangle$ that are coupled to qubits in quantum computer Q. The unitary interaction U to couple ancillas to the quantum computer is used to introduce the following effect:

$$U[|a\rangle\{E_s|q\rangle\}] = |s\rangle\{E_s|q\rangle\}, \tag{4.50}$$

indicating that the final ancilla state $|s\rangle$ depends on the error operator induced by the environment E_s, but not on the quantum state of the quantum computer $|q\rangle$. Furthermore, the final ancilla states form the orthonormal set. The operator U produces the following transformation of the tripartite system environment–ancilla–quantum computer:

$$U\left[\sum_{s}|e_s\rangle|a\rangle\{E_s|q\rangle\}\right] = \sum_{s}|e_s\rangle|s\rangle\{E_s|q\rangle\}. \tag{4.51}$$

Given the orthonormality of states $\{|s\rangle\}$ they can be distinguished and by measuring the ancilla on this basis, we obtain the nonentangled pure state as a post-measurement state:

$$|\psi_{\text{post measurement}}\rangle = |e_s\rangle|S\rangle\{E_s|q\rangle\}, \tag{4.52}$$

where S is the *syndrome error*, identifying the error operator E_s. This process of determination of the error syndrome is commonly referred to as *syndrome extraction*. Because the error operator E_s is identified from the syndrome S, we can apply E_s^{-1} to the quantum computer Q to undo the action of the environment:

$$|\psi_{\text{final}}\rangle = |e_s\rangle|S\rangle|q\rangle. \tag{4.53}$$

Therefore the unwanted entanglement between the quantum computer Q and its environment E has been undone by further entangling Q with the corresponding set of ancillas, together with appropriate measurement on ancillas to deentangle the Q and E and identify the most probable error operator.

Entanglement can also find application in so-called *superdense coding* [40], in which Alice sends two classical bits even though she has only one qubit to work with. Let the Bell states $|B_{ab}\rangle$ correspond to the two classical bits ab to be transmitted, wherein $ab = 00$, 01, 10, or 11. The starting state is the Bell state $|B_{00}\rangle$:

$$|\Phi\rangle = |B_{00}\rangle = 2^{-1/2}(|00\rangle + |11\rangle), \tag{4.54}$$

and we assume that Alice is in possession of the first qubit and Bob of the second one. Depending on which two classical bits she wants to send she will act on her qubit by performing the corresponding single-qubit gate. If she wants to send 00 bits, she leaves her qubit untouched. If she applies the X gate on initial qubit $|\Phi\rangle$, she will get:

$$(X \otimes I)|\Phi\rangle = 2^{-1/2}(|10\rangle + |01\rangle) = |B_{01}\rangle, \tag{4.55}$$

and thus will be able to transmit classical bits $ab = 01$. On the other hand, if she applies single-qubit gate Z to her qubit before sending it to Bob, she will transform the $|\Phi\rangle$ state to:

$$(Z \otimes I)|\Phi\rangle = 2^{-1/2}(|00\rangle - |11\rangle) = |B_{10}\rangle, \tag{4.56}$$

and thus will be able to transmit classical bits $ab = 10$. Finally, if she applies single-qubit gate jY to her qubit before sending to Bob, she will transform the $|\Phi\rangle$ state to:

$$(jY \otimes I)|\Phi\rangle = 2^{-1/2}(|01\rangle - |10\rangle) = |B_{11}\rangle, \tag{4.57}$$

and thus will be able to transmit classical bits $ab = 11$. On the receiver side, once Bob receives the qubit from Alice he will perform the measurements in Bell basis and will be able to find one of the Bell states $\{|B_{00}\rangle, |B_{01}\rangle, |B_{10}\rangle, |B_{11}\rangle\}$, representing two classical bits 00, 01, 10, or 11.

Entanglement has applications in quantum teleportation, which was discussed in the previous chapter. Here we discuss *entanglement swapping* instead. Entanglement swapping starts with two EPR pairs, labeled $|B_{00}\rangle_{12}$ and $|B_{00}\rangle_{34}$, that is:

$$|B_{00}\rangle_{12} = 2^{-1/2}(|00\rangle_{12} + |11\rangle_{12}),\ |B_{00}\rangle_{34} = 2^{-1/2}(|00\rangle_{34} + |11\rangle_{34}). \tag{4.58}$$

Alice is in possession of qubits 1 and 4, while Bob is in position of qubits 2 and 3. The product state will be:

$$|B_{00}\rangle_{12}|B_{00}\rangle_{34} = 2^{-1}(|00\rangle_{12}|00\rangle_{34} + |00\rangle_{12}|11\rangle_{34} + |11\rangle_{12}|00\rangle_{34} + |11\rangle_{12}|11\rangle_{34}). \tag{4.59}$$

This product state can be rearranged as follows:

$$|B_{00}\rangle_{12}|B_{00}\rangle_{34} = 2^{-1}(|B_{00}\rangle_{14}|B_{00}\rangle_{23} + |B_{01}\rangle_{14}|B_{01}\rangle_{23} + |B_{10}\rangle_{14}|B_{10}\rangle_{23} + |B_{11}\rangle_{14}|B_{11}\rangle_{23}). \tag{4.60}$$

Alice performs the Bell-state measurement on qubits 1 and 4 with possible outcomes being $\{|B_{00}\rangle_{14}, |B_{01}\rangle_{14}, |B_{10}\rangle_{14}, |B_{11}\rangle_{14}\}$, each occurring with probability 1/4. Depending on Alice's measurement result, Bob's subsystem collapses into one of the Bell states $|B_{00}\rangle_{23}$, $|B_{01}\rangle_{23}$, $|B_{10}\rangle_{23}$, or $|B_{11}\rangle_{23}$ indicating that qubits 2 and 3 are entangled even though they have been independent initially.

4.6 OPERATOR-SUM REPRESENTATION

Let the composite system C be composed of quantum register Q and environment E. This kind of system can be modeled as a closed quantum system. Because the composite system is closed, its dynamic is unitary, and the final state is specified by a unitary operator U as follows $U(\rho \otimes \varepsilon_0)U^\dagger$, where ρ is a density operator of the initial state of the quantum register Q, and ε_0 is the initial density operator of the environment E. The reduced density operator of Q upon interaction ρ_f can be obtained by tracing out the environment:

$$\rho_f = \mathrm{Tr}_E\left[U(\rho \otimes \varepsilon_0)U^\dagger\right] \equiv \xi(\rho). \tag{4.61}$$

The transformation (mapping) of initial density operator ρ to the final density operator ρ_f, denoted as $\xi : \rho \to \rho_f$, given by Eq. (4.61), is often called the *superoperator* or *quantum operation*. The final density operator can be expressed in so-called *operator-sum representation* as follows:

$$\rho_f = \sum_k E_k \rho E_k^\dagger, \tag{4.62}$$

where E_k are the operation elements for the superoperator. Clearly, in the absence of the environment, the superoperator becomes: $U\rho U^\dagger$, which is nothing else but a conventional time-evolution quantum operation.

The operator-sum representation can be used in the classification of quantum operations into two categories: (1) *trace preserving* when $\mathrm{Tr}\xi(\rho) = \mathrm{Tr}\ \rho = 1$ and (2) *nontrace preserving* when $\mathrm{Tr}\xi(\rho) < 1$. Starting from the trace-preserving condition:

$$\mathrm{Tr}\rho = \mathrm{Tr}\xi(\rho) = \mathrm{Tr}\left[\sum_k E_k \rho E_k^\dagger\right] = \mathrm{Tr}\left[\rho \sum_k E_k E_k^\dagger\right] = 1, \tag{4.63}$$

we obtain:

$$\sum_k E_k E_k^\dagger = I. \tag{4.64}$$

For nontrace-preserving quantum operation, Eq. (4.64) is not satisfied, and informally we can write $\sum_k E_k E_k^\dagger < I$.

If the *environment dimensionality is large enough* it can be found in pure state, $\varepsilon_0 = |\phi_0\rangle\langle\phi_0|$, and the corresponding superoperator becomes:

$$\begin{aligned}
\xi(\rho) &= \mathrm{Tr}_E\left[U(\rho\otimes\varepsilon)U^\dagger\right] = \sum_k \langle\phi_k|\left(U\rho\otimes\varepsilon U^\dagger\right)|\phi_k\rangle \\
&= \sum_k \langle\phi_k|\left(U\rho\otimes\left(\underbrace{|\phi_0\rangle\langle\phi_0|}_{\varepsilon}\right)U^\dagger\right)|\phi_k\rangle \\
&= \sum_k \underbrace{\langle\phi_k|U|\phi_0\rangle}_{E_k}\ \rho\underbrace{\langle\phi_0|U^\dagger|\phi_k\rangle}_{E_k^\dagger} = \\
&= \sum_k E_k \rho E_k^\dagger,\ E_k = \langle\phi_k|U|\phi_0\rangle.
\end{aligned} \tag{4.65}$$

The E_k operators in operator-sum representation are known as *Kraus operators*. As an illustration, let us consider *bit-flip* and *phase-flip* channels, which are illustrated in Fig. 4.5. (The simultaneous bit-and-phase-flip operation is described by the Pauli-Y operator.) Let the composite system be given by: $|\phi_E\rangle|\psi_Q\rangle$, wherein the initial state of the environment is $|\phi_E\rangle = |0_E\rangle$. Furthermore, let the quantum subsystem Q interact with the environment E by a Pauli-X operator:

$$U = \sqrt{1-p}I\otimes I + \sqrt{p}X\otimes X,\ 0 \le p \le 1. \tag{4.66}$$

Therefore, with probability $1 - p$, we leave the quantum system untouched, while with probability p we apply the Pauli-X operator to both quantum subsystem

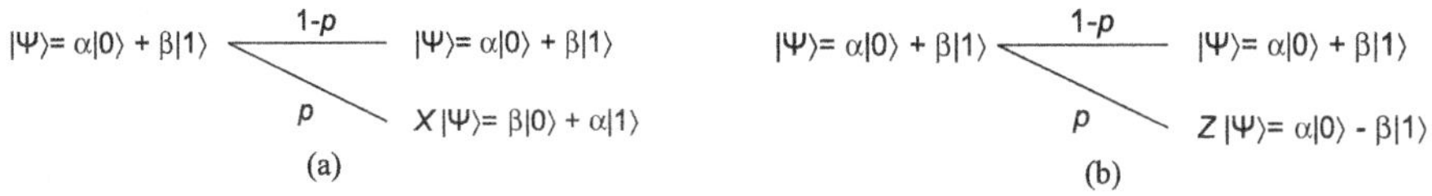

FIGURE 4.5

(A) The bit-flip channel model. (B) The phase-flip channel model.

and the environment. By applying the operator U on the environment state we obtain:

$$U|\phi_E\rangle = \sqrt{1-p}I\otimes I|0_E\rangle + \sqrt{p}X\otimes X|0_E\rangle = \sqrt{1-p}|0_E\rangle I + \sqrt{p}|1_E\rangle X. \quad (4.67)$$

The corresponding Kraus operators are given by:

$$E_0 = \langle 0_E|U|\phi_E\rangle = \sqrt{1-p}I,\ E_1 = \langle 1_E|U|\phi_E\rangle = \sqrt{p}X. \quad (4.68)$$

Finally, the operator-sum representation is given by:

$$\xi(\rho) = E_0\rho E_0^\dagger + E_1\rho E_1^\dagger = (1-p)\rho + pX\rho X. \quad (4.69)$$

In similar fashion, the Kraus operators for the phase-flip channel are given by:

$$E_0 = \langle 0_E|U|\phi_E\rangle = \sqrt{1-p}I,\ E_1 = \langle 1_E|U|\phi_E\rangle = \sqrt{p}Z, \quad (4.70)$$

and the corresponding operator-sum representation is:

$$\xi(\rho) = E_0\rho E_0^\dagger + E_1\rho E_1^\dagger = (1-p)\rho + pZ\rho Z. \quad (4.71)$$

The Pauli operators play a key role in quantum computing and quantum communication applications. Let us prove that Pauli matrices along with the identity operator form an orthogonal basis for a complex Hilbert space of 2 × 2 matrices, denoted as $H_{2\times 2}$. To that end we employ the *Hilbert–Schmidt inner product*, defined as:

$$\langle \boldsymbol{A}, \boldsymbol{B}\rangle_{HS} = \mathrm{Tr}(\boldsymbol{A}^\dagger \boldsymbol{B}), \quad (4.72)$$

where $\boldsymbol{A},\boldsymbol{B} \in H_{2\times 2}$, to first show that all the four matrices are orthogonal:

$$\frac{1}{2}\mathrm{Tr}(\boldsymbol{I}^\dagger \boldsymbol{X}) = \frac{1}{2}\mathrm{Tr}\left[\begin{pmatrix}1 & 0\\0 & 1\end{pmatrix}\begin{pmatrix}0 & 1\\1 & 0\end{pmatrix}\right] = \frac{1}{2}\mathrm{Tr}\begin{pmatrix}0 & 1\\1 & 0\end{pmatrix} = 0 = \frac{1}{2}\mathrm{Tr}(\boldsymbol{I}^\dagger \boldsymbol{Z}) = \frac{1}{2}\mathrm{Tr}(\boldsymbol{I}^\dagger \boldsymbol{Y}), \quad (4.73a)$$

$$\frac{1}{2}\mathrm{Tr}(\boldsymbol{X}^\dagger \boldsymbol{Y}) = \frac{1}{2}\mathrm{Tr}\left[\begin{pmatrix}0 & 1\\1 & 0\end{pmatrix}\begin{pmatrix}0 & -j\\j & 0\end{pmatrix}\right] = \frac{1}{2}\mathrm{Tr}\begin{pmatrix}j & 0\\0 & -j\end{pmatrix} = 0 = \frac{1}{2}\mathrm{Tr}(\boldsymbol{Y}^\dagger \boldsymbol{X}), \quad (4.73b)$$

$$\frac{1}{2}\mathrm{Tr}(\boldsymbol{X}^\dagger \boldsymbol{Z}) = \frac{1}{2}\mathrm{Tr}\left[\begin{pmatrix}0 & 1\\1 & 0\end{pmatrix}\begin{pmatrix}1 & 0\\0 & -1\end{pmatrix}\right] = \frac{1}{2}\mathrm{Tr}\begin{pmatrix}0 & -1\\1 & 0\end{pmatrix} = 0 = \frac{1}{2}\mathrm{Tr}(\boldsymbol{Z}^\dagger \boldsymbol{X}), \quad (4.73c)$$

$$\frac{1}{2}\mathrm{Tr}(\boldsymbol{Y}^\dagger \boldsymbol{Z}) = \frac{1}{2}\mathrm{Tr}\left[\begin{pmatrix}0 & -j\\j & 0\end{pmatrix}\begin{pmatrix}1 & 0\\0 & -1\end{pmatrix}\right] = \frac{1}{2}\mathrm{Tr}\begin{pmatrix}0 & j\\j & 0\end{pmatrix} = 0 = \frac{1}{2}\mathrm{Tr}(\boldsymbol{Z}^\dagger \boldsymbol{Y}). \quad (4.73d)$$

Based on the Hilbert–Schmidt inner product we can go a step further and show that $\boldsymbol{I}$, $\boldsymbol{X}$, $\boldsymbol{Y}$, $\boldsymbol{Z}$ are not only orthogonal but also orthonormal, which is proved in an exercise for readers later. Next, we need to prove that $\boldsymbol{I}$, $\boldsymbol{X}$, $\boldsymbol{Y}$, $\boldsymbol{Z}$ span $H_2 \times 2H_{2\times2}$, that is, we can express any matrix $\boldsymbol{P} \in H_{2\times2}$ as follows:

$$\boldsymbol{P} = c_1\boldsymbol{I} + c_2\boldsymbol{X} + c_3\boldsymbol{Y} + c_4\boldsymbol{Z}, \tag{4.74}$$

such that $c_i \in C$. To show this, let us first define $\boldsymbol{P}$ as:

$$\boldsymbol{P} = \begin{pmatrix} r & s \\ t & u \end{pmatrix}, \tag{4.75}$$

where $r, s, t, u \in C$. Due to the orthonormality of $\boldsymbol{I}$, $\boldsymbol{X}$, $\boldsymbol{Y}$, $\boldsymbol{Z}$, the coefficients can be easily calculated using the Hilbert–Schmidt inner product by:

$$c_1 = \frac{1}{2}\mathrm{Tr}(\boldsymbol{I}^\dagger\boldsymbol{P}) = \frac{1}{2}\mathrm{Tr}\left[\begin{pmatrix} 1 & 0 \\ 0 & 1 \end{pmatrix}\begin{pmatrix} r & s \\ t & u \end{pmatrix}\right] = \frac{1}{2}(r+u) \in C, \tag{4.76a}$$

$$c_2 = \frac{1}{2}\mathrm{Tr}(\boldsymbol{X}^\dagger\boldsymbol{P}) = \frac{1}{2}\mathrm{Tr}\left[\begin{pmatrix} 0 & 1 \\ 1 & 0 \end{pmatrix}\begin{pmatrix} r & s \\ t & u \end{pmatrix}\right] = \frac{1}{2}(t+s) \in C, \tag{4.76b}$$

$$c_3 = \frac{1}{2}\mathrm{Tr}(\boldsymbol{Y}^\dagger\boldsymbol{P}) = \frac{1}{2}\mathrm{Tr}\left[\begin{pmatrix} 0 & -j \\ j & 0 \end{pmatrix}\begin{pmatrix} r & s \\ t & u \end{pmatrix}\right] = \frac{j}{2}(s-t) \in C, \tag{4.76c}$$

$$c_4 = \frac{1}{2}\mathrm{Tr}(\boldsymbol{Z}^\dagger\boldsymbol{P}) = \frac{1}{2}\mathrm{Tr}\left[\begin{pmatrix} 1 & 0 \\ 0 & -1 \end{pmatrix}\begin{pmatrix} r & s \\ t & u \end{pmatrix}\right] = \frac{1}{2}(r-u) \in C. \tag{4.76d}$$

Substituting Eq. (4.76a) through Eq. (4.76d) in Eq. (4.74) verifies our claim, thereby completing the proof.

4.7 DECOHERENCE EFFECTS, DEPOLARIZATION, AND AMPLITUDE-DAMPING CHANNEL MODELS

Quantum computation works by manipulating the quantum interference effect. Quantum interference, a manifestation of coherent superposition of quantum states, is the cornerstone behind all quantum information tasks such as quantum computation and quantum communication. A major source of the problem is our inability to prevent our quantum system of interest from interacting with the surrounding environment. This interaction results in an entanglement between the quantum system and the environment leading to decoherence. To understand this system–environment entanglement and decoherence better, let us consider a qubit described by density-state (matrix) $\rho = \begin{bmatrix} a & b \\ c & d \end{bmatrix}$ interacting with the environment, described by the following three states: $|0_E\rangle$, $|1_E\rangle$, and $|2_E\rangle$. Without loss of generality, we assume that the environment was initially in state $|0_E\rangle$. The unitary operator introducing the entanglement between the quantum system and the environment is defined as:

$$\begin{aligned} U|0\rangle|0_E\rangle &= \sqrt{1-p}|0\rangle|0_E\rangle + \sqrt{p}|0\rangle|1_E\rangle \\ U|0\rangle|0_E\rangle &= \sqrt{1-p}|1\rangle|0_E\rangle + \sqrt{p}|1\rangle|2_E\rangle. \end{aligned} \tag{4.77}$$

The corresponding Kraus operators are given by:

$$E_0 = \sqrt{1-p}I,\ E_1 = \sqrt{p}|0\rangle\langle 0|,\ E_2 = \sqrt{p}|1\rangle\langle 1|. \tag{4.78}$$

The operator-sum representation is given by:

$$\xi(\rho) = E_0\rho E_0^\dagger + E_1\rho E_1^\dagger + E_2\rho E_2^\dagger = \begin{bmatrix} a & (1-p)\,b \\ (1-p)\,c & d \end{bmatrix}. \tag{4.79}$$

By applying these quantum operations n times, the corresponding final state would be:

$$\rho_f = \begin{bmatrix} a & (1-p)^n\,b \\ (1-p)^n\,c & d \end{bmatrix}. \tag{4.80}$$

If the probability p is expressed as $p = \gamma\Delta t$ we can write $n = t/\Delta t$ and in limit we obtain:

$$\lim_{\Delta t \to 0} (1-p)^n = (1-\gamma\Delta t)^{t/\Delta t} = e^{-\gamma t}. \tag{4.81}$$

Therefore the corresponding operator-sum representation as n tends to plus infinity is given by:

$$\xi(\rho) = \begin{bmatrix} a & e^{-\gamma t}b \\ e^{-\gamma t}c & d \end{bmatrix}. \tag{4.82}$$

Clearly, the terms b and c go to zero as t increases, indicating that the relative phase in the original state of the quantum system is lost, and the corresponding channel model is known as the *phase-damping* channel model.

In this example we have considered the coupling between a single-qubit quantum system and the environment and discussed the resulting loss of interference or coherent superposition. In general, for multiple qubit systems, decoherence also results in loss of coupling between the qubits. In fact, with increasing complexity and size of the quantum computer the decoherence effect becomes worse. Additionally, the quantum system lies in some complex Hilbert space where there are infinite variations of errors that can cause decoherence.

A more general example of dephasing is depolarization. The *depolarizing channel*, as shown in Fig. 4.6, with probability $1-p$ leaves the qubit as it is, while with probability p moves the initial state into $\rho_f = I/2$ that maximizes the von Neumann entropy $S(\rho) = -\text{Tr}\rho \log \rho = 1$. The properties describing the model can be summarized as follows:

1. Qubit errors are independent.
2. Single-qubit errors (X, Y, Z) are equally likely.
3. All qubits have the same single-error probability $p/4$.

The Kraus operators E_i of the channel should be selected as follows:

$$E_0 = \sqrt{1-3p/4}I;\ E_1 = \sqrt{p/4}X;\ E_2 = \sqrt{p/4}Y;\ E_3 = \sqrt{p/4}Z. \tag{4.83}$$

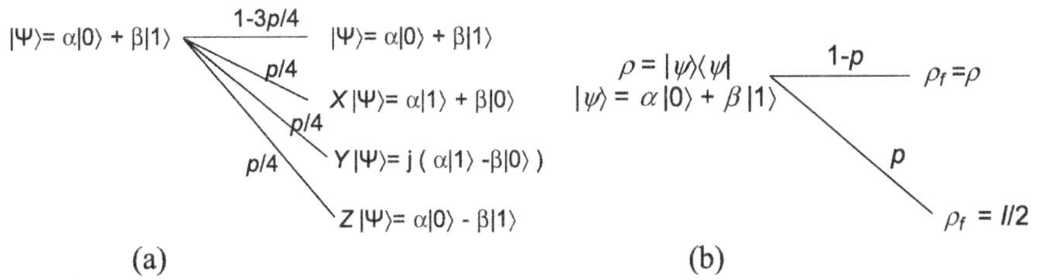

FIGURE 4.6

Depolarizing channel model: (A) Pauli operator description and (B) density operator description.

The action of the depolarizing channel is to perform the following mapping: $\rho \rightarrow \xi(\rho) = \sum E_i \rho E_i^{\dagger}$, where ρ is the initial density operator. Without loss of generality we will assume that the initial state was pure $|\psi\rangle = a|0\rangle + b|1\rangle$ so that:

$$\rho = |\psi\rangle\langle\psi| = (a|0\rangle + b|1\rangle)(\langle 0|a^* + \langle 1|b^*)$$
$$= |a|^2 \begin{bmatrix} 1 & 0 \\ 0 & 0 \end{bmatrix} + ab^* \begin{bmatrix} 0 & 1 \\ 0 & 0 \end{bmatrix} + a^* b \begin{bmatrix} 0 & 0 \\ 1 & 0 \end{bmatrix} + |b|^2 \begin{bmatrix} 0 & 0 \\ 0 & 1 \end{bmatrix}. \tag{4.84}$$
$$= \begin{bmatrix} |a|^2 & ab^* \\ a^* b & |b|^2 \end{bmatrix}$$

The resulting quantum operation can be represented using operator-sum representation as follows:

$$\xi(\rho) = \sum_i E_i \rho E_i^{\dagger} = \left(1 - \frac{3p}{4}\right)\rho + \frac{p}{4}(X\rho X + Y\rho Y + Z\rho Z)$$
$$= \left(1 - \frac{3p}{4}\right)\begin{bmatrix} |a|^2 & ab^* \\ a^* b & |b|^2 \end{bmatrix} + \frac{p}{4}\begin{bmatrix} 0 & 1 \\ 1 & 0 \end{bmatrix}\begin{bmatrix} |a|^2 & ab^* \\ a^* b & |b|^2 \end{bmatrix}\begin{bmatrix} 0 & 1 \\ 1 & 0 \end{bmatrix}$$
$$+\frac{p}{4}\begin{bmatrix} 1 & 0 \\ 0 & -1 \end{bmatrix}\begin{bmatrix} |a|^2 & ab^* \\ a^* b & |b|^2 \end{bmatrix}\begin{bmatrix} 1 & 0 \\ 0 & -1 \end{bmatrix} + \frac{p}{4}\begin{bmatrix} 0 & -j \\ j & 0 \end{bmatrix}\begin{bmatrix} |a|^2 & ab^* \\ a^* b & |b|^2 \end{bmatrix}\begin{bmatrix} 0 & -j \\ j & 0 \end{bmatrix}$$
$$= \left(1 - \frac{3p}{4}\right)\begin{bmatrix} |a|^2 & ab^* \\ a^* b & |b|^2 \end{bmatrix} + \frac{p}{4}\begin{bmatrix} |b|^2 & a^* b \\ ab^* & |a|^2 \end{bmatrix} + \frac{p}{4}\begin{bmatrix} |a|^2 & -ab^* \\ -a^* b & |b|^2 \end{bmatrix}$$
$$+\frac{p}{4}\begin{bmatrix} |b|^2 & -a^* b \\ -ab^* & |a|^2 \end{bmatrix}$$

$$= \begin{bmatrix} \left(1-\frac{p}{2}\right)|a|^2 + \frac{p}{2}|b|^2 & (1-p)ab^* \\ (1-p)a^*b & \frac{p}{2}|a|^2 + \left(1-\frac{p}{2}\right)|b|^2 \end{bmatrix}$$

$$= \begin{bmatrix} (1-p)|a|^2 + \frac{p}{2}\left(|a|^2+|b|^2\right) & (1-p)ab^* \\ (1-p)a^*b & \frac{p}{2}\left(|a|^2+|b|^2\right) + (1-p)|b|^2 \end{bmatrix}. \tag{4.85a}$$

Since $|a|^2 + |b|^2 = 1$, the operator-sum representation cab be written as:

$$\xi(\rho) = \sum_i E_i \rho E_i^\dagger = (1-p)\begin{bmatrix} |a|^2 & ab^* \\ a^*b & |b|^2 \end{bmatrix} + \frac{p}{2}I = (1-p)\rho + \frac{p}{2}I. \tag{4.85b}$$

The first line in Eq. (4.85a) corresponds to the model shown in Fig. 4.6A and Eq. (4.85b) corresponds to model shown in Fig. 4.6B. It is clear from Eq. (4.85a) and (4.85b) that $\mathrm{Tr}\xi(\rho) = \mathrm{Tr}(\rho) = 1$ meaning that the superoperator is trace preserving. Notice that the depolarizing channel model in some other books/papers can be slightly different from the one described here.

We now describe the *amplitude-damping channel* model. In certain quantum channels the errors X, Y, and Z do not occur with the same probability. In the amplitude-damping channel, the operation elements are given by:

$$E_0 = \begin{pmatrix} 1 & 0 \\ 0 & \sqrt{1-\varepsilon^2} \end{pmatrix}, \quad E_1 = \begin{pmatrix} 0 & \varepsilon \\ 0 & 0 \end{pmatrix}. \tag{4.86}$$

Spontaneous emission is an example of a physical process that can be modeled using the amplitude-damping channel model. If $|\psi\rangle = a\,|0\rangle + b\,|1\rangle$ is the initial qubit state $\left(\rho = \begin{bmatrix} |a|^2 & ab^* \\ a^*b & |b|^2 \end{bmatrix}\right)$, the effect of the amplitude-damping channel is to perform the following mapping:

$$\rho \to \xi(\rho) = E_0 \rho E_0^\dagger + E_1 \rho E_1^\dagger = \begin{pmatrix} 1 & 0 \\ 0 & \sqrt{1-\varepsilon^2} \end{pmatrix}\begin{bmatrix} |a|^2 & ab^* \\ a^*b & |b|^2 \end{bmatrix}\begin{pmatrix} 1 & 0 \\ 0 & \sqrt{1-\varepsilon^2} \end{pmatrix}$$

$$+ \begin{pmatrix} 0 & \varepsilon \\ 0 & 0 \end{pmatrix}\begin{bmatrix} |a|^2 & ab^* \\ a^*b & |b|^2 \end{bmatrix}\begin{pmatrix} 0 & 0 \\ \varepsilon & 0 \end{pmatrix}$$

$$= \begin{pmatrix} |a|^2 & ab^*\sqrt{1-\varepsilon^2} \\ a^*b\sqrt{1-\varepsilon^2} & |b|^2\left(1-\varepsilon^2\right) \end{pmatrix} + \begin{pmatrix} |b|^2\varepsilon^2 & 0 \\ 0 & 0 \end{pmatrix}$$

$$= \begin{pmatrix} |a|^2 + \varepsilon^2|b|^2 & ab^*\sqrt{1-\varepsilon^2} \\ a^*b\sqrt{1-\varepsilon^2} & |b|^2\left(1-\varepsilon^2\right) \end{pmatrix}. \tag{4.87}$$

Probabilities $P(0)$ and $P(1)$ that E_0 and E_1 occur are given by:

$$P(0) = \mathrm{Tr}\left(E_0 \rho E_0^\dagger\right) = \mathrm{Tr}\begin{pmatrix} |a|^2 & ab^* \sqrt{1-\varepsilon^2} \\ a^* b\sqrt{1-\varepsilon^2} & |b|^2 (1-\varepsilon^2) \end{pmatrix} = 1 - \varepsilon^2 |b|^2$$

$$P(1) = \mathrm{Tr}\left(E_1 \rho E_1^\dagger\right) = \mathrm{Tr}\begin{pmatrix} |b|^2 \varepsilon^2 & 0 \\ 0 & 0 \end{pmatrix} = \varepsilon^2 |b|^2. \tag{4.88}$$

The corresponding amplitude-damping channel model is shown in Fig. 4.7.

To verify how much the entanglement is preserved during interaction with the quantum channel, *entanglement fidelity* is typically used. Let us assume that the bipartite system $H_A \otimes H_B$ is prepared in the general pure state $\rho_{AB} = |\psi_{AB}\rangle\langle\psi_{AB}|$. Under the assumption that the second subsystem undergoes an interaction with the channel, *entanglement fidelity* can be defined as:

$$F_e = \langle \rho_{AB}, \rho_{AB,\,\mathrm{final}} \rangle_{HS} = \mathrm{Tr}\left(\rho_{AB}\rho_{AB,\,\mathrm{final}}\right) = \langle \psi_{AB} | \rho_{AB,\,\mathrm{final}} | \psi_{AB} \rangle, \tag{4.89}$$

which represents the Hilbert–Schmidt inner product (see Eq. (4.72)) between initial and final density operators, thus describing the variation in entanglement due to interaction with the quantum channel. The values of F_e close to 1 indicate that entanglement is well preserved, while values close to 0 imply that the entanglement is destroyed during interaction with the quantum channel. If we trace out the first subsystem, the reduced density operator for subsystem H_B will be:

$$\rho_{B,\,\mathrm{final}} = \mathrm{Tr}_A \rho_{AB,\,\mathrm{final}}. \tag{4.90}$$

Now we can define the fidelity between initial and final states of subsystem H_B as follows:

$$F\left(\rho_B, \rho_{B,\,\mathrm{final}}\right) = \langle \rho_B, \rho_{B,\,\mathrm{final}} \rangle_{HS} = \mathrm{Tr}\left(\rho_B \rho_{B,\,\mathrm{final}}\right),\ \rho_B = \mathrm{Tr}_A \rho_{AB}. \tag{4.91}$$

This definition of fidelity is an upper bound for entanglement fidelity:

$$F_e \le F\left(\rho_B, \rho_{B,\,\mathrm{final}}\right), \tag{4.92}$$

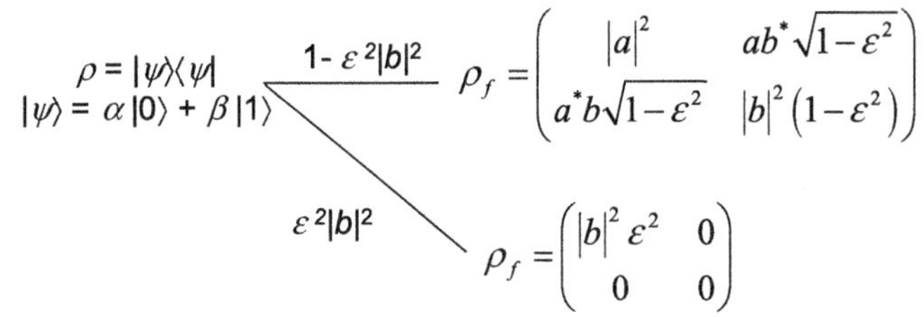

FIGURE 4.7

Amplitude-damping channel model.

because it is a measure of similarity (distance) between two states, while entanglement fidelity is the measure of preservation of entanglement after the interaction with the quantum channel is over. Mathematically speaking, the trace operation and the superoperator do not commute in general. A given Bell state can be converted to a different Bell state by the quantum channel, which will result in 0 entanglement fidelity, while the final state is maximally entangled. This is the reason why other entanglement figures of merit should be considered as well, such as the concurrence and the entanglement of formation introduced in Section 4.5.

Before concluding this section, we provide another definition of fidelity. Let ρ and σ be two density operators. Then, the fidelity can be defined as:

$$F(\rho, \sigma) = \mathrm{Tr}\left(\sqrt{\rho^{1/2}\sigma\rho^{1/2}}\right). \tag{4.93}$$

For two pure states with density operators $\rho = |\psi\rangle\langle\psi|$, $\sigma = |\phi\rangle\langle\phi|$ the corresponding fidelity will be:

$$\begin{aligned} F(\rho, \sigma) &= \mathrm{Tr}\left(\sqrt{\rho^{1/2}\sigma\rho^{1/2}}\right) \overset{\rho^2=\rho}{=} \mathrm{Tr}\left(\sqrt{|\psi\rangle\langle\psi|(|\phi\rangle\langle\phi|)|\psi\rangle\langle\psi|}\right) \\ &= \mathrm{Tr}\Big(|\psi\rangle\underbrace{\langle\psi|\phi\rangle\langle\phi|\psi\rangle}_{|\langle\phi|\psi\rangle|^2}\langle\psi|\Big)^{1/2} = |\langle\phi|\psi\rangle| \underbrace{Tr(|\psi\rangle\langle\psi|)}_{1} = |\langle\phi|\psi\rangle|. \end{aligned} \tag{4.94}$$

where we used the property of the density operators for pure states $\rho^2 = \rho$ (or equivalently $\rho = \rho^{1/2}$). Clearly, for pure states the fidelity corresponds to the square root of the probability of finding the system in state $|\phi\rangle$ if it is known to be prepared in state $|\psi\rangle$ (and vice versa). Since fidelity is related to the probability, it ranges between 0 and 1, with 0 indicating there is no overlap and 1 meaning that the states are identical. The following *properties* of fidelity hold:

1. The symmetry property:

$$F(\rho, \sigma) = F(\sigma, \rho). \tag{4.95}$$

2. The fidelity is invariant under unitary operations:

$$F\left(U\rho U^{\dagger}, U\sigma U^{\dagger}\right) = F(\rho, \sigma). \tag{4.96}$$

3. If ρ and σ commute, the fidelity can be expressed in terms of eigenvalues of ρ, denoted as r_i, and σ, denoted as s_i, as follows:

$$F(\rho, \sigma) = \sum_i (r_i s_i)^{1/2}. \tag{4.97}$$

4.8 QUANTUM COMPUTING SIMULATORS

The purpose of a quantum computing simulator is to test the syntax and flow of an algorithm before actually implementing it in the quantum hardware, and this simulation can be done locally or on cloud. Unfortunately, to classically simulate the N-qubit system would require storing 2^N complex numbers in classical memory. The quantum circuitry performs the unitary transformation U on the N-qubit state, that is, $|\psi'\rangle = U|\psi\rangle$, which would require $2^N \times 2^N$ memory units to store the whole quantum circuit, which is not feasible in practice. To avoid this problem, we should apply one-qubit and two-qubit gates on a given qubit location. For instance, for single-qubit gates, this will result in applying a 2×2 matrix to the given qubit location. The popular quantum computing development libraries include [41]: Qiskit, Forest, ProjectQ, Quantum Development Kit (QDK), and Cirq.

The *quantum information science kit* (*Qiskit*) is the IBM-created quantum computing dev library, which provides a framework to program quantum computers. This open-source software platform allows one to work with the quantum language known as Open Quantum Assembly Language (OpenQASM), used in quantum computers in the IBM Q Experience. The Qiskit open-source development library is composed of four main (core) modules [42]:

- Qiskit Terra module, which contains core elements to create quantum programs at the circuit/pulses level as well as to optimize these core elements by considering the particular physical quantum processor constraints.
- Qiskit Aer module, which provides a C++ simulator framework and tools to develop noise models to conduct realistic simulations in the presence of noise and errors occurring during execution on real devices.
- Qiskit Ignis module, which provides a framework to understand the noise sources in quantum circuits/devices as well as to develop different noise mitigation/reduction strategies.
- Qiskit Aqua module, which contains a library of cross-domain quantum algorithms suitable for application in near-term quantum computing.

Qiskit is available in Python (version 3.5 or higher), Swift, and JavaScript; however, given the high popularity of Python, we will only discuss Python. To install Qiskit we can use the Python package manager pip by simply typing in the command line the following:

```
pip install qiskit
```

Regarding Qiskit syntax, we have to apply the following high-level steps: (1) a building step in which we design a quantum circuit that represents the problem at hand, (2) an execution step in which we run experiments on different backends (including both actual quantum hardware and simulators), and (3) an analysis step in which we summarize the statistics and visualize the results. As an illustration, let us consider the *quantum random number generator* (QRNG) implemented by applying the Hadamard gate on the $|0\rangle$state to obtain the superposition of CB states, each occurring with the same probability amplitude:

$$H|0\rangle = \frac{1}{\sqrt{2}}(|0\rangle + |1\rangle).$$

By measuring this state the probabilities of obtaining 0 and 1 are equally likely. Let us implement this QRNG using Qiskit. We need first to import the packages to create the quantum and classical registers, quantum circuit, function to execute the circuit, and simulator backend as follows:

```
import numpy as np
from qiskit import(
    QuantumCircuit
    execute
    Aer)
from qiskit.visualization import plot_histogram
```

After calling the Aer simulator, we need to initialize quantum and classical registers as well as the quantum circuitry:

```
simulator = Aer.get_backend('qasm_simulator')
qr = QuantumRegister (1)
cr = ClassicalRegister (1)
qc = QuantumCircuit(qr, cr)
```

The next step would be to create the gates required, namely the Hadamard gate and measurement circuit:

```
qc.h(qr[0])
qc.measure(qr[0], cr[0])
```

It might be a good idea to draw our quantum circuit by:

```
print(qc.draw())
```

to get the following circuitry:

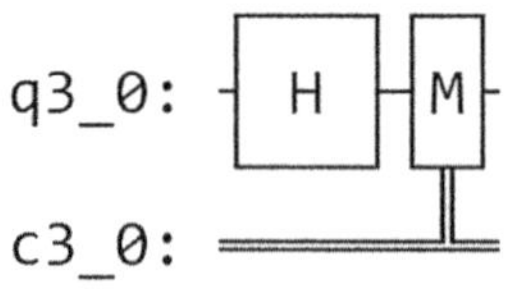

To execute the quantum circuit we need to type:

```
res = execute(qc, simulator, shots = 10,000).result()
```

To visualize the results we can create the histogram by:

```
plot_histogram(res.get_counts())
```

to get the following histogram:

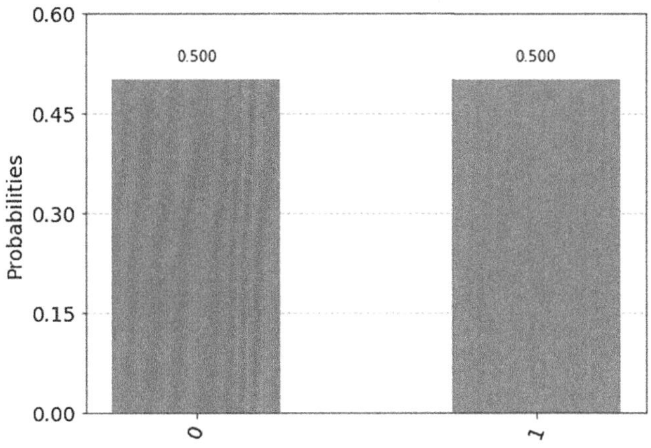

Clearly, for 10,000 runs the probability of both 0 and 1 is 0.5 each. To get these results we used Python 3.7 (Spyder).

OpenQASM is the quantum language to provide instructions for an actual quantum computer, and Qiskit has a feature to generate the corresponding code. For instance, for the QRNG example by typing:

```
print(qc.qasm())
```

we obtain the following QASM code for the quantum circuit:

```
OPENQASM 2.0;
include "qelib1.inc";
qreg q3[1];
creg c3[1];
h q3[0];
measure q3[0] -> c3[0];
```

The first two lines are common and can be found in any QASM file. The next two lines define the quantum and classical registers, with one item each. The subsequent line applies the Hadamard gate to the 0th qubit in the quantum register q3. The last line describes the measurement on qubit q3[0] and the result of measurement is placed in the 0th position of the classical register.

As another example, let us simulate the Bell state $|B_{00}\rangle = 2^{-1/2}(|00\rangle + |11\rangle)$ preparation. We first need to import the packages in the same fashion as in the previous example. We then define the quantum circuit composed of two qubits and apply the Hadamard gate on qubit 0 followed by a CNOT gate with qubit 0 being the control qubit and qubit 1 the target qubit:

```
qc = QuantumCircuit(2)
# Apply Hadamard gate to the qubit 01:
qc.h(0)
# Apply the CNOT^{0,1} gate:
qc.cx(0,1)
```

To draw the quantum circuit we type:

```
qc.draw()
```

to get:

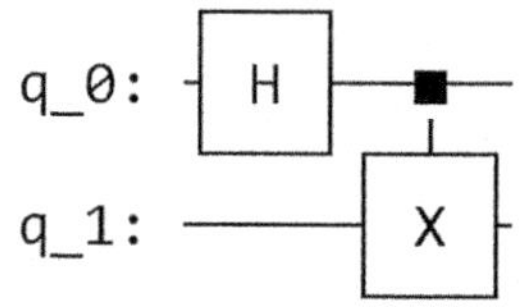

To collect the result we need to select the appropriate simulator:

```
backend = Aer.get_backend('statevector_simulator')
final_state = execute(qc,backend).result().get_statevector()
```

Let us finally verify what is the probability of getting $|00\rangle$ and $|11\rangle$ by typing:

```
results = execute(qc,backend).result().get_counts()
plot_histogram(results)
```

to get:

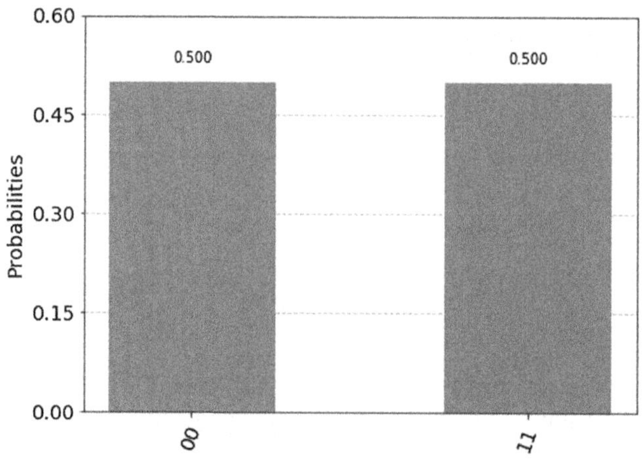

Qiskit documentation can be found at https://qiskit.org/documentation/. It contains information on installation and setup, Qiskit elements, developing strategy, various tutorials and libraries, references, and troubleshooting. The tutorials are grouped into the following categories: circuits, advanced circuits, high-performance simulators, quantum system error analysis, and optimization. The web page also provides documentation for each of four backends, related to connectivity, coherence times, and gate application time. Information about the IBM Quantum Experience can be found at https://quantum-computing.ibm.com/. Qiskit documentation also offers a course in quantum computing: https://qiskit.org/textbook/preface.html. The quantum backends supported by Qiskit include

IBMQX2 (5 qubits), IBMQX4 (5 qubits), IBMQX5 (16 qubits), and QS1_1 (20 qubits). The last quantum computer is available only to members of the IBM Q network. Details of the IBMAX5 computer can be found in [43].

Cirq represents Google's quantum computing development library, available at https://cirq.readthedocs.io/en/latest/index.html, which is specifically designed for noisy intermediate scale quantum computers. To install it we need to type the following text in the command line:

```
pip install cirq
```

Cirq has certain similarities to Qiskit; however, the documentation is not that comprehensive compared to the Qiskit documentation. To become familiarized with the Cirq syntax, let us observe the same QRNG.

We need first to import the package by typing:

```
import cirq
```

Now we need to allocate the quantum register and the quantum circuit by:

```
qubits = [cirq.LineQubit(0)]
crq = cirq.Circuit()
```

We then apply the Hadamard gate to the qubit, followed by the measurement in the CB:

```
crq.append(cirq.H(qubits[0]))
crq.append(cirq.measure(qubits[0],key="z"))
```

Finally, we activate the quantum simulator, let it run for 10 receptions, and print the results:

```
simulator=cirq.Simulator()
result=simulator.run(crq,repetitions=10)
print(result)
```

Forest is a Python-based quantum software platform developed by Rigetti Computing, which includes:

- *PyQuil*, the open source to develop, analyze, and run quantum programs;
- *Quil*, short for quantum instruction language, representing the open low-level assembly/instruction language suitable for near-term quantum computers employing a shared classical/quantum memory model [44].

Once Anaconda is installed, to install PyQuil we need to type the following text in the command line:

```
pip install -c rigetti pyquil
```

Forest has comprehensive documentation containing installation details, various programs, a gate operation description, a quantum virtual machine (QVM) simulator, a quantum computer with eight qubits, Quil language and complier [45], as well as some material on quantum computing fundamentals.

To become familiarized with the Forest syntax, let us observe the same QRNG program. We need first to import the package, declare the quantum circuit/program, perform gate operations, and execute the quantum with the following three lines:

```
from pyquil.quil import Program
import pyquil.gates as gates
from pyquil import api
```

Furthermore, we instantiate the quantum program and specify to apply the Hadamard gate to qubit 0, followed by measurement of the same qubit, and store the result into a classical qubit 0:

```
qpr = Program()
qpr += [gates.H(0),gates.MEASURE(0,0)]
```

We then establish the connection to the QVM, run the program, and display the output of the quantum circuit for 2 trials. (Notice that sometimes you might experience a problem connecting to the QVM server.) PyQuil has the option to generate the Quil code. By simply typing:

```
print(qpr)
```

we get:

H 0

MEASURE 0 ro[0]

To access Rigetti's quantum computer the potential user must place the formal request by filling in the form at https://qcs.rigetti.com/request-access.

QDK is the quantum computing library developed by Microsoft. QDK is not based on Python, but contains its own language called Q# (pronounced "Q sharp"). Unlike the superconducting quantum computer by IBM and Rigetti, Microsoft relies on topological qubits based on Majorana fermions [46]. Given that the quantum computer is not available yet and the syntax is completely different from previous languages, we refer interested readers to [41,42].

4.9 SUMMARY

This chapter was devoted to QIP fundamentals. The basic QIP features were discussed in Section 4.1. The superposition principle and quantum parallelism concept were introduced in Section 4.2, together with QIP basics. The noncloning theorem was formulated and proved in Section 4.3. Furthermore, the theorem of the inability to distinguish nonorthogonal quantum states was formulated and proved in Section 4.4. The next Section 4.5 was devoted to various aspects of entanglement, including the Schmidt decomposition theorem, purification, superdense coding, and entanglement swapping. Various measures of entanglement were introduced, including the entanglement of formation and concurrence. How entanglement can be used in detecting and correcting quantum errors was described as well. Furthermore, in Section 4.6 the operator-sum representation of a quantum operation was introduced and applied to the bit-flip and phase-flip channels. In Section 4.7 quantum errors and decoherence effects were discussed and the phase-damping, depolarizing, and amplitude-damping channels were described. Different quantum computing libraries, including Qiskit, Cirq, Forest, and QDK were described in Section 4.8. For a detailed description of various QIP realizations, interested readers are referred to Chapter 13. For a deeper understanding of QIP concepts, a set of problems is provided in the following section.

4.10 PROBLEMS

1. Which of the following are legitimate quantum states?
   (a) $|\psi\rangle = 0.5|0\rangle + 0.5|1\rangle$;
   (b) $|\psi\rangle = \cos\varphi\,|0\rangle + \sin\varphi\,|1\rangle$;
   (c) $|\psi\rangle = (3^{-1/2}|0\rangle + j|1\rangle)/2$;
   (d) $|\psi\rangle = \cos^2\varphi\,|0\rangle + \sin^2\varphi\,|1\rangle$.
2. Let the qubit be given by $|\psi\rangle = 2^{-1/2}(|0\rangle + |1\rangle)$. Verify that the action of the phase-flip channel is given by:

$$|\psi_{out}\rangle = 2^{-1/2}\left[p\begin{pmatrix}1\\-1\end{pmatrix} + (1-p)\begin{pmatrix}1\\-1\end{pmatrix}\right].$$

   Employ the operator-sum notation to verify this.
3. The ket vectors and the corresponding density matrices in the previous problem are expressed in the CB $\{|0\rangle, |1\rangle\}$. Express each of them in the diagonal basis DB $= \{|+\rangle = 2^{-1/2}(|0\rangle + |1\rangle, |-\rangle)$ instead.
4.
   (a) Show that for some quantum state ρ the following is valid:

$$I = \frac{I\rho I + X\rho X + Y\rho Y + Z\rho Z}{2}.$$

   (b) Based on (a) derive the operator-sum relation for:

$$\xi(\rho) = p\frac{I}{2} + (1-p)\rho.$$

5. Derive the operator-sum representation (4.62) by assuming that the environment state is the mixed state.
6. Consider a qubit initially in the quantum state $|\psi\rangle = \alpha|0\rangle + \beta|1\rangle$, wherein $|\alpha|^2 + |\beta|^2 = 1$, which interacts with a generalized damping channel with the Kraus operators given by:

$$E_1 = \sqrt{p}\begin{pmatrix}1 & 0\\ 0 & \sqrt{1-\gamma}\end{pmatrix},\ E_2 = \sqrt{p}\begin{pmatrix}0 & \sqrt{\gamma}\\ 0 & 0\end{pmatrix},\ E_3 = \sqrt{1-p}\begin{pmatrix}\sqrt{1-\gamma} & 0\\ 0 & 1\end{pmatrix},\ E_4 = \sqrt{1-p}\begin{pmatrix}0 & 0\\ \sqrt{\gamma} & 0\end{pmatrix}$$

   (a) Show that the quantum operation associated with this channel is trace preserving.
   (b) Determine operator-sum representation.
   (c) Determine the probabilities $P(i)$ that processes E_i $(i = 1,2,3,4)$ will occur.
7.
   (a) Show that the Pauli matrix X performs bit-flip operation.
   (b) Show that the Pauli matrix Z performs phase-flip operation.
   (c) Show that the Pauli matrix Y performs bit-phase-flip operation.

8. If no-cloning theorem could be violated, would it be possible to reliably distinguish two nonorthogonal quantum states?

9. Determine the Schmidt decomposition of the following quantum states:

(a) All the Bell states and

(b) $|\psi\rangle = 3^{-1/2}[|0\rangle|1\rangle + |1\rangle|0\rangle + |1\rangle|1\rangle]$

10. Suppose we have a quantum state $|\psi\rangle$ that belongs to the bipartite quantum system $H_A \otimes H_B$. Assuming that both Hilbert spaces have the same dimension, show that the partial traces $\text{Tr}_A(|\psi\rangle\langle\psi|)$ and $\text{Tr}_B(|\psi\rangle\langle\psi|)$ have the same eigenvalues.

11. Show that the EPR pairs form an orthonormal basis for the Hilbert space of any bipartite quantum system.

12. Determine:

(a) The trace of the Pauli matrices;

(b) The determinant of the Pauli matrices;

(c) The square of the Pauli matrices.

13. Using Qiskit simulate the quantum teleportation circuit.

14. Using Cirq simulate the Bell preparation state and quantum teleportation circuits.

15. Using Forest simulate the Bell preparation state and quantum teleportation circuits.

References

[1] F. Gaitan, Quantum Error Correction and Fault Tolerant Quantum Computing, CRC Press, 2008.

[2] I.B. Djordjevic, Quantum Information Processing and Quantum Error Correction: An Engineering Approach, Elsevier/Academic Press, 2012.

[3] D. McMahon, Quantum Computing Explained, John Wiley & Sons, 2008.

[4] I.B. Djordjevic, Physical-Layer Security and Quantum Key Distribution, Springer International Publishing, Switzerland, 2019.

[5] C.E. Shannon, A mathematical theory of communication, Bell Syst. Tech. J. 27 (1948), 379–423, 623–656.

[6] C.E. Shannon, W. Weaver, The Mathematical Theory of Communication, The University of Illinois Press, Urbana, Illinois, 1949.

[7] T. Cover, J. Thomas, Elements of Information Theory, Wiley-Interscience, 1991.

[8] R. Landauer, Irreversibility and heat generation in the computing process, IBM J. Res. Dev. 5 (3) (1961).

[9] A.N. Kolmogorov, Three approaches to the quantitative definition of information, Probl. Inf. Transm. 1 (1) (1965) 1–7.

[10] A. Turing, On computable numbers, with an application to the Entscheidungs problem, Proc. Lond. Math. Soc. 2 (42) (1936) 230–265.

[11] A. Turing, Computability and lambda-definability, J. Symbolic Logic 2 (1937) 153–163.

[12] A. Turing, Systems of logic based on ordinals, Proc. Lond. Math. Soc. 3 (45) (1939) 161–228.
[13] M. Planck, On the law of distribution of energy in the normal spectrum, Ann. Phys. 4 (1901) 553.
[14] A. Einstein, Concerning an heuristic point of view toward the emission and transformation of light, Ann. Phys. 17 (1905) 132.
[15] L. de Broglie, Waves and quanta, Comput. Ren. 177 (1923) 507.
[16] W. Pauli, On the connection between the completion of electron groups in an atom with the complex structure of spectra, Z. Phys. 31 (1925) 765.
[17] W. Heisenberg, Quantum-theoretical re-interpretation of kinematic and mechanical relations, Z. Phys. 33 (1925) 879.
[18] M. Born, P. Jordan, On quantum mechanics, Z. Phys. 34 (1925) 858.
[19] E. Schrödinger, Quantization as a problem of proper values. Part I, Ann. Phys. 79 (1926) 361.
[20] E. Schrödinger, On the relation between the quantum mechanics of Heisenberg, Born, and Jordan, and that of schrödinger, Ann. Phys. 79 (1926) 734.
[21] W. Heisenberg, The actual content of quantum theoretical kinematics and mechanics, Z. Phys. 43 (1927) 172.
[22] J.J. Sakurai, Modern Quantum Mechanics, Addison Wesley, 1993.
[23] R.P. Feynman, Simulating physics with computers, Int. J. Theor. Phys. 21 (1982) 467–488.
[24] D. Deutsch, The Fabric of Reality, Penguin Press, New York, 1997.
[25] D. Deutsch, Quantum theory, the Church-Turing principle and the universal quantum computer, Proc. Roy. Soc. Lond. A 400 (1985) 97–117.
[26] D. Deutsch, R. Jozsa, Rapid solution of problems by quantum computation, Proc. Math. Phys. Sci. 439 (1907) (1992) 553–558.
[27] P. Shor, Algorithms for quantum computation: discrete logarithms and factoring, IEEE Comput. Soc. Press (1994) 124–134.
[28] L.K. Grover, A fast quantum mechanical algorithm for database search, in: Proceedings 28th Annual ACM Symposium on the Theory of Computing, May 1996, p. 212.
[29] A. Einstein, B. Podolsky, N. Rosen, Can quantum-mechanical description of physical reality Be considered complete? Phys. Rev. 47 (1935) 777–780.
[30] J.S. Bell, On the Einstein-Podolsky-Rosen paradox, Physics 1 (1964) 195–200.
[31] J.S. Bell, Speakable and Unspeakable in Quantum Mechanics, *Cambridge University Press*, 1987.
[32] J.F. Clauser, A. Shimony, Bell's theorem experimental tests and implications, Rep. Prog. Phys. 41 (1978) 1981.
[33] A. Aspect, P. Grangier, G. Roger, Experimental tasks of realistic local theories by Bell's theorem, Phys. Rev. Lett. 47 (1981) 460.
[34] P.G. Kwiat, et al., New high-intensity source of polarization-entangled photon pairs, Phys. Rev. Lett. 75 (1995) 4337.
[35] R. Horodecki, P. Horodecki, M. Horodecki, K. Horodecki, Quantum entanglement, Rev. Mod. Phys. 81 (2009) p865.
[36] V. A. Mousolou, Entanglement Fidelity and Measure of Entanglement, (2020). Available at: https://arxiv.org/abs/1911.10854.
[37] C.H. Bennett, D.P. DiVincenzo, J. Smolin, W.K. Wootters, Mixed-state entanglement and quantum error correction, Phys. Rev. A 54 (1996) 3824.

[38] S. Hill, W.K. Wootters, Entanglement of a pair of quantum bits, Phys. Rev. Lett. 78 (1997) 5022.
[39] W.K. Wootters, Entanglement of formation of an arbitrary state of two qubits, Phys. Rev. Lett. 80 (1998) 2245.
[40] A. Harrow, Superdense coding of quantum states, Phys. Rev. Lett. 92 (18) (2004) 187901-1−187901-4.
[41] R. LaRose, Overview and comparison of gate level quantum software platforms, Quantum 3 (2019) 130, available at: https://arxiv.org/abs/1807.02500v2.
[42] J.D. Hidary, Quantum Computing: An Applied Approach, Springer Nature Switzerland AG, Cham, Switzerland, 2019.
[43] B. Abdo, et al., IBM Q 16 Rueschlikon V1.x.x, September 2018. Available at: https://ibm.biz/qiskit-ibmqx5.
[44] Rigetti Forest: An API for Quantum Computing in the Cloud, (2020). Available at: https://www.welcome.ai/tech/technology/rigetti-forest .
[45] R.S. Smith, M.J. Curtis, W.J. Zeng, A Practical Quantum Instruction Set Architecture, 2016 available at, https://arxiv.org/abs/1608.03355.
[46] H. Zhang, et al., Quantized majorana conductance, Nature 556 (2018) 74−79.

CHAPTER

Quantum Algorithms and Methods

5

CHAPTER OUTLINE

This chapter is devoted to basic quantum algorithms and methods. The chapter starts by revisiting the quantum parallelism concept and describing its power in calculating a global property of a certain function by performing only one evaluation of that function, namely Deutsch's and the Deutsch–Jozsa algorithms. We also describe the Bernstein–Vazirani algorithm, which is able to determine a string encoded in a function in only one computational step. Furthermore, the Grover search algorithm to perform a search for an entry in an unstructured database is described. Next, the quantum Fourier transform (FT) is described, which is a basic algorithm used in many other quantum algorithms. In addition, an algorithm to evaluate the period of a function is provided. How to can crack the Rivest–Shamir–

https://doi.org/10.1016/B978-0-12-821982-9.00006-X

Adleman encryption protocol is also described. Then, Shor's factorization algorithm is provided. Furthermore, Simon's algorithm is described. Next, the quantum phase estimation algorithm is described. Quantum computational complexity and Turing machine representation are discussed as well. The quantum circuits are imperfect, which prevents us from running well-known quantum algorithms using the gates-based quantum computing approach. To solve this problem, adiabatic quantum computing and variational (quantum) circuits are introduced. After summarizing the chapter, a set of problems is offered to provide a deeper understanding of the material presented in this chapter.

5.1 QUANTUM PARALLELISM (REVISITED)

We start our discussion on basic quantum computing algorithms [1–11] by revisiting the quantum parallelism concept. The quantum computation C implemented on a quantum register maps the input string $i_1 \ldots i_N$ to the output string $O_1(i),\ldots,O_N(i)$:

$$\begin{pmatrix} O_1(i) \\ \vdots \\ O_N(i) \end{pmatrix} = U(C)\begin{pmatrix} i_1 \\ \vdots \\ i_N \end{pmatrix}; \quad (i)_{10} = (i_1 \cdots i_N)_2. \tag{5.1}$$

The computation basis (CB) states are denoted by:

$$|i_1 \cdots i_N\rangle = |i_1\rangle \otimes \cdots \otimes |i_N\rangle; \quad i_1, \cdots, i_N \in \{0,1\}. \tag{5.2}$$

The linear superposition allows us to form the following $2N$-qubit state:

$$|\psi_{in}\rangle = \left[\frac{1}{\sqrt{2^N}}\sum_i |i_1 \cdots i_N\rangle\right] \otimes |0 \cdots 0\rangle, \tag{5.3}$$

and upon the application of quantum operation $U(C)$, the output can be represented by:

$$|\psi_{\text{out}}\rangle = U(C)|\psi_{\text{in}}\rangle = \frac{1}{\sqrt{2^N}}\sum_i |i_1 \cdots i_N\rangle \otimes |O_1(i) \cdots O_N(i)\rangle. \tag{5.4}$$

The quantum computer has been able to encode all input strings generated by C into $|\psi_{\text{out}}\rangle$; in other words, it has simultaneously pursued 2^N classical paths. This ability of a quantum computer to encode multiple computational results into a quantum state in a single quantum computational step is known as *quantum parallelism*.

The quantum circuit to simultaneously evaluate $f(0)$ and $f(1)$ of the mapping $f(x)$: $\{0,1\} \rightarrow \{0,1\}$ is shown in Fig. 5.1. Its output state contains information about both $f(0)$ and $f(1)$.

We can show that the quantum operation U_f is unitary as follows:

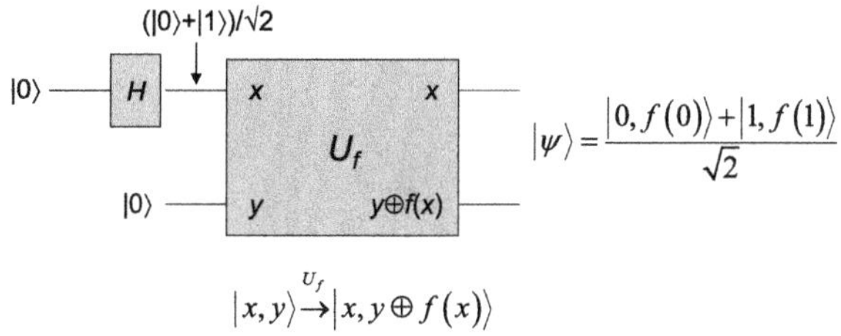

FIGURE 5.1

Quantum circuit to simultaneously compute $f(0)$ and $f(1)$.

$$|x, y \oplus f(x)\rangle \xrightarrow{U_f} |x, [y \oplus f(x)] \oplus f(x)\rangle = |x, y\rangle \Rightarrow U_f^2 = I. \tag{5.5}$$

The action of operation U_f in operator notation can be represented by:

$$U_f|x, 0\rangle = |x, f(x)\rangle \quad U_f|x, y\rangle = |x, y \oplus f(x)\rangle. \tag{5.6}$$

The output of the circuit shown in Fig. 5.1 can be obtained now as:

$$|\psi\rangle = U_f|H0 \otimes 0\rangle = U_f\frac{1}{\sqrt{2}}(|00\rangle + |10\rangle) = \frac{1}{\sqrt{2}}(|0,f(0)\rangle + |1,f(1)\rangle), \tag{5.7}$$

proving that the circuit from Fig. 5.1 can indeed simultaneously evaluate $f(0)$ and $f(1)$.

We would like now to generalize the computation of $f(x)$ on $n + m$ qubits. Before we provide the quantum circuit to perform this computation, we need the quantum circuit to create the superposition state from Eq. (5.3). The Walsh–Hadamard transform on two ancilla qubits in state $|00\rangle$ can be implemented by applying the Hadamard gates on ancilla qubits as shown in Fig. 5.2A.

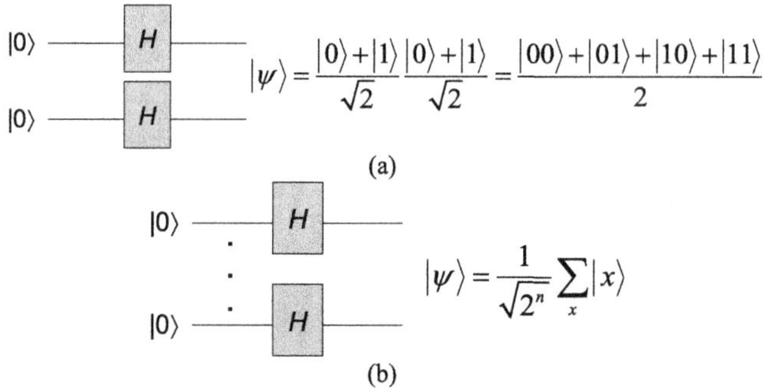

FIGURE 5.2

Walsh–Hadamard transform: (a) on two qubits, and (b) on n qubits.

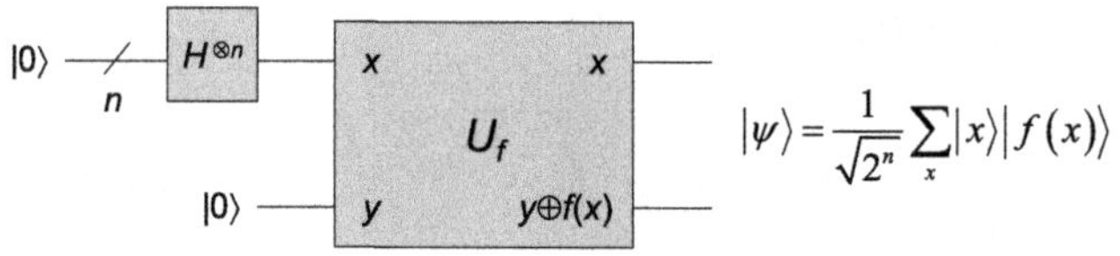

FIGURE 5.3

Quantum circuit to evaluate $f(x)$: $\{0,1\}^n \rightarrow \{0,1\}$.

The Walsh–Hadamard transform on n qubits can be implemented as shown in Fig. 5.2B. Now, by using the Walsh–Hadamard transform circuit, the quantum circuit to evaluate $f(x)$: $\{0,1\}^n \rightarrow \{0,1\}$ can be implemented as shown in Fig. 5.3.

The quantum circuit performing U_f on $n + m$ qubits is shown in Fig. 5.4. Its operation can be described as follows:

$$U_f|x, z\rangle = |x, z \oplus f(x)\rangle, \quad |z\rangle = |z_{m-1} \cdots z_0\rangle, \quad z_i = 0, 1. \tag{5.8}$$

By setting $z = 0 \ldots 0$, the action of this gate is:

$$U_f\left|x, 0^{\otimes m}\right\rangle = \left|x, f(x)\right\rangle, \tag{5.9}$$

which is the same as the action of the gate shown in Fig. 5.3. By using this circuit, we can generalize the evaluation $f(x)$ to $n + m$ qubits as illustrated in Fig. 5.5. The output state of the quantum circuit shown in Fig. 5.5 can be determined as:

$$|\psi\rangle = U_f\left|\left(H^{\otimes n} 0^{\otimes n}\right) \otimes 0^{\otimes m}\right\rangle = \frac{1}{2^{n/2}} U_f\left(\sum_{x=0}^{2^n-1} |x\rangle \otimes \left|0^{\otimes m}\right\rangle\right) = \frac{1}{2^{n/2}} \sum_{x=0}^{2^n-1} |x, f(x)\rangle. \tag{5.10}$$

5.2 DEUTSCH'S, DEUTSCH–JOZSA, AND BERNSTEIN–VAZIRANI ALGORITHMS

5.2.1 Deutsch's Algorithm

Although $|\psi\rangle = (|0, f(0)\rangle + |1, f(1)\rangle)/\sqrt{2}$ contains information about both $f(0)$ and $f(1)$, there is no any advantage to classical computation if the table of $f(x)$ is desired. On the other hand, quantum computation allows us to evaluate some global property,

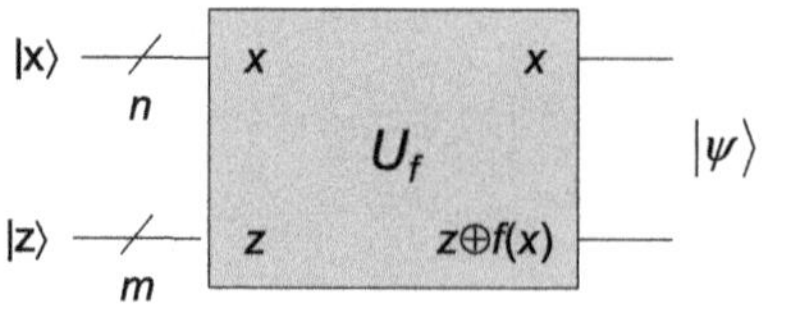

FIGURE 5.4

Generalization of quantum operation U_f on $n + m$ qubits.

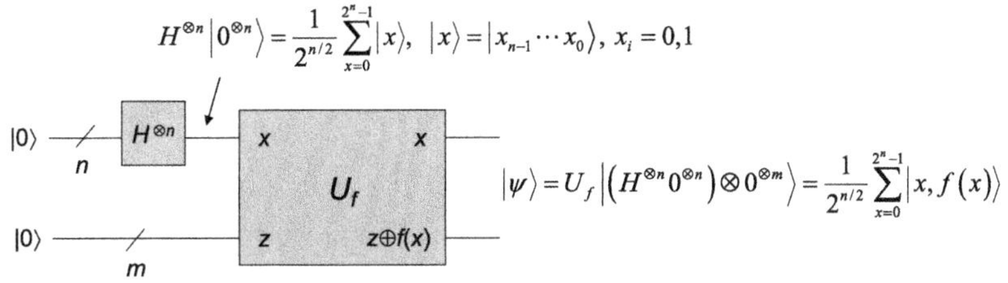

FIGURE 5.5

Quantum circuit to evaluate $f(x)$ on $n + m$ qubits.

say $f(0) \oplus f(1)$, by performing only one evaluation of $f(x)$. Although this example is trivial and not very practical, it allows us to describe the basic concepts of quantum computing. The quantum circuit implementing Deutsch's algorithm is shown in Fig. 5.6. The state $|\psi\rangle$ after the Hadamard gates stage is given by:

$$|\psi\rangle = H|0\rangle \otimes H|1\rangle = \frac{1}{2}(|0\rangle + |1\rangle)(|0\rangle - |1\rangle) = \frac{1}{2}\left(\sum_{x=0}^{1}|x\rangle\right)(|0\rangle - |1\rangle). \tag{5.11}$$

The application of operator U_f to $|\psi\rangle$ performs the following mapping:

- Option 1: $f(x) = 0 \Rightarrow (|0\rangle - |1\rangle) \rightarrow (|0\rangle - |1\rangle)$.
- Option 2: $f(x) = 1 \Rightarrow (|0\rangle - |1\rangle) \rightarrow (|1\rangle - |0\rangle) = -(|0\rangle - |1\rangle)$.

By combining these two options, the same mapping can be represented by $(|0\rangle - |1\rangle) \rightarrow (-1)^{f(x)}(|0\rangle - |1\rangle)$ so that the action of U_f to $|\psi\rangle$ is as follows:

$$U_f|\psi\rangle = U_f H|0\rangle \otimes H|1\rangle = \frac{1}{2}\left(\sum_{x=0}^{1}(-1)^{f(x)}|x\rangle\right)(|0\rangle - |1\rangle). \tag{5.12}$$

The operator that performs the following mapping $|x\rangle \xrightarrow{U_f} (-1)^{f(x)}|x\rangle$ is known as an *oracle* operator. The upper and lower registers on the right side of the circuit shown in Fig. 5.6 are unentangled after the oracle operator. This operator is important in the Grover search algorithm, which will be described later. The output state in

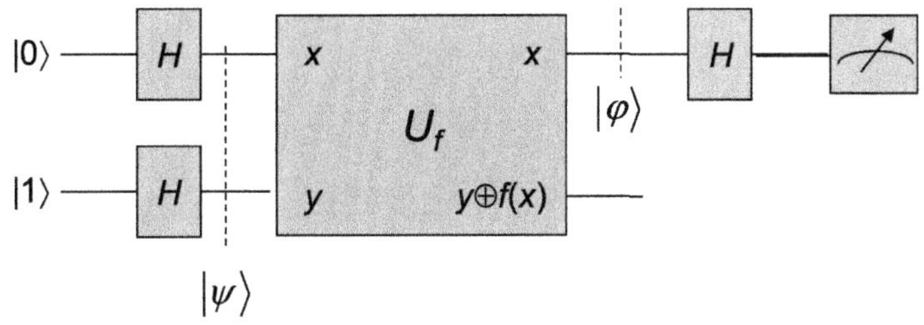

FIGURE 5.6

Quantum circuit implementing Deutsch's algorithm.

the upper branch of Fig. 5.6 is given by $|\varphi\rangle = \left[(-1)^{f(0)}|0\rangle + (-1)^{f(1)}|1\rangle\right]/\sqrt{2}$, and by applying the Hadamard gate we obtain:

$$H|\varphi\rangle = \frac{1}{2}\left[(-1)^{f(0)} + (-1)^{f(1)}\right]|0\rangle + \frac{1}{2}\left[(-1)^{f(0)} - (-1)^{f(1)}\right]|1\rangle. \quad (5.13)$$

By performing the measurement we can obtain two possible states:

- Option 1: qubit $|0\rangle$ is the final state upon the measurement, meaning that $f(0) = f(1)$; in other words the function $f(x)$ is constant.
- Option 2: qubit $|1\rangle$ is obtained, meaning that $f(0) \neq f(1)$, indicating that the function $f(x)$ is balanced (the outputs are opposite for half of the inputs).

Therefore, in this example, we employed the quantum parallelism concept to bypass the explicit calculation of $f(x)$.

5.2.2 Deutsch–Jozsa Algorithm

The Deutsch–Jozsa algorithm represents a generalization of Deutsch's algorithm. Namely, it verifies if the mapping $f(x)$: $\{0,1\}^n \rightarrow \{0,1\}$ is constant or balanced for all values of $x \in \{0, 2^n - 1\}$. The quantum circuit implementing the Deutsch–Jozsa algorithm is shown in Fig. 5.7. We will describe the Deutsch–Jozsa algorithm for $n = 2$ (Fig. 5.8), while the description for arbitrary n is left as a homework problem. Based on Fig. 5.8 we conclude that the action of U_f on $|\psi\rangle$ is given by:

$$U_f|\psi\rangle = U_f\left[\frac{1}{2}\sum_{x=0}^{3}|x\rangle\right] \otimes \frac{1}{\sqrt{2}}(|0\rangle - |1\rangle) = \left[\frac{1}{2}\sum_{x=0}^{3}(-1)^{f(x)}|x\rangle\right] \otimes \frac{1}{\sqrt{2}}(|0\rangle - |1\rangle). \quad (5.14)$$

Clearly, the upper and lower registers on the right side of gate U_f are unentangled.

By applying the Hadamard gates on upper qubits $|\psi'\rangle$ and performing measurement we obtain the following options:

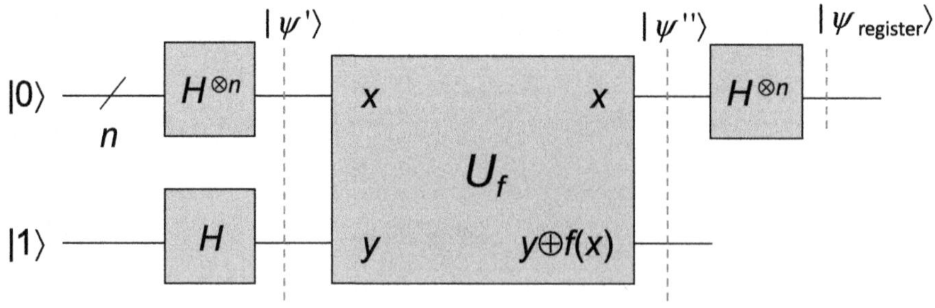

FIGURE 5.7

Deutsch–Jozsa algorithm implementation.

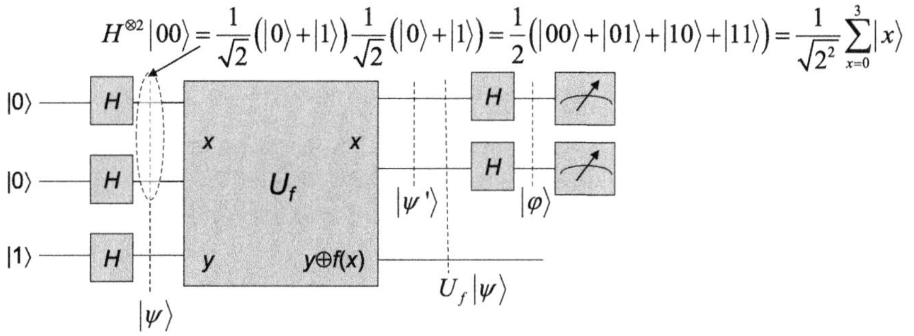

FIGURE 5.8

Deutsch–Jozsa algorithm implementation for $n = 2$.

- Option 1: $f(x) = \text{const}$, in which case the upper qubits $|\psi'\rangle$ can be represented as:

$$|\psi'\rangle = \pm\frac{1}{2}(|00\rangle + |01\rangle + |10\rangle + |11\rangle) = \pm\frac{1}{\sqrt{2}}(|0\rangle + |1\rangle)\frac{1}{\sqrt{2}}(|0\rangle + |1\rangle). \quad (5.15a)$$

By applying the corresponding Hadamard gates, the upper qubits $|\psi'\rangle$ are transformed to:

$$|\varphi\rangle = H^{\otimes 2}|\psi'\rangle = \pm|00\rangle. \quad (5.15b)$$

- Option 2: $f(x) = x \bmod 2$, in which case the upper qubits (based on Eq. (5.14)) are in fact:

$$|\psi'\rangle = \frac{1}{2}(|00\rangle - |01\rangle + |10\rangle - |11\rangle) = \frac{1}{2}[|0\rangle(|0\rangle - |1\rangle) + |1\rangle(|0\rangle - |1\rangle)]. \quad (5.16a)$$

Now, by applying the Hadamard gates, the upper qubits $|\psi'\rangle$ are transformed to:

$$H^{\otimes 2}|\psi'\rangle = H^{\otimes 2}\frac{1}{2}[(|0\rangle + |1\rangle) \otimes (|0\rangle - |1\rangle)] = |01\rangle. \quad (5.16b)$$

The result in upper qubits is unambiguous only if $|\psi'\rangle$ is a nonentangled state, indicating that it must be:

$$(-1)^{f(0)+f(1)} = (-1)^{f(1)+f(2)} \quad (5.17)$$

5.2.3 Bernstein–Vazirani Algorithm

This algorithm was introduced by Bernstein and Vazirani [12] and was experimentally demonstrated using trapped ions technology in [13]. The function $f(x)$ calculates the dot product of x and a secret string $s \in \{0,1\}^n$; that is, $f(x) = x \cdot s = x_1 s_1 + \cdots + x_n s_n \bmod 2$. The goal of the Bernstein–Vazirani (BV) algorithm is to determine the string encoded in this function in only one computational step. The corresponding classical algorithm will require n computational steps:

$$f(10\cdots0) = s_1, f(010\cdots0) = s_2, \cdots, f(00\cdots01) = s_n.$$

Implementation of the BV algorithm is based on the same quantum circuitry already used in the Deutsch–Jozsa algorithm, shown in Fig. 5.7, but the function $f(x)$ is different. The input state $|\psi'\rangle$ to the U_f-unitary and the corresponding output state can be written as:

$$|\psi'\rangle = 2^{-n/2} \sum_{x\in\{0,1\}^n} |x\rangle \frac{|0\rangle - |1\rangle}{\sqrt{2}}, \; |\psi''\rangle = 2^{-n/2} \sum_{x\in\{0,1\}^n} (-1)^{x\cdot s}|x\rangle \frac{|0\rangle - |1\rangle}{\sqrt{2}}, \tag{5.18}$$

respectively. The action of Hadamard gates on the n-qubit state $|z\rangle$ is given by $H^{\otimes n}|z\rangle = 2^{-n/2} \sum_{x\in\{0,1\}^n} (-1)^{z\cdot x}|x\rangle$. Because the Hamdard gate is the second-order operator $H^2 = I$, by reapplying $H^{\otimes n}$ on $H^{\otimes n}|z\rangle$ we obtain back $|z\rangle$, in other words we can write:

$$H^{\otimes n}\left(H^{\otimes n}|z\rangle\right) = H^{\otimes n} 2^{-n/2} \sum_{x\in\{0,1\}^n} (-1)^{z\cdot x}|x\rangle = |z\rangle. \tag{5.19}$$

By applying this property to Eq. (5.18) on the top register, the output of the register will be $|\psi_{\text{register}}\rangle = |s\rangle$. Now, by performing the measurements on the register states we determine the content of the secret sequence s.

5.3 GROVER SEARCH ALGORITHM

The Grover search algorithm performs a *search* for an entry in an *unstructured database*. If N is the number of entries, a classical algorithm will take on average $N/2$ attempts, while the Grover search algorithm solves this problem in $\sim\sqrt{N}$ operations. To efficiently describe this algorithm we use an interpretation similar to that provided in [1]. The search is conducted under the following assumptions: (1) the database address contains n qubits and (2) the *search function* is defined by:

$$f_y(x) = \begin{cases} 0, x \neq y \\ 1, x = y \end{cases} = \delta_{xy}, x \in \{0, 1, \cdots, 2^n - 1\}, \tag{5.20}$$

meaning that the search function result is one if the correct item $|y\rangle$ is found. In the previous section, we defined the *oracle* operator as follows:

$$O_y|x\rangle = (-1)^{f_y(x)}|x\rangle. \tag{5.21a}$$

The oracle operator can also be represented as:

$$O_y = I - 2|y\rangle\langle y|, \tag{5.21b}$$

and the action of oracle operator on the CB state $|x\rangle$ will then be:

$$O_y|x\rangle = (I - 2|y\rangle\langle y|)|x\rangle = |x\rangle - 2|y\rangle\langle y|x\rangle = \begin{cases} -|y\rangle, & x = y \\ |x\rangle, & x \neq y \end{cases}. \tag{5.21c}$$

Based on the oracle operator, we define the *Grover* operator by:

$$G = H^{\otimes n} X H^{\otimes n} O, \tag{5.22}$$

where the action of operator X (not related to the Pauli X operator) on state $|x\rangle$ is given by:

$$X|x\rangle = -(-1)^{\delta_{x0}}|x\rangle, \tag{5.23}$$

while the corresponding matrix representation of the operator X is given by:

$$X \doteq \begin{bmatrix} 1 & 0 & 0 & 0 & \cdots & 0 \\ 0 & -1 & 0 & 0 & \cdots & 0 \\ 0 & 0 & -1 & 0 & \cdots & 0 \\ \cdots & & \vdots & & & \\ 0 & 0 & 0 & 0 & \cdots & -1 \end{bmatrix}. \tag{5.24}$$

From Eq. (5.24) it is clear that the X operator can be represented as $X = 2|0\rangle\langle 0| - I$, and upon substitution into Eq. (5.22), the Grover search operator becomes:

$$G = H^{\otimes n}(2|0\rangle\langle 0| - I)H^{\otimes n}O. \tag{5.25}$$

To simplify further the representation of the Grover operator, we represent it in terms of the superposition state:

$$|\Psi\rangle = H^{\otimes n}|0^{\otimes n}\rangle = \frac{1}{2^{n/2}}\sum_{x=0}^{2^n-1}|x\rangle, \tag{5.26}$$

as follows:

$$G = H^{\otimes n}(2|0\rangle\langle 0| - I)H^{\otimes n}O = \left(2\underbrace{H^{\otimes n}|0\rangle}_{|\Psi\rangle}\underbrace{\langle 0|H^{\otimes n}}_{\langle\Psi|} - \underbrace{H^{\otimes n}H^{\otimes n}}_{I} \right) O \overset{H^2=I}{=} (2|\Psi\rangle\langle\Psi| - I)O. \tag{5.27}$$

The Grover operator implementation circuit for $n = 3$ is shown in Fig. 5.9. The geometric interpretation of the Grover search algorithm, in terms of reflection operator O, is essentially the rotation in a 2D plane as shown in Fig. 5.10. The key idea of the Grover search algorithm is to determine the desired entry in an iterative fashion by rotating a current state $|\alpha\rangle$ in small angles θ until we reach the $|y\rangle$ state. The rotation of superposition state $|\Psi\rangle$ for an angle θ in counterclockwise fashion can be implemented by two reflections (or equivalently by applying the oracle operator

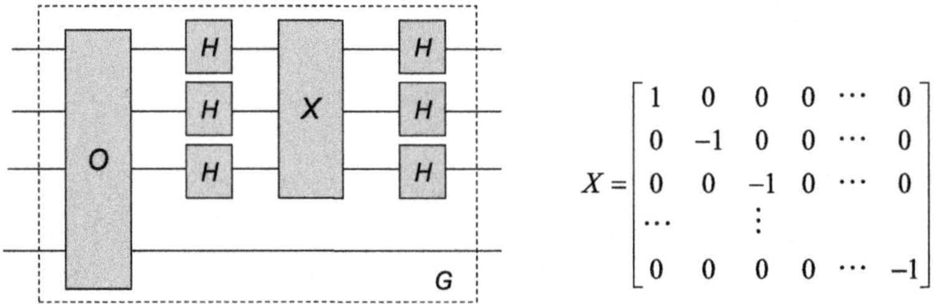

FIGURE 5.9

Quantum circuit to perform the Grover search.

twice): the first reflection of $|\Psi\rangle$ with respect to the state $|\alpha\rangle$ to get $O|\Psi\rangle$, and the second reflection of $O|\Psi\rangle$ with respect to the superposition state $|\Psi\rangle$ to get the resulting state $G|\Psi\rangle$, which is just the counterclockwise rotation of $|\Psi\rangle$ for θ.

The state $|\alpha\rangle$ is essentially the superposition of all other database entries $|x\rangle$ different from $|y\rangle$; in other words we can write:

$$|\alpha\rangle = \frac{1}{\sqrt{N-1}}\sum_{x \neq y}|x\rangle. \tag{5.28}$$

The superposition state, given by Eq. (5.26), can now be expressed in terms of Eq. (5.28) as follows:

$$|\Psi\rangle = \sqrt{1-\frac{1}{N}}|\alpha\rangle + \sqrt{\frac{1}{N}}|y\rangle. \tag{5.29}$$

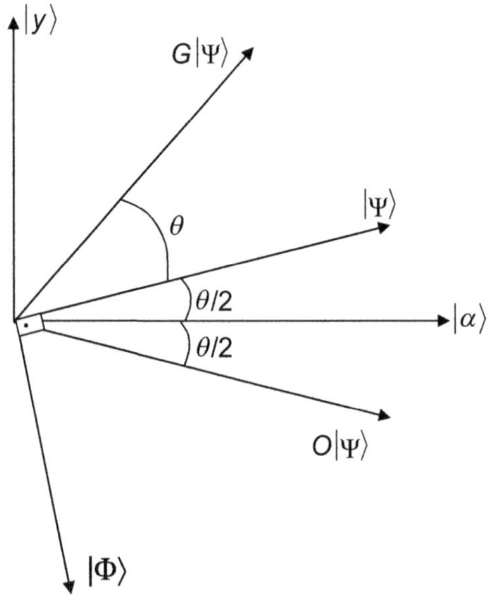

FIGURE 5.10

Geometric interpretation of the Grover search algorithm.

By substituting $\sqrt{1-1/N}=\cos(\theta/2)$, we can rewrite the superposition state $|\Psi\rangle$ as:

$$|\Psi\rangle=\cos\left(\frac{\theta}{2}\right)|\alpha\rangle+\sin\left(\frac{\theta}{2}\right)|y\rangle. \tag{5.30}$$

Let us now apply the oracle operator on $\lambda|\alpha\rangle+\mu|y\rangle$ to obtain:

$$O_y(\lambda|\alpha\rangle+\mu|y\rangle)=\lambda(-1)^{\delta_{\alpha y}}|\alpha\rangle+\mu(-1)^{\delta_{yy}}\left|y\right\rangle=\lambda|\alpha\rangle-\mu|y\rangle, \tag{5.31}$$

proving therefore that the action of operator O is a reflection with respect to the direction of $|\alpha\rangle$. The Grover operator action on state $\lambda|\Psi\rangle+\mu|\Phi\rangle$ is given by:

$$\begin{aligned}(2|\Psi\rangle\langle\Psi|-I)(\lambda|\Psi\rangle+\mu|\Phi\rangle)&=2\lambda|\Psi\rangle\underbrace{\langle\Psi|\Psi\rangle}_{=1}-\lambda|\Psi\rangle+2\mu|\Psi\rangle\underbrace{\langle\Psi|\Phi\rangle}_{=0}-\mu|\Phi\rangle\\&=\lambda|\Psi\rangle-\mu|\Phi\rangle,\end{aligned} \tag{5.32}$$

which is just the reflection with respect to the direction of $|\Psi\rangle$. Therefore the product of two reflections is a rotation, and from Fig. 5.10 we conclude that the angle between $G|\Psi\rangle$ and $|\alpha\rangle$ is $\theta+\theta/2=3\theta/2$ so that we can write:

$$G|\Psi\rangle=\cos\left(\frac{3\theta}{2}\right)|\alpha\rangle+\sin\left(\frac{3\theta}{2}\right)|y\rangle. \tag{5.33}$$

In similar fashion, since the angle between $G|\Psi\rangle$ and $|\Psi\rangle$ is θ, $G^2|\Psi\rangle$ can be obtained from $|\Psi\rangle$ by rotation by an angle 2θ or equivalently from $|\alpha\rangle$ by rotation by $(2\theta+\theta/2)$. After k iterations, $G^k|\Psi\rangle$ is obtained from $|\alpha\rangle$ by rotation by $(k\theta+\theta/2)=(2k+1)\theta/2$ and we can write:

$$G^k|\Psi\rangle=\cos\left(\frac{(2k+1)\theta}{2}\right)|\alpha\rangle+\sin\left(\frac{(2k+1)\theta}{2}\right)|y\rangle. \tag{5.34}$$

The optimum number of iterations can be obtained by aligning $G^k|\Psi\rangle$ with $|y\rangle$ or equivalently by setting:

$$\cos\left(\frac{(2k+1)\theta}{2}\right)=0\Leftrightarrow\cos k\theta\cos\frac{\theta}{2}-\sin k\theta\sin\frac{\theta}{2}=0. \tag{5.35}$$

By using the substitution $\cos(\theta/2)=\sqrt{1-1/N}$, we obtain that $\cos k\theta=1/\sqrt{N}$, and the optimum number of iteration steps is then:

$$k_o=\left[\frac{1}{\theta}\cos^{-1}\sqrt{\frac{1}{N}}\right]+1. \tag{5.36}$$

If $N>>1$ we can use Taylor expansion and keep the first term to obtain $\theta\simeq 2/\sqrt{N}$ and by substituting it in Eq. (5.36) we obtain:

$$k_o\simeq\frac{\sqrt{N}}{2}\cos^{-1}\sqrt{\frac{1}{N}}\simeq\frac{\pi\sqrt{N}}{4}. \tag{5.37}$$

Therefore to complete successfully the Grover search, we need to apply the oracle operator $\sim\sqrt{N}$ times. The quantum circuit to perform a Grover search for $n=3$ is shown in Fig. 5.11.

After the optimum number of steps k_o, the angle between $G^{ko}|\Psi\rangle$ and $|y\rangle$ is less than $\theta/2$, so that the probability of error is less than $O(1/N)$. It can be shown that the Grover algorithm is optimal. By taking all gates into account, the total number of operations in the Grover search algorithm is proportional to $\sqrt{N}\log N$.

5.4 QUANTUM FOURIER TRANSFORM

In this section we are concerned with quantum FT. The unitary FT U_{FT} can be represented by the following matrix elements:

$$\langle y|U_{\mathrm{FT}}|x\rangle = (U_{\mathrm{FT}})_{yx} = \frac{1}{2^{n/2}} e^{j2\pi xy/2^n}; \quad |x\rangle = |x_{n-1}\cdots x_0\rangle, \quad x_i = 0, 1. \tag{5.38}$$

We can show that the U_{FT} operator is unitary as follows:

$$\sum_{y=0}^{2^n-1}\left(U_{\mathrm{FT}}^{\dagger}\right)_{x'y}(U_{\mathrm{FT}})_{yx} = \sum_{y=0}^{2^n-1}\left(U_{\mathrm{FT}}^{*}\right)_{x'y}(U_{\mathrm{FT}})_{yx} = \frac{1}{2^{n/2}}\sum_{y=0}^{2^n-1} e^{j2\pi(x-x')y/2^n} = \delta_{x'x}. \tag{5.39}$$

The quantum FT notation is provided in Fig. 5.12.

Let $|\Psi\rangle$ be a linear superposition of vectors $|x\rangle$:

$$|\Psi\rangle = \sum_{x=0}^{2^n-1} f(x)|x\rangle, \tag{5.40}$$

where $f(x) = \langle x|\Psi\rangle$ is a projection of $|\Psi\rangle$ along $|x\rangle$, while the probability amplitudes satisfy the following normalization condition: $\sum_{x=0}^{2^n-1}|f(x)|^2 = 1$. By applying the U_{FT} operator on the superposition state we obtain:

$$|\Phi\rangle = U_{\mathrm{FT}}|\Psi\rangle = U_{\mathrm{FT}}\sum_{x=0}^{2^n-1}\langle x|\Psi\rangle|x\rangle. \tag{5.41}$$

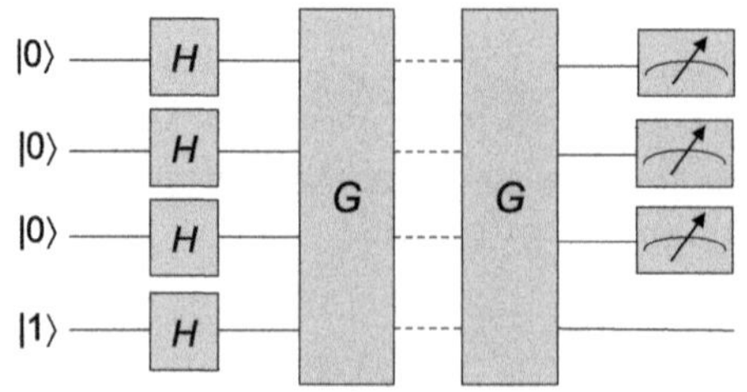

FIGURE 5.11

Quantum circuit to perform the Grover search for $n=3$.

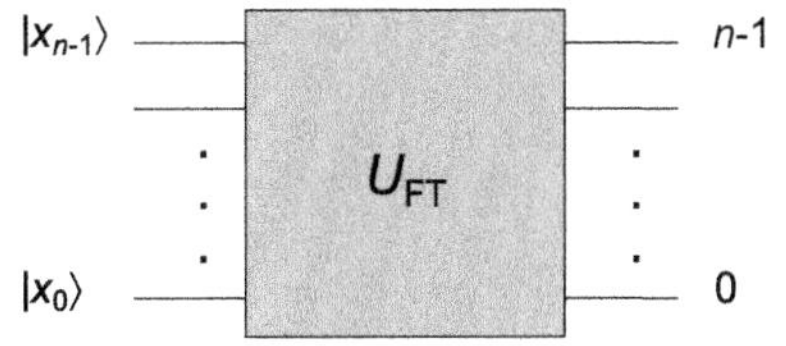

FIGURE 5.12

Notation of the quantum Fourier transform circuit.

The probability amplitude of finding the state $|y\rangle$ of CB at the output of the FT block is given by:

$$\langle y|\Phi\rangle = \sum_{x=0}^{2^n-1}\langle y|U_{\mathrm{FT}}|x\rangle\langle x|\Psi\rangle = \frac{1}{2^{n/2}}\sum_{x=0}^{2^n-1} e^{j2\pi xy/2^n} f(x), \tag{5.42}$$

which is just the FT of $f(x)$:

$$\widetilde{f}(y) = \frac{1}{2^{n/2}}\sum_{x=0}^{2^n-1} f(x)e^{j2\pi xy/2^n}. \tag{5.43}$$

The action of U_{FT} on $|x\rangle$ in operator notation is given by:

$$U_{\mathrm{FT}}|x\rangle = \sum_{y=0}^{2^n-1}|y\rangle\langle y|U_{\mathrm{FT}}|x\rangle = \frac{1}{2^{n/2}}\sum_{y=0}^{2^n-1} e^{j2\pi xy/2^n}|y\rangle, \tag{5.44}$$

where:

$$x = x_0 + 2x_1 + 2^2x_2 + \cdots + 2^{n-1}x_{n-1} \quad y = y_0 + 2y_1 + 2^2y_2 + \cdots + 2^{n-1}y_{n-1}.$$

For $n = 3$, $N = 2^n = 8$ we can represent xy as:

$$\frac{xy}{8} = y_0\left(\frac{x_2}{2}+\frac{x_1}{4}+\frac{x_0}{8}\right) + y_1\left(\frac{x_1}{2}+\frac{x_0}{4}\right) + y_2\frac{x_0}{2}. \tag{5.45}$$

We also know that:

$$0.x_p x_{p-1}\cdots x_1 x_0 = \frac{x_p}{2}+\frac{x_{p-1}}{2^2}+\cdots+\frac{x_0}{2^p}, \tag{5.46}$$

and $\exp(j2\pi p) = 1$, when p is an integer. By representing $|y\rangle = |y_{n-1}\cdots y_0\rangle$, from Eq. (5.44) we obtain:

$$\begin{aligned} U_{\mathrm{FT}}|x\rangle &= \frac{1}{2^{n/2}}\sum_{y_0,\cdots,y_{n-1}} e^{j2\pi y_{n-1}.x_0}\cdots e^{j2\pi y_0.x_{n-1}\cdots x_0}|y_{n-1}\cdots y_0\rangle \\ &= \frac{1}{2^{n/2}}\left(\sum_{y_{n-1}} e^{j2\pi y_{n-1}.x_0}|y_{n-1}\rangle\right)\cdots\left(\sum_{y_1} e^{j2\pi y_1.x_{n-2}\cdots x_0}|y_1\rangle\right)\left(\sum_{y_0} e^{j2\pi y_0.x_{n-1}\cdots x_0}|y_0\rangle\right). \end{aligned} \tag{5.47}$$

By representing Eq. (5.47) in developed form we obtain:

$$U_{FT}|x\rangle = \frac{1}{2^{n/2}}\left(|0_{n-1}\rangle + e^{j2\pi .x_0}|1_{n-1}\rangle\right)\cdots\left(|0_1\rangle + e^{j2\pi .x_{n-2}\cdots x_0}|1_1\rangle\right)\left(|0_0\rangle + e^{j2\pi .x_{n-1}\cdots x_0}|1_0\rangle\right). \tag{5.48}$$

By following a similar procedure, for $n=2$, $N=2^n=4$, the application of U_{FT} on $|x\rangle$ yields to:

$$\begin{aligned} U_{FT}|x\rangle = U_{FT}|x_1x_0\rangle &= \frac{1}{2}\left(|0_1\rangle + e^{j2\pi .x_0}|1_1\rangle\right)\left(|0_0\rangle + e^{j2\pi .x_1x_0}|1_0\rangle\right) \\ &= \frac{1}{2}\left(|00\rangle + e^{j2\pi .x_1x_0}|01\rangle + e^{j2\pi .x_0}|10\rangle + e^{j2\pi(.x_1x_0 + .x_0)}|11\rangle\right). \end{aligned} \tag{5.49}$$

The quantum circuit to perform FT for $n=3$ is shown in Fig. 5.13.

Since $H|0\rangle_2 = (|0\rangle_2 + |1\rangle_2)/\sqrt{2}, H|1\rangle_2 = (|0\rangle_2 - |1\rangle_2)/\sqrt{2}$, the action of the Hadamard gate on qubit $|x_2\rangle$ (Fig. 5.13) is given by:

$$H|x\rangle_2 = \frac{1}{\sqrt{2}}\left(|0\rangle_2 + e^{j2\pi .x_2}|1\rangle_2\right). \tag{5.50}$$

In Fig. 5.13 with R_d we denoted the *discrete phase shift gate*, whose action in matrix representation is given by:

$$R_d \doteq \begin{bmatrix} 1 & 0 \\ 0 & e^{j\pi/2^d} \end{bmatrix}. \tag{5.51}$$

Let the $C_{ij}(R_d)$ denote the control-R_d gate, where i is the control qubit and j is the target qubit. If $x_1 = 0$, the action of $C_{12}(R_d)$ is given by:

$$C_{12}(R_1)H|x_2\rangle = \frac{1}{\sqrt{2}}\left(|0\rangle_2 + e^{j2\pi .x_2}|1\rangle_2\right). \tag{5.52}$$

On the other hand, if $x_1 = 1$, then the action of $C_{12}(R_d)$ is given by:

$$C_{12}(R_1)H|x_2\rangle = \frac{1}{\sqrt{2}}\left(|0\rangle_2 + e^{j\pi/2}e^{j2\pi .x_2}|1\rangle_2\right). \tag{5.53}$$

Eqs. (5.52) and (5.53) can be jointly combined into:

$$C_{12}(R_1)H|x_2\rangle = \frac{1}{\sqrt{2}}\left(|0\rangle_2 + e^{j2\pi .x_2x_1}|1\rangle_2\right). \tag{5.54}$$

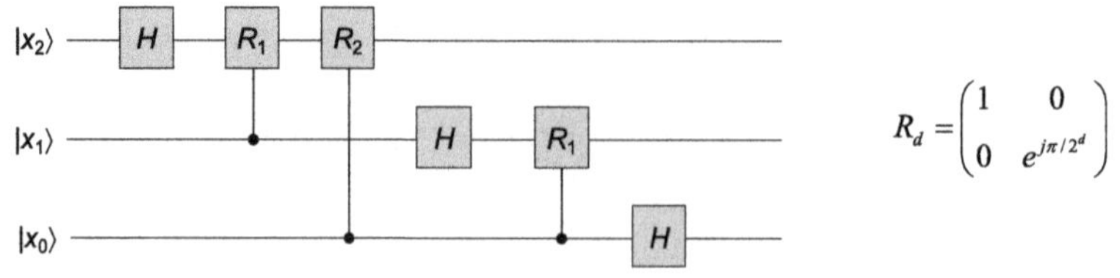

FIGURE 5.13

Quantum circuit to perform the Fourier transform for $n=3$.

To determine the final state in the top branch of Fig. 5.13, we further apply $C_{02}(R_2)$ on $C_{12}(R_1)|x_2\rangle$ and obtain:

$$C_{02}(R_2)C_{12}(R_1)H|x_2\rangle = \frac{1}{\sqrt{2}}\left(|0\rangle_2 + e^{j2\pi.x_2x_1x_0}|1\rangle_2\right). \tag{5.55}$$

The final state of the FT circuit from Fig. 5.13 is then given by:

$$|\Psi'\rangle = \frac{1}{\sqrt{8}}\left(|0\rangle_0 + e^{j2\pi.x_0}|1\rangle_0\right)\left(|0\rangle_1 + e^{j2\pi.x_1x_0}|1\rangle_1\right)\left(|0\rangle_2 + e^{j2\pi.x_2x_1x_0}|1\rangle_2\right). \tag{5.56}$$

We can see that the qubits are put in the wrong order so that the additional swap gates are needed. Another option would be to redefine the CB states as follows: $|x\rangle = |x_0x_1\cdots x_{n-1}\rangle$.

The quantum circuit to perform FT for arbitrary n can be obtained by generalization of the corresponding circuit from Fig. 5.13, which is shown in Fig. 5.14.

Regarding the complexity of the quantum FT circuit, the number of required Hadamard gates is n, while the number of controlled gates is given by:

$$n + (n-1) + \cdots + 1 \simeq \frac{n^2}{2}. \tag{5.57}$$

Therefore the overall complexity of the FT computation circuit is $O(n^2)$.

5.5 THE PERIOD OF A FUNCTION AND SHOR'S FACTORING ALGORITHM

Shor's factorization algorithm [14, 15] is based on the possibility of determining in polynomial time the period of a function $f(x)$. The function $f(x)$ in Shor's algorithm is $f(x) = b^x \bmod N$. This function has a period r: $f(x) = f(x + r)$, $x \in \{0,1,\ldots,2^{n-1}\}$. For the algorithm to be successful it is important that $2^n > N^2$. A classical algorithm requires $O(N)$ elementary operations. The corresponding quantum algorithm requires $O(n^3)$ elementary operations. The function $b^x \bmod N$ behaves as random noise over a period in a classical case. Let us first describe the algorithm to find the period of this function.

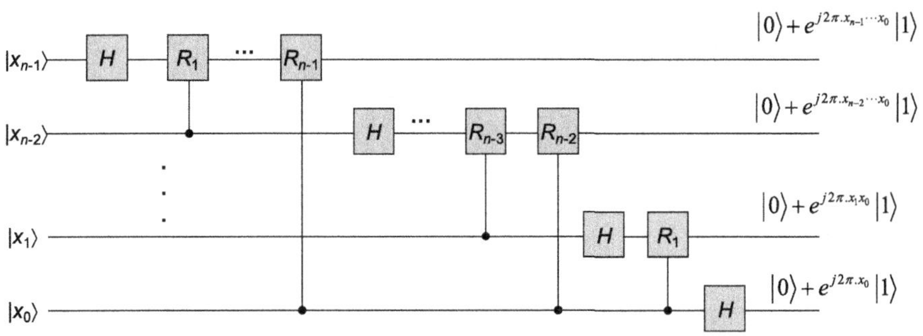

FIGURE 5.14

Quantum circuit to perform the Fourier transform for arbitrary n.

5.5.1 Period of Function $f(x) = b^x \bmod N$

To efficiently describe this algorithm we use an interpretation similar to that provided in [1]. The initial state of $n + m$ qubits is given by:

$$|\Phi\rangle = \frac{1}{2^{n/2}} \left(\sum_{x=0}^{2^n-1} |x\rangle \right) \otimes |0\cdots0\rangle. \tag{5.58}$$

By applying the function operator U_f on initial state $|\Phi\rangle$, the final state is:

$$|\Phi_f\rangle = U_f|\Phi\rangle = \frac{1}{2^{n/2}} \sum_{x=0}^{2^n-1} |x, f(x)\rangle. \tag{5.59}$$

If we now perform the measurement of the output register and obtain the result f_0, the corresponding final state of the first register will be "the state vector collapse":

$$|\Psi_0\rangle = \frac{1}{\mathcal{N}} \sum_{x:f(x)=f_0} |x\rangle, \tag{5.60}$$

where the summation is performed over all x's for which $f(x) = f_0$, and $\mathcal{N}$ is the normalization factor. Since $f(x + r) = f(x)$, we expect that pr, where p is an integer, will be periodic as well: $f(x + pr) = f(x)$. Let x_0 denote the minimum x to satisfy $f(x_0) = f_0$, then Eq. (5.60) can be rewritten as follows:

$$|\Psi_0\rangle = \frac{1}{\sqrt{K}} \sum_{k=0}^{K-1} |x_0 + kr\rangle; K = [2^n/r] \text{or} [2^n/r] + 1, \tag{5.61}$$

where with $[z]$ we denote the integer part of z. The schematic description of the quantum circuit to calculate the period is shown in Fig. 5.15.

The state of the input register is an incoherent superposition of vectors $|\Psi_i\rangle$:

$$|\Psi_i\rangle = \frac{1}{\sqrt{K_i}} \sum_{k=0}^{K_i-1} |x_i + kr\rangle, \tag{5.62}$$

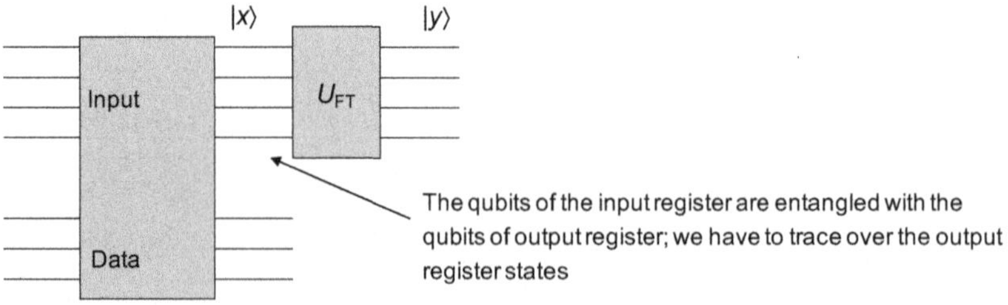

FIGURE 5.15

Quantum circuit to determine the period of $f(x)$.

where the minimum x to satisfy $f(x_i) = f_i$ is denoted by x_i. The input register density operator can be written, based on Eq. (5.59), as:

$$\rho_{\text{tot}} = |\Psi_f\rangle\langle\Psi_f| = \frac{1}{2^n}\sum_{x,z}|x,f(x)\rangle\langle z,f(z)|, \tag{5.63}$$

and the density operator of the input register can be obtained by tracing over the output register states:

$$\rho_{\text{in}} = \text{Tr}_{\text{out}}\rho_{\text{tot}} = \frac{1}{2^n}\sum_{x,z}|x\rangle\langle z|\langle f(x)|f(z)\rangle. \tag{5.64}$$

As an illustration, let the function $f(x)$ take the value f_0 exactly N_0 times and the value f_1 exactly N_1 times so that $N_0 + N_1 = 2^n$, then the following is valid:

$$\langle f(x)|f(z)\rangle = \begin{cases} 0, & f(x) \neq f(z) \\ 1, & f(x) = f(z) \end{cases}. \tag{5.65}$$

Upon substitution of Eq. (5.65) into Eq. (5.64), we obtain:

$$\rho_{\text{in}} = \frac{1}{2^n}\left(\sum_{x,z:f(x)=f(z)=f_0}|x\rangle\langle z| + \sum_{x,z:f(x)=f(z)=f_1}|x\rangle\langle z|\right). \tag{5.66}$$

Therefore the density operator ρ_{in} represents the incoherent superposition of states $|\Psi_0\rangle$ and $|\Psi_1\rangle$, each occurring with probability $p_i = N_0/2^n$ $(i = 0,1)$, defined as:

$$|\Psi_0\rangle = \frac{1}{\sqrt{N_0}}\sum_{x:f(x)=f_0}|x\rangle \quad |\Psi_1\rangle = \frac{1}{\sqrt{N_1}}\sum_{x:f(x)=f_1}|x\rangle. \tag{5.67}$$

Based on Eq. (5.61), we can rewrite Eq. (5.66) as follows:

$$\rho_{\text{in}} = \frac{1}{2^n}\sum_{i=0}^{r-1}\sum_{k_i,k_j=0}^{K_i-1}|x_i + k_i r\rangle\langle x_i + k_j r| \tag{5.68}$$

To determine the probability amplitude of finding the CB state $|y\rangle$ we can employ the FT of state:

$$|\Psi\rangle = \sum_{x=0}^{2^n-1}f(x)|x\rangle, \; f(x) = \langle x|\Psi\rangle, \; \sum_{x=0}^{2^n-1}|f(x)|^2 = 1, \tag{5.69}$$

which becomes $|\Phi_0\rangle$ for $f(x) = 1/\sqrt{K}, \; x = x_0 + kr$. Since $|\Phi_0\rangle = U_{\text{FT}}|\Psi_0\rangle$, the probability amplitude of finding the CB state $|y\rangle$ is given by:

$$a(\Phi_0 \to y) = \langle y|\Phi_0\rangle = \frac{1}{2^{n/2}}\frac{1}{\sqrt{K}}\sum_{k=0}^{K-1}e^{j2\pi y(x_0+kr)/2^n}, \tag{5.70}$$

and the corresponding probability is:

$$p(y) = |a(\Phi_0 \to y)|^2 = \frac{1}{2^n K}\left|\sum_{k=0}^{K-1} e^{j2\pi y(x_0+kr)/2^n}\right|^2. \tag{5.71}$$

By using the geometric series formula we obtain:

$$\sum_{k=0}^{K-1} e^{j2\pi ykr/2^n} = \frac{1-e^{j2\pi yKr/2^n}}{1-e^{j2\pi yr/2^n}} = e^{j\pi(K-1)r/2^n}\frac{\sin(\pi yKr/2^n)}{\sin(\pi yr/2^n)}. \tag{5.72}$$

By using Eq. (5.72), the probability $p(y)$ for $2^n/r = K$ can be written as:

$$p(y) = \frac{1}{2^n K}\frac{\sin^2(\pi y)}{\sin^2(\pi y/K)} = \begin{cases} 1/r, & y = iK \\ 0, & \text{otherwise} \end{cases}. \tag{5.73}$$

By substituting $y_i = i2^n/r + \delta_i$, we can rewrite Eq. (5.73) as:

$$p(y_i) = \frac{1}{2^n K}\frac{\sin^2(\pi\delta_i Kr/2^n)}{\sin^2(\pi\delta_i r/2^n)}. \tag{5.74}$$

It can be shown that δ_i satisfies the following inequality:

$$|\delta_i| = \left|y_i - i\frac{2^n}{r}\right| < 1/2, \tag{5.75}$$

so that the probability $p(y)$ can be lower bounded by:

$$p(y_i) \geq \frac{4}{\pi^2}\frac{K}{2^n} \simeq \frac{4}{\pi^2}\frac{1}{r}. \tag{5.76}$$

It can be shown that the function $p(y)$ has sharp maxima when the value of y is close to $i2^n/r$. The value i/r can be determined by developing $y/2^n$ in continued fractions by repeatedly applying the split and invert method. For example, 5/13 can be represented by using continued fractions expansion as:

$$\frac{1}{2 + \frac{1}{1 + \frac{1}{1+\frac{1}{2}}}}$$

The i/r is then obtained as an irreducible fraction i_0/r_0. If i and r do not have a common factor, then we can immediately obtain $r = r_0$. The probability that two large numbers do not have a common factor is larger than 60%. Therefore with probability 0.4×0.6 (0.4 comes from the probability of finding y_i close to $i2^n/r$) the foregoing protocol will directly give the period $r = r_0$. If $f(x) \neq f(x + r_0)$, we try to see whether the periodicity condition $f(x) = f(x + kr_0)$ can be satisfied with small multiples of r_0 (such as $2r_0$, $3r_0$, ...). If all these trials are unsuccessful, it means $|\delta_i| >$ 1/2 and the protocol needs to be repeated. This procedure requires $O(n^3)$ elementary operations: $O(n^2)$ for FT and $O(n)$ for the calculation of b^x.

Once we determine the period r, we can crack the Rivest–Shamir–Adleman (RSA) encryption protocol. Before describing how to do so, let us provide a brief overview of the RSA encryption protocol.

5.5.2 Solving the RSA Encryption Protocol

The RSA encryption protocol is summarized in Fig. 5.16. Bob chooses two primes p and q, and determines $N = pq$ and a number c that does not have a common divisor with the product $(p - 1)(q - 1)$. He further calculates d, the inverse for $\mathrm{mod}(p - 1)(q - 1)$ multiplication, by:

$$cd \equiv 1 \mathrm{mod}(p-1)(q-1). \tag{5.77}$$

By a nonsecure path, by using a classical communication channel, he sends Alice the values of N and c (but not p and q of course). Alice then sends Bob an encrypted message, which must be represented as a number $a < N$. Alice calculates:

$$b \equiv a^c \mathrm{mod} N, \tag{5.78}$$

and sends the result b over the nonsecure channel. Bob upon receiving b calculates:

$$b^d \mathrm{mod} N = a, \tag{5.79}$$

and recovers the encrypted number.

Example. Let $p = 3$ and $q = 7$, meaning that $N = 21$ and $(p - 1)(q - 1) = 12$. We further choose $c = 5$, which does not have a common divisor with 12. Next, d is determined as the inverse for mod 12 as 5, namely from c^d mod 12 = 25 mod 12 = 1. Let us assume that Alice wants to send $a = 4$. She first computes a^c mod 21 = 4^5 mod 21 = (1024 = 21 × 48 + 16) = 16 mod 21 = b, and sends it to Bob. Bob computes b^5 mod 21 = 16^5 mod 21 = (49,932 × 21 + 4)mod 21 = 4 mod 21 and therefore recovers the encrypted number a.

Cracking the RSA protocol. After this short overview of RSA protocol, we now describe how to crack it. Eve obtains b, N, c from the public channel, and calculates d' from:

$$cd' \equiv \mathrm{mod} r. \tag{5.80}$$

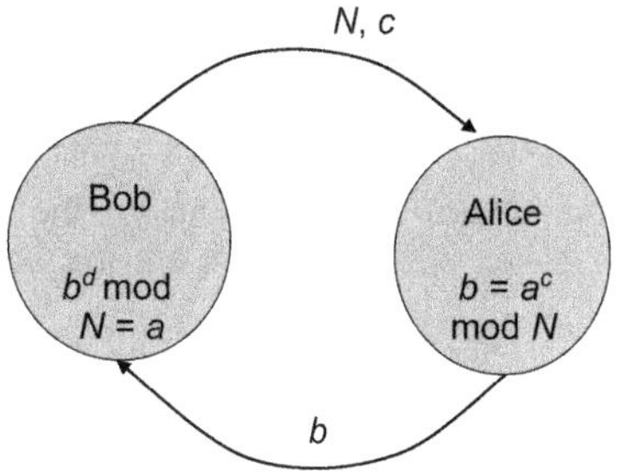

FIGURE 5.16

Illustration of Rivest–Shamir–Adleman encryption protocol.

She further calculates b^{dl} mod N as follows:

$$b^{d'}\text{mod}N = a^{cd'}\text{mod}N = a^{1+mr}\text{mod}N = a(a^r)^m\text{mod}N = a\text{mod}N, \quad a^r \equiv 1\text{mod}N. \tag{5.81}$$

5.5.3 Shor's Factorization Algorithm

To *factorize* N we must write $a^r - 1$ as:

$$a^r - 1 = \left(a^{r/2} - 1\right)\left(a^{r/2} + 1\right), \tag{5.82}$$

while $a^{r/2} \neq \pm 1$ mod N. If the two factors are integers (and $a^{r/2} \neq \pm 1$ mod N), then the product of integers $(a^{r/2} - 1)$ $(a^{r/2} + 1)$ is divisible by $N = pq$. The values of p and q are obtained by:

$$p = \gcd\left(N, a^{r/2} - 1\right) \text{ and } q = \gcd\left(N, a^{r/2} + 1\right). \tag{5.83}$$

From fundamental arithmetic theorem we know that any positive integer N can be factorized as:

$$N = \prod_{i=1}^{k} p_i^{l_i}, \tag{5.84}$$

where p_i are the primes and l_i are their corresponding powers. To factor N we can generate an integer b between 1 and $N - 1$ randomly, and apply the classical Euclidean algorithm to determine if $\gcd(b,N) = 1$, which requires $O[(\log_2 N)^3]$ (classical) steps. If $\gcd(b,N) > 1$, then the b is a nontrivial factor of N. If, on the other hand, $\gcd(b,N) = 1$, we choose the order of b to be r, that is b^r mod $N = 1$ and N divides $b^r - 1$. Furthermore, if r is even we can factor $b^r - 1$ as $(b^{r/2} - 1)(b^{r/2} + 1)$, and we can find the factors of N by Eq. (5.83). Based on the previous discussion, we can summarize *Shor's factorization algorithm* as follows:

1. Randomly generate an integer b between 0 and $N - 1$, determine the $\gcd(b,N)$. If $\gcd(b,N) > 1$, then b is a nontrivial factor of N; otherwise go to next step.
2. Prepare two quantum registers initialized to the $|0\rangle$state as follows:

$$|\Psi_{\text{in}}\rangle = |0\rangle^{\otimes m}|0\rangle^{\otimes n}, \quad n = \lceil \log_2 N \rceil. \tag{5.85}$$

3. By applying the m-qubit Hadamard transform on the first register, create the superposition state:

$$|\Psi_1\rangle = \frac{1}{2^{m/2}} \sum_{x=0}^{2^m - 1} |x\rangle|0\rangle^{\otimes n}. \tag{5.86}$$

4. Apply the unitary transformation performing the following mapping:

$$|\Psi_2\rangle = C\sum_{x=0}^{2^m-1}|x\rangle|b^x \mathrm{mod} N\rangle, \tag{5.87}$$

where C is a normalization constant.

5. Perform the measurement on the second register in CB to obtain f_0, while leaving the first register in uniform superposition of all states for which $f(x) = f_0$:

$$|\Psi_3\rangle = C'\sum_{k=0}^{K-1}|x_0 + kr\rangle|f_0\rangle, \quad K = \lceil 2^m / r\rceil. \tag{5.88}$$

6. Apply the inverse quantum FT on the first register to obtain:

$$|\Psi_4\rangle = C''\sum_{l=0}^{2^m-1}\sum_{k=0}^{K-1} e^{j2\pi(x_0+kr)l/2^m}|l\rangle|f_0\rangle. \tag{5.89}$$

7. Perform the measurement on the first register to obtain i/r, as described in the discussion for inequality Eq. (5.75).
8. Apply the continued fractions expansion algorithm to determine an irreducible fraction i_0/r_0, followed by the procedure described in the paragraph following Eq. (5.76) that determines the period r.
9. If r is even and $b^{r/2} \neq 1 \bmod N$, then compute $\gcd(b^{r/2} - 1, N) = p$ and $\gcd(b^{r/2} + 1, N) = q$ to check if either p or q is a nontrivial factor.

Repeat this procedure until full prime factorization in the form of Eq. (5.84) is obtained.

5.6 SIMON'S ALGORITHM

Shor's factorization algorithm requires determination of the period of a function. An alternative to this algorithm is Simon's algorithm [16], which represents a particular instance of a more general problem known as the *Abelian hidden subgroup* problem. Let us observe the function f from $(F_2)^N$ to itself, represented by the following unitary map:

$$|x\rangle|0\rangle \xrightarrow{U_f} |x\rangle|f(x)\rangle \quad \forall x. \tag{5.90}$$

Let H' be a subgroup of $(F_2)^N$ for which the function f is a unique constant for every right coset of H'. Simon's algorithm performs $O(N)$ evaluations of the oracle f in combination with classical computation to determine the *generators* of subgroup H'. Simon's algorithm can be formulated as follows:

1. *Initialization.* Prepare two N-qubit quantum registers initialized to the $|0\rangle$state:

$$|\Psi_{\text{in}}\rangle = |0\rangle^{\otimes N}|0\rangle^{\otimes N}. \tag{5.91}$$

2. *Superposition state.* By applying the N-qubit Hadamard transform on the first register, create the superposition state:

$$|\Psi_1\rangle = \frac{1}{2^{N/2}}\sum_{x\in F_2^N}|x\rangle|0\rangle^{\otimes N}. \tag{5.92}$$

3. *Unitary operation U_f application.* Upon applying the U_f operator on the second register we obtain:

$$|\Psi_2\rangle = \frac{1}{2^{N/2}}\sum_{x\in F_2^N}|x\rangle|f(x)\rangle. \tag{5.93}$$

4. *Measurement step.* Perform the measurement on the second register to obtain a value y, which yields to the coset of the hidden subgroup in the first register:

$$|\Psi_3\rangle = \frac{1}{|H'|^{1/2}}\sum_{x:f(x)=y}|x\rangle|y\rangle. \tag{5.94}$$

5. *Hadamard transform followed by measurement.* Apply the N-qubit Hadamard transform to the first register, followed by the measurement on the first register in CB.
6. Repeat steps (1)–(4) approximately N times. The result of this algorithm will be the orthogonal complement of H' with respect to the scalar product in $(F_2)^N$.
7. Solve the system of linear equations by *Gauss elimination* to obtain the kernel of a square matrix, which contains the generators of H'.

5.7 QUANTUM PHASE ESTIMATION

In quantum phase estimation (QPE) we are concerned with determining the eigenvalues of a certain unitary operator U:

$$U|\psi\rangle = \omega|\psi\rangle = e^{j2\pi\theta}|\psi\rangle, \tag{5.95}$$

where $|\psi\rangle$ is an eigenvector, while ω is a corresponding eigenvalue. Given that the operator U is unitary, all eigenvalues are located on a unit circle and can be represented as $\omega = \exp(j2\pi\theta)$. This phase (angle) θ is what we are going to estimate in the QPE algorithm with some finite precision. The corresponding quantum circuit to perform the QPE is provided in Fig. 5.17. We use N ancilla qubits in the top register, and the input to the bottom register is the eigenvector whose eigenvalue we are concerned about. We apply the controlled-U operation for different durations,

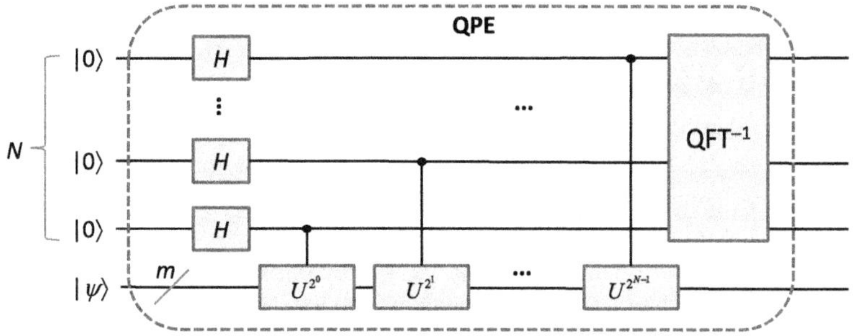

FIGURE 5.17

Quantum circuit to perform the quantum phase estimation.

ranging from 2^0 to $2^N - 1$ time intervals. With the help of N-Hadamard gates we create the superposition state $H^{\otimes N}|0\rangle^{\otimes N} = 2^{-N/2} \sum_{x\in\{0,1\}^N} |x\rangle$. We use the superposition states to perform controlled-U operations. At the end of this process, the overall state just before the inverse quantum FT (QFT^{-1}) gate will be:

$$2^{-N/2}\sum_{x=0}^{2^N-1}|x\rangle\underbrace{(U^x|\psi\rangle)}_{e^{j2\pi\theta x}|\psi\rangle} = 2^{-N/2}\sum_{x=0}^{2^N-1}e^{j2\pi\theta x}|x\rangle|\psi\rangle = \left(2^{-N/2}\sum_{x=0}^{2^N-1}e^{j2\pi\theta x}|x\rangle\right)|\psi\rangle. \tag{5.96}$$

By performing the QFT^{-1} on the top register, based on Eq. (5.44), we obtain:

$$U_{\text{FT}}^{\dagger}\left(2^{-N/2}\sum_{x=0}^{2^N-1}e^{j2\pi\theta x}|x\rangle\right) \overset{U_{\text{FT}}^{\dagger}|x\rangle = \frac{1}{2^{N/2}}\sum_{y=0}^{2^N-1}e^{-j2\pi xy/2^N}|y\rangle}{=} 2^{-N}\sum_{x,y=0}^{2^N-1}e^{j2\pi x(\theta-y/2^N)}|y\rangle. \tag{5.97}$$

The probability of finding the subsystem in state $|y\rangle$ is given by:

$$\begin{aligned}\Pr(y) &= \left|2^{-N}\sum_{x=0}^{2^N-1}e^{j2\pi x(\theta-y/2^N)}\right|^2 = 2^{-2N}\left|\sum_{x=0}^{2^N-1}\underbrace{\left[e^{j2\pi(\theta-y/2^N)}\right]}_{r}^{x}\right|^2 \\ &= 2^{-2N}\left|\underbrace{\sum_{x=0}^{2^N-1}r^x}_{\frac{r^{2^N}-1}{r-1}}\right|^2 = 2^{-2N}\left|\frac{r^{2^N}-1}{r-1}\right|^2.\end{aligned} \tag{5.98}$$

As shown in [17], the $\Pr(y) > 0.4$ indicating that the probability of measuring θ with N-bits precision is greater than 0.4.

5.8 CLASSICAL/QUANTUM COMPUTING COMPLEXITIES AND TURING MACHINES

Quantum algorithms have raised a number of questions on the accuracy of statements from classical algorithm theory when algorithm complexity is a subject of interest. Namely, some problems that appear to be "intractable" (the complexity is exponential in the number of bits n) for classical computing can be solved polynomially in n by means of quantum computers. For example, if prime factorization is an intractable problem, which was suggested by practice but not proved, then Shor's algorithm discussed earlier contradicts this observation, as it can represent a composite number in terms of primes in a number of steps that is polynomial in n. The key advantage of quantum algorithms is that they can explore all branches of a nondeterministic algorithm in parallel, through the concept of quantum parallelism.

The classical algorithms can be classified as *effective* when the number of steps is a *polynomial* function of size n. The computational complexity of these algorithms is typically denoted as **P**. The class of problems that are solvable *nondeterministically* in *polynomial* time are called **NP**. The subclass of these problems, which are the most difficult ones, are **NP-complete** problems. For example, the traveling salesman problem belongs to this subclass. If one these problems can be solved efficiently, then all of them can be solved. The class of problems that can be solved with the amount of *memory polynomial* in input size is called **PSPACE**. Furthermore, the class of problems that can be solved with high probability with the help of a random generator is known as **BPP**, originating from "bounded error probability polynomial in time." Finally, the class of problems that can be solved in polynomial time, if sums of exponentially many contributors are themselves computable in polynomial time, is denoted as $\mathbf{P^{\#P}}$. These different classes of problems can be related as follows:

$$\mathbf{P} \subset \mathbf{BPP}, \mathbf{P} \subset \mathbf{NP} \subset \mathbf{P}^{\#\mathbf{P}} \subset \mathbf{PSPACE}. \tag{5.95}$$

Turing defined a class of machines, known as *Turing machines*, which can be used to study the complexity of a computational algorithm. In particular, there exist so-called *universal machines*, which can be used to simulate any other Turing machine. Interestingly enough, the Turing machine can be used to simulate all operations performed on a modern computer. We can formulate the *Church–Turing* thesis as follows: The class of functions that can be computed by a Turing machine corresponds exactly to the class of functions that one would naturally consider to be computable by an algorithm. This thesis essentially establishes the equivalence between rigorous mathematical description (Turing machine) and an intuitive concept. There exist problems that are not calculable; there is no known algorithm to solve them. For example, the halting problem of a Turing machine belongs [18,19] to this class. The Church–Turing thesis is also applicable to quantum algorithms. In the previous paragraph we said that an algorithm is effective if it can be solved in a polynomial number of steps. We also learned that Turing machines can be used to describe these effective algorithms. These two observations can be used to formulate the *strong* version of Church–Turing thesis: Any computational model can be simulated on a probabilistic Turing machine with a polynomial number of computational steps. There exist several types of Turing machines depending on the type of computation. The deterministic, probabilistic, and multitape Turing machines are described briefly next.

The *deterministic Turing machine* is described by an *alphabet* A, a set of *control states* Q, and by a *transition function* δ:

$$\delta: Q\times A\rightarrow Q\times A\times D, D=\{-1,0,1\}. \tag{5.96}$$

The elements of the alphabet are called *letters*, and by concatenating the letters we obtain *words*. The set D is related to the *read/write* head, with the elements -1, $+1$, and 0 denoting the movement of the head to the left, right, and standing, respectively. The deterministic Turing machine can be *defined* by (Q,A,δ,q_0,q_a,q_r), where the *state* of the machine is specified by $q\in Q$. In particular, $q_0,q_a,q_r\in Q$ denote the initial state, the accepting state, and rejecting state, respectively. *Configuration* of the Turing machine is given by $c=(q,x,y)$, where $x,y\in A'$, with A' being the set of all words obtained by concatenating the letters from A. The Turing machine has a *tape* (memory), specified by xy with x being the *scanning* (*reading*). A *computation* is a sequence of configurations beginning with an *initial configuration* c_0, until we reach the *halting configuration*. The computation halts after t computation steps when either one of the configurations does not have a successor or if its state is q_a or q_r.

The *probabilistic Turing machine* is more general, as the transition function assigns *probabilities* to possible operations:

$$\delta: Q\times A\times Q\times A\times D\rightarrow[0,1]. \tag{5.97}$$

In other words, a probabilistic Turing machine is a nondeterministic Turing machine that randomly selects possible transitions according to some probability distribution. As a consequence, the machine-state transitions can be described by a *stochastic matrix*. A given configuration is a successor configuration with probability δ. A terminal configuration can be computed from an initial configuration with a probability given by the product of probabilities of intermediate configurations leading to it by a particular computation, defined by a sequence of states. The deterministic Turing machine is just a particular instance of a probabilistic Turing machine.

An *m-type deterministic Turing machine* is characterized by m tapes, an alphabet A, a finite state of control states Q, and the following transition function:

$$\delta: Q\times A^m\rightarrow Q\times(A\times D)^{\times m}, \tag{5.98}$$

and it is defined by (Q,A,δ,q_0,q_a,q_r). A configuration of an m-type machine is given by $(q,x_1,y_1,\ldots,x_m,y_m)$, where q is the current state of the machine, $(x_i,y_i)\in A'\times A'$, and x_iy_i denotes the content of the ith type. The m-type machines are suitable for problems involving parallelism. If the computational time needed for a one-type machine is t, then the computation time needed for an m-type machine is $O(t^{1/2})$.

The *quantum Turing machine* is characterized by the following transition function:

$$\delta: Q \times A \times Q \times A \times D \rightarrow C, \tag{5.99}$$

which moves a given configuration to a range of successor configurations, each occurring with quantum *probability amplitude*, which corresponds to a unitary transformation of a quantum state. (With C we denote the set of complex numbers.) Bennet has shown that m-type Turing machines can be simulated by *reversible Turing machines*, with certain reduction in efficiency [20]. Furthermore, Tofolli has shown that arbitrary finite mapping can be computed reversibly by padding strings with zeros, permuting them, and projecting some of the bit strings to other bit strings [21]. The elementary reversible gates can be used to implement permutations of bit strings. Finally, Benioff has shown that unitary quantum state evolution (which is reversible) is at least as powerful as the Turing machine [22,23]. The probabilistic classical process can be represented by a tree, which grows exponentially with possible outcomes. The key difference of quantum computing is that we assign the quantum probability amplitudes to the branches in a tree, which can interfere with each other.

5.9 ADIABATIC QUANTUM COMPUTING

Many problems in quantum computing and quantum machine learning can be mapped to the Ising problem formalism, therefore we will first briefly describe the *Ising model* [24]. The magnetic dipole moments of the atomic spins can be in one of two states: either $+1$ or -1. The spins, represented as the discrete variables, are arranged in a graph, typically the *lattice* graph, and this model allows each spin to interact with its closest neighbors. The spins in the neighborhood having the same value tend to have lower energy; however, heat disturbs this tendency so that there exist different phase transitions. Let the individual spin on the lattice be denoted by $\sigma_i \in \{-1,1\}$, $i = 1,2,\ldots,L$ with L being the lattice size. To represent the energy of the Ising model the following Hamiltonian is typically used:

$$H(\boldsymbol{\sigma}) = -\sum_{\langle i,j \rangle} J_{ij}\sigma_i\sigma_j - \mu\sum_i h_i\sigma_i, \tag{5.100}$$

where J_{ij} denotes the interaction strength between the ith and jth sites, while h_i donates the strength of an external magnetic field interacting with the ith spin site in the lattice, and μ is the magnetic moment. We use the notation $\langle i,j \rangle$ to denote the direct nearest neighbors. By replacing the spin variables with corresponding Pauli operators (matrices), we can express the Ising model using quantum mechanical interpretation:

$$\hat{H} = -\sum_{\langle i,j \rangle} J_{ij}\sigma_i^Z\sigma_j^Z - \sum_i h_i\sigma_i^Z, \tag{5.101}$$

where the Pauli $\sigma^Z = Z$ matrix is defined by $\sigma^Z = Z = \begin{bmatrix} 1 & 0 \\ 0 & -1 \end{bmatrix}$, and the action on the CB vector will be $\sigma^Z|0\rangle = |0\rangle$, $\sigma^Z|1\rangle = -|1\rangle$. When the external field is transversal, the *transverse-field Hamiltonian* will be:

$$\hat{H} = -\sum_{\langle i,j \rangle} J_{ij}\sigma_i^Z\sigma_j^Z - \sum_i h_i\sigma_i^X, \tag{5.102}$$

where the Pauli X-operator is defined by $\sigma^X = X = \begin{bmatrix} 0 & 1 \\ 1 & 0 \end{bmatrix}$.

An *adiabatic process* is the process that changes slowly (gradually) so that the system can adapt its configuration accordingly. If the system was initially in the ground state of an initial Hamiltonian H_0, it will after the adiabatic change end up in the ground state of the final Hamiltonian H_1. This claim is commonly referred to as the *adiabatic theorem*, proved by Born and Fock in 1928 [25]. For initial Hamiltonian H_0 we typically choose one whose ground state is easy to implement, while for Hamiltonian H_1 one for which ground state we are interested in. To determine its ground state we consider the following Hamiltonian:

$$H(t) = (1-t)H_0 + tH_1, \quad t \in [0,1], \tag{5.103}$$

and start with the ground state of $H(0) = H_0$ and then gradually increase the parameter t, representing the time normalized with the duration of the whole process. According to the quantum adiabatic theorem, the system will gradually evolve to the ground state of H_1, providing that there is no degeneracy for the ground state energy. This transition depends on the time Δt during which the adiabatic change takes place and depends on the minimum energy gap between ground state and the first excited state of the adiabatic process. When applied to the Ising model, we select the initial Hamiltonian as $H_0 = \sum_i X_i$, describing the interaction of transverse external field with the spin sites. On the other hand, the final Hamiltonian is given by $H_1 = -\sum_{\langle i,j \rangle} J_{ij}Z_iZ_j - \sum_i h_iX_i$, representing the classical Ising model. The time-dependent Hamiltonian combines two models by Eq. (5.103). We gradually change the parameter t from zero to 1, so that the adiabatic pathway is chosen, and the final state will be the ground state of Hamiltonian H_1, encoding the solution of the problem we are trying to solve. The gap ΔE between the ground sate and the first excited state is parameter t dependent. The time scale (the runtime of the algorithm) is reversely proportional to the $\min[\Delta E(t)]^2$ [26]. So, if the minimum gap, during adiabatic pathway, is too small the runtime of the entire algorithm could be large.

In *adiabatic quantum computing* the Hamiltonian is defined as follows:

$$H = \underbrace{-\sum_{\langle i,j \rangle} J_{ij}Z_iZ_j - \sum_i h_iZ_i}_{\text{classical Ising model}} - \underbrace{\sum_{\langle i,j \rangle} g_{ij}X_iX_j}_{\text{interaction between transverse fields}}, \tag{5.104}$$

and it can be shown that this model can simulate *universal* quantum computation.

5.10 VARIATIONAL QUANTUM EIGENSOLVER

Quantum circuits are imperfect, which prevents us from running well-known quantum algorithms using the gates-based quantum computing approach. To overcome this problem, a new breed of quantum algorithms has been introduced [27–29], employing the parametrized shallow quantum circuits, which can be called *variational (quantum) circuits*. The *variational circuits* are suitable for noisy and imperfect quantum computers, which can also be called noisy intermediate-scale quantum devices [30]. They employ the variational method to find the approximation to the lowest energy eigenket or ground state, which is illustrated in Fig. 5.18. The quantum processing unit (QPU) is embedded in the environment (bath), and there is an interaction between the QPU and the environment causing the decoherence. The decoherence problem limits the ability to control the circuit and therefore the quantum circuit depth. To avoid this problem we can run a short burst of calculation on QPU, extract the results for the classical processor unit, which will optimize the parameters of the QPU to reduce the energy of the system, and feed back a new set of parameters to the QPU for the next iteration. We iterate this hybrid quantum classical algorithm until global minimum energy of the system is achieved. Any Hamiltonian H can be decomposed in terms of Hamiltonians of corresponding subsystems, that is, $H=\sum_i H_i$, where H_i is a Hamiltonian of the ith subsystem. The expected value of the Hamiltonian can then be determined as:

$$\langle H\rangle = \langle\psi|H|\psi\rangle = \langle\psi|\sum_i H_i|\psi\rangle = \sum_i \langle\psi|H_i|\psi\rangle = \sum_i \langle H_i\rangle, \tag{5.105}$$

where we exploited the linearity of quantum observables and $|\psi\rangle$ denotes the arbitrary input state. By applying this approach we can effectively replace the long coherent evolution of QPU with multiple short coherent evolutions of the subsystems. The expected value of the Hamiltonian can also be decomposed in terms of corresponding eigenkets of Hamiltonian $|\lambda_i\rangle$ as follows:

$$\begin{aligned}\langle H\rangle = \langle\psi|H|\psi\rangle &= \sum_{\lambda_i,\lambda_j\in \text{Spectrum}(H)} \langle\psi|(|\lambda_i\rangle\langle\lambda_i|)H(|\lambda_j\rangle\langle\lambda_j|)|\psi\rangle = \sum_{\lambda_i,\lambda_j}\langle\psi|\lambda_i\rangle\langle\lambda_i|H|\lambda_j\rangle\langle\lambda_j|\psi\rangle \\ &= \sum_{\lambda_i}|\langle\psi|\lambda_i\rangle|^2\underbrace{\langle\lambda_i|H|\lambda_i\rangle}_{\lambda_i} \geq \sum_{\lambda_i}|\langle\psi|\lambda_i\rangle|^2 E_0 = E_0,\end{aligned} \tag{5.106}$$

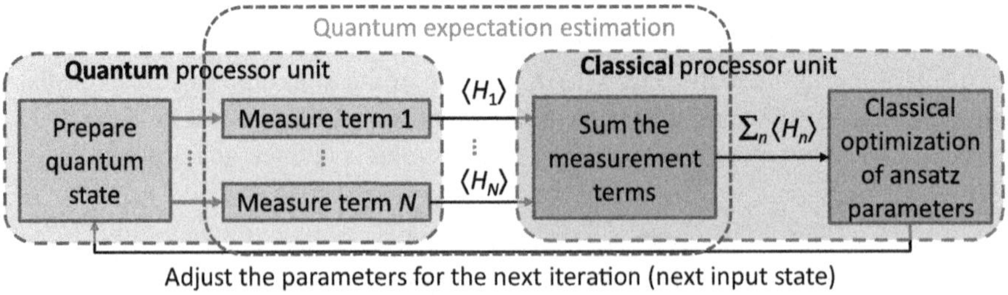

FIGURE 5.18

Illustrating the variational quantum eigensolver principle.

where E_0 is the lowest energy eigenket of the Hamiltonian. Varying over the entire Hilbert space is too complicated, so we can choose a subspace of the entire Hilbert space, parametrized by some real-valued parameters p_i ($i = 1,2,\ldots,N$), which is known as the ansatz. The choice of ansatz is relevant, as some choices of ansatzes can lead to better approximations.

5.11 SUMMARY

This chapter was devoted to various quantum algorithms and methods: (1) Deutsch's, Deutsch–Jozsa, and Bernstein–Vazirani algorithms, presented in Section 5.2; (2) Grover search algorithm, presented in Section 5.3; (3) quantum FT, presented in Section 5.4; (4) period-finding and Shor's factorization algorithms, presented in Section 5.5; (5) Simon's algorithm, presented in Section 5.6, and (6) quantum phase estimation algorithm, presented in Section 5.7. Section 5.1 reviewed the quantum parallelism concept. In Section 5.5 we also described how to crack the RSA encryption protocol. Section 5.8 was devoted to classical/quantum computing complexities and Turing machines. Adiabatic quantum computing was described in Section 5.9, while the variational quantum eigensolver was introduced in Section 5.10. In the next section we provide a set of problems that will help readers achieve a deeper understanding. In addition, a couple of problems have been devoted to the quantum discrete logarithm algorithm, Kitaev's algorithm, and quantum simulation. Additional quantum algorithms relevant in quantum machine learning are provided in Chapter 14.

5.12 PROBLEMS

1. By using mathematical induction, prove that the following circuit can be used to implement the Deutsch–Jozsa algorithm, that is, to verify whether the mapping $\{0,1\}^n \rightarrow \{0,1\}$ is constant or balanced.

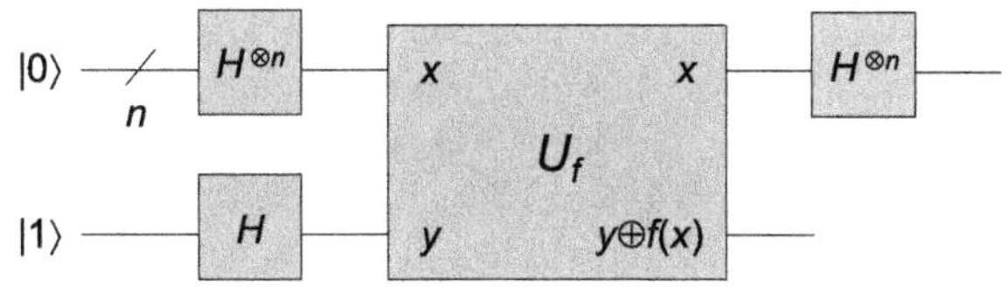

2. By using mathematical induction, prove that the following circuit can be used to calculate the FT for arbitrary n:

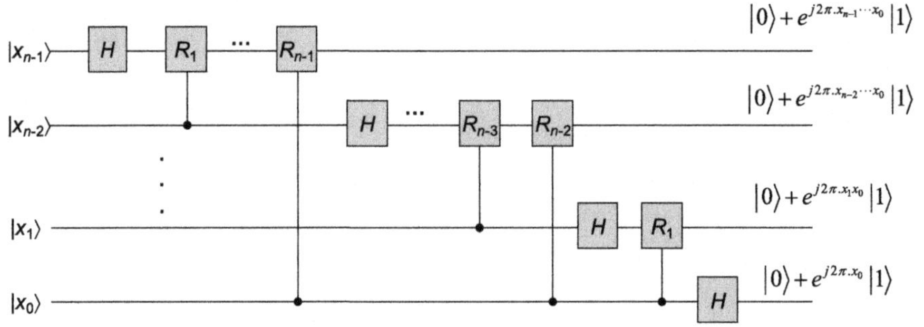

3. Assume that a unitary operator U has an eigneket $|u\rangle$ with eigenvalue $\exp(j2\pi\varphi_u)$. The goal of phase estimation is to estimate the phase φ_u. Describe how the quantum FT can be used for phase estimation.
4. This problem is related to the *shift-invariance property* of quantum FT. Let G be a group and H be a subgroup of G. If a function f on G is constant on cosets of H, then the FT of f is invariant over cosets of H. Prove the claim.
5. The quantum circuit to perform the Grover search for $n = 3$ was shown in Fig. 5.11. Provide a quantum circuit that can be used to perform the Grover search for arbitrary n. Explain the operation principle and prove that this circuit can indeed be used for any n.
6. Provide a quantum circuit to implement Shor's factorization algorithm for $n = 15$. Describe the operation principle of this circuit.
7. This problem is devoted to *quantum discrete logarithms*. Let us consider the following function: $f(x_1, x_2) = a^{bx_1+x_2} \text{mod} N$, where all variables are integers. Let r be the smallest positive integer for which $a^r \bmod N = 1$; this integer can be determined by using the order-finding algorithm. Clearly, this function is periodic as $f(x_1 + i, x_2 - ib) = f(x_1, x_2)$, where i is an integer. The discrete logarithm problem can be formulated as follows: given a and $c = a^b$, determine b. This problem is important in cracking the RSA encryption protocol as discussed in Section 5.5. Your task is to provide a quantum algorithm that can solve this problem by using one query of a quantum block U that performs the following unitary mapping: $U|x_1\rangle|x_2\rangle|y\rangle \rightarrow |x_1\rangle|x_2\rangle|y \oplus f(x_1, x_2)\rangle$.
8. In Problem 6 you were asked to provide a quantum circuit to implement Shor's factorization algorithm for $n = 15$. By using Simon's algorithm, describe how to perform the same task. Provide the corresponding quantum circuit. Analyze the complexity of both algorithms.

9. Suppose that the list of numbers $x_1,\ldots,x_n$ is stored in quantum memory. How many memory accesses are needed to determine the smallest number in the list with a success probability of $\geq 1/2$?
10. Provide the quantum circuit to perform the following mapping:

$$|m\rangle \rightarrow \frac{1}{\sqrt{p}} \sum_{n=0}^{p-1} e^{j2\pi mn/p} |n\rangle,$$

where p is a prime.

11. Design a quantum circuit to perform the following mapping: $|x\rangle \rightarrow |x + c \bmod 2^n\rangle$, where $x \in [0, 2^{n-1}]$ and c is the constant, by using quantum FT.
12. The following circuit can be used to perform addition:

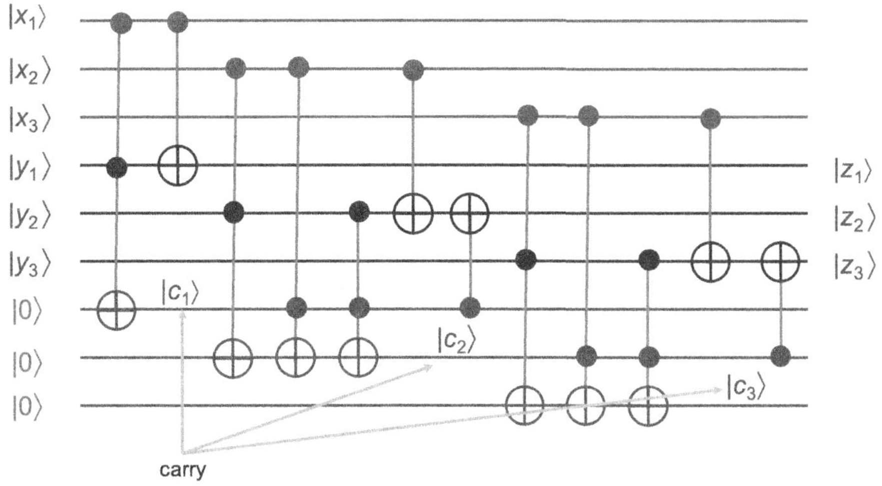

Describe the operation principle, and provide the result of addition. Can you generalize this addition problem?

13. The following problem is related to *Kitaev's algorithm*, which represents an alternative way to estimate the phase. Let us observe the following quantum circuit:

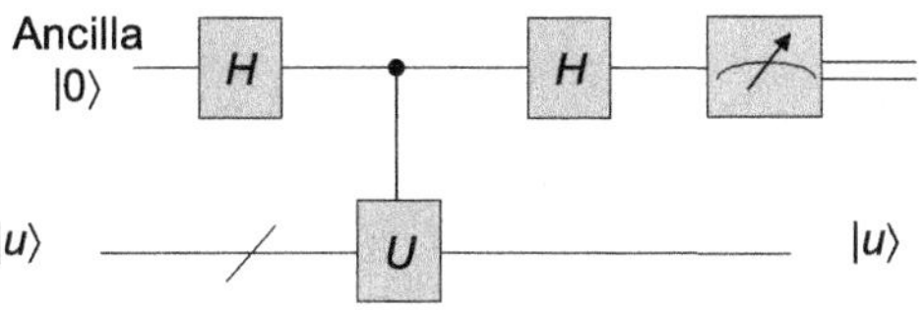

where $|u\rangle$ is an eigneket of U with eigenvalue $\exp(j2\pi\varphi_u)$. Show that the result of the measurement 0 appears with a probability of $p = \cos^2(\pi\varphi)$. Since the eigenket is insensitive to measurement the operator U can be replaced with U^m, where m is an arbitrary positive integer. Show that by repeating this circuit appropriately we can obtain the arbitrary precision of p and consequently estimate the phase φ with desired precision. Compare the complexity of Kitaev's algorithm with respect to that from Problem 3.

14. The state ket after application of Hadamard gates of the following circuit:

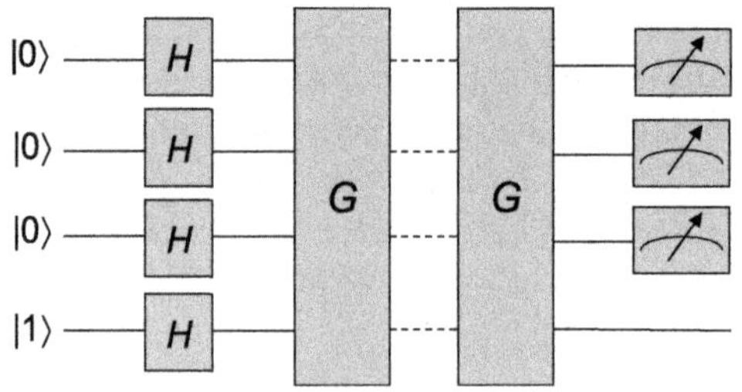

can be written in the following form:

$$|\psi\rangle = \frac{1}{N^{1/2}} \sum_x a_x^{(0)} |x\rangle, \quad a_x^{(0)} = 1.$$

The application of operator GO on $|\psi\rangle$ leads to the following ket:

$$|\psi\rangle = \frac{1}{N^{1/2}} \sum_x a_x^{(1)} |x\rangle.$$

Establish the connection between $a_x^{(1)}$ and $a_x^{(0)}$.

15. This problem is related to *quantum simulation*. When the Hamiltonian H that can be represented as the sum of polynomial many terms H_m, namely $H = \sum_m H_m$, each of which can be efficiently implemented, we can efficiently simulate the evolution operator $\exp\left(-\frac{j}{\hbar}Ht\right)$ and approximate the evolution of state $|\psi(t)\rangle = \exp\left(-\frac{j}{\hbar}Ht\right)\left|\psi(0)\right\rangle$. If for all m,n $[H_m,H_n] = 0$, then we can write $\exp\left(-\frac{j}{\hbar}Ht\right) = \prod_k e^{-\frac{j}{\hbar}H_k t}$. However, if $[H_m,H_n] \neq 0$, then the previous equation is not valid. The following formula, known as the Trotter formula, can be used for approximations leading to quantum simulation algorithms: $\lim_{n\to\infty} \left(e^{jU_1 t/n} e^{jU_2 t/n}\right) = e^{j(U_1+U_2)t}$, where U_1 and U_2 are Hermitian operators. Prove the Trotter formula. Prove also the following useful approximations:

$e^{j(U_1+U_2)\Delta t} = e^{jU_1\Delta t}e^{jU_2\Delta t} + O(\Delta t^2)$, $e^{j(U_1+U_2)\Delta t} = e^{jU_1\Delta t/2}e^{jU_2\Delta t}e^{jU_1\Delta t/2} + O(\Delta t^3)$. Finally, the following approximation, known as the Baker–Campbell–Hausdorf formula, is also useful in quantum simulation: $e^{(U_1+U_2)\Delta t} = e^{U_1\Delta t}e^{U_2\Delta t}e^{-[U_1,U_2]\Delta t^2/2} + O(\Delta t^3)$. Prove it. Consider now a single particle living in 1D potential $V(x)$, governed by Hamiltonian: $H = p^2/(2m) + V(x)$. Perform the computation $|\psi(t)\rangle = exp(-jHt/\hbar)|\psi(0)\rangle$ by using the foregoing approximations.

16. Construct the quantum circuit to simulate the Hamiltonian $H = Z_1Z_2 \ldots Z_n$, performing the unitary transform $|\psi(t)\rangle = exp(-jHt/\hbar)|\psi(0)\rangle$for arbitrary Δt.

References

[1] M. Le Bellac, An Introduction to Quantum Information and Quantum Computation, Cambridge University Press, 2006.

[2] M.A. Neilsen, I.L. Chuang, Quantum Computation and Quantum Information, Cambridge University Press, 2000.

[3] F. Gaitan, Quantum Error Correction and Fault Tolerant Quantum Computing, CRC Press, 2008.

[4] G. Jaeger, Quantum Information: An Overview, Springer, 2007.

[5] D. Petz, Quantum Information Theory and Quantum Statistics, Theoretical and Mathematical Physics, Springer, Berlin, 2008.

[6] P. Lambropoulos, D. Petrosyan, Fundamentals of Quantum Optics and Quantum Information, Springer-Verlag, Berlin, 2007.

[7] G. Johnson, A Shortcut Thought Time: The Path to the Quantum Computer, Knopf, N. York, 2003.

[8] J. Preskill, Quantum Computing, 1999. Available at: http://www.theory.caltech.edu/~preskill/.

[9] J. Stolze, D. Suter, Quantum Computing, Wiley, N. York, 2004.

[10] R. Landauer, Information is physical, Phys. Today 44 (5) (May 1991) 23–29.

[11] R. Landauer, The physical nature of information, Phys. Lett. A 217 (1991) 188–193.

[12] E. Bernstein, U. Vazirani, Quantum complexity theory, SIAM J. Comput. 26 (5) (1997) 1411–1473.

[13] S.D. Fallek, C.D. Herold, B.J. McMahon, K.M. Maller, K.R. Brown, J.M. Amini, Transport implementation of the Bernstein–Vazirani algorithm with ion qubits, New J. Phys. 18 (2016) 083030.

[14] P.W. Shor, Algorithms for quantum computation: discrete logarithm and factoring, p. 124, in: Proc. IEEE 35th Annual Symposium on Foundations of Computer Science, 1994.

[15] P.W. Shor, Polynomial-time algorithms for prime number factorization and discrete logarithms on a quantum computer, SIAM J. Comput. 26 (1997) 1484.

[16] A. Ekert, R. Josza, Shor's factoring algorithm, Rev. Mod. Phys. 68 (1996) 733–753.

[17] D. McMahon, Quantum Computing Explained, John Wiley & Sons, 2008.

[18] P. Benioff, The computer as a physical system: a microscopic quantum mechanical Hamiltonian model of computers as represented by turing machines, J. Stat. Phys. 22 (5) (1980) 563–591.

[19] P. Benioff, Models of quantum turing machines, Fortschr. Phys. 46 (1998) 423.
[20] T. Beth, M. Rötteler, Quantum algorithms: applicable algebra and quantum physics, in: G. Alber, T. Beth, M. Horodečki, P. Horodečki, R. Horodečki, M. Rötteler, H. Weinfurter, R. Werner, A. Zeilinger (Eds.), Quantum Information, Springer-Verlag, Berlin, 2001, pp. 96–150.
[21] D.R. Simon, On the power of quantum computation, in: S. Goldwasser (Ed.), Proc. IEEE 35th Annual Symposium on the Foundations of Computer Science, 1994, p. 116. Los Alamitos CA.
[22] C.H. Bennett, Time/space trade-offs for reversible computation, SIAM J. Comput. 18 (1989) 766–776.
[23] T. Toffoli, Reversible computing, p. 632, in: G. Goos, J. Hartmanis (Eds.), Automata, Languages and Programming, Lecture Notes in Computer Science, vol. 85Springer-Verlag, Berlin, 1980.
[24] G.H. Wannier, Statistical Physics, Dover Publications, Inc., New York, 1987.
[25] M. Born, V. Fock, Beweis des adiabatensatzes, Z. Phys. 51 (1–3) (1928) 165–180.
[26] W. van Dam, M. Mosca, U. Vazirani, How powerful is adiabatic quantum computation?, in: Proc. The 42nd Annual Symposium on Foundations of Computer Science Newport Beach, , CA, USA, 2001, pp. 279–287.
[27] E. Farhi, J. Goldstone, S. Gutmann, A Quantum Approximate Optimization Algorithm, 2014, pp. 1–16. Available at: https://arxiv.org/abs/1411.4028.
[28] A. Peruzzo, et al., A variational eigenvalue solver on a photonic quantum processor, Nat. Commun. 5 (2014) 4213.
[29] J. R McClean, J. Romero, R. Babbush, A. Aspuru-Guzik, The theory of variational hybrid quantum-classical algorithms, New J. Phys. 18 (2016) 023023.
[30] J. Preskill, Quantum computing in the NISQ era and beyond, Quantum 2 (2018) 79.

CHAPTER

Information Theory and Classical Error Correcting Codes

6

CHAPTER OUTLINE

Quantum Information Processing, Quantum Computing, and Quantum Error Correction
https://doi.org/10.1016/B978-0-12-821982-9.00009-5

In this Chapter, we provide basic concepts of information theory and coding theory. These topics are described to the level needed to easier understand in later chapters various topics such as quantum information theory, quantum error correction, fault-tolerant error correction, fault-tolerant computing, and QKD. The chapter starts with definitions of entropy, joint entropy, conditional entropy, relative entropy, mutual information, and channel capacity; followed by information capacity theorem. We also briefly describe the source coding and data compaction concepts. We then discuss the channel capacity of discrete memoryless channels, continuous channels, and some optical channels. After that, the fundamentals of blocks codes are introduced, such as linear block codes (LBCs), definition of generator and parity-check matrices, syndrome decoding, distance properties of LBCs, and some important coding bounds. Further, the cyclic codes are introduced. The BCH codes are described next. The RS codes, concatenated, and product codes are then described. After short summary section, a set of problems is provided for readers to get a deeper understanding of information theory and classical error correction.

6.1 CLASSICAL INFORMATION THEORY FUNDAMENTALS

In this section we try to answer two fundamental questions [1–13]: (1) what is the lowest, irreducible complexity below which the signal cannot be compressed and (2) what is the highest reliable transmission rate over a given communication noisy channel? To answer the first question we introduce the concept of *entropy*, which represents the basic measure of information, and is defined in terms of the probabilistic behavior of the source. It will be shown later in the text that average number of bits per symbol to represent a discrete source can be made as small as possible but not smaller than the entropy of the source, which represents the claim of so-called *source-coding theorem*. To answer the second question we introduce the concept of *channel capacity*, as the largest possible amount of information that can be reliably transmitted over the channel. As long as the transmission rate is lower than the channel capacity we can design a channel coding scheme that can allow for reliable transmission over the channel and to reconstruct the transmitted information with arbitrary small probability of error.

6.1.1 Entropy, Conditional Entropy, Relative Entropy, and Mutual Information

Let us observe a *discrete memoryless source* (DMS), characterized by a finite alphabet $S = \{s_0, s_1, \ldots, s_{K-1}\}$, wherein each symbol is generated with probability $P(S = s_k) = p_k$, $k = 0,1,\ldots,K-1$. At a given time instance, the DMS must generate one symbol from the alphabet, so that we can write $\sum_{k=0}^{K-1} p_k = 1$. The generation of a symbol at a given time instance is independent of previously generated symbols. The amount of information that a given symbol carries is related to the surprise when it occurs, and is therefore reversely proportional to the probability of its occurrence. Since there is uncertainty about which symbol will be generated by the source, it appears that the terms uncertainty, surprise, and amount of information are interrelated. Given that certain symbols can occur with very low probability, the amount of information value will be huge if reverse probability is used to determine the amount of information. In practice, we use the logarithm of reverse probability of occurrence as the amount of information to solve this problem:

$$I(s_k) = \log\left(\frac{1}{p_k}\right) = -\log p_k;\ k = 0, 1, \ldots, K-1. \tag{6.1}$$

The most common base of a logarithm is the base 2, and the unit for the amount of information is *bi*nary uni*t* (bit). When $p_k = 1/2$, the amount of information is $I(s_k) = 1$ bit, indicating that 1 bit is the amount of information gained when one out of two equally likely events occurs. It is straightforward to show that the amount of information is nonnegative, that is, $I(s_k) \geq 0$. Furthermore, when $p_k > p_i$, then $I(s_i) > I(s_k)$. Finally, when two symbols s_i and s_k are independent, then the joint amount of information is additive, that is, $I(s_k s_i) = I(s_k) + I(s_i)$.

The average information content per symbol is commonly referred as the *entropy*:

$$H(S) = E_P\, I(s_k) = E\left[\log\left(\frac{1}{p_k}\right)\right] = \sum_{k=0}^{K-1} p_k \log\left(\frac{1}{p_k}\right). \tag{6.2}$$

It can be easily be shown that the entropy is upper and lower bounded as follows:

$$0 \leq H(S) \leq |S|. \tag{6.3}$$

The entropy of binary source $S = \{0,1\}$ is given by:

$$H(S) = p_0 \log\left(\frac{1}{p_0}\right) + p_1 \log\left(\frac{1}{p_1}\right) = -p_0 \log p_0 - (1-p_0)\log(1-p_0) = H(p_0), \tag{6.4}$$

where $H(p_0)$ is known as the *binary entropy function.*

The entropy definition is also applicable to any random variable X, namely we can write: $H(X) = -\sum_k p_k \log p_k$. The *joint entropy* of a pair of random variables (X,Y), denoted as $H(X,Y)$, is defined as:

$$H(X,Y) = -E_{p(x,y)} \log p(X,Y) = -\sum_{x\in X}\sum_{x\in Y} p(x,y)\log p(x,y). \tag{6.5}$$

When a pair of random variables (X,Y) has the joint distribution $p(x,y)$, the *conditional entropy* $H(Y|X)$ is defined as:

$$\begin{aligned} H(Y|X) &= -E_{p(x,y)} \log p(Y|X) = -\sum_{x\in X}\sum_{y\in Y} \underbrace{p(x,y)}_{p(y|x)p(x)} \log p(y|x) \\ &= -\sum_{x\in X} p(x) \sum_{y\in Y} p(y|x)\log p(y|x). \end{aligned} \tag{6.6}$$

Since $p(x,y) = p(x)p(y|x)$, by taking the logarithm we obtain:

$$\log p(X,Y) = \log p(X) + \log p(Y|X), \tag{6.7}$$

and now by applying the expectation operator we obtain:

$$\underbrace{E\log p(X,Y)}_{H(X,Y)} = \underbrace{E\log p(X)}_{H(X)} + \underbrace{E\log p(Y|X)}_{H(Y|X)} \Leftrightarrow H(X,Y) = H(X) + H(Y|X). \tag{6.8}$$

Eq. (6.8) is commonly referred to as the *chain rule*.

The *relative entropy* is a measure of distance between two distributions $p(x)$ and $q(x)$, and is defined as follows:

$$D(p||q) = E_p \log\left[\frac{p(X)}{q(X)}\right] = \sum_{x\in X} p(x)\log\left[\frac{p(X)}{q(X)}\right]. \tag{6.9}$$

The relative entropy is also known as the Kullback–Leibler distance, and can be interpreted as the measure of inefficiency of assuming that distribution is $q(x)$, when true distribution is $p(x)$. Now, by replacing $p(X)$ with $p(X,Y)$ and $q(X)$ with $p(X)q(Y)$, the corresponding relative entropy between the joint distribution and the product of distribution is well known as *mutual information*:

$$\begin{aligned} D(p(X,Y)||p(X)q(Y)) &= E_{p(X,Y)} \log\left[\frac{p(X,Y)}{p(X)q(Y)}\right] \\ &= \sum_{x\in X}\sum_{y\in Y} p(x,y)\log\left[\frac{p(X,Y)}{p(X)q(Y)}\right] \doteq I(X,Y). \end{aligned} \tag{6.10}$$

6.1.2 Source Coding and Data Compaction

An important problem in communication systems and data storage is to efficiently represent the data generated by a source. Source coding represents the process by which this goal is achieved. In particular, when the statistics of the source is known, we can assign short codewords to frequently occurring symbols, and long codewords to rarely occurring symbols, which results in a *variable-length source code*. There are two basic requirements to be satisfied: (1) the codewords generated by a source encoder are binary and (2) the source code must be *uniquely decodable*, that is, the original sequence can be perfectly recovered from a binary encoded sequence.

The average number of bits per symbol can be determined by:

$$L = \sum_i p_k L_k, \; L_k = \text{length}(s_k), \tag{6.11}$$

where L_k is the length of the kth source symbol codeword s_k, which occurs with probability p_k. The source coding efficiency can be defined by $\eta = L_{\min}/L$, $L_{\min} = \min L$.

The *source coding theorem*, also known as Shannon's first theorem, can be formulated as follows [1–6]: the average source codeword length for distortionless transmission, for a given memoryless source S of entropy $H(S)$, is lower bounded by:

$$L \geq H(S). \tag{6.12}$$

In other words, we can make the average number of bits per symbol for a DMS as small as possible but not smaller than the entropy of the source. Now we can redefine the *source coding efficiency* as follows:

$$\eta = \frac{H(S)}{L}. \tag{6.13}$$

For efficient transmission, the redundant information should be removed before the transmission takes place. This can be done by quantization (lossy compression) or *data compaction* (lossless data compression), considered here. An efficient way to do so is by prefix coding. The *prefix code* is defined as a source code for which no codeword can be used as a prefix of a longer codeword. It can be shown that the prefix code is uniquely decodable, but the converse is not necessarily true. Decoding of the prefix codes can be performed as soon as a source symbol codeword is fully received, and as such it can be called instantaneous code. To verify whether a given source code is uniquely decodable we can use the *Kraft–McMillan inequality* [1–6]:

$$\sum_{k=0}^{K-1} 2^{-L_k} \leq 1, \tag{6.14}$$

where K is the cardinality of the source.

Because the entropy of the DMS is $H(S)$, we can construct the prefix code such that:

$$p_k = 2^{-l_k} \Rightarrow \sum_{k=0}^{K-1} 2^{-l_k} = \sum_{k=0}^{K-1} p_k = 1. \tag{6.15}$$

The average codeword length will then be:

$$L = \sum_{k=0}^{K-1} p_k L_k = \sum_{k=0}^{K-1} 2^{-L_k} L_k. \tag{6.16}$$

The corresponding entropy of this prefix code will be:

$$H(S) = \sum_{k=0}^{K-1} p_k \log\left(\frac{1}{p_k}\right) = \sum_{k=0}^{K-1} 2^{-L_k} \log_2\left(2^{L_k}\right) = \sum_{k=0}^{K-1} \frac{L_k}{2^{L_k}} = L. \tag{6.17}$$

Clearly, this prefix code is matched to the source since $L = H(S)$ and thus the code efficiency is $\eta = 1$. The key question now arises: how can the prefix code be matched to an arbitrary DMS? The answer to this question requires the help of so-called *extended source*. In extended code, we consider the blocks of symbols rather than individual symbols, obtained by the Cartesian product of symbols from the original source. If the symbols from the source are independent, then the probability of an nth order extended source will be n times larger than the entropy of the original source; in other words we can write:

$$H(S^n) = nH(S). \tag{6.18}$$

The average codeword length of the extended source, denoted as $\overline{L}_n$, will be limited as:

$$H(S^n) \leq \overline{L}_n \leq H(S^n) + 1 \Rightarrow nH(S) \leq \overline{L}_n \leq nH(S) + 1. \tag{6.19}$$

By dividing both sides of the inequalities by n we obtain:

$$H(S) \leq \frac{\overline{L}_n}{n} \leq H(S) + \frac{1}{n}. \tag{6.20}$$

In limit, as n tends to infinity we obtain:

$$\lim_{n->\infty} \frac{\overline{L}_n}{n} = H(S), \tag{6.21}$$

indicating that we can faithfully represent the source by sufficiently extending the order of the source.

6.1.2.1 Huffman Coding

An important class of prefix codes is the class of *Huffman codes* [14]. The key idea behind the Huffman code is to represent a symbol from a source alphabet by a sequence of bits of length being proportional to the amount of information conveyed by the symbol under consideration, that is, $L_k \cong -\log(p_k)$. Clearly, the Huffman code requires knowledge of the source statistics and attempts to represent the DMS statistics by a simpler one. The *Huffman encoding algorithm* can be summarized as follows:

1. List the symbols of the source in decreasing order of probability of occurrence. The two symbols with a lowest probability are assigned to a 0 or 1.
2. Two symbols with a lowest probability are combined into a new symbol (super-symbol) with a probability being the sum of individual symbols within the super-symbol. Location of the super-symbol in the list in the next stage is according to the combined probability.
3. The procedure is repeated until we are left with only two symbols to which we assign bits 0 and 1.

By reading out the bits assigned until we reach the root we get the codeword assigned to each symbol from the source. The Huffman code is applicable not only to binary source code, but also to nonbinary codes.

Example. A DMS has an alphabet of eight symbols whose probabilities of occurrence are as follows:

Symbols:	s_1	s_2	s_3	s_4	s_5	s_6	s_7	s_8
Probabilities:	0.28	0.18	0.15	0.13	0.10	0.07	0.05	0.04

We are concerned with the designing of the Huffman code for this source by moving a super-symbol as high as possible and assuming ternary transmission with symbols {0,1,2}. The Huffman procedure for this ternary code is summarized in Fig. 6.1.

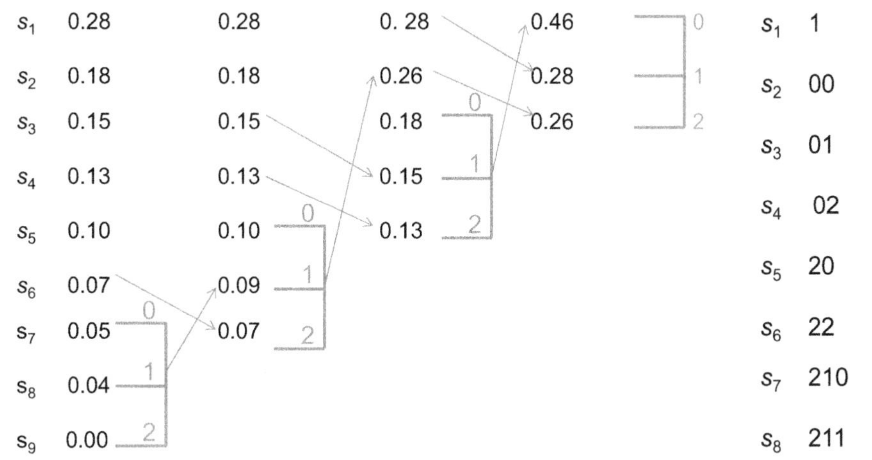

FIGURE 6.1

Huffman procedure (*left*) and corresponding ternary code (*right*).

6.1.2.2 Ziv–Lempel Algorithm

The Ziv–Lempel algorithm, on the other hand, does not require knowledge of the source statistics and is therefore easier to implement. The key idea behind this algorithm is to parse the source data stream into the shortest segments not encountered before. As an illustration, let us consider the following binary sequence: 110100111011110.... We would like to use the Ziv–Lempel algorithm to encode this sequence. We assume that the binary symbols 0 and 1 are already in the code book at addresses 1 and 2, respectively. By parsing this sequence into segments not encountered before we obtain 11 | 01| 00| 111| 011| 110 .... The corresponding Ziv–Lempel code is provided in Fig. 6.2. The first row represents the numerical positions of segments, assuming that symbols 0 and 1 are already in the code book. The second row represents the segments of the parsed sequence. The third row provides the numerical representation of segments, with each segment represented with the address of the segment already encountered previously and the bit in the last position representing the new bit from the sequence known as the innovation bit. The final row provides binary-encoded blocks, wherein each address is represented in binary form followed by the innovation bit.

Addresses:	1	2	3	4	5	6	7	8
Segments:	0	1	11	01	00	111	011	110
Numerical representations:			21	11	10	31	41	30
Binary encoded blocks:			**0101**	**0011**	**0010**	**0111**	**1001**	**0110**

FIGURE 6.2

Ziv–Lempel algorithm on the binary sequence 110100111011110....

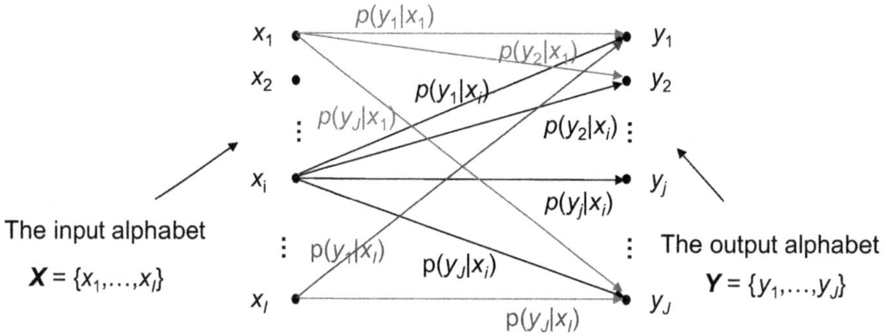

FIGURE 6.3

Discrete memoryless channel.

6.1.3 Mutual Information, Channel Capacity, Channel Coding Theorem, and Information Capacity Theorem

6.1.3.1 Mutual Information and Information Capacity

Fig. 6.3 shows an example of a discrete memoryless channel (DMC), which is characterized by channel (transition) probabilities. If $X = \{x_0,x_1,\ldots,x_{I-1}\}$ and $Y = \{y_0,y_1,\ldots,y_{J-1}\}$ denote the channel input alphabet and the channel output alphabet, respectively, the channel is completely characterized by the following set of transition probabilities:

$$p(y_j|x_i) = P(Y = y_j|X = x_i),\ 0 \le p(y_j|x_i) \le 1, \tag{6.22}$$

where $i \in \{0,1,\ldots,I-1\}$, $j \in \{0,1,\ldots,J-1\}$, while I and J denote the sizes of input and output alphabets, respectively. The transition probability $p(y_j|x_i)$ represents the conditional probability that $Y = y_j$ for a given input $X = x_i$.

One of the most important characteristics of the transmission channel is the information capacity, which is obtained by maximization of mutual information $I(\boldsymbol{X},\boldsymbol{Y})$ over all possible input distributions:

$$C = \max_{\{p(x_i)\}} I(\boldsymbol{X},\boldsymbol{Y}),\ I(\boldsymbol{X},\boldsymbol{Y}) = H(\boldsymbol{X}) - H(\boldsymbol{X}|\boldsymbol{Y}), \tag{6.23}$$

where $H(\boldsymbol{U}) = E\{-\log_2 P(\boldsymbol{U})\}$ denotes the entropy of a random variable $\boldsymbol{U}$.

For DMC, the mutual information can be determined as:

$$I(\boldsymbol{X};\boldsymbol{Y}) = H(\boldsymbol{X}) - H(\boldsymbol{X}|\boldsymbol{Y})$$
$$= \sum_{i=1}^{M} p(x_i)\log_2\left[\frac{1}{p(x_i)}\right] - \sum_{j=1}^{N} p(y_j)\sum_{i=1}^{M} p(x_i|y_j)\log_2\left[\frac{1}{p(x_i|y_j)}\right]. \tag{6.24}$$

In Eq. (6.24), $H(\boldsymbol{X})$ represents the uncertainty about the channel input before observing the channel output, also known as entropy, while $H(\boldsymbol{X}|\boldsymbol{Y})$ denotes the conditional entropy or the amount of uncertainty remaining about the channel input after the channel output has been received. (The log function from Eq. (6.24) relates to the base 2, and it will be like that throughout this chapter). Therefore the mutual information represents the amount of information (per symbol) that is conveyed by the channel, which represents the uncertainty about the channel input that is resolved by observing the channel output. The mutual information can be interpreted by means of a Venn diagram [1–13] as shown in Fig. 6.4A. The left and right circles represent the entropy of the channel input and channel output, respectively, while the mutual information is obtained as the intersection area of these two circles. Another interpretation is illustrated in Fig. 6.4B [3]. The mutual information, i.e., the information conveyed by the channel, is obtained as the output information minus unwanted information introduced by the channel.

Since for M-ary input and M-ary output symmetric channels we have that $p(y_j|x_i) = P_s/(M-1)$ and $p(y_j|x_j) = 1 - P_s$, where P_s is symbol error probability, the channel capacity, in bits/symbol, can be found as:

$$C = \log_2 M + (1 - P_s)\log_2(1 - P_s) + P_s \log_2\left(\frac{P_s}{M-1}\right). \tag{6.25}$$

The channel capacity represents an important bound on data rates achievable by any modulation and coding scheme. It can also be used in comparison of different coded modulation schemes in terms of their distance to the maximum channel capacity curve. In Fig. 6.5, we show the channel capacity for a Poisson M-ary pulse-position modulation (PPM) channel against the average number of signal photons per slot, expressed in dB scale, for the average number of background photons of $K_b = 1$.

6.1.3.2 Channel Coding Theorem

We have now built enough knowledge to introduce a very important theorem, the channel coding theorem [1–13], which can be formulated as follows. Let a DMS with an alphabet $\boldsymbol{S}$ have the entropy $H(\boldsymbol{S})$ and emit the symbols every T_s seconds. Let a DMC have capacity C and be used once in T_c seconds. Then, if:

$$H(\boldsymbol{S})/T_s \leq C/T_c, \tag{6.26}$$

there exists a coding scheme for which the source output can be transmitted over the channel and reconstructed with an arbitrary small probability of error. The parameter $H(\boldsymbol{S})/T_s$ is related to the average information rate, while the parameter C/T_c is related to the channel capacity per unit time.

For binary symmetric channel ($N = M = 2$) the inequality is reduced to $R \leq C$, where R is the code rate. Since the proof of this theorem can be found in any textbook on information theory, such as [2–5], the proof of this theorem will be omitted.

6.1.3.3 Capacity of Continuous Channels

In this section we will discuss the channel capacity of continuous channels. Let $\boldsymbol{X} = [X_1, X_2, \ldots, X_n]$ denote an n-dimensional multivariate, with a probability

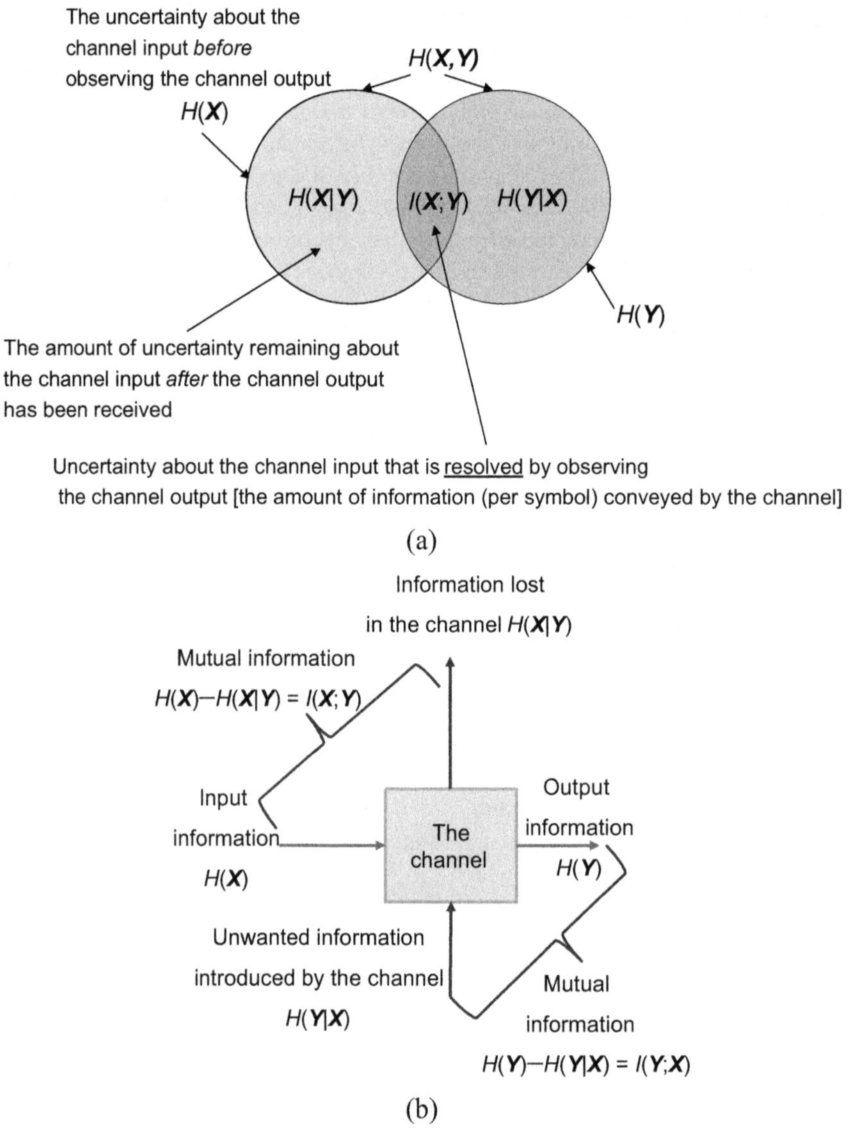

FIGURE 6.4

Interpretation of the mutual information by using: (a) Venn diagrams and (b) the approach due to Ingels.

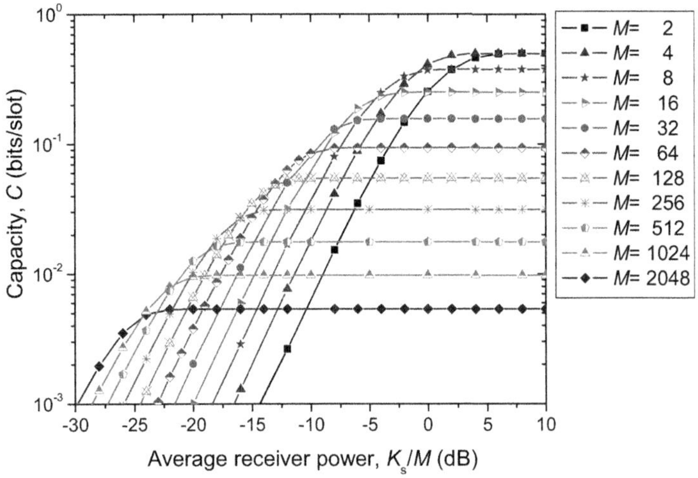

FIGURE 6.5

Channel capacity for M-ary pulse-position modulation on a Poisson channel for $K_b = 1$.

density function (PDF) $p_1(x_1,x_2,\ldots,x_n)$, representing the channel input. The corresponding differential entropy is defined by [2–6]:

$$h(X_1, X_2, \ldots, X_n) = -\underbrace{\int_{-\infty}^{\infty} \cdots \int_{-\infty}^{\infty}}_{n} p_1(x_1, x_2, \ldots, x_n) \log p_1(x_1, x_2, \ldots, x_n) dx_1 dx_2 \ldots dx_n$$
$$= \langle -\log p_1(x_1, x_2, \ldots, x_n) \rangle, \tag{6.27}$$

where for brevity we use $\langle \cdot \rangle$ to denote the expectation operator. To simplify explanations we will use the compact form of Eq. (6.27), namely $h(\boldsymbol{X}) = \langle -\log p_1(\boldsymbol{X}) \rangle$, which was introduced in [4]. In similar fashion, the channel output can be represented as m-dimensional random variable $Y = [Y_1,Y_2,\ldots,Y_m]$ with a PDF $p_2(y_1,y_2,\ldots,y_m)$, while corresponding differential entropy is defined by:

$$h(Y_1, Y_2, \ldots, Y_m) = -\underbrace{\int_{-\infty}^{\infty} \cdots \int_{-\infty}^{\infty}}_{m} p_2(y_1, y_2, \ldots, y_m) \log p_1(y_1, y_2, \ldots, y_m) dy_1 dy_2 \ldots dy_m$$
$$= \langle -\log p_2(y_1, y_2, \ldots, y_m) \rangle. \tag{6.28}$$

In compact form, the differential entropy of output can be written as $h(\boldsymbol{Y}) = \langle -\log p_1(\boldsymbol{Y}) \rangle$.

Example. Let an n-dimensional multivariate $\boldsymbol{X} = [X_1,X_2,\ldots,X_n]$ with a PDF $p_1(x_1,x_2,\ldots,x_n)$ be applied to the nonlinear channel with the following nonlinear characteristic $\boldsymbol{Y} = g(\boldsymbol{X})$, where $\boldsymbol{Y} = [Y_1,Y_2,\ldots,Y_n]$ represents the channel output with PDF $p_2(y_1,y_2,\ldots,y_n)$. Since the corresponding PDFs are related by the Jacobian symbol as follows: $p_2(y_1,\ldots,y_n)=p_1(x_1,\ldots,x_n)\left|J\left(\frac{X_1,\ldots,X_n}{Y_1,\ldots,Y_m}\right)\right|$, the output entropy can be determined as:

$$h(Y_1,\ldots,Y_m)\cong h(X_1,\ldots,X_n)-\left\langle \log\left|J\left(\frac{X_1,\ldots,X_n}{Y_1,\ldots,Y_m}\right)\right|\right\rangle.$$

To account for the channel distortions and additive noise influence we can observe the corresponding conditional and joint PDFs:

$$\begin{aligned}&P(y_1<Y_1<y_1+dy_1,\ldots,y_m<Y_m<y_m+dy_m|X_1=x_1,\ldots,X_n=x_n)=p(\tilde{y}|\tilde{x})d\tilde{y}\\&P(y_1<Y_1<y_1+dy_1,\ldots,x_n<X_n<x_n+dx_n)=p(\tilde{x},\tilde{y})d\tilde{x}d\tilde{y}.\end{aligned} \tag{6.29}$$

The mutual information (also known as information rate) can be written in compact form as follows [4]:

$$I(\boldsymbol{X};\boldsymbol{Y})=\left\langle \log\frac{p(\boldsymbol{X},\boldsymbol{Y})}{p(\boldsymbol{X})P(\boldsymbol{Y})}\right\rangle. \tag{6.30}$$

Notice that various differential entropies $h(\boldsymbol{X})$, $h(\boldsymbol{Y})$, $h(\boldsymbol{Y}|\boldsymbol{X})$ do not have direct interpretation as far as the information processing in the channel is concerned, as compared to their discrete counterparts, from previous Subsection 6.1.3.1. Some authors, such as Gallager in [13], prefer to define the mutual information directly by Eq. (6.30) without considering the differential entropies at all. The mutual information, however, has the theoretical meaning and represents the average information processed in the channel (or amount of information conveyed by the channel). The mutual information has the following important properties [2–6]: (1) it is symmetric: $I(X;Y)=I(Y;X)$; (2) it is nonnegative; (3) it is finite; (4) it is invariant under linear transformation; (5) it can be expressed in terms of the differential entropy of channel output by $I(X;Y)=h(Y)-h(Y|X)$; and (6) it is related to the channel input differential entropy by $I(X;Y)=h(X)-h(X|Y)$.

The information capacity can be obtained by maximization of Eq. (6.30) under all possible input distributions, which is:

$$C=\max I(\boldsymbol{X};\boldsymbol{Y}). \tag{6.31}$$

Let us now determine the mutual information of two random vectors $\boldsymbol{X}=[X_1,X_2,\ldots,X_n]$ and $\boldsymbol{Y}=[Y_1,Y_2,\ldots,Y_m]$, which are normally distributed. Let $\boldsymbol{Z}=[\boldsymbol{X};\boldsymbol{Y}]$ be the random vector describing the joint behavior. Without loss of generality, we further assume that $\overline{X}_k=0\,\forall k$ and $\overline{Y}_k=0\,\forall k$ (the mean values are zero). The corresponding PDF for $\boldsymbol{X}$, $\boldsymbol{Y}$, and $\boldsymbol{Z}$ are, respectively, given as [4]:

$$\begin{aligned}p_1(\boldsymbol{x})&=\frac{1}{(2\pi)^{n/2}(\det\boldsymbol{A})^{1/2}}\exp\left(-0.5\left(\boldsymbol{A}^{-1}\boldsymbol{x},\boldsymbol{x}\right)\right),\\\boldsymbol{A}&=[a_{ij}],\ a_{ij}=\int x_ix_jp_1(\boldsymbol{x})d\boldsymbol{x}\end{aligned} \tag{6.32}$$

$$p_2(\mathbf{y}) = \frac{1}{(2\pi)^{n/2}(\det \mathbf{B})^{1/2}} \exp\left(-0.5\left(\mathbf{B}^{-1}\mathbf{y},\mathbf{y}\right)\right),$$
$$\mathbf{B} = [b_{ij}],\ b_{ij} = \int y_i y_j p_2(\mathbf{y}) d\mathbf{y} \tag{6.33}$$

$$p_3(\mathbf{z}) = \frac{1}{(2\pi)^{(n+m)/2}(\det \mathbf{C})^{1/2}} \exp\left(-0.5\left(\mathbf{C}^{-1}\mathbf{z},\mathbf{z}\right)\right),$$
$$\mathbf{C} = [c_{ij}],\ c_{ij} = \int z_i z_j p_3(\mathbf{z}) d\mathbf{z} \tag{6.34}$$

where $(\bullet,\bullet)$ denotes the dot product of two vectors. By substitution of Eqs. (6.32)−(6.34) into Eq. (6.30) we obtain [4]:

$$I(\mathbf{X};\mathbf{Y}) = \frac{1}{2}\log\frac{\det\mathbf{A}\det\mathbf{B}}{\det\mathbf{C}}. \tag{6.35}$$

The mutual information between two Gaussian random vectors can also be expressed in terms of their correlation coefficients [4]:

$$I(\mathbf{X};\mathbf{Y}) = -\frac{1}{2}\log\left[\left(1-\rho_1^2\right)\ldots\left(1-\rho_l^2\right)\right],\ l = \min(m,n), \tag{6.36}$$

where ρ_j is the correlation coefficient between X_j and Y_j.

To obtain the information capacity for additive Gaussian noise, we make the following assumptions: (1) the input X, output Y, and noise Z are n-dimensional random variables; (2) $\overline{X}_k = 0$, $\overline{X_k^2} = \sigma_{x_k}^2\ \forall k$ and $\overline{Z}_k = 0$, $\overline{Z_k^2} = \sigma_{z_k}^2\ \forall k$; and (3) the noise is additive: $Y = X + Z$. Since we have that:

$$p_{\mathbf{x}}(\mathbf{y}|\mathbf{x}) = p_{\mathbf{x}}(\mathbf{x}+\mathbf{z}|\mathbf{x}) = \prod_{k=1}^{n}\left[\frac{1}{(2\pi)^{1/2}\sigma_{z_k}}e^{-z_k^2/2\sigma_{z_k}^2}\right] = p(\mathbf{z}), \tag{6.37}$$

the conditional differential entropy can be obtained as:

$$H(\mathbf{Y}|\mathbf{X}) = H(\mathbf{Z}) = -\int_{-\infty}^{\infty} p(\mathbf{z})\log p(\mathbf{z})d\mathbf{z}. \tag{6.38}$$

The mutual information is then:

$$I(\mathbf{X};\mathbf{Y}) = h(\mathbf{Y}) - h(\mathbf{Y}|\mathbf{X}) = h(\mathbf{Y}) - h(\mathbf{Z}) = h(\mathbf{Y}) - \frac{1}{2}\sum_{k=1}^{n}\log 2\,\pi e\sigma_{z_k}^2. \tag{6.39}$$

The information capacity, expressed in bits per channel use, is therefore obtained by maximizing $h(\mathbf{Y})$. Because the distribution maximizing the differential entropy is Gaussian, the information capacity is obtained as:

$$
\begin{aligned}
C(\boldsymbol{X};\boldsymbol{Y}) &= \frac{1}{2}\sum_{k=1}^{n}\log 2\,\pi e\sigma_{y_k}^2 - \frac{1}{2}\sum_{k=1}^{n}\log 2\,\pi e\sigma_{z_k}^2 = \frac{1}{2}\sum_{k=1}^{n}\log\left(\frac{\sigma_{y_k}^2}{\sigma_{z_k}^2}\right) \\
&= \frac{1}{2}\sum_{k=1}^{n}\log\left(\frac{\sigma_{x_k}^2+\sigma_{z_k}^2}{\sigma_{z_k}^2}\right) = \frac{1}{2}\sum_{k=1}^{n}\log\left(1+\frac{\sigma_x^2}{\sigma_z^2}\right).
\end{aligned}
\tag{6.40}
$$

For $\sigma_{x_k}^2 = \sigma_x^2$, $\sigma_{z_k}^2 = \sigma_z^2$ we obtain the following expression for information capacity:

$$
C(\boldsymbol{X};\boldsymbol{Y}) = \frac{n}{2}\log\left(1+\frac{\sigma_x^2}{\sigma_z^2}\right),
\tag{6.41}
$$

where σ_x^2/σ_z^2 presents the signal-to-noise ratio (SNR). Expression (6.41) represents the maximum amount of information that can be transmitted per symbol.

From a practical point of view it is important to determine the amount of information conveyed by the channel per second, which is the information capacity per unit time, also known as the channel capacity. For bandwidth-limited channels and Nyquist signaling employed, there will be 2BW samples per second (BW is the channel bandwidth) and the corresponding channel capacity becomes:

$$
C = W\log\left(1+\frac{P}{N_o BW}\right)\ [\text{bits / s}],
\tag{6.42}
$$

where P is the average transmitted power and $N_0/2$ is the noise power spectral density. Eq. (6.42) represents the well-known information capacity theorem, commonly referred to as Shannon's third theorem [15]. Since the Gaussian source has the maximum entropy clearly, it will maximize the mutual information. Therefore Eq. (6.42) can be derived as follows. Let the n-dimensional multivariate $\boldsymbol{X} = [X_1,\ldots,X_n]$ represent the Gaussian channel input with samples generated from zero-mean Gaussian distribution with variance σ_x^2. Let the n-dimensional multivariate $\boldsymbol{Y} = [Y_1,\ldots,Y_n]$ represent the Gaussian channel output, with samples spaced 1/(2BW) apart. The channel is additive with noise samples generated from zero-mean Gaussian distribution with variance σ_z^2. Let the PDFs of input and output be denoted by $p_1(\boldsymbol{x})$ and $p_2(\boldsymbol{y})$, respectively. Finally, let the joint PDF of input and output of the channel be denoted by $p(\boldsymbol{x},\boldsymbol{y})$. The maximum mutual information can be calculated from:

$$
I(\boldsymbol{X};\boldsymbol{Y}) = \iint p(\boldsymbol{x},\boldsymbol{y})\log\frac{p(\boldsymbol{x},\boldsymbol{y})}{p(\boldsymbol{x})P(\boldsymbol{y})}\,d\boldsymbol{x}d\boldsymbol{y}.
\tag{6.43}
$$

By following the similar procedure to that just used we obtain the corresponding expression for Gaussian channel capacity given by Eq. (6.42).

By using Eq. (6.42) and Fano's inequality [2]:

$$
H(\boldsymbol{X}|\boldsymbol{Y}) \le H(P_e) + P_e\log_2(M-1),\ H(P_e) = -P_e\log_2 P_e - (1-P_e)\log(1-P_e)
\tag{6.44}
$$

For an optical channels amplified spontaneous emission, noise-dominated scenario, and binary phase-shift keying (BPSK) at 40 Gb/s in Fig. 6.6A we report the minimum bit error rates (BERs) against optical SNR for different code rates. In Fig. 6.6B, we show the minimum channel capacity BERs for different code rates for 64 PPM over the Poisson channel in the presence of background radiation with average number of background photons of $K_b = 0.2$. For example, it has been shown in [16] that for 64 PPM over the Poisson channel with serial concatenated turbo code (8184,4092) we are about 1 dB away from channel capacity (for $K_b = 0.2$ at BER of 10^{-5}), while with low-density parity-check (LDPC) (8192,4096) code, of fixed column weight 6 and irregular row weight, we are 1.4 dB away from channel capacity.

6.2 CHANNEL CODING PRELIMINARIES

Two key system parameters are transmitted: power and channel bandwidth, which together with additive noise sources determine the SNR and correspondingly BER. In practice, we very often come across a situation when the target BER cannot be achieved with a given modulation format. For the fixed SNR, the only practical option to change the data quality transmission from unacceptable to acceptable is through the use of channel coding. Another practical motivation of introducing channel coding is to reduce the required SNR for a given target BER. The amount of energy that can be saved by coding is commonly described by coding gain. Coding gain refers to the savings attainable in the energy per information bit-to-noise spectral density ratio (E_b/N_0) required to achieve a given bit error probability when coding is used compared to that with no coding. A typical digital communication system employing channel coding is shown in Fig. 6.7. The discrete source generates the information in the form of a sequence of symbols. The channel encoder accepts the message symbols and adds redundant symbols according to a corresponding prescribed rule. Channel coding is the act of transforming a length-k sequence into a length-n codeword. The set of rules specifying this transformation is called the channel code, which can be represented as the following mapping:

$$\mathscr{C}: \mathscr{M} \rightarrow \mathscr{X}, \tag{6.45}$$

where $\mathscr{C}$ is the channel code, $\mathscr{M}$ is the set of information sequences of length k, and $\mathscr{X}$ is the set of codewords of length n. The decoder exploits these redundant symbols to determine which message symbol was actually transmitted. Encoder and decoder consider the whole digital transmission system as a discrete channel. Different classes of channel codes can be categorized into three broad categories: (1) error detection in which we are concerned only with detecting the errors occurring during transmission (examples include automatic request for transmission), (2) forward error correction (FEC), where we are interested in correcting the errors occurring

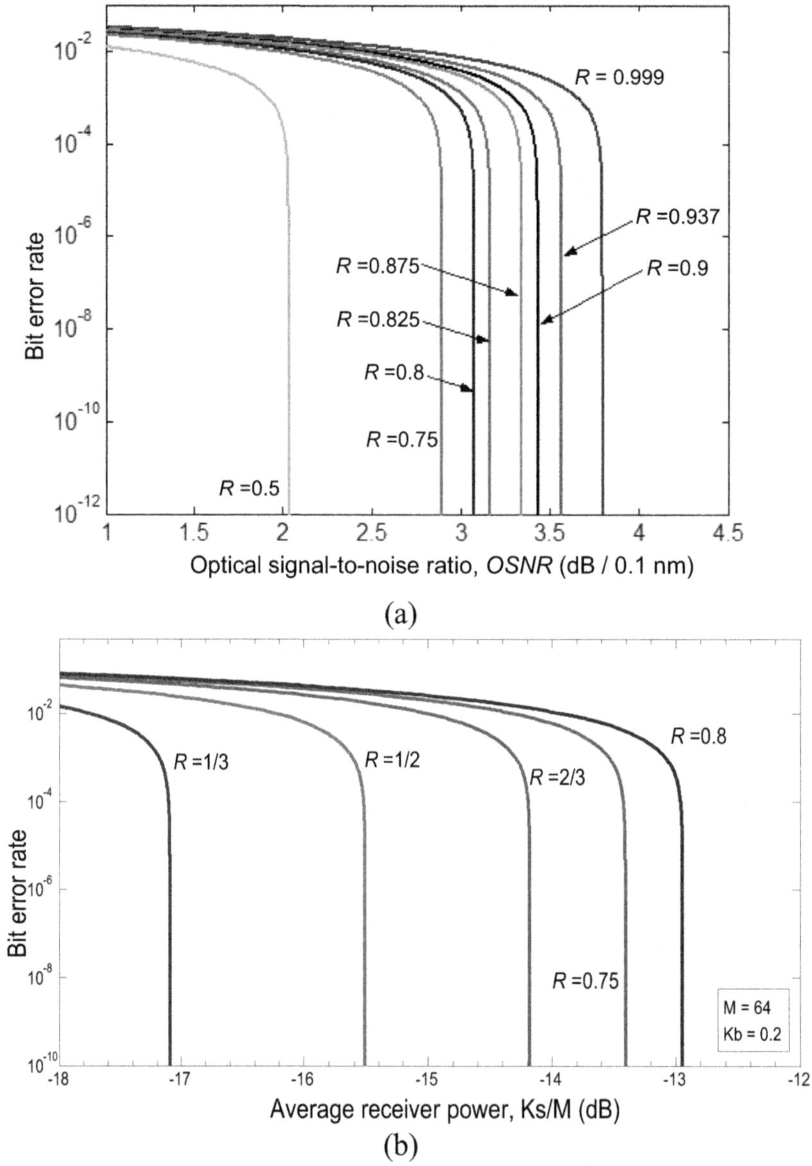

FIGURE 6.6

(a) Minimum bit error rate (BER against optical signal-to-noise ratio for different code rate values (for binary phase-shift keying at 40 Gb/s). (b) Channel capacity BER) curves against average receiver power for 64 pulse-position modulation over a Poisson channel with background radiation of $K_b = 0.2$ and different code rates.

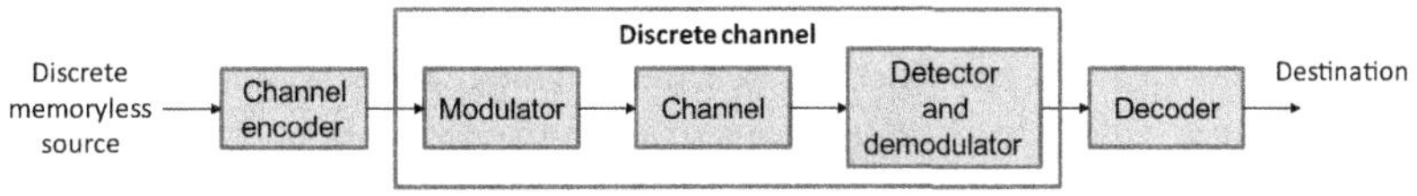

FIGURE 6.7

Block diagram of a point-to-point digital communication system.

during transmission, and (3) hybrid channel codes that combine the previous two approaches. In this chapter we are concerned only with FEC.

The key idea behind the forward error correcting codes is to add extra redundant symbols to the message to be transmitted, and use those redundant symbols in the decoding procedure to correct the errors introduced by the channel. Redundancy can be introduced in time, frequency, or space domains. For example, redundancy in the time domain is introduced if the same message is transmitted at least twice; the technique is known as the repetition code. Space redundancy is used as a means to achieve high spectrally efficient transmission, in which the modulation is combined with error control.

The codes commonly considered in digital communications and storage applications belong either to the class of block codes or to the class of convolutional codes. In an (n,k) block code the channel encoder accepts information in successive k-symbol blocks, and adds $n - k$ redundant symbols that are algebraically related to the k message symbols, thereby producing an overall encoded block of n symbols $(n > k)$, known as a codeword. If the block code is systematic, the information symbols stay unchanged during the encoding operation, and the encoding operation may be considered as adding the $n - k$ generalized parity checks to k information symbols. Since the information symbols are statistically independent (a consequence of source coding or scrambling), the next codeword is independent of the content of the current codeword. The code rate of an (n,k) block code is defined as $R = k/n$, and overhead by $OH = (1/R - 1) \cdot 100\%$. In convolutional code, however, the encoding operation may be considered as the discrete-time convolution of the input sequence with the impulse response of the encoder. Therefore the $n - k$ generalized parity checks are functions of not only k information symbols but also the functions of m previous k-tuples, with $m + 1$ being the encoder impulse response length. Statistical dependence is introduced to the window of length $n(m + 1)$, the parameter known as the constraint length of convolutional codes.

In the rest of this chapter an elementary introduction to linear block codes, cyclic codes, Reed–Solomon (RS) codes, concatenated codes, and product codes is given. For a detailed treatment of different error-control coding schemes, interested readers are referred to [17–22].

6.3 LINEAR BLOCK CODES

The linear block code (n,k), using the language of vector spaces, can be defined as a subspace of a vector space over finite field GF(q), with q being the prime power.

Every space is described by its basis—a set of linearly independent vectors. The number of vectors in the basis determines the dimension of the space. Therefore for an (n,k) linear block code the dimension of the space is n, and the dimension of the code subspace is k.

Example. $(n,1)$ repetition code. The repetition code has two codewords $\boldsymbol{x}_0 = (00 \ldots 0)$ and $\boldsymbol{x}_1 = (11 \ldots 1)$. Any linear combination of these two codewords is another codeword:

$$\boldsymbol{x}_0 + \boldsymbol{x}_0 = \boldsymbol{x}_0$$

$$\boldsymbol{x}_0 + \boldsymbol{x}_1 = \boldsymbol{x}_1 + \boldsymbol{x}_0 = \boldsymbol{x}_1$$

$$\boldsymbol{x}_1 + \boldsymbol{x}_1 = \boldsymbol{x}_0$$

The set of codewords from a linear block code forms a group under the addition operation, because an all-zero codeword serves as the identity element, and the codeword itself serves as the inverse element. This is the reason why the linear block codes are also called group codes. The linear block code (n,k) can be observed as a k-dimensional subspace of the vector space of all n-tuples over the binary filed GF(2) $= \{0,1\}$, with addition and multiplication rules given in Table 6.1. All n-tuples over GF(2) form the vector space. The sum of two n-tuples $\boldsymbol{a} = (a_1\ a_2 \ldots a_n)$ and $\boldsymbol{b} = (b_1\ b_2 \ldots b_n)$ is clearly an n-tuple and the commutative rule is valid because $\boldsymbol{c} = \boldsymbol{a} + \boldsymbol{b} = (a_1 + b_1\ a_2 + b_2 \ldots a_n + b_n) = (b_1 + a_1\ b_2 + a_2 \ldots b_n + a_n) = \boldsymbol{b} + \boldsymbol{a}$. The all-zero vector $\boldsymbol{0} = (0\ 0 \ldots 0)$ is the identity element, while n-tuple $\boldsymbol{a}$ itself is the inverse element $\boldsymbol{a} + \boldsymbol{a} = 0$. Therefore the n-tuples form the Abelian group with respect to the addition operation. The scalar multiplication is defined by: $\alpha\boldsymbol{a} = (\alpha a_1\ \alpha a_2 \ldots \alpha a_n)$, $\alpha \in$ GF(2). The distributive laws:

$$\alpha(\boldsymbol{a} + \boldsymbol{b}) = \alpha\boldsymbol{a} + \alpha\boldsymbol{b}$$

$$(\alpha + \beta)\boldsymbol{a} = \alpha\boldsymbol{a} + \beta\boldsymbol{a},\ \forall\ \alpha,\ \beta \in \text{GF}(2)$$

are also valid. The associate law $(\alpha \cdot \beta)\boldsymbol{a} = \alpha \cdot (\beta\boldsymbol{a})$ is clearly satisfied. Therefore the set of all n-tuples is a vector space over GF(2). The set of all code words from an (n,k) linear block code forms an Abelian group under the addition operation. It can be shown, in a fashion similar to the foregoing, that all codewords of an (n,k) linear block code form the vector space of dimensionality k. There exist k basis vectors (codewords) such that every codeword is a linear combination of these codewords.

Table 6.1 Addition (+) and multiplication (•) rules.

+	0	1	•	0	1
0	0	1	0	0	0
1	1	0	1	0	1

Example. $(n,1)$ repetition code: $C = \{(0\,0 \ldots 0), (1\,1 \ldots 1)\}$. Two codewords in C can be represented as a linear combination of an all-ones basis vector: $(11 \ldots 1) = 1 \cdot (11 \ldots 1)$, $(00 \ldots 0) = 1 \cdot (11 \ldots 1) + 1 \cdot (11 \ldots 1)$.

6.3.1 Generator Matrix for Linear Block Code

Any codeword $\boldsymbol{x}$ from the (n,k) linear block code can be represented as a linear combination of k basis vectors $\boldsymbol{g}_i$ $(i = 0,1,\ldots,k-1)$:

$$\boldsymbol{x} = m_0\boldsymbol{g}_0 + m_1\boldsymbol{g}_1 + \ldots + m_{k-1}\boldsymbol{g}_{k-1} = \boldsymbol{m}\begin{bmatrix} \boldsymbol{g}_0 \\ \boldsymbol{g}_1 \\ \ldots \\ \boldsymbol{g}_{k-1} \end{bmatrix} = \boldsymbol{mG};\ \boldsymbol{G} = \begin{bmatrix} \boldsymbol{g}_0 \\ \boldsymbol{g}_1 \\ \ldots \\ \boldsymbol{g}_{k-1} \end{bmatrix},$$
$$\boldsymbol{m} = \begin{pmatrix} m_0 & m_1 & \ldots & m_{k-1} \end{pmatrix}, \tag{6.46}$$

where $\boldsymbol{m}$ is the message vector and $\boldsymbol{G}$ is the generator matrix (of dimensions $k \times n$), in which every row represents a vector from the coding subspace. Therefore to encode, the message vector $\boldsymbol{m}(m_0,m_1,\ldots,m_{k-1})$ has to be multiplied with a generator matrix $\boldsymbol{G}$ to get $\boldsymbol{x} = \boldsymbol{mG}$, where $\boldsymbol{x}(x_0,x_1,\ldots,x_{n-1})$ is a codeword.

Example. Generator matrices for repetition $(n,1)$ code $\boldsymbol{G}_{\text{rep}}$ and $(n,n-1)$ single-parity-check code $\boldsymbol{G}_{\text{par}}$ are given, respectively, as:

$$\boldsymbol{G}_{\text{rep}} = [11\ldots1] \quad \boldsymbol{G}_{\text{par}} = \begin{bmatrix} 100\ldots01 \\ 010\ldots01 \\ \ldots \\ 000\ldots11 \end{bmatrix}.$$

By elementary operations on rows in the generator matrix, the code may be transformed into systematic form:

$$\boldsymbol{G}_s = [\boldsymbol{I}_k|\boldsymbol{P}], \tag{6.47}$$

where $\boldsymbol{I}_k$ is unity matrix of dimensions $k \times k$ and $\boldsymbol{P}$ is the coefficient matrix of dimensions $k \times (n-k)$ with columns denoting the positions of parity checks:

$$\boldsymbol{P} = \begin{bmatrix} p_{00} & p_{01} & \cdots & p_{0,n-k-1} \\ p_{10} & p_{11} & \cdots & p_{1,n-k-1} \\ \ldots & \ldots & & \ldots \\ p_{k-1,0} & p_{k-1,1} & \cdots & p_{k-1,n-k-1} \end{bmatrix}.$$

The codeword of a systematic code is obtained by:

$$\boldsymbol{x} = [\boldsymbol{m}|\boldsymbol{b}] = \boldsymbol{m}[\boldsymbol{I}_k|\boldsymbol{P}] = \boldsymbol{mG}, \qquad \boldsymbol{G} = [\boldsymbol{I}_k|\boldsymbol{P}], \tag{6.48}$$

and the structure of the systematic codeword is shown in Fig. 6.8.

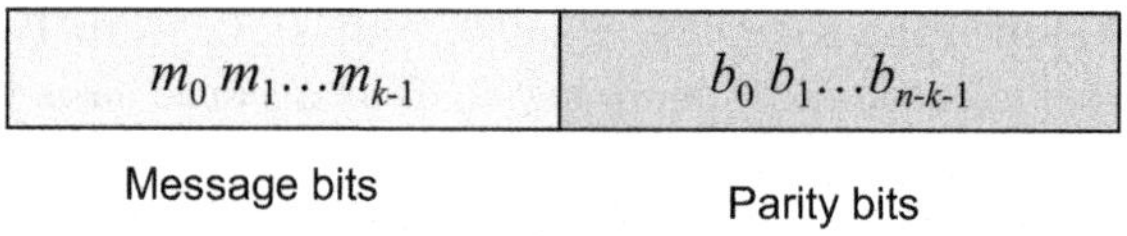

FIGURE 6.8

Structure of the systematic codeword.

Therefore during encoding the message vector stays unchanged and the elements of a vector of parity checks $\boldsymbol{b}$ are obtained by:

$$b_i = p_{0i}m_0 + p_{1i}m_1 + \ldots + p_{k-1,i}m_{k-1}, \tag{6.49}$$

where:

$$p_{ij} = \begin{cases} 1, & b_i \quad \text{depends} \quad m_j \\ 0, & \text{otherwise} \end{cases}$$

During transmission the channel introduces the errors so that the received vector $\boldsymbol{r}$ can be written as $\boldsymbol{r} = \boldsymbol{x} + \boldsymbol{e}$, where $\boldsymbol{e}$ is the error vector (pattern) with components determined by:

$$e_i = \begin{cases} 1 & \text{if an error occurred in the } i\text{th location} \\ 0 & \text{otherwise.} \end{cases}$$

To determine whether the received vector $\boldsymbol{r}$ is a codeword vector, we introduce the concept of a parity-check matrix.

6.3.2 Parity-Check Matrix for Linear Block Code

Another useful matrix associated with linear block codes is the parity-check matrix. Let us expand the matrix equation $\boldsymbol{x} = \boldsymbol{mG}$ in scalar form as follows:

$$\begin{aligned}
x_0 &= m_0 \\
x_1 &= m_1 \\
&\ldots \\
x_{k-1} &= m_{k-1} \\
x_k &= m_0p_{00} + m_1p_{10} + \ldots + m_{k-1}p_{k-1,0} \\
x_{k+1} &= m_0p_{01} + m_1p_{11} + \ldots + m_{k-1}p_{k-1,1} \\
&\ldots \\
x_{n-1} &= m_0p_{0,n-k-1} + m_1p_{1,n-k-1} + \ldots + m_{k-1}p_{k-1,n-k-1}.
\end{aligned} \tag{6.50}$$

By using the first k equalities, the last $n - k$ equations can be rewritten as follows:

$$\begin{aligned}
&x_0p_{00} + x_1p_{10} + \ldots + x_{k-1}p_{k-1,0} + x_k = 0\\
&x_0p_{01} + x_1p_{11} + \ldots + x_{k-1}p_{k-1,0} + x_{k+1} = 0\\
&\ldots\\
&x_0p_{0,\,n-k+1} + x_1p_{1,\,n-k-1} + \ldots + x_{k-1}p_{k-1,\,n-k+1} + x_{n-1} = 0.
\end{aligned} \tag{6.51}$$

The matrix representation of Eq. (6.51) is:

$$\begin{bmatrix} x_0 & x_1 & \ldots & x_{n-1} \end{bmatrix} \begin{bmatrix} p_{00} & p_{10} & \ldots & p_{k-1,\,0} & 1 & 0 & \ldots & 0\\ p_{01} & p_{11} & \ldots & p_{k-1,\,1} & 0 & 1 & \ldots & 0\\ \ldots & & \ldots & & \ldots & & & \ldots\\ p_{0,\,n-k-1} & p_{1,\,n-k-1} & \ldots & p_{k-1,\,n-k-1} & 0 & 0 & \ldots & 1 \end{bmatrix}^T$$

$$= \boldsymbol{x}\begin{bmatrix} \boldsymbol{P}^T & \boldsymbol{I}_{n-k} \end{bmatrix} = \boldsymbol{x}\boldsymbol{H}^T = \boldsymbol{0}, \quad \boldsymbol{H} = \begin{bmatrix} \boldsymbol{P}^T & \boldsymbol{I}_{n-k} \end{bmatrix}_{(n-k)xn}. \tag{6.52}$$

The $\boldsymbol{H}$-matrix in Eq. (6.52) is known as the parity-check matrix. We can easily verify that:

$$\boldsymbol{G}\boldsymbol{H}^T = \begin{bmatrix} \boldsymbol{I}_k & \boldsymbol{P} \end{bmatrix}\begin{bmatrix} \boldsymbol{P} \\ \boldsymbol{I}_{n-k} \end{bmatrix} = \boldsymbol{P} + \boldsymbol{P} = \boldsymbol{0}, \tag{6.53}$$

meaning that the parity-check matrix of an (n,k) linear block code $\boldsymbol{H}$ is a matrix of rank $n - k$ and dimensions $(n - k) \times n$ whose null space is a k-dimensional vector with basis forming the generator matrix $\boldsymbol{G}$.

Example. Parity-check matrices for $(n,1)$ repetition code $\boldsymbol{H}_{\text{rep}}$ and $(n,n - 1)$ single-parity check code $\boldsymbol{H}_{\text{par}}$ are given, respectively, as:

$$\boldsymbol{H}_{\text{rep}} = \begin{bmatrix} 100\ldots01 \\ 010\ldots01 \\ \ldots \\ 000\ldots11 \end{bmatrix} \quad \boldsymbol{H}_{\text{par}} = [11\ldots1].$$

Example. For Hamming (7,4) code the generator $\boldsymbol{G}$ and parity-check $\boldsymbol{H}$ matrices are given, respectively, as:

$$\boldsymbol{G} = \begin{bmatrix} 110|1000 \\ 011|0100 \\ 111|0010 \\ 101|0001 \end{bmatrix}, \quad \boldsymbol{H} = \begin{bmatrix} 100|1011 \\ 010|1110 \\ 001|0111 \end{bmatrix}$$

Every (n,k) linear block code with generator matrix $\boldsymbol{G}$ and parity-check matrix $\boldsymbol{H}$ has a dual code with generator matrix $\boldsymbol{H}$ and parity-check matrix $\boldsymbol{G}$. For example, $(n,1)$ repetition and $(n,n - 1)$ single-parity-check codes are dual.

6.3.3 Distance Properties of Linear Block Codes

To determine the error correction capability of the code we have to introduce the concept of Hamming distance and Hamming weight. Hamming distance between two codewords $\boldsymbol{x}_1$ and $\boldsymbol{x}_2$, $d(\boldsymbol{x}_1,\boldsymbol{x}_2)$, is defined as the number of locations in which their respective elements differ. Hamming weight, $w(\boldsymbol{x})$, of a codeword vector $\boldsymbol{x}$ is defined as the number of nonzero elements in the vectors. The minimum distance, $d_{\min}$, of a linear block code is defined as the smallest Hamming distance between any pair of code vectors in the code. Since the zero vector is a codeword, the minimum distance of a linear block code can be determined simply as the smallest Hamming weight of the nonzero code vectors in the code. Let the parity-check matrix be written as $\boldsymbol{H} = [\boldsymbol{h}_1\ \boldsymbol{h}_2 \ldots \boldsymbol{h}_n]$, where $\boldsymbol{h}_i$ is the ith column in $\boldsymbol{H}$. Since every codeword $\boldsymbol{x}$ must satisfy the syndrome equation, $\boldsymbol{x}\boldsymbol{H}^T = \boldsymbol{0}$ (see Eq. 6.52), the minimum distance of a linear block code is determined by the minimum number of columns of the $\boldsymbol{H}$-matrix whose sum is equal to the zero vector. For example, (7,4) Hamming code in the foregoing example has the minimum distance $d_{\min} = 3$ since the addition of first, fifth, and sixth columns leads to zero vector. The codewords can be represented as points in n-dimensional space, as shown in Fig. 6.9. The decoding process can be visualized by creating the spheres of radius t around codeword points. The received word vector $\boldsymbol{r}$ in Fig. 6.9A will be decoded as a codeword $\boldsymbol{x}_i$ because its Hamming distance $d(\boldsymbol{x}_i,\boldsymbol{r}) \leq t$ is closest to the codeword $\boldsymbol{x}_i$. On the other hand, in the example shown in Fig. 6.9B the Hamming distance $d(\boldsymbol{x}_i,\boldsymbol{x}_j) \leq 2t$ and the received vector $\boldsymbol{r}$ that falls in the intersection area of the two spheres cannot be uniquely decoded.

Therefore an (n,k) linear block code of minimum distance $d_{\min}$ can correct up to t errors if, and only if, $t \leq \rfloor 1/2(d_{\min} - 1) \lfloor$ (where $\lfloor\rfloor$ denotes the largest integer less than or equal to the enclosed quantity) or equivalently $d_{\min} \geq 2t + 1$. If we are only interested in detecting e_d errors, then $d_{\min} \geq e_d + 1$. Finally, if we are interested in detecting e_d errors and correcting e_c errors, then $d_{\min} \geq e_d + e_c + 1$. The Hamming (7,4) code is therefore a single error-correcting and double error-detecting code. More generally, a family of (n,k) linear block codes with the following parameters:

- Block length: $n = 2^m - 1$
- Number of message bits: $k = 2^m - m - 1$
- Number of parity bits: $n - k = m$
- $d_{\min} = 3$

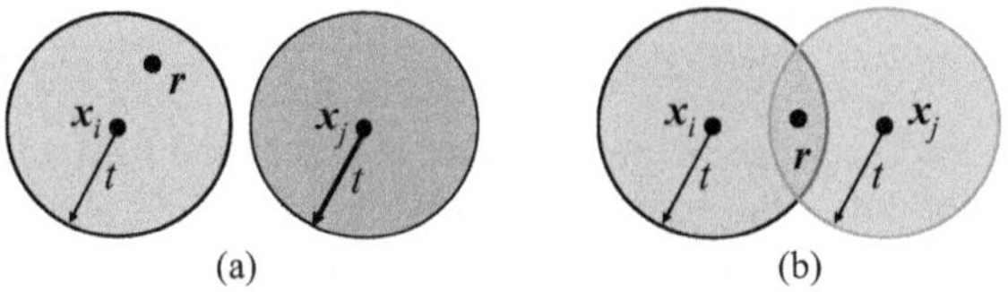

FIGURE 6.9

Hamming distance: (a) $d(x_i,x_j) \geq 2t + 1$ and (b) $d(x_i,x_j) < 2t + 1$.

where $m \geq 3$, are known as Hamming codes. Hamming codes belong to the class of perfect codes, the codes that satisfy the following Hamming inequality with equality sign bib11[2,11]:

$$2^{n-k} \geq \sum_{i=0}^{t} \binom{n}{i}. \tag{6.54}$$

This bound gives how many errors t can be corrected with an (n,k) linear block code by using syndrome decoding (described later).

6.3.4 Coding Gain

A very important characteristic of an (n,k) linear block code is so-called coding gain, which was introduced earlier as being the savings attainable in the energy per information bit-to-noise spectral density ratio (E_b/N_0) required to achieve a given bit error probability when coding is used compared to that with no coding. Let E_c denote the transmitted bit energy and E_b denote the information bit energy. Since the total information word energy kE_b must be the same as the total codeword energy nE_c, we obtain the following relationship between E_c and E_b:

$$E_c = (k\,/\,n)E_b = RE_b. \tag{6.55}$$

The probability of error for BPSK on an additive white Gaussian noise (AWGN) channel, when a coherent hard decision (bit-by-bit) demodulator is used, can be obtained as follows:

$$p = \frac{1}{2}\mathrm{erfc}\left(\sqrt{\frac{E_c}{N_0}}\right) = \frac{1}{2}\mathrm{erfc}\left(\sqrt{\frac{RE_b}{N_0}}\right), \tag{6.56}$$

where the erfc(x) function is defined by:

$$\mathrm{erfc}(x) = \frac{2}{\sqrt{\pi}} \int_{x}^{+\infty} e^{-z^2} dz$$

For high SNRs the word error probability (remaining upon decoding) of a t-error correcting code is dominated by a $t+1$-error event:

$$P_w(e) \approx \binom{n}{t+1} p^{t+1}(1-p)^{n-t+1} \approx \binom{n}{t+1} p^{t+1}. \tag{6.57}$$

The bit error probability P_b is related to the word error probability by:

$$P_b \approx \frac{2t+1}{n} P_w(e) \approx c(n,t)p^{t+1}, \tag{6.58}$$

because $2t+1$ and more errors per codeword cannot be corrected, they can be located anywhere on n codeword locations, and $c(n,t)$ is a parameter dependent on error correcting capability t and codeword length n. By using the upper bound on $\mathrm{erfc}(x)$ we obtain:

$$P_b \approx \frac{c(n,t)}{2}\left[\exp\left(\frac{-RE_b}{N_0}\right)\right]^{t+1}. \tag{6.59}$$

The corresponding approximation for the uncoded case is:

$$P_{b,\,\text{uncoded}} \approx \frac{1}{2}\exp\left(-\frac{E_b}{N_0}\right). \tag{6.60}$$

By equating Eqs. (6.59) and (6.60) and ignoring the parameter $c(n,t)$ we obtain the following expression for hard decision decoding coding gain:

$$\frac{(E_b/N_0)_{\text{uncoded}}}{(E_b/N_0)_{\text{coded}}} \approx R(t+1). \tag{6.61}$$

The corresponding soft decision asymptotic coding gain of convolutional codes is [19–21]:

$$\frac{(E_b/N_0)_{\text{uncoded}}}{(E_b/N_0)_{\text{coded}}} \approx Rd_{\min}, \tag{6.62}$$

and is about 3 dB better than hard decision decoding (because $d_{\min} \geq 2t + 1$).

In optical communications it is very common to use the Q-factor as the figure of merit instead of SNR, which is related to the BER on an AWGN as follows:

$$\mathrm{BER} = \frac{1}{2}\mathrm{erfc}\left(\frac{Q}{\sqrt{2}}\right). \tag{6.63}$$

Let $\mathrm{BER_{in}}$ denote the BER at the input of an FEC decoder, let $\mathrm{BER_{out}}$ denote the BER at the output of an FEC decoder, and let $\mathrm{BER_{ref}}$ denote target BER (such as either 10^{-12} or 10^{-15}). The corresponding coding gain (CG) and net coding gain (NCG) are, respectively, defined as [23]:

$$\mathrm{CG} = 20\log_{10}\left[\mathrm{erfc}^{-1}(2\mathrm{BER_{ref}})\right] - 20\log_{10}\left[\mathrm{erfc}^{-1}(2\mathrm{BER_{in}})\right] \quad [\mathrm{dB}] \tag{6.64}$$

$$\mathrm{NCG} = 20\log_{10}\left[\mathrm{erfc}^{-1}(2\mathrm{BER_{ref}})\right] - 20\log_{10}\left[\mathrm{erfc}^{-1}(2\mathrm{BER_{in}})\right] + 10\log_{10}R \quad [\mathrm{dB}]. \tag{6.65}$$

All coding gains reported in this chapter are in fact NCG, although they are sometimes called coding gains only, because this is the common practice in coding theory literature [19–21].

6.3.5 Syndrome Decoding and Standard Array

The received vector $\boldsymbol{r} = \boldsymbol{x} + \boldsymbol{e}$ ($\boldsymbol{x}$ is the codeword and $\boldsymbol{e}$ is the error pattern introduced earlier) is a codeword if the following syndrome equation is satisfied: $\boldsymbol{s} = \boldsymbol{r}\boldsymbol{H}^T = \boldsymbol{0}$. The syndrome has the following important properties:

1. The syndrome is the only function of the error pattern. This property can easily be proved from the definition of syndrome as follows: $\boldsymbol{s} = \boldsymbol{r}\boldsymbol{H}^T = (\boldsymbol{x} + \boldsymbol{e})\boldsymbol{H}^T = \boldsymbol{x}\boldsymbol{H}^T + \boldsymbol{e}\boldsymbol{H}^T = \boldsymbol{e}\boldsymbol{H}^T$.
2. All error patterns that differ by a codeword have the same syndrome. This property can also be proved from syndrome definition. Let $\boldsymbol{x}_i$ be the ith $(i = 0,1,\ldots,2^{k-1})$ codeword. The set of error patterns that differ by a codeword is known as a coset: $\{\boldsymbol{e}_i = \boldsymbol{e} + \mathbf{x}_i;\ i = 0,1,\ldots,2^{k-1}\}$. The syndrome corresponding to ith error pattern from this set $\boldsymbol{s}_i = \boldsymbol{r}_i\boldsymbol{H}^T = (\boldsymbol{x}_i + \boldsymbol{e})\boldsymbol{H}^T = \boldsymbol{x}_i\boldsymbol{H}^T + \boldsymbol{e}\boldsymbol{H}^T = \boldsymbol{e}\boldsymbol{H}^T$ is the only function of the error pattern, and therefore all error patterns from the coset have the same syndrome.
3. The syndrome is a function of only those columns of a parity-check matrix corresponding to the error locations. The parity-check matrix can be written in the following form: $\boldsymbol{H} = [\boldsymbol{h}_1 \ldots \boldsymbol{h}_n]$, where the ith element $\boldsymbol{h}_i$ denotes the ith column of $\boldsymbol{H}$. Based on syndrome definition for an error pattern $\boldsymbol{e} = [e_1 \ldots e_n]$ the following is valid:

$$\boldsymbol{s} = \boldsymbol{e}\boldsymbol{H}^T = \begin{bmatrix} e_1 & e_2 & \ldots & e_n \end{bmatrix} \begin{bmatrix} \boldsymbol{h}_1^T \\ \boldsymbol{h}_2^T \\ \ldots \\ \boldsymbol{h}_n^T \end{bmatrix} = \sum_{i=1}^{n} e_i \boldsymbol{h}_i^T, \tag{6.66}$$

which proves the claim of property 3.

4. With syndrome decoding an (n,k) linear block code can correct up to t errors, providing that Hamming bound (6.21) is satisfied. (This property will be proved in the next subsection.)

By using property 2, 2^k codewords partition the space of all received words into 2^k disjoint subsets. Any received word within the subsets will be decoded as the unique codeword. A standard array is a technique by which this partition can be achieved, and can be constructed using the following two steps [19–21]:

1. Write down 2^k codewords as elements of the first row, with the all-zero codeword as the leading element.
2. Repeat steps 2(a) and 2(b) until all 2^n words are exhausted.
   - **(a)** Out of the remaining unused n-tuples, select one with the least weight for the leading element of the next row.
   - **(b)** Complete the current row by adding the leading element to each nonzero codeword appearing in the first row and writing down the resulting sum in the corresponding column.

The standard array for an (n,k) block code obtained by this algorithm is illustrated in Fig. 6.10. The columns represent 2^k disjoint sets, and every row represents the coset of the code with leading elements being called the coset leaders:

$$
\begin{array}{ccccccccc}
x_1 = 0 & x_2 & x_3 & \cdots & x_i & \cdots & x_{2^k} \\
e_2 & x_2 + e_2 & x_3 + e_2 & \cdots & x_i + e_2 & \cdots & x_{2^k} + e_2 \\
e_3 & x_2 + e_3 & x_3 + e_3 & \cdots & x_i + e_2 & \cdots & x_{2^k} + e_3 \\
\cdots & \cdots & & \cdots & & \cdots & \cdots \\
e_j & x_2 + e_j & x_3 + e_j & \cdots & x_i + e_j & \cdots & x_{2^k} + e_j \\
\cdots & \cdots & & \cdots & & & \cdots \\
e_{2^{n-k}} & x_2 + e_{2^{n-k}} & x_3 + e_{2^{n-k}} & \cdots & x_i + e_{2^{n-k}} & \cdots & x_{2^k} + e_{2^{n-k}}
\end{array}
$$

FIGURE 6.10

Standard array architecture.

Example. A standard array of (5,2) code $C = \{(00000),(11010),(10101),(01111)\}$ is given in Table 6.2. The parity-check matrix of this code is given by:

$$
H = \begin{bmatrix} 1 & 0 & 0 & 1 & 1 \\ 0 & 1 & 0 & 1 & 0 \\ 0 & 0 & 1 & 0 & 1 \end{bmatrix}
$$

Because the minimum distance of this code is 3 (first, second, and fourth columns add to zero), this code is able to correct all single errors. For example, if the word 01010 is received it will be decoded to the top-most codeword 11010 of the column in which it lies. In the same table, corresponding syndromes are provided as well.

The syndrome decoding procedure is a three-step procedure [19–21]:

1. For the received vector $\boldsymbol{r}$, compute the syndrome $\boldsymbol{s} = \boldsymbol{rH}^T$. From property 3 we can establish one-to-one correspondence with the syndromes and error patterns (Table 6.2), leading to the lookup table (LUT) containing the syndrome and corresponding error pattern (the coset leader).

Table 6.2 Standard array of (5,2) code and corresponding decoding table.

	Codewords				**Syndrome s**	**Error pattern**
Coset leader	00000	11010	10101	01111	000	00000
	00001	11011	10100	01110	101	00001
	00010	11000	10111	01101	110	00010
	00100	11110	10001	01011	001	00100
	01000	10010	11101	00111	010	01000
	10000	01010	00101	11111	100	10000
	00011	11001	10110	01100	011	00011
	00110	11100	10011	01001	111	00110

2. Within the coset characterized by the syndrome s, identify the coset leader, say e_0. The coset leader corresponds to the error pattern with the largest probability of occurrence.
3. Decode the received vector as $x = r + e_0$.

Example. Let the received vector for the foregoing (5,2) code example be $r =$ (01010). The syndrome can be computed as $s = rH^T = (100)$, and the corresponding error pattern from the LUT is found to be $e_0 = (10000)$. The decoded word is obtained by adding the error pattern to the received word $x = r + e_0 =$ (11010), and the error on the first bit position is corrected.

The standard array can be used to determine the probability of word error as follows:

$$P_w(e) = 1 - \sum_{i=0}^{n} \alpha_i p^i (1-p)^{n-i}, \tag{6.67}$$

where α_i is the number of coset leaders of weight i (distribution of weights is also known as weight distribution of coset leaders) and p is the crossover probability of the binary symmetric channel (BSC). Any error pattern that is not a coset leader will result in decoding error. For example, the weight distributions of coset leaders in (5,2) code are $\alpha_0 = 1$, $\alpha_1 = 5$, $\alpha_2 = 2$, $\alpha_i = 0$, $i = 3,4,5$, which leads to the following word error probability:

$$P_w(e) = 1 - (1-p)^5 - 5p(1-p)^4 - 2p^2(1-p)^3\big|_{p=10^{-3}} = 7.986 \cdot 10^{-6}.$$

We can use Eq. (6.67) to estimate the coding gain of a given linear block code. For example, the word error probability for Hamming (7,4) code is:

$$P_w(e) = 1 - (1-p)^7 - 7p(1-p)^6 = \sum_{i=2}^{7} \binom{7}{i} p^i (1-p)^{7-i} \approx 21p^2.$$

In the previous section, we established the following relationship between bit and word error probabilities: $P_b \approx P_w(e)(2t+1)/n = (3/7)P_w(e) \approx (3/7)21p^2 = 9p^2$. Therefore the crossover probability can be evaluated as:

$$p = \sqrt{P_b}/3 = (1/2)\operatorname{erfc}\left(\sqrt{\frac{RE_b}{N_0}}\right)$$

From this expression we can easily calculate the required SNR to achieve target P_b. By comparing such obtained SNR with corresponding SNR for uncoded BPSK we can evaluate the corresponding coding gain.

To evaluate the probability of undetected error, we have to determine the other codeword weights as well. Because the undetected errors are caused by error patterns being identical to the nonzero codewords, the undetected error probability can be evaluated by:

$$P_u(e) = \sum_{i=1}^{n} A_i p^i (1-p)^{n-i} = (1-p)^n \sum_{i=1}^{n} A_i \left(\frac{p}{1-p}\right), \tag{6.68}$$

where p is the crossover probability and A_i denotes the number of codewords of weight i. The codeword weight can be determined by McWilliams identity that establishes the connection between codeword weights A_i and the codeword weights of the corresponding dual code B_i by [17]:

$$A(z) = 2^{-(n-k)}(1+z)^n B\left(\frac{1-z}{1+z}\right), \quad A(z) = \sum_{i=0}^{n} A_i z^i, \quad B(z) = \sum_{i=0}^{n} B_i z^i, \tag{6.69}$$

where $A(z)$ ($B(z)$) represents the polynomial representation of codeword weights (dual codeword weights). By substituting $z = p/(1-p)$ in Eq. (6.69) and knowing that $A_0 = 1$ we obtain:

$$A\left(\frac{p}{1-p}\right) - 1 = \sum_{i=1}^{n} A\left(\frac{p}{1-p}\right)^i. \tag{6.70}$$

Substituting Eq. (6.70) into Eq. (6.68) we obtain:

$$P_u(e) = (1-p)^n \left[A\left(\frac{p}{1-p}\right) - 1\right]. \tag{6.71}$$

An alternative expression for $P_u(e)$ in terms of $B(z)$ can be obtained from Eq. (6.69), which is more suitable for use when $n - k < k$, as follows:

$$P_u(e) = 2^{-(n-k)} B(1-2p) - (1-p)^n. \tag{6.72}$$

For large n, k, $n-k$ the use of McWilliams identity is impractical; in this case an upper bound on average probability of undetected error of an (n,k) systematic code should be used instead:

$$\overline{P_u}(e) \leq 2^{-(n-k)}[1-(1-p)^n]. \tag{6.73}$$

For a q-ary maximum distance separable code, which satisfies the Singleton bound introduced in the next section with equality, we can determine a closed formula for weight distribution [17]:

$$A_i = \binom{n}{i}(q-1)\sum_{j=0}^{i-d_{\min}} (-1)^j \binom{i-1}{j} q^{i-d_{\min}-j}, \tag{6.74}$$

where $d_{\min}$ is the minimum distance of the code, $A_0 = 1$, and $A_i = 0$ for $i \in [1, d_{\min}-1]$.

6.3.6 Important Coding Bounds

In this section we describe several important coding bounds, including Hamming, Plotkin, Gilbert–Varshamov, and Singleton bounds [19–21]. The Hamming bound has already been introduced for binary linear block codes by Eq. (6.54). The Hamming bound for q-ary (n,k) linear block code is given by:

$$\left[1+(q-1)\binom{n}{1}+(q-1)^2\binom{n}{2}+\ldots+(q-1)^i\binom{n}{i}+\ldots+(q-1)^t\binom{n}{t}\right]q^k \le q^n, \tag{6.75}$$

where t is the error correction capability and $(q-1)^i\binom{n}{i}$ is the number of received words that differ from a given codeword in i symbols. Namely, there are n chooses i ways in which symbols can be chosen out of n and there are $(q-1)^i$ possible choices for symbols. The codes satisfying the Hamming bound with an equality sign are known as perfect codes. Hamming codes are perfect codes because $n = 2^{n-k} - 1$, which is equivalent to $(1+n)2^k = 2^n$ so that the foregoing inequality is satisfied with equality. $(n,1)$ repetition code is also a perfect code. The three-error correcting (23,12) Golay code is another example of a perfect code because:

$$\left[1+\binom{23}{1}+\binom{23}{2}+\binom{23}{3}\right]2^{12}=2^{23}$$

The Plotkin bound is the bound on the minimum distance of a code:

$$d_{\min} \le \frac{n2^{k-1}}{2^k-1}. \tag{6.76}$$

Namely, if all codewords are written as the rows of a $2^k \times n$ matrix, each column will contain 2^{k-1} 0s and 2^{k-1} 1s, with the total weight of all codewords being $n2^{k-1}$.

The Gilbert–Varshamov bound is based on the property that the minimum distance $d_{\min}$ of a linear (n,k) block code can be determined as the minimum number of columns in an $\boldsymbol{H}$-matrix that sum to zero:

$$\binom{n-1}{1}+\binom{n-1}{2}+\ldots+\binom{n-1}{d_{\min}-2} < 2^{n-k}-1. \tag{6.77}$$

Another important bound is the Singleton bound:

$$d_{\min} \le n-k+1. \tag{6.78}$$

This bound is straightforward to prove. Let only 1 bit of value 1 be present in the information vector. If it is involved in $n-k$ parity checks, then the total number of 1s in the codeword cannot be larger than $n-k+1$. The codes satisfying the Singleton bound with an equality sign are known as the maximum distance separable (MDS) codes (e.g., RS codes are MDS codes).

6.4 CYCLIC CODES

The most commonly used class of linear block codes is the class of cyclic codes. Examples of cyclic codes include Bose–Chaudhuri–Hocquenghem (BCH) codes, Hamming codes, and Golay codes. RS codes are also cyclic but nonbinary codes. Even LDPC codes can be designed in cyclic or quasi-cyclic fashion.

Let us observe the vector space of dimension n. The subspace of this space is cyclic code if for any codeword $\boldsymbol{c}(c_0,c_1,\ldots,c_{n-1})$ arbitrary cyclic shift $\boldsymbol{c}_j(c_{n-j},c_{n-j+1},\ldots,c_{n-1},c_0,c_1,\ldots,c_{n-j-1})$ is another codeword. With every codeword $\boldsymbol{c}(c_0,c_1,\ldots,c_{n-1})$ from a cyclic code, we associate the codeword polynomial:

$$c(x) = c_0 + c_1x + c_2x^2 + \cdots + c_{n-1}x^{n-1}. \tag{6.79}$$

The jth cyclic shift, observed mod$(x^n - 1)$, is also a codeword polynomial:

$$c^{(j)} = x^j c(x)\text{mod}(x^n - 1). \tag{6.80}$$

It is straightforward to show that observed subspace is cyclic if composed from polynomials divisible by a polynomial $g(x) = g_0 + g_1x + \ldots + g_{n-k}x^{n-k}$ that divides $x^n - 1$ at the same time. The polynomial $g(x)$, of degree $n - k$, is called the generator polynomial of the code. If $x^n - 1 = g(x)h(x)$, then the polynomial of degree k is called the parity-check polynomial. The generator polynomial has the following three important properties [19–21]:

1. The generator polynomial of an (n,k) cyclic code is unique (usually proved by contradiction);
2. Any multiple of the generator polynomial is a codeword polynomial; and
3. The generator polynomial and parity-check polynomial are factors of $x^n - 1$.

The generator polynomial $g(x)$ and the parity-check polynomial $h(x)$ serve the same role as the generator matrix $\boldsymbol{G}$ and parity-check matrix $\boldsymbol{H}$ of a linear block code. n-Tuples related to the k polynomials $g(x),xg(x),\ldots,x^{k-1}g(x)$ may be used in rows of the $k \times n$ generator matrix $\boldsymbol{G}$, while n-tuples related to the $(n - k)$ polynomials $x^kh(x^{-1}),x^{k+1}h(x^{-1}),\ldots,x^{n-1}h(x^{-1})$ may be used in rows of the $(n - k) \times n$ parity-check matrix $\boldsymbol{H}$.

To encode we have simply to multiply the message polynomial $m(x) = m_0 + m_1x + \ldots + m_{k-1}x^{k-1}$ with the generator polynomial $g(x)$, i.e., $c(x) = m(x)g(x)$ mod$(x^n - 1)$, where $c(x)$ is the codeword polynomial. To encode in systematic form we have to find the remainder of $x^{n-k}m(x)/g(x)$ and add it to the shifted version of the message polynomial $x^{n-k}m(x)$, i.e., $c(x) = x^{n-k}m(x) + \text{rem}[x^{n-k}m(x)/g(x)]$, where rem[ ] denoted as the remainder of a given entity. The general circuit for generating the codeword polynomial in systematic form is given in Fig. 6.11A. The encoder operates as follows. When the switch S is in position 1 and the gate is closed (on), the information bits are shifted into the shift register and at the same time transmitted onto the channel. Once all information bits are shifted into the register in k shifts, with the gate being open (off), the switch S is moved to position 2, and the content of the $(n - k)$-shift register is transmitted onto the channel.

To check if the received word polynomial is the codeword polynomial $r(x) = r_0 + r_1x + \ldots + r_{n-1}x^{n-1}$ we have simply to determine the syndrome polynomial

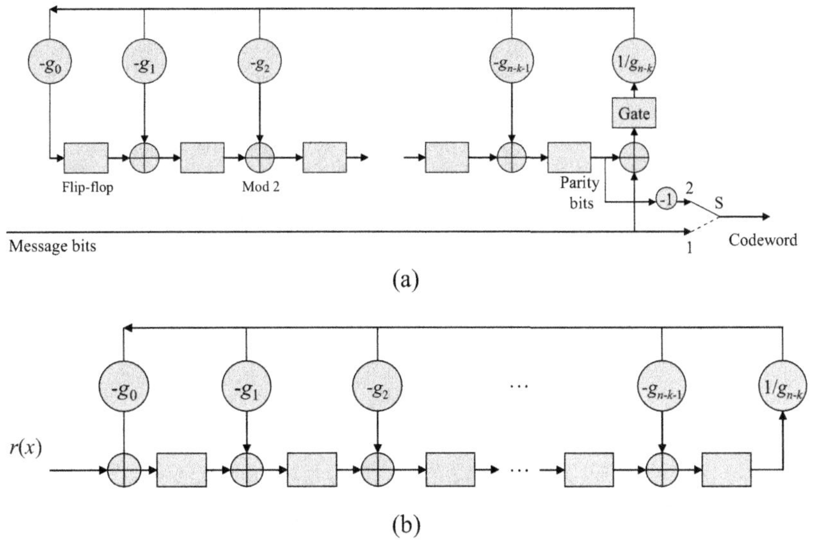

FIGURE 6.11

(a) Systematic cyclic encoder and (b) syndrome calculator.

$s(x) = \text{rem}[r(x)/g(x)]$. If $s(x)$ is zero, then there is no error introduced during transmission. The corresponding circuit is shown in Fig. 6.11B.

For example, the encoder and syndrome calculator for (7,4) Hamming code are given in Fig. 6.12A and B, respectively. The generating polynomial is given by $g(x) = 1 + x + x^3$. The polynomial $x^7 + 1$ can be factorized as follows: $x^7 + 1 = (1 + x)(1 + x^2 + x^3)(1 + x + x^3)$. If we select $g(x) = 1 + x + x^3$ as the generator polynomial, based on property 3 of the generator polynomial, the corresponding parity-check polynomial will be $h(x) = (1 + x)(1 + x^2 + x^3) = 1 +$

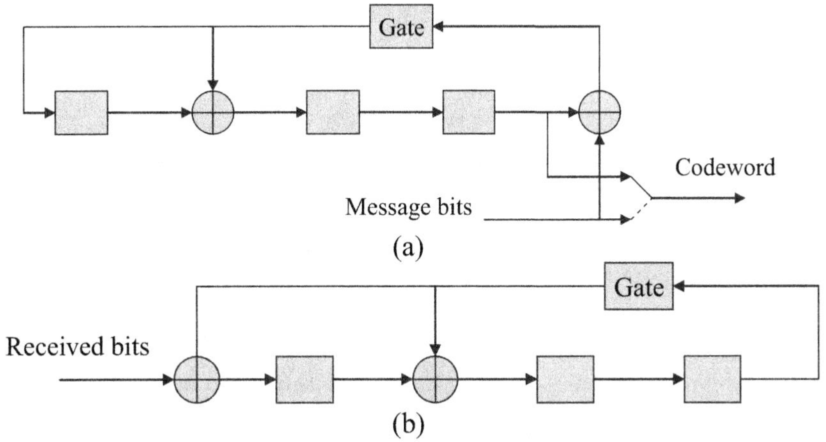

FIGURE 6.12

Hamming (7,4) encoder (a) and syndrome calculator (b).

$x + x^2 + x^4$. The message sequence 1001 can be represented in polynomial form by $m(x) = 1 + x^3$. For representation in systematic form, we have to multiply $m(x)$ by x^{n-k} to obtain: $x^{n-k}m(x) = x^3m(x) = x^3 + x^6$. The codeword polynomial is obtained by $c(x) = x^{n-k}m(x) + \text{rem}[x^{n-k}m(x)/g(x)] = x + x^3 + \text{rem}[(x + x^3)/(1 + x + x^3)] = x + x^2 + x^3 + x^6$. The corresponding codeword is 0111,001. To obtain the generator matrix of this code we can use the following polynomials: $g(x) = 1 + x + x^3$, $xg(x) = x + x^2 + x^4$, $x^2g(x) = x^2 + x^3 + x^5$, and $x^3g(x) = x^3 + x^4 + x^6$, and write the corresponding n-tuples in the form of a matrix as follows:

$$G' = \begin{bmatrix} 1101000 \\ 0110100 \\ 0011010 \\ 0001101 \end{bmatrix}.$$

By Gaussian elimination we can put the generator matrix in systematic form:

$$G = \begin{bmatrix} 1101000 \\ 0110100 \\ 1110010 \\ 1010001 \end{bmatrix}.$$

The parity-check matrix can be obtained from the following polynomials:

$$x^4h(x^{-1}) = 1 + x^2 + x^3 + x^4,\ x^5h(x^{-1}) = x + x^3 + x^4 + x^5,\ x^6h(x^{-1}) = x^2 + x^3 + x^5 + x^6,$$

by writing down the corresponding n-tuples in the form of a matrix:

$$H' = \begin{bmatrix} 1011100 \\ 0101110 \\ 0010111 \end{bmatrix}.$$

The $\boldsymbol{H}$-matrix can be put in systematic form by Gaussian elimination:

$$H = \begin{bmatrix} 1001011 \\ 0101110 \\ 0010111 \end{bmatrix}.$$

The syndrome polynomial $s(x)$ has the following three important properties, which can be used to simplify the implementation of decoders [19–21]:

1. The syndrome of received word polynomial $r(x)$ is also the syndrome of corresponding error polynomial $e(x)$;
2. The syndrome of a cyclic shift of $r(x)$, $xr(x)$, is determined by $xs(x)$; and

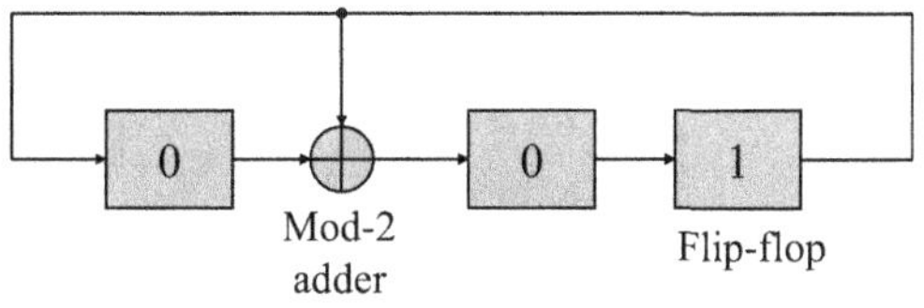

FIGURE 6.13

Encoder for the (7, 3) maximal-length code.

3. The syndrome polynomial $s(x)$ is identical to the error polynomial $e(x)$, if the errors are confined to the $(n - k)$ parity-check bits of the received word polynomial $r(x)$.

Maximal-length codes $(n = 2^m - 1, m)$ $(m \geq 3)$ are dual Hamming codes and have a minimum distance $d_{\min} = 2^m - 1$. The parity-check polynomial for (7,3) maximal-length codes is therefore $h(x) = 1 + x + x^3$. The encoder for (7,3) maximum-length code is given in Fig. 6.13. The generator polynomial gives one period of maximum-length code, providing that the encoder is initialized to 0 … 01. For example, the generator polynomial for the foregoing (7,3) maximum-length code is $g(x) = 1 + x + x^2 + x^4$, and the output sequence is given by:

$$\underbrace{1\,0\,0}_{\text{initial state}} \quad \underbrace{1\,1\,1\,0\,1\,0\,0}_{g(x)=1+x+x^2+x^4}$$

Cyclic redundancy check (CRC) codes are very popular codes for error detection. (n,k) CRC codes are capable of detecting [19–21]:

1. All error bursts of length $n - k$, with an error burst of length $n - k$ being defined as a contiguous sequence of $n - k$ bits in which the first and last bits or any other intermediate bits are received in error;
2. A fraction of error burst of length equal to $n - k + 1$; the fraction equals $1 - 2^{-(n-k-1)}$;
3. A fraction of error of length greater than $n - k + 1$; the fraction equals $1 - 2^{-(n-k-1)}$;
4. All combinations of $d_{\min} - 1$ (or fewer) errors;
5. All error patterns with an odd number of errors if the generator polynomial $g(x)$ for the code has an even number of nonzero coefficients.

In Table 6.3 we list the generator polynomials of several CRC codes, which are currently used in various communication systems.

Decoding of cyclic codes is composed of the same three steps used in decoding linear block codes, namely syndrome computation, error pattern identification, and error correction [19–21]. The *Meggit decoder* configuration, which is implemented based on syndrome property 2 (also known as Meggit theorem), is shown in Fig. 6.14. The syndrome is calculated by dividing the received word by generating polynomial $g(x)$, and at the same time the received word is shifted into the buffer

Table 6.3 Generator polynomials of several cyclic redundancy check (CRC) codes.

CRC codes	Generator polynomial	$n-k$
CRC-8 code (IEEE 802.16, WiMax)	$1+x^2+x^8$	8
CRC-16 code (IBM CRC-16, ANSI, USB, SDLC)	$1+x^2+x^{15}+x^{16}$	16
CRC-ITU (X25, V41, CDMA, Bluetooth, HDLC, PPP)	$1+x^5+x^{12}+x^{16}$	16
CRC-24 (WLAN, UMTS)	$1+x+x^5+x^6+x^{23}+x^{24}$	24
CRC-32 (Ethernet)	$1+x+x^2+x^4+x^5+x^7+x^8+x^{10}+x^{11}+x^{12}+x^{16}+x^{22}+x^{23}+x^{26}+x^{32}$	32

register. Once the last bit of the received word enters the decoder, the gate is turned off. The syndrome is further read into the error pattern detection circuit, and implemented as the combinational logic circuit, which generates 1 if and only if the content of the syndrome register corresponds to a correctible error pattern at the highest-order position x^{n-1}. By adding the output of the error pattern detector to the bit in error, the error can be corrected, and at the same time the syndrome has to be modified. If an error occurred on position x^1, by cyclically shifting the received word $n-1-1$ times, the erroneous bit will appear in position x^{n-1}, and can be corrected. The decoder therefore corrects the errors in a bit-by-bit fashion until the entire received word is read out from the buffer register.

The Hamming (7,4) cyclic decoder configuration, for generating polynomial $g(x) = 1+x+x^3$, is shown in Fig. 6.15. Because we expect only single errors, the error polynomial corresponding to the highest-order position is $e(x) = x^6$ and

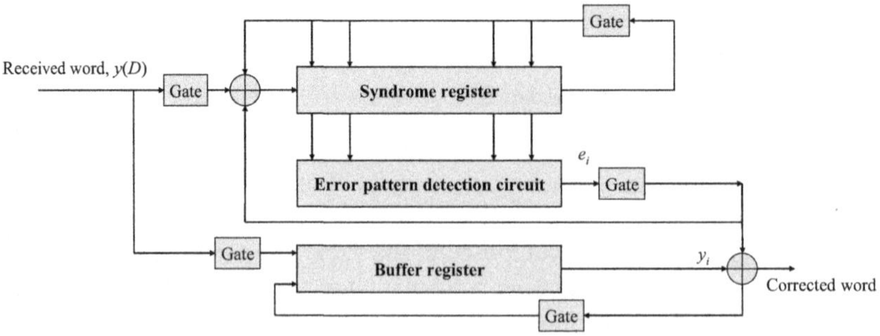

FIGURE 6.14

A Meggit decoder configuration.

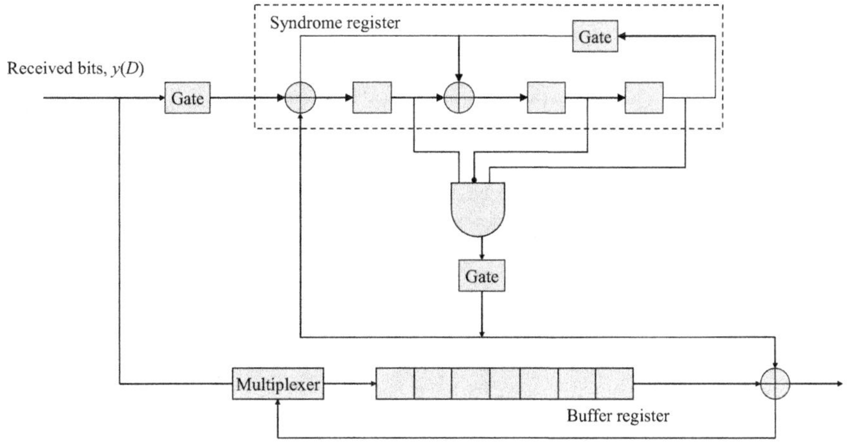

FIGURE 6.15

Hamming (7,4) decoder configuration.

the corresponding syndrome is $s(x) = 1 + x^2$. Once this syndrome is detected, the erroneous bit at position x^6 is to be corrected. Let us now assume that an error occurred on position x^i, with corresponding error pattern $e(x) = x^i$. Once the entire received word is shifted into the syndrome register, the error syndrome is not 101. But after $6 - i$ additional shifts, the content of the syndrome register will become 101, and the error at position x^6 (initially at position x^i) will be corrected.

There exist several versions of Meggit decoders; however, the basic idea is essentially similar to that just described. The complexity of this decoder increases very quickly as the number of errors to be corrected increases, and it is rarely used for the correction of more than 3 single errors or one burst of errors. The Meggit decoder can be simplified under certain assumptions. Let us assume that errors are confined to only highest-order information bits of the received polynomial $r(x)$: $x^k, x^{k+1}, \ldots, x^{n-1}$, so that syndrome polynomial $s(x)$ is given by:

$$s(x) = r(x)\text{mod}g(x) = [c(x) + e(x)]\text{mod}g(x) = e(x)\text{mod}g(x), \tag{6.81}$$

where $r(x)$ is the received polynomial, $c(x)$ is the codeword polynomial, and $g(x)$ is the generator polynomial. The corresponding error polynomial can be estimated by:

$$\begin{aligned} e'(x) &= e'(x)\text{mod}g(x) = s'(x) = x^{n-k}s(x)\text{mod}g(x) \\ &= e_k + e_{k+1}x + \ldots + e_{n-2}x^{n-k-2} + e_{n-1}x^{n-k-1}. \end{aligned} \tag{6.82}$$

The error polynomial will be at most of degree $n - k - 1$ because $\deg[g(x)] = n - k$. Therefore the syndrome register content is identical to the error pattern, and we say that the error pattern is trapped in the syndrome register, and the corresponding decoder, known as an error-trapping decoder, is shown in Fig. 6.16 [17,19]. If t or fewer errors occur in $n - k$ consecutive locations, it can be shown that the error pattern

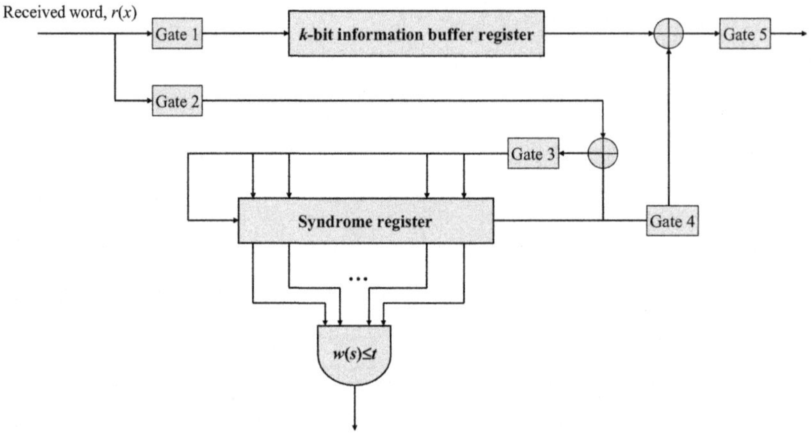

FIGURE 6.16

Error-trapping decoding architecture.

is trapped in the syndrome register only when the weight of the syndrome $w(s)$ is less than or equal to t [17,19]. Therefore the test for the error-trapping condition is to check if the weight of the syndrome is t or less. With gates 1, 2, and 3 closed (4 and 5 open), the received word is shifted into the syndrome register from the right end, which is equivalent to premultiplying the received polynomial $r(x)$ with x^{n-k}. Once the highest-order k bits (corresponding to information bits) are shifted into the information buffer, gate 1 is opened. Gate 2 is opened once all n bits from the received word are shifted into the syndrome register. The syndrome register at this point contains the syndrome corresponding to $x^{n-k}r(x)$. If its weight is t or less, gates 4 and 5 are closed (all others are opened), and the corrected information is shifted out. If $w(s) > t$, the errors are not confined to the $n - k$ higher-order positions of $s(x)$, and we keep shifting with gate 3 being on (other gates are switched off) until $w(s) \leq t$. If $w(s)$ is never $\leq t$ and the syndrome register is shifted k times, either an error pattern with errors confined to $n - k$ consecutive end-around locations has occurred or an uncorrectable error pattern has occurred. For additional explanation of this decoder and other types of cyclic decoders, interested readers are referred to [17].

6.5 BOSE–CHAUDHURI–HOCQUENGHEM CODES

BCH codes, the most famous cyclic codes, were discovered by Hocquenghem in 1959 and by Bose and Chaudhuri in 1960 [19–21]. Among many different decoding algorithms, the most important are the Massey–Berlekamp algorithm and Chien's search algorithm. An important subclass of BCH is a class of RS codes proposed in 1960. Before we continue with study of BCH codes, we have to introduce some properties of finite fields [17,21,24,25].

6.5.1 Galois Fields

Before we can proceed with finite fields, we introduce the concepts of ring, field, and congruencies from abstract algebra [24,25]. A ring is defined as a set of elements R with two operations: addition "+" and multiplication "·", satisfying the following three properties: (1) R is an Abelian (commutative) group under addition, (2) multiplication operation is associative, and (3) multiplication is associative over addition. A field is defined as a set of elements F with two operations: addition "+" and multiplication "·", satisfying the following properties: (1) F is an Abelian group under addition operation, with 0 being the identity element, (2) the nonzero elements of F form an Abelian group under multiplication, with 1 being the identity element, and (3) the multiplication operation is distributive over the addition operation.

The quantity a is said to be congruent to quantity b observed per modulus n, denoted as $a \equiv b \pmod n$, if $a - b$ is divisible by n. If $x \equiv a \pmod n$, then a is called a residue to x to modulus n. A class of residues to modulus n is the class of all integers congruent to a given residue (mod n), and every member of the class is called a representative of the class. There are n classes, represented by $(0),(1),(2),\ldots,(n-1)$, and the representatives of these classes are called a complete system of incongruent residues to modulus n. If i and j are two members of a complete system of incongruent residues to modulus n, then addition and multiplication between i and j can be introduced by:

1. $i + j = (i + j) \pmod n$
2. $i \cdot j = (i \cdot j) \pmod n$

A complete system of residues (mod n) forms a commutative ring with unity element. Let s be a nonzero element of these residues. Then s possesses an inverse element if and only if n is a prime, p. When p is a prime, a complete system of residues (mod p) forms a Galois (finite) field, and is commonly denoted by GF(p).

Let $P(x)$ be any given polynomial in x of degree m with coefficients belonging to GF(p), and let $F(x)$ be any polynomial in x with integral coefficients. Then, $F(x)$ may be expressed as [24]:

$$F(x) = f(x) + p \cdot q(\mathrm{x}) + P(x) \cdot Q(x),$$

where $f(x) = a_0 + a_1x + a_2x^2 + \ldots + a_{m-1}x^{m-1}$, $a_i \in$ GF(p). This relationship may be written as $F(x) \equiv f(x) \bmod \{p, P(x)\}$, and we say that $f(x)$ is the residue of $F(x)$ modulus p and $P(x)$. If p and $P(x)$ are kept fixed but $f(x)$ varies, p^m classes can be formed (because each coefficient of $f(x)$ may take p values of GF(p)). The classes defined by $f(x)$ form a commutative (Abelian) ring, which will be a field if and only if $P(x)$ is irreducible over GF(p) (not divisible with any other polynomial of degree $m - 1$ or less) [17,19,21]. The finite field formed by p^m classes of residues is called a Galois field of order p^m and is denoted by GF(p^m). Two important properties of GF(q), $q = p^m$, are [17,19,21]:

1. The roots of polynomial $x^{q-1} - 1$ are all nonzero elements of GF(q).
2. Let $P(x)$ be an irreducible polynomial of degree m with coefficients from GF(p) and β be a root from the extended field GF($q = p^m$). Then all the m roots of $P(x)$ are $\beta, \beta^p, \beta^{p^2}, \ldots, \beta^{p^{m-1}}$.

The nonzero elements of GF(p^m) can be represented as polynomials of degree at most $m - 1$ or as powers of a primitive root α such that [17,19,21]:

$$\alpha^{p^m - 1} = 1, \quad \alpha^d \neq 1 (\text{for } d \text{ dividing } p^m - 1)$$

Therefore the primitive element (root) is a field element that generates all nonzero field elements as its successive powers. An irreducible polynomial that has a primitive element as its root is called a primitive polynomial.

The function $P(x)$ is said to be a minimum polynomial for generating the elements of GF(p^m) and represents the smallest degree polynomial over GF(p) having a field element $\beta \in$ GF(p^m) as a root. To obtain a minimum polynomial we have to divide $x^q - 1$ ($q = p^m$) by the least common multiple (LCM) of all factors of the form $x^d - 1$, where d is a divisor of $p^m - 1$, and obtain a so-called cyclotomic equation (that is, the equation having for its roots all primitive roots of equation $x^{q-1} - 1 = 0$). The order of this equation is $O(p^m - 1)$, where $O(k)$ is the number of all positive integers less than k and relatively prime to it. By substituting each coefficient in this equation by least nonzero residue to modulus p, we get the cyclotomic polynomial of order $O(p^m - 1)$. Let $P(x)$ be an irreducible factor of this polynomial, then $P(x)$ is a minimum polynomial, which is in general not the unique one.

Example. Let us determine the minimum polynomial for generating the elements of GF(2^3). The cyclotomic polynomial is: $(x^7 - 1)/(x - 1) = x^6 + x^5 + x^4 + x^3 + x^2 + x + 1 = (x^3 + x^2 + 1)(x^3 + x + 1)$. Hence, $P(x)$ can be either $x^3 + x^2 + 1$ or $x^3 + x + 1$. Let us choose $P(x) = x^3 + x^2 + 1$. The degree of this polynomial is deg $[P(x)] = 3$. Now, we explain how we can construct GF(2^3) using the $P(x)$. The construction always starts with elements from the basic field (in this case GF(2) $= \{0,1\}$). All nonzero elements can be obtained as successive powers of primitive root α, until no new element is generated, which is given in the first column of Table 6.4. The second column is obtained exploiting the primitive polynomial $P(x) = x^3 + x^2 + 1$. α is the root of $P(x)$ and therefore $P(\alpha) = \alpha^3 + \alpha^2 + 1 = 0$, and α^3 can be expressed as $\alpha^2 + 1$. α^4 can be expressed as $\alpha\alpha^3 = \alpha(\alpha^2 + 1) = \alpha^3 + \alpha = \alpha^2 + 1 + \alpha$. The third column in Table 6.4 is obtained by reading off coefficients in the second column, with the leading coefficient multiplying the α^2.

6.5.2 The Structure and Encoding of BCH Codes

Equipped we this knowledge of Galois fields we can continue our description of the structure of BCH codes. Let the finite field GF(q) (symbol field) and the extension field GF(q^m) (locator field), $m \geq 1$, be given. For every $m_0 \geq 1$ and Hamming distance d there exists a BCH code with the generating polynomial $g(x)$, if and

Table 6.4 Three different representations of GF(2^3) generated by $x^3 + x^2 + 1$.

Power of α	Polynomial	3-tuple
0	0	000
α^0	1	001
α^1	α	010
α^2	α^2	100
α^3	$\alpha^2 + 1$	101
α^4	$\alpha^2 + \alpha + 1$	111
α^5	$\alpha + 1$	011
α^6	$\alpha^2 + \alpha$	110
α^7	1	001

only if it is of the smallest degree with coefficients from GF(q) and with roots from the extension field GF(q^m) as follows [17]:

$$\alpha^{m_0},\ \alpha^{m_0+1},\ \cdots,\alpha^{m_0+d-2}, \tag{6.83}$$

where α is from GF(q^m). The codeword length is determined as the LCM of orders of roots. (The order of an element β from the finite field is the smallest positive integer j such that $\beta^j = 1$.)

It can be shown that for any positive integer m ($m \geq 3$) and t ($t < 2^{m-1}$) there exists a binary BCH code having the following properties [17,19]:

- Codeword length: $n = 2^m - 1$
- Number of parity bits: $n - k \leq mt$
- Minimum Hamming distance: $d \geq 2t + 1$.

This code is able to correct up to t errors. The generator polynomial can be found as the LCM of the minimal polynomials of α^i [17,19]:

$$g(x) = \text{LCM}[P_{\alpha^1}(x), P_{\alpha^3}(x), \ldots, P_{\alpha^{2t-1}}(x)], \tag{6.84}$$

where α is a primitive element in GF(2^m), and $P_{\alpha^i}(x)$ is the minimal polynomial of α^i.

Let $c(x) = c_0 + c_1x + c_2x^2 + \ldots + c_{n-1}x^{n-1}$ be the codeword polynomial, and let the roots of the generator polynomial be $\alpha, \alpha^2, \ldots, \alpha^{2t}$, where t is the error correction capability of the BCH code. Because the generator polynomial $g(x)$ is the factor of codeword polynomial $c(x)$, the roots of $g(x)$ must also be the roots of $c(x)$:

$$c(\alpha^i) = c_0 + c_1\alpha^i + \ldots + c_{n-1}\alpha^{(n-1)i} = 0;\ \ 1 \leq i \leq 2t. \tag{6.85}$$

This equation can also be written as an inner (scalar) product of codeword vector $\boldsymbol{c} = [c_0\ c_1\ \ldots\ c_{n-1}]$ and the following vector $[1\ \alpha^i\ \alpha^{2i}\ \ldots\ \alpha^{2(n-1)i}]$:

$$[c_0\ c_1 \ldots c_{n-1}]\begin{bmatrix} 1 \\ \alpha^i \\ \cdots \\ \alpha^{(n-1)i} \end{bmatrix} = 0;\ \ 1 \leq i \leq 2t. \tag{6.86}$$

Eq. (6.86) can also be written as the following matrix:

$$[c_0\ c_1 \ldots c_{n-1}]\begin{bmatrix} \alpha^{n-1} & \alpha^{n-2} & \ldots & \alpha & 1 \\ (\alpha^2)^{n-1} & (\alpha^2)^{n-2} & \ldots & \alpha^2 & 1 \\ (\alpha^3)^{n-1} & (\alpha^3)^{n-2} & \ldots & \alpha^3 & 1 \\ \ldots & \ldots & \ldots & \ldots & \ldots \\ (\alpha^{2t})^{n-1} & (\alpha^{2t})^{n-2} & \ldots & \alpha^{2t} & 1 \end{bmatrix}^T = \boldsymbol{c}\boldsymbol{H}^T = \boldsymbol{0},$$

$$\boldsymbol{H} = \begin{bmatrix} \alpha^{n-1} & \alpha^{n-2} & \ldots & \alpha & 1 \\ (\alpha^2)^{n-1} & (\alpha^2)^{n-2} & \ldots & \alpha^2 & 1 \\ (\alpha^3)^{n-1} & (\alpha^3)^{n-2} & \ldots & \alpha^3 & 1 \\ \ldots & \ldots & \ldots & \ldots & \ldots \\ (\alpha^{2t})^{n-1} & (\alpha^{2t})^{n-2} & \ldots & \alpha^{2t} & 1 \end{bmatrix}, \tag{6.87}$$

where $\boldsymbol{H}$ is the parity-check matrix of the BCH code. Using property 2 of GF(q) from the previous section, we conclude that α^i and α^{2i} are the roots of the same minimum polynomial, so that the even rows in $\boldsymbol{H}$ can be omitted to get the final version of the parity-check matrix of the BCH codes:

$$\boldsymbol{H} = \begin{bmatrix} \alpha^{n-1} & \alpha^{n-2} & \ldots & \alpha & 1 \\ (\alpha^3)^{n-1} & (\alpha^3)^{n-2} & \ldots & \alpha^3 & 1 \\ (\alpha^5)^{n-1} & (\alpha^5)^{n-2} & \ldots & \alpha^5 & 1 \\ \ldots & \ldots & \ldots & \ldots & \ldots \\ (\alpha^{2t-1})^{n-1} & (\alpha^{2t-1})^{n-2} & \ldots & \alpha^{2t-1} & 1 \end{bmatrix}. \tag{6.88}$$

For example, (15,7) two-error correcting BCH code has the generator polynomial [26]:

$$\begin{aligned} g(x) &= LCM[P_\alpha(x), P_{\alpha^3}(x)] \\ &= LCM\left[x^4 + x + 1,\ (x+\alpha^3)(x+\alpha^6)(x+\alpha^9)(x+\alpha^{12})\right] \\ &= x^8 + x^7 + x^6 + x^4 + 1, \end{aligned}$$

and the parity-check matrix [26]:

$$\begin{aligned} \boldsymbol{H} &= \begin{bmatrix} \alpha^{14} & \alpha^{13} & \alpha^{12} & \cdots & \alpha & 1 \\ \alpha^{42} & \alpha^{39} & \alpha^{36} & \cdots & \alpha^3 & 1 \end{bmatrix} \\ &= \begin{bmatrix} \alpha^{14} & \alpha^{13} & \alpha^{12} & \alpha^{11} & \alpha^{10} & \alpha^9 & \alpha^8 & \alpha^7 & \alpha^6 & \alpha^5 & \alpha^4 & \alpha^3 & \alpha^2 & \alpha & 1 \\ \alpha^{12} & \alpha^9 & \alpha^6 & \alpha^3 & 1 & \alpha^{12} & \alpha^9 & \alpha^6 & \alpha^3 & 1 & \alpha^{12} & \alpha^9 & \alpha^6 & \alpha^3 & 1 \end{bmatrix}. \end{aligned}$$

In the previous expression we used the fact that in GF(2^4) $\alpha^{15} = 1$. The primitive polynomial used to design this code was $p(x) = x^4 + x + 1$. Every element in GF(2^4) can be represented as a 4-tuple, as shown in Table 6.5.

Table 6.5 GF(2^4) generated by $x^4 + x + 1$.

Power of α	Polynomial of α	4-tuple
0	0	0000
α^0	1	0001
α^1	α	0010
α^2	α^2	0100
α^3	α^3	1000
α^4	$\alpha + 1$	0011
α^5	$\alpha^2 + \alpha$	0110
α^6	$\alpha^3 + \alpha^2$	1100
α^7	$\alpha^3 + \alpha + 1$	1011
α^8	$\alpha^2 + 1$	0101
α^9	$\alpha^3 + \alpha$	1010
α^{10}	$\alpha^2 + \alpha + 1$	0111
α^{11}	$\alpha^3 + \alpha^2 + \alpha$	1110
α^{12}	$\alpha^3 + \alpha^2 + \alpha + 1$	1111
α^{13}	$\alpha^3 + \alpha^2 + 1$	1101
α^{14}	$\alpha^3 + 1$	1001

To create the second column we have used the relation $\alpha^4 = \alpha + 1$, and the 4-tuples are obtained reading off coefficients in the second column. By replacing the powers of α in the foregoing parity-check matrix by corresponding 4-tuples, the parity-check matrix can be written in the following binary form:

$$H = \begin{bmatrix} 1 & 1 & 1 & 1 & 0 & 1 & 0 & 1 & 1 & 0 & 0 & 1 & 0 & 0 & 0 \\ 0 & 1 & 1 & 1 & 1 & 0 & 1 & 0 & 1 & 1 & 0 & 0 & 1 & 0 & 0 \\ 0 & 0 & 1 & 1 & 1 & 1 & 0 & 1 & 0 & 1 & 1 & 0 & 0 & 1 & 0 \\ 1 & 1 & 1 & 0 & 1 & 0 & 1 & 1 & 0 & 0 & 1 & 0 & 0 & 0 & 1 \\ 1 & 1 & 1 & 1 & 0 & 1 & 1 & 1 & 1 & 0 & 1 & 1 & 1 & 1 & 0 \\ 1 & 0 & 1 & 0 & 0 & 1 & 0 & 1 & 0 & 0 & 1 & 0 & 1 & 0 & 0 \\ 1 & 1 & 0 & 0 & 0 & 1 & 1 & 0 & 0 & 0 & 1 & 1 & 0 & 0 & 0 \\ 1 & 0 & 0 & 0 & 1 & 1 & 0 & 0 & 0 & 1 & 1 & 0 & 0 & 0 & 1 \end{bmatrix}$$

In Table 6.6 we listed the parameters of several BCH codes generated by primitive elements of order less than $2^5 - 1$ that are of interest for optical communications. The complete list can be found in Appendix C of [17].

Generally speaking there is no need for q to be a prime; it could be a prime power. However, the symbols must be taken from GF(q) and the roots from GF(q^m). From nonbinary BCH codes, RS codes are the most well known and these codes are briefly explained in the next section.

Table 6.6 A set of primitive binary Bose–Chaudhuri–Hocquenghem codes.

n	*K*	*t*	Generator polynomial (in octal form)
15	11	1	23
63	57	1	103
63	51	2	12471
63	45	3	1701317
127	120	1	211
127	113	2	41567
127	106	3	11554743
127	99	4	3447023271
255	247	1	435
255	239	2	267,543
255	231	3	156720665
255	223	4	75626641375
255	215	5	23157564726421
255	207	6	16176560567636227
255	199	7	7633031270420722341
255	191	8	2663470176115333714567
255	187	9	52755313540001322236351
255	179	10	22624710717340432416300455

6.5.3 Decoding of BCH Codes

BCH codes can be decoded as any other cyclic codes class. For example, in Fig. 6.17 we provide the error-trapping decoder for BCH (15,7) double-error correcting code. The operation principle of this circuit was already explained in the previous section. Here, we explain the decoding process by employing the algorithms especially

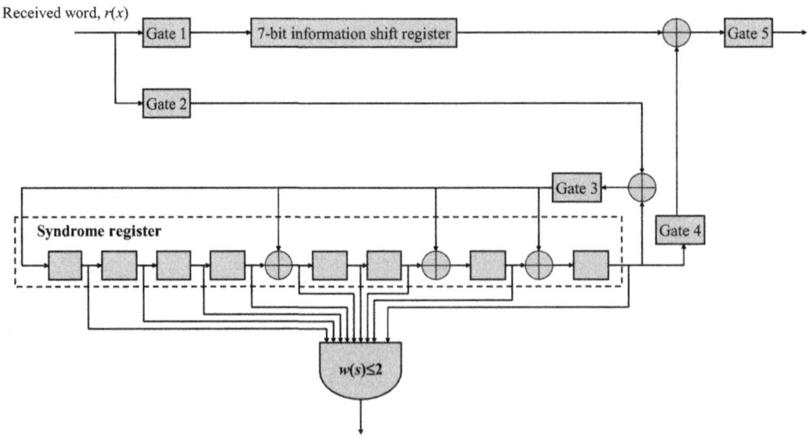

FIGURE 6.17

Error-trapping decoder for (15,7) Bose–Chaudhuri–Hocquenghem code generated by $g(x) = 1 + x^4 + x^6 + x^7 + x^8$.

developed for the decoding of BCH codes. Let $g(x)$ be the generator polynomial with corresponding roots $\alpha,\alpha^2,\ldots,\alpha^{2t}$. Let $c(x) = c_0 + c_1x + c_2x^2 + \ldots + c_{n-1}x^{n-1}$ be the codeword polynomial, $r(x) = r_0 + r_1x + r_2x^2 + \ldots + r_{n-1}x^{n-1}$ be the received word polynomial, and $e(x) = e_0 + e_1x + e_2x^2 + \ldots + e_{n-1}x^{n-1}$ be the error polynomial. The roots of the generator polynomial are also roots of the codeword polynomial, that is:

$$c(\alpha^i) = 0;\ i = 0,\ 1,\ \ldots,\ 2t. \tag{6.89}$$

For binary BCH codes the only nonzero element is 1, therefore the indices i of coefficients $e_i \neq 0$ (or $e_i = 1$) determine the error locations. For nonbinary BCH codes the error magnitudes are also important in addition to error locations. By evaluating the received word polynomial $r(x)$ for α^i we obtain:

$$r(\alpha^i) = c(\alpha^i) + e(\alpha^i) = e(\alpha^i) = S_i, \tag{6.90}$$

where S_i is the ith component of the syndrome vector defined by:

$$\mathbf{S} = [S_1\ S_2\ \ldots\ S_{2t}] = \boldsymbol{rH}^{\mathrm{T}}. \tag{6.91}$$

BCH code is able to correct up to t errors. Let us assume that error polynomial $e(x)$ does not have more than t errors, which can then be written as:

$$e(x) = e_{j_1}x^{j_1} + e_{j_2}x^{j_2} + \ldots + e_{j_l}x^{j_l} + \ldots + e_{j_v}x^{j_v}; \qquad 0 \leq v \leq t, \tag{6.92}$$

where e_{j_l} is the error magnitude and j_l is the error location. The corresponding syndrome components can be obtained from Eqs. (6.90) and (6.92) as follows:

$$S_i = e_{j_1}(\alpha^i)^{j_1} + e_{j_2}(\alpha^i)^{j_2} + \ldots + e_{j_l}(\alpha^i)^{j_l} + \ldots + e_{j_v}(\alpha^i)^{j_v}; \qquad 0 \leq v \leq t, \tag{6.93}$$

where α^{j_l} is the error location number. Notice that the error magnitudes are from the symbol field, while the error location numbers are from the extension field. To avoid double indexing, let us introduce the following notation [6]: $X_l = \alpha^{j_l}$, $Y_l = e_{j_l}$. The pairs (X_l,Y_l) completely identify the errors ($l \in [1,v]$). We have to solve the following set of equations [6]:

$$\begin{aligned} S_1 &= Y_1X_1 + Y_2X_2 + \ldots + Y_vX_v \\ S_2 &= Y_1X_1^2 + Y_2X_2^2 + \ldots + Y_vX_v^2 \\ &\ldots \\ S_{2t} &= Y_1X_1^{2t} + Y_2X_2^{2t} + \ldots + Y_vX_v^{2t}. \end{aligned} \tag{6.94}$$

The procedure to solve this system of equations represents the corresponding decoding algorithm. Direct solution of this system of equations is impractical. There exist many different algorithms to solve the system of Eqs. (6.94), ranging from iterative to Euclidean algorithms [19–21]. The very popular decoding algorithm of BCH codes is the Massey–Berlekamp algorithm [19–21]. In this algorithm the BCH decoding is observed as a shift register synthesis problem: given the syndromes S_i we have to find the minimal length shift register that generates the syndromes.

Once we determine the coefficients of this shift register, we construct the error locator polynomial [6,19,21]:

$$\sigma(x) = \prod_{i=1}^{\nu}(1 + X_i x) = \sigma_\nu x^\nu + \sigma_{\nu-1}x^{\nu-1} + \ldots + \sigma_1 x + 1, \tag{6.95}$$

where the σ_i's, also known as elementary symmetric functions, are given by Viète's formulas:

$$\begin{aligned}\sigma_1 &= X_1 + X_2 + \ldots + X_\nu \\ \sigma_2 &= \sum_{i<j} X_i X_j \\ \sigma_3 &= \sum_{i<j<k} X_i X_j X_k \\ &\ldots \\ \sigma_\nu &= X_1 X_2 \ldots X_\nu.\end{aligned} \tag{6.96}$$

Because $\{X_l\}$ are the inverses of the roots of $\sigma(x)$, $\sigma(1/X_l) = 0\ \forall\ l$, and we can write [6,19,21]:

$$X_l^\nu \sigma\left(X_l^{-1}\right) = X_l^\nu + \sigma_1 X_l^{\nu-1} + \ldots + \sigma_\nu. \tag{6.97}$$

By multiplying the previous equation by X_l^j and performing summation over l for fixed j we obtain:

$$S_{\nu+j} + \sigma_1 S_{\nu+j-1} + \ldots + \sigma_\nu S_j = 0;\ j = 1, 2, \ldots, \nu. \tag{6.98}$$

This equation can be rewritten as follows:

$$S_{\nu+j} = -\sum_{i=1}^{\nu} \sigma_i S_{\nu+j-i};\ j = 1, 2, \ldots, \nu. \tag{6.99}$$

$S_{\nu+j}$ represents the output of the shift register shown in Fig. 6.18. The Massey–Berlekamp algorithm is summarized by the flowchart shown in Fig. 6.19, which is self-explanatory. Once the locator polynomial is determined, we have to find the roots and invert them to obtain the error locators.

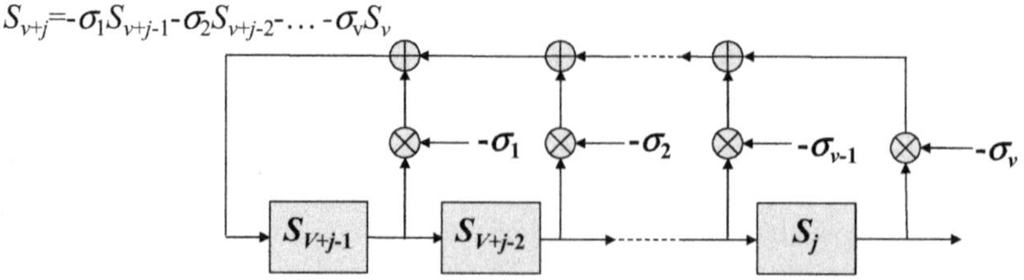

FIGURE 6.18

Shift register that generates the syndromes S_j.

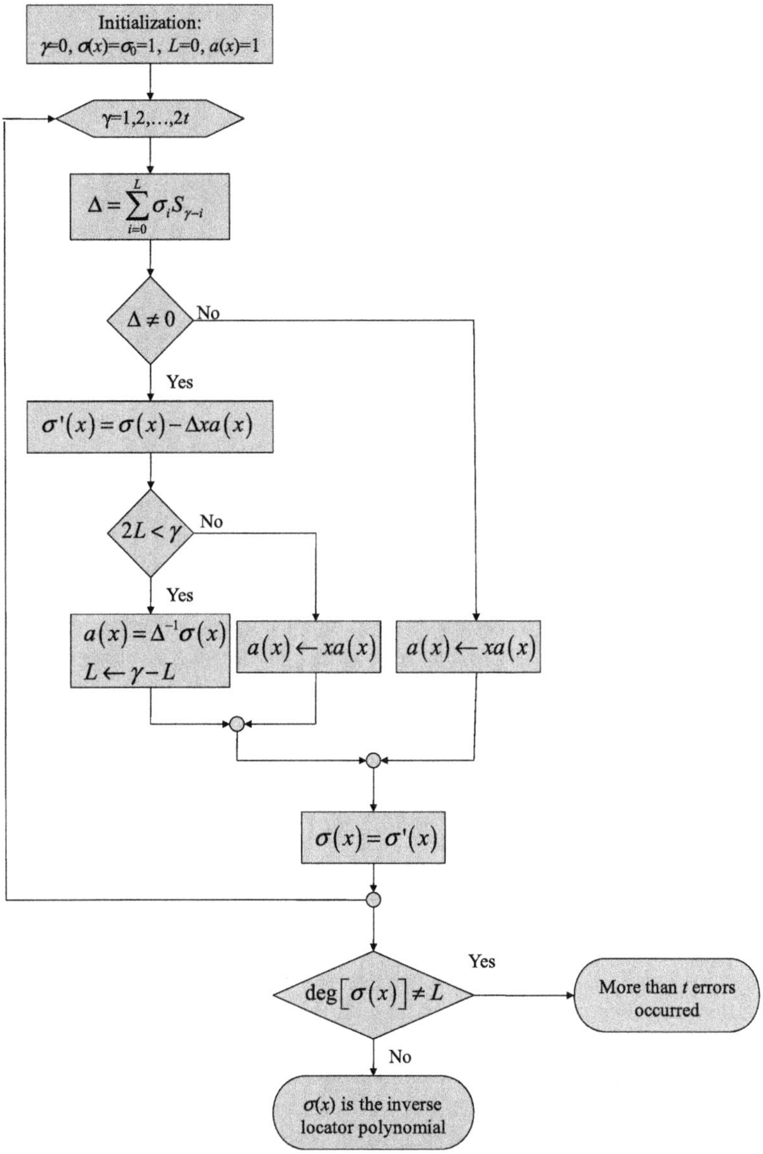

FIGURE 6.19

Flowchart of the Massey–Berlekamp algorithm. Δ denotes the error (discrepancy) between the syndrome and the shift register output, and $a(x)$ stores the content of the shift register (normalized by Δ^{-1}) prior to lengthening.

To determine the error magnitudes, we have to define another polynomial, known as the error-evaluator polynomial, defined as follows [6,19]:

$$\varsigma(x) = 1 + (S_1 + \sigma_1)x + (S_2 + \sigma_1 S_1 + \sigma_2)x^2 + \ldots + (S_\nu + \sigma_1 S_{\nu-1} + \ldots + \sigma_\nu)x^\nu. \tag{6.100}$$

The error magnitudes are then obtained from [6,19]:

$$Y_l = \frac{\varsigma(X_l^{-1})}{\prod_{i=1,\, i\neq l}^{\nu} (1 + X_i X_l^{-1})}. \tag{6.101}$$

For binary BCH codes we set $Y_l = 1$ so that we do not need to evaluate the error magnitudes.

6.6 REED–SOLOMON CODES, CONCATENATED CODES, AND PRODUCT CODES

RS codes were discovered in 1960, and represent a special class of nonbinary BCH codes [27,28]. RS codes represent the most commonly used nonbinary codes. Both the code symbols and the roots of the generating polynomial are from the locator field. In other words, the symbol field and locator field are the same ($m = 1$) for RS codes. The codeword length of RS codes is determined by $n = q^m - 1 = q - 1$, so that RS codes are relatively short codes. The minimum polynomial for some element β is $P_\beta(x) = x - \beta$. If α is the primitive element of GF(q) (q is a prime or prime power), the generator polynomial for t-error correcting RS code is given by [19–21]:

$$g(x) = (x - \alpha)(x - \alpha^2)\cdots(x - \alpha^{2t}). \tag{6.102}$$

The generator polynomial degree is $2t$ and is the same as the number of parity symbols $n - k = 2t$, while the block length of the code is $n = q - 1$. Since the minimum distance of BCH codes is $2t + 1$, the minimum distance of RS codes is $d_{\min} = n - k + 1$, satisfying therefore the Singleton bound ($d_{\min} \leq n - k + 1$) with equality and belonging to the class of MDS codes. When $q = 2^m$, the RS code parameters are: $n = m(2^m - 1)$, $n - k = 2mt$, and $d_{\min} = 2mt + 1$. Therefore the minimum distance of RS codes, when observed as binary codes, is large. The RS codes may be considered as burst error correcting codes, and as such are suitable for bursty-error prone channels. This binary code is able to correct up to t bursts of length m. Equivalently, this binary code is able to correct a single burst of length $(t - 1)m + 1$.

The weight distribution of RS codes can be determined by [17]:

$$A_i = \binom{n}{i}(q-1)\sum_{j=0}^{i-d_{\min}} (-1)^j \binom{i-1}{j} q^{i-d_{\min}-j}, \tag{6.103}$$

and by using this expression we can evaluate the undetected error probability by Eq. (6.68).

Example. Let GF(4) be generated by $1 + x + x^2$, as we explained in Section 6.5.1. The symbols of GF(4) are 0, 1, α, and α^2. The generator polynomial for RS(3,2) code is given by $g(x) = x - \alpha$. The corresponding codewords are: 000, 101, $\alpha 0 \alpha$, $\alpha^2 0 \alpha^2$, 011, 110, $\alpha 1 \alpha^2$, $\alpha^2 1 \alpha$, $0 \alpha \alpha$, $1 \alpha \alpha^2$, $\alpha \alpha 0$, $\alpha^2 \alpha 1$, $0 \alpha^2 \alpha^2$, $1 \alpha^2 \alpha$, $\alpha \alpha^2 1$, and $\alpha^2 \alpha^2 0$. This code is essentially the even parity check code ($\alpha^2 + \alpha + 1 = 0$). The generator polynomial for RS(3,1) is $g(x) = (x - \alpha)(x - \alpha^2) = x^2 + x + 1$, while the corresponding codewords are: 000, 111, $\alpha\alpha\alpha$, and $\alpha^2\alpha^2\alpha^2$. Therefore this code is in fact the repetition code.

Since RS codes are a special class of nonbinary BCH codes, they can be decoded using the same decoding algorithm already explained in the previous section.

To improve the burst error correction capability of RS codes, RS codes can be combined with an inner binary block code in a concatenation scheme as shown in Fig. 6.20. The key idea behind the concatenation scheme can be explained as follows [19]. Consider the codeword generated by inner (n,k,d) code (with d being the minimum distance of the code), and transmitted over the bursty channel. The decoder processes the erroneously received codeword and decodes it correctly. However, occasionally the received codeword is decoded incorrectly. Therefore the inner encoder, the channel, and the inner decoder may be considered as a super channel whose input and output alphabets belong to GF(2^k). The outer encoder (N,K,D) (D is the minimum distance of the outer code) encodes input K symbols and generates output N symbols transmitted over the super channel. The length of each symbol is k information digits. The resulting scheme, known as concatenated code and proposed initially by Forney [22], is an $(Nn,Kk,\geq Dd)$ code with the minimum distance of at least Dd. For example, RS(255,239,8) code can be combined with the (12,8,3) single-parity-check code in the concatenation scheme $(12{\cdot}255,239{\cdot}8,\geq 24)$. The concatenated scheme from Fig. 6.20 can be generalized to q-ary channels, the inner code operating over GF(q) and the outer over GF(q^k).

Two RS codes can be combined in a concatenated scheme by interleaving. An interleaved code is obtained by taking L codewords (of length N) of a given code $\mathbf{x}_j = (x_{j1},x_{j2},\ldots,x_{jN})$ $(j = 1,2,\ldots,L)$, and forming the new codeword by interleaving the L codewords as follows $\mathbf{y}_i = (x_{11},x_{21},..,x_{L1}, x_{12},x_{22},..,x_{L2},\ldots,x_{1N},x_{2N},..,x_{LN})$. The process of interleaving can be visualized as the process of forming an $L \times N$ matrix of L codewords written row by row and transmitting the matrix column by column:

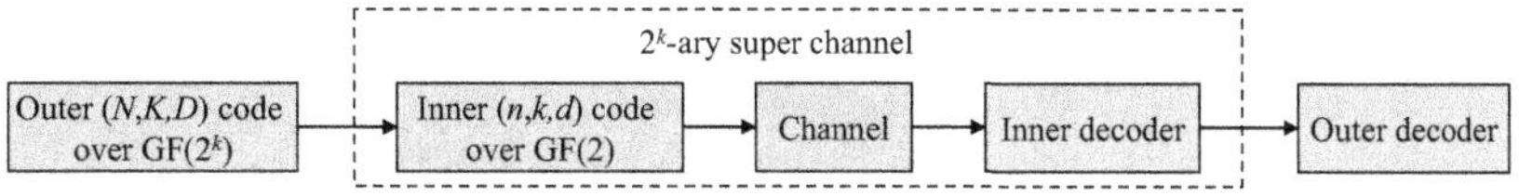

FIGURE 6.20

The concatenated $(Nn,Kk,\geq Dd)$ code.

$$
\begin{array}{l}
x_{11}\ x_{12}\ \ldots x_{1N} \\
x_{21}\ x_{22}\ \ldots x_{2N} \\
\ldots \\
x_{L1}\ x_{L2}\ \ldots x_{LN}
\end{array}
$$

The parameter L is known as the interleaving degree. The transmission must be postponed until L codewords are collected. To be able to transmit a column whenever a new codeword becomes available, the codewords should be arranged down diagonals as follows, and the interleaving scheme is known as the delayed interleaving (1-frame delayed interleaving):

$$
\begin{array}{cccccccc}
x_{i-(N-1),1} & \cdots & x_{i-2,1} & & x_{i-1,1} & x_{i,1} & & \\
& x_{i-(N-1),2} & \cdots & & x_{i-2,2} & x_{i-1,2} & & x_{i,2} \\
& & \cdots\cdots & \cdots\cdots & \cdots\cdots & \cdots\cdots & & \\
& & & & x_{i-(N-1),N-1} & x_{i-(N-2),N-1} & & \\
& & & & & x_{i-(N-1),N} & & x_{i-(N-2),N}
\end{array}
$$

Each new codeword completes one column of this array. In the foregoing example the codeword $\boldsymbol{x}_i$ completes the column (frame) $x_{i,1}, x_{i-1,2}, \ldots, x_{i-(N-1),N}$. A generalization of this scheme, in which the components of the ith codeword $\boldsymbol{x}_i$, say $x_{i,j}$ and $x_{i,j+1}$, are spaced λ frames apart, is known as λ-frame delayed interleaved.

Another way to deal with burst errors is to arrange two RS codes in product manner as shown in Fig. 6.21. A product code [6,9,19] is an (n_1n_2, k_1k_2, d_1d_2) code in which codewords form an $n_1 x n_2$ array such that each row is a codeword from an (n_1, k_1, d_1) code C_1, and each column is a codeword from an (n_2, k_2, d_2) code C_2; with n_i, k_i, and d_i ($i = 1,2$) being the codeword length, dimension, and minimum distance, respectively, of the ith component code. The product codes were proposed

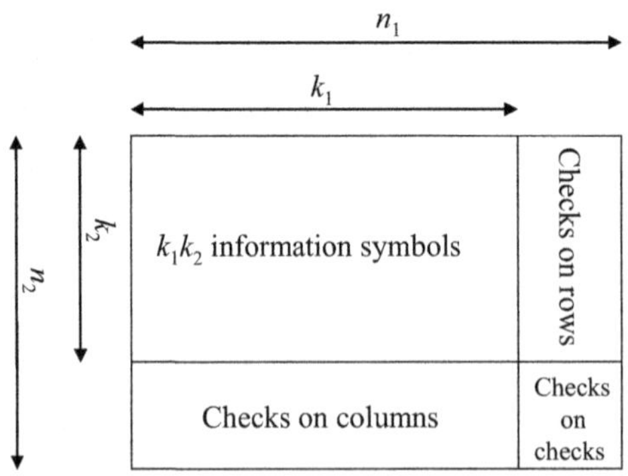

FIGURE 6.21

The structure of a codeword of a product code.

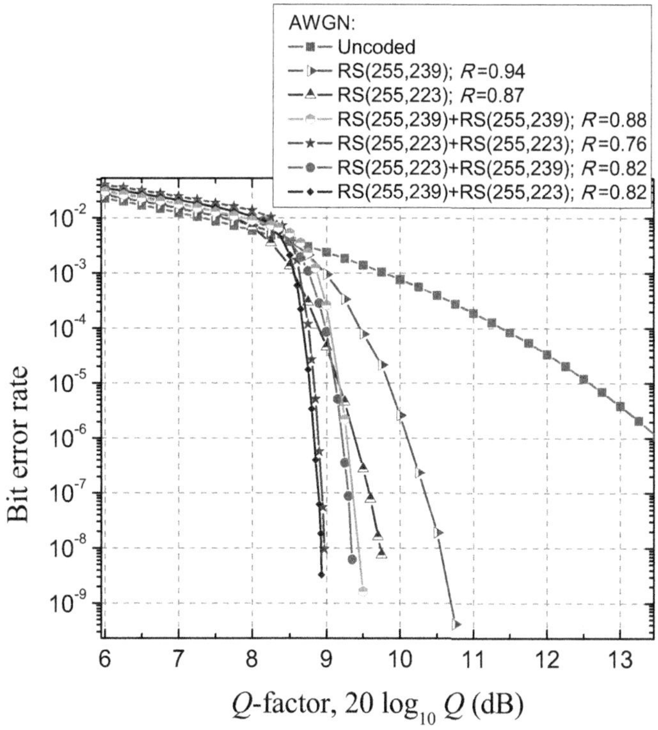

FIGURE 6.22

Bit error rate performance of concatenated Reed–Solomon codes on an optical on-off keying channel. The *Q*-factor is defined by $Q = (\mu_1 - \mu_0)/(\sigma_1 + \sigma_0)$, where μ_i and σ_i are the mean value and standard deviation corresponding to symbol i ($i = 0,1$). *AWGN*, Additive white Gaussian noise.

by Elias [18]. Both binary (such as binary BCH codes) and nonbinary codes (such as RS codes) may be arranged in the product code manner. It is possible to show [19] that the minimum distance of a product code is the product of minimum distances of component codes. It is straightforward to show that the product code is able to correct the burst error of length $b = \max(n_1b_2, n_2b_1)$, where b_i is the burst error capability of component code $i = 1,2$.

The results of Monte Carlo simulations for different RS concatenation schemes on an optical on/off keying AWGN channel are shown in Fig. 6.22. Interestingly, the concatenation scheme RS(255,239) + RS(255,223) of code rate $R = 0.82$ outperforms the concatenation scheme RS(255,223) + RS(255,223) of lower code rate $R = 0.76$, as well as the concatenation scheme RS(255,223) + RS(255,239) of the same code rate.

More powerful FEC schemes belong to the class of iteratively decodable codes [29–31], including turbo, turbo product, and LDPC codes.

6.7 CONCLUDING REMARKS

This chapter was devoted to the basic concepts of information theory and coding theory. In Section 6.1, we provided the definitions of entropy, joint entropy, conditional entropy, relative entropy, mutual information, and channel capacity, followed by information capacity theorem. We also briefly described the source coding and data compaction concepts. Furthermore, we discussed the channel capacity of DMCs, continuous channels, and some optical channels.

The standard FEC schemes that belong to the class of hard decision codes were described in the rest of this chapter. More powerful FEC schemes belong to the class of soft iteratively decodable codes, but their description is out of the scope of this chapter (see, for instance, [9,10,32]). Iteratively decodable codes, such as LDPC codes, are described in Chapter 10. In Section 6.2 we introduced the classical channel coding preliminaries. Section 6.3 was devoted to the basics of linear block codes, such as definition of generator and parity-check matrices, syndrome decoding, distance properties of LBCs, and some important coding bounds. In Section 6.4 cyclic codes were introduced. The BCH codes were described in Section 6.5. The RS, concatenated, and product codes were described in Section 6.6. In the following section, we provide the set of problems for readers to obtain a deeper understanding of classical error correction concepts.

6.8 PROBLEMS

1. Prove the following properties of mutual information:
   - **(a)** The mutual information is symmetric: $I(X,Y)= I(Y,X)$;
   - **(b)** The mutual information of a channel is nonnegative: $I(X,Y) \geq 0$;
   - **(c)** The mutual information of a channel is related to the joint entropy $H(X,Y)$ as follows:

$$I(X;Y)=H(X)+H(Y)-H(X,Y),\ H(X,Y)=\sum_j\sum_k p(x_j;y_k)\log_2\left[\frac{1}{p(x_j;y_k)}\right].$$

2. Let us observe the M-ary input M-ary output symmetric channel, such as M-ary PPM, for which $p(y_j|x_i) = P_s/(M-1)$ and $p(y_j|x_j) = 1-P_s$, where P_s is the symbol error probability. Derive the following expression for channel capacity in bits/symbol:

$$C=\log_2 M+(1-P_s)\log_2(1-P_s)+P_s\log_2\left(\frac{P_s}{M-1}\right).$$

   Plot the channel capacity as a function of P_s by using the M as a parameter.

3. By using the information capacity theorem and Fano's inequality, generate the minimum BER versus SNR plots for different code rates. Observe the BPSK transmission over the zero-mean additive Gaussian noise channel.
4. The binary erasure channel has two inputs and three outputs as depicted in Fig. 6.P4. The inputs are labeled x_0 and x_1, and the outputs y_0, y_1, and e. A fraction p of the incoming bits is erased by the channel. Determine the capacity of this channel.

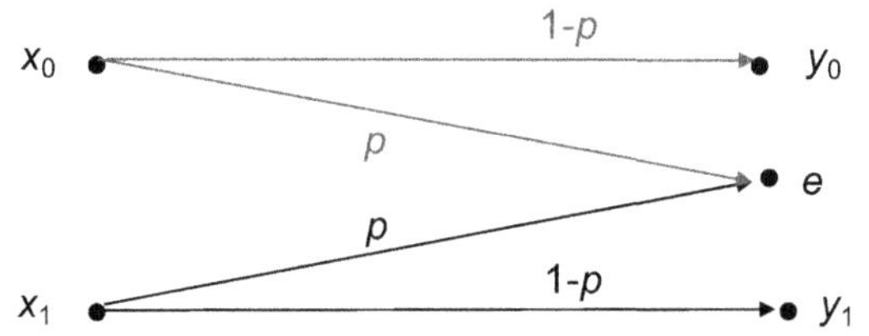

FIGURE 6.P4

Binary erasure channel model.

5. The BSC is described by the following channel matrix:

$$\boldsymbol{P}_{\mathrm{BSC}} = \begin{bmatrix} q & p \\ p & q \end{bmatrix}, \; p + q = 1.$$

Determine the channel capacity of this channel and plot its dependence against the crossover probability p.

Observe now the channel that is obtained by cascading the BSC channel. Determine the channel matrix after cascading two and three BSC channels. Plot the channel capacity of cascaded channels against p. Discuss the results.

6. The binary asymmetric channel (BAC) is described by the following channel matrix:

$$\boldsymbol{P}_{\mathrm{BAC}} = \begin{bmatrix} q_1 & p_1 \\ p_2 & q_2 \end{bmatrix}, \; p_i + q_i = 1 \; (i = 1, 2).$$

Determine the channel capacity of this channel and plot its dependence against p_1 by using p_2 as the parameter. Discuss BAC and BSC results.

7. An (n,k) linear block code is described by the following parity-check matrix:

$$\boldsymbol{H} = \begin{bmatrix} 0 & 0 & 0 & 0 & 0 & 0 & 0 & 1 & 1 & 1 & 1 & 1 & 1 & 1 & 1 \\ 0 & 0 & 0 & 1 & 1 & 1 & 1 & 0 & 0 & 0 & 0 & 1 & 1 & 1 & 1 \\ 0 & 1 & 1 & 0 & 0 & 1 & 1 & 0 & 0 & 1 & 1 & 0 & 0 & 1 & 1 \\ 1 & 0 & 1 & 0 & 1 & 0 & 1 & 0 & 1 & 0 & 1 & 0 & 1 & 0 & 1 \end{bmatrix}.$$

(a) Determine the code parameters: codeword length, number of information bits, code rate, overhead, minimum distance, and error correction capability.
(b) Represent the $\boldsymbol{H}$-matrix in systematic form and determine the generator matrix $\boldsymbol{G}$ of the corresponding systematic code.

8. Consider a (7,4) code with generator matrix:

$$\boldsymbol{G} = \begin{bmatrix} 0 & 1 & 0 & 1 & 1 & 0 & 0 \\ 1 & 0 & 1 & 0 & 1 & 0 & 0 \\ 0 & 1 & 1 & 0 & 0 & 1 & 0 \\ 1 & 1 & 0 & 0 & 0 & 0 & 1 \end{bmatrix}$$

(a) Find all the codewords of the code.
(b) What is the minimum distance of the code?
(c) Determine the parity-check matrix of the code.
(d) Determine the syndrome for the received vector [1101011].
(e) Assuming that an information bit sequence of all 0s is transmitted, find all minimum weight error patterns $\boldsymbol{e}$ that result in a valid codeword that is not the all-zero codeword.
(f) Use row and column operations to transform $\boldsymbol{G}$ to systematic form and find its corresponding parity-check matrix. Sketch a shift register implementation of this systematic code.
(g) All Hamming codes have a minimum distance of 3. What is the error correction and error detection capability of a Hamming code?

9. Consider the following set:

$$C = \{(000000),\ (001011),\ (010101),\ (011110),\ (100111),\ (101100),\ (110010),\ (111001)\}$$

(a) Show that C is vector space over GF(2).
(b) Determine the dimension of this set.
(c) Determine the set of basis vectors.

If the set C is considered as the code book of an LBC:

(d) Determine the parameters of this code.
(e) Determine the generator and parity-check matrices in systematic form.
(f) Determine the minimum distance and error correction capability of this code.

10. Let C be an (n,k) LBC of minimum Hamming distance d. Define a new code C_e by adding an additional overall parity-check equation. Such obtained code is known as extended code.
(a) If the minimum distance of the original code is d, determine the minimum distance of the extended code.
(b) Show that extended code is an $(n+1,k)$ code.

11. The maximum-length $C^m{}_{\text{dual}}$ is the dual of $(2^m - 1, 2^m - 1 - m, 3)$ of the corresponding Hamming code. Let $m = 3$:
   (a) List all codewords of this code.
   (b) Show that $C^3{}_{\text{dual}}$ is a (7,3,4) LBC.

12. Determine an (8,4) code obtained by extending the (7,4) Hamming code. Determine its minimum distance and error correction capability. Show that (8,4) code and its dual are identical. Such codes are called self-dual and are of high importance in quantum error correction.

13. Let $\boldsymbol{G} = [\boldsymbol{I}_k|\boldsymbol{P}]$ be the generator matrix of a systematic (n,k) LBC. Show that this code is self-dual if and only if $\boldsymbol{P}$ is a square and $\boldsymbol{P}\boldsymbol{P}^T = \boldsymbol{I}$. Show that for dual codes $n = 2k$. Finally, design self-dual codes for $n = 4$, 6, and 8.

14. Prove the following properties of a standard array:
   (a) All n-tuples of a row are distinct.
   (b) Each n-tuple appears exactly once in the standard array.
   (c) There are exactly 2^{n-k} rows in the standard array.
   (d) For perfect codes (satisfying the Hamming bound with equality sign) all n-tuples of weight $t = \text{int}[(d_{\min} - 1)/2]$ or less appear as coset leaders (int[x] is the integer part of x).
   (e) For quasi-perfect codes, in addition to all n-tuples of weight t or less, some but not all n-tuples of weight $t + 1$ appear as coset leaders.
   (f) All elements in the same row (coset) have the same syndrome.
   (g) Elements in different rows have different syndromes.
   (h) There are 2^{n-k} different syndromes corresponding to 2^{n-k} rows.

15. For the extended code from Problem 12 determine the standard array and syndrome decoding table. If the received word was (10101010) what would be the result of standard array decoding? Determine the weight distribution of coset leaders and word error probability for the crossover probability of BSC being 10^{-4}. Finally, determine the coding gain at a target BER of 10^{-6}.

16. The (n,k) LBC can be shortened by deleting the information symbols, which is equivalent to the deleting of parity columns in the parity-check matrix $\boldsymbol{H} = [\boldsymbol{P}^T|\boldsymbol{I}_{n-k}]$. Let us remove all columns of even weight in $\boldsymbol{P}^T$ of the $(2^m - 1, 2^m - 1 - m)$ Hamming code. Derive the new code parameters in terms of the original code and determine the minimum distance of the shortened code. What is the error correction capability of this shortened code?

17. Reed–Muller (RM) codes have been used in deep-space communications such as Mariner 9, Voyager 1, and Voyager 2. The rth order RM code has the following generator matrix:

$$\boldsymbol{G} = \begin{bmatrix} \boldsymbol{G}_0 \\ \boldsymbol{G}_1 \\ \vdots \\ \boldsymbol{G}_r \end{bmatrix},$$

where G_0 is an all-one row vector of length $n = 2^m$. The submatrix G_1 has dimensions $m \times 2^m$ in which columns represent all possible m-tuples. The submatrix G_2 is obtained from G_1 by taking two row products of G_1, with only indices in increasing order taken into account. The product of two rows $\boldsymbol{a} = (a_1,\ldots,a_n)$ and $\boldsymbol{b} = (b_1,\ldots,b_n)$ is defined by $\boldsymbol{ab} = (a_1b_1,\ldots,a_nb_n)$. The submatrix G_3 is obtained from G_2 by taking three row products of G_1, and so on:

(a) For $m = 4$ and $r = 3$ provide the corresponding generator matrix and determine the code parameters.

(b) Show that the 0th-order RM code is in fact the $(n,1)$ repetition code.

(c) Determine the minimum distance and parameters of the first-order RM codes.

(d) Prove that for any RM code the following is valid:

$$n = \sum_{i=0}^{r} \binom{m}{i}, \quad n - k = \sum_{i=0}^{m-1-i} \binom{m}{i}$$

(e) Determine the first-order RM code for $m = 3$ and the second-order RM code for $m = 4$. How are these related to the Hamming codes?

18. Prove that an LBC can correct all error patterns having e_c or fewer errors and simultaneously detect all error patterns containing e_d ($e_d \geq e_c$) or fewer errors if the minimum distance $d_{\min}$ satisfies the following inequality: $d_{\min} \geq e_d + e_c + 1$.

19. Let the polynomial $x^7 + 1$ over GF(2) be given. Determine the generator and parity-check polynomials of cyclic (7,4) code. By using this polynomial create the corresponding generator and parity-check matrices. Encode the information sequence (1101) using this cyclic code. Provide the encoder and syndrome circuits.

20. Design a (15,11) cyclic code that is capable of correcting single errors. Provide the generator and parity-check polynomials. Provide the corresponding Meggit decoder. Finally, provide the corresponding error-trapping decoder.

21. Consider a three-error correcting (15,5) BCH code with generator polynomial $g(x) = 1 + x + x^2 + x^4 + x^5 + x^8 + x^{10}$.

(a) Decode $r(x) = 1 + x^5 + x^6 + x^7 + x^9$ using the error-trapping decoder.

(b) Decode $r(x) = 1 + x + x^3 + x^7$ using the Massey—Berlekamp algorithm.

22. In GSM systems, Fire codes are used to correct the burst of errors. The error burst of length B is defined as a contiguous sequence of B bits in which the first and last bits or any other intermediate bits are received in error. The (n,k) codes for burst error correction should satisfy the Reiger bound: $n - k \geq 2B$. The Fire code over GF(q) is constructed by the following generator polynomial:

$$g(x) = \left(x^{2B-1} - 1\right)p(x), \ \deg(p(x)) = m \geq B, \ \text{rem}\left[\left(x^{2B-1} - 1\right) / p(x)\right] \neq 0.$$

Determine the block length of this code. For $n - k = 40$ determine the parameter B and parameters of the corresponding Fire code so that the Reiger bound is satisfied.

23. This problem is also related to Fire codes. Let us consider the primitive polynomial $p(x) = 1 + x + x^6$ over GF(2). Use this primitive polynomial to determine the Fire code that can correct bursts of length 7 or less.

24. Let α be the primitive element in GF(2^4). Determine the generator polynomial and parity-check matrix of binary BCH code with error correction capability of 2 and codeword length of 15.

25. Let us observe a two-error correcting binary BCH (1023,1003) code. By using Table 6.P1 and the McWilliams identities determine the weight enumerator of this code. Plot $\log_{10}(A_i)$ as a function of Hamming weight.

26. Consider RS (31,15) code.

(a) Determine the number of bits per symbol in the code and determine the block length in bits.

(b) Determine the minimum distance of the code and the error correction capability.

Table 6.P1 Weight distribution of the dual two-error corrective primitive binary Bose–Chaudhuri–Hocquenghem codes of length $2^m - 1$.

Weight w	**Number of vectors with weight w, B_w**
Odd $m \geq 3$	
Odd $m \geq 3$	
0	1
$2^{m-1} - 2^{(m+1)/2-1}$	$(2^{m-2} + 2^{(m-1)/2-1})(2^m - 1)$
2^{m-1}	$(2^m - 2^{m-1} + 1)(2^m - 1)$
$2^{m-1} + 2^{(m+1)/2-1}$	$(2^{m-2} - 2^{(m-1)/2-1})(2^m - 1)$
Even $m \geq 4$	
0	1
$2^{m-1} - 2^{(m+2)/2-1}$	$2^{(m-2)/2-1}(2^{(m-2)/2} + 1)(2^m - 1)/3$
$2^{m-1} - 2^{m/2-1}$	$2^{(m+2)/2-1}(2^{m/2} + 1)(2^m - 1)/3$
2^{m-1}	$(2^{m-2} + 1)(2^m - 1)$
$2^{m-1} + 2^{m/2-1}$	$2^{(m+2)/2-1}(2^{m/2} - 1)(2^m - 1)/3$
$2^{m-1} + 2^{(m+2)/2-1}$	$2^{(m-2)/2-1}(2^{(m-2)//2} - 1)(2^m - 1)/3$

Modified from S. Lin, D. J. Costello, Error Control Coding: Fundamentals and Applications, Prentice-Hall, Inc., USA, 1983.

27. Let us observe RS (n,k) code:

(a) Prove that the minimum distance is given by $d_{\min} = n - k + 1$.

(b) Prove that the weight distribution of RS codes can be determined by:

$$A_i = \binom{n}{i}(q-1)\sum_{j=0}^{i-d_{\min}}(-1)^j\binom{i-1}{j}q^{i-d_{\min}-j}.$$

28. Let GF(2^3) be obtained as shown in Table 6.4.

(a) Determine the generator and parity-check polynomials of RS code capable of correcting two errors. If the binary message (011000011) is to be transmitted, determine the codeword polynomial.

(b) Determine the generator polynomial of systematic RS code capable of correcting two errors. If the binary message (011000011) is to be transmitted, determine the codeword polynomial.

29. Design the RS code over GF(2^4), given in Table 6.5, capable of correcting three errors. If the received word $\boldsymbol{r} = (000\alpha^7 00\alpha^3 00000\alpha^4 00)$ was obtained, by using Massey–Berlekamp algorithm, determine the error polynomial.

30. Design a concatenated code using (15,9) three-symbol error correcting RS code as the outer code and (7,4) binary Hamming code as the inner code.

(a) Determine the overall codeword length and the number of binary information bits contained in the codeword.

(b) Determine the error correction capability of this code.

References

[1] S. Haykin, Digital Communication Systems, John Wile & Sons, 2014.

[2] T.M. Cover, J.A. Thomas, Elements of Information Theory, John Wiley & Sons, Inc., New York, 1991.

[3] F.M. Ingels, Information and Coding Theory, Intext Educational Publishers, Scranton, 1971.

[4] F.M. Reza, An Introduction to Information Theory, Dover Publications, New York, 1994.

[5] R.E. Blahut, Principles and Practice of Information Theory, Addison-Wesley Publishing Company, Reading, 1990.

[6] D. Drajic, P. Ivanis, Introduction to Information Theory and Coding, third ed., Academic Misao, Belgrade, 2009 (In Serbian.).

[7] I.B. Djordjevic, Physical-Layer Security and Quantum Key Distribution, Springer International Publishing, Switzerland, 2019.

[8] I.B. Djordjevic, Quantum Biological Information Theory, Springer, Nov, 2015.

[9] I.B. Djordjevic, Advanced Optical and Wireless Communications Systems, Springer International Publishing, Switzerland, Dec. 2017.

[10] I.B. Djordjevic, W. Ryan, B. Vasic, Coding for Optical Channels, Springer, March 2010.

[11] S. Haykin, Communication Systems, John Wiley & Sons, Inc., 2004.

[12] J.G. Proakis, Digital Communications, McGaw-Hill, Boston, MA, 2001.

[13] R.G. Gallager, Information Theory and Reliable Communication, John Wiley and Sons, New York, 1968.

[14] D. Huffman, A method for the construction of minimum-redundancy codes, Proc. IRE 40 (9) (1952) 1098–1101.

[15] C.E. Shannon, A mathematical theory of communication, Bell Syst. Tech. J. 27 (1948), 379–423 and 623–656.
[16] M.F. Barsoum, B. Moision, M. Fitz, D. Divsalar, J. Hamkins, Itrative coded pulse-position-modulation for deep-space optical communications, in: Proc. ITW 2007, September 2–6, 2007, pp. 66–71. Lake Tahoe, California.
[17] S. Lin, D.J. Costello, Error Control Coding: Fundamentals and Applications, Prentice-Hall, Inc., USA, 1983.
[18] P. Elias, Error-free coding', IRE Trans. Inf. Theory IT-4 (1954) 29–37.
[19] J.B. Anderson, S. Mohan, Source and Channel Coding: An Algorithmic Approach, Kluwer Academic Publishers, Boston, MA, 1991.
[20] F.J. MacWilliams, N.J.A. Sloane, The Theory of Error-Correcting Codes, North Holland, Amsterdam, the Netherlands, 1977.
[21] S.B. Wicker, Error Control Systems for Digital Communication and Storage, Prentice-Hall, Inc., Englewood Cliffs, NJ, 1995.
[22] G.D. Forney Jr., Concatenated Codes, MIT Press, Cambridge, MA, 1966.
[23] T. Mizuochi, Recent progress in forward error correction and its interplay with transmission impairments, IEEE J. Sel. Top. Quant. Electron. 12 (4) (2006) 544–554.
[24] D. Raghavarao, Constructions and Combinatorial Problems in Design of Experiments, Dover Publications, Inc., New York, 1988 (reprint).
[25] C.C. Pinter, A Book of Abstract Algebra, Dover Publications, Inc., New York, 2010 (reprint).
[26] J.K. Wolf, Efficient maximum likelihood decoding of linear block codes using a trellis, IEEE Trans. Inf. Theor. IT-24 (1) (1978) 76–80.
[27] I.S. Reed, G. Solomon, Polynomial codes over certain finite fields, SIAM J. Appl. Math. 8 (1960) 300–304.
[28] S.B. Wicker, V.K. Bhargva (Eds.), Reed-Solomon Codes and Their Applications, IEEE Press, New York, 1994.
[29] L.R. Bahl, J. Cocke, F. Jelinek, J. Raviv, Optimal decoding of linear codes for minimizing symbol error rate, IEEE Trans. Inf. Theor. IT-20 (2) (1974) 284–287.
[30] W.E. Ryan, Concatenated convolutional codes and iterative decoding, in: J.G. Proakis (Ed.), Wiley Encyclopedia in Telecommunications, John Wiley and Sons, 2003.
[31] R.H. Morelos-Zaragoza, The Art of Error Correcting Coding, John Wiley & Sons, Boston, MA, 2002.
[32] W. Shieh, I. Djordjevic, OFDM for Optical Communications, Elsevier, October 2009.

Further reading

[1] C. Berrou, A. Glavieux, P. Thitimajshima, Near Shannon limit error-correcting coding and decoding: turbo codes, in: Proc. 1993 Int. Conf. Comm. (ICC 1993), 1993, pp. 1064–1070.
[2] C. Berrou, A. Glavieux, Near optimum error correcting coding and decoding: turbo codes, IEEE Trans. Commun. (1996) 1261–1271.
[3] R.M. Pyndiah, Near optimum decoding of product codes, IEEE Trans. Commun. 46 (1998) 1003–1010.
[4] R.G. Gallager, Low Density Parity Check Codes, MIT Press, Cambridge, MA, 1963.
[5] I.B. Djordjevic, S. Sankaranarayanan, S.K. Chilappagari, B. Vasic, Low-density parity-check codes for 40 Gb/s optical transmission systems, IEEE/LEOS J. Sel. Top. Quantum Electron. 12 (4) (2006) 555–562.

[6] I.B. Djordjevic, O. Milenkovic, B. Vasic, Generalized low-density parity-check codes for optical communication systems, IEEE/OSA J. Lightwave Technol. 23 (2005) 1939–1946.
[7] B. Vasic, I.B. Djordjevic, R. Kostuk, Low-density parity check codes and iterative decoding for long haul optical communication systems, IEEE/OSA J. Lightwave Technol. 21 (2003) 438–446.
[8] I.B. Djordjevic, et al., Projective plane iteratively decodable block codes for WDM high-speed long-haul transmission systems, IEEE/OSA J. Lightwave Technol. 22 (2004) 695–702.
[9] O. Milenkovic, I.B. Djordjevic, B. Vasic, Block-circulant low-density parity-check codes for optical communication systems, IEEE/LEOS J. Selected Top. Quantum Electron. 10 (2004) 294–299.
[10] B. Vasic, I.B. Djordjevic, Low-density parity check codes for long haul optical communications systems, IEEE Photon. Technol. Lett. 14 (2002) 1208–1210.
[11] I. B. Djordjevic, M. Arabaci, L. Minkov, Next generation FEC for high-capacity communication in optical transport networks, IEEE/OSA J. Lightwave Technol., accepted for publication. (Invited Paper.).
[12] I.B. Djordjevic, On advanced FEC and coded modulation for ultra-high-speed optical transmission, IEEE Commun. Surveys Tutorials 18 (3) (2016) 1920–1951.
[13] S. Chung, et al., On the design of low-density parity-check codes within 0.0045 dB of the Shannon limit, IEEE Commun. Lett. 5 (2001) 58–60.
[14] I.B. Djordjevic, L.L. Minkov, H.G. Batshon, Mitigation of linear and nonlinear impairments in high-speed optical networks by using LDPC-coded turbo equalization, IEEE J. Sel. Areas Commun., Opt. Commun. Netw. 26 (6) (2008) 73–83.
[15] M.E. van Valkenburg, Network Analysis, third ed., Prentice-Hall, Englewood Cliffs, 1974.
[16] B. Vucetic, J. Yuan, Turbo Codes-Principles and Applications, Kluwer Academic Publishers, Boston, 2000.
[17] M. Ivkovic, I.B. Djordjevic, B. Vasic, Calculation of achievable information rates of long-haul optical transmission systems using instanton approach, IEEE/OSA J. Lightwave Technol. 25 (2007) 1163–1168.
[18] D. Divsalar, F. Pollara, Turbo codes for deep-space communications, in: TDA Progress Report 42-120, February 15, 1995, pp. 29–39.

CHAPTER

Quantum Information Theory Fundamentals

7

CHAPTER OUTLINE

This chapter is devoted quantum information theory fundamentals. The chapter starts with classical and von Neumann entropies definitions and properties, followed by quantum representation of classical information. Next, we introduce the accessible information as the maximum mutual information over all possible measurement schemes. We further introduce an important bound, Holevo bound and prove it. Next we introduce typical sequence, typical state, typical subspace and projector on the typical subspace. Then formulate and prove the Schumacher's source coding theorem, the quantum equivalent of Shannon's source coding theorem. We further introduce the concept of quantum compression, decompression, and introduce the concept of reliable compression. We then describe various quantum channel models including bit-flip, phase-flip, depolarizing, and amplitude damping channel models. We then formulate and prove the Holevo-Schumacher-Westmoreland (HSW) theorem, the quantum equivalent of Shannon's channel capacity theorem.

Quantum Information Processing, Quantum Computing, and Quantum Error Correction
https://doi.org/10.1016/B978-0-12-821982-9.00012-5

7.1 INTRODUCTORY REMARKS

Quantum information theory, similar to its classical counterpart, studies the meaning and limits of communicating classical and quantum information over quantum channels. In this chapter we introduce the basic concepts underlying this vast and fascinating area that is currently a subject of intense research [1,2]. In the classical context, information theory provides answers to two fundamental questions [3–5]: To what extent can information be compressed? and What is the maximum rate for reliable communication over a noisy channel? The former is the basis of source coding and data compression in communication theory, while the latter forms the basis for channel coding. Information theory also has overlap with many areas in other disciplines. Since its inception—with the 1948 seminal paper by Claude Shannon titled "A Mathematical Theory of Communication" [3]—information theory has established links with many areas, including, but not limited to, Kolmogorov complexity in computer science [6], large deviation theory within probability theory [7], hypothesis testing (e.g., Fisher information [8]) in statistics, and Kelly's rule in investment and portfolio theory [9].

In the last several decades, classical information theory has provided the foundation for the development of its quantum analog. This is not to say that both are the same. In fact, as we will see in this chapter, quantum information theory has revealed consequences for the quantum world that cannot be predicted by its classical information theory. Nevertheless, classical information theory provides the logical structure to coherently introduce the ideas of quantum information theory and we will exploit this link throughout this chapter. It is challenging to do justice to quantum information theory in a single chapter. It is a comprehensive area with many consequences and a whole book can be, and has been, devoted to it. Since our space is limited we focus on telling a coherent story to readers that, hopefully, will provide them with a helpful introduction to the subject. The references listed at the end of this chapter provide resources for readers who decide to pursue this field further. References [3–9] provide the classical information theory background along with its relation to a variety of other fields in computer science, statistics, probability, estimation and detection, and finance. References [10–25] list important papers that led to breakthroughs in quantum information theory. Finally, references [14–25] detail interesting theoretical work done in quantifying the capacity of quantum optical channels.

7.2 VON NEUMANN ENTROPY

Let us observe a classical discrete memoryless source with the alphabet $X = \{x_1,x_2,\ldots,x_N\}$. The symbols from the alphabet are emitted by the source with probabilities $P(X = x_n) = p_n$, $n = 1,2,\ldots,N$. The amount of information carried by the kth symbol is related to the uncertainty that is resolved when this symbol occurs, and it is defined as $I(x_n) = \log(1/p_n) = -\log(p_n)$, where the logarithm is to the base 2. The classical (Shannon) entropy is defined as the measure of the average amount of information per source symbol:

$$H(X) = E[I(x_n)] = \sum_{n=1}^{N} p_n I(x_n) = \sum_{n=1}^{N} p_n \log_2\left(\frac{1}{p_n}\right). \tag{7.1}$$

The Shannon entropy satisfies the following inequalities:

$$0 \leq H(X) \leq \log N = \log|X|. \tag{7.2}$$

In research literature, $H(X)$ is interchangeably denoted as $H(P)$ to highlight the important fact that Shannon entropy is a functional of the probability mass function.

Let Q be a quantum system with the state described by the density operator $\boldsymbol{\rho}_x$. The probability that the output $\boldsymbol{\rho}_x$ is obtained is given by $p_x = P(x)$. The quantum source is therefore described by the ensemble $\{\boldsymbol{\rho}_x, p_x\}$, characterized by the mixed density operator $\boldsymbol{\rho} = \sum_{x \in X} p_x \boldsymbol{\rho}_x$. Then, the *quantum (von Neumann) entropy* is defined as:

$$S(\boldsymbol{\rho}) = -Tr(\boldsymbol{\rho} \log \boldsymbol{\rho}) = -\sum_{\lambda_i} \lambda_i \log \lambda_i, \; \boldsymbol{\rho} = \sum_i \lambda_i |\lambda_i\rangle\langle\lambda_i|, \tag{7.3}$$

where λ_i are eigenvalues of $\boldsymbol{\rho}$ and $|\lambda_i\rangle$ are corresponding eigenkets. When all quantum states are pure and mutually orthogonal, then von Neumann entropy is equal to the Shannon entropy as in that case $p_i = \lambda_i$. As an illustration, in Fig. 7.1 we provide an interpretation of the quantum representation of the classical information. Two different preparations $P^{(i)}$ and $P^{(j)}$, with $H(X^{(i)}) \neq H(X^{(j)})$ $(i \neq j)$, can generate the same $\boldsymbol{\rho}$ and hence have the same von Neumann entropy $S(\boldsymbol{\rho})$ because the states of the two preparations may not be physically distinguishable from each other.

We should mention that, although $S(\boldsymbol{\rho}) = H(P)$ for pure and mutually orthogonal states, they do not have the same meaning. $S(\boldsymbol{\rho})$ represents the measure of the quantum uncertainty before any measurement takes place, while $H(P)$ is the measure of the classical uncertainty of the outcomes after the quantum measurements. To clarify this point, let us assume that we measure the observable A denoted by:

$$A = \sum_i a_i |a_i\rangle\langle a_i|. \tag{7.4}$$

This measurement for the quantum system Q in a state described by $\boldsymbol{\rho}$ yields a_i with probability given by $p_i = \text{Tr}(|a_i\rangle\langle a_i|\boldsymbol{\rho})) = \langle a_i|\boldsymbol{\rho}|a_i\rangle$ If the observable A commutes with $\boldsymbol{\rho}$, that is, $[A,\boldsymbol{\rho}] = 0$, then the same basis diagonalizes both A and $\boldsymbol{\rho}$, and both the classical and quantum entropies are equal. This is the case for pure orthogonal states. In general, however, the two do not commute and $S(\boldsymbol{\rho}) < H(P_{A,\boldsymbol{\rho}})$, where $P_{A,\boldsymbol{\rho}}$ denotes the probability of the measurement outcomes when the observable is A and the density operator is $\boldsymbol{\rho}$. Given that Shannon entropy is upper limited by $\log N$, while $S(\rho)$ is upper limited by Shannon entropy,

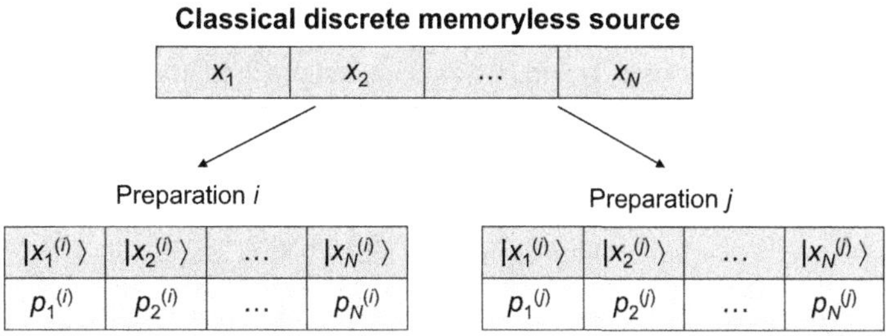

FIGURE 7.1

Quantum representation of the classical information.

we would expect $S(\rho) \leq \log d$, where d is the dimensionality of the Hilbert space. This situation corresponds to the completely mixed state for which $p_i = 1/d$. Because any density operator is nonnegative Hermitian with trace 1, we have that:

$$\sum_i \lambda_i = 1, \quad 0 \leq \lambda_i \leq 1. \tag{7.5}$$

Clearly, when $\boldsymbol{\rho}$ is the pure state, 1 of λ_i is equal to 1 so that $S(\boldsymbol{\rho}) = 0$. Otherwise, based on Eqs. (7.3) and (7.5) we conclude that $S(\boldsymbol{\rho}) > 0$. To summarize, von Neumann entropy is bounded as follows:

$$0 \leq S(\boldsymbol{\rho}) \leq \log d. \tag{7.6}$$

Another relevant property of the von Neumann entropy is *invariancy under the unitary transformation*; in other words, for any unitary transformation U the following is valid:

$$S(U\boldsymbol{\rho}U^{\dagger}) = S(\boldsymbol{\rho}). \tag{7.7}$$

If we choose the U to diagonalize $\boldsymbol{\rho}$ we can simplify the calculation of the von Neumann entropy.

The von Neumann entropy is *concave*, that is, given the set of positive numbers p_i summing up to 1 and the density operators $\boldsymbol{\rho}_i$ we have that:

$$S\left(\sum_i p_i \boldsymbol{\rho}_i\right) \geq \sum_i p_i S(\boldsymbol{\rho}_i). \tag{7.8}$$

In other words, the von Neumann entropy is higher if we are ignorant as to how the state was prepared.

Given the set of positive numbers p_i summing up to 1 and the density operators $\boldsymbol{\rho}_i$ the von Neumann entropy satisfies the following bound:

$$S\left(\sum_i p_i \boldsymbol{\rho}_i\right) \leq \sum_i p_i S(\boldsymbol{\rho}_i) - \sum_i p_i \log p_i, \tag{7.9}$$

wherein the equality is achieved when $\boldsymbol{\rho}_i$ have orthogonal support.

Given two quantum systems in states specified by ρ and $\boldsymbol{\sigma}$, the *relative entropy* of $\boldsymbol{\rho}$ to $\boldsymbol{\sigma}$ is given by:

$$S(\boldsymbol{\rho} \parallel \boldsymbol{\sigma}) = \mathrm{Tr}(\boldsymbol{\rho} \log \boldsymbol{\rho}) - \mathrm{Tr}(\boldsymbol{\rho} \log \boldsymbol{\sigma}). \tag{7.10}$$

This definition is analogous to the definition of classical relative entropy between two probability distributions $p(x)$ and $q(x)$ introduced in Chapter 6. Classical relative entropy is a measure of distance between two distributions $p(x)$ and $q(x)$, and is defined as follows:

$$D(p \,||\, q) = E_p \log\left[\frac{p(X)}{q(X)}\right] = \sum_{x \in X} p(x) \log\left[\frac{p(X)}{q(X)}\right]. \tag{7.11}$$

Relative entropy is also known as Kullback–Leibler (KL) distance and can be interpreted as the measure of inefficiency of assuming that distribution is $q(x)$, when true distribution is $p(x)$. From the definition of classical relative entropy it is clear that in certain cases it can be infinite, for example, when the support of $p(x)$ exceeds that of $q(x)$. A similar scenario also arises for quantum relative entropy. Let us define the kernel of any density matrix as being the eigenspace spanned by the eigenkets of the density operator corresponding to the eigenvalue 0. Additionally, let the support of any density operator refer to the eigenspace spanned by the eigenvectors of the density operator corresponding to nonzero eigenvalues. Then, for the cases when the nontrivial intersection between the support of $\boldsymbol{\rho}$ and the kernel of $\boldsymbol{\sigma}$ is empty, and the support of $\boldsymbol{\rho}$ is a subspace of the support of $\boldsymbol{\sigma}$, the relative entropy $S(\boldsymbol{\rho} \parallel \boldsymbol{\sigma})$ is finite. Otherwise it is infinite. An important property of the relative entropy, known as the *Klein inequality*, states that:

$$S(\boldsymbol{\rho} \parallel \boldsymbol{\sigma}) \geq 0, \tag{7.12}$$

with equality being satisfied if and only if $\boldsymbol{\rho} = \boldsymbol{\sigma}$. By setting $\boldsymbol{\sigma} = \boldsymbol{I}/d$, from Eq. (7.12) we obtain:

$$S(\boldsymbol{\rho} \parallel \boldsymbol{I}/d) = \mathrm{Tr}(\boldsymbol{\rho} \log \boldsymbol{\rho}) - \mathrm{Tr}(\boldsymbol{\rho} \log(\boldsymbol{I}/d)) \geq 0 \Rightarrow S(\boldsymbol{\rho}) = -\mathrm{Tr}(\boldsymbol{\rho} \log \boldsymbol{\rho}) \leq \log d, \tag{7.13}$$

thus proving the upper bound in Eq. (7.6).

To prove the Klein inequality, we can perform the eigen decompositions of $\boldsymbol{\rho}$ and $\boldsymbol{\sigma}$:

$$\boldsymbol{\rho} = \sum_i \rho_i |\rho_i\rangle\langle\rho_i|, \ \sigma = \sum_i \sigma_i |\sigma_i\rangle\langle\sigma_i|; \ 0 \leq \rho_i, \ \sigma_i \leq 1, \tag{7.14}$$

and apply the definition formula for the relative entropy:

$$\begin{aligned}
S(\boldsymbol{\rho} \parallel \boldsymbol{\sigma}) &= \mathrm{Tr}(\boldsymbol{\rho} \log \boldsymbol{\rho}) - \mathrm{Tr}(\boldsymbol{\rho} \log \boldsymbol{\sigma}) = \sum_i \rho_i \log \rho_i - \mathrm{Tr}\left[\sum_i \rho_i |\rho_i\rangle\langle\rho_i| \log \boldsymbol{\sigma}\right] \\
&= \sum_i \rho_i \log \rho_i - \sum_i \rho_i \langle\rho_i| \underbrace{\log \boldsymbol{\sigma}}_{\sum_j |\sigma_j\rangle \log \sigma_j \langle\sigma_j|} |\rho_i\rangle \\
&= \sum_i \rho_i \left[\log \rho_i - \sum_j \log \sigma_j \underbrace{\langle p_i|\sigma_j\rangle\langle\sigma_j|p_i\rangle}_{|\langle p_i|\sigma_j\rangle|^2}\right].
\end{aligned} \tag{7.15}$$

Since:

$$\sum_j \log \sigma_j |\langle p_i|\sigma_j\rangle|^2 \leq \log \left[\underbrace{\sum_i \sigma_j |\langle p_i|\sigma_j\rangle|^2}_{\gamma_i}\right], \tag{7.16}$$

from Eq. (7.15) we obtain:

$$S(\boldsymbol{\rho} \parallel \boldsymbol{\sigma}) \geq \sum_i \rho_i \log \frac{\rho_i}{\gamma_i} = D(\{\rho_i\} \parallel \{\gamma_i\}), \tag{7.17}$$

and since classical relative entropy is nonnegative we obtain that the quantum relative entropy is nonnegative too, which proves the Klein inequality (7.12). In the following subsection we turn our attention to bipartite and multipartite quantum systems.

7.2.1 Composite Systems

Let us consider the bipartite quantum system Q being composed of uncorrelated subsystems A and B, described by the product state density operator $\boldsymbol{\rho}_{AB} = \boldsymbol{\rho}_A \otimes \boldsymbol{\rho}_B$. This system has the interesting property that uncorrelated subsystems have the same quantum entropy, that is, we can write:

$$S(\boldsymbol{\rho}_A) = S(\boldsymbol{\rho}_B). \tag{7.18}$$

To prove this claim, we employ the Schmidt decomposition theorem (proved in Chapter 4), in which a pure state $|\psi\rangle$ of a bipartite system AB can be decomposed as follows:

$$|\psi\rangle = \sum_i c_i |i_A\rangle |i_B\rangle, \tag{7.19}$$

where $|i_A\rangle$ and $|i_B\rangle$ are orthonormal bases of the subsystems H_A and H_B, respectively, and $c_i \in R^+$ (R^+ is the set of nonnegative real numbers) are *Schmidt coefficients* that satisfy the following condition $\sum_i c_i^2 = 1$. By employing the Schmidt decomposition theorem we can express the density operator $\boldsymbol{\rho}_{AB}$ as follows:

$$\rho_{AB} = |\psi\rangle\langle\psi| = \sum_i \sum_j c_i c_j |i_A\rangle\langle j_A| \otimes |i_B\rangle\langle j_B|. \tag{7.20}$$

By taking the partial trace of $\boldsymbol{\rho}_{AB}$ over subsystem A we obtain:

$$\mathrm{Tr}_A(\rho_{AB}) = \sum_i \sum_j c_i c_j \underbrace{\langle j_A | i_A\rangle}_{\delta_{ji}} |i_B\rangle\langle j_B| = \sum_j c_j^2 |j_B\rangle\langle j_B| = \rho_B. \tag{7.21}$$

In similar fashion, by taking the partial trace of $\boldsymbol{\rho}_{AB}$ over subsystem B we obtain:

$$\mathrm{Tr}_B(\rho_{AB}) = \sum_i \sum_j c_i c_j \underbrace{\langle i_B | j_B\rangle}_{\delta_{ij}} |i_A\rangle\langle j_A| = \sum_i c_i^2 |i_A\rangle\langle i_A| = \rho_A. \tag{7.22}$$

From Eqs. (7.21) and (7.22) we conclude that for a pure joint state given by $\boldsymbol{\rho}_{AB} = \boldsymbol{\rho}_A \otimes \boldsymbol{\rho}_B$, the partial traces $\boldsymbol{\rho}_A$ and $\boldsymbol{\rho}_B$ have the same expansion coefficients c_i^2 and consequently the same von Neumann entropy, i.e., $S(\boldsymbol{\rho}_A) = S(\boldsymbol{\rho}_B)$.

For a general bipartite system AB, however, the relation between the quantum entropies of the subsystem and composite system is described by the *subadditivity inequality*:

$$S(\rho_{AB}) \leq S(\rho_A) + S(\rho_B), \tag{7.23}$$

with the equality sign being satisfied when $\rho_{AB} = \rho_A \otimes \rho_B$. Thus the quantum entropy of the composite system is bounded from above by the sum of the entropies of the constituent subsystems, except when the subsystems are uncorrelated, in which case it is purely additive. The subadditivity inequality can be thought of as the quantum analog of the Shannon inequality for classical sources X and Y:

$$H(X,Y) \leq H(X) + H(Y), \tag{7.24}$$

indicating that the joint entropy of XY with a joint distribution $p(x,y)$ is lower than the sum of the entropy of sources X and Y with marginal distributions $p(x)$ and $p(y)$, respectively, except for the case when the outputs of the two sources are independent of each other, when the equality sign is satisfied. This result, however, should not make the reader draw a false equivalence between the classical and the quantum worlds. In fact, the difference between them is starkly demonstrated by considering the lower bound on $S(\rho_{AB})$ given by the *Araki–Lieb triangle inequality*:

$$|S(\rho_A) - S(\rho_B)| \leq S(\rho_{AB}), \tag{7.25}$$

with the equality satisfied when ρ_{AB} is a pure state. Let us look at a consequence of this inequality through an example. The bounds Eqs. (7.23) and (7.25) can be combined as follows:

$$|S(\rho_A) - S(\rho_B)| \leq S(\rho_{AB}) \leq S(\rho_A) + S(\rho_B). \tag{7.26}$$

Example. We consider here subsystems of a bipartite system that are entangled. Entanglement implies that the subsystems have nonlocal quantum correlations. Let us consider one such system whose state is specified by the Bell state:

$$|B_{10}\rangle = 2^{-1/2}(|00\rangle - |11\rangle).$$

The corresponding density operator is given by:

$$\rho_{AB} = |B_{10}\rangle\langle B_{10}| = 2^{-1}(|00\rangle - |11\rangle)(\langle 00| - \langle 11|) = \frac{1}{2}\begin{bmatrix} 1 & 0 & 0 & -1 \\ 0 & 0 & 0 & 0 \\ 0 & 0 & 0 & 0 \\ -1 & 0 & 0 & 1 \end{bmatrix}.$$

Clearly, $S(\rho_{AB}) = -\log 1 = 0$. The reduced density matrix for Bob will be:

$$\rho_B = \mathrm{Tr}_A(\rho_{AB}) = \frac{1}{2}\begin{bmatrix} 1 & 0 \\ 0 & 1 \end{bmatrix},$$

while the corresponding von Neumann entropy for Bob will be $S(\rho_B) = -\log_2(1/2) = 1$, and similarly for Alice. Clearly, both inequalities in Eq. (7.26) are satisfied. The result says that the quantum entropy of a composite system with entangled subsystems is *less* than the quantum entropy of either subsystem. Thus, even though the composite system is in a pure state, measurements of the observables of the subsystems result in outcomes that have uncertainty associated with them. This result has considerable application in quantum cryptography and quantum error correction codes. It is important to note that there is no classical counterpart for this observation. In fact, classical considerations provide exactly the opposite result, $H(X,Y) \geq H(X)$ and $H(X,Y) \geq H(Y)$, implying that joint Shannon entropy of the classical composite system is lower bounded by the Shannon entropy of either subsystems.

To prove the *Araki–Lieb triangle inequality*, let us consider a purifying quantum system C, such that the resulting tripartite system ABC is in a pure state. Employing the subadditivity inequality for a composite system we have:

$$S(\rho_{AC}) \leq S(\rho_A) + S(\rho_C),\ S(\rho_{BC}) \leq S(\rho_B) + S(\rho_C). \tag{7.27}$$

Furthermore, because ABC is in a pure state, using Eq. (7.18) we have that:

$$S(\rho_{AB}) = S(\rho_C),\ S(\rho_{AC}) = S(\rho_B),\ S(\rho_{BC}) = S(\rho_A). \tag{7.28}$$

By using Eqs. (7.27) and (7.28) we obtain that:

$$S(\rho_A) - S(\rho_B) = \underbrace{S(\rho_{BC})}_{\leq S(\rho_B)+S(\rho_C)} - S(\rho_B) \leq S(\rho_C) = S(\rho_{AB}), \tag{7.29}$$

which proves the Araki–Lieb triangle inequality. The equality is satisfied only when the bipartite system AB is in a pure state.

Before concluding this section, let us state without proof a very important inequality in quantum information theory known as *strong subadditivity*: For a tripartite system ABC in a state specified by $\boldsymbol{\rho}_{ABC}$ the following inequalities are satisfied:

$$S(\rho_A) + S(\rho_B) \leq S(\rho_{AC}) + S(\rho_{BC}), \tag{7.30a}$$

$$S(\rho_{ABC}) + S(\rho_B) \leq S(\rho_{AB}) + S(\rho_{BC}). \tag{7.30b}$$

Strong subadditivity has important consequences in quantum information theory related to the conditioning of quantum entropy, bounds on quantum mutual information, and the effect of quantum operations on quantum mutual information.

7.3 HOLEVO INFORMATION, ACCESSIBLE INFORMATION, AND HOLEVO BOUND

Let us go back to the KL distance. By replacing $p(X)$ with $p(X,Y)$ and $q(X)$ with $p(X)$ $q(Y)$, the corresponding classical relative entropy between the joint distribution $p(X,Y)$ and product of distributions $p(X)$ and $q(X)$ is well known as *mutual information*:

$$\begin{aligned} D(p(X,Y)\|p(X)q(Y)) &= E_{p(X,Y)} \log\left[\frac{p(X,Y)}{p(X)q(Y)}\right] \\ &= \sum_{x\in X}\sum_{y\in Y} p(x,y)\log\left[\frac{p(X,Y)}{p(X)q(Y)}\right] \doteq I(X,Y). \end{aligned} \tag{7.31}$$

In this case, the mutual information between random variables X and Y is related to the amount of information that Y has about X on average, namely $I(X;Y) = H(X) - H(X|Y)$. In other words, it represents the amount of uncertainty about X that is resolved given that we know Y.

Let $\{\boldsymbol{\rho}_x\}$ be a set of mutually orthogonal mixed states occurring with probabilities $\{p_x\}$, and let Alice prepare a quantum state given by the density operator:

$$\boldsymbol{\rho} = \sum_x p_x \boldsymbol{\rho}_x. \tag{7.32}$$

If all that is available to Bob is $\boldsymbol{\rho}$, then the associated uncertainty is given by $S(\boldsymbol{\rho})$. If, however, Bob also knows how Alice prepared the state, that is, he knows the probability p_x associated with each mixed state $\boldsymbol{\rho}_x$, then the uncertainty is given by $\sum_x p_x S(\boldsymbol{\rho}_x)$. We would intuitively expect that:

$$S\left(\sum_x p_x \boldsymbol{\rho}_x\right) \geq \sum_x p_x S(\boldsymbol{\rho}_x), \tag{7.33}$$

because ignorance of how $\boldsymbol{\rho}$ was prepared is reduced in the latter case. This inequality represents the special case of the *concavity property* of the von Neumann entropy given by Eq. (7.8). The difference between the two represents the average reduction in quantum entropy given that we know how $\boldsymbol{\rho}$ was prepared, namely $\boldsymbol{\rho} = \Sigma_x p_x \boldsymbol{\rho}_x$, and is commonly referred to as the *Holevo information*:

$$\chi = S(\boldsymbol{\rho}) - \sum_x p_x S(\boldsymbol{\rho}_x). \tag{7.34}$$

Let us now consider the ensemble $\{\boldsymbol{\rho}_x, p_x\}$ of mutually orthogonal mixed states, for which $\mathrm{Tr}(\boldsymbol{\rho}_{x_1}\boldsymbol{\rho}_{x_2}) = 0,\ x_1 \neq x_2$. By choosing the basis for which $\boldsymbol{\rho} = \sum_x p_x \boldsymbol{\rho}_x$ is block diagonal we have that:

$$\begin{aligned}
S(\boldsymbol{\rho}) &= -\mathrm{Tr}(\boldsymbol{\rho}\log\boldsymbol{\rho}) = -\sum_x \mathrm{Tr}(p_x\boldsymbol{\rho}_x)\log(p_x\boldsymbol{\rho}_x) \\
&= -\sum_x p_x\left[\underbrace{\mathrm{Tr}(\boldsymbol{\rho}_x)}_{=1}\log(p_x) + \mathrm{Tr}\boldsymbol{\rho}_x\log(\boldsymbol{\rho}_x)\right] \\
&= \underbrace{-\sum_x p_x\log(p_x)}_{H(X)} + \sum_x p_x\underbrace{[-\mathrm{Tr}\boldsymbol{\rho}_x\log(\boldsymbol{\rho}_x)]}_{S(\boldsymbol{\rho}_x)} = H(X) + \sum_x p_x S(\boldsymbol{\rho}_x).
\end{aligned} \tag{7.35}$$

By rearranging we obtain:

$$\underbrace{S(\boldsymbol{\rho}) - \sum_x p_x S(\boldsymbol{\rho}_x)}_{\chi} = H(X), \tag{7.36}$$

indicating that as long as $\boldsymbol{\rho}_x$ have mutually orthogonal support, the Holevo information is equal to the Shannon entropy of a classical source X with probability distribution given by $P(X=x)=p_x$. In general, however, when the ensemble of mixed states does not satisfy the mutually orthogonal condition, Shannon entropy upper bounds the Holevo information. To summarize, the Holevo information is upper bounded by Shannon entropy, that is:

$$\chi \leq H(X). \tag{7.37}$$

Suppose now that Alice wants to send classical information to Bob over a quantum channel. Alice has a classical source X that outputs the character x drawn from an alphabet X with probability p_x. For each x Alice prepares the state $\boldsymbol{\rho}_x$ and sends it over the channel. For the sake of simplicity we assume that the quantum operation of the channel is characterized by an identity operator. When Bob receives $\boldsymbol{\rho}_x$, he performs the generalized positive operator-valued measure (POVM) $\boldsymbol{M}_y$, resulting in outcome y with probability:

$$p(y|x)=\mathrm{Tr}(\boldsymbol{M}_y\boldsymbol{\rho}_x). \tag{7.38}$$

Bob's goal in performing this measurement is to maximize the mutual information $I(X,Y)$. This maximized mutual information is the best that Bob can hope to do, and it represents the information accessible to Bob. Since Bob can perform the optimization over all possible measurement strategies to maximize the mutual information, the *accessible information* is defined as:

$$H(X:Y)=\max_{\boldsymbol{M}_y} I(X;Y). \tag{7.39}$$

If the quantum states are pure and mutually orthogonal, instead of POVM we consider projective measurements such that $p(y|x)=\mathrm{Tr}(\boldsymbol{M}_y\boldsymbol{\rho}_x)=\mathrm{Tr}(\boldsymbol{P}_y\boldsymbol{\rho}_x)=1$, if $x=y$ and zero otherwise. In that case $H(X:Y)=H(X)$ and Bob (receiver) is able accurately to estimate the information sent by Alice (transmitter). If the states are nonorthogonal, the accessible information is bounded by:

$$H(X:Y)\leq S(\boldsymbol{\rho})\leq H(X),\ \rho=\sum_x p_x\boldsymbol{\rho}_x, \tag{7.40}$$

and this result is a consequence of the fact that nonorthogonal states are not completely distinguishable. An important interpretation of Eq. (7.40) is as a consequence of the no-cloning theorem. To see this, suppose two nonorthogonal states $|\psi_1\rangle$ and $|\psi_2\rangle$ can be cloned. We can then perform repeated cloning of the states to obtain the states $|\psi_1\rangle\otimes\ \ldots\ \otimes|\psi_1\rangle$ and $|\psi_2\rangle\otimes\ \ldots\ \otimes|\psi_2\rangle$. As the number of clones tends to infinity these new states start becoming orthogonal, so that Bob can unambiguously determine them resulting in $H(X:Y)=H(X)$. Since the arbitrary quantum states cannot be cloned we reach the desired result Eq. (7.40).

When $\boldsymbol{\rho}_x$ states are mixed states, the accessible information is bounded by the Holevo information:

$$H(X:Y)\leq \chi, \tag{7.41}$$

and this generalized result is known as the *Holevo bound*. The equality holds when the mixed states have mutually orthogonal support. Since the Holevo information is upper bounded by the Shannon entropy we can write:

$$H(X:Y) \leq \chi \leq H(X), \tag{7.42}$$

implying that apart from the case when $\{\rho_x\}$ have mutually orthogonal support there is no possibility for Bob to completely recover the classical information, characterized by *H*(*X*), that Alice sent him over the quantum channel!

Example. Let us consider two orthogonal pure states $|0\rangle$ and $|1\rangle$ with the corresponding density operators being:

$$\rho_0 = \begin{bmatrix} 1 & 0 \\ 0 & 0 \end{bmatrix}, \ \rho_1 = \begin{bmatrix} 0 & 0 \\ 0 & 1 \end{bmatrix}.$$

Clearly, $S(\rho_0) = S(\rho_1) = -\log 1 = 0$. Furthermore, let us assume that Alice selects each state with probability (1/2), resulting in:

$$\rho = \frac{1}{2}\begin{bmatrix} 1 & 0 \\ 0 & 1 \end{bmatrix}.$$

As a result $S(\rho) = -2 \times (1/2) \log_2(1/2) = 1$, while $\sum_x p_x S(\rho_x) = 0.5S(\rho_0) + 0.5S(\rho_1) = 0$. Consequently, $\chi = 1 - 0 = 1$, and as expected is equal to the Shannon entropy $H(X) = -2 \times (1/2) \log_2(1/2) = 1$. Now let us suppose that we still have the same pure orthogonal states, but Alice is uncertain whether a given preparation is $|0\rangle$ and $|1\rangle$ resulting in mixed density operators:

$$\rho_0 = \rho_1 = \frac{1}{2}\begin{bmatrix} 1 & 0 \\ 0 & 1 \end{bmatrix}.$$

As in the previous case, Alice selects each with probability 1/2, resulting in:

$$\rho = \frac{1}{2}\begin{bmatrix} 1 & 0 \\ 0 & 1 \end{bmatrix},$$

and

$$\chi = S(\rho) - \sum_x p_x S(\rho_x) = 1 - 1 = 0.$$

Thus the Holevo information is zero and there is no hope for Bob to recover any information Alice sent him. Note that although ρ_0 and ρ_1 are mixed states they are not mutually orthogonal. In fact, they are identical and therefore Bob's measurement outcomes are completely uncertain.

7.4 DATA COMPRESSION AND SCHUMACHER'S NOISELESS QUANTUM CODING THEOREM

Having studied the basic building blocks of quantum information theory we now turn to the quantum version of the first central issue in information theory: To what extent can a quantum message be compressed without significantly affecting the quantum message fidelity? The answer to this question, given by Schumacher's quantum noiseless channel coding theorem, is the topic of this section. Schumacher's theorem is the quantum analog of Shannon's noiseless channel coding theorem that establishes the condition under which data compression of classical information can be done reliably. We therefore begin by discussing it and outlining its main ideas. This discussion will form the basis of our development of Schumacher's theorem and quantum data compression.

7.4.1 Shannon's Noiseless Source Coding Theorem

Consider a classical source X that generates symbols 0 and 1 with probabilities p and $1-p$, respectively. The probability of output sequence $x_1,\ldots,x_N$ is given by:

$$p(x_1,\ldots,x_N) = p^{\sum_i x_i}(1-p)^{\sum_i (1-x_i)} \xrightarrow[N\to\infty]{} p^{Np}(1-p)^{N(1-p)}. \tag{7.43}$$

By taking the logarithm of this probability we obtain:

$$\log p(x_1,\ldots,x_N) \approx Np\log p + N(1-p)\log(1-p) = -NH(X). \tag{7.44}$$

Therefore the probability of occurrence of the so-called *typical sequence* is $p(x_1,\ldots,x_N) \approx 2^{-NH(X)}$, and on average $NH(X)$ bits are needed to represent any typical sequence, which is illustrated in Fig. 7.2A. We denoted the typical set by $T(N,\varepsilon)$. Clearly, not much of the information will be lost if we consider $2^{NH(X)}$ typical sequences instead of 2^N possible sequences, in particular for sufficiently large N. This observation can be used in *data compression*. Namely, with $NH(X)$ bits we can enumerate all typical sequences. If the source output is a typical sequence it will be represented with $NH(X)$ bits. On the other hand, when an atypical sequence

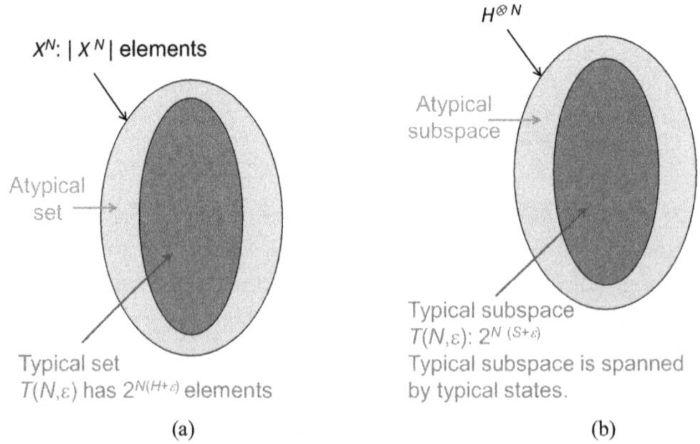

FIGURE 7.2

Typical set (a) and typical subspace (b).

is generated by the source we will assign a fixed index to it resulting in compression loss. However, as $N \to \infty$, the probability of occurrence of atypical sequence will tend to zero resulting in arbitrarily small information loss.

Let us now extend this concept to nonbinary sources. The law of large numbers applied to a nonbinary source with i.i.d. outputs $X_1,\ldots,X_N$ claims that as $N \to \infty$, the expected value of the source approaches the true value $E(X)$ in probability sense. In other words:

$$P\left(\left|\frac{1}{N}\sum_i X_i - E(X)\right| \le \varepsilon\right) > 1 - \delta, \tag{7.45}$$

for sufficiently large N and ε, $\delta > 0$. By replacing the random variable X_i with the function of the random variable, namely $-\log P(X_i)$, we obtain:

$$-N[H(X) + \varepsilon] \le \log P(x_1, \ldots, x_N) \le -N[H(X) - \varepsilon]. \tag{7.46}$$

The sequences $\{x_1,x_2,\ldots,x_N\}$ that satisfy Eq. (7.46) are the typical sequences, and the set of these typical sequences is the typical set $T(N,\varepsilon)$. Thus as a consequence of Eq. (7.45), for sufficiently large N, we have:

$$P(T(N, \varepsilon)) > 1 - \delta. \tag{7.47}$$

Since $P[T(N,\varepsilon)] > 1 - \delta$ and given Eq. (7.46), the size of a typical set, denoted as $|T(N,\varepsilon)|$, will be bounded by:

$$(1 - \delta)2^{N[H(X)-\varepsilon]} \le |T(N, \varepsilon)| \le 2^{N[H(X)+\varepsilon]}. \tag{7.48}$$

To prove the right side of inequality, we note that the sum of probabilities of all typical sequences cannot be more than 1, so that the following is valid:

$$\begin{aligned} 1 \ge \sum_{(x_1,\ldots,x_n)\in T(N,\varepsilon)} P(x_1, \ldots, x_n) &\ge \sum_{(x_1,\ldots,x_n)\in T(N,\varepsilon)} 2^{-N[H(X)+\varepsilon]} \\ &= |T(N, \varepsilon)| 2^{-N[H(X)+\varepsilon]}. \end{aligned} \tag{7.49}$$

Resulting in:

$$|T(N, \varepsilon)| \le 2^{N[H(X)+\varepsilon]}. \tag{7.50}$$

To prove the left side of inequality (7.48), we invoke Eq. (7.47) to obtain:

$$\begin{aligned} 1 - \delta < P(T(N, \varepsilon)) &\le \sum_{(x_1,\ldots,x_n)\in T(N,\varepsilon)} P(x_1, \ldots, x_n) \\ &\le \sum_{(x_1,\ldots,x_n)\in T(N,\varepsilon)} 2^{-N[H(X)-\varepsilon]} = |T(N, \varepsilon)| 2^{-N[H(X)-\varepsilon]}, \end{aligned} \tag{7.51}$$

resulting in:

$$|T(N, \varepsilon)| > (1 - \delta)2^{N[H(X)-\varepsilon]}. \tag{7.52}$$

Therefore we have a couple of powerful results. First, as $N \to \infty$, the probability that an output sequence will be a typical sequence approaches 1 and the probability of occurrence of any one particular typical sequence is approximately given by $2^{-NH(X)}$. Second, the total number of such typical sequences, that is, the size of the typical set, is approximately $2^{NH(X)}$.

Suppose now that a compression rate R is larger than $H(X)$, say $R > H(X) + \varepsilon$. Then, based on the foregoing discussion the total number of typical sequences, for sufficiently large N, is bounded by:

$$|T(N,\varepsilon)| \le 2^{N[H(X)+\varepsilon]} \le 2^{NR}, \tag{7.53}$$

and the probability of occurrence of a typical sequence is lower bounded by $1 - \delta$, where δ is arbitrarily small.

The source encoding strategy can be described as follows. Let us divide the set of all sequences generated by the source into typical and atypical sets as shown in Fig. 7.2A. From Eq. (7.53) it clear that sequences in the typical set can be represented by at most NR bits. If an atypical sequence occurs, we assign to it a fixed index. As N tends to infinity, the probability for this event to occur tends to zero, indicating that we can encode and decode typical sequences reliably.

On the other hand, when $R < H(X)$ we can encode maximum 2^{NR} typical sequences, labeled as $T_R(N,\varepsilon)$. Since the probability of occurrence of typical sequences is upper bounded by $2^{-N[H(X)\ -\varepsilon]}$, the typical sequences in $T_R(N,\varepsilon)$ will occur with probability $2^{N[R-H(X)+\varepsilon]}$, which tends to zero as N tends to infinity, indicating that a reliable compression scheme does not exist in this case. With this we have just proved Shannon's source coding theorem, which can be formulated as follows.

Shannon's Source Coding Theorem. For an i.i.d. source X, the reliable compression method exists for $R > H(X)$. If, on the other hand, $R < H(X)$, then no reliable compression scheme exists.

7.4.2 Schumacher's Noiseless Quantum Source Coding Theorem

Consider now a quantum source emitting a pure state $|\psi_x\rangle$ with probability p_x, described in terms of the mixed density operator:

$$\rho = \sum_x p_x |\psi_x\rangle\langle\psi_x|. \tag{7.54}$$

The quantum message comprises N quantum source outputs, independent of each other, so that:

$$\rho_{\otimes N} = \rho \otimes \cdots \otimes \rho. \tag{7.55}$$

Schumacher's theorem gives us the condition under which there exists a reliable compression scheme that can compress and decompress the quantum message with high *fidelity*. We will define fidelity later in this section. The development of Schumacher's noiseless quantum channel coding theorem closely parallels Shannon's classical version.

For this purpose, let us express the mixed density operator in terms of its orthonormal eigenkets as:

$$\boldsymbol{\rho} = \sum_{\alpha} \lambda_\alpha |\lambda_\alpha\rangle\langle\lambda_\alpha|. \tag{7.56}$$

Then, the quantum source can effectively be thought of as a classical source, and we say that the sequence $\{\lambda_1 \ldots \lambda_N\}$ is the typical sequence if it satisfies the following inequality for sufficiently large N:

$$P\left(\left|\frac{1}{N}\log\left(\frac{1}{P(\lambda_1)P(\lambda_2)\cdots P(\lambda_N)}\right) - S(\boldsymbol{\rho})\right| \leq \varepsilon\right) > 1 - \delta, \tag{7.57}$$

and this inequality is a direct consequence of the law of large numbers in a manner identical to Eqs. (7.45) and (7.46). We can now define the *typical state* as a state $|\lambda_1\rangle \ldots |\lambda_N\rangle$ for which a sequence of eigenvalues $\boldsymbol{\lambda} = \{\lambda_1,\ldots,\lambda_N\}$ is a typical sequence. Thus similarly to the concept of the set of typical sequences—the typical set—for a classical information source, we introduce the notion of a typical subspace spanned by the typical basis states $|\lambda_1\rangle \ldots |\lambda_N\rangle$. We can then define the projection operator onto the typical subspace as:

$$\boldsymbol{P}_T = \sum_{\lambda} |\lambda_1\rangle\langle\lambda_1| \otimes \cdots \otimes |\lambda_N\rangle\langle\lambda_N|. \tag{7.58}$$

The *projection* of the quantum message on the typical subspace is determined by $\boldsymbol{P}_T\boldsymbol{\rho}_{\otimes N}$. The probability of the projection is given by its trace. By using the projection operator we can separate the total Hilbert space into typical and atypical subspaces, as illustrated in Fig. 7.2B. The probability that the state of the quantum message lies in the typical subspace is given by:

$$P(\boldsymbol{P}_T\boldsymbol{\rho}_{\otimes N}) > 1 - \delta. \tag{7.59}$$

Earlier we showed that the probability of the typical set approaches unity for sufficiently large N. Analogously, we now have the result that the probability that the quantum message lies in the typical subspace approaches unity for sufficiently large N. In the last section we were able to bound the size of the typical set. Following a similar procedure we can bound the dimensionality of the typical subspace, characterized by $\mathrm{Tr}(\boldsymbol{P}_T)$. Denoting the typical subspace by $T_\lambda(N,\varepsilon)$, the bounds are given by:

$$(1-\delta)2^{N[S(\boldsymbol{\rho})-\varepsilon]} \leq |T_\lambda(N,\varepsilon)| \leq 2^{N[S(\boldsymbol{\rho})+\varepsilon]}. \tag{7.60}$$

Analogous to the classical development of Shannon's noiseless coding theorem we have a couple of powerful results. As $N \to \infty$, the probability that the state of the quantum message will lie in the typical subspace approaches unity and the dimensionality of the typical subspace is approximately $2^{NS(\boldsymbol{\rho})}$, with the probability of the quantum message being any one typical state approaching $2^{-NS(\boldsymbol{\rho})}$.

Now we are in a position to formulate the corresponding *compression procedure*. Define the projector on the typical subspace $\boldsymbol{P}_T$ and its complement projecting on orthogonal subspace $\boldsymbol{P}_T^{\perp} = \boldsymbol{I} - \boldsymbol{P}_T$ ($\boldsymbol{I}$ is the identity operator) with corresponding outcomes 0 and 1. For compression purposes, we perform the measurements using two orthogonal operators with outputs denoted as 1 and 0, respectively. If the outcome is 1 we know that the message is in a typical state and we do nothing further. If, on the other hand, the outcome is 0 we know that it belongs to an atypical subspace and numerate it as a *fixed* state from the typical subspace. Given that the probability of this happening can be made as small as possible for large N, we can compress the quantum message without the loss of information.

Schumacher showed that when the rate $R > S(|\psi_x\rangle)$ the foregoing compression scheme was reliable. Reliability implies that the compressed quantum message can be decompressed with high fidelity. In other words, we say that the *compression is reliable* if the corresponding *entanglement fidelity* tends to 1 for large N:

$$F_e(\boldsymbol{\rho}_{\otimes N}; D \circ C) \xrightarrow[N\to\infty]{} 1;\ \boldsymbol{\rho}_{\otimes N} = |\psi_{\otimes N}\rangle\langle\psi_{\otimes N}|,\ |\psi_{\otimes N}\rangle = |\psi_{x_1}\rangle \otimes \ldots \otimes |\psi_{x_N}\rangle, \tag{7.61}$$

where we used D and C to denote the decompression and compression operations, respectively, defined as the following mappings:

$$D:\ H_c^N \to H^N,\ C:\ H^N \to H_c^N, \tag{7.62}$$

with H_c^N being the 2^{NR}-dimensional subspace of H^N.

Let us reconsider our i.i.d. quantum source defined by the ensemble $\{|\psi_x\rangle, p_x\}$ where $|\psi_x\rangle$ are pure states. Given this ensemble, the kth instance of a length N quantum message is given by:

$$\left|\psi_{\otimes N}^k\right\rangle = \left|\psi_{x_1}^k\right\rangle \otimes \cdots \otimes \left|\psi_{x_N}^k\right\rangle, \tag{7.63}$$

and the corresponding density operator is $\boldsymbol{\rho}_{\otimes N}^k = \left|\psi_{\otimes N}^k\right\rangle\left\langle\psi_{\otimes N}^k\right|$. Alice encodes the message by performing a fuzzy measurement whose outcome will be 0_{typical} with probability $\text{Tr}\left(\boldsymbol{P}_T\boldsymbol{\rho}_{\otimes N}^k\right)\rangle 1-\delta$ and 1_{atypical} with probability $\text{Tr}\left(\boldsymbol{P}_T^{\perp}\boldsymbol{\rho}_{\otimes N}^k\right) \le \delta$. As already mentioned, the atypical state is coded as some fixed state $|\varphi\rangle$ in the typical subspace, with:

$$E = \sum_j \underbrace{|\varphi\rangle\langle j|}_{E_j} = \sum_j E_j,\ E_j = |\varphi\rangle\langle j|, \tag{7.64}$$

where $|j\rangle$ is the orthonormal basis of the atypical subspace. The compressed state is therefore given by:

$$\rho_C = \boldsymbol{P}_T\boldsymbol{\rho}_{\otimes N}^k\boldsymbol{P}_T + \sum_j E_j\boldsymbol{\rho}_{\otimes N}^k E_j^{\dagger}. \tag{7.65}$$

On substituting Eq. (7.64) into Eq. (7.65) we obtain:

$$\begin{aligned}
\rho_C &= \boldsymbol{P}_T\boldsymbol{\rho}_{\otimes N}^k\boldsymbol{P}_T + \sum_j |\varphi\rangle\left\langle j|\psi_{\otimes N}^k\right\rangle\left\langle\psi_{\otimes N}^k|j\right\rangle\langle\varphi| \\
&= \boldsymbol{P}_T\boldsymbol{\rho}_{\otimes N}^k\boldsymbol{P}_T + |\varphi\rangle\langle\varphi|\sum_j\left\langle\psi_{\otimes N}^k|j\right\rangle\left\langle j|\psi_{\otimes N}^k\right\rangle \\
&= \boldsymbol{P}_T\boldsymbol{\rho}_{\otimes N}^k\boldsymbol{P}_T + |\varphi\rangle\langle\varphi|\left\langle\psi_{\otimes N}^k\right|\boldsymbol{P}_T^{\perp}\left|\psi_{\otimes N}^k\right\rangle.
\end{aligned} \tag{7.66}$$

The decoding operator for the compressed Hilbert space of the typical subspace is the identity operator. Therefore the decoding operation results in the state $\boldsymbol{\rho}_{D\circ C}$ given by Eq. (7.66). The question we need to answer is what the fidelity of the decoding is when N is sufficiently large. For the kth quantum message we have considered, the fidelity $F_k = \langle \psi^k_{\otimes N} | \boldsymbol{\rho}_{D\circ C} | \psi^k_{\otimes N}\rangle$ is equal to unity if $\boldsymbol{\rho}_{D\circ C} = \boldsymbol{\rho}^k_{\otimes N}$. Thus fidelity is a measure of how close the decompressed quantum message is to the original quantum message. Because the quantum state of the source output is derived from an ensemble of quantum messages we compute the average fidelity as:

$$F = \sum_k p_k F_k = \sum_k p_k \left\langle \psi^k_{\otimes N} | \boldsymbol{\rho}_{D\circ C} | \psi^k_{\otimes N} \right\rangle, \; p_k = P\left(\boldsymbol{\rho}^k_{\otimes N}\right). \tag{7.67}$$

By substituting Eq. (7.66) into Eq. (7.67) we obtain:

$$\begin{aligned}
F &= \sum_k p_k F_k = \sum_k p_k \left\langle \psi^k_{\otimes N} | \boldsymbol{P}_T \boldsymbol{\rho}^k_{\otimes N} \boldsymbol{P}_T \middle| \psi^k_{\otimes N} \right\rangle \\
&\quad + \underbrace{\sum_k p_k \left\langle \psi^k_{\otimes N} | \varphi \rangle \langle \varphi | \left\langle \psi^k_{\otimes N} \middle| \boldsymbol{P}_T^{\perp} \middle| \psi^k_{\otimes N} \right\rangle \middle| \psi^k_{\otimes N} \right\rangle}_{F_{atypical}} \\
&= \sum_k p_k \left\langle \psi^k_{\otimes N} \middle| \boldsymbol{P}_T \middle| \psi^k_{\otimes N} \right\rangle \left\langle \psi^k_{\otimes N} \middle| \boldsymbol{P}_T \middle| \psi^k_{\otimes N} \right\rangle \\
&\quad + F_{atypical} = \sum_k p_k \underbrace{\mathrm{Tr}\left(\boldsymbol{P}_T \boldsymbol{\rho}^k_{\otimes N}\right) \mathrm{Tr}\left(\boldsymbol{P}_T \boldsymbol{\rho}^k_{\otimes N}\right)}_{\mathrm{Tr}\left(\boldsymbol{P}_T \boldsymbol{\rho}^k_{\otimes N}\right)^2} + F_{atypical} \\
&\geq \sum_k p_k \mathrm{Tr}\left(\boldsymbol{P}_T \boldsymbol{\rho}^k_{\otimes N}\right)^2 = \mathrm{Tr}(\boldsymbol{P}_T \boldsymbol{\rho}_{\otimes N})^2.
\end{aligned} \tag{7.68}$$

Given Eq. (7.59), from Eq. (7.68) we obtain the following inequality:

$$F > (1 - \delta)^2 \geq 1 - 2\delta. \tag{7.69}$$

Thus we have just shown that we can compress the quantum message to $N(S(\boldsymbol{\rho}) + \varepsilon)$ qubits with arbitrarily small loss in fidelity for sufficiently large N.

What happens if the quantum message is encoded to $R < S(\boldsymbol{\rho})$ qubits? In this case the dimension of the Hilbert subspace Ω in which the compressed quantum message lies is 2^{NR}. Let the projection operator onto this subspace be $\boldsymbol{V}$ and let the decompressed state be $\boldsymbol{\rho}_{D\circ C}$. This decompressed state lies in Ω and consequently can be expanded in the basis of this subspace as follows:

$$\boldsymbol{\rho}_{D\circ C} = \sum_i |v_i\rangle \langle v_i | \boldsymbol{\rho}_{D\circ C} | v_i \rangle \langle v_i |. \tag{7.70}$$

The average fidelity of reconstruction can now be calculated as:

$$\begin{aligned}F &= \sum_k p_k F_k = \sum_k p_k \left\langle \psi^k_{\otimes N} | \boldsymbol{\rho}_{D\circ C} | \psi^k_{\otimes N} \right\rangle \\ &= \sum_{k,i} p_k \left\langle \psi^k_{\otimes N} \| v_i \right\rangle \underbrace{\langle v_i | \boldsymbol{\rho}_{D\circ C} | v_i \rangle}_{\geq 0} \left\langle v_i \| \psi^k_{\otimes N} \right\rangle \\ &\geq \sum_{k,i} p_k \left\langle \psi^k_{\otimes N} \| v_i \right\rangle \left\langle v_i \| \psi^k_{\otimes N} \right\rangle = \sum_k p_k \mathrm{Tr}\left(\boldsymbol{V} \boldsymbol{\rho}^k_{\otimes N} \right).\end{aligned} \tag{7.71}$$

Following similar discussion to that related to Shannon's noiseless coding theorem, the probability that a typical state lies in Ω is $2^{N[R-S(\boldsymbol{\rho})+\varepsilon]}$. For $R < S(\boldsymbol{\rho})$, as $N \to \infty$ this probability approaches 0. Thus the fidelity of a quantum message reconstructed after being compressed to $R < S(\boldsymbol{\rho})$ qubits is not good at all. We therefore have the following statement for Schumacher's quantum noiseless coding theorem.

Schumacher's Source Coding Theorem. Let $\{|\psi_x\rangle, p_x\}$ be an i.i.d. quantum source. If $R > S(\rho)$, then there exists a reliable compression scheme of rate R for this source. Otherwise, if $R < S(\rho)$, then no reliable compression scheme of rate R exists.

Example [1]. Let us consider a quantum source that produces states $|\psi_1\rangle$ and $|\psi_2\rangle$ with probabilities p_1 and p_2, respectively. Let us define the states to be:

$$|\psi_1\rangle = \frac{1}{\sqrt{2}}(|0\rangle + |1\rangle),\ |\psi_2\rangle = \frac{\sqrt{3}}{2}|0\rangle + \frac{1}{2}|1\rangle,$$

with $p_1 = p_2 = 0.5$. Thus the resulting density matrix is given by:

$$\boldsymbol{\rho} = p_1|\psi_1\rangle\langle\psi_1| + p_2|\psi_2\rangle\langle\psi_2| = \begin{pmatrix} 0.6250 & 0.4665 \\ 0.4665 & 0.3750 \end{pmatrix}.$$

The eigen decomposition of $\boldsymbol{\rho}$ is:

$$\boldsymbol{\rho} = \begin{pmatrix} -0.7934 & -0.6088 \\ -0.6088 & 0.7934 \end{pmatrix} \begin{pmatrix} 0.9830 & 0 \\ 0 & 0.0170 \end{pmatrix} \begin{pmatrix} -0.7934 & -0.6088 \\ -0.6088 & 0.7934 \end{pmatrix}^{\dagger},$$

and therefore the eigenvectors are:

$$|\lambda_1\rangle = \begin{pmatrix} -0.7934 \\ -0.6088 \end{pmatrix},\ |\lambda_2\rangle = \begin{pmatrix} -0.6088 \\ 0.7934 \end{pmatrix},$$

with the corresponding eigenvalues given by $\lambda_1 = 0.9830$ and $\lambda_2 = 0.0170$. With this setup in mind, let us consider the compression of a three-qubit quantum message. With our source as defined earlier, the possible quantum messages are:

$$\begin{aligned}|\psi_{111}\rangle &= |\psi_1\rangle \otimes |\psi_1\rangle \otimes |\psi_1\rangle, & |\psi_{112}\rangle &= |\psi_1\rangle \otimes |\psi_1\rangle \otimes |\psi_2\rangle. \\ |\psi_{121}\rangle &= |\psi_1\rangle \otimes |\psi_2\rangle \otimes |\psi_1\rangle, & |\psi_{122}\rangle &= |\psi_1\rangle \otimes |\psi_2\rangle \otimes |\psi_2\rangle. \\ |\psi_{211}\rangle &= |\psi_2\rangle \otimes |\psi_1\rangle \otimes |\psi_1\rangle, & |\psi_{212}\rangle &= |\psi_2\rangle \otimes |\psi_1\rangle \otimes |\psi_2\rangle. \\ |\psi_{221}\rangle &= |\psi_2\rangle \otimes |\psi_2\rangle \otimes |\psi_1\rangle, & |\psi_{222}\rangle &= |\psi_2\rangle \otimes |\psi_2\rangle \otimes |\psi_2\rangle.\end{aligned}$$

Let us also define the eigenvector-based basis states:

$$\begin{aligned}
|\lambda_{111}\rangle &= |\lambda_1\rangle \otimes |\lambda_1\rangle \otimes |\lambda_1\rangle, & |\lambda_{112}\rangle &= |\lambda_1\rangle \otimes |\lambda_1\rangle \otimes |\lambda_2\rangle. \\
|\lambda_{121}\rangle &= |\lambda_1\rangle \otimes |\lambda_2\rangle \otimes |\lambda_1\rangle, & |\lambda_{122}\rangle &= |\lambda_1\rangle \otimes |\lambda_2\rangle \otimes |\lambda_2\rangle. \\
|\lambda_{211}\rangle &= |\lambda_2\rangle \otimes |\lambda_1\rangle \otimes |\lambda_1\rangle, & |\lambda_{212}\rangle &= |\lambda_2\rangle \otimes |\lambda_1\rangle \otimes |\lambda_2\rangle. \\
|\lambda_{221}\rangle &= |\lambda_2\rangle \otimes |\lambda_2\rangle \otimes |\lambda_1\rangle, & |\lambda_{222}\rangle &= |\lambda_2\rangle \otimes |\lambda_2\rangle \otimes |\lambda_2\rangle.
\end{aligned}$$

These states can be used to define the projective operators onto the Hilbert space $\mathbb{H}^{\otimes 3}$ spanned by $|\lambda_{i,j,k}\rangle, i,j = 1,2$. Using these measurement states, let us see if there exists a subspace of the Hilbert space where the quantum messages lie. In other words, can we find the typical subspace? To do this we first note that:

$$\mathrm{Tr}((|\psi_i\rangle\langle\psi_i|)(|\lambda_j\rangle\langle\lambda_j|)) = |\langle\psi_i|\lambda_j\rangle|^2 = \lambda_j;\ i,j = 1,2.$$

Furthermore:

$$\mathrm{Tr}((|\psi_{ijk}\rangle\langle\psi_{ijk}|)(|\lambda_{i'j'k'}\rangle\langle\lambda_{i'j'k'}|)) = |\langle\psi_i|\lambda_{i'}\rangle|^2 |\langle\psi_j|\lambda_{j'}\rangle|^2 |\langle\psi_k|\lambda_{k'}\rangle|^2 = \lambda_{i'}\lambda_{j'}\lambda_{k'},$$

where $i,j,k = 1,2; i',j',k' = 1,2$. We now compute $\lambda_{i'}, \lambda_{j'}, \lambda_{k'}$:

$$\lambda_1\lambda_1\lambda_1 = 0.9499,\ \lambda_1\lambda_1\lambda_2 = 0.0164,\ \lambda_1\lambda_2\lambda_1 = 0.0164,\ \lambda_1\lambda_2\lambda_2 = 2.8409\mathrm{E}(-4),$$

$$\lambda_2\lambda_1\lambda_1 = 0.0164,\ \lambda_2\lambda_1\lambda_2 = 2.8409\mathrm{E}(-4),\ \lambda_2\lambda_2\lambda_1 = 2.8409\mathrm{E}(-4),$$

$$\lambda_2\lambda_2\lambda_2 = 4.9130\mathrm{E}(-6).$$

From these values we can see that the probability that the quantum message lies in the subspace spanned by $|\lambda_{111}\rangle, |\lambda_{112}\rangle, |\lambda_{121}\rangle, |\lambda_{211}\rangle$ is $0.9499 + 0.0164 + 0.0164 + 0.0164 = 0.9991$, while the probability that the quantum message lies in the subspace spanned by $|\lambda_{122}\rangle, |\lambda_{212}\rangle, |\lambda_{221}\rangle, |\lambda_{212}\rangle$ is $2.8409\mathrm{E}(-4) + 2.8409\mathrm{E}(-4) + 2.8409\mathrm{E}(-4) + 4.9130\mathrm{E}(-6) = 8.5717\mathrm{E}(-4)$. Thus with probability close to 1, the quantum message will lie in the subspace spanned by the first set of quantum states, while the probability δ of the quantum message lying in subspace spanned by the second set of quantum states lies close to 0. This becomes even more apparent if n, which is equal to 3 right now, increases. Thus $|\lambda_{111}\rangle, |\lambda_{112}\rangle, |\lambda_{121}\rangle, |\lambda_{211}\rangle$ span the typical subspace, while $|\lambda_{122}\rangle, |\lambda_{212}\rangle, |\lambda_{221}\rangle, |\lambda_{212}\rangle$ span the complement subspace.

We can now perform data compression using unitary encoding. Unitary encoding implies that a unitary operator $\boldsymbol{U}$ acts on the Hilbert space such that any quantum message state $|\psi_{ijk}\rangle$ that lies in the typical subspace goes to $|\xi_1\rangle \otimes |\xi_2\rangle \otimes |fixed_{typical}\rangle$, while the message that lies in the atypical subspace goes to $|\xi_1\rangle \otimes |\xi_2\rangle \otimes |fixed_{atypical}\rangle$. The unitary operator can be chosen such that, say for example, $|fixed_{typical}\rangle = |0\rangle$ and $|fixed_{atypical}\rangle = |1\rangle$. Thus for encoding the message, Alice measures the third qubit. If the output is 0, then the quantum message is in the typical subspace and Alice sends $|\xi_1\rangle \otimes |\xi_2\rangle$ to Bob. If the output is 1, then the quantum message is in the atypical subspace. In this case Alice assigns a fixed quantum state $|\varphi\rangle$ in the typical subspace whose unitary transformation is the state $|a\rangle \otimes |b\rangle \otimes |0\rangle$. Alice then sends $|a\rangle \otimes |b\rangle$ to Bob.

To decompress the message, Bob first appends |0 to the quantum state received by him, and applies the transform U^{-1}. The possible resulting states are $|\psi_{ijk}\rangle$ or $|\varphi\rangle$. Thus Bob is able to decompress the message as desired. For the cases where we get $|\varphi\rangle$ we lose information. However, as shown earlier the probability of this happening is very small, and can be made smaller for larger n.

Let us also look at the fidelity of the retrieved message. For a given quantum message m the fidelity is given by:

$$F_m = \langle \psi^{\mathrm{m}}_{\otimes \mathrm{n}} | \boldsymbol{P}_T \boldsymbol{\rho}^{\mathrm{m}}_{\otimes \mathrm{n}} \boldsymbol{P}_T | \psi^{\mathrm{m}}_{\otimes \mathrm{n}} \rangle + \textit{atypical contribution}.$$

Here, $\boldsymbol{P}_T$ is the projection onto the typical subspace. For our example for any quantum message, fidelity reduces to:

$$F = (1-\delta)^2 + \delta \left| \langle \varphi | \psi_{ijk} \rangle \right|^2.$$

We have already computed $\delta = 8.5717\mathrm{E}(-4)$, and this gives us the fidelity greater than $(1-\delta)^2 = 0.9983$. We thus have achieved a rate of $R = \frac{2}{3} = 0.667$ with high reconstruction fidelity. Let us compare this rate with the quantum entropy of the source. The quantum entropy turns out to be:

$$S(\boldsymbol{\rho}) = \lambda_1 \log\lambda_1 + \lambda_2 \log\lambda_2 = 0.1242.$$

We therefore have $R > S$, therefore satisfying Schumacher's condition. The foregoing equation also hints at another interesting fact: we can do better than two qubits and go down to a single qubit. For this case, $R = \frac{1}{3} = 0.333$ is still greater than $S(\boldsymbol{\rho})$. We note that here, quantum entropy represents the incompressible information in the message generated by a quantum source in a manner similar to Shannon's entropy representing the incompressible information of a classical source.

7.5 QUANTUM CHANNELS

The quantum channel performs the transformation (mapping) of initial density operator ρ to the final density operator ρ_f, denoted as $\xi : \rho \rightarrow \rho_f$, which is known as the *superoperator* or *quantum operation*. The final density operator can be expressed in the *operator-sum representation* as follows:

$$\rho_f = \sum_k E_k \rho E_k^\dagger, \tag{7.72}$$

where E_k are the operation elements for the superoperator. To describe the actions of the quantum channel we need to determine the Kraus operators E_k.

The operator-sum representation can be used in the classification of quantum channels into two categories: (1) *trace preserving* when $\mathrm{Tr}\xi(\rho) = \mathrm{Tr}\rho = 1$ and (2) *nontrace preserving* when $\mathrm{Tr}\xi(\rho) < 1$. Starting from the trace-preserving condition:

$$\mathrm{Tr}\rho = \mathrm{Tr}\xi(\rho) = \mathrm{Tr}\left[\sum_k E_k \rho E_k^\dagger\right] = \mathrm{Tr}\left[\rho \sum_k E_k E_k^\dagger\right] = 1, \tag{7.73}$$

we obtain:

$$\sum_k E_k E_k^\dagger = I. \tag{7.74}$$

The corresponding proof has been provided already in Chapter 4. For nontrace-preserving quantum operation, Eq. (7.75) is not satisfied, and informally we can write $\sum_k E_k E_k^\dagger < I$.

As an illustration, let us consider *bit-flip* (b.f.) and *phase-flip* (p.f.) channels, which are illustrated in Fig. 7.3. The corresponding Kraus operators for the b.f. channel are given by:

$$E_0^{(\text{b.f.})} = \sqrt{1-p}I,\ E_1^{(\text{b.f.})} = \sqrt{p}X, \tag{7.75}$$

and the operator sum representation is given by:

$$\xi^{(\text{b.f.})}(\rho) = E_0^{(\text{b.f.})}\rho\left(E_0^{(\text{b.f.})}\right)^\dagger + E_1^{(\text{b.f.})}\rho\left(E_1^{(\text{b.f.})}\right)^\dagger = (1-p)\rho + pX\rho X. \tag{7.76}$$

In similar fashion, the Kraus operators for the p.f. channel are given by:

$$E_0^{(\text{p.f.})} = \sqrt{1-p}I,\ E_1^{(\text{p.f.})} = \sqrt{p}Z, \tag{7.77}$$

and the corresponding operator-sum representation is:

$$\xi^{(\text{p.f.})}(\rho) = E_0^{(\text{p.f.})}\rho\left(E_0^{(\text{p.f.})}\right)^\dagger + E_1^{(\text{p.f.})}\rho\left(E_1^{(\text{p.f.})}\right)^\dagger = (1-p)\rho + pZ\rho Z. \tag{7.78}$$

Another relevant quantum channel model is the *depolarizing channel*. The depolarizing channel, as shown in Fig. 7.4, with probability $1-p$ leaves the qubit as it is, while with probability p moves the initial state into $\rho_f = I/2$ that maximizes the von

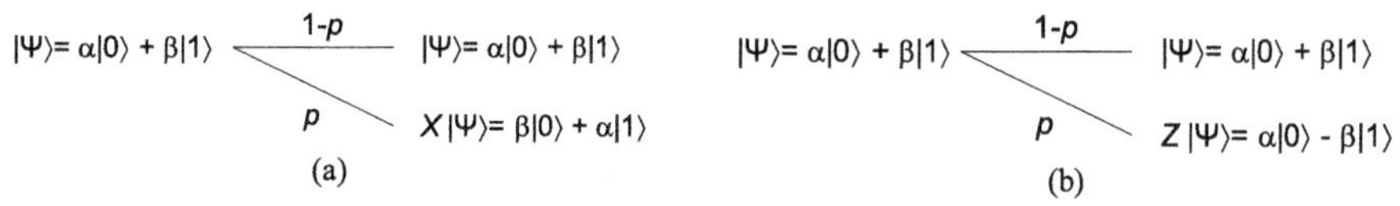

FIGURE 7.3

(a) Bit-flip channel model. (b) Phase-flip channel model.

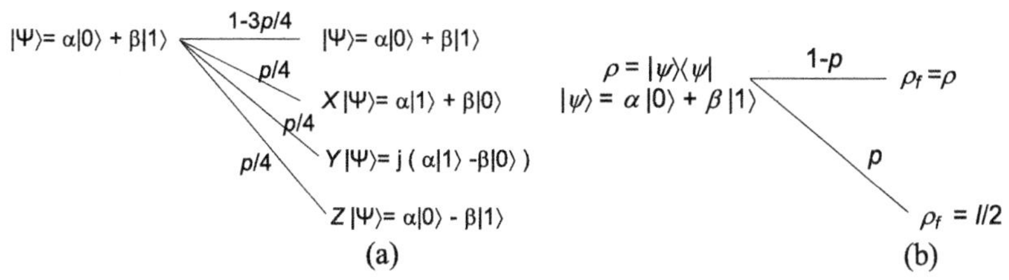

FIGURE 7.4

Depolarizing channel model: (a) Pauli operator description and (b) density operator description.

Neumann entropy $S(\rho) = -\text{Tr}\rho \log \rho = 1$. The properties describing the model can be summarized as follows:

1. Qubit errors are independent.
2. Single-qubit errors (X, Y, Z) are equally likely.
3. All qubits have the same single-error probability $p/4$.

The Kraus operators E_i of the channel should be selected as follows:

$$E_0 = \sqrt{1-3p/4}I;\ E_1 = \sqrt{p/4}X;\ E_1 = \sqrt{p/4}X;\ E_1 = \sqrt{p/4}Z. \tag{7.79}$$

The action of the depolarizing channel is to perform the following mapping: $\rho \rightarrow \xi(\rho) = \sum_i E_i \rho E_i^\dagger$, where ρ is the initial density operator. The operator-sum representation for the depolarization channel can be written as:

$$\begin{aligned}\xi(\rho) &= \sum_i E_i \rho E_i^\dagger = \left(1-\frac{3p}{4}\right)\rho + \frac{p}{4}(X\rho X + Y\rho Y + Z\rho Z) \\ &= (1-p)\rho + \frac{p}{2}I.\end{aligned} \tag{7.80}$$

The first line in Eq. (7.80) corresponds to the model shown in Fig. 7.4A and the second line in Eq. (7.80) corresponds to the model shown in Fig. 7.4B. It is clear from Eq. (7.80) that $\text{Tr}\xi(\rho) = \text{Tr}(\rho) = 1$ meaning that the superoperator is trace preserving. Notice that the depolarizing channel model in some other books/papers can be slightly different from the one described here.

We now describe the *amplitude-damping channel* model. In certain quantum channels the errors X, Y, and Z do not occur with the same probability. In the amplitude-damping channel, the Kraus operators are given by:

$$E_0 = \begin{pmatrix} 1 & 0 \\ 0 & \sqrt{1-\varepsilon^2} \end{pmatrix},\ E_1 = \begin{pmatrix} 0 & \varepsilon \\ 0 & 0 \end{pmatrix}. \tag{7.81}$$

Spontaneous emission is an example of a physical process that can be modeled using the amplitude-damping channel model. If $|\psi\rangle = a\,|0\rangle + b\,|1\rangle$ is the initial qubit sate, the effect of the amplitude-damping channel is to perform the following mapping:

$$\rho \rightarrow \xi(\rho) = E_0 \rho E_0^\dagger + E_1 \rho E_1^\dagger = \begin{pmatrix} |a|^2 + \varepsilon^2|b|^2 & ab^*\sqrt{1-\varepsilon^2} \\ a^*b\sqrt{1-\varepsilon^2} & |b|^2(1-\varepsilon^2) \end{pmatrix}. \tag{7.82}$$

Probabilities $P(0)$ and $P(1)$ that E_0 and E_1 occur are given by:

$$\begin{aligned} P(0) &= \text{Tr}\left(E_0\rho E_0^\dagger\right) = \text{Tr}\begin{pmatrix} |a|^2 & ab^*\sqrt{1-\varepsilon^2} \\ a^*b\sqrt{1-\varepsilon^2} & |b|^2(1-\varepsilon^2) \end{pmatrix} = 1-\varepsilon^2|b|^2 \\ P(1) &= \text{Tr}\left(E_1\rho E_1^\dagger\right) = \text{Tr}\begin{pmatrix} |b|^2\varepsilon^2 & 0 \\ 0 & 0 \end{pmatrix} = \varepsilon^2|b|^2. \end{aligned} \tag{7.83}$$

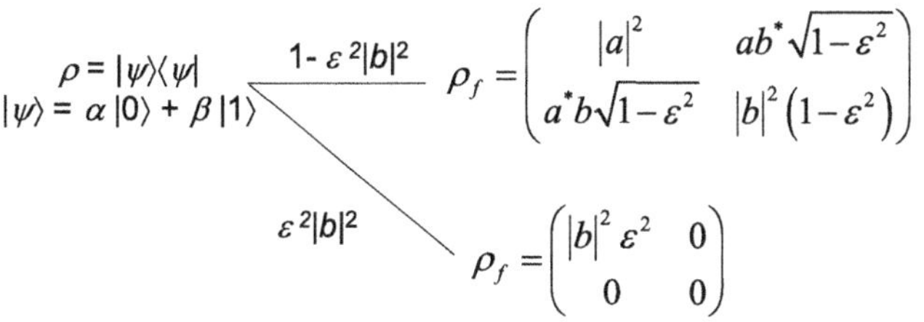
$\rho = |\psi\rangle\langle\psi|$
$|\psi\rangle = \alpha|0\rangle + \beta|1\rangle$
$1-\varepsilon^2|b|^2$
$\rho_f = \begin{pmatrix} |a|^2 & ab^*\sqrt{1-\varepsilon^2} \\ a^*b\sqrt{1-\varepsilon^2} & |b|^2(1-\varepsilon^2) \end{pmatrix}$
$\varepsilon^2|b|^2$
$\rho_f = \begin{pmatrix} |b|^2\varepsilon^2 & 0 \\ 0 & 0 \end{pmatrix}$

FIGURE 7.5

Amplitude-damping channel model.

The corresponding amplitude-damping channel model is shown in Fig. 7.5.

Every quantum channel error E (whether discrete or continuous) can be described as a linear combination of elements from the following discrete set $\{I,X,Y,Z\}$, given by $E = e_1I + e_2X + e_3Y + e_4Z$, which was proved in Chapter 4. For example, the following error $E = \begin{bmatrix} 1 & 0 \\ 0 & 0 \end{bmatrix}$ can be represented in terms of Pauli operators as $E = (I + Z)/2$. An error operator that affects several qubits can be written as a weighted sum of Pauli operators $\sum c_iP_i$ acting on ith qubits.

7.6 QUANTUM CHANNEL CODING AND HOLEVO–SCHUMACHER–WESTMORELAND THEOREM

We now turn to the second question we raised at the beginning of this chapter: What is the maximum rate at which we can transfer information over a noisy communication channel? Specifically, we will try to answer this question for classical information over quantum channels. As in the previous section, we begin by looking at the classical counterpart, given similarities in derivation.

7.6.1 Classical Error Correction and Shannon's Channel Coding Theorem

The classical discrete memoryless channel, for input X and output Y, is described by the set of transition probabilities:

$$p(y_k|x_j) = P(Y = y_k|X = x_j);\ 0 \le p(y_k|x_j) \le 1, \tag{7.84}$$

satisfying the condition $\sum_k p(y_k|x_j) = 1$. The conditional entropy of X, given $Y = y_k$, is related to the uncertainty of the channel input X by observing the channel output y_k, and is defined as:

$$H(X|Y = y_k) = \sum_j p(x_j|y_k)\log\left[\frac{1}{p(x_j|y_k)}\right], \tag{7.85}$$

where:

$$p(x_j|y_k)=p(y_k|x_j)p(x_j)/p(y_k),\ p(y_k)=\sum_j p(y_k|x_j)p(x_j). \tag{7.86}$$

The amount of uncertainty remaining about the channel input X after the observing channel output Y can then be determined by:

$$H(X|Y)=\sum_k H(X|Y=y_k)p(y_k)=\sum_k p(y_k)\sum_j p(x_j|y_k)\log\left[\frac{1}{p(x_j|y_k)}\right]. \tag{7.87}$$

Therefore the amount of uncertainty about the channel input X that is resolved by observing the channel input would be the difference $H(X) - H(X|Y)$, and this difference is known as the *mutual information* $I(X,Y)$. In other words:

$$I(X,Y)=H(X)-H(X|Y)=\sum_j p(x_j)\sum_k p(y_k|x_j)\log\left[\frac{p(y_k|x_j)}{p(y_k)}\right]. \tag{7.88}$$

The $H(X)$ represents the uncertainty about the channel input X before observing the channel output Y, while $H(X|Y)$ denotes the conditional entropy or the amount of uncertainty remaining about the channel input after the channel output has been received. Therefore the mutual information represents the amount of information (per symbol) that is conveyed by the channel. In other words, it represents the uncertainty about the channel input that is resolved by observing the channel output. The mutual information can be interpreted by means of the Venn diagram shown in Fig. 7.6A. The left circle represents the entropy of channel input, the right circle represents the entropy of channel output, and the mutual information is obtained in the intersection of these two circles. Another interpretation due to Ingels [26] is shown in Fig. 7.6B. The mutual information, i.e., the information conveyed by the channel, is obtained as the output information minus information lost in the channel. One important figure of merit of the classical channel is *channel capacity*, which is obtained by maximization of mutual information $I(X,Y)$ over all possible input distributions:

$$C=\max_{\{p(x_i)\}} I(X;Y). \tag{7.89}$$

The classical channel encoder accepts the message symbols and adds redundant symbols according to a corresponding prescribed rule. Channel coding is the act of transforming a length-k sequence into a length-n codeword. The set of rules specifying this transformation is called the channel code, which can be represented as the following mapping: C: $M \rightarrow X$, where C is the channel code, M is the set of information sequences of length k, and X is the set of codewords of length n. The decoder exploits these redundant symbols to determine which message symbol was actually transmitted. The concept of classical channel coding is introduced in Fig. 7.7. An important class of channel codes is the class of *block codes*. In an (n,k) block code, the channel encoder accepts information in successive k-symbol blocks and

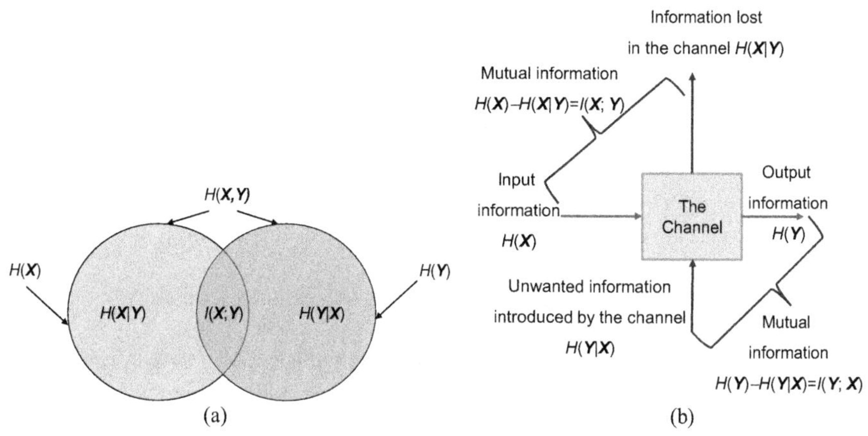

FIGURE 7.6

Interpretation of the mutual information: (a) using Venn diagrams and (b) using the approach due to Ingels.

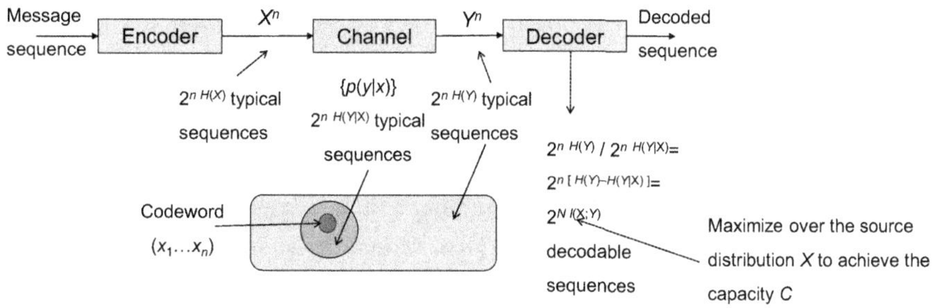

FIGURE 7.7

Classical error correction and Shannon's capacity theorem.

adds $n - k$ redundant symbols that are algebraically related to the k message symbols, thus producing an overall encoded block of n symbols ($n > k$), known as a *codeword*. If the block code is *systematic*, the information symbols stay unchanged during the encoding operation, and the encoding operation may be considered as adding the $n - k$ generalized parity checks to k information symbols. The code rate of the code is defined as $R = k/n$.

Shannon has shown that if $R < C$, we can construct 2^{NR} length-N codewords that can be sent over the (classical) channel with maximum probability of error approaching zero for large N. To prove this claim, we introduce the concept of jointly typical sequences. Two length-n sequences $\boldsymbol{x}$ and $\boldsymbol{y}$ are jointly typical sequences if they satisfy the following set of inequalities:

$$P\left(\left|\frac{1}{n}\log\left(\frac{1}{P(\boldsymbol{x})}\right)-H(X)\right|\leq\varepsilon\right)>1-\delta, \tag{7.90a}$$

$$P\left(\left|\frac{1}{n}\log\left(\frac{1}{P(\boldsymbol{y})}\right)-H(Y)\right|\leq\varepsilon\right)>1-\delta, \tag{7.90b}$$

$$P\left(\left|\frac{1}{n}\log\left(\frac{1}{P(\boldsymbol{x},\boldsymbol{y})}\right)-H(X,Y)\right|\leq\varepsilon\right)>1-\delta, \tag{7.90c}$$

where $P(\boldsymbol{x},\boldsymbol{y})$ denotes the joint probability of the two sequences and $H(X,Y)$ is their joint entropy.

For an n-length input codeword randomly generated according to the probability distribution of a source X, the number of random input sequences is approximately $2^{nH(X)}$ and the number of output typical sequences is approximately $2^{nH(Y)}$. Furthermore, the total number of input and output sequences that are jointly typical is $2^{nH(X,Y)}$. Therefore the total pairs of sequences that are simultaneously $\boldsymbol{x}$-typical, $\boldsymbol{y}$-typical, and also jointly typical are $2^{n\underbrace{[H(X)+H(Y)-H(X,Y)]}_{I(X,Y)}}=2^{nI(X,Y)}$, where $I(X,Y)$ is the mutual information between X and Y, as illustrated in Fig. 7.7. These are the maximum number of codeword sequences that can be distinguished. One way of seeing this is to consider a single codeword. For this codeword, the action of the channel, characterized by the conditional probability $P(\boldsymbol{y}|\boldsymbol{x})$, defines the Hamming sphere in which this codeword can lie after the action of the channel. The size of this Hamming sphere is approximately $2^{nH(Y|X)}$. Given that the total number of output typical sequences is approximately $2^{nH(Y)}$, if we desire to have no overlap between two Hamming spheres the maximum number of codewords we can consider is given by $2^{nH(X)}/2^{nH(Y|X)}=2^{nI(X,Y)}$. To increase this number we need to maximize $I(X,Y)$ over the distribution of X, as we do not have control over the channel. This maximal mutual information is referred to as the capacity of the channel. If we have a rate $R<C$, then Shannon's noisy channel coding theorem tells us that we can construct 2^{nR} length-n codewords that can be sent over the channel with maximum probability of error approaching zero for large n. Now we can formally formulate Shannon's channel coding theorem as follows.

Shannon's Channel Coding Theorem. Let us consider the transmission of $2^{N(C-\varepsilon)}$ equiprobable messages. Then, there exists a classical channel coding scheme of rate $R<C$ in which the codewords are selected from all 2^N possible words such that decoding error probability can be made arbitrarily small for sufficiently large N.

7.6.2 Quantum Error Correction and Holevo–Schumacher–Westmoreland Theorem

Now we consider communication over the quantum channel, as illustrated in Fig. 7.8. The quantum encoder for each message m out of $M = 2^{NR}$ messages generates a product state codeword $\boldsymbol{\rho}^m$ drawn from the ensemble $\{p_x, \boldsymbol{\rho}_x\}$ as follows:

$$\boldsymbol{\rho}^m = \boldsymbol{\rho}_{m_1} \otimes \cdots \otimes \boldsymbol{\rho}_{m_N} = \boldsymbol{\rho}_{\otimes N}. \tag{7.91}$$

This quantum codeword is sent over the quantum channel described by the trace-preserving quantum operation ξ, resulting in the received quantum word:

$$\boldsymbol{\sigma}_m = \xi(\boldsymbol{\rho}_{m_1}) \otimes \xi(\boldsymbol{\rho}_{m_2}) \otimes \cdots \otimes \xi(\boldsymbol{\rho}_{m_N}) = \boldsymbol{\sigma}_{\otimes N}. \tag{7.92}$$

Bob performs measurements on $\boldsymbol{\sigma}_m$ to decode Alice's message. Our goal is to find the maximum rate at which Alice can communicate over the quantum channel such that the probability of decoding error is negligibly small. This maximum rate is the capacity $C(\xi)$ of the quantum channel for transmitting classical information. The key idea behind the Holevo–Schumacher–Westmoreland (HSW) theorem is that for $R < C(\xi)$ there exist quantum codes with codewords of length N such that the probability of incorrectly decoding the quantum codeword can be made arbitrarily small.

The von Neumann entropy associated with the quantum encoder would be:

$$S(\boldsymbol{\rho}) = S\left(\sum_x p_x \boldsymbol{\rho}_x\right), \tag{7.93}$$

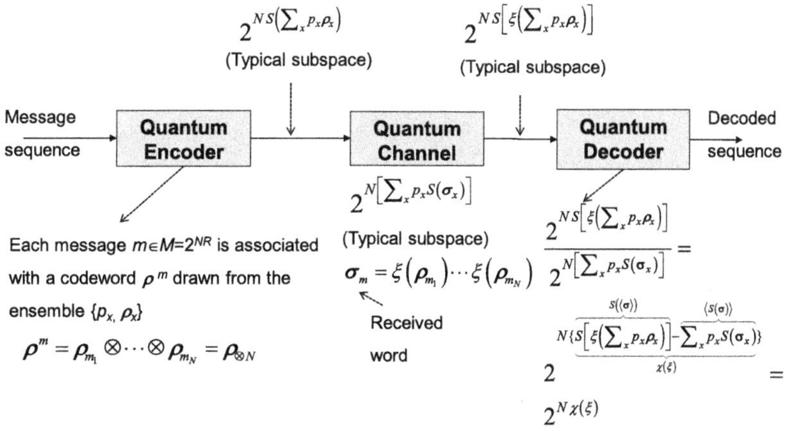

FIGURE 7.8

Quantum error correction and Holevo–Schumacher–Westmoreland theorem.

while the dimensionality of the typical subspace of the quantum encoder is given by:

$$2^{NS(\boldsymbol{\rho})} = 2^{NS\left(\sum_x p_x \boldsymbol{\rho}_x\right)}. \tag{7.94}$$

On the other hand, the entropy associated with $\boldsymbol{\sigma}$ will be:

$$\langle S(\boldsymbol{\sigma})\rangle = \sum_x p_x S(\boldsymbol{\sigma}_x), \tag{7.95}$$

while the dimensionality of the quantum subspace characterizing the quantum channel is given by:

$$2^{N\langle S(\boldsymbol{\sigma})\rangle} = 2^{N\sum_x p_x S(\boldsymbol{\sigma}_x)}. \tag{7.96}$$

The quantum channel perturbs the quantum codeword transmitted over the quantum channel by performing the trace-preserving quantum operation so that the entropy at the channel output can be written as:

$$S(\langle \boldsymbol{\sigma}\rangle) = S\left(\xi\left(\sum_x p_x \boldsymbol{\rho}_x\right)\right), \tag{7.97}$$

while the dimensionality of the corresponding subspace is given by:

$$2^{NS(\langle \boldsymbol{\sigma}\rangle)} = 2^{NS\left(\xi\left(\sum_x p_x \boldsymbol{\rho}_x\right)\right)}. \tag{7.98}$$

The number of decodable codewords would then be:

$$\frac{2^{NS(\langle \boldsymbol{\sigma}\rangle)}}{2^{N\langle S(\boldsymbol{\sigma})\rangle}} = 2^{N\left[\underbrace{S(\langle \boldsymbol{\sigma}\rangle) - \langle S(\boldsymbol{\sigma})\rangle}_{\chi(\xi)}\right]} = 2^{N\chi(\xi)}. \tag{7.99}$$

To maximize the number of decodable codewords we need to perform the optimization of $\chi(\xi)$ over p_x and $\boldsymbol{\rho}_x$, which represents the simplified derivation of the HSW theorem.

Holevo–Schumacher–Westmoreland Theorem. Let us consider the transmission of a codeword $\boldsymbol{\rho}_m$ drawn from the ensemble $\{p_x, \boldsymbol{\rho}_x\}$ over the quantum channel, characterized by the trace-preserving quantum operation ξ, with the rate $R < C(\xi)$, where $C(\xi)$ is the product state capacity defined as:

$$\begin{aligned} C(\xi) &= \max_{\{p_x,\boldsymbol{\rho}_x\}} \chi(\xi) = \max_{\{p_x,\boldsymbol{\rho}_x\}} \left[S(\langle \boldsymbol{\sigma}\rangle) - \langle S(\boldsymbol{\sigma})\rangle\right] \\ &= \max_{\{p_x,\boldsymbol{\rho}_x\}} \left[S\left[\xi\left(\sum_x p_x \boldsymbol{\rho}_x\right)\right] - \sum_x p_x S(\xi(\boldsymbol{\rho}_x))\right], \end{aligned} \tag{7.100}$$

where the maximization of $\chi(\xi)$ is performed over p_x and $\boldsymbol{\rho}_x$. Then, there exists a coding scheme that allows reliable error-free transmission over the quantum channel.

To make the idea of capacity concrete we compute the capacity of two example quantum channels [1]: the *depolarizing channel* and the *phase-flip* channel. We assume quantum source outputs $\boldsymbol{\rho}_0$ with probability p_0, and $\boldsymbol{\rho}_1$ with probability p_1. We further assume that both the states are orthogonal pure states given by:

$$\boldsymbol{\rho}_0 = \begin{pmatrix} 1 & 0 \\ 0 & 0 \end{pmatrix}, \ \boldsymbol{\rho}_1 = \begin{pmatrix} 0 & 0 \\ 0 & 1 \end{pmatrix}.$$

Depolarizing channel. The depolarizing channel, defined by the relation:

$$\varepsilon(\boldsymbol{\rho}) = q\frac{\boldsymbol{I}}{2} + (1-q)\boldsymbol{\rho},$$

where ε denotes the superoperator, either transforms the input state to a uniform mixed state—denoted by the identity operator $\boldsymbol{I}$—with probability q, or leaves the state unchanged with probability $1 - q$. The transformation to the mixed state is referred to as the depolarization action of the channel. Applying the foregoing quantum operation to both $\boldsymbol{\rho}_0$ and $\boldsymbol{\rho}_1$ we obtain:

$$\boldsymbol{\sigma}_0 = \varepsilon(\boldsymbol{\rho}_0) = \frac{q}{2}\begin{pmatrix} 1 & 0 \\ 0 & 1 \end{pmatrix} + (1-q)\begin{pmatrix} 1 & 0 \\ 0 & 0 \end{pmatrix} = \begin{pmatrix} 1-\frac{q}{2} & 0 \\ 0 & \frac{q}{2} \end{pmatrix},$$

$$\boldsymbol{\sigma}_1 = \varepsilon(\boldsymbol{\rho}_1) = \frac{q}{2}\begin{pmatrix} 1 & 0 \\ 0 & 1 \end{pmatrix} + (1-q)\begin{pmatrix} 0 & 0 \\ 0 & 1 \end{pmatrix} = \begin{pmatrix} \frac{q}{2} & 0 \\ 0 & 1-\frac{q}{2} \end{pmatrix}.$$

The corresponding quantum entropies are easily seen to be:

$$\begin{aligned} S(\boldsymbol{\sigma}_0) &= \left(1-\frac{q}{2}\right)\log\left(1-\frac{q}{2}\right) + \frac{q}{2}\log\frac{q}{2}, \\ S(\boldsymbol{\sigma}_1) &= \frac{q}{2}\log\frac{q}{2} + \left(1-\frac{q}{2}\right)\log\left(1-\frac{q}{2}\right) = S(\boldsymbol{\sigma}_0). \end{aligned}$$

Consequently, we have:

$$\begin{aligned} S(\boldsymbol{\sigma}) &= p_0\, S(\boldsymbol{\sigma}_0) + p_1 S(\boldsymbol{\sigma}_1) \\ &= (p_0 + p_1) S(\boldsymbol{\sigma}_1) \\ &= \frac{q}{2}\log\frac{q}{2} + \left(1-\frac{q}{2}\right)\log\left(1-\frac{q}{2}\right). \end{aligned}$$

This gives us the second term of Holevo information χ. To find the first term we first compute:

$$\sigma = \varepsilon\left(\sum_i p_i \rho_i\right) = \frac{q}{2}\begin{pmatrix} 1 & 0 \\ 0 & 1 \end{pmatrix} + (1-q)\begin{pmatrix} p_0 & 0 \\ 0 & p_1 \end{pmatrix}$$
$$= \begin{pmatrix} \frac{q}{2} + (1-q)p_0 & 0 \\ 0 & \frac{q}{2} + (1-q)p_1 \end{pmatrix},$$

and therefore have:

$$S(\sigma) = \left(\frac{q}{2} + (1-q)p_0\right)\log\left(\frac{q}{2} + (1-q)p_0\right) + \left(\frac{q}{2} + (1-q)p_1\right)\log\left(\frac{q}{2} + (1-q)p_1\right).$$

Under the constraint that $p_0 + p_1 = 1$, $S(\langle\sigma\rangle)$ is maximized for $p_0 = p_1 = \frac{1}{2}$, and this maximum value is 1 bit. Therefore the Holevo information reduces to:

$$\chi = 1 - \left(1 - \frac{q}{2}\right)\log\left(1 - \frac{q}{2}\right) + \frac{q}{2}\log\frac{q}{2}.$$

We can see that χ is equal to 1 for $q = 0$ and is 0 when $q = 1$. Thus the capacity $C(\varepsilon) = 1$ bit. It is interesting to note that for $q = 1$, χ vanishes, and the channel has no use. The reason can be deduced from the definition of the depolarization channel. When we set $q = 1$, the output of the channel is a mixed state $\frac{I}{2}$, no matter what the input. Consequently, we deterministically know that the channel output is a mixed state (maximum uncertainty) and the channel has no use!

Phase-flip channel. As the next example, let us consider the phase-flip channel. If the input to the channel is a state given by $\alpha|0\rangle + \beta|1\rangle$, the action of the phase-flip channel results in $\alpha|0\rangle - \beta|1\rangle$ with probability q, or leaves the state unchanged with probability $1 - q$. The phase-flip action is performed using the Pauli matrix:

$$\mathbf{Z} = \begin{pmatrix} 1 & 0 \\ 0 & -1 \end{pmatrix}.$$

We therefore define the phase-flip channel mathematically by:

$$\varepsilon(\rho) = q\mathbf{Z}\rho\mathbf{Z} + (1-q)\rho.$$

Following the steps in the previous example we have:

$$\sigma_0 = \varepsilon(\rho_0) = q\begin{pmatrix} 1 & 0 \\ 0 & -1 \end{pmatrix}\begin{pmatrix} 1 & 0 \\ 0 & 0 \end{pmatrix}\begin{pmatrix} 1 & 0 \\ 0 & -1 \end{pmatrix} + (1-q)\begin{pmatrix} 1 & 0 \\ 0 & 0 \end{pmatrix} = \begin{pmatrix} 1 & 0 \\ 0 & 0 \end{pmatrix},$$

$$\sigma_1 = \varepsilon(\rho_1) = q\begin{pmatrix} 1 & 0 \\ 0 & -1 \end{pmatrix}\begin{pmatrix} 0 & 0 \\ 0 & 1 \end{pmatrix}\begin{pmatrix} 1 & 0 \\ 0 & -1 \end{pmatrix} + (1-q)\begin{pmatrix} 0 & 0 \\ 0 & 1 \end{pmatrix} = \begin{pmatrix} 0 & 0 \\ 0 & 1 \end{pmatrix}.$$

The corresponding quantum entropies are:

$$S(\sigma_0) = (1)\log(1) + 0\log 0 = 0,\ S(\sigma_1) = 0\log 0 + (1)\log(1) = 0 = S(\sigma_0),$$

and therefore the second term of the Holevo information is given by:

$$\langle S(\boldsymbol{\sigma})\rangle = p_0\, S(\boldsymbol{\sigma}_0) + p_1 S(\boldsymbol{\sigma}_1) = 0.$$

To calculate the first term, as in the previous example, we first find the expression for:

$$\langle \boldsymbol{\sigma} \rangle = \varepsilon\left(\sum_i p_i \boldsymbol{\rho}_i\right) = q\begin{pmatrix} 1 & 0 \\ 0 & -1 \end{pmatrix}\begin{pmatrix} p_0 & 0 \\ 0 & p_1 \end{pmatrix}\begin{pmatrix} 1 & 0 \\ 0 & -1 \end{pmatrix} + (1-q)\begin{pmatrix} p_0 & 0 \\ 0 & p_1 \end{pmatrix} = \begin{pmatrix} p_0 & 0 \\ 0 & p_1 \end{pmatrix}.$$

This gives us:

$$S(\langle \boldsymbol{\sigma} \rangle) = p_0 \, \log p_0 + p_1 \, \log p_1.$$

Therefore:

$$\chi = S(\langle \boldsymbol{\sigma} \rangle) - \langle S(\boldsymbol{\sigma})\rangle = p_0 \, \log p_0 + p_1 \, \log p_1.$$

Holevo information is maximized over $\{p_0, p_1\}$, and this maximization can be done by noting that the capacity of any channel is upper bounded by the entropy of the source:

$$S(\langle \boldsymbol{\rho} \rangle) = S\begin{pmatrix} p_0 & 0 \\ 0 & p_1 \end{pmatrix} = p_0 \, \log p_0 + p_1 \, \log p_1,$$

which is the same as χ! Under the constraint $p_0 + p_1 = 1$, the maximum χ is 1 bit for $p_0 = p_1 = \frac{1}{2}$. Therefore the capacity of the phase-flip channel is 1 bit!

Lossy bosonic channel. As a last example we consider a continuous variable quantum channel, the lossy bosonic channel. In quantum optical communication, bosonic channels generally refer to photon-based communication channels like free-space optical and fiber-optics transmission links. They are an important area of research because they allow us to systematically explain and study quantum optical communication by exploiting system linearity coupled with certain nonlinear devices such as parametric down converters and squeezers [20]. For a lossy channel the effect of the channel on the input quantum states can be modeled by a quantum noise source. Taking this noise source into account, the evolution of a multimode lossy bosonic channel can be described as:

$$A_{o_k} = \sqrt{T_k} A_{i_k} + \sqrt{1 - T_k} B_k, \tag{7.101}$$

where A_{i_k} and A_{o_k} are the annihilation ladder operators of the kth mode of the input and output electromagnetic field, B_k is the annihilation operator associated with the kth environmental noise mode, and T_k is the kth mode channel transmissivity or quantum efficiency. For such a channel the capacity is given by:

$$C(\xi) = \sum_k g(T_k N_k(\beta)), \tag{7.102}$$

with $N_k(\beta)$ being the optimal photon-count distribution given by:

$$N_k(\beta) = \frac{T_k^{-1}}{e^{\beta T_k^{-1} \hbar \omega_k} - 1}, \tag{7.103}$$

where $\hbar$ is the reduced Planck constant, ω_k is the frequency of the kth mode, and β is the Lagrange multiplier determined by the average transmitted energy. A detailed derivation of this result is given in [21] and the references therein. The function $g(\bullet)$:

$$g(x) = (x+1)\log(x+1) - x \log x, \tag{7.104}$$

is known as the Shannon entropy of the Bose–Einstein probability distribution [24].

An interesting consequence of Eq. (7.102) is that by constructing the encoded product state:

$$\rho_{\otimes k} = \otimes_k \int p_k(u) |u\rangle_k \langle u| du, \tag{7.105}$$

where the kth modal state $\int p_k(u) |u\rangle_k \langle u| du$ is a mix of coherent states $|u\rangle_k$ and $p_k(u)$ is the Gaussian probability distribution:

$$p_k(u) = \frac{1}{\pi N_k} e^{-|u|^2/N_k}, \tag{7.106}$$

with N_k being the modal photon count and u being the continuous variable, the channel capacity can be realized with a single of use of the channel without employing entanglement. More details can be found in [20–25].

7.7 SUMMARY

This chapter was devoted to quantum information theory fundamentals. After introductory remarks in Section 7.1, the chapter introduced classical and von Neumann entropies definitions and properties, followed by quantum representation of classical information in Section 7.2. Next, in Section 7.3, we introduced the accessible information as the maximum mutual information over all possible measurement schemes. In the same section, we further introduced an important bound, the Holevo bound, and proved it. In Section 7.4 we introduced typical sequence, typical state, typical subspace, and projector on the typical subspace. Furthermore, we formulated and proved Schumacher's source coding theorem, the quantum equivalent of Shannon's source coding theorem. In the same section we introduced the concept of quantum compression and decompression as well as the concept of reliable compression. In Section 7.5 we described various quantum channel models, including bit-flip, phase-flip, depolarizing, and amplitude-damping channel models. In Section 7.6 we formulated and proved the HSW theorem, the quantum equivalent of Shannon's channel capacity theorem. For a deeper understanding of underlying concepts, we have provided a set of problems in the following section.

7.8 PROBLEMS

1. Consider a source $X = \{x_1, x_2\}$ wherein the symbols x_i $(i = 1,2)$ appear with probabilities p and $1 - p$, respectively. Determine the Shannon entropy $H(P)$ of this source. Show that $H(P)$ is a concave function of p. Determine the p maximizing the $H(P)$.
2. Given two distributions P and Q on X, prove that the relative entropy $D(P||Q) \geq 0$, with equality if and only if $P = Q$.
3. For two random variables X and Y, the classical mutual information is:

$$I(X;Y) = H(X) - H(X|Y),$$

where $H(X|Y)$ is the conditional entropy defined as:

$$H(X|Y) = \sum_{x \in X} \sum_{y \in Y} P(x,y) \log(y|x).$$

Show that mutual information is symmetric:

$$I(X;Y) = I(Y;X),$$

and nonnegative:

$$I(X;Y) \geq 0,$$

with equality if and only if X and Y are independent.

4. Consider the following state of a bipartite quantum system AB:

$$|\psi = \sqrt{\frac{1}{3}}(|1> \otimes |1> + |1> \otimes |0> + |0> \otimes |1).$$

   (a) Compute the quantum entropy of the joint system.
   (b) Compute the quantum entropy of the subsystems.
   (c) Verify the triangle inequality and the Araki–Lieb triangle inequality.
   (d) Is the composite system in a pure state?

5. For the product state $\rho_x \otimes \rho_y$, show that $S(\rho_x \otimes \rho_y) = S(\rho_x) + S(\rho_y)$.
6. Given a quantum source characterized by the density matrix:

$$\rho = \frac{1}{4}\begin{pmatrix} 3 & -1 \\ -1 & 1 \end{pmatrix}.$$

   (a) Find the maximum achievable rate for this quantum source.
   (b) Given that we have three qubit quantum messages, what is the corresponding typical subspace? Assume $\delta < 0.1$.
   (c) What will be the typical subspace if the message length is four qubits?

7. Repeat Problem 6, but with the input state given by:

$$|\psi> = \frac{1}{\sqrt{2}}\begin{pmatrix} 1 \\ 1 \end{pmatrix} + \frac{1}{\sqrt{2}}\begin{pmatrix} 1 \\ -1 \end{pmatrix}.$$

8. Let $\{P_m\}$ be a set of projective measurements that satisfy the following properties:

$$P_m^\dagger = P_m,$$
$$\sum_m P_m = I, \text{ and}$$
$$P_{m_2} = \delta_{m_1 m_2} P_{m_1}.$$

For $\boldsymbol{\sigma} = \sum_m P_m \boldsymbol{\rho} P_m$, show that $S(\boldsymbol{\sigma}) \geq S(\boldsymbol{\rho})$, with the necessary and sufficient condition for the equality being $\boldsymbol{\sigma} = \boldsymbol{\rho}$.

9. Prove the concavity property of the von Neumann entropy.

10. Prove the strong subadditivity property of the von Neumann entropy.

11. Consider the depolarizing channel:

$$\varepsilon(\boldsymbol{\rho}_\iota) = (1-q)\boldsymbol{\rho} + \frac{q}{3}\boldsymbol{X}\boldsymbol{\rho}\boldsymbol{X}^\dagger + \frac{q}{3}\boldsymbol{Y}\boldsymbol{\rho}\boldsymbol{Y}^\dagger + \frac{q}{3}\boldsymbol{Z}\boldsymbol{\rho}\boldsymbol{Z}^\dagger,$$

where $\boldsymbol{X}$, $\boldsymbol{Y}$, $\boldsymbol{Z}$ are the Pauli matrices, and the source emits two equally likely states:

$$|\psi_1\rangle = \frac{1}{\sqrt{2}}(|0\rangle + |1\rangle), |\psi_2\rangle = \frac{\sqrt{3}}{2}|0\rangle + \frac{1}{2}|1\rangle.$$

This channel leaves the state invariant with probability $1-q$ or results in a bit flip ($\boldsymbol{X}$), phase flip ($\boldsymbol{Z}$), or a bit and phase flip ($\boldsymbol{Y}$) with equal probability $\frac{q}{3}$.

(a) Show ε is a trace-preserving quantum channel operation.

(b) Determine the capacity of this quantum channel.

(c) We write the quantum channel in the operator-sum representation:

$$\varepsilon(\boldsymbol{\rho}_i) = \boldsymbol{E}_1\boldsymbol{\rho}_i\boldsymbol{E}_1^\dagger + \boldsymbol{E}_2\boldsymbol{\rho}_i\boldsymbol{E}_2^\dagger + \boldsymbol{E}_3\boldsymbol{\rho}_i\boldsymbol{E}_3^\dagger + \frac{q}{3}\boldsymbol{E}_4\boldsymbol{\rho}_i\boldsymbol{E}_4^\dagger,$$

where:

$$\boldsymbol{E}_1 = \sqrt{1-q}\boldsymbol{I},\ \boldsymbol{E}_2 = \sqrt{\frac{q}{3}}\boldsymbol{X},$$
$$\boldsymbol{E}_3 = -i\sqrt{\frac{q}{3}}\boldsymbol{Y},\ \boldsymbol{E}_4 = \sqrt{\frac{q}{3}}\boldsymbol{Z}.$$

Determine the quality of the transmitted bit given by the channel fidelity:

$$F = \sum_i (\mathrm{Tr}(\boldsymbol{\rho}\boldsymbol{E}_i))(\mathrm{Tr}(\boldsymbol{\rho}\boldsymbol{E}_i)).$$

12. Consider the bit-flip channel:

$$\varepsilon(\boldsymbol{\rho}) = q\boldsymbol{X}\boldsymbol{\rho}\boldsymbol{X} + (1-q)\boldsymbol{\rho},$$

that takes as input the qubit $\alpha|0\rangle + \beta|1\rangle$ and performs the bit-flip operation with probability q while leaving the qubit alone with probability $1 - q$.

(a) Compute its Holevo information.

(b) What can you say about the capacity of this channel?

(c) Is it possible that under certain conditions the capacity is 0?

References

[1] I.B. Djordjevic, Quantum Information Processing and Quantum Error Correction: An Engineering Approach, Elsevier/Academic Press, 2012.

[2] I.B. Djordjevic, Physical-Layer Security and Quantum Key Distribution, Springer International Publishing, Switzerland, 2019.

[3] C.E. Shannon, A mathematical theory of communication, Bell Syst. Tech. J. 27 (1948), 379–423, 623–656.

[4] C.E. Shannon, W. Weaver, The Mathematical Theory of Communication, The University of Illinois Press, Urbana, Illinois, 1949.

[5] T. Cover, J. Thomas, Elements of Information Theory, Wiley-Interscience, 1991.

[6] A.N. Kolmogorov, Three approaches to the quantitative definition of information, Probl. Inf. Transm. 1 (1) (1965) 1–7.

[7] H. Touchette, The large deviation approach to statistical mechanics, Phys. Rep. 478 (1–3) (2009) 1–69.

[8] B.R. Frieden, Science from Fisher Information: A Unification, Cambridge Univ. Press, 2004.

[9] J.L. Kelly Jr., A new interpretation of information rate, Bell Syst. Tech. J. 35 (1956) 917–926.

[10] M.A. Nielsen, I.L. Chuang, Quantum Computation and Quantum Information, Cambridge Univ. Press, 2000.

[11] J. Preskill, Reliable quantum computers, Proc. Roy. Soc. Lond. A 454 (1998) 385–410.

[12] A.S. Holevo, Problems in the mathematical theory of quantum communication channels, Rep. Math. Phys. 12 (1977) 273–278.

[13] A.S. Holevo, Capacity of a quantum communications channel, Probl. Inf. Transm. 15 (1979) 247–253.

[14] R. Jozsa, B. Schumacher, A new proof of the quantum noiseless coding theorem, J. Mod. Optic. 41 (1994) 2343–2349.

[15] B. Schumacher, M. Westmoreland, W.K. Wootters, Limitation on the amount of accessible information in a quantum channel, Phys. Rev. Lett. 76 (1997) 3452–3455.

[16] A.S. Holevo, The capacity of the quantum channel with general signal states, IEEE Trans. Inf. Theor. 44 (1998) 269–273.

[17] M. Sassaki, K. Kato, M. Izutsu, O. Hirota, Quantum channels showing superadditivity in channel capacity, Phys. Rev. A 58 (1998) 146–158.

[18] C.A. Fuchs, Nonorthogonal quantum states maximize classical information capacity, Phys. Rev. Lett. 79 (1997) 1162–1165.

[19] C.H. Bennett, P.W. Shor, J.A. Smolin, A.V. Thapliyal, Entanglement enhanced classical capacity of noisy quantum channels, Phys. Rev. Lett. 83 (1999) 3081–3084.

[20] A.S. Holevo, R.F. Werner, Evaluating capacities of boson Gaussian channels, Phys. Rev. A 63 03 (2001) 2312.

[21] V. Giovannetti, S. Lloyd, L. Maccone, P.W. Shor, Broadband channel capacities, Phys. Rev. A 68 (2003) 062323.

[22] V. Giovannetti, S. Lloyd, L. Maccone, P.W. Shor, Entanglement assisted capacity of the broadband lossy channel, Phys. Rev. Lett. 91 (2003) 047901.

[23] V. Giovannetti, S. Guha, S. Lloyd, L. Maccone, J.H. Shapiro, H.P. Yuen, Classical capacity of the lossy bosonic channel: the exact solution, Phys. Rev. Lett. 92 (2004) 027902.

[24] J.H. Shapiro, S. Guha, B.I. Erkmen, Ultimate channel capacity of free-space optical communications, J. Opt. Netw. 4 (2005) 501−516.

[25] S. Guha, J.H. Shapiro, B.I. Erkmen, Classical capacity of bosonic broadcast communication and a minimum output entropy conjecture, Phys. Rev. A 76 (2007) 032303.

[26] F.M. Ingels, Information and Coding Theory, Intext Educational Publishers, Scranton, 1971.

CHAPTER

Quantum Error Correction

8

CHAPTER OUTLINE

Quantum Information Processing, Quantum Computing, and Quantum Error Correction
https://doi.org/10.1016/B978-0-12-821982-9.00013-7

The concept of this chapter is to gradually introduce readers to the quantum error correction coding principles, starting from an intuitive description to a rigorous mathematical description. The chapter starts with Pauli operators, basic definitions, and representation of quantum errors. Although the Pauli operators were introduced in Chapter 2, they are put here in the context of quantum errors and quantum error correction. Next, basic quantum codes, such as three-qubit flip code, three-qubit phase-flip code, and Shor's nine-qubit code, are presented. The projection measurements are used to determine the error syndrome and perform corresponding error correction action. Furthermore, the stabilizer formalism and the stabilizer group are introduced. The basic stabilizer codes are described as well. The whole of the next chapter is devoted to stabilizer codes; here only the basic concepts are introduced. An important class of codes, the class of Calderbank–Shor–Steane (CSS) codes, is described next. The connection between classical and quantum codes is then established, and two classes of CSS codes, dual-containing and quantum codes derived from classical codes over GF(4), are described. The concept of quantum error correction is then formally introduced, followed by the necessary and sufficient conditions for quantum code to correct a given set of errors. Then, the minimum distance of a quantum code is defined and used to relate it to the error correction capability of a quantum code. The CSS codes are then revisited by using this mathematical framework. The next section is devoted to important quantum coding bounds, such as the Hamming quantum bound, quantum Gilbert–Varshamov bound, and quantum Singleton bound (also known as the Knill–Laflamme bound). Next, the concept of operator-sum representation is introduced, and used to provide physical interpretation and describe the measurement of the environment. Finally, several important quantum channel models are introduced, such as depolarizing channel, amplitude-damping channel, and generalized amplitude-damping channel. After the summary section, the set of problems for self-study is provided, which enables readers to better understand the underlying concepts of quantum error correction.

8.1 PAULI OPERATORS (REVISITED)

Pauli operators $G = \{I,X,Y,Z\}$ [1–25], as already introduced in Chapter 2, can be represented in matrix form as follows:

$$X=\sigma_x \doteq \begin{bmatrix} 0 & 1 \\ 1 & 0 \end{bmatrix},\ Y=\sigma_y \doteq \begin{bmatrix} 0 & -j \\ j & 0 \end{bmatrix},\ Z=\sigma_z \doteq \begin{bmatrix} 1 & 0 \\ 0 & -1 \end{bmatrix}. \tag{8.1}$$

Their action on qubit $|\psi\rangle = a|0\rangle + b|1\rangle$ can be described as follows:

$$X(a|0\rangle + b|1\rangle) = a|1\rangle + b|0\rangle \quad Y(a|0\rangle + b|1\rangle) = j(a|1\rangle - b|0\rangle)$$

$$Z(a|0\rangle + b|1\rangle) = a|0\rangle - b|1\rangle. \tag{8.2}$$

Therefore the action of an X-operator is to introduce the bit flip, the action of the Z-operator is to introduce the phase flip, and the action of the Y-operator is to introduce simultaneously the bit and phase flips. The properties of Pauli operators can be summarized as:

$$\begin{aligned} &X^2 = I,\ Y^2 = I,\ Z^2 = I \\ &XY = jZ \quad YX = -jZ \quad YZ = jX \quad ZY = -jX \quad ZX = jY \quad XZ = -jY. \end{aligned} \tag{8.3}$$

The properties of Pauli operators given by Eq. (8.3) can also be summarized by the following multiplication table:

×	I	X	Y	Z
I	I	X	Y	Z
X	X	I	jZ	$-jY$
Y	Y	$-jZ$	Y	jX
Z	Z	jY	$-jX$	I

It is interesting to notice that two Pauli operators commute only if they are identical or one of them is an identity operator, otherwise they anticommute. Because the set G is not closed under multiplication, it is not a multiplicative group. However, from Chapter 2 we know that two states $|\psi\rangle$ and $e^{j\theta}\,|\psi\rangle$ are equal up to the global phase shift because the results of a measurement are the same. Therefore often in quantum error correction we can simply omit the imaginary unit j. The corresponding multiplication table in which the imaginary unit is ignored is as follows:

×	I	X	Y	Z
I	I	X	Y	Z
X	X	I	Z	Y
Y	Y	Z	Y	X
Z	Z	Y	X	I

Clearly, all operators now commute and the corresponding set G' with such defined multiplication is a commutative (Abelian) group. Such a group can be called a "projective" Pauli group [15], and is isomorphic to the group of binary 2-tuples $(Z_2)^2 = \{00,01,10,11\}$ with an addition table given by:

+	00	01	11	10
00	00	01	11	10
01	01	00	10	11
11	11	10	00	01
10	10	11	01	00

The projective Pauli group is also isomorphic to the quaternary group $F_4 = \{0, 1, \omega, \overline{\omega}\}$, $\overline{\omega} = 1 + \omega$, with a corresponding addition table given by:

+	0	$\overline{\omega}$	1	ω
0	0	$\overline{\omega}$	1	ω
$\overline{\omega}$	$\overline{\omega}$	0	ω	1
1	1	ω	0	$\overline{\omega}$
ω	ω	1	$\overline{\omega}$	0

Based on these tables we can establish the following correspondence among the projective Pauli group G', $(Z_2)^2$, and F_4:

G'	$(Z_2)^2$	F_4
I	00	0
X	01	$\overline{\omega}$
Y	11	1
Z	10	ω

This correspondence can be used to relate the quantum codes to classical codes over $(Z_2)^2$ and F_4.

In addition to Pauli operators, the Hadamard operator (gate) will also be used a lot in this chapter. Its action can be described as:

$$H(a|0\rangle + b|1\rangle) = a\frac{1}{\sqrt{2}}(|0\rangle + |1\rangle) + b\frac{1}{\sqrt{2}}(|0\rangle - |1\rangle)$$

$$= \frac{1}{\sqrt{2}}[(a+b)|0\rangle + (a-b)|1\rangle]. \tag{8.4}$$

The quantum error correction code (QECC) can be defined as mapping from K-qubit space to N-qubit space. To facilitate its definition, we introduce the concept of Pauli operators. A *Pauli operator on N qubits* has the following form: $cO_1O_2 \ldots O_N$, where each operator $O_i \in \{I,X,Y,Z\}$, and $c = j^l$ ($l = 1,2,3,4$). This operator takes $|i_1 i_2 \ldots i_N\rangle$ to $cO_1|i_1\rangle \otimes O_2|i_2\rangle \ldots \otimes O_N|i_N\rangle$. For example, the action of $IXZ(|000\rangle + |111\rangle) = |010\rangle - |101\rangle$ is to bit flip the second qubit and phase flip the third qubit if it was $|1\rangle$. For convenience, we will also often use a shorthand notation for representing Pauli operators, in which only the nonidentity operators O_i are written, while the identity operators are assumed. For example, $IXIZI$ can be denoted as X_2Z_4, meaning that the operator X acts on the second qubit and the operator Z acts on the fourth qubit (the action of identity operators I to other qubits is simply omitted since it does not cause any change). Two Pauli operators *commute* if and only if there is an even number of places where they have different Pauli matrices neither of which is the identity I. For example, XXI and IYZ do not commute, whereas XXI and ZYX do commute. If two Pauli operators do not commute, they anticommute, since their individual Pauli matrices either commute or anticommute.

The set of Pauli operators on N qubits forms the *multiplicative Pauli group* G_N. For the multiplicative group we can define the *Clifford operator* [8] U as the operator that preserves the elements of the Pauli group under conjugation, namely $\forall\ O \in G_N$: $UOU^\dagger \in G_N$. The encoded operator for quantum error correction typically belongs to the Clifford group. To implement any unitary operator from the Clifford group, the use of CNOT U_{CNOT}, Hadamard H, and phase gate P is sufficient. The operation of the H gate is already given by Eq. (8.4), the action of the P gate can be described as $P(a|0\rangle + b|1\rangle) = a|0\rangle + jb|1\rangle$, and the action of the CNOT gate can be described as $U_{\text{CNOT}}|i\rangle|j\rangle = |i\rangle|i \oplus j\rangle$.

Every quantum channel error E (be it discrete or continuous) can be described as a linear combination of elements from the following discrete set $\{I,X,Y,Z\}$, given by $E = e_1 I + e_2 X + e_3 Y + e_4 Z$. For example, the following error $E = \begin{bmatrix} 1 & 0 \\ 0 & 0 \end{bmatrix}$ can be represented in terms of Pauli operators as $E = (I + Z)/2$. An error operator that affects several qubits can be written as a weighted sum of Pauli operators $\sum c_i P_i$ acting on ith qubits. An error may act not only on the code qubits but also on the environment. Given an initial state $|\psi\rangle\ |\phi\rangle^e$, which is a tensor product of code qubits $|\psi\rangle$ and environmental states $|\phi\rangle^e$, any error acting on both the code and the environment can be written as a weighted sum $\sum c_{i,j}\ P_i P_j^e$ of Pauli operators that act on both code and environment qubits. If S_i are syndrome operators that identify the error term $P_{\alpha'}|\psi\rangle$, then the operators $S_i I^e$ will pick up terms of the form $\sum_{\beta'} c_{\alpha',\beta'} P_{\alpha'}|\psi\rangle\ P^e{}_{\beta'}|\phi\rangle^e$ from the quantum noise-affected state, and these terms can be written as $P_{\alpha'}|\psi\rangle\ |\mu\rangle^e$ for some new environmental state $|\mu\rangle^e$. Therefore the measuring of the syndrome restores a tensor product state of qubits and environment, suggesting that the code and the environment evolve independently of each other.

8.2 QUANTUM ERROR CORRECTION CONCEPTS

Quantum error correction is essentially more complicated than classical error correction. Difficulties for quantum error correction can be summarized as follows: (1) The no-cloning theorem indicates that it is impossible to make a copy of an arbitrary quantum state $|\psi\rangle = \alpha|0\rangle + \beta|0\rangle$. If this would be possible, then $|\psi\rangle|\psi\rangle = \alpha^2|00\rangle + \alpha\beta|01\rangle + \alpha\beta|10\rangle + \beta^2|11\rangle$ must be $\alpha\beta = 0$, so that the state $|\psi\rangle$ is not arbitrary any more. (2) Quantum errors are continuous and a qubit can be in any superposition of the two bases states. (3) The measurements destroy the quantum information. A quantum error correction consists of four major steps: encoding, error detection, error recovery, and decoding. The elements of QECCs are shown in Fig. 8.1. Based on discussion of quantum errors and Pauli operators on N qubits, the QECC can be defined as follows. The $[N,K]$ QECC performs encoding of the quantum state of K qubits, specified by 2^K complex coefficients α_s ($s = 0,1,\ldots,2^K - 1$), into a quantum state of N qubits, in such a way that errors can be detected and corrected, and all 2^K complex coefficients can be perfectly restored, up to the global phase shift. The sender (Alice) encodes quantum information in state $|\psi\rangle$ with the help of local ancilla qubits $|0\rangle$, and then sends the encoded qubits over a noisy quantum channel (say free-space optical channel or optical fiber). The receiver (Bob) performs multiqubit measurement on all qubits to diagnose the channel error and performs a recovery unitary operation R to reverse the action of the channel. The quantum error correction principles will be more evident after several simple quantum codes provided next.

8.2.1 Three-Qubit Flip Code

Assume we want to send a single qubit $|\Psi\rangle = \alpha|0\rangle + \beta|1\rangle$ through the quantum channel in which during transmission the transmitted qubit can be flipped to $X|\Psi\rangle = \beta|0\rangle + \alpha|1\rangle$ with probability p. Such a quantum channel is called a *bit-flip channel* and can be described as shown in Fig. 8.2A.

The three-qubit flip code sends the same qubit three times, and therefore represents the repetition code equivalent. The corresponding codewords in this code are $|\bar{0}\rangle = |000\rangle$ and $|\bar{1}\rangle = |111\rangle$. The three-qubit flip code encoder is shown in Fig. 8.2B. One input qubit and two ancillas are used at the encoder inputs, which can be represented by $|\psi_{123}\rangle = \alpha|000\rangle + \beta|100\rangle$. The first ancilla qubit (the second qubit at the encoder input) is controlled by the information qubit (the first qubit at the

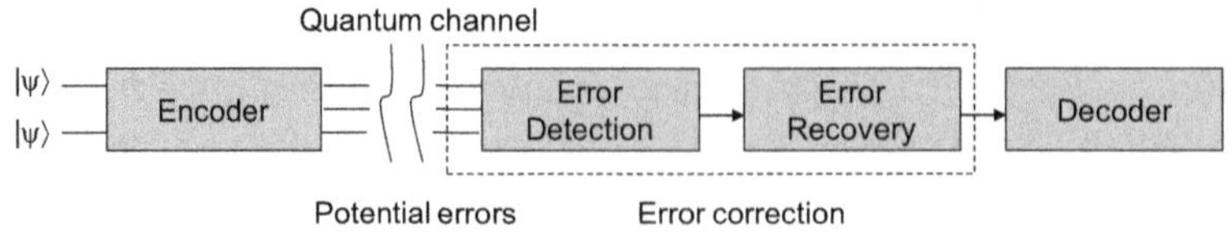

FIGURE 8.1

Quantum error correction principle.

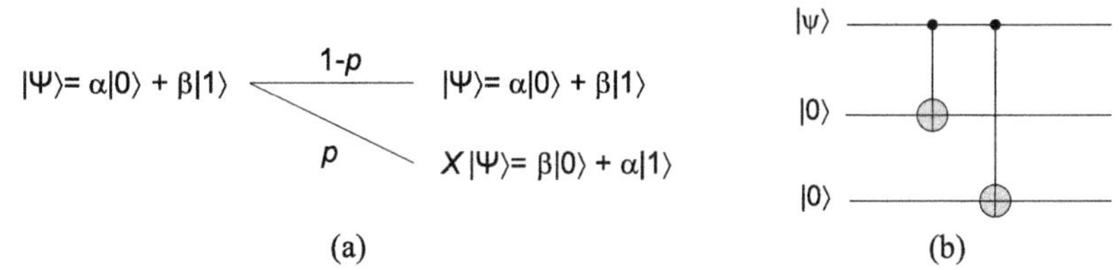

FIGURE 8.2

(a) Bit-flipping channel model. (b) Three-qubit flip code encoder.

encoder input) so that its output can be represented by $\text{CNOT}_{12}(\alpha|000\rangle + \beta|100\rangle) = \alpha|000\rangle + \beta|110\rangle$ (if the control qubit is $|1\rangle$ the target qubit is flipped, otherwise it stays unchanged). The output of the first CNOT gate is used as input to the second CNOT gate in which the second ancilla qubit (the third qubit) is controlled by the information qubit (the first qubit) so that the corresponding encoder output is obtained as $\text{CNOT}_{13}(\alpha|000\rangle + \beta|110\rangle) = \alpha|000\rangle + \beta|111\rangle$, which indicates that basis codewords are indeed $|\overline{0}\rangle$ and $|\overline{1}\rangle$.

With this code, we are capable of correcting a single qubit flip, which occurs with probability $(1-p)^3 + 3p(1-p)^2 = 1-3p^2 + 2p^3$. Therefore the probability of an error remaining uncorrected or wrongly corrected with this code is $3p^2 - 2p^3$. It is clear from Fig. 8.2B that a three-qubit bit-flip encoder is a *systematic encoder* in which the information qubit is unchanged, and the ancilla qubits are used to impose the encoding operation and create the parity qubits (the output qubits 2 and 3).

Example. Let us assume that a qubit flip occurred on the first qubit leading to the received quantum word $|\Psi_r\rangle = \alpha|100\rangle + \beta|011\rangle$. The error correction, as indicated previously, consists of two steps. The first step is the error detection, which performs the measurement with projection operators P_i ($i = 0,1,2,3$) defined as:

$$P_0 = |000\rangle\langle 000| + |111\rangle\langle 111| \quad P_1 = |100\rangle\langle 100| + |011\rangle\langle 011|$$
$$P_2 = |010\rangle\langle 010| + |101\rangle\langle 101| \quad P_3 = |001\rangle\langle 001| + |110\rangle\langle 110|$$

The projection operator P_i determines whether bit-flip error occurred on the ith qubit location, and P_0 means there is no bit-flip error at all. The syndrome measurements give the following result: $\langle\psi_r|P_0|\psi_r\rangle = 0$, $\langle\psi_r|P_1|\psi_r\rangle = 1$, $\langle\psi_r|P_2|\psi_r\rangle = 0$, $\langle\psi_r|P_3|\psi_r\rangle = 0$, which can be represented as syndrome vector $S = [0\ 1\ 0\ 0]$, indicating that the error occurred in the first qubit. The second step is error recovery in which we flip the first qubit back to the original one by applying the X_1 operator to it. Another approach would be to perform measurements on the following observables Z_1Z_2 and Z_2Z_3. The result of measurement is the eigenvalue ± 1 and corresponding eigenvectors are two valid codewords, namely $|000\rangle$ and $|111\rangle$. The observables can be represented as follows:

$$Z_1Z_2 = (|00\rangle\langle 11| + |11\rangle\langle 11|) \otimes I - (|01\rangle\langle 01| + |10\rangle\langle 10|) \otimes I$$
$$Z_2Z_3 = I \otimes (|00\rangle\langle 11| + |11\rangle\langle 11|) - I \otimes (|01\rangle\langle 01| + |10\rangle\langle 10|).$$

It can be shown that $\langle\psi_r|Z_1Z_2|\psi_r\rangle = -1$, $\langle\psi_r|Z_2Z_3|\psi_r\rangle = +1$, indicating that an error occurred on either the first or second qubit, but not on the second or third qubit. The intersection reveals that the first qubit was in error. By using this approach we can create the three-qubit look-up table (LUT), given as Table 8.1.

A three-qubit flip code error detection and error correction circuit is shown in Fig. 8.3.

We learned in Chapter 3 how to perform the measurements of an observable without completely destroying the quantum information with the help of an ancilla and two Hadamard gates by performing the measurement on the ancilla instead. Because for error detection we need two observables (Z_1Z_2 and Z_2Z_3) we need two ancillas and four Hadamard gates (Fig. 8.3A). The results of measurements on ancillas determine the error syndrome $[\pm1 \ \ \pm1]$, and based on the LUT given by Table 8.1 we identify the error event and apply the corresponding X_i gate on the ith qubit being in error, and the error is corrected since $X^2 = I$. The control logic operation is described in Table 8.1. For example, if both outputs at the measurements' circuits are -1, the operator X_2 is activated. The last step is to perform decoding as shown in Fig. 8.3B by simply reversing the order of elements in the corresponding encoder.

Table 8.1 Three-qubit flip code look-up table.

Z_1Z_2	Z_2Z_3	Error
+1	+1	I
+1	−1	X_3
−1	+1	X_1
−1	−1	X_2

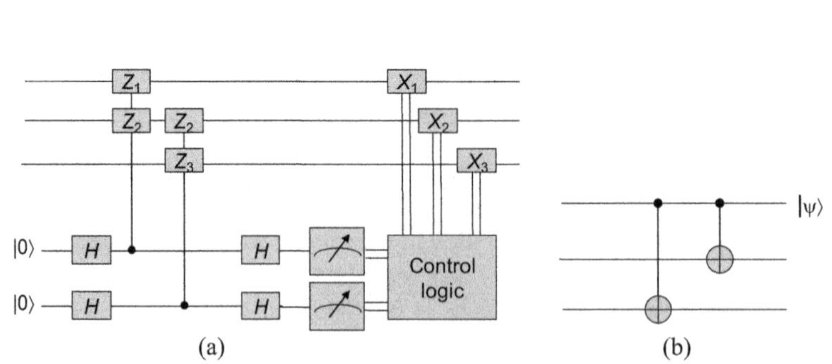

FIGURE 8.3

(a) Three-qubit flip code error detection and error correction circuit. (b) Decoder circuit configuration.

From the universal quantum gates section 3.6 we know that CNOT and Hadamard gates are sufficient for many QECCs. Therefore we can use the equivalent circuits for measuring the Z and X operators shown in Fig. 8.4A and B, respectively, to implement the three-qubit flip code error detection and correction circuits as shown in Fig. 8.5. The corresponding LUT for this representation is given as Table 8.2. The control logic operates as described in this table. For example, if $M_1 = M_2 = 1$ the operator X_2 is activated.

8.2.2 Three-Qubit Phase-Flip Code

Assume we want to send a single qubit $|\Psi\rangle = \alpha|0\rangle + \beta|1\rangle$ through the quantum channel in which during transmission the qubit can be phase flipped as follows: $Z|\Psi\rangle = \alpha|0\rangle - \beta|1\rangle$ with certain probability p. Such a quantum channel is known as the quantum *phase-flip channel*, and the corresponding channel model is shown in Fig. 8.6.

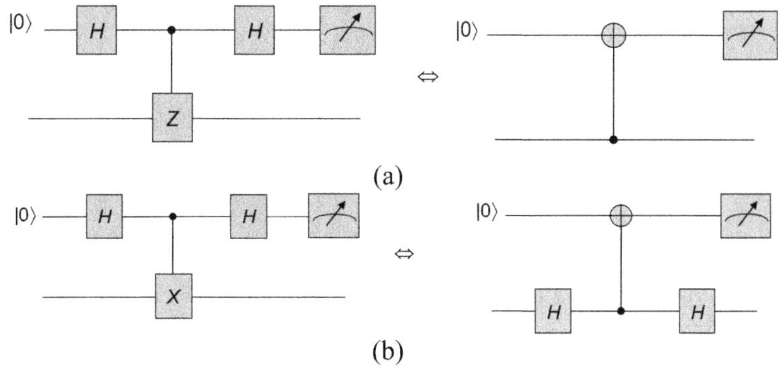

FIGURE 8.4

Equivalent circuits for measuring the: (a) Z-operator and (b) X-operator.

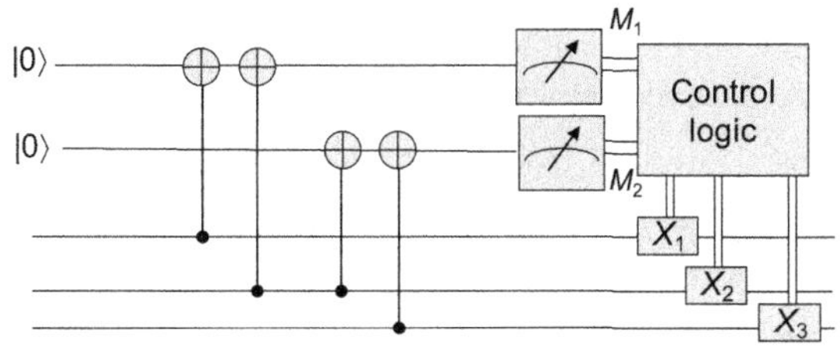

FIGURE 8.5

Equivalent three-qubit flip code error detection and error correction circuit.

Table 8.2 Three-qubit code syndrome look-up table for the equivalent circuit shown in Fig. 8.5.

M_1	M_2	Error
0	0	I
1	0	X_1
1	1	X_2
0	1	X_3

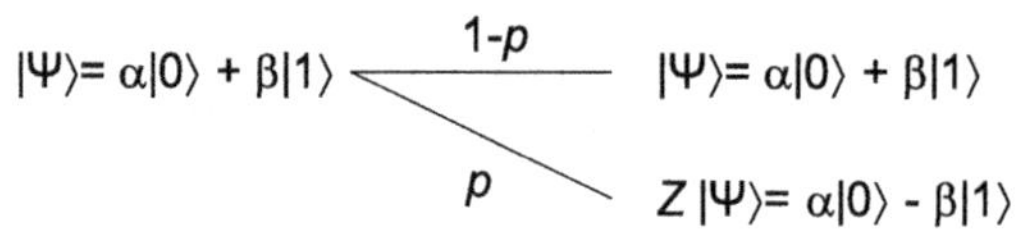

FIGURE 8.6

Phase-flipping channel model.

To protect against the phase-flip errors we work in a different computational basis (CB) instead, known as the diagonal basis:

$$|+\rangle = \frac{|0\rangle + |1\rangle}{\sqrt{2}}, \quad \left|-\right\rangle = \frac{|0\rangle - |1\rangle}{\sqrt{2}}. \tag{8.5}$$

In this basis, the phase-flip operator Z acts as an ordinary qubit-flip operator because the action on new CB states is:

$$Z|+\rangle = \frac{Z|0\rangle + Z|1\rangle}{\sqrt{2}} = \frac{|0\rangle - |1\rangle}{\sqrt{2}} = \left|-\right\rangle, \; Z|-\rangle = \frac{Z|0\rangle - Z|1\rangle}{\sqrt{2}} = \frac{|0\rangle + |1\rangle}{\sqrt{2}} = \left|+\right\rangle. \tag{8.6}$$

The alternative syndrome measurement for this code can therefore be described as: $H^{\otimes 3}Z_1Z_2H^{\otimes 3} = X_1X_2$ and $H^{\otimes 3}Z_2Z_3H^{\otimes 3} = X_2X_3$. For base conversion we can use the Hadamard gate, whose action is described by Eq. (8.4). The three-qubit phase-flip encoding circuit, shown in Fig. 8.7A, can be implemented using the three-qubit flip encoder shown in Fig. 8.3B followed by three Hadamard gates to perform the basis conversion. Corresponding error detection and recovery circuits are provided in Fig. 8.7B.

8.2.3 Shor's Nine-Qubit Code

The key idea of Shor's nine-qubit code is to encode first the qubit using the phase-flip encoder and then to encode each of the three resulting qubits again with the bit-flip encoder, providing therefore protection against both qubit-flip and

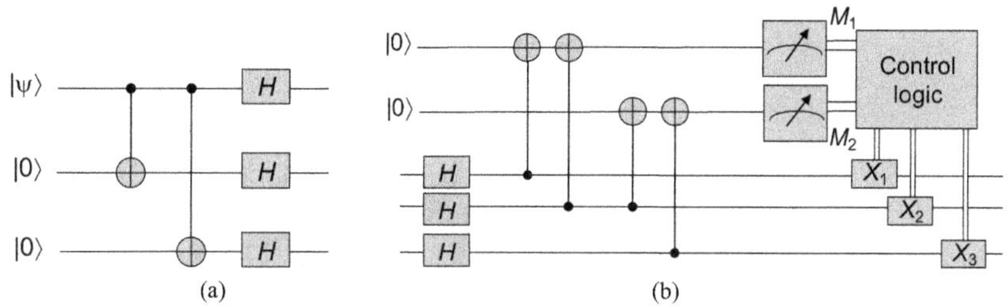

FIGURE 8.7

Three-qubit phase-flip: (a) encoder circuit and (b) error detection and recovery circuits.

phase-flip errors. The two base codewords can be obtained, based on this description, as follows:

$$\begin{aligned}|\bar{0}\rangle &= \frac{1}{2\sqrt{2}}(|000\rangle + |111\rangle)(|000\rangle + |111\rangle)(|000\rangle + |111\rangle) \\ &= \frac{1}{2\sqrt{2}}(|000000000\rangle + |000000111\rangle + |000111000\rangle + |111000000\rangle \\ &\quad + |000111111\rangle + |111111000\rangle + |111111111\rangle) \\ |\bar{1}\rangle &= \frac{1}{2\sqrt{2}}(|000\rangle - |111\rangle)(|000\rangle - |111\rangle)(|000\rangle - |111\rangle) \\ &= \frac{1}{2\sqrt{2}}(|000000000\rangle - |000000111\rangle - |000111000\rangle - |111000000\rangle \\ &\quad + |000111111\rangle + |111111000\rangle - |111111111\rangle).\end{aligned} \tag{8.7}$$

Shor's encoder, shown in Fig. 8.8, is composed of two sections: (1) the phase-flip encoder and (2) every qubit from the phase-flip encoder is further encoded using the qubit-flip encoder. Therefore this code can be considered as a *concatenated code*. Shor's nine-qubit error correction and decoder circuit, shown in Fig. 8.9, is composed of two stages: (1) the first stage is composed of three three-qubit bit-flip detector, recovery, and decoder circuits followed by three Hadamard gates to perform CB conversion, and (2) an additional three-qubit bit-flip error detector, recovery, and decoder circuit.

Because the component codes in Shor's code are systematic, while the overall code is not systematic, the first, fourth, and seventh qubits are used, after CB conversion, as inputs to the second stage three-qubit bit-flip code detector, recovery, and detector circuit.

After this introductory treatment of quantum error correction, in the next section we describe a very important class of QECCs, namely stabilizer codes. Stabilizer codes are described in this section at a conceptual level, while in Section 8.3 they are treated in a more formal way. Because this is an important class of quantum codes, the whole of the next chapter is devoted to stabilizer codes.

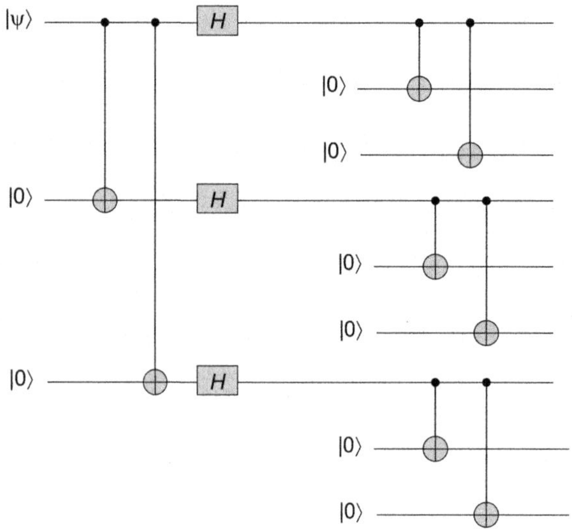

FIGURE 8.8

Nine-qubit Shor's code encoding circuit.

8.2.4 Stabilizer Codes Concepts

To facilitate the description of stabilizer codes we introduce several definitions. The Pauli group over one qubit is defined as $G_1 = \{\pm I, \pm jI, \pm X, \pm jX, \pm Y, \pm jY, \pm Z, \pm jZ\}$. The Pauli group G_N on N qubits consists of all Pauli operators of the form

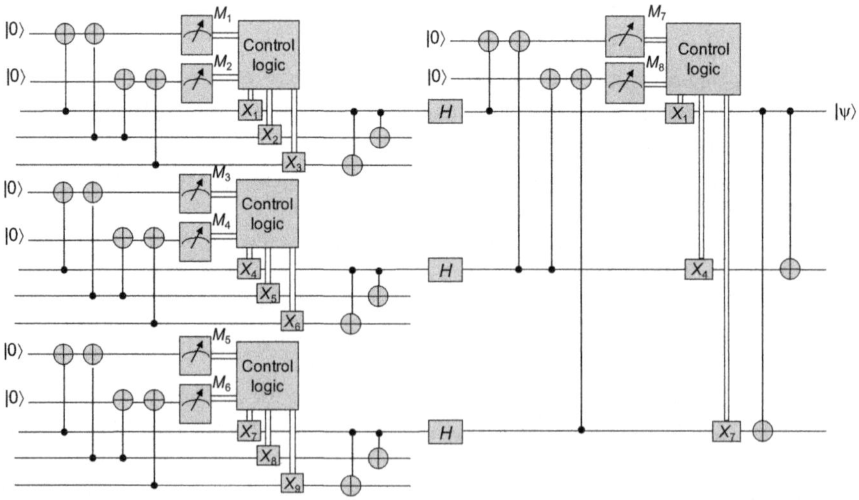

FIGURE 8.9

Error correction and decoding circuit for nine-qubit Shor's code.

$cO_1O_2 \ldots O_N$, where $O_i \in \{I, X, Y, \text{or } Z\}$, and $c \in \{1, -1, j, -j\}$. Therefore any Pauli vector on N qubits can be written uniquely as the product of X- and Z-containing operators together with the phase factor (± 1 or $\pm j$). In a stabilizer framework, a codeword is defined to be a ket $|\psi\rangle$, that is, $+1$ eigenket of all stabilizers s_i, so that $s_i|\psi\rangle = |\psi\rangle$ for all i. In other words, the set of operators s_i that "stabilizes" ("fixes") the codeword states forms the group S that is called the stabilizer group. A *stabilizer group* S consists of a set of Pauli matrices (X, Y, Z together with multiplicative factors ± 1, $\pm j$), with a property that any two operators in the group S commute so that they can be simultaneously measured. We say that an operator fixes the state if the state is an eigenket with eigenvalue $+1$ for this operator.

Example. An alternative syndrome measurement basis for Shor's code is given by Z_1Z_2; Z_2Z_3; Z_3Z_4; Z_4Z_5; Z_5Z_6; Z_7Z_8; Z_8Z_9; $X_1X_2X_3X_4X_5X_6$; and $X_4X_5X_6X_7X_8X_9$. The Z-syndromes are used for bit-flip error detection, and X-syndromes for phase-flip error detection. The key idea of the stabilizer formalism is that quantum states can be more efficiently described by working on the operators that stabilize them than by working explicitly on the state. All operators that fix the base codewords of Shor's code $|\overline{0}\rangle$ and $|\overline{1}\rangle$, introduced earlier, can be written as the product of eight operators M_i, as shown in Table 8.3.

The stabilizer S is an Abelian subgroup of G_N. Let V_S denote the vector space stabilized by S, that is, the set of states on N qubits that are fixed with every element from S. The stabilizer S is described by its generators g_i. Any subgroup of G_N can be used as a stabilizer, providing that the following two conditions are satisfied: (1) the elements of S commute with each other, and (2) $-I$ is not an element of S.

An $[N,K]$ *stabilizer code* is defined as a vector subspace V_S stabilized by $S = \langle g_1, g_2, \ldots, g_{N-K} \rangle$. Any element s from S can be written as a unique product of powers of the generators, as described in Section 8.3. This interpretation is very similar to classical error correction, in which the basis vectors g_i in code space have a similar role as stabilizer generators in quantum error correction.

Table 8.3 Stabilizer table for Shor's code.

	1	2	3	4	5	6	7	8	9	
M_1	Z	Z	I	I	I	I	I	I	I	Z_1Z_2
M_2	I	Z	Z	I	I	I	I	I	I	Z_2Z_3
M_3	I	I	I	Z	Z	I	I	I	I	Z_4Z_5
M_4	I	I	I	I	Z	Z	I	I	I	Z_5Z_6
M_5	I	I	I	I	I	I	Z	Z	I	Z_7Z_8
M_6	I	I	I	I	I	I	I	Z	Z	Z_8Z_9
M_7	X	X	X	X	X	X	I	I	I	$X_1X_2X_3X_4X_5X_6$
M_8	I	I	I	X	X	X	X	X	X	$X_4X_5X_6X_7X_8X_9$

Example. Let us observe the Bell (Einstein–Podolsky–Rosen, EPR) state: $|\psi\rangle = (|00\rangle + |11\rangle)/\sqrt{2}$. It can be easily shown that $X_1X_2|\psi\rangle = |\psi\rangle$ and $Z_1Z_2|\psi\rangle = |\psi\rangle$. Therefore the EPR state is stabilized by X_1X_2 and Z_1Z_2. It can be shown that $|\psi\rangle$ is the unique state that is stabilized by the two operators.

Let us analyze what is going to happen when a unitary gate U is applied to a vector subspace V_S stabilized by S:

$$s \in S: U|\psi\rangle = Us|\psi\rangle = UsU^{+}U|\psi\rangle = \left(UsU^{+}\right)U|\psi\rangle. \tag{8.8}$$

Therefore the state $U|\psi\rangle$ is stabilized by UsU^{+}. We say the set of U such that $UG_NU^{+} = G_N$ is the *normalizer* of G_N denoted by $N(G_N)$. With this formalism we can define the *distance* of a stabilizer code to be the minimum weight of an element of $N(S)$-S. More details of stabilizer group and stabilizer codes can be found in Section 8.3. In the next section, instead we turn our attention to establishing the relationship between quantum and classical codes.

8.2.5 Relationship Between Quantum and Classical Codes

As was indicated earlier, given any Pauli operator on N qubits, we can write it uniquely as a product of X-containing operators and Z-containing operators and a phase factor (± 1, $\pm j$). For instance, XIYZYI = −(XIXIXI) · (IIZZZI). We can now express the X-operator as a binary string of length N, with "1" standing for X and "0" for I, and do the same for the Z-operator. Thus each stabilizer can be written as the X string followed by the Z string, giving a matrix of width $2N$. We mark the boundary between the two types of strings with vertical bars, so, for instance, the set of generators of Shor's code appears as the *quantum check matrix A*. The quantum check matrix of Shor's code (Table 8.3) is given by:

$$\mathbf{A} = \begin{bmatrix} \quad X \qquad\qquad Z \\ 111111000|000000000 \\ 000111111|000000000 \\ 000000000|110000000 \\ 000000000|011000000 \\ 000000000|000110000 \\ 000000000|000011000 \\ 000000000|000000110 \\ 000000000|000000011 \end{bmatrix}.$$

The commutativity of stabilizers now appears as an *orthogonality of rows* with respect to a *twisted* (*sympletic*) *product*, formulated as follows: if the kth row is

$r_k = (x_k; z_k)$, where x_k is the X binary string and z_k the Z string, then the twisted product of rows k and l is defined by:

$$r_k \odot r_l = x_k \cdot z_l + x_l \cdot z_k \text{ mod2}, \tag{8.9}$$

where $x_k \cdot z_l$ is a dot (scalar) product defined by $x_k \cdot z_l = \sum_j x_{kj} z_{lj}$. The twisted product is zero if and only if there is an even number of places where the operators corresponding to rows k and l differ (and neither are the identity), i.e., if the operators commute. If we write the quantum-check $\boldsymbol{A}$ as $\boldsymbol{A} = (\boldsymbol{A}_1|\boldsymbol{A}_2)$, then the condition that the twisted product is zero for all k and l can be written compactly as:

$$\boldsymbol{A}_l \boldsymbol{A}_2^T + \boldsymbol{A}_2 \boldsymbol{A}_1^T = \boldsymbol{0}. \tag{8.10}$$

A *Pauli error operator* E can be interpreted as a binary string $\boldsymbol{e}$ of length $2N$, in which we reverse the order of the X and Z strings. For example, $E = Z_1X_2Y_9$ can be written as $\boldsymbol{e} = [100000001|010000001]$. Thus the quantum syndrome for the noise is exactly the classical syndrome $\boldsymbol{eA}^T$, considering $\boldsymbol{A}$ as a parity-check matrix and $\boldsymbol{e}$ as binary noise vector. In this formalism, we perform the ordinary dot product (mod 2) of error vector $\boldsymbol{e}$ with the corresponding row of the quantum-check matrix, if the result is zero the error operator and corresponding stabilizer of the row commute, otherwise the result is 1 and they do not commute.

In conclusion, the properties of stabilizer codes can be inferred from those of a special class of classical codes. Given any binary matrix of size $M \times 2N$ that has the property that the twisted product of any two rows is zero, an equivalent quantum code can be constructed that encodes $N - M$ qubits in N qubits. Several examples are provided in the following three subsections to illustrate this representation.

8.2.6 Quantum Cyclic Codes

An $(N;K) = (5;1)$ quantum code is generated by the following four stabilizers: XZZXI, IXZZX, XIXZZ, and ZXIXZ. The corresponding quantum-check matrix is given by:

$$\boldsymbol{A} = \begin{bmatrix} X \quad Z \\ 10010|01100 \\ 01001|00110 \\ 10100|00011 \\ 01010|10001 \end{bmatrix}.$$

A correctable set of errors consists of all operators with one nonidentity term, e.g., XIIII, IIIYI, or IZIII. These correspond to binary strings such as 00000|10000 for XIIII, 00010|00010 for IIIYI, and so on. There are 15 of these, and each has a distinct syndrome, thereby using 2^4 1 possible nonzero syndromes. Because this code satisfies the quantum Hamming inequality (see Section 8.4) with an equality sign it belongs to the class of *perfect quantum codes*.

8.2.7 Calderbank–Shor–Steane Codes

An important class of codes, invented by Calderbank, Shor, and Steane, known as the class of CSS codes [1,2], has the form:

$$A = \begin{bmatrix} \boldsymbol{H} & | & \boldsymbol{0} \\ \boldsymbol{0} & | & \boldsymbol{G} \end{bmatrix}, \quad \boldsymbol{H}\boldsymbol{G}^T = \boldsymbol{0}, \tag{8.11}$$

where $\boldsymbol{H}$ and $\boldsymbol{G}$ are $M \times N$ matrices. The condition $\boldsymbol{HG}^T = \boldsymbol{0}$ ensures that the twisted product condition is satisfied. As there are $2M$ stabilizer conditions applying to N qubit states, $N - 2M$ qubits are encoded in N qubits. The special case of CSS codes are dial-containing codes, also known as weakly self-dual codes, in which $\boldsymbol{H} = \boldsymbol{G}$, so that the quantum check matrix $\boldsymbol{A}$ has the following form:

$$A = \begin{bmatrix} \boldsymbol{H} & | & \boldsymbol{0} \\ \boldsymbol{0} & | & \boldsymbol{H} \end{bmatrix}, \quad \boldsymbol{H}\boldsymbol{H}^T = \boldsymbol{0}. \tag{8.12}$$

The $\boldsymbol{HH}^T = 0$ condition is equivalent to $C^{\perp}(\boldsymbol{H}) \subset C(\boldsymbol{H})$, where $C(\boldsymbol{H})$ is the code having $\boldsymbol{H}$ as its parity check matrix and $C^{\perp}(\boldsymbol{H})$ is its dual code. An example of a dual-containing code is *Steane's* 7-qubit code, defined by the Hamming (7,4) code:

$$\boldsymbol{H} = \begin{bmatrix} 0001111 \\ 0110011 \\ 1010101 \end{bmatrix}.$$

The rows have an even number of 1s, and any two of them overlap by an even number of 1s, so that $C^{\perp}(\boldsymbol{H}) \subset C(\boldsymbol{H})$. Here, $M = 3$, $N = 7$, and so $N - 2M = 1$, and thus 1 qubit is encoded in 7 qubits, representing the [7,1] code.

8.2.8 Quantum Codes Over GF(4)

Let the elements of GF(4) be 0, 1, w, and $w^2 = w' = 1 + w$. One can write a row $r_1 = (a_1 a_2 \ldots a_n | b_1 b_2 \ldots b_n)$ of quantum-check matrix in the binary string representation as a vector over GF(4) as follows: $\rho_1 = (a_1 + b_1 w;\ a_2 + b_2 w;\ \ldots;\ a_n + b_n w)$. Given a second row $r_2 = (c_1 c_2 \ldots c_n | d_1 d_2 \ldots d_n)$, with $\rho_2 = (c_1 + d_1 w;\ c_2 + d_2 w;\ \ldots;\ c_n + d_n w)$, the *Hermitian inner product* is defined by:

$$\rho_1 \cdot \rho_2 = \sum_i (a_i + b_i w')(c_i + d_i w) = \sum_i [(a_i c_i + b_i d_i + b_i c_i) + (a_i d_i + b_i c_i) w]. \tag{8.13}$$

Because $a_i d_i + b_i c_i = r_1 \odot r_2$, the orthogonality in the Hermitian sense ($\rho_1 \cdot \rho_2 = 0$) leads to orthogonality in the simpletic sense too ($r_1 \odot r_2 = 0$). The opposite is not true because the term $a_i c_i + b_i d_i + b_i c_i$ is not necessarily zero when $a_i d_i + b_i c_i = 0$. Therefore the codes over GF(4) satisfying the property that any two rows are orthogonal in the Hermitian inner product sense can be used as quantum codes.

8.3 QUANTUM ERROR CORRECTION

In the previous section, quantum error correction was introduced on a conceptual level. Here, we describe quantum codes more formally. The section starts with redundancy, followed by the stabilizer group, and quantum syndrome decoding. We also formally establish the connection between classical and quantum codes. We then discuss the necessary and sufficient conditions for quantum error correction and provide a detailed quantum error correction example. Furthermore, the distance properties of quantum codes are discussed, and distance is related to error correction capability. Finally, we describe quantum encoder and decoder implementations.

8.3.1 Redundancy and Quantum Error Correction

A QECC that encodes K qubits into N qubits, denoted as $[N,K]$, is defined by an encoding mapping U from the K-qubit Hilbert space H_2^K onto a 2^K-dimensional subspace C_q of the N-qubit Hilbert space H_2^N. The subspace C_q is called the *code space*; the states belonging to C_q are called the *codewords*, and the encoded CB kets are called *basis codewords*. The single-qubit CB states are typically chosen to be the eigenkets of Z_j:

$$Z_j|b_j\rangle = (-1)^{b_j}|b_j\rangle;\ b_j = 0, 1;\ j = 1, 2, ..., K. \tag{8.14}$$

The CB states of $H^k{}_2$ are given by:

$$|\boldsymbol{b}\rangle \equiv |b_1...b_k\rangle = |b_1\rangle \otimes ... \otimes |b_k\rangle. \tag{8.15}$$

The encoded CB states are obtained by applying the encoding mapping to the CB states as follows:

$$\left|\overline{\boldsymbol{b}}\right\rangle \equiv \left|\overline{b_1...b_k}\right\rangle = U|b_1...b_k\rangle, \tag{8.16}$$

where the action of encoding mapping is given by $Z_j \xrightarrow{U} \overline{Z}_j = UZ_jU^\dagger$. The encoded CB states are simultaneous eigenkets of $\{Z_j\text{: } j = 1, \ldots, K\}$ because:

$$\overline{Z}_j\left|\overline{\boldsymbol{b}}\right\rangle = UZ_jU^\dagger\left|\overline{b_1...b_k}\right\rangle = UZ_jU^\dagger(U|b_1...b_k\rangle) = UZ_j|\boldsymbol{b}\rangle = U(-1)^{b_j}|\boldsymbol{b}\rangle = (-1)^{b_j}\left|\overline{\boldsymbol{b}}\right\rangle. \tag{8.17}$$

Therefore encoding mapping preserves the eigenvalues.

The simplest QECCs are *canonical codes* [16,22], which can be introduced by the following trivial encoding mapping U_c:

$$U_c\text{: } |\psi\rangle \rightarrow |0\rangle|\psi\rangle. \tag{8.18}$$

In canonical codes, see Fig. 8.10, the quantum register containing $N-K$ ancillas $|\mathbf{0}\rangle_{N-K} = \underbrace{|0\rangle \otimes \cdots \otimes |0\rangle}_{N-K}$ is appended to the information quantum register containing

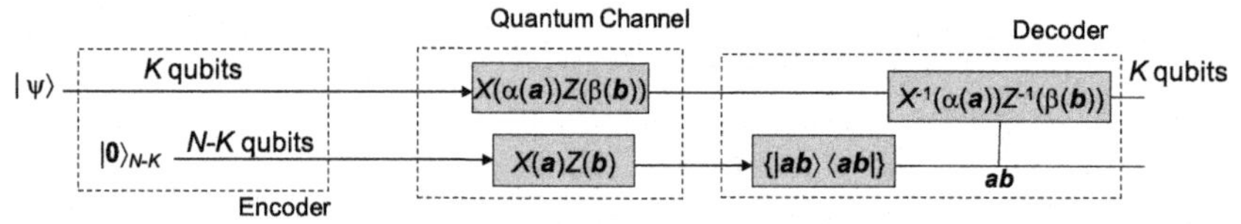

FIGURE 8.10

Canonical quantum error correction coding.

K qubits. The basis for single-qubit errors is given by $\{I,X,Y,Z\}$. The basis for N-qubit errors is obtained by forming all possible direct products:

$$E = j^l O_1 \otimes \cdots \otimes O_N; \qquad O_i \in \{I, X, Y, Z\}, l = 0, 1, 2, 3. \tag{8.19}$$

The N-qubit error basis can be transformed into multiplicative *Pauli group* G_N if we allow E to be premultiplied by $j^{l'} = 1, j, -1, -j$ (for $l' = 0,1,2,3$). By noticing that $Y = -jXZ$, any error in the Pauli group of N-qubit errors can be represented by:

$$\begin{aligned} &E = j^{l'} X(\boldsymbol{a})Z(\boldsymbol{b}); \boldsymbol{a} = a_1 \cdots a_N; \boldsymbol{b} = b_1 \cdots b_N; a_i, b_i = 0, 1; l' = 0, 1, 2, 3 \\ &X(\boldsymbol{a}) \equiv X_1^{a_1} \otimes \cdots \otimes X_N^{a_N}; Z(\boldsymbol{b}) \equiv Z_1^{b_1} \otimes \cdots \otimes Z_N^{b_N}. \end{aligned} \tag{8.20}$$

In Eq. (8.20), the subscript i ($i = 0,1,\ldots,N$) denotes the location of the qubit to which operator X_i (Z_i) is applied; a_i (b_i) takes the value 1 if the ith X (Z) operator is to be included and value 0 if the same operator is to be excluded. Therefore to uniquely determine the error operator (up to the phase constant $j^{l'} = 1, j, -1, -j$ for $l' = 0,1,2,3$) it is sufficient to specify vectors $\boldsymbol{a}$ and $\boldsymbol{b}$. For example, the error $E = Z_1 X_2 Y_9 = -jZ_1 X_2 X_9 Z_9$ can be identified (up to the phase constant) by specifying $\boldsymbol{a} = (010000001)$ and $\boldsymbol{b} = (100000001)$. Notice this representation is equivalent to classical representation of error E from Section 8.2 as follows: $\boldsymbol{e} = [100000001|010000001]$, by reversing the positions of vectors $\boldsymbol{a}$ and $\boldsymbol{b}$.

In Fig. 8.10, we employ this simplified notation and represent the quantum error as the product of X- and Z-containing operators. We observe separately possible errors introduced on information qubits, denoted as $X(\alpha(\boldsymbol{a}))$ and $Z(\beta(\boldsymbol{b}))$, and errors introduced on ancillas, denoted as $X(\boldsymbol{a})$ and $Z(\boldsymbol{b})$, where α and β are functions of $\boldsymbol{a}$ and $\boldsymbol{b}$, respectively. The action of a correctable quantum error E:

$$\begin{aligned} &E \in \boldsymbol{E}_c = \{X(\boldsymbol{a})Z(\boldsymbol{b}) \otimes X(\alpha(\boldsymbol{a}))Z(\beta(\boldsymbol{b})) : \boldsymbol{a}, \boldsymbol{b} \in F_2^{N-K}; \\ &F_2 = \{0, 1\}\}; \qquad \alpha, \beta : F_2^{N-K} \to F_2^{N-K} \end{aligned} \tag{8.21}$$

can be described as follows:

$$E(|\boldsymbol{0}\rangle|\psi\rangle) = X(\boldsymbol{a})Z(\boldsymbol{b})|0\rangle \otimes X(\alpha(\boldsymbol{a}))Z(\beta(\boldsymbol{b}))|\psi\rangle. \tag{8.22}$$

Since $X(\boldsymbol{a})Z(\boldsymbol{b})|\boldsymbol{0}\rangle = X(\boldsymbol{a})|\boldsymbol{0}\rangle = |\boldsymbol{a}\rangle$ we obtain:

$$E(|\boldsymbol{0}\rangle|\psi\rangle) = |\boldsymbol{a}\rangle \otimes \overbrace{X(\alpha(\boldsymbol{a}))Z(\beta(\boldsymbol{b}))|\psi\rangle}^{|\psi'\rangle} = |\boldsymbol{a}\rangle|\psi'\rangle. \tag{8.23}$$

On the receiver side, we perform measurements on ancillas to determine the syndrome $S = (\boldsymbol{a},\boldsymbol{b})$ without affecting the information qubits. Once the syndromes are determined, we perform reverse recovery operator actions, denoted in Fig. 8.10 as $X^{-1}(\alpha(\boldsymbol{a}))$ and $Z^{-1}(\beta(\boldsymbol{b}))$, and the proper information state is recovered. Notice that by employing the entanglement between the source and destination we can simplify the decoding process, which is the subject of the section on entanglement-assisted quantum error correction in the next chapter (see also [16,22]).

An arbitrary quantum error correcting code can be observed as a generalization of the canonical code, as shown in Fig. 8.11, where U is the corresponding encoding operator. The set of errors introduced by the channel, which can be corrected by this code, can be represented now by:

$$E = \{U[X(\boldsymbol{a})Z(\boldsymbol{b}) \otimes X(\alpha(\boldsymbol{a}))Z(\beta(\boldsymbol{b}))]U^{-1}:\ \boldsymbol{a},\boldsymbol{b} \in F_2^{N-K};$$
$$F_2 = \{0,1\}\}; \qquad \alpha,\beta:\ F_2^{N-K} \rightarrow F_2^{N-K}. \tag{8.24}$$

8.3.2 Stabilizer Group S

The quantum stabilizer code C_q, with parameters $[N,K]$, can be defined as the unique subspace of H_2^N that is fixed by the elements from stabilizer S of C_q as given by:

$$\forall\ s \in S,\ |c\rangle \in C_q:\ s|c\rangle = |c\rangle. \tag{8.25}$$

The stabilizer group S is constructed from a set of $N-K$ operators $g_1,\ldots,g_{N-K}$, also known as the *generators* of S. Generators have the following properties: (1) they commute among each other, (2) they are unitary and Hermitian, and (3) they have the order of 2 because $g_i^2 = I$. Any element $s \in S$ can be written as a unique product of powers of the generators:

$$s = g_1^{a_1} \cdots g_{N-K}^{a_{N-K}},\ a_i \in \{0,1\};\ i = 1,\cdots,N-K. \tag{8.26}$$

When a_i is set to 1 (0) the ith generator is included (excluded) from the product of generators. Therefore the elements from S can be labeled by simple binary strings of length $N-K$, namely $\boldsymbol{a} = a_1 \ldots a_{N-K}$. Here, we can draw the parallel with classical error correction, in which any codeword in a linear block code can be represented as a linear combination of basis codewords. The key difference is that the

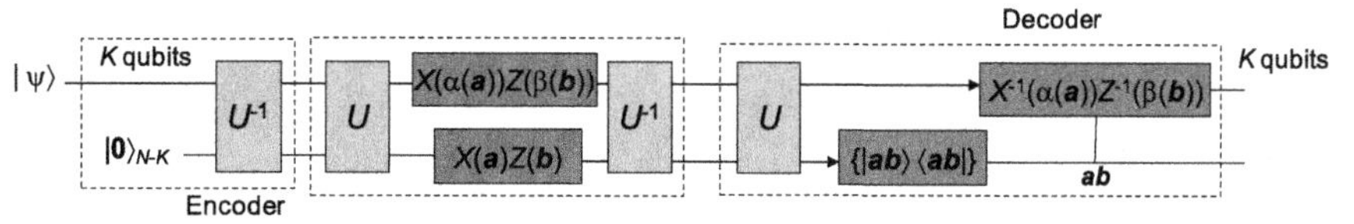

FIGURE 8.11

Arbitrary quantum error correcting code represented as a generalization of the canonical code.

corresponding group in quantum error correction is multiplicative rather than additive. The eigenvalues of generators g_i, $\{\mathrm{eig}(g_i)\}$, can be found starting from property (3):

$$|\mathrm{eig}(g_i)\rangle = I|\mathrm{eig}(g_i)\rangle = g_i^2|\mathrm{eig}(g_i)\rangle = \mathrm{eig}^2(g_i)|\mathrm{eig}(g_i)\rangle. \tag{8.27}$$

From Eq. (8.27) it is clear that $\mathrm{eig}^2(g_i) = 1$ meaning that eigenvalues are $\mathrm{eig}(g_i) = \pm 1 = (-1)^{\lambda_i}$; $\lambda_i = 0, 1$. Because the "parent" space for C_q is 2^N-dimensional space, we need to determine N-commuting operators that specify a unique state $|\psi\rangle \in H_2^N$. The following 2^N simultaneous eigenstates of $\{g_1, g_2, \cdots, g_{N-K}; \overline{Z}_1, \overline{Z}_2, \cdots, \overline{Z}_K\}$ can be used as a basis for H_2^N. Namely, we have chosen the subset $\{\overline{Z}_i\}$ in such a way that each element for this subset commutes with all generators $\{g_i\}$ as well as among all elements of the subset itself. The eigenkets can be labeled by $\boldsymbol{\lambda} = \lambda_1 \ldots \lambda_{N-K}$ and encoded CBs will be:

$$g_i\left|\boldsymbol{\lambda};\overline{\boldsymbol{b}}\right\rangle = (-1)^{\lambda_i}\left|\boldsymbol{\lambda};\overline{\boldsymbol{b}}\right\rangle \quad \overline{Z}_j\left|\boldsymbol{\lambda};\overline{\boldsymbol{b}}\right\rangle = (-1)^{b_j}\left|\boldsymbol{\lambda};\overline{\boldsymbol{b}}\right\rangle;\ i = 1, \ldots, N-K;$$

$$j = 1, \ldots, K;\ \lambda_i, b_j = 0, 1. \tag{8.28}$$

For every $s(\boldsymbol{a}) \in S$ the following is valid:

$$\begin{aligned} s(\boldsymbol{a})\left|\boldsymbol{\lambda};\overline{\boldsymbol{b}}\right\rangle &= g_1^{a_1}\ldots g_{n-k}^{a_{n-k}}\left|\boldsymbol{\lambda};\overline{\boldsymbol{b}}\right\rangle = (-1)^{\overbrace{\sum_{i=1}^{N-K}\lambda_i a_i \bmod 2}^{\boldsymbol{\lambda}\cdot\boldsymbol{a}}}\left|\boldsymbol{\lambda};\overline{\boldsymbol{b}}\right\rangle \\ &= (-1)^{\boldsymbol{\lambda}\cdot\boldsymbol{a}}\left|\boldsymbol{\lambda};\overline{\boldsymbol{b}}\right\rangle;\quad \boldsymbol{\lambda}\cdot\boldsymbol{a} = \sum_{i=1}^{N-K}\lambda_i a_i \bmod 2. \end{aligned} \tag{8.29}$$

From Eq. (8.25) is clear that:

$$s(\boldsymbol{a})\left|\boldsymbol{\lambda};\overline{\boldsymbol{b}}\right\rangle = \left|\boldsymbol{\lambda};\overline{\boldsymbol{b}}\right\rangle, \tag{8.30}$$

when $\boldsymbol{\lambda}\cdot\boldsymbol{a} = 0 \bmod 2$. The simplest way to satisfy Eq. (8.30) is for $\boldsymbol{\lambda} = 0 \ldots 0$, meaning that eigenkets $\left\{\left|\boldsymbol{\lambda} = 0\ldots0;\overline{\boldsymbol{b}}\right\rangle\right\}$ can be used as encoded CB states.

The following interpretation of the same result is due to Gaitan [14]. The set $C(\boldsymbol{\lambda}) = \left\{\left|\boldsymbol{\lambda};\overline{\boldsymbol{b}}\right\rangle : \overline{\boldsymbol{b}} \in F_2^K\right\}$ (where $F_2 = \{0,1\}$) is clearly the subspace of H_N^2 whose elements are simultaneous eigenvectors of the generators $g_1, \ldots, g_{N-K}$ with corresponding eigenvalues $(-1)^{\lambda_1}, \cdots, (-1)^{\lambda_{N-K}}$. It can be shown that the set of 2^{N-K} subspaces $\{C(\boldsymbol{\lambda})$: $\boldsymbol{\lambda} \in F_2^{N-K}\}$ partitions H_2^N. Because the quantum stabilizer code C_q is fixed by stabilizer S, the states $|c\rangle \in C_q$ are fixed by $g_1, \ldots, g_{N-K}$ with corresponding eigenvalues $\mathrm{eig}(g_i) = 1$ ($i = 1,2,\ldots,N\text{-}K$), meaning that $C_q \subset C(\boldsymbol{\lambda} = 0 \ldots 0)$. Since both C_q and $C(\boldsymbol{\lambda} = 0 \ldots 0)$ are 2^K dimensional, it must be $C_q = C(\boldsymbol{\lambda} = 0 \ldots 0)$. Therefore the encoded CB states are the eigenkets $\left\{\left|\boldsymbol{\lambda} = 0\ldots0;\overline{\boldsymbol{b}}\right\rangle\right\}$.

8.3.3 Quantum-Check Matrix and Syndrome Equation

In the previous section, we already established the connection between quantum and classical codes, but we did not explain how this connection was actually derived, which is the topic of this subsection. The following two theorems are important in establishing the connection between classical and quantum codes.

Theorem 1. Let E be an error and S the stabilizer group for a quantum stabilizer code C_q. If S contains an element that anticommutes with E, then for all $|c\rangle, |c'\rangle \in C_q$, $E|c\rangle$ is orthogonal to $|c'\rangle$:

$$\langle c'|E|c\rangle = 0. \tag{8.31}$$

This theorem is quite straightforward to prove. Since $|c\rangle, |c'\rangle \in C_q$ and the error E anticommutes with $s \in S$ ($\{E, s\} = 0$) we can write:

$$E|c\rangle = Es|c\rangle = -sE|c\rangle. \tag{8.32}$$

By multiplying with $\langle c'|$ from the left side we obtain:

$$\langle c'|E|c\rangle = -\langle c'|sE|c\rangle = -\langle c'|E|c\rangle. \tag{8.33}$$

By solving Eq. (8.33) for $\langle c'|E|c\rangle$ we obtain $\langle c'|E|c\rangle = 0$.

Theorem 2. Let E be an error and C_q be a quantum stabilizer code with generators $g_1,\ldots,g_{N-K}$. The image $E(C_q)$ under E is $C(\boldsymbol{\lambda})$, where $\boldsymbol{\lambda} = \lambda_1 \ldots \lambda_{N-K}$ such that:

$$\lambda_i = \begin{cases} 0, & [E, g_i] = 0 \\ 1, & \{E, g_i\} = 0 \end{cases} \quad (i = 1, \ldots, N-K). \tag{8.34}$$

This theorem can be proved by observing the action of generator g_i and an error operator E on $|c\rangle \in C_q$:

$$g_i E|c\rangle = (-1)^{\lambda_i} E g_i|c\rangle = (-1)^{\lambda_i} E|c\rangle. \tag{8.35}$$

It is clear from Eq. (8.35) that for $\lambda_i = 0$, E and g_i commute since $g_i E = E g_i$. On the other hand, for $\lambda_i = 1$, from Eq. (8.35) we obtain $g_i E = -E g_i$, which means that E and g_i anticommute ($\{E, g_i\} = 0$). These two cases can be written in one equation as follows:

$$\lambda_i = \begin{cases} 0, & [E, g_i] = 0 \\ 1, & \{E, g_i\} = 0 \end{cases} \quad (i = 1, \ldots, N-K),$$

proving therefore Eq. (8.34). What remains is to prove that the image $E(C_q)$ equals $C(\boldsymbol{\lambda}) = \{|\boldsymbol{\lambda}; \overline{\boldsymbol{\delta}}\rangle : \overline{\boldsymbol{\delta}} \in F_2^K\}$. The ket $E|c\rangle$ can be represented in terms of eigenkets $|\boldsymbol{\lambda}; \overline{\boldsymbol{a}}\rangle$ as follows:

$$E|c\rangle = \sum_{\boldsymbol{\lambda}'} \sum_{\boldsymbol{a}} \alpha(\boldsymbol{\lambda}'; \overline{\boldsymbol{\delta}}) |\boldsymbol{\lambda}'; \overline{\boldsymbol{a}}\rangle, \tag{8.36}$$

where $\alpha(\boldsymbol{\lambda}'; \overline{\boldsymbol{a}})$ are projections along basis kets $|\boldsymbol{\lambda}'; \overline{\boldsymbol{a}}\rangle$. Let us now employ the property we just derived and observe the simultaneous action of g_i and E on $|c\rangle$:

$$g_i E|c\rangle = \sum_{\boldsymbol{\lambda}'} \sum_{\overline{\boldsymbol{a}}} \alpha(\boldsymbol{\lambda}'; \overline{\boldsymbol{a}}) g_i |\boldsymbol{\lambda}'; \overline{\boldsymbol{a}}\rangle = \sum_{\boldsymbol{\lambda}} \sum_{\overline{\boldsymbol{a}}} (-1)^{\lambda_i} \alpha(\boldsymbol{\lambda}; \overline{\boldsymbol{a}}) |\boldsymbol{\lambda}; \overline{\boldsymbol{a}}\rangle. \tag{8.37}$$

It is clear from Eqs. (8.37) and (8.35), in which the right side of the equation is only a function of eigenvalues $(-1)^{\lambda_i}$, that $E|c\rangle$ is equal to:

$$E|c\rangle = \sum_{\bar{a}} \alpha(\boldsymbol{\lambda};\bar{\boldsymbol{a}})|\boldsymbol{\lambda};\bar{\boldsymbol{a}}\rangle, \tag{8.38}$$

and therefore $E(C_q) = C(\boldsymbol{\lambda})$.

Because $I(C_q) = C(0 \ldots. 0)$, where I is an identity operator, it is clear that $\boldsymbol{\lambda} = \lambda_1 \ldots \lambda_{N-K}$ can be used as an error syndrome of E, denoted as $S(E)$. Therefore the *error syndrome* is defined as $S(E) = \lambda_1 \ldots \lambda_{N-K}$, where λ_i is defined by Eq. (8.34). Based on the discussion from Section 8.2 we can represent the generator g_i as a binary vector of length $2N$: $\boldsymbol{g}_i = (\boldsymbol{a}_i|\boldsymbol{b}_i)$, where the row vector $\boldsymbol{a}_i$ $(\boldsymbol{b}_i)$ is obtained from operator representation by replacing the X-operators (Z-operators) by 1s and identity operators by 0s. The binary representations of all generators can be used to create the *quantum-check matrix* $\boldsymbol{A}$ obtained by writing the binary representation of g_i as the ith row of $(\boldsymbol{A})_i$ as follows:

$$\boldsymbol{A} = \begin{pmatrix} (\boldsymbol{A})_1 \\ \vdots \\ (\boldsymbol{A})_{N-K} \end{pmatrix}, (\boldsymbol{A})_i = \boldsymbol{g}_i = (\boldsymbol{a}_i|\boldsymbol{b}_i). \tag{8.39}$$

By representing the error operator E as binary error row vector $\boldsymbol{e} = (\boldsymbol{c}|\boldsymbol{d})$, but now with $\boldsymbol{c}$ $(\boldsymbol{d})$ obtained from error operator E by replacing the Z-operators (X-operators) with 1s and identity operators by 0s, the syndrome equation can be written as:

$$S(E) = \boldsymbol{e}\boldsymbol{A}^{\mathrm{T}}, \tag{8.40}$$

which is very similar to classical error correction.

8.3.4 Necessary and Sufficient Conditions for Quantum Error Correction Coding

A quantum error correction circuit determines the error syndrome with the help of ancilla qubits. The measured value of syndrome S determines a set of errors $\boldsymbol{E}_S = (E_1, E_2, \ldots)$ with syndromes being equal to the measured value: $S(E_i) = S$, $\forall E_i \in \boldsymbol{E}_S$. In relation to classical error correction, this set of errors $\boldsymbol{E}_S$ having the same syndrome S can be called the *coset*. Among the different candidate errors, we choose for the expected error the most probable one, say E_S. We then perform a recovery operation by simply applying $E^{\dagger}{}_S$. If the actual error differs from true error E_S, the resulting state $E^{\dagger}{}_S\, E|s\rangle$ will be from C_q but different from the uncorrupted state $|s\rangle$. The following two conditions should be satisfied during the design of a quantum code: (1) the encoded CB states must be chosen carefully so that the environment is not able to distinguish among different CBs, and (2) the corrupted images of codewords must be orthogonal among each other. These two conditions can be formulated as a theorem that provides the *necessary and sufficient conditions* to be satisfied for a quantum code C_q to be able to correct a given set of errors $\{\boldsymbol{E}\}$.

Theorem 3. (Necessary and sufficient conditions for QECCs) [14]: The code C_q is an $\boldsymbol{E}$ error correcting code if and only if $\forall\ |\bar{i}\rangle, |\bar{j}\rangle\ (\bar{i} \neq \bar{j})$ and $\forall\ E_a,\ E_b \in \boldsymbol{E}$ and the following two conditions are simultaneously satisfied:

$$\langle\bar{i}|E_a^{\dagger}E_b|\bar{i}\rangle = \langle\bar{j}|E_a^{\dagger}E_b|\bar{j}\rangle \quad \langle\bar{i}|E_a^{\dagger}E_b|\bar{j}\rangle = 0. \tag{8.41}$$

The first condition indicates that the action of the environment is similar to all CBs so that the channel is not able to distinguish among different CBs. The second condition indicates that the images of codewords are orthogonal to each other. The proof of this theorem is left as a homework problem.

8.3.5 A Quantum Stabilizer Code for Phase-Flip Channel (Revisited)

The different quantum error models considered so far assume: (1) errors on different qubits are independent, (2) single-qubit errors are equally likely, and (3) single qubit-error probability is the same for all qubits. Several such models, including the phase-flip channel model, have already been introduced in the previous section. The phase-flip channel generates eight possible errors on the three qubits as shown in Table 8.4.

At least one out of two generators must anticommute with each of the single-qubit errors $\{E_1, E_2, E_3\}$, justifying the following selection of generators:

$$g_1 = X_1X_2 \qquad g_2 = X_1X_3.$$

The error syndrome can easily be calculated from Theorem 2 by:

$$S(E) = \lambda_1 \ldots \lambda_{N-K};\ \lambda_i = \begin{cases} 0, [E, g_i] = 0 \\ 1, \{E, g_i\} = 0 \end{cases},$$

and is given in the third column of Table 8.4. The expected error determined from the syndrome and corresponding recovery operator are provided in Table 8.5, in which out of several candidate error operators having the same syndrome, the error with the lowest weight is selected.

Table 8.4 Syndrome look-up table for the phase-flip channel.

Error E	Error probability $P_e(E)$	Error syndrome $S(E)$
$E_0 = I$	$(1-p)^3$	00
$E_1 = Z_1$	$p(1-p)^2$	11
$E_2 = Z_2$	$p(1-p)^2$	10
$E_3 = Z_3$	$p(1-p)^2$	01
$E_4 = Z_1, Z_2$	$p^2(1-p)$	01
$E_5 = Z_1, Z_3$	$p^2(1-p)$	10
$E_6 = Z_2Z_3$	$p^2(1-p)$	11
$E_7 = Z_1, Z_2Z_3$	p^3	00

Table 8.5 Syndrome look-up table and recovery operators for the phase-flip channel.

Most probable error E_s	Error syndrome $S(E)$	Recovery operator R_s
$E_{00} = I$	00	$R_{00} = I$
$E_{01} = Z_3$	01	$R_{01} = Z_3$
$E_{10} = Z_2$	10	$R_{10} = Z_2$
$E_{11} = Z_1$	11	$R_{11} = Z_1$

The stabilizer group is the set of elements of the form $s(\boldsymbol{p}) = g_1^{p1} g_2^{p2}$, where $p_1,p_2 = 0,1$. By simply varying p_i (0 or 1) we obtain the following stabilizer:

$$S = \{I, X_1X_2, X_1X_3, X_2X_3\}.$$

The single-qubit CB states can be chosen to be the eigenkets of X:

$$|0\rangle \equiv |\text{eig}(X) = +1\rangle \qquad |1\rangle \equiv |\text{eig}(X) = -1\rangle.$$

The stabilizer S must fix the code space C_q, including the encoded CB states:

$$|0\rangle \rightarrow |\bar{0}\rangle = |000\rangle \qquad |1\rangle \rightarrow |\bar{1}\rangle = |111\rangle.$$

This mapping is consistent with no-cloning theorem claiming that arbitrary kets cannot be cloned, which does not prevent the cloning of orthogonal kets.

The *failure probability* for error correction P_f is the probability that one of the nonexpected errors $\{E_4,E_5,E_6,E_7\}$ occurs, which is:

$$P_f = 3p^2(1-p) + p^3$$

Example. Let the error syndrome be $S = 10$, while the actual error was $E_5 = Z_1Z_3$. Based on the LUT (Table 8.5) we will choose that $E_{10} = Z_2$ so that the corresponding recovery operator will be $R_{10} = E_{10}^\dagger = Z_2$. The transmitted codeword can be represented as $|c\rangle = a|\bar{0}\rangle + b|\bar{1}\rangle$. The action of the channel leads to the state $E_5|c\rangle$, while the simultaneous action of error E_5 and recovery operator R_{10} leads to $|\psi\rangle = E_{10}^\dagger E_5|c\rangle = Z_1Z_2Z_3|c\rangle$. Because the action of the resulting operator is as follows:

$$Z\left\{\begin{array}{c}|+1\rangle\\|-1\rangle\end{array}\right. = \left\{\begin{array}{c}|-1\rangle\\|+1\rangle\end{array}\right. \Rightarrow E_{10}^\dagger E_5\left\{\begin{array}{c}|\bar{0}\rangle\\|\bar{1}\rangle\end{array}\right. = Z_1Z_2Z_3\left\{\begin{array}{c}|\bar{0}\rangle\\|\bar{1}\rangle\end{array}\right. = \left\{\begin{array}{c}|\bar{1}\rangle\\|\bar{0}\rangle\end{array}\right.,$$

the final state $|\psi_f\rangle = a|\bar{1}\rangle + b|\bar{0}\rangle$ is different from the original state, and so the error correction has failed to correct the error operator, but it was able to return the received state back to the code space.

8.3.6 Distance Properties of Quantum Error Correction Codes

A QECC is defined by the encoding operation of mapping K qubits into N qubits. The QECC $[N,K]$ can be interpreted as a 2^K-dimensional subspace C_q of N-qubit Hilbert space H_2^N, together with corresponding recovery operation R. As already mentioned, this subspace (C_q) is called the code space; the kets belonging to C_q are known as codewords; and the encoded CB kets are called the basis codewords. The basis for single-qubit errors is given by $\{I,X,Y,Z\}$, as already described. Because any error, be it discrete or continuous, can be represented as a linear combination of the basis errors, the linear combination of correctable errors will also be a correctable error. The basis for N-qubit quantum errors is obtained by forming all possible direct products:

$$E = j^l O_1 \otimes \cdots \otimes O_N = j^{l'} X(\boldsymbol{a}) Z(\boldsymbol{b}); \boldsymbol{a} = a_1 \cdots a_N; \boldsymbol{b} = b_1 \cdots b_N; a_i, b_i = 0, 1; l,$$

$$l' = 0, 1, 2, 3 \quad X(\boldsymbol{a}) \equiv X_1^{a_1} \otimes \cdots \otimes X_N^{a_N}; Z(\boldsymbol{b}) \equiv Z_1^{b_1} \otimes \cdots \otimes Z_N^{b_N}; \quad O_i \in \{I, X, Y, Z\}. \tag{8.42}$$

The weight of an error operator $E(\boldsymbol{a},\boldsymbol{b})$ is defined to be a number of qubits different from the identity operator. Necessary and sufficient conditions for a QECC to correct a set of errors $E = \{E_p\}$, given by Eq. (8.41), can be compressed as follows:

$$\left\langle \bar{i} \middle| E_p^\dagger E_q \middle| \bar{j} \right\rangle = C_{pq} \delta_{ij}, \tag{8.43}$$

where the matrix elements C_{pq} satisfy the condition $C_{pq} = C^*{}_{qp}$, so that the square matrix $\boldsymbol{C} = (C_{pq})$ is a Hermitian. A QECC for which matrix $\boldsymbol{C}$ is singular is said to be degenerate. If we interpret $E_p^\dagger E_q$ as a new error operator E, Eq. (8.43) can be rewritten as:

$$\langle \bar{i} | E | \bar{j} \rangle = C_E \delta_{ij}. \tag{8.44}$$

We say that a QECC has a distance D if all errors of weight less than D satisfy Eq. (8.44), and there exists at least one error of weight D to violate it. In other words, the distance of QECC is the weight of the smallest weight D of error E that cannot be detected by the code. Similarly to classical codes, we can relate the distance D to the error correction capability t as follows: $D \geq 2t + 1$. Namely, since $E = E_p^\dagger E_q$ the weight of error operator E will be $\mathrm{wt}\left(E_p^\dagger E_q\right) = 2t$. If we are only interested in detecting the errors but not correcting them, the error detection capability d is related to the distance D by $D \geq d + 1$. Since we are interested only in the detection of errors, we can set $E_p = I$ to obtain $\mathrm{wt}\left(E_p^\dagger E_q\right) = \mathrm{wt}(E_q) = d$. If we are interested in simultaneously detecting d errors and correcting t errors, the distance of the code must be $D \geq d + t + 1$. The following theorem can be used to determine if a given quantum code C_q of quantum distance D is degenerate or not.

Theorem 4. The quantum code C_q of distance D is a degenerate code if and only if its stabilizer S contains an element with weight less than D (excluding the identity element).

The theorem can be proved as follows. If the code C_q is a degenerate code of distance D, there will exist two correctable errors E_1, E_2 such that their action on a CB codeword is the same: $E_1|\bar{i}\rangle = E_2|\bar{i}\rangle$. By multiplying with $E_2^\dagger$ from the left we obtain: $E_2^\dagger E_1|\bar{i}\rangle = |\bar{i}\rangle$, which indicates that the error $E_2^\dagger E_1 \in S$. Since any correctable error satisfies $\langle\bar{i}|E|\bar{j}\rangle = C_E\delta_{ij}$ we obtain $\left\langle\bar{i}\middle|E_2^\dagger E_1\middle|\bar{j}\right\rangle = C_{12}\delta_{ij}$. Because the C_q code has a distance D it is clear that $\mathrm{wt}\left(E_2^\dagger E_1\right) < D$. On the other hand, if there exists an $s \in S$ with $\mathrm{wt}(s) < D$ we can find another $s_a \in S$ so that $s_a s = s_b \in S$. By multiplying both sides with $s_a^\dagger$ we obtain $s = s_a^\dagger s_b$. From the definition of stabilizer codes we know that $s|\bar{i}\rangle = |\bar{i}\rangle$, $\forall s \in S$. By expressing $s = s_a^\dagger s_b$ we obtain $s_a^\dagger s_b|\bar{i}\rangle = |\bar{i}\rangle$, which by multiplying with s_a from the left becomes $s_b|\bar{i}\rangle = s_a|\bar{i}\rangle$, which is equivalent to $(s_a - s_b)|\bar{i}\rangle = 0$. Since the matrix S_{ab} is singular, the code C_q is degenerate.

8.3.7 Calderbank–Shor–Steane Codes (Revisited)

CSS codes are constructed from two classical binary codes C and C' satisfying the following three properties:

1. C and C' are (n,k,d) and (n,k',d') codes, respectively;
2. $C' \subset C$; and
3. C and $C'^{\perp}$ are both t error correcting codes.

The code construction partitions C into the cosets of C':

$C = C' \cup (c_1 + C') \cup \ldots \cup (c_N + C')$, where $c_1, c_2, \ldots, c_N \in C$, and N is the number of cosets determined by (from Lagrange's theorem): $N = 2^k/2^{k'} = 2^{k-k'}$. The basis codewords are obtained by identifying each one with the corresponding coset $|\bar{c}_i\rangle \Leftrightarrow c_i + C'$, so that:

$$|\bar{c}_i\rangle = \frac{1}{\sqrt{2^{k'}}}\sum_{c' \in C'} |c_i + c'\rangle; i = 1, \cdots, N. \tag{8.45}$$

If $v, w \in C$ so that $v - w = d \in C'$ (belonging to the same coset of C'), then $|\bar{v}\rangle = |\bar{w}\rangle$ because:

$$\begin{aligned}|\bar{v}\rangle &= \frac{1}{\sqrt{2^{k'}}}\sum_{c' \in C'} |v + c'\rangle \overset{v-w=d}{=} \frac{1}{\sqrt{2^{k'}}}\sum_{c' \in C'} |w + d + c'\rangle \\ &\overset{d+c'=d'}{=} \frac{1}{\sqrt{2^{k'}}}\sum_{d' \in C'} |w + d'\rangle = |\bar{w}\rangle.\end{aligned} \tag{8.46}$$

The code space C_q is a subspace of $H^n{}_2$ spanned by the basis codewords and is therefore $N = 2^{k-k'}$ dimensional.

Example. CSS [7,3,1] code. For Steane [7,3,3] code, C is the Hamming (7,4,3) and C' is (7,3,4) maximum-length code:

$$H(C) = H(C'_{\perp}) = \begin{bmatrix} 0 & 0 & 0 & 1 & 1 & 1 & 1 \\ 0 & 1 & 1 & 0 & 0 & 1 & 1 \\ 1 & 0 & 1 & 0 & 1 & 0 & 1 \end{bmatrix}.$$

Based on Eqs. (8.12) and (8.39), and the foregoing H-matrix we obtain the following generators:

$$g_1 = X_4X_5X_6X_7 \quad g_2 = X_2X_3X_6X_7 \quad g_3 = X_1X_3X_5X_7$$
$$g_4 = Z_4Z_5Z_6Z_7 \quad g_5 = Z_2Z_3Z_6Z_7 \quad g_6 = Z_1Z_3Z_5Z_7.$$

The number of cosets is $N = 2^{k-k'} = 2$, and the corresponding cosets are:

$(0000000 + C') = \{000\ 0\ 000, 011\ 0\ 011, 101\ 0\ 101, 110\ 0\ 110, 000\ 1\ 111, 011\ 1\ 100, 101\ 1\ 010, 110\ 1\ 001\}$
$(1111111 + C') = \{111\ 1\ 111, 100\ 1\ 100, 010\ 1\ 010, 001\ 1\ 001, 111\ 0\ 000, 100\ 0\ 011, 010\ 0\ 101, 001\ 0\ 110\}$.

Therefore based on Eq. (8.45) we can represent the basis codewords by:

$$|\overline{0}\rangle = \frac{1}{\sqrt{2^3}}[|0000000\rangle + |0110011\rangle + |1010101\rangle + |1100110\rangle + |0001111\rangle$$
$$+ |0111100\rangle + |1011010\rangle + |1101001\rangle]$$

$$|\overline{1}\rangle = \frac{1}{\sqrt{2^3}}[|1111111\rangle + |1001100\rangle + |0101010\rangle + |0011001\rangle + |1110000\rangle$$
$$+ |1000011\rangle + |0100101\rangle + |0010110\rangle].$$

It can clearly been shown that the generators $g_1,\ldots,g_6$ fix the basis codewords.

By definition, a distance D of the QECC is the weight of the smallest error not satisfying Eq. (8.44). We can easily find that $\langle\overline{0}|X_1X_2X_3|\overline{1}\rangle = 1$. Since the weight of error $X_1X_2X_3$ is 3, the distance of the code is 3 and error correction capability is 1. It can be shown that Steane [7,3,1] code is nondegenerate since the stabilizer S does not contain any element of weight smaller than 3.

8.3.8 Encoding and Decoding Circuits of Quantum Stabilizer Codes

In this section, we briefly describe the implementation of encoders and decoders for quantum stabilizer codes. An efficient encoding and decoding of quantum stabilizer codes is described in the next chapter, which is devoted to stabilizer codes. Here, we provide very simple descriptions of encoders and decoders, which is provided for completeness of presentation. In particular, decoders are quite easy to implement by using the stabilizer formalism. The error detector can be obtained by concatenation of circuits corresponding to different stabilizers. The transmission error can be identified by an intersection of corresponding syndrome measurements. Let us observe the implementation of stabilizer $S_a = X_1X_4X_5X_6$. The syndrome quantum circuit for measurement of stabilizer S_a is shown in Fig. 8.12A and B. In Fig. 8.12A the quantum syndrome implementation circuit is based on Hadamard (*H*) and controlled-*X* gates, while in Fig. 8.12B the corresponding implementation is based on *H* and CNOT gates only. The initial and final Hadamard gates are applied on the ancilla. We perform the measurement on the ancilla qubit only in $\{|0\rangle, |1\rangle\}$ basis, which gives the outcome 0 if S_a has the outcome +1, and the outcome 1 when S_a has the outcome −1. This can be explained as follows [12]:

$$\begin{aligned}(I\otimes H)S_a^c(I\otimes H)|\psi\rangle|0\rangle &= (I\otimes H)S_a^c|\psi\rangle(|0\rangle+|1\rangle)/\sqrt{2}\\ &= (I\otimes H)(|\psi\rangle|0\rangle+S_a|\psi\rangle|1\rangle)/\sqrt{2}\\ &= [|\psi\rangle(|0\rangle+|1\rangle)+S_a|\psi\rangle(|0\rangle-|1\rangle)]/2\\ &= [(I+S_a)|\psi\rangle|0\rangle+(I-S_a)|\psi\rangle|1\rangle]/2. \end{aligned} \tag{8.47}$$

Because the measuring of the ancilla projects $|\psi\rangle$ on the eigenkets of S_a, we will obtain outcome 0 when S_a has outcome +1 and 1 when S_a has outcome −1. (In Eq. (8.47) we use superscript *c* to denote the control operation; for $c = 0$ the action S_a is excluded.)

We now describe how to implement the encoders for a dual-containing code defined by a full rank matrix ***H*** with $N > 2M$, based on a proposal due to MacKay et al [12]. We first need to transform the ***H***-matrix by Gauss elimination into the

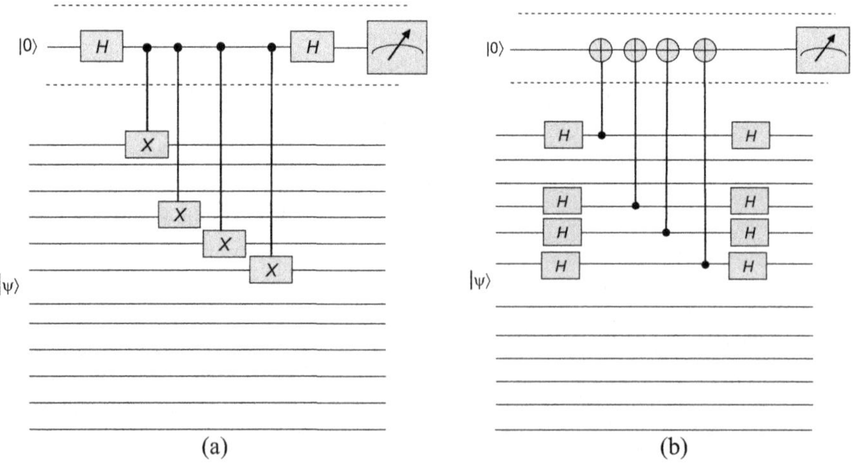

FIGURE 8.12

Syndrome quantum circuit for stabilizer $S_a = X_1X_4X_5X_6$: (a) based on *H* and controlled-*X* gates and (b) based on *H* and CNOT gates.

following form: $\widetilde{\boldsymbol{H}} = \left[\boldsymbol{I}_M \middle| \boldsymbol{P}_{M\times(N-M)}\right]$, where $\boldsymbol{I}$ is the identity matrix of size $M \times M$ and $\boldsymbol{P}$ is the binary matrix of size $M \times (N - M)$ of full rank. We further transform the $\boldsymbol{P}$-matrix also by Gaussian elimination into the following form: $\widetilde{\boldsymbol{P}} = \left[\boldsymbol{I}_M \middle| \boldsymbol{Q}_{M\times(N-2M)}\right]$, where $\boldsymbol{Q}$ is the binary matrix of size $M \times (N - 2M)$. Clearly, for arbitrary string $\boldsymbol{f}$ of length $K = N - 2M$, $[0|\boldsymbol{Qf}|\boldsymbol{f}]$ is a codeword of $C(\boldsymbol{H})$. The encoder can now be implemented in two stages as shown in Fig. 8.13, which is modified from [12].

The first stage performs the following mapping:

$$|\mathbf{0}\rangle_M|\mathbf{0}\rangle_M|\boldsymbol{s}\rangle_K \rightarrow |\mathbf{0}\rangle_M|\boldsymbol{Qs}\rangle_M|\boldsymbol{s}\rangle_K, \tag{8.48}$$

where $|\boldsymbol{s}\rangle_K$ is the K-qubit information state, $\boldsymbol{s}$ is the binary string of length K, and $|\mathbf{0}\rangle_M = \underbrace{|0\rangle\cdots|0\rangle}_{M\text{ times}}$. In other words, the information qubits serve as control qubits, while the middle M ancilla qubits serve as target qubits. The control operations

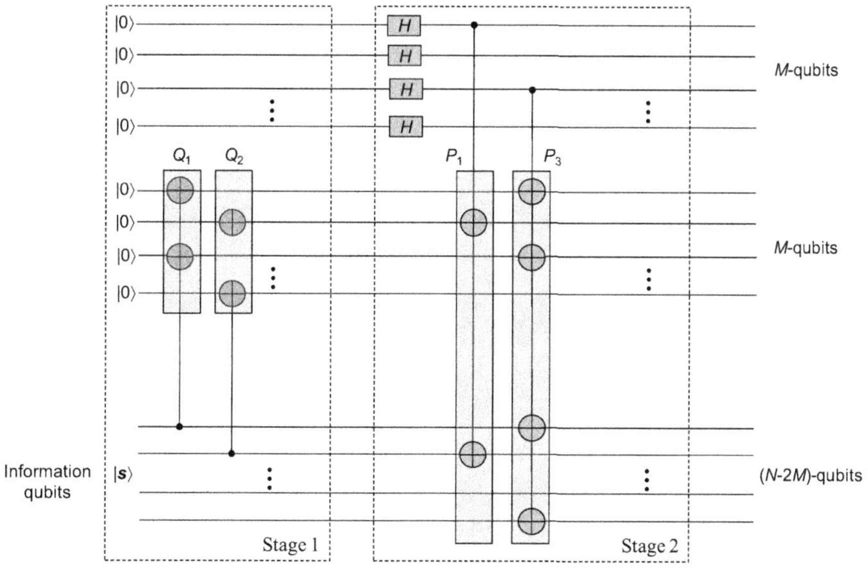

FIGURE 8.13

Encoder implementation for dual-containing codes. The block Q_k in the first stage corresponds to the kth column of submatrix Q. This block is controlled by the kth information qubit, and executed only when the kth information qubit is $|1\rangle$. The CNOT gates are placed according to the nonzero positions in Q_k. In the second stage, the rows of the P-matrix are conditionally executed based on the content of the H gate's output. The kth Hadamard gate controls the set of gates corresponding to the kth row of the P-matrix. The CNOT gates are placed in accordance with nonzero positions in P_k.

are dictated by columns in the $\boldsymbol{Q}$-matrix. In the second stage, the first M-qubits from stage 1 (ancillas) are processed by Hadamard gates mapping the state $|\boldsymbol{0}\rangle_M$ to:

$$|\boldsymbol{0}\rangle_M \rightarrow \underbrace{\frac{|0\rangle+|0\rangle}{\sqrt{2}} \otimes \cdots \otimes \frac{|0\rangle+|0\rangle}{\sqrt{2}}}_{M \text{ times}} = \frac{1}{2^{M/2}} \sum_{\boldsymbol{v}} |\boldsymbol{v}\rangle_M, \tag{8.49}$$

where vector $\boldsymbol{v}$ runs over all possible binary M-tuples. Let us now introduce the binary vector $\boldsymbol{y}$ to denote the concatenation of binary strings $\boldsymbol{Qs}$ and $\boldsymbol{s}$ in Eq. (8.48). From Fig. 8.13 it is clear that the purpose of stage 2 is to conditionally execute the row operators of $\widetilde{\boldsymbol{H}} = \left[\boldsymbol{I}_M \middle| \boldsymbol{P}_{M\times(N-M)}\right]$ on target qubits $|\boldsymbol{y}\rangle_{N-M}$. The qubits $|\boldsymbol{v}\rangle_M$ serve as control qubits. Notice that this stage is very similar to classical encoders of systematic linear block codes (LBCs). Therefore the operation of the final stage is to perform the following mapping:

$$\frac{1}{2^{M/2}} \sum_{\boldsymbol{v}} |\boldsymbol{v}\rangle_M |\boldsymbol{y}\rangle_{N-M} \rightarrow \frac{1}{2^{M/2}} \sum_{\boldsymbol{v}} \left(\prod_{m=1}^{M} P_m^c \right) |\boldsymbol{v}\rangle_M |\boldsymbol{y}\rangle_{N-M}$$
$$= \frac{1}{2^{M/2}} \sum_{\boldsymbol{x}\in C^{\perp}(\boldsymbol{H})} |\boldsymbol{x}+\boldsymbol{y}\rangle_N, \boldsymbol{y}\in C(\boldsymbol{H}). \tag{8.50}$$

The most general codeword of the form:

$$|\psi\rangle = \sum_{\boldsymbol{y}\in C(\boldsymbol{H})} \alpha_{\boldsymbol{y}} \sum_{\boldsymbol{x}\in C^{\perp}(\boldsymbol{H})} |\boldsymbol{x}+\boldsymbol{y}\rangle_N, \tag{8.51}$$

can be obtained if we start from $\sum_{\boldsymbol{s}} \alpha_{\boldsymbol{s}} |\boldsymbol{s}\rangle_K$ instead of $|\boldsymbol{s}\rangle_K$.

8.4 IMPORTANT QUANTUM CODING BOUNDS

In this section we describe several important quantum coding bounds, including Hamming, Gilbert–Varshamov, and Singleton bounds. We also discuss quantum weight enumerators and quantum MacWilliams identity.

8.4.1 Quantum Hamming Bound

The quantum Hamming bound for an $[N,K]$ QECC of error correction capability t is given by:

$$\sum_{j=0}^{t} \binom{N}{j} 3^j 2^K \leq 2^N, \tag{8.52}$$

where t is the error correction capability, K is the information word length, and N is the codeword length. This inequality is quite straightforward to prove. The number of information words can be found as 2^K, the number of error locations is N chooses j, and the number of possible errors $\{X,Y,Z\}$ at every location is 3^j. The total number of errors for all codewords that can be corrected cannot be larger than the code space that is 2^N dimensional. Similarly to classical codes, the QECCs satisfying the Hamming inequality with an equality sign can be called the perfect codes. For $t = 1$, the quantum Hamming bound becomes: $(1 + 3N)2^K \leq 2^N$. For an $[N,1]$ quantum code with $t = 1$, the quantum Hamming bound is simply $2(1 + 3N) \leq 2^N$. The smallest possible N to satisfy the Hamming bound is $N = 5$, which represents the perfect code. This cyclic code has already been introduced in Section 8.2. It is interesting to notice that the quantum Hamming bound is identical to the classical Hamming bound for q-ary LBCs (see Section 6.3.6) by setting $q = 4$, corresponding to the cardinality of a set of errors $\{I,X,Y,Z\}$. This is consistent with the connection we established between QECCs and classical codes over GF(4).

The asymptotic quantum Hamming bound can be obtained by letting $N \to \infty$. For very large N, the last term in summation (8.52) dominates and we can write: $\binom{N}{t}3^j2^K \leq 2^N$. By taking the $\log_2(\bullet)$ from both sides of the inequality we obtain: $\log_2\binom{N}{t} + t\log_2 3 + K \leq N$. By using the following approximation: $\log_2\binom{N}{t} \simeq NH_2(t/N)$, where $H_2(p)$ is the binary entropy function $H_2(p) = -p\log p - (1-p)\log(1-p)$, we obtain the following asymptotic quantum Hamming bound:

$$\frac{K}{N} \leq 1 - H_2(t/N) - \frac{t}{N}\log_2 3. \tag{8.53}$$

8.4.2 Quantum Gilbert–Varshamov Bound

Let us consider an $[N,K]$ QECC with distance D. Following the analogy with classical q-ary LBC we established in the previous subsection, we expect that the quantum Gilbert–Varshamov bound is going to be the same as the classical one for $q = 4$:

$$\sum_{j=0}^{D-1}\binom{N}{j}3^j2^K \geq 2^N. \tag{8.54}$$

Because all basis codewords and all errors E from Pauli group G_N with $\mathrm{wt}(E) < D$ must satisfy the equation:

$$\langle \bar{i}|E|\bar{j}\rangle = C_E\delta_{ij}, \tag{8.55}$$

the number of such errors is: $N_E = \sum_{j=0}^{D-1}\binom{N}{j}3^j$. The following derivation is due to Gottesman [8] (see also [14]). Consider the state $|\psi_1\rangle$ satisfying Eq. (8.55) for all possible errors E, which can be used as a basis codeword. Let us observe now the space orthogonal to both $|\psi_1\rangle$ and $E|\psi_1\rangle$. The dimensionality of this subspace is $2^N - N_E$. Consider now the state $|\psi_2\rangle$ from this subspace satisfying Eq. (8.55) and determine the subspace orthogonal to both $E|\psi_1\rangle$ and $E|\psi_2\rangle$. The dimensionality of this subspace is $2^N - 2N_E$. We can iterate this procedure i ($<2^K$) times. In the ith step the corresponding subspace will have dimensionality $2^N - iN_E > 0$, where $i < 2^K$. The inequality $2^N - iN_E > 0$ can be rewritten by substituting the expression N_E as follows:

$$\sum_{j=0}^{D-1} \binom{N}{j} 3^j i < 2^N; i < 2^K. \tag{8.56}$$

Since i is strictly smaller than 2^K for $i = 2^K$ we obtain the inequality Eq. (8.54). Similarly to the quantum Hamming bound, the quantum Gilbert–Varshamov bound is identical to the classical Gilbert–Varshamov bound for q-ary LBCs by setting $q = 4$ (corresponding to the cardinality of the set of single-qubit errors $\{I,X,Y,Z\}$).

The asymptotic quantum Gilbert–Varshamov bound can be determined by following the similar procedure as for asymptotic quantum Hamming bound to obtain:

$$\frac{K}{N} \geq 1 - H_2(d/N) - \frac{d}{N}\log_2 3. \tag{8.57}$$

By combining the asymptotic quantum Hamming and Gilbert–Varshamov bounds we obtain the following upper and lower bound for quantum code rate $R = K/N$:

$$1 - H_2(D/N) - \frac{D}{N}\log_2 3 \leq \frac{K}{N} \leq 1 - H_2(t/N) - \frac{t}{N}\log_2 3. \tag{8.58}$$

8.4.3 Quantum Singleton Bound (Knill–Laflamme Bound [17])

In Chapter 6 (see Section 6.3.6), we derived the classical Singleton bound for (n,k,d) LBCs as follows:

$$d \leq n - k + 1 \Leftrightarrow n - k \geq d - 1 = 2t + 1 - 1 = 2t. \tag{8.59}$$

For the quantum $[N,K]$ codes obtained by CSS construction, we have $2M = 2(n - k)$ stabilizer conditions applied to N-qubit states so that the corresponding Singleton bound will be $2M = 2(n - k) \geq 4t$. Because in CSS construction $N - 2M$ qubits are encoded into N qubits, the quantum Singleton bound becomes:

$$N - K \geq 4t \Leftrightarrow N - K \geq 2(D - 1) \Leftrightarrow \frac{K}{N} \leq 1 - \frac{2}{N}(D - 1). \tag{8.60}$$

Similarly to classical codes, the quantum codes satisfying the quantum Singleton bound with an equality sign can be called the maximum distance separable quantum codes. By combining the asymptotic Gilbert–Varshamov and Singleton bounds we obtain the following bounds for the quantum code rate:

$$1 - H_2(D/N) - \frac{D}{N}\log_2 3 \leq \frac{K}{N} \leq 1 - \frac{2}{N}(D-1). \tag{8.61}$$

For entanglement-assisted-like QECCs [16,22], the Singleton bound can be derived as follows. Consider $[N,K,D]$ QECC and assume that K Bell states are shared between the source and destination. We further encode the transmitter portion of the Bell states only. The corresponding codeword can be represented in the following form:

$$\overbrace{\boldsymbol{d}}^{K}\ \overbrace{\boldsymbol{a}}^{D-1}\ \overbrace{\boldsymbol{b}}^{D-1}\ \overbrace{\boldsymbol{c}}^{N-2(D-1)},\ K = 2(D-1), \tag{8.62}$$

where $\boldsymbol{d}$ is the receiver portion of the Bell states, which stays unencoded. The von Neumann entropy of density operator ρ, introduced in Chapter 2, is defined by $S(\rho) = -\text{Tr}\rho\log\rho$. We know from the quantum information theory chapter that $S(\rho)$ vanishes for the nonentangled pure state, and it can be as large as $\log N_D$ for the maximally entangled state for two N_D-state systems $|\psi\rangle = \frac{1}{\sqrt{N_D}}\sum_i |ii\rangle$, where $|ij\rangle$ is the CB of the composite system. An important property of the von Neumann entropy is the subadditivity property. Let a composite system S be a bipartite system composed of component subsystems S_1 and S_2. The subadditivity property can be stated as follows: $S(\rho(S_1S_2)) \leq S(\rho(S_1)) + S(\rho(S_2))$, with the equality sign being satisfied if component subsystems are independent of each other: $\rho(S_1S_2) = \rho(S_1) \otimes \rho(S_2)$. It is clear from Eq. (8.62) that:

$$S(\rho(\boldsymbol{da})) = S(\rho(\boldsymbol{bc})) \text{ and } S(\rho(\boldsymbol{db})) = S(\rho(\boldsymbol{ac})). \tag{8.63}$$

From the subadditivity property it follows that:

$$S(\rho(\boldsymbol{BC} = \boldsymbol{bc})) \leq S(\rho(\boldsymbol{b})) + S(\rho(\boldsymbol{c})) \text{ and } S(\rho(\boldsymbol{ac})) \leq S(\rho(\boldsymbol{a})) + S(\rho(\boldsymbol{c})). \tag{8.64}$$

Since $\boldsymbol{d}$ and $\boldsymbol{a}$ ($\boldsymbol{d}$ and $\boldsymbol{b}$) independent of each other we obtain:

$$S(\rho(\boldsymbol{da})) = S(\rho(\boldsymbol{d})) + S(\rho(\boldsymbol{a})) \text{ and } S(\rho(\boldsymbol{db})) = S(\rho(\boldsymbol{d})) + S(\rho(\boldsymbol{b})). \tag{8.65}$$

By properly combining Eqs. (8.63)–(8.65) we obtain:

$$\begin{aligned} &S(\rho(\boldsymbol{d})) \leq S(\rho(\boldsymbol{b})) + S(\rho(\boldsymbol{c})) - S(\rho(\boldsymbol{a})) \text{ and} \\ &S(\rho(\boldsymbol{d})) \leq S(\rho(\boldsymbol{a})) + S(\rho(\boldsymbol{c})) - S(\rho(\boldsymbol{b})). \end{aligned} \tag{8.66}$$

By summing two inequalities in Eq. (8.66) and dividing by 2 we obtain the Singleton bound:

$$\log_2 2^K \leq \log_2 2^{N-2(D-1)} \Leftrightarrow K \leq N - 2(D-1) \Leftrightarrow N - K \geq 2(D-1). \tag{8.67}$$

8.4.4 Quantum Weight Enumerators and Quantum MacWilliams Identity

We have shown in Chapter 6 that in classical error correction, the codeword weight distribution can be used to determine word error probability and some other properties of the code. Similar ideas can be adopted in quantum error correction. Let A_w be the number of stabilizer S elements of weight w, and B_w be the number of elements of the same weight in the centralizer of S, denoted as $C(S)$. The polynomials $A(z) = \Sigma_w A_w z^w$ and $B(z) = \Sigma_w B_w z^w$ ($A_0 = B_0 = 1$) can be used as the weight enumerators of S and $C(S)$, respectively.

Let F be a finite field GF(q), where q is a prime power, and let F^N be a vector space of dimension N over F. A linear code C_F of length N over F is a subspace of F^N. Let $C_F^{\perp}$ denote a dual code of C_F and let the elements of F be denoted by $\omega_0 = 0, \omega_1,\ldots,\omega_{q-1}$. The composition of a vector $\boldsymbol{v} \in F^N$ is defined to be $\text{comp}(\boldsymbol{v}) = \boldsymbol{s} = (s_0,s_1,\ldots,s_{q-1})$, where s_i is the number of coordinates of $\boldsymbol{v}$ equal to ω_i. Clearly, $\sum_{i=0}^{q-1} s_i = N$. The Hamming weight of vector $\boldsymbol{v}$, denoted as wt($\boldsymbol{v}$), is the number of nonzero coordinates, namely $\text{wt}(\boldsymbol{v}) = \sum_{i=1}^{q-1} s_i(\boldsymbol{v})$. Let A_i be the number of vectors in C_F having $\text{wt}(\boldsymbol{v}) = i$. Then, $\{A_i\}$ will be the weight enumerator of C_F. Similarly, let $\{B_i\}$ denote the weight numerator of dual code. The weight enumerators are related by the following MacWilliams identity [19,20]:

$$\sum_{i=0}^{n} B_i z^i = \frac{1}{|C_F|}\sum_{i=0}^{n} A_i(1+(q-1)z)^{n-i}(1-z)^i$$

$$= \frac{1}{|C_F|}(1+(q-1)z)^n \sum_{i=0}^{n} A_i\left(\frac{1-z}{1+(q-1)z}\right)^i. \tag{8.68}$$

Expression (8.68) can be rewritten in terms of polynomials introduced earlier as follows:

$$B(z) = \frac{1}{|C_F|}(1+(q-1)z)^n A\left(\frac{1-z}{1+(q-1)z}\right). \tag{8.69}$$

We have already shown that both quantum Hamming and Gilbert–Varshamov bounds can be obtained as a special case of classical nonbinary codes for $q = 4$ (corresponding to the cardinality of set $\{I,X,Y,Z\}$). Therefore the quantum MacWilliams identity for $[N,K]$ quantum code can be obtained from Eq. (8.69) by setting $q = 4$:

$$B(z) = \frac{1}{2^{N-K}}(1+3z)^n A\left(\frac{1-z}{1+3z}\right). \tag{8.70}$$

By matching the coefficients of z^i in Eq. (8.68) and by setting $q = 4$ we establish the relationship between B_i and A_i:

$$B_i = \frac{1}{2^{N-K}}\sum_{w=0}^{N}\left[\sum_{j=0}^{i}(-1)^j 3^{i-j}\binom{w}{j}\binom{N-w}{i-j}\right]A_w. \tag{8.71}$$

The weight distribution of classical Reed–Solomon (RS) (n,k,d) code with symbols from GF(q) is given by:

$$A_w = \binom{n}{w}(q-1)\sum_{j=0}^{w-d}(-1)^j\binom{w-1}{j}q^{w-d-j}. \tag{8.72}$$

If the quantum code is derived from RS code, then the weight distribution A_w can be found in closed form by setting $q = 4$ in Eq. (8.72).

Readers interested in rigorous derivation of Eqs. (8.70) and (8.71) and for some other quantum bounds are referred to [8,14]. In the next section, we introduce the operator-sum representation that is a powerful tool for studying different quantum channels. Namely, we can describe the action of a quantum channel by so-called superoperator or quantum operation, expressed in terms of channel operation elements. Several important quantum channel models will then be discussed using this framework. This study is useful in quantum information theory to determine the quantum channel capacity and for quantum codes design. It is also relevant in the study of the performance of QECCs over various quantum channels.

8.5 QUANTUM OPERATIONS (SUPEROPERATORS) AND QUANTUM CHANNEL MODELS

In this section we are concerned with the operator-sum representation of a quantum operation (also known as superoperator), and with various quantum channel models, including depolarizing channel, amplitude-damping channel, and generalized amplitude-damping channel.

8.5.1 Operator-Sum Representation

Let the composite system C be composed of quantum register Q and environment E. This kind of system can be modeled as a closed quantum system. Because the composite system is closed, its dynamic is unitary, and final state is specified by a unitary operator U as follows: $U(\rho \otimes \varepsilon_0)U^\dagger$, where ρ is a density operator of the initial state of the quantum register Q, and ε_0 is the initial density operator of the environment E. The reduced density operator of Q upon interaction with ρ_f can be obtained by tracing out the environment:

$$\rho_f = \mathrm{Tr}_E\left[U(\rho \otimes \varepsilon_0)U^\dagger\right] \equiv \xi(\rho). \tag{8.73}$$

The transformation (mapping) of initial density operator ρ to the final density operator ρ_f, denoted as $\xi : \rho \rightarrow \rho_f$, given by Eq. (8.73), is often called the superoperator or quantum operation. The final density operator can be expressed in so-called operator-sum representation as follows:

$$\rho_f = \sum_k E_k \rho E_k^\dagger, \tag{8.74}$$

where E_k are the operation elements for the superoperator. The operator-sum representation is suitable to represent the action of the channel, evaluate performance of QECCs, and study the quantum channel capacity, which is the subject of the quantum information theory chapter. To derive Eq. (8.74) we perform spectral decomposition of the environment initial state:

$$\varepsilon_0 = \sum_l \lambda_l |\phi_l\rangle\langle\phi_l|;\ |\phi_l\rangle = \text{eigenkets}(\varepsilon_0), \tag{8.75}$$

and use the definition of the trace operation from Chapter 2, namely $\text{Tr}(X) = \sum_i \langle a^{(i)}|X|a^{(i)}\rangle$. From properties of trace operation we know that it is independent on the basis, and we choose for the basis the orthonormal basis of environmental Hilbert space H_E: $\{|e_m\rangle\}$. Therefore by using the spectral decomposition and trace definition, the final density operator becomes:

$$\begin{aligned} \rho_f &= \sum_m \langle e_m|U\{\rho\otimes\varepsilon_0\}U^\dagger|e_m\rangle = \sum_{l,m}\lambda_l\langle e_m|U\{\rho\otimes|\phi_l\rangle\langle\phi_l|\}U^\dagger|e_m\rangle \\ &= \sum_{l,m} E_{lm}\rho E_{lm}^\dagger,\ E_{lm} = \sqrt{\lambda_l}\langle e_m|U|\phi_l\rangle, \end{aligned} \tag{8.76}$$

where E_{lm} are the operation elements of the superoperator. By using single subscript k instead of double subscript lm we obtain the operator-sum representation Eq. (8.74). The operator-sum representation can be used to classify quantum operations into two categories: (1) trace preserving when $\text{Tr}\xi(\rho) = \text{Tr}\rho = 1$ and (2) nontrace preserving when $\text{Tr}\xi(\rho) < 1$.

Starting from the trace-preserving condition:

$$\text{Tr}\rho = \text{Tr}\xi(\rho) = \text{Tr}\left[\sum_k E_k\rho E_k^\dagger\right] = \text{Tr}\left[\rho\sum_k E_k E_k^\dagger\right] = 1,$$

we obtain:

$$\sum_k E_k E_k^\dagger = I. \tag{8.77}$$

For nontrace-preserving quantum operation, Eq. (8.77) is not satisfied, and informally we can write $\sum_k E_k E_k^\dagger < I$.

Notice that in the foregoing derivation we assumed that the quantum register and environment are initially nonentangled and that the environment was in a pure state, which does not reduce the generality of discussion. Observe again the situation in which the quantum register Q and its environment E are initially nonentangled, with corresponding density operators being ρ and $\varepsilon_0 = |e_0\rangle\langle e_0|$, respectively. Upon interaction we perform the measurement on observable $M = \sum_m \mu_m P_m$ of E, where $\{\mu_m\}$ are the eigenvalues of M and P_m is the projection operator onto the subspace of states of E with eigenvalues μ_m. For the outcome μ_m, the final (normalized) state of the composite system is given by:

$$\rho_{\text{tot}}^k = \frac{P_k U(\rho\otimes|e_0\rangle\langle e_0|)U^\dagger P_k}{\text{Tr}(P_k U(\rho\otimes|e_0\rangle\langle e_0|)U^\dagger P_k)}. \tag{8.78}$$

The denominator can be simplified by applying the tracing per environment first:

$$\begin{aligned}\mathrm{Tr}\left(P_k U(\rho \otimes |e_0\rangle\langle e_0|)U^\dagger P_k\right) &= \mathrm{Tr}_Q \mathrm{Tr}_E\left(P_k U(\rho \otimes |e_0\rangle\langle e_0|)U^\dagger P_k\right) \\ &= \mathrm{Tr}_Q\left(E_k \rho E_k^\dagger\right), E_k = \langle \mu_k|U|e_0\rangle,\end{aligned} \tag{8.79}$$

and by substituting this result in Eq. (8.78) to obtain:

$$\rho_{\mathrm{tot}}^k = \frac{P_k U(\rho \otimes |e_0\rangle\langle e_0|)U^\dagger P_k}{\mathrm{Tr}_Q\left(E_k \rho E_k^\dagger\right)}. \tag{8.80}$$

In practice, we are interested in the (normalized) reduced density operator for the quantum register Q, which can be obtained by tracing out the environment:

$$\rho^k = \frac{\mathrm{Tr}_E\left(P_k U(\rho \otimes |e_0\rangle\langle e_0|)U^\dagger P_k\right)}{\mathrm{Tr}_Q\left(E_k \rho E_k^\dagger\right)} = \frac{E_k \rho E_k^\dagger}{\mathrm{Tr}_Q\left(E_k \rho E_k^\dagger\right)}. \tag{8.81}$$

In Chapter 2 we learned that the probability of obtaining μ_k from the measurement of M can be determined by $\Pr(\mu_k) = \mathrm{Tr}(\rho_f P_k)$. Therefore the probability of measurement being μ_k can be obtained from:

$$\begin{aligned}P(k) = \Pr(\mu_k) = \mathrm{Tr}[P_k U(\rho \otimes |e_0\rangle\langle e_0|)U^\dagger] &\underset{\mathrm{Tr}AB=\mathrm{Tr}BA}{\overset{P_k^2=P_k}{=}} \mathrm{Tr}[P_k U(\rho \otimes |e_0\rangle\langle e_0|)U^\dagger P_k] \\ &= \mathrm{Tr}_Q(E_k \rho E_k^\dagger).\end{aligned} \tag{8.82}$$

If the measurement outcome is not observed, the composite system will be in a mixed state:

$$\rho_{tot} = \sum_k P(k)\rho_{tot}^k. \tag{8.83}$$

The reduced density operator upon measurement will then be:

$$\rho_Q = \sum_k P(k)\mathrm{Tr}_E\left(\rho_{tot}^k\right) = \sum_k P(k)\rho_k = \sum_k E_k \rho E_k^\dagger. \tag{8.84}$$

Therefore the representation of ρ_Q is the same as that of the operator-sum representation. In the derivation of Eq. (8.74) we did not perform any measurement; nevertheless, we obtained the same result. This conclusion can be formulated as the *principle of implicit measurement*: Once a subsystem E of composite system C has finished interacting with the rest of composite system $Q = C - E$, the reduced density operator for Q will not be affected by any operations that we carry out solely on E. As a consequence of this principle we can interpret the quantum operation of the environment to the quantum register as performing the mapping $\rho \to \rho_k$ with probability $P(k)$.

Without loss of generality, let us assume that the quantum register Q and the environment E are initially in the product state $\rho \otimes \varepsilon_0$, and ε_0 is in the mixed state

$\varepsilon_0 = \sum_l \lambda_l \, |\phi_l\rangle \, \langle\phi_l|$ ($\mathrm{tr}(\varepsilon_0) = \sum_l \lambda_l = 1$). Furthermore, Q and E undergo the interaction described by U, and upon interaction we perform the measurement, which leaves the state of E in subspace $S \subset H_E$, where H_E is the Hilbert space of the environment. The projection operator P_E associated with S and the superoperator can be represented in terms of the orthonormal basis for $S\{|g_m\rangle\}$ as follows:

$$P_E = \sum_m |g_m\rangle\langle g_m|. \tag{8.85}$$

Let $S^\perp$ denote the dual space with orthonormal basis $\{|h_n\rangle\}$. Since the Hilbert space of environment H_E can be written as $S \oplus S^\perp = H_E$, the combined basis sets $\{|g_m\rangle\}$ and $\{|h_n\rangle\}$ can be used to create the basis for H_E. The reduced density operator for the quantum register after the measurement will be:

$$\xi(\rho) = \mathrm{Tr}_E\left(P_E U(\rho \otimes \varepsilon_0)U^\dagger P_E\right) = \sum_{l,m} E_{l,m}\rho E_{l,m}^\dagger, \; E_{lm} = \sqrt{\lambda_l}\langle g_m|U|\phi_l\rangle. \tag{8.86}$$

The probability that after the measurement the state of E is found in the subspace S is then as follows:

$$P(S) = \mathrm{Tr}_Q(\xi(\rho)). \tag{8.87}$$

because $P(S)$ is the probability $0 \le \mathrm{Tr}_Q(\xi(\rho)) \le 1$, and the quantum operation is nontrace preserving. If we do not observe the outcome of the measurement the corresponding setting is called nonselective dynamics, otherwise it is called selective dynamics. For nonselective dynamics it is clear that $P_E = I_E$ and the set $\{|g_m\rangle\}$ spans H_E (meaning that $H_E = S$). Because of this fact we expect that the completeness relationship involving the operation elements E_{lm} from Eq. (8.84) is satisfied:

$$\begin{aligned}
\sum_{lm} E_{lm}^\dagger E_{lm} &= \sum_{lm} \lambda_l \langle\phi_l|U^\dagger|g_m\rangle\langle g_m|U|\phi_l\rangle \\
&= \sum_l \lambda_l \langle\phi_l|U^\dagger U|\phi_l\rangle \overset{U^\dagger U = I_Q \otimes I_E}{=} \sum_l \lambda_l \langle\phi_l|\phi_l\rangle I_Q \\
&= I_Q \sum_l \lambda_l \overset{\sum_l \lambda_l = \mathrm{Tr}\varepsilon_0 = 1}{=} I_Q.
\end{aligned} \tag{8.88}$$

The quantum operation now is clearly trace preserving and nothing has been learned about the final state of the environment. Therefore the nonselective dynamics corresponds to the trace-preserving quantum operations.

In selective dynamics we observe the outcome of the measurement of E so that now $P_E \neq I_E$ and $S \subset H_E$. Since $P(S) + P(S^\perp) = 1$, the trace of the quantum operation is $\mathrm{Tr}(\xi(\rho)) = P(S) = 1 - P(S^\perp) < 1$ and the quantum operation is not trace preserving. Therefore the selective dynamics of quantum operations is nontrace preserving and we gain some knowledge about the final state of the environment. We have already mentioned that the Hilbert space of environment H_E can be written as $S \oplus S^\perp = H_E$, and the combined basis set $\{|g_m\rangle\}$ of S and basis set $\{|h_n\rangle\}$ of $S^\perp$ can be used as the basis for H_E. We also defined earlier the elementary operations corresponding to S as $E_{lm} = \sqrt{\lambda_l}\langle g_m|U|\phi_l\rangle$. In a similar fashion we define the element operations corresponding to $S_\perp$ as $H_{ln} = \sqrt{\lambda_l}\langle h_n|U|\phi_l\rangle$. We can show that the completeness relation is now satisfied when both element operations E_{lm} and H_{ln} are involved:

$$\begin{aligned}\sum_{lm} E_{lm}^\dagger E_{lm} + \sum_{\ln} H_{\ln}^\dagger H_{\ln} &= \sum_{lm} \lambda_l\langle\phi_l|U^\dagger|g_m\rangle\langle g_m|U|\phi_l\rangle + \sum_{ln}\lambda_l\langle\phi_l|U^\dagger|h_n\rangle\langle h_n|U|\phi_l\rangle \\ &= \sum_l \lambda_l\left\langle \phi_l\left|U^\dagger\left\{\sum_m |g_m\rangle\langle g_m| + \sum_n |h_n\rangle\langle h_n|\right\}U\right|\phi_l\right\rangle \\ &= \sum_l \lambda_l\langle\phi_l|U^\dagger U|\phi_l\rangle = I_Q. \end{aligned} \tag{8.89}$$

This derivation justifies our previous claim that for nonpreserving operations we can write informally that $\sum_{lm} E_{lm}E_{lm}^\dagger < I$.

It is clear from this discussion that the operator-sum representation is base dependent. By choosing a different basis for the environment and a different basis for S from those discussed, we will obtain a different operator-sum representation. Therefore the operator-sum representation is not a unique one. We also expect that we should be able to transform one set of elementary operations used in the operator-sum representation to another set of elementary operations by properly choosing the unitary transformation. In the remainder of this section we describe several important quantum channels by employing, just introduced, operator-sum representation.

8.5.2 Depolarizing Channel

The depolarizing channel (Fig. 8.14) with probability $1-p$ leaves the qubit as it is, while with probability p moves the initial state into $\rho_f = I/2$ that maximizes the von Neumann entropy $S(\rho) = -\text{Tr}\rho \log \rho = 1$. The properties describing the model can be summarized as follows:

1. Qubit errors are independent.
2. Single-qubit errors (X, Y, Z) are equally likely.
3. All qubits have the same single-error probability $p/4$.

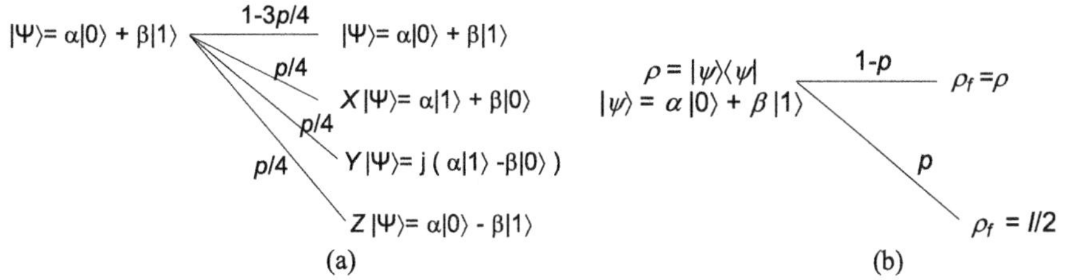

FIGURE 8.14

Depolarizing channel model: (a) Pauli operator description and (b) density operator description.

Operation elements E_i of the channel model should be selected as follows:

$$E_0 = \sqrt{1-3p/4I};\ E_1 = \sqrt{p/4X};\ E_1 = \sqrt{p/4X};\ E_1 = \sqrt{p/4Z}. \tag{8.90}$$

The action of the depolarizing channel is to perform the following mapping: $\rho \to \xi(\rho) = \sum_i E_i \rho E_i^\dagger$, where ρ is the initial density operator. Without loss of generality we will assume that the initial state was pure $|\psi\rangle = a|0\rangle + b|1\rangle$ so that:

$$\begin{aligned}
\rho = |\psi\rangle\langle\psi| &= (a|0\rangle + b|1\rangle)(\langle 0|a^* + \langle 1|b^*) \\
&= |a|^2 \begin{bmatrix} 1 & 0 \\ 0 & 0 \end{bmatrix} + ab^* \begin{bmatrix} 0 & 1 \\ 0 & 0 \end{bmatrix} + a^*b \begin{bmatrix} 0 & 0 \\ 1 & 0 \end{bmatrix} + |b|^2 \begin{bmatrix} 0 & 0 \\ 0 & 1 \end{bmatrix} \\
&= \begin{bmatrix} |a|^2 & ab^* \\ a^*b & |b|^2 \end{bmatrix}.
\end{aligned} \tag{8.91}$$

The resulting quantum operation can be represented using operator-sum representation as follows:

$$\begin{aligned}
\xi(\rho) &= \sum_i E_i \rho E_i^\dagger = \left(1 - \frac{3p}{4}\right)\rho + \frac{p}{4}(X\rho X + Y\rho Y + Z\rho Z) \\
&= \left(1 - \frac{3p}{4}\right)\begin{bmatrix} |a|^2 & ab^* \\ a^*b & |b|^2 \end{bmatrix} + \frac{p}{4}\begin{bmatrix} 0 & 1 \\ 1 & 0 \end{bmatrix}\begin{bmatrix} |a|^2 & ab^* \\ a^*b & |b|^2 \end{bmatrix}\begin{bmatrix} 0 & 1 \\ 1 & 0 \end{bmatrix} + \frac{p}{4}\begin{bmatrix} 1 & 0 \\ 0 & -1 \end{bmatrix}\begin{bmatrix} |a|^2 & ab^* \\ a^*b & |b|^2 \end{bmatrix} \\
&\quad \times \begin{bmatrix} 1 & 0 \\ 0 & -1 \end{bmatrix} \\
&\quad + \frac{p}{4}\begin{bmatrix} 0 & -j \\ j & 0 \end{bmatrix}\begin{bmatrix} |a|^2 & ab^* \\ a^*b & |b|^2 \end{bmatrix}\begin{bmatrix} 0 & -j \\ j & 0 \end{bmatrix} = \left(1 - \frac{3p}{4}\right)\begin{bmatrix} |a|^2 & ab^* \\ a^*b & |b|^2 \end{bmatrix} \\
&\quad + \frac{p}{4}\begin{bmatrix} |b|^2 & a^*b \\ ab^* & |a|^2 \end{bmatrix} + \frac{p}{4}\begin{bmatrix} |a|^2 & -ab^* \\ -a^*b & |b|^2 \end{bmatrix} + \frac{p}{4}\begin{bmatrix} |b|^2 & -a^*b \\ -ab^* & |a|^2 \end{bmatrix} \\
&= \begin{bmatrix} \left(1 - \frac{p}{2}\right)|a|^2 + \frac{p}{2}|b|^2 & (1-p)ab^* \\ (1-p)a^*b & \frac{p}{2}|a|^2 + \left(1 - \frac{p}{2}\right)|b|^2 \end{bmatrix} \\
&= \begin{bmatrix} (1-p)|a|^2 + \frac{p}{2}\left(|a|^2 + |b|^2\right) & (1-p)ab^* \\ (1-p)a^*b & \frac{p}{2}\left(|a|^2 + |b|^2\right) + (1-p)|b|^2 \end{bmatrix} \\
&\overset{|a|^2+|b|^2=1}{=} (1-p)\begin{bmatrix} |a|^2 & ab^* \\ a^*b & |b|^2 \end{bmatrix} + \frac{p}{2}I = (1-p)\rho + \frac{p}{2}I.
\end{aligned} \tag{8.92}$$

The first line in Eq. (8.92) corresponds to the model shown in Fig. 8.14A and the last line corresponds to the model shown in Fig. 8.14B. It is clear from Eqs. (8.91) and (8.92) that $\mathrm{Tr}\xi(\rho) = \mathrm{Tr}(\rho) = 1$ meaning that this superoperator is trace preserving. Notice that the depolarizing channel model in some other books/papers can be slightly different from the one described here; one such model is given in homework problem no. 15.

8.5.3 Amplitude-Damping Channel

In certain quantum channels the errors X, Y, and Z do not occur with the same probability. In the amplitude-damping channel, the operation elements are given by:

$$E_0 = \begin{pmatrix} 1 & 0 \\ 0 & \sqrt{1-\varepsilon^2} \end{pmatrix}, \quad E_1 = \begin{pmatrix} 0 & \varepsilon \\ 0 & 0 \end{pmatrix}. \tag{8.93}$$

Spontaneous emission is an example of a physical process that can be modeled using the amplitude-damping channel model. If $|\psi\rangle = a|0\rangle + b|1\rangle$ is the initial qubit state $\left(\rho = \begin{bmatrix} |a|^2 & ab^* \\ a^*b & |b|^2 \end{bmatrix}\right)$, the effect of the amplitude-damping channel is to perform the following mapping:

$$\begin{aligned}
\rho \to \xi(\rho) = E_0\rho E_0^\dagger + E_1\rho E_1^\dagger &= \begin{pmatrix} 1 & 0 \\ 0 & \sqrt{1-\varepsilon^2} \end{pmatrix} \begin{bmatrix} |a|^2 & ab^* \\ a^*b & |b|^2 \end{bmatrix} \begin{pmatrix} 1 & 0 \\ 0 & \sqrt{1-\varepsilon^2} \end{pmatrix} \\
&+ \begin{pmatrix} 0 & \varepsilon \\ 0 & 0 \end{pmatrix} \begin{bmatrix} |a|^2 & ab^* \\ a^*b & |b|^2 \end{bmatrix} \begin{pmatrix} 0 & 0 \\ \varepsilon & 0 \end{pmatrix} \\
&= \begin{pmatrix} |a|^2 & ab^*\sqrt{1-\varepsilon^2} \\ a^*b\sqrt{1-\varepsilon^2} & |b|^2(1-\varepsilon^2) \end{pmatrix} \\
&+ \begin{pmatrix} |b|^2\varepsilon^2 & 0 \\ 0 & 0 \end{pmatrix} = \begin{pmatrix} |a|^2 + \varepsilon^2|b|^2 & ab^*\sqrt{1-\varepsilon^2} \\ a^*b\sqrt{1-\varepsilon^2} & |b|^2(1-\varepsilon^2) \end{pmatrix}.
\end{aligned} \tag{8.94}$$

Probabilities $P(0)$ and $P(1)$ that E_0 and E_1 occur are given by:

$$\begin{aligned}
P(0) &= \mathrm{Tr}\left(E_0\rho E_0^\dagger\right) = \mathrm{Tr}\begin{pmatrix} |a|^2 & ab^*\sqrt{1-\varepsilon^2} \\ a^*b\sqrt{1-\varepsilon^2} & |b|^2(1-\varepsilon^2) \end{pmatrix} = 1 - \varepsilon^2|b|^2 \\
P(1) &= \mathrm{Tr}\left(E_1\rho E_1^\dagger\right) = \mathrm{Tr}\begin{pmatrix} |b|^2\varepsilon^2 & 0 \\ 0 & 0 \end{pmatrix} = \varepsilon^2|b|^2.
\end{aligned} \tag{8.95}$$

The corresponding amplitude-damping channel model is shown in Fig. 8.15.

We have shown in Chapter 2 that for a two-level system the density operator can be written as $\rho = (I + \boldsymbol{R}\cdot\boldsymbol{\sigma})/2$, where $\boldsymbol{\sigma} = [X \;\; Y \;\; Z]$ and $\boldsymbol{R} = [R_x \;\; R_y \;\; R_z]$ is the Bloch vector, whose components can be determined by

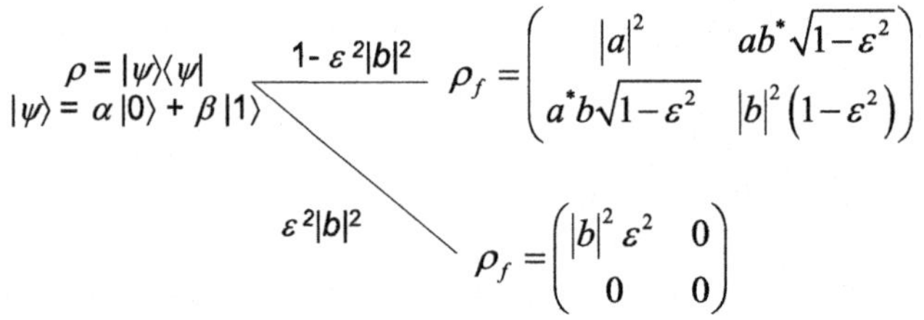

FIGURE 8.15

Amplitude-damping channel model.

$R_x = \mathrm{Tr}(\rho X), R_y = \mathrm{Tr}(\rho Y), R_z = \mathrm{Tr}(\rho Z)$. Using this result we can determine the Bloch vector $\boldsymbol{R}_\rho$ for ρ by:

$$R_{\rho,x} = \mathrm{Tr}(\rho X) = \mathrm{Tr}\left(\begin{bmatrix} |a|^2 & ab^* \\ a^*b & |b|^2 \end{bmatrix}\begin{bmatrix} 0 & 1 \\ 1 & 0 \end{bmatrix}\right) = ab^* + a^*b = 2\mathrm{Re}(a^*b)$$

$$R_{\rho,y} = 2\mathrm{Im}(a^*b)\ R_{\rho,z} = |a|^2 - |b|^2, \tag{8.96}$$

and $\boldsymbol{R}_\xi$ for $\xi(\rho)$ by:

$$R_{\xi,x} = \mathrm{Tr}(\rho_f X) = \mathrm{Tr}\left(\begin{pmatrix} |a|^2 + \varepsilon^2|b|^2 & ab^*\sqrt{1-\varepsilon^2} \\ a^*b\sqrt{1-\varepsilon^2} & |b|^2(1-\varepsilon^2) \end{pmatrix}\begin{bmatrix} 0 & 1 \\ 1 & 0 \end{bmatrix}\right)$$

$$= (ab^* + a^*b)\sqrt{1-\varepsilon^2} = R_{\rho,x}\sqrt{1-\varepsilon^2}, \quad R_{\xi,y} = R_{\rho,y}\sqrt{1-\varepsilon^2},$$

$$R_{\xi,z} = R_{\rho,z}(1-\varepsilon^2) + \varepsilon^2. \tag{8.97}$$

Example 1. If the initial state was $|1\rangle$, the Bloch vectors $\boldsymbol{R}_\rho$ and $\boldsymbol{R}_\xi$ for ρ and $\xi(\rho)$ are given, respectively, by:

$$\boldsymbol{R}_\rho = -\hat{z}\quad \boldsymbol{R}_\xi = (-1 + 2\varepsilon^2)\hat{z}.$$

Clearly, the initial amplitude of the Bloch vector is reduced. The error introduced by the channel can be interpreted as $X + jY$.

Example 2. On the other hand, if the initial state was $|0\rangle$, the $\xi(\rho)$ is given by $\xi(\rho) = |0\rangle\langle 0|$, suggesting that the initial state was not changed at all. Therefore $\Pr(|1\rangle \rightarrow |0\rangle) = \varepsilon^2$.

8.5.4 Generalized Amplitude-Damping Channel

In the generalized amplitude-damping channel, the operation elements are given by:

$$E_0 = \sqrt{p}\begin{pmatrix} 1 & 0 \\ 0 & \sqrt{1-\gamma} \end{pmatrix}, \quad E_1 = \sqrt{p}\begin{pmatrix} 0 & \sqrt{\gamma} \\ 0 & 0 \end{pmatrix},$$
$$E_2 = \sqrt{1-p}\begin{pmatrix} \sqrt{1-\gamma} & 0 \\ 0 & 1 \end{pmatrix}, \quad E_3 = \sqrt{1-p}\begin{pmatrix} 0 & 0 \\ \sqrt{\gamma} & 0 \end{pmatrix}. \tag{8.98}$$

If $|\psi\rangle = a|0\rangle + b|1\rangle$ is the initial qubit sate, the effect of the amplitude-damping channel is to perform the following mapping:

$$\begin{aligned} \rho \rightarrow \xi(\rho) &= \xi(\rho) \\ &= \frac{I}{2} + \frac{2p-1}{2}Z + \frac{1}{2}\left[R_{\rho,x}\sqrt{1-\gamma}X + R_{\rho,y}\sqrt{1-\gamma}Y + R_{\rho,z}(1-\gamma)Z\right], \end{aligned} \tag{8.99}$$

where the initial Bloch vector is given by $R_\rho = (R_{\rho,x}, R_{\rho,y}, R_{\rho,z}) = \left(2\text{Re}(a^*b), 2\text{Im}(a^*b), |a|^2 - |b|^2\right)$ The operation elements E_i $(i = 0,1,2,3)$ occur with probabilities:

$$P(0) = \text{Tr}\left(E_0\rho E_0^\dagger\right) = p\left[1 - \frac{\gamma}{2}(1 - R_{\rho,z})\right], \quad P(1) = \text{Tr}\left(E_1\rho E_1^\dagger\right) = p\frac{\gamma}{2}(1 - R_{\rho,z}),$$

$$P(3) = \text{Tr}\left(E_3\rho E_3^\dagger\right) = (1-p)\left[1 - \frac{\gamma}{2}(1 + R_{\rho,z})\right],$$

$$P(4) = \text{Tr}\left(E_4\rho E_4^\dagger\right) = (1-p)\frac{\gamma}{2}(1 + R_{\rho,z}). \tag{8.100}$$

The quantum errors considered so far are uncorrelated. If the error probability of correlated errors drops sufficiently rapidly with the number of errors so that the correlated errors with more than t errors are unlikely, we can use quantum error correction that can fix up to t errors. Errors that "knock" the qubit state out of this 2D Hilbert space are known as leakage errors.

8.6 SUMMARY

This chapter was devoted to quantum error correction concepts, ranging from an intuitive description to a rigorous mathematical framework. After the Pauli operators were introduced we described basic quantum codes, such as three-qubit flip code, three-qubit phase-flip code, Shor's nine-qubit code, stabilizer codes, and CSS codes. We then formally introduced quantum error correction, including quantum error correction mapping, quantum error representation, stabilizer group definition, quantum-check matrix representation, and the quantum syndrome equation. We further provided the necessary and sufficient conditions for quantum error correction, discussed the distance properties and error correction capability, and revisited the CSS codes. Section 8.4 was devoted to important quantum coding bounds, including the quantum Hamming bound, Gilbert–Varshamov bound, and Singleton bound. In the same section, we also discussed quantum weight enumerators and quantum MacWilliams identities. We further discussed quantum superoperators and various quantum channels, including the quantum depolarizing, amplitude-damping, and generalized amplitude-damping channels. For a deeper understanding of the presented quantum error correction material, the following section provides a set of homework problems.

8.7 Problems

1. The set of elements from the Pauli group G_N that commute with all elements from stabilizer S is called the centralizer of S, and is denoted by $C(S)$. On the other hand, the set of elements from the Pauli group G_N that fixes S under conjugation is called the normalizer of S, and is denoted as $N(S)$. In other words, $N(S) = \{e \in G_N | eSe\dagger = S\}$.
   (a) Show that the centralizer of S is the subgroup of G_N.
   (b) Show that $N(S) = C(S)$.
2. Let C_q be a quantum error correcting code with distance D. Prove that this code is degenerate code if and only if its stabilizer S has an element with weight less than D (excluding the identity element).
3. To ensure that the set of N-qubit errors from G_N, defined by Eq. (8.20), is the group, commonly we need to work with the quotient group G_N/C, where $C = \{\pm I, \pm jI\}$, so that the coset $eC = \{\pm e, \pm je\}$ is considered as a single error since all errors from this coset have the same syndrome.
   (a) Show that the orders of G_N and G_N/C are 2^{2N+2} and 2^{2N}, respectively.
   (b) Show that $\forall\, e_1, e_2 \in G_N$: $[e_1, e_2] = 0$ or $\{e_1, e_2\} = 0$. Therefore any two errors from Pauli group G_N either commute or anticommute.
   (c) Show that $\forall\, e \in G_N$ the following is valid: (1) $e^2 = \pm I$, (2) $e^+ = \pm e$, (3) $e^{-1} = e^+$.
4. Let us observe Shor's [9,1] code, whose basis codewords are given by:

$$
\begin{aligned}
|\bar{0}\rangle &= \frac{1}{2\sqrt{2}}(|000\rangle + |111\rangle)(|000\rangle + |111\rangle)(|000\rangle + |111\rangle) \\
&= \frac{1}{2\sqrt{2}}(|000000000\rangle + |000000111\rangle + |000111000\rangle + |111000000\rangle \\
&\quad + |000111111\rangle + |111111000\rangle + |111000111\rangle + |111111111\rangle)
\end{aligned}
$$

$$
\begin{aligned}
|\bar{1}\rangle &= \frac{1}{2\sqrt{2}}(|000\rangle - |111\rangle)(|000\rangle - |111\rangle)(|000\rangle - |111\rangle) \\
&= \frac{1}{2\sqrt{2}}(|000000000\rangle - |000000111\rangle - |000111000\rangle - |111000000\rangle \\
&\quad + |000111111\rangle + |111111000\rangle + |111000111\rangle - |111111111\rangle).
\end{aligned}
$$

   (a) By using the fact that this code can be obtained by concatenating two [3,1,1] codes, its generators can be written as:

$$g_1 = Z_1Z_2 \quad g_2 = Z_1Z_3 \quad g_3 = Z_4Z_5$$
$$g_4 = Z_4Z_6 \quad g_5 = Z_7Z_8 \quad g_6 = Z_7Z_9$$
$$g_7 = X_1X_2X_3X_4X_5X_6 \quad g_8 = X_1X_2X_3X_7X_8X_9$$

Show that these generators fix the basis codewords.

(b) Find the distance of this code.

(c) Determine if this code is degenerate or nondegenerate.

5. The quantum-check matrix is given by:

$$A = \begin{bmatrix} 1001 & 0110 \\ 1111 & 1001 \end{bmatrix}.$$

(a) Determine the generators of the corresponding stabilizer code.

(b) Find the parameters $[N,K]$ of the code.

(c) Determine the stabilizer of this code.

(d) Find the distance of the code.

(e) Determine if this code is degenerate or nondegenerate.

(f) Provide the syndrome decoder implementation.

6. The quantum-check matrix is given by:

$$A = \begin{bmatrix} 11111111 & 00000000 \\ 00111100 & 00010111 \\ 01011010 & 00101011 \\ 00101011 & 01001101 \\ 01001101 & 10001110 \end{bmatrix}.$$

(a) Show that the twisted product between any two rows is zero.

(b) Determine the generators of the corresponding stabilizer code.

(c) Find the parameters $[N,K,D]$ of the code.

(d) Provide the syndrome decoder implementation.

7. The quantum-check matrix $\boldsymbol{A}$ of a CSS code is given by:

$$\boldsymbol{A} = \begin{bmatrix} \boldsymbol{H} & \boldsymbol{0} \\ \boldsymbol{0} & \boldsymbol{H} \end{bmatrix}.$$

Can the classical code with the following $\boldsymbol{H}$-matrix be used in CSS code? If not, why? Can you modify the $\boldsymbol{H}$-matrix so that it can be used to construct the corresponding CSS code? Determine the parameters of such obtained code.

$$\boldsymbol{H} = \begin{bmatrix} 0 & 0 & 1 & 0 & 1 & 0 & 1 \\ 1 & 0 & 0 & 1 & 1 & 0 & 0 \\ 0 & 1 & 0 & 1 & 0 & 0 & 1 \\ 1 & 0 & 0 & 0 & 0 & 1 & 1 \\ 0 & 0 & 1 & 1 & 0 & 1 & 0 \\ 1 & 1 & 1 & 0 & 0 & 0 & 0 \\ 0 & 1 & 0 & 0 & 1 & 1 & 0 \end{bmatrix}.$$

8. Prove that the quantum code C_q is an $\boldsymbol{E}$ error correcting code if and only if $\forall\ |\bar{i}\rangle, |\bar{j}\rangle\ (\bar{i} \neq \bar{j})$ and $\forall\ E_a,\ E_b \in \boldsymbol{E}$ the following two conditions are valid:

$$\langle\bar{i}|E_a^{\dagger}E_b|\bar{i}\rangle = \langle\bar{j}|E_a^{\dagger}E_b|\bar{j}\rangle \quad \langle\bar{i}|E_a^{\dagger}E_b|\bar{j}\rangle = 0.$$

9. For the decoder shown in Fig. 8.P1 provide the corresponding generators and determine the code parameters. Determine also the quantum-check matrix. Can this code be related to CSS codes? If yes, how?

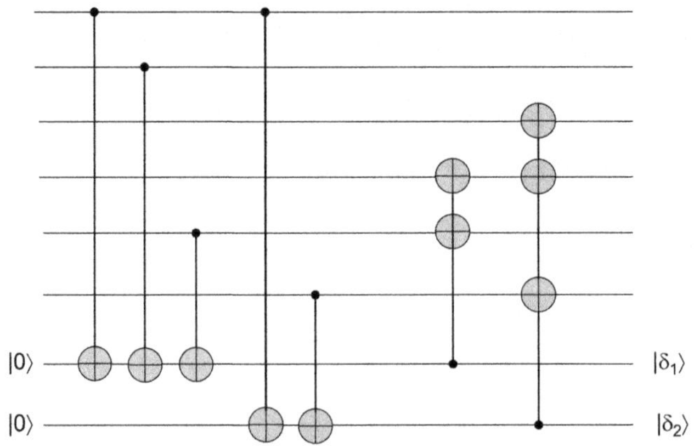

FIGURE 8.P1

Decoder circuit under study. The information qubits are denoted by $|\delta_1\rangle$ and $|\delta_2\rangle$.

10. For the encoder shown in Fig. 8.P2 provide the corresponding decoder. For such obtained decoder provide the corresponding generators and determine the code parameters. Finally, provide the quantum-check matrix.

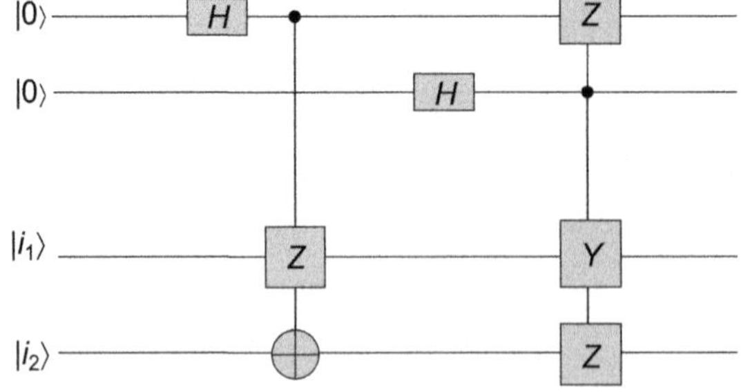

FIGURE 8.P2

Encoder circuit under study. The information qubits are denoted by $|i_1\rangle$ and $|i_2\rangle$.

11. Let the dual-containing CSS code be given by the following quantum-check matrix:

$$A = \begin{bmatrix} H & | & \mathbf{0} \\ \mathbf{0} & | & H \end{bmatrix} \qquad H = \begin{bmatrix} 1 & 0 & 0 & 1 & 1 & 1 \\ 1 & 1 & 1 & 0 & 0 & 1 \\ 0 & 1 & 1 & 1 & 1 & 0 \end{bmatrix}.$$

(a) By using MacKay's encoder implementation shown in Fig. 8.13, provide the corresponding encoder implementation.

(b) Based on the foregoing quantum-check matrix and syndrome decoder approach implementation shown in Fig. 8.12, provide the corresponding decoder implementation.

12. The classical quaternary (4,2,3) code is given by:

$$H_4 = \begin{bmatrix} 1 & \omega & 1 & 0 \\ 1 & 1 & 0 & 1 \end{bmatrix}.$$

How can we establish the connection between elements from GF(4) and Pauli operators? Can this code be used to design a quantum code? If not, explain why. Can you modify it by extension so that it can be used as a quantum code?

13. Consider a qubit, initially in the normalized state: $|\psi\rangle = a|0\rangle + b|1\rangle$, that interacts with a damping channel.

(a) Show that $\xi(\rho)$ is given by:

$$\xi(\rho) = \begin{pmatrix} |a|^2 + \varepsilon^2 |b|^2 & ab^* \sqrt{1-\varepsilon^2} \\ ab^* \sqrt{1-\varepsilon^2} & |b|^2 \left(1-\varepsilon^2\right) \end{pmatrix}.$$

(b) Derive the probabilities $P(0)$ and $P(1)$ that errors E_0 and E_1 will occur. Derive also initial and final Bloch vectors $\boldsymbol{R}_\rho$ and $\boldsymbol{R}_\xi$.

(c) Let $\rho = (1/2)(|0\rangle + |1\rangle)(\langle 0| + \langle 1|)$, corresponding to Bloch vector $\boldsymbol{R}_\rho = \hat{\boldsymbol{x}}$. Show that:
$P(0) = 1 - \varepsilon^2/2$, $P(1) = \varepsilon^2/2$ and $\boldsymbol{R}_\xi = \sqrt{1-\varepsilon^2}\hat{\boldsymbol{x}} + \varepsilon^2\hat{\boldsymbol{z}}$.

14. Consider a qubit, initially in the normalized state: $|\psi\rangle = a|0\rangle + b|1\rangle$, that interacts with a generalized damping channel, with operation elements given by:

$$E_0 = \sqrt{p}\begin{pmatrix} 1 & 0 \\ 0 & \sqrt{1-\gamma} \end{pmatrix}, \quad E_1 = \sqrt{p}\begin{pmatrix} 0 & \sqrt{\gamma} \\ 0 & 0 \end{pmatrix}$$

$$E_2 = \sqrt{1-p}\begin{pmatrix} \sqrt{1-\gamma} & 0 \\ 0 & 1 \end{pmatrix}, \quad E_3 = \sqrt{1-p}\begin{pmatrix} 0 & 0 \\ \sqrt{\gamma} & 0 \end{pmatrix}.$$

The Bloch vector $\boldsymbol{R}_\rho$ for initial $\rho = |\psi\rangle\langle\psi|$ is given by:

$$R_{\rho,x} = 2\mathrm{Re}(a^* b) \qquad R_{\rho,y} = 2\mathrm{Im}(a^* b) \qquad R_{\rho,z} = |a|^2 - |b|^2.$$

(a) Show that the quantum operation associated with this channel is trace preserving.
(b) Determine $\xi(\rho)$ and the final Bloch vector $\boldsymbol{R}_\xi$.
(c) Determine the probabilities $P(i)$ that operators E_i $(i = 0,1,2,3)$ will occur.
(d) Show that density matrix $\rho_{fp} = (1/2)(I + \boldsymbol{R}_{fp} \cdot \boldsymbol{\sigma}), \boldsymbol{R}_{fp} = (2p-1)\hat{z}$ is a fixed point for this channel: $\xi(\rho_{fp}) = \rho_{fp}$.

15. Consider a qubit, initially in the normalized state: $|\psi\rangle = a|0\rangle + b|1\rangle$, that interacts with a depolarizing channel, with operation elements given now by:

$$E_1 = \sqrt{1-p}I, \quad E_2 = \sqrt{\frac{p}{3}}X$$

$$E_3 = \sqrt{\frac{p}{3}}Z, \quad E_4 = -j\sqrt{\frac{p}{3}}Y,$$

where X, Y, and Z are the Pauli matrices. The depolarizing channel leaves the qubit intact with probability $1 - p$; with probability $p/3$ it performs E_2, E_3, or E_4.

(a) Is the quantum operation associated with this channel trace preserving or nontrace preserving?
(b) Determine the superoperator $\xi(\rho)$, where ρ is the initial state density operator.
(c) Determine the fidelity F, which represents the "quality of transmitted qubit," defined by:

$$F = \sum_i (\mathrm{Tr}\rho E_i)\left(\mathrm{Tr}\rho E_i^\dagger\right).$$

References

[1] A.R. Calderbank, P.W. Shor, Good quantum error-correcting codes exist, Phys. Rev. A 54 (1996) 1098–1105.

[2] A.M. Steane, Error correcting codes in quantum theory, Phys. Rev. Lett. 77 (1996) 793.

[3] A.M. Steane, Simple quantum error-correcting codes,, Phys. Rev. A 54 (6) (1996) 4741–4751.

[4] A.R. Calderbank, E.M. Rains, P.W. Shor, N.J.A. Sloane, Quantum error correction and orthogonal geometry, Phys. Rev. Lett. 78 (3) (1997) 405–408.

[5] E. Knill, R. Laflamme, Concatenated quantum codes, 1996. Available at: http://arxiv.org/abs/quant-ph/9608012.

[6] R. Laflamme, C. Miquel, J.P. Paz, W.H. Zurek, Perfect quantum error correcting code, Phys. Rev. Lett. 77 (1) (1996) 198–201.

[7] D. Gottesman, Class of quantum error correcting codes saturating the quantum hamming bound, Phys. Rev. A 54 (1996) 1862–1868.
[8] D. Gottesman, Stabilizer Codes and Quantum Error Correction (Ph.D. Dissertation), California Institute of Technology, Pasadena, CA, 1997.
[9] R. Cleve, D. Gottesman, Efficient computations of encoding for quantum error correction, Phys. Rev. A 56 (1997) 76–82.
[10] A.Y. Kitaev, Quantum error correction with imperfect gates, in: Quantum Communication, Computing, and Measurement, Plenum Press, New York, 1997, pp. 181–188.
[11] C.H. Bennett, D.P. DiVincenzo, J.A. Smolin, W.K. Wootters, Mixed-state entanglement and quantum error correction, Phys. Rev. A 54 (6) (1996) 3824–3851.
[12] D.J.C. MacKay, G. Mitchison, P.L. McFadden, Sparse-graph codes for quantum error correction, IEEE Trans. Inf. Theor. 50 (2004) 2315–2330.
[13] M.A. Neilsen, I.L. Chuang, Quantum Computation and Quantum Information, Cambridge University Press, Cambridge, 2000.
[14] F. Gaitan, Quantum Error Correction and Fault Tolerant Quantum Computing, CRC Press, 2008.
[15] G.D. Forney Jr., M. Grassl, S. Guha, Convolutional and tail-biting quantum error-correcting codes, IEEE Trans. Inf. Theor. 53 (2007) 865–880.
[16] M.-H. Hsieh, Entanglement-Assisted Coding Theory (Ph.D. Dissertation), University of Southern California, August 2008.
[17] E. Knill, R. Laflamme, Theory of quantum error-correcting codes, Phys. Rev. A 55 (1997) 900.
[18] J. Preskill, Ph219/CS219 Quantum Computing (Lecture Notes), Caltech, 2009. Available at: http://theory.caltech.edu/people/preskill/ph229/.
[19] F.J. MacWilliams, N.J.A. Sloane, J.-M. Goethals, The MacWilliams identities for nonlinear codes, Bell Syst. Tech. J. 51 (4) (1972) 803–819.
[20] F.J. MacWilliams, N.J.A. Sloane, The Theory of Error Correcting Codes, North-Holland Mathematical Library, New York, 1977.
[21] A.R. Calderbank, E.M. Rains, P.W. Shor, N.J.A. Sloane, Quantum error correction via codes over GF(4), IEEE Trans. Inf. Theor. 44 (1998) 1369–1387.
[22] I. Devetak, T.A. Brun, M.-H. Hsieh, Entanglement-assisted quantum error-correcting codes, in: V. Sidoravičius (Ed.), New Trends in Mathematical Physics, Selected Contributions of the XVth International Congress on Mathematical Physics, Springer, 2009, pp. 161–172.
[23] I.B. Djordjevic, Quantum LDPC codes from balanced incomplete block designs, IEEE Commun. Lett. 12 (2008) 389–391.
[24] I.B. Djordjevic, Photonic quantum dual-containing LDPC encoders and decoders, IEEE Photon. Technol. Lett. 21 (13) (2009) 842–844.
[25] I.B. Djordjevic, Photonic entanglement-assisted quantum low-density parity-check encoders and decoders, Opt. Lett. 35 (9) (2010) 1464–1466.

CHAPTER

Quantum Stabilizer Codes and Beyond

9

CHAPTER OUTLINE

In this chapter, a well-studied class of quantum codes is described, namely stabilizer codes and related quantum codes. The basic concept of stabilizer codes was already introduced in Chapter 8. Here the stabilizer codes are described in a more rigorous way. After rigorous introduction of stabilizer codes (Section 9.1), their properties and the encoded operations are introduced in Section 9.2. Furthermore, stabilizer codes are described by using the finite geometry interpretation in Section 9.3. The standard form of stabilizer codes is introduced in Section 9.4, which is further used to provide

Quantum Information Processing, Quantum Computing, and Quantum Error Correction
https://doi.org/10.1016/B978-0-12-821982-9.00011-3

efficient encoder and decoder implementations in Section 9.5. Section 9.6 is devoted to nonbinary stabilizer codes. Next, subsystem codes are described in Section 9.7, followed by the topological codes in Section 9.8. Section 9.9 is related to surface codes. Entanglement-assisted codes are described in Section 9.10. For a better understanding of this material, a set of problems is provided at the end of the chapter.

9.1 STABILIZER CODES

Quantum stabilizer codes [1–50] are based on the stabilizer concept, introduced in the Appendix. Before the stabilizer code is formally introduced, the Pauli group definition from the previous chapter is revisited. A *Pauli operator on N qubits* has the following form $cO_1O_2 \ldots O_N$, where each $O_i \in \{I, X, Y, Z\}$, and $c = j^l$ $(l = 1,2,3,4)$. The set of Pauli operators on N qubits forms the *multiplicative Pauli group* G_N. It was shown earlier that the basis for single-qubit errors is given by $\{I,X,Y,Z\}$, while the basis for N-qubit errors is obtained by forming all possible direct products as follows:

$$E = O_1 \otimes \ldots \otimes O_N; \quad O_i \in \{I, X, Y, Z\}. \tag{9.1}$$

The N-qubit error basis can be transformed into multiplicative Pauli group G_N if it is allowed to premultiply E by the j^l $(l = 0,1,2,3)$ factor. By noticing that $Y = -jXZ$, any error in the Pauli group of N-qubit errors can be represented by:

$$\begin{aligned} &E = j^{\lambda} X(\boldsymbol{a}) Z(\boldsymbol{b}); \ \boldsymbol{a} = a_1 \ldots a_N; \ \boldsymbol{b} = b_1 \ldots b_N; \ a_i, b_i = 0, 1; \ \lambda = 0, 1, 2, 3 \\ &X(\boldsymbol{a}) \equiv X_1^{a_1} \otimes \ldots \otimes X_N^{a_N}; \quad Z(\boldsymbol{b}) \equiv Z_1^{b_1} \otimes \ldots \otimes Z_N^{b_N}. \end{aligned} \tag{9.2}$$

In Eq. (9.2), the subscript i $(i = 0,1,\ldots,N)$ is used to denote the location of the qubit to which operator X_i (Z_i) is applied; a_i (b_i) takes the value 1 (0) if the ith X (Z) operator is to be included (excluded). Therefore to uniquely determine the error operator (up to the phase constant $j^{\lambda} = 1, j, -1, -j$ for $l' = 0,1,2,3$), it is sufficient to specify the binary vectors $\boldsymbol{a}$ and $\boldsymbol{b}$.

The action of Pauli group G_N on a set W is further studied. The *stabilizer* of an element $w \in W$, denoted as S_w, is the set of elements from G_N that fix w: $S_w = \{g \in G_N | gw = w\}$. It can simply be shown that S_w forms the group. Let S be the largest Abelian subgroup of G_N that fixes all elements from quantum code C_Q, which is commonly called the stabilizer group. The *stabilizer group* $S \subseteq G_N$ is constructed from a set of $N - K$ operators $g_1,\ldots,g_{N-K}$, also known as *generators* of S. Generators have the following properties: (1) they commute among each other, (2) they are unitary and Hermitian, and (3) they have the order of 2 because $g^2{}_i = I$. Any element $s \in S$ can be written as a unique product of powers of the generators:

$$s = g_1^{c_1} \ldots g_{N-K}^{c_{N-K}}, \quad c_i \in \{0, 1\}; \quad i = 1, \ldots, N - K. \tag{9.3}$$

When c_i equals 1 (0), the ith generator is included (excluded) from the product of the foregoing generators. Therefore the elements from S can be labeled by simple binary strings of length $N - K$, namely $\boldsymbol{c} = c_1 \ldots c_{N-K}$. The eigenvalues of generators g_i, denoted as $\{\text{eig}(g_i)\}$, can be found starting from property (3) as follows:

$$|\text{eig}(g_i)\rangle = I|\text{eig}(g_i)\rangle = g_i^2|\text{eig}(g_i)\rangle = \text{eig}^2(g_i)|\text{eig}(g_i)\rangle. \tag{9.4}$$

From Eq. (9.4), it is clear that $\text{eig}^2(g_i) = 1$ meaning that eigenvalues are given by $\text{eig}(g_i) = \pm 1 = (-1)^{\lambda_i};\ \lambda_i = 0, 1$.

The quantum stabilizer code C_Q, with parameters $[N,K]$, can now be defined as the unique subspace of Hilbert space H_2^N that is fixed by the elements from stabilizer S of C_Q as follows:

$$C_Q = \bigcap_{s \in S} \{|c\rangle \in H_2^N \,|\, s\,|c\rangle = |c\rangle\}. \tag{9.5}$$

Notice that S cannot include $-I$ and $\text{j}I$ because the corresponding code space C_Q will be trivial. Namely, if $-I$ is included in S, then it must fix arbitrary codeword $-I|c\rangle = |c\rangle$, which can be satisfied only when $|c\rangle = 0$. Since $(\text{j}I)^2 = -I$, the same conclusion can be made about $\text{j}I$. Because the original space for C_Q is 2^N-dimensional space, it is necessary to determine N commuting operators that specify a unique state ket $|\psi\rangle \in H_2^N$. The following 2^N simultaneous eigenkets of $\{g_1, g_2, \ldots, g_{N-K}; \overline{Z}_1, \overline{Z}_2, \ldots, \overline{Z}_K\}$ can be used as a basis for H_2^N. Namely, if the subset $\{\overline{Z}_i\}$ is chosen in such a way that each element from this subset commutes with all generators $\{g_i\}$ as well as among all elements of the subset itself, the elements from the foregoing set can be used as basis kets. This consideration will be the subject of interest of the next section. In the rest of this section, instead, some important properties of the stabilizer group and stabilizer codes are studied.

In most of the cases for quantum error correction code (QECC), it is sufficient to observe the *quotient group* G_N/C, $C = \{\pm I, \pm \text{j}I\}$ (see Appendix for the definition of quotient group). Namely, since in quantum mechanics it is not possible to distinguish unambiguously two states that differ only in a global phase constant (see Chapter 2), the coset $EC = \{\pm E, \pm \text{j}E\}$ can be considered as a single error. From Eq. (9.2), it is clear that any error from Pauli group G_N can be uniquely described by two binary vectors $\boldsymbol{a}$ and $\boldsymbol{b}$ of length N each. The number of such errors is therefore $2^{2N} \times 4$, where the factor 4 originates from global phase constant j^l $(l = 0,1,2,3)$. Therefore the order of G_N is 2^{2N+2}, while the order of quotient group G_N/C is $2^{2N+2}/4 = 2^{2N}$.

Two important *properties* of N-qubit errors are:

1. $\forall\, E_1, E_2 \in G_N$ the following is valid: $[E_1,E_2] = 0$ or $\{E_1,E_2\} = 0$. In other words, the elements from the Pauli group either commute or anticommute.
2. For $\forall\, E \in G_N$ the following is valid: (1) $E^2 = \pm I$, (2) $E^\dagger = \pm E$, and (3) $E^{-1} = E^\dagger$.

To prove property (1), the fact that operators X_i and Z_j commute if $i \neq j$ and anticommute when $i = j$ can be used. By using error representation given by Eq. (9.2), the error E_1E_2 can be represented as:

$$E_1E_2 = j^{l_1+l_2}X(\boldsymbol{a}_1)Z(\boldsymbol{b}_1)X(\boldsymbol{a}_2)Z(\boldsymbol{b}_2) = j^{l_1+l_2}(-1)^{\boldsymbol{a}_1\boldsymbol{b}_2+\boldsymbol{b}_1\boldsymbol{a}_2}X(\boldsymbol{a}_2)Z(\boldsymbol{b}_2)X(\boldsymbol{a}_1)Z(\boldsymbol{b}_1)$$

$$= (-1)^{\boldsymbol{a}_1\boldsymbol{b}_2+\boldsymbol{b}_1\boldsymbol{a}_2}E_2E_1 = \begin{cases} E_2E_1, & \boldsymbol{a}_1\boldsymbol{b}_2+\boldsymbol{b}_1\boldsymbol{a}_2 = 0 \bmod 2 \\ -E_2E_1, & \boldsymbol{a}_1\boldsymbol{b}_2+\boldsymbol{b}_1\boldsymbol{a}_2 = 1 \bmod 2 \end{cases}, \tag{9.6}$$

where the notation $\boldsymbol{ab} = \sum_i a_i b_i$ is used to denote the dot product. The first claim of property (2) can easily be proved from Eq. (9.2) as follows:

$$E^2 = j^{2l'} X(\boldsymbol{a})Z(\boldsymbol{b})X(\boldsymbol{a})Z(\boldsymbol{b}) = (-1)^{l'+\boldsymbol{ab}} X(\boldsymbol{a})X(\boldsymbol{a})Z(\boldsymbol{b})Z(\boldsymbol{b}) = (-1)^{l'+\boldsymbol{ab}} I = \pm I, \tag{9.7}$$

where the following property of Pauli operators $Z^2 = X^2 = I$ is used. The second claim of property (2) can also be proved by using again definition Eq. (9.2) by:

$$\begin{aligned} E^\dagger = \left(j^{l'} X(\boldsymbol{a})Z(\boldsymbol{b})\right)^\dagger &= (-j)^{l'} Z^\dagger(\boldsymbol{b})X^\dagger(\boldsymbol{a}) = (-j)^{l'} Z(\boldsymbol{b})X(\boldsymbol{a}) \\ &= (-j)^{l'} (-1)^{\boldsymbol{ab}} X(\boldsymbol{a})Z(\boldsymbol{b}) = \pm E, \end{aligned} \tag{9.8}$$

where the Hermitian property of Pauli matrices $X^\dagger = X$, $Z^\dagger = Z$ is used. Finally, the claim (3) of property (2) can be proved by using Eq. (9.8) as follows:

$$EE^\dagger = E(\pm E) = \pm I, \tag{9.9}$$

indicating that $E^{-1} = E^\dagger$.

In the previous chapter, the concept of syndrome of an error was introduced as follows. Let C_Q be a quantum stabilizer code with generators $g_1, g_2, \ldots, g_{N-K}$ and let $E \in G_N$ be an error. The *error syndrome* for error E is defined by the bit string $S(E) = [\lambda_1 \lambda_2 \ldots \lambda_{N-K}]^T$ with component bits being determined by:

$$\lambda_i = \begin{cases} 0, & [E, g_i] = 0 \\ 1, & \{E, g_i\} = 0 \end{cases} \quad (i = 1, \ldots, N-K). \tag{9.10}$$

Similarly as in syndrome decoding for classical error correction, it is expected that errors with a nonvanishing syndrome are detectable. In Chapter 8 (Section 8.3, Theorem 1), it was proved that if an error E anticommutes with some element from stabilizer group S, then the image of computational basis (CB) codeword is orthogonal to another CB codeword so that following can be written:

$$\langle \bar{i} | E | \bar{j} \rangle = 0. \tag{9.11}$$

It is also known from Chapter 8 (see Eq. (8.44)) that:

$$\langle \bar{i} | E | \bar{j} \rangle = C_E \delta_{\bar{i}\bar{j}}. \tag{9.12}$$

Therefore from Eqs. (9.11) and (9.12), it is clear that $C_E = 0$, which indicates that errors with a nonvanishing syndrome are correctable.

Let the set of errors $E = \{E_m\}$ from G_N for which $S(E^+{}_m E_n) \neq 0$ be observed. From Theorem 1 from Section 8.3 it is clear that:

$$\langle \bar{i} | E_m^\dagger E_n | \bar{j} \rangle = 0. \tag{9.13}$$

From Eq. (9.12), it can be concluded that $C_{mn} = 0$, indicating that this set of errors can be corrected.

The third case of interest are errors E with vanishing syndrome $S(E) = 0$. Clearly, these errors commute with all generators from S. Because the stabilizer is Abelian, $S \subseteq C(S)$, where $C(S)$ is the *centralizer* of S introduced in the Appendix, that is, the set of errors $E \in G_N$ that commute with all elements from S. There are two options for error E: (1) $E \in S$, in which case there is no need to correct the error since it fixes all codewords already, and (2) $E \in C(S) - S$, in which case the error is not detectable. This observation can formally be proved as follows. Given that $E \in C(S) - S$, $s \in S$, $|c\rangle \in C_Q$, the following is valid:

$$sE|c\rangle = Es|c\rangle = E|c\rangle. \tag{9.14}$$

Because $E \notin S$, the following is valid $E|c\rangle = |c'\rangle \neq |c\rangle$ and the error E results in a codeword different from the original codeword. Without loss of generality, it is assumed that the erroness codeword is a basis codeword $|c'\rangle = |\bar{i}\rangle$, which can be expanded in terms of basis codewords, as any other codeword, as follows:

$$E|\bar{i}\rangle = \sum_j p_j|\bar{j}\rangle \neq |\bar{i}\rangle. \tag{9.15}$$

By multiplying with dual of basis codeword $\langle\bar{k}|$ from the left, the following is obtained:

$$\langle\bar{k}|E|\bar{i}\rangle = p_k \neq 0 \, (k \neq i). \tag{9.16}$$

For error E to be detectable, Eq. (9.12) requires p_k to be 0 when k is different from i, while Eq. (9.16) claims that $p_k \neq 0$, which indicates that error E is not detectable.

Because $C(S)$ is the subgroup of G_N, from the Appendix (see Theorem T5) we know that its cosets can be used to partition G_N as follows:

$$EC(S) = \{Ec | c \in C(S)\}. \tag{9.17}$$

From Lagrange theorem the following is true:

$$|G_N| = |C(S)|[G_N : C(S)] = 2^{2N+2}. \tag{9.18}$$

Based on syndrome definition Eq. (9.10), it is clear that with an error syndrome binary vector of length $N - K$, the 2^{N-K} different errors can be identified, which means that the index of $C(S)$ in G_N is $[G_N{:}C(S)] = 2^{N-K}$. From Eq. (9.18), the following is obtained:

$$|C(S)| = |G_N|/[G_N : C(S)] = 2^{N+K+2}. \tag{9.19}$$

Finally, since $C = \{\pm I, \pm jI\}$, the quotient group $C(S)/C$ order is:

$$|C(S) / C| = 2^{N+K}. \tag{9.20}$$

Similarly as in classical error correction, where the set of errors that differ in a codeword have the same syndrome, it is expected that errors $E_1, E_2 \in G_N$, belonging to the same coset $EC(S)$, will have the same syndrome $S(E) = \boldsymbol{\lambda} = [\lambda_1 \ldots \lambda_{N-K}]^{\mathrm{T}}$. This observation can be expressed as the following theorem.

Theorem 1: The set of errors $\{E_i|E_i \in G_N\}$ have the same syndrome $\boldsymbol{\lambda} = [\lambda_1 \ldots \lambda_{N-K}]^{\mathrm{T}}$ if an only if they belong to the same coset $EC(S)$ given by Eq. (9.17).

Proof. To prove the theorem, without loss of generality, two arbitrary errors E_1 and E_2 will be observed. First, by assuming that these two errors have the same syndrome, namely $S(E_1) = S(E_2) = \boldsymbol{\lambda}$, it will be proved that they belong to the same coset. For each generator g_i, the following is valid:

$$E_1E_2g_i = (-1)^{\lambda_i}E_1g_iE_2 = (-1)^{2\lambda_i}g_iE_1E_2 = g_iE_1E_2, \tag{9.21}$$

which means that E_1E_2 commutes with g_i and therefore $E_1E_2 \in C(S)$. From claim (2) of property (ii) on page 339, we know that $E_1^{\dagger} = \pm E_1$ and since $E_1E_2 \in C(S)$, then $E_1^{\dagger}E_2 = c \in C(S)$. By multiplying $E_1^{\dagger}E_2 = c$ with E_1 from the left, the following is obtained: $E_2 = E_1c$, which means that $E_2 \in E_1C(S)$. Since E_2 belongs to both $E_2C(S)$and $E_1C(S)$ (and there is no intersection between equivalence classes), it can be concluded that $E_1C(S) = E_2C(S)$, meaning that errors E_1 and E_2 belong to the same coset. To complete the proof of Theorem 1, the opposite side of the claim will be proved. Namely, by assuming that two errors E_1 and E_2 belong to the same coset, it will be proved that they have the same syndrome. Since $E_1,\ E_2 \in E_1C(S)$ from the definition of coset (9.17), it is clear that $E_2 = E_1c$ and therefore $E_1^{\dagger}E_2 = c \in C(S)$, indicating that $E_1^{\dagger}E_2$ commutes with all generators from S. From claim (2) of property (ii) on page 339, it is clear that $E_1^{\dagger} = \pm E_1$ meaning that E_1E_2 must commute with all generators from S, which can be written as:

$$[E_1E_2,\ g_i] = 0,\quad \forall g_i \in S. \tag{9.22}$$

Let the syndromes corresponding to E_1 and E_2 be denoted by $S(E_1) = [\lambda_1 \ldots \lambda_{N-K}]^{\mathrm{T}}$ and $S(E_2) = [\lambda_1' \ldots \lambda_{N-K}']^{\mathrm{T}}$, respectively. Furthermore, the commutation relations between E_1E_2 and generator g_i are established as:

$$E_1E_2g_i = (-1)^{\lambda_i'}E_1g_iE_2 = (-1)^{\lambda_i+\lambda_i'}g_iE_1E_2. \tag{9.23}$$

Because of commutation relation (9.22), clearly, $\lambda_i + \lambda_i' = 0 \bmod 2$, which is equivalent to $\lambda_i' = \lambda_i \bmod 2$, proving therefore that errors E_1 and E_2 have the same syndrome.

From the previous discussion at the beginning of the section, it is clear that nondetectable errors belong to $C(S) - S$, and the *distance* D of a QECC can be defined as the lowest weight among weights of all elements from $C(S) - S$. The distance of QECC is related to the error correction capability of the codes, as discussed in the section on distance properties of QECCs in the previous chapter. In the same section, the concept of *degenerate* codes is introduced as codes for which the matrix $\boldsymbol{C}$ with elements from Eq. (9.12) is singular. In the same chapter, a useful theorem is proved, which can be used as a relatively simple check on degeneracy of a QECC. Namely, this theorem claims that a quantum stabilizer code with distance D is a degenerate code if and only if its stabilizer S contains at least one element (not counting the identity) of weight less than D. Therefore two errors E_1 and E_2 are degenerate if $E_1E_2 \in S$. It can be shown that for nondegenerate quantum stabilizer codes, linearly independent correctable errors have different syndromes. Let E_1 and E_2 be arbitrary, two linearly independent correctable errors with corresponding syndromes being $S(E_1)$ and $S(E_2)$, respectively. Because these two errors are correctable, their weight in $E^{\dagger}{}_1E_2$ must be smaller than the distance of the code and since the code is nondegenerate, clearly $E_1^{\dagger}E_2 \notin S,\ E_1^{\dagger}E_2 \notin C(S) - S$. Since $E_1^{\dagger}E_2$ does not belong either to S or to $C(S) - S$, it must belong to $G_N - C(S)$ and anticommute with at least one element, say g, from S. If E_1 anticommutes with g, E_2 must commute with it, and vice versa, and so $S(E_1) \neq S(E_2)$. From Eq. (9.11), it is clear that $\left\langle \bar{i}\middle|E_1^{\dagger}E_2\middle|\bar{j}\right\rangle = 0$ and therefore because of Eq. (9.12) $C_{12} = 0$. In conclusion, only degenerate codes can have $\left\langle \bar{i}\middle|E_1^{\dagger}E_2\middle|\bar{j}\right\rangle \neq 0$.

In some books and papers, the description of stabilizer quantum codes is based on the normalizer concept introduced in the Appendix instead of the centralizer concept employed in this section. Namely, the *normalizer* $N_G(S)$ of the stabilizer S is defined as the set of errors $E \in G_N$ that fix S under conjugation; in other words, $N_G(S) = \{E \in G_N \mid ESE^{\dagger} = S\}$. These two concepts are equivalent to each other since it can be proved that $N_G(S)$= C(S). To prove this claim, let us observe an element c from $C(S)$. Then, for every element $s \in S$, the following is valid: $csc^{\dagger} = scc^{\dagger} = s$, which indicates that $C(S) \subseteq N(S)$. Let us now observe an element n from $N(S)$. By using property (1), that the elements n and s either commute or anticommute, we have: $nsn^{\dagger} = \pm snn^{\dagger} = \pm s$. Since $-I$ cannot be an element from S, clearly, $nsn^{\dagger} = s$ or equivalently $ns = sn$, meaning that s and n commute and therefore $N(S) \subseteq C(S)$. Since the only way to satisfy both $C(S) \subseteq N(S)$ and $N(S) \subseteq C(S)$ requirements is by setting $N(S) = C(S)$, proving therefore that these two representations of quantum stabilizer codes are equivalent.

9.2 ENCODED OPERATORS

A $[N,K]$ QECC introduces an encoding map U from 2^K-dimensional Hilbert space, denoted by H_2^K, to 2^N-dimensional Hilbert space, denoted by H_2^N, as follows:

$$|\delta\rangle \in H_2^K \rightarrow |c\rangle = U|\delta\rangle \quad O \in G_K \rightarrow \overline{O} = UOU^{\dagger}. \tag{9.24}$$

In particular, Pauli operators X and Z are mapped to:

$$X_i \rightarrow \overline{X}_i = UX_iU^{\dagger} \quad \text{and} \quad Z_i \rightarrow \overline{Z}_i = UZ_iU^{\dagger}, \tag{9.25}$$

respectively. We have shown in the previous section that arbitrary N-qubit error can be written in terms of X- and Z-containing operators as given by Eq. (9.2). In similar fashion, the unencoded operator O can also be represented in terms of X- and Z-containing operators as follows:

$$O = j^{\lambda}X(\boldsymbol{a})Z(\boldsymbol{b});\ \boldsymbol{a} = a_1\cdots a_K;\ \boldsymbol{b} = b_1\cdots b_K;\ a_i,\ b_i \in \{0,\ 1\}. \tag{9.26}$$

The encoding given by Eq. (9.24) maps the operator O to encoded operator $\overline{O}$ as follows:

$$\overline{O} = U\left[j^{\lambda}X(\boldsymbol{a})Z(\boldsymbol{b})\right]U^{\dagger} = j^{\lambda}UX_1^{a_1}U^{\dagger}UX_2^{a_2}\ldots U^{\dagger}UX_N^{a_N}U^{\dagger}UZ_1^{b_1}U^{\dagger}UZ_2^{b_2}\ldots U^{\dagger}UZ_N^{b_N}U^{\dagger}$$

$$= j^{\lambda}\left(UX_1^{a_1}U^{\dagger}\right)\left(UX_2^{a_2}U^{\dagger}\right)\ldots U^{\dagger}\left(UX_N^{a_N}U^{\dagger}\right)\left(UZ_1^{b_1}U^{\dagger}\right)UZ_2^{b_2}\ldots U^{\dagger}\left(UZ_N^{b_N}U^{\dagger}\right) =$$

$$= j^{\lambda}\underbrace{(\overline{X}_1)^{a_1}(\overline{X}_2)^{a_2}\ldots(\overline{X}_N)^{a_N}}_{\overline{X}(\boldsymbol{a})}\underbrace{(\overline{Z}_1)^{b_1}\ldots(\overline{Z}_N)^{b_N}}_{\overline{Z}(\boldsymbol{b})} = j^{\lambda}\overline{X}(\boldsymbol{a})\overline{Z}(\boldsymbol{b}). \tag{9.27}$$

By using Eq. (9.27) we can easily prove the following commutativity properties:

$$\begin{aligned}&\left[\overline{X}_i, \overline{X}_j\right] = \left[\overline{Z}_i, \overline{Z}_j\right] = 0\\&\left[\overline{X}_i, \overline{Z}_j\right] = 0(i \neq j)\\&\{\overline{X}_i, \overline{Z}_i\} = 0.\end{aligned} \tag{9.28}$$

It can also be proved that $\overline{X}_i$, $\overline{Z}_i$ commute with the generators g_i from S, indicating that *Pauli-encoded operators belong to C(S).*

The *decoding map* $U^{\dagger}$ returns the codewords to unencoded kets, namely $U^{\dagger}|c\rangle = |\delta\rangle;\ |c\rangle \in H_2^N,\ |\delta\rangle \in H_2^K$. To demonstrate this, let us start with the definition of a stabilizer:

$$s|c\rangle = |c\rangle, \ \forall s \in S, \tag{9.29}$$

and apply $U^{\dagger}$ on both sides of Eq. (9.29) to obtain:

$$U^{\dagger}s|c\rangle = U^{\dagger}|c\rangle. \tag{9.30}$$

By inserting $UU^{\dagger} = I$ between s and $|c\rangle$ we obtain:

$$U^{\dagger}s\left(UU^{\dagger}\right)|c\rangle = U^{\dagger}|c\rangle \Leftrightarrow \left(U^{\dagger}sU\right)|\delta\rangle = |\delta\rangle. \tag{9.31}$$

Namely, since $U^{\dagger} : |c\rangle \in H_2^N \rightarrow |\delta\rangle \in H_2^K$, it is clear that stabilizer elements must be mapped to identity operators $I_K \in G_K$, meaning that $U^{\dagger}sU = I_k$. To be able to better characterize the decoding mapping we will prove that *centralizer C(S) is generated by encoded Pauli operators* $\{\overline{X}_i, \overline{Z}_i | i = 1, \ldots, K\}$ *and generators of S*, $\{g_i\!: i = 1,\ldots,N - K\}$. To prove this claim, let us consider the set W of 2^{N+K+2} operators generated by all possible powers of g_k, X_l, and Z_m as follows:

$$\begin{aligned}W &= \left\{\widehat{O}(\boldsymbol{a}, \boldsymbol{b}, \boldsymbol{c}) = j^{\lambda}(\overline{X}_1)^{a_1}\ldots(\overline{X}_K)^{a_K}(\overline{Z}_1)^{b_1}\ldots(\overline{Z}_K)^{b_K}(g_1)^{c_1}\ldots(g_{N-K})^{c_{N-K}}\right\}\\&= \{j^{\lambda}\overline{X}(\boldsymbol{a})\overline{Z}(\boldsymbol{b})g(\boldsymbol{c})\};\\ \boldsymbol{a} &= a_1\ldots a_K;\ \boldsymbol{b} = b_1\ldots b_K;\\ \boldsymbol{c} &= c_1\ldots c_{N-K};\ g(\boldsymbol{c}) = (g_1)^{c_1}\ldots(g_{N-K})^{c_{N-K}}.\end{aligned} \tag{9.32}$$

From commutation relations and from Eq. (9.27) we can conclude that encoded operators $\overline{O}$ commute with generators of stabilizer S, and therefore $\overline{O} \in C(S)$. From Eq. (9.32) we see that an operator $\widehat{O}(\boldsymbol{a}, \boldsymbol{b}, \boldsymbol{c})$ can be represented as a product of operators $\overline{O}$ and $g(\boldsymbol{c})$. Since $\overline{O} \in C(S)$ and $g(\boldsymbol{c})$ contains generators of stabilizer S, clearly, $\widehat{O}(\boldsymbol{a}, \boldsymbol{b}, \boldsymbol{c}) \in C(S)$. The cardinality of set W is $|W| = 2^{K+K+N-K+2} = 2^{N+K+2}$ and it is the same as $|C(S)|$ (see Eq. (9.20)), which leads us to the conclusion that $C(S) = W$. Let us now observe an error E from $C(S)$; the decoding mapping will perform the following action:

$$\begin{aligned}\overline{E} \to U^\dagger \overline{E} U &= U^\dagger \left[j^\lambda \overline{X}(\boldsymbol{a}) \overline{Z}(\boldsymbol{b}) \overline{g}(\boldsymbol{c}) \right] U \\ &= j^\lambda U^\dagger (\overline{X}_1)^{a_1} U U^\dagger \ldots (\overline{X}_K)^{a_K} U U^\dagger (\overline{Z}_1)^{b_1} U U^\dagger \ldots (\overline{Z}_K)^{b_K} U U^\dagger \\ &\quad (g_1)^{c_1} U U^\dagger \ldots U U^\dagger (g_{N-K})^{c_{N-K}} U^\dagger \\ &= j^\lambda (U^\dagger U X_1^{a_1} U^\dagger U) \ldots (U^\dagger U X_K^{a_K} U^\dagger U) \left(U^\dagger U Z_1^{b_1} U^\dagger U \right) \ldots \left(U^\dagger U Z_K^{b_K} U^\dagger U \right) \\ &\quad (U^\dagger g_1^{c_1} U) \ldots (U^\dagger g_{N-k}^{c_{N-K}} U).\end{aligned} \tag{9.33}$$

By using the fact that stabilizer elements are mapped to identity during demapping, that is, $U^\dagger s U = I_k$, Eq. (9.33) becomes:

$$U^\dagger \overline{E} U = j^\lambda (X_1^{a_1}) \ldots (X_K^{a_K}) \left(Z_1^{b_1} \right) \ldots \left(Z_K^{b_K} \right) = j^\lambda X(\boldsymbol{a}) Z(\boldsymbol{b}). \tag{9.34}$$

Therefore decoding mapping is essentially mapping from $C(S)$ to G_K. The image of this mapping is G_K and based on Eq. (9.31) we know that the kernel of mapping is S; in other words we can write:

$$\mathrm{Im}(U^\dagger) = G_K \quad \text{and} \quad \mathrm{Ker}(U^\dagger) = S. \tag{9.35}$$

It can also be shown that decoding mapping $U^\dagger$ is a *homomorphism* from $C(S) \to G_K$. Namely, from Eq. (9.34), it is clear that decoding mapping is onto (surjective). Let us now observe two operators from $C(S)$, denoted as $\overline{O}_1$ and $\overline{O}_2$. Decoding will lead to:

$$U^\dagger : \overline{O}_1 \to U^\dagger \overline{O}_1 U = U^\dagger U O_1 U^\dagger U = O_1, \quad U^\dagger : \overline{O}_2 \to O_2. \tag{9.36}$$

The decoding mapping of the product of $\overline{O}_1$ and $\overline{O}_2$ is clearly:

$$U^\dagger : \overline{O}_1 \overline{O}_2 \to U^\dagger \overline{O}_1 \overline{O}_2 U = U^\dagger U O_1 U^\dagger U O_2 U^\dagger U = O_1 O_2. \tag{9.37}$$

Since the image of the product of operators equals the product of the images of corresponding operators, the decoding mapping is clearly a homomorphism. Let us now apply the fundamental homomorphism theorem from the Appendix on decoding homomorphism $U^{\dagger}: C(S) \rightarrow G_K$, which claims that the quotient group $C(S)/\mathrm{Ker}(U^{\dagger}) = C(S)/S$ is isomorphic to G_K so that we can write:

$$C(S)/S \cong G_K. \tag{9.38}$$

Because two isomorphic groups have the same order and from the previous section we know that the order of the unencoded group is $|G_K| = 2^{2K+2}$ we conclude that:

$$|C(S)/S| = |G_K| = 2^{2K+2}. \tag{9.39}$$

In the previous section we have shown that an error $E \in C(S) - S$ is not detectable by the QECC scheme because these errors perform mapping from one codeword to another codeword and can therefore be used to implement encoded operations. The errors E and Es ($s \in S$) essentially perform the same encoded operation, which means that encoded operations can be identified with cosets of S in $C(S)$. This observation is consistent with the isomorphism given by Eq. (9.38). Since the same vectors $\boldsymbol{a}$ and $\boldsymbol{b}$ uniquely determine both O and $\overline{O}$ (see Eqs. (9.26) and (9.27)), and because of isomorphism (9.38), we can identify encoded operators $\overline{O}$ with cosets of quotient group $C(S)/S$. Therefore we have established bijection between the encoded operations and the cosets of $C(S)/S$. In conclusion, the Pauli-encoded operators $\{\overline{X}_i, \overline{Z}_j | i, j = 1, \ldots, K\}$ must belong to different cosets. Finally, we are going to show that the quotient group can be represented by:

$$C(S)/S = \cup \{\overline{X}_m S, \overline{Z}_n S: m, n = 1, \ldots, K\} \wedge \{j^l: l = 0, 1, 2, 3\}. \tag{9.40}$$

This representation can be established by labeling the cosets with the help of Eq. (9.27) as follows:

$$\overline{O}S = j^l \overline{X}(\boldsymbol{a}) \overline{Z}(\boldsymbol{b}) S = j^l (\overline{X}_1 S)^{a_1} \ldots (\overline{X}_k S)^{a_K} (\overline{Z}_1 S)^{b_1} \ldots (\overline{Z}_1 S)^{b_K}. \tag{9.41}$$

Indeed, the cosets have similar forms as given by Eq. (9.40).

Example. [5,1,3] code (cyclic and quantum perfect code). If the generators are given by:

$$g_1 = X_1 Z_2 Z_3 X_4, \quad g_2 = X_2 Z_3 Z_4 X_5, \quad g_3 = X_1 X_3 Z_4 Z_5, \quad g_4 = Z_1 X_2 X_4 Z_5,$$

then stabilizer elements can be generated as follows $s = g_1^{c_1} \ldots g_4^{c_4},\ c_i \in \{0, 1\}$. The encoded Pauli operators are given by: $\overline{X} = X_1 X_2 X_3 X_4 X_5$, $\overline{Z} = Z_1 Z_2 Z_3 Z_4 Z_5$. Since $g_1 \overline{X} = Y_2 Y_3 X_5 \in C(S) - S$, $\mathrm{wt}(g_1 \overline{X}) = 3$ and the distance is $D = 3$. Since the smallest weight of stabilizer elements is 4, which is larger than D, clearly the code is nondegenerate. Because the Hamming bound is satisfied with the equality sign, the code is perfect. The basis codewords are given by:

$$|\overline{0}\rangle = \sum_{s \in S} s|0000\rangle$$

$$|\overline{1}\rangle = \overline{X}|\overline{0}\rangle.$$

9.3 FINITE GEOMETRY INTERPRETATION

The quotient group G_N/C, where $C = \{\pm I, \pm jI\}$, as we discussed earlier, is the normal subgroup of G_N, and therefore it can be constructed from its cosets with coset multiplication (see the Appendix) being defined as $(E_1C)(E_2C) = (E_1E_2)C$, where E_1 and E_2 are two error operators. We have shown in the previous chapter and Section 8.2 that bijection between G_N/C and $2N$-dimensional Hilbert space F_2^{2N} can be established by Eq. (9.2). We have also shown, see Eq. (9.6), that two error operators either commute or anticommute. We can modify the commutativity relation (9.6) as follows:

$$E_1E_2 = \begin{cases} E_2E_1, & \boldsymbol{a}_1\boldsymbol{b}_2 + \boldsymbol{b}_1\boldsymbol{a}_2 = 0 \bmod 2 \\ -E_2E_1, & \boldsymbol{a}_1\boldsymbol{b}_2 + \boldsymbol{b}_1\boldsymbol{a}_2 = 1 \bmod 2 \end{cases} = \begin{cases} E_2E_1, & \boldsymbol{a}_1\boldsymbol{b}_2 + \boldsymbol{b}_1\boldsymbol{a}_2 = 0 \\ -E_2E_1, & \boldsymbol{a}_1\boldsymbol{b}_2 + \boldsymbol{b}_1\boldsymbol{a}_2 = 1 \end{cases}, \tag{9.42}$$

where the dot product is now observed per mod 2: $\boldsymbol{a}_1\boldsymbol{b}_2 = a_{11}b_{21} + \cdots a_{1N}b_{2N} \bmod 2$. The inner product of the type $\boldsymbol{a}_1\boldsymbol{b}_2 + \boldsymbol{b}_1\boldsymbol{a}_2$ in equation above is known as a *sympletic (twisted) inner product* and can be denoted by $v(E_1) \odot v(E_2)$, where $v(E_i) = [\boldsymbol{a}_i|\boldsymbol{b}_i]$ $(i = 1,2)$ is the vector representation of E_i. The addition operation in the sympletic inner product is also per mod 2, meaning that it performs mapping of two vectors from F_2^{2N} into $F_2 = \{0,1\}$. Based on Eq. (9.42) and sympletic inner product notation, the commutativity relation can be written as:

$$v(E_1) \odot v(E_2) = \begin{cases} 0, & [E_1, E_2] = 0 \\ 1, & \{E_1, E_2\} = 0 \end{cases}. \tag{9.43}$$

Important *properties* of the sympletic inner product are:

1. Self-orthogonality: $\boldsymbol{v} \odot \boldsymbol{v} = 0$,
2. Symmetry: $\boldsymbol{u} \odot \boldsymbol{v} = \boldsymbol{v} \odot \boldsymbol{u}$, and
3. Bilinearity property:

$$(\boldsymbol{u} + \boldsymbol{v}) \odot \boldsymbol{w} = \boldsymbol{u} \odot \boldsymbol{w} + \boldsymbol{v} \odot \boldsymbol{w} \quad \boldsymbol{u} \odot (\boldsymbol{v} + \boldsymbol{w}) = \boldsymbol{u} \odot \boldsymbol{v} + \boldsymbol{u} \odot \boldsymbol{w}, \quad \forall \boldsymbol{u}, \boldsymbol{v}, \boldsymbol{w} \in F_2^{2N}.$$

We defined earlier the weight of an N-dimensional operator as the number of nonidentity operators. Given the fact that an N-dimensional operator E can be represented in vector (finite geometry) representation by $v(E) = [a_1 \ldots a_N|b_1b_2 \ldots b_N]$, the *weight* of a vector $v(E)$, denoted as wt($v(E)$), sometimes called *sympletic weight* and denoted by swt($v(E)$), is equal to the number of components i for which $a_i = 1$ and/or $b_i = 1$; in other words, $\text{wt}([\boldsymbol{a}|\boldsymbol{b}]) = |\{i|(a_i,b_i) \neq (0,0)\}|$. The weight in finite geometry representation is equal to the weight in operator form representation, defined as the number of nonidentity tensor product components in $E = cO_1 \ldots O_N$ (where $O_i \in \{I,X,Y,Z\}$ and $c = j^\lambda$, $\lambda = 0,1,2,3$), that is, $\text{wt}(E) = |\{O_i \neq I\}|$. In similar fashion, the (sympletic) *distance* between two vectors $v(E_1) = (\boldsymbol{a}_1|\boldsymbol{b}_1)$ and $v(E_2) = (\boldsymbol{a}_2|\boldsymbol{b}_2)$ can be defined as the weight of their sum, that is, $D(v(E_1),v(E_2)) = \text{wt}(v(E_1) + v(E_2))$. For example, let us observe the following operators: $E_1 = X_1Z_2Z_3X_4$ and $E_2 = X_2Z_3Z_4$. The corresponding finite geometry representations are $v(E_1) = (1001|0110)$ and $v(E_2) = (0100|0011)$, the weights are 4 and 3, respectively, and the distance between them is:

$$D(v(E_1),\ v(E_2)) = \mathrm{wt}(v(E_1) + v(E_2)) = \mathrm{wt}(1101|0101) = 3.$$

The *dual* QECC can be introduced in similar fashion as was done in classical error correction from Chapter 6. The QECC, as described in Section 9.1, can be described in terms of stabilizer S. Any element s from S can be represented as a vector $v(s) \in F_2^{2N}$. The stabilizer generators:

$$g_i = j^l X(\boldsymbol{a}_i) Z(\boldsymbol{b}_i);\ i = 1, \ldots N - K \tag{9.44}$$

can be represented, using finite geometry formalism, by:

$$\boldsymbol{g}_i \doteq v(g_i) = (\boldsymbol{a}_i|\boldsymbol{b}_i). \tag{9.45}$$

The *dual* (null) space $S_\perp$ can then be defined as the set of vectors orthogonal to stabilizer generator vectors; in other words, $\{v_d \in F_2^{2n} : v_d \odot v(s) = 0\}$. Because of Eq. (9.43), $S_\perp$ is in fact the image of $C(S)$.

We turn our attention now to *syndrome* representation using finite geometry representation. An error operator E from G_N can be represented using finite geometry interpretation as follows:

$$E \in G_N \rightarrow \boldsymbol{v}_E = v(E) = (\boldsymbol{a}_E|\boldsymbol{b}_E) \in F_2^{2N}. \tag{9.46}$$

The syndrome components can be determined, based on Eqs. (9.10) and (9.43), by:

$$S(E) = [\lambda_1 \lambda_2 \ldots \lambda_{N-K}]^{\mathrm{T}};\ \lambda_i = \boldsymbol{v}_E \odot \boldsymbol{g}_i,\ i = 1, \ldots, N - K. \tag{9.47}$$

Notice that the ith component of syndrome λ_i can be calculated by matrix multiplication, instead of sympletic product, as follows:

$$\begin{aligned} &\lambda_i = \boldsymbol{v}_E \odot \boldsymbol{g}_i = \boldsymbol{b}_E \boldsymbol{a}_i + \boldsymbol{a}_E \boldsymbol{b}_i = (\boldsymbol{b}_E|\boldsymbol{a}_E)\begin{pmatrix}\boldsymbol{a}_i \\ \hline \boldsymbol{b}_i\end{pmatrix} = \boldsymbol{v}_E \boldsymbol{J} \boldsymbol{g}_i^T = \boldsymbol{v}_E \boldsymbol{J}(\boldsymbol{A})_i^{\mathrm{T}}; \\ &(\boldsymbol{A})_i = \boldsymbol{g}_i = (\boldsymbol{a}_i|\boldsymbol{b}_i);\ i = 1, \ldots, N - K, \end{aligned} \tag{9.48}$$

where $\boldsymbol{J}$ is the matrix used to permute the location of X- and Z-containing operators in $\boldsymbol{v}_E$, namely:

$$\boldsymbol{v}_E \boldsymbol{J} = (\boldsymbol{a}_E|\boldsymbol{b}_E)\begin{pmatrix}\boldsymbol{0}_N & \boldsymbol{I}_N \\ \boldsymbol{I}_N & \boldsymbol{0}_N\end{pmatrix} = (\boldsymbol{b}_E|\boldsymbol{a}_E),\ \boldsymbol{J} = \begin{pmatrix}\boldsymbol{0}_N & \boldsymbol{I}_N \\ \boldsymbol{I}_N & \boldsymbol{0}_N\end{pmatrix},$$

with $\boldsymbol{0}_N$ being the $N \times N$ all-zero matrix and $\boldsymbol{I}_N$ being the $N \times N$ identity matrix. Eq. (9.48) allows us to represent the sympletic inner product as the matrix multiplication. By defining the (quantum)-check matrix as follows:

$$A = \begin{pmatrix} (A)_1 \\ (A)_2 \\ \vdots \\ (A)_{N-K} \end{pmatrix}, \tag{9.49}$$

where each row is the finite geometry representation of generator g_i, and we can represent Eqs. (9.47) and (9.48) as follows:

$$S(E) = [\lambda_1 \lambda_2 \ldots \lambda_{N-K}]^{\mathrm{T}} = \boldsymbol{v}_E \boldsymbol{J} \boldsymbol{A}^T. \tag{9.50}$$

The syndrome Eq. (9.50) is very similar to the classical error correction syndrome equation (see Chapter 6).

The *encoded Pauli operators* can be represented using finite geometry formalism as follows:

$$v(\overline{X}_i) = (\boldsymbol{a}(\overline{X}_i)|\boldsymbol{b}(\overline{X}_i)) \quad v(\overline{Z}_i) = (\boldsymbol{a}(\overline{Z}_i)|\boldsymbol{b}(\overline{Z}_i)). \tag{9.51}$$

The commutativity properties (9.28) can be now written as:

$$\begin{aligned} &v(\overline{X}_m) \odot \boldsymbol{g}_n = 0; \quad n = 1, \ldots, N-K \\ &v(\overline{X}_m) \odot v(\overline{X}_n) = 0; \quad n = 1, \ldots, K \\ &v(\overline{X}_m) \odot v(\overline{Z}_n) = 0; \quad n \neq m \\ &v(\overline{X}_m) \odot v(\overline{Z}_m) = 1. \end{aligned} \tag{9.52}$$

From Eq. (9.52), it is clear that to completely characterize the encoded Pauli operators in finite geometry representation we have to solve the system of $N-K$ equations (given by Eq. (9.52)). Since there are $2N$ unknowns, we have $N-K$ degrees of freedom, which can be used to simplify encoder and decoder implementations as described in the following sections. Since there are $N-K$ degrees of freedom, there are 2^{N-K} ways in which the Pauli-encoded operators $\overline{X}_i$ can be chosen, which is the same as the number of cosets in $C(S)$.

Example: [4,2,2] code. The generators and encoded Pauli operators are given by:

$$\begin{aligned} &g_1 = X_1 Z_2 Z_3 X_4 \quad && g_2 = Y_1 X_2 X_3 Y_4 \\ &\overline{X}_1 = X_1 Y_3 Y_4 && \overline{Z}_1 = Y_1 Z_2 Y_3 \ . \\ &\overline{X}_2 = X_1 X_3 Z_4 && \overline{Z}_2 = X_2 Z_3 Z_4 \end{aligned}$$

Corresponding finite geometry representations are given by:

$$\begin{aligned} &v(g_1) = (1001|0110) \quad && v(g_2) = (1111|1001) \\ &v(\overline{X}_1) = (1011|0011) && v(\overline{X}_2) = (1010|0001). \\ &v(\overline{Z}_1) = (1010|1110) && v(\overline{Z}_2) = (0100|0011) \end{aligned}$$

Finally, the quantum-check matrix is given by:

$$A = \begin{pmatrix} 1001|0110 \\ 1111|1001 \end{pmatrix}.$$

Before concluding this section, we show how to represent an arbitrary $s \in S$ by using finite geometry formalism. Based on generator representation (9.44) we can express an arbitrary element s from S by:

$$\begin{aligned} s &= (g_1)^{c_1}...(g_{N-K})^{c_{N-K}} = \prod_{i=1}^{N-K} g_i^{c_i} = \prod_{i=1}^{N-K} \left(j^{\lambda} X(\boldsymbol{a}_i) Z(\boldsymbol{b}_i)\right)^{c_i} \\ &= j^{\lambda'} \prod_{i=1}^{N-K} \prod_k (X_k)^{c_i a_k} (Z_k)^{c_i b_k} = j^{\lambda'} X\left(\sum_{i=1}^{N-K} c_i \boldsymbol{a}_i\right) Z\left(\sum_{i=1}^{N-K} c_i \boldsymbol{b}_i\right), \end{aligned} \tag{9.53}$$

The corresponding finite geometry representation, based on Eq. (9.45), is as follows:

$$v(s) = \boldsymbol{v}_s = \sum_{i=1}^{N-K} c_i \boldsymbol{g}_i. \tag{9.54}$$

Therefore the set of vectors $\{\boldsymbol{g}_i: i = 1,...,N-K\}$ span a linear subspace of F_2^{2N}, which represents the image of the stabilizer S. Eq. (9.54) is very similar to classical linear block encoding from Chapter 6.

9.4 STANDARD FORM OF STABILIZER CODES

The quantum-check matrix of QECC C_q, denoted by A_q, can be written, based on Eq. (9.49), as follows:

$$\boldsymbol{A} = \begin{pmatrix} \boldsymbol{g}_1 \\ \vdots \\ \boldsymbol{g}_{N-K} \end{pmatrix} = \left(\begin{array}{ccc|ccc} a_{1,1} & \dots & a_{1,N} & b_{1,1} & \dots & b_{1,N} \\ & \vdots & & & \vdots & \\ a_{N-K,1} & \dots & a_{N-K,N} & b_{N-K,1} & \dots & b_{N-K,N} \end{array}\right) = (\boldsymbol{A}'|\boldsymbol{B}'), \tag{9.55}$$

where the ith row corresponds to finite geometry representation of the generator g_i (see Eq. (9.45)). Based on linear algebra we can make the following two observations:

- *Observation 1*: swapping the mth and nth qubits in the quantum register corresponds to the mth and nth column swap within both submatrices $\boldsymbol{A}$ and $\boldsymbol{B}$.
- *Observation 2*: adding the nth row of $\boldsymbol{A}_q$ to the mth row ($m \neq n$) maps $\boldsymbol{g}_m \to \boldsymbol{g}_m + \boldsymbol{g}_n$ and generator $g_m \to g_m g_n$, so that the codewords and stabilizer are invariant to this change.

Therefore we can perform the *Gauss–Jordan elimination* to put the quantum-check matrix into the following form:

$$\boldsymbol{A}_q = \left(\begin{array}{cc|cc} \overbrace{\boldsymbol{I}}^{r} & \overbrace{\boldsymbol{A}}^{N-r} & \overbrace{\boldsymbol{B}}^{r} & \overbrace{\boldsymbol{C}}^{N-r} \\ \boldsymbol{0} & \boldsymbol{0} & \boldsymbol{D} & \boldsymbol{E} \end{array}\right) \begin{matrix} \}r \\ \}N-K-r \end{matrix}; \quad r = \text{rank}(\boldsymbol{A}_q). \tag{9.56}$$

In the second step we further perform Gauss–Jordan elimination on submatrix $\boldsymbol{E}$ to obtain:

$$\boldsymbol{A}_q = \left(\begin{array}{ccc|ccc} \overbrace{\boldsymbol{I}}^{r} & \overbrace{\boldsymbol{A}_1}^{N-K-r-s} & \overbrace{\boldsymbol{A}_2}^{K+s} & \overbrace{\boldsymbol{B}}^{r} & \overbrace{\boldsymbol{C}_1}^{N-K-r} & \overbrace{\boldsymbol{C}_2}^{K+s} \\ \boldsymbol{0} & \boldsymbol{0} & \boldsymbol{0} & \boldsymbol{D}_1 & \boldsymbol{I} & \boldsymbol{E}_2 \\ \boldsymbol{0} & \boldsymbol{0} & \boldsymbol{0} & \boldsymbol{D}_2 & \boldsymbol{0} & \boldsymbol{0} \end{array} \right) \begin{array}{l} \}r \\ \}N-K-r-s. \\ \}s \end{array} \tag{9.57}$$

Since the last s generators do not commute with the first r operators we have to set $s = 0$, which leads to the standard form of quantum-check matrix:

$$\boldsymbol{A}_q = \left(\begin{array}{ccc|ccc} \overbrace{\boldsymbol{I}_r}^{r} & \overbrace{\boldsymbol{A}_1}^{N-K-r} & \overbrace{\boldsymbol{A}_2}^{K} & \overbrace{\boldsymbol{B}}^{r} & \overbrace{\boldsymbol{C}_1}^{N-K-r} & \overbrace{\boldsymbol{C}_2}^{K} \\ \boldsymbol{0} & \boldsymbol{0} & \boldsymbol{0} & \boldsymbol{D} & \boldsymbol{I}_{N-K-r} & \boldsymbol{E} \end{array} \right) \begin{array}{l} \}r \\ \}N-K-r \end{array}. \tag{9.58}$$

Example. The standard form for Steane's code can be obtained as follows:

$$\boldsymbol{A}_q = \left(\begin{array}{c|c} 0001111 & 0000000 \\ 0110011 & 0000000 \\ 1010101 & 0000000 \\ 0000000 & 0001111 \\ 0000000 & 0110011 \\ 0000000 & 1010101 \end{array} \right)$$

$$\sim \left[\begin{array}{ccccccc|ccccccc} 1 & 0 & 0 & 0 & 1 & 1 & 1 & 0 & 0 & 0 & 0 & 0 & 0 & 0 \\ 0 & 1 & 0 & 1 & 0 & 1 & 1 & 0 & 0 & 0 & 0 & 0 & 0 & 0 \\ 0 & 0 & 1 & 1 & 1 & 1 & 0 & 0 & 0 & 0 & 0 & 0 & 0 & 0 \\ 0 & 0 & 0 & 0 & 0 & 0 & 0 & 1 & 0 & 1 & 1 & 0 & 0 & 1 \\ 0 & 0 & 0 & 0 & 0 & 0 & 0 & 0 & 1 & 1 & 0 & 1 & 0 & 1 \\ 0 & 0 & 0 & 0 & 0 & 0 & 0 & 1 & 1 & 1 & 0 & 0 & 1 & 0 \end{array} \right].$$

We turn our attention now to the *standard form* representation for the *encoded Pauli operators*. Based on the previous section, Pauli-encoded operators $\overline{X}_i$ can be represented using finite geometry representation as follows:

$$v(\overline{X}_i) = \left(\begin{array}{ccc|ccc} \overbrace{\boldsymbol{u}_1(i)}^{r} & \overbrace{\boldsymbol{u}_2(i)}^{N-K-r} & \overbrace{\boldsymbol{u}_3(i)}^{K} & \overbrace{\boldsymbol{v}_1(i)}^{r} & \overbrace{\boldsymbol{v}_2(i)}^{N-K-r} & \overbrace{\boldsymbol{v}_3(i)}^{K} \end{array} \right). \tag{9.59}$$

Since encoded Pauli operators have $N-K$ degrees of freedom, as shown in the previous section, we can set arbitrary $N-K$ components of Eq. (9.59) to zero:

$$v(\overline{X}_i)=\left(\mathbf{0}\quad \overbrace{\mathbf{u}_2(i)}^{N-K-r}\quad \overbrace{\mathbf{u}_3(i)}^{K}\ \middle|\ \overbrace{\mathbf{v}_1(i)}^{r}\quad \mathbf{0}\quad \overbrace{\mathbf{v}_3(i)}^{K}\right). \tag{9.60}$$

Because $\overline{X}_i \in C(S)$, it must commute with all generators of S. Based on Eq. (9.50), for matrix multiplication $v(\overline{X}_i)\boldsymbol{A}_q^T$ we have to change the positions of X- and Z-containing operators so that the commutativity can be expressed as:

$$\left(\frac{\mathbf{0}}{\mathbf{0}}\right)=(\mathbf{v}_1(i)\mathbf{0}\mathbf{v}_3(i)|\mathbf{0}\mathbf{u}_2(i)\mathbf{u}_3(i))\begin{pmatrix} \boldsymbol{I} & \boldsymbol{A}_1 & \boldsymbol{A}_2 & \boldsymbol{B} & \boldsymbol{C}_1 & \boldsymbol{C}_2 \\ \mathbf{0} & \mathbf{0} & \mathbf{0} & \boldsymbol{D} & \boldsymbol{I} & \boldsymbol{E}\end{pmatrix}^{\mathrm{T}}. \tag{9.61}$$

After matrix multiplication in Eq. (9.61) we obtain the following two sets of linear equations:

$$\begin{aligned} \mathbf{v}_1(i)+\boldsymbol{A}_2\mathbf{v}_3(i)+\boldsymbol{C}_1\mathbf{u}_2(i)+\boldsymbol{C}_2\mathbf{u}_3(i)&=0 \\ \mathbf{u}_2(i)+\boldsymbol{E}\mathbf{u}_3(i)&=0. \end{aligned} \tag{9.62}$$

For convenience, let us write down the Pauli-encoded operators as the following matrix:

$$\begin{aligned}
&\overline{\boldsymbol{X}}=\begin{pmatrix} v(\overline{X}_1) \\ \vdots \\ v(\overline{X}_K)\end{pmatrix}=(\mathbf{0}\quad \mathbf{u}_2\quad \mathbf{u}_3\,|\,\mathbf{v}_1\quad \mathbf{0}\quad \mathbf{v}_3\,);\\
&\mathbf{u}_2=\begin{pmatrix} u_{2,1}(1) & \cdots & u_{2,N-K-r}(1) \\ \vdots & & \vdots \\ u_{2,1}(K) & \cdots & u_{2,N-K-r}(K)\end{pmatrix}\\
&\mathbf{u}_3=\begin{pmatrix} u_{3,1}(1) & \cdots & u_{3,K}(1) \\ \vdots & & \vdots \\ u_{3,1}(K) & \cdots & u_{3,K}(K)\end{pmatrix};\ \mathbf{v}_1=\begin{pmatrix} v_{1,1}(1) & \cdots & v_{1,r}(1) \\ \vdots & & \vdots \\ v_{1,1}(K) & \cdots & v_{1,r}(K)\end{pmatrix},\\
&\mathbf{v}_3=\begin{pmatrix} v_{3,1}(1) & \cdots & v_{3,K}(1) \\ \vdots & & \vdots \\ v_{3,1}(K) & \cdots & v_{3,K}(K)\end{pmatrix},
\end{aligned} \tag{9.63}$$

where each row represents one of the encoded Pauli operators. Based on Eq. (9.28) we conclude that the Pauli-encoded operators $\overline{X}_i$ must commute among each other, which in matrix representation can be written as:

$$\mathbf{0}=(\mathbf{v}_1\quad \mathbf{0}\quad \mathbf{v}_3\,|\,\mathbf{0}\quad \mathbf{u}_2\quad \mathbf{u}_3\,)(\mathbf{0}\quad \mathbf{u}_2\quad \mathbf{u}_3\,|\,\mathbf{v}_1\quad \mathbf{0}\quad \mathbf{v}_3\,)^{\mathrm{T}}=\mathbf{v}_3\mathbf{u}_3^{\mathrm{T}}+\mathbf{u}_3\mathbf{v}_3^{\mathrm{T}}. \tag{9.64}$$

By solving the system of Eqs. (9.62) and (9.64) we obtain the following solution:

$$u_3 = I,\ \ v_3 = \mathbf{0},\ \ u_2 = E,\ \ v_1 = C_1E + C_2. \tag{9.65}$$

Based on Eq. (9.63) we derive the standard form of encoded Pauli operators $\overline{X}_i$:

$$\overline{X} = \begin{pmatrix} \mathbf{0} & E^T & I \,|\, (E^T C_1^T + C_2^T) & \mathbf{0} & \mathbf{0} \end{pmatrix}. \tag{9.66}$$

The standard form for the encoded Pauli operators $\overline{Z}_i$ can be derived in similar fashion. Let us first introduce the matrix representation of encoded Pauli operators $\overline{Z}_i$, similarly to Eq. (9.63):

$$\overline{Z} = \begin{pmatrix} v(\overline{Z}_1) \\ \vdots \\ v(\overline{Z}_K) \end{pmatrix} = \begin{pmatrix} \mathbf{0} & u_2' & u_3' \,|\, v_1' & \mathbf{0} & v_3' \end{pmatrix}. \tag{9.67}$$

The encoded Pauli operators $\overline{Z}_i$ must commute with generators g_i, which in matrix form can be represented by:

$$\begin{pmatrix} \mathbf{0} \\ \hline \mathbf{0} \end{pmatrix} = \begin{pmatrix} v_1' & \mathbf{0} & v_3' | \mathbf{0} & u_2' & u_3' \end{pmatrix} \left(\begin{array}{ccc|ccc} I & A_1 & A_2 & B & C_1 & C_2 \\ \mathbf{0} & \mathbf{0} & \mathbf{0} & D & I & E \end{array}\right)^{\mathrm{T}}, \tag{9.68}$$

which leads to the following set of linear equations:

$$\begin{aligned} v_1' + A_2 v_3' + C_1 u_2' + C_2 u_3' &= \mathbf{0} \\ u_2' + E u_3' &= \mathbf{0} \end{aligned} \tag{9.69}$$

Furthermore, the encoded Pauli operators $\overline{Z}_i$ commute with $\overline{X}_i$ ($j \neq i$), but anticommute for $j = i$ (see Eq. (9.28)), which in matrix form can be represented by:

$$I = \begin{pmatrix} \mathbf{0} & u_2' & u_3' \,|\, v_1' & \mathbf{0} & v_3' \end{pmatrix} \begin{pmatrix} v_1 & \mathbf{0} & \mathbf{0} \,|\, \mathbf{0} & u_2 & I \end{pmatrix}^T = v_3', \tag{9.70}$$

and the corresponding solution is $v'_3 = I$. Finally, the $\overline{Z}_i$ commute among themselves, which in matrix form can be written by:

$$\mathbf{0} = \begin{pmatrix} \mathbf{0} & u_2' & u_3' \,|\, v_1' & \mathbf{0} & I \end{pmatrix} \begin{pmatrix} v_1' & \mathbf{0} & I \,|\, \mathbf{0} & u_2' & u_3' \end{pmatrix}^{\mathrm{T}} = u_3' + (u_3')^{\mathrm{T}}, \tag{9.71}$$

whose solution is $u'_3 = \mathbf{0}$. By substituting solutions for Eqs. (9.70) and (9.71) we obtain:

$$v_1' = A_2,\ u_2' = \mathbf{0}. \tag{9.72}$$

Based on solutions of Eqs. (9.70)–(9.72), from Eq. (9.67) we derive the standard form of the encoded Pauli operators $\overline{Z}_i$ as:

$$\overline{Z} = \begin{pmatrix} \mathbf{0} & \mathbf{0} & \mathbf{0} \,|\, A_2^T & \mathbf{0} & I \end{pmatrix}. \tag{9.73}$$

Example. *Standard form of [5,1,3] code*. The generators of this code are:

$g_1 = X_1Z_2Z_3X_4$, $g_2 = X_2Z_3Z_4X_5$, $g_3 = X_1X_3Z_4Z_5$, and $g_4 = Z_1X_2X_4Z_5$.

The standard form of the quantum-check matrix can be obtained by Gaussian elimination:

$$A_q = \left(\begin{array}{ccccc|ccccc} 1 & 0 & 0 & 1 & 0 & 0 & 1 & 1 & 0 & 0 \\ 0 & 1 & 0 & 0 & 1 & 0 & 0 & 1 & 1 & 0 \\ 1 & 0 & 1 & 0 & 0 & 0 & 0 & 0 & 1 & 1 \\ 0 & 1 & 0 & 1 & 0 & 1 & 0 & 0 & 0 & 1 \end{array}\right) \sim \left(\begin{array}{ccccc|ccccc} 1 & 0 & 0 & 0 & 1 & 1 & 1 & 0 & 1 & 1 \\ 0 & 1 & 0 & 0 & 1 & 0 & 0 & 1 & 1 & 0 \\ 0 & 0 & 1 & 0 & 1 & 1 & 1 & 0 & 0 & 0 \\ 0 & 0 & 0 & 1 & 1 & 1 & 0 & 1 & 1 & 1 \end{array}\right).$$

From Eq. (9.58) we conclude that:

$$r=4,\ \ A_1 = 0,\ \ C_1 = 0,\ \ E = 0,\ \ B = \begin{pmatrix} 1 & 1 & 0 & 1 \\ 0 & 0 & 1 & 1 \\ 1 & 1 & 0 & 0 \\ 1 & 0 & 1 & 1 \end{pmatrix},$$

$$A_2 = \begin{pmatrix} 1 \\ 1 \\ 1 \\ 1 \end{pmatrix},\ \ C_2 = \begin{pmatrix} 1 \\ 0 \\ 0 \\ 1 \end{pmatrix}.$$

From Eq. (9.66), the standard form for encoded Pauli X-operators is:

$$\overline{X} = \left(\mathbf{0}\ \ E^T\ \ I \,\middle|\, \left(E^T C_1^T + C_2^T\right)\ \ \mathbf{0}\ \ \mathbf{0}\right) = \left(0\ \ 0\ \ 0\ \ 0\ \ 1 \,\middle|\, 1\ \ 0\ \ 0\ \ 1\ \ 0\right).$$

From Eq. (9.73), the standard form for encoded Pauli Z-operators is:

$$\overline{Z} = \left(\mathbf{0}\ \ \mathbf{0}\ \ \mathbf{0} \,\middle|\, A_2^T\ \ \mathbf{0}\ \ I\right) = \left(0\ \ 0\ \ 0\ \ 0\ \ 0 \,\middle|\, 1\ \ 1\ \ 1\ \ 1\ \ 1\right).$$

By using $XZ = -jY$ we obtain from the standard form of A_q the following generators (by ignoring the global phase constant):

$$g_1 = Y_1 Z_2 Z_4 Y_5,\ \ g_2 = X_2 Z_3 Z_4 X_5,\ \ g_3 = Z_1 Z_2 X_3 X_5 \ \text{and}\ g_4 = Z_1 Z_3 Y_4 Y_5.$$

The encoded Pauli operators, based on the foregoing, can be written in operator form as:

$$\overline{X} = Z_1 Z_4 X_5,\ \ \overline{Z} = Z_1 Z_2 Z_3 Z_4 Z_5.$$

9.5 EFFICIENT ENCODING AND DECODING

This section is devoted to efficient encoder and decoder implementations, initially introduced by Gottesman and Cleve [7–9] (see also [14]).

9.5.1 Efficient Encoding

Here we study an efficient implementation of an encoder for an [*N*,*K*] quantum stabilizer code. The unencoded *K*-qubit CB states can be obtained as simultaneous eigenkets of Pauli Z_i-operators from:

$$\begin{aligned} Z_i|\delta_1...\delta_K\rangle &= (-1)^{\delta_i}|\delta_1...\delta_k\rangle;\ \ \delta_i = 0,\ 1;\ \ i = 1, ..., K \\ |\delta_1...\delta_k\rangle &= X_1^{\delta_1}...X_K^{\delta_K}|0...0\rangle_K. \end{aligned} \tag{9.74}$$

The encoding operator U maps:

- The unencoded K-qubit CB ket $|\delta_1...\delta_K\rangle$ to N-qubit basis codeword $\left|\overline{\delta_1...\delta_K}\right\rangle = U\left|\delta_1...\delta_K\right\rangle$, and
- The single-qubit Pauli operator X_i to the N-qubit encoded operator $\overline{X}_i$, that is, $X_i \rightarrow \overline{X}_i = UX_iU^\dagger$.

It follows from this and Eq. (9.74) that:

$$\begin{aligned} \left|\overline{\delta_1...\delta_k}\right\rangle &= U|\delta_1...\delta_k\rangle = U\left(X_1^{\delta_1}...X_K^{\delta_K}\right)\left(U^\dagger U\right)|0...0\rangle_K \\ &= \left[U\left(X_1^{\delta_1}...X_K^{\delta_K}\right)U^\dagger\right]\underbrace{U|0...0\rangle_K}_{|\overline{0...0}\rangle} = (\overline{X}_1)^{\delta_1}...(\overline{X}_K)^{\delta_K}|\overline{0...0}\rangle. \end{aligned} \tag{9.75}$$

By defining the basis codeword by:

$$|\overline{0...0}\rangle = \sum_{s\in S} s|0...0\rangle_N, \tag{9.76}$$

we can simplify the implementation of the encoder. Namely, it can be shown by mathematical induction that:

$$\sum_{s\in S} s = \prod_{i=1}^{N-K}(I_N + g_i). \tag{9.77}$$

For $N - K = 1$ we have only one generator g so that the LHS of Eq. (9.77) becomes:

$$\sum_{s\in S} s = \sum_{i=0}^{1} g^i = I + g. \tag{9.78}$$

Let us assume that Eq. (9.77) is correct for $N-K=M$, and we can prove that the same equation is also valid for $M+1$ as follows:

$$\sum_{s\in S} s = \sum_{c_1=0;\ \ldots;\ c_M=0;\ c_{M+1}=0}^{1} g_1^{c_1}\ldots g_M^{c_M} g_{M+1}^{c_{M+1}} = \prod_{i=1}^{M}(I_N+g_i)\sum_{c_{M+1}=0}^{1} g_{M+1}^{c_{M+1}}$$
$$= \prod_{i=1}^{M}(I_N+g_i)\cdot(I_N+g_{M+1}) = \prod_{i=1}^{M+1}(I_N+g_i). \tag{9.79}$$

By using Eq. (9.77), Eq. (9.76) can be written as:

$$|\overline{0\ldots0}\rangle = \prod_{i=1}^{N-K}(I_N+g_i)|0\ldots0\rangle_N. \tag{9.80}$$

Since $\left|\overline{\delta_1\ldots\delta_K}\right\rangle = (\overline{X}_1)^{\delta_1}\ldots(X_K)^{\delta_K}\left|\overline{0\ldots0}\right\rangle$ by using Eq. (9.80), the N-qubit basis codeword can be obtained as:

$$\left|\overline{\delta_1\ldots\delta_K}\right\rangle = \prod_{i=1}^{N-K}(I_N+g_i)\overline{X}_1^{\delta_1}\ldots\overline{X}_K^{\delta_K}|0\ldots0\rangle_N. \tag{9.81}$$

We will use Eq. (9.81) to efficiently implement the encoder with the help of ancilla qubits. Namely, the encoding operation can be represented as the following mapping:

$$|0\ldots0\delta_1\ldots\delta_K\rangle \equiv |0\ldots0\rangle_{N-K}\otimes|\delta_1\ldots\delta_K\rangle \rightarrow \left|\overline{\delta_1\ldots\delta_K}\right\rangle. \tag{9.82}$$

Based on Eq. (9.81) we have first to determine the *action of encoded Pauli operators* $\overline{X}_i$:

$$\overline{X}_1^{\delta_1}\ldots\overline{X}_K^{\delta_K}|0\ldots0\rangle_N = \breve{U}_1\ldots\breve{U}_K|0\ldots0\delta_1\ldots\delta_k\rangle, \tag{9.83}$$

where $\breve{U}_i$ are controlled operations, whose action is going to be determined as follows, based on the standard form of $\overline{X}_i$:

$$v(\overline{X}_i) = \left(\overbrace{\mathbf{0}}^{r}\quad \overbrace{\mathbf{u}_2(i)}^{N-K-r}\quad \overbrace{\mathbf{u}_3(i)}^{K}\ \middle|\ \overbrace{\mathbf{v}_1(i)}^{r}\quad \mathbf{0}\quad \mathbf{0}\right);\ \mathbf{u}_3(i) = (0\ldots1_i\ldots0). \tag{9.84}$$

By introducing the following two definitions:

$$S_x[\mathbf{u}_2(i)] \equiv X_{r+1}^{u_{2,\,1}(i)}\ldots X_{N-K}^{u_{2,\,N-K-r}(i)};\ S_z[\mathbf{v}_1(i)] \equiv Z_1^{v_{1,\,1}(i)}\ldots Z_r^{v_{1,\,r}(i)}, \tag{9.85}$$

we can express the action of encoded Pauli operator $\overline{X}_i$, based on Eqs. (9.84) and (9.85), as follows:

$$\overline{X}_i = S_x[\mathbf{u}_2(i)]S_z[\mathbf{v}_1(i)]X_{N-K+i}. \tag{9.86}$$

In special cases, for $i = K$, Eq. (9.86) becomes:

$$\overline{X}_K = S_x[\boldsymbol{u}_2(K)]S_z[\boldsymbol{v}_1(K)]X_N, \tag{9.87}$$

so that the action of $\overline{X}_K^{\delta_K}$ on $|0\ldots0\rangle_N$ is as follows:

$$\overline{X}_K^{\delta_k}|0\ldots0\rangle_N \overbrace{\underbrace{=}_{S_z[\boldsymbol{v}_1(K)]|0\ldots1\rangle_N=|0\ldots1\rangle_N}}^{X_N|0\ldots0\rangle_N=|0\ldots1\rangle_N} \begin{cases} |0\ldots0\rangle_N, \delta_K = 0 \\ S_x[\boldsymbol{u}_2(K)]|0\ldots1\rangle_N, \delta_K = 1 \end{cases} = \{S_x[\boldsymbol{u}_2(K)]\}^{\delta_K}|0\ldots\delta_K\rangle_N. \tag{9.88}$$

The overall action of $\overline{X}_1^{\delta_1}\ldots\overline{X}_K^{\delta_K}$ on $|0\ldots0\rangle_N$ can be obtained in an iterative fashion:

$$\overline{X}_1^{\delta_1}\ldots\overline{X}_K^{\delta_K}|0\ldots0\rangle_N = \prod_{i=1}^{K}\{S_x[\boldsymbol{u}_2(i)]\}^{\delta_i}|0\ldots0\delta_1\ldots\delta_K\rangle_N. \tag{9.89}$$

Clearly, the action of operators in Eq. (9.83) is:

$$\breve{U}_i = \{S_x[\boldsymbol{u}_2(i)]\}^{\delta_i}. \tag{9.90}$$

Therefore the action of $\breve{U}_i$ is controlled-$S_x[\boldsymbol{u}_2(i)]$ operation, controlled by information qubit δ_i, associated with the $(N - K + i)$th qubit. The qubits affected by $\breve{U}_i$ operation are the target qubits at positions $r + 1$ to $N - K$. The number of required two-qubit gates is $\leq K(N - K - r)$.

To determine N-qubit basis codewords, based on Eq. (9.81), we now have to apply the operator $G = \prod_{i=1}^{N-K}(I_N + g_i)$ on $\overline{X}_1^{\delta_1}\ldots\overline{X}_K^{\delta_K}|0\ldots0\rangle_N$. The action of G can simply be determined by the following factorization:

$$G = \prod_{i=1}^{N-K}(I_N + g_i) = G_1G_2;\ G_1 = \prod_{i=1}^{r}(I_N + g_i),\ G_2 = \prod_{i=r+1}^{N-K}(I_N + g_i). \tag{9.91}$$

Since the qubits at positions $r + 1$ to $N - K$ are affected by $\breve{U}_i$, G_1 and $\breve{U}_i$ operations commute (it does not matter in which order we apply them) and we can write:

$$\left|\overline{\delta_1\ldots\delta_K}\right\rangle = G_1G_2\overline{X}_1^{\delta_1}\ldots\overline{X}_K^{\delta_K}|0\ldots0\rangle_N = G_1\overline{X}_1^{\delta_1}\ldots\overline{X}_K^{\delta_K}G_2|0\ldots0\rangle_N. \tag{9.92}$$

From the standard form of quantum-check matrix (9.58), it is clear that G_2 is composed of Z-operators only and since $Z_i|0\ldots0\rangle_N = |0\ldots0\rangle_N$, the state $|0\cdots0\rangle_N$ is invariant with respect to operation G_2 and we can write:

$$\left|\overline{\delta_1\ldots\delta_K}\right\rangle = G_1\overline{X}_1^{\delta_1}\ldots\overline{X}_K^{\delta_K}|0\ldots0\rangle_N = G_1\breve{U}_T|0\ldots0\delta_1\ldots\delta_K\rangle,\ \breve{U}_T = \breve{U}_1\ldots\breve{U}_K. \tag{9.93}$$

To complete the encoder implementation we need to determine the *action of* G_1, which contains factors like $(I+g_i)$. From standard form (9.58) we know that finite geometry representation of generator g_i is given by:

$$g_i = (0\ldots01_i0\ldots0A_1(i)A_2(i)|B(i)C_1(i)C_2(i)). \tag{9.94}$$

From Eq. (9.94), it is clear that the ith position of operator X is always present, while the presence of operator Z is dependent on the content of $B_i(i)$ so that we can express g_i as follows:

$$\underline{g_i = T_i X_i Z_i^{B_i(i)}};\quad i=1,\ldots,r, \tag{9.95}$$

where T_i is the operator derived from g_i by removing all operators associated with the ith qubit.

Example. Generator g_1 for [5,1,3] code is given by $g_1 = Y_1Z_2Z_4Y_5$. By removing all operators associated with the first qubit we obtain $T_1 = Z_2Z_4Y_5$.

The action of $(I+g_i)$ on $|\psi\rangle = \overline{X}_1^{\delta_1}\ldots\overline{X}_K^{\delta_K}|0\ldots0\rangle_N$, based on Eq. (9.95), is given by:

$$(I+g_i)|\psi\rangle = \breve{U}_T|0\ldots0\delta_1\ldots\delta_K\rangle + T_iX_iZ_i^{B_i(i)}\breve{U}_T|0\ldots0\delta_1\ldots\delta_K\rangle. \tag{9.96}$$

From Eq. (9.95), it is clear that the action of g_i affects the qubits on positions 1 to r, while $\breve{U}_T$ affect qubits from $r+1$ to $N-K$ (see Eqs. (9.85) and (9.90)), which indicates that these operators commute $Z_i^{B_i(i)}\breve{U}_T = \breve{U}_TZ_i^{B_i(i)}$. Because the state $|0\rangle$ is invariant on the action of Z, the state $|0\ldots0\delta_1\ldots\delta_K\rangle$ is also invariant on the action of $Z_i^{B_i(i)}$, that is, $Z_i^{B_i(i)}|0\ldots0\delta_1\ldots\delta_K\rangle = |0\ldots0\delta_1\ldots\delta_K\rangle$. On the other hand, the action of the X-operator on state $|0\rangle$ is to perform qubit flipping to $|1\rangle$ so that we can write $X_i|0\ldots0\delta_1\ldots\delta_K\rangle = |0\ldots01_i0\ldots0\delta_1\ldots\delta_K\rangle$. The overall action of $(I+g_i)$ on $|\psi\rangle$ is therefore:

$$(I+g_i)|\psi\rangle = \breve{U}_T|0\ldots0\delta_1\ldots\delta_K\rangle + T_i\breve{U}_T|0\ldots1_i\ldots0\delta_1\ldots\delta_K\rangle. \tag{9.97}$$

Eq. (9.97) can be expressed in terms of Hadamard gate action on the ith qubit:

$$H_i|\psi\rangle = \breve{U}_TH_i|0\ldots0_i\ldots0\delta_1\ldots\delta_K\rangle \overset{H_i|\delta_i\rangle=\frac{1}{\sqrt{2}}[|0\rangle+(-1)^{\delta_i}|1\rangle],\ \delta_i=0,\,1}{=} \breve{U}_T\{|0\ldots0_i\ldots0\delta_1\ldots\delta_K\rangle$$
$$+|0\ldots1_i\ldots0\delta_1\ldots\delta_K\rangle\}. \tag{9.98}$$

Let us further apply the operator controlled-T_i, that is, $T_i^{\alpha_i}$ $(\alpha_i = 0,\,1)$, on Eq. (9.98) from the left to obtain:

$$T_i^{\alpha_i}H_i|\psi\rangle = \breve{U}_T|0\ldots0_i\ldots0\delta_1\ldots\delta_K\rangle + T_i\breve{U}_T|0\ldots1_i\ldots0\delta_1\ldots\delta_k\rangle, \tag{9.99}$$

which is the same as the action of $I + g_i$ given by Eq. (9.97). We conclude therefore that:

$$(I+g_i)|\psi\rangle = T_i^{\alpha_i} H_i|\psi\rangle, \tag{9.100}$$

meaning that *the action of $(I + g_i)$ on $|\psi>$ is to first apply Hadamard gate H_i on $|\psi>$ followed by the controlled-T_j gate* (T_i gate is a controlled qubit at position i). Since T_i acts on no more than $N - 1$ qubits, and controlled-T_i gates act on no more than $N - 1$ two-qubit gates, plus one-qubit gate Hadamard gate H_i is applied, the number of gates needed to perform operation (9.100) is $(N - 1) + 1$ gates.

Based on the definition of G_1 operation (see Eq. (9.91)) we have to iterate the action $(I + g_i)$ per $i = 1,\ldots,r$ to obtain the *basis codewords*:

$$\left|\overline{\delta_1\ldots\delta_k}\right\rangle = \underbrace{\prod_{i=1}^{N-K}(I+g_i)}_{\prod_{i=1}^{r} T_i^{\alpha_i} H_i}\ \underbrace{\overline{X}_1^{\delta_1}\ldots\overline{X}_k^{\delta_k}}_{\prod_{m=1}^{K}\breve{U}_m}|0\ldots0\rangle_N = \left(\prod_{i=1}^{r} T_i^{\alpha_i} H_i\right)\left(\prod_{m=1}^{K}\breve{U}_m\right)|0\ldots0\delta_1\ldots\delta_K\rangle. \tag{9.101}$$

From Eq. (9.101), it is clear that the number of required Hadamard gates is r, the number of two-qubit gates implementing controlled-T_i gates is $r(N - 1)$, and the number of controlled-$S_x[\boldsymbol{u}_2(i)]$ gates (operating on qubits $r + 1$ to $N - K$) is $K(N - K - r)$, so that the number of total gates is no larger than:

$$N_{\text{gates}} = r[(N-1)+1] + K(N-K-r) = (K+r)(N-K). \tag{9.102}$$

Since the number of gates is a linear function of codeword length N, this implementation can be called "efficient." Based on Eqs. (9.90) and (9.85) we can rewrite Eq. (9.101) in a form suitable for implementation:

$$\left|\overline{\delta_1\ldots\delta_k}\right\rangle = \left(\prod_{i=1}^{r} T_i^{\alpha_i} H_i\right)\left(\prod_{m=1}^{K}\{S_x[\boldsymbol{u}_2(m)]\}^{\delta_m}\right)\left|0\ldots0\delta_1\ldots\delta_K\right\rangle, \tag{9.103}$$
$$S_x[\boldsymbol{u}_2(m)] = X_{r+1}^{u_{2,1}(m)}\ldots X_{N-K}^{u_{2,N-K-r}(m)}.$$

Example. Let us now study the implementation of the encoding circuit for [5,1,3] quantum stabilizer code; the same code was observed in the previous section. The generators derived from standard form representation were found to be:

$$g_1 = Y_1Z_2Z_4Y_5,\ g_2 = X_2Z_3Z_4X_5,\ g_3 = Z_1Z_2X_3X_5 \text{ and } g_4 = Z_1Z_3Y_4Y_5.$$

The encoded Pauli X-operator was found as:

$$\overline{X} = Z_1Z_4X_5,$$

and corresponding finite geometry representation is:

$$v(\overline{X}) = (0\ \ 0\ \ 0\ \ 0\ \ 1|1\ \ 0\ \ 0\ \ 1\ \ 0).$$

Since $r=4$, $N=5$, and $K=1$, from Eq. (9.84) it is clear that $N-K-r=0$ and so $\boldsymbol{u}_2=0$, meaning that $S_x[u_2]=I$. From Eq. (9.103) we see that we need to determine the T_i operators, which can be simply determined from generators g_i by removing the ith qubit term:

$$T_1=Z_2Z_4Y_5,\ T_2=Z_3Z_4X_5,\ T_3=Z_1Z_2X_5 \text{ and } T_4=Z_1Z_3Y_5.$$

Based on Eq. (9.103), the basis codewords can be obtained as:

$$|\bar{\delta}\rangle=\left[\prod_{i=1}^{4}(T_i)^{\delta}H_i\right]|0000\delta\rangle,$$

and the corresponding encoder implementation is shown in Fig. 9.1.

Example. The generators for [8,3,3] quantum stabilizer code are given by:

$$g_1=X_1X_2X_3X_4X_5X_6X_7X_8,\ g_2=Z_1Z_2Z_3Z_4Z_5Z_6Z_7Z_8,\ g_3=X_2X_4Y_5Z_6Y_7Z_8,$$
$$g_4=X_2Z_3Y_4X_6Z_7Y_8 \text{ and } g_5=Y_2X_3Z_4X_5Z_6Y_8.$$

The corresponding quantum-check matrix, based on the foregoing generators, is given by:

$$A=\left(\begin{array}{cccccccc|cccccccc}
1&1&1&1&1&1&1&1&0&0&0&0&0&0&0&0\\
0&0&0&0&0&0&0&0&1&1&1&1&1&1&1&1\\
0&1&0&1&1&0&1&0&0&0&0&0&1&1&1&1\\
0&1&0&1&0&1&0&1&0&0&1&1&0&0&1&1\\
0&1&1&0&1&0&0&1&0&1&0&1&0&1&0&1
\end{array}\right).$$

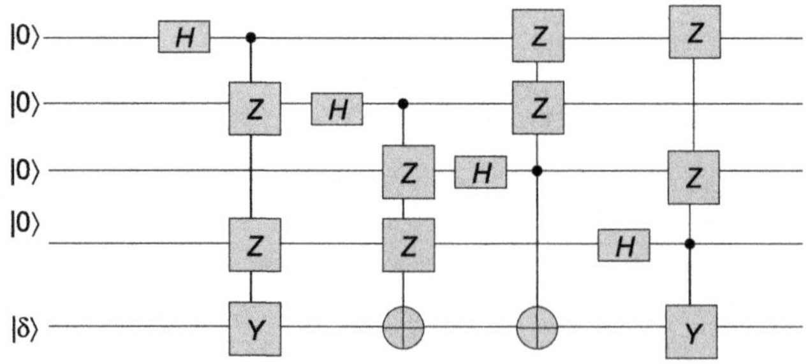

FIGURE 9.1

Efficient implementation of encoding circuit for [5,1,3] quantum stabilizer code. Since the action of Z in T_i^{δ} ($i=1,2$) is trivial, it can be omitted.

By Gaussian elimination we can put the foregoing quantum-check matrix in standard form (9.58) as follows:

$$A=\left(\begin{array}{cccc|c|ccc|cccc|c|ccc} 1&0&0&0&1&1&1&0&0&1&0&0&1&1&0&1\\ 0&1&0&0&1&1&0&1&0&0&1&0&1&0&1&1\\ 0&0&1&0&1&0&1&1&0&1&0&1&1&0&1&0\\ 0&0&0&1&0&1&1&1&0&0&1&1&1&1&0&0\\ 0&0&0&0&0&0&0&0&1&1&1&1&1&1&1&1 \end{array}\right).$$

From code parameters and Eq. (9.58), it is clear that $r=4$, $K=3$, $N-K-r=1$, and:

$$A_1=\begin{pmatrix}1\\1\\1\\0\end{pmatrix},\quad A_2=\begin{pmatrix}1&1&0\\1&0&1\\0&1&1\\1&1&1\end{pmatrix}$$

$$B=\begin{pmatrix}0&1&0&0\\0&0&1&0\\0&1&0&1\\0&0&1&1\end{pmatrix},\quad C_1=\begin{pmatrix}1\\1\\1\\1\end{pmatrix},\quad C_2=\begin{pmatrix}1&0&1\\0&1&1\\0&1&0\\1&0&0\end{pmatrix},$$

$$D=(1\ \ 1\ \ 1\ \ 1),\quad E=(1\ \ 1\ \ 1).$$

Based on standard form (9.66) of encoded Pauli X-operators we obtain:

$$\overline{X}=\left(\mathbf{0}\quad E^T\quad I\,\middle|\,\left(E^TC_1^T+C_2^T\right)\quad \mathbf{0}\quad \mathbf{0}\right)$$

$$=\left(\begin{array}{cccc|c|ccc|cccc|c|ccc} 0&0&0&0&1&1&0&0&0&1&1&0&0&0&0&0\\ 0&0&0&0&1&0&1&0&1&0&0&1&0&0&0&0\\ 0&0&0&0&1&0&0&1&0&0&1&1&0&0&0&0 \end{array}\right).$$

Based on standard form (9.73) of encoded Pauli Z-operators we obtain:

$$\overline{Z}=\left(\mathbf{0}\quad \mathbf{0}\quad \mathbf{0}\,\middle|\,A_2^T\quad \mathbf{0}\quad I\right)$$

$$=\left(\begin{array}{cccccccc|cccccccc} 0&0&0&0&0&0&0&0&1&1&0&1&0&1&0&0\\ 0&0&0&0&0&0&0&0&1&0&1&1&0&0&1&0\\ 0&0&0&0&0&0&0&0&0&1&1&1&0&0&0&1 \end{array}\right).$$

From the encoded Pauli X-matrix, Eqs. (9.84) and (9.103), it is clear that:

$$u_2 = \begin{pmatrix} 1 \\ 1 \\ 1 \end{pmatrix}, \quad S_x[u_2(1)] = X_5, \quad S_x[u_2(2)] = X_5, \quad S_x[u_2(3)] = X_5.$$

The generators derived from the standard form quantum-check matrix are obtained as:

$$g_1 = X_1 Z_2 Y_5 Y_6 X_7 Z_8, \quad g_2 = X_2 Z_3 Y_5 X_6 Z_7 Y_8, \quad g_3 = Z_2 X_3 Z_4 Y_5 Y_7 X_8,$$
$$g_4 = Z_3 Y_4 Z_5 Y_6 X_7 X_8, \quad \text{and} \quad g_5 = Z_1 Z_2 Z_3 Z_4 Z_5 Z_6 Z_7 Z_8.$$

The corresponding T_i operators are obtained from generators g_i by omitting the ith term:

$$T_1 = Z_2 Y_5 Y_6 X_7 Z_8, \quad T_2 = Z_3 Y_5 X_6 Z_7 Y_8, \quad T_3 = Z_2 Z_4 Y_5 Y_7 X_8, \quad T_4 = Z_3 Z_5 Y_6 X_7 X_8.$$

Based on Eq. (9.103) we obtain the efficient implementation of the encoder shown in Fig. 9.2.

To simplify further the encoder implementation we can modify the standard form (9.58) by Gaussian elimination to obtain:

$$A_q = \begin{pmatrix} \overset{r}{I_r} & \overset{N-K-r}{A_1} & \overset{K}{A_2} \Big| \overset{r}{B} & \overset{N-K-r}{0} & \overset{K}{C} \\ 0 & 0 & 0 \Big| D & I_{N-K-r} & E \end{pmatrix} \begin{matrix} \}r \\ \}N-K-r \end{matrix} \tag{9.104}$$

The corresponding encoded Pauli operators can be represented by:

$$\bar{X} = \begin{pmatrix} 0 & E^T & I_K \Big| C^T & 0 & 0 \end{pmatrix}, \quad \bar{Z} = \begin{pmatrix} 0 & 0 & 0 \Big| A_2^T & 0 & I_K \end{pmatrix}. \tag{9.105}$$

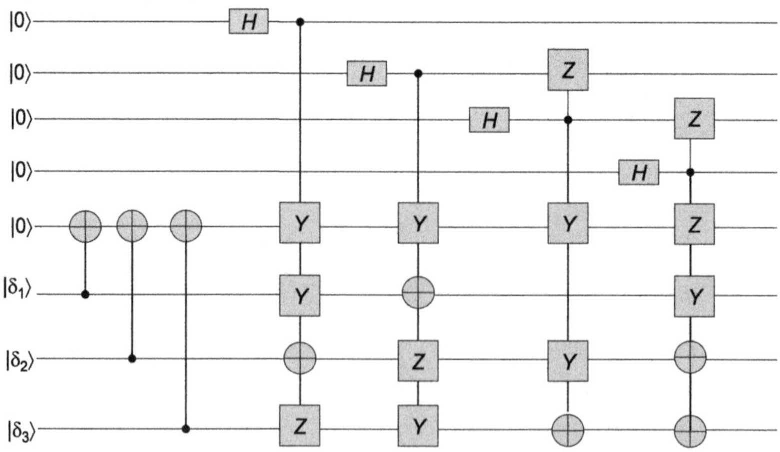

FIGURE 9.2

Efficient implementation of encoding circuit for [8,3,3] quantum stabilizer code.

From Eq. (9.104), it is clear that generators can be represented, using finite geometry representation, by:

$$g_i = (0 \quad \ldots \quad 0 \quad 1_i \quad 0\ldots a_{r+1} \quad \ldots \quad a_N| \quad b_1 \quad \ldots \quad b_r \quad 0 \quad \ldots \quad 0 \quad b_{N-K+1} \quad \ldots \quad b_N). \tag{9.106}$$

Based on Eq. (9.105), the ith encoded Pauli X-operator can be represented by:

$$v(\overline{X}_i) = \left(0 \quad \ldots \quad 0 \quad a'_{r+1} \quad \ldots \quad a'_{N-K} \quad 0 \quad \ldots 0 \quad 1_{N-K+i} \quad 0 \quad \ldots \quad 0| \quad b'_1 \quad \ldots \quad b'_r \quad 0\ldots 0\right). \tag{9.107}$$

Based on Eqs. (9.103), (9.106), and (9.107) we show in Fig. 9.3 the two-stage encoder configuration. The first stage is related to controlled-X operations in which the ith information qubit (or equivalently the $(N-K+i)$th codeword qubit) is used to control ancilla qubits at positions $r+1$ to $N-K$, based on Eq. (9.107). In the second stage, the ith ancilla qubit upon application of the Hadamard gate is used to control qubits at codeword locations $i+1$ to N, based on Eq. (9.106).

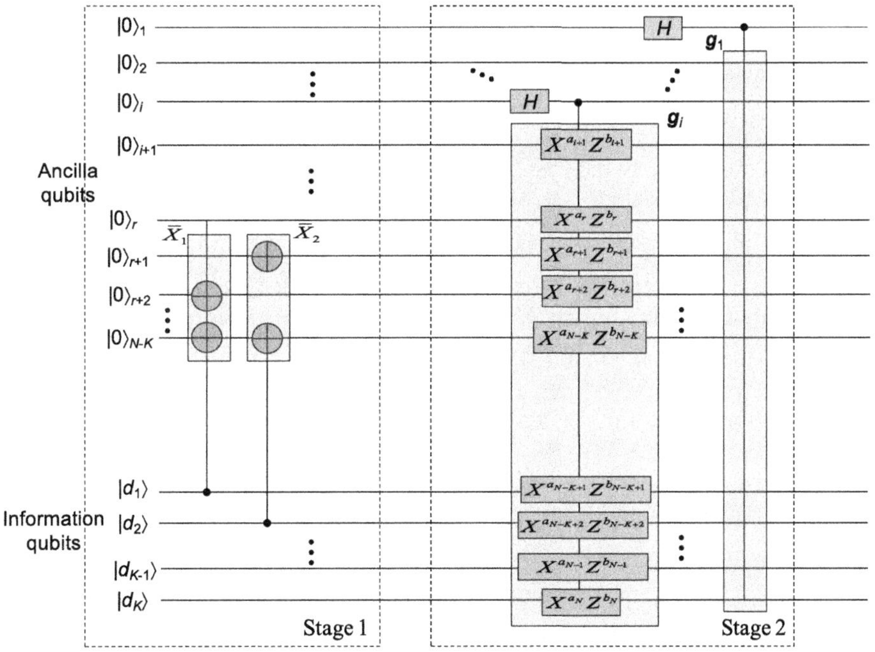

FIGURE 9.3

Efficient encoder implementation based on the standard form given by Eq. (9.104).

9.5.2 Efficient Decoding

The straightforward approach to decoding is to implement the decoding circuit as an encoding circuit with gates put in reverse order. However, this approach is not necessarily optimal in terms of gate utilization. Here we describe an efficient implementation due to Gottesman [8] (see also [14]). Gottesman's approach is to introduce K-ancilla qubits being in state $|0 \ldots 0>_K$, and perform decoding as the following mapping:

$$|\Psi_{\text{in}}\rangle = \left|\overline{\delta_1 \ldots \delta_K}\right\rangle \otimes |0\ldots0\rangle_K \rightarrow |\Psi_{\text{out}}\rangle = U_{\text{decode}}|\Psi_{\text{in}}\rangle = U_{\text{decode}}\left|\overline{\delta_1 \ldots \delta_K}\right\rangle \otimes |0\ldots0\rangle_K$$
$$= \left|\overline{0\ldots0}\right\rangle \otimes |\delta_1 \cdots \delta_K\rangle. \tag{9.108}$$

With respect to Eq. (9.108), the following two remarks are important:

- *Remark 1.* U_{decode} is a linear operator; it is sufficient to concentrate the decoding of basis codewords.
- *Remark 2.* The decoding procedure in addition to the decoded state $|\delta_1 \ldots \delta_K>$ also returns N qubits in state $\left|\overline{0\ldots0}\right\rangle$, which is needed in the encoding process. It occurs as a result of applying the operator G on ket $|0 \ldots 0>_N$. Therefore by implementing encoder and decoder in a transceiver by applying Gottesman's approach we can save Nr gates in the encoding process.

Decoding can be performed in *two stages*:

1. By applying properly the CNOT gates to the ancilla qubits we put the ancillas into the decoded state $|\delta_1 \ldots \delta_K>$, that is, we perform the following mapping: $|\Psi_{\text{in}}\rangle = \left|\overline{\delta_1 \ldots \delta_K}\right\rangle \otimes |0\ldots0\rangle_K \rightarrow \left|\overline{\delta_1 \ldots \delta_K}\right\rangle \otimes |\delta_1 \ldots \delta_K\rangle$, and
2. By applying a controlled-X_i operation to each encoded qubit i conditioned on the ith ancilla qubit we perform the following mapping: $\left|\overline{\delta_1 \ldots \delta_K}\right\rangle \otimes \left|\delta_1 \ldots \delta_K\right\rangle \rightarrow \left|\overline{0\ldots0}\right\rangle \otimes |\delta_1 \ldots \delta_K\rangle$.

The starting point for the first stage is the standard form of encoded Pauli operators $\overline{Z}_i$:

$$\overline{\mathbf{Z}} = \begin{pmatrix} \mathbf{0} & \mathbf{0} & \mathbf{0} \,|\, \mathbf{A}_2^T & \mathbf{0} & \boldsymbol{I} \end{pmatrix}, \; v(\overline{Z}_i) = (v_l),$$
$$v_l = \begin{cases} A_{2,\,l}(i); \; l = 1, \ldots, r \\ \delta_{l,\,N-K+i}; \; l = r+1, \ldots, N \end{cases}, \tag{9.109}$$

which are given in operator form by:

$$\overline{Z}_i = Z_1^{A_{2,1}(i)} \ldots Z_r^{A_{2,r}(i)} Z_{N-K+i}. \tag{9.110}$$

From the previous chapter we know that the basis codeword $\left|\overline{\delta_1 \ldots \delta_K}\right\rangle$ is the eigenket of $\overline{Z}_i$:

$$\overline{Z}_i\left|\overline{\delta_1 \ldots \delta_K}\right\rangle = (-1)^{\delta_i}\left|\overline{\delta_1 \ldots \delta_K}\right\rangle. \tag{9.111}$$

The basis codeword $|\overline{\boldsymbol{\delta}}\rangle = \left|\overline{\delta_1 \ldots \delta_K}\right\rangle$ can be expanded in terms of N-qubit CB kets $\{|\boldsymbol{d}\rangle = |d_1 \ldots d_N\rangle|\, d_i = 0,1;\, i = 1,\ldots,N\}$ as follows:

$$|\overline{\boldsymbol{\delta}}\rangle = \left|\overline{\delta_1 \ldots \delta_K}\right\rangle = \sum_{\boldsymbol{d} \in F_2^N} C_{\overline{\boldsymbol{\delta}}}(\boldsymbol{d})|d_1 \ldots d_N\rangle. \tag{9.112}$$

Let us now apply Eq. (9.111) on Eq. (9.112) to obtain:

$$(-1)^{\delta_i}|\overline{\boldsymbol{\delta}}\rangle = \overline{Z}_i|\overline{\boldsymbol{\delta}}\rangle = \sum_{\boldsymbol{d} \in F_2^n} C_{\overline{\boldsymbol{\delta}}}(\boldsymbol{d})\overline{Z}_i|d_1 \ldots d_N\rangle = \sum_{\boldsymbol{d} \in F_2^N} C_{\overline{\boldsymbol{\delta}}}(\boldsymbol{d})(-1)^{v(\overline{Z}_i)\cdot \boldsymbol{d}}|d_1 \ldots d_N\rangle$$

$$= (-1)^{v(\overline{Z}_i)\cdot \boldsymbol{d}} \sum_{\boldsymbol{d} \in F_2^N} C_{\overline{\boldsymbol{\delta}}}(\boldsymbol{d})|d_1 \ldots d_N\rangle = (-1)^{v(\overline{Z}_i)\cdot \boldsymbol{d}}|\overline{\boldsymbol{\delta}}\rangle. \tag{9.113}$$

From Eq. (9.113), it is obvious that:

$$\delta_i = v(\overline{Z}_i)\cdot \boldsymbol{d}. \tag{9.114}$$

We indicated earlier that in the *first stage* of the decoding procedure we need to properly apply CNOT gates on ancillas, and it is therefore important to determine the action of the X_{a_i} gate on $\left|\overline{\delta_1 \ldots \delta_K}\right\rangle \otimes |0\ldots 0\rangle_K$. By applying the expansion (9.113) we obtain:

$$X_{a_1}^{\delta_1}\left|\overline{\delta_1 \ldots \delta_K}\right\rangle \otimes |0\ldots 0\rangle_K = \sum_{\boldsymbol{d} \in F_2^N} C_{\overline{\boldsymbol{\delta}}}(\boldsymbol{d})|d_1 \ldots d_N\rangle \otimes \left[X_{a_1}^{\delta_1}|0\ldots 0\rangle_K\right] \overset{\delta_i = v(\overline{Z}_1)\cdot \boldsymbol{d}}{=}$$
$$\sum_{\boldsymbol{d} \in F_2^N} C_{\overline{\boldsymbol{\delta}}}(\boldsymbol{d})|d_1 \ldots d_N\rangle \otimes X_{a_1}^{v(\overline{Z}_1)\cdot \boldsymbol{d}}|0\ldots 0\rangle_K. \tag{9.115}$$

From Eq. (9.115), it is clear that we need to determine the action of $X_{a_1}^{v(\overline{Z}_1)\cdot \boldsymbol{d}}$ to $|0\rangle$:

$$X_{a_1}^{v(\overline{Z}_1)\cdot \boldsymbol{d}}|0\rangle = X_{a_1}^{v_1(\overline{Z}_1)d_1} \ldots X_{a_1}^{v_N(\overline{Z}_1)d_N}|0\rangle = |v(\overline{Z}_1)\cdot \boldsymbol{d}\rangle = |\delta_1\rangle. \tag{9.116}$$

Therefore the action of the $X_{a_1}^{\delta_1}$ gate on $\left|\overline{\delta_1 \ldots \delta_K}\right\rangle \otimes |0\ldots 0\rangle_K$ is:

$$X_{a_1}^{\delta_1}\left|\overline{\delta_1 \ldots \delta_k}\right\rangle \otimes |0\ldots 0\rangle_K = \left|\overline{\delta_1 \ldots \delta_K}\right\rangle \otimes |\delta_1 \ldots 0\rangle_K. \tag{9.117}$$

By applying a similar procedure on the remaining ancilla qubits we obtain:

$$\prod_{i=1}^{K} X_{a_i}^{\delta_i} \left|\overline{\delta_1 \ldots \delta_K}\right\rangle \otimes |0\ldots0\rangle_K = \left|\overline{\delta_1 \ldots \delta_K}\right\rangle \otimes |\delta_1 \ldots \delta_K\rangle. \tag{9.118}$$

In the *second stage* of decoding, as indicated earlier, we have to perform the following mapping: $\left|\overline{\delta_1 \ldots \delta_K}\right\rangle \otimes \left|\delta_1 \ldots \delta_K\right\rangle \rightarrow \left|\overline{0\ldots0}\right\rangle \otimes |\delta_1 \ldots \delta_K\rangle$. This mapping can be implemented by applying a controlled-X_i operation to the ith encoded qubit, which serves as a target qubit. The ith ancilla qubit serves as the control qubit. This action can be described as follows:

$$\begin{aligned}(\overline{X}_1)^{\delta_1} \left|\overline{\delta_1 \delta_2 \ldots \delta_K}\right\rangle \otimes \left|\delta_1 \delta_2 \ldots \delta_K\right\rangle &= \left|\overline{(\delta_1 \oplus \delta_1)\delta_2 \ldots \delta_K}\right\rangle \otimes \left|\delta_1 \delta_2 \ldots \delta_K\right\rangle \\ &= \left|\overline{0\delta_2 \ldots \delta_K}\right\rangle \otimes \left|\delta_1 \ldots \delta_K\right\rangle.\end{aligned} \tag{9.119}$$

Furthermore we perform a similar procedure on the remaining encoded qubits:

$$\prod_{i=1}^{K} (\overline{X}_i)^{\delta_i} \left|\overline{\delta_1 \ldots \delta_K}\right\rangle \otimes \left|\delta_1 \ldots \delta_K\right\rangle = \left|\overline{0\cdots0}\right\rangle \otimes |\delta_1 \ldots \delta_K\rangle. \tag{9.120}$$

The *overall decoding process*, by applying Eqs. (9.118) and (9.120), can be represented by:

$$\prod_{i=1}^{K} (\overline{X}_i)^{\delta_i} \prod_{l=1}^{K} X_{a_l}^{\delta_l} \left|\overline{\delta_1 \ldots \delta_K}\right\rangle \otimes |0\ldots0\rangle_K = \left|\overline{0\ldots0}\right\rangle \otimes |\delta_1 \ldots \delta_K\rangle. \tag{9.121}$$

The *total number of gates for decoding*, based on Eq. (9.121), can be found as:

$$K(r+1) + K(N-K+1) = K(N-K+r+2) \leq (K+r)(N-K), \tag{9.122}$$

because $\overline{Z}_i$ acts on $r+1$ qubits (see Eq. (9.109)) and there are K such actions based on Eq. (9.121); controlled-X_i operations require no more than $N-K+1$ two-qubit gates and there are K such actions (see Eq. (9.121)). On the other hand, the number of gates required in reverse encoding is $(K+r)(N-K)$ (see Eq. (9.102)). Therefore the decoder implemented based on Eq. (9.121) is more efficient than decoding based on reverse efficient encoding.

Example. A decoding circuit for [4,2,2] quantum stabilizer code can be obtained based on Eq. (9.121). The starting point are generators of the code:

$$g_1 = X_1 Z_2 Z_3 X_4 \text{ and } g_2 = Y_1 X_2 X_3 Y_4.$$

The corresponding quantum-check matrix and its standard form (obtained by Gaussian elimination) are given by:

$$\mathbf{A} = \left(\begin{array}{cccc|cccc} 1 & 0 & 0 & 1 & 0 & 1 & 1 & 0 \\ 1 & 1 & 1 & 1 & 1 & 0 & 0 & 1 \end{array}\right) \sim \left(\begin{array}{cccc|cccc} 1 & 0 & 0 & 1 & 0 & 1 & 1 & 0 \\ 0 & 1 & 1 & 0 & 1 & 1 & 1 & 1 \end{array}\right).$$

The encoded Pauli operators in finite geometry form are given by:

$$\overline{X} = \begin{pmatrix} \mathbf{0} & E^T & I \,|\, (E^T C_1^T + C_2^T) & \mathbf{0} & \mathbf{0} \end{pmatrix} = \left(\begin{array}{cccc|cccc} 0 & 0 & 1 & 0 & 1 & 1 & 0 & 0 \\ 0 & 0 & 0 & 1 & 0 & 1 & 0 & 0 \end{array}\right)$$

$$\overline{Z} = \begin{pmatrix} \mathbf{0} & \mathbf{0} & \mathbf{0} \,|\, A_2^T & \mathbf{0} & I \end{pmatrix} = \left(\begin{array}{cccc|cccc} 0 & 0 & 0 & 0 & 0 & 1 & 1 & 0 \\ 0 & 0 & 0 & 0 & 1 & 0 & 0 & 1 \end{array}\right).$$

The corresponding operator representation of Pauli-encoded operators is:

$$\overline{X}_1 = Z_1Z_2X_3, \ \overline{X}_2 = Z_2X_4.$$
$$\overline{Z}_1 = Z_2Z_3, \ \overline{Z}_2 = Z_1Z_4.$$

The decoding circuit obtained by using Eq. (9.121) is shown in Fig. 9.4. The decoding circuit has two stages. In the first stage we apply CNOT gates on ancillas, with control states being the encoded states. The Pauli-encoded state $\overline{Z}_1 = Z_2Z_3$ indicates that the first ancilla qubit is controlled by encoded kets at positions 2 and 3. The Pauli-encoded state $\overline{Z}_2 = Z_1Z_4$ indicates that the second ancilla qubit is controlled by encoded kets at positions 1 and 4. The second stage is implemented based on Pauli-encoded operators $\overline{X}_i$. The ancilla qubits now serve as control qubits, and encoded qubits as target qubits. The Pauli-encoded state $\overline{X}_1 = Z_1Z_2X_3$ indicates that the first ancilla qubit controls the encoded kets at positions 1, 2, and 3 and corresponding controlled gates are Z at positions 1 and 2 and X at position 3. The Pauli-encoded state $\overline{X}_2 = Z_2X_4$ indicates that the second ancilla qubit controls the encoded kets at positions 2 and 4, while the corresponding controlled gates are Z at position 2 and X at position 4.

Example. The decoding circuit for [5,1,3] quantum stabilizer code can be obtain in similar fashion to the previous example, and the starting point are generators of the code:

$$g_1 = Y_1Z_2Z_4Y_5, \ g_2 = X_2Z_3Z_4Z_5, \ g_3 = Z_1Z_2X_3X_5 \text{ and } g_4 = Z_1Z_3Y_4Y_5.$$

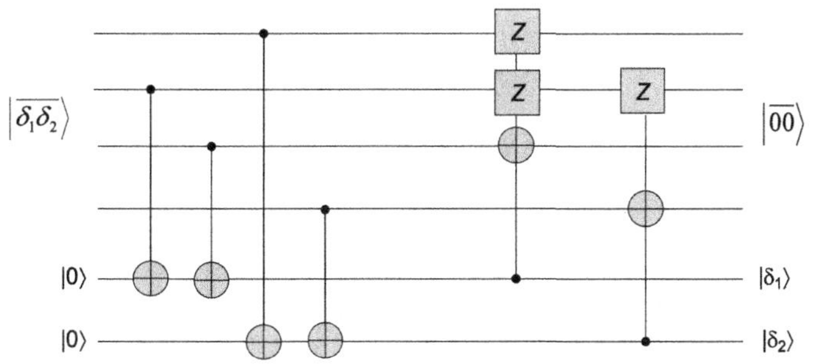

FIGURE 9.4

Efficient implementation of decoding circuit for [4,2,2] quantum stabilizer code.

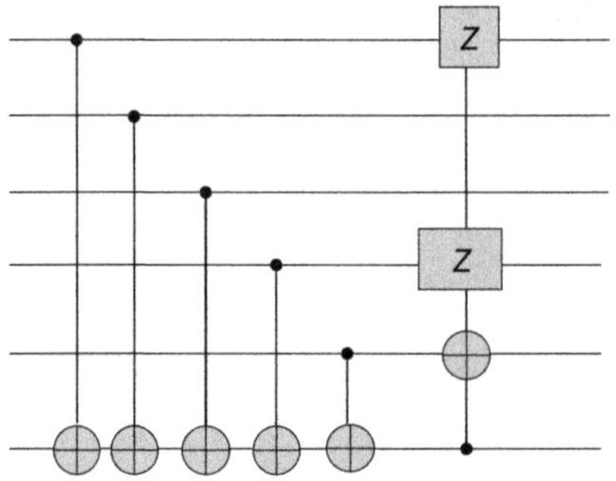

FIGURE 9.5

Efficient implementation of decoding circuit for [5,1,3] quantum stabilizer code.

Following a similar procedure as in the foregoing example, the standard form Pauli-encoded operators are given by:

$$\overline{Z} = Z_1Z_2Z_3Z_4Z_5 \text{ and } \overline{X} = Z_1Z_4X_5.$$

The corresponding decoding circuit is shown in Fig. 9.5.

9.6 NONBINARY STABILIZER CODES

In previous sections we considered quantum stabilizer codes with finite geometry representation over F_2^{2N}, $F_2 = \{0,1\}$. Because these codes are defined over F_2^{2N}, they can be called "binary" stabilizer codes. In this section we are concerned with stabilizer codes defined over F_q^{2N}, where $q = p^l$ is a prime power (p is a prime and $l \geq 1$ is an integer) [22–27]. This class of codes can be called "nonbinary" stabilizer codes. Although many definitions and properties from previous sections are applicable here, certain modifications are needed as described next. First of all we operate on q-ary quantum digits, which in analogy with qubits can be called "qudits." Second, we need to extend the definitions of quantum gates to qudits. In the previous chapter we have seen that arbitrary qubit error can be represented in terms of Pauli operators $\{I,X,Y,Z\}$. We have also seen that the Y-operator can be expressed in terms of X- and Z-operators. A similar strategy can be applied here. We need to extend definitions of X- and Z-operators to qudits as follows:

$$X(a)|x\rangle = |x + a\rangle, \quad Z(b) = \omega^{\mathrm{tr}(bx)}|x\rangle; \; x, \; a, \; b \in F_q, \tag{9.123}$$

where tr(•) denotes the trace operation from F_q to F_p and ω is a pth root of unity, namely $\omega = \exp(\mathrm{j}2\pi/p)$. The trace operation from F_{q^m} to F_q is defined as:

$$\mathrm{tr}_{q^m}(x) = \sum_{i=0}^{m-1} x^{q^i}. \tag{9.124}$$

If F_q is the prime field, the subscript can be omitted, as was done in Eq. (9.123).

Before we proceed further with nonbinary quantum stabilizer codes we believe it is convenient to introduce several additional *nonbinary quantum gates*, which will be useful in describing arbitrary quantum computation on qudits. The nonbinary quantum gates are shown in Fig. 9.6, in which the action of the gates is described as well. The F gate corresponds to the discrete Fourier transform gate. Its action on ket $|0\rangle$ is the superposition of all basis kets with the same probability amplitude:

$$F|0\rangle = \frac{1}{\sqrt{q}} \sum_{u \in F_q} |u\rangle. \tag{9.125}$$

From Chapter 2 we know that an operator A can be represented in the following form: $A = \sum_{i,j} A_{ij} |i\rangle\langle j|$. Based on Fig. 9.6 we determine the action of operator $FX(a)F^\dagger$ as follows:

$$\begin{aligned} FX(a)F^\dagger &= \frac{1}{\sqrt{q}} \sum_{i,j \in F_q} \omega^{\mathrm{tr}(ij)} |i\rangle\langle j| \sum_{x \in F_q} \omega |x+a\rangle\langle x| \frac{1}{\sqrt{q}} \sum_{l,k \in F_q} \omega^{-\mathrm{tr}(lk)} |l\rangle\langle k| \\ &= \frac{1}{q} \sum_{i,j \in F_q} \omega^{\mathrm{tr}(ia)} \underbrace{\sum_{x \in F_q} \omega^{\mathrm{tr}(ix-xl)}}_{1,\ i=l\ \ 0,\ i \neq l} |i\rangle\langle l| \\ &= \sum_{i \in F_q} \omega^{\mathrm{tr}(ia)} |i\rangle\langle i| = Z(a). \end{aligned} \tag{9.126}$$

By using the basic nonbinary gates shown in Fig. 9.6 and Eq. (9.123) we can perform more complicated operations, as illustrated in Fig. 9.7. These circuits will be used later in the section on efficient implementation of quantum nonbinary encoders and decoders. Notice that the circuit on the bottom of Fig. 9.7 is obtained by concatenation of the circuits on the top.

It can be shown that the set of errors $\varepsilon = \{X(a)X(b) | a, b \in F_q\}$ satisfies the following properties: (1) it contains the identity operator, (2) $\mathrm{tr}\left(E_1^\dagger E_2\right) = 0 \quad \forall E_1, E_2 \in \varepsilon$, and (3) $\forall E_1, E_2 \in \varepsilon : E_1 E_2 = cE_3, E_3 \in \varepsilon, c \in F_q$.

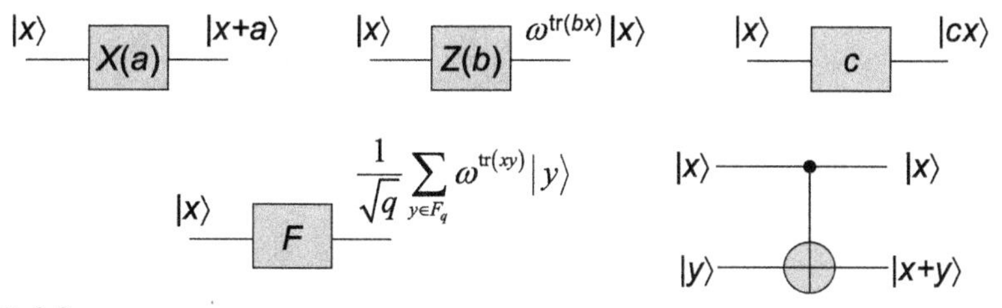

FIGURE 9.6

Basic nonbinary quantum gates.

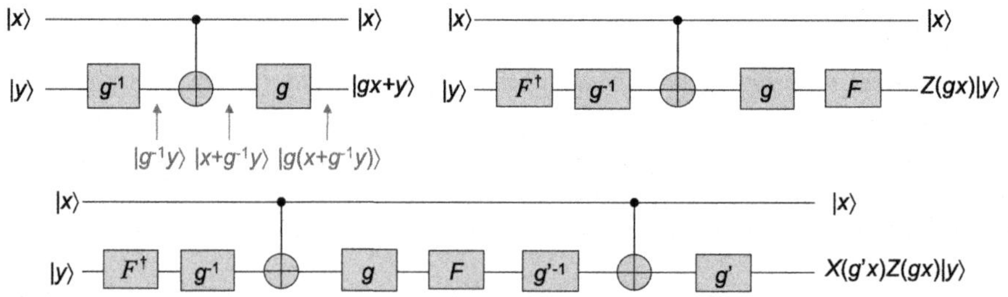

FIGURE 9.7

Nonbinary quantum circuits derived from basic nonbinary gates.

This set forms an error basis for the set of $q \times q$ matrices, and sometimes is called the "nice error basis." It can also be shown that: (1) $X(a)Z(b) = \omega^{-tr(ab)}Z(b)X(a)$, and (2) $X(a+a')Z(b+b') = \omega^{-tr(a'b)}X(a)Z(b)X(a')Z(b')$.

Example. The 4-ary nice error basis over $F_4 = \{0, 1, \alpha, \overline{\alpha}\}$ can be obtained as a tensor product of the Pauli basis:

$$X(0) = II, \ X(1) = IX, \ X(\alpha) = XI, \ X(\overline{\alpha}) = XX,$$
$$Z(0) = II, \ Z(1) = ZI, \ Z(\alpha) = ZZ, \ Z(\overline{\alpha}) = IZ.$$

To determine the nice basis error on N qudits we introduce the following notation: $X(\boldsymbol{a}) = X(a_1) \otimes \ldots \otimes X(a_N)$ and $Z(\boldsymbol{b}) = Z(b_1) \otimes \ldots \otimes Z(b_N)$, where $\boldsymbol{a} = (a_1,\ldots,a_N)$, $\boldsymbol{b} = (b_1,\ldots,b_N)$, and $a_i, b_i \in F_q$. The set $\varepsilon_N = \left\{X(\boldsymbol{a})X(\boldsymbol{b})|\boldsymbol{a}, \boldsymbol{b} \in F_q^N\right\}$ is the nice error basis defined over F_q^{2N}. Similarly to the Pauli multiplicative group we can define the *error group* by:

$$G_N = \left\{\omega^c X(\boldsymbol{a})Z(\boldsymbol{b})|\boldsymbol{a}, \boldsymbol{b} \in F_q^N, \ c \in F_q\right\}. \tag{9.127}$$

Let S be the largest Abelian subgroup of G_N that fixes all elements from quantum code C_Q, called the stabilizer group. The $[N,K]$ nonbinary stabilizer code C_Q is defined as the K-dimensional subspace of the N-qudit Hilbert space H_q^N as follows:

$$C_Q = \bigcap_{s \in S} \left\{|c\rangle \in H_q^N \middle| s|c\rangle = |c\rangle\right\}. \tag{9.128}$$

Clearly, the definition of nonbinary stabilizer codes is a straightforward generalization of quantum stabilizer code over H_2^N. Therefore similar properties, definitions, and theorems introduced in previous sections are applicable here. For example, two errors $E_1 = \omega^{c_1}X(\boldsymbol{a}_1)Z(\boldsymbol{b}_1)$, $E_2 = \omega^{c_2}X(\boldsymbol{a}_2)Z(\boldsymbol{b}_2) \in G_N$ commute if and only if their trace sympletic product vanishes, that is:

$$\text{tr}(\boldsymbol{a}_1\boldsymbol{b}_2 - \boldsymbol{a}_2\boldsymbol{b}_1) = 0. \tag{9.129}$$

From property (2) of errors, namely $X(a)Z(b)X(a')Z(b') = \omega^{\mathrm{tr}(a'b)}X(a+a')Z(b+b')$, it can be simply verified that $E_1E_2 = \omega^{\mathrm{tr}(\boldsymbol{a}_2\boldsymbol{b}_1)}X(\boldsymbol{a}_1+\boldsymbol{a}_2)Z(\boldsymbol{b}_1+\boldsymbol{b}_2)$ and $E_2E_1 = \omega^{\mathrm{tr}(\boldsymbol{a}_1\boldsymbol{b}_2)}X(\boldsymbol{a}_1+\boldsymbol{a}_2)Z(\boldsymbol{b}_1+\boldsymbol{b}_2)$. Clearly, $E_1E_2 = E_2E_1$ only when $\omega^{\mathrm{tr}(\boldsymbol{a}_2\boldsymbol{b}_1)} = \omega^{\mathrm{tr}(\boldsymbol{a}_1\boldsymbol{b}_2)}$, which is equivalent to Eq. (9.129).

The (sympletic) weight of a qudit error $E = \omega^c X(\boldsymbol{a})Z(\boldsymbol{b})$ can be defined, in similar fashion to the weight of a qubit error, as the number of components i for which $(a_i,b_i) \neq (0,0)$. We say that nonbinary quantum stabilizer code has distance D if it can detect all errors of weight less than D, but none of weight D. The error correction capability t of nonbinary quantum code is related to minimum distance by $t = \lfloor (D-1)/2 \rfloor$. We say that nonbinary quantum stabilizer code is nondegenerate if its stabilizer group S does not contain an element of weight smaller than t. From the foregoing definitions we can see that nonbinary quantum stabilizer codes are straightforward generalizations of corresponding qubit stabilizer codes. The key difference is that instead of the sympletic product we need to use the trace-sympletic product. Similar theorems can be proved, and properties can be derived by following similar procedures, by considering differences outlined earlier. On the other hand, quantum hardware implementation is more challenging. For example, instead of using Pauli X and Z gates we must use the $X(a)$ and $Z(b)$ gates shown in Fig. 9.6. We can also define the standard form, but instead of using F_2 during Gaussian elimination, we must perform Gaussian elimination in F_q instead. For example, we can determine the syndrome based on syndrome measurements, as illustrated in previous sections. Let the generator $\boldsymbol{g}_i$ be given as follows:

$$\boldsymbol{g}_i = [\boldsymbol{a}_i|\boldsymbol{b}_i] = [0\ldots 0a_i\ldots a_N|0\ldots 0b_i\ldots b_N] \in F_q^{2N},\ a_i,\ b_i \in F_q. \tag{9.130}$$

The quantum circuit shown in Fig. 9.8 will provide the nonzero measurement if a detectable error does not commute with a multiple of $\boldsymbol{g}_i$. By comparison with

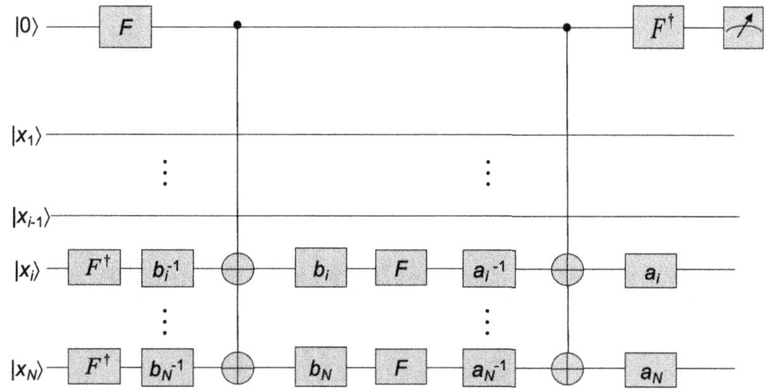

FIGURE 9.8

Nonbinary syndrome decoding circuit corresponding to generator $\boldsymbol{g}_i$.

corresponding qubit syndrome decoding circuits from the previous chapter we conclude that the F gate for nonbinary quantum codes has a similar role to the Hadamard gate for codes over F_2. Notice that the syndrome circuit is a generalization of the bottom circuit shown in Fig. 9.7. From Fig. 9.8, it is clear that if the input to the stabilizer circuit is $E|\psi\rangle$, where E is the qudit error and $|\psi\rangle$ is the codeword, then the stabilizer circuit performs the following mapping:

$$|0\rangle E|\psi\rangle \to \sum_{u\in F_q} F^\dagger|u\rangle X(u\boldsymbol{a}_i)Z(u\boldsymbol{b}_i)E|\psi\rangle. \tag{9.131}$$

The qudit error E, by ignoring the complex phase constant, can be represented as $E = X(\boldsymbol{a})Z(\boldsymbol{b})$. Based on the text following Eq. (9.129), it is clear that $X(\boldsymbol{a}_i)Z(\boldsymbol{b}_i)E = \omega^{\mathrm{tr}(\boldsymbol{ab}_i-\boldsymbol{a}_i\boldsymbol{b})}EX(\boldsymbol{a}_i)Z(\boldsymbol{b}_i)$. Based on Fig. 9.6 we can show that $X(u\boldsymbol{a}_i)Z(u\boldsymbol{b}_i)E = \omega^{\mathrm{tr}(u(\boldsymbol{ab}_i-\boldsymbol{a}_i\boldsymbol{b}))}EX(\boldsymbol{a}_i)Z(\boldsymbol{b}_i)$, and by substituting in Eq. (9.131) we obtain:

$$|0\rangle E|\psi\rangle \to \sum_{u\in F_q} F^\dagger|u\rangle X(u\boldsymbol{a}_i)Z(u\boldsymbol{b}_i)E|\psi\rangle = \sum_{u\in F_q} F^\dagger|u\rangle \omega^{\mathrm{tr}(u(\boldsymbol{ab}_i-\boldsymbol{a}_i\boldsymbol{b}))}EX(\boldsymbol{a}_i)Z(\boldsymbol{b}_i)|\psi\rangle. \tag{9.132}$$

Since $X(\boldsymbol{a}_i)Z(\boldsymbol{b}_i)|\psi\rangle = |\psi\rangle$, because $X(\boldsymbol{a}_i)Z(\boldsymbol{b}_i)\in S$, Eq. (9.132) becomes:

$$\begin{aligned}|0\rangle E|\psi\rangle &\to \left[\sum_{u\in F_q} F^\dagger|u\rangle \omega^{\mathrm{tr}(u(\boldsymbol{ab}_i-\boldsymbol{a}_i\boldsymbol{b}))}\right]E|\psi\rangle\\ &= \left[\sum_{u\in F_q}\sum_{v\in F_q} \omega^{-\mathrm{tr}(uv)}\omega^{\mathrm{tr}(u(\boldsymbol{ab}_i-\boldsymbol{a}_i\boldsymbol{b}))}|v\rangle\right]E|\psi\rangle\\ &= \sum_{v\in F_q}|v\rangle \underbrace{\sum_{u\in F_q}\omega^{\mathrm{tr}(u(\boldsymbol{ab}_i-\boldsymbol{a}_i\boldsymbol{b})-uv)}}_{1,\ \text{for } v=\boldsymbol{ab}_i-\boldsymbol{a}_i\boldsymbol{b}} E|\psi\rangle = |\boldsymbol{ab}_i-\boldsymbol{a}_i\boldsymbol{b}\rangle E|\psi\rangle.\end{aligned} \tag{9.133}$$

From Eq. (9.133), it is evident that the result of the measurement gives us the ith syndrome component $\lambda_i = \mathrm{tr}(\boldsymbol{ab}_i - \boldsymbol{a}_i\boldsymbol{b})$. By $N-K$ measurements on corresponding generators $\boldsymbol{g}_i$ ($i = 1,\ldots,N-K$) we obtain the following syndrome $S(E)$:

$$S(E) = [\lambda_1\lambda_2\ldots\lambda_{N-K}]^{\mathrm{T}};\ \lambda_i = \mathrm{tr}(\boldsymbol{ab}_i - \boldsymbol{a}_i\boldsymbol{b}),\ i = 1,\ldots,N-K. \tag{9.134}$$

Clearly, the syndrome Eq. (9.134) is very similar to Eq. (9.47), and similar conclusions can be made. The correctable qudit error maps the code space to q^K-dimensional subspace of q^N-dimensional Hilbert space. Since there are $N-K$ generators, or equivalently syndrome positions, there are q^{N-K} different cosets. All qudit errors belonging to the same coset have the same syndrome. By selecting the most probable qudit error for the coset representative, typically lowest weight error, we can uniquely identify the qudit error and consequently perform the error correction action.

We can define the projector P onto the code space C_Q as follows:

$$P = \frac{1}{|S|}\sum_{s\in S} s. \tag{9.135}$$

Clearly, $Ps = s$ for every $s \in S$. Furthermore:

$$P^2 = \frac{1}{|S|}\sum_{s\in S} Ps = \frac{1}{|S|}\sum_{s\in S} s = P. \tag{9.136}$$

Since $s^\dagger \in S$, the following is valid: $P^\dagger = P$, which indicates that P is an orthogonal projector. The dimensionality of C_Q is $\dim C_Q = q^N/|S|$.

Quantum bounds similar to those described in the previous chapter can be derived for nonbinary stabilizer codes as well. For example, the Hamming bound for nonbinary quantum stabilizer code $[N,K,D]$ (where K is the number of information qudits and N is the codeword length) over F_q is given by:

$$\sum_{i=0}^{\lfloor (D-1)/2 \rfloor} \binom{N}{i} (q^2-1)^i q^K \le q^N. \tag{9.137}$$

This inequality is obtained from inequality (8.52) in Chapter 8 by substituting term 3^i with $(q^2-1)^i$, and it can be proved using similar methodology. The number of information qudits is now q^K and the number of possible codewords is q^N.

9.7 SUBSYSTEM CODES

Subsystem codes [26–31] represent a generalization of nonbinary stabilizer codes from the previous section. They can also be considered as a generalization of decoherence-free subspaces [32,33] and noiseless subsystems [34]. The key idea behind subsystem codes is to decompose the quantum code C_Q as the tensor product of two subsystems A and B as follows $C_Q = A \otimes B$ The information qudits belong to subsystem A and noninformation qudits, also known as gauge qudits, belong to subsystem B. We are only concerned with errors introduced in subsystem A.

It was discussed in the previous chapter that both noise $\mathcal{E}$ and recovery $\mathcal{R}$ processes can described as quantum operations $\mathcal{E},\mathcal{R}$: $L(H) \rightarrow L(H)$, where $L(H)$ is the space of linear operators on Hilbert space H. These mappings can be represented in terms of operator-sum representation, $\mathcal{E}(\rho) = \sum_i E_i \rho E_i^\dagger,\ E_i \in L(H),\ \mathcal{E} = \{E_i\}$. Given the quantum code C_Q that is the subspace of H we say that the set of errors $\mathcal{E}$ is correctable if there exists a recovery operation $\mathcal{R}$ such that $\mathcal{R}\mathcal{E}(\rho) = \rho$ from any state

ρ from $L(C_Q)$. In terms of subsystem codes we say that there exists a recovery operation such that for any ρ^A from $L(A)$ and ρ^B from $L(B)$, the following is valid: $\mathcal{RE}(\rho^A \otimes \rho^B) = \rho^A \otimes \rho'^B$, where the state ρ'^B is not relevant. The necessary and sufficient condition for the set of errors $\mathcal{E} = \{E_m\}$ to be correctable is that $PE_m^\dagger E_n P = I^A \otimes g_{mn}^B$, $\forall m, n$, where P is the projector on code subspace. Such set of errors is called a correctable set of errors. Clearly, the linear combination of errors from $\mathcal{E}$ is also correctable so that it makes sense to observe the correctable set of errors as linear space with properly chosen operator basis. If we are concerned with quantum registers/systems composed of N qubits, the corresponding error operators are Pauli operators. The Pauli group on N qubits, G_N, was introduced in Section 9.1. The stabilizer formalism can also be used in describing the stabilizer subsystem codes. Stabilizer subsystem codes are determined by a subgroup of G_N that contains the element jI, called the *gauge group* G, and by stabilizer group S that is properly chosen such that $S' = j^\lambda S$ is the center of G, denoted as $Z(G)$ (that is, the set of elements from G that commute with all elements from G). We are concerned with the following decomposition of H: $H = C \oplus C^\perp = (H_A \otimes H_B) \oplus C^\perp$, where $C^\perp$ is the dual of $C = H_A \otimes H_B$. The gauge operators are chosen in such a way that they act trivially on subsystem A, but generate full algebra of subsystem B. The information is encoded on subsystem A, while subsystem B is used to absorb the effects of gauge operations. The Pauli operators for K logical qubits are obtained from the isomorphism $N(\mathcal{G})/S \simeq G_K$, where $N(\mathcal{G})$ is the normalizer of $\mathcal{G}$ On the other hand, subsystem B consists of R gauge qubits recovered from isomorphism $\mathcal{G}/S \simeq G_N$, where N $K + R + s$, $s \geq 0$. If $\widetilde{X}_1$ and $\widetilde{Z}_1$ represent the images of X_i and Z_i under automorphism U of G_N, the stabilizer can be described by $S = \langle \widetilde{Z}_1, \ldots, \widetilde{Z}_s \rangle$; $R + s \leq N$; R, $s \geq 0$. The gauge group can be specified by: $\mathcal{G} = \langle jI, \widetilde{Z}_1, \ldots, \widetilde{Z}_{s+R}, \widetilde{X}_{s+1}, \ldots, \widetilde{X}_R \rangle$. The images must satisfy the commutative relations (9.28). The logical (encoded) Pauli operators will be then $\overline{X}_1 = \widetilde{X}_{s+R+1}$, $\overline{Z}_1 = \widetilde{Z}_{s+R+1}, \ldots,$ $\overline{X}_K = \widetilde{X}_N$, $\overline{Z}_K = \widetilde{Z}_N$. The detectable errors are the elements of $G_N - N(S)$ and undetectable errors in $N(S) - G$. Undetectable errors are related to the logical Pauli operators since $N(S)/\mathcal{G} \simeq N(\mathcal{G})/S'$. Namely, if $n \in N(S)$, then there exists $g \in G$ such that $ng \in N(G)$ and if $g' \in G$ such that $ng' \in N(G)$ the $gg' \in G \cap N(G) = S'$. The distance D of this code is defined as the minimum weight among undetectable errors. The subsystem code encodes K qubits into N-qubit codewords and has R gauge qubits, and can therefore be denoted as $[N,K,R,D]$ code.

We turn our attention now to subsystem codes defined as the subspace of q^N-dimensional Hilbert space H_q^N. For a and b from F_q we define unitary operators (qudit errors) $X(a)$ and $Z(b)$ as follows:

$$X(a)|x\rangle = |x + a\rangle, \; Z(b) = \omega^{\mathrm{tr}(bx)}|x\rangle; \; x, \; a, \; b \in F_q; \; \omega = e^{j2\pi/p}. \tag{9.138}$$

As expected, since nonbinary stabilizer subsystem codes are the generalization of nonbinary stabilizer codes, similar qudit errors occur. The trace operation is defined by Eq. (9.124). The qudit *error group* is defined by:

$$G_N = \left\{\omega^c X(\boldsymbol{a})Z(\boldsymbol{b})|\boldsymbol{a},\ \boldsymbol{b} \in F_q^N,\ c \in F_p\right\};$$
$$\boldsymbol{a} = (a_1 \ldots a_N),\ \boldsymbol{b} = (b_1 \ldots b_N);\ a_i,\ b_i \in F_q. \tag{9.139}$$

Let C_Q be a quantum code such that $H = C_Q \oplus C_Q^{\perp}$. The $[N,K,R,D]$ subsystem code over F_q is defined as the decomposition of code space C_Q into a tensor product of two subsystems A and B such that $C_Q = A \otimes B$, where dimensionality of A equals $\dim A = q^K$, $\dim B = q^R$ and all errors of weight less than $D_{\min}$ on subsystem A can be detected. What is interesting about this class of codes is that when constructed from classical codes, the corresponding classical code does not need to be dual containing (self-orthogonal). Notice that subsystem codes can also be defined over F_{q^2}.

For $[N,K,R,D_{\min}]$ stabilizer subsystem codes over F_2, it can be shown that the centralizer of S is given by (see Problem 17):

$$C_{G_N}(S) = \langle G,\ \overline{X}_1,\ \overline{Z}_1, \ldots, \overline{X}_K,\ \overline{Z}_K\rangle. \tag{9.140}$$

Since $C_Q = A \otimes B$, then $\dim A = 2^K$ and $\dim B = 2^R$. From Eq. (9.135) we know that stabilizer S can be used as the projector to C_Q. The dimensionality of quantum code defined by stabilizer S is 2^{K+R}. The stabilizer S therefore defines an $[N,K+R,D]$ stabilizer code. Based on Eq. (9.140) we conclude that image operators $\widetilde{Z}_i,\ \widetilde{X}_i;\ i = s+1, \ldots, R$ behave as encoded operators on gauge qubits, while $\overline{Z}_i,\ \overline{X}_i$ act on information qubits. In total we have the set of $2(K+R)$ encoded operators of $[N,K+R,D]$ stabilizer codes given by:

$$\{\overline{X}_1,\ \overline{Z}_1, \ldots, \overline{X}_K,\ \overline{Z}_K,\ \widetilde{X}_{s+1},\ \widetilde{Z}_{s+1}, \ldots, \widetilde{X}_{s+R},\ \widetilde{Z}_{s+R}\}. \tag{9.141}$$

Therefore with Eq. (9.141) we have just established the connection between stabilizer codes and subsystem codes. Since subsystem codes are more flexible for design, they can be used to design new classes of stabilizer codes. More importantly, Eq. (9.141) gives us an opportunity to implement encoders and decoders for subsystem codes using similar approaches already developed for stabilizer codes, such as standard form formalism. For example, subsystem code [4,1,1,2] is described by stabilizer group $S = \langle \widetilde{Z}_1,\ \widetilde{Z}_2\rangle$ and gauge group $\mathcal{G} = \langle S,\ \widetilde{X}_3,\ \widetilde{Z}_3,\ jI\rangle$, where $\widetilde{Z}_1 = X_1X_2X_3X_4$, $\widetilde{Z}_2 = Z_1Z_2Z_3Z_4$, $\widetilde{Z}_3 = Z_1Z_2$, and $\widetilde{X}_3 = X_2X_4$. The encoded operators are given by: $\overline{X}_1 = X_2X_3$ and $\overline{Z}_1 = Z_2Z_4$. Corresponding [4,2] stabilizer code, based on Eq. (9.141), is described by generators: $g_1 = \widetilde{X}_3$, $g_2 = \widetilde{Z}_3$, $g_3 = \overline{X}_1$, and $g_4 = \overline{Z}_1$. In Fig. 9.9 we show two possible encoder versions of the same [4,1,1,2] subsystem code. Notice that the state of gauge qubits is irrelevant as long as information qubits are concerned. They can be randomly chosen. In Fig. 9.9b we show that by proper setting of gauge qubits we can simplify encoder implementation.

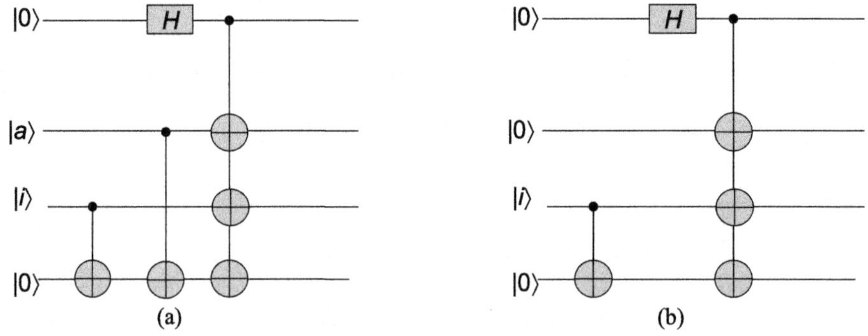

FIGURE 9.9

Encoder implementation of [4,1,1,2] subsystem code: (a) gauge qubit $|a\rangle$ has arbitrary value and (b) gauge qubit is set to $|0\rangle$.

Based on Eq. (9.141) and the text around it we can conclude that with the help of stabilizer S and gauge G groups of subsystem code, we are able to create a stabilizer S_A, whose encoder can be used as an encoder for subsystem code providing that gauge qubits are initialized to $|0\rangle$ qubit. Clearly, $S_A \supseteq S$, which indicates that the space stabilized by S_A is the subspace of $C_Q = A \otimes B$, which is, on the other hand, stabilized by S. Since $|S_A|/|S| = 2^R$, the dimension of the subspace stabilized by S_A is $2^{K+R-R} = 2^K$. If the subsystem stabilizer and gauge groups are given by $S = \langle \widetilde{Z}_1, \ldots, \widetilde{Z}_s \rangle; s = N - K - R$ and $\mathcal{G} = \langle jI, S, \widetilde{Z}_{s+1}, \ldots, \widetilde{Z}_{s+R}, \widetilde{X}_{s+1}, \ldots, \widetilde{X}_{s+R} \rangle$, respectively, then the stabilizer group of S_A will be given by: $S_A = \langle S, \widetilde{Z}_{s+1}, \ldots, \widetilde{Z}_{s+R} \rangle$. The quantum-check matrix for S_A can be put in standard form (9.104) as follows:

$$A(S_A) = \left(\begin{array}{ccc|ccc} \overset{r}{I_r} & \overset{N-K-r}{A_1} & \overset{K}{A_2} & \overset{r}{B} & \overset{N-K-r}{0} & \overset{K}{C} \\ 0 & 0 & 0 & D & I_{N-K-r} & E \end{array} \right) \begin{array}{l} \}r \\ \}N-K-r \end{array}, \quad r = \mathrm{rank}(A). \tag{9.142}$$

The encoded Pauli operators of S_A are given by:

$$\bar{X} = \left(0 \;\; E^T \;\; I_K \,|\, C^T \;\; 0 \;\; 0 \right), \; \bar{Z} = \left(0 \;\; 0 \;\; 0 \,|\, A_2^T \;\; 0 \;\; I_K \right). \tag{9.143}$$

The encoder circuit for [N,K,R,D] subsystem code can be implemented as shown in Fig. 9.3, wherein all ancilla qubits are initialized to $|0\rangle$ qubit.

As an illustration, let us study the encoder implementation of [9,1,4,3] Bacon−Shor code [35,36] (see also [26]). The stabilizer S and gauge G groups of this code are given by:

$$\begin{aligned} &S = \langle \widetilde{Z}_1 = X_1X_2X_3X_7X_8X_9, \widetilde{Z}_2 = X_4X_5X_6X_7X_8X_9, \widetilde{Z}_3 = Z_1Z_3Z_4Z_6Z_7Z_9, \widetilde{Z}_4 = Z_2Z_3Z_5Z_6Z_8Z_9 \rangle \\ &\mathcal{G} = \langle S, \mathcal{G}_X, \mathcal{G}_Z \rangle, \mathcal{G}_X = \langle \widetilde{X}_5 = X_2X_5, \widetilde{X}_6 = X_3X_6, \widetilde{X}_7 = X_5X_8, \widetilde{X}_8 = X_6X_9 \rangle, \\ &\mathcal{G}_Z = \langle \widetilde{Z}_5 = Z_1Z_3, \widetilde{Z}_6 = Z_4Z_6, \widetilde{Z}_7 = Z_2Z_3, \widetilde{Z}_8 = Z_5Z_6 \rangle. \end{aligned}$$

The stabilizer S_A can be formed by adding the images from G_Z to S to get $S_A = \langle S, \mathcal{G}_Z \rangle$, and the corresponding encoder is shown in Fig. 9.10a. The stabilizer S_A can also be obtained by G_X to S to get $S_A = \langle S, \mathcal{G}_X \rangle$, with the corresponding encoder being shown in Fig. 9.10b. Clearly, the encoder from Fig. 9.10b has a larger number of Hadamard gates, but a smaller number of CNOT gates, which are much more difficult for implementation. Therefore the gauge qubits provide flexibility in the implementation of encoders.

We further describe another encoder implementation method due to Grassl et al. [37] (see also [26]), also known as the *conjugation method*. The key idea of this method is to start with quantum-check matrix $\boldsymbol{A}$, given by Eq. (9.142), and transform it into the following form:

$$\boldsymbol{A}' = [\mathbf{0}\ \mathbf{0} | \boldsymbol{I}_{N-K}\ \mathbf{0}]. \tag{9.144}$$

During this transformation we have to memorize the actions we performed, which is equivalent to the gates being employed on information and ancilla qubits to perform this transformation. Notice that quantum-check matrix (9.144) corresponds to canonical quantum codes described in the previous chapter, which simply perform the following mapping:

$$|\psi\rangle \to \underbrace{|0\rangle \otimes \ldots \otimes |0\rangle}_{N-K \text{ times}}\ |\psi\rangle = |\mathbf{0}\rangle_{N-K}|\psi\rangle. \tag{9.145}$$

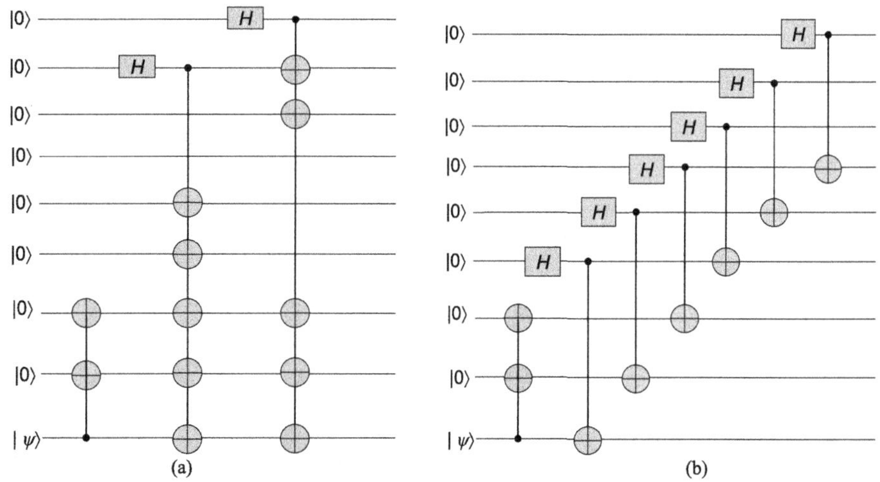

FIGURE 9.10

Encoder for [9,1,4,3] subsystem code (or equivalently [9,1,3] stabilizer code) for S_A given by: (a) $S_A = \langle S, G_Z \rangle$ and (b) $S_A = \langle S, G_X \rangle$.

In this encoding mapping, $N-K$ ancilla qubits are appended to K information qubits. Corresponding encoded Pauli operators are given by:

$$\overline{X}=(\mathbf{0}\ \mathbf{I}_K|\mathbf{0}\ \mathbf{0}),\quad \overline{Z}=(\mathbf{0}\ \mathbf{0}|\mathbf{0}\ \mathbf{I}_K). \tag{9.146}$$

Let us observe one particular generator of the quantum-check matrix, say $\mathbf{g}=(a_1,\ldots,a_N|b_1,\ldots,b_N)$. The action of the Hadamard gate on the ith qubit is to change the position of a_i and b_i as follows, where we underlined the affected qubits for convenience:

$$\left(a_1\ldots\underline{a_i}\ldots a_N|b_1\ldots\underline{b_i}\ldots b_N\right)\xrightarrow{H}\left(a_1\ldots\underline{b_i}\ldots a_N|b_1\ldots\underline{a_i}\ldots b_N\right). \tag{9.147}$$

The action of the CNOT gate on $\mathbf{g}$, where the ith qubit is the control qubit and the jth qubit is the target qubit, the action denoted as $\text{CNOT}^{i,j}$, can be described as:

$$\left(a_1\ldots\underline{a_j}\ldots a_N|b_1\ldots\underline{b_i}\ldots b_N\right)\xrightarrow{\text{CNOT}^{i,j}}\left(a_1\ldots a_{j-1}\underline{a_j+a_i}\,a_{j+1}\ldots a_N|b_1\ldots b_{i-1}\underline{b_i+b_j}\,b_{i+1}\ldots b_N\right). \tag{9.148}$$

Therefore the jth entry in the X-portion and the ith entry in the Z-portion are affected. Finally, the action of phase gate P on the ith qubit is to perform the following mapping:

$$\left(a_1\ldots\underline{a_i}\ldots a_N|b_1\ldots\underline{b_i}\ldots b_N\right)\xrightarrow{P}\left(a_1\ldots\underline{a_i}\ldots a_N|b_1\ldots\underline{a_i+b_i}\ldots b_N\right). \tag{9.149}$$

Based on Eqs. (9.147–9.149) we can create the following look-up table (LUT):

Based on Table 9.1, the transformation of $\mathbf{g}=(a_1,\ldots,a_N|b_1,\ldots,b_N)$ to $\mathbf{g}'=(a'_1,\ldots,a'_N|0,\ldots,0)$ can be achieved by application of Hadamard H and phase P gates as follows:

$$\bigotimes_{i=1}^{N} H^{\overline{a}_i b_i} P^{a_i b_i},\ \overline{a}_i=a_i+1 \bmod 2. \tag{9.150}$$

For example, $\mathbf{g}=(10010|01110)$ can be transformed to $\mathbf{g}'=(11110|00000)$ by application of the following sequence of gates: $H_2H_3P_4$, based on Table 9.1 or Eq. (9.150).

Table 9.1 Look-up-table of actions of gates I, H, and P on the ith qubit. The ith qubit (a_i,b_i) is through the action of the corresponding gate transformed into (c,d).

(a_i,b_i)	Gate	(c,d)
(0,0)	I	(0,0)
(1,0)	I	(1,0)
(0,1)	H	(1,0)
(1,1)	P	(1,0)

Based on Eq. (9.148), the error operator $\boldsymbol{e} = (a_1,\ldots,a_i = 1,\ldots, a_N|0,\ldots,0)$ can be converted to $\boldsymbol{e}' = (0,\ldots,a'_i = 1,0\ldots0|0\ldots0)$ by the application of the following sequence of gates:

$$\prod_{m=1,m\neq n}^{N} (CNOT^{m,n})^{a_n}. \tag{9.151}$$

For example, the error operator $\boldsymbol{e} = (11110|00000)$ can be transformed to $e' = (0100|0000)$ by the application sequence of CNOT gates: $\mathrm{CNOT}^{2,1}\mathrm{CNOT}^{2,3}\mathrm{CNOT}^{2,4}$.

Therefore the first stage is to convert the Z-portion of the stabilizer matrix to an all-zero submatrix by application of H and P gates by means of Eq. (9.150). In the second stage we convert every row of the stabilizer matrix upon the first stage $(\boldsymbol{a}'|\mathbf{0})$ to $(0\ldots a'_i = 10\ldots0|0\ldots0)$ by means of Eq. (9.151). After the second stage, the stabilizer matrix has the form $(\boldsymbol{I}_{N-K}0|\mathbf{00})$. In the third stage we transform such obtained stabilizer matrix to $(\mathbf{00}|\boldsymbol{I}_{N-K}\mathbf{0})$ by applying the H gates to the first $N-K$ qubits. The final form of stabilizer matrix is that of canonical code, which simply appends $N-K$ ancilla qubits in the $|0\rangle$ state to K information qubits, that is, it performs the mapping (9.145). During this three-stage transformation, encoded Pauli operators are transformed into the form given by Eq. (9.146). Let us denote the sequence of gates being applied during three-stage transformation as $\{U\}$. By applying this sequence of gates in opposite order we obtain the encoding circuit of the corresponding stabilizer code.

Let us now apply the conjugation method on subsystem codes. The key difference with respect to stabilizer codes is that we need to observe the whole gauge group instead. One option is to create the stabilizer group S_A based on gauge group $G = <S,G_Z,G_X>$ as by either $S_A = <S,G_Z>$ or $S_A = <S,G_X>$ and then apply the conjugation method. The second option is to transform the gauge group, represented using finite geometry formalism, as follows:

$$G = \begin{bmatrix} \underline{S} \\ \underline{\mathcal{G}_Z} \\ \mathcal{G}_X \end{bmatrix} \xrightarrow{\{U\}} \mathcal{G}' = \begin{bmatrix} \mathbf{0} & \mathbf{0} & \mathbf{0} | & \boldsymbol{I}_s & \mathbf{0} & \mathbf{0} \\ \mathbf{0} & \mathbf{0} & \mathbf{0} | & \mathbf{0} & \boldsymbol{I}_R & \mathbf{0} \\ \mathbf{0} & \boldsymbol{I}_R & \mathbf{0} | & \mathbf{0} & \mathbf{0} & \mathbf{0} \end{bmatrix}. \tag{9.152}$$

With transformation (9.152) we perform the following mapping:

$$|\phi\rangle|\psi\rangle \rightarrow \underbrace{|0\rangle\otimes\ldots\otimes|0\rangle}_{s=N-K-R \text{ times}} |\phi\rangle|\psi\rangle = |\mathbf{0}\rangle_s|\phi\rangle|\psi\rangle, \tag{9.153}$$

where quantum state $|\psi\rangle$ corresponds to information qubits, while the quantum state $|\phi\rangle$ represents the gauge qubits. Therefore with transformation (9.152) we obtain the *canonical subsystem code*, that is, the subsystem code in which $s = N-K-R$ ancilla qubits in the $|0\rangle$ state are appended to information and gauge qubits. By applying the sequence of gates $\{U\}$ in Eq. (9.152) in opposite order we obtain the encoder of the subsystem code.

As an illustrative example, let us observe again the subsystem [4,1,1,2] code, whose gauge group can be represented using the finite geometry representation as follows:

$$\mathcal{G}=\left[\begin{array}{cccc|cccc} 1 & 1 & 1 & 1 & 0 & 0 & 0 & 0 \\ 0 & 0 & 0 & 0 & 1 & 1 & 1 & 1 \\ \hline 0 & 0 & 0 & 0 & 0 & 0 & 1 & 1 \\ 0 & 1 & 0 & 1 & 0 & 0 & 0 & 0 \end{array}\right].$$

By applying a sequence of transformations we can represent the [4,1,1,2] subsystem code in canonical form:

$$\mathcal{G}\xrightarrow{U_1=\mathrm{CNOT}^{1,2}\mathrm{CNOT}^{1,3}\mathrm{CNOT}^{1,4}}\left[\begin{array}{cccc|cccc} 1 & 0 & 0 & 0 & 0 & 0 & 0 & 0 \\ 0 & 0 & 0 & 0 & 0 & 1 & 1 & 1 \\ \hline 0 & 0 & 0 & 0 & 0 & 0 & 1 & 1 \\ 0 & 1 & 0 & 1 & 0 & 0 & 0 & 0 \end{array}\right]\xrightarrow{U_2=H_2H_3H_4}\left[\begin{array}{cccc|cccc} 1 & 0 & 0 & 0 & 0 & 0 & 0 & 0 \\ 0 & 1 & 1 & 1 & 0 & 0 & 0 & 0 \\ \hline 0 & 0 & 1 & 1 & 0 & 0 & 0 & 0 \\ 0 & 0 & 0 & 0 & 0 & 1 & 0 & 1 \end{array}\right]$$

$$\xrightarrow{U_3=\mathrm{CNOT}^{2,3}\mathrm{CNOT}^{2,4}}\left[\begin{array}{cccc|cccc} 1 & 0 & 0 & 0 & 0 & 0 & 0 & 0 \\ 0 & 1 & 0 & 0 & 0 & 0 & 0 & 0 \\ \hline 0 & 0 & 1 & 1 & 0 & 0 & 0 & 0 \\ 0 & 0 & 0 & 0 & 0 & 0 & 0 & 1 \end{array}\right]\xrightarrow{U_4=\mathrm{CNOT}^{4,3}}\left[\begin{array}{cccc|cccc} 1 & 0 & 0 & 0 & 0 & 0 & 0 & 0 \\ 0 & 1 & 0 & 0 & 0 & 0 & 0 & 0 \\ \hline 0 & 0 & 0 & 1 & 0 & 0 & 0 & 0 \\ 0 & 0 & 0 & 0 & 0 & 0 & 0 & 1 \end{array}\right]$$

$$\xrightarrow{U_5=H_1H_2}\left[\begin{array}{cccc|cccc} 0 & 0 & 0 & 0 & 1 & 0 & 0 & 0 \\ 0 & 0 & 0 & 0 & 0 & 1 & 0 & 0 \\ \hline 0 & 0 & 0 & 1 & 0 & 0 & 0 & 0 \\ 0 & 0 & 0 & 0 & 0 & 0 & 0 & 1 \end{array}\right].$$

By starting now from canonical codeword $|0\rangle|0\rangle|\psi\rangle|\phi\rangle$ and applying the sequence of transformations $\{U_i\}$ in opposite order we obtain the encoder shown in Fig. 9.11a. An alternative encoder can be obtained by swapping control and target qubits of U_3 and U_4 to obtain the encoder shown in Fig. 9.11b. By setting the gauge qubit to $|0\rangle$ and by several simple modifications we obtain the optimized encoder circuit shown in Fig. 9.11c.

Before concluding this section we provide the following theorem, whose proof has been left as a homework problem, which allows us to establish the connection between classical codes and quantum subsystem codes.

Theorem 2. Let C be a classical linear subcode of F_2^{2N} and let D denote its subcode $D=C\cap C^{\perp}$. If $x=|C|$ and $y=|D|$, then there exits subsystem code

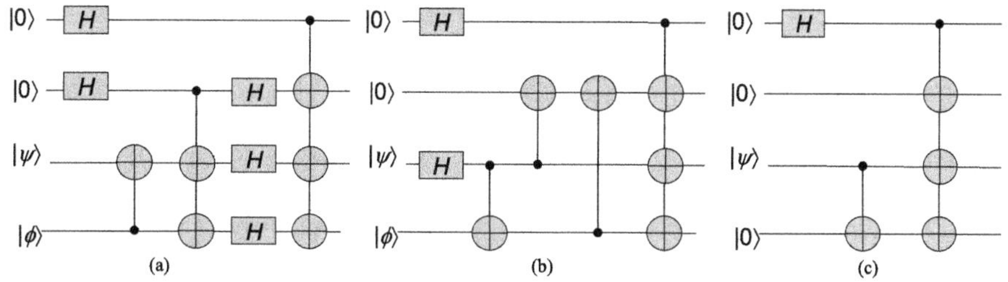

FIGURE 9.11

Encoder for [4,1,1,2] subsystem code by the conjugation method: (a) first version, (b) second version, and (c) optimized version.

$C_Q = A \otimes B$, where dim$A = q^N/(xy)^{1/2}$ and dim$B = (x/y)^{1/2}$. The minimum distance of subsystem A is given by:

$$D_{\min} = \begin{cases} \mathrm{wt}\left(\left(C + C^{\perp}\right) - C\right) = \mathrm{wt}\left(D^{\perp} - C\right), D^{\perp} \neq C \\ \mathrm{wt}\left(D^{\perp}\right), D^{\perp} = C \end{cases},$$
$$C^{\perp} = \left\{ u \in F_q^{2N} \middle| u \odot v = 0 \ \ \forall v \in C \right\}. \tag{9.154}$$

The minimum distance of $[N,K,R]$ subsystem code $D_{\min}$, given by Eq. (9.154), is more challenging to determine than that of stabilizer code because of the set difference. The definition can be relaxed and we can define minimum distance as $D_m = \mathrm{wt}(C)$. Furthermore, we say that subsystem code is nondegenerate (pure) if $D_m \geq D_{\min}$. The Hamming and Singleton bounds can now be simply determined by observing the corresponding bounds of nonbinary stabilizer codes. Since dim$A = q^K$, dim$B = q^R$, the Singleton bound of $[N,K,R,D_{\min}]$ subsystem code is given by:

$$K + R \leq N - 2D_{\min} + 2. \tag{9.155}$$

The corresponding Hamming quantum bound of subsystem code is given, based on Eq. (9.137), by:

$$\sum_{i=0}^{\lfloor (D_{\min}-1)/2 \rfloor} \binom{N}{i} \left(q^2 - 1\right)^i q^{K+R} \leq q^N. \tag{9.156}$$

9.8 TOPOLOGICAL CODES

Topological QECCs [40–50] are typically defined on a 2D lattice, with quantum parity check being geometrically local. The locality of parity check is of crucial importance since the syndrome measurements are easy to implement when the qubits involved in syndrome verification are in close proximity to each other. In addition, the possibility to implement the universal quantum gates topologically yields to the increased interest in topological quantum codes. On the other hand, someone may argue that quantum low-density parity-check (LDPC) codes can be designed on such a way that quantum parity checks are only local. In addition, through the concept of subsystem codes, the portion of qubits can be converted to qubits not carrying any encoded information, but instead serving as the gauge qubits. The gauge qubits can be used to "absorb" the effect of errors. Moreover, the subsystem codes allow syndrome measurements with a smaller number of qubit interactions. The combination of locality of topological codes and small number of interactions leads us to another generation of QECCs, known as topological subsystem codes due to Bombin [43].

Topological codes on a square lattice such as Kitaev's toric code [40,41], quantum lattice code with a boundary [42], surface codes, and planar codes [43] are easier to design and implement [44–46]. These basic codes on qubits can be generalized to higher alphabets [47,48] resulting in quantum topological codes on qudits. In this section we are concerned only with quantum topological codes on qubits, the corresponding topological codes on qudits can be obtained following a similar analogy provided in Section 9.6 (see also [47,48]).

Kitaev's toric code [40,41] is defined on a square lattice with periodic boundary conditions, meaning that the lattice has a topology of torus as shown in Fig. 9.12.

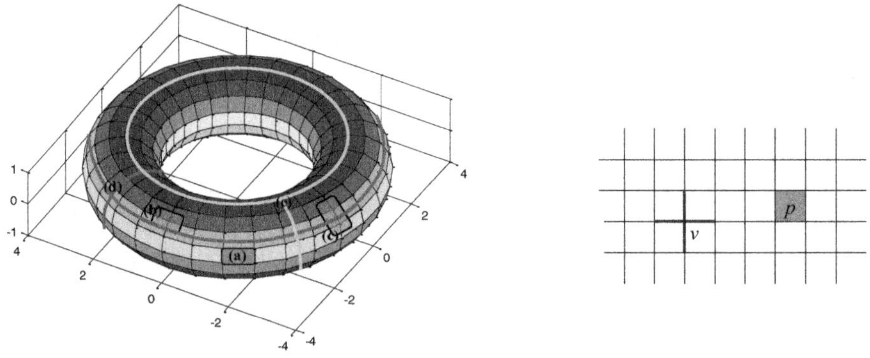

FIGURE 9.12

Square lattice on the torus. The qubits are associated with the edges. The figure on the right shows an enlarged portion of square lattice on the torus. The *Z*-containing operators represent string on the lattice, while the *X*-containing operators represent strings on the dual lattice. The elementary trivial loop (cycle) denoted with (a) corresponds to a certain B_p operator, while the elementary trivial cycle denoted with (b) corresponds to a certain A_v operator on the lattice dual. The trivial loops can be obtained as the product of individual elementary loops, see, for example, the cycle denoted with (c). The nontrivial loops on the dual lattice, such as (d), or on the lattice, such as (e), correspond to the generators from the set of encoded (logical) Pauli operators.

The qubits are associated with the edges of the lattice. For an $m \times m$ square lattice on the torus, there are $N = 2m^2$ qubits. For each vertex v and plaquette (or face) p (Fig. 9.13(right)) we associate the stabilizer operators as follows:

$$A_v = \bigotimes_{i \in n(v)} X_i \quad \text{and} \quad B_p = \bigotimes_{i \in n(p)} Z_i, \tag{9.157}$$

where X_i and Z_i denote the corresponding Pauli X- and Z-operators on position i. With $n(v)$ we denote the neighborhood of vertex v, that is, the set of edges incident to vertex v, while with $n(p)$ we denote the neighborhood of plaquette p, that is, the set of edges encircling the face p. (We use the symbol $\otimes$ to denote the tensor product, as we have done earlier.) Based on Pauli operator properties (see Chapter 2), it is clear that operators A_v and B_p mutually commute. Namely, the commutation of operators A_v among themselves is trivial to prove. Since A_v and B_p operators have either zero or two common edges (Fig. 9.12 (right)), they commute (an even number of anti-commuting Pauli operators results in commuting stabilizer operators). The operators A_v and B_p are Hermitian and have eigenvalues 1 and -1. Let H_2^N denote the Hilbert space, where $N = 2m^2$. The toric code space C_Q can be defined as follows:

$$C_Q = \left\{ |c\rangle \in H_2^N \middle| A_v|c\rangle = |c\rangle, B_p|c\rangle = |c\rangle; \forall v, p \right\}. \tag{9.158}$$

Eq. (9.158) (see the content inside the brackets) represents the eigenvalue equation, and since A_v and B_p mutually commute, they have common eigenkets. The stabilizer group is defined by $S = \langle A_v, B_p \rangle$.

Since $\prod_v A_v = I$, $\prod_p B_p = I$, there are $M = 2m^2 - 2$ independent stabilizer operators. The number of information qubits can be determined by $KN - M = 2$ and the code space dimensionality is $\dim C_Q = 2^{N-K} = 4$. Clearly, this toric code has a low quantum code rate. Another interesting property to notice is that stabilizers A_v contain only X-operators, while stabilizers B_p contain only Z-operators. Clearly, this toric code represents a particular instance of Calderbank, Shor, and Steane (CSS) codes, described in Chapter 8. The encoder and decoder can therefore be equivalently implemented as described in Section 9.5. We saw earlier that commutation of A_v and B_p arises from the fact that these two operators have either zero or two common edges, which is equivalent to the classical code with the property that any two rows overlap in an even number of positions, representing a dual-containing code.

Let us now observe the error E acting on codeword $|c\rangle$, which results in state $E|c\rangle$, that is, not necessarily the eigenket with eigenvalue $+1$ of vertex and plaquette operators. If E and A_p commute, then the ket $E|c\rangle$ will be the $+1$ eigenket; otherwise, it will be the -1 eigenket. A similar conclusion applies to B_p. By performing the measurements on each vertex A_v and plaquette B_p operators we obtain the syndrome pair $s = (s_v, s_p)$, where $s_i = \pm 1$, $i \in \{v, p\}$. The intersection of syndrome pairs will give us the set of errors having the same syndrome, namely the coset (the set of errors that differ in an element from stabilizer S). Out of all possible errors from the coset we choose the most probable one, typically the lowest weight one. We can also consider stabilizer elements individually instead of in pairs, and the corresponding error syndrome can be obtained as explained in the previous chapter or in Sections 9.1 and 9.3. Another approach would be to select the most likely error compatible with the syndrome, namely:

$$E_{\mathrm{ML}} = \arg \max_{E \in \mathscr{E}} P(E), \quad \mathscr{E} = \{E | EA_v = s_v A_v E,\ EB_p = s_p B_p E\}, \tag{9.159}$$

where $P(E)$ is the probability of occurrence of a given error E, and subscript ML is used to denote maximum-likelihood decision. For example, for the depolarizing channel from the previous chapter, $P(E)$ is given by:

$$P(E) = (1-p)^{N-\mathrm{wt}(E)} \frac{p^{\mathrm{wt}(E)}}{3}, \tag{9.160}$$

where with probability $1 - p$ we leave the qubit unaffected and apply the Pauli operators X, Y, and Z each with probability $p/3$. The error correction action is to apply the error E_{ML} selected according to Eq. (9.159). If the proper error was selected, the overall action of the quantum channel and error correction would be $E^2 = \pm I$. Notice that this approach is not necessarily the optimum approach as it will become apparent soon. From Fig. 9.12, it is clear that only Z-containing operators correspond to the strings on the lattice, while only X-containing operators correspond to the strings on the dual lattice. The plaquette operators are represented as elementary cycles (loops) on the lattice, while the vertex operators are represented by elementary loops on the dual lattice. Trivial loops can be obtained as the product of elementary loops. Therefore the plaquette operators generate the group of homologically trivial loops on the torus, while the vertex operators generate homologically trivial loops on the dual lattice. Two error operators E_1 and E_2 have the same effect on a codeword $|c\rangle$ if E_1E_2 contains only homologically trivial loops, and we say that these two errors are homologically equivalent. This concept is very similar to the error coset concept introduced in the previous chapter. There exist four independent operators that perform the mapping of stabilizer S to itself. These operators commute with all operators from S (both A_v's and B_p's), but do not belong to S and are known as encoded (logical) Pauli operators. The encoded Pauli operators correspond to the homologically nontrivial loops of Z and X type, as shown in Fig. 9.12. Let the set of logical Pauli operators be denoted by L. Because the operators $l \in L$ and ls for every $s \in S$ are equivalent, only homology classes (cosets) are important. The corresponding maximum-likelihood error operator selection can be performed as follows:

$$l_{\mathrm{ML}} = \underset{l \in L}{\mathrm{argmax}} \sum_{s \in S} P(E = lsR(\boldsymbol{s})), \tag{9.161}$$

where $R(\boldsymbol{s})$ is the representative error from the coset corresponding to syndrome $\boldsymbol{s}$. The error correction action is performed by applying the operator $l_{\mathrm{ML}}R(\boldsymbol{s})$. Notice that the foregoing algorithm can be computationally extensive. An interesting algorithm of complexity of $O(m^2 \log m)$ can be found in[44,45]. Since we have already established the connection between quantum LDPC codes and torus codes we can use the LDPC decoding algorithms described in Chapter 10 instead. We now determine encoded (logical) Pauli operators following a description due to Bravyi and Kitaev [42]. Let us consider the operator of the form:

$$\overline{Y}(c, c') = \prod_{i \in c} Z_i \prod_{j \in c'} X_j, \tag{9.162}$$

where c is an I-cycle (loop) on the lattice and c' is an I-cycle on the dual lattice. It can be simply been shown that $\overline{Y}(c,\ c')$ commutes with all stabilizers and thus it performs mapping of code subspace C_Q to itself. Since this mapping depends on homology classes (cosets) of c and c' we can denote the encoded Pauli operators by $\overline{Y}([c],\ [c'])$. It can be shown that $\overline{Y}([c],\ [c'])$ forms a linear basis for $L(C_Q)$. The logical operators $\overline{X}_1(0,\ [c_1'])$, $\overline{X}_2(0,\ [c_2'])$, $\overline{Z}_1([c_1], 0)$, and $\overline{Z}_2([c_2], 0)$, where c_1,c_2 are the cycles on the original lattice and c'_1,c'_2 are the cycles on the dual lattice, can be used as generators of $L(C_Q)$.

We turn our attention now to the quantum topological codes on a *lattice with boundary*, introduced by Bravyi and Kitaev [42]. Instead of dealing with lattices on the torus, here we consider the finite square lattice on the plane. Two types of boundary can be introduced: z-type boundary shown in Fig. 9.13a and x-type boundary shown in Fig. 9.13b. In Fig. 9.13c we provide an illustrative example [42] for a 2×3 lattice.

The simplest form of quantum boundary code can be obtained by simply alternating x-type and z-type boundaries, as illustrated in Fig. 9.13c. Since an $n \times m$ lattice has $(n + 1)(m + 1)$ horizontal edges and nm vertical edges, by associating the edges with qubits, the corresponding codeword length will be $N = 2nm + n + m + 1$. The stabilizers can be formed in a very similar fashion to toric codes. The vertex

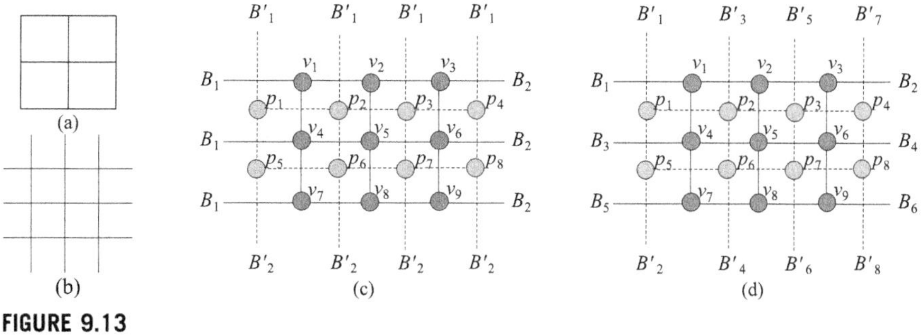

FIGURE 9.13

Quantum codes derived from a square lattice on the plane: (a) z-type boundary, (b) x-type boundary, (c) 2×3 lattice-based code, and (d) generalization of topological code (c). With B_1 and B_2 we denote the boundaries of the original lattice, while with B'_1 and B'_2 we denote the boundaries of the dual lattice.

Modified from A.Y. Kitaev, Fault-tolerant quantum computation by anyons, Ann. Phys. 303 (1) (2003) 2–30. Available at: http://arxiv.org/abs/quant-ph/9707021.

and plaquette operators can be defined in a fashion similar to Eq. (9.157), and the code subspace can be introduced by Eq. (9.158). The free ends of edges do not contribute to stabilizers. Notice that inner vertex and plaquette stabilizers are of weight 4, while outer stabilizers are of weight 3. This quantum code is an example of irregular quantum stabilizer code. The number of plaquette and vertex stabilizers is determined by $n(m+1)$ and $(n+1)m$, respectively. For example, as shown in Fig. 9.13c, the vertex stabilizers are given by:

$$A_{v_1} = X_{B_1v_1}X_{v_1v_2}X_{v_1v_4}, \quad A_{v_2} = X_{v_1v_2}X_{v_2v_3}X_{v_2v_5}, \quad A_{v_3} = X_{v_2v_3}X_{v_3B_2}X_{v_3v_6},$$
$$A_{v_4} = X_{B_1v_4}X_{v_4v_5}X_{v_1v_4}X_{v_4v_7}, \quad A_{v_5} = X_{v_4v_5}X_{v_5v_6}X_{v_2v_5}X_{v_5v_8}, \quad A_{v_6} = X_{v_5v_6}X_{v_6B_2}X_{v_3v_6}X_{v_6v_9},$$
$$A_{v_7} = X_{B_1v_7}X_{v_7v_8}X_{v_4v_7}, \quad A_{v_8} = X_{v_7v_8}X_{v_8v_9}X_{v_5v_8}, \quad A_{v_9} = X_{v_8v_9}X_{v_9B_2}X_{v_6v_9}.$$

On the other hand, the plaquette stabilizers are given by:

$$B_{p_1} = Z_{B'_1p_1}Z_{p_1p_5}Z_{p_1p_2}, \quad B_{p_2} = Z_{B'_1p_2}Z_{p_2p_6}Z_{p_1p_2}Z_{p_2p_3}, \quad B_{p_3} = Z_{B'_1p_3}Z_{p_3p_7}Z_{p_2p_3}Z_{p_3p_4},$$
$$B_{p_4} = Z_{B'_1p_4}Z_{p_4p_8}Z_{p_3p_4}, \quad B_{p_5} = Z_{p_1p_5}Z_{p_5B'_2}Z_{p_5p_6}, \quad B_{p_6} = Z_{p_2p_6}Z_{p_6B'_2}Z_{p_5p_6}Z_{p_6p_7},$$
$$B_{p_7} = Z_{p_3p_7}Z_{p_7B'_2}Z_{p_6p_7}Z_{p_7p_8}, \quad B_{p_8} = Z_{p_4p_8}Z_{p_8B'_2}Z_{p_7p_8}.$$

From this example, it is clear that the number of independent stabilizers is given by $N-K=2nm+n+m$, so that the number of information qubits of this code is only $K=1$. The encoded (logical) Pauli operators can also be obtained by Eq. (9.162), but now with the *I*-cycle c (c') being defined as a string that ends on the boundary of the lattice (dual lattice). An error E is undetectable if it commutes with stabilizers, that is, it can be represented as a linear combination of operators given by Eq. (9.162). Since the distance is defined as the minimum weight of error that cannot be detected, the *I*-cycle c (c') has the length $n+1$ ($m+1$), and the minimum distance is determined by $D=\min(n+1,m+1)$. The parameters of this quantum code are $[2nm+n+m+1, 1, \min(n+1,m+1)]$. Clearly, by letting one dimension go to infinity, the minimum distance will be infinitely large, but quantum code rate will tend to zero.

This class of codes can be generalized in many different ways: (1) by setting $B_1=B_2=B$ we obtain the square lattice on the *cylinder*, (2) by setting $B_1=B_2=B'_1=B'_2=B$ we obtain the square lattice on the *sphere*, and (3) by assuming that free edges end at different boundaries we obtain a stronger code, since the number of cycles of length 4 in the corresponding bipartite graph representation (see Chapter 10 for definition) is smaller. The 2×3-based lattice code belonging to class (3) is shown in Fig. 9.13d, and its properties are studied in Problem 28. Notice that the lattice does not need to be square/rectangular, which can further be used as another generalization. Several examples of such codes are provided in [42,43]. By employing the good properties of topological codes and subsystem codes we can design the so-called subsystem topological codes [44].

9.9 SURFACE CODES

Surface codes are closely related to the quantum topological codes on the boundary, introduced by Bravyi and Kitaev [42,43]. Given their popularity and possible application in quantum computing [51–57], this section is devoted to this class of quantum codes. There exist various versions of surface codes; here we describe two very popular ones. The surface code is defined on a 2D lattice, such as the one shown in Fig. 9.14, with qubits being clearly indicated in Fig. 9.14a. The stabilizers of plaquette type can be defined as provided in Fig. 9.14b. Each plaquette stabilizer denoted by X (Z) is composed of Pauli X (Z)-operators on qubits located in the intersection of edges of the corresponding plaquette. As an illustration, the plaquette stabilizer denoted by X related to qubits 1 and 2 will be X_1X_2. On the other hand, the plaquette stabilizer denoted by Z related to qubits 5, 6, 8, and 9 will be $Z_5Z_6Z_8Z_9$. To simplify the notation we can use the representation provided in Fig. 9.14c, where the shaded plaquettes correspond to all X-containing operators' stabilizers, while the white plaquettes correspond to all Z-containing operators' stabilizers. The weight-2 stabilizers are allocated around the perimeter, while the weight-4 stabilizers are located in the interior. If we denote the set of all blue (white) plaquettes with $\{B_p\}$ ($\{W_p\}$) we can define the stabilizer on plaquette p as follows:

$$S_p = \underset{i \in p}{\otimes} \sigma_i, \sigma_i = \begin{cases} X_i, \text{for } p \in B_p \\ Z_i, \text{for } p \in W_p \end{cases}. \tag{9.163}$$

All stabilizers commute and have the eigenvalues +1 and −1. The surface code (SC) can now be defined in similar fashion as given by Eq. (9.158), namely $C_{SC} = \{|c\rangle \in H_2^N | S_p|c\rangle = |c\rangle; \forall p\}$. Given that the surface code space is 2D, it encodes a single logical qubit.

The logical operators preserve the code space and as such are used to manipulate the logical qubit state. For surface code, the logical operators are run over both sides of the lattice as shown in Fig. 9.14d and can be represented as

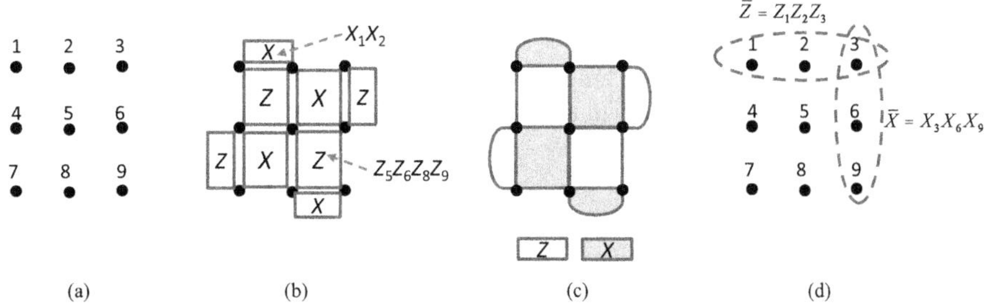

FIGURE 9.14

Surface code: (a) the qubits are located in the lattice positions, (b) all-X and all-Z plaquette operators, (c) popular representation of surface codes in which X- and Z-plaquettes are clearly indicated, and (d) logical operators.

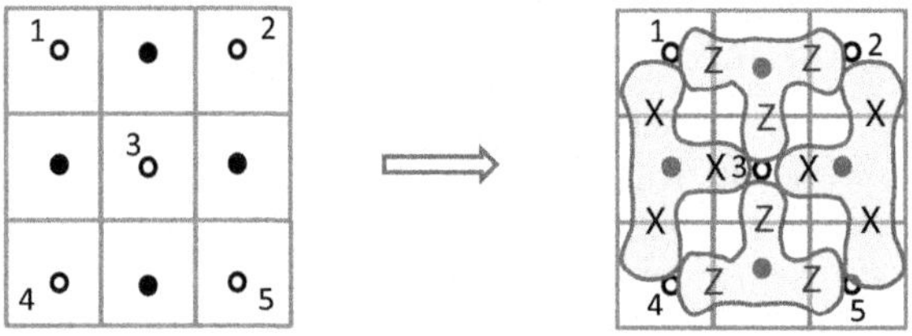

FIGURE 9.15

Another interpretation of surface code.

$\overline{Z} = Z_1Z_2Z_3$, $\overline{X} = X_3X_6X_9$. Clearly, the logical operator spans two boundaries of the same type. The codeword length is determined as the product of side lengths, expressed in number of qubits, and for the surface code from Fig. 9.14 we have that $N = L_X \cdot L_Z = 3 \cdot 3 = 9$. On the other hand, the number of information qubits is $K = 1$. The minimum distance of this code is determined as the minimum side length, that is, $d = \min(L_X, L_Z) = 3$, indicating that this code can correct a single-qubit error. Let E denote the error operator applied on codeword $|c\rangle$, that is, we can write $E|c\rangle$. Let us assume that this error operator represents the single physical qubit Pauli flip. The $E|c\rangle$ will anticommute with all codewords $|c'\rangle$, that is, $\langle c'|E|c\rangle = -1$. By placing ancilla qubits in the middle of each plaquette we can perform the syndrome measurements. The intersection of syndromes can help identify the most probable error operator. Given that the surface codes are topological codes, the decoding algorithm described in the previous section is applicable here as well.

Let us provide another description of surface code. Let us now place the qubits on the dual lattice, as illustrated in Fig. 9.15. There are two types of qubits: quiescent states, denoted by open (white) circles, and measurement qubits, denoted by filled (black) circles. We manipulate the quiescent states to preform desired quantum computing. The measurement qubits serve the role of ancilla qubits and are used to facilitate syndrome measurements. In this particular example, there are five data qubits (quiescent states) and four measurement qubits. Clearly, the measurement qubits are associated with either all-X or all-Z stabilizers. From Fig. 9.15 (right) we can read out the following stabilizers for this surface code: $Z_1Z_2Z_3$, $X_1X_3X_4$, $X_2X_3X_5$, and $Z_3Z_4Z_5$. Since the codeword length is $N = 5$, the number of stabilizer generators $N - K = 4$, we conclude that the number of information qubits are $N - (N - K) = 1$ and surface code dimensionality is $2^K = 2$.

9.10 ENTANGLEMENT-ASSISTED QUANTUM CODES

In this section, the entanglement-assisted (EA) QECCs [19,58–62] are described, which make use of preexisting entanglement between transmitter and receiver to improve the reliability of transmission. EA quantum codes can be considered as a generalization of superdense coding [38,58]. They can also be considered as particular instances of subsystem codes [28–30], described in Section 9.7. The key advantage of EA quantum codes compared to CSS codes is that EA quantum codes do not require that corresponding classical codes, from which they are derived, be dual containing. Namely, arbitrary classical code can be used to design EA quantum codes, providing that there exists a preentanglement between source and destination. In this chapter, the notation due to Brun, Devetak, Hsieh, and Wilde [19,58–62] is used, who can be considered as inventors of this class of quantum codes.

9.10.1 Introduction to Entanglement-Assisted Quantum Codes

The EA-QECCs make use of preexisting entanglement between transmitter and receiver to improve the reliability of transmission [19,58–62]. In CSS codes only dual-containing classical codes can be used. In the EA-QECC concept, arbitrary classical code can be used as quantum code, providing that there exists entanglement between sender and receiver. Namely, if classical codes are not dual containing, they correspond to a set of stabilizer generators that do not commute. By providing the entanglement between source and destination, the corresponding generators can be embedded into a larger set of commuting generators, which gives a well-defined code space. We start our discussion with *superdense coding* [38]. In this scheme, transmitter (Alice) and receiver (Bob) share the entangled state, also known as entanglement qubit (ebit) state [19,58,60]:

$$|\Phi\rangle = \frac{1}{\sqrt{2}}(|0\rangle \otimes |0\rangle + |1\rangle \otimes |1\rangle).$$

The first half of entangled pair belongs to Alice and the second half to Bob. The state $|\Phi\rangle$ is the simultaneous $(+1,+1)$ eigenstate of the commuting operators $Z \otimes Z$ and $X \otimes X$. Alice sends a 2-bit message $(a_1,a_2) \in F_2^2$ to Bob as follows. Alice performs the encoding operation on half of the ebit based on the message (a_1,a_2):

$$\left(Z^{a_1}X^{a_2} \otimes I^B\right)|\Phi\rangle = |a_1, a_2\rangle. \tag{9.164}$$

Alice sends the encoded state to Bob over the "perfect" qubit channel. Bob performs decoding based on measurement in

$$\left\{\left(Z^{a_1}X^{a_2} \otimes I^B\right)|\Phi\rangle : (a_1, a_2) \in F_2^2\right\} \tag{9.165}$$

basis, that is, by simultaneously measuring the $Z \otimes Z$ and $X \otimes X$ observables. This scheme can be generalized as follows. Alice and Bob share the state $|\Phi\rangle^{\otimes m}$ that is the simultaneous $+1$ eigenstate of:

$$\begin{aligned} &Z(e_1) \otimes Z(e_1)\ldots Z(e_m) \otimes Z(e_m) \wedge X(e_1) \otimes X(e_1)\ldots X(e_m) \otimes X(e_m); \\ &e_i = (0\ldots 01_i 0\ldots 0). \end{aligned} \tag{9.166}$$

Alice encodes the message $(a_1,a_2) \in F_2^{2m}$, producing the encoded state:

$$\left(Z(a_1)X(a_2) \otimes I^B\right)|\Phi\rangle = |a_1,\ a_2\rangle. \tag{9.167}$$

Bob performs decoding by simultaneously measuring the following observables:

$$Z(e_1) \otimes Z(e_1)\ldots Z(e_m)\otimes Z(e_m)\wedge X(e_1)\otimes X(e_1)\ldots X(e_m)\otimes X(e_m). \tag{9.168}$$

Let G_N be the Pauli group on N qubits. For any non-Abelian subgroup $S \subset G_N$ of size 2^{N-K}, there exists the sets of generators for S $\{\underline{Z}_1,\ldots,\underline{Z}_{s+e}, \underline{X}_{s+1},\ldots,\underline{X}_{s+e}\}$ $(s + 2e = N - K)$ with the following commutation properties:

$$[\underline{Z}_i, \underline{Z}_j] = 0, \forall i,j;\ [\underline{X}_i, \underline{X}_j] = 0, \forall i,j;\ [\underline{X}_i, \underline{Z}_j] = 0, \forall i \neq j;\ \{\underline{Z}_i, \underline{X}_i\} = 0, \forall i. \tag{9.169}$$

The non-Abelian group can be partitioned into:

1. A commuting subgroup, *isotropic subgroup*, $S_I = \{Z_1,\ldots,Z_a\}$ and
2. An *entanglement subgroup* $S_E = \{Z_{e+1}, Z_{e+c}, X_{e+1},\ldots,X_{e+c}\}$ with anticommuting pairs. The anticommuting pairs (Z_i, X_i) are shared between source A and destination B.

For example, the state shared between A and B, the ebit,

$$|\phi^{AB}\rangle = \frac{1}{\sqrt{2}}\left(|00\rangle^{AB} + |11\rangle^{AB}\right),$$

has two operators to fix it $X^A X^B$ and $Z^A Z^B$. These two operators commute $[X^A X^B, Z^A Z^B] = 0$, while local operators anticommute $\{X^A, X^B\} = \{Z^A, Z^B\} = 0$. We can therefore use the entanglement to resolve anticommutativity of two generators in non-Abelian group S. From the previous chapter we know that two basic properties of Pauli operators are: (1) two Pauli operators *commute* if and only if there is an even number of places where they have different Pauli matrices neither of which is the identity I, and (2) if two Pauli operators do not commute, they anticommute, since their individual Pauli matrices either commute or anticommute. Let us observe the following four-qubit operators:

$$\begin{array}{cccc|c} Z & X & Z & I & X \\ Z & Z & I & Z & Z \\ Y & X & X & Z & I \\ Z & Y & Y & X & I \end{array}.$$

Clearly, they do not commute. However, if we extend them by adding an additional qubit as given earlier, they can be embedded in the larger space in which they commute.

The operation principle of EA quantum error correction is illustrated in Fig. 9.16. The sender A (Alice) encodes quantum information in state $|\psi\rangle$ with the help of local ancilla qubits $|0\rangle$ and her half of shared ebits $|\phi^{AB}\rangle$, and then sends the encoded qubits over a noisy quantum channel (say free-space optical channel or optical fiber). The receiver B (Bob) performs multiqubit measurement on all qubits to diagnose the channel error and performs a recovery unitary operation R to reverse the action of the

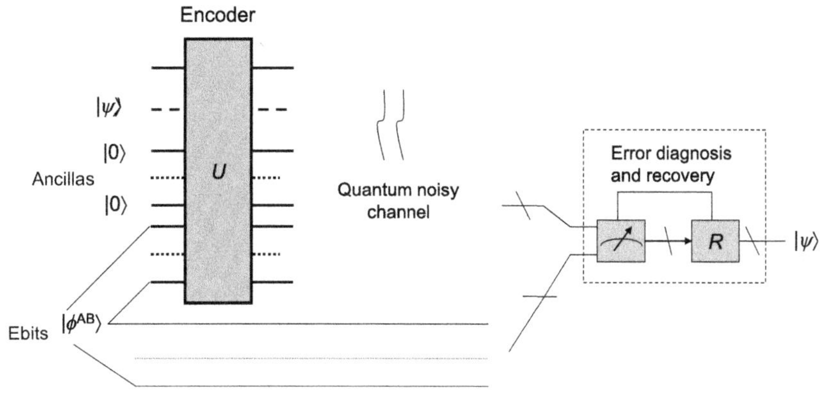

FIGURE 9.16

Entanglement-assisted quantum error correction operation principle.

channel. Notice that the channel does not affect at all the receiver's half of shared ebits. The *operation steps* of an EA quantum error correction scheme:

1. The sender and receiver share e ebits before quantum communication starts, and senders employs a ancilla qubits. The *unencoded state* is a simultaneous $+1$ eigenstate of the following operators:

$$\{Z_{a+1}|Z_1, \ldots, Z_{a+e}|Z_e, X_{a+1}|X_1, \ldots, X_{a+e}|X_e, Z_1, \ldots, Z_a\}, \tag{170.1}$$

where the first half of ebits correspond to the sender, and the second half to the receiver. The sender encodes K information qubits with the help of a ancilla qubits and the sender's half of the e ebits. The encoding unitary transforms the unencoded operators to the following encoded operators:

$$\{\underline{Z}_{a+1}|Z_1, \ldots, \underline{Z}_{a+e}|Z_e, \underline{X}_{a+1}|X_1, \ldots, \underline{X}_{a+e}|X_e, \underline{Z}_1, \ldots, \underline{Z}_a\}. \tag{170.2}$$

2. The sender transmits N qubits over a noisy quantum channel, which affects only these N qubits, not the receiver's half of e ebits.
3. The receiver combines the received qubits with c ebits and performs the measurement on all $N + e$ qubits.
4. Upon identification of the error, the receiver performs a recovery operation to reverse the quantum channel error.

9.10.2 Entanglement-Assisted Quantum Error Correction Principles

The block scheme of EA quantum code, which requires a certain number of ebits to be shared between the source and destination, is shown in Fig. 9.17. The source encodes quantum information in the K-qubit state $|\psi\rangle$ with the help of local ancilla

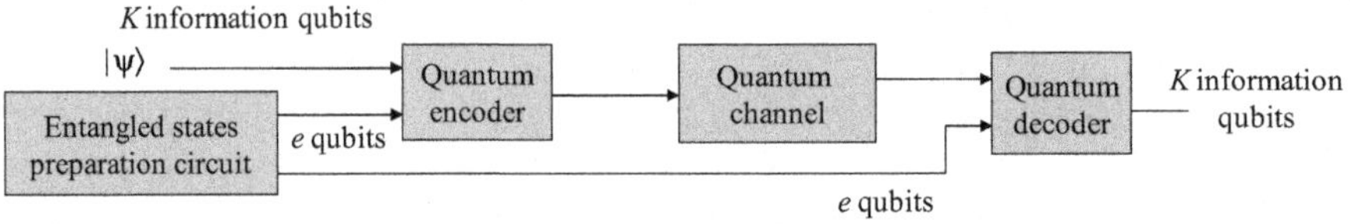

FIGURE 9.17

Generic entanglement-assisted quantum error correction scheme.

qubits $|0\rangle$ and the source half of shared ebits (e ebits) into N qubits, and then sends the encoded qubits over a noisy quantum channel (such as a free-space or fiber-optic channel). The receiver performs decoding on all qubits ($N+e$ qubits) to diagnose the channel error and performs a recovery unitary operation to reverse the action of the channel. Notice that the channel does not affect at all the receiver's half of the shared ebits. By omitting the ebits, the conventional quantum coding scheme is obtained.

As was discussed in previous sections, both noise $\mathcal{N}$ and recovery $\mathcal{R}$ processes can be described as quantum operations $\mathcal{N},\mathcal{R}$: $L(H) \rightarrow L(H)$, where $L(H)$ is the space of linear operators on Hilbert space H. These mappings can be represented in terms of operator-sum representation:

$$\mathcal{N}(\rho)=\sum_i E_i\rho E_i^\dagger,\quad E_i\in L(H),\quad \mathcal{N}=\{E_i\}. \tag{9.171}$$

Given the quantum code C_Q that is a subspace of H we say that the set of errors $\mathcal{E}$ is correctable if there exists a recovery operation $\mathcal{R}$ such that $\mathcal{R}\mathcal{N}(\rho)=\rho$ from any state ρ from $L(C_Q)$. Furthermore, each error operator E_i can be expanded in terms of Pauli operators:

$$\begin{aligned}
&E_i=\sum_{[\boldsymbol{a}|\boldsymbol{b}]\in F_2^{2N}}\alpha_{i,[\boldsymbol{a}|\boldsymbol{b}]}E_{[\boldsymbol{a}|\boldsymbol{b}]},\quad E_{[\boldsymbol{a}|\boldsymbol{b}]}=j^\lambda X(\boldsymbol{a})Z(\boldsymbol{b});\ \ \boldsymbol{a}=a_1\ldots a_N;\ \ \boldsymbol{b}=b_1\ldots b_N;\\
&\qquad a_i,\ b_i=0,\ 1;\ \ \lambda=0,\ 1,\ 2,\ 3\\
&X(\boldsymbol{a})\equiv X_1^{a_1}\otimes\ldots\otimes X_N^{a_N};Z(\boldsymbol{b})\equiv Z_1^{b_1}\otimes\ldots\otimes Z_N^{b_N}.
\end{aligned} \tag{9.172}$$

An $[N,K,e;D]$ EA-QECC consists of: (1) an encoding map:

$$\widehat{U}_{\text{enc}}: L^{\otimes K}\otimes L^{\otimes e}\rightarrow L^{\otimes N}, \tag{9.173}$$

and (2) a decoding trace-preserving mapping:

$$\mathcal{D}: L^{\otimes N}\otimes L^{\otimes e}\rightarrow L^{\otimes K}, \tag{9.174}$$

such that the composition of encoding mapping, channel transformation, and decoding mapping is the identity mapping:

$$\mathcal{D}\circ\mathcal{N}\circ\breve{U}_{\text{enc}}\circ\breve{U}_{\text{app}}=I^{\otimes K}; I: L\rightarrow L;\ U_{\text{app}}|\psi\rangle=|\psi\rangle|\Phi^{\otimes e}\rangle, \tag{9.175}$$

where U_{app} is an operation that simply appends the transmitter half of the maximum-entangled state $|\Phi^{\otimes e}\rangle$.

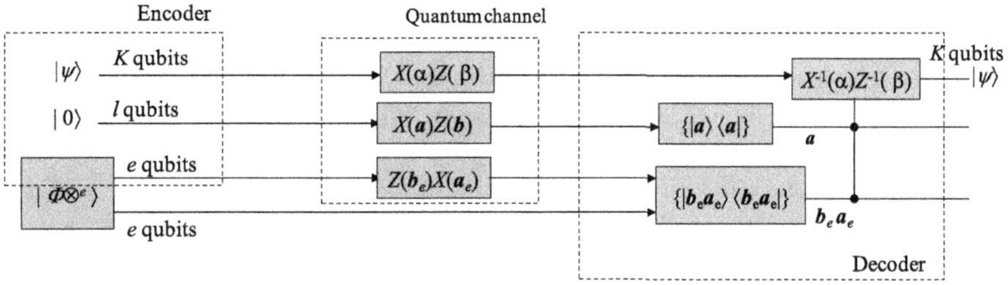

FIGURE 9.18

Entanglement-assisted canonical code.

9.10.3 Entanglement-Assisted Canonical Quantum Codes

The simplest EA quantum code, EA canonical code, performs the following trivial encoding operation:

$$\widehat{U}_c: |\psi\rangle \rightarrow |\mathbf{0}\rangle_l \otimes |\Phi^{\otimes e}\rangle \otimes |\psi\rangle_K; l = N - K - e. \tag{9.176}$$

The canonical EA code, shown in Fig. 9.18, is therefore an extension of the canonical code presented in Chapter 8, in which e ancilla qubits are replaced by e maximally entangled kets shared between transmitter and receiver. Since the number of information qubits is K and codeword length is N, the number of remaining ancilla qubits in the $|0\rangle$ state is $l = N - K - e$.

The EA canonical code can correct the following set of errors:

$$\begin{aligned}N_{\text{canonical}} = &\left\{X(\boldsymbol{a})Z(\boldsymbol{b}) \otimes Z(\boldsymbol{b}_{\text{eb}})X(\boldsymbol{a}_{\text{eb}}) \otimes X(\alpha(\boldsymbol{a},\, \boldsymbol{b}_{\text{eb}},\, \boldsymbol{a}_{\text{eb}}))Z(\beta(\boldsymbol{a},\, \boldsymbol{b}_{\text{eb}},\, \boldsymbol{a}_{\text{eb}})):\right.\\ &\left.\boldsymbol{a},\, \boldsymbol{b} \in F_2^l; \boldsymbol{b}_{\text{eb}},\, \boldsymbol{a}_{\text{eb}} \in F_2^e\right\}\\ \alpha,\, \beta: &F_2^l \times F_2^e \times F_2^e \rightarrow F_2^K,\end{aligned} \tag{9.177}$$

where the Pauli operators $X(\boldsymbol{c})$ ($\boldsymbol{c} \in \{\boldsymbol{a}, \boldsymbol{a}_{\text{eb}}, \alpha(\boldsymbol{a},\boldsymbol{b}_{\text{eb}},\boldsymbol{a}_{\text{eb}})\}$) and $Z(\boldsymbol{d})$ ($\boldsymbol{d} \in \{\boldsymbol{b}, \boldsymbol{b}_{\text{eb}}, \beta(\boldsymbol{a},\boldsymbol{b}_{\text{eb}},\boldsymbol{a}_{\text{eb}})\}$) are introduced in Eq. (9.172).

To prove claim (9.177), let us observe one particular error E_c from the set of errors (9.177) and study its action on transmitted codeword $|\mathbf{0}\rangle_l \otimes |\Phi^{\otimes e}\rangle \otimes |\psi\rangle_K$:

$$\begin{aligned}&X(\boldsymbol{a})Z(\boldsymbol{b})|\mathbf{0}\rangle_l \otimes \left(Z(\boldsymbol{b}_{\text{eb}})X(\boldsymbol{a}_{\text{eb}}) \otimes I^{R_x}\right)|\Phi^{\otimes e}\rangle \otimes X(\alpha(\boldsymbol{a},\boldsymbol{b}_{\text{eb}},\boldsymbol{a}_{\text{eb}}))\\ &Z(\beta(\boldsymbol{a},\boldsymbol{b}_{\text{eb}},\boldsymbol{a}_{\text{eb}}))|\psi\rangle.\end{aligned} \tag{9.178}$$

The action of operator $X(\boldsymbol{a})Z(\boldsymbol{b})$ on state $|\mathbf{0}\rangle_l$ is given by $X(\boldsymbol{a})Z(\boldsymbol{b})|\mathbf{0}\rangle_l \overset{Z(\boldsymbol{b})|\mathbf{0}\rangle_l = |\mathbf{0}\rangle_l}{=} X(\boldsymbol{a})|\mathbf{0}\rangle_l = |\boldsymbol{a}\rangle$. On the other hand, the action of $Z(\boldsymbol{b}_{\text{eb}})X(\boldsymbol{a}_{\text{eb}})$ on the maximum-entangled state is $(Z(\boldsymbol{b}_{\text{eb}})X(\boldsymbol{a}_{\text{eb}}) \otimes I^{R_x})|\Phi^{\otimes e}\rangle = |\boldsymbol{b}_{\text{eb}}, \boldsymbol{a}_{\text{eb}}\rangle$, where I^{Rx} denotes the identity operators applied on the receiver half of the maximum-entangled state. Finally, the action of the channel on the information portion of the codeword is given by $X(\alpha(\boldsymbol{a},\, \boldsymbol{b}_{\text{eb}},\, \boldsymbol{a}_{\text{eb}}))Z(\beta(\boldsymbol{a},\, \boldsymbol{b}_{\text{eb}},\, \boldsymbol{a}_{\text{eb}}))|\psi\rangle = |\psi'\rangle$. Clearly, the vector ($\boldsymbol{a}\ \boldsymbol{b}\ \boldsymbol{b}_{\text{eb}}\ \boldsymbol{a}_{\text{eb}}$) uniquely specifies the error operator E_c, and can be called the syndrome vector. Since the state $|\mathbf{0}\rangle_l$ is invariant on the action of $Z(\boldsymbol{b})$, the vector $\boldsymbol{b}$ can be omitted from the syndrome vector. To obtain the $\boldsymbol{a}$-portion of the syndrome we have to perform simultaneous measurements on $Z(\boldsymbol{d}_i)$, where $\boldsymbol{d}_i = (0\cdots01_i0\cdots0)$. On the other hand, to obtain the $(\boldsymbol{b}_{\text{eb}}\ \boldsymbol{a}_{\text{eb}})$-portion of the syndrome vector we have to perform measurements on $Z(\boldsymbol{d}_1)Z(\boldsymbol{d}_1)$, ..., $Z(\boldsymbol{d}_e)Z(\boldsymbol{d}_e)$ for $\boldsymbol{b}_{\text{eb}}$ and $X(\boldsymbol{d}_1)X(\boldsymbol{d}_1)$, ..., $X(\boldsymbol{d}_e)X(\boldsymbol{d}_e)$ for $\boldsymbol{a}_{\text{eb}}$. Once the syndrome vector is determined we apply $X^{-1}(\alpha(\boldsymbol{a},\ \boldsymbol{b}_e,\ \boldsymbol{a}_e))Z^{-1}(\beta(\boldsymbol{a},\ \boldsymbol{b}_e,\ \boldsymbol{a}_e))$ to undo the action of the channel. From syndrome measurements, it is clear that we applied exactly the same measurements as in the superdense coding, indicating that EA codes can indeed be interpreted as a generalization of superdense coding (see [38] and Section 8.8 of [60]). To avoid the need for measurement we can perform the following controlled unitary decoding:

$$U_{\text{canocnical, dec}} = \sum_{\boldsymbol{a},\boldsymbol{b}_{\text{eb}},\boldsymbol{a}_{\text{eb}}} |\boldsymbol{a}\rangle\langle\boldsymbol{a}| \otimes |\boldsymbol{b}_{\text{eb}},\ \boldsymbol{a}_{\text{eb}}\rangle\langle\boldsymbol{b}_{\text{eb}},\ \boldsymbol{a}_{\text{eb}}| \otimes X^{-1}(\alpha(\boldsymbol{a},\ \boldsymbol{b}_{\text{eb}},\ \boldsymbol{a}_{\text{eb}}))$$
$$Z^{-1}(\beta(\boldsymbol{a},\ \boldsymbol{b}_{\text{eb}},\ \boldsymbol{a}_{\text{eb}})). \tag{9.179}$$

The EA error correction can also be described by using the stabilizer formalism interpretation. Let $S_{\text{canonical}}$ be the non-Abelian group generated by:

$$\begin{aligned} S_{\text{canonical}} &= \langle S_{\text{canonical},I}, S_{\text{canonical},E}\rangle;\ S_{\text{canonical},I} = \langle Z_1, \ldots, Z_l\rangle, \\ S_{\text{canonical},E} &= \langle Z_{l+1}, \ldots, Z_{l+e}, X_{l+1}, \ldots, X_{l+e}\rangle, \end{aligned} \tag{9.180}$$

where with $S_{\text{canonical},I}$ we denote the isotropic subgroup, and with $S_{\text{canonical},E}$ we denote the sympletic subgroup of anticommuting operators. The representation (9.180) has certain similarities with subsystem codes from Section 9.7. Namely, by interpreting $S_{\text{canonical},I}$ as the stabilizer group and $S_{\text{canonical},E}$ as the gauge group, it turns out that EA codes represent a generalization of subsystem codes. Moreover, the subsystem codes do not require the entanglement to be established between transmitter and receiver. On the other hand, EA codes with few ebits have reasonable complexity of encoder and decoder, compared dual-containing codes, and as such represent promising candidates for fault-tolerant quantum computing and quantum teleportation applications.

We perform an Abelian extension of non-Abelian group $S_{\text{canonical}}$, denoted as $S_{\text{canonical,ext}}$, which now acts on $N+e$ qubits as follows:

$$\begin{aligned} &Z_1 \otimes I, \ldots, Z_l \otimes I \\ &Z_{l+1} \otimes Z_1, X_{l+1} \otimes X_1, \ldots, Z_{l+e} \otimes Z_e, X_{l+e} \otimes X_e. \end{aligned} \tag{9.181}$$

Out of $N+e$ qubits, the first N qubits correspond to the transmitter and the additional e qubits correspond to the receiver. The operators on the right side of Eq. (9.181) act on the receiver qubits. This extended group is Abelian and can be used as a stabilizer group that fixes the code subspace $C_{\text{canonical,EA}}$. Such obtained stabilizer group can be called the EA stabilizer group. The number of information qubits of this stabilizer can be determined by $K\ N-l-e$, so that the code can be denoted as $[N,K,e,D]$, where D is the distance of the code.

Before concluding this subsection, it will be useful to identify the correctable set of errors. In analogy with stabilizer codes we expect the correctable set of errors of EA-QECC, defined by $S_{\text{canonical}} = <S_{\text{canonical},I}, S_{\text{canonical},E}>$, to be given by:

$$N_{\text{canonical}} = \left\{E_m | \forall E_1, \ E_2 \Rightarrow E_2^{\dagger} E_1 \in S_{\text{canonical}, \ I} \cup (G_N - C(S_{\text{canonical}}))\right\}, \quad (9.182)$$

where $C(S_{\text{canonical}})$ is the centralizer of $S_{\text{canonical}}$ (that is, the set of errors that commute with all elements from $S_{\text{canonical}}$). We have already shown (see Eq. (9.177) and corresponding text) that every error can be specified by the following syndrome ($\boldsymbol{a}\ \boldsymbol{b}_{\text{eb}}\ \boldsymbol{a}_{\text{eb}}$). If two errors E_1 and E_2 have the same syndrome ($\boldsymbol{a}\ \boldsymbol{b}_{\text{eb}}\ \boldsymbol{a}_{\text{eb}}$), then $E_2^{\dagger}E_1$ will have an all-zeros syndrome and clearly such an error belongs to $S_{\text{canonical},I}$. Such an error $E_2^{\dagger}E_1$ is trivial, as it fixes all codewords. If two errors have different syndromes, then the syndrome for $E_2^{\dagger}E_1$, which has the form ($\boldsymbol{a}\ \boldsymbol{b}_{\text{eb}}\ \boldsymbol{a}_{\text{eb}}$), indicates that such an error can be corrected provided that it does not belong to $C(S_{\text{canonical}})$. Namely, when an error belongs to $C(S_{\text{canonical}})$, then its action results in another codeword different from the original codeword, so that such an error is undetectable.

9.10.4 General Entanglement-Assisted Quantum Codes

From group theory, it is known that if V is an arbitrary subgroup of G_N of order 2^M, then there exists a set of generators $\{\overline{X}_{p+1}, \ldots, \overline{X}_{p+q}; \overline{Z}_1, \ldots, \overline{Z}_{p+q}\}$ satisfying similar commutation properties to Eq. (9.28), namely:

$$[\overline{X}_m, \overline{X}_n] = [\overline{Z}_m, \overline{Z}_n] = 0 \ \forall m, \ n; \ \ [\overline{X}_m, \overline{Z}_n] = 0 \ \ \forall m \neq n; \ \ \{\overline{X}_m, \overline{Z}_n\} = 0 \ \forall m = n. \quad (9.183)$$

We also know from the theorem provided in Section 2.8 that the unitary equivalent observables, A and UAU^{-1}, have identical spectra. This means that we can establish bijection between two groups V and S, which is going to preserve their commutation relations, so that $\forall \ v \in V$ there exists corresponding $s \in S$ such that $v = UsU^{-1}$ (up to a general phase constant). By using these results, we can establish bijection between canonical EA quantum code described by $S_{\text{canonical}} = \langle S_{\text{canonical}, \ I}, \ S_{\text{canonical}, \ E}\rangle$ and a general EA code characterized by $S = \langle S_I, S_E\rangle$ with $S = US_{\text{canonical}}U^{-1}$ (or, equivalently, $S_{\text{canonical}} = U-^{1}SU$), where U is a corresponding unitary operator.

Based on the foregoing discussion we can generalize the results from the previous subsection as follows. Given a general group $S = <S_I, S_E>$ ($|S_I| = 2^{N-K-e}$ and $|S_E| = 2^{2e}$), there exits an $[N,K,e;D]$ EA-QECC, illustrated in Fig. 9.19, denoted as C_{EA}, defined by encoding–decoding pair (E,D) with the following properties:

1. The EA code C_{EA} can correct any error from the following set of errors N:

$$\mathcal{N} = \left\{E_m | \forall E_1, \ E_2 \Rightarrow E_2^{\dagger} E_1 \in S_I \cup (G_N - C(S))\right\}. \quad (9.184)$$

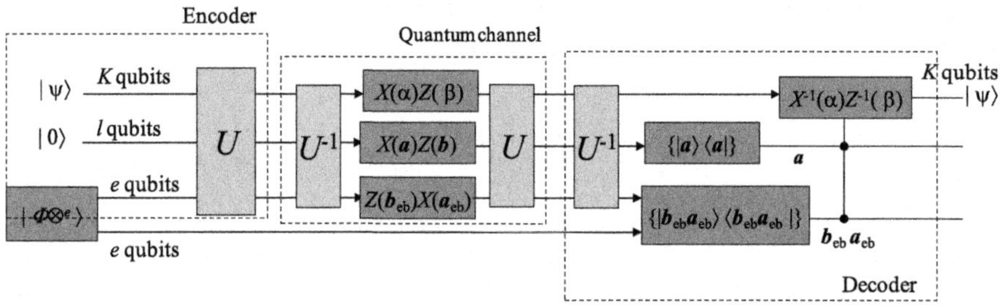

FIGURE 9.19

Entanglement-assisted code obtained as a generalization of the canonical entanglement-assisted code.

2. The code space C_{EA} is a simultaneous eigenspace of the Abelian extension of S, denoted as S_{ext}.
3. To decode, the error syndrome is obtained by simultaneously measuring the observables from S_{ext}.

Because the commutation relations of S are the same as that of EA canonical code $C_{canonical,EA}$ we can find a unitary operator U such that $S_{canonical} = USU^{-1}$. Based on Fig. 9.19, it is clear that encoding mapping can be represented as a composition of unitary operator U and encoding mapping of canonical code $E_{canonical}$, namely $\mathscr{E} = U \circ \mathscr{E}_{canonical}$. In similar fashion, decoding mapping can be represented by $\mathscr{D} = \mathscr{D}_{canonical} \circ \widehat{U}^{-1}$, $\widehat{U} = \text{extension}(U)$. Furthermore, the composition of encoding, error, and decoding mapping would result in identity operator:

$$\mathscr{D} \circ \mathscr{N} \circ \mathscr{E} = \mathscr{D}_{canonical} \circ \widehat{U}^{-1} \circ \widehat{U} \mathscr{N}_{canonical} \widehat{U}^{-1} \circ \widehat{U} \mathscr{E} = I^{\otimes K}. \tag{9.185}$$

The set of errors given by Eq. (9.184) is the correctable set of errors because $C_{canonical,EA}$ is the simultaneous eigenspace of $S_{canonical,ext}$, and $S_{ext} = US_{canonical,ext}U^{-1}$, and by definition $C_{EA} = U(C_{canonical,EA})$, so that we conclude that C_{ext} is a simultaneous eigenspace of S_{ext}. We learned in the previous subsection that decoding operation of $D_{canonical}$ involves: (1) measuring the set of generators of $S_{canonical,ext}$ yielding the error syndrome according to the error E_c, and (2) performing the recovery operation to undo the action of the channel error E_c. The decoding action of C_{EA} is equivalent to the measurement of $S_{ext} = US_{canonical,ext}U^{-1}$, followed by recovery operation UE_cU^{-1} and U^{-1} to undo the encoding.

The distance D of EA code can be introduced in a similar fashion to that of a stabilizer code. The distance D of EA code is defined as the minimum weight among undetectable errors. In other words, we say that EA quantum code has distance D if it can detect all errors of weight less than D, but none of weight D. The error correction capability t of EA quantum code is related to minimum distance D by $t = \lfloor (D-1)/2 \rfloor$. The weight of a qubit error $[\mathbf{u}|\mathbf{v}]$ ($\mathbf{u},\mathbf{v} \in F_2^N$) can be defined in similar fashion to that for stabilizer codes, as the number of components i for which $(u_i,b_i) \neq (0,0)$ or equivalently $\mathrm{wt}([\mathbf{u}|\mathbf{v}]) = \mathrm{wt}(\mathbf{u}+\mathbf{v})$, where "+" denotes bitwise addition mod 2.

The EA quantum bounds are similar to the quantum coding bounds presented in Chapter 8. For example, the EA quantum Singleton bound is given by:

$$N + e - K \geq 2(D-1).$$

9.10.5 EA-QECCs Derived From Classical Quaternary and Binary Codes

We turn our attention now to establishing the connection between EA codes and quaternary classical codes, which are defined over $F_4 = \mathrm{GF}(4) = \{0,1,\omega,\omega^2 = \varpi\}$. As a reminder, the addition and multiplication tables for F_4 are given as follows.

+	0	ϖ	1	ω
0	0	ϖ	1	ω
ϖ	ϖ	0	ω	1
1	1	ω	0	ϖ
ω	ω	1	ϖ	0

×	0	ϖ	1	ω
0	0	0	0	0
ϖ	0	ω	ϖ	1
1	0	ϖ	1	ω
ω	0	1	ω	ϖ

On the other hand, the addition table for binary 2-tuples, $F_2^2 = \{00,01,10,11\}$, is given by:

+	00	01	11	10
00	00	01	11	10
01	01	00	10	11
11	11	10	00	01
10	10	11	01	00

Furthermore, from Chapter 8 we know that the multiplication table of Pauli matrices is given by:

×	*I*	X	Y	Z
I	*I*	X	Y	Z
X	X	*I*	jZ	-jY
Y	Y	-jZ	*I*	jX
Z	Z	jY	-jX	*I*

If we ignore the phase factor in Pauli matrices multiplication we can establish the following correspondences between G_N, F_2^2, and F_4:

G	$F_2{}^2$	F_4
I	00	0
X	10	ω
Y	11	1
Z	01	$\overline{\omega}$

It can easily be shown that the mapping f: $F_4 \rightarrow F_2^2$ is an isomorphism. Therefore if a classical code (n,k,d) C_4 over GF(4) exists, then an $[N,K,e,D]=[n,2k-n+e,e;D]$ EA-QECC exists for some nonnegative integer e, which can be described by the following quantum-check matrix:

$$A_{\mathrm{EA}}=f(\widetilde{H}_4);\widetilde{H}_4=\begin{pmatrix}\omega H_4\\ \overline{\omega}H_4\end{pmatrix}, \tag{9.186}$$

where with H_4 we denote the parity-check matrix of classical code (n,k,d) over F_4. The sympletic product of binary vectors from Eq. (9.186) is equal to the trace of the product of their GF(4) representations as follows:

$$h_m \odot h_n=\mathrm{tr}\left(f^{-1}(h_m)\overline{f^{-1}(h_n)}\right), \mathrm{tr}(w)=w+\overline{w}, \tag{9.187}$$

where h_m (h_n) is the mth (nth) row of A_{EA} and the overbar denotes the conjugation. The sympletic product of A_{EA} is given by:

$$\begin{aligned}\mathrm{tr}\left\{\begin{pmatrix}\omega H_4\\ \overline{\omega}H_4\end{pmatrix}\begin{pmatrix}\omega H_4\\ \overline{\omega}H_4\end{pmatrix}^{\dagger}\right\}&=\mathrm{tr}\left\{\begin{pmatrix}\omega H_4\\ \overline{\omega}H_4\end{pmatrix}\begin{pmatrix}\overline{\omega}H_4^{\dagger} & \omega H_4^{\dagger}\end{pmatrix}\right\}\\ &=\mathrm{tr}\left\{\begin{pmatrix}H_4H_4^{\dagger} & \overline{\omega}H_4H_4^{\dagger}\\ \omega H_4H_4^{\dagger} & H_4H_4^{\dagger}\end{pmatrix}\right\}=\mathrm{tr}\left\{\begin{pmatrix}1 & \overline{\omega}\\ \omega & 1\end{pmatrix}\otimes H_4H_4^{\dagger}\right\}\\ &=\begin{pmatrix}1 & \overline{\omega}\\ \omega & 1\end{pmatrix}\otimes H_4H_4^{\dagger}+\begin{pmatrix}1 & \omega\\ \overline{\omega} & 1\end{pmatrix}\otimes \overline{H}_4H_4^{\mathrm{T}}.\end{aligned} \tag{9.188}$$

The rank of $\widetilde{H}_4\widetilde{H}_4^{\dagger}$ is not going to change if we perform the following similarity transformation $BA\widetilde{H}_4H_4^{\dagger}A^{\dagger}B^{\dagger}$, where: $A=\begin{bmatrix}1 & \overline{\omega}\\ \omega & 1\end{bmatrix}\otimes I, B=\begin{bmatrix}1 & 0\\ 1 & 1\end{bmatrix}\otimes I$ to obtain:

$$BA\widetilde{H}_4H_4^{\dagger}A^{\dagger}B^{\dagger}=\begin{bmatrix}\overline{H}_4H_4^{\mathrm{T}} & \mathbf{0}\\ \mathbf{0} & H_4H_4^{\dagger}\end{bmatrix}=\overline{H}_4H_4^{\mathrm{T}}\otimes H_4H_4^{\dagger}. \tag{9.189}$$

The rank of $B A \widetilde{H}_4 H_4^{\dagger} A^{\dagger} B^{\dagger}$ is then obtained by:

$$\begin{aligned}\operatorname{rank}\left(B A \widetilde{H}_4 H_4^{\dagger} A^{\dagger} B^{\dagger}\right) &= \operatorname{rank}\left(\overline{H}_4 H_4^{\mathrm{T}} \otimes H_4 H_4^{\dagger}\right) = \operatorname{rank}\left(\overline{H}_4 H_4^{\mathrm{T}}\right) + \operatorname{rank}\left(H_4 H_4^{\dagger}\right) \\ &= 2\operatorname{rank}\left(H_4 H_4^{\dagger}\right) = 2e,\ e = \operatorname{rank}\left(H_4 H_4^{\dagger}\right).\end{aligned} \tag{9.190}$$

Eq. (9.190) indicates that the number of required ebits is $e = \operatorname{rank}\left(H_4 H_4^{\dagger}\right)$, so that the parameters of this EA code, based on Eq. (9.186), are $[N,K,e] = [n,n+e-2(n-k),e] = [n,2k-n+e,e]$.

Example. The EA code derived from (4,2,3) 4-ary classical code described by parity-check matrix

$$H_4 = \begin{pmatrix} 1 & 1 & 0 & 1 \\ 1 & \omega & 1 & 0 \end{pmatrix}$$

has parameters [4,1,1,3], and based on Eq. (9.186), the corresponding quantum-check matrix is given by:

$$A_{EA} = f(\widetilde{H}_4) = f\left(\begin{bmatrix} \omega H_4 \\ \overline{\omega} H_4 \end{bmatrix}\right) = f\left(\begin{bmatrix} \omega & \omega & 0 & \omega \\ \omega & \overline{\omega} & \omega & 0 \\ \overline{\omega} & \overline{\omega} & 0 & \overline{\omega} \\ \overline{\omega} & 1 & \overline{\omega} & 0 \end{bmatrix}\right) = \begin{pmatrix} X & X & I & X \\ X & Z & X & I \\ Z & Z & I & Z \\ Z & Y & Z & I \end{pmatrix}.$$

In the remainder of this subsection we describe how to relate EA quantum codes with binary classical codes. Let H be a binary parity-check matrix of dimension $(n-k) \times n$. We will show that the corresponding EA-QECC has the parameters $[n,2k-n+e;e]$, where $e = \operatorname{rank}(HH^{\mathrm{T}})$. The quantum-check matrix of CSS-like EA quantum code can be represented as follows:

$$A_{\mathrm{EA}} = \left(\begin{array}{c|c} H & 0 \\ 0 & H \end{array}\right). \tag{9.191}$$

The rank of sympletic product of A_{AE}, based on Eq. (9.187), is given by:

$$\begin{aligned}\operatorname{rank}\{A_{\mathrm{EA}} \odot A_{EA}^{\mathrm{T}}\} &= \operatorname{rank}\left\{\begin{pmatrix} H \\ 0 \end{pmatrix}\begin{pmatrix} 0 & H^{\mathrm{T}} \end{pmatrix} + \begin{pmatrix} 0 \\ H \end{pmatrix}\begin{pmatrix} H^{\mathrm{T}} & 0 \end{pmatrix}\right\} \\ &= \operatorname{rank}\left\{\begin{pmatrix} 0 & HH^{\mathrm{T}} \\ 0 & 0 \end{pmatrix} + \begin{pmatrix} 0 & 0 \\ HH^{\mathrm{T}} & 0 \end{pmatrix}\right\} = \operatorname{rank}\{HH^{\mathrm{T}} + HH^{\mathrm{T}}\} \\ &= 2\operatorname{rank}\{HH^{\mathrm{T}}\} = 2e,\ e = \operatorname{rank}\{HH^{\mathrm{T}}\}.\end{aligned} \tag{9.192}$$

From Eq. (9.192), it is clear that the number of required qubits is $e = \operatorname{rank}(H\ \ H^{\mathrm{T}})$, so that the parameters of this EA code are $[N,K,e] = [n,n+e-2(n-k),e] = [n,2k-n+e,e]$. Dual-containing codes can therefore be considered as EA codes for which $e = 0$.

The EA code given by Eq. (9.191) can be generalized as follows:

$$A_{\mathrm{EA}} = (A_x | A_z), \tag{9.193}$$

where (n- − k) × n submatrix A_x (A_z) contains only X-operators (Z-operators). The rank of sympletic product of A_{AE} can be obtained as:

$$\mathrm{rank}\{A_{\mathrm{EA}} \odot A_{EA}^{\mathrm{T}}\} = \mathrm{rank}\{A_x A_z^{\mathrm{T}} + A_z A_x^{\mathrm{T}}\} = 2e, e = \mathrm{rank}\{A_x A_z^{\mathrm{T}} + A_z A_x^{\mathrm{T}}\}/2. \tag{9.194}$$

The parameters of this EA code are $[N,K,e] = [n,n + e - (n - k),e] = [n,k + e,e]$.

Another generalization of Eq. (9.191) is given by:

$$A_{\mathrm{EA}} = \left(\begin{array}{c|c} H_x & 0 \\ 0 & H_z \end{array} \right), \tag{9.195}$$

where (n − k_x) × n submatrix H_x contains only X-operators, while (n−k_z) × n submatrix H_z contains only Z-operators. The rank of sympletic product of A_{AE} is given by:

$$\begin{aligned}\mathrm{rank}\{A_{\mathrm{EA}} \odot A_{EA}^{\mathrm{T}}\} &= \mathrm{rank}\left\{ \begin{pmatrix} H_x \\ 0 \end{pmatrix} \begin{pmatrix} 0 & H_z^{\mathrm{T}} \end{pmatrix} + \begin{pmatrix} 0 \\ H_z \end{pmatrix} \begin{pmatrix} H_x^{\mathrm{T}} & 0 \end{pmatrix} \right\} \\ &= \mathrm{rank}\left\{ \begin{pmatrix} 0 & H_x H_z^{\mathrm{T}} \\ 0 & 0 \end{pmatrix} + \begin{pmatrix} 0 & 0 \\ H_z H_x^{\mathrm{T}} & 0 \end{pmatrix} \right\} \qquad (9.196) \\ &= \mathrm{rank}\{H_x H_z^{\mathrm{T}} + H_z H_x^{\mathrm{T}}\} = 2e, e = \mathrm{rank}\{H_x H_z^{\mathrm{T}}\}.\end{aligned}$$

The EA code parameters are $[N,K,e;D] = [n,n + e - (n - k_x) - (n - k_y), e; \min(d_x,d_z)] = [n,k_x + k_y - n + e, e; \min(d_x,d_z)]$.

9.10.6 Encoding and Decoding for Entanglement-Assisted Quantum Codes

Encoders and decoders for EA codes can be implemented in similar fashion to that of subsystem codes, described in Section 9.7: (1) by using standard format formalism [8,9] and (2) by using the conjugation method due to Grassl et al. [37] (see also [26,63]).

In the previous subsection we defined an EA $[N,K,e]$ code by using a non-Ableian subgroup S of G_N (multiplicative Pauli group on N qubits) with $2e + l$ generators, where $l = N - K - e$. The subgroup S can be described by the following set of independent generators $\{\overline{X}_{e+1}, \ldots, \overline{X}_{e+l}; \overline{Z}_1, \ldots, \overline{Z}_{e+l}\}$, which satisfy the commutation relations (9.183). The non-Abelian subgroup S can be decomposed into two subgroups: commuting isotropic group $S_I = \{\overline{Z}_1, \ldots, \overline{Z}_l\}$, and entanglement subgroup with anticommuting pairs $\{\overline{X}_{l+1}, \ldots, \overline{X}_{l+e}; \overline{Z}_{l+1}, \ldots, \overline{Z}_{l+e}\}$. Such decomposition allows us to determine the EA code parameters. Because the isotropic subgroup is the commuting group, it corresponds to ancilla qubits. On the other hand, since the entanglement subgroup is composed of anticommuting pairs, its elements correspond to transmitter and receiver halves of ebits. This procedure can be performed by the sympletic Gram–Schmidt procedure [63]. Let us assume that the following generators create the subgroup S: $g_1, \ldots, g_M$. We start the procedure with generator g_1 and check if it commutes with all other generators. If it commutes we remove it from further consideration. If g_1 anticommutes with a generator g_i we relabel g_2 as g_i and vice versa. For the remaining generators we perform the following manipulation:

$$g_m = g_m g_1^{f(g_2 g_m)} g_2^{f(g_1 g_m)}, \quad f(a, b) = \begin{cases} 0, & [a, b] = 0 \\ 1, & \{a, b\} = 0 \end{cases}; m = 3, 4, \ldots, M. \quad (9.197)$$

The generators g_1 and g_2 are then removed from further discussion, and the same algorithm is applied on the remaining generators. At the end of this procedure, the generators being removed from consideration create the code generators satisfying the commutation relations (9.183).

We can perform the extension of the non-Abelian subgroup S into Abelian and then apply the standard form procedure for encoding and decoding, which was fully explained earlier (Section 9.7). Here, instead, we will concentrate to the conjugation method for EA encoder implementation. The key idea is very similar to that of subsystem codes. Namely, we represent the non-Abelian group S using finite geometry interpretation. We then perform Gaussian elimination to transform the EA code into canonical EA code:

$$|c\rangle \rightarrow |\Phi^{\otimes e}\rangle \underbrace{|0\rangle \otimes \ldots \otimes |0\rangle}_{l=N-K-e \text{ times}} |\psi\rangle_K = |\Phi^{\otimes e}\rangle |\mathbf{0}\rangle_l |\psi\rangle_K, \quad (9.198)$$

where $|c\rangle$ is the EA codeword, $|\Phi^{\otimes e}\rangle$ is the ebit state, and $|\mathbf{0}\rangle_l$ are l ancilla states prepared into the $|0\rangle$ state. In other words, we have to apply the sequence of gates $\{U\}$ to perform the following transformation of the quantum-check matrix of EA code:

$$\mathbf{A}_{\text{EA}} = [\mathbf{A}_x|\mathbf{A}_z] \xrightarrow{\{U\}} \mathbf{A}_{\text{EA,canonical}} = \left[\begin{array}{ccc|ccc} \overbrace{\mathbf{0}}^{e} & \overbrace{\mathbf{0}}^{l} & \overbrace{\mathbf{0}}^{K} & \mathbf{I}_e & \mathbf{0} & \mathbf{0} \\ \mathbf{I}_e & \mathbf{0} & \mathbf{0} & \mathbf{0} & \mathbf{0} & \mathbf{0} \\ \mathbf{0} & \mathbf{0} & \mathbf{0} & \mathbf{0} & \mathbf{I}_l & \mathbf{0} \end{array}\right]. \quad (9.199)$$

The RHS of Eq. (9.199) is obtained from Eq. (9.181). We apply the same set of rules as for subsystem codes. The action of the Hadamard gate on the ith qubit is to change the position of a_i and b_i as follows, where we underline the affected qubit for convenience:

$$\left(a_1 \ldots \underline{a_i} \ldots a_N | b_1 \ldots \underline{b_i} \ldots b_N\right) \xrightarrow{H_i} \left(a_1 \ldots \underline{b_i} \ldots a_N | b_1 \ldots \underline{a_i} \ldots b_N\right). \tag{9.200}$$

The action of the CNOT gate on generator $\mathbf{g}$, where the ith qubit is the control qubit and the jth qubit is the target qubit, the action denoted as $\text{CNOT}_{i,j}$, can be described as:

$$\left(a_1 \ldots \underline{a_j} \ldots a_N | b_1 \ldots \underline{b_i} \ldots b_N\right) \xrightarrow{\text{CNOT}_{i,j}} \left(a_1 \ldots a_{j-1} \underline{a_j + a_i}\, a_{j+1} \ldots a_N | b_1 \ldots b_{i-1} \underline{b_i + b_j}\, b_{i+1} \ldots b_N\right). \tag{9.201}$$

Therefore the jth entry in the X-portion and the ith entry in the Z-portion are affected. The action of phase gate P on the ith qubit is to perform the following mapping:

$$\left(a_1 \ldots \underline{a_i} \ldots a_N | b_1 \ldots \underline{b_i} \ldots b_N\right) \xrightarrow{P_i} \left(a_1 \ldots \underline{a_i} \ldots a_N | b_1 \ldots \underline{a_i + b_i} \ldots b_N\right). \tag{9.202}$$

Finally, the application of the CNOT gate between the ith and jth qubit in an alternative fashion three times leads to the swapping of the ith and jth columns in both X- and Z-portions of the quantum-check matrix, $\text{SWAP}_{i,j} = \text{CNOT}_{i,j}\text{CNOT}_{j,i}\text{CNOT}_{i,j}$, as shown in Chapter 3:

$$\left(a_1 \ldots \underline{a_i} \ldots \underline{a_j} \ldots a_N | b_1 \ldots \underline{b_i} \ldots \underline{b_j} \ldots b_N\right) \xrightarrow{\text{SWAP}_{i,j}} \left(a_1 \ldots \underline{a_j} \ldots \underline{a_i} \ldots a_N | b_1 \ldots \underline{b_j} \ldots \underline{b_i} \ldots b_N\right). \tag{9.203}$$

Let us denote the sequence of gates being applied during this transformation to canonical EA code as $\{U\}$. By applying this sequence of gates in opposite order we obtain the encoding circuit of corresponding EA code. Notice that adding the nth row to the mth row maps $g_m \rightarrow g_m g_n$ so that codewords are invariant to this change.

Since this method is very similar to that used for subsystem codes from Section 9.7 we turn our attention now to the sympletic Gram–Schmidt method due to Wilde [63], which has already been discussed at the operator level. Here we describe this method using finite geometry interpretation. Let us observe the following example due to Wilde [63]:

$$
\boldsymbol{A}_{\mathrm{EA}}=\left[\begin{array}{cccc|cccc}0&1&0&0&1&0&1&0\\0&0&0&0&1&1&0&1\\1&1&1&0&0&1&0&0\\1&1&0&1&0&0&0&0\end{array}\right]\xrightarrow{SWAP_{1,2}}\left[\begin{array}{cccc|cccc}1&0&0&0&0&1&1&0\\0&0&0&0&1&1&0&1\\1&1&1&0&1&0&0&0\\1&1&0&1&0&0&0&0\end{array}\right]\xrightarrow{H_2,H_3}\left[\begin{array}{cccc|cccc}1&1&1&0&0&0&0&0\\0&1&0&0&1&0&0&1\\1&0&0&0&1&1&1&0\\1&0&0&1&0&1&0&0\end{array}\right]
$$

$$
\xrightarrow{CNOT_{1,2},CNOT_{1,3}}\left[\begin{array}{cccc|cccc}1&0&0&0&0&0&0&0\\0&1&0&0&1&0&0&1\\1&1&1&0&1&1&1&0\\1&1&1&1&1&1&0&0\end{array}\right]\xrightarrow{H_1,H_4}\left[\begin{array}{cccc|cccc}0&0&0&0&1&0&0&0\\1&1&0&1&0&0&0&0\\1&1&1&0&1&1&1&0\\1&1&1&0&1&1&0&1\end{array}\right]\xrightarrow{CNOT_{1,2},CNOT_{1,4}}\left[\begin{array}{cccc|cccc}0&0&0&0&1&0&0&0\\1&0&0&0&0&0&0&0\\1&0&1&1&0&1&1&0\\1&0&1&1&1&1&0&1\end{array}\right]
$$

$$
\xrightarrow[row_4\leftarrow row_4+row_2]{row_4\leftarrow row_4+row_1,\ row_3\leftarrow row_3+row_2}\left[\begin{array}{cccc|cccc}0&0&0&0&1&0&0&0\\1&0&0&0&0&0&0&0\\0&0&1&1&0&1&1&0\\0&0&1&1&0&1&0&1\end{array}\right]\xrightarrow{H_2}\left[\begin{array}{cccc|cccc}0&0&0&0&1&0&0&0\\1&0&0&0&0&0&0&0\\0&1&1&1&0&0&1&0\\0&1&1&1&0&0&0&1\end{array}\right]\xrightarrow{CNOT_{2,3},CNOT_{2,4}}\left[\begin{array}{cccc|cccc}0&0&0&0&1&0&0&0\\1&0&0&0&0&0&0&0\\0&1&0&0&0&1&1&0\\0&1&0&0&0&1&0&1\end{array}\right]
$$

$$
\xrightarrow{P_2,H_3}\left[\begin{array}{cccc|cccc}0&0&0&0&1&0&0&0\\1&0&0&0&0&0&0&0\\0&1&1&0&0&0&0&0\\0&1&0&0&0&0&0&1\end{array}\right]\xrightarrow{CNOT_{2,3}}\left[\begin{array}{cccc|cccc}0&0&0&0&1&0&0&0\\1&0&0&0&0&0&0&0\\0&1&0&0&0&0&0&0\\0&1&1&0&0&0&0&1\end{array}\right]\xrightarrow{row_4\leftarrow row_4+row_3}\left[\begin{array}{cccc|cccc}0&0&0&0&1&0&0&0\\1&0&0&0&0&0&0&0\\0&1&0&0&0&0&0&0\\0&0&1&0&0&0&0&1\end{array}\right]
$$

$$
\xrightarrow{H_2,H_4}\left[\begin{array}{cccc|cccc}0&0&0&0&1&0&0&0\\1&0&0&0&0&0&0&0\\0&0&0&0&0&1&0&0\\0&0&1&1&0&0&0&0\end{array}\right]\xrightarrow{CNOT_{3,4}}\left[\begin{array}{cccc|cccc}0&0&0&0&1&0&0&0\\1&0&0&0&0&0&0&0\\0&0&0&0&0&1&0&0\\0&0&1&0&0&0&0&0\end{array}\right]\xrightarrow{H_3}\left[\begin{array}{cccc|cccc}0&0&0&0&1&0&0&0\\1&0&0&0&0&0&0&0\\0&0&0&0&0&1&0&0\\0&0&0&0&0&0&1&0\end{array}\right]
$$

Clearly, we have transformed the initial EA into canonical form (9.199). The corresponding encoding circuits can be obtained by applying the set of gates to the canonical codeword, used to transform the quantum-check matrix into canonical form, but now in the opposite order. The encoding circuit for this EA code is shown in Fig. 9.20.

9.10.7 Operator Quantum Error Correction Codes (Subsystem Codes)

The operator quantum error correcting codes (OQECCs) [26,35,59–72] represent a very important class of quantum codes, which are also known as subsystem codes. This class of codes has already been discussed in Section 9.7. Nevertheless, the following interpretation due to Hsieh, Devetak, and Brun [59,60] is useful for a deeper understanding of the underlying concept.

We start our description with canonical OQECCs. An $[N,K,R]$ canonical OQECC can be represented by the following mapping:

$$
|\phi\rangle|\psi\rangle\rightarrow\underbrace{|0\rangle\otimes\ldots\otimes|0\rangle}_{s=N-K-R\text{ times}}|\phi\rangle_R|\psi\rangle_K=|\mathbf{0}\rangle_s|\phi\rangle_R|\psi\rangle_K, \tag{9.204}
$$

where $|\psi\rangle$ represents K-qubit information state, $|\phi\rangle$ represents R-qubit gauge state, and $|\mathbf{0}\rangle_s$ represent s ancilla qubits. So the key difference with respect to canonical

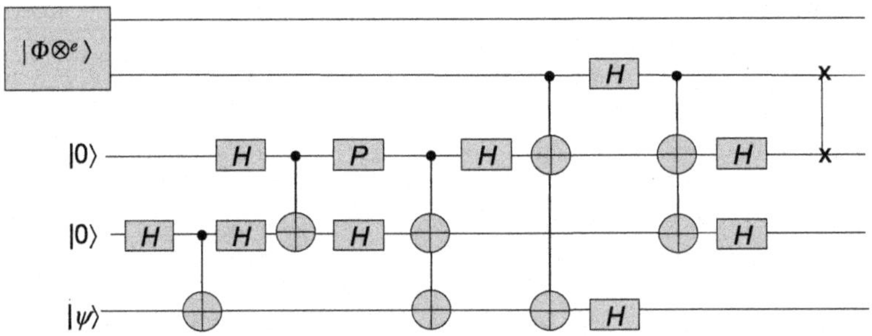

FIGURE 9.20

The encoding circuit for entanglement assisted code from the foregoing example.

stabilizer code is that R ancilla $|0\rangle$ qubits are converted into gauge qubits. This canonical OQECC can correct the following set of errors:

$$N_{\text{canonical, OQECC}} = \{X(\boldsymbol{a})Z(\boldsymbol{b}) \otimes X(\boldsymbol{c})Z(\boldsymbol{d}) \otimes X(\alpha(\boldsymbol{a}))Z(\beta(\boldsymbol{a})) \\ : \boldsymbol{a}, \boldsymbol{b} \in F_2^s; \boldsymbol{c}, \boldsymbol{d} \in F_2^R\}; \alpha, \beta: F_2^s \to F_2^K. \tag{9.205}$$

To prove claim (9.205), let us observe one particular error E_c from the set of errors (9.205) and study its action on transmitted codeword $|\mathbf{0}\rangle_s|\phi\rangle_R|\psi\rangle_K$:

$$X(\boldsymbol{a})Z(\boldsymbol{b})|\mathbf{0}\rangle_s \otimes (X(\boldsymbol{c})Z(\boldsymbol{d}))|\phi\rangle \otimes X(\alpha(\boldsymbol{a}))Z(\beta(\boldsymbol{a}))|\psi\rangle. \tag{9.206}$$

The action of operator $X(\boldsymbol{a})Z(\boldsymbol{b})$ on state $|\mathbf{0}\rangle_s$ is given by $X(\boldsymbol{a})Z(\boldsymbol{b})|\mathbf{0}\rangle_s \overset{Z(\boldsymbol{b})|\mathbf{0}\rangle_s=|\mathbf{0}\rangle_s}{\widehat{=}} X(\boldsymbol{a})|\mathbf{0}\rangle_s = |\boldsymbol{a}\rangle$. On the other hand, the action of $X(\boldsymbol{c})Z(\boldsymbol{d})$ on the gauge state is $(X(\boldsymbol{c})Z(\boldsymbol{d}))|\phi\rangle = |\phi'\rangle$. Finally, the action of the channel on the information portion of the codeword is given by $X(\alpha(\boldsymbol{a}))Z(\beta(\boldsymbol{a}))|\psi\rangle = |\psi'\rangle$. Clearly, the vector $(\boldsymbol{a}\ \boldsymbol{b}\ \boldsymbol{c}\ \boldsymbol{d})$ uniquely specifies the error operator E_c, and can be called the syndrome vector. Since the state $|\mathbf{0}\rangle_s$ is invariant on the action of $Z(\boldsymbol{b})$, the vector $\boldsymbol{b}$ can be omitted from the syndrome vector. Also since the final state of the gauge subsystem is irrelevant to the information qubit states, the vectors $\boldsymbol{c}$ and $\boldsymbol{d}$ can also be omitted from consideration. To obtain the $\boldsymbol{a}$-portion of the syndrome we have to perform simultaneous measurements on $Z(\boldsymbol{d}_i)$, where $\boldsymbol{d}_i = (0\ldots01_i0\ldots0)$ on ancilla qubits. Once the syndrome vector $\boldsymbol{a}$ is determined we apply the operator $X^{-1}(\alpha(\boldsymbol{a}))Z^{-1}(\beta(\boldsymbol{a}))$ to the information portion of the codeword to undo the action of the channel, as illustrated in Fig. 9.21. To avoid the need for measurement we can perform the following controlled unitary decoding operation:

$$U_{\text{canocnical, OECC, dec}} = \sum_{\boldsymbol{a}} |\boldsymbol{a}\rangle\langle\boldsymbol{a}| \otimes \boldsymbol{I} \otimes X^{-1}(\alpha(\boldsymbol{a}))Z^{-1}(\beta(\boldsymbol{a})), \tag{9.207}$$

and discard the unimportant portion of the received quantum word.

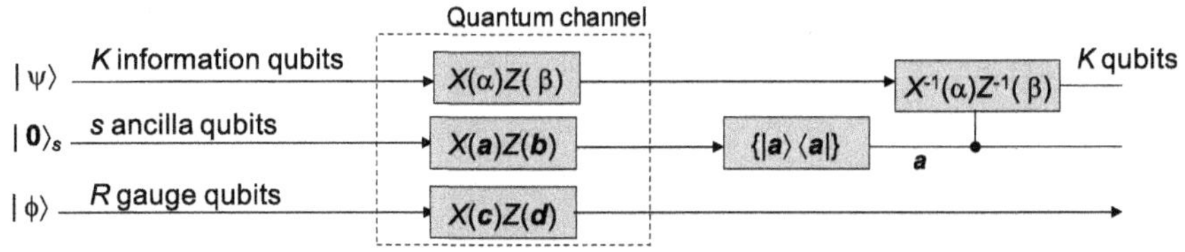

FIGURE 9.21

Canonical operator quantum error correcting code encoder and decoder principles.

The OQECCs can also be described by using the stabilizer formalism interpretation as follows:

$$\begin{aligned} S_{\text{canonical, OQECC}} &= \langle S_{\text{canonical, OQECC},I}, S_{\text{canonical, OQECC},E}\rangle \\ S_{\text{canonical, OQECC},I} &= \langle Z_1, \dots, Z_s\rangle, S_{\text{canonical, OQECC},E} \\ &= \langle Z_{s+1}, \dots, Z_{s+R}, X_{s+1}, \dots, X_{s+R}\rangle, \end{aligned} \tag{9.208}$$

where with $S_{\text{canonical, OQECC},I}$ we denote the isotropic subgroup of size 2^s, and with $S_{\text{canonical, OQECC},E}$ we denote the sympletic subgroup of size 2^{2R}. The isotropic subgroup defines a 2^{K+R}-dimensional code space, while the sympletic subgroup defines all possible operations on gauge qubits. The OQECC can correct the following set of errors:

$$\mathcal{N} = \left\{E_m \middle| \forall E_1, E_2 \Rightarrow E_2^\dagger E_1 \in S_{\text{canonical,OQECC}} \cup \left(G_N - C\left(S_{\text{canonical,OQECC},I}\right)\right)\right\}. \tag{9.209}$$

By a similar approach as in Section 9.3 we can establish bijection between canonical OQECCs described by $S_{\text{canonical, OQECC}} = \langle S_{\text{canonical, OQECC},I}, S_{\text{canonical, OQECC},E}\rangle$ and a general OQECC characterized by $S_{\text{OQECC}} = \langle S_{\text{OQECC},I}, S_{\text{OQECC},E}\rangle$ with $S_{\text{OQECC}} = US_{\text{canonical,OQECC}}U^{-1}$ (or equivalently $S_{\text{canonical, OQECC}} = U^{-1}S_{\text{OQECC}}U$), where U is a corresponding unitary operator. This $[N,K,R,D]$ OQECC, illustrated in Fig. 9.22, denoted as C_{OQECC}, is defined by encoding–decoding pair (E,D) with the following properties:

1. The OQECC C_{OQECC} can correct any error for the following set of errors:

$$\mathcal{N} = \left\{E_m | \forall E_1,\ E_2 \Rightarrow E_2^\dagger E_1 \in S_{\text{OQECC}} \cup \left(G_N - C\left(S_{\text{OQECC},\ I}\right)\right)\right\}. \tag{9.210}$$

2. The code space C_{OQECC} is a simultaneous eigenspace of $S_{\text{OQECC},I}$.
3. To decode, the error syndrome is obtained by simultaneously measuring the observables from $S_{\text{OQECC},I}$.

The encoding and decoding of OQECC (subsystem) codes is already described in Section 9.7.

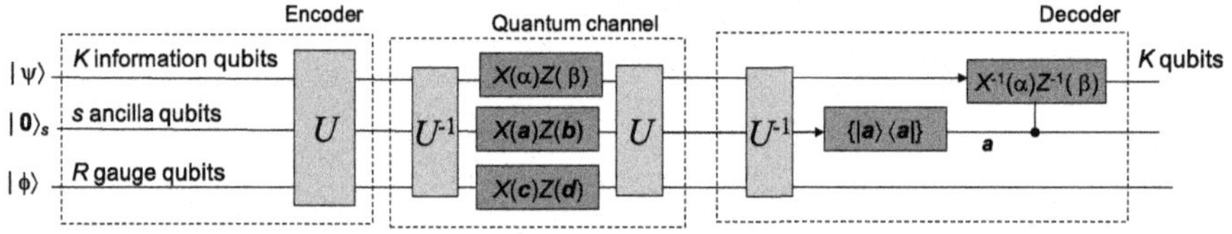

FIGURE 9.22

Generalized operator quantum error correcting code encoder and decoder principles.

9.10.8 Entanglement-Assisted Operator Quantum Error Correction Coding

EA-OQECCs represent the generalization of EA codes and OQECCs (also known as subsystem codes). Namely, in addition to information and gauge qubits used in OQECCs, several entanglement qubits (ebits) are shared between the source and destination. The simplest EA-QECC, the canonical code, performs the following trivial encoding operation:

$$\hat{U}_c : |\psi\rangle_K \rightarrow |\psi\rangle_K \otimes |\mathbf{0}\rangle_s \otimes |\phi\rangle_R \otimes |\Phi^{\otimes e}\rangle; s = N - K - R - e. \qquad (9.211)$$

The codeword in canonical $[N,K,R,e]$ EA-OQECC, shown in Fig. 9.23a, is composed of K information qubits, R gauge qubits, and e maximally entangled kets shared between transmitter and receiver. Since the number of information qubits is K and codeword length is N, the number of remained ancilla qubits in the $|0\rangle$ state is $s = N - K - R - e$.

The canonical EA-OQECC can correct the following set of errors:

$$\mathcal{N}_{\text{canonical}} = \{X(\alpha(\boldsymbol{a}, \boldsymbol{a}_{\text{eb}}, \boldsymbol{b}_{\text{eb}}))Z(\beta(\boldsymbol{a}, \boldsymbol{a}_{\text{eb}}, \boldsymbol{b}_{\text{eb}})) \otimes X(\boldsymbol{a})Z(\boldsymbol{b}) \otimes X(\boldsymbol{c})Z(\boldsymbol{d}) \otimes X(\boldsymbol{a}_{\text{eb}})Z(\boldsymbol{b}_{\text{eb}}) :$$
$$\boldsymbol{a}, \boldsymbol{b} \in F_2^s; \boldsymbol{c}, \boldsymbol{d} \in F_2^R; \boldsymbol{b}_{\text{eb}}, \boldsymbol{a}_{\text{eb}} \in F_2^e\}; \alpha, \beta : F_2^s \times F_2^e \times F_2^e \rightarrow F_2^K, \qquad (9.212)$$

where the Pauli operators $X(\boldsymbol{e})$ ($\boldsymbol{e} \in \{\boldsymbol{a}, \boldsymbol{a}_{\text{eb}}, \boldsymbol{c}, \alpha(\boldsymbol{a},\boldsymbol{b}_{\text{eb}},\boldsymbol{a}_{\text{eb}})\}$) and $Z(\boldsymbol{e})$ ($\boldsymbol{e} \in \{\boldsymbol{b}, \boldsymbol{b}_{\text{eb}}, \boldsymbol{d}, \beta(\boldsymbol{a},\boldsymbol{b}_{\text{eb}},\boldsymbol{a}_{\text{eb}})\}$) are introduced by Eq. (9.172).

To prove claim (9.212), let us observe one particular error E_c from the set of errors (9.212) and study its action on transmitted codeword $|\psi\rangle_K \otimes |\mathbf{0}\rangle_s \otimes |\phi\rangle_R \otimes |\Phi^{\otimes e}\rangle$:

$$X(\alpha(\boldsymbol{a}, \boldsymbol{a}_{\text{eb}}, \boldsymbol{b}_{\text{eb}}))Z(\beta(\boldsymbol{a}, \boldsymbol{a}_{\text{eb}}, \boldsymbol{b}_{\text{eb}}))|\psi\rangle \otimes X(\boldsymbol{a})Z(\boldsymbol{b})|\mathbf{0}\rangle_s \otimes X(\boldsymbol{c})Z(\boldsymbol{d})|\phi\rangle_R \otimes \left(X(\boldsymbol{a}_{\text{eb}}) Z(\boldsymbol{b}_{\text{eb}}) \otimes I^{R_x}\right)|\Phi^{\otimes e}\rangle. \qquad (9.213)$$

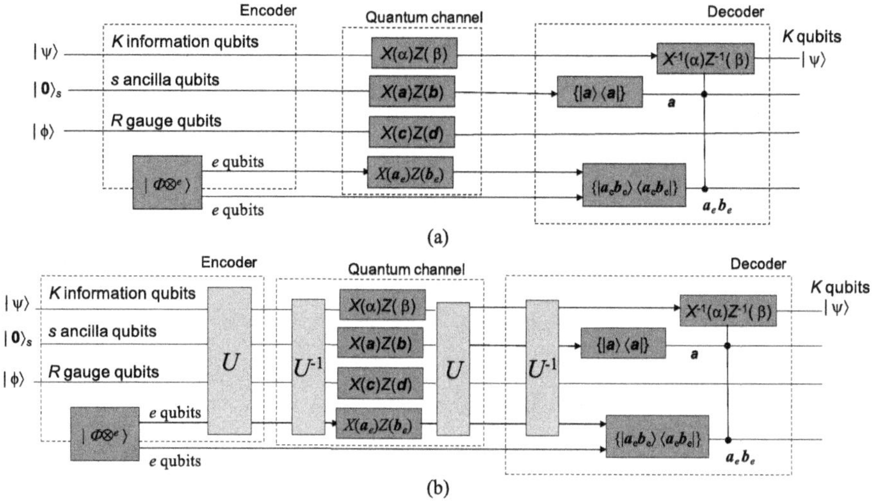

FIGURE 9.23

Entanglement-assisted operator quantum error correcting codes: (a) canonical code and (b) general code.

The action of operator $X(\boldsymbol{a})Z(\boldsymbol{b})$ on state $|\mathbf{0}\rangle_s$ is given by $X(\boldsymbol{a})Z(\boldsymbol{b})|\mathbf{0}\rangle_s \overset{Z(\boldsymbol{b})|\mathbf{0}\rangle_s=|\mathbf{0}\rangle_s}{=} X(\boldsymbol{a})|\mathbf{0}\rangle_s = |\boldsymbol{a}\rangle$. On the other hand, the action of $X(\boldsymbol{a}_{\text{eb}})$ $Z(\boldsymbol{b}_{\text{eb}})$ on the maximum-entangled state is $(X(\boldsymbol{a}_{\text{eb}})Z(\boldsymbol{b}_{\text{eb}}) \otimes I^{R_x})|\Phi^{\otimes e}\rangle = |\boldsymbol{a}_{\text{eb}}, \boldsymbol{b}_{\text{eb}}\rangle$, where I^{Rx} denotes the identity operators applied on the receiver half of the maximum-entangled state. The action of the channel on gauge qubits is described by $(X(\boldsymbol{c})Z(\boldsymbol{d}))|\phi\rangle = |\phi'\rangle$. Finally, the action of the channel on the information portion of the codeword is given by $X(\alpha(\boldsymbol{a}, \boldsymbol{a}_{\text{eb}}, \boldsymbol{b}_{\text{eb}}))\,Z(\beta(\boldsymbol{a}, \boldsymbol{a}_{\text{eb}}, \boldsymbol{b}_{\text{eb}}))|\psi\rangle = |\psi'\rangle$. Clearly, the vector $(\boldsymbol{a}\ \boldsymbol{b}\ \boldsymbol{c}\ \boldsymbol{d}\ \boldsymbol{a}_{\text{eb}}\ \boldsymbol{b}_{\text{eb}})$ uniquely specifies the error operator E_c, and can be called the syndrome vector. Since the state $|\mathbf{0}\rangle_s$ is invariant on the action of $Z(\boldsymbol{b})$, the vector $\boldsymbol{b}$ can be omitted from the syndrome vector. Similarly, since the state of gauge qubits is irrelevant to the information qubits state, the vectors $\boldsymbol{c}$ and $\boldsymbol{d}$ can also be dropped from the syndrome vector. To obtain the $\boldsymbol{a}$-portion of the syndrome we have to perform simultaneous measurements on $Z(\boldsymbol{d}_i)$, where $\boldsymbol{d}_i = (0\ldots01_i0\ldots0)$. On the other hand, to obtain the $(\boldsymbol{a}_{\text{eb}}\ \boldsymbol{b}_{\text{eb}})$-portion of the syndrome vector we have to perform measurements on $Z(\boldsymbol{d}_1)Z(\boldsymbol{d}_1), \ldots, Z(\boldsymbol{d}_e)Z(\boldsymbol{d}_e)$ for $\boldsymbol{b}_{\text{eb}}$ and $X(\boldsymbol{d}_1)X(\boldsymbol{d}_1), \ldots, X(\boldsymbol{d}_e)X(\boldsymbol{d}_e)$ for $\boldsymbol{a}_{\text{eb}}$. Once the syndrome vector is determined we apply $X^{-1}(\alpha(\boldsymbol{a}, \boldsymbol{a}_e, \boldsymbol{b}_e))$

$Z^{-1}(\beta(\boldsymbol{a}, \boldsymbol{b}_e, \boldsymbol{a}_e))$ to undo the action of the channel. To avoid the need for measurement we can perform the following controlled unitary decoding:

$$U_{\text{canocnical, dec}} = \sum_{\boldsymbol{a}, \boldsymbol{b}_{\text{eb}}, \boldsymbol{a}_{\text{eb}}} X^{-1}(\alpha(\boldsymbol{a}, \boldsymbol{a}_{\text{eb}}, \boldsymbol{b}_{\text{eb}}))Z^{-1}(\beta(\boldsymbol{a}, \boldsymbol{a}_{\text{eb}}, \boldsymbol{b}_{\text{eb}}))$$

$$\otimes |\boldsymbol{a}\rangle\langle \boldsymbol{a}| \otimes I \otimes |\boldsymbol{a}_{\text{eb}}, \boldsymbol{b}_{\text{eb}}\rangle\langle \boldsymbol{a}_{\text{eb}}, \boldsymbol{b}_{\text{eb}}|. \tag{9.214}$$

The canonical EA-OQECC can also be described by using the stabilizer formalism interpretation. Let $S_{\text{canonical}}$ be the non-Abelian group generated by;

$$S_{\text{canonical}} = \langle S_{\text{canonical},I}, S_{\text{canonical},E}, S_{\text{canonical},G} \rangle, \tag{9.215}$$

where with $S_{\text{canonical},I}$ we denote the isotropic subgroup, with $S_{\text{canonical},E}$ we denote the entanglement subgroup, and $S_{\text{canonical}, G}$ is the gauge subgroup given respectively by:

$$S_{\text{canonical},I} = \langle Z_1, \ldots, Z_s \rangle, S_{\text{canonical},E} = \langle Z_{s+1}, \ldots, Z_{s+e}, X_{s+1}, \ldots, X_{s+e} \rangle, S_{\text{canonical},G}$$
$$= \langle Z_{s+e+1}, \ldots, Z_{s+e+R}, X_{s+e+1}, \ldots, X_{s+e+R} \rangle. \tag{9.216}$$

By using finite geometry representation, the generators of Eqs. (9.215) and (9.216) can be arranged as follows:

$$A_{\text{EA-OQECC,canonical}} = \left[\begin{array}{cccc|cccc} \overbrace{\mathbf{0}}^{K} & \overbrace{\mathbf{0}}^{s} & \overbrace{\mathbf{0}}^{R} & \overbrace{\mathbf{0}}^{e} & \overbrace{\mathbf{0}}^{K} & \overbrace{I_s}^{s} & \overbrace{\mathbf{0}}^{R} & \overbrace{\mathbf{0}}^{e} \\ \mathbf{0} & \mathbf{0} & \mathbf{0} & \mathbf{0} & \mathbf{0} & \mathbf{0} & I_R & \mathbf{0} \\ \mathbf{0} & \mathbf{0} & I_R & \mathbf{0} & \mathbf{0} & \mathbf{0} & \mathbf{0} & \mathbf{0} \\ \mathbf{0} & \mathbf{0} & \mathbf{0} & \mathbf{0} & \mathbf{0} & \mathbf{0} & \mathbf{0} & I_e \\ \mathbf{0} & \mathbf{0} & \mathbf{0} & I_e & \mathbf{0} & \mathbf{0} & \mathbf{0} & \mathbf{0} \end{array}\right]. \tag{9.217}$$

The EA-OQECC can correct the following set of errors:

$$\mathcal{N} = \left\{ E_m \middle| \forall E_1, E_2 \Rightarrow E_2^\dagger E_1 \in \langle S_{\text{canonical},I}, S_{\text{canonical},G} \rangle \right.$$
$$\left. \cup \left(G_N - C(\langle S_{\text{canonical},I}, S_{\text{canonical},G} \rangle) \right) \right\}. \tag{9.218}$$

By a similar approach as in Section 9.3 we can establish the following bijection between canonical EA-OQECC described by $S_{\text{canonical}} = \langle S_{\text{canonical},I}, S_{\text{canonical},E}, S_{\text{canonical},G} \rangle$, and a general EA-OQECC characterized by $S = \langle S_I, S_E, S_G \rangle$, with $S = US_{\text{canonical}}U^{-1}$ (or equivalently $S_{\text{canonical}} = U^{-1}SU$), where U is a corresponding unitary operator. This $[N,K,R,e]$ EA-OQECC, illustrated in Fig. 9.23b, is defined by encoding–decoding pair (E,D) with the property that it can correct any error from the following set of errors:

$$\mathcal{N} = \left\{ E_m | \forall E_1,\ E_2 \Rightarrow E_2^\dagger E_1 \in \langle S_I,\ S_G \rangle \cup (G_N - C(\langle S_I,\ S_G \rangle)) \right\}. \tag{9.219}$$

The EA-OQECC is very flexible, as it can be transformed into another code by adding or removing certain stabilizer elements. It can be shown that an $[N,K,R,e;D_1]$ can be transformed into an $[N,K+R,0,e;D_2]$ or an $[N,K,0,e;D_3]$ code. The minimum distance of EQ-OQECCs is defined in a fashion like that in previous sections. For encoder implementation we can use the conjugation method, in which we employ the approach described in Section 9.10.6 to transform the EA-OQECC into corresponding canonical form (9.217). By applying this sequence of gates in opposite order we obtain the encoding circuit of the corresponding EA-OQECC.

9.10.8.1 EA-OQECCs Derived From Classical Binary and Quaternary Codes

EA-OQECCs can be derived from EA codes described in Section 9.10.4 by properly rearranging and combing the stabilizers.

Example. Let us observe the [8,1,1; 3] EA code described by the following group:

$$S_{EA}=\langle S_I,\ S_E\rangle,\ \ S_I=\langle Z_1Z_2,\ Z_1Z_3,\ Z_4Z_5,\ Z_4Z_6,\ Z_7Z_8,\ X_1X_2X_3X_4X_5X_6\rangle,\ \ S_E=\langle Z_8,\ X_1X_2X_3X_7X_8\rangle.$$

The corresponding Pauli-encoded operators are given by: $\{Z_1Z_4Z_8, X_4X_5X_6\}$. By using this EA code we can derive the following $[8,1,R=2,e=1;\ 3]$ EA-OQECC:

$$S_{EA}=\langle S_I,\ S_E,\ S_G\rangle,\ \ S_I=\langle Z_1Z_2Z_4Z_5,\ Z_1Z_3Z_4Z_6,\ Z_7Z_8,\ X_1X_2X_3X_4X_5X_6\rangle,$$
$$S_E=\langle Z_8,\ X_1X_2X_3X_7X_8\rangle,\ \ S_G=\langle Z_1Z_2,\ X_2X_5,\ Z_4Z_6,\ X_3X_6\rangle.$$

For CSS-like codes, the number of ebits can be determined by $e=\text{rank}(\boldsymbol{H}\boldsymbol{H}^{\text{T}})$, where $\boldsymbol{H}$ is the parity-check matrix of classical code. For example, the Bose, Chaudhuri, and Hocquenghem (BCH) codes described in Chapter 6 can be used in the design of EA codes. The BCH code over GF(2^m) has the codeword length $n=2^m-1$, minimum distance $d\geq 2t+1$, number of parity bits $n-k\leq mt$, and its parity-check matrix is given by:

$$\boldsymbol{H}=\begin{bmatrix} \alpha^{n-1} & \alpha^{n-2} & \cdots & \alpha & 1 \\ \alpha^{3(n-1)} & \alpha^{3(n-2)} & \cdots & \alpha^3 & 1 \\ \alpha^{5(n-1)} & \alpha^{5(n-2)} & \cdots & \alpha^5 & 1 \\ \cdots & \cdots & \cdots & \cdots & \cdots \\ \alpha^{(2t-1)(n-1)} & \alpha^{(2t-1)(n-2)} & \cdots & \alpha^{2t-1} & 1 \end{bmatrix}. \tag{9.220}$$

In Chapter 6 we described how to obtain binary representation of this parity-check matrix. For example, for $m=6$, the number of ebits required is $e=\text{rank}(\boldsymbol{H}\boldsymbol{H}^{\text{T}})=6$. This BCH (63,39; 9) code therefore can be used to design $[n=63, 2k-n+e=21, e=6]$ EA code. By inspection we can find that the last six rows of the binary $\boldsymbol{H}$-matrix are sympletic pairs that form the entanglement group. The gauge group can be formed by removing one sympletic pair at a time from the entanglement group and adding it to the gauge group. Clearly, the following set of quantum codes can be obtained using this approach: $\{[63,21,R=0,e=6],\ [63,21,R=1,e=5],$

$[63,21,R=2,e=4]$, $[63,21,R=3,e=3]$, $[63,21,R=4,e=2]$, $[63,21,R=5,e=1]$, $[63,21,R=6,e=0]\}$. The first code from the set is a pure EA code, and the last code from the set is a pure subsystem code, while the quantum codes in between are EA-OQECCs.

The EA-OQECCs can also be derived from classical LDPC codes, and the whole of the next chapter is devoted to different classes of quantum codes derived from classical LDPC codes [18,19,69,73–75].

The quaternary classical codes can also be used in the design of EA-QEECCs. The first step is to start with corresponding EA code, using the concepts outlined in Section 9.10.4. The second step is to carefully move some of the sympletic pairs from the entanglement group to the gauge group so that the minimum distance is not affected that much. Let us consider the following (15,10,4) quaternary code [60]:

$$\boldsymbol{H}_4=\begin{bmatrix} 1 & 0 & 0 & 0 & 1 & 1 & \alpha^2 & 0 & 1 & \alpha^2 & 0 & \alpha & \alpha^2 & 1 & 0 \\ 0 & 1 & 0 & 0 & 1 & 0 & \alpha & \alpha^2 & 1 & \alpha & 0 & 0 & 1 & \alpha & 1 \\ 0 & 0 & 1 & 0 & \alpha & \alpha^2 & 1 & \alpha & 1 & 0 & 0 & \alpha & 1 & \alpha^2 & \alpha \\ 0 & 0 & 0 & 1 & 1 & \alpha^2 & 0 & 1 & \alpha^2 & \alpha & 0 & \alpha^2 & 1 & 0 & \alpha^2 \\ 0 & 0 & 0 & 0 & 0 & 0 & 0 & 0 & 0 & 0 & 1 & 0 & 0 & 0 & 0 \end{bmatrix}, \tag{9.221}$$

where the elements of GF(4) are $\{0,1,\alpha,\alpha^2\}$, generated by primitive polynomial $p(x)=x^2+x+1$. By using Eq. (9.186) we obtain the following quantum-check matrix of $[15,9,e=4;4]$ EA code:

$$A_{EA}=f(\widetilde{\boldsymbol{H}}_4)=f\left(\begin{bmatrix}\alpha\boldsymbol{H}_4\\ \alpha^2\boldsymbol{H}_4\end{bmatrix}\right)$$
$$=\begin{bmatrix} X & I & I & I & X & X & Y & I & Z & Y & I & Z & Y & X & I \\ I & X & I & I & X & I & Z & Y & X & Z & I & I & X & Z & X \\ I & I & X & I & Z & Y & X & Z & X & I & I & Z & X & Y & Z \\ I & I & I & X & X & Y & I & X & Y & Z & I & Y & X & I & Y \\ I & I & I & I & I & I & I & I & I & I & X & I & I & I & I \\ Z & I & I & I & Z & Z & X & I & Z & X & I & Y & X & Z & I \\ I & Z & I & I & Z & I & Y & X & Z & Y & I & I & Z & Y & Z \\ I & I & Z & I & Y & X & Z & Y & Z & I & I & Y & Z & X & Y \\ I & I & I & Z & Z & X & I & Z & X & Y & I & X & Z & I & X \\ I & I & I & I & I & I & I & I & I & I & Z & I & I & I & I \end{bmatrix}. \tag{9.222}$$

From Eq. (9.222), it is clear that stabilizers in the third and fourth rows commute, and therefore they represent the isotropic subgroup. The remaining eight anti-commute and represent the entanglement subgroup. By removing any two anti-commuting pairs from the entanglement subgroup, say the first two, and moving them to the gauge subgroup we obtain the [15,9,$e=3$,$R=1$] EA-OQECC. By using the MAGMA approach [76] we can identify an EA-QECC derived from Eq. (9.222) having the largest possible minimum distance.

9.11 SUMMARY

This chapter was devoted to stabilizer codes and their relatives. The stabilizer codes were introduced in Section 9.1. Their basic properties were discussed in Section 9.2 The encoded operations were introduced in the same section (Section 9.2). In Section 9.3, finite geometry interpretation of stabilizer codes was introduced. This representation was used in Section 9.4 to introduce the so-called standard form of stabilizer code. The standard form is the basic representation for the efficient encoder and decoder implementations, which were discussed in Section 9.5. Section 9.6 was devoted to nonbinary stabilizer codes, which generalizes the previous sections. Subsystem codes were introduced in Section 9.7. In the same section, efficient encoding and decoding of subsystem codes was discussed as well. An important class of quantum codes, namely topological codes, was discussed in Section 9.8. Section 9.9 was devoted to surface codes. Entanglement-assisted codes were introduced in Section 9.10. Finally, for a better understanding of the material presented in this chapter, a set of problems has been provided in the next section.

9.12 PROBLEMS

1. By using Eq. (9.28), prove the following commutative properties:

$$\begin{aligned}&[\overline{X}_i,\overline{X}_j]=[\overline{Z}_i,\overline{Z}_j]=0\\&[\overline{X}_i,\overline{Z}_j]=0(i\neq j)\\&\{\overline{X}_i,\overline{Z}_i\}=0.\end{aligned}$$

Prove also that $\overline{X}_i,\overline{Z}_i$ commute with the generators g_i from S.

2. The [4,2] QECC is given by the following generators: $g_1=X_1Z_2Z_3X_4$, $g_2=Y_1X_2X_3Y_4$. Determine appropriately the Pauli-encoded operators so that the commutation relations from Problem 1 are satisfied. What is the distance of this code? Is this code degenerate? Determine the basis codewords.

3. The generators for [8,3,3] code are given by:

$$g_1 = X_1X_2X_3X_4X_5X_6X_7X_8,\ \ g_2 = Z_1Z_2Z_3Z_4Z_5Z_6Z_7Z_8,\ \ g_3 = X_2X_4X_5X_6X_7X_8,$$
$$g_4 = X_2X_3X_4X_6X_7X_8,\ \text{ and }\ g_5 = Y_2X_3Z_4X_5Z_6Y_8.$$

- **(a)** Determine the standard form of quantum-check matrix, and determine corresponding generators in standard form.
- **(b)** Determine encoded Pauli operators in standard form.
- **(c)** Determine the efficient encoder implementation.
- **(d)** Determine the efficient decoder implementation.
- **(e)** Prove that the distance of this code is $D = 3$.
- **(f)** Determine whether this code is nondegenerate.

4. The quantum check matrix $\boldsymbol{A}$ of a CSS code is given by:

$$\boldsymbol{A} = \begin{bmatrix} \boldsymbol{H} & | & \boldsymbol{0} \\ \boldsymbol{0} & | & \boldsymbol{H} \end{bmatrix} \quad \boldsymbol{H} = \begin{bmatrix} 1 & 0 & 0 & 1 & 1 & 1 \\ 1 & 1 & 1 & 0 & 0 & 1 \\ 0 & 1 & 1 & 1 & 1 & 0 \end{bmatrix}.$$

- **(a)** Determine the standard form of quantum-check matrix, and determine corresponding generators in standard form.
- **(b)** Determine encoded Pauli operators in standard form.
- **(c)** Given the results of (a) and (b), determine the efficient encoding circuit for the corresponding quantum stabilizer code.
- **(d)** Given the results of (a) and (b), determine the efficient decoding circuit for the corresponding quantum stabilizer code.

5. The quantum-check matrix of a stabilizer code is given below as matrix $\boldsymbol{A}$.

- **(a)** Determine the standard form of quantum-check matrix and determine corresponding generators in standard form.
- **(b)** Determine encoded Pauli operators in standard form.
- **(c)** Determine the code parameters.
- **(d)** Given the results of (a) and (b), determine the efficient encoding circuit for the corresponding quantum stabilizer code.
- **(e)** Given the results of (a) and (b), determine the efficient decoding circuit for the corresponding quantum stabilizer code.

$$A = \begin{bmatrix} 1 & 1 & 1 & 1 & 1 & 1 & 1 & 1 & 1 & 0 & 0 & 0 & 0 & 0 & 0 & 0 & 0 & 0 \\ 0 & 0 & 0 & 1 & 1 & 1 & 1 & 1 & 1 & 0 & 0 & 0 & 0 & 0 & 0 & 0 & 0 & 0 \\ 0 & 0 & 0 & 0 & 0 & 0 & 0 & 0 & 0 & 1 & 0 & 1 & 1 & 0 & 1 & 1 & 0 & 1 \\ 0 & 0 & 0 & 0 & 0 & 0 & 0 & 0 & 0 & 0 & 1 & 1 & 0 & 1 & 1 & 0 & 1 & 1 \\ 0 & 0 & 0 & 0 & 0 & 0 & 0 & 0 & 0 & 1 & 0 & 1 & 0 & 0 & 0 & 0 & 0 & 0 \\ 0 & 0 & 0 & 0 & 0 & 0 & 0 & 0 & 0 & 0 & 0 & 0 & 1 & 0 & 1 & 0 & 0 & 0 \\ 0 & 0 & 0 & 0 & 0 & 0 & 0 & 0 & 0 & 0 & 1 & 1 & 0 & 0 & 0 & 0 & 0 & 0 \\ 0 & 0 & 0 & 0 & 0 & 0 & 0 & 0 & 0 & 0 & 0 & 0 & 0 & 1 & 1 & 0 & 0 & 0 \end{bmatrix}.$$

6. We would like to design a CSS code based on the following parity-check matrix of classical code:

$$\boldsymbol{H} = \begin{bmatrix} 0 & 0 & 1 & 0 & 1 & 0 & 1 \\ 1 & 0 & 0 & 1 & 1 & 0 & 0 \\ 0 & 1 & 0 & 1 & 0 & 0 & 1 \\ 1 & 0 & 0 & 0 & 0 & 1 & 1 \\ 0 & 0 & 1 & 1 & 0 & 1 & 0 \\ 1 & 1 & 1 & 0 & 0 & 0 & 0 \\ 0 & 1 & 0 & 0 & 1 & 1 & 0 \end{bmatrix}.$$

(a) Can this $\boldsymbol{H}$-matrix be used to design a CSS code? If not, how it can be modified so that it can be used to design a corresponding CSS code.

(b) Determine the standard form of a quantum-check matrix, and determine corresponding generators in standard form.

(c) Determine encoded Pauli operators in standard form.

(d) Determine the code parameters.

(e) Given the results of (a) and (b), determine the efficient encoding circuit for the corresponding quantum stabilizer code.

(f) Given the results of (a) and (b), determine the efficient decoding circuit for the corresponding quantum stabilizer code.

7. For nonbinary quantum gates from Fig. 9.6, show that: (1) $F^{-1}Z(b)F = X(b)$ and (2) $F^{-1}X(a)F = Z(-a)$.

8. Show that the set of errors from Section 9.6 $\varepsilon = \{X(a)X(b)|a,\ b \in F_q\}$ satisfies the following properties: (1) it contains the identity operator, (2) $\mathrm{tr}\left(E_1^{\dagger}E_2\right) = 0\ \forall E_1,\ E_2 \in \varepsilon$, and (3) $\forall E_1,\ E_2 \in \varepsilon : E_1E_2 = cE_3,\ E_3 \in \varepsilon,\ c \in F_q$.

9. If the operators $X(a)$ and $Z(b)$ are defined as $X(a)|x\rangle = |x+a\rangle,\ Z(b) = \omega^{\mathrm{tr}(bx)}|x\rangle; x,\ a,\ b \in F_q$, show that: (1) $X(a)Z(b) = \omega^{-tr(ab)}Z(b)X(a)$, and (2) $X(a+a')Z(b+b') = \omega^{-tr(a'b)}X(a)Z(b)X(a')Z(b')$.

10. Show that the syndrome circuit from Fig. 9.8 for input $E|\psi\rangle$, where E is the qudit error and $|\psi\rangle$ is the codeword, performs the following mapping: $|0\rangle E|\psi\rangle \rightarrow \sum_{u \in F_q} F^{\dagger}|u\rangle X(u\boldsymbol{a}_i)Z(u\boldsymbol{b}_i)E|\psi\rangle$.

11. Prove that the Hamming bound for nonbinary quantum stabilizer code $[N,K,D]$ (where K is the number of information qudits and N is the codeword length) over F_q is given by:

$$\sum_{i=0}^{\lfloor (D-1)/2 \rfloor} \binom{N}{i} (q^2-1)^i q^K \leq q^N.$$

12. Prove that the Singleton bound for nonbinary quantum stabilizer code $[N,K,D]$ (where K is the number of information qudits and N is the codeword length) over F_q is given by:

$$K \leq N - 2D + 2.$$

13. We say that the nonbinary quantum stabilizer code $[N,K,D]$ (where K is the number of information qudits and N is the codeword length) is nondegenerate (or pure) if and only if its stabilizer S does not contain any nonscalar error operator of weight less than D. Prove that there does not exist any *perfect* nondegenerate code of distance greater than 3.

14. A quantum stabilizer code $[N,K,D]$ (where K is the number of information qudits and N is the codeword length) exists only if there exists a classical linear code $C \subseteq F_q^{2N}$ of size $|C| = q^{N-K}$ such that $C \subseteq C^{\perp}$ and $\text{wt}(C^{\perp} - C) = D$. Prove the claim.

15. In CSS code construction, let C_1 and C_2 denote two classical linear codes with parameters (n_1,k_1,d_1) and (n_2,k_2,d_2) such that $C_2^{\perp} \subseteq C_1$. Then there exists an $[n,k_1 + k_2 - n,d]$ quantum stabilizer code over F_q such that $d = \min\{\text{wt}(c) | c \in (C_1 - C_2^{\perp}) \cup (C_2 - C_1^{\perp})\}$ that is nondegenerate to min $\{d_1,d_2\}$. Prove the claim.

16. Let C be a classical linear code over F_q with parameters (n,k,d) that contain its dual, $C^{\perp} \subseteq C$. Then there exists an $[n,2k - n, \geq d]$ quantum stabilizer code over F_q that is nondegenerate to d. Prove the claim.

17. Let C_Q be an $[N,K,R,D]$ stabilizer subsystem code over F_2 with stabilizer S and gauge group G. Denote the encode operators $\overline{X}_i, \overline{Z}_i$; $i = 1,\ldots,K$ satisfying commutative relations (9.28). Then there exist operators $\widetilde{X}_i, \widetilde{Z}_i \in G_N; i = 1,\ldots,N$ such that:

(a) $S = \langle \widetilde{Z}_1,\ldots,\widetilde{Z}_s \rangle; R + s \leq N; R, s \geq 0$,

(b) $\mathcal{G} = \langle jI, \widetilde{Z}_1,\ldots,\widetilde{Z}_{s+R}, \widetilde{X}_{s+1},\ldots,\widetilde{X}_R \rangle$,

(c) $\overline{X}_1 = \widetilde{X}_{s+R+1}, \overline{Z}_1 = \widetilde{Z}_{s+R+1},\ldots,\overline{X}_K = \widetilde{X}_N, \overline{Z}_K = \widetilde{Z}_N$, and

(d) $C_{G_N}(S) = \langle G, \overline{X}_1, \overline{Z}_1,\ldots,\overline{X}_K, \overline{Z}_K \rangle$.

Prove claims (a)–(d).

18. For [9,1,4,3] Bacon–Shore code described in Section 9.7, describe how encoder architectures shown in Fig. 9.10 are obtained.

19. Let us observe one particular generator of a quantum-check matrix, say $\boldsymbol{g} = (a_1,\ldots,a_N \,|b_1,\ldots,b_N)$. Show that the action of H and P gates on the ith qubit is given by Eqs. (9.147) and (9.149), respectively. Show that the action of the CNOT gate on $\boldsymbol{g}$, where the ith qubit is the control qubit and the jth qubit is the target qubit, the action denoted as $\text{CNOT}^{i,j}$, is given by Eq. (9.148).

20. Based on Table 9.1, show that the transformation of $\boldsymbol{g} = (a_1,\ldots,a_N|b_1,\ldots,b_N)$ to $\boldsymbol{g}' = (a'_1,\ldots,a'_N|0,\ldots,0)$ can be achieved by application of Hadamard and phase gates as follows:

$$\bigotimes_{i=1}^{N} H^{\overline{a}_i b_i} P^{a_i b_i}, \overline{a}_i = a_i + 1 \text{mod} 2.$$

21. Based on Eq. (9.148), show that the error operator $\boldsymbol{e} = (a_1,\ldots,a_i = 1,\ldots,a_N|0,\ldots,0)$ can be converted to $\boldsymbol{e}' = (0,\ldots,a'_i = 1,0\ldots0|0\ldots0)$ by the application of the following sequence of gates:

$$\prod_{m=1, m\neq n}^{N} (\text{CNOT}^{m,n})^{a_n}.$$

22. For quantum stabilizer codes from Problems P3–P6, determine the corresponding encoding circuits by using the conjugation method.

23. For [9,1,4,3] Bacon–Shore code described in Section 9.7, determine the encoding circuit by using the conjugation method.

24. Show that by transforming the gauge group, represented using finite geometry formalism, as follows:

$$\mathcal{G} = \begin{bmatrix} \underline{S} \\ \underline{\mathcal{G}_Z} \\ \mathcal{G}_X \end{bmatrix} \xrightarrow{\{U\}} \mathcal{G}' = \begin{bmatrix} \mathbf{0} & \mathbf{0} & \mathbf{0} & I_s & \mathbf{0} & \mathbf{0} \\ \mathbf{0} & \mathbf{0} & \mathbf{0} & \mathbf{0} & I_R & \mathbf{0} \\ \mathbf{0} & I_R & \mathbf{0} & \mathbf{0} & \mathbf{0} & \mathbf{0} \end{bmatrix},$$

we essentially implement the following mapping:

$$|\phi\rangle|\psi\rangle \rightarrow \underbrace{|0\rangle \otimes \ldots \otimes |0\rangle}_{s=N-K-R \text{ times}} |\phi\rangle|\psi\rangle = |\mathbf{0}\rangle_s|\phi\rangle|\psi\rangle.$$

Which sequence of gates should be applied on the gauge group to perform the foregoing transformation?

25. Explain how the encoder circuit shown in Fig. 9.11c has been derived from Fig. 9.11b.

26. Prove Theorem 2.

27. Let us observe the torus code with $m \times m$ lattice. Let us further consider the operator of the form:

$$\overline{Y}(c,\ c') = \prod_{i \in c} Z_i \prod_{j \in c'} X_j,$$

where c is an I-cycle (loop) on the lattice and c' is an I-cycle on the dual lattice.

(a) Show that $\overline{Y}(c,\ c')$ operators commute with all stabilizers and thus perform mapping of code subspace C_Q to itself.

(b) Show that $\overline{Y}([c],\ [c'])$ forms a linear basis for $L(C_Q)$.

(c) Show that the logical operators $\overline{X}_1(0,\ [c_1'])$, $\overline{X}_2(0,\ [c_2'])$, $\overline{Z}_1([c_1], 0)$, and $\overline{Z}_2([c_2], 0)$, where c_1,c_2 are the cycles on the original lattice and c'_1,c'_2 are the cycles on the dual lattice, can be used as generators of $L(C_Q)$.

28. Let us observe the plane code with an $n \times m$ lattice with boundaries as illustrated in Fig. 9.13d. Determine the quantum code parameters [N,K,D]. For the 2×3 example shown in Fig. 9.13d, determine the stabilizer and encoded Pauli operators. Determine the corresponding quantum-check matrix and represent it in standard form. Finally, provide efficient encoder and decoder implementations.

29. If V is an arbitrary subgroup of G_N of order 2^M, then there exists a set of generators $\{\overline{X}_{p+1}, \ldots, \overline{X}_{p+q}; \overline{Z}_1, \ldots, \overline{Z}_{p+q}\}$ satisfying similar commutation properties to Eq. (9.28) at the beginning of the chapter, namely:

$$[\overline{X}_m, \overline{X}_n] = [\overline{Z}_m, \overline{Z}_n] = 0\ \forall m, n;\ [\overline{X}_m, \overline{Z}_n] = 0\ \forall m \neq n;\ \{\overline{X}_m, \overline{Z}_n\} = 0\ \forall m = n.$$

Prove this claim.

30. If there exists bijection between two groups V and S that preserves their commutation relations, prove that $\forall\ v \in V$ there exists corresponding $s \in S$ such that $v = UsU^{-1}$ (up to a general phase constant).

31. As a generalization of the EA quantum codes represented earlier we can design a *continuous variable* EA quantum code given by:

$$\mathbf{A}_{\text{EA}} = (\mathbf{A}_x | \mathbf{A}_z),$$

where A_{EA} is an $(n - k) \times 2n$ real matrix, and both $\mathbf{A}_x$ and $\mathbf{A}_z$ are $(n - k) \times n$ real matrices. The number of required entangled states is:

$$e = \text{rank}\{\mathbf{A}_z\mathbf{A}_x^{\text{T}} - \mathbf{A}_x\mathbf{A}_z^{\text{T}}\}/2.$$

Prove this claim.

32. A formula like that from the previous example can be derived for EA qudit codes, namely by replacing the subtraction operation by subtraction mod p (where p is a prime) as follows:

$$e = \text{rank}\{(\boldsymbol{A}_z\boldsymbol{A}_x^{\mathrm{T}}\text{-}\boldsymbol{A}_x\boldsymbol{A}_z^{\mathrm{T}})\text{mod}p\}/2.$$

The quantum-check matrix of this qudit code is given by: $\boldsymbol{A}_{\mathrm{EA}} = (\boldsymbol{A}_x|\boldsymbol{A}_z)$, with matrix elements being from finite field $F_p = \{0,1,\ldots,p-1\}$. Notice that error operators need to be redefined as was done in Section 9.6. The corresponding entanglement qudits (edits) have the form $\sum_{m=0}^{p-1}|m\rangle|m\rangle/\sqrt{p}$. Prove the claims. Can you generalize the EA qudit code design over GF(q) ($q = p^m$, $m > 1$)?

33. This problem is related to the extended EA codes. If an EA $[N,K,e,D]$ code exists, prove that an extended $[N+1,K-1,e',D']$ also exists for some e' and $D' \geq D$. Provide the quantum-check matrix of extended EA code, assuming that the quantum-check matrix of the original EA code is known.

34. In certain quantum technologies, the CNOT gate is challenging to implement. For EA code given by the following quantum-check matrix, determine the encoding circuit without using any SWAP operators because they require the use of three CNOT gates.

$$\boldsymbol{A}_{\mathrm{EA}} = \left[\begin{array}{cccc|cccc} 0 & 1 & 0 & 0 & 1 & 0 & 1 & 0 \\ 0 & 0 & 0 & 0 & 1 & 1 & 0 & 1 \\ 1 & 1 & 1 & 0 & 0 & 1 & 0 & 0 \\ 1 & 1 & 0 & 1 & 0 & 0 & 0 & 0 \end{array}\right].$$

35. By using the standard form method, conjugation method, and sympletic Gram–Schmidt method for EA code encoder implementation, described by the following quantum-check matrix, provide the corresponding encoding circuits. Discuss the complexity of corresponding realizations.

$$\boldsymbol{A}_{EA} = \begin{bmatrix} 0 & 0 & 1 & 0 & 1 & 0 & 1 \\ 1 & 0 & 0 & 1 & 1 & 0 & 0 \\ 0 & 1 & 0 & 1 & 0 & 0 & 1 \\ 1 & 0 & 0 & 0 & 0 & 1 & 1 \\ 0 & 0 & 1 & 1 & 0 & 1 & 0 \\ 1 & 1 & 1 & 0 & 0 & 0 & 0 \\ 0 & 1 & 0 & 0 & 1 & 1 & 0 \end{bmatrix}.$$

36. Prove that canonical OQECC can correct the following set of errors:

$$N = \left\{ E_m | \forall E_1, E_2 \Rightarrow E_2^{\dagger} E_1 \in S_{\text{canonical,OQECC}} \cup \left(G_N - C\left(S_{\text{canonical,OQECC},I} \right) \right) \right\}.$$

37. Prove that $[N,K,R,D]$ OQECC has the following properties:

(a) It can correct any error for the following set of errors:

$$\mathcal{N} = \left\{ E_m | \forall E_1, E_2 \Rightarrow E_2^{\dagger} E_1 \in S_{\text{OQECC}} \cup \left(G_N - C\left(S_{\text{OQECC},I} \right) \right) \right\}.$$

(b) The code space C_{OQECC} is a simultaneous eigenspace of $S_{\text{OQECC},I}$.

(c) To decode, the error syndrome is obtained by simultaneously measuring the observables from $S_{\text{OQECC},I}$.

38. Prove that canonical $[N,K,R,e]$ EA-OQECC can correct the following set of errors:

$$\mathcal{N} = \left\{ E_m | \forall E_1,\ E_2 \Rightarrow E_2^{\dagger} E_1 \in \left\langle S_{\text{canonical},I},\ S_{\text{canonical},G} \right\rangle \cup \left(G_N - C\left(\left\langle S_{\text{canonical},I},\ S_{\text{canonical},G} \right\rangle \right) \right) \right\}.$$

Prove also that $[N,K,R,e]$ EA-OQECC can correct the following set of errors:

$$\mathcal{N} = \left\{ E_m | \forall E_1,\ E_2 \Rightarrow E_2^{\dagger} E_1 \in \langle S_I,\ S_G \rangle \cup \left(G_N - C(\langle S_I,\ S_G \rangle) \right) \right\}.$$

39. Prove that an $[N,K,R,e;D_1]$ EA-OQECC can be transformed into an $[N,K+R,0,e;D_2]$ or an $[N,K,0,e;D_3]$ code. Show that minimum distances satisfy the following inequalities: $D_2 \le D_1 \le D_3$.

40. The (15,7) BCH code has the following parity-check matrix:

$$H = \begin{bmatrix} 1 & 1 & 1 & 1 & 0 & 1 & 0 & 1 & 1 & 0 & 0 & 1 & 0 & 0 & 0 \\ 0 & 1 & 1 & 1 & 1 & 0 & 1 & 0 & 1 & 1 & 0 & 0 & 1 & 0 & 0 \\ 0 & 0 & 1 & 1 & 1 & 1 & 0 & 1 & 0 & 1 & 1 & 0 & 0 & 1 & 0 \\ 1 & 1 & 1 & 0 & 1 & 0 & 1 & 1 & 0 & 0 & 1 & 0 & 0 & 0 & 1 \\ 1 & 1 & 1 & 1 & 0 & 1 & 1 & 1 & 1 & 0 & 1 & 1 & 1 & 1 & 0 \\ 1 & 0 & 1 & 0 & 0 & 1 & 0 & 1 & 0 & 0 & 1 & 0 & 1 & 0 & 0 \\ 1 & 1 & 0 & 0 & 0 & 1 & 1 & 0 & 0 & 0 & 1 & 1 & 0 & 0 & 0 \\ 1 & 0 & 0 & 0 & 1 & 1 & 0 & 0 & 0 & 1 & 1 & 0 & 0 & 0 & 1 \end{bmatrix}.$$

Determine the parameters of corresponding EA code. Describe all possible EA-OQECCs that can be derived from this code.

41. Let us observe an EA code described by the following quantum-check matrix:

$$A_{\mathrm{EA}} = \begin{bmatrix} X & I & I & I & X & X & Y & I & Z & Y & I & Z & Y & X & I \\ I & X & I & I & X & I & Z & Y & X & Z & I & I & X & Z & X \\ I & I & X & I & Z & Y & X & Z & X & I & I & Z & X & Y & Z \\ I & I & I & X & X & Y & I & X & Y & Z & I & Y & X & I & Y \\ I & I & I & I & I & I & I & I & I & I & X & I & I & I & I \\ Z & I & I & I & Z & Z & X & I & Z & X & I & Y & X & Z & I \\ I & Z & I & I & Z & I & Y & X & Z & Y & I & I & Z & Y & Z \\ I & I & Z & I & Y & X & Z & Y & Z & I & I & Y & Z & X & Y \\ I & I & I & Z & Z & X & I & Z & X & Y & I & X & Z & I & X \\ I & I & I & I & I & I & I & I & I & I & Z & I & I & I & I \end{bmatrix}.$$

By using this quantum-check matrix, determine an EA-OQECC of largest possible minimum distance. What are the parameters of this code? By using the conjugation method, provide the corresponding encoding circuit.

References

[1] A.R. Calderbank, P.W. Shor, Good quantum error-correcting codes exist, Phys. Rev. A 54 (1996) 1098–1105.

[2] A.M. Steane, Error correcting codes in quantum theory, Phys. Rev. Lett. 77 (1996) 793.

[3] A.M. Steane, Simple quantum error-correcting codes, Phys. Rev. A 54 (6) (December 1996) 4741–4751.

[4] A.R. Calderbank, E.M. Rains, P.W. Shor, N.J.A. Sloane, Quantum error correction and orthogonal geometry, Phy. Rev. Lett. 78 (3) (January 1997) 405–408.

[5] E. Knill, R. Laflamme, Concatenated Quantum Codes, 1996. Available at: http://arxiv.org/abs/quant-ph/9608012.

[6] R. Laflamme, C. Miquel, J.P. Paz, Wojciech H. Zurek, Perfect quantum error correcting code, Phys. Rev. Lett. 77 (1) (July 1996) 198–201.

[7] D. Gottesman, Class of quantum error correcting codes saturating the quantum Hamming bound, Phys. Rev. A 54 (September 1996) 1862–1868.

[8] D. Gottesman, Stabilizer Codes and Quantum Error Correction, PhD Dissertation, California Institute of Technology, Pasadena, CA, 1997.

[9] R. Cleve, D. Gottesman, Efficient computations of encoding for quantum error correction, Phys. Rev. A 56 (July 1997) 76–82.

[10] A.Y. Kitaev, Quantum error correction with imperfect gates, in: Quantum Communication, Computing, and Measurement, Plenum Press, New York, 1997, pp. 181–188.

[11] C.H. Bennett, D.P. DiVincenzo, J.A. Smolin, W.K. Wootters, Mixed-state entanglement and quantum error correction, Phys. Rev. A 54 (6) (November 1996) 3824–3851.
[12] D.J.C. MacKay, G. Mitchison, P.L. McFadden, Sparse-graph codes for quantum error correction, IEEE Trans. Inf. Theory 50 (2004) 2315–2330.
[13] M.A. Neilsen, I.L. Chuang, Quantum Computation and Quantum Information, Cambridge University Press, Cambridge, 2000.
[14] F. Gaitan, Quantum Error Correction and Fault Tolerant Quantum Computing, CRC Press, 2008.
[15] G.D. Forney Jr., M. Grassl, S. Guha, Convolutional and tail-biting quantum error-correcting codes, IEEE Trans. Inf. Theory 53 (March 2007) 865–880.
[16] A.R. Calderbank, E.M. Rains, P.W. Shor, N.J.A. Sloane, Quantum error correction via codes over GF(4), IEEE Trans. Inf. Theory 44 (1998) 1369–1387.
[17] I.B. Djordjevic, Quantum LDPC codes from balanced incomplete block designs, IEEE Comm. Lett. 12 (May 2008) 389–391.
[18] I.B. Djordjevic, Photonic quantum dual-containing LDPC encoders and decoders, IEEE Photon. Technol. Lett. 21 (13) (July 1, 2009) 842–844.
[19] I.B. Djordjevic, Photonic entanglement-assisted quantum low-density parity-check encoders and decoders, Opt. Lett. 35 (9) (May 1, 2010) 1464–1466.
[20] S.A. Aly, A. Klappenecker, P.K. Sarvepalli, On quantum and classical BCH codes, IEEE Trans. Inf. Theory 53 (3) (2007) 1183–1188.
[21] S.A. Aly, A. Klappenecker, P.K. Sarvepalli, Primitive quantum BCH codes over finite fields, in: Proc. Int. Symp. Inform. Theory 2006 (ISIT 2006), 2006, pp. 1114–1118.
[22] A. Ashikhmin, E. Knill, Nonbinary quantum stabilizer codes, IEEE Trans. Inf. Theory 47 (7) (November 2001) 3065–3072.
[23] A. Ketkar, A. Klappenecker, S. Kumar, P.K. Sarvepalli, Nonbinary stabilizer codes over finite fields, IEEE Trans. Inf. Theory 52 (11) (November 2006) 4892–4914.
[24] J.-L. Kim, J. Walker, Nonbinary quantum error-correcting codes from algebraic curves, Discrete Math. 308 (14) (2008) 3115–3124.
[25] P.K. Sarvepalli, A. Klappenecker, Nonbinary quantum Reed-Muller codes, in: Proc. Int. Symp. Inform. Theory 2005 (ISIT 2005), 2005, pp. 1023–1027.
[26] P.K. Sarvepalli, Quantum Stabilizer Codes and Beyond, PhD Dissertation, Texas A&M University, August 2008.
[27] S.A.A.A. Ahmed, Quantum Error Control Codes, PhD Dissertation, Texas A&M University, May 2008.
[28] S.A. Aly, A. Klappenecker, Subsystem code constructions, in: Proc. ISIT 2008, July 6–11, 2008, pp. 369–373. Toronto, Canada.
[29] S. Bravyi, Subsystem codes with spatially local generators, Phys. Rev. A 83 (2011), 012320–012321–012320-9.
[30] A. Klappenecker, P.K. Sarvepalli, On subsystem codes beating the quantum Hamming or Singleton Bound, Proc. Math. Phys. Eng. Sci. 463 (2087) (November 8, 2007) 2887–2905.
[31] A. Klappenecker, P.K. Sarvepalli, Clifford code construction of operator quantum error correcting codes, quant-ph/0604161 (April 2006).
[32] D.A. Lidar, I.L. Chuang, K.B. Whaley, Decoherence-free subspaces for quantum computation, Phys. Rev. Lett. 81 (12) (1998) 2594–2597.
[33] P.G. Kwiat, A.J. Berglund, J.B. Altepeter, A.G. White, Experimental verification of decoherence-free subspaces, Science 290 (October 20, 2000) 498–501.

[34] L. Viola, E.M. Fortunato, M.A. Pravia, E. Knill, R. Laflamme, D.G. Cory, Experimental realization of noiseless subsystems for quantum information processing, Science 293 (September 14, 2001) 2059–2063.
[35] D. Bacon, Operator quantum error correcting subsystems for self-correcting quantum memories, Phys. Rev. A 73 (1) (2006), 012340–012341–012340-13.
[36] D. Bacon, A. Casaccino, Quantum error correcting subsystem codes from two classical linear codes, in: Proc. Forty-fourth Annual Allerton Conference on Communication, Control, and Computing, 2006, pp. 520–527. Monticello, Illinois.
[37] M. Grassl, M. Rötteler, T. Beth, Efficient quantum circuits for non-qubit quantum error-correcting codes, Int. J. Found. Comput. Sci. 14 (5) (2003) 757–775.
[38] A. Harrow, Superdense coding of quantum states, Phys. Rev. Lett. 92 (18) (May 7, 2004) 187901-1–187901-4.
[39] E. Knill, R. Laflamme, Theory of quantum error-correcting codes, Phys. Rev. A 55 (1997) 900.
[40] J. Preskill, Ph219/CS219 quantum computing (lecture notes), Caltech (2009). Available at: http://theory.caltech.edu/people/preskill/ph229/.
[41] A. Kitaev, Topological quantum codes and anyons, in: Quantum computation: a grand mathematical challenge for the twenty-first century and the millennium (Washington, DC, 2000), volume 58 of Proc. Sympos. Appl. Math, Amer. Math. Soc., Providence, RI, 2002, pp. 267–272.
[42] A.Y. Kitaev, Fault-tolerant quantum computation by anyons, Ann. Phys. 303 (1) (2003) 2–30. Available at: http://arxiv.org/abs/quant-ph/9707021.
[43] S.B. Bravyi, A.Y. Kitaev, Quantum Codes on a Lattice With Boundary, 1998. Available at, http://arxiv.org/abs/quant-ph/9811052.
[44] H. Bombin, Topological subsystem codes, Phys. Rev. A 81 (March 3, 2010) 032301-1–032301-15.
[45] G. Duclos-Cianci, D. Poulin, Fast decoders for topological subsystem codes, Phys. Rev. Lett. 104 (2010) 050504-1–050504-4.
[46] G. Duclos-Cianci, D. Poulin, A renormalization group decoding algorithm for topological quantum codes, in: Proc. 2010 IEEE Information Theory Workshop-ITW 2010, Aug. 30 2010-Sept. 3 2010, Dublin, Ireland, 2010.
[47] M. Suchara, S. Bravyi, B. Terhal, Constructions and noise threshold of topological subsystem codes, J. Phys. A: Math. Theory 44 (2011) 155301 (26 pages).
[48] P. Sarvepalli, Topological color codes over higher alphabet, in: Proc. 2010 IEEE Information Theory Workshop-ITW 2010, 2010, Dublin, Ireland, 2010.
[49] F. Hernando, M.E. O'Sullivan, E. Popovici, S. Srivastava, Subfield-subcodes of generalized toric codes, in: Proc. ISIT 2010, June 13–18, 2010, pp. 1125–1129. Austin, Texas, U.S.A.
[50] H. Bombin, Topological order with a twist: Ising anyons from an abelian model, Phys. Rev. Lett. 105 (2010) 030403-1–030403-4.
[51] A.G. Fowler, M. Mariantoni, J.M. Martinis, A.N. Cleland, Surface codes: Towards practical large-scale quantum computation, Phys. Rev. A 86 (3) (18 September 2012) 032324.
[52] D. Litinski, A Game of Surface Codes: Large-Scale Quantum Computing with Lattice Surgery, Quantum 3 (2019) 128.
[53] A.J. Landahl, C. Ryan-Anderson, Quantum Computing by Color-code Lattice Surgery, 2014. Available at, https://arxiv.org/abs/1407.5103.

[54] C. Horsman, A.G. Fowler, S. Devitt, R.V. Meter, Surface code quantum computing by lattice surgery, New J. Phys. 14 (2012) 123011.

[55] A.G. Fowler, C. Gidney, Low Overhead Quantum Computation Using Lattice Surgery, 2018. Available at: https://arxiv.org/abs/1808.06709.

[56] R. Raussendorf, D.E. Browne, H.J. Briegel, Measurement-based quantum computation with cluster states, Phys. Rev. A 68 (2003) 022312.

[57] S. Bravyi, A. Kitaev, Universal quantum computation with ideal Clifford gates and noisy ancillas, Phys. Rev. A 71 (2) (2005) 022316.

[58] T. Brun, I. Devetak, M.H. Hsieh, Correcting quantum errors with entanglement, Science 314 (October 20, 2006) 436−439.

[59] M.-H. Hsieh, I. Devetak, T. Brun, General entanglement-assisted quantum error correcting codes, Phys. Rev. A 76 (December 19, 2007) 062313-1−062313-7.

[60] M.-H. Hsieh, Entanglement-Assisted Coding Theory, PhD Dissertation, University of Southern California, August 2008.

[61] I. Devetak, T.A. Brun, M.-H. Hsieh, Entanglement-Assisted Quantum Error-Correcting Codes, in: V. Sidoravičius (Ed.), New Trends in Mathematical Physics, Selected Contributions of the XVth International Congress on Mathematical Physics, Springer, 2009, pp. 161−172.

[62] M.-H. Hsieh, W.-T. Yen, L.-Y. Hsu, High Performance Entanglement-Assisted Quantum LDPC Codes Need Little Entanglement, IEEE Trans. Inf. Theory 57 (3) (March 2011) 1761−1769.

[63] M.M. Wilde, Quantum Coding with Entanglement, PhD Dissertation, University of Southern California, August 2008.

[64] M.M. Wilde, T.A. Brun, Optimal entanglement formulas for entanglement-assisted quantum coding, Phys. Rev. A 77 (2008) 064302.

[65] M.M. Wilde, H. Krovi, T.A. Brun, Entanglement-assisted quantum error correction with linear optics, Phys. Rev. A 76 (2007) 052308.

[66] M.M. Wilde, T.A. Brun, Protecting quantum information with entanglement and noisy optical modes, Quantum Inf. Process. 8 (2009) 401−413.

[67] M.M. Wilde, T.A. Brun, Entanglement-assisted quantum convolutional coding, Phys. Rev. A 81 (2010) 042333.

[68] M.M. Wilde, D. Fattal, Nonlocal quantum information in bipartite quantum error correction, Quantum Inf. Process. 9 (2010) 591−610.

[69] Y. Fujiwara, D. Clark, P. Vandendriessche, M. De Boeck, V.D. Tonchev, Entanglement-assisted quantum low-density parity-check codes, Phys. Rev. A 82 (4) (2010) 042338.

[70] D. Bacon, A. Casaccino, Quantum error correcting subsystem codes from two classical linear codes, in: Proc. Forty-Fourth Annual Allerton Conference, September 27−29, 2006, pp. 520−527. Allerton House, UIUC, Illinois, USA.

[71] D. Kribs, R. Laflamme, D. Poulin, Unified and generalized approach to quantum error correction, Phys Rev. Lett. 94 (18) (May 9, 2005) 180501.

[72] M.A. Nielsen, D. Poulin, Algebraic and information-theoretic conditions for operator quantum error correction, Phys. Rev. A 75 (June 21, 2007) 064304.

[73] I.B. Djordjevic, Quantum LDPC codes from balanced incomplete block designs, IEEE Comm. Lett. 12 (May 2008) 389−391.

[74] T. Camara, H. Ollivier, J.-P. Tillich, A class of quantum LDPC codes: construction and performances under iterative decoding, in: Proc. ISIT 2007, 24-29 June 2007, pp. 811–815.
[75] S. Lin, S. Zhao, A class of quantum irregular LDPC codes constructed from difference family, in: Proc. ICSP, October 24–28, 2010, pp. 1585–1588.
[76] W. Bosma, J.J. Cannon, C. Playoust, The magma algebra system I: The user language, J. Symb. Comp. 4 (1997) 235–266.

CHAPTER

Quantum LDPC Codes

10

CHAPTER OUTLINE

This chapter is devoted to quantum low-density parity-check (LDPC) codes, which have many advantages compared to other classes of quantum codes, thanks to the sparseness of their quantum-check matrices. Both semirandom and structured quantum LDPC codes are described. Key advantages of structured quantum LDPC codes compared to other codes include: (1) regular structure in corresponding parity-check (***H***-) matrices leads to low complexity encoders/decoders, and (2) their sparse ***H***-matrices require a small number of interactions per qubit to determine the error location. The chapter begins with the introduction of classical LDPC codes in Section 10.1, their design and decoding algorithms. Furthermore, dual-containing quantum LDPC codes are described in Section 10.2. The next section (Section 10.3) is devoted to entanglement-assisted quantum LDPC codes. Furthermore, in Section 10.4, the probabilistic sum-product algorithm based on the quantum-check matrix instead of the classical parity-check matrix is described. Notice that encoders for

Quantum Information Processing, Quantum Computing, and Quantum Error Correction
https://doi.org/10.1016/B978-0-12-821982-9.00015-0

dual-containing and entanglement-assisted quantum LDPC codes can be implemented based on either the standard form method or the conjugation method (described in Chapter 9). Since there is no difference in encoder implementation of quantum LDPC codes compared to other classes of quantum block codes described in Chapter 9 we omit discussion on encoder implementation and concentrate instead on designing and decoding algorithms for quantum LDPC codes. Quantum spatially coupled LDPC codes are briefly introduced in Section 10.5.

10.1 CLASSICAL LDPC CODES

Codes on graphs, such as turbo codes [1] and LDPC codes [2–14], have revolutionized communications, and are becoming standard in many applications. LDPC codes, invented by Gallager in the 1960s, are linear block codes for which the parity-check matrix has a low density of 1s [4]. LDPC codes have generated great interest in the coding community recently [2–21], and this has resulted in a great deal of understanding of the different aspects of LDPC codes and their decoding process. An iterative LDPC decoder based on the sum-product algorithm (SPA) has been shown to achieve a performance as close as 0.0045 dB to the Shannon limit [20]. The inherent low complexity of this decoder opens up avenues for its use in different high-speed applications. Because of their low decoding complexity, LDPC codes are intensively studied for quantum applications as well.

If the parity-check matrix has a low density of 1s and the number of 1s per row and per column are both constant, the code is said to be a *regular LDPC* code. To facilitate implementation at high speed we prefer the use of regular rather than irregular LDPC codes. The graphical representation of LDPC codes, known as bipartite (Tanner) graph representation, is helpful in the efficient description of LDPC decoding algorithms. A *bipartite (Tanner) graph* is a graph whose nodes may be separated into two classes (*variable* and *check* nodes), and where *undirected edges* may only connect two nodes not residing in the same class. The Tanner graph of a code is drawn according to the following rule: check (function) node c is connected to variable (bit) node v whenever element h_{cv} in a parity-check matrix $\boldsymbol{H}$ is a 1. In an $m \times n$ parity-check matrix, there are $m = n - k$ check nodes and n variable nodes.

Example 1. As an illustrative example, consider the $\boldsymbol{H}$-matrix of the following code:

$$H = \begin{bmatrix} 1 & 0 & 1 & 0 & 1 & 0 \\ 1 & 0 & 0 & 1 & 0 & 1 \\ 0 & 1 & 1 & 0 & 0 & 1 \\ 0 & 1 & 0 & 1 & 1 & 0 \end{bmatrix}.$$

For any valid codeword $\boldsymbol{x} = [x_0\, x_1\, \ldots\, x_{n-1}]$, the checks used to decode the codeword are written as:

- Eq. (c_0): $x_0 + x_2 + x_4 = 0 \pmod 2$
- Eq. (c_1): $x_0 + x_3 + x_5 = 0 \pmod 2$
- Eq. (c_2): $x_1 + x_2 + x_5 = 0 \pmod 2$
- Eq. (c_3): $x_1 + x_3 + x_4 = 0 \pmod 2$

The bipartite graph (Tanner graph) representation of this code is given in Fig. 10.1a. The circles represent the bit (variable) nodes, while squares represent the check (function) nodes. For example, the variable nodes x_0, x_2, and x_4 are involved in Eq. (c_0), and therefore connected to the check node c_0. A closed path in a bipartite graph comprising l edges that closes back on itself is called a cycle of length l. The shortest cycle in the bipartite graph is called the girth. The girth influences the minimum distance of LDPC codes, correlates the extrinsic log-likelihood ratios (LLRs), and therefore affects the decoding performance. The use

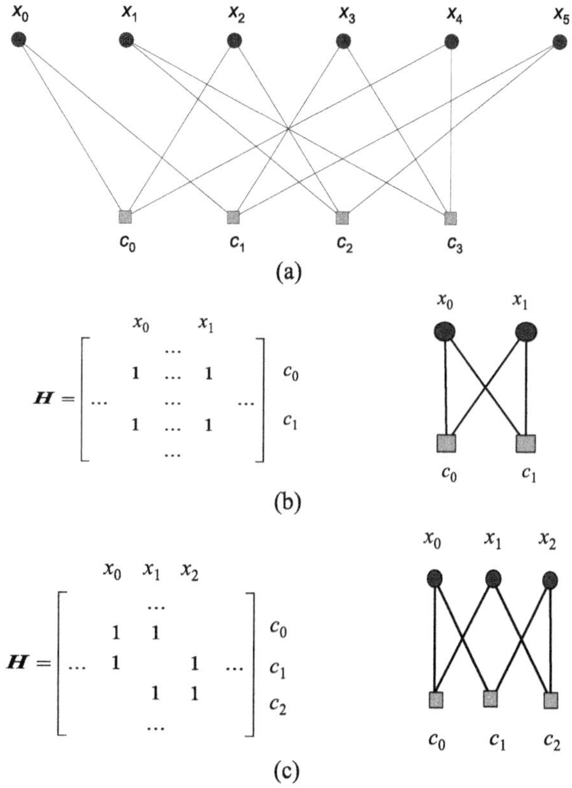

FIGURE 10.1

(a) Bipartite graph of (6,2) code described by the H-matrix. Cycles in a Tanner graph: (b) cycle of length 4 and (c) cycle of length 6.

of large-girth LDPC codes is preferable because the large girth increases the minimum distance and decorrelates the extrinsic info in the decoding process. To improve the iterative decoding performance, we have to avoid cycles of length 4, and preferably 6 as well. To check for the existence of short cycles, one has to search over the **H**-matrix for the patterns shown in Fig. 10.1b and c.

In the rest of this section we describe a method for designing large-girth quasi-cyclic (QC)-LDPC codes (Subsection 10.1.1), and an efficient and simple variant of SPA suitable for use in optical communications, namely the min-sum-with-correction-term algorithm (Subsection 10.1.2). We also evaluate bit error rate (BER) performance of QC-LDPC codes against concatenated codes and turbo-product codes (TPCs) (Subsection 10.1.3).

10.1.1 Large-Girth Quasi-Cyclic Binary LDPC Codes

Based on Tanner's bound for the minimum distance of an LDPC code [22]:

$$d \geq \begin{cases} 1+\dfrac{w_c}{w_c-2}\left((w_c-1)^{\lfloor(g-2)/4\rfloor}-1\right), g/2=2\text{m}+1 \\ 1+\dfrac{w_c}{w_c-2}\left((w_c-1)^{\lfloor(g-2)/4\rfloor}-1\right)+(w_c-1)^{\lfloor(g-2)/4\rfloor},\ g/2=2\text{m} \end{cases}, \tag{10.1}$$

(where g and w_c denote the girth of the code graph and the column weight, respectively, and where d stands for the minimum distance of the code), it follows that large girth leads to an exponential increase in the minimum distance, provided that the column weight is at least 3. ($\lfloor\rfloor$ denotes the largest integer less than or equal to the enclosed quantity.) For example, the minimum distance of girth-10 codes with column weight $r=3$ is at least 10. The parity-check matrix of regular QC-LDPC codes [23,24] can be represented by:

$$\boldsymbol{H}=\begin{bmatrix} I & I & I & \ldots & I \\ I & P^{S[1]} & P^{S[2]} & \ldots & P^{S[c-1]} \\ I & P^{2S[1]} & P^{2S[2]} & \ldots & P^{2S[c-1]} \\ \ldots & \ldots & \ldots & \ldots & \ldots \\ I & P^{(r-1)S[1]} & P^{(r-1)S[2]} & \ldots & P^{(r-1)S[c-1]} \end{bmatrix}, \tag{10.2}$$

where I is the $B \times B$ (B is a prime number) identity matrix, P is the $B \times B$ permutation matrix given by $P=(p_{ij})_{B\times B}$, $p_{i,i+1}=p_{B,1}=1$ (zero otherwise), and r and c represent the number of block rows and block columns in Eq. (10.2), respectively. The set of integers S is to be carefully chosen from the set $\{0,1,\ldots,B-1\}$ so that the cycles of short length, in the corresponding Tanner (bipartite) graph representation of Eq. (10.2), are avoided. According to Theorem 2.1 in Ref. [24], we have to avoid the cycles of length $2k$ ($k=3$ or 4) defined by the following equation:

$$S[i_1]j_1+S[i_2]j_2+\ldots+S[i_k]j_k=S[i_1]j_2+S[i_2]j_3+\ldots+S[i_k]j_1 \text{mod} p, \tag{10.3}$$

where the closed path is defined by (i_1, j_1), (i_1, j_2), (i_2, j_2), (i_2, j_3), …, (i_k, j_k), (i_k, j_1) with the pair of indices denoting row-column indices of permutation blocks in Eq. (10.2) such that $l_m \neq l_{m+1}$, $l_k \neq l_1$ ($m = 1,2,...,k$; $l \in \{i,j\}$). Therefore we have to identify the sequence of integers $S[i] \in \{0,1,\ldots,B-1\}$ ($i = 0,1,\ldots,r-1$; $r < B$) not satisfying Eq. (10.3), which can be done either by computer search or in a combinatorial fashion. For example, to design the QC-LDPC codes in Ref. [18], we introduced the concept of the cyclic-invariant difference set (CIDS). The CIDS-based codes come naturally as girth-6 codes, and to increase the girth we had to selectively remove certain elements from a CIDS. The design of LDPC codes of rate above 0.8, column weight 3, and girth 10 using the CIDS approach is very challenging and is still an open problem. Instead, in our recent paper [23], we solved this problem by developing an efficient computer search algorithm. We add an integer at a time from the set $\{0,1,\ldots,B-1\}$ (not used before) to the initial set S and check if Eq. (10.3) is satisfied. If Eq. (10.3) is satisfied, we remove that integer from the set S and continue our search with another integer from set $\{0,1,\ldots,B-1\}$ until we exploit all the elements from $\{0,1,\ldots,B-1\}$. The code rate of these QC codes, R, is lower bounded by:

$$R \geq \frac{|S|B - rB}{|S|B} = 1 - r/|S|, \tag{10.4}$$

and the codeword length is $|S|B$, where $|S|$ denotes the cardinality of set S. For a given code rate R_0, the number of elements from S to be used is $\lfloor r/(1-R_0) \rfloor$. With this algorithm, LDPC codes of arbitrary rate can be designed.

Example 2. By setting $B = 2311$, the set of integers to be used in Eq. (10.2) is obtained as $S = \{1, 2, 7, 14, 30, 51, 78, 104, 129, 212, 223, 318, 427, 600, 808\}$. The corresponding LDPC code has rate $R_0 = 1 - 3/15 = 0.8$, column weight 3, girth 10, and length $|S|B = 15 \cdot 2311 = 34{,}665$. In Example 1, the initial set of integers was $S = \{1, 2, 7\}$, and the set of rows to be used in (6.25) was $\{1, 3, 6\}$. The use of a different initial set will result in a different set from that obtained earlier.

Example 3. By setting $B = 269$, the set S is obtained as $S = \{0, 2, 3, 5, 9, 11, 12, 14, 27, 29, 30, 32, 36, 38, 39, 41, 81, 83, 84, 86, 90, 92, 93, 95, 108, 110, 111, 113, 117, 119, 120, 122\}$. If 30 integers are used, the corresponding LDPC code has rate $R_0 = 1 - 3/30 = 0.9$, column weight 3, girth 8, and length $30 \cdot 269 = 8070$.

10.1.2 Decoding of Binary LDPC Codes

In this subsection we describe the min-sum-with-correction-term decoding algorithm [25,26]. It is a simplified version of the original algorithm proposed by Gallager [4]. Gallager proposed a near-optimal iterative decoding algorithm for LDPC codes that computes the distributions of the variables to calculate the a posteriori *probability* (APP) of a bit v_i of a codeword $\boldsymbol{v} = [v_0\ v_1\ \ldots\ v_{n-1}]$ to be equal to 1, given a received vector $\boldsymbol{y} = [y_0\ y_1\ \ldots\ y_{n-1}]$. This iterative decoding scheme engages passing the extrinsic info back and forth among the c-nodes and the v-nodes over the edges to update the distribution estimation. Each iteration in this scheme is composed of two half-iterations. In Fig. 10.2, we illustrate both the first and second

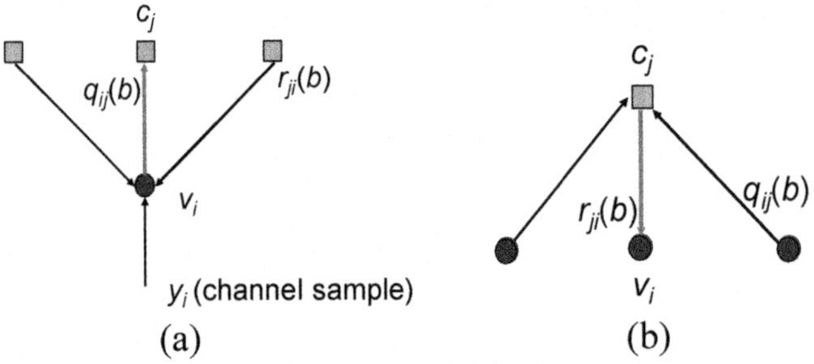

FIGURE 10.2

Half-iterations of the sum-product algorithm: (a) first half-iteration: extrinsic info sent from v-nodes to c-nodes, and (b) second half-iteration: extrinsic info sent from c-nodes to v-nodes.

halves of an iteration of the algorithm. As an example, in Fig. 10.2a we show the message sent from the v-node v_i to the c-node c_j. The v_i-node collects the information from the channel (y_i sample), in addition to extrinsic info from other c-nodes connected to the v_i-node, processes them, and sends the extrinsic info (not already available info) to c_j. This extrinsic info contains the information about the probability $\Pr(c_i = b|y_0)$, where $b \in \{0,1\}$. This is performed in all c-nodes connected to the v_i-node. On the other hand, Fig. 10.2b shows the extrinsic info sent from the c-node c_i to the v-node v_j, which contains the information about $\Pr(c_i$ equation is satisfied $|\, \mathbf{y})$. This is done repeatedly to all the c-nodes connected to the v_i-node.

After this intuitive description, we describe the min-sum-with-correction-term algorithm in more detail [25] because of its simplicity and suitability for high-speed implementation. Generally, we can either compute APP $\Pr(v_i|\mathbf{y})$ or the APP ratio $l(v_i) = \Pr(v_i = 0|\mathbf{y})/\Pr(v_i = 1|\mathbf{y})$, which is also referred to as the likelihood ratio. In the log-domain version of the sum-product algorithm we replace these likelihood ratios with LLRs because the probability domain includes many multiplications that lead to numerical instabilities, whereas the computation using LLR computation involves addition only. Moreover, the log-domain representation is more suitable for finite precision representation. Thus we compute the LLRs by $L(v_i) = \log[\Pr(v_i = 0|\mathbf{y})/\Pr(v_i = 1|\mathbf{y})]$. For the final decision, if $L(v_i) > 0$ we decide in favor of 0 and if $L(v_i) < 0$ we decide in favor of 1. To further explain the algorithm, we introduce the following notations due to MacKay [27] (see also [25]):

$$
\begin{aligned}
V_j &= \{v\text{-nodes connected to } c\text{-node } c_j\} \\
V_j \backslash i &= \{v\text{-nodes connected to } c\text{-node } c_j\} \backslash \{v\text{-node } v_i\} \\
C_i &= \{c\text{-nodes connected to } v\text{-node } v_i\} \\
C_i \backslash j &= \{c\text{-nodes connected to } v\text{-node } v_i\} \backslash \{c\text{-node } c_j\} \\
M_v(\sim i) &= \{\text{messages from all } v\text{-nodes except node } v_i\} \\
M_c(\sim j) &= \{\text{messages from all } c\text{-nodes except node } c_j\} \\
P_i &= \Pr(v_i = 1|y_i)
\end{aligned}
$$

S_i = event that the check equations involving c_i are satisfied
$q_{ij}(b) = \Pr(v_i = b|S_i, y_i, M_c(\sim j))$
$r_{ji}(b) = \Pr(\text{check equation } c_j \text{ is satisfied}|v_i = b, M_v(\sim i))$

In the log-domain version of the sum-product algorithm, all the calculations are performed in the log domain as follows:

$$L(v_i) = \log\left[\frac{\Pr(v_i = 0|y_i)}{\Pr(v_i = 1|y_i)}\right], \quad L(r_{ji}) = \log\left[\frac{r_{ji}(0)}{r_{ji}(1)}\right], \quad L(q_{ji}) = \log\left[\frac{q_{ji}(0)}{q_{ji}(1)}\right]. \quad (10.5)$$

The algorithm starts with the initialization step where we set $L(v_i)$ as follows:

$$\begin{aligned}
L(v_i) &= (-1)^{y_i}\log\left(\frac{1-\varepsilon}{\varepsilon}\right), \quad \text{for BSC} \\
L(v_i) &= 2y_i/\sigma^2, \quad \text{for binary input AWGN} \\
L(v_i) &= \log\left(\frac{\sigma_1}{\sigma_0}\right) - \frac{(y_i-\mu_0)^2}{2\sigma_0^2} + \frac{(y_i-\mu_1)^2}{2\sigma_1^2}, \quad \text{for BA} - \text{AWGN} \\
L(v_i) &= \log\left(\frac{\Pr(v_i = 0|y_i)}{\Pr(v_i = 1|y_i)}\right), \quad \text{for abritrary channel,}
\end{aligned} \quad (10.6)$$

where ε is the probability of error in the binary symmetric channel (BSC), σ^2 is the variance of the Gaussian distribution of the additive white Gaussian noise (AWGN), and μ_j and σ_j^2 ($j = 0,1$) represent the mean and variance of the Gaussian process corresponding to the bits $j = 0,1$ of a binary asymmetric AWGN channel. After initialization of $L(q_{ij})$ we calculate $L(r_{ji})$ as follows:

$$\begin{aligned}
L(r_{ji}) &= L\left(\sum_{i' \in V_j \backslash i} b_{i'}\right) = L(\ldots \oplus b_k \oplus b_l \oplus b_m \oplus b_n \ldots) \\
&= \ldots L_k \boxplus L_l \boxplus L_m \boxplus L_n \boxplus \ldots,
\end{aligned} \quad (10.7)$$

where $\oplus$ denotes the modulo-2 addition and $\boxplus$ denotes a pairwise computation defined by:

$$L_1 \boxplus L_2 = \prod_{k=1}^{2} \text{sign}(L_k) \cdot \phi\left(\sum_{k=1}^{2} \phi(|L_k|)\right), \quad \phi(x) = -\log\tanh(x/2). \quad (10.8)$$

Upon calculation of $L(r_{ji})$ we update:

$$L(q_{ij}) = L(v_i) + \sum_{j' \in C_{i \setminus j}} L(r_{j'i}), \quad L(Q_i) = L(v_i) + \sum_{j \in C_i} L(r_{ji}). \tag{10.9}$$

Finally, the decision step is as follows:

$$\hat{v}_i = \begin{cases} 1, & L(Q_i) < 0 \\ 0, & \text{otherwise} \end{cases}. \tag{10.10}$$

If the syndrome equation $\hat{\boldsymbol{v}}\boldsymbol{H}^T = \boldsymbol{0}$ is satisfied or the maximum number of iterations is reached, we stop, otherwise we recalculate $L(r_{ji})$ and update $L(q_{ij})$ and $L(Q_i)$ and check again. It is important to set the number of iterations high enough to ensure that most of the codewords are decoded correctly and low enough not to affect the processing time. It is important to mention that the decoder for good LDPC codes requires fewer iterations to guarantee successful decoding. The *Gallager log-domain SPA* can be formulated as follows:

1. Initialization: For $j = 0,1,\ldots,n-1$, initialize the messages to be sent from v-node i to c-node j to channel LLRs, namely $L(q_{ij}) = L(c_i)$.
2. c-node update rule: For $j = 0,1,\ldots,n-k-1$, compute $L(r_{ji}) = \left(\prod_{i' \in R_j \setminus i} \alpha_{i'j}\right)$ $\phi\left[\sum_{i' \in R_j \setminus i} \phi(\beta_{i'j})\right]$, where $\alpha_{ij} = \text{sign}[L(q_{ij})]$, $\beta_{ij} = |L(q_{ij})|$, and $\phi(x) = -\log \tanh(x/2) = \log[(e^x+1)/(e^x-1)]$
3. v-node update rule: For $i = 0,1,\ldots,n-1$, set $L(q_{ij}) = L(c_i) + \sum_{j' \in C_i \setminus j} L(r_{j'i})$ for all c-nodes for which $h_{ji} = 1$.
4. Bit decisions: Update $L(Q_i)$ $(i = 0,\ldots,n-1)$ by $L(Q_i) = L(c_i) + \sum_{j \in C_i} L(r_{ji})$ and set $\hat{c}_i = 1$ when $L(Q_i) < 0$ (otherwise, $\hat{c}_i = 0$). If $\hat{\boldsymbol{c}}\boldsymbol{H}^T = \boldsymbol{0}$ or a predetermined number of iterations has been reached, then stop, otherwise go to step 1.

Because the c-node update rule involves log and tanh functions, it is computationally intensive, and there exist many approximations. Very popular is the min-sum-plus-correction-term approximation [6, 26]. Namely, it can be shown that "box-plus" operator $\boxplus$ can also be calculated by:

$$L_1 \boxplus L_2 = \prod_{k=1}^{2} \text{sign}(L_k) \cdot \min(|L_1|, |L_2|) + c(x, y), \tag{10.11}$$

where $c(x,y)$ denotes the correction factor defined by:

$$c(x, y) = \log[1 + \exp(-|x+y|)] - \log[1 + \exp(-|x-y|)], \tag{10.12}$$

commonly implemented as a look-up table.

10.1.3 BER Performance of Binary LDPC Codes

The results of simulations for an AWGN channel model are given in Fig. 10.3, where we compare the large-girth LDPC codes (Fig. 10.3a) against Reed–Solomon (RS) codes, concatenated RS codes, TPCs, and other classes of LDPC codes. In all simulation results in this section, we maintained the double precision. For the LDPC(16935,13550) code we also provided 3- and 4-bit fixed-point simulation results (Fig. 10.3a). Our results indicate that the 4-bit representation performs comparable to the double-precision representation, whereas the 3-bit representation performs 0.27 dB worse than the double-precision representation at the BER of $2 \cdot 10^{-8}$. The girth-10 LDPC(24015,19212) code of rate 0.8 outperforms the concatenation RS(255,239) + RS(255,223) (of rate 0.82) by 3.35 dB and RS(255,239) by 4.75 dB both at a BER of 10^{-7}. The same LDPC code outperforms projective geometry (PG) $(2,2^6)$-based LDPC(4161,3431) (of rate 0.825) of girth 6 by 1.49 dB at a BER of 10^{-7}, and outperforms CIDS-based LDPC(4320,3242) of rate 0.75 and girth-8 LDPC codes by 0.25 dB. At a BER of 10^{-10}, it outperforms lattice-based LDPC(8547,6922) of rate 0.81 and girth-8 LDPC code by 0.44 dB, and Bose–Chaudhuri–Hocquenghem (BCH)(128,113) × BCH(256,239) TPC of rate 0.82 by 0.95 dB. The net coding gain at a BER of 10^{-12} is 10.95 dB. In Fig. 10.3b, different LDPC codes are compared against RS(255,223) code, concatenated RS code of rate 0.82, and convolutional code (CC) (of constraint length 5). It can be seen that LDPC codes, both regular and irregular, offer much better performance than hard-decision codes. It should be noticed that pairwised balanced design [58]-based irregular LDPC code of rate 0.75 is only 0.4 dB away from the concatenation of convolutional-RS codes (denoted in Fig. 10.3b as RS + CC) with significantly lower code rate $R = 0.44$ at a BER of 10^{-6}. As expected, irregular LDPC codes (black-colored curves) outperform regular LDPC codes.

The main problem in decoder implementation for large-girth *binary* LDPC codes is the excessive codeword length. To solve this problem, we will consider *nonbinary* LDPC codes over GF(2^m). By designing codes over higher-order fields we aim to achieve coding gains comparable to binary LDPC codes but for shorter codeword lengths. Notice also that 4-ary LDPC codes (over GF(4)) can be simply related to entanglement-assisted quantum LDPC codes given by Eq. (9.16) in Chapter 9. Moreover, properly designed classical nonbinary LDPC codes can be used as nonbinary stabilizer codes, described in Section 8.6.

10.1.4 Nonbinary LDPC Codes

The parity-check matrix $\boldsymbol{H}$ of a nonbinary QC-LDPC code can be organized as an array of submatrices of equal size as in Eq. (10.13), where $\boldsymbol{H}_{i,j}$, $0 \le i < \gamma$, $0 \le j < \rho$, is a $B \times B$ submatrix in which each row is a cyclic shift of the row preceding it. This modular structure can be exploited to facilitate hardware implementation of the decoders of QC-LDPC codes [7,8]. Furthermore, the QC nature of their

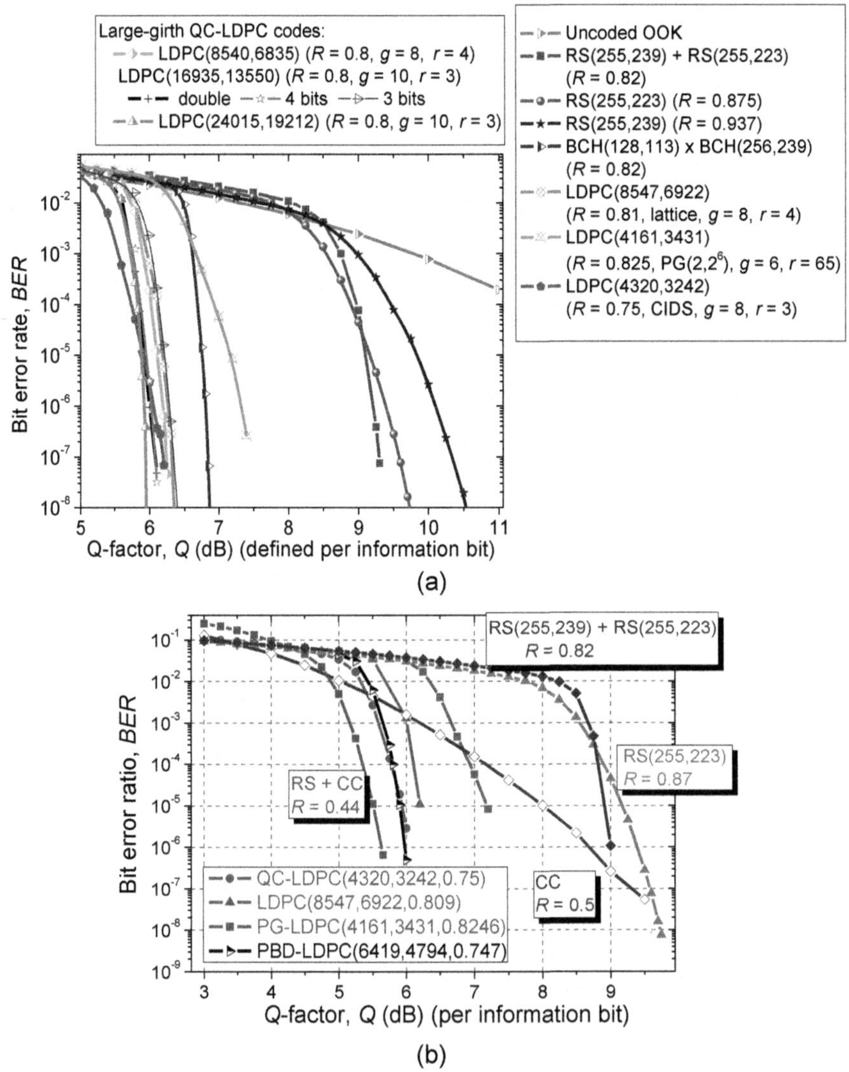

FIGURE 10.3

(a) Large-girth quasi-cyclic low-density parity-check (QC-LDPC) codes against Reed–Solomon (RS) codes, concatenated RS codes, turbo-product codes, and girth-6 LDPC codes on an additive white Gaussian noise (AWGN) channel model, and (b) LDPC codes versus convolutional, concatenated RS, and concatenation of convolutional and RS codes on an AWGN channel. Number of iterations in sum-product-with-correction-term algorithm was set to 25. The *Q*-factor is defined as $Q = (\mu_1 - \mu_0)/(\sigma_1 + \sigma_0)$, where μ_i is the mean value corresponding to bit i ($i = 0,1$) and σ_i is the corresponding standard deviation. *BCH*, Bose–Chaudhuri–Hocquenghem; *CC*, convolutional code; *CIDS*, cyclic-invariant difference set; *PBD*, pairwise balanced design; *PG*, projective geometry.

generator matrices enables encoding of QC-LDPC codes to be performed in linear time using simple shift register-based architectures [8,11]:

$$H = \begin{bmatrix} H_{0,0} & H_{0,1} & \cdots & H_{0,\rho-1} \\ H_{1,0} & H_{1,1} & \cdots & H_{1,\rho-1} \\ \vdots & \vdots & \ddots & \vdots \\ H_{\gamma-1,0} & H_{\gamma-1,1} & \cdots & H_{\gamma-1,\rho-1} \end{bmatrix}. \tag{10.13}$$

If we select the entries of H from the binary field GF(2), as was done in the previous section, then the resulting QC-LDPC code is a binary LDPC code. On the other hand, if the selection is made from the Galois field of q elements denoted by GF(q), then we obtain a q-ary QC-LDPC code. To decode binary LDPC codes, as described earlier, an iterative message-passing algorithm is referred to as SPA. For nonbinary LDPC codes, a variant of the SPA known as the q-ary SPA (QSPA) is used [9]. When the field order is a power of 2, i.e., $q = 2^m$, where m is an integer and $m \geq 2$, a fast Fourier transform (FFT)-based implementation of QSPA, referred to as FFT-QSPA, significantly reduces the computational complexity of QSPA. FFT-QSPA is further analyzed and improved in Ref. [10]. A mixed-domain FFT-QSPA implementation, in short, MD-FFT-QSPA, aims to reduce the hardware implementation complexity by transforming multiplications in the probability domain into additions in the log domain whenever possible. It also avoids instability issues commonly faced in probability-domain implementations.

Following the code design we discussed earlier, we generated (3,15)-regular, girth-8 LDPC codes over the fields GF(2^p), where $0 \leq p \leq 7$. All the codes had a code rate (R) of at least 0.8 and hence an overhead $=$ ($1/R - 1$) of 25% or less. We compared the BER performances of these codes against each other and against some other well-known codes, such as RS(255,239), RS(255,223) and concatenated RS(255,239) + RS(255,223) codes, and BCH(128,113) × BCH(256,239) TPC. We used the binary AWGN channel model in our simulations and set the maximum number of iterations to 50. In Fig. 10.4a we present the BER performances of the set of nonbinary LDPC codes discussed earlier. Using the figure we can conclude that when we fix the girth of a nonbinary regular, rate-0.8 LDPC code at 8, increasing the field order above 8 exacerbates the BER performance. In addition to having better BER performance than codes over higher-order fields, codes over GF(4) have smaller decoding complexities when decoded using the MD-FFT-QSPA algorithm since the complexity of this algorithm is proportional to the field order. Thus we focus our attention on nonbinary, regular, rate-0.8, girth-8 LDPC codes over GF(4) in the rest of the section.

In Fig. 10.4b we compare the BER performance of the LDPC(8430,6744) code over GF(4) discussed in Fig. 10.4a against that of the RS(255,239) code, RS(255,223) code, RS(255,239) + RS(255,223) concatenation code, and BCH(128,113) × BCH(256,239) TPC. We observe that the LDPC code over GF(4) outperforms all of these codes with a significant margin. In particular, it

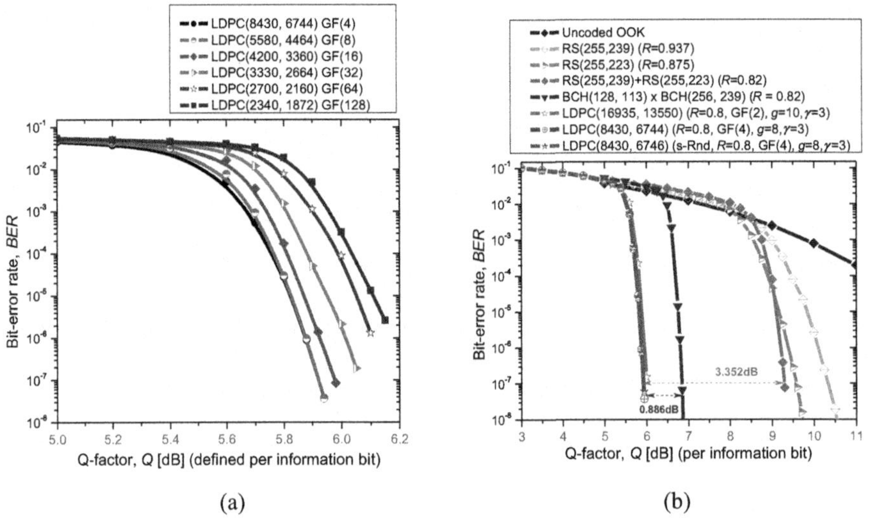

FIGURE 10.4

(a) Comparison of nonbinary, (3,15)-regular, girth-8 low-density parity-check (LDPC) codes over the binary additive white Gaussian noise channel. (b) Comparison of 4-ary (3,15)-regular, girth-8 LDPC codes, a binary, girth-10 LDPC code, three Reed–Solomon (RS) codes, and a turbo-product code. *BCH*, Bose–Chaudhuri–Hocquenghem; *GF*, Galois field.

provides an additional coding gain of 3.363 and 4.401 dB at a BER of 10^{-7} when compared to the concatenation code RS(255,239) + RS(255,223) and the RS(255,239) code, respectively. Its coding gain improvement over BCH(128,113) × BCH(256,239) TPC is 0.886 dB at a BER of 4×10^{-8}. We also presented in Fig. 10.4b a competitive, binary, (3,15)-regular, LDPC(16935,13550) code proposed in [23]. We can see that the 4-ary, (3,15)-regular, girth-8 LDPC(8430,6744) code beats the bit-length-matched binary LDPC code with a margin of 0.089 dB at a BER of 10^{-7}. More importantly, the complexity of the MD-FFT-QSPA used for decoding the nonbinary LDPC code is lower than the min-sum-with-correction-term algorithm [6] used for decoding the corresponding binary LDPC code. When the MD-FFT-QSPA is used for decoding a (γ,ρ)-regular q-ary LDPC($N/\log q$, $K/\log q$) code, which is bit length matched to a (γ,ρ)-regular binary LDPC(N,K) code, the complexity is given by $(M/\log q)2\rho q(\log q + 1 - 1/(2\rho))$ additions, where $M = N - K$ is the number of check nodes in the binary code. On the other hand, to decode the bit-length-matched binary counterpart using the min-sum-with-correction-term algorithm [6], one needs $15M(\rho - 2)$ additions. Thus a (3,15)-regular 4-ary LDPC code requires 91.28% of the computational resources required for decoding a (3,15)-regular binary LDPC code of the same rate and bit length.

10.1.5 LDPC Codes Design

The most obvious way to design LDPC codes is to construct an LDPC matrix with prescribed properties. Some important designs among others include: (1) Gallager codes (semirandom construction) [4], (2) MacKay codes (semirandom construction) [26], (3) finite geometry-based LDPC codes [17,30,31], (4) combinatorial design-based LDPC codes [19], (5) QC (array, block-circulant) LDPC codes [2,3,18,23,24] (and references therein), (6) irregular LDPC codes [31], and (7) Tanner codes [22]. The QC-LDPC codes of large girth have already been discussed in Section 10.1.1. In the rest of this section we describe briefly several classes of LDPC codes, namely Gallager, Tanner, and MacKay codes. Other classes of codes, commonly referred to as structured (because they have a regular structure in their parity-check matrices, which can be exploited to facilitate hardware implementation of encoders and decoders), will be described in corresponding sections related to quantum LDPC codes. Notice the basic design principles for both classical and quantum LDPC codes are very similar, and the same designs already used for classical LDPC code design can be used to design quantum LDPC codes. Therefore various code designs from Chapter 17 of Ref. [21], with certain modifications, can be used for quantum LDPC code design as well.

10.1.5.1 Gallager Codes

The $\boldsymbol{H}$-matrix for Gallager code has the following general form:

$$\boldsymbol{H}=[\boldsymbol{H}_1 \quad \boldsymbol{H}_2 \quad \cdots \quad \boldsymbol{H}_{w_c}]^T, \tag{10.14}$$

where $\boldsymbol{H}_1$ is $p \times p \cdot w_r$ matrix of row weight w_r and $\boldsymbol{H}_i$ are column-permuted versions of the $\boldsymbol{H}_1$-submatrix. The row weight of $\boldsymbol{H}$ is w_r, and column weight is w_c. The permutations are carefully chosen to avoid cycles of length 4. The $\boldsymbol{H}$-matrix is obtained by computer search.

10.1.5.2 Tanner Codes and Generalized LDPC Codes

In Tanner codes [22], each bit node is associated with a code bit, and each check node is associated with a subcode whose length is equal to the degree of the node, which is illustrated in Fig. 10.5. Notice that so-called generalized LDPC (GLDPC) codes [15,32–35] were inspired by Tanner codes.

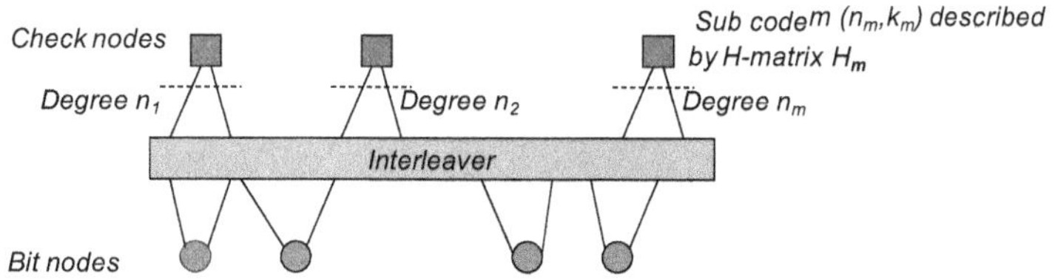

FIGURE 10.5

Tanner codes.

Example 4. Let us consider the following example:

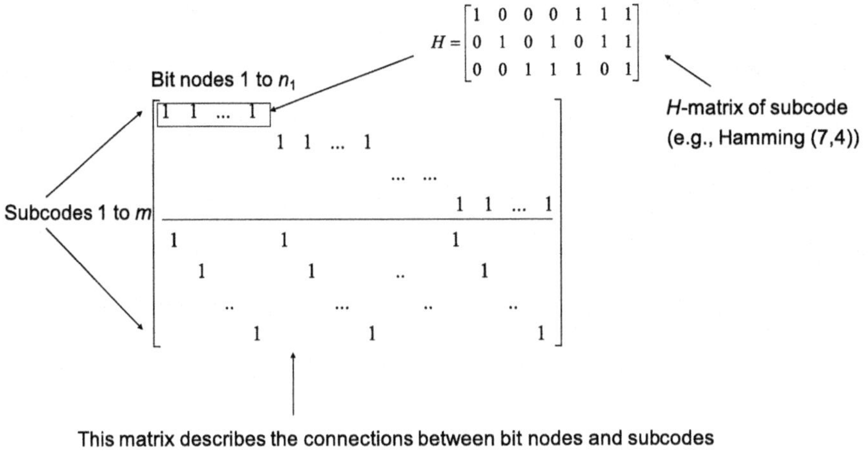

Tanner code design in this example is performed in two stages. We first design the global code by starting from an identity matrix $I_{m/2}$ and replace every nonzero element with n_1 ones, and every zero element by n_1 zeros. The lower submatrix is obtained by concatenating the identity matrices In_1. In the second stage we substitute all-one row vectors (of length n_1) by the parity-check matrix of local linear block code (n_1,k_1), such as Hamming, BCH, or Reed–Muller (RM) code. The resulting parity-check matrix is used as the parity-check matrix of an LDPC code.

The GLDPC codes can be constructed in similar fashion. To construct a GLDPC code one can replace each single parity-check equation of a global LDPC code by the parity-check matrix of a simple linear block code, known as the constituent (local) code, and this construction is proposed by Lentmaier and Zigangirov [33], and we will refer to this construction as LZ-GLDPC code construction. An illustrative example is shown in Fig. 10.6. In another construction proposed by Boutros et al. in Ref. [32], referred to here as B-GLDPC code construction, the

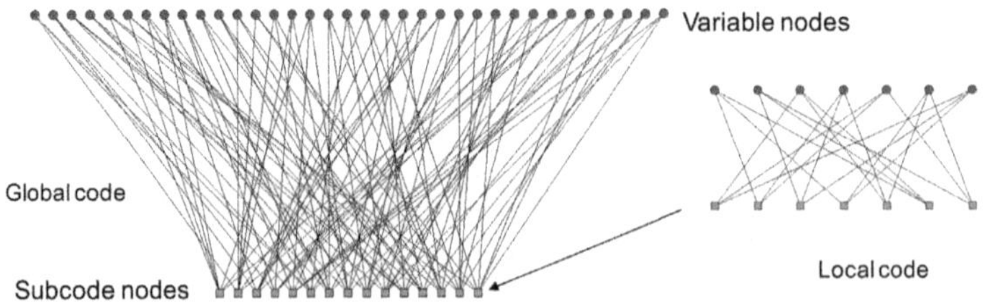

FIGURE 10.6

Illustration of construction of Lentmaier and Zigangirov generalized low-density parity-check codes.

parity-check matrix, $\boldsymbol{H}$, is a sparse matrix partitioned into W submatrices $\boldsymbol{H}_1,\ldots,\boldsymbol{H}_W$. The $\boldsymbol{H}_1$-submatrix is a block-diagonal matrix generated from an identity matrix by replacing the 1s by a parity-check matrix $\boldsymbol{H}_0$ of a local code of codeword length n and dimension k:

$$\boldsymbol{H} = \begin{bmatrix} \boldsymbol{H}_1^T & \boldsymbol{H}_2^T & \ldots & \boldsymbol{H}_W^T \end{bmatrix}^T \boldsymbol{H}_1 = \begin{bmatrix} \boldsymbol{H}_0 & & & 0 \\ & \boldsymbol{H}_0 & & \\ & & \cdots & \\ 0 & & & \boldsymbol{H}_0 \end{bmatrix}. \tag{10.15}$$

Each submatrix $\boldsymbol{H}_j$ in Eq. (10.15) is derived from $\boldsymbol{H}_1$ by random column permutations. The code rate of a GLDPC code is lower bounded by:

$$R = K/N \geq 1 - W(1 - k/n), \tag{10.16}$$

where K and N denote the dimension and the codeword length of a GLDPC code, W is the column weight of a global LDPC code, and k/n is the code rate of a local code (k and n denote the dimension and codeword length of a local code). The GLDPC codes can be classified as follows [15]: (1) GLDPC codes with algebraic local codes of short length, such as Hamming codes, BCH codes, RS codes, or RM codes, (2) GLDPC codes for which the local codes are high-rate regular or irregular LDPC codes with large minimum distance, and (3) fractal GLDPC codes in which the local code is in fact another GLDPC code. For more details on LZ-GLDPC- and B-GLDPC-like codes and their generalization-fractal GLDPC codes (a local code is another GLDPC code), interested readers are referred to [15,35] (and references therein).

10.1.5.3 MacKay Codes

Following MacKay (1999) [27] the following are several ways to generate the sparse matrices in order of increasing algorithm complexity (not necessarily improved performance):

1. $\boldsymbol{H}$-matrix is generated by starting from an all-zero matrix and randomly inverting w_c (not necessarily distinct bits) in each column. The resulting LDPC code is an irregular code.
2. $\boldsymbol{H}$-matrix is generated by randomly creating weight-w_c columns.
3. $\boldsymbol{H}$-matrix is generated with weight-w_c columns and uniform row weight (as near as possible).
4. $\boldsymbol{H}$-matrix is generated with weight-w_c columns, weight-w_r rows, and no two columns have an overlap larger than 1.
5. $\boldsymbol{H}$-matrix is generated as in (4), and short cycles are avoided.
6. $\boldsymbol{H}$-matrix is generated as in (5), and can be represented by $\boldsymbol{H} = [\boldsymbol{H}_1|\boldsymbol{H}_2]$, where $\boldsymbol{H}_2$ is invertible or at least has a full rank.

The construction via (5) may lead to an $\boldsymbol{H}$-matrix that is not of full rank. Nevertheless, it can be put in the following form by column swapping and Gauss–Jordan elimination:

$$\boldsymbol{H} = \begin{bmatrix} \boldsymbol{P}^T & \boldsymbol{I} \\ \boldsymbol{0} & \boldsymbol{0} \end{bmatrix}, \qquad (10.17)$$

and by eliminating the all-zero submatrix we obtain the parity-check matrix of systematic LDPC code:

$$\widetilde{\boldsymbol{H}} = [\boldsymbol{P}^T \quad \boldsymbol{I}]. \qquad (10.18)$$

Among the various construction algorithms listed earlier, construction (5) will be described in more detail.

Outline of construction algorithm (5):

1. Choose code parameters n, k, w_c, w_r, and g (minimum cycle length girth). The resulting $\boldsymbol{H}$-matrix will be an $m \times n$ ($m = n - k$) matrix with w_c ones per column and w_r ones per row.
2. Set the column counter to $i_c = 0$.
3. Generate a weight-w_c column vector and place it in the i_cth column of the $\boldsymbol{H}$-matrix.
4. If the weight of each row is $\leq w_r$, the overlap between any two columns is ≤ 1, and if all cycle lengths are $\geq g$, then increment the counter $i_c = i_c + 1$.
5. If $i_c = n$, stop; else go to step 3.

This algorithm could take hours to run with no guarantee of regularity of the $\boldsymbol{H}$-matrix. Moreover, it may not finish at all, and we need to restart the search with another set of parameters. Richard and Urbanke (in 2001) proposed a linear complexity in length technique based on the $\boldsymbol{H}$-matrix [28].

An alternative approach to simplify encoding is to design the codes via algebraic, geometric, or combinatoric methods [17,24,29,30]. The $\boldsymbol{H}$-matrix of those designs can be put in cyclic or QC form, leading to implementations based on shift registers and mod-2 adders. Since these classes of codes can also be used for the design of quantum LDPC codes we postpone their description until later sections.

10.2 DUAL-CONTAINING QUANTUM LDPC CODES

It has been shown by Calderbank, Shor, and Steane [36,37] that the quantum codes, now known as CSS codes, can be designed using a pair of conventional linear codes satisfying the *twisted property*, that is, one of the codes includes the dual of another code. This class of quantum codes has already been studied in Chapter 7. Among CSS codes particularly are simple CSS codes based on dual-containing codes [38], whose (quantum) check matrix can be represented by [38]:

$$\boldsymbol{A} = \left[\begin{array}{c|c} \boldsymbol{H} & \boldsymbol{0} \\ \boldsymbol{0} & \boldsymbol{H} \end{array}\right], \qquad (10.19)$$

where $\boldsymbol{HH}^T = \mathbf{0}$, which is equivalent to $C^{\perp}(\boldsymbol{H}) \subset C(\boldsymbol{H})$, where $C(\boldsymbol{H})$ is the code having $\boldsymbol{H}$ as the parity-check matrix, and $C^{\perp}(\boldsymbol{H})$ is its corresponding dual code. It has been shown in Ref. [38] that the requirement $\boldsymbol{HH}^T = \mathbf{0}$ is satisfied when rows of $\boldsymbol{H}$ have an even number of 1s, and any two of them overlap by an even number of 1s. The LDPC codes satisfying these two requirements in Ref. [38] were designed by exhaustive computer search, in Ref. [39] they were designed as codes over GF(4) by identifying the Pauli operators I,X,Y,Z with elements from GF(4), while in Ref. [40] they were designed in QC fashion. In what follows we will show how to design the dual-containing LDPC codes using the combinatorial objects known as balanced incomplete block designs (BIBDs) [46–50]. Notice that the theory behind BIBDs is well known (see Ref. [41]), and BIBDs of unity index have already been used to design LDPC codes of girth 6 [42]. Notice, however, that dual-containing LDPC codes are girth-4 LDPC codes, and they can be designed based on BIBDs with even index.

A balanced incomplete block design, denoted as BIBD(v,b,r,k,λ), is a collection of subsets (also known as *blocks*) of a set $V$of size v, with a size of each subset being k, so that: (1) each pair of elements (also known as *points*) occurs in *exactly* λ of the subsets, and (2) every element occurs in exactly r subsets. The BIBD parameters satisfy the following two conditions [41]: (1) $vr = bk$ and (2) $\lambda(v-1) = r(k-1)$. Because the BIBD parameters are related (conditions 1 and 2) it is sufficient to identify only three of them: v, k, and λ. It can be easily verified [42] that a point-block incident matrix represents a parity-check matrix H of an LDPC code of the code rate R lower bounded by $R \geq [b - \text{rank}(\boldsymbol{H})]/b$, where b is the codeword length, and with rank( ) we denoted the rank of the parity-check matrix. The parameter k corresponds to the column weight, r to the row weight, and v to the number of parity checks. The corresponding quantum code rate is lower bounded by $R_Q \geq [b - 2 \cdot \text{rank}(\boldsymbol{H})]/b$. By selecting the index of BIBD $\lambda = 1$, the parity-check matrix has a girth of at least 6. For a classical LDPC code to be applicable in quantum error correction the following two conditions are to be satisfied [38]: (1) the LDPC code must contain its dual or equivalently any two rows of the parity-check matrix must have even overlap and the row weight must be even ($\boldsymbol{HH}^T = \mathbf{0}$), and (2) the code must have a rate greater than 1/2. The BIBDs with even index λ satisfy condition 1. The parameter λ corresponds to the number of ones in which two rows overlap.

Example 5. The parity-check matrix from BIBD(7,7,4,4,2) = {{1,2,4,6}, {2,6,3,7}, {3,5,6,1}, {4,3,2,5}, {5,1,7,2}, {6,7,5,4}, {7,4,1,3}}, given as the following $\boldsymbol{H}_1$-matrix, has the rank 3, an even overlap between any two rows, and a row weight that is even as well (or equivalently $\boldsymbol{H}_1\boldsymbol{H}_1^T = \mathbf{0}$). In Fig. 10.7 we show the equivalence between this BIBD, parity-check matrix $\boldsymbol{H}$, and the Tanner graph. Namely, the blocks correspond to the bit nodes and provide the position of nonzero elements in corresponding columns. For example, block $B_1 = \{1,2,4,6\}$ corresponds to the bit node v_1 and it is involved in parity checks c_1, c_2, c_4, and c_6.

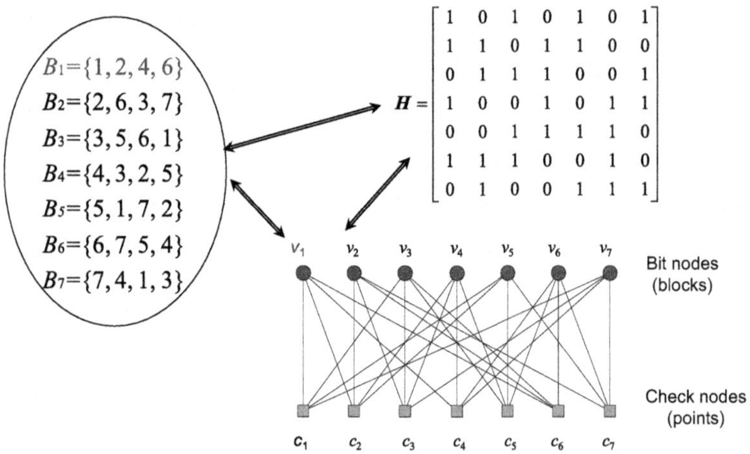

FIGURE 10.7

Equivalence between balanced incomplete block design, parity-check matrix *H*, and Tanner graph.

Therefore we establish edges between variable node v_1 and set check nodes c_1, c_2, c_4, and c_6. The B_1 block also gives the positions of nonzero elements in the first column of the parity-check matrix.

Notice that so-called λ-configurations (in the definition of BIBD the word *exactly* from condition 1 is replaced by *at most*) are not applicable here. However, the $\boldsymbol{H}$-matrix from the BIBD of the unity index can be converted to an $\boldsymbol{H}$-matrix satisfying condition 1 by adding a column with all ones, which is equivalent to adding an additional block to unity index BIBD having all elements from *V*. For example, the parity-check matrix from BIBD(7,7,3,3,1), after the addition of a column with all ones, is given as the following $\boldsymbol{H}_2$-matrix, and satisfies the condition $\boldsymbol{H}_2\boldsymbol{H}_2^T = \boldsymbol{0}$. The following method due to Bose [43] is a powerful method to design many different BIBDs with desired index λ. Let S be a set of elements. Associate to each element u from the set S the following symbols u_1, u_2,...,u_n. Let the subsets of elements from S (also called blocks) S_1,...,S_t satisfy the following three conditions: (1) every subset S_i ($i = 1$,...,t) contains k symbols (the symbols from the same block are different from one another); (2) among kt symbols in t blocks exactly r symbols belong to each of n classes ($nr = kt$); and (3) the differences from t blocks are symmetrically repeated so that each repeats λ times. If s is an element from S, from each S_i set we are able to form another subset $S_{i,s}$ by adding s to S_i keeping the class number (subscript) unchanged; then, sets $S_{i,s}$ ($i = 1$,...,t; $s \in S$) represent an (mn, nt, r, k, λ) BIBD. By observing the elements from BIBD blocks as a position of ones in corresponding columns of a parity-check matrix, the code rate of an LDPC code such obtained is lower bounded by $R \geq 1 - m/t$ ($R_Q \geq 1 - 2m/t$), and the codeword length is determined by $b = nt$.

$$
\boldsymbol{H}_1 = \begin{bmatrix} 1 & 0 & 1 & 0 & 1 & 0 & 1 \\ 1 & 1 & 0 & 1 & 1 & 0 & 0 \\ 0 & 1 & 1 & 1 & 0 & 0 & 1 \\ 1 & 0 & 0 & 1 & 0 & 1 & 1 \\ 0 & 0 & 1 & 1 & 1 & 1 & 0 \\ 1 & 1 & 1 & 0 & 0 & 1 & 0 \\ 0 & 1 & 0 & 0 & 1 & 1 & 1 \end{bmatrix}
$$

$$
\boldsymbol{H}_2 = \begin{bmatrix} 1 & 1 & 1 & 0 & 0 & 0 & 0 & 1 \\ 1 & 0 & 0 & 1 & 1 & 0 & 0 & 1 \\ 1 & 0 & 0 & 0 & 0 & 1 & 1 & 1 \\ 0 & 1 & 0 & 1 & 0 & 1 & 0 & 1 \\ 0 & 1 & 0 & 0 & 1 & 0 & 1 & 1 \\ 0 & 0 & 1 & 1 & 0 & 0 & 1 & 1 \\ 0 & 0 & 1 & 0 & 1 & 1 & 0 & 1 \end{bmatrix}.
$$

In the rest of this section we introduce several constructions employing the method due to Bose.

Construction 1: If $6t+1$ is a prime or prime power and θ is a primitive root of GF($6t+1$), then the following t initial sets $S_i = (0,\theta^i,\theta^{2t+i},\theta^{4t+i})$ $(i=0,1,\ldots,t-1)$ form a BIBD($6t+1$,$t(6t+1)$,$4t$,4,2). The BIBD is formed by adding the elements from GF($6t+1$) to the initial blocks S_i. Because the index of BIBD is even ($\lambda=2$), and row weight $r=4t$ is even, the corresponding LDPC code is a dual-containing code ($\boldsymbol{HH}^T=\boldsymbol{0}$). The quantum code rate for this construction, and constructions 2 and 3 as well, is lower bounded by $R_Q \geq (1-2/t)$. The BIBD(7,7,4,4,2) given earlier is obtained using this construction method. For $t=30$ a dual-containing LDPC(5430,5249) code is obtained. The corresponding quantum LDPC code has the rate 0.934, which is significantly higher than any of the codes introduced in Ref. [38] ($R_Q=1/4$).

Construction 2: If $10t+1$ is a prime or prime power and θ is a primitive root of GF($10t+1$), then the following t initial sets $S_i = (\theta^i,\theta^{2t+i},\theta^{4t+i},\theta^{6t+i})$ form a BIBD($10t+1$,$t(10t+1)$,$4t$,4,2). For example, for $t=24$ a dual-containing LDPC(5784,5543) code is obtained, and corresponding CSS code has the rate 0.917.

Construction 3: If $5t+1$ is a prime or prime power and θ is a primitive root of GF($5t+1$), then the following t initial sets $(\theta^i,\theta^{2t+i},\theta^{4t+i},\theta^{6t+i},\theta^{8t+i})$ form a BIBD($5t+1$,$t(5t+1)$,$5t$,5,4). Notice that parameter t is to be even for LDPC code to satisfy condition 1. For example, for $t=30$, the dual-containing LDPC(4530,4379) code is obtained, and corresponding CSS LDPC code has the rate 0.934.

Construction 4: If $2t + 1$ is a prime or prime power and θ is a primitive root of GF($2t + 1$), then the following $5t + 2$ initial sets:

$$\left(\theta_1^i, \theta_1^{t+i}, \theta_3^{i+a}, \theta_3^{t+i+a}, 0_2\right) \quad (i = 0, 1, ..., t-1)$$
$$\left(\theta_2^i, \theta_2^{t+i}, \theta_4^{i+a}, \theta_4^{t+i+a}, 0_3\right) \quad (i = 0, 1, ..., t-1)$$
$$\left(\theta_3^i, \theta_3^{t+i}, \theta_5^{i+a}, \theta_5^{t+i+a}, 0_4\right) \quad (i = 0, 1, ..., t-1)$$
$$\left(\theta_4^i, \theta_4^{t+i}, \theta_1^{i+a}, \theta_1^{t+i+a}, 0_5\right) \quad (i = 0, 1, ..., t-1)$$
$$\left(\theta_5^i, \theta_5^{t+i}, \theta_2^{i+a}, \theta_2^{t+i+a}, 0_1\right) \quad (i = 0, 1, ..., t-1)$$
$$(0_1, 0_2, 0_3, 0_4, 0_5), (0_1, 0_2, 0_3, 0_4, 0_5)$$

form a BIBD($10t + 5$,$(5t + 2)(2t + 1)$,$5t + 2$,5,2). Similarly as in the previous construction, the parameter t is to be even. The quantum code rate is lower bounded by $R_Q \geq [1 - 2 \cdot 5/(5t + 2)]$, and the codeword length is determined by $(5t + 2)(2t + 1)$. For $t = 30$, the dual-containing LDPC(5490,5307) is obtained, and corresponding quantum LDPC code has the rate 0.934. The following two constructions are obtained by converting unity index BIBD into $\lambda = 2$BIBD.

Construction 5: If $12t + 1$ is a prime or prime power and θ is a primitive root of GF($12t + 1$), then the following t initial sets $(0,\theta^i,\theta^{4t+i},\theta^{8t+i})$ $(i = 0,2,\ldots,2t - 2)$ form a BIBD($12t + 1$,$t(12t + 1)$,$4t$,4,1). To convert this unity index BIBD into $\lambda = 2$BIBD we have to add an additional block $(1,\ldots,12t + 1)$.

Construction 6: If $20t + 1$ is a prime or prime power and θ is a primitive root of GF($20t + 1$), then the following t initial sets $(\theta^i,\theta^{4t+i},\theta^{8t+i},\theta^{12t+i},\theta^{16t+i})$ $(i = 0,2,\ldots,2t - 2)$ form a BIBD($20t + 1$,$t(20t + 1)$,$5t$,5,1). Similarly as we did in the previous construction, to convert this unity index BIBD into $\lambda = 2$BIBD we have to add an additional block $(1,\ldots,20t + 1)$.

Construction 7: If $2(2\lambda + 1)t + 1$ is a prime power and θ is a primitive root of GF[$2(2\lambda + 1)t + 1$], then the following t initial sets $S_i = (\theta^i,\theta^{2t+i},\theta^{4t+i},\ldots, \theta^{4\lambda t+i})$ $(i = 0,1,\ldots,t - 1)$ form a BIBD($2(2\lambda + 1)t + 1$,$t[2(2\lambda + 1)t + 1]$,$(2\lambda + 1)t$,$2\lambda + 1$,λ). The BIBD is formed by adding the elements from GF[$2(2\lambda + 1)t + 1$] to the initial blocks S_i. For any even index λ, and even parameter t (the row weight is even), the corresponding LDPC code is a dual-containing code ($\boldsymbol{HH}^T = 0$). The quantum code rate for this construction is lower bounded by $R_Q \geq (1 - 2/t)$, and the minimum distance is lower bounded by $d_{\min} \geq 2\lambda + 2$. For any odd index λ design we have to add an additional block $(1,2,\ldots,2(2\lambda + 1)t + 1)$, so that the row weight of $\boldsymbol{H}$ becomes even.

Construction 8: If $2(2\lambda - 1)t + 1$ is a prime power and θ is a primitive root of GF[$2(2\lambda - 1)t + 1$], then the following t initial sets $S_i = (0,\theta^i,\theta^{2t+i},\ldots,\theta^{(4\lambda-1)t+i})$ form a BIBD($2(2\lambda - 1)t + 1$,$[2(2\lambda - 1)t + 1]t$,$2\lambda t$,2λ,λ). For any even index λ, the corresponding LDPC code is a dual-containing code. The quantum LPDC code rate is lower bounded by $R_Q \geq (1 - 2/t)$, and the minimum distance is lower bounded by $d_{\min} \geq 2\lambda + 1$.

Construction 9: If $(\lambda-1)t$ is a prime power and θ is a primitive root of GF[$(\lambda-1)t+1$], then the following t initial sets $(0,\theta^i,\theta^{t+i},\ldots,\theta^{(\lambda-2)t+i})$ form a BIBD[$(\lambda-1)t+1,((\lambda-1)t+1)t,\lambda t,\lambda,\lambda$]. Again for an even index λ the quantum LDPC code of rate $R_Q \geq (1-2/t)$ is obtained, whose minimum distance is lower bounded by $d_{\min} \geq \lambda + 1$. Similarly as in the previous two constructions, for any odd index λ design we have to add an additional block $(1,2,\ldots,(\lambda-1)t)$.

Construction 10: If $2k-1$ is a prime power and θ is a primitive root of GF($2k-1$), then the following initial sets:

$$\left(0,\theta^i,\theta^{i+2},\ldots,0^{i+2k-4}\right),\quad \left(\infty,\theta^{i+1},\theta^{i+3},\ldots,0^{i+2k-3}\right)\quad (i=0,1)$$

form a BIBD($2k,4(2k-1),2(2k-1),k,2(k-1)$). For even k the quantum code rate is lower bounded by $R_Q \geq [1-1/(2k-1)]$, the codeword length is determined by $4(2k-1)$, and the minimum distance is lower bounded by $d_{\min} \geq k+1$.

We performed the simulations for error correcting performance of BIBD-based dual-containing codes described earlier as the function of noise level by Monte Carlo simulations. We simulated the classical BSC to be compatible with current literature [38–40]. The results of simulations are shown in Fig. 10.7 for 30 iterations in a sum-product-with-correction-term algorithm described in the previous section. We simulated dual-containing LDPC codes of high-rate and moderate lengths, so that corresponding quantum LDPC code has a rate around 0.9. BER curves correspond to the $C/C^{\perp}$ case, and are obtained by counting the errors only on those codewords from C not belonging to $C^{\perp}$. The codes from BIBD with index $\lambda=2$ outperform the codes with index $\lambda=4$. The codes derived from unity index BIBDs by adding the all-ones column outperform the codes derived from BIBDs of the even index. In simulations presented in Fig. 10.7, the codes with parity-check matrices with column weight $k=4$ or 5 are observed. The code from PG is an exception. It is based on BIBD($s^2+s+1,s+1,1$), where s is a prime power; in our example, parameter s was set to 64. The LDPC(4162,3432) code based on this BIBD outperforms other codes; however, the code rate is lower ($R=0.825$), and the column weight is large (65). For more details on PG codes and secondary structures developed from them, interested readers are referred to the next section.

To improve the BER performance of proposed high-rate sparse dual-containing codes we employed an efficient algorithm due to Sankaranarayanan and Vasic [44] for removing the cycles of length 4 in the corresponding bipartite graph. As shown in Fig. 10.8, this algorithm can significantly improve the BER performance, especially for weak (index 4 BIBD-based) codes. For example, with LDPC(3406,3275) code the BER of 10^{-5} can be achieved at crossover probability $1.14\cdot 10^{-4}$ if a sum-product-with-correction-term algorithm is employed ($g=4$ curve), while the same BER can be achieved at crossover probability $6.765\cdot 10^{-4}$ when the algorithm proposed in Ref. [44] is employed ($g=6$ curve). Notice that this algorithm modifies the parity-check matrix by adding the auxiliary variables and checks, so that the four cycles are removed in the modified parity-check matrix. The algorithm attempts to minimize the required number of auxiliary variable/check nodes while removing the four cycles. The modified parity-check matrix is used only in the decoding phase, while the encoder remains the same.

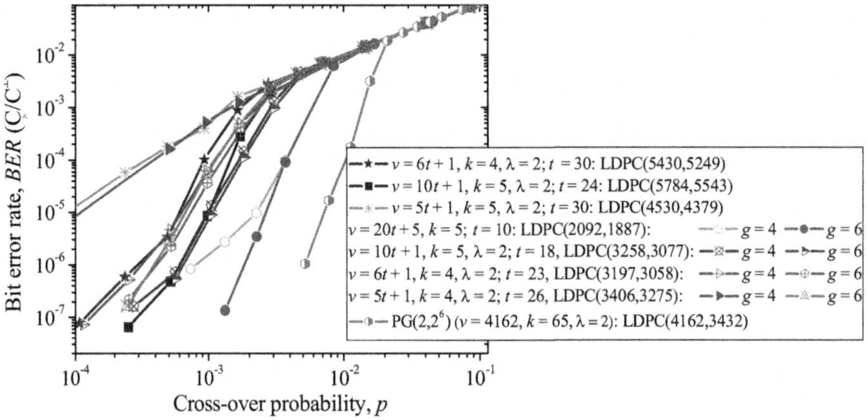

FIGURE 10.8

Bit error rates against crossover probability on a binary symmetric channel. *LDPC*, Low-density parity-check; *PG*, projective geometry.

In Fig. 10.9 we further compare the BER performance of three quantum LDPC codes of quantum rate above 0.9, which are designed by employing constructions 7 and 8: (1) quantum LDPC(7868,7306,0.9285,≥6) code from construction 7 by setting $t = 28$ and $\lambda = 2$, (2) quantum LDPC(8702,8246,0.946,≥5) code from construction 8 by setting $t = 38$ and $\lambda = 2$, and (3) quantum LDPC (8088,7416,0.917,≥9) from construction 8 by setting $t = 24$ and $\lambda = 4$. For comparison purposes, two curves for quantum LDPC codes from Fig. 10.8 are plotted as well. The BIBD codes from constructions 7 and 8 outperform the codes from previous constructions for BERs around 10^{-7}. The code from BIBD with index $\lambda = 4$ and

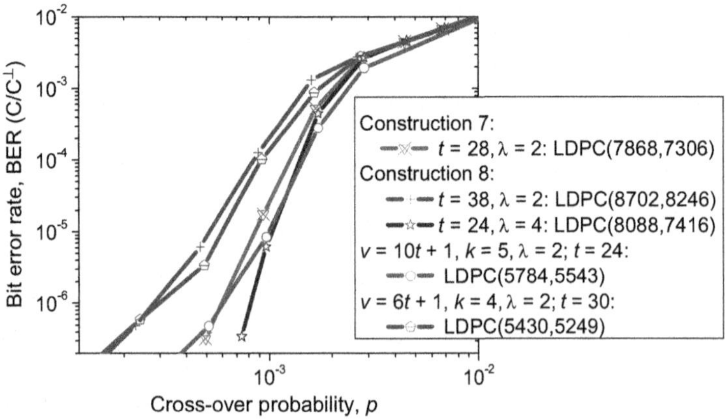

FIGURE 10.9

Bit error rate s against crossover probability on a binary symmetric channel. *LDPC*, Low-density parity-check.

construction 8 outperforms the codes with index $\lambda = 2$ from both constructions because it has larger minimum distance. The code with index $\lambda = 2$ from construction 7 outperforms the corresponding code from construction 8.

10.3 ENTANGLEMENT-ASSISTED QUANTUM LDPC CODES

Entanglement-assisted quantum error correction codes (EA-QECCs) [51–57] make use of preexisting entanglement between transmitter and receiver to improve the reliability of transmission. Dual-containing LDPC codes themselves have a number of advantages compared to other classes of quantum codes because of the sparseness of their quantum parity-check matrices, which leads to a small number of interactions required to perform decoding. However, the dual-containing LDPC codes are in fact girth-4 codes, and the existence of short cycles deteriorates the error performance by correlating the extrinsic information in the decoding process. By using the EA-QECC concept, arbitrary classical code can be used as quantum code. Namely, if classical codes are not dual containing they correspond to a set of stabilizer generators that do not commute. By providing the entanglement between source and destination, the corresponding generators can be embedded into larger set of commuting generators, which gives a well-defined code space.

By using the LDPC codes of girth at least 6 we can dramatically improve the performance and reliability of current quantum key distribution schemes. The number of ebits required for EA-QECC is determined by $e = \text{rank}\,(\boldsymbol{H}\boldsymbol{H}^T)$, as we described in the previous chapter. Notice that for dual-containing codes, $\boldsymbol{H}\boldsymbol{H}^T = \boldsymbol{0}$ meaning that the number of required ebits is zero. Therefore the minimum number of ebits in an EA-QECC is 1, and here we are concerned with the design of LDPC codes of girth $g \geq 6$ with $e = 1$ or reasonably small e. For example, an EA quantum LDPC code of quantum-check matrix $\boldsymbol{A}$, given in the following, has rank $(\boldsymbol{H}\boldsymbol{H}^T) = 1$ and girth 6:

$$\boldsymbol{A} = \begin{pmatrix} \boldsymbol{H} & \boldsymbol{0} \\ \boldsymbol{0} & \boldsymbol{H} \end{pmatrix} \quad \boldsymbol{H} = \begin{bmatrix} 1 & 0 & 1 & 1 & 1 \\ 1 & 1 & 0 & 0 & 1 \\ 0 & 1 & 1 & 1 & 0 \end{bmatrix}. \tag{10.20}$$

Because two Pauli operators on n qubits commute if and only if there is an even number of places in which they differ (neither of which is the identity I operator) we can extend the generators in $\boldsymbol{A}$ by adding the $e = 1$ column so that they can be embedded into a larger Abelian group; the procedure is known as Abelianization in abstract algebra. For example, by adding the $(0\ 1\ 1)^T$ column to the $\boldsymbol{H}$-matrix in Eq. (10.20), the newly obtained code is dual-containing quantum code, and the corresponding quantum-check matrix is given by:

$$\boldsymbol{A}' = \begin{pmatrix} \boldsymbol{H}' & \boldsymbol{0} \\ \boldsymbol{0} & \boldsymbol{H}' \end{pmatrix} \quad \boldsymbol{H}' = \left(\begin{array}{ccccc|c} 1 & 0 & 1 & 1 & 1 & 0 \\ 1 & 1 & 0 & 0 & 1 & 1 \\ 0 & 1 & 1 & 1 & 0 & 1 \end{array}\right). \tag{10.21}$$

The last column in $\boldsymbol{H}'$ corresponds to the ebit, the qubit that is shared between the source and destination, which is illustrated in Fig. 10.10. The source encodes quantum information in state $|\psi\rangle$ with the help of local ancilla qubits $|0\rangle$ and the source-half of shared ebits, and then sends the encoded qubits over a noisy quantum channel (say free-space optical channel or optical fiber). The receiver performs decoding on all qubits to diagnose the channel error and performs a recovery unitary operation to reverse the action of the channel. Notice that the channel does not affect at all the receiver's half of shared ebits. The encoding of quantum LDPC codes is very similar to the different classes of EA block codes as described in the previous chapter. Since the encoding of EA-LDPC is no different to that of any other EA block code, in the rest of this section we deal with the design of EA-LDPC codes instead. The following theorem can be used for the design of EA codes from BIBDs that require only one ebit to be shared between source and destination.

Theorem 1. Let $\boldsymbol{H}$ be a $b \times v$ parity-check matrix of an LDPC code derived from a BIBD(v,k,1) of odd r. The rank of $\boldsymbol{HH}^T$ is equal to 1, while the corresponding EA-LDPC code of CSS type has the parameters $[v,v-2b+1]$ and requires one ebit to be shared between source and destination.

Proof: Because any two rows or columns in $\boldsymbol{H}$ of size $b \times v$, derived from BIBD(v,k,1), overlap in *exactly* one position, by providing that row weight r is odd, the matrix $\boldsymbol{HH}^T$ is an all-one matrix. The rank of all-one matrix $\boldsymbol{HH}^T$ is 1. Therefore LDPC codes from BIBDs of unity index ($\lambda = 1$) and odd r have $e =$ rank $(\boldsymbol{HH}^T) = 1$, while the number of required ebits to be shared between source and destination is $e = 1$. If the EA code is put in CSS form, then the parameters of EA codes will be $[v,v-2b+1]$.

Because the girth of codes derived by employing Theorem 1 is 6, they can significantly outperform quantum dual-containing LDPC codes. Notice that certain blocks in so-called λ-configurations [41] have the overleap of zero and therefore cannot be used for the design of EA codes with $e = 1$. Shrikhande [66] has shown that the generalized Hadamard matrices, affine-resolvable BIBDs, group-divisible designs, and orthogonal arrays of strength 2 are all related, and because for these combinatorial objects $\lambda = 0$ or 1, they are not suitable for the design of EA codes of $e = 1$. Notice that if the quantum-check matrix is put into the following format $\boldsymbol{A} = [\boldsymbol{H}|\boldsymbol{H}]$, then the corresponding EA code has parameters $[v,v-b+1]$. This design can be generalized by employing two BIBDs having the same parameter v, as explained in Theorem 2.

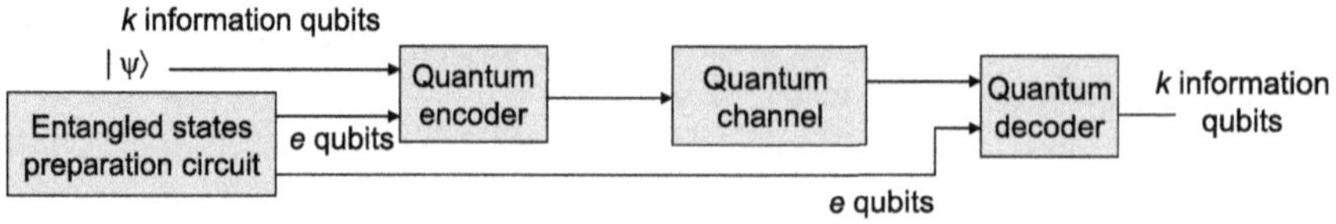

FIGURE 10.10

Operation principle of entanglement-assisted quantum codes.

Theorem 2. Let $\boldsymbol{H}_1$ be a $b_1 \times v$ parity-check matrix of an LDPC code derived from a BIBD(v,k_1,1) of odd r_1 and $\boldsymbol{H}_2$ be a $b_2 \times v$ parity-check matrix of an LDPC code derived from a BIBD(v,k_2,1) of odd r_2. The number of ebits required in the corresponding EA code with quantum-check matrix $\boldsymbol{A} = [\boldsymbol{H}_1|\boldsymbol{H}_2]$ is $e =$ rank $(\boldsymbol{H}_1\boldsymbol{H}_2^T + \boldsymbol{H}_2\boldsymbol{H}_1^T)/2 = 1$, while the corresponding EA-LDPC code has the parameters $[v,v - b + 1]$.

The proof of this theorem is a straightforward generalization of the proof of Theorem 1. Next, we describe three designs belonging to the class of codes from Theorem 1.

Design 1 (Steiner triple system): If $6t + 1$ is a prime power and θ is a primitive root of GF($6t + 1$), then the following t initial blocks $(\theta^i,\theta^{2t+i},\theta^{4t+i})$ $(i = 0,1, \ldots,t - 1)$ form BIBD($6t + 1$,3,1) with $r = 3t$. The BIBD is formed by adding the elements from GF($6t + 1$) to the initial blocks. The corresponding LDPC code has rank $(\boldsymbol{HH}^T) = 1$ and a girth of 6.

Design 2 (projective planes): A finite projective plane of order q, say PG(2, q), is a $(q^2 + q + 1, q + 1, 1)$ BIBD, $q \geq 2$ and q is a power of prime [17,17,21,30,53]. The finite projective plane-based geometries are summarized in Table 10.1. The point set of the design consists of all the *points* on PG(2, q), and the block set of the design consists of all the *lines* on PG(2, q). The points and lines of a finite projective plane satisfy the following set of axioms: (1) every line consists of the same number of points, (2) any two points on the plane are connected by a unique line, (3) any two lines on the plane intersect at a unique point, and (4) a fixed number of lines pass through any point on the plane. Since any two lines on a projective plane intersect at a unique point, there are no parallel lines on the plane ($\lambda = 1$). The incidence matrix (parity-check matrix of corresponding LDPC codes) of such a design, for $q = 2^s$, is cyclic and hence any row of the matrix can be obtained by cyclic shifting (right or left) another row of the matrix. Since the row weight $q + 1 = 2^s+1$ is odd, the corresponding LDPC code has rank $(\boldsymbol{HH}^T) = 1$. The minimum distance of codes from projective planes, affine planes, oval designs, and unitals [45] is at least 2^s+2, 2^s+1, $2^{s-1} + 1$ and $2^{s/2} + 2$, respectively. Corresponding ranks of $\boldsymbol{H}$-matrices are given, respectively, by Ref. [45] 3^s+1, 3^s, 3^s-2^s, and $2^{3s/2}$. Notice, however, that affine plane-based codes have rank $(\boldsymbol{HH}^T) > 1$ because they contain parallel lines. The code rate of classical projective plane codes is given by

Table 10.1 Finite projective plane $2 - (v, k, 1)$-based geometries.

k	v	Parameter	Name
$q + 1$	$q^2 + q + 1$	q-a prime power	Projective planes
q	q^2	q-a prime power	Affine planes
$q/2$	$q(q - 1)/2$	q-a prime power and even	Oval designs
$\sqrt{q} + 1$	$q\sqrt{q} + 1$	q-a prime power and square	Unitals

$R = (q^2 + q + 1 - \text{rank}(\boldsymbol{H}))/(q^2 + q + 1) = (q^2 + q - 3^s)/(q^2 + q + 1)$, while the quantum code rate of corresponding quantum code is given by:

$$\begin{aligned} R_Q &= \left(q^2 + q + 1 - 2\text{rank}(\boldsymbol{H}) + \text{rank}\left(\boldsymbol{H}\boldsymbol{H}^{\mathrm{T}}\right)\right)/\left(q^2 + q + 1\right) \\ &= \left(q^2 + q - 2 \cdot 3^s\right)/\left(q^2 + q + 1\right). \end{aligned} \tag{10.22}$$

Design 3 (m-dimensional projective geometries): The finite projective geometries [41,58] PG(m, p^s) are constructed using (m + 1)-tuples $\boldsymbol{x} = (x_0, x_1, \cdots, x_m)$ of elements x_i from GF(p^s) (p is a prime, s is a positive integer), not all simultaneously equal to zero, called points. Two (m + 1)-tuples $\boldsymbol{x}$ and $\boldsymbol{y} = (y_0, y_1, \cdots, y_m)$ represent the same point if $\boldsymbol{y} = \lambda\boldsymbol{x}$, where λ is a nonzero element from GF(p^s). Therefore each point can be represented on $p^s - 1$ ways (an equivalence class). The number of points in PG(m, p^s) is given by:

$$v = \left[p^{(m+1)s} - 1\right]/(p^s - 1). \tag{10.23}$$

The points (equivalence classes) can be represented by $\left[\alpha^i\right] = \left\{\alpha^i, \beta\alpha^i, \cdots, \beta^{p^s-2}\alpha^i\right\}$ $(0 \le i \le v)$, where $\beta = \alpha^v$. Let $[\alpha^i]$ and $[\alpha^j]$ be two distinct points in PG(m, p^s), then the line passing through them consists of points of the form $[\lambda_1\alpha^i + \lambda_2\alpha^j]$, where $\lambda_1, \lambda_2 \in$ GF(p^s). Because $[\lambda_1\alpha^i + \lambda_2\alpha^j]$ and $[\beta^k\lambda_1\alpha^i + \beta^k\lambda_2\alpha^j]$ represent the same point, each line in PG(m, p^s) consists of $k = (p^{ms} - 1)/(p^s - 1)$ points. The number of lines intersecting at a given point is given by k, and the number of lines in m-dimensional PG is given by:

$$b = \left[p^{s(m+1)} - 1\right](p^{sm} - 1)/\left[\left(p^{2s} - 1\right)(p^s - 1)\right]. \tag{10.24}$$

The parity-check matrix is obtained as the incidence matrix $\boldsymbol{H} = (h_{ij})_{b \times v}$ with rows corresponding to the lines and columns to the points, and columns being arranged in the following order: $[\alpha^0]$,...,$[\alpha^v]$. This class of codes are sometimes called type I projective geometry codes [21]. The $h_{ij} = 1$ if the jth point belongs to the ith line, and zero otherwise. Each row has weight $p^s + 1$ and each column has weight k. Any two rows or columns have exactly one "1" in common, and providing that p^s is even, then the row weight ($p^s + 1$) will be odd, and corresponding LDPC code will have rank($\boldsymbol{H}\boldsymbol{H}^T$) = 1. The quantum code rate of corresponding codes will be given by:

$$R_Q = \left(v - 2\text{rank}(\boldsymbol{H}) + \text{rank}\left(\boldsymbol{H}\boldsymbol{H}^T\right)\right)/v. \tag{10.25}$$

By defining a parity-check matrix as an incidence matrix with lines corresponding to columns and points to rows, the corresponding LDPC code will have rank($\boldsymbol{H}\boldsymbol{H}^T$) = 1 by providing that a row weight k is odd. This type of PG code is sometimes called type II PG code [21].

Design 4 (finite geometry codes on m-flats) [45]: In projective geometric terminology, an (m + 1)-dimensional element of the set is referred to as an *m-flat*. For example, *points* are 0-flats, *lines* are 1-flats, *planes* are 2-flats, and *m-dimensional* spaces are (m − 1) flats. An l-dimensional flat is said to be contained in an

m-dimensional flat if it satisfies a set-theoretic containment. Since any m-flat is a finite vector subspace, it is justified to talk about a basis set composed of basis elements independent in GF(p) (p is a prime). In the following discussion, the basis element of a 0-flat is referred to as a point and generally this should cause no confusion because this basis element is representative of the 0-flat. Using the foregoing definitions and linear algebraic concepts, it is straightforward to count the number of m-flats, defined by a basis with $m + 1$ points, in PG(n, q). The number of such flats, say $N_{PG}(m, n, q)$, is the number of ways of choosing $m + 1$-independent points in PG(n, q) divided by the number of ways of choosing $m + 1$-independent points in any m-flat:

$$\begin{aligned} N_{PG}(m,n,q) &= \frac{(q^{n+1}-1)(q^{n+1}-q)(q^{n+1}-q^2)\cdots(q^{n+1}-q^m)}{(q^{m+1}-1)(q^{m+1}-q)(q^{m+1}-q^2)\cdots(q^{m+1}-q^m)} \\ &= \prod_{i=0}^{m}\frac{(q^{n+1-i}-1)}{(q^{m+1-i}-1)}. \end{aligned} \quad (10.26)$$

Hence, the number of 0-flats in PG(n, q) is $N_{PG}(0,n,q) = \frac{q^{n+1}-1}{q-1}$, and the number of 1-flats is:

$$N_{PG}(1,n,q) = \frac{(q^{n+1}-1)(q^n-1)}{(q^2-1)(q-1)}. \quad (10.27)$$

When $n = 2$, the number of points is equal to the number of lines, and this agrees with dimensions of point-line incidence matrices of projective plane codes introduced earlier.

An algorithm to construct an m-flat in PG(n, q) begins by recognizing that the elements of GF(q^{n+1}) can be used to represent points of PG(n, q) in the following manner [45,59]. If α is the primitive element of GF(q^{n+1}), then α^v, where $v = (q^{n+1}-1)/(q-1)$, is a primitive element of GF(q). Each one of the first v powers of α can be taken as the basis of one of the 0-flats in PG(n, q). In other words, if α^i is the basis of a 0-flat in PG(n, q), then every α^k, such that $i \equiv k \mathrm{mod} v$, is contained in the subspace. In a similar fashion, a set of $m + 1$ powers of α that are independent in GF(q) forms a basis of an m-flat. If $\alpha^{s_1}, \alpha^{s_2}, \ldots, \alpha^{s_{m+1}}$ is a set of $m + 1$ basis elements of an m-flat, then any point in the flat can be written as:

$$\zeta_i = \sum_{j=1}^{m+1} \varepsilon_{ij}\alpha^{s_j}, \quad (10.28)$$

where the ε_{ij} are chosen such that no two vectors $< \varepsilon_{i1}, \varepsilon_{i2}, \ldots, \varepsilon_{i(m+1)} >$ are linear multiples over GF(q). Now, every ζ_i can be equivalently written as a power of the primitive element α.

To construct an LDPC code, we generate all m-flats and l-flats in PG(n, q) using the method just described. An incidence matrix of m-flats and l-flats in PG(n, q) is a binary matrix with $N_{PG}(m, n, q)$ rows and $N_{PG}(l, n, q)$ columns. The (i, j)th element of the incidence matrix $\mathbf{A}_{PG}(m, l, n, q)$ is a one if and only if the jth l-flat is contained in the ith m-flat of the geometry. An LDPC code is constructed by considering the incidence matrix or its transpose as the parity-check matrix of the code. It is widely accepted that the performance of an LDPC code under an iterative decoder is deteriorated by the presence of four cycles in the Tanner graph of the code. To avoid four cycles, we impose an additional constraint that l should be one less than m. If l is one less than m, then no two distinct m-flats have more than one l-dimensional subspace in common. This guarantees that the girth (length of the shortest cycle) of the graph is 6.

In Fig. 10.11 we show the code length as a function of quantum code rate for EA-LDPC codes derived from designs 2 and 3, obtained for different values of parameter $s = 1-9$. The results of simulations are shown in Fig. 10.12 for 30 iterations in sum-product algorithm. The EA quantum LDPC codes from design 3 are compared against dual-containing quantum LDPC codes discussed in the previous chapter. We can see that EA-LDPC codes significantly outperform dual-containing LDPC codes.

Before concluding this section, we provide a very general design method [29], which is based on the theory of mutually orthogonal Latin squares (MOLS) constructed using MacNeish–Mann theorem [58]. This design allows us to determine a corresponding code for an arbitrary composite number q.

Design 5 (iterative decodable codes based on MacNeish–Mann theorem) [29]: Let the prime decomposition of an integer q be:

$$q = p_1^{s_1} p_2^{s_2} \cdots p_m^{s_m}, \tag{10.29}$$

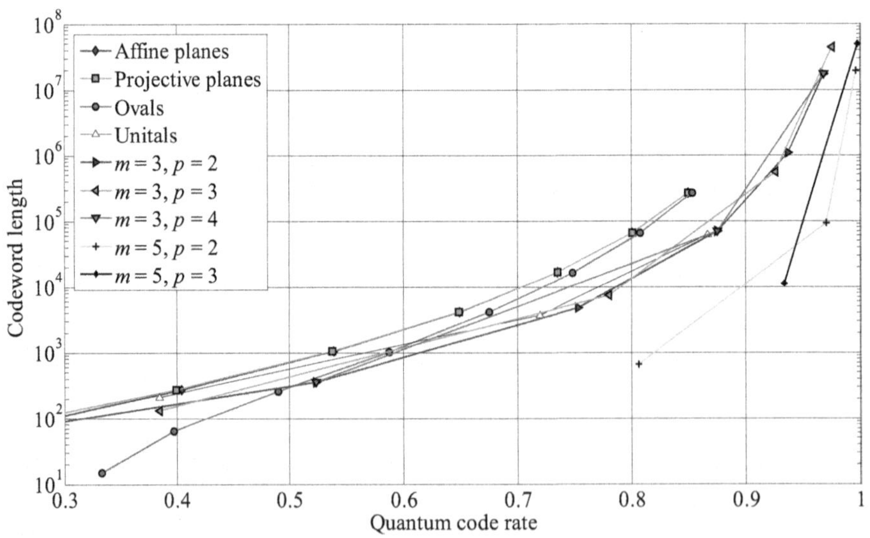

FIGURE 10.11

Entanglement-assisted quantum low-density parity-check codes from m-dimensional projective geometries.

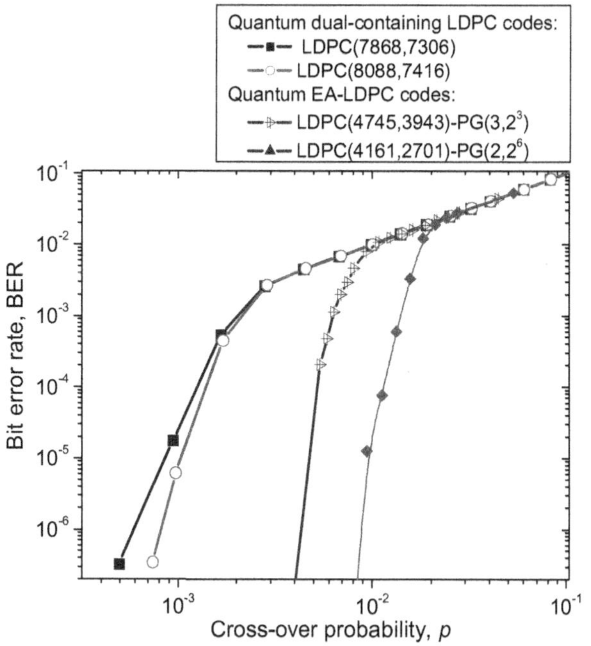

FIGURE 10.12

Bit error rates against crossover probability on a binary symmetric channel.

with p_i $(i = 1,2,\ldots,m)$ being the prime numbers. MacNeish and Mann showed [58] that a complete set of:

$$n(q) = \min\left(p_1^{s_1}, p_2^{s_2}, \cdots, p_m^{s_m}\right) - 1 \tag{10.30}$$

MOLS of order q can always be constructed. The construction algorithm is given next. Consider m-tuples:

$$\gamma = (g_1, g_2, \ldots, g_m), \tag{10.31}$$

where $g_i \in \mathrm{GF}\left(p_i^{s_i}\right)$, $i = 1,2,\ldots,m$, with addition and multiplication operations defined by:

$$\gamma_1 + \gamma_2 = (g_1, g_2, \ldots, g_m) + (h_1, h_2, \ldots, h_m) = (g_1 + h_1, g_2 + h_2, \ldots, g_m + h_m), \tag{10.32}$$

and:

$$\gamma_1 \gamma_2 = (g_1 h_1, g_2 h_2, \ldots, g_m h_m), \tag{10.33}$$

wherein the operations in brackets are performed in the corresponding Galois fields (e.g., $g_1 h_1$ and $g_1 + h_1$ are performed in $\mathrm{GF}\left(p_1^{s_1}\right)$).

Denote the elements from $\mathrm{GF}(p_i^{s_i})$, $i = 1, 2, .., m$ as $0, g_i^{(1)} = 1, g_i^{(2)}, ..., g_i^{(p_i^{s_i}-1)}$. The following elements possess multiplicative inverses:

$$\gamma_i = \left(g_1^{(i)}, g_2^{(i)}, ..., g_m^{(i)}\right), \quad i = 1, 2, ..., n(q), \tag{10.34}$$

since they are composed of nonzero elements of corresponding Galois fields, and $\gamma_i - \gamma_j$ for $i \neq j$ as well. Denote the first element as $\gamma_0 = (0, 0, ..., 0)$, the next $n(q)$ as in Eq. (10.34), and the rest $q - n(q) - 1$ elements as:

$$\begin{aligned}&\gamma_j = \left(g_1^{(i_1)}, g_2^{(i_2)}, ..., g_m^{(i_m)}\right), \; j = n(q) + 1, ..., q - 1; \; i_1 = 0, 1, ..., p_1^{s_1} - 1;\\&i_2 = 0, 1, ..., p_2^{s_2} - 1; ...; i_m = 0, 1, ..., p_m^{s_m} - 1,\end{aligned} \tag{10.35}$$

with combinations of components leading to Eq. (10.32) being excluded. The $n(q)$ arrays of elements, MOLS, L_i, $i = 1,2,\ldots,n(q)$ are further formed with the (j,k)th cell being filled by $\gamma_j + \gamma_i\gamma_k$; $i = 1, 2, ..., n(q)$; $0 \leq j, k \leq q - 1$. (The multiplication $\gamma_i\gamma_k$ is performed using Eq. (10.31), and the addition $\gamma_j + \gamma_i\gamma_k$ using Eq. (10.30).) To establish the connection between MOLS and BIBD, we introduce the following one-to-one mapping l: $L > V$, with L being the MOLS and V being the integer set of q^2 elements (*points*). The simplest example of such mapping is linear mapping: $l(x, y) = q(x-1) + y$; $1 \leq x, y \leq q$ (therefore $1 \leq l(x,y) \leq q^2$). (The numbers $l(x,y)$ are referred to as the cell *labels*.) Each L_i ($i = 1,\ldots,n(q)$) defines a set of q *blocks* (*lines*) $B_i = \{l(x, y)|L_i(x, y) = s, \; 1 \leq s \leq q\}$. Each line s ($1 \leq s \leq q$) in B_i contains the labels of element s from L_i. These blocks are equivalent to the sets of lines of slopes $0 \leq s \leq q - 1$ in the lattice design we introduced in Ref. [19]. Labels of every line of a design specify the positions of "1"s in a row of the parity-check matrix. Namely, the *incidence matrix* of a design is a $qn(q) \times q^2$ matrix $H = (h_{ij})$ defined by:

$$h_{ij} = \begin{cases} 1 & \text{if the ith block contains the jth point} \\ 0 & \text{otherwise} \end{cases} \tag{10.36}$$

and represents the parity-check matrix. The code length is q^2 and the code rate is determined by $R = \left(q^2 - \mathrm{rank}(\boldsymbol{H})\right)/q^2$, and may be estimated as $R \cong \left(q^2 - n(q)q\right)/q^2 = 1 - n(q)/q$. The quantum code rate of the corresponding quantum code is given by:

$$R_Q = \left(q^2 - 2\mathrm{rank}(\boldsymbol{H}) + \mathrm{rank}\left(\boldsymbol{H}\boldsymbol{H}^T\right)\right)/q^2 \simeq \left(q^2 - 2n(q)q + \mathrm{rank}\left(\boldsymbol{H}\boldsymbol{H}^T\right)\right)/q^2. \tag{10.37}$$

10.4 ITERATIVE DECODING OF QUANTUM LDPC CODES

In Chapters 7 and 8 we defined the QECC, which encodes K qubits into N qubits, denoted as $[N,K]$, by an encoding mapping U from the K-qubit Hilbert space H_2^K onto a 2^K-dimensional subspace C_q of the N-qubit Hilbert space H_2^N. The $[N,K]$ quantum stabilizer code C_Q has been defined as the unique subspace of Hilbert space H_2^N that is fixed by the elements from stabilizer S of C_Q as follows:

$$C_Q = \bigcap_{s \in S} \{ |c\rangle \in H_2^N \,|\, s|c\rangle = |c\rangle \}. \tag{10.38}$$

Any error in Pauli group G_N of N-qubit errors can be represented by:

$$\begin{aligned} &E = j^{\lambda} X(\boldsymbol{a}) Z(\boldsymbol{b}); \ \boldsymbol{a} = a_1 \cdots a_N; \ \boldsymbol{b} = b_1 \cdots b_N; \ a_i, b_i = 0, 1; \ \lambda = 0, 1, 2, 3 \\ &X(\boldsymbol{a}) \equiv X_1^{a_1} \otimes \cdots \otimes X_N^{a_N}; \ Z(\boldsymbol{b}) \equiv Z_1^{b_1} \otimes \cdots \otimes Z_N^{b_N}. \end{aligned} \tag{10.39}$$

An important *property* of N-qubit errors is: $\forall\ E_1, E_2 \in G_N$ and the following is valid: $[E_1,E_2] = 0$ or $\{E_1,E_2\} = 0$. In other words, the elements from the Pauli group either commute or anticommute:

$$\begin{aligned} E_1 E_2 &= j^{l_1+l_2} X(\boldsymbol{a}_1) Z(\boldsymbol{b}_1) X(\boldsymbol{a}_2) Z(\boldsymbol{b}_2) = j^{l_1+l_2} (-1)^{\boldsymbol{a}_1 \boldsymbol{b}_2 + \boldsymbol{b}_1 \boldsymbol{a}_2} X(\boldsymbol{a}_2) Z(\boldsymbol{b}_2) X(\boldsymbol{a}_1) Z(\boldsymbol{b}_1) \\ &= (-1)^{\boldsymbol{a}_1 \boldsymbol{b}_2 + \boldsymbol{b}_1 \boldsymbol{a}_2} E_2 E_1 = \begin{cases} E_2 E_1, & \boldsymbol{a}_1 \boldsymbol{b}_2 + \boldsymbol{b}_1 \boldsymbol{a}_2 = 0 \bmod 2 \\ -E_2 E_1, & \boldsymbol{a}_1 \boldsymbol{b}_2 + \boldsymbol{b}_1 \boldsymbol{a}_2 = 1 \bmod 2 \end{cases}. \end{aligned} \tag{10.40}$$

In Chapters 8 and 9 we introduced the concept of syndrome of an error as follows. Let C_Q be a quantum stabilizer code with generators $g_1, g_2, \ldots, g_{N-K}$ and let $E \in G_N$ be an error. The error syndrome for error E is defined by the bit string $S(E) = [\lambda_1 \lambda_2 \ldots \lambda_{N-K}]^{\mathrm{T}}$ with component bits being determined by:

$$\lambda_i = \begin{cases} 0, & [E, g_i] = 0 \\ 1, & \{E, g_i\} = 0 \end{cases} \quad (i = 1, \ldots, N-K). \tag{10.41}$$

We have shown in Chapter 8 that the set of errors $\{E_i |\, E_i \in G_N\}$ have the same syndrome $\boldsymbol{\lambda} = [\lambda_1 \ldots \lambda_{N-K}]^{\mathrm{T}}$ if and only if they belong to the same coset $EC(S) = \{Ec | c \in C(S)\}$, where $C(S)$ is the *centralizer* of S (that is, the set of errors $E \in G_N$ that commute with all elements from S). The Pauli-encoded operators have been introduced by the following mappings:

$$X_i \to \overline{X}_i = U X_{N-K+i} U^{\dagger} \quad \text{and} \quad Z_i \to \overline{Z}_i = U Z_{N-K+i} U^{\dagger}; \ i = 1, \cdots, K, \tag{10.42}$$

where U is a Clifford operator acting on N qubits (an element of Clifford group-$N(G_N)$, where N is the normalizer of G_N). By using this interpretation, the encoding operation is represented by:

$$U(|\boldsymbol{0}\rangle_{N-K} \otimes |\psi\rangle), \quad |\psi\rangle \in H_2^K. \tag{10.43}$$

An equivalent definition of syndrome elements is given by Ref. [59]:

$$\lambda_i = \begin{cases} 1, & Eg_i = g_iE \\ -1, & Eg_i = -g_iE \end{cases} \quad (i=1,...,N-K). \tag{10.44}$$

This definition can simplify verification of commutativity of two errors $E = E_1 \otimes \cdots \otimes E_N$ and $F = F_1 \otimes \cdots \otimes F_N$, where $E_i, F_i \in \{I,X,Y,Z\}$, as follows:

$$EF = \prod_{i=1}^{N} E_iF_i. \tag{10.45}$$

The Pauli channels can be described using the following error model, which has already been discussed in Section 7.5:

$$\xi(\rho) = \sum_{E \in G_N} p(E)E\rho E^{\dagger}, \quad p(E) \geq 0, \quad \sum_{E \in G_N} p(E) = 1. \tag{10.46}$$

A memoryless channel model can be defined in similar fashion to the discrete memoryless channel models for classical codes (see Chapter 6) as follows [59]:

$$p(E = E_1 \cdots E_N) = p(E_1) \cdots p(E_N), \quad E \in G_N. \tag{10.47}$$

The depolarizing channel is a very popular channel model for which $p(I) = 1 - p$, $p(X) = p(Y) = p(Z) = p/3$ for certain depolarizing strength $0 \leq p \leq 1$.

The probability of E error given syndrome $\boldsymbol{\lambda}$ can be calculated by:

$$p(E|\boldsymbol{\lambda}) \sim p(E) \prod_{i=1}^{N-K} \delta_{\lambda_i, Eg_i}. \tag{10.48}$$

The most likely error E_{opt} given the syndrome can be determined by:

$$E_{\text{opt}}(\lambda) = \underset{E \in G_N}{\text{argmax}} p(E|\boldsymbol{\lambda}). \tag{10.49}$$

Unfortunately, this approach is an NP-hard problem. We can use instead the qubit-wise most likely error evaluation as follows [59]:

$$E_{\text{q,opt}}(\lambda) = \underset{E_q \in G_1}{\text{argmax}} p_q(E_q|\boldsymbol{\lambda}), \quad p_q(E_q|\boldsymbol{\lambda}) = \sum_{E_1,\cdots,E_{q-1},E_{q+1},\cdots,E_N \in G_1} p(E_1 \cdots E_N|\boldsymbol{\lambda}). \tag{10.50}$$

Notice that this marginal optimum does not necessarily coincide with the global optimum.

It is clear from Eq. (10.50) that this marginal optimization involves summation of many terms exponential in N. The SPA similar to that discussed in Section 10.1 can be used here. For this purpose we can also use *bipartite graph* representation of stabilizers. Namely, we can identify stabilizers with function nodes and qubits with variable nodes. As an illustrative example, let us consider (5,1) quantum cyclic code generated by the following four stabilizers:

$g_1 = X_1Z_2Z_3X_4$, $g_2 = X_2Z_3Z_4X_5$, $g_3 = X_1X_3Z_4X_5$, $g_4 = Z_1X_2X_4Z_5$. The corresponding quantum-check matrix is given by:

$$A = \begin{bmatrix} \overset{X}{10010} | \overset{Z}{01100} \\ 01001|00110 \\ 10100|00011 \\ 01010|10001 \end{bmatrix}.$$

The corresponding bipartite graph is shown in Fig. 10.13, where we used dashed lines to denote the X-operator action on particular qubits and solid lines to denote the action of Z-operators. This is an example of a *regular* quantum stabilizer code, with qubit node degree 3 and stabilizer node degree 4. For example, the qubit q_1 is involved in stabilizers g_1, g_3, and g_4 and therefore there exist the edges between this qubit and corresponding stabilizers. The weight of the edges is determined by the action of the corresponding Pauli operator on the observed qubit:

Let us denote the messages to be sent from qubit q to generator (check) node c as $m_{q\to c}$, which are concerned probability distributions over $E_q \in G_1 = \{I,X,Y,Z\}$ and represent an array of four positive numbers, one for each value of E_q. Let us denote the neighborhood of q by $n(q)$ and neighborhood of c by $n(c)$. The initialization is performed as follows: $m_{q\to c}(E_q) = p_q(E_q)$, where $p_q(E_q)$ is given by Eq. (10.50). Each generator ("check") node c sends the message to its neighbor qubit node q by properly combining the extrinsic information from other neighbors as follows:

$$m_{c\to q}(E_q) \sim \sum_{E_{q'},\ q'\in n(c)\backslash q} \left(\delta_{\lambda_c, E_c g_c} \prod_{q'\in n(c)\backslash q} m_{q'\to c}(E_{q'}) \right), \tag{10.51}$$

where $n(c)\backslash q$ denotes all neighbors of generator (check) node c except q. The proportionality factor can be determined from the normalization condition: $\sum_{E_i} m_{c\to q} = 1$. Clearly, the message $m_{c\to q}$ is related to the syndrome component λ_c associated with generator g_c and extrinsic information collected from all neighbors of c except q. This step can be called, in analogy with classical LDPC codes, the *generator (check)-node update rule*.

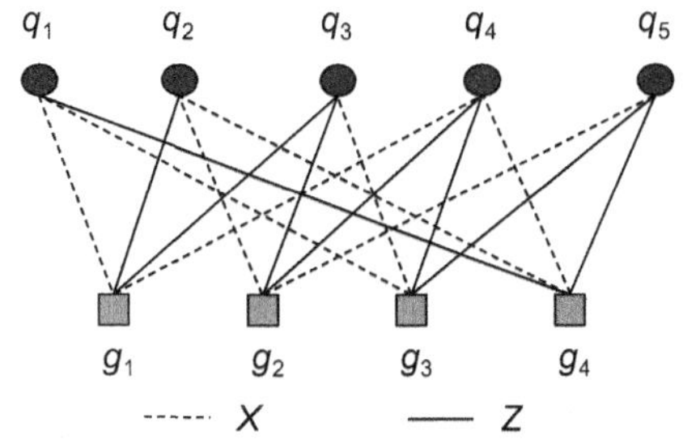

FIGURE 10.13

The bipartite graph of (5,1) quantum cyclic code with generators:

$$g_1 = X_1Z_2Z_3X_4,\ g_2 = X_2Z_3Z_4X_5,\ g_3 = X_1X_3Z_4X_5,\ g_4 = Z_1X_2X_4Z_5.$$

The message to be sent from qubit q to generator (check) c is obtained by properly combining extrinsic information (message) from all generator neighbors of q, say c', except generator node c:

$$m_{q\to c}(E_q) \sim p_q(E_q) \prod_{c'\in n(q)\backslash c} m_{c'\to q}(E_q), \tag{10.52}$$

where $n(q)\backslash c$ denotes all neighbors of qubit node q except c. Clearly, $m_{q\to c}$ is a function of prior probability $p(E_q)$ and extrinsic messages collected from all neighbors of q except c. The proportionality constant in Eq. (10.52) can also be obtain from the normalization condition. Again, in analogy with classical LDPC codes, this step can be called the *qubit-node update rule*.

Steps (10.51) and (10.52) are iterated until a valid codeword is obtained or predetermined number of iteration is reached. Now we have to calculate the beliefs $b_q(E_q)$, that is, the marginal conditional probability estimation $p_q(E_q|\lambda)$, as follows:

$$b_q(E_q) \sim p_q(E_q) \prod_{c'\in n(q)} m_{c'\to q}(E_q). \tag{10.53}$$

Clearly, the beliefs are determined by taking the extrinsic messages of all neighbors of qubit q into account and the prior probability $p_q(E_q)$. The recovery operator can be obtained as tensor product of qubit-wise maximum-belief Pauli operators:

$$E_{\text{SPA}} = \otimes_{q=1}^{N} E_{\text{SPA},q}, \quad E_{\text{SPA},q} = \underset{E_q\in G_1}{\text{argmax}}\{b_q(E_q)\}. \tag{10.54}$$

If the bipartite graph is a tree, then the beliefs will converge to true conditional marginal probabilities $b_q(E_q) = p_q(E_q|\lambda)$ in a number of steps equal to the tree's depth. However, in the presence of cycles the beliefs will not converge to the correct conditional marginal probabilities in general. We will perform iterations until trivial error syndrome is obtained, that is, until $E_{\text{SPA}} g_i = \lambda_i$ ($\forall\, i = 1,\ldots,N-K$) or until the predetermined number of iterations has been reached. This algorithm is applicable to both dual-containing and EA quantum codes.

The foregoing quantum SPA algorithm is essentially very similar to the classical probability-domain SPA. In Section 10.1 we described the log-domain version of this algorithm, which is more suitable for hardware implementation as the multiplication of probabilities is replaced by addition of reliabilities.

The cycles of length 4, which exist in dual-containing codes, deteriorate the decoding BER performance. There exist various methods to improve decoding performance, such as freezing, random perturbation, and collision methods [59]. Some may use method [44], which was already described in Section 10.2. In addition, by avoiding the so-called trapping sets the BER performance of LDPC decoders can be improved [60,62].

10.5 SPATIALLY COUPLED QUANTUM LDPC CODES

The starting point can be the QC-LPDC code with the template parity-check matrix given by:

$$\boldsymbol{H}_{\mathrm{QC}} = \begin{bmatrix} \boldsymbol{I} & \boldsymbol{I} & \boldsymbol{I} & \cdots & \boldsymbol{I} \\ \boldsymbol{I} & \boldsymbol{B}^{S[1]} & \boldsymbol{B}^{S[2]} & \cdots & \boldsymbol{B}^{S[c-1]} \\ \boldsymbol{I} & \boldsymbol{B}^{2S[1]} & \boldsymbol{B}^{2S[2]} & \cdots & \boldsymbol{B}^{2S[c-1]} \\ \cdots & \cdots & \cdots & \cdots & \cdots \\ \boldsymbol{I} & \boldsymbol{B}^{(r-1)S[1]} & \boldsymbol{B}^{(r-1)S[2]} & \cdots & \boldsymbol{B}^{(r-1)S[c-1]} \end{bmatrix},$$

$$\boldsymbol{B} = \begin{bmatrix} 0 & 1 & 0 & \cdots & 0 \\ 0 & 0 & 1 & \cdots & 0 \\ \cdots & \cdots & \cdots & \cdots & \cdots \\ 0 & 0 & 0 & \cdots & 1 \\ 1 & 0 & 0 & \cdots & 0 \end{bmatrix}_{b\times b}, \tag{10.55}$$

where $\boldsymbol{I}$ and $\boldsymbol{B}$ are identity and permutation matrices of size $b \times b$, and integers $S[i] \in \{0,1,\ldots,b-1\}$ $(i = 0,1,\ldots,r-1;\ r < b)$ are properly chosen to satisfy the girth (the largest cycle in the corresponding bipartite graph representation of $\boldsymbol{H}_{\mathrm{QC}}$) constraints, as described in Ref. [61]. To memorize the parity-check matrix of this QC-LDPC code, for each block row we need to layer just the position of the first nonzero entry in the first row of the corresponding permutation submatrix, all other rows are block-cyclic permutations of the first row. By using this QC-LDPC code as a template design we create a spatially coupled (SC)-LDPC code as illustrated in Fig. 10.14 left. The codeword length of this SC-LDPC code will be $b \times (l \times c - m \times (l-1))$, where l is the number of coupled template QC-LDPC codes and m is the coupling length expressed in terms of number of blocks. Because there are $r \times c \times l$ nonempty submatrices in the parity-check matrix of the SC-LDPC

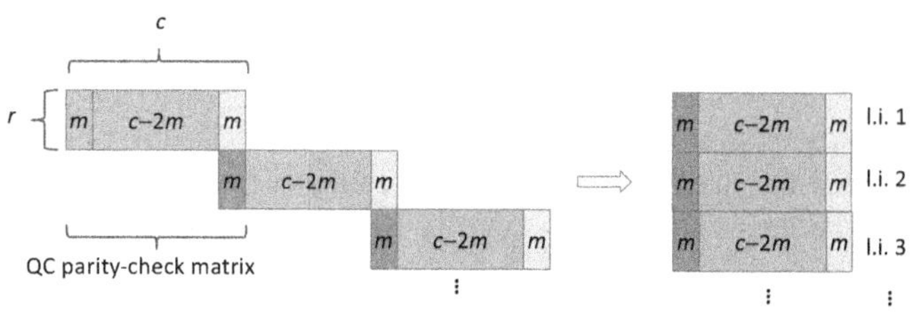

FIGURE 10.14

Illustrating the parity-check matrix of a spatially coupled low-density parity-check code design derived from the template quasi-cyclic low-density parity-check code. *l.i.*, Layer index.

code, we can introduce the layer index and thus reduce the memory requirements as illustrated in Fig. 10.14 right. In such a way we do not need to memorize the all-zeros submatrices. For full-rank parity-check matrix of the template QC-LDPC code, the code rate of the SC-LDPC code is given by:

$$R = 1 - \frac{rl}{lc - m(l-1)}. \tag{10.56}$$

Therefore for fixed l, by increasing the coupling length m we can reduce the code rate and thus improve the error correction capability of the code. For the field-programmable gate array implementation of decoders for SC-LDPC codes, interested readers are referred to our recent paper [62]. In a more general SC design, where the coupled code layers are not the same, which is known as *time-variant* SC-LDPC code, the memory size will be c times larger than that of the template QC-LDPC code.

By ensuring that the following condition is satisfied for the SC-LDPC code:

$$\boldsymbol{H}_{SC}\boldsymbol{H}_{SC}^T = \boldsymbol{0}, \tag{10.57}$$

the corresponding quantum LDPC code will belong to the class of CSS codes. The quantum-check matrix of the quantum SC-LDPC code will be:

$$\boldsymbol{A}_{SC} = \left[\begin{array}{c|c} \boldsymbol{H}_{SC} & \boldsymbol{0} \\ \boldsymbol{0} & \boldsymbol{H}_{SC} \end{array}\right], \tag{10.58}$$

and can be decoded using the SPA similar to that provided in the previous section.

SC-LDPC code design can simply be generalized to the convolutional LDPC code, such as the one introduced in [63]. Similarly to the classical counterparts [64], quantum SC-LDPC codes outperform corresponding quantum block LDPC codes as shown in Ref. [65].

10.6 SUMMARY

This chapter was devoted to quantum LDPC codes. In Section 10.1 we described classical LDPC codes, their design, and decoding algorithms. In Section 10.2 we described dual-containing quantum LDPC codes. Various design algorithms have been described. Section 10.3 was devoted to entanglement-assisted quantum LDPC codes. Various classes of finite geometry codes were described. In Section 10.4 we described the probabilistic sum-product algorithm based on the quantum-check matrix. The quantum spatially coupled LDPC codes were introduced briefly in Section 10.5. In the following section we provide a set of problems that will help readers to better understand the material from this chapter.

10.7 PROBLEMS

1. Consider an (8,4) product code composed of a (3,2) single-parity-check code:

c_0	c_1	c_2
c_3	c_4	c_5
c_6	c_7	

The parity-check equations of this code are given by:

$$c_2 = c_0 + c_1, \quad c_5 = c_3 + c_4, \quad c_6 = c_0 + c_3, \quad c_7 = c_1 + c_4.$$

Write down the corresponding parity-check matrix and provide its Tanner graph. What is the minimum distance of this code? Can this classical code be used as a dual-containing quantum code? Explain your answer.

2. Consider a sequence of m-independent binary digits $\boldsymbol{a} = (a_1, a_2, \ldots, a_m)$ in which $\Pr(a_k = 1) = p_k$.
   (a) Determine the probability that $\boldsymbol{a}$ contains an even number of 1s.
   (b) Determine the probability that $\boldsymbol{a}$ contains an odd number of 1s.
3. Let us use the notation introduced in Section 10.1. Let $q_{ij}(b)$ denote the extrinsic information (message) to be passed from variable node v_i to function node f_j regarding the probability that $c_i = b$, $b \in \{0,1\}$, as illustrated in Fig. 10.P3 on the left. This message is concerned with the probability that $c_i = b$ given extrinsic information from all check nodes, except node f_j, and given channel sample y_j. Let $r_{ji}(b)$ denote extrinsic information to be passed from node f_j to node v_i, as illustrated in Fig. 10.P3 on the right. This message is concerned about the probability that the jth parity-check equation is satisfied given $c_i = b$, and other bits have separable distribution given by $\{q_{ij'}\}_{j' \neq j}$.
   (a) Show that $r_{ji}(0)$ and $r_{ji}(1)$ can be calculated from $q_{i'j}(0)$ and $q_{i'j}(1)$ by using Problem 2 as follows:

$$r_{ji}(0) = \frac{1}{2} + \frac{1}{2} \prod_{i' \in V_j \backslash i} \left(1 - 2q_{i'j}(1)\right), \quad r_{ji}(1) = \frac{1}{2} - \frac{1}{2} \prod_{i' \in V_j \backslash i} \left(1 - 2q_{i'j}(1)\right).$$

(b) Show that q_{ji} (0) and q_{ji} (1) can be calculated as the functions of r_{ji} (0) and r_{ji} (1) as follows:

$$q_{ij}(0) = (1 - P_i) \prod_{j' \in C_i \backslash j} r_{j'i}(0), \quad q_{ij}(1) = P_i \prod_{j' \in C_i \backslash j} r_{j'i}(1).$$

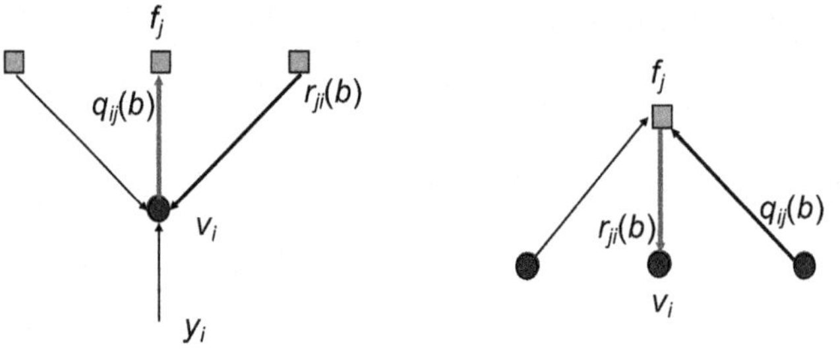

FIGURE 10.P3

Sum-product algorithm.

4. The probability-domain sum-product algorithm "suffers" because: (1) multiplications are involved (additions are less costly to implement), and (2) in the calculation of the product, many probabilities could become numerically unstable (especially for long codes with more than 50 iterations). The log-domain version of the sum-product algorithm is a preferable option. Instead of beliefs in the log domain we use LLRs, which for codeword bits c_i ($i = 2,\ldots,n$) are defined by:

$$L(c_i) = \log\left[\frac{\Pr(c_i = 0|y_i)}{\Pr(c_i = 1|y_i)}\right],$$

where y_i is a noisy sample corresponding to c_i. The LLRs corresponding to r_{ji} and q_{ij} are defined as follows:

$$L(r_{ji}) = \log\left[\frac{r_{ji}(0)}{r_{ji}(1)}\right], \quad L(q_{ij}) = \log\left[\frac{q_{ij}(0)}{q_{ij}(1)}\right].$$

By using $\tanh[\log(p_0/p_1)/2] = p_0 - p_1 = 1 - 2p_1$ show that:

(a) $L(r_{ji}) = 2\tanh^{-1}\left\{\prod_{i' \in V_j \backslash i} \tanh\left[\frac{1}{2}L(q_{i'j})\right]\right\},$

(b) $L(q_{ij}) = L(c_i) + \sum_{j' \in C_i \backslash j} L(r_{j'i})$.

5. By using the following function:

$$\phi(x) = -\log \tanh\left(\frac{x}{2}\right) = \log\left[\frac{e^x + 1}{e^x - 1}\right],$$

show that $L(r_{ij})$ from Problem 4(a) can be written as:

$$L(r_{ji}) = \left(\prod_{i'} \alpha_{i'j}\right) \phi\left[\sum_{i' \in R_j \backslash i} \phi(\beta_{i'j})\right],$$

where $\alpha_{ij} = \text{sign}\left[L(q_{ij})\right]$, $\beta_{ij} = \left|L(q_{ij})\right|$

6. Show that $L(a_1 \oplus a_2) = L_1 \boxplus L_2$, $L_i = L(a_i)$ can be written as:

$$L_1 \boxplus L_2 = \log\left(\frac{1 + e^{L_1 + L_2}}{e^{L_1} + e^{L_2}}\right).$$

7. Show that $L_1 \boxplus L_2$ from the previous problem can also be written as:

$$L_1 \boxplus L_2 = \text{sign}(L_1)\,\text{sign}(L_2)\min(|L_1|, |L_2|) + s(L_1, L_2),$$

where $s(L_1, L_2) = \log(1 + e^{-|x+y|}) - \log(1 + e^{-|x-y|})$

8. This problem is related to the LDPC decoder implementation using the min-sum-with-correction-term algorithm. Implement the LDPC decoder using the min-sum-with-correction-term algorithm as described in Section 10.1.2.

(a) By assuming that zero codeword was transmitted over an AWGN channel for PG(2,2^6)-based LDPC code, show BER dependence against signal-to-noise ratio in dB scale for binary phase-shift keying. Determine the coding gain at a BER of 10^{-6}.

(b) Implement now a corresponding encoder, generate sufficiently long pseudorandom binary sequence, encode it, transmit over the AWGN channel, and repeat the simulation from (a). Discuss the differences.

(c) Repeat the procedure for a BSC.

9. This problem is related to the LDPC decoder implementation using the probabilistic-domain algorithm described in Section 10.5. Implement this algorithm as described in Section 10.5. Repeat the steps (a)–(c) from the previous problem. Compare the results.

10. Let us now design an EA code based on Steiner triple systems from Section 10.3 having the length comparable to PG(2,2^6)-based EA code. Which one is performing better at a BER of 10^{-6}? Justify the answer. Discuss the decoder complexity for these two codes.

11. Let us again observe EA code based on Steiner triple systems and PG(2,2^6)-based EA code. Evaluate the performance of these two EA codes (of comparable lengths) for the following channel models:
(a) Depolarizing channel; and
(b) Amplitude-damping channel.

For channel descriptions please refer to Chapter 7 (and Section 7.5).

References

[1] W.E. Ryan, Concatenated convolutional codes and iterative decoding, in: J.G. Proakis (Ed.), Wiley Encyclopedia of Telecommunications, John Wiley & Sons, 2003.

[2] I.B. Djordjevic, M. Arabaci, L. Minkov, Next generation FEC for high-capacity communication in optical transport networks, IEEE/OSA J. Lightw. Technol. 27 (2009) 3518–3530 (Invited Paper).

[3] I. Djordjevic, W. Ryan, B. Vasic, Coding for Optical Channels, Springer, 2010.

[4] R.G. Gallager, Low Density Parity Check Codes, MIT Press, Cambridge, 1963.

[5] I.B. Djordjevic, S. Sankaranarayanan, S.K. Chilappagari, B. Vasic, Low-density parity-check codes for 40 Gb/s optical transmission systems, IEEE/LEOS J. Sel. Top. Quantum Electron. 12 (4) (July/Aug. 2006) 555–562.

[6] J. Chen, A. Dholakia, E. Eleftheriou, M. Fossorier, X.-Y. Hu, Reduced-complexity decoding of LDPC codes, IEEE Trans. Commun. 53 (2005) 1288–1299.

[7] I.B. Djordjevic, B. Vasic, Nonbinary LDPC codes for optical communication systems, IEEE Photon. Technol. Lett. 17 (10) (Oct. 2005) 2224–2226.

[8] M. Arabaci, I.B. Djordjevic, R. Saunders, R.M. Marcoccia, Non-binary quasi-cyclic LDPC based coded modulation for beyond 100-Gb/s transmission, IEEE Photon. Technol. Lett. 22 (6) (March 2010) 434–436.

[9] M.C. Davey, Error-correction Using Low-Density Parity-Check Codes (Ph.D. dissertation), Univ. Cambridge, Cambridge, U.K., 1999.

[10] D. Declercq, M. Fossorier, Decoding algorithms for nonbinary LDPC codes over GF(q), IEEE Trans. Commun. 55 (4) (2007) 633–643.

[11] M. Arabaci, I.B. Djordjevic, R. Saunders, R.M. Marcoccia, Polarization-multiplexed rate-adaptive non-binary-LDPC-coded multilevel modulation with coherent detection for optical transport networks, Optic Express 18 (3) (2010) 1820–1832.

[12] M. Arabaci, I.B. Djordjevic, An alternative FPGA implementation of decoders for quasi-cyclic LDPC codes, in: Proc. TELFOR 2008, November 2008, pp. 351–354.

[13] M. Arabaci, I.B. Djordjevic, R. Saunders, R. Marcoccia, Non-binary LDPC-coded modulation for high-speed optical metro networks with back propagation, in: Proc. SPIE Photonics West 2010, OPTO: Optical Communications: Systems and Subsystems,

Optical Metro Networks and Short-Haul Systems II, January 2010, pp. 23–28. San Francisco, California, USA, Paper no. 7621-17.
[14] M. Arabaci, Nonbinary-LDPC-Coded Modulation Schemes for High-Speed Optical Communication Networks (Ph.D. dissertation), University of Arizona, November 2010.
[15] I.B. Djordjevic, O. Milenkovic, B. Vasic, Generalized low-density parity-check codes for optical communication systems, IEEE/OSA J. Lightw. Technol. 23 (May 2005) 1939–1946.
[16] B. Vasic, I.B. Djordjevic, R. Kostuk, Low-density parity check codes and iterative decoding for long haul optical communication systems, IEEE/OSA J. Lightw. Technol. 21 (February 2003) 438–446.
[17] I.B. Djordjevic, et al., Projective plane iteratively decodable block codes for WDM high-speed long-haul transmission systems, IEEE/OSA J. Lightw. Technol. 22 (3) (March 2004) 695–702.
[18] O. Milenkovic, I.B. Djordjevic, B. Vasic, Block-circulant low-density parity-check codes for optical communication systems, IEEE/LEOS J. Sel. Top. Quantum Electron. 10 (March/April 2004) 294–299.
[19] B. Vasic, I.B. Djordjevic, Low-density parity check codes for long haul optical communications systems, IEEE Photon. Technol. Lett. 14 (8) (August 2002) 1208–1210.
[20] S. Chung, et al., On the design of low-density parity-check codes within 0.0045 dB of the Shannon Limit, IEEE Commun. Lett. 5 (February 2001) 58–60.
[21] S. Lin, D.J. Costello, Error Control Coding: Fundamentals and Applications, second ed., Prentice-Hall, Inc., USA, 2004.
[22] R.M. Tanner, A recursive approach to low complexity codes,, IEEE Tans. Information Theory IT-27 (September 1981) 533–547.
[23] I.B. Djordjevic, L. Xu, T. Wang, M. Cvijetic, Large girth low-density parity-check codes for long-haul high-speed optical communications, in: Proc. OFC/NFOEC, IEEE/OSA, San Diego, CA, 2008. Paper no. JWA53.
[24] M.P.C. Fossorier, Quasi-cyclic low-density parity-check codes from circulant permutation matrices, IEEE Trans. Inf. Theor. 50 (2004) 1788–1793.
[25] W.E. Ryan, An introduction to LDPC codes,, in: B. Vasic (Ed.), CRC Handbook for Coding and Signal Processing for Recording Systems, CRC Press, 2004.
[26] H. Xiao-Yu, E. Eleftheriou, D.-M. Arnold, A. Dholakia, Efficient implementations of the sum-product algorithm for decoding of LDPC codes, Proc. IEEE Globecom 2 (November 2001), 1036-1036E.
[27] D.J.C. MacKay, Good error correcting codes based on very sparse matrices, IEEE Trans. Inf. Theor. 45 (1999) 399–431.
[28] T. Richardson, A. Shokrollahi, R. Urbanke, Design of capacity approaching irregular low-density parity-check codes, IEEE Trans. Inf. Theor. 47 (2) (February 2001) 619–637.
[29] I.B. Djordjevic, B. Vasic, MacNeish-Mann theorem based iteratively decodable codes for optical communication systems, IEEE Commun. Lett. 8 (8) (August 2004) 538–540.
[30] Y. Kou, S. Lin, M.P.C. Fossorier, Low-density parity-check codes based on finite geometries: a rediscovery and new results, IEEE Trans. Inf. Theor. 47 (7) (November 2001) 2711–2736.
[31] I.B. Djordjevic, S. Sankaranarayanan, B. Vasic, Irregular low-density parity-check codes for long haul optical communications, IEEE Photon. Technol. Lett. 16 (1) (January 2004) 338–340.

[32] J. Boutros, O. Pothier, G. Zemor, Generalized low density (Tanner) codes, in: Proc. 1999 IEEE Int. Conf. Communication (ICC 1999), vol. 1, 1999, pp. 441–445.

[33] M. Lentmaier, K.S. Zigangirov, On generalized low-density parity check codes based on Hamming component codes, IEEE Commun. Lett. 3 (8) (August 1999) 248–250.

[34] T. Zhang, K.K. Parhi, High-performance, low-complexity decoding of generalized low-density parity-check codes, in: Proc. IEEE Global Telecommunications Conf. 2001 (IEEE Globecom), vol. 1, 2001, pp. 181–185.

[35] I.B. Djordjevic, L. Xu, T. Wang, M. Cvijetic, GLDPC codes with Reed-Muller component codes suitable for optical communications, IEEE Commun. Lett. 12 (September 2008) 684–686.

[36] A.R. Calderbank, P.W. Shor, Good quantum error-correcting codes exist, Phys. Rev. A 54 (1996) 1098–1105.

[37] A.M. Steane, Error correcting codes in quantum theory, Phys. Rev. Lett. 77 (1996) 793.

[38] D.J.C. MacKay, G. Mitchison, P.L. McFadden, Sparse-graph codes for quantum error correction, IEEE Trans. Inf. Theor. 50 (2004) 2315–2330.

[39] T. Camara, H. Ollivier, J.-P. Tillich, Constructions and Performance of Classes of Quantum LDPC Codes, 2005 quant-ph/0502086.

[40] M. Hagiwara, H. Imai, Quantum Quasi-Cyclic LDPC Codes, 2007 quant-ph/0701020.

[41] I. Anderson, Combinatorial Designs and Tournaments, Oxford University Press, Oxford, 1997.

[42] B. Vasic, O. Milenkovic, Combinatorial constructions of low-density parity-check codes for iterative decoding, IEEE Trans. Inf. Theor. 50 (June 2004) 1156–1176.

[43] R.C. Bose, On the construction of balanced incomplete block designs, Ann. Eugen. 9 (1939) 353–399.

[44] S. Sankaranarayanan, B. Vasic, Iterative decoding of linear block codes: a parity-check orthogonalization approach, IEEE Trans. Inf. Theor. 51 (September 2005) 3347–3353.

[45] S. Sankaranarayanan, I.B. Djordjevic, B. Vasic, Iteratively decodable codes on m-flats for WDM high-speed long-haul transmission, J. Lightwave Technol. 23 (November 2005) 3696–3701.

[46] I.B. Djordjevic, Quantum LDPC codes from balanced incomplete block designs, IEEE Commun. Lett. 12 (May 2008) 389–391.

[47] I.B. Djordjevic, Photonic quantum dual-containing LDPC encoders and decoders, IEEE Photon. Technol. Lett. 21 (13) (July 1, 2009) 842–844.

[48] I.B. Djordjevic, On the photonic implementation of universal quantum gates, Bell states preparation circuit, quantum relay and quantum LDPC encoders and decoders, IEEE Photon. J. 2 (1) (February 2010) 81–91.

[49] I.B. Djordjevic, Photonic implementation of quantum relay and encoders/decoders for sparse-graph quantum codes based on optical hybrid, IEEE Photon. Technol. Lett. 22 (19) (October 1, 2010) 1449–1451.

[50] I.B. Djordjevic, Cavity quantum electrodynamics based quantum low-density parity-check encoders and decoders, in: SPIE Photonics West 2011, Advances in Photonics of Quantum Computing, Memory, and Communication IV, The Moscone Center, San Francisco, California, USA, January 2011, pp. 22–27. Paper no. 7948-38.

[51] T. Brun, I. Devetak, M.-H. Hsieh, Correcting quantum errors with entanglement, Science 314 (2006) 436–439.

[52] M.-H. Hsieh, I. Devetak, T. Brun, General entanglement-assisted quantum error correcting codes, Phys. Rev. A 76 (December 19, 2007), 062313-1 - 062313-7.

[53] I.B. Djordjevic, Photonic entanglement-assisted quantum low-density parity-check encoders and decoders, Opt. Lett. 35 (9) (May 1, 2010) 1464–1466.

[54] M.-H. Hsieh, Entanglement-Assisted Coding Theory, PhD Dissertation, University of Southern California, August 2008.

[55] I. Devetak, T.A. Brun, M.-H. Hsieh, Entanglement-assisted quantum error-correcting codes, in: V. Sidoravičius (Ed.), New Trends in Mathematical Physics, Selected Contributions of the XVth International Congress on Mathematical Physics, Springer, 2009, pp. 161–172.

[56] M.-H. Hsieh, W.-T. Yen, L.-Y. Hsu, High performance entanglement-assisted quantum LDPC codes need little entanglement 57 (3) (March 2011) 1761–1769.

[57] M.M. Wilde, Quantum Coding with Entanglement, PhD Dissertation, University of Southern California, August 2008.

[58] D. Raghavarao, Constructions and Combinatorial Problems in Design of Experiments, Dover Publications, Inc., New York, 1988.

[59] D. Poulin, Y. Chung, On the iterative decoding of sparce quantum codes, Quant. Inf. Comput. 8 (10) (2008) 987–1000.

[60] Z. Zhang, L. Dolecek, B. Nikolic, V. Anantharam, M. Wainwright, Design of LDPC decoders for improved low error rate performance: quantization and algorithm choices, IEEE Trans. Commun. 57 (11) (November 2009) 3258–3268.

[61] I.B. Djordjevic, L. Xu, T. Wang, M. Cvijetic, Large girth low-density parity-check codes for long-haul high-speed optical communications, in: Optical Fiber Communication Conference/National Fiber Optic Engineers Conference, OSA Technical Digest (CD), Optical Society of America, 2008 paper JWA53.

[62] X. Sun, I.B. Djordjevic, FPGA implementation of rate-adaptive spatially-coupled LDPC codes suitable for optical communications, Optics Express 27 (3) (2019) 3422–3428.

[63] J. Felstrom, K.S. Zigangirov, Time-varying periodic convolutional codes with low-density parity-check matrix, IEEE Trans. Inf. Theor. 45 (6) (1999) 2181–2191.

[64] D.J. Costello, et al., A Comparison between LDPC Block and Convolutional Codes, IEEE Information Theory and Applications Workshop, 2006.

[65] I. Andriyanova, D. Mauricey, J.-P. Tillich, Spatially coupled quantum LDPC codes, in: Proc. 2012 Information Theory Workshop, 2012, pp. 327–331.

[66] S.S. Shrikhande, Generalized Hadamard matrices and orthogonal arrays of strength two, Can. J. Math. 16 (1964) 736–740.

CHAPTER

11 Fault-Tolerant Quantum Error Correction and Fault-Tolerant Quantum Computing

CHAPTER OUTLINE

This chapter considers one of the most important applications of quantum error correction, namely the protection of quantum information as it dynamically

Quantum Information Processing, Quantum Computing, and Quantum Error Correction
https://doi.org/10.1016/B978-0-12-821982-9.00017-4

undergoes computation through so-called fault-tolerant quantum computing [1–7]. The chapter starts by introducing fault tolerance basics and traversal operations (Section 11.1). It continues with fault-tolerant quantum computation concepts and procedures (Section 11.2). In Section 11.2, the universal set of fault-tolerant quantum gates is introduced, followed by fault-tolerant measurement, fault-tolerant state preparation, and fault-tolerant encoded state preparation using Steane's code as an illustrative example. Section 11.3 provides a rigorous description of fault-tolerant quantum error correction. The section starts with a short review of some basic concepts from stabilizer codes. In subsection 11.3.1 fault-tolerant syndrome extraction is described, followed by a description of fault-tolerant encoded operations in subsection 11.3.2. Subsection 11.3.3 is concerned with the application of the quantum gates on a quantum register by means of a measurement protocol. This method, when applied transversally, is used in subsection 11.3.4 to enable fault-tolerant error correction based on an arbitrary quantum stabilizer code. The fault-tolerant stabilizer codes are described in subsection 11.3.4. In subsection 11.3.5, the [5,1,3] fault-tolerant stabilizer code is described as an illustrative example. Section 11.4 covers fault-tolerant computing, in particular the fault-tolerant implementation of the Tofolli gate. Finally, in Section 11.5, the accuracy threshold theorem is formulated and proved. After a summary in Section 11.6, in the final section (Section 11.7) a set of problems is provided to enable readers to gain a deeper understanding of fault-tolerance theory.

11.1 FAULT-TOLERANCE BASICS

One of the most powerful applications of quantum error correction is the protection of quantum information as it dynamically undergoes quantum computation [1–5]. Imperfect quantum gates affect quantum computation by introducing errors in computed data. Moreover, the imperfect control gates introduce errors in processed sequence since wrong operations are applied. The quantum error correction coding (QECC) scheme now needs to deal not only with errors introduced by the quantum channel by also with errors introduced by imperfect quantum gates during the encoding/decoding process. Because of this fact, the reliability of data processed by quantum computer is not a priori guaranteed by QECC. The reason is threefold: (1) the gates used for encoders and decoders are composed of imperfect gates, including controlled imperfect gates; (2) syndrome extraction applies unitary operators to entangle ancilla qubits with code block; and (3) the error recovery action requires the use of a controlled operation to correct for the errors. Nevertheless, it can be shown that arbitrary good quantum error protection can be achieved even with imperfect gates, providing that the error probability per gate is below a certain *threshold*, this claim is known as accuracy threshold theorem and will be discussed later in the chapter.

From this discussion it is clear that QECC does not a priori improve reliability of a quantum computer. So, the purpose of *fault-tolerant design* is to ensure the reliability of the quantum computer given the threshold by properly implementing the fault-tolerant quantum gates. Similarly as in previous chapters we observe the following error model: (1) qubit errors occur independently, (2) *X*-, *Y*-, and *Z*-qubit errors occur with the same probability, (3) the error probability per qubit is the same for all qubits, and (4) the errors introduced by an imperfect quantum gate affect only the qubits acted on by that gate. Condition (4) is added to ensure that errors introduced by imperfect gates do not propagate and cause catastrophic errors. Based on this error mode we can define fault-tolerant operation as follows.

Definition 1. We say that a quantum operation is fault tolerant if the occurrence of single gate/storage error during the operation does not produce more than one error on each encoded qubit block.

In other words, an operation is said to be fault tolerant if, up to the first order in single-error probability p (i.e., $O(p)$), the operation generates no more than one error per block. The key idea of fault-tolerant computing is to construct fault-tolerant operations to perform logic on encoded states. Notice that some may argue that if we use strong enough QECC we might allow for more than one error per code block. However, the QECC scheme is typically designed to deal with random errors introduced by the quantum channel or storage device. By allowing for too many errors due to imperfect gates may lead to exceeding the error correction capability of QECC when both random and imperfect gate errors are present. By restricting the number of faulty gate errors to one we can simplify the fault-tolerant design.

Example. Let us observe the SWAP gate introduced in Chapter 3, which interchanges the states of two qubits: $U_{\mathrm{SWAP}}^{(12)}|\psi_1\rangle|\psi_2\rangle = |\psi_2\rangle|\psi_1\rangle$ If the SWAP gate is imperfect it can introduce errors on both qubits, and based on Definition 1, this gate is faulty. However, if we introduce an additional auxiliary state we can perform the SWAP operation fault tolerantly as follows: $U_{\mathrm{SWAP}}^{(23)}U_{\mathrm{SWAP}}^{(12)}U_{\mathrm{SWAP}}^{(13)}|\psi_1\rangle|\psi_2\rangle|\psi_3\rangle = |\psi_2\rangle|\psi_1\rangle|\psi_3\rangle$ For example, let us assume that a storage error occurred on qubit 1, leading to $|\psi_1\rangle \rightarrow |\psi_1'\rangle$, before the SWAP operation takes place. The resulting state will be $|\psi_2\rangle|\psi_1'\rangle|\psi_3\rangle$, meaning that a single error occurring during SWAP operation results in only one error per block, meaning that this modified gate operates fault tolerantly. Another related definition, commonly used in fault-tolerant quantum computing, namely *transversal operation*, can be shown to be fault tolerant.

Definition 2. An operation that satisfies one of the following two conditions is a transversal operation: (1) it only employs one-qubit gates to the qubits in a code block, and (2) the *i*th qubit in one code block interacts only with the *i*th qubit in a different code block or block of ancilla qubits.

Condition (1) is obviously consistent with the fault-tolerance definition. Let us now observe two-qubit operations, in particular the operation of the gate involving *i*th qubits of two different blocks. Based on the foregoing error model, only these two qubits can be affected by this faulty gate. Since these two interacting qubits belong to two different blocks, the transversal operation is fault tolerant. The transversal operation definition is quite useful as it is much easier to verify than the fault-tolerant definition. In the following section, to familiarize readers with fault-tolerant quantum computing concepts we will develop fault-tolerant gates assuming that the QECC scheme is based on Steane code [5].

11.2 FAULT-TOLERANT QUANTUM COMPUTATION CONCEPTS

In this section we introduce basic fault-tolerant computation concepts. The purpose of this section is to gradually introduce readers to fault-tolerant concepts before moving on to rigorous exposition in Section 11.3.

11.2.1 Fault-Tolerant Pauli Gates

The key idea of fault-tolerant computing is to construct fault-tolerant operations to perform logic on encoded states. The encoded Pauli operators for Pauli X- and Z-operators for Steane [7,1] code are given as follows:

$$\overline{Z} = Z_1 Z_2 Z_3 Z_4 Z_5 Z_6 Z_7 \quad \overline{X} = X_1 X_2 X_3 X_4 X_5 X_6 X_7. \tag{11.1}$$

The encoded Pauli gate Y can easily be implemented as:

$$\overline{Y} = \overline{X}\,\overline{Z}. \tag{11.2}$$

Let us now check if the encoded Pauli gates are transversal gates. Based on the quantum channel model from the previous section, it is clear that single-qubit error does not propagate. Because every encoded gate is implemented in bitwise fashion, the encoded Pauli gates are transversal and consequently fault tolerant. The implementation of fault-tolerant Pauli gates is shown in Fig. 11.1.

11.2.2 Fault-Tolerant Hadamard Gate

An encoded Hadamard gate H should interchange X and Z under conjugation the same way an unencoded Hadamard gate interchanges Z and X:

$$\overline{H} = H_1 H_2 H_3 H_4 H_5 H_6 H_7. \tag{11.3}$$

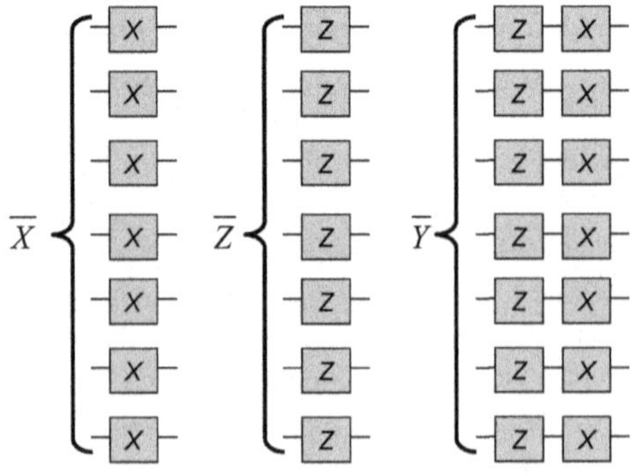

FIGURE 11.1

Transversal Pauli gates on a qubit encoded using Steane [7,1] code.

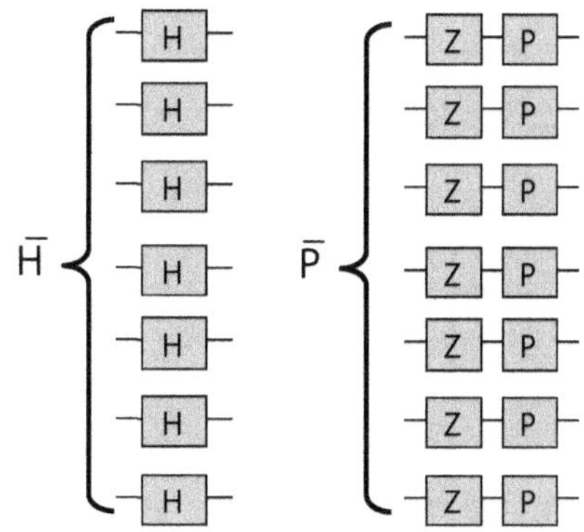

FIGURE 11.2

Transversal Hadamard and phase gates on a qubit encoded using Steane [7,1] code.

Fault-tolerant Hadamard gate implementation is shown in Fig. 11.2 (left). A single error, occurring on one qubit, does not propagate any further, indicating that an encoded Hadamard gate is transversal and consequently fault tolerant.

11.2.3 Fault-Tolerant Phase Gate (*P*)

Under conjugation the phase gate P maps Z to Z and X to Y as follows:

$$PZP^{\dagger} = Z \quad PXP^{\dagger} = Y. \tag{11.4}$$

The fault-tolerant phase gate $\overline{P}$ should perform the same action on encoded Pauli gates. The bitwise operation of $\overline{P} = P_1P_2P_3P_4P_5P_6P_7$ on encoded Pauli Z and X gates gives:

$$\overline{P}\,\overline{Z}\,\overline{P}^{\dagger} = \overline{Z} \quad \overline{P}\,\overline{X}\,\overline{P}^{\dagger} = -\overline{Y}, \tag{11.5}$$

indicating that the sign in conjugation operation of an encoded Pauli-X gate is incorrect, which can be fixed by inserting a Z-operator in front of the single P-operator, so that the fault-tolerant phase gates can be represented as shown in Fig. 11.2 (right). A single error, occurring on one qubit, does not propagate any further, indicating that the operation is transversal and fault tolerant.

11.2.4 Fault-Tolerant CNOT Gate

A fault-tolerant CNOT gate is shown in Fig. 11.3. Let us assume that the ith qubit in the upper block interacts with the ith qubit of the lower block. From Fig. 11.3 is clear that an error in the ith qubit of the upper block will only affect the ith qubit of the lower block, so that condition (2) of Definition 2 is satisfied and the gate from Fig. 11.3 is transversal and consequently (based on Definition 1) fault tolerant. The fault-tolerant quantum gates discussed so far are sufficient to implement an arbitrary encoder and decoder fault tolerantly.

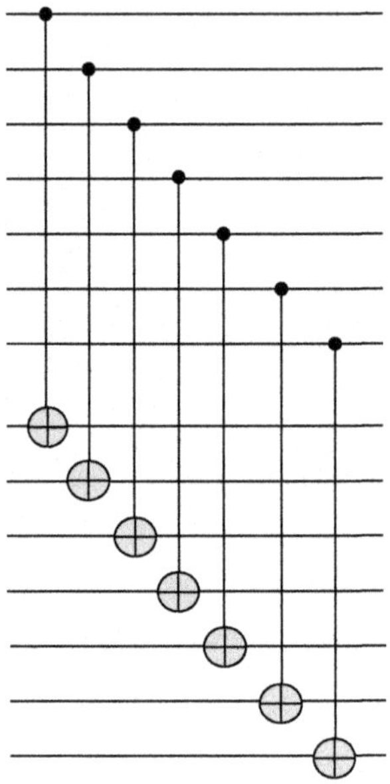

FIGURE 11.3

Fault-tolerant CNOT gate implementation.

11.2.5 Fault-Tolerant $\pi/8$ (T) Gate

For the complete set of universal gates, the implementation of a nontransversal gate such as the $\pi/8$ (T) gate is needed. To implement the T gate fault tolerantly we need to apply the following three-step procedure [3,5]:

1. Prepare an ancilla, in base state, and apply a Hadamard followed by a nonfault-tolerant T gate:

$$|\phi\rangle = TH|0\rangle = T\frac{|0\rangle + |1\rangle}{\sqrt{2}} = \frac{|0\rangle + e^{j\pi/4}|1\rangle}{\sqrt{2}}. \tag{11.6}$$

2. Apply a CNOT gate on the prepared state $|\phi\rangle$ as control qubit and on the state we want to transform $|\psi\rangle$ as target qubit:

$$\begin{aligned} U_{\text{CNOT}}^{(12)}|\phi\rangle|\psi\rangle &= U_{\text{CNOT}}^{(12)}\frac{|0\rangle + e^{j\pi/4}\left|1\right\rangle}{\sqrt{2}}\otimes(a|0\rangle + b|1\rangle) \\ &= \frac{1}{\sqrt{2}}U_{\text{CNOT}}^{(12)}\left[|0\rangle(a|0\rangle + b|1\rangle) + e^{j\pi/4}|1\rangle(a|0\rangle + b|1\rangle)\right] \\ &= \frac{1}{\sqrt{2}}(a|0\rangle|0\rangle + b|0\rangle|1\rangle) + e^{j\pi/4}(a|1\rangle|1\rangle + b|1\rangle|0\rangle) \\ &= \frac{1}{\sqrt{2}}\left[\left(a|0\rangle + be^{j\pi/4}|1\rangle\right)|0\rangle + \left(b|0\rangle + ae^{j\pi/4}|1\rangle\right)|1\rangle\right], \end{aligned} \tag{11.7}$$

where we used superscript (12) to denote the action of the CNOT gate from qubit 1 (control qubit) to qubit 2 (target qubit).

3. In the third step we need to measure the original qubit $|\psi\rangle$: if the result of a measurement is zero, the final state is $a|0\rangle + be^{j\pi/4}|1\rangle$; otherwise we have to apply a PX gate on the original ancilla qubit as follows:

$$\begin{aligned} PX\begin{bmatrix} b \\ ae^{j\pi/4} \end{bmatrix} &= \begin{bmatrix} 1 & 0 \\ 0 & j \end{bmatrix}\begin{bmatrix} 0 & 1 \\ 1 & 0 \end{bmatrix}\begin{bmatrix} b \\ ae^{j\pi/4} \end{bmatrix} = \begin{bmatrix} 0 & 1 \\ j & 0 \end{bmatrix}\begin{bmatrix} b \\ ae^{j\pi/4} \end{bmatrix} \\ &= \begin{bmatrix} ae^{j\pi/4} \\ be^{j\pi/2} \end{bmatrix} = e^{j\pi/4}\begin{bmatrix} a \\ be^{j\pi/4} \end{bmatrix} = e^{j\pi/4}T|\psi\rangle. \end{aligned} \tag{11.8}$$

Because $THZHT^{+} = TXT^{+} = e^{-j\pi/4}PX$, the state $|\phi\rangle$ can also be prepared from $|0\rangle$ fault tolerantly by measuring PX: if the result is $+1$, preparation was successful; otherwise, the procedure needs to be repeated. The fault-tolerant T-gate implementation is shown in Fig. 11.4. In this implementation we assumed that the measurement procedure is perfect. The fault-tolerant measurement procedure is discussed in the following section.

11.2.6 Fault-Tolerant Measurement

A quantum circuit for performing the measurement on a single-qubit operator U is shown in Fig. 11.5A. Its operation was already described in Chapter 3. This circuit is clearly faulty. The first idea of performing the fault-tolerant measurement is to put U in a transversal form U', as illustrated in Fig. 11.5B, and then perform

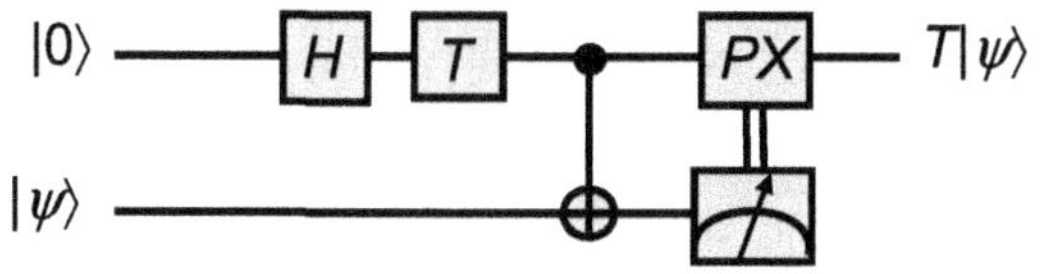

FIGURE 11.4

Fault-tolerant *T* gate implementation.

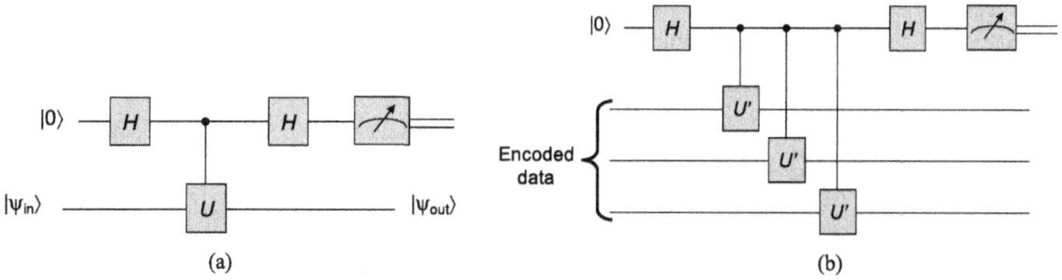

FIGURE 11.5

Faulty measurement circuits: (a) original circuit and (b) modified circuit obtained by putting *U* in transversal form.

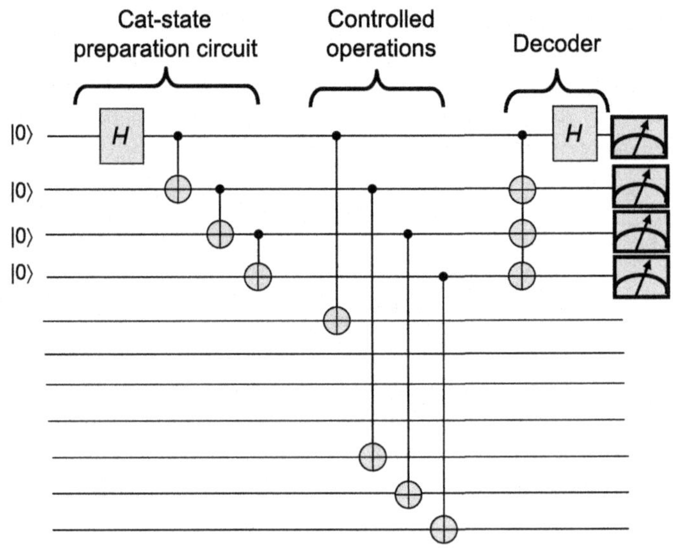

FIGURE 11.6

Fail-tolerant measurement circuit.

measurement in bitwise fashion with the same control ancilla. Unfortunately, such a measurement is not fault tolerant because an error in the ancilla qubit would affect several data qubits. We can convert the circuit shown in Fig. 11.5B into a fault-tolerant circuit by using three ancilla qubits as shown in Fig. 11.6. This new circuit has several stages. The first stage is the cat-state $(|000\rangle + |111\rangle)/\sqrt{2}$ preparation circuit. The encoder in this stage is very similar to the three-qubit flip code encoder described in Chapter 8. The second stage is the verification stage, which has certain similarities with the three-qubit flip code error correction circuit. Namely, this stage is used to verify if the state after the first stage is indeed the cat state. The third stage is the controlled-U stage, similarly as in Fig. 11.5B. The fourth stage is the decoder stage, which is used to return ancilla states back to the original state. As expected, this circuit has similarities with the three-qubit code decoder. The final measurement stage can be faulty. To reduce its importance on qubit error probability, we can perform the whole measurement procedure several times and apply the majority rule for the final measurement result.

11.2.7 Fault-Tolerant State Preparation

The fault-tolerant state preparation procedure can be described as follows. We prepare the desired state and apply the foregoing verification procedure measurement section. If the verification step is successful, that state is used for further quantum computation; otherwise we start the verification procedure again with fresh ancilla qubits.

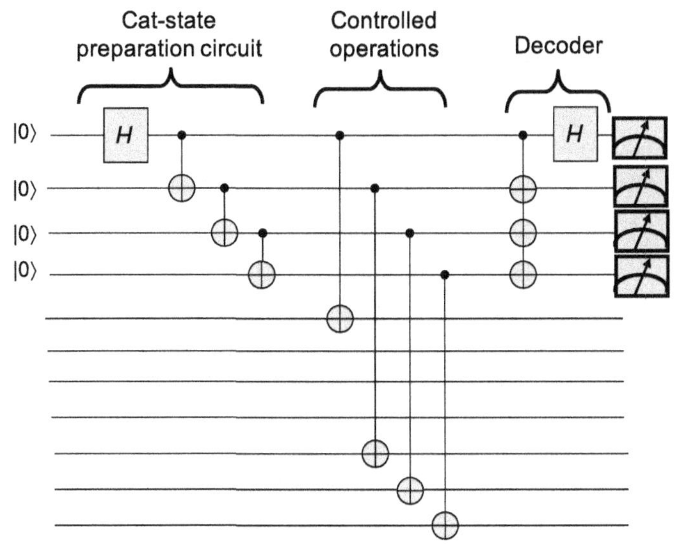

FIGURE 11.7

Fault-tolerant measurement for the generator $X_1X_5X_6X_7$.

11.2.8 Fault-Tolerant Measurement of Stabilizer Generators

A fault-tolerant measurement procedure has been introduced for an observable corresponding to a single qubit. A similar procedure can be applied to an arbitrary stabilizer generator. For example, the fault-tolerant measurement circuit of the first stabilizer generator of Steane code $g_1 = X_1X_5X_6X_7$, from Section 8.2, is shown in Fig. 11.7. After this elementary introduction of fault-tolerant concepts, gates, and procedures we turn our attention to a more rigorous description.

11.3 FAULT-TOLERANT QUANTUM ERROR CORRECTION

This section is devoted to fault-tolerant quantum error correction. Before we proceed with this topic we are going to review some basic concepts from stabilizer codes introduced in Chapters 7 and 8. Let S be the largest Abelian subgroup of G_N (the N-qubit Pauli multiplicative error group) that fixes all elements $|c\rangle$ from quantum code C_Q, called the stabilizer group. The $[N,K]$ stabilizer code C_Q is defined as the K-dimensional subspace of the N-qubit Hilbert space H_2^N as follows:

$$C_Q = \bigcap_{s \in S} \{ |c\rangle \in H_2^N \, | s|c\rangle = |c\rangle \}. \tag{11.9}$$

The very important concept of quantum stabilizer codes is the concept of syndrome of an error E from G_N. Let C_Q be a quantum stabilizer code with generators

$g_1, g_2, \ldots, g_{N-K}$, then the error syndrome of $E \in G_N$ can be defined as the bit string $S(E) = [\lambda_1 \lambda_2 \ldots \lambda_{N-K}]^T$ with component bits being determined by:

$$\lambda_i = \begin{cases} 0, & [E, g_i] = 0 \\ 1, & \{E, g_i\} = 0 \end{cases} \quad (i = 1, \ldots, N-K). \tag{11.10}$$

It has been shown in Chapter 8 that the corrupted codeword $E|c\rangle$ is a simultaneous eigenket of generators g_i $(i = 1,\ldots,N - K)$ so that we can write:

$$g_i(E|c\rangle) = (-1)^{\lambda_i} E(g_i|c\rangle) = (-1)^{\lambda_i} E|c\rangle. \tag{11.11}$$

By inspection of Eqs. (11.10) and (11.11) it can be concluded that eigenvalues $(-1)^{\lambda_i}$ (and consequently the syndrome components λ_i) can be determined by measuring g_i when the codeword $|c\rangle$ is corrupted by an error E. By representing the error E and generator g_i as products of corresponding X- and Z-containing operators as was done in Chapter 8, namely by $E = X(\boldsymbol{a}_E)Z(\boldsymbol{b}_E),\ g_i = X(\boldsymbol{a}_i)Z(\boldsymbol{b}_i)$, when nonzero positions in $\boldsymbol{a}$ ($\boldsymbol{b}$) give the locations of X- (Z-) operators, and based on Eq. (11.11), we can write:

$$g_i E = (-1)^{\boldsymbol{a}_E \cdot \boldsymbol{b}_i + \boldsymbol{b}_E \cdot \boldsymbol{a}_i} E g_i, \tag{11.12}$$

where $\boldsymbol{a}_E \cdot \boldsymbol{b}_i + \boldsymbol{b}_E \cdot \boldsymbol{a}_i$ denotes the sympletic product introduced in Section 8.3 (the bitwise addition is mod 2). From Eqs. (11.11) and (11.12) it is obvious that the ith component of the syndrome can be obtained by:

$$\lambda_i = \boldsymbol{a}_E \cdot \boldsymbol{b}_i + \boldsymbol{b}_E \cdot \boldsymbol{a}_i \bmod 2; \quad i = 1, \ldots, N-K. \tag{11.13}$$

By inserting identity operators, represented in terms of a unitary operator U as follows $I = U^\dagger U$, into Eq. (11.11), the following is obtained:

$$\begin{aligned} g_i \overset{I}{U^\dagger U} E \overset{I}{U^\dagger U}|c\rangle = (-1)^{\lambda_i} E \overset{I}{U^\dagger U}|c\rangle \overset{\cdot U \text{ (from the left)}}{\Leftrightarrow} (U g_i U^\dagger)(U E U^\dagger) U|c\rangle \\ = (-1)^{\lambda_i} (U E U^\dagger) U|c\rangle. \end{aligned} \tag{11.14}$$

By introducing the following notation: $\overline{g}_i = U g_i U^\dagger, \overline{E} = U E U^\dagger$, and $|\overline{c}\rangle = U|c\rangle$, the right side of Eq. (11.14) can be rewritten as:

$$\overline{g}_i \overline{E}|\overline{c}\rangle = (-1)^{\lambda_i} \overline{E}\left|\overline{c}\right\rangle. \tag{11.15}$$

Therefore the corrupted image of the codeword $\overline{E}|\overline{c}\rangle$ is a simultaneous eigenket of the images of generators $\overline{g}_i$. Similarly to Eq. (11.14) we can show that:

$$g_i|c\rangle = |c\rangle \Leftrightarrow g_i \overset{I}{U^\dagger U}|c\rangle = |c\rangle \overset{\cdot U \text{ (from the left)}}{\Leftrightarrow} U g_i \overset{I}{U^\dagger U}|c\rangle = U|c\rangle \Leftrightarrow \overline{g}_i|\overline{c}\rangle = |\overline{c}\rangle, \tag{11.16}$$

which will be used in the next section.

11.3.1 Fault-Tolerant Syndrome Extraction

To simplify further exposition of fault-tolerant quantum error correction, for the moment we will assume that generators $g_i = X(\boldsymbol{a}_i)Z(\boldsymbol{b}_i)$ do not contain the *Y*-operators; in other words, $a_{ij}b_{ij} = 0\,\forall j$. In syndrome extraction, the ancilla qubits interact with a codeword in such a way to encode the syndrome *S*(*E*) information into ancilla qubits and then perform appropriate measurements on ancilla qubits to reveal the syndrome vector *S*(*E*). We will show that the syndrome information *S*(*E*) can be incorporated by applying the transversal Hadamard $\overline{H}$ and transversal CNOT gate $\overline{U}_{\text{CNOT}}$ on corrupted codeword $E|c\rangle$ as follows:

$$\overline{H}_i\overline{U}_{\text{CNOT}}(\boldsymbol{a}_i+\boldsymbol{b}_i)\overline{H}_iE|c\rangle \otimes |\boldsymbol{A}\rangle = E|c\rangle \otimes \sum_{\boldsymbol{A}_{\text{even}}} |\boldsymbol{A}_{\text{even}} + \boldsymbol{b}_E\cdot\boldsymbol{a}_i + \boldsymbol{a}_E\cdot\boldsymbol{b}_i\rangle, \tag{11.17}$$

where $|\boldsymbol{A}\rangle$ is the *Shore state* defined by:

$$|\boldsymbol{A}\rangle = \frac{1}{2^{(\text{wt}(g_i)-1)/2}}\sum_{\boldsymbol{A}_{\text{even}}}|\boldsymbol{A}_{\text{even}}\rangle, \tag{11.18}$$

with summation being performed over all even-parity bit strings of length wt(g_i), denoted as $\boldsymbol{A}_{\text{even}}$. The Shore state is obtained by first applying the cat-state preparation circuit on $|\boldsymbol{0}\rangle_{\text{wt}(g_i)} = \underbrace{|0\rangle\otimes\ldots\otimes|0\rangle}_{\text{wt}(g_i)\text{times}}$ ancilla qubits, as explained in the corresponding text of Fig. 11.7, followed by the set of wt(g_i) Hadamard gates, as shown in Fig. 11.8. Therefore we can represent the Shore state in terms of cat state $|\psi_{\text{cat}}\rangle$ as follows:

$$|\boldsymbol{A}\rangle = \prod_{j=1}^{\text{wt}(g_i)} H_j|\psi_{\text{cat}}\rangle,\ |\psi_{\text{cat}}\rangle = \frac{1}{\sqrt{2}}[|0\ldots0\rangle + |1\ldots1\rangle]. \tag{11.19}$$

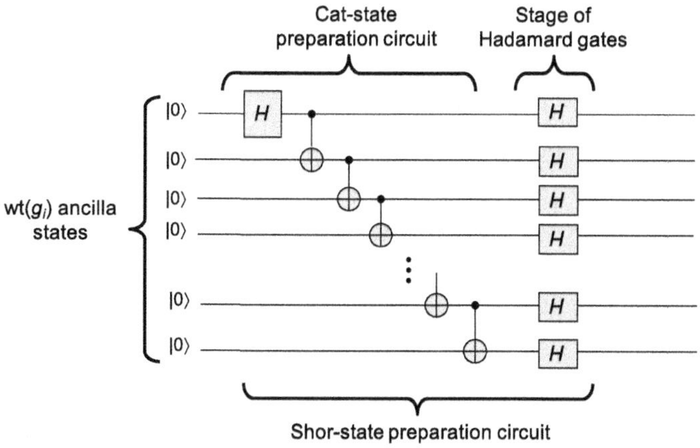

FIGURE 11.8

Shore-state preparation circuit.

It is evident from Fig. 11.8 that the stage of Hadamard gates is transversal as the ith error in the observed block introduces an error only on the ith qubit of the corresponding interacting block. On the other hand, the error introduced in the arbitrary inner CNOT gate of the cat-state preparation circuit propagates and qubits below are affected, indicating that the cat-state preparation circuit is not fault tolerant. We will return to this problem and Shor state verification later in this section, once we introduce the transversal Hadamard gate $\overline{H}$ and transversal CNOT gate $\overline{U}_{\text{CNOT}}$. The transversal Hadamard gate $\overline{H}$ and transversal CNOT gate $\overline{U}_{\text{CNOT}}$, related to generator:

$$g_i = X(\boldsymbol{a}_i)Z(\boldsymbol{b}_i);\;\; \boldsymbol{a}_i = [a_{i1} \ldots a_{iN}]^{\mathrm{T}}, \boldsymbol{a}_i = [b_{i1} \ldots b_{iN}]^{\mathrm{T}} (a_{ij} b_{ij} = 0 \forall j),$$

are defined, respectively, as:

$$\overline{H}_i \doteq \prod_{j=1}^{N} (H_j)^{a_{ij}}, \overline{U}_{\text{CNOT}}(\boldsymbol{a}_i + \boldsymbol{b}_i) \doteq \prod_{j=1}^{N} (U_{\text{CNOT}})^{a_{ij} + b_{ij}}. \tag{11.20}$$

In Eq. (11.20), the transversal CNOT gate $\overline{U}_{\text{CNOT}}$ is controlled by $\text{wt}(g_i)$ qubits for which $a_{ij} + b_{ij} = 1$. From Eq. (11.17) it is evident that encoded block and ancilla block are in a nonentangled state, and by performing measurement on the ancilla block we are not going to affect the encoded block (to collapse into one of eigenkets). By conducting a measurement on the ancilla qubits, the modified Shor state will collapse into one of eigenkets, say $|\boldsymbol{A}'_{\text{even}} + \boldsymbol{b}_E \cdot \boldsymbol{a}_i + \boldsymbol{a}_E \cdot \boldsymbol{b}_i\rangle$. Since $\text{wt}(\boldsymbol{A}'_{\text{even}}) \bmod 2 = 0$, the result of the measurement will be:

$$(-1)^{\boldsymbol{b}_E \cdot \boldsymbol{a}_i + \boldsymbol{a}_E \cdot \boldsymbol{b}_i} = (-1)^{\lambda_i}, \tag{11.21}$$

and the ith syndrome component $\lambda_i = \boldsymbol{b}_E \cdot \boldsymbol{a}_i + \boldsymbol{a}_E \cdot \boldsymbol{b}_i \bmod 2$ will be revealed. By repeating a similar procedure for all components of the syndrome we can identify the most probable error E based on syndrome $S(E) = [\lambda_1 \ldots \lambda_{N-K}]^T$ and perform the corresponding error recovery action $\boldsymbol{R} = R_1 \ldots R_N$.

As a reminder, the action of the Hadamard gate on the jth qubit is to perform the following transformation: $H_j X_j H_j = Z_j$. The application of the transversal H gate on generator g_i yields to:

$$\begin{aligned}\overline{g}_i &= \overline{H}_i g_i \overline{H}_i^{\dagger} \overset{\overline{H}_i^{\dagger} = \overline{H}_i}{=} \overline{H}_i X(\boldsymbol{a}_i) Z(\boldsymbol{b}_i) \overline{H}_i = \overline{H}_i X(\boldsymbol{a}_i) \overline{H}_i Z(\boldsymbol{b}_i) \overset{H_j X_j H_j = Z_j}{=} Z(\boldsymbol{a}_i) Z(\boldsymbol{b}_i) \\ &= Z(\boldsymbol{a}_i + \boldsymbol{b}_i).\end{aligned} \tag{11.22}$$

In similar fashion we can show that the action of transversal H gate on error E gives:

$$\begin{aligned}
\overline{E} &= \overline{H}_i E \overline{H}_i^{\dagger} \overset{\overline{H}_i^{\dagger}=\overline{H}_i}{=} \overline{H}_i X(\boldsymbol{a}_E) Z(\boldsymbol{b}_E) \overline{H}_i \overset{\overline{\boldsymbol{a}}_i+\boldsymbol{a}_i=(1\cdots1)}{=} \overline{H}_i X(\boldsymbol{a}_E\cdot(\overline{\boldsymbol{a}}_i+\boldsymbol{a}_i)) Z(\boldsymbol{b}_E\cdot(\overline{\boldsymbol{a}}_i+\boldsymbol{a}_i)) \overline{H}_i \\
&= \overline{H}_i \underbrace{X(\boldsymbol{a}_E\cdot\overline{\boldsymbol{a}}_i+\boldsymbol{a}_E\cdot\boldsymbol{a}_i)}_{X(\boldsymbol{a}_E\cdot\overline{\boldsymbol{a}}_i)X(\boldsymbol{a}_E\cdot\boldsymbol{a}_i)} \underbrace{Z(\boldsymbol{b}_E\cdot\overline{\boldsymbol{a}}_i+\boldsymbol{b}_E\cdot\boldsymbol{a}_i)}_{Z(\boldsymbol{b}_E\cdot\overline{\boldsymbol{a}}_i)Z(\boldsymbol{b}_E\cdot\boldsymbol{a}_i)} \overline{H}_i \\
&= \overline{H}_i X(\boldsymbol{a}_E\cdot\overline{\boldsymbol{a}}_i)\overline{H}_i \underbrace{\overline{H}_i X(\boldsymbol{a}_E\cdot\boldsymbol{a}_i)\overline{H}_i}_{Z(\boldsymbol{a}_E\cdot\boldsymbol{a}_i)} \overline{H}_i Z(\boldsymbol{b}_E\cdot\overline{\boldsymbol{a}}_i)\overline{H}_i \underbrace{\overline{H}_i Z(\boldsymbol{b}_E\cdot\boldsymbol{a}_i)\overline{H}_i}_{X(\boldsymbol{b}_E\cdot\boldsymbol{a}_i)} \\
&= X(\boldsymbol{a}_E\cdot\overline{\boldsymbol{a}}_i) Z(\boldsymbol{a}_E\cdot\boldsymbol{a}_i) Z(\boldsymbol{b}_E\cdot\overline{\boldsymbol{a}}_i) X(\boldsymbol{b}_E\cdot\boldsymbol{a}_i),
\end{aligned} \tag{11.23}$$

where the "$\cdot$" operation is defined by $\boldsymbol{a}\cdot\boldsymbol{b} = (a_1b_1, \ldots, a_Nb_N)$ and $\overline{\boldsymbol{a}}_i$ is the bitwise complement $\overline{\boldsymbol{a}}_i = (a_{i1}+1, \ldots, a_{iN}+1)$ In the foregoing derivation we used the following property: $\overline{\boldsymbol{a}}_i + \boldsymbol{a}_i = (1, \ldots, 1)$. The LHS of (11.17) can be rewritten, based on Eq. (11.23), as follows:

$$\begin{aligned}
&\overline{H}_i \overline{U}_{\mathrm{CNOT}}(\boldsymbol{a}_i+\boldsymbol{b}_i)\overline{H}_i E|c\rangle \otimes |\boldsymbol{A}\rangle \\
&= \overline{H}_i \overline{U}_{\mathrm{CNOT}}(\boldsymbol{a}_i+\boldsymbol{b}_i) \overbrace{\overline{H}_i E \overline{H}_i^{\dagger}}^{\overline{E}} \overbrace{\overline{H}_i g_i \overline{H}_i g_i \overline{H}_i |c\rangle}^{g_i^2|\overline{c}\rangle=|\overline{c}\rangle} \otimes |\boldsymbol{A}\rangle \\
&= \overline{H}_i \overline{U}_{\mathrm{CNOT}}(\boldsymbol{a}_i+\boldsymbol{b}_i)\overline{E}|\overline{c}\rangle \otimes |\boldsymbol{A}\rangle.
\end{aligned} \tag{11.24}$$

By substituting the last line of Eq. (11.23) into Eq. (11.24) we obtain:

$$\begin{aligned}
&\overline{H}_i \overline{U}_{\mathrm{CNOT}}(\boldsymbol{a}_i+\boldsymbol{b}_i)\overline{H}_i E|c\rangle \otimes |\boldsymbol{A}\rangle \overset{|\overline{c}\rangle=\sum_m \overline{c}(\boldsymbol{m})|\boldsymbol{m}\rangle}{=} \\
&\overline{H}_i \overline{U}_{\mathrm{CNOT}}(\boldsymbol{a}_i+\boldsymbol{b}_i) X(\boldsymbol{a}_E\cdot\overline{\boldsymbol{a}}_i) Z(\boldsymbol{a}_E\cdot\boldsymbol{a}_i+\boldsymbol{b}_E\cdot\overline{\boldsymbol{a}}_i) X(\boldsymbol{b}_E\cdot\boldsymbol{a}_i) \sum_{\boldsymbol{m}} \overline{c}(\boldsymbol{m})|\boldsymbol{m}\rangle \otimes \big|\boldsymbol{A}\big\rangle,
\end{aligned} \tag{11.25}$$

where we used the expansion $|\overline{c}\rangle = \sum_{\boldsymbol{m}} \overline{c}(\boldsymbol{m})|\boldsymbol{m}\rangle$ in terms of computational basis (CB) kets. Upon applying the X-containing operators we obtain:

$$\begin{aligned}
&\overline{H}_i \overline{U}_{\mathrm{CNOT}}(\boldsymbol{a}_i+\boldsymbol{b}_i)\overline{H}_i E|c\rangle \otimes |\boldsymbol{A}\rangle \\
&= \overline{H}_i \overline{U}_{\mathrm{CNOT}}(\boldsymbol{a}_i+\boldsymbol{b}_i) X(\boldsymbol{a}_E\cdot\overline{\boldsymbol{a}}_i) Z(\boldsymbol{a}_E\cdot\boldsymbol{a}_i+\boldsymbol{b}_E\cdot\overline{\boldsymbol{a}}_i) \sum_{\boldsymbol{m}} \overline{c}(\boldsymbol{m})|\boldsymbol{m}+\boldsymbol{b}_E\cdot\boldsymbol{a}_i\rangle \otimes |\boldsymbol{A}\rangle \\
&= (-1)^{\lambda} \overline{H}_i \overline{U}_{\mathrm{CNOT}}(\boldsymbol{a}_i+\boldsymbol{b}_i) Z(\boldsymbol{a}_E\cdot\boldsymbol{a}_i+\boldsymbol{b}_E\cdot\overline{\boldsymbol{a}}_i) X(\boldsymbol{a}_E\cdot\overline{\boldsymbol{a}}_i) \sum_{\boldsymbol{m}} \overline{c}(\boldsymbol{m})|\boldsymbol{m}+\boldsymbol{b}_E\cdot\boldsymbol{a}_i\rangle \otimes |\boldsymbol{A}\rangle \\
&= (-1)^{\lambda} \overline{H}_i \overline{U}_{\mathrm{CNOT}}(\boldsymbol{a}_i+\boldsymbol{b}_i) Z(\boldsymbol{a}_E\cdot\boldsymbol{a}_i+\boldsymbol{b}_E\cdot\overline{\boldsymbol{a}}_i) \sum_{\boldsymbol{m}} \overline{c}(\boldsymbol{m})|\boldsymbol{m}+\boldsymbol{b}_E\cdot\boldsymbol{a}_i+\boldsymbol{a}_E\cdot\overline{\boldsymbol{a}}_i\rangle \otimes |\boldsymbol{A}\rangle,
\end{aligned} \tag{11.26}$$

where $\lambda = (\boldsymbol{b}_E\cdot\overline{\boldsymbol{a}}_i)\cdot(\boldsymbol{a}_E\cdot\overline{\boldsymbol{a}}_i)$ (Since $X(\boldsymbol{a}_E\cdot\overline{\boldsymbol{a}}_i)$ and $Z(\boldsymbol{a}_E\cdot\boldsymbol{a}_i)$ act on different qubits, they commute.) By the application of a transversal CNOT gate on the Shor state from Eq. (11.26), we obtain:

$$\begin{aligned}\overline{H}_i\overline{U}_{\text{CNOT}}(\boldsymbol{a}_i+\boldsymbol{b}_i)\overline{H}_iE|c\rangle\otimes|\boldsymbol{A}\rangle &=(-1)^{\lambda}\overline{H}_iZ(\boldsymbol{a}_E\cdot\boldsymbol{a}_i+\boldsymbol{b}_E\cdot\overline{\boldsymbol{a}}_i)\sum_{\boldsymbol{m}}\overline{c}(\boldsymbol{m})|\boldsymbol{m}+\boldsymbol{b}_E\cdot\boldsymbol{a}_i\\ &\quad+\boldsymbol{a}_E\cdot\overline{\boldsymbol{a}}_i\rangle\otimes\sum_{\boldsymbol{A}_{\text{even}}}\Big|\boldsymbol{A}_{\text{even}}+\underbrace{(\boldsymbol{a}_i+\boldsymbol{b}_i)(\boldsymbol{b}_E\cdot\boldsymbol{a}_i+\boldsymbol{a}_E\cdot\boldsymbol{b}_i)}_{\boldsymbol{a}_E\cdot\boldsymbol{b}_i+\boldsymbol{b}_E\cdot\boldsymbol{a}_i}\Big\rangle\\ &=E|c\rangle\otimes\sum_{\boldsymbol{A}_{\text{even}}}|\boldsymbol{A}_{\text{even}}+\boldsymbol{b}_E\cdot\boldsymbol{a}_i+\boldsymbol{a}_E\cdot\boldsymbol{b}_i\rangle,\end{aligned} \tag{11.27}$$

proving therefore Eq. (11.17).

Example. *The [5,1,3] code (revisited).* The generators of this code are:

$$g_1=X_1Z_2Z_3X_4,\quad g_2=X_2Z_3Z_4X_5,\quad g_3=X_1X_3Z_4Z_5,\ \text{and}\ g_4=Z_1X_2X_4Z_5.$$

Based on Eq. (11.20) and the foregoing generators we obtain the transversal Hadamard gates as follows:

$$\overline{H}_1=H_1H_4,\quad \overline{H}_2=H_2H_5,\quad \overline{H}_3=H_1H_3,\quad \overline{H}_4=H_2H_4.$$

Based on Eq. (11.22), the application of the transversal H gate on generators g_i ($i=1,\ldots,4$) yields to:

$$\begin{aligned}\overline{g}_1&=Z(\boldsymbol{a}_1+\boldsymbol{b}_1)=Z_1Z_2Z_3Z_4,\quad \overline{g}_2=Z(\boldsymbol{a}_2+\boldsymbol{b}_2)=Z_2Z_3Z_4Z_5,\\ \overline{g}_3&=Z(\boldsymbol{a}_3+\boldsymbol{b}_3)=Z_1Z_3Z_4Z_5,\quad \overline{g}_4=Z_1Z_2Z_4Z_5.\end{aligned}$$

The corresponding circuit for syndrome extraction is shown in Fig. 11.9. The upper block corresponds to the action on codeword qubits, and the lower block corresponds to the action on ancilla qubits (Shor state). It is clear that single-qubit error in a codeword block can affect only one qubit in the ancilla block suggesting that this implementation is transversal and consequently fault tolerant.

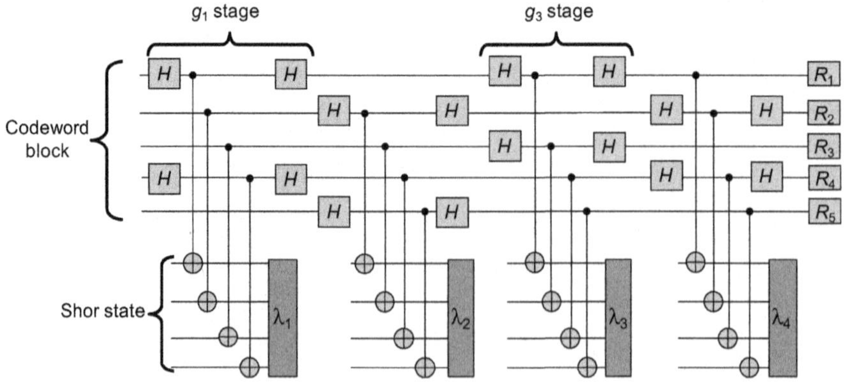

FIGURE 11.9

Syndrome extraction circuit for [5,1,3] stabilizer code.

Let us observe the stage corresponding to generator g_1. Based on transparent Hadamard gate $\overline{H}_1 = H_1H_4$ we have to place two Hadamard gates taking action on codeword block qubits H_1 and H_4, respectively. The corresponding generator image $\overline{g}_1 = Z_1Z_2Z_3Z_4$ indicates that qubits 1, 2, 3, and 4 in the codeword block are control qubits, while qubits in the Shor state block are target qubits. The purpose of this manipulation is to encode the syndrome information into the ancilla qubits. We further perform the measurements on the modified Shor state and the result of the measurements provides information on syndrome component λ_1. The other g_i ($i = 2,3,4$) stages operate in a similar fashion. Once the syndrome vector is determined, from the corresponding look-up table we determine the most probable error and apply corresponding recovery action $\boldsymbol{R} = R_1R_2R_3R_4R_5$ to undo the action of imperfect gates.

As an alternative, Plenio et al. [6] proposed to rearrange the sequence of CNOT gates in Eq. (11.17) as follows:

$$\overline{H}_i\overline{U}_{\mathrm{CNOT}}(\boldsymbol{a}_i)\overline{H}_i\overline{U}_{\mathrm{CNOT}}(\boldsymbol{b}_i)E|c\rangle \otimes |\boldsymbol{A}\rangle, \tag{11.28}$$

where the transversal Hadamard gate is now applied on all qubits:

$$\overline{H}_i = \prod_{j=1}^{N} H_j. \tag{11.29}$$

Based on these generators and Eqs. (11.28) and (11.29), the corresponding syndrome extraction circuit is shown in Fig. 11.10. The operation principle of this circuit version is very similar to that shown in Fig. 11.9.

In the foregoing discussion we assumed that the generators do not contain Y-operators. In case the generators contain Y-operators, when constructing the images of generators $\overline{g}_i$ instead of applying the Hadamard gate H_k to the kth qubit, we need to find the gate that performs the following transformation: $\widetilde{H}_kY_k\widetilde{H}_k^{\dagger} = Z_k$. It can

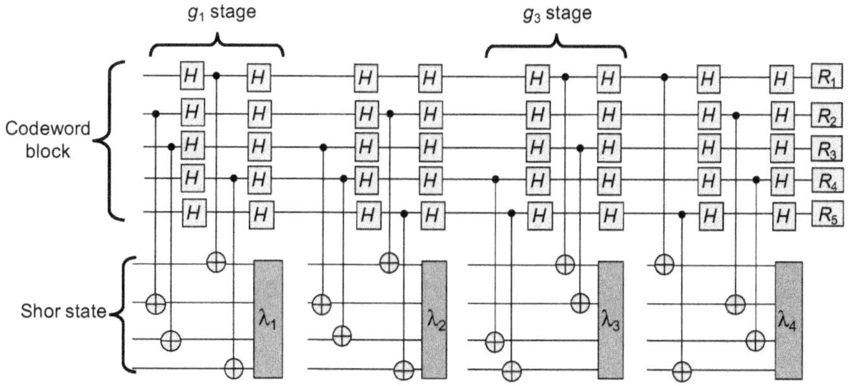

FIGURE 11.10

Syndrome extraction circuit for [5,1,3] stabilizer code based on the approach due to Plenio et al.

simply be verified by matrix multiplication that the following gate performs this transformation:

$$\widetilde{H}_k = \frac{1}{\sqrt{2}}\begin{bmatrix} 1 & -j \\ -j & 1 \end{bmatrix}. \tag{11.30}$$

The transversal Hadamard gate is now, based on Eqs. (11.20) and (11.30), defined as follows:

$$\overline{H}_i \doteq \prod_{j=1}^{N} \left(\widetilde{H}_j\right)^{(\boldsymbol{a}_i\cdot\boldsymbol{b}_i)_j} (H_j)^{(\boldsymbol{a}_i\cdot\overline{\boldsymbol{b}}_i)_j}, \overline{b}_{ij} = b_{ij} + 1, \tag{11.31}$$

where we used the notation $(\boldsymbol{a}_i \cdot \boldsymbol{b}_i)_j$ to denote the jth component of $\boldsymbol{a}_i \cdot \boldsymbol{b}_i$. The images of generators now become:

$$\begin{aligned} \overline{g}_i &= \overline{H}_i g_i \overline{H}_i^\dagger = \overline{H}_i X(\boldsymbol{a}_i) Z(\boldsymbol{b}_i) \overline{H}_i^\dagger = \overline{H}_i Y(\boldsymbol{a}_i\cdot\boldsymbol{b}_i) X(\boldsymbol{a}_i\cdot\overline{\boldsymbol{b}}_i) Z(\overline{\boldsymbol{a}}_i\cdot\boldsymbol{b}_i)\overline{H}_i^\dagger \\ &= \underbrace{\prod_{j=1}^{N} \left(\widetilde{H}_j\right)^{(\boldsymbol{a}_i\cdot\boldsymbol{b}_i)_j} Y(\boldsymbol{a}_i\cdot\boldsymbol{b}_i)\left(\widetilde{H}_j^\dagger\right)^{(\boldsymbol{a}_i\cdot\boldsymbol{b}_i)_j}}_{Z(\boldsymbol{a}_i\cdot\boldsymbol{b}_i)} \underbrace{\prod_{j=1}^{N} (H_j)^{(\boldsymbol{a}_i\cdot\overline{\boldsymbol{b}})_j} X(\boldsymbol{a}_i\cdot\overline{\boldsymbol{b}}_i)(H_j)^{(\boldsymbol{a}_i\cdot\overline{\boldsymbol{b}}_i)_j}}_{Z(\boldsymbol{a}_i\cdot\overline{\boldsymbol{b}})} Z(\overline{\boldsymbol{a}}_i\cdot\boldsymbol{b}_i) \\ &= Z(\boldsymbol{a}_i\cdot\boldsymbol{b}_i + \boldsymbol{a}_i\cdot\overline{\boldsymbol{b}}_i + \overline{\boldsymbol{a}}_i\cdot\boldsymbol{b}_i) = Z(\boldsymbol{a}_i\cdot\boldsymbol{b}_i + \boldsymbol{a}_i + \boldsymbol{a}_i\cdot\boldsymbol{b}_i + \boldsymbol{b}_i + \boldsymbol{a}_i\cdot\boldsymbol{b}_i) = Z(\boldsymbol{a}_i + \boldsymbol{b}_i + \boldsymbol{a}_i\cdot\boldsymbol{b}_i). \end{aligned} \tag{11.32}$$

In a similar fashion, the image of the error E can be found by:

$$\overline{E} = \overline{H}_i E \overline{H}_i^\dagger = (-1)^\lambda Z\left(\boldsymbol{a}_E\cdot(\boldsymbol{a}_i\cdot\overline{\boldsymbol{b}}_i) + \boldsymbol{b}_E\cdot\overline{\boldsymbol{a}_i\cdot\overline{\boldsymbol{b}}_i}\right) X\left(\boldsymbol{a}_E\cdot\overline{\boldsymbol{a}_i\cdot\overline{\boldsymbol{b}}_i} + \boldsymbol{b}_E\cdot\boldsymbol{a}_i\right), \tag{11.33}$$

where $\lambda = \left(\boldsymbol{a}_E\cdot\overline{\boldsymbol{a}_i\cdot\overline{\boldsymbol{b}}_i}\right)\cdot\left(\boldsymbol{b}_E\cdot\overline{\boldsymbol{a}_i\cdot\overline{\boldsymbol{b}}_i}\right)$ By the application of the transversal CNOT gate on $E|c\rangle \otimes |\boldsymbol{A}\rangle$, we obtain:

$$\overline{U}_{\text{CNOT}}(\boldsymbol{a}_i + \boldsymbol{b}_i + \boldsymbol{a}_i\cdot\boldsymbol{b}_i)E|c\rangle \otimes |\boldsymbol{A}\rangle = \overline{E}|\overline{c}\rangle \otimes \sum_{\boldsymbol{A}_{\text{even}}} |\boldsymbol{A}_{\text{even}} + \boldsymbol{b}_E\cdot\boldsymbol{a}_i + \boldsymbol{a}_E\cdot\boldsymbol{b}_i\rangle. \tag{11.34}$$

Finally, by applying the transversal Hadamard gate given by Eqs. (11.31) on (11.34) we derive:

$$\overline{H}_i\overline{U}_{\text{CNOT}}(\boldsymbol{a}_i + \boldsymbol{b}_i + \boldsymbol{a}_i\cdot\boldsymbol{b}_i)\overline{H}_i E|c\rangle \otimes |\boldsymbol{A}\rangle = E|c\rangle \otimes \sum_{\boldsymbol{A}_{\text{even}}} |\boldsymbol{A}_{\text{even}} + \boldsymbol{b}_E\cdot\boldsymbol{a}_i + \boldsymbol{a}_E\cdot\boldsymbol{b}_i\rangle, \tag{11.35}$$

which has a similar form to that of Eq. (11.17).

Example. *The [4,2,2] code (revisited).* The starting point in this example are generators of the code:

$$g_1 = X_1Z_2Z_3X_4 \text{ and } g_2 = Y_1X_2X_3Y_4.$$

Based on Eq. (11.31) we obtain the following transversal Hadamard gates corresponding to generators g_i $(i = 1,2)$:

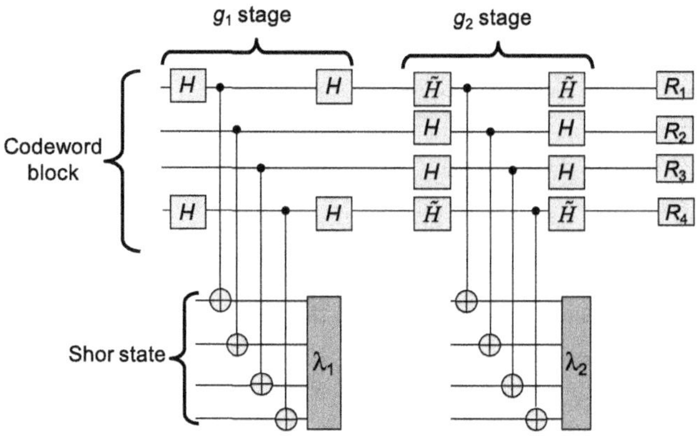

FIGURE 11.11

Syndrome extraction circuit for [4,2,2] stabilizer code.

$$\overline{H}_1 = H_1 H_4 \text{ and } H_2 = \widetilde{H}_1 H_2 H_3 \widetilde{H}_4.$$

Based on the last line of Eq. (11.32) we obtain the images of the generators as follows:

$$\overline{g}_1 = Z_1 Z_2 Z_3 Z_4 \text{ and } \overline{g}_2 = Z_1 Z_2 Z_3 Z_4.$$

Finally, based on Eq. (11.35), the syndrome extraction circuit for [4,2,2] code is shown in Fig. 11.11.

The syndrome extraction circuits discussed so far employ only transversal operations providing that the Shor state was prepared fault tolerantly. Closer inspection of the Shor preparation circuit (Fig. 11.8) reveals that the error introduced by a wrong qubit or a faulty CNOT gate can cause the propagation of errors. From Fig. 11.8 we can see that the first and last qubits have an even parity. Therefore by adding an additional parity check on the first and last qubit we can verify if the even parity condition is satisfied. If satisfied, with know that up to $O(p)$ no flip error has occurred. If the measurement of Z on the addition ancilla qubit is 1 we know that an odd number of errors occurred, and we have to repeat the verification procedure. The modified Shor preparation circuit enabling this verification is shown in Fig. 11.12.

We learned earlier that if we perform the measurement on a modified Shor state, the modified Shor state will collapse into one of its eigenkets, say $|\boldsymbol{A}'_{\text{even}} + \boldsymbol{b}_E \cdot \boldsymbol{a}_i + \boldsymbol{a}_E \cdot \boldsymbol{b}_i\rangle$, and the result of the measurement will be $(-1)^{\boldsymbol{b}_E \cdot \boldsymbol{a}_i + \boldsymbol{a}_E \cdot \boldsymbol{b}_i + \text{wt}(\boldsymbol{A}'_{\text{even}})} = (-1)^{\lambda_i + \text{wt}(\boldsymbol{A}'_{\text{even}})}$. If an odd number of bit-flip errors has occurred during Shor-state preparation, the $\text{wt}(\boldsymbol{A}'_{\text{even}}) = 1$ and the syndrome value will be $\lambda_i + 1$. The other sources of errors are storage errors and imperfect CNOT gates used in syndrome calculation. The syndrome $S(E)$ can be used to identify the error introduced by various sources of imperfections discussed earlier. In an ideal world, the most probable error can be detected based on the syndrome and used to perform error correction action. However, the syndrome

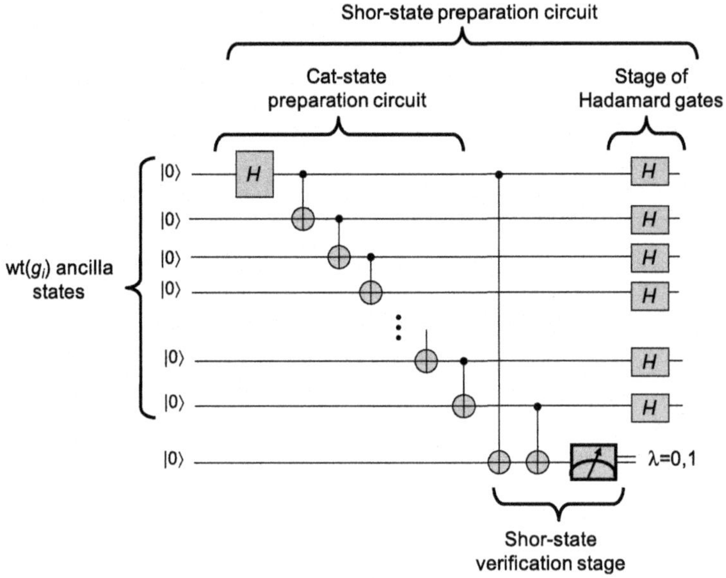

FIGURE 11.12

Modified Shor-state preparation circuit with verification stage.

extraction procedure, as we mentioned, can also be faulty. The wrongly determined syndrome during the syndrome extraction procedure can introduce additional errors in the error correction action. This problem can be solved by the following *syndrome verification protocol* due to Plenio et al. [6]:

1. If the result of measurements returns $S(E) = 0$ we accept the measurement result as correct and assume that no recovery operation is needed.
2. On the other hand, if the result of measurements returns $S(E) \neq 0$, instead of using the syndrome to identify the most probable error followed by error correction action, we repeat the whole procedure until the measurement returns $S(E) = 0$.

This protocol might be time consuming, but ensures that a wrongly determined syndrome cannot introduce the propagation of errors. Notice that even an $S(E) = 0$ result does not guarantee that storage error did not occur after the syndrome was determined, for example, in the last transversal gates of Figs. 11.9–11.11. However, this undetectable error will be present in the codeword block after the error correction, indicating that the protocol is still fault tolerant.

11.3.2 Fault-Tolerant Encoded Operations

We defined the encoded gates in Chapter 9 as unitary operators that perform the mapping of one codeword to another codeword as given here:

$$\left(UsU^{\dagger}\right)U|c\rangle = Us|c\rangle \overset{s|c\rangle=|c\rangle}{=} U|c\rangle; \quad \forall |c\rangle \in C_Q, \quad s \in S. \tag{11.36}$$

If $UsU^{\dagger} = s \quad \forall s \in S$, that is, if stabilizer S is fixed by U under conjugation ($USU^{\dagger} = S$), then $U|c\rangle$ will be a codeword as, based on Eq. (11.36), it is fixed for any element from S. The operator U therefore represents an *encoded operation* (the operation that maps one codeword to another codeword). As shown in Chapter 8, the set of operators U that fix S represent the subgroup of the *unitary group* $U(N)$ (the set of all unitary matrices of dimensions $N \times N$). This subgroup is called the *normalizer* of S with respect to $U(N)$, and it is denoted here as $N_U(S)$. Namely, the normalizer $N_U(S)$ of the stabilizer S is defined as the set of operators $U \in U(N)$ that fixes S under conjugation; in other words, $N_U(S) = \{U \in U(N) \mid USU^{\dagger} = S\}$. In the Appendix (see also Chapter 8) we introduce an equivalent concept, namely the centralizer of S with respect to $U(N)$, denoted as $C_U(S)$, as the set of all operators U (from $U(N)$) that commute for all stabilizer elements; in other words, $C_U(S) = \{U \in U(N) \mid US = SU\}$. Clearly, the centralizer condition $US = SU$ can be converted to the normalizer by multiplying from the right by $U^{\dagger}$ to obtain $USU^{\dagger} = S$. (For a more formal proof, refer to Section 8.1). Notice that the key difference between this section and Section 8.1 is that here we are concerned with the action of U from $U(N)$ on S, while in Section 8.1 we were concerned with the action of an operator from Pauli multiplicative group G_N. As $G_N \subset U(N)$, it is obvious that $N_G(S) \subset N_U(S)$.

An important group for fault-tolerant quantum error correction is the *Clifford* group, which is defined as the normalizer of the Pauli multiplicative group G_N with respect to $U(N)$, and denoted as $N_U(G_N)$. The *Clifford operator* U is therefore an operator that preserves the elements of the Pauli group under conjugation, namely $\forall O \in G_N : UOU^{\dagger} \in G_N$. The Clifford group elements can be generated based on a simple set of gates: Hadamard, phase, and CNOT gates, or equivalently Hadamard, phase, and controlled-Z gates. In this section, we are concerned with encoded gates $U \in N(G_N) \cap N(S)$ that can be constructed using these gates and encoded operations that are implemented in transversal fashion, and which are consequently fault tolerant.

We learned in Chapter 8 that the quotient group $C(S)/S$ can be represented by:

$$C(S)/S = \cup\{\overline{X}_m S, \overline{Z}_n S: m, n = 1, ..., K\} \wedge \{j^l : l = 0, 1, 2, 3\}, \tag{11.37}$$

where $\{\overline{X}_m, \overline{Z}_n | m, n = 1, \ldots, K\}$ denote the encoded Pauli operators. The elements of $C(S)/S$ (cosets), denoted as $\overline{O}S$, can be represented as follows:

$$\overline{O}S = j^l \overline{X}(\boldsymbol{a})\overline{Z}(\boldsymbol{b})S = j^l (\overline{X}_1 S)^{a_1} \ldots (\overline{X}_k S)^{a_K} (\overline{Z}_1 S)^{b_1} \ldots (\overline{Z}_1 S)^{b_K}, \tag{11.38}$$

as proved in Chapter 9. By denoting the conjugation mapping image of encoded Pauli operators by $\widetilde{X}_m = U\overline{X}_m U^{\dagger}, \widetilde{Z}_n = U\overline{Z}_n U^{\dagger}$ we can show that:

$$\begin{aligned}\left[\widetilde{X}_m, g_n\right] &= \widetilde{X}_m g_n - g_n \widetilde{X}_m = U\overline{X}_m U^{\dagger} g_n - g_n U\overline{X}_m U^{\dagger} \overset{g_n = Us_n U^{\dagger}}{=} U\overline{X}_m U^{\dagger} U s_n U^{\dagger} - U s_n U^{\dagger} U\overline{X}_m U^{\dagger} \\ &= U\overline{X}_m s_n U^{\dagger} - U s_n \overline{X}_m U^{\dagger} \overset{s_n \overline{X}_m = \overline{X}_m s_n}{=} U\overline{X}_m s_n U^{\dagger} - U\overline{X}_m s_n U^{\dagger} = 0,\end{aligned} \tag{11.39}$$

where we represented the generators g_n in terms some stabilizer element s_n by $g_n = Us_nU^\dagger$ (namely, as U fixes the stabilizer under conjugation, any element $s \in S$, including generators, can be represented in the form $UsU^\dagger$). Clearly, the image of Pauli operators also commutes with generators g_n and therefore belongs to $C(S)$ as well. Because the operator U fixes the stabilizer S under conjugation, based on Eq. (11.38) and commutation relation Eq. (11.39), the operator U maps $C(S)/S$ to $C(S)/S$, as the coset $\overline{O}S$ is mapped to:

$$\tilde{O}S = j^l(\tilde{X}_1S)^{a_1}\dots(\tilde{X}_kS)^{a_K}(\tilde{Z}_1S)^{b_1}\dots(\tilde{Z}_1S)^{b_K}. \tag{11.40}$$

In the Appendix we have defined the *homomorphism* as the mapping f from group G onto group H if the image of a product ab equals the product of images of a and b. If f is a homomorphism and it is bijective (1−1 and onto), then f is said to be an *isomorphism* from G to H, and commonly is denoted by $G \cong H$. Finally, if $G = H$ we call the isomorphism an *automorphism*. Let us now observe the mapping f from G_N to G_N that acts on elements from G_N by conjugation, namely: $f(g) = UgU^\dagger\ \forall\ g \in G_N,\ U \in N(G_N)$. It will be shown later that this mapping is an automorphism. We first need to show that this mapping is 1−1. Let us observe two *different* elements $a,b \in G_N$ and assume that $f(a) = f(b)$. The conjugation definition ($f(g) = UgU^\dagger$) function indicates that $f(a) = UaU^\dagger = UbU^\dagger = f(b)$. By multiplying by U from the right we obtain $Ua = Ub$, and by multiplying by $U^\dagger$ from the left we obtain $a = b$, which is a contradiction. Therefore the mapping $f(g)$ is 1−1 mapping. In the second step we need to prove that the mapping $f(g)$ is an onto mapping. We have shown in the previous paragraph that the range of mapping $f(g)$, denoted as $f(G_N)$, is contained in G_N. Since the mapping $f(g)$ is 1−1 it must be $f(G_N) = G_N$, indicating that the conjugation mapping is indeed onto mapping. In the final stage we need to prove that the image of the product ab, where $a,b \in G_N$, must be equal to the product of images $f(a)f(b)$. This claim can easily be proved from the definition of conjugation mapping as follows:

$$f(ab) = UabU^\dagger = Ua\overset{I}{U^\dagger U}bU^\dagger = \overset{f(a)}{(UaU^\dagger)}\ \overset{f(b)}{(UbU^\dagger)} = f(a)f(b). \tag{11.41}$$

Therefore the mapping $f(g) = UgU^\dagger$ is an automorphism of G_N.

Because every Clifford operator U from $N(G_N)$ is an automorphism of G_N, to determine the action of U on G_N it is sufficient to determine the action of generators of $N(G_N)$ on generators of G_N. We also know that every Clifford generator can be represented in terms of Hadamard, phase, and CNOT (or controlled-Z) gates, and it is essential to study the action of single-qubit Hadamard and phase gates on generators of G_1 and the action of two-qubit CNOT gates on generators of G_2.

The matrix representation of the Hadamard gate in the CBs, based on Chapter 2, is given by:

$$H = \frac{1}{\sqrt{2}}\begin{bmatrix} 1 & 1 \\ 1 & -1 \end{bmatrix}. \tag{11.42}$$

The Pauli gates X, Y, and Z are under conjugation operation $HOH^{\dagger}$ ($O \in \{X,Y,Z\}$) mapped to:

$$
\begin{aligned}
X \rightarrow HXH^{\dagger} &= \frac{1}{\sqrt{2}}\begin{bmatrix} 1 & 1 \\ 1 & -1 \end{bmatrix}\begin{bmatrix} 0 & 1 \\ 1 & 0 \end{bmatrix}\frac{1}{\sqrt{2}}\begin{bmatrix} 1 & 1 \\ 1 & -1 \end{bmatrix} = Z, \\
Z \rightarrow HZH^{\dagger} &= \frac{1}{\sqrt{2}}\begin{bmatrix} 1 & 1 \\ 1 & -1 \end{bmatrix}\begin{bmatrix} 1 & 0 \\ 0 & -1 \end{bmatrix}\frac{1}{\sqrt{2}}\begin{bmatrix} 1 & 1 \\ 1 & -1 \end{bmatrix} = X, \\
Y \rightarrow HYH^{\dagger} &= \frac{1}{\sqrt{2}}\begin{bmatrix} 1 & 1 \\ 1 & -1 \end{bmatrix}\begin{bmatrix} 0 & -j \\ j & 0 \end{bmatrix}\frac{1}{\sqrt{2}}\begin{bmatrix} 1 & 1 \\ 1 & -1 \end{bmatrix} = -Y.
\end{aligned}
\tag{11.43}
$$

The matrix representation of the phase gate in the CB is given by:

$$
P = \begin{bmatrix} 1 & 0 \\ 0 & j \end{bmatrix}. \tag{11.44}
$$

The Pauli gates X, Y, and Z are under conjugation operation $POP^{\dagger}$ ($O \in \{X,Y,Z\}$) mapped to:

$$
\begin{aligned}
X \rightarrow PXP^{\dagger} &= \begin{bmatrix} 1 & 0 \\ 0 & j \end{bmatrix}\begin{bmatrix} 0 & 1 \\ 1 & 0 \end{bmatrix}\begin{bmatrix} 1 & 0 \\ 0 & -j \end{bmatrix} = Y, \\
Z \rightarrow PZP^{\dagger} &= \begin{bmatrix} 1 & 0 \\ 0 & j \end{bmatrix}\begin{bmatrix} 1 & 0 \\ 0 & -1 \end{bmatrix}\begin{bmatrix} 1 & 0 \\ 0 & -j \end{bmatrix} = Z, \\
Y \rightarrow PYP^{\dagger} &= \begin{bmatrix} 1 & 0 \\ 0 & j \end{bmatrix}\begin{bmatrix} 0 & -j \\ j & 0 \end{bmatrix}\begin{bmatrix} 1 & 0 \\ 0 & -j \end{bmatrix} = -X.
\end{aligned}
\tag{11.45}
$$

From Eqs. (11.43) and (11.45) is clear that both operators (H and P) permute the generators of G_1, and that the phase operator performs the rotation around the z-axis for 90 degrees.

The last gate needed to implement any Clifford operator is the CNOT gate. Since the CNOT gate is a two-qubit gate we need to study its action on generators from G_2; namely $X_1 = XI$, X_2, Z_1, and Z_2. In CB, the CNOT gate U_{CNOT} and generators from G_2 can be represented as:

$$
\begin{aligned}
U_{\text{CNOT}} &= \begin{bmatrix} I & 0 \\ 0 & X \end{bmatrix}, X_1 = X \otimes I = \begin{bmatrix} 0 & 1 \\ 1 & 0 \end{bmatrix} \otimes I = \begin{bmatrix} 0 & I \\ I & 0 \end{bmatrix}, \\
X_2 &= I \otimes X = \begin{bmatrix} 1 & 0 \\ 0 & 1 \end{bmatrix} \otimes X = \begin{bmatrix} X & 0 \\ 0 & X \end{bmatrix}, \\
Z_1 &= Z \otimes I = \begin{bmatrix} 1 & 0 \\ 0 & -1 \end{bmatrix} \otimes I = \begin{bmatrix} I & 0 \\ 0 & -I \end{bmatrix}, \\
Z_2 &= I \otimes Z = \begin{bmatrix} 1 & 0 \\ 0 & 1 \end{bmatrix} \otimes Z = \begin{bmatrix} Z & 0 \\ 0 & Z \end{bmatrix}.
\end{aligned}
\tag{11.46}
$$

The generators of G_2, X_1, X_2, Z_1, and Z_2, are under conjugation operation $U_{\text{CNOT}} O U_{\text{CNOT}}^{\dagger}$ ($O \in \{X_1, X_2, Z_1, Z_2\}$) mapped to:

$$
\begin{aligned}
X_1 &\rightarrow U_{\text{CNOT}} X_1 U_{\text{CNOT}}^{\dagger} = \begin{bmatrix} I & 0 \\ 0 & X \end{bmatrix} \begin{bmatrix} 0 & I \\ I & 0 \end{bmatrix} \begin{bmatrix} I & 0 \\ 0 & X \end{bmatrix} = \begin{bmatrix} 0 & X \\ X & 0 \end{bmatrix} = X \otimes X = X_1 X_2, \\
X_2 &\rightarrow U_{\text{CNOT}} X_2 U_{\text{CNOT}}^{\dagger} = \begin{bmatrix} I & 0 \\ 0 & X \end{bmatrix} \begin{bmatrix} X & 0 \\ 0 & X \end{bmatrix} \begin{bmatrix} I & 0 \\ 0 & X \end{bmatrix} = I \otimes X = X_2, \\
Z_1 &\rightarrow U_{\text{CNOT}} Z_1 U_{\text{CNOT}}^{\dagger} = \begin{bmatrix} I & 0 \\ 0 & X \end{bmatrix} \begin{bmatrix} I & 0 \\ 0 & -I \end{bmatrix} \begin{bmatrix} I & 0 \\ 0 & X \end{bmatrix} = Z \otimes I = Z_1, \\
Z_2 &\rightarrow U_{\text{CNOT}} Z_2 U_{\text{CNOT}}^{\dagger} = \begin{bmatrix} I & 0 \\ 0 & X \end{bmatrix} \begin{bmatrix} Z & 0 \\ 0 & Z \end{bmatrix} \begin{bmatrix} I & 0 \\ 0 & X \end{bmatrix} = Z \otimes Z = Z_1 Z_2.
\end{aligned}
\tag{11.47}
$$

From Eq. (11.47) it is clear that an error introduced on the control qubit affects the target qubit, while the phase error introduced on the target qubit affects the control qubit.

For certain implementations, such as cavity quantum electrodynamics implementation, the use of a controlled-*Z* U_{CZ} gate is more appropriate than the use of a CNOT gate [8]. Namely, we have shown in Chapter 3 that the CNOT gate can be implemented as shown in Fig. 11.13. The gate shown in Fig. 11.13 in CB can be expressed as:

$$
\begin{bmatrix} I & 0 \\ 0 & HZH \end{bmatrix} = \begin{bmatrix} I & 0 \\ 0 & X \end{bmatrix} = U_{\text{CNOT}}, \tag{11.48}
$$

indicating that the controlled-*Z* gate can indeed be used instead of the CNOT gate. We now study the action of controlled-*Z* gate U_{CZ} on generators of G_2. The

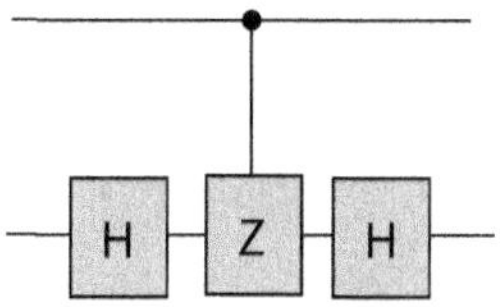

FIGURE 11.13

CNOT gate expressed in terms of the controlled-Z gate.

generators of G_2, X_1, X_2, Z_1, and Z_2, are under conjugation operation $U_{CZ}OU_{CZ}^{\dagger}$ ($O \in \{X_1,X_2,Z_1,Z_2\}$) mapped to:

$$\begin{aligned}
X_1 &\rightarrow U_{CZ}X_1U_{CZ}^{\dagger} = \begin{bmatrix} I & 0 \\ 0 & Z \end{bmatrix}\begin{bmatrix} 0 & I \\ I & 0 \end{bmatrix}\begin{bmatrix} I & 0 \\ 0 & Z \end{bmatrix} = X \otimes Z = X_1Z_2, \\
X_2 &\rightarrow U_{CZ}X_2U_{CZ}^{\dagger} = \begin{bmatrix} I & 0 \\ 0 & Z \end{bmatrix}\begin{bmatrix} X & 0 \\ 0 & X \end{bmatrix}\begin{bmatrix} I & 0 \\ 0 & Z \end{bmatrix} = Z \otimes X = Z_1X_2, \\
Z_1 &\rightarrow U_{CZ}Z_1U_{CZ}^{\dagger} = \begin{bmatrix} I & 0 \\ 0 & Z \end{bmatrix}\begin{bmatrix} I & 0 \\ 0 & -I \end{bmatrix}\begin{bmatrix} I & 0 \\ 0 & Z \end{bmatrix} = Z \otimes I = Z_1, \\
Z_2 &\rightarrow U_{CNOT}Z_2U_{CNOT}^{\dagger} = \begin{bmatrix} I & 0 \\ 0 & X \end{bmatrix}\begin{bmatrix} Z & 0 \\ 0 & Z \end{bmatrix}\begin{bmatrix} I & 0 \\ 0 & X \end{bmatrix} = I \otimes Z = Z_2.
\end{aligned} \tag{11.49}$$

From Eq. (11.49), it is evident that the bit-flip error on one qubit introduces the phase error on the other qubit, while the phase error on either qubit does not propagate.

We turn our attention now to the determination of a unitary operator when the corresponding automorphism is known. We will use the example due to Gottesman [9]. The automorphism on G_1 in this example is given by cyclic permutation of Pauli gates [9]:

$$f_U: X \rightarrow Y \rightarrow Z \rightarrow X. \tag{11.50}$$

We know from Chapter 2 that CB ket $|0\rangle$ is an eigenket of Pauli operator Z with an eigenvalue of $+1$. Let us now apply the unknown gate U on basis ket $|0\rangle$:

$$U|0\rangle \overset{Z|0\rangle=|0\rangle}{=} UZ|0\rangle = UZ\overset{I}{U^{\dagger}U}|0\rangle = (UZU^{\dagger})U|0\rangle = X(U|0\rangle), \tag{11.51}$$

where we use the mapping property (11.50). By comparing the first and last portions of Eq. (11.51) we conclude that $U|0\rangle$ is an eigenket of Pauli operator X with eigenvalue $+1$, which is given by (see Chapter 2):

$$U|0\rangle = \frac{1}{\sqrt{2}}(|0\rangle + |1\rangle). \tag{11.52}$$

The elements of the first column in matrix representation of U:

$$U = \begin{bmatrix} U_{00} & U_{01} \\ U_{10} & U_{11} \end{bmatrix}, \tag{11.53}$$

can be determined from Eq. (11.52) as follows: $U_{00} = \langle 0|U|0\rangle = 1/\sqrt{2}$, and $U_{10} = \langle 1|U|0\rangle = 1/\sqrt{2}$. The computational ket $|1\rangle$ can be obtained as $X|0\rangle = |1\rangle$, and by using a similar approach as in Eq. (11.51) we obtain:

$$\begin{aligned} U|1\rangle &\overset{|1\rangle=X|0\rangle}{=} UX|0\rangle = UX\overbrace{U^\dagger U}^{I}|0\rangle = (UXU^\dagger)U|0\rangle = Y(U|0\rangle) \\ &\overset{U|0\rangle=\frac{1}{\sqrt{2}}(|0\rangle+|1\rangle)}{=} \frac{1}{\sqrt{2}}Y(|0\rangle + |1\rangle) = \frac{-j}{\sqrt{2}}(|0\rangle - |1\rangle). \end{aligned} \tag{11.54}$$

The elements of the second column of Eq. (11.53) can be found from Eq. (11.54) as follows: $U_{01} = \langle 0|U|1\rangle = -j/\sqrt{2}$, and $U_{11} = \langle 1|U|1\rangle = j/\sqrt{2}$, so that the matrix representation of U is given by:

$$U_T = \frac{1}{\sqrt{2}}\begin{bmatrix} 1 & -j \\ 1 & j \end{bmatrix}, \tag{11.55}$$

The generators of G_1, X, Y, and Z, are under conjugation operation $U_T O U_T^\dagger$ ($O \in \{X,Y,Z\}$) mapped to:

$$\begin{aligned} X \to U_T X U_T^\dagger &= \begin{bmatrix} 1 & -j \\ 1 & j \end{bmatrix}\begin{bmatrix} 1 & 0 \\ 0 & 1 \end{bmatrix}\begin{bmatrix} 1 & 1 \\ j & -j \end{bmatrix} = Y, \\ Y \to U_T Y U_T^\dagger &= \begin{bmatrix} 1 & -j \\ 1 & j \end{bmatrix}\begin{bmatrix} 0 & -j \\ j & 0 \end{bmatrix}\begin{bmatrix} 1 & 1 \\ j & -j \end{bmatrix} = Z, \\ Z \to U_T Z U_T^\dagger &= \begin{bmatrix} 1 & -j \\ 1 & j \end{bmatrix}\begin{bmatrix} 1 & 0 \\ 0 & -1 \end{bmatrix}\begin{bmatrix} 1 & 1 \\ j & -j \end{bmatrix} = X, \end{aligned} \tag{11.56}$$

which is consistent with Eq. (11.50).

11.3.3 Measurement Protocol

In this section we are concerned with applying the quantum gate on a quantum register by means of a measurement. This method, applied transversally, will be used in the following sections to enable fault-tolerant quantum computation by using arbitrary quantum stabilizer code. DiVincenzo and Shor [10] described the method that can be used to perform the measurements using any operator O from multiplicative Pauli group G_N (see also [1,9]). There are three possible options involving the relationship of this operator O and stabilizer group S: (1) it can belong to S, when the measuring of O will not provide any useful information of the state of the system as the result will always be $+1$ for a valid codeword; (2) it can commute with all elements from S, indicating that O belongs to $N(S)/S$; and (3) it can anticommute with certain elements from S, which represents the case useful in practice, as discussed in Chapters 7 and 8. Let us create a set of generators $S_0 = \{g_1, g_2, \ldots, g_M\}$ such that the operator O anticommutes with g_1 and commutes with the remaining generators. If it happens that the mth generator g_m anticommutes with O we can replace it with $g_1 g_m$, which will clearly commute with O. Because the operator O commutes with generators g_m ($m = 2,3,\ldots,M$), the measurement of O will not affect them (they can be simultaneously measured with complete precision). On the other hand, since g_1 anticommutes with O, the measurement of O will affect g_1. The measurement of O will perform a projection of an arbitrary quantum state $|\psi_0\rangle$ as follows: $P_\pm|\psi_0\rangle$, with projection operators ($P_\pm^2 = P_\pm$) given by:

$$P_\pm = \frac{1}{2}(I \pm O), \tag{11.57}$$

see Chapter 2 for justification of this representation. Because the operator O has second order ($O^2 = I$, see Chapter 8), its eigenvalues are ± 1. Therefore the projection operators $P_\pm$ will project state $|\psi_0\rangle$ onto ± 1 eigenkets, respectively. The action of generators g_m on state $|\psi_0\rangle$ is given by $g_m|\psi_0\rangle = |\psi_0\rangle$, as expected. We now come to the point where we can introduce the *measurement protocol*, which consists of three steps:

1. Perform the measurement of O fault tolerantly, as described in Section 11.2.
2. If the measurement outcome is $+1$, do not take any further action. The final state will be $P_+|\psi_0\rangle = |\psi_+\rangle$.
3. If the measurement outcome is -1, apply the g_1-operator on the postmeasurement state to obtain:

$$\begin{aligned} g_1 P_- |\psi_0\rangle &= g_1 \frac{1}{2}(I - O)|\psi_0\rangle = \frac{1}{2}(I + O) g_1 |\psi_0\rangle = P_+ g_1 |\psi_0\rangle \\ &= P_+|\psi_0\rangle = |\psi_+\rangle. \end{aligned} \tag{11.58}$$

In both cases, the measurement of O performs the mapping $|\psi_0\rangle \to |\psi_+\rangle$ The final state is fixed by stabilizer $S_1 = \{O, g_1', \ldots, g_m'\}$, with generators given by:

$$g_m' = \begin{cases} g_m, & [g_m, O] = 0 \\ g_1 g_m, & \{g_m, O\} = 0 \end{cases}; \quad m = 2, \ldots, M. \tag{11.59}$$

It can be shown, by observing the generators from S_0, that the foregoing measurement procedure also performs the following $S_0 \to S_1$. In the Appendix, we defined the centralizer of S_i ($i = 0,1$) as the set of operators from G_N that commute with all elements from S_i. We will show later that the measurement protocol performs the following mapping: $C(S_0)/S_0 \to C(S_1)/S_1$. To prove this claim, let us observe an element n from $C(S_0)$ and a state $|\psi_0\rangle$ that is fixed by S_0. The action of operator n on state $P_+|\psi_0\rangle$, during the measurement of operator O, must be the same as $P_+ n'|\psi_0\rangle$, where n' is the image of n (under measurement of O). If n commutes with O, then $n' = n$. On the other hand, if n anticommutes with O, $g_1 n$ will commute with O, and the image of n will be $n' = g_1 n$. Therefore the measurement of O in the first case maps $n \to n$, and in the second case it maps $n \to g_1 n$. Since in the first case n commutes with O and $n' = n$, then n' must belong to the coset nS_1. In the second case, $g_1 n$ commutes with O and n' belongs to the coset $(g_1 n)S_1$. In conclusion, in the first case the measurement of O performs the mapping $nS_0 \rightarrow nS_1$, while in the second case $nS_0 \rightarrow g_1 nS_1$, proving the foregoing claim. Let us denote the encoded Pauli operators for S_0 by $\widehat{X}_i$, $\widehat{Z}_i$. To summarize, to determine the action of measurement of O on initial state $|\psi_0\rangle$ we need to follow the following procedure [2]:

1. Identify a generator g_1 from S_0 that anticommutes with O. Substitute any generator g_m $(m > 1)$ that anticommutes with O with $g_1 g_m$.
2. Create the stabilizer S_1 from S_0 by replacing g_1 (from S_0) with operator O. The remaining generators for S_1 are created according to the following rule:

$$g'_m = \begin{cases} g_m, & [g_m, O] = 0 \\ g_1 g_m, & \{g_m, O\} = 0 \end{cases}; \quad m = 2, \ldots, M.$$

3. As the final stage, substitute any $\widehat{X}_i(\widehat{Z}_i)$ that anticommutes with operator O by $g_1\widehat{X}_i(g_1\widehat{Z}_i)$, to ensure that $\widehat{X}_i(\widehat{Z}_i)$ belongs to $C(S_1)$.

The following example due to Gottesman [9] (see also [2]) will be used to illustrate this measurement procedure. Consider a two-qubit system $|\psi\rangle \otimes |0\rangle$, where $|\psi\rangle$ is an arbitrary state of qubit 1. Let us now apply the CNOT gate using the rules we derived by Eq. (11.47). The stabilizer for $|\psi\rangle \otimes |0\rangle$ has a generator g given by $I \otimes Z$. The operators $\widehat{X} = X \otimes I, \widehat{Z} = Z \otimes I$ both commute with generator g and mutually anticommute, and satisfy properties of encoded Pauli operators. The CNOT gate with qubit 1 serving as control qubit and qubit 2 as target qubit will perform the following mapping based on (11.47):

$$\begin{aligned} \widehat{X} &= X_1 \rightarrow \widehat{X}' = X \otimes X = X_1 X_2, \\ \widehat{Z} &= Z_1 \rightarrow \widehat{Z}' = Z \otimes I = Z_1, \\ g &= Z_2 \rightarrow g' = Z \otimes Z = Z_1 Z_2. \end{aligned} \tag{11.60}$$

We now study the measurement of operator $O = I \otimes Y$. Clearly, the operators $\widehat{X}$ and g anticommute with O (as they both differ in one position with O). The measurement of O will perform the following mapping, based on the previous three-step procedure:

$$\begin{aligned} \widehat{X}' &= X \otimes X = X_1 X_2 \rightarrow \widehat{X}'' = g'\widehat{X}' = (Z \otimes Z)(X \otimes X) = -Y \otimes Y = -Y_1 Y_2, \\ \widehat{Z}' &= Z \otimes I = Z_1 \rightarrow \widehat{Z}'' = \widehat{Z}' = Z_1, \\ g' &= Z \otimes Z = Z_1 Z_2 \rightarrow g'' = O = Y_2. \end{aligned} \tag{11.61}$$

The measurement protocol leaves the second qubit in $+1$ eigenket of Y_2, and the second portions of g'', $\widehat{X}''$, and $\widehat{Z}''$ fix it. As qubits 1 and 2 stay nonentangled after the measurement, qubit 2 can be discarded from further consideration. The left portions of operators $\widehat{X}, \widehat{X}', \widehat{Z}, \widehat{Z}'$ will determine the operation carried out on qubit 1 during the measurement. From Eq. (11.61) it is evident that during the measurement of O, the X is mapped to $-Y$, and the Z is mapped to Z, which is the same as the action of $P^{\dagger}$. Therefore the measurement of O and CNOT gate can be used to implement the phase gate.

The Hadamard gate can be represented in terms of the phase gate as follows:

$$PR_x(-\pi/2)^{\dagger}P = \begin{bmatrix} 1 & 0 \\ 0 & j \end{bmatrix} \frac{1}{\sqrt{2}} \begin{bmatrix} 1 & j \\ j & 1 \end{bmatrix}^{\dagger} \begin{bmatrix} 1 & 0 \\ 0 & j \end{bmatrix} = \frac{1}{\sqrt{2}} \begin{bmatrix} 1 & 1 \\ 1 & -1 \end{bmatrix} = H, \quad (11.62)$$

where with $R_x(-\pi/2)$ we denoted the rotation about the x-axis by $-\pi/2$, which is an operator introduced in Chapter 3, with matrix representation given by:

$$R_x(-\pi/2) = \frac{1}{\sqrt{2}} \begin{bmatrix} 1 & j \\ j & 1 \end{bmatrix}.$$

It maps the Pauli operators X and Z to X and Y, respectively. What remains is to show that this rotation operator can be implemented by using the measurement protocol and CNOT gate. The initial two-qubit system will now be $|\psi\rangle \otimes (|0\rangle + |1\rangle)$, and the corresponding generator $g = X_2$. The encoded Pauli operator analogs are given by $\widehat{X} = X_1, \widehat{Z} = Z_1$, and the operator O as X_2. Let us now apply the CNOT gate where qubit 2 is the control qubit and qubit 1 is the target qubit. The CNOT gate will perform the following mapping based on Eq. (11.47):

$$\begin{aligned} \widehat{X} &= X_1 \rightarrow \widehat{X}' = X \otimes I = X_1, \\ \widehat{Z} &= Z_1 \rightarrow \widehat{Z}' = Z \otimes Z = Z_1 Z_2, \\ g &= X_2 \rightarrow g' = X \otimes X = X_1 X_2. \end{aligned} \quad (11.63)$$

We then apply the phase gate on qubit 2, which performs the following mapping:

$$\begin{aligned} \widehat{X}' &= X_1 \rightarrow \widehat{X}'' = X_1, \\ \widehat{Z}' &= Z_1 Z_2 \rightarrow \widehat{Z}'' = Z_1 Z_2, \\ g' &= X_1 X_2 \rightarrow g'' = X_1 Y_2. \end{aligned} \quad (11.64)$$

In the final stage we perform the measurement on operator O. It is clear from Eq. (11.64) that $\widehat{Z}'' = Z_1 Z_2$ and g'' anticommute with operator O (because they both differ from O in one position). By applying the previous three-step measurement procedure we obtain the following mapping:

$$\begin{aligned} \widehat{X}'' &\rightarrow \widehat{X}''' = X_1, \\ \widehat{Z}'' &\rightarrow \widehat{Z}''' = g''\widehat{Z}'' = (X_1 Y_2)(Z_1 Z_2) = Y_1 X_2, \\ g'' &\rightarrow g''' = O = X_2. \end{aligned} \quad (11.65)$$

By disregarding qubit 2 we conclude that the measurement procedure maps X_1 to X_1 and Z_1 to Y_1, which is the same as the action of rotation operator $R_x(-\pi/2)$. In conclusion, we have just shown that any gate from the Clifford group can be implemented based on the CNOT gate, the measurement procedure described earlier, and a properly prepared initial state.

11.3.4 Fault-Tolerant Stabilizer Codes

We start our description of fault-tolerant stabilizer codes with a four-qubit unitary operation due to Gottesman [9], which when applied transversally can generate an encoded version of itself for all $[N,1]$ stabilizer codes. This four-qubit operation performs the following mapping [9]:

$$\begin{aligned}
&X\otimes I\otimes I\otimes I\rightarrow X\otimes X\otimes X\otimes I\\
&I\otimes X\otimes I\otimes I\rightarrow I\otimes X\otimes X\otimes X\\
&I\otimes I\otimes X\otimes I\rightarrow X\otimes I\otimes X\otimes X\\
&I\otimes I\otimes I\otimes X\rightarrow X\otimes X\otimes I\otimes X\\
&Z\otimes I\otimes I\otimes I\rightarrow Z\otimes Z\otimes Z\otimes I\\
&I\otimes Z\otimes I\otimes I\rightarrow I\otimes Z\otimes Z\otimes Z\\
&I\otimes I\otimes Z\otimes I\rightarrow Z\otimes I\otimes Z\otimes Z\\
&I\otimes I\otimes I\otimes Z\rightarrow Z\otimes Z\otimes I\otimes Z.
\end{aligned} \tag{11.66}$$

Clearly, the ith row ($i=2,3,4$) is obtained by cyclic permutation of the $(i-1)$th row, one position to the right. Similarly, the jth row ($j=6,7,8$) is obtained by cyclic permutation of the $(j-1)$th row, one position to the right. The corresponding quantum circuit to perform this manipulation is shown in Fig. 11.14. For any element $s=e^{jl\pi/2}X(\boldsymbol{a}_s)X(\boldsymbol{b}_s)\in S$, where the phase factor $\exp(jl\pi/2)$ is employed to ensure the Hermitian properties of s, this quantum circuit maps $s\otimes I\otimes I\otimes I\rightarrow s\otimes s\otimes s\otimes I$ and its cyclic permutations as given here:

$$\begin{aligned}
&s\otimes I\otimes I\otimes I\rightarrow s\otimes s\otimes s\otimes I\\
&I\otimes s\otimes I\otimes I\rightarrow I\otimes s\otimes s\otimes s\\
&I\otimes I\otimes s\otimes I\rightarrow s\otimes I\otimes X\otimes s\\
&I\otimes I\otimes I\otimes s\rightarrow s\otimes s\otimes I\otimes s.
\end{aligned} \tag{11.67}$$

It can easily be verified that images of generators are themselves the generators of stabilizer $S\otimes S\otimes S\otimes S$. The encoded Pauli operators $\overline{X}_m=e^{jl_m\pi/2}X(\boldsymbol{a}_m)Z(\boldsymbol{b}_m)$, $\overline{Z}_m=e^{jl'_m\pi/2}X(\boldsymbol{c}_m)Z(\boldsymbol{d}_m)$ are mapped in a similar fashion:

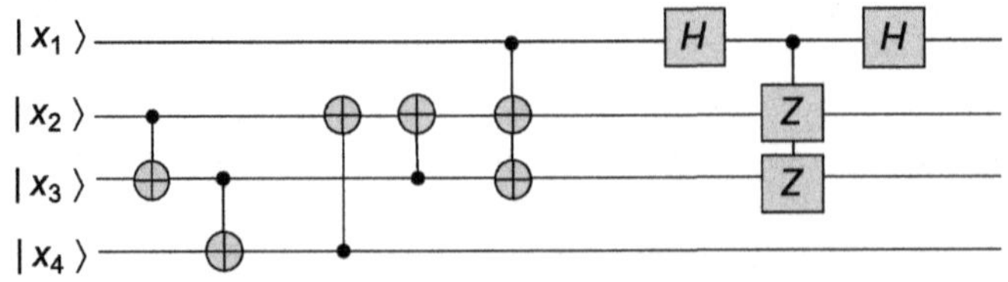

FIGURE 11.14

Quantum circuit to perform the mapping due to Gottesman [9] given by Eq. (11.66).

$$\begin{aligned}
&\overline{X}_m \otimes I \otimes I \otimes I \to \overline{X}_m \otimes \overline{X}_m \otimes \overline{X}_m \otimes I \\
&I \otimes \overline{X}_m \otimes I \otimes I \to I \otimes \overline{X}_m \otimes \overline{X}_m \otimes \overline{X}_m \\
&I \otimes I \otimes \overline{X}_m \otimes I \to \overline{X}_m \otimes I \otimes \overline{X}_m \otimes \overline{X}_m \\
&I \otimes I \otimes I \otimes \overline{X}_m \to \overline{X}_m \otimes \overline{X}_m \otimes I \otimes \overline{X}_m \\
&\overline{Z}_m \otimes I \otimes I \otimes I \to \overline{Z}_m \otimes \overline{Z}_m \otimes \overline{Z}_m \otimes I \\
&I \otimes \overline{Z}_m \otimes I \otimes I \to I \otimes \overline{Z}_m \otimes \overline{Z}_m \otimes \overline{Z}_m \\
&I \otimes I \otimes \overline{Z}_m \otimes I \to \overline{Z}_m \otimes I \otimes \overline{Z}_m \otimes \overline{Z}_m \\
&I \otimes I \otimes I \otimes \overline{Z}_m \to \overline{Z}_m \otimes \overline{Z}_m \otimes I \otimes \overline{Z}_m .
\end{aligned} \tag{11.68}$$

The transversal application of four-qubit operation provides the encoded version of the operation itself, which is true for all $[N,1]$ stabilizer codes. On the other hand, for $[N,K>1]$ stabilizer codes, the four-qubit operation maps the encoded Pauli operators $\overline{X}_m$, $\overline{Z}_m$ for all m-positions simultaneously. To be suitable to map a subset of encoded Pauli operators, certain modifications are needed, which will be described later in this section.

Let us now provide a simple but illustrative example due to Gottesman [9] (see also [2]). We are concerned with the mapping of the following four-qubit state $|\psi_1\psi_2\rangle \otimes |00\rangle$, where $|\psi_1\psi_2\rangle$ is an arbitrary two-qubit state. The stabilizer S_0 is described by the following two generators: $g_1 = Z_3, g_2 = Z_4$. The encoded Pauli operators are given by:

$$\widetilde{X}_1 = X_1,\ \widetilde{X}_2 = X_2,\ \widetilde{Z}_1 = Z_1,\ \widetilde{Z}_2 = Z_2. \tag{11.69}$$

Applying the four-qubit operation given by Eq. (11.66) we obtain the following mapping:

$$\begin{aligned}
&g_1 \to g_1' = Z_1Z_3Z_4, \quad g_2 \to g_1' = Z_1Z_2Z_4, \quad \widetilde{X}_1 \to \widetilde{X}_1' = X_1X_2X_3, \\
&\widetilde{X}_2 \to \widetilde{X}_2' = X_2X_3X_4, \quad \widetilde{Z}_1 \to \widetilde{Z}_1' = Z_1Z_2Z_3, \quad \widetilde{Z}_2 \to \widetilde{Z}_2' = Z_2Z_3Z_4.
\end{aligned}$$

We now perform the measurements using operators $O_1 = X_3$ and $O_2 = X_4$. By applying the measurement protocol from the previous section we obtain the following mapping as the result of measurements:

$$\begin{aligned}
&g_1'' \to O_1,\ g_1' = Z_1Z_3Z_4,\ g_2 \to O_2,\ g_2' = Z_1Z_2Z_4,\ \widetilde{X}_1' \to \widetilde{X}_1'' = X_1X_2X_3, \\
&\widetilde{X}_2' \to \widetilde{X}_2'' = X_2X_3X_4,\ \widetilde{Z}_1' \to \widetilde{Z}_1'' = g_1'g_2'\widetilde{Z}_1' = Z_1,\ \widetilde{Z}_2' \to \widetilde{Z}_2'' = g_1'g_2'\widetilde{Z}_2' = Z_4.
\end{aligned} \tag{11.70}$$

By disregarding the measured ancilla qubits in Eqs. (11.69) and (11.70) (qubits 3 and 4) we conclude that effectively we performed the following mapping on the first two qubits:

$$\begin{aligned}
&X_1 \to X_1X_2,\ X_2 \to X_2, \\
&Z_1 \to Z_1,\ Z_2 \to Z_1Z_2,
\end{aligned} \tag{11.71}$$

which is the same as the CNOT gate given by Eq. (11.47). Therefore by using the four-qubit operation and measurements we can implement the CNOT gate. In the previous section we learned how to implement an arbitrary Clifford operator by using the CNOT gate and measurements. If we apply a similar procedure to four code blocks and disregard the ancilla code blocks we can perform the fault-tolerant encoded operation that maps (on the first two qubits):

$$\begin{aligned}&\overline{X}_m \otimes I \rightarrow \overline{X}_m \otimes \overline{X}_m,\ \ I \otimes \overline{X}_m \rightarrow I \otimes \overline{X}_m,\\ &\overline{Z}_m \otimes I \rightarrow \overline{Z}_m \otimes I,\ \ I \otimes \overline{Z}_m \rightarrow \overline{Z}_m \otimes \overline{Z}_m.\end{aligned} \tag{11.72}$$

Eq. (11.72) is nothing else but a block-encoded CNOT gate. For $[N,K>1]$ stabilizer codes, with this procedure the encoded CNOT gate is applied to all mth-encoded qubits in two code blocks simultaneously. Another difficulty that arises is the fact that this procedure cannot be applied to two encoded qubits lying in the same code block or one lying in the first block and the other one in the second block.

We turn our attention to the description of the procedure that can be applied to $[N,K>1]$ stabilizer codes as well. The first problem can be solved by moving the mth qubit to two ancilla blocks, by employing the procedure described later. During this transfer the mth-encoded qubit has been transferred to two ancilla qubit blocks, while other qubits in ancilla blocks are initialized to $|\overline{0}\rangle$ We then apply the block-encoded CNOT gate given by Eq. (11.71) and the description above it. Because we transferred the mth-encoded qubit to ancilla blocks, the other coded blocks and other ancilla block will not be affected. Once we apply the desired operation to the mth-encoded qubit we can transfer it back to the original code block. Let us assume that the data qubit is in an arbitrary state $|\psi\rangle$ and ancilla qubit prepared in the $+1$ eigenket of $X^{(a)}$ $((|0\rangle + |1\rangle)/\sqrt{2}$ state in CB) where we used superscript (a) to denote the ancilla qubit. The stabilizer for this two-qubit state is described by the following generator: $g = I^{(d)}X^{(a)}$, where we now use the superscript (d) to denote the data qubit. The encoded Pauli operator analogs are given by: $\tilde{X} = X^{(d)}I^{(a)}, \tilde{Z} = Z^{(d)}I^{(a)}$. The CNOT gate is further applied assuming that the ancilla qubit is a control qubit and the data qubit is a target qubit. The CNOT gate performs the following mapping (see Eq. (11.47)):

$$\begin{aligned}&g = I^{(d)}X^{(a)} \rightarrow g' = X^{(d)}X^{(a)},\ \ \tilde{X} = X^{(d)}I^{(a)} \rightarrow \tilde{X}' = X^{(d)}I^{(a)},\\ &\tilde{Z} = Z^{(d)}I^{(a)} \rightarrow \tilde{Z}' = Z^{(d)}Z^{(a)}.\end{aligned} \tag{11.73}$$

We then perform the measurement on operator $O = Z^{(d)}I^{(a)}$, and by applying the measurement procedure from the previous section we perform the following mapping:

$$\begin{aligned}&g' = X^{(d)}X^{(a)} \rightarrow g'' = O = Z^{(d)}I^{(a)},\ \ \tilde{X}' \rightarrow \tilde{X}'' = g'\tilde{X}' = I^{(d)}X^{(a)},\\ &\tilde{Z}' = Z^{(d)}I^{(a)} \rightarrow \tilde{Z}'' = Z^{(d)}Z^{(a)}.\end{aligned} \tag{11.74}$$

By disregarding the measurement (data) qubit we conclude that effectively we performed the mapping:

$$X^{(d)} \rightarrow X^{(a)},\ \ Z^{(d)} \rightarrow Z^{(a)}, \tag{11.75}$$

indicating that the data qubit is transferred to the ancilla qubit. If, on the other hand, the initial ancilla qubit was in the $|0\rangle$ state, with the CNOT gate application with the ancilla qubit serving as the control qubit, the target (data) qubit would be unaffected. The procedure discussed in this paragraph can therefore be used to transfer the encoded qubit to an ancilla block, with other qubits being initialized to $|\overline{0}\rangle$ To summarize, the procedure can be described as follows. The ancilla block is prepared in such a way that all encoded qubits n, different from the mth qubit, are initialized to $+1$ eigenket $|\overline{0}\rangle$ of $\overline{Z}_n^{(a)}$. The encoded qubit m is, on the other hand, initialized to the $+1$ eigenket of $\overline{X}_m^{(a)}$. The block-encoded CNOT gate, described by Eq. (11.72), is then applied by using the ancilla block as the control block. We further perform the measurement on $\overline{Z}_m^{(d)}$. With this procedure, the mth-encoded qubit is transferred to the ancilla block, while other encoded ancilla qubits stay unaffected. Upon measurement, the mth qubit is left $+1$ eigentket $|\overline{0}\rangle$ of $\overline{Z}_m^{(d)}$, wherein the remaining encoded data qubits stay unaffected. Clearly, with the just described procedure we are capable of transferring an encoded qubit from one block to another. What remains now is to transfer the mth-encoded qubit from the ancilla-encoded block back to the original data block, which can be done by the procedure described in the next paragraph.

Let us assume now that our two-qubit data-ancilla system is initialized as follows: $|0\rangle^{(d)} \otimes |\psi\rangle^{(a)}$, with the generator of the stabilizer given by $g = Z^{(d)}I^{(a)}$. The corresponding encoded Pauli generator analogs are given by $\widetilde{X} = I^{(d)}X^{(a)}, \widetilde{Z} = I^{(d)}Z^{(a)}$. We then apply the CNOT gate using the ancilla qubit as a control qubit and data qubit as a target qubit, which performs the following mapping (see again Eq. (11.47)):

$$\begin{aligned} &g = Z^{(d)}I^{(a)} \rightarrow g' = Z^{(d)}Z^{(a)}, \quad \widetilde{X} = I^{(d)}X^{(a)} \rightarrow \widetilde{X}' = X^{(d)}X^{(a)}, \\ &\widetilde{Z} = I^{(d)}Z^{(a)} \rightarrow \widetilde{Z}' = I^{(d)}Z^{(a)}. \end{aligned} \tag{11.76}$$

We further perform the measurement on operator $O = I^{(d)}X^{(a)}$, which performs the mapping given by:

$$\begin{aligned} &g' = Z^{(d)}Z^{(a)} \rightarrow g'' = O = I^{(d)}X^{(a)}, \quad \widetilde{X}' = X^{(d)}X^{(a)} \rightarrow \widetilde{X}'' = X^{(d)}X^{(a)}, \\ &\widetilde{Z}' = I^{(d)}Z^{(a)} \rightarrow \widetilde{Z}'' = g'\widetilde{Z}' = Z^{(d)}I^{(a)}. \end{aligned} \tag{11.77}$$

By disregarding the measurement (ancilla) qubit we conclude that effectively we performed the mapping:

$$X^{(a)} \rightarrow X^{(d)}, \quad Z^{(a)} \rightarrow Z^{(d)}, \tag{11.78}$$

which is just the opposite of the mapping given by Eq. (11.75). With this procedure therefore we can move the mth-encoded qubit back to the original encoded data block.

We turn our attention now to the second problem of applying the CNOT gate from the mth to the nth qubit within the same block or between two different blocks. This problem can be solved by transferring qubits under consideration to two ancilla

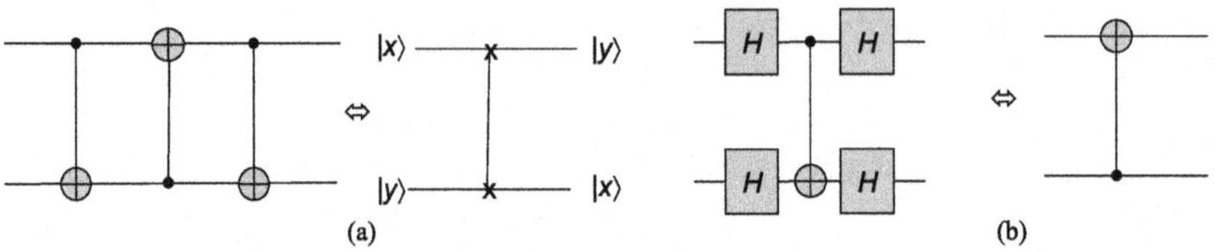

FIGURE 11.15

Quantum circuits to perform: (a) SWAP operation and (b) CNOT gate action from qubit 2 to qubit 1.

blocks with all other positions being in the $|\overline{0}\rangle$ state. We further apply the block-encoded CNOT gate to the corresponding ancilla blocks in a procedure that will leave other ancilla qubits unaffected. After performing the desired CNOT operation we will need to transfer the corresponding qubits back to original data positions. The key device for this manipulation is the fault-tolerant SWAP gate. The SWAP gate has already been described in Chapter 3, and the corresponding quantum circuit is provided in Fig. 11.15A to facilitate a description of fault-tolerant SWAP gate implementation. In Fig. 11.15B we also describe a convenient method to implement CNOT gate operation from qubit 2 to qubit 1, which has also been described in Chapter 3. Although we have already proved the correctness of equivalences given in Fig. 11.15, it is interesting to prove the equivalencies by using the procedure we introduced in Section 11.3.2. To prove the equivalence given in Fig. 11.15A we apply the CNOT gate from qubit 1 to qubit 2, which performs the following mapping:

$$XI \rightarrow XX, IX \rightarrow IX, \quad ZI \rightarrow ZI, IZ \rightarrow ZZ. \tag{11.79}$$

In the second stage we apply the CNOT gate using qubit 2 as the control qubit, and based on Eq. (11.47) we perform the following mapping:

$$\begin{aligned} &XX = (XI)(IX) \rightarrow (XI)(XX) \overset{X^2=I}{=} IX, \quad IX \rightarrow XX, \quad ZI \rightarrow ZZ, \\ &ZZ = (ZI)(IZ) \rightarrow (ZZ)(IZ) \overset{Z^2=I}{=} ZI. \end{aligned} \tag{11.80}$$

In the third stage we apply again the CNOT gate by using qubit 1 as the control qubit:

$$\begin{aligned} &IX \rightarrow IX, \quad XX = (XI)(IX) \rightarrow (XX)(IX) = XI, \\ &ZZ = (ZI)(IZ) \rightarrow (ZI)(ZZ) = IZ, \quad ZI \rightarrow ZI. \end{aligned} \tag{11.81}$$

The overall mapping can be written as:

$$XI \rightarrow IX, \quad IX \rightarrow XI, \quad ZI \rightarrow IZ, \quad IZ \rightarrow ZI, \tag{11.82}$$

which is the SWAP-ing action of qubit 1 and qubit 2. To prove the equivalence given in Fig. 11.15B we first apply the Hadamard gates on both qubits, and based on Eq. (11.43) we can write:

$$XI \rightarrow ZI,\ \ IX \rightarrow IZ,\ \ ZI \rightarrow XI,\ \ IZ \rightarrow IX. \tag{11.83}$$

We further apply the CNOT gate using qubit 1 as the control qubit, and based on Eq. (11.47) we perform the following mapping:

$$ZI \rightarrow ZI,\ \ IZ \rightarrow ZZ,\ \ XI \rightarrow XX,\ \ IX \rightarrow IX. \tag{11.84}$$

Finally, we apply again the Hadamard gates on both qubits to obtain:

$$ZI \rightarrow XI,\ \ ZZ \rightarrow XX,\ \ XX \rightarrow ZZ,\ \ IX \rightarrow IZ. \tag{11.85}$$

The overall mapping can be written as:

$$XI \rightarrow XI,\ IX \rightarrow XX,\ ZI \rightarrow ZZ,\ \ IZ \rightarrow IZ, \tag{11.86}$$

which is the same as the action of the CNOT gate from qubit 2 to qubit 1.

In certain implementations, it is difficult to implement the *automorphism group* of the stabilizer S, denoted as Aut(S), and defined as the group of functions performing the mapping $G_N \rightarrow G_N$. (Note that the automorphism group of the group G_N preserves the multiplication table.) In these situations, it is instructive to design a SWAP gate based on one of its unitary elements, say U, which is possible to implement. This operator is an automorphism of stabilizer S; it fixes it, and is an encoded operator that maps codewords to codewords and maps $C(S)/S$ to $C(S)/S$. Based on fundamental isomorphism theorem (see Appendix and Section 8.2, Eq. (11.38)), we know that $C(S)/S \cong G_K$, and $U \in N(S)$, indicating that the action of U can be determined by studying its action on the generators of G_K. The automorphism U must act on the first and nth-encoded qubits, and by measuring other encoded qubits upon action of U we come up with an encoded operation in $N(G_2)$ that acts only on the first and nth qubits. Gottesman has shown in [10] that as far as the single-qubit operations are concerned, the only two-qubit gates in $N(G_2)$ are: identity, CNOT gate, SWAP gate, and CNOT gate, followed by SWAP gate. If the automorphism U yields to the action of the SWAP gate, our design is completed. On the other hand, if the automorphism U yields to the action of the CNOT gate, by using the equivalency in Fig. 11.15A we can implement the SWAP gate. Finally, if the automorphism U yields to the action of the CNOT gate followed by the SWAP gate, the desired SWAP gate that swaps the first qubit being in the encoded state $|\bar{0}\rangle$ and nth-encoded qubit can be implemented based on the circuit shown in Fig. 11.16A. Notice that the SWAP of

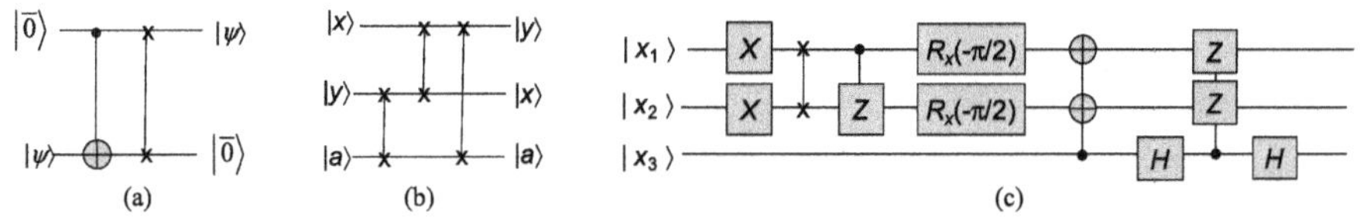

FIGURE 11.16

(a) Quantum circuit to swap encoded qubits 1 and n, (b) quantum circuit to swap kets $|x\rangle$ and $|y\rangle$ by means of auxiliary ket $|a\rangle$, and (c) quantum circuit to perform the mapping due to Gottesman [9], given by Eq. (11.96).

two qubits in the block is not a transversal operation. An error introduced during the swapping operation can cause errors on both qubits being swapped. This problem can be solved by using the circuit shown in Fig. 11.16B, in which two qubits to be swapped do not interact directly but by means of an ancilla qubit instead.

There are certain implementations, such as linear optics, in which the implementation of the CNOT gate is challenging, so that we should keep the number of CNOT gates low. Typically, the CNOT gate in linear optics is implemented as a probabilistic gate [12–14], which performs a desired action with certain probability. For such technologies we can implement the SWAP gate by means of *quantum teleportation* [15] introduced in Chapter 3. Alternatively, the highly nonlinear optical fibers can be used in combination with a four-wave mixing effect for quantum teleportation. Namely, in quantum teleportation, the entanglement in the Bell state, also known as the Einstein–Podolsky–Rosen pair, is used to transport arbitrary quantum state $|\psi\rangle$ between two distant observers A and B (often called Alice and Bob). The quantum teleportation system employs three qubits, qubit 1 is an arbitrary state to be transferred, while qubits 2 and 3 are in Bell state $(|00\rangle + |11\rangle)/\sqrt{2}$. Observer A has access to qubits 1 and 2, while observer B has access to qubit 3. The observers also share an auxiliary classical communication channel. Therefore the initial three-qubit state is $|\psi\rangle \otimes (|00\rangle + |11\rangle)/\sqrt{2}$. The corresponding generators of the stabilizer are $g_1 = IXX$ and $g_2 = IZZ$, while the encoded Pauli operator analogs are $\widetilde{X} = XII$, $\widetilde{Z} = ZII$. Observer A first applies the CNOT gate by using qubit 1 as the control qubit and qubit 2 as the target qubit. Based on Eq. (11.47) we know that generators and encoded Pauli operator analogs will be mapped to:

$$\begin{aligned} g_1 &= IXX \rightarrow g_1' = IXX, \quad g_2 = IZZ \rightarrow g_2' = ZZZ, \\ \widetilde{X} &= XII \rightarrow \widetilde{X}' = XXI, \quad \widetilde{Z} = ZII \rightarrow \widetilde{Z}' = ZII. \end{aligned} \tag{11.87}$$

Observer A further applies the Hadamard gate on qubit 1, which performs the following mapping:

$$\begin{aligned} g_1' &= IXX \rightarrow g_1'' = IXX, \quad g_2' = ZZZ \rightarrow g_2'' = XZZ, \\ \widetilde{X}' &= XXI \rightarrow \widetilde{X}'' = ZXI, \quad \widetilde{Z}' = ZII \rightarrow \widetilde{Z}'' = XII. \end{aligned} \tag{11.88}$$

Observer A then performs measurement on operators $O_1 = Z_1$ and $O_2 = Z_2$ by using the measurement procedure (protocol) described in Section 11.3.3. This measurement protocol performs the following mapping:

$$\begin{aligned} g_1'' &= IXX \rightarrow g_1''' = O_2, \quad g_2'' = XZZ \rightarrow g_2''' = O_1, \\ \widetilde{X}'' &= ZXI \rightarrow \widetilde{X}''' = g_1''\widetilde{X}'' = ZIX, \quad \widetilde{Z}'' = XII \rightarrow \widetilde{Z}''' = g_2''\widetilde{Z}'' = IZZ. \end{aligned} \tag{11.89}$$

By discarding the measured qubits we conclude that we have effectively performed the following mapping: $X_1 \rightarrow X_3, Z_1 \rightarrow Z_3$, meaning that qubit 1 has been transferred to qubit 3, which is equivalent to the SWAP operation. We now explain how to use the quantum teleportation procedure to perform encoded qubit swapping. We first prepare an encoded block in which all qubits are in state $|\overline{0}\rangle$ except the nth-encoded qubit that is an arbitrary state $|\psi\rangle$. We further prepare another block in which all qubits are in state $|\overline{0}\rangle$ except the first and nth-encoded qubits that are prepared into encoded Bell state $(|\overline{00}\rangle+|\overline{11}\rangle)/\sqrt{2}$. Quantum teleportation is then performed between the nth-encoded qubits in two blocks, which finishes by transferring the nth-encoded qubit from the first block into the first encoded qubit in the second block, while leaving the nth-encoded qubit in the second block in state $|\overline{0}\rangle$. By discarding the first coded block we effectively performed an encoded SWAP operation. This procedure requires: block-encoded CNOT gate, block-Hadamard gate, and $O_n = Z_n$ measurements, which is doable for the arbitrary stabilizer code. Therefore we have just completed a description of how to fault tolerantly apply encoded operations in the Clifford group $N(G_N)$ for arbitrary stabilizer code. Before concluding this section, in the following subsection we provide an illustrative example of [5,1,3] stabilizer code, which is due to Gottesman [9] (see also [2]).

11.3.5 The [5,1,3] Fault-Tolerant Stabilizer Code

This is a stabilizer code that can correct a single error (because the minimum distance is 3), and it is also a perfect quantum code, as it satisfies the quantum Hamming bound with equality sign (see Section 7.4). This quantum code is generated by the following four generators of the stabilizer: $g_1 = XZZXI$, $g_2 = IXZZX$, $g_3 = XIXZZ$, and $g_4 = ZXIXZ$, and the corresponding quantum-check matrix is given by:

$$A_Q = \begin{bmatrix} X \quad Z \\ 10010|01100 \\ 01001|00110 \\ 10100|00011 \\ 01010|10001 \end{bmatrix}. \tag{11.90}$$

Since every next row is the cyclic shift of the previous row, this code is a cyclic quantum code. The corresponding encoded Pauli operators, by using standard form formalism, are obtained as follows:

$$\overline{X} = X_1X_2X_3X_4X_5, \quad \overline{Z} = Z_1Z_2Z_3Z_4Z_5. \tag{11.91}$$

In Section 11.3.2 we introduced the following Pauli gates cyclic permutation operator:

$$U: X \rightarrow Y \rightarrow Z \rightarrow X. \tag{11.92}$$

When this operator is applied transversally to the codeword of [5,1,3] code, it leaves the stabilizer S invariant as shown here:

$$g_1 = XZZXI \to YXXYI = g_3g_4,\ \ g_2 = IXZZX \to IYXXY = g_1g_2g_3,$$
$$g_3 = XIXZZ \to YIYXX = g_2g_3g_4,\ \ g_4 = ZXIXZ \to XYIYX = g_1g_2. \tag{11.93}$$

Since the operator U fixes the stabilizer S, when applied transversally, it can serve as an encoded fault-tolerant operation. To verify this claim, let us apply U to encoded Pauli operators:

$$\overline{X} = X_1X_2X_3X_4X_5 \to Y_1Y_2Y_3Y_4Y_5 = \overline{Y},\ \ \overline{Z} = Z_1Z_2Z_3Z_4Z_5 \to X_1X_2X_3X_4X_5 = \overline{X},$$
$$\overline{Y} = Y_1Y_2Y_3Y_4Y_5 \to Z_1Z_2Z_3Z_4Z_5 = \overline{Z}. \tag{11.94}$$

In Section 11.3.4 we introduced the four-qubit operation (see Eq. (11.66)) that can be used, when applied transversally, to generate an encoded version of itself for all stabilizer codes. Alternatively, the following three-qubit operation can serve the same role for [5,1,3] code [9]:

$$T_3 \doteq \begin{bmatrix} 1 & 0 & j & 0 & j & 0 & 1 & 0 \\ 0 & -1 & 0 & j & 0 & j & 0 & -1 \\ 0 & j & 0 & 1 & 0 & -1 & 0 & -j \\ j & 0 & -1 & 0 & 1 & 0 & -j & 0 \\ 0 & j & 0 & -1 & 0 & 1 & 0 & -j \\ j & 0 & 1 & 0 & -1 & 0 & -j & 0 \\ -1 & 0 & j & 0 & j & 0 & -1 & 0 \\ 0 & 1 & 0 & j & 0 & j & 0 & 1 \end{bmatrix}. \tag{11.95}$$

This three-qubit operation performs, with the corresponding quantum circuit shown in Fig. 11.16C, the following mapping on unencoded qubits:

$$XII \to XYZ,\ \ ZII \to ZXY,\ \ IXI \to YXZ,\ \ IZI \to XZY,\ \ IIX \to XXX,\ \ IIZ \to ZZZ. \tag{11.96}$$

The transversal application of T_3 performs the following mapping on the generators of [5,1,3] code:

$$\begin{aligned}
g_1 \otimes I \otimes I &= XZZXI \otimes IIIII \otimes IIIII \to XZZXI \otimes \underbrace{YXXYI}_{g_3g_4} \otimes \underbrace{ZYYZI}_{g_1g_3g_4} \\
&= g_1 \otimes g_3g_4 \otimes g_1g_3g_4 \\
g_2 \otimes I \otimes I &= IXZZX \otimes IIIII \otimes IIIII \to IXZZX \otimes \underbrace{IYXXY}_{g_1g_2g_3} \otimes \underbrace{IZYYZ}_{g_1g_3} \\
&= g_2 \otimes g_1g_2g_3 \otimes g_1g_3 \\
g_3 \otimes I \otimes I &\to g_3 \otimes g_2g_3g_4 \otimes g_2g_4,\ \ g_4 \otimes I \otimes I \to g_4 \otimes g_1g_2 \otimes g_1g_2g_4 \\
I \otimes g_1 \otimes I &\to g_3g_4 \otimes g_1 \otimes g_1g_3g_4,\ \ I \otimes g_2 \otimes I \to g_1g_2g_3 \otimes g_2 \otimes g_1g_3 \\
I \otimes g_3 \otimes I &\to g_2g_3g_4 \otimes g_3 \otimes g_2g_4,\ \ I \otimes g_4 \otimes I \to g_1g_2 \otimes g_4 \otimes g_1g_2g_4 \\
I \otimes I \otimes g_n &\to g_n \otimes g_n \otimes g_n;\ \ \ n = 1,2,3,4.
\end{aligned} \tag{11.97}$$

By closer inspection of Eq. (11.97) we conclude that the T_3-operator applies the U and U^2 on the other two blocks, except the last line. Since both of these operators fixes S, it turns out that T_3 fixes $S \otimes S \otimes S$ as well for all lines in Eq. (11.97) except the last one. On the other hand, by inspection of the last line in Eq. (11.97) we conclude that the T_3-operator also fixes $S \otimes S \otimes S$ for this case. Therefore the T_3-operator applied transversally represents an encoded fault-tolerant operation for the [5,1,3] code. In similar fashion, by applying the U and U^2 on the other two blocks in encoded Pauli operators we obtain the following mapping:

$$\begin{aligned}&\overline{X}\otimes I\otimes I\rightarrow\overline{X}\otimes\overline{Y}\otimes\overline{Z},\ \overline{Z}\otimes I\otimes I\rightarrow\overline{Z}\otimes\overline{X}\otimes\overline{Y},\ I\otimes\overline{X}\otimes I\rightarrow\overline{Y}\otimes\overline{X}\otimes\overline{Z},\\&I\otimes\overline{Z}\otimes I\rightarrow\overline{X}\otimes\overline{Z}\otimes\overline{Y},\ I\otimes I\otimes\overline{X}\rightarrow\overline{X}\otimes\overline{X}\otimes\overline{X},\ I\otimes I\otimes\overline{Z}\rightarrow\overline{Z}\otimes\overline{Z}\otimes\overline{Z}.\end{aligned} \tag{11.98}$$

Therefore the T_3-operator, applied transversally, has performed the mapping to the encoded version of the operation itself. Now, by combining the operations U and T_3 together with the measurement protocol we are in a position to implement fault tolerantly all operators in $N(G_N)$ for [5,1,3] stabilizer code. Because of the equivalence of Eqs. (11.96) and (11.98), it is sufficient to describe the unencoded design procedures. We start our description with the phase gate implementation description. Our starting point is the following three-qubit system: $|00\rangle \otimes |\psi\rangle$ The generators of the stabilizer are $g_1 = ZII$ and $g_2 = IZI$, while the corresponding encoded Pauli operator analogs are $\widetilde{X} = IIX$, $\widetilde{Z} = IIZ$. The application of T_3 operation results in the following mapping:

$$\begin{aligned}&g_1 = ZII \rightarrow g'_1 = ZXY,\quad g_2 = IZI \rightarrow g'_2 = XZY,\\&\widetilde{X} = IIX \rightarrow \widetilde{X}' = XXX,\quad \widetilde{Z} = IIZ \rightarrow \widetilde{Z}' = ZZZ.\end{aligned} \tag{11.99}$$

Next, we perform the measurement of operators $O_1 = IZI$ and $O_2 = IIZ$. We see that g'_1 and $\widetilde{X}'$ anticommute with O_1, while g'_1, g'_2, and $\widetilde{X}'$ anticommute with O_2. Based on measurement protocol we perform the following mapping:

$$\begin{aligned}&g'_1 = ZXY \rightarrow g''_1 = O_1,\quad g'_2 \rightarrow g''_2 = O_2,\\&\widetilde{X}' = XXX \rightarrow \widetilde{X}'' = g'_1\widetilde{X}' = YIZ,\quad \widetilde{Z}' \rightarrow \widetilde{Z}'' = ZZZ.\end{aligned} \tag{11.100}$$

By disregarding the measurement qubits (qubits 2 and 3) we effectively performed the following mapping $X_3 \rightarrow Y_1$, and $Z_3 \rightarrow Z_1$, which is in fact the phase gate combined with transfer from qubit 3 to qubit 1. Since:

$$U^{\dagger}P = \frac{1}{\sqrt{2}}\begin{bmatrix}1 & j\\ j & 1\end{bmatrix} = R_x(-\pi/2),$$

and the Hadamard gate can be represented as $H = PR_x(-\pi/2)P^{\dagger}$, we can implement an arbitrary single-qubit gate in the Clifford group. For arbitrary Clifford operator implementation, the CNOT gate is needed. The starting point is the following three-qubit system: $|\psi_1\rangle \otimes |\psi_2\rangle \otimes |0\rangle$. The generator of the stabilizer is given by $g = IIZ$, while the encoded Pauli operator analogs are given by $\widetilde{X}_1 = XII, \widetilde{X}_2 = IXI, \widetilde{Z}_1 = ZII, \widetilde{Z}_2 = IZI$. We then apply the T_3-operator, which performs the following mapping:

$$g = IIZ \to g' = ZZZ,\ \tilde{X}_1 = XII \to \tilde{X}'_1 = XYZ,\ \tilde{X}_2 = IXI \to \tilde{X}'_2 = YXZ,$$
$$\tilde{Z}_1 = ZII \to \tilde{Z}'_1 = ZXY,\ \tilde{Z}_2 = IZI \to \tilde{Z}'_2 = XZY. \tag{11.101}$$

We further perform the measurement of $O = IXI$. Clearly, g', $\tilde{X}'_1$ and $\tilde{Z}_2$ anticommute with O, and the measurement protocol performs the mapping:

$$\begin{aligned} &g' \to g'' = O,\ \tilde{X}'_1 = XYZ \to \tilde{X}''_1 = g'\tilde{X}'_1 = YXI,\ \tilde{X}'_2 \to \tilde{X}''_2 = YXZ, \\ &\tilde{Z}'_1 \to \tilde{Z}''_1 = ZXY,\ \tilde{Z}'_2 = XZY \to \tilde{Z}''_2 = g'\tilde{Z}'_2 = YIX. \end{aligned} \tag{11.102}$$

By disregarding the measurement qubit (namely, qubit 2) we effectively performed the following mapping:

$$X_1 I_2 \to Y_1 I_3,\ \tilde{X}_2 = I_1 X_2 \to Y_1 Z_3,\ \tilde{Z}_1 = Z_1 I_2 \to Z_1 Y_3,\ I_1 Z_2 \to Y_1 X_3. \tag{11.103}$$

By relabeling qubit 3 in the images with qubit 2, the mapping (11.103) can be described by the following equivalent gate U_e [9] (see also [1]):

$$U_e = \left(I \otimes U^2\right)(P \otimes I) U_{\mathrm{CNOT}}^{(21)} (I \otimes R_x(-\pi/2)), \tag{11.104}$$

where with $U_{\mathrm{CNOT}}^{(21)}$ we denoted the CNOT gate with qubit 2 as the control qubit and qubit 1 as the target qubit. By using the fact that the U gate is of order 3 ($U^3 = I \Leftrightarrow \left(U^2\right)^{\dagger} = U$), the $U_{\mathrm{CNOT}}^{(21)}$ gate can be expressed in terms of the U_e gate as:

$$U_{\mathrm{CNOT}}^{(21)} = \left(P^{\dagger} \otimes I\right)(I \otimes U) U_e \left(I \otimes (R_x(-\pi/2))^{\dagger}\right). \tag{11.105}$$

The CNOT gate $U_{\mathrm{CNOT}}^{(21)}$ can be implemented based on $U_{\mathrm{CNOT}}^{(21)}$ and four Hadamard gates in a fashion similar to that shown in Fig. 11.15B. This description of CNOT gate implementation completes the description of all gates needed for implementation of an arbitrary Clifford operator. For the implementation of a universal set of gates required for quantum computation we have to study the design of the Toffoli gate, which is the subject of the next section.

11.4 FAULT-TOLERANT QUANTUM COMPUTATION

As already discussed in Chapter 3, there exist various sets of universal quantum gates: (1) {Hadamard, phase, $\pi/8$ gate, CNOT} gates, (2) {Hadamard, phase, CNOT, Toffoli} gates, (3) the {Barenco} gate [16], and (4) the {Deutsch} gate [17]. To complete set (1), the fault-tolerant $\pi/8$ gate is needed, which was discussed in Section 11.2.5 (see also [5]). To complete set (2), the Toffoli gate [2,3,11] is needed, which is the subject of interest in this section. The following description of the fault-tolerant Toffoli gate is based on Gottesman's proposal [11] (see also [2]).

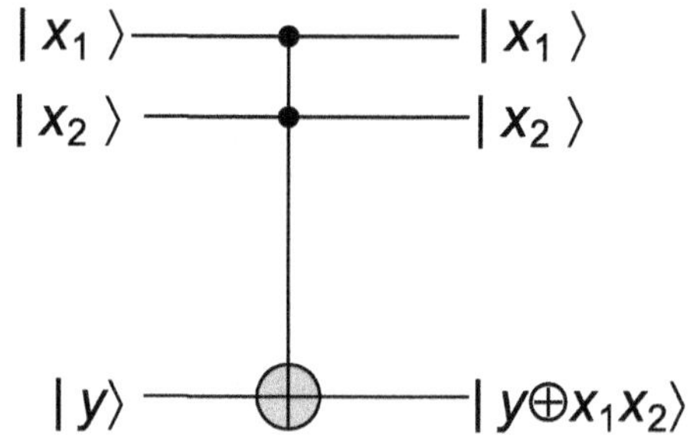

FIGURE 11.17

Toffoli gate description.

11.4.1 Fault-Tolerant Toffoli Gate

The Toffoli gate has already been introduced in Chapter 2, and its operation principle is described in Fig. 11.17. It can be considered as a generalization of the CNOT gate, which has two control qubits ($|x_1\rangle$ and $|x_2\rangle$) and one target qubit $|y\rangle$. The NOT gate operation is applied to the target qubit only when $x_1 = x_2 = 1$:

$$U_{\text{Toffoli}}|x_1x_2\rangle|y\rangle = |x_1x_2\rangle|y \oplus x_1x_2\rangle, \tag{11.106}$$

where with $\oplus$ we denote mod 2 addition, as usual. For convenience we can express the Toffoli gate action in terms of projection operators. Namely, we have shown earlier (see Chapter 2), that for arbitrary Hermitian operator O of order 2 ($O^2 = I$, with eigenvalues ± 1), the projection operators are defined as:

$$P_{\pm}(O) = \frac{1}{2}(I \pm O). \tag{11.107}$$

By using the projection operators we can express the Toffoli operator as follows [11] (see also [2]):

$$\begin{aligned} U_{\text{Toffoli}} &= \frac{1}{2}\{P_+(Z_1) + P_+(Z_2) + P_-(Z_1Z_2)\} + P_-(Z_1)P_-(Z_2)X_3 \\ &= \frac{1}{4}[3I + Z_1 + Z_2 - Z_1Z_2 + (I - Z_1)(I - Z_2)X_3]. \end{aligned} \tag{11.108}$$

Namely, Eq. (11.108) can be interpreted as:

$$\begin{aligned} U_{\text{Toffoli}}|x_1x_2y\rangle &= \frac{1}{2}\left(\delta_{x_1,0} + \delta_{x_2,0} + \delta_{x_1+x_2,1}\right)|x_1x_2y\rangle + \delta_{x_1x_2,1}|x_1x_2y\rangle \\ &= \begin{cases} |00y\rangle; & x_1 = x_2 = 0 \\ |10y\rangle; & x_1 = 1, x_2 = 0 \\ |01y\rangle; & x_1 = 0, x_2 = 1 \\ |11(y\oplus 1)\rangle; & x_1 = 1, x_2 = 1 \end{cases}, \end{aligned} \tag{11.109}$$

proving therefore its correctness. We are interested to see the action of the Toffoli operator on elements from the Pauli multiplicative group G_3, which is essentially the conjugation operation:

$$f(a) = U_{\text{Toffoli}} a U_{\text{Toffoli}}^{\dagger}, \quad \forall a \in G_3. \tag{11.110}$$

If this mapping is a homomorphism, then it will be sufficient to study the action of the Toffoli operator on generator elements from G_3. Let us first verify whether the multiplication table is preserved by conjugation:

$$\begin{aligned} f(ab) &= U_{\text{Toffoli}} ab U_{\text{Toffoli}}^{\dagger} = U_{\text{Toffoli}} a \overbrace{U_{\text{Toffoli}} U_{\text{Toffoli}}^{\dagger}}^{I} b U_{\text{Toffoli}}^{\dagger} \\ &= \underbrace{U_{\text{Toffoli}} a U_{\text{Toffoli}}}_{f(a)} \underbrace{U_{\text{Toffoli}}^{\dagger} b U_{\text{Toffoli}}^{\dagger}}_{f(b)} = f(a) f(b), \quad \forall a, b \in G_3. \end{aligned} \tag{11.111}$$

Since the conjugation function of the product of two elements from G_3 equals the product of images, clearly the mapping is a homomorphism. It is evident that the image of arbitrary element from G_3 under conjugation by the Toffoli operator is contained in $N(G_3)$, which indicates that the Toffoli operator does not fix G_3, and consequently it cannot belong to the normalizer of G_3, $N(G_3)$. Since $G_3 \subset G_N$, the Toffoli operator cannot fix G_N and consequently the Toffoli operator does not belong to the Clifford group $N(G_N)$. Let us now study the action of the Toffoli operator on X_i, Z_i $(i = 1,2,3)$. We start the study with mapping $f(X_1)$ [11]:

$$\begin{aligned} X_1 \rightarrow U_{\text{Toffoli}} X_1 U^{\dagger} &= \frac{1}{4}[3I + Z_1 + Z_2 - Z_1 Z_2 + (I - Z_1)(I - Z_2) X_3] X_1 \\ &\frac{1}{4}[3I + Z_1 + Z_2 - Z_1 Z_2 + (I - Z_1)(I - Z_2) X_3] \\ &= \underbrace{\frac{1}{2}(I + Z_2 + (I - Z_2) X_3)}_{U_{\text{CNOT}}^{(23)}} X_1 \\ &= U_{\text{CNOT}}^{(23)} X_1. \end{aligned} \tag{11.112}$$

The mapping $f(X_2)$ can be obtained in analogy to mapping (11.112) as follows:

$$\begin{aligned} X_2 \rightarrow U_{\text{Toffoli}} X_2 U_{\text{Toffoli}}^{\dagger} &= \frac{1}{4}[3I + Z_1 + Z_2 - Z_1 Z_2 + (I - Z_1)(I - Z_2) X_3] X_2 \\ &\frac{1}{4}[3I + Z_1 + Z_2 - Z_1 Z_2 + (I - Z_1)(I - Z_2) X_3] \\ &= \underbrace{\frac{1}{2}(I + Z_1 + (I - Z_1) X_3)}_{U_{\text{CNOT}}^{(13)}} X_2 \\ &= U_{\text{CNOT}}^{(13)} X_2. \end{aligned} \tag{11.113}$$

The operators X_3, Z_1, and Z_2 do not change during conjugation mapping, while the mapping $f(Z_3)$ is given by:

$$
\begin{aligned}
Z_3 \rightarrow U_{\text{Toffoli}} Z_3 U_{\text{Toffoli}}^{\dagger} &= \frac{1}{4}[3I + Z_1 + Z_2 - Z_1 Z_2 + (I - Z_1)(I - Z_2)X_3]Z_3 \\
& \frac{1}{4}[3I + Z_1 + Z_2 - Z_1 Z_2 + (I - Z_1)(I - Z_2)X_3] \\
&= \underbrace{\frac{1}{2}(I + Z_1 + (I - Z_1)Z_2)\, Z_3}_{U_{\text{CP}}^{(12)}} \\
&= U_{\text{CP}}^{(12)} Z_3,
\end{aligned}
\tag{11.114}
$$

where U_{CP} is the control-phase (or control-Z) gate. Based on Eqs. (11.112) −(11.114), the following stabilizer generators:

$$
g_1 = U_{\text{CNOT}}^{(23)} X_1,\ \ g_2 = U_{\text{CNOT}}^{(13)} X_2,\ \ g_3 = U_{\text{CP}}^{(12)} Z_3 \tag{11.115}
$$

fix the state:

$$
A = \frac{1}{2}(|000\rangle + |010\rangle + |100\rangle + |111\rangle). \tag{11.116}
$$

We now create the composite state $|A\rangle \otimes \underbrace{|\psi_1\rangle \otimes |\psi_2\rangle \otimes |\psi_3\rangle}_{|\psi\rangle}$. The corresponding generators of stabilizer for this composite state are:

$$
G_1 = U_{\text{CNOT}}^{(23)} X_1 III,\ \ G_2 = U_{\text{CNOT}}^{(13)} X_2 III,\ \ G_3 = U_{\text{CP}}^{(12)} Z_3 III, \tag{11.117}
$$

while the corresponding encoded Pauli operator analogs are:

$$
\tilde{X}_1 = X_4, \tilde{X}_2 = X_5,\ \ \tilde{X}_3 = X_6,\ \ \tilde{Z}_1 = Z_4,\ \ \tilde{Z}_2 = Z_5, \tilde{Z}_3 = Z_6. \tag{11.118}
$$

By applying the CNOT gates $U_{\text{CNOT}}^{(14)}, U_{\text{CNOT}}^{(25)}, U_{\text{CNOT}}^{(63)}$ we perform the following mapping:

$$
\begin{aligned}
& G_1 \rightarrow G_1' = U_{\text{CNOT}}^{(23)} X_1 X_4,\ \ G_2 \rightarrow G_2' = U_{\text{CNOT}}^{(13)} X_2 X_5,\ \ G_3 \rightarrow G_3' = U_{\text{CP}}^{(12)} Z_3 Z_6, \\
& \tilde{X}_1 \rightarrow \tilde{X}_1' = X_4, \tilde{X}_2 \rightarrow \tilde{X}_2' = X_5, \tilde{X}_3 \rightarrow \tilde{X}_3' = X_3 X_6,\ \ \tilde{Z}_1 \rightarrow \tilde{Z}_1' = Z_1 Z_4, \\
& \tilde{Z}_2 \rightarrow \tilde{Z}_2' = Z_2 Z_5,\ \ \tilde{Z}_3 \rightarrow \tilde{Z}_3' = Z_6.
\end{aligned}
\tag{11.119}
$$

By measuring the operators $O_1 = Z_4$, $O_2 = Z_5$, and $O_3 = X_6$, since the operator O_1 (O_2) anticommutes with $G_1', \tilde{X}_1' (G_2', \tilde{X}_2')$ and operator O_3 anticommutes with $G_3', \tilde{Z}_3'$ the measurement protocol performs the mapping:

$$G'_1 \to G''_1 = O_1,\ G'_2 \to G''_2 = O_2,\ G'_3 \to G''_3 = O_3,$$
$$\widetilde{X}'_1 \to \widetilde{X}''_1 = G'_1\widetilde{X}'_1 = U^{(23)}_{\text{CNOT}}X_1 = G_1,\ \widetilde{X}'_2 \to \widetilde{X}''_2 = G'_2\widetilde{X}_2 = U^{(13)}_{\text{CNOT}}X_2 = G_2,$$
$$\widetilde{X}'_3 \to \widetilde{X}''_3 = X_3X_6,$$
$$\widetilde{Z}'_1 \to \widetilde{Z}''_1 = Z_1Z_4,\ \widetilde{Z}'_2 \to \widetilde{Z}''_2 = Z_2Z_5,\ \widetilde{Z}'_3 \to \widetilde{Z}''_3 = G'_3\widetilde{Z}'_3 = U^{(12)}_{\text{CP}}Z_3 = G_3. \tag{11.120}$$

By closer inspection of Eqs. (11.118) and (11.120), and by disregarding the measured qubits (1, 2, and 3), we conclude that the foregoing procedure has performed the following mapping:

$$X_4 \to g_1,\ X_5 \to g_2,\ X_6 \to X_3,\ Z_4 \to Z_1,\ Z_5 \to Z_2,\ Z_6 \to g_3, \tag{11.121}$$

which is the same as the homomorphism introduced by the Toffoli operator given by Eqs. (11.112)–(11.114), in combination with the transfer of data from qubits 1–3 to qubits 4–6.

The encoded Toffoli gate can now be implemented by using the encoded CNOT gate and measurement. The encoded g_3 operation, denoted as $\overline{g}_3$, can be represented as the product of the encoded controlled-phase gate on encoded qubits 1 and 3 and encoded Pauli operator $\overline{Z}_3$, namely $\overline{g}_3 = \overline{U}^{(12)}_{\text{CP}}\overline{Z}_3$ (see Eq. (11.115)). From Fig. 11.13 we know that the encoded controlled-phase (controlled-*Z*) gate can be expressed in terms of the encoded CNOT gate as follows:

$$\overline{U}^{(12)}_{\text{CP}} = (I \otimes \overline{H})\overline{U}^{(12)}_{\text{CNOT}}(I \otimes \overline{H}). \tag{11.122}$$

This equation allows us to implement $\overline{U}^{(12)}_{\text{CP}}$ transversally since we have already learned how to implement $\overline{U}^{(12)}_{\text{CNOT}}$ and $\overline{H}$ transversally. Because the encoded Z_3-operator can be represented by $\overline{Z}_3 = j^{l_3}X(\boldsymbol{a}_3)Z(\boldsymbol{b}_3)$, it contains only single-qubit operators and its action is by default transversal.

The previous description and discussion are correct providing that ancilla state $|A\rangle$ can be prepared fault tolerantly. To be able to perform the Toffoli gate on encoded states we must find a way to generate the encoded version of $|A\rangle$, denoted as $|\overline{A}\rangle$ To facilitate implementation we introduce another auxiliary encoded state $|\overline{B}\rangle$ that is related to $|\overline{A}\rangle$ by $|\overline{B}\rangle = \overline{X}_3|\overline{A}\rangle$, where $\overline{X}_3$ is the encoded X_3-operator. We will assume that the corresponding coding blocks required are obtained from an [*N*,1,*D*] stabilizer code. For other [*N*,*K*,*D*] stabilizer codes we will need to start with three ancilla blocks with encoded qubits being in the $|\overline{0}\rangle$ state. We then further need to apply the procedure leading to the encoded state $|\overline{A}\rangle$. The first portion in these blocks will be in the $|\overline{A}\rangle$ state, while the remaining qubits will stay in the $|\overline{0}\rangle$ state. The three encoded data qubits to be involved in the Toffoli gate will further be moved into three ancilla blocks initialized into the $|\overline{0}\rangle$ state. Furthermore, the Toffoli gate procedure will be applied to the six ancilla blocks by using block-encoded CNOT gates and the measurement protocol. Finally, the encoded data qubits will be transferred back to the original code blocks. Therefore the study of encoding procedures by using [N,1,D] stabilizer code will not affect the generality of procedures. By ignoring the normalization constants, the encoded $|\overline{A}\rangle$ and $|\overline{B}\rangle(=\overline{X}_3|\overline{A}\rangle)$ states can be represented as:

$$\begin{aligned}|\overline{A}\rangle &= |\overline{000}\rangle + |\overline{010}\rangle + |\overline{100}\rangle + |\overline{111}\rangle,\\ |\overline{B}\rangle &= \overline{X}_3|\overline{A}\rangle = |\overline{001}\rangle + |\overline{011}\rangle + |\overline{101}\rangle + |\overline{110}\rangle.\end{aligned} \tag{11.123}$$

It is interesting to notice that the action of the encoded g_3-operator on encoded states $|\overline{A}\rangle$ and $|\overline{B}\rangle$ is as follows:

$$\overline{g}_3|\overline{A}\rangle = |\overline{A}\rangle, \quad \overline{g}_3|\overline{B}\rangle = -|\overline{B}\rangle. \tag{11.124}$$

Therefore the transversal application of the encoded g_3-operator maps $|\overline{A}\rangle \to |\overline{A}\rangle$ and $|\overline{B}\rangle \to -|\overline{B}\rangle$ From Eq. (11.123) it is clear that:

$$\begin{aligned}|\overline{A}\rangle + |\overline{B}\rangle &= |\overline{000}\rangle + |\overline{010}\rangle + |\overline{100}\rangle + |\overline{111}\rangle + |\overline{001}\rangle + |\overline{011}\rangle + |\overline{101}\rangle + |\overline{110}\rangle\\ &= \left(|\overline{0}\rangle + |\overline{1}\rangle\right)^3 = \sum_{d_1,d_2,d_3=0}^{1} |\overline{d_1 d_2 d_3}\rangle,\end{aligned} \tag{11.125}$$

which indicates that the state $|\overline{A}\rangle + |\overline{B}\rangle$ can be obtained by measuring $\overline{X}$ for each code block. Let the ancilla block be prepared in cat state $|0\ldots0\rangle + |1\ldots1\rangle$ and the code block in state $|\overline{A}\rangle + |\overline{B}\rangle$. By executing the encoded g_3-operator only when the ancilla is in state $|1\ldots1\rangle$ we will effectively perform the following mapping on the composite state:

$$(|0\ldots0\rangle + |1\ldots1\rangle)(|\overline{A}\rangle + |\overline{B}\rangle) \to (|0\ldots0\rangle + |1\ldots1\rangle)|\overline{A}\rangle + (|0\ldots0\rangle - |1\ldots1\rangle)|\overline{B}\rangle. \tag{11.126}$$

Now, we perform the measurement of $O = X_1 \ldots X_N$ on the ancilla cat-state block. We know that:

$$X_1 \ldots X_N(|0\ldots0\rangle \pm |1\ldots1\rangle) = \pm(|0\ldots0\rangle \pm |1\ldots1\rangle), \tag{11.127}$$

so that if the result of the measurement is +1, the code block will be in state $|\overline{A}\rangle$, otherwise it will be in state $|\overline{B}\rangle$ When in state $|\overline{B}\rangle$ we need to apply the encoded operator $\overline{X}_3$ to obtain $\overline{X}_3|\overline{B}\rangle = |\overline{A}\rangle$ The only problem that has remained is the cat-state preparation circuit that is not fault tolerant, as we discussed in Section 11.3.1. The quantum circuit from Fig. 11.12 (when Hadamard gates are omitted) generates the cat state where there is no error. It also verifies that up to the order of p (p is the single-qubit error probability) there is no bit-flip error present. However, the phase error in cat-state generation will introduce an error in the implementation of state $|\overline{A}\rangle$. Since:

$$Z_i(|0\ldots0\rangle + |1\ldots1\rangle)(|\overline{A}\rangle + |\overline{B}\rangle) = (|0\ldots0\rangle - |1\ldots1\rangle)(|\overline{A}\rangle + |\overline{B}\rangle), \tag{11.128}$$

the application of mapping Eq. (11.126) in the presence of the phase-flip error yields to:

$$Z_i(|0\ldots0\rangle + |1\ldots1\rangle)(|\overline{A}\rangle + |\overline{B}\rangle) \rightarrow (|0\ldots0\rangle - |1\ldots1\rangle)|\overline{A}\rangle + (|0\ldots0\rangle + |1\ldots1\rangle)|\overline{B}\rangle. \tag{11.129}$$

By measuring the operator $O = X_1 \ldots X_N$ on the ancilla cat-state block, the result $+1$ leads to $|\overline{B}\rangle$, while the result -1 leads to $|\overline{A}\rangle$. This problem can be solved by repeating the cat-state preparation procedure and measurement of O every time with a fresh ancilla block until we get the same result twice in a row. It can be shown that in this case up to the order of p, the code block will be in state $|\overline{A}\rangle$.

11.5 ACCURACY THRESHOLD THEOREM

In our discussion about the Shor-state preparation circuit we learned that up to $O(P_e)$ no flip error has occurred, where P_e is the qubit-error probability. Moreover, in our discussion of the fault-tolerant Toffoli gate we learned that up to the order of P_e we can prepare state $|\overline{A}\rangle$ fault tolerantly. Therefore the encoding data using QECCs and applying fault-tolerant operations can also introduce new errors if the qubit-error probability P_e is not sufficiently low. It can be shown that the probability for a fault-tolerant circuit to introduce two or more errors into a single code block is proportional to P_e^2 for some constant of proportionality C. So the encoded procedure will be successful with probability $1 - CP_e^2$. If the probability of qubit-error P_e is small enough, e.g., $P_e < 10^{-4}$, there is a real benefit of using fault-tolerant technology. For an arbitrary long quantum computation with the number of qubits involved in the order of millions, the failure probability per block CP_e^2 is still too high. Namely, the success probability for an N-qubit computation is approximately $(1 - CP_e^{2N})^N$. Even with millions of qubits being involved with a failure probability $P_e < 10^{-4}$ we cannot make the computation error arbitrary small. Knill and Laflamme [18] (see also [19–21]) solved this problem by introducing concatenated codes. We have already met the concept of concatenated codes in Shor's nine-qubit code discussion from Chapter 7. Moreover, the concept of quantum concatenated codes is very similar to the concept of classical concatenated codes introduced in Chapter 6. A concatenated code first encodes a set of K qubits by using an $[N,K,D]$ code. Every qubit from the codeword is encoded again by using an $[N_1,1,D_1]$ code. Such obtained codeword qubits are further encoded using an $[N_2,1,D_2]$ code, and so on. This procedure is illustrated in Fig. 11.18. After k concatenation stages, the resulting quantum code has the parameters $[NN_1N_2 \ldots N_k,K,DD_1D_2 \ldots D_k]$. Such obtained concatenated code has a failure probability of order $(CP_e)^{2^k}/C$. Therefore the improvement with quantum concatenated code is remarkable. With quantum concatenated code we are in a position to design QECCs with an arbitrarily small probability of error. Unfortunately, the complexity increases as the number of concatenation stages increases. In practice, careful engineering is needed to satisfy these conflicting requirements. We are now in a position to formulate the famous accuracy threshold theorem.

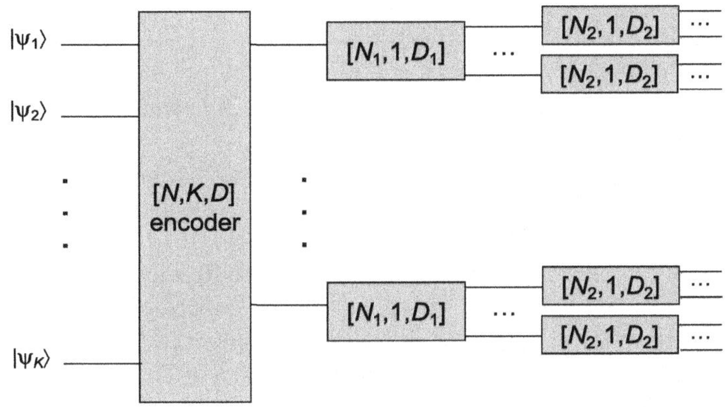

FIGURE 11.18

Principle of quantum concatenated code (two-stages example).

Accuracy threshold theorem [2]**:** A quantum computation operation of arbitrary duration can be performed with arbitrarily small failure error probability in the presence of quantum noise and by using imperfect (faulty) quantum gates providing that the following conditions are satisfied:

1. Computational operation is protected by using a concatenated QECC.
2. Fault-tolerant procedures are used for encoded quantum gates, error correction, and measurement protocols.
3. Storage quantum registers and a universal set of quantum gates are available that have failure error probabilities P_e, which are smaller than a threshold value P_{tsh}, known as threshold accuracy: $P_e < P_{tsh}$.

For the proof of the theorem, interested readers are referred to Chapter 6 of [2] (see also [11]). Another interesting formulation is due to Neilsen and Chuang [3].

Threshold theorem for quantum computation [3]**:** A quantum circuit containing $p(N)$ gates, where $p(\bullet)$ is the polynomial function, can be simulated with a failure probability of at most P_f by using:

$$O\left(\text{poly}\left(\frac{\log p(N)}{P_f}\right)p(N)\right) \tag{11.130}$$

quantum gates with quantum hardware whose components fail with probability at most P_e, provided P_e is below a constant threshold, $P_e < P_{tsh}$, and given reasonable assumptions about the quantum noise in the underlying quantum hardware. In Eq. (11.130), with poly($\bullet$) we denoted the polynomial complexity.

Namely, if the desired failure probability of a quantum computation circuit is P_f, the error probability per gate must be smaller than $P_f/p(N)$, where $p(N)$ is the number of gates. The number of stages needed in concatenated code for this accuracy can be determined from the following inequality:

$$\frac{(CP_e)^{2^k}}{C} \leq \frac{P_f}{p(N)}, \tag{11.131}$$

where $P_e < P_{tsh}$, $C = 1/P_{tsh}$. By solving Eq. (11.131) we obtain that the number of stages is lower bounded by:

$$k \geq \log_2\left\{1 + \frac{\log[p(N)/P_f]}{\log(P_{tsh}/P_e)}\right\}. \tag{11.132}$$

The size of the corresponding quantum circuit is given by d^k, where d is the maximum number of gates used in a fault-tolerant procedure to implement an encoded gate. From Eq. (11.132) it is clear that the quantum circuit size per encoded gate is of the order poly(log($p(N)/P_f$)), proving the claim of Eq. (11.130). This theorem is very important because it dispels the doubts about capabilities of quantum computers to perform arbitrarily long calculations due to the error accumulation phenomenon. For more rigorous proof, interested readers are referred to [2] (see also [11]).

11.6 SURFACE CODES AND LARGE-SCALE QUANTUM COMPUTING

Surface codes are closely related to quantum topological codes on the boundary, introduced by Bravyi and Kitaev [22,23]. This class of codes is highly popular in quantum computing [24,25]. The surface code is defined on a 2D lattice, such as the one shown in Fig. 11.19, with qubits being clearly indicated in Fig. 11.19A. Stabilizers of the plaquette type can be defined as provided in Fig. 11.19B. Each plaquette stabilizer denoted by $X(Z)$ is composed of Pauli $X(Z)$-operators on qubits located in the intersection of edges of the corresponding plaquette. As an illustration, the plaquette stabilizer denoted by X related to qubits 1 and 2 will be X_1X_2. On the other hand, the plaquette stabilizer denoted by Z related to qubits 5, 6, 8, and 9 will

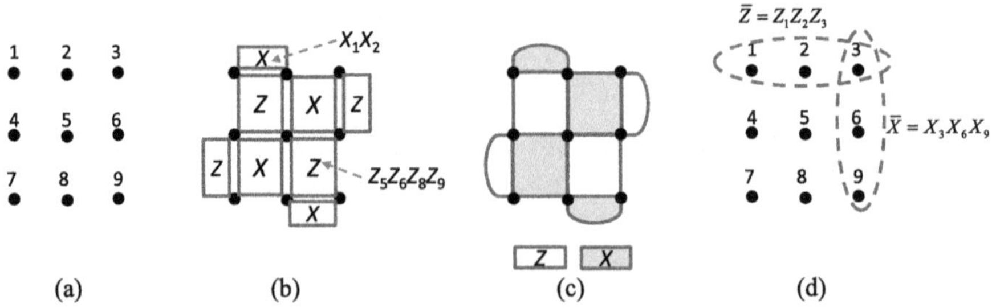

FIGURE 11.19

Surface code: (a) the qubits are located in the lattice positions, (b) all-X and all-Z plaquette operators, (c) popular representation of surface codes in which stabilizers are clearly indicated, and (d) logical operators.

be $Z_5Z_6Z_8Z_9$. To simplify the notation we can use the representation provided in Fig. 11.19C, where the shaded plaquettes correspond to all-X-containing operators' stabilizers, while the white plaquettes correspond to all-Z-containing operators' stabilizers. The weight-2 stabilizers are allocated around the perimeter, while weight-4 stabilizers are located in the interior. The logical operators for this code are run over both sides of the lattice as shown in Fig. 11.19D and can be represented as $\overline{Z} = Z_1Z_2Z_3$, $\overline{X} = X_3X_6X_9$. Clearly, the logical operator spans two boundaries of the same type.

The codeword length is determined as the product of side lengths, expressed in number of qubits, and for the surface code from Fig. 11.19 we have that $N = L_x \cdot L_z = 3 \cdot 3 = 9$. One the other hand, the number of information qubits is $K = 1$. The minimum distance of this code is determined as the minimum side length, that is, $d = \min(L_x, L_z) = 3$, indicating that this code can correct a single-qubit error. Given that the surface codes are topological codes, the decoding algorithm described in Chapter 9 (see Section 9.8) is applicable here as well.

Let us now provide another interpretation due to Litinski [25], which is suitable for large-scale quantum computing applications. In *Litinski's framework*, the surface code is represented as a game played on a board partitioned into a certain number of tiles, like in the example provided in Fig. 11.20A. The corresponding physical implementation is provided in Fig. 11.20B, with dots representing the physical qubits. On each tile we can place a logical qubit, represented as a *patch*. The edges of qubits represent the logical Pauli operators. The solid (dashed) edges represent the logical Pauli $Z(X)$-operator. Let us consider four-corner and six-corner patches. Four-corner patches have four edges, two dashed edges representing logical X qubits and two solid edges representing logical Z qubits. The logical qubits correspond to surface code patches, as shown in Fig. 11.20B. The patches can be of different shapes and can occupy more than one tile, see, for example, the logical qubit $|q_3\rangle$ (refer to Fig. 11.20). The six-corner patches occupy two tiles and represent two

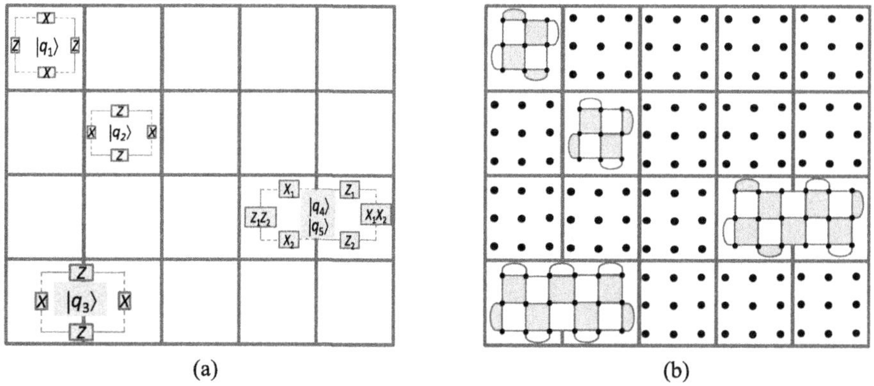

FIGURE 11.20

One-qubit and two-qubit patches: (a) notation and (b) actual physical implementation.

qubits $|q_4\rangle$ and $|q_5\rangle$, one on the top and the other on the bottom. The one on the top $|q_4\rangle$ has two edges representing X_1 and Z_1 logical Pauli operators, while the one on the bottom $|q_5\rangle$ also has two edges representing X_2 and Z_2 logical Pauli operators. The remaining edges represent the product operators X_1X_2 and Z_1Z_2. The surface code will now be represented by three dashed edges (*smooth (X) boundaries*) and three solid edges (*rough (Z) boundaries*).

Let us now specify the *rules of the game*, that is, the operations that can be applied to the patches (qubits), which can be categorized as: (1) *initialization*, (2) *qubit measurements*, and (3) *patch deformations*. With each of these operations we associate the cost, expressed in terms of time-steps, with each time-step (t.s.) corresponding to $\sim d$ code cycles (related to the measuring of all stabilizers d times), with d being the distance of underlying surface code per tile. One-qubit patches can be initialized to $|0\rangle$ or $|+\rangle$ states, while two-qubit patches can be initialized to $|00\rangle$ or $|++\rangle$ states, with associated cost being 0 t.s. (The logic $|+\rangle$ state indicates that all physical qubits are initialized into the $|+\rangle$ state.) In principle, one-qubit patches can be initialized to the arbitrary states, such as the *magic state* $|m\rangle = |0\rangle + \exp(j\pi/4)|1\rangle$; however, an undetected Pauli error can spoil the initialized state. As an illustration, by applying the Pauli error with probability p, the magic state can be converted to one of the following states: $|m_1\rangle = |0\rangle - \exp(j\pi/4)|1\rangle$, $|m_2\rangle = |1\rangle + \exp(j\pi/4)|0\rangle$, and $|m_2\rangle = |1\rangle - \exp(j\pi/4)|0\rangle$. When related to the surface codes, this corresponds to the *state injection* [26], which is illustrated in Fig. 11.21. In state injection we start with the physical magic state surrounded by the stabilizer states, and we switch the stabilizer configuration to include the magic state and create the patch, followed by corresponding measurements. *Single-patch measurements* can be performed in X or Z bases, and after the measurement the corresponding patches are removed from the board, thus freeing up the occupied tiles for future use. The cost associated with single-patch measurements is 0 t.s. For *two-patch measurements*, when the edges in neighboring tiles are different, we can perform the product measurements. As an illustration, the product $Z \otimes Z$ between adjacent patches can be measured as illustrated in Fig. 11.22 (left). In surface codes this corresponds to the *lattice surgery* [27], in which we change the configuration as shown in Fig. 11.22 (right) by introducing the patches with highlited edges;

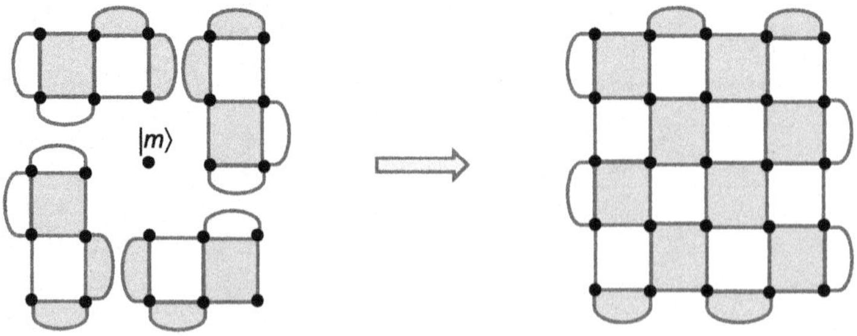

FIGURE 11.21

State injection procedure.

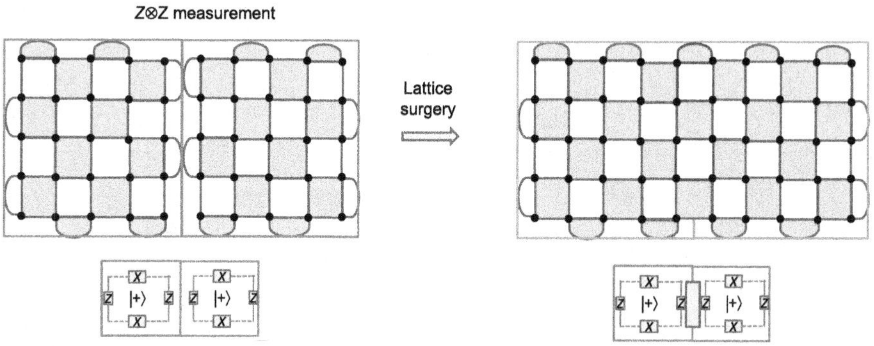

FIGURE 11.22

Lattice surgery procedure for the measurement on product $Z \otimes Z$.

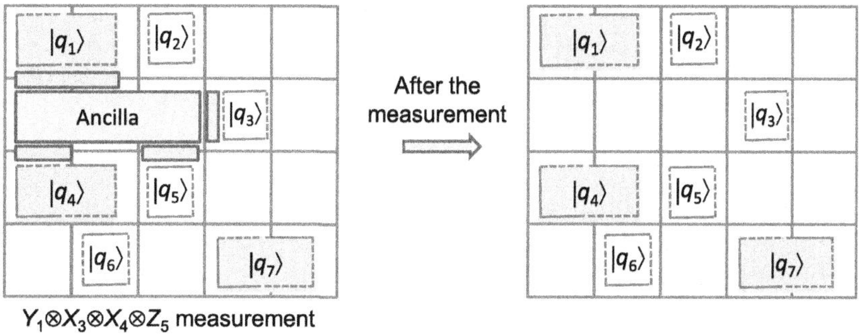

FIGURE 11.23

Multipatch measurement in 1 t.s.

after that we measure the stabilizers for d cycles to get the outcome of measurements, and split again. The cost associated with the lattice surgery is 1 t.s. This represents the way to introduce the entanglement between two adjacent patches. Namely, we start with the state $|++\rangle = 0.5(|00\rangle + |11\rangle + |01\rangle + |10\rangle)$ and perform the measurement on the $Z \otimes Z$-operator. If the result of the measurement is $+1$ the qubits end up in state $2^{-1/2}(|00\rangle + |11\rangle)$; otherwise, in state $2^{-1/2}(|01\rangle + |10\rangle)$. In either case, the qubits are maximally entangled. We can apply a similar procedure for the $X \otimes Z$ product operator. Of course, it is also possible to measure the product operator involving the Y-operator. What is even more interesting is that it is possible to measure the product for more than two encoded Pauli operators through *multipatch measurements*. The key idea here is to initialize the ancilla that exists for the duration of the multipatch measurement (i.e., for 1 t.s.). As an illustration, let us observe the $Y_{|q_1\rangle}X_{|q_3\rangle}X_{|q_4\rangle}Z_{|q_5\rangle}$ measurement example shown in Fig. 11.23. We perform the measurements on the desired product of Pauli operators next to the ancilla patch. Once the measurements are done, the ancilla patch goes. This approach is enabled by multiple lattice surgeries as described in [28].

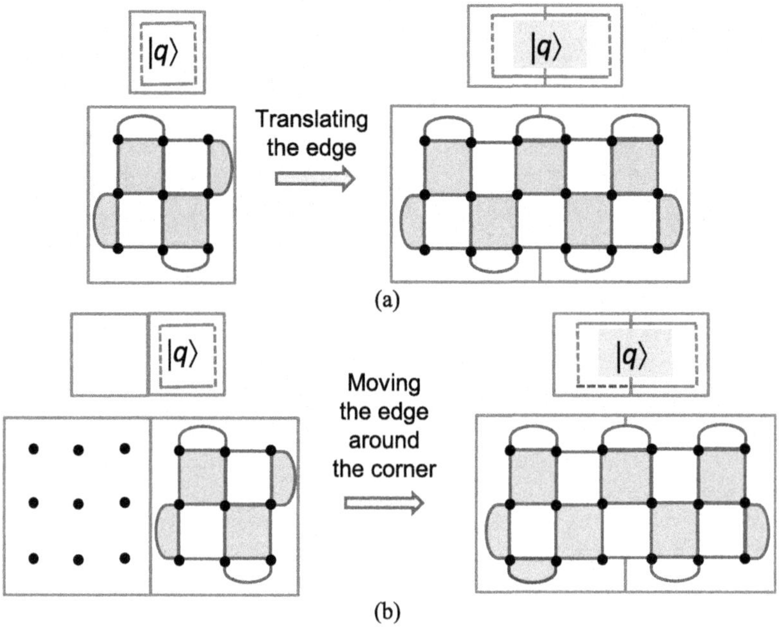

FIGURE 11.24

Patch deformation operations: (a) translating the edge to expand the patch and (b) moving the edge around the corner.

The *patch deformation* procedures are rather simple, and their purpose is to change the shape of the patch by moving the edges and corners as needed, which is illustrated in Fig. 11.24A. If we move the edge of the patch to expand it to an additional tile, this process will require 1 t.s. since we need to measure new stabilizers for decode cycles. On the other hand, if we reduce the size of the patch, it requires 0 t.s. since it involves just a bunch of measurements on physical qubits. We can move the edge around the corner as shown in Fig. 11.24B. Finally, we can move the corners of the patch along the patch boundary to change its shape as illustrated in Fig. 11.25. Given that this process involves new stabilizers, it will consume 1 t.s.

By using these patch operations we can implement a desired quantum circuit. The popular gates for quantum circuits are S gate, Hadamard gate, T gate, and CNOT gate. They can be expressed in terms of the Pauli product rotations $P_\varphi = \exp(-jP\varphi)$, where P is the Pauli product operator and φ is the rotation angle, which is illustrated in Fig. 11.26. The single-qubit rotations are straightforward to represent. The CNOT gate, on the other hand, can be represented as the $\pi/4$ rotation of two operators Z and X, and two single-qubit, Z and X, $-\pi/4$ rotations. The generic quantum circuit can be represented in terms of $\pi/4$ rotations, $\pi/8$ rotations, and measurements.

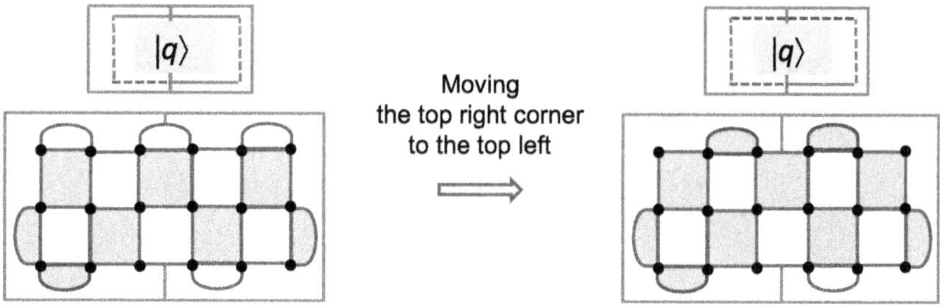

FIGURE 11.25

Moving the corner operation.

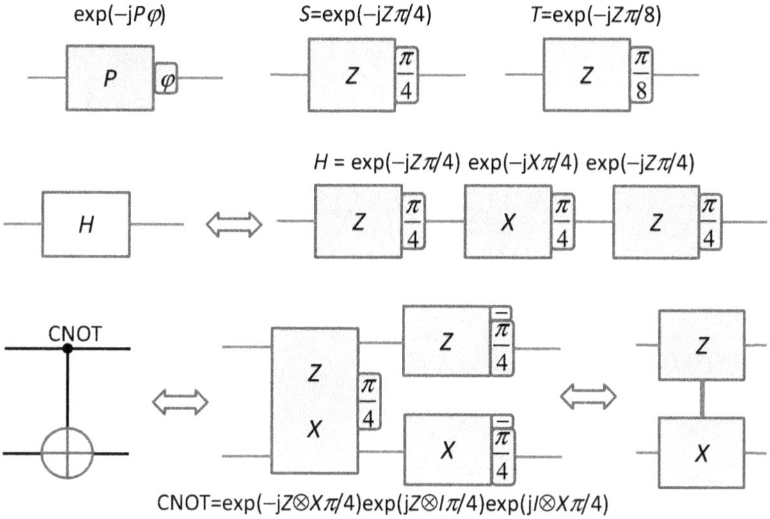

FIGURE 11.26

Single-qubit rotations and CNOT gate in terms of Pauli product rotations P_φ.

The single-qubit commutation rules, derived in Chapter 3, can be summarized as shown in Fig. 11.27 (top). In the same figure we provide the "absorption" rule of the measurements (Fig. 11.27 (bottom)), which can simplify significantly the computation complexity. Similarly to one-way quantum computing, it might be a good idea to propagate the one-qubit gates through CNOT gates and have them absorbed by applying the measurement's absorption rule. The CNOT gate propagation rules can be summarized as follows [29]:

$$\begin{aligned}&\mathrm{CNOT}^{(c,t)}X_t = X_t C_X^{(c,t)},\quad \mathrm{CNOT}^{(c,t)}X_c = X_c X_t \mathrm{CNOT}^{(c,t)},\\&\mathrm{CNOT}^{(c,t)}Z_c = Z_c \mathrm{CNOT}^{(c,t)},\quad \mathrm{CNOT}^{(c,t)}Z_t = Z_c Z_t \mathrm{CNOT}^{(c,t)},\end{aligned}\tag{11.133}$$

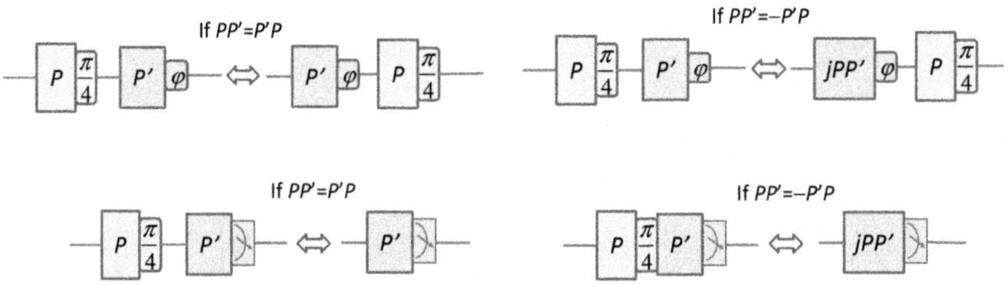

FIGURE 11.27

Single-qubit commutation rules and the measurement absorption rules.

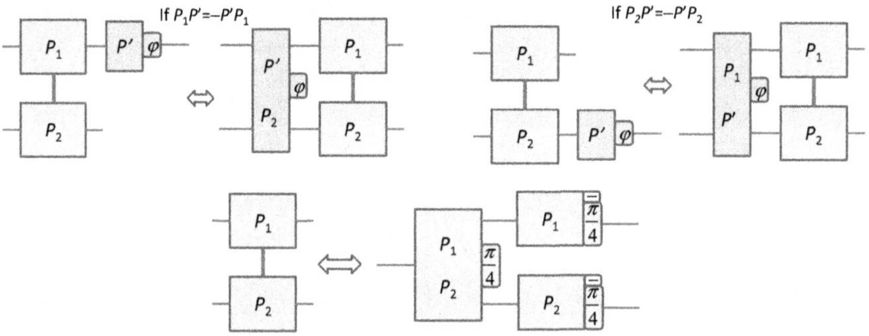

FIGURE 11.28

Controlled gate propagation rules.

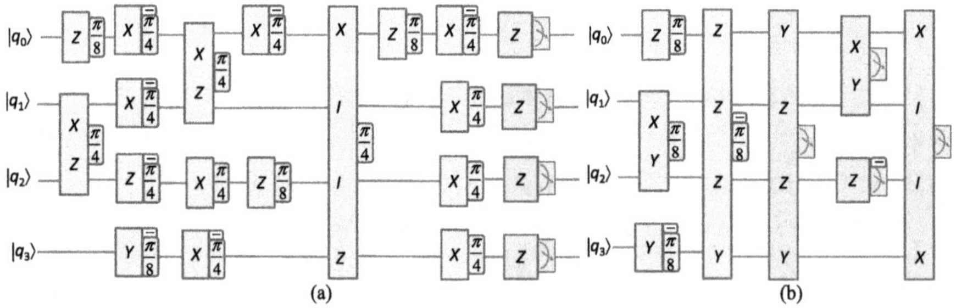

FIGURE 11.29

Quantum computation: (a) a generic quantum circuit and (b) equivalent quantum circuit.

with c and t being the control and target qubits, respectively. Alternatively, we can apply the approach due to Litinski to propagate all gates toward the end except for the T gates, which are used to create the magic states, so that the equivalent quantum circuit can be decomposed into two stages: the stage employing T gates, in both single-qubit and two-qubit operations, and the measurement stage. The CNOT

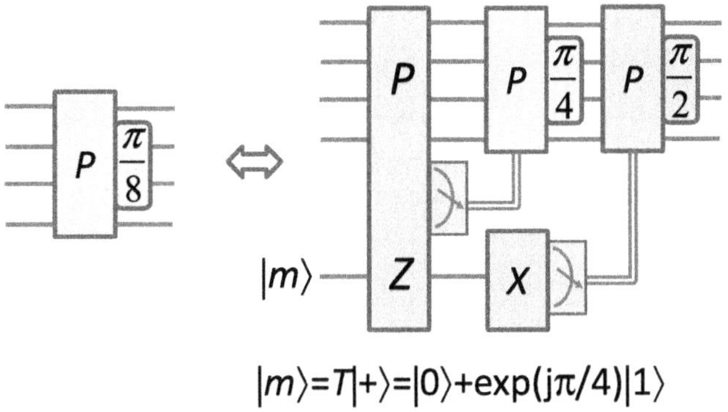

FIGURE 11.30

Magic state-based quantum circuit to perform the $\pi/8$ rotation.

gate propagation rules given by Eq. (11.133) can be generalized as shown in Fig. 11.28.

As an illustration, by applying the commutation rules from Figs. 11.27 and 11.28, the quantum circuit provided in Fig. 11.29A can be represented in equivalent form provided in Fig. 11.29B. Clearly, by employing the commutation rules, the computation complexity can be simplified. The equivalent circuit involves only $\pi/8$ rotations and measurements. In principle, any quantum circuit can be represented by employing only $\pi/8$ rotations and measurements. So far we have shown how to perform the measurements, but not $\pi/8$ rotations. If we want to perform $P\pi/8 = \exp(-jP\pi/8)$ rotation by employing the magic state $|m\rangle$, we can use the circuit provided in Fig. 11.30. To do so we can measure the Pauli product $P \otimes Z$ between the qubit and the magic state and depending on the outcome of this measurement the Clifford correction might be needed, which can be commuted to the end of the circuit. Unfortunately, the process of magic state creation is not fault tolerant. To solve this problem we can apply the *magic state distillation* procedure [25]. An intuitive approach to perform quantum computing would be to split the quantum circuit into two stages: (1) the *data block* (stage) that stores n qubits by employing $f(n)$ tiles, while consuming the magic state every m t.s., and (2) the *magic state distillation block* (stage) that employs x tiles and generates the magic state every y t.s.

As an illustration, let us consider the example provided in Fig. 11.31, where we are interested in implementing the gate $(X \otimes I \otimes Z \otimes Y)_{\pi/8}$ by employing the

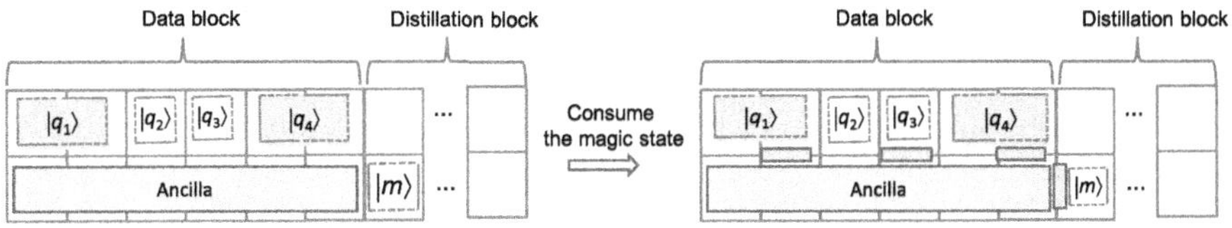

FIGURE 11.31

How the measurement of $X_1Z_3Y_4Z_m$ can implement the gate $(X\otimes I\otimes Z\otimes Y)_{\pi/8}$.

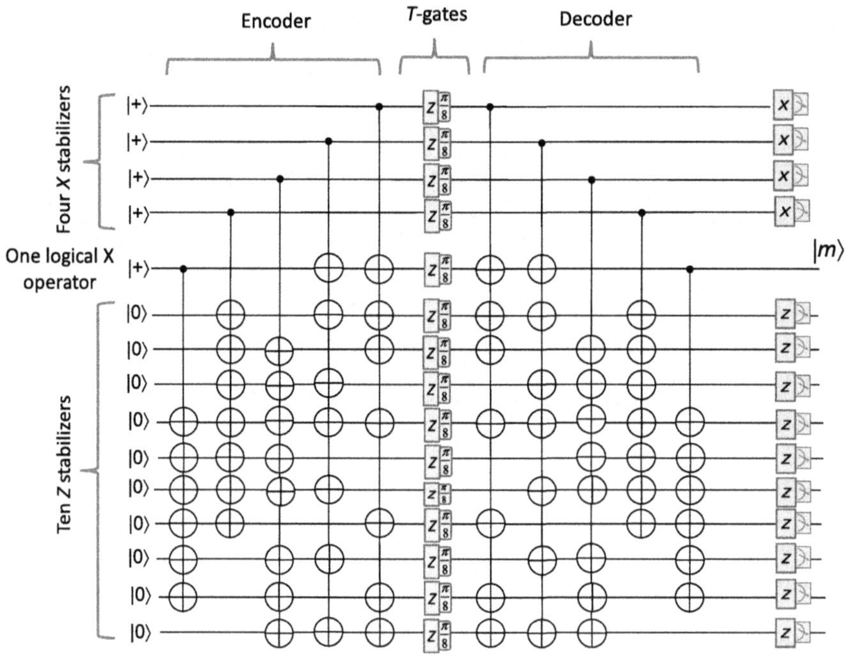

FIGURE 11.32

15-to-1 magic state distillation circuit. The output magic state error rate is $35p^3$, compared to the input error rate $p << 1$.

$X_1Z_3Y_4Z_m$ measurement. We initialize the ancilla, then perform the product measurements, and finally consume the magic state. So this data block consumes one magic state every 1 t.s. Once we are done with this $\pi/8$ rotation gate we move to the next one. Once we are done with all $\pi/8$ rotation gates we move to the measurement stage of the quantum computing circuit. More efficient implementations of the date blocks are discussed in [25]. What remains to explain is how to implement the magic state distillation stage (block).

Magic state distillation is the short computation step to generate the magic state, and given that we need to repeat this procedure many times, it is worth the effort to optimize it. The starting point could be the 15-to-1 distillation circuit due to Bravyi and Kitaev [30] (see also [25]), shown in Fig. 11.32. Clearly, this distillation circuit is based on 15-qubit quantum error correction code with transversal T gate, meaning that the logical T gate corresponds to the faulty T gates on each qubit. In this code there are four X stabilizers, 10 Z stabilizers, and one logical X-operator corresponding to the magic state location. Even though the T gates are faulty we perform the syndrome measurements and repeat the procedure until all syndromes are satisfied (all measured values are $+1$). At that point we can output the correct magic state. By applying the commutation rules from Figs. 11.27 and 11.28 we can simplify the magic state distillation circuit as shown in Fig. 11.33, which is suitable for

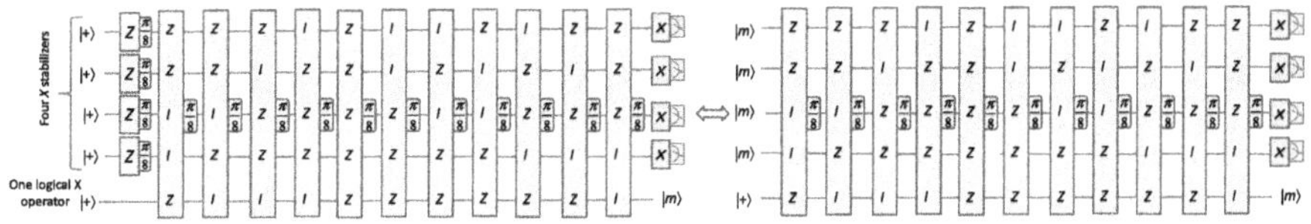

FIGURE 11.33

Equivalent 5-to-1 magic state distillation circuit employing 11 $\pi/8$ rotations.

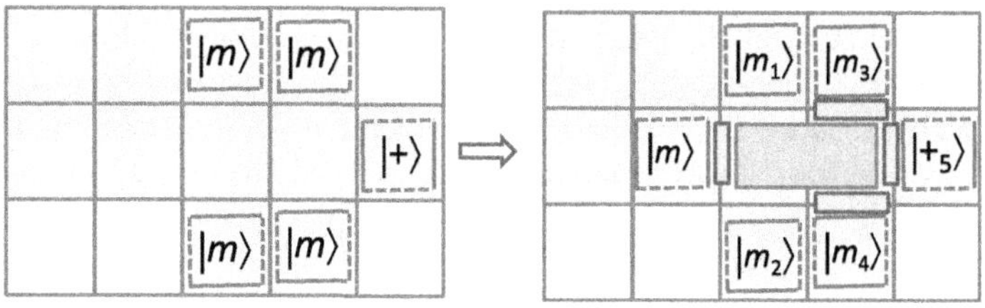

FIGURE 11.34

Toward the implementation of the distillation circuit shown in Fig. 11.33 by employing the $\pi/8$ rotation circuit from Fig. 11.30.

implementation by using the surface codes, by employing 11 tiles as shown in Fig. 11.34. Namely, by employing the $\pi/8$ rotation circuit from Fig. 11.30 we can consume an extra faulty magic state and perform the product measurements as shown in Fig. 11.34 (right). Unfortunately, with probability 50% the $\pi/8$ rotation will be converted to the $\pi/4$ rotation. For instance, with probability 50%, the $(I\otimes I\otimes Z\otimes Z\otimes Z)_{\pi/8}$ gate will be converted to the $(I\otimes I\otimes Z\otimes Z\otimes Z)_{\pi/4}$ gate. In principle, we can commute this to the end of the circuit; however, we will end up with the product of Pauli measurements instead of Pauli measurements, which are more difficult to implement.

Alternatively, we can use the autocorrected $\pi/8$ rotation circuit, proposed by Litinski [25] and provided in Fig. 11.35, which automatically applies the Clifford correction when needed. The corresponding implementation of the 5-to-1 distillation circuit from Fig. 11.33 (right) is illustrated in Fig. 11.36. To perform the $\pi/8$

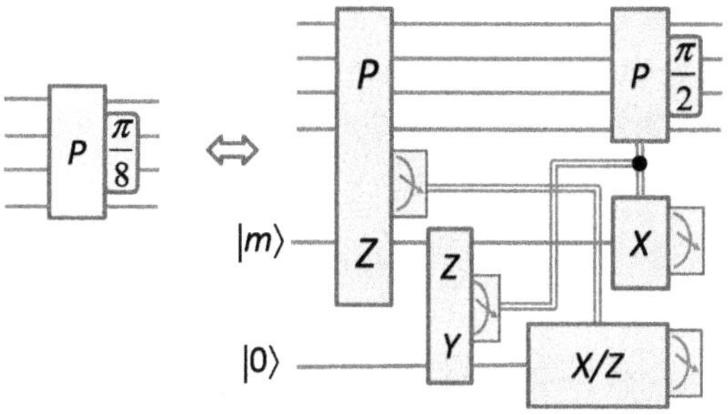

FIGURE 11.35

Autocorrected $\pi/8$ rotation circuit due to Litinski.

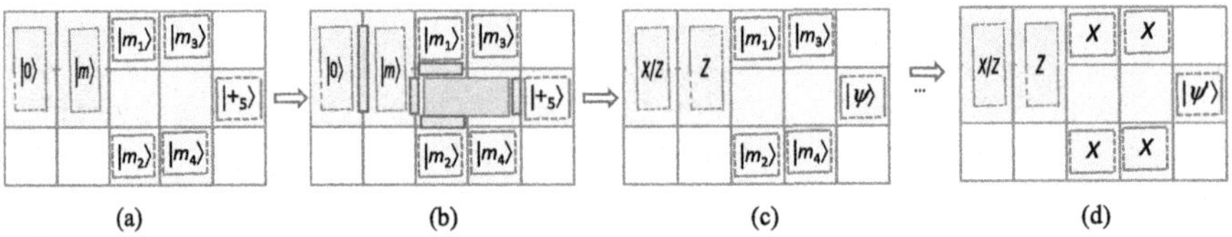

FIGURE 11.36

Implementation of the 5-to-1 distillation circuit from Fig. 11.28 (right).

rotation we initialize the faulty magic state and $|0\rangle$ state as shown in Fig. 11.36A. We then measure $P \otimes Z$, in this case $P = ZZIIZ$, and at the same time we measure $Z \otimes Y$ on magic and $|0\rangle$ states. Depending on the outcome of $P \otimes Z$ measurement, see Fig. 11.36B, we perform the measurement on the $|0\rangle$ state in either X basis or Z basis (either performing correction or not) as shown in Fig. 11.36C. This procedure is applied for all 11 $\pi/8$ rotations until we reach the end of the circuit, when we perform X measurements on all one to four qubits, see Fig. 11.36D. If all outcomes are $+1$ we have just distilled the magic state. We have done so by employing 11 tiles or $11d^2$ physical qubits; in other words, $11d + O(1)$ code cycles.

11.7 SUMMARY

This chapter was devoted to fault-tolerant quantum computation concepts. After the introduction of fault-tolerance basics and transversal operations (Section 11.1) we provided the basics of fault-tolerant quantum computation concepts and procedures (Section 11.2) by using Steane's code as an illustrative example. Section 11.3 was devoted to a rigorous description of fault-tolerant QECC. Section 11.4 was devoted to fault-tolerant computing, in particular to the fault-tolerant implementation of the Toffoli gate. In Section 11.5 we formulated and proved the accuracy threshold theorem. Section 11.6 was devoted to surface code-based large-scale computing. In the next section we provide a set of problems for interested readers to get a deeper understanding of fault-tolerance theory.

11.8 PROBLEMS

1. This problem is related to fault-tolerant T gate implementation. As the first step in derivation, show that the following three circuits are equivalent to each other:

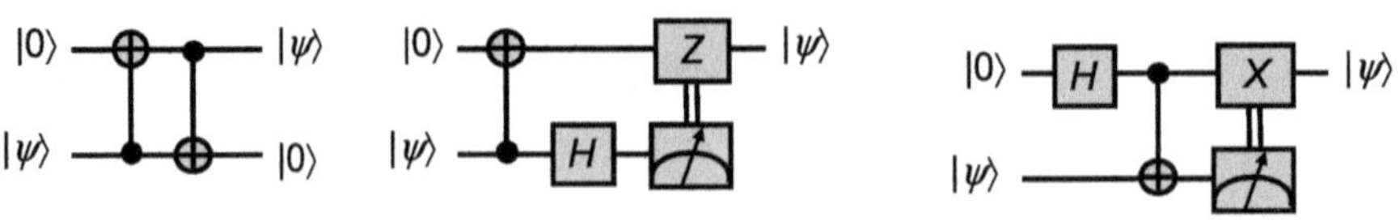

Now prove that the following circuit is in fact the fault-tolerant T gate:

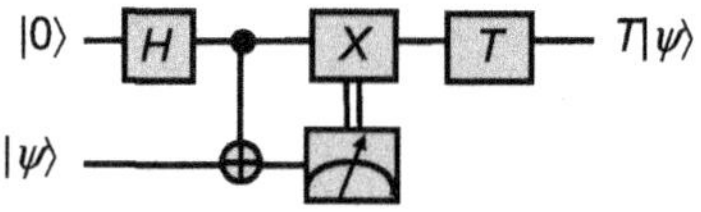

2. Determine the fault-tolerant Toffoli gate using the [7,1] Steane code.
3. In Section 11.3 we studied the fault-tolerant syndrome extraction circuit implementation of [5,1,3] code. The corresponding generators did not contain Y-operators. For an efficient implementation, the standard form described in Chapter 9 is desirable. However, the corresponding generators obtained from standard representation contain Y-operators:

   $$g_1 = Y_1Z_2Z_4Y_5,\ g_2 = X_2Z_3Z_4Z_5,\ g_3 = Z_1Z_2X_3X_5 \text{ and } g_4 = Z_1Z_3Y_4Y_5.$$

   Determine the quantum circuit for syndrome extraction based on the foregoing generators. Compare the complexity of the corresponding schemes. Determine also the syndrome extraction circuit using Plenio's approach. Discuss the complexity of this circuit with respect to previous ones.
4. The generators for [8,3,3] quantum stabilizer code are given by:

   $$g_1 = X_1X_2X_3X_4X_5X_6X_7X_8,\ g_2 = Z_1Z_2Z_3Z_4Z_5Z_6Z_7Z_8,$$
   $$g_3 = X_2X_4Y_5Z_6Y_7Z_8,\ g_4 = X_2Z_3Y_4X_6Z_7Y_8 \text{ and } g_5 = Y_2X_3Z_4X_5Z_6Y_8.$$

   Determine the corresponding transversal syndrome extraction circuit. Provide the configuration of the Shore-state preparation circuit as well. Discuss the Shor-state verification procedure. Finally, describe the syndrome verification procedure.
5. Prove that the following quantum circuit:

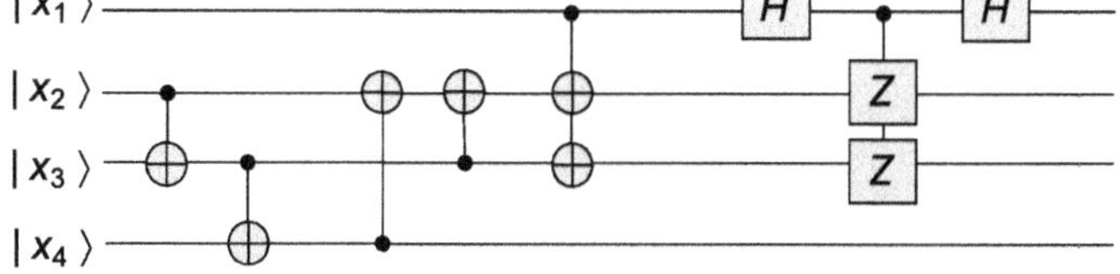

implements the following mapping:

$$X\otimes I\otimes I\otimes I\rightarrow X\otimes X\otimes X\otimes I$$
$$I\otimes X\otimes I\otimes I\rightarrow I\otimes X\otimes X\otimes X$$
$$I\otimes I\otimes X\otimes I\rightarrow X\otimes I\otimes X\otimes X$$
$$I\otimes I\otimes I\otimes X\rightarrow X\otimes X\otimes I\otimes X$$
$$Z\otimes I\otimes I\otimes I\rightarrow Z\otimes Z\otimes Z\otimes I$$
$$I\otimes Z\otimes I\otimes I\rightarrow I\otimes Z\otimes Z\otimes Z$$
$$I\otimes I\otimes Z\otimes I\rightarrow Z\otimes I\otimes Z\otimes Z$$
$$I\otimes I\otimes I\otimes Z\rightarrow Z\otimes Z\otimes I\otimes Z.$$

This circuit, when applied transversally, can generate an encoded version of itself for any $[N,1]$ stabilizer code as discussed earlier.

6. The one-qubit operator U performs the following Pauli operator cyclic shift:

$$U: X\rightarrow Y\rightarrow Z\rightarrow X.$$

Let us consider a two-qubit system: $|\psi\rangle\otimes|0\rangle_Y=|\psi\rangle\otimes(|0\rangle+j|1\rangle)/\sqrt{2}$, where $|\psi\rangle$ is an arbitrary state. Let us now apply the CNOT gate by using qubit 2 as the control qubit and qubit 1 as the target qubit, followed by measurement of $O=Y_1$. Prove that this procedure implements the cyclic shift of Pauli operators introduced earlier.

The arbitrary state can be represented by $|\psi\rangle=a|0\rangle+b|1\rangle$. Determine the state of qubit 2 upon application of the measurement protocol described in the previous paragraph. Finally, determine the matrix representation of U.

7. Prove that the following quantum circuit:

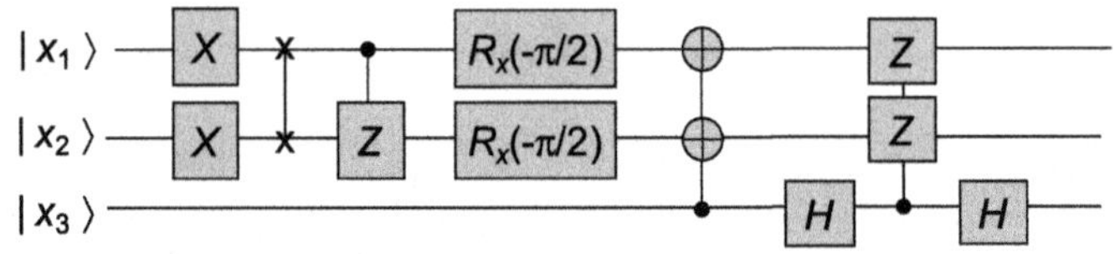

implements the following mapping:

$$XII\rightarrow XYZ,\ ZII\rightarrow ZXY,\ IXI\rightarrow YXZ,\ IZI\rightarrow XZY,\ IIX\rightarrow XXX,\ IIZ\rightarrow ZZZ.$$

This circuit, when applied transversally, can generate an encoded version of itself for the [5,1,3] stabilizer code as discussed earlier.

8. This problem is devoted to the fault-tolerant encoded operations [4,2,2] code and fault-tolerant computing based on this code. The generators of this code are given by $g_1=X_1X_2X_3X_4$ and $g_2=Z_1Z_2Z_3Z_4$. The corresponding encoded Pauli operators are given by

$$\overline{X}_1=X_1X_2,\ \overline{X}_2=X_1Z_3,\ \overline{Z}_1=Z_2Z_3,\ \overline{Z}_2=Z_3Z_4.$$

(a) Determine the mapping caused by transversal application of CNOT gates using the qubits in code block 1 as control qubits.
(b) Prove that encoded operation is a block-encoded CNOT gate.
(c) Describe how to implement a fault-tolerant SWAP gate.
(d) Describe how to implement a fault-tolerant Toffoli gate.

9. Let the quantum $[N,K,D]$ code be given. Consider now two encoded blocks. Prove that transversal application of the CNOT gate to these two encoded blocks yields to an encoded operation only if the quantum code is CSS code.

10. The generators for [8,3,3] quantum stabilizer code are given by:

$$g_1 = X_1X_2X_3X_4X_5X_6X_7X_8,\ g_2 = Z_1Z_2Z_3Z_4Z_5Z_6Z_7Z_8,\ g_3 = X_2X_4Y_5Z_6Y_7Z_8,$$
$$g_4 = X_2Z_3Y_4X_6Z_7Y_8 \text{ and } g_5 = Y_2X_3Z_4X_5Z_6Y_8.$$

(a) Determine the corresponding encoded Pauli operators.
(b) Determine the mapping caused by transversal application of CNOT gates using the qubits in code block 1 as control qubits.
(c) Prove that encoded operation is a block-encoded CNOT gate.
(d) Describe how to implement a fault-tolerant SWAP gate.
(e) Describe how to implement a fault-tolerant Toffoli gate.

11. Show that in the absence of phase errors in the cat state, the following circuit can be used to generate the state:

$$|A\rangle = [|000\rangle + |010\rangle + |100\rangle + |111\rangle]/2.$$

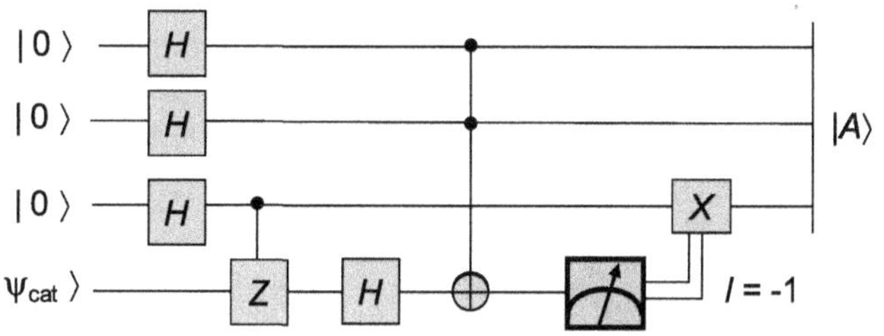

In the presence of phase error, the measurement result must be verified. This can be done by replacing the foregoing cat state with three cat states and then applying majority voting rule. Each cat state interacts with three qubits as in the foregoing figure, and upon measurements the X gate is applied if the majority of measurements results is -1. Provide the corresponding circuit. Justify that the measurement result will be accurate up to the order of P_e (qubit error probability). Prove that the following circuit applies the Toffoli gate to the state $|A\rangle$:

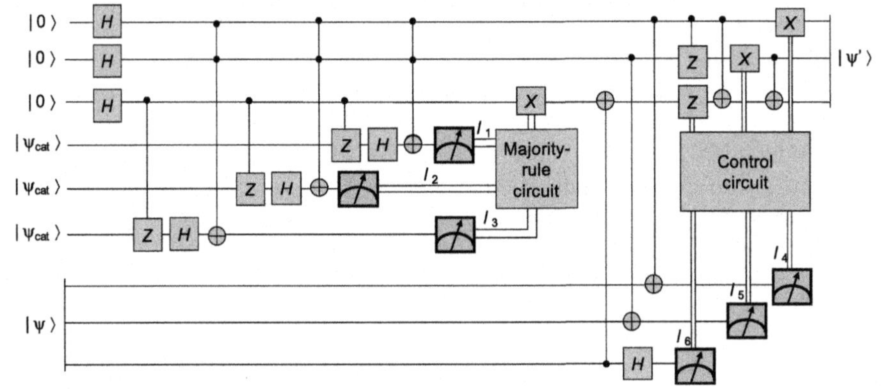

Describe the operation principle of this quantum circuit.

12. The quantum-check matrix $\boldsymbol{A}$ of a CSS code is given by:

$$\boldsymbol{A}=\begin{bmatrix}\boldsymbol{H} & | & \boldsymbol{0}\\ \boldsymbol{0} & | & \boldsymbol{H}\end{bmatrix}\quad \boldsymbol{H}=\begin{bmatrix}1 & 0 & 0 & 1 & 1 & 1\\ 1 & 1 & 1 & 0 & 0 & 1\\ 0 & 1 & 1 & 1 & 1 & 0\end{bmatrix}.$$

(a) Determine the mapping caused by transversal application of CNOT gates using the qubits in code block 1 as control qubits.
(b) Prove that encoded operation is a block-encoded CNOT gate.
(c) Describe how to implement a fault-tolerant SWAP gate.
(d) Describe how to implement a fault-tolerant Toffoli gate.

13. Let us observe the 5-to-1 magic state distillation circuit shown in Fig. 11.P13. By using the approach similar to one in Fig. 11.31, describe different steps in the distillation process.

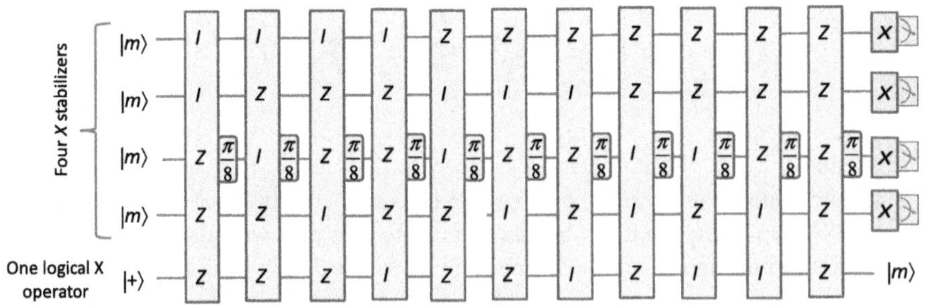

FIGURE 11.P13

5-to-1 magic state distillation circuit employing 11 $\pi/8$ rotations.

14. Prove the equivalence of the magic state-based $\pi/8$ rotation circuits shown in Figs. 11.30 and 11.35.

References

[1] D. Gottesman, I.L. Chuang, Demonstrating the viability of universal quantum computation using teleportation and single-qubit operations, Nature 402 (November 25, 1999) 390−393.

[2] F. Gaitan, Quantum Error Correction and Fault Tolerant Quantum Computing, CRC Press, 2008.

[3] M.A. Neilsen, I.L. Chuang, Quantum Computation and Quantum Information, Cambridge University Press, Cambridge, 2000.

[4] P.W. Shor, Fault-tolerant quantum computation, in: Proc. 37nd Annual Symposium on Foundations of Computer Science, IEEE Computer Society Press, 1996, pp. 56−65.

[5] X. Zhou, D.W. Leung, I.L. Chuang, Methodology for quantum logic gate construction, Phys. Rev. A 62 (5) (2000) paper 052316.

[6] M.B. Plenio, V. Vedral, P.L. Knight, Conditional generation of error syndromes in fault-tolerant error correction, Phys. Rev. A 55 (6) (1997) 4593−4596.

[7] A.Y. Kitaev, Quantum computations: algorithms and error correction, Russ. Math. Surv. 52 (6) (1997) 1191−1249.

[8] I.B. Djordjevic, Cavity quantum electrodynamics based quantum low-density parity-check encoders and decoders, in: Proc. SPIE Photonics West 2011, Advances in Photonics of Quantum Computing, Memory, and Communication IV, The Moscone Center, San Francisco, California, USA, 22 - 27 January 2011, vol. 7948, paper 794813.

[9] D. Gottesman, Theory of fault-tolerant quantum computation, Phys. Rev. A 57 (1) (1998) 127−137.

[10] D.P. DiVincenzo, P.W. Shor, Fault-tolerant error correction with efficient quantum codes,, Phys. Rev. Lett. 77 (15) (1996) 3260−3263.

[11] D. Gottesman, Stabilizer Codes and Quantum Error Correction, PhD Dissertation, California Institute of Technology, Pasadena, CA, 1997.

[12] T.C. Ralph, N.K. Langford, T.B. Bell, A.G. White, Linear optical Controlled-NOT gate in the coincidence basis, Phys. Rev. A 65 (2002), 062324.

[13] E. Knill, R. Laflamme, G.J. Milburn, A scheme for efficient quantum computation with linear optics, Nature 409 (2001) 46−52.

[14] A. Politi, M. Cryan, J. Rarity, S. Yu, J.L. O'Brien, Silica-on-silicon waveguide quantum circuits, Science 320 (5876) (May 2008) 646−649.

[15] C.H. Bennett, G. Brassard, C. Crépeau, R. Jozsa, A. Peres, W.K. Wootters, Teleporting an unknown quantum state via dual classical and Einstein-Podolsky-Rosen channels, Phys. Rev. Lett. 70 (13) (1993) 1895−1899.

[16] A. Barenco, A universal two-bit quantum computation, Proc. R. Soc. Lond. A, Math. Phys. Sci. 449 (1937) (June 1995) 679−683.

[17] D. Deutsch, Quantum computational networks, Proc. R. Soc. Lond. A, Math. Phys. Sci. 425 (1868) (September 1989) 73−90.

[18] E. Knill, R. Laflamme, Concatenated Quantum Codes, 1996. Available at: http://arxiv.org/abs/quant-ph/9608012.

[19] E. Knill, R. Laflamme, W.H. Zurek, Resilient quantum computation, Science 279 (5349) (January 16 , 1998) 342−345.

[20] C. Zalka, Threshold Estimate for Fault Tolerant Quantum Computation, 1996. Available at: http://arxiv.org/abs/quant-ph/9612028.

[21] M. Grassl, P. Shor, G. Smith, J. Smolin, B. Zeng, Generalized concatenated quantum codes,, Phys. Rev. A 79 (5) (2009) paper 050306(R).

[22] A.Y. Kitaev, Fault-tolerant quantum computation by anyons, Ann. Phys. 303 (1) (2003) 2−30. Available at: http://arxiv.org/abs/quant-ph/9707021.

[23] S.B. Bravyi, A.Y. Kitaev, Quantum Codes on a Lattice with Boundary, Available at: http://arxiv.org/abs/quant-ph/9811052.

[24] A.G. Fowler, M. Mariantoni, J.M. Martinis, A.N. Cleland, Surface codes: towards practical large-scale quantum computation, Phys. Rev. A 86 (3) (September 18 , 2012) 032324.

[25] D. Litinski, A game of surface codes: large-scale quantum computing with lattice surgery, Quantum 3 (2019) 128.

[26] A.J. Landahl, C. Ryan-Anderson, Quantum Computing by Color-Code Lattice Surgery, 2014. Available at: https://arxiv.org/abs/1407.5103.

[27] C. Horsman, A.G. Fowler, S. Devitt, R.V. Meter, Surface code quantum computing by lattice surgery, New J. Phys. 14 (2012) 123011.

[28] A.G. Fowler, C. Gidney, Low Overhead Quantum Computation Using Lattice Surgery, 2018. Available at: https://arxiv.org/abs/1808.06709.

[29] R. Raussendorf, D.E. Browne, H.J. Briegel, Measurement-based quantum computation with cluster states, Phys. Rev. A 68 (2003) 022312.

[30] S. Bravyi, A. Kitaev, Universal quantum computation with ideal Clifford gates and noisy ancillas, Phys. Rev. A 71 (2) (2005) 022316.

CHAPTER

Cluster State-based Quantum Computing

12

CHAPTER OUTLINE

12.1 CLUSTER STATES

The cluster state is a highly entangled state relevant in measurement-based quantum computing and one-way quantum computing applications [1−5]. By employing the cluster state and properly measuring the qubits in specific order and basis we can effectively perform arbitrary quantum computation [1−3]. By employing local adaptive measurements on qubits within the cluster state we exploit the quantum correlations to perform universal quantum computing. With cluster state-based quantum computing we avoid the need for coherent control of quantum states and reduce the problem of the creation of a highly entangled state at the initial stage, typically with the help of Ising-type interaction. The introduction of cluster states has been motivated by the development of qubits arranged in a 2D array [6,7] such as cold atoms in optical lattices and ion traps. Cluster state-based quantum computing can be summarized as follows [8,9]:

- Initialize N qubits in the same state, say $|+\rangle = (|0\rangle + |1\rangle)/\sqrt{2}$.
- Apply the controlled-phase (C_Z) gates to qubits being connected.
- Measure the qubits on the basis dictated by the previous measurement result, thus creating feedback.

Cluster state-based quantum computing has attracted significant research attention [10−15] and represents an active field of research even today [16−18]. The cluster states belong to the class of graph states [15,19], which also include Bell states, Greenberger−Horne−Zeilinger (GHZ) states, W states, and various entangled states used in quantum error correction. When the cluster C is defined

Quantum Information Processing, Quantum Computing, and Quantum Error Correction
https://doi.org/10.1016/B978-0-12-821982-9.00004-6

as a connected subset on a d-dimensional lattice, it obeys the set of eigenvalue equations:

$$K_a|\phi\rangle_C = (-1)^{k_a}|\phi\rangle_C,\ K_a = X_a \bigotimes_{b\in N(a)} Z_b, \tag{12.1}$$

where K_a are *correlation operators* with $N(a)$ denoting the neighborhood of $a \in C$, and $k_a = 0,1$. In practice, by setting $k_a = 0$ for all a's, we can relate correlation operators to the stabilizer formalism introduced in previous chapters. For the graph $G = (V,E)$, where V is the set of vertices (corresponding to qubits) and E is the set of edges (corresponding to the controlled-phase gates), we can define the *graph state* $|G\rangle$ as follows:

$$g_a|G\rangle = |G\rangle,\ g_a = X_a \bigotimes_{b\in N(a)} Z_b, \tag{12.2}$$

where g_a are the *stabilizing operators*. We can define the subset of stabilizing operators satisfying the following properties: (1) they commute among each other, (2) they are unitary and Hermitian, and (3) they are of the second order ($g_i^2 = I$), which are commonly referred to as the *generators* of stabilizer group S, denoted as $\{g_1,\ldots,g_{N-K}\}$. Any element of the stabilizer group, $s\in S$, can be represented in terms of a unique product of generators:

$$s = g_1^{c_1}\cdots g_{N-K}^{c_{N-K}},\ c_i\in\{0,1\};\ i = 1,\cdots,N-K. \tag{12.3}$$

Clearly, when $c_i = 1$ (0) the corresponding generator is included (excluded) from the product introduced above. The graph state $|G\rangle$ can be expanded using the computational basis as follows:

$$|G\rangle = \frac{1}{2^N}\sum_{\boldsymbol{x}\in\{0,1\}^N}(-1)^{\boldsymbol{x}^{\mathrm{T}}\boldsymbol{A}\boldsymbol{x}}|\boldsymbol{x}\rangle,\ \boldsymbol{x} = x_1\cdots x_N \in \{0,1\}^N, \tag{12.4}$$

where $\boldsymbol{A}$ is the adjacency matrix. For vertices $a,b \in V$, representing the end points of an edge, the corresponding element in the adjacency matrix is given by:

$$A_{a,b} = \begin{cases} 1, & \{a,b\}\in E \\ 0, & \text{otherwise} \end{cases}. \tag{12.5}$$

For weighted graphs, we also need to associate the weight with edge $\{a,b\}$, that is, $A_{a,b} = w_{a,b}$. The neighborhood of vertex a, denoted as $N(a)$, is defined as the set of vertices adjacent to it:

$$N(a) = \{b\in V|\{a,b\}\in E\}. \tag{12.6}$$

The number of neighbors related to a is called the degree of the vertex a. An ordered list of vertices $a_1 = a, a_2,\ldots,a_n = b$, wherein a_i and a_{i+1} are adjacent (for all i), is called the $\{a,b\}$ path. The graph that has an $\{a,b\}$ path for every $a,b \in V$ is called the *connected graph*.

As an illustration, in Fig. 12.1 we provided three examples of graph states: (1) three-vertex path, (2) triangle of three vertices, and (3) the star graph corresponding

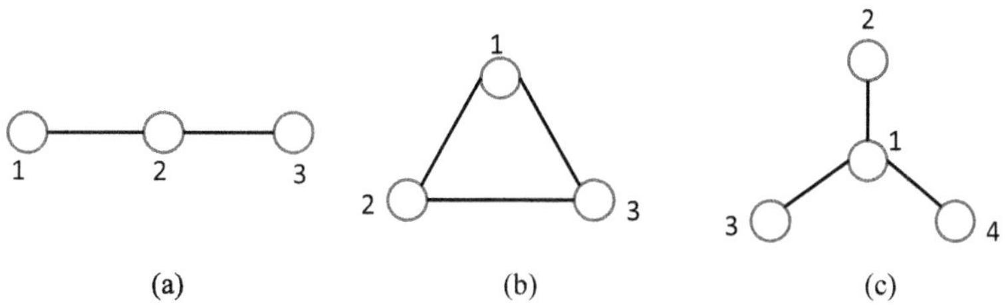

FIGURE 12.1

Graph states: (a) three-vertex path, (b) triangle of three vertices, and (c) the star graph.

to the $|\text{GHZ}\rangle$ state on $N=4$ qubits. The neighbors of vertex 1 in the star graph are $\{2,3,4\}$. The corresponding adjacency matrices are provided as follows:

$$\boldsymbol{A}_a = \begin{bmatrix} 0 & 1 & 0 \\ 1 & 0 & 1 \\ 0 & 1 & 0 \end{bmatrix}, \quad \boldsymbol{A}_b = \begin{bmatrix} 0 & 1 & 1 \\ 1 & 0 & 1 \\ 1 & 1 & 0 \end{bmatrix}, \quad \boldsymbol{A}_c = \begin{bmatrix} 0 & 1 & 1 & 1 \\ 1 & 0 & 0 & 0 \\ 1 & 0 & 0 & 0 \\ 1 & 0 & 0 & 0 \end{bmatrix}.$$

Based on Eq. (12.2) the corresponding stabilizers for the three-vertex path are *XZI*, *ZXZ*, and *IZX*, while for the triangle graph the stabilizers are *XZZ*, *ZXZ*, and *ZZX*. A two-qubit cluster state has stabilizers *XZ* and *ZX*, and the corresponding state can be represented by:

$$|\phi\rangle_2 = 2^{-1/2}(|0+\rangle + |1-\rangle), \tag{12.7}$$

and is related to the Bell state (up to corresponding local transformations). On the other hand, the cluster state corresponding to the three-vertex path is given by:

$$|\phi\rangle_3 = 2^{-1/2}(|{+}0{+}\rangle + |{-}1{-}\rangle), \tag{12.8}$$

and is related to the GHZ state (up to corresponding local transformations). The vertices in the graph represent $|+\rangle$ states, while the edges represent the controlled-phase gates. The corresponding quantum gates-based model for a three-vertex path graph is provided in Fig. 12.2. The action of a controlled-*Z* gate on initial state $|+\rangle|+\rangle$ can be described by:

$$C_Z^{(1,2)}|+\rangle|+\rangle = 2^{-1/2}(|0\rangle|+\rangle + |1\rangle|-\rangle). \tag{12.9}$$

The output of this circuit is the cluster state $|\phi\rangle_3$ since:

$$\begin{aligned} |\Psi_{\text{out}}\rangle &= 2^{-1/2} C_Z^{(2,3)}(|0\rangle_1|+\rangle_2 + |1\rangle_1|-\rangle_2)|+\rangle_3 \\ &= 2^{-1/2} C_Z^{(2,3)}(|+\rangle_1|0\rangle_2 + |-\rangle_1|1\rangle_2)|+\rangle_3 \\ &= 2^{-1/2}(|+\rangle_1|0\rangle_2|+\rangle_3 + |-\rangle_1|1\rangle_2|-\rangle_3) = |\phi_3\rangle. \end{aligned} \tag{12.10}$$

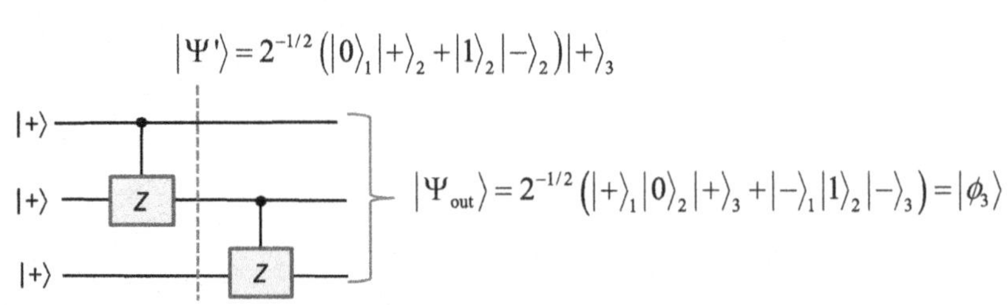

FIGURE 12.2

Quantum circuit corresponding to the three-qubit linear cluster state (three-vertex path graph).

The initial state does not need to be an all $|+\rangle$ state, and the gate connecting two neighboring nodes in the graph does not need to be the C_Z gate. For an arbitrary initial state $|\Psi\rangle$, any two-qubit unitary U_{ij} satisfying the following constraints can be used [19]:

1. Any two-qubit unitaries must commute since the graph is undirected:

$$[U_{ij}, U_{jk}] = 0, \ \forall \ i,j,k \in V. \tag{12.11}$$

2. Since the graph is undirected the unitaries must be symmetric:

$$U_{ij} = U_{ji}, \ \forall \ i,j \in V. \tag{12.12}$$

3. When the edges are not weighted, the same unitary U can be applied to any two neighboring qubits:

$$U_{ij} = U, \ \forall \ i,j \in V \tag{12.13}$$

The unitary U_{ij} can be related to the corresponding interaction Hamiltonian H_{ij} by:

$$U_{ij} = \exp(-j\varphi_{ij}H_{ij}), \tag{12.14}$$

where φ_{ij} is the interaction (coupling) strength. At this point it is convenient to introduce the Ising model [20]. The *Ising model*, named after German physicist Ernst Ising, was introduced to study the magnetic dipole moments of atoms in statistical physics. The magnetic dipole moments of the atomic spins can be in one of two states, either $+1$ or -1. The spins, represented as the discrete variables, are arranged in a graph, typically the *lattice* graph, and this model allows each spin to interact with the closest neighbors. The spins in the neighborhood having the same value tend to have lower energy; however, the heat disturbs this tendency so that there exist different phase transitions. Let the individual spin on the lattice be denoted by $\sigma_i \in$

$\{-1,1\}$; $i=1,2,\ldots,L$ with L being the lattice size. To represent the energy of the Ising model the following Hamiltonian is typically used:

$$H(\boldsymbol{\sigma}) = -\sum_{\langle i,j \rangle} J_{ij}\sigma_i\sigma_j - \mu\sum_i h_i\sigma_i, \tag{12.15}$$

where J_{ij} denotes the interaction strength between the ith and jth sites, while h_i donates the strength of the external magnetic field interacting with the ith spin site in the lattice, and μ is the magnetic moment. We use the notation $\langle i,j \rangle$ to denote the direct nearest neighbors. By replacing the spin variables with corresponding Pauli operators (matrices), we can express the Ising module using quantum mechanical interpretation:

$$\hat{H} = -\sum_{\langle i,j \rangle} J_{ij}Z_iZ_j - \sum_i h_iZ_i. \tag{12.16}$$

Now we can represent any set of commuting two-qubit unitaries, satisfying Eq. (12.11), with the help of a Hamiltonian of the form [19]:

$$\varphi_{ij}\hat{H}_{ij} = \varphi_{ij}Z_iZ_j + h_iZ_i + h_jZ_j. \tag{12.17}$$

The corresponding unitaries can be represented in terms of evolution operators by:

$$\begin{aligned} U_{ij} &= \exp\left(-j\varphi_{ij}\hat{H}_{ij}\right) = \exp\left(-j\varphi_{ij}Z_iZ_j - jh_iZ_i - jh_jZ_j\right) \\ &= \underbrace{\exp\left(-j\varphi_{ij}Z_iZ_j\right)}_{U_{ij}^{(I)}} \exp\left(-jh_iZ_i - jh_jZ_j\right) \\ &= U_{ij}^{(I)} \exp\left(-jh_iZ_i - jh_jZ_j\right),\ U_{ij}^{(I)} = \exp\left(-j\varphi_{ij}Z_iZ_j\right). \end{aligned} \tag{12.18}$$

By setting the initial state to $|\Psi\rangle = |+\rangle \otimes \ldots \otimes |+\rangle$, to ensure that the state $U_{ij}^{(I)}|+\rangle|+\rangle$ is maximally entangled, we set $\varphi_{ij}=\pi$ and for the controlled-phase gate we can write:

$$U_{ij}(\pi) = \exp\left(-j\pi\hat{H}_{ij}\right), \hat{H}_{ij} = |1\rangle\langle 1| \otimes |1\rangle\langle 1|. \tag{12.19}$$

Given that projectors onto the eigenkets of Z_i with eigenvalues ± 1 are given by:

$$P_i^{(z,\pm 1)} = \frac{I \pm Z_i}{2}, \tag{12.20}$$

where I_2 is the identity operator (with corresponding matrix representation being an identity of size 2×2), we can represent the Hamiltonian H_{ij} as follows:

$$H_{ij} = \frac{I_2 - Z_i}{2}\frac{I_2 - Z_j}{2} = \frac{1}{4}\left(I_4 - Z_i - Z_j + Z_iZ_j\right). \tag{12.21}$$

The corresponding evolution operator will be:

$$U_{ij}(\varphi_{ij}) = e^{-j\frac{\varphi_{ij}}{4}I_4} e^{j\frac{\varphi_{ij}}{4}Z_i} e^{j\frac{\varphi_{ij}}{4}Z_j} \underbrace{e^{-j\frac{\varphi_{ij}}{4}Z_iZ_j}}_{U_{ij}^{(I)}(\varphi_{ij})}, \quad U_{ij}^{(I)}(\varphi_{ij}) = e^{-j\frac{\varphi_{ij}}{4}Z_iZ_j}. \tag{12.22}$$

The terms in front of $U_{ij}^{(I)}$ represent the rotations around the z-axis for each qubit, while the term $U_{ij}^{(I)}$ for $\varphi_{ij} = \pi$ can be represented by:

$$U_{ij}^{(I)}(\pi) = e^{-j\frac{\pi}{4}Z_iZ_j} = \cos\left(\frac{\pi}{4}\right)I_4 - j\sin\left(\frac{\pi}{4}\right)Z_iZ_j. \tag{12.23}$$

If we ignore the phase rotation term $\exp(-j\pi I_4/4)$, we can rewrite Eq. (12.22) as follows:

$$\begin{aligned} U_{ij}(\pi) &\cong \underbrace{e^{j\frac{\pi}{4}Z_i}}_{\cos\left(\frac{\pi}{4}\right)I + j\sin\left(\frac{\pi}{4}\right)Z_i} \underbrace{e^{j\frac{\pi}{4}Z_j}}_{\cos\left(\frac{\pi}{4}\right)I + j\sin\left(\frac{\pi}{4}\right)Z_j} \left[\cos\left(\frac{\pi}{4}\right)I - j\sin\left(\frac{\pi}{4}\right)Z_iZ_j\right] \\ &= \frac{1+j}{2\sqrt{2}}(I + Z_iI + IZ_j - Z_iZ_j) = \frac{1+j}{2\sqrt{2}}\left(\begin{bmatrix}1&0\\0&1\end{bmatrix}\otimes\begin{bmatrix}1&0\\0&1\end{bmatrix} + \begin{bmatrix}1&0\\0&-1\end{bmatrix}\otimes\begin{bmatrix}1&0\\0&1\end{bmatrix}\right. \\ &\quad \left. + \begin{bmatrix}1&0\\0&1\end{bmatrix}\otimes\begin{bmatrix}1&0\\0&-1\end{bmatrix} - \begin{bmatrix}1&0\\0&-1\end{bmatrix}\otimes\begin{bmatrix}1&0\\0&-1\end{bmatrix}\right) \\ &= \frac{1+j}{\sqrt{2}} \underbrace{\begin{bmatrix}1&0&0&0\\0&1&0&0\\0&0&1&0\\0&0&0&-1\end{bmatrix}}_{C_Z} = \frac{1+j}{\sqrt{2}}C_Z = e^{j\pi/4}C_Z, \end{aligned} \tag{12.24}$$

which clearly represents the controlled-phase gate up to a global phase constant.

The cluster state for a *linear chain (LC) of qubits* can be represented by:

$$|\phi\rangle_{LC,N} = 2^{-N/2} \bigotimes_{n=1}^{N} (|0\rangle_n Z_{n+1} + |0\rangle_n), Z_{N+1} = I. \tag{12.25}$$

As an illustration, the cluster state for a four-qubit LC is given by:

$$|\phi\rangle_{LC,4} = \frac{1}{2}(|+\rangle|0\rangle|+\rangle|0\rangle + |+\rangle|0\rangle|-\rangle|1\rangle + |-\rangle|1\rangle|-\rangle|0\rangle + |-\rangle|1\rangle|+\rangle|1\rangle). \tag{12.26}$$

Let L specify the lattice sites filled by qubits. The cluster $C \subset L$ is specified by the connected lattice sites and the corresponding cluster state can be represented by:

$$|\phi\rangle_C = \underset{c \in C}{\otimes}\left(|0\rangle_c \underset{b \in N(c)}{\otimes} Z_b + |0\rangle_c\right). \tag{12.27}$$

12.2 UNIVERSALITY OF CLUSTER STATE-BASED QUANTUM COMPUTING

One of the driving forces for one-way quantum computing and measurement-based quantum computing is 1-bit teleportation [8,16,21], with the corresponding quantum circuit provided in Fig. 12.3, in which the first qubit is arbitrary state $|\psi\rangle$, while the second qubit is the $|+\rangle$ state. When we perform the measurement on the first qubit in the diagonal basis $\{|+\rangle,|-\rangle\}$, the state of the second qubit will be $X^m H|\psi\rangle$, where m is the result (outcome) of the measurement (on the first qubit). Therefore by performing the $C_Z(|\psi\rangle \otimes |+\rangle)$ and measuring the first qubit in the X-basis, we effectively teleported the quantum state from physical qubit 1 to qubit 2. Let us now apply this concept to the graph shown in Fig. 12.4. By applying the C_Z gate between qubits 0 and 1, followed by the X-measurements on qubits 0, 1, and 2, the resulting quantum state of the third qubit will be:

$$X^{m_2} \underbrace{HX^{m_1}H}_{Z^{m_1}} X^{m_0} H|\psi\rangle_3 = X^{m_2} Z^{m_1} X^{m_0} H|\psi\rangle_3, \tag{12.28}$$

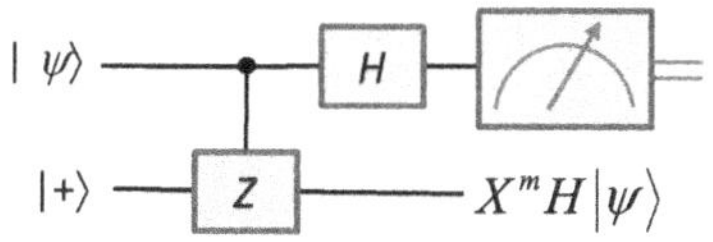

FIGURE 12.3

Quantum circuit to perform 1-bit teleportation.

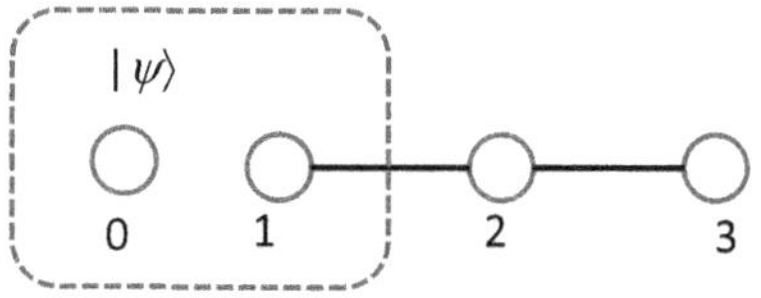

FIGURE 12.4

Teleportation of the quantum state though a linear chain of qubits. We apply the C_Z gate between qubits 0 and 1, followed by the X-measurements on qubits 0, 1, and 2.

where m_i are outcomes of measurements on qubits 0, 1, and 2, and we employed the following identity: $HXH=Z$.

A 1-bit teleportation circuit can be generalized as shown in Fig. 12.5, where $R_Z(\theta)$ denotes the Z-rotation by an angle θ, that is, $R_Z(\theta)=\exp(-j\theta Z/2)$. The action of the Z-rotation operator on diagonal states will be:

$$R_Z(\theta)|\pm\rangle=\exp(-j\theta Z/2)|\pm\rangle=2^{-1/2}[\cos(\theta/2)I-jZ\sin(\theta/2)](|0\rangle\pm|1\rangle)$$
$$=e^{-j\theta/2}\underbrace{2^{-1/2}(|0\rangle\pm e^{j\theta}|1\rangle)}_{|\pm_\theta\rangle}=e^{-j\theta/2}|\pm_\theta\rangle, \quad (12.29)$$

wherein the global phase term can be ignored. Given that $R_Z(\theta)$ and Z operators commute with the C_Z gate, the equivalency on the right side of Fig. 12.5 is valid, which helps us to teleport the quantum state $|\psi\rangle$, after the Z-rotation, to physical qubit 2. The basis $\{|+_\theta\rangle,|-_\theta\rangle\}$ lies in the (x,y) plane of the Bloch sphere and can be called the *equatorial* basis. By using this equivalency, let us observe the linear chain of qubits in Fig. 12.6, in which the qubits are initialized to the $|+\rangle$ state and then we apply the single-qubit gates $HR_Z(\theta_i)$ $(i=1,2)$ to physical qubits 1 and 2. By performing the X-measurements on qubits 1 and 2, and applying the 1-bit teleportation rule from Fig. 12.5, the state of the third qubit will be:

$$X^{m_2}HR_Z(\pm\theta_2)X^{m_1}HR_Z(\theta_1)|+\rangle\overbrace{\underset{R_Z(\pm\theta_2)X^{m_1}=X^{m_1}R_Z(\theta_2)}{=}}^{HX^{m_1}=Z^{m_1}H}X^{m_2}Z^{m_1}HR_Z(\theta_2)HR_Z(\theta_1)|+\rangle, \quad (12.30)$$

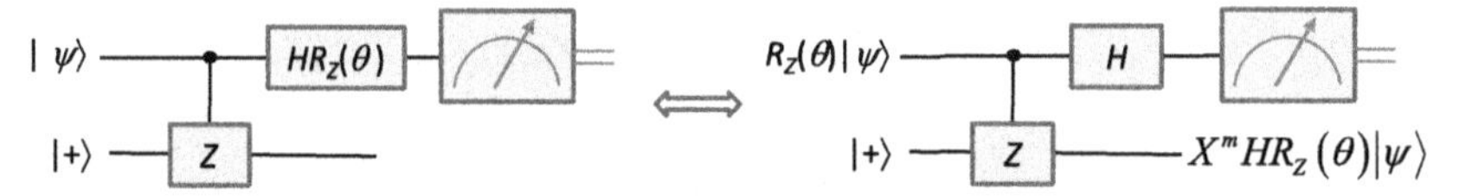

FIGURE 12.5

Generalized 1-bit teleportation quantum circuit.

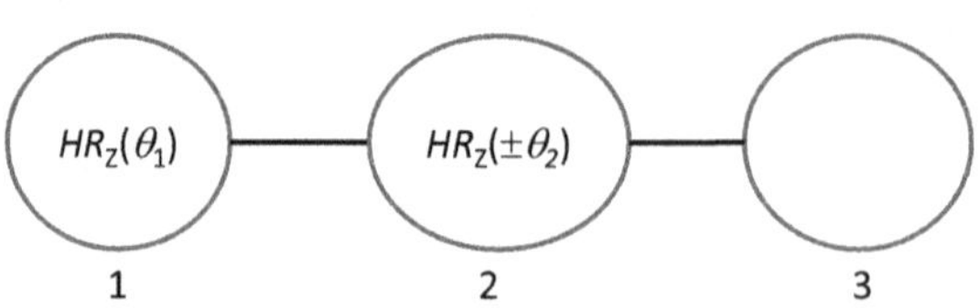

FIGURE 12.6

Cluster state-based processing describing the following operation: $C_Z^{(12)}C_Z^{(23)}\,HR_Z(\theta_1)|+\rangle_1 HR_Z(\theta_2)|+\rangle_2|+\rangle_3$.

and the corresponding equivalent circuit is provided in Fig. 12.7. The sign $\pm$ in front of θ_2 is properly chosen such that the following identity is valid: $R_Z(\pm\theta_2)X^{m_1} = X^{m_1}R_Z(\pm\theta_2)$. This equivalency can be generalized to the multiqubit equivalency provided in Fig. 12.8. When we omit the top HR_Z gates and bottom R_Z gates, we effectively simulated the CNOT gate, since $HZH = X$.

The Euler decomposition of the arbitrary single-qubit operation is given by Ref. [22,23]:

$$U(\theta_1, \theta_2, \theta_3) = R_X(\theta_3)R_Z(\theta_2)R_X(\theta_1), \tag{12.31}$$

where the X-rotation for angle θ_i $(i = 1,3)$ is defined by $R_X(\theta_i) = \exp(-j\theta_i\, X/2)$, while the Z-rotation operator is introduced above. Alternatively, we can also represent an arbitrary single-qubit operator U as follows [5]:

$$U(\alpha, \beta, \gamma) = HR_Z(\gamma)HR_Z(\beta)HR_Z(\alpha)HR_Z(0), \tag{12.32}$$

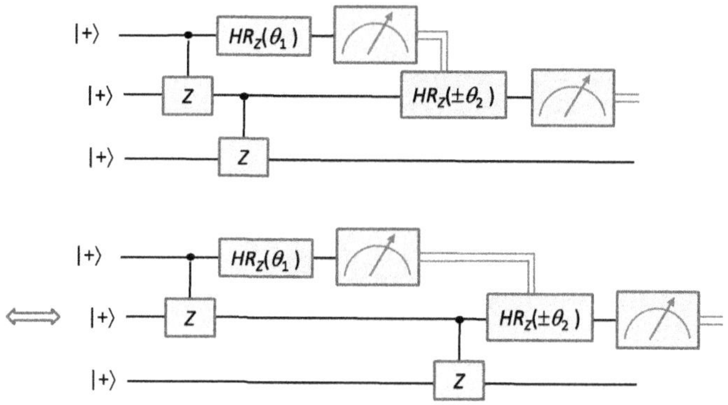

FIGURE 12.7

Equivalent quantum circuit for the cluster state from Fig. 12.6.

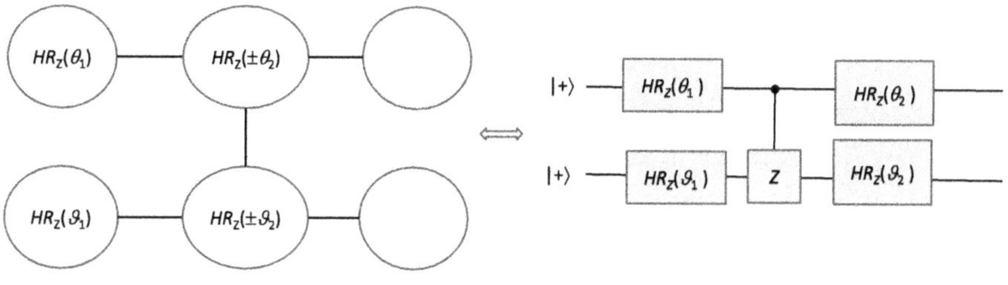

FIGURE 12.8

Cluster state-based simulation of controlled-Z quantum circuit combined with HR_Z single-qubit operations.

for some properly chosen angles α, β, γ. To simulate arbitrary single-qubit rotation, we can use the graph state shown in Fig. 12.9, wherein the first qubit is initialized to the input state, while the last qubit represents the final state. This problem represents generalization of the problem from Fig. 12.4. However, we perform the measurements in the equatorial basis introduced by Eq. (12.29), and one-step operation is provided in Fig. 12.4. So the output state can be described by:

$$|\psi\rangle_5 = X^{m_4} HR_Z(\gamma) X^{m_3} HR_Z(\beta) X^{m_2} HR_Z(\alpha) X^{m_1} HR_Z(0)|\psi\rangle. \tag{12.33}$$

Now by applying the *Pauli propagation relations*:

$$R_Z(\vartheta)Z = ZR_Z(\vartheta), \; R_Z(\vartheta)X = XR_Z(-\vartheta), \tag{12.34}$$

and the $HX = ZH$ equality, we can rewrite the previous state as follows:

$$|\psi\rangle_5 = X^{m_2+m_4} Z^{m_1+m_3} HR_Z\left[(-1)^{m_1+m_3}\gamma\right] HR_Z[(-1)^{m_2}\beta] HR_Z[(-1)^{m_1}\alpha] HR_Z(0)|\psi\rangle, \tag{12.35}$$

indicating that the action of the unitary operator is different from Eq. (12.32) in sign of angles, in addition to a Pauli operator. To solve this problem we need to update the angles to be used for equatorial measurements as shown in Fig. 12.9. Therefore the arbitrary single-qubit operator can indeed be implemented by the five-qubit linear chain cluster state, providing that the measurement pattern from Fig. 12.9 is used.

The *procedure to implement an arbitrary one-qubit gate* can also be interpreted as follows [22]:

- Initialize the subsystem in Fig. 12.9 as follows:

$$|\Psi\rangle_{C',5} = |\psi\rangle_1 \bigotimes_{n=2}^{5} |+\rangle_n.$$

- Entangle the qubits by applying the controlled-phase operation to neighboring qubits.
- Measure the qubits 1, 2, 3, and 4 by applying the measurements in the equatorial basis $\{|+_\theta\rangle, |-_\theta\rangle\}$ with corresponding angles in the basis being 0, $-(-1)^{m_1}\alpha$, $-(-1)^{m_2}\beta$, and $-(-1)^{m_1+m_3}\gamma$, respectively.

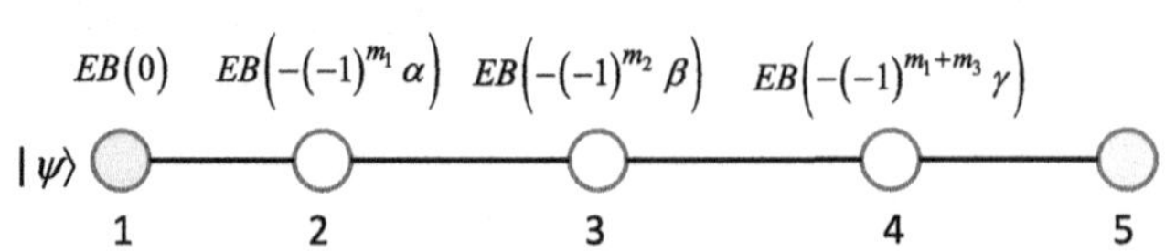

FIGURE 12.9

Cluster state-based arbitrary single-qubit rotation. Rotation angles correspond to the Euler decomposition angles as described in the text.

- The foregoing procedure implements the gate $U(\alpha,\beta,\gamma)=HR_Z(\gamma)HR_Z(\beta)HR_Z(\alpha)HR_Z(0)$ up to the Pauli operators $X^{m_2+m_4}Z^{m_1+m_3}$.

The CNOT gate can also be implemented by employing the cluster on 15 qubits [22], as shown in Fig. 12.10, in which the measurement pattern is provided as well. The corresponding measurements can be performed simultaneously. The *procedure to implement the CNOT gate* can be summarized as follows:

- Initialize the subsystem in Fig. 12.10 as follows:

$$|\Psi\rangle_{C',15}=|\psi\rangle_1|\psi\rangle_9\bigotimes_{n=2}^{7}|+\rangle_n\bigotimes_{n=10}^{15}|+\rangle_n.$$

- Entangle the qubits by applying the controlled-phase operation to neighboring qubits.
- Measure the qubits 1, 9, 10, 11, 13, and 14 in the diagonal basis (X-basis). On the other hand, measure the qubits 2–6, 8, and 12 in the Y-basis. All the measurements can be performed at the same time.
- The foregoing procedure implements the CNOT gate up to the Pauli gates, wherein qubit 7 is the control qubit and qubit 15 is the target qubit.

Given that the set of gates {CNOT, $U(\alpha,\beta,\gamma)$} represent the universal set of quantum gates, the cluster state-based quantum computation is universal. We learned from the foregoing that when the result of the measurement is not zero, then an extra Pauli operator will occur. The Pauli-Z operators commute with the controlled-phase gate and can be thus pushed through the gate. However, the Pauli-X gate does not commute with the controlled-phase gate, but we can apply the following identity:

$$C_Z^{(i,j)}X_i=X_iZ_jC_Z^{(i,j)}. \tag{12.36}$$

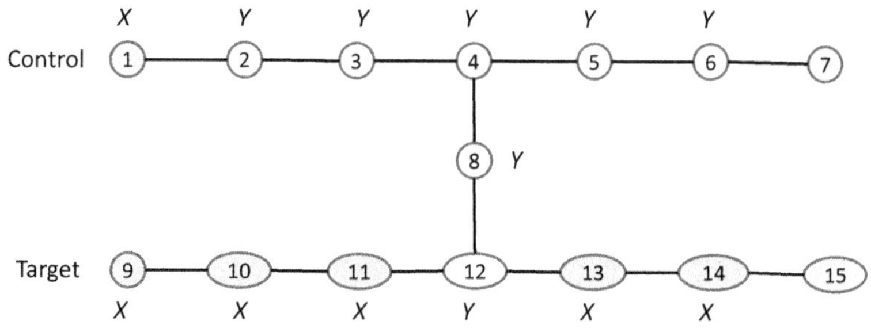

FIGURE 12.10

Cluster state-based implementation of the CNOT gate.

12.3 CLUSTER STATE PROCESSING AND ONE-WAY QUANTUM COMPUTATION

We have already discussed the role of X and equatorial measurements in Section 12.2. For the linear chain of qubits $|\psi\rangle_1|+\rangle_2\ldots|+\rangle_N$, which are pairwise entangled by the C_Z gates, we transfer $|\psi\rangle$ to the Nth physical qubit by a series of X-measurements, and the overall state after $N-1$ X-measurements can be represented by:

$$|m_1\rangle|m_2\rangle\cdots|m_{N-1}\rangle\otimes U_N|\psi\rangle,\ U_N\in\{I,X_N,Z_N,X_NZ_N\}, \qquad (12.37)$$

where m_n $(n=1,\ldots,N-1)$ is the outcome of the nth X-measurement and $|m_n\rangle$ is the corresponding eigenket. Let us now observe the linear chain cluster state, and we perform the measurement on a qubit i, located between $i-1$ and $i+1$ qubits. The X-measurement on the $i$thqubit will have the following two effects: (1) removal of the ith qubit from the cluster and (2) joining of the neighboring qubits $i-1$ and $i+1$ into a single joint logical qubit, as illustrated in Fig. 12.11. The single logic qubit will represent 0 as $|{+}{+}\rangle$ and 1 as $|{-}{-}\rangle$, which is similar to the repetition coding.

We now study the role of Z-measurements. Suppose that we want to eliminate the qubit a in the cluster provided in Fig. 12.12. Let us apply the Z-measurement on qubit a and denote the outcome of the measurement as m_a. Given that the measurement Z commutes with C_Z gates, we can first apply the C_Z gate between a and 2 qubits, and then perform the Z-measurement on qubit a, before applying C_Z gates

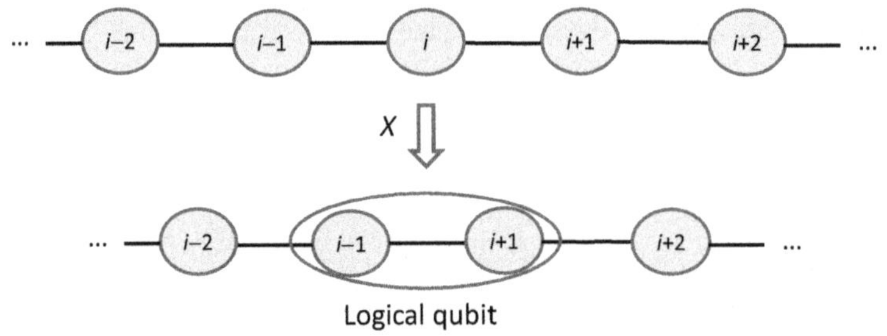

FIGURE 12.11

Role of X-measurement in a linear chain cluster.

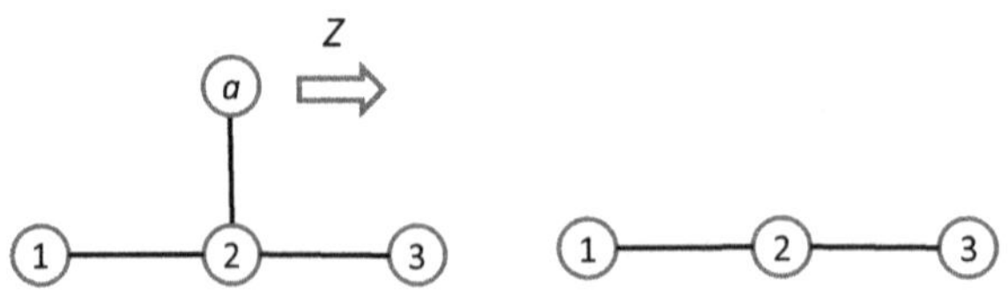

FIGURE 12.12

The role of Z-measurements is to remove the unwanted qubits in the cluster, such as qubit a.

between 1 and 2 as well as 2 and 3. So the action of the C_Z gate between qubits a and 2 will be $C_Z^{(1,2)}|+\rangle_a|+\rangle_2 = |0\rangle_a|+\rangle_2 + |1\rangle_a|-\rangle_2$. Now by performing the Z-measurement on qubit a, qubit a will be removed and the state of qubit 2 will be $|(-1)^{ma}\rangle$. The remaining qubits in the cluster will be in the state $|+\rangle_1|(-1)^{ma}\rangle_2|+\rangle_3$. Let us now observe the linear chain cluster state as in Fig. 12.13 and let us perform the Z-measurement on a qubit i, located between $i-1$ and $i+1$ qubits. The Z-measurement on the ith qubit will have the following two effects: (1) breaking of the connections between the ith qubit and its neighbors and (2) removal of the ith qubit from the cluster. The Z-measurements should also be applied in the final stage of one-way quantum computing on remaining qubits to obtain the classical output results.

Finally, let us consider the role of Y-measurement on the linear chain cluster state shown in Fig. 12.14. We are interested in the effect of the Y-measurement on qubit i, located between $i-1$ and $i+1$ qubits. The Y-measurement on the ith qubit will have the following two effects: (1) removal of the ith qubit from the cluster and (2) linking of the neighboring qubits $i-1$ and $i+1$, as illustrated in Fig. 12.14.

In a conventional quantum network model for quantum computation, we prepare the input state in a quantum register, and then apply a sequence of quantum gates corresponding to the desired unitary transformation of the states yielding to the output state. We then perform the measurements on the output state to obtain a classical readout. In one-way quantum computing, on the other hand, we perform the single-qubit measurements on the cluster state. Now the pattern of measurement

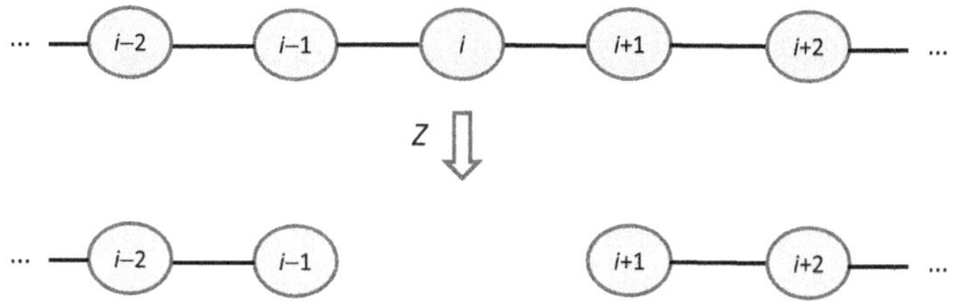

FIGURE 12.13

Role of Z-measurement in the linear chain cluster.

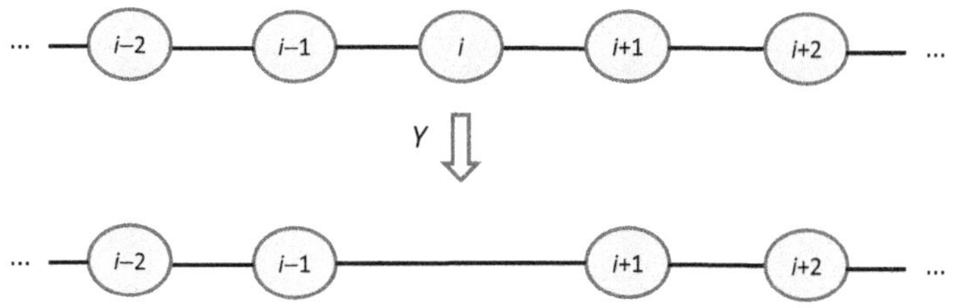

FIGURE 12.14

Role of Y-measurement in the linear chain cluster.

directions together with the classical processing of the measurement results represents the quantum algorithm [22,24]. As shown in Ref. [16], the $M \times N$ 2D cluster array, such as the one shown in Fig. 12.15, can be used to execute an arbitrary quantum computing algorithm. Even though the result of the measurement is random, this randomness can be compensated for by properly selecting the basis for subsequent measurements, so that the quantum computing is deterministic. The cluster state is split into three disjoint sections: input state (the first column of qubits in the 2D array, lattice), the output register (the last column in the 2D lattice), and operational (body) qubits on which proper measurements will be performed. As long as the arbitrary input state is known, it is sufficient to consider the all-$|+\rangle$ state as the input state. Namely, we can always transform the all-$|+\rangle$ state by a proper unitary transformation U_{in} to obtain $|\psi_{\text{in}}\rangle = U_{\text{in}}|+\rangle^M$. Now instead of implementing the unitary transformation U on input state $|\psi_{\text{in}}\rangle$, we instead need to implement UU_{in} on $|+\rangle^M$ by using the cluster state with index n running from 1 to $N - 1$. To implement UU_{in}, based on the generic cluster state from Fig. 12.15, it is sufficient to perform the equatorial measurements [16]. Together with the equatorial measurement the unitary transformation can be represented by Ref. [16]:

$$\hat{U} = \prod_{n=1}^{N-1} \left(\bigotimes_{m=1}^{M} H_m \right)_n \left(\bigotimes_{m=1}^{M-1} C_Z^{(m,m+1)} \right)_n . \tag{12.38}$$

As an illustration, by employing a 2×4 cluster state with the help of equatorial measurements we can implement the identity gate, single-qubit Z- and X-rotations, as well as two-qubit control gate as shown in Fig. 12.16, based on Ref. [16].

The basic buildings blocks in one-way quantum computing (1WQC) are single-qubit measurements rather than quantum gates [22–24]. The order of gates applied in the quantum circuit model is not necessarily the optimum one in 1WQC, but rather determined by the set of rules to be established for conditional adaptive measurements. The key idea behind the cluster state-based computational model is to track the outcomes of the previous measurements to determine the basis for

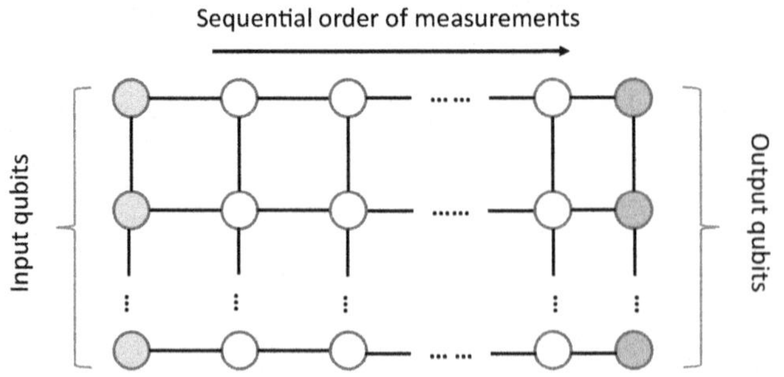

FIGURE 12.15

Generic (2D) cluster state to be used in one-way quantum computing.

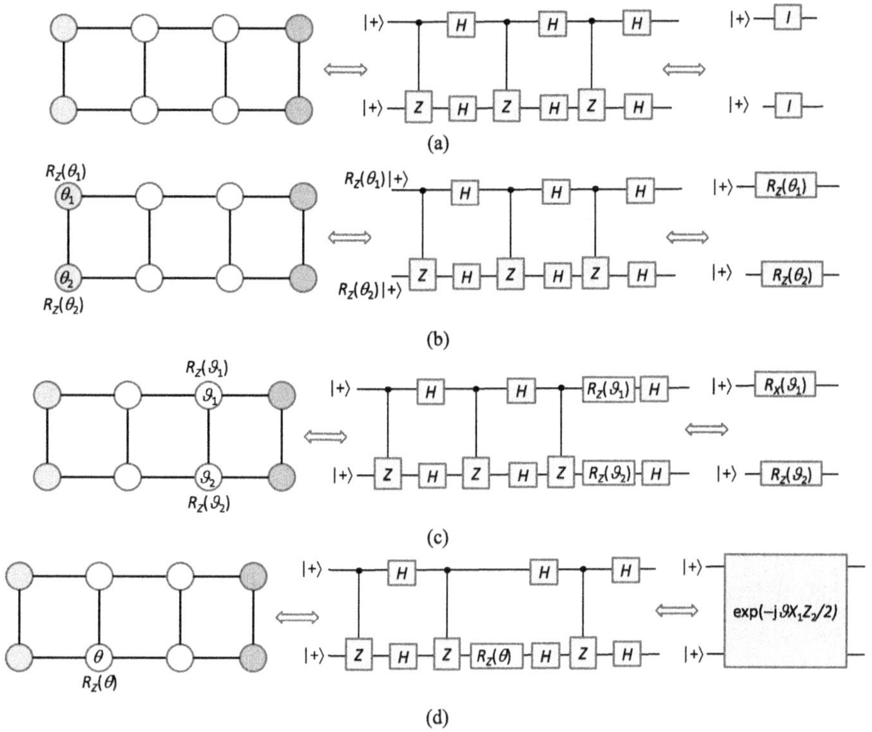

FIGURE 12.16

Employing the 2 × 4 cluster state to implement: (a) identity gate, (b) $R_Z(\theta)$, (c) $R_X(\vartheta)$, and two-qubit control gate with the help of equatorial measurements.

incoming measurements, and this information is stored in the so-called *information flow vector* $\boldsymbol{I}(t)$ [22,24]. The cluster state C can be divided into temporally ordered disjoint subsets of qubits $Q_t \subset C$ $(0 \le t \le t_{\max})$ such that the union of subsets represents the whole cluster; in other words, we can write [22,24]:

$$\bigcup_{t=0}^{t_{\max}} Q_t = C,\ Q_t \cap Q_{t'} = \varnothing\ \forall\ t \neq t'. \tag{12.39}$$

The qubits within each cluster can be measured at the same time. However, the measurements in higher-order clusters are dependent on the outcomes in previous clusters. The information needed to move the quantum computation forward is stored in the information flow vector. The qubits in subset Q_0 are those with fixed bases; in other words, they do not depend on the outcomes of measurements in other subsets. The measurements in Q_0 are those in the bases that correspond to the Pauli operators and should be used to implement the Clifford portion of the quantum circuit as well as the output state. As a reminder, the *Clifford group* $\mathscr{C}_N$ *is composed of unitaries for which the elements of the Pauli group on* N *qubits, denoted as* $\mathscr{P}_N$, *are invariant under conjugation, that is, we can write* $c\mathcal{P}_N c^\dagger = \mathcal{P}_N \ \forall \ c \in \mathcal{C}_N$ *[25]. The bases to be used in the* Q_t *round* ($t > 0$) *are dependent on measurement outcomes in all previous rounds. The result of the quantum computation is derived from the measurement outcomes of all rounds. The measurement operators in the* Q_t *round* ($t > 0$) *have the following form [22,24]:*

$$\cos(\theta)X \pm j\sin(\theta)Y,\ 0 < |\theta| < \pi/2, \tag{12.40}$$

where the sign depends on some of the outcomes from previous measurements. As an illustration, the computational model for 1WQC is provided in Fig. 12.17. The cluster is divided into disjoint subsets Q_t labeled according to the temporal order, wherein the bases in the Q_t stage are determined based on previous measurement outcomes stored in the information flow vector. The information flow vector $\boldsymbol{I}(t)$ is a $2N$-dimensional binary vector, wherein N is the number of qubits involved in the corresponding quantum circuit model, and t is the index of iteration (round). The Pauli operator on N qubits can be represented as a product of X- and Z-containing operators, and this vector stores the results of measurements corresponding to X and Z bases. The information flow vector is related to the by-products of the form (Eq. 12.35) as explained in Ref. [24].

The *overall by-product operator* U_Σ is a Pauli operator and can therefore also be represented as a product of X- and Z-containing terms [22]:

$$U_\Sigma = \prod_{n=1}^{N} X_i^{x_i} Z_i^{z_i};\ x_i, z_i \in \{0,1\}, \tag{12.41}$$

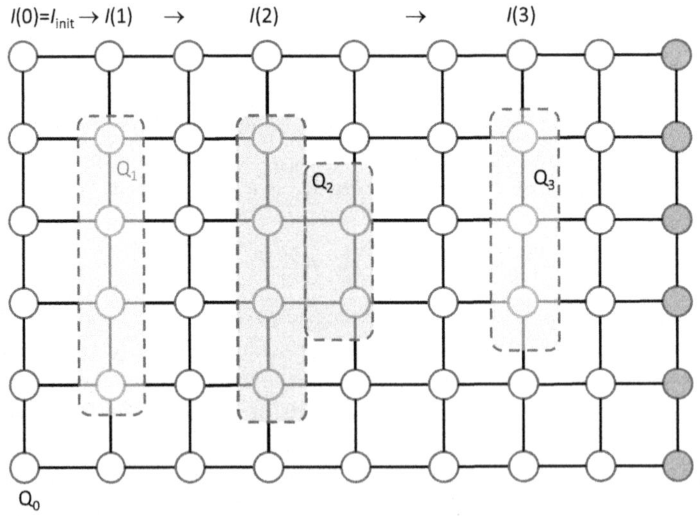

FIGURE 12.17

Illustrative computational model for one-way quantum computing.

so that the state of the output register, before the reading-out takes place, can be represented by:

$$|\Psi\rangle = U_\Sigma|\psi_{\text{out}}\rangle, \tag{12.42}$$

where $|\psi_{\text{out}}\rangle$ will be the content of the output register in the quantum circuit (network) model. Let the corresponding outputs of measurements on $|\psi_{\text{out}}\rangle$ be denoted by $\{m'_i\}$, while the corresponding measurement outcomes on $U_\Sigma|\psi_{\text{out}}\rangle$ be denoted by $\{m_i\}$. Based on Eq. (12.20) we conclude that upon measurement the qubits after readout are projected to the following state:

$$\begin{aligned} |M\rangle &= \prod_{n=1}^{N}\frac{I+(-1)^{m_n}Z_n}{2}U_\Sigma|\psi_{\text{out}}\rangle = U_\Sigma \underbrace{U_\Sigma^\dagger\prod_{n=1}^{N}\frac{I+(-1)^{m_n}Z_n}{2}U_\Sigma}_{\prod_{n=1}^{N}\frac{I+(-1)^{m'_n}Z_n}{2}}|\psi_{\text{out}}\rangle \\ &= \prod_{n=1}^{N}\frac{I+(-1)^{m'_n}Z_n}{2}|\psi_{\text{out}}\rangle, \end{aligned} \tag{12.43}$$

where m'_i is the desired computation result related to m_i by $m'_i = m_i \oplus x_i$, where x_i is dictated by the overall by-product operator U_Σ. Therefore the performance of 1WQC is not affected by random measurements, since the outcomes before and after measurements correspond to one-qubit operations that can be tracked and corrected in the last step, just before the reading-out stage.

The 1WQC can be represented by the sequence of simulated gates U_{gn} as follows:

$$|\Psi\rangle = \prod_{n=1}^{|N|}U_{\Sigma_{gn}}U_{g_n}\Big|\psi_{\text{in}}\Big\rangle, \tag{12.44}$$

wherein $U_{\Sigma_{gn}}$ represents the by-product term and $|N|$ denotes the number of simulated gates. To obtain the overall by-product term we need to propagate forward the individual by-product gates by applying the Pauli X-gate propagation rule through the C_Z gate, given by Eq. (12.36). The Pauli-Z gate commutes with the C_Z gate. Additionally, we need to apply the identities we applied already: $HZ = XH$ and $HX = ZX$. If the quantum computation is expressed in terms of CNOT gates, denoted as C_X, we need to apply the following propagation rules [22]:

$$\begin{aligned} C_X^{(c,t)}X_t &= X_tC_X^{(c,t)}, C_X^{(c,t)}X_c = X_cX_tC_X^{(c,t)}, \\ C_X^{(c,t)}Z_c &= Z_cC_X^{(c,t)}, C_X^{(c,t)}Z_t = Z_cZ_tC_X^{(c,t)}, \end{aligned} \tag{12.45}$$

with c and t being the control and target qubits, respectively. Moreover, in 1WQC the following identities are relevant:

$$U(\alpha,\beta,\gamma)X = XU(\alpha,-\beta,\gamma),\ U(\alpha,\beta,\gamma)Z = ZU(-\alpha,\beta,-\gamma). \tag{12.46}$$

For gates c from Clifford group C the following is valid:

$$cU_{\Sigma} = \underbrace{cU_{\Sigma}c^{-1}}_{U_{\Sigma'}}\, c = U_{\Sigma'}c, \tag{12.47}$$

indicating that the Clifford gates are unaffected by propagation of the by-product and they can be measured simultaneously in a single time step.

12.4 PHYSICAL IMPLEMENTATIONS

Although the inspiration for cluster state-based quantum computing originates from quantum teleportation-based quantum gates, another relevant contribution is the realization of physical qubits arranged in a 2D array or lattice [6,7]. The qubits arranged in a lattice, in close proximity, could be interacted by external electric or magnetic fields [26–28]. As an illustration, two quantum dots with one valence electron each can be coupled via a transverse electrical and magnetic field to implement the desired two-qubit gate [26]. Let the spins for these electrons in valence bands be denoted by $\boldsymbol{S}_1$ and $\boldsymbol{S}_2$. By applying the magnetic field B along the z-axis and electric field E along the x-axis, we can introduce the coupling that can be described by the interaction Hamiltonian $H_I(t) = J(t)\boldsymbol{S}_1\boldsymbol{S}_2$, wherein the coupling factor $J(t)$ is a function of B, E, and the distance between quantum dots [26]. By arranging the quantum dots in a 2D lattice as shown in Fig. 12.15 and ensuring that the interaction between neighboring quantum dots implements the C_Z gate, we can create the desired cluster state to be used in 1WQC. Various schemes to generate cluster states in solid-state double quantum dots have been studied in Ref. [27–31]. Other implementations to generate the cluster states include linear optics [14], optical lattices [32], trapped ions [33,34], cavity quantum electrodynamics [35], and superconducting quantum circuits [36], to mention a few.

For trapped ion- and neutral atom-based two-qubit gates, the internal degrees-of-freedom (typically electron energy levels) of qubits are coupled with the multiqubit vibrational states [34,37]. For trapped ions, the interaction is enabled by the Coulomb potential, while for neutral atoms the interaction is mediated by the dipole–dipole interaction between atoms excited to low-order Rydberg states in a constant electric field. State manipulation and read-out are performed by coherent optical pulses of properly selected power and duration. Trapped ion- and neutral atom-based cluster states suffer from decoherence effects.

The photons are less sensitive to the decoherence effects but are more challenging to interact to create the cluster state. Knill, Lafflamme, and Milburn (KLM) have shown in Ref. [38] that optical quantum computing is possible by employing beam splitters (BSs), phase shifters, photodetectors, and single photon sources. Unfortunately, the corresponding CNOT gate is probabilistic (nondeterministic) in nature. The corresponding Knill C_Z gate, introduced in Ref. [39], is provided in Fig. 12.18, which performs the controlled-phase operation with 2/27 probability. As a reminder, the BS unitary matrix is given by Ref. [40]:

$$BS(\theta,\phi)=\begin{bmatrix}\cos\theta & -e^{j\phi}\sin\theta\\ e^{j\phi}\sin\theta & \cos\theta\end{bmatrix}, \tag{12.48}$$

where θ and ϕ are related to transmissivity t and reflectivity r. For symmetric BS we set $\phi=\pi/2$, so that the unitarity transformation conditions are satisfied ($|t|^2+|r|^2=1$, $t^*r+tr^*=0$), and the previous transformation simplifies to:

$$BS(\theta)=\begin{bmatrix}\cos\theta & -j\sin\theta\\ j\sin\theta & \cos\theta\end{bmatrix}=\begin{bmatrix}t & r\\ r & t\end{bmatrix}=\cos\theta I-j\sin\theta X=e^{-j\theta X}=R_X(2\theta), \tag{12.49}$$

and this BS was used in Fig. 12.18. If the average gate success probability is denoted by p, the quantum circuit performing the quantum computation with N gates will perform the desired computation with small probability p^N. Therefore p^{-N} such systems must be run in parallel. By using teleportation [41], the success probability of KLM gates can be increased to $N^2/(N+1)^2$. Alternatively, quantum error correction can be employed to reduce the failure probability [42,43]. Nevertheless, the complexity of this approach is still prohibitively high.

To solve this problem, Browne and Rudolph proposed in Ref. [14] a cluster state-based approach to linear optics quantum computation. Instead of implementing the C_Z gates between neighboring qubits, the authors proposed to probabilistically fuse the existing clusters, as illustrated in Fig. 12.19, with the *fusion operator* being defined by:

$$F=|0\rangle\langle 00|+|1\rangle\langle 11|. \tag{12.50}$$

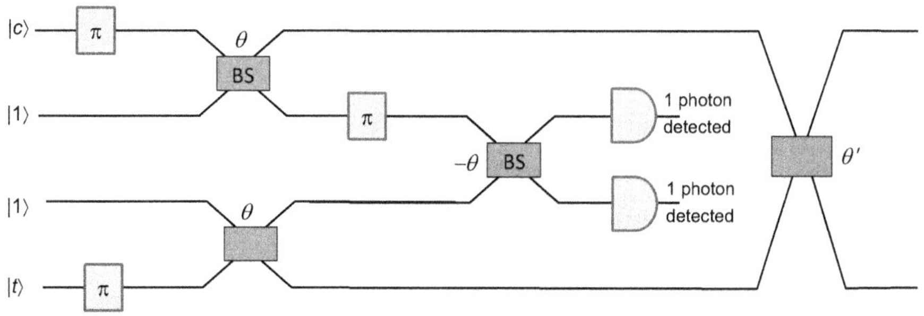

FIGURE 12.18

Knill's probabilistic C_Z gate. The beam splitter (BS) parameters are $\theta=54.74$ degrees and $\theta'=17.63$ degrees.

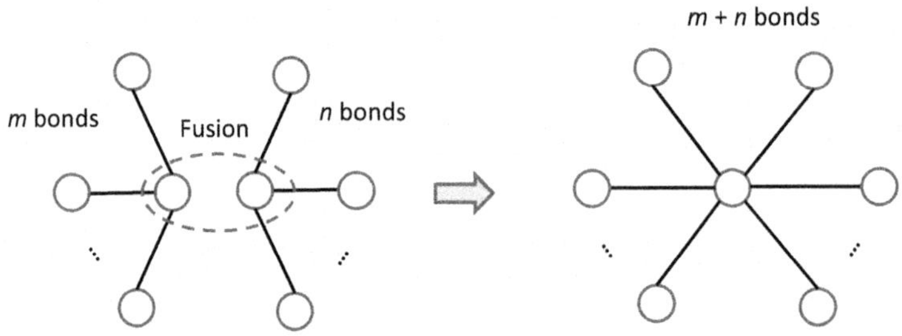

FIGURE 12.19

Fusion operator.

In other words, the fusion operator merges two qubits while preserving the existing cluster state bonds, which is illustrated in Fig. 12.19. The fusion operation can be described as follows:

$$\begin{aligned}&\left(|0\rangle\prod_{i=1}^{m}I_i|\psi\rangle+|1\rangle\prod_{i=1}^{m}Z_i|\psi\rangle\right)\otimes\left(|0\rangle\prod_{i=1}^{n}I_i|\psi'\rangle+|1\rangle\prod_{i=1}^{n}Z_i|\psi'\rangle\right)\\&\xrightarrow{F}|0\rangle\prod_{i=1}^{m}I_i|\psi\rangle\prod_{i=1}^{n}I_i|\psi'\rangle+|1\rangle\prod_{i=1}^{n}Z_i|\psi\rangle\prod_{i=1}^{n}Z_i|\psi'\rangle,\end{aligned}\tag{12.51}$$

indicating that we start with two cluster states $|\psi\rangle$ and $|\psi'\rangle$ and we merge them together.

The initial cluster states to start with are photon Bell states $|0\rangle|+\rangle + |1\rangle|-\rangle$, where $|0\rangle$ is the horizontally polarized photon $|H\rangle$, while $|1\rangle$ is the vertically polarized photon $|V\rangle$. The basic block to perform the fusion operation is the polarization beam splitter (PBS), as shown in Fig. 12.20. Similarly to Ref. [40], the PBS operation can be described as:

$$\begin{aligned}U_{PBS}=&A_r\left|V_l^{(\mathrm{out})}\right\rangle\left\langle V_l^{(\mathrm{in})}\right|+A_r\left|V_r^{(\mathrm{out})}\right\rangle\left\langle V_r^{(\mathrm{in})}\right|\\&+A_t\left|H_l^{(\mathrm{out})}\right\rangle\left\langle H_r^{(\mathrm{in})}\right|+A_t\left|H_r^{(\mathrm{out})}\right\rangle\left\langle H_l^{(\mathrm{in})}\right|,\end{aligned}\tag{12.52}$$

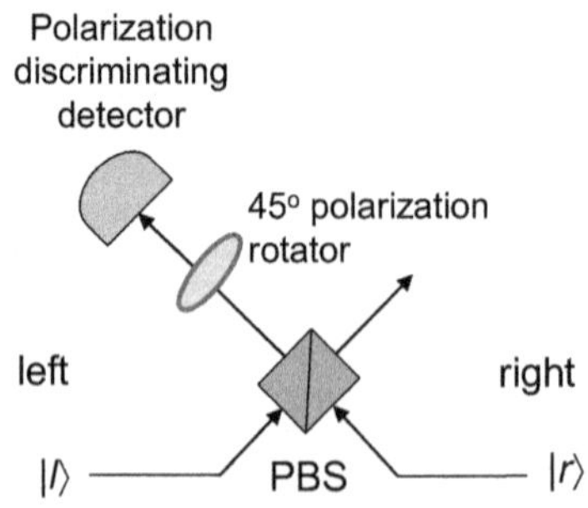

FIGURE 12.20

Type I fusion. *PBS,* Polarization beam splitter.

where A_r and A_t are reflection and transmission amplitude coefficients, respectively. In other words, the vertical photon is reflected by the PBS, while the horizontal photon is transmitted through the PBS. With the photons present at both left and right input ports, there are four possible outcomes, each occurring with probability 0.25. Two outcomes correspond to the desired fusion operators $|0\rangle\langle 00| \pm |1\rangle\langle 11|$, and the success probability of the fusion is 0.5. When a single photon is detected by the detector, the successful fusion is declared. On the other hand, when 0 or two photons are detected we know that the fusion resulted in failure, and both qubits are cut from their respective clusters. Therefore fusion failure is similar to measuring both qubits in the Z-basis (after which the measured qubits are removed from the cluster).

Two illustrative examples related to the fusion process are provided in Fig. 12.21. In the example at the top of Fig. 12.21, on average there is no increase in the cluster length. On the other hand, in the example at the bottom of the same figure, on average the cluster length is increased by 0.5 qubits. If we employ five-qubit clusters, we will need on average 6.5 Bell states per single qubit added to the cluster.

1D clusters are not sufficient for universal quantum computing, but we need to implement the 2D cluster states. This can be done by appropriately fusing several linear clusters as shown in Fig. 12.22. Unfortunately, the success rate is 0.5, and the fusion failure is similar to the Z-measurement, resulting in the breaking up of the bonds. To avoid this problem, we can introduce the fusion similar to the X-measurement, which does not cut the qubit bonds but merges the neighboring qubits into an encoded logic qubit.

Let us modify the fusion operation by introducing the additional 45-degree rotation to both qubits, with failure corresponding to the X-measurements, which is illustrated in Fig. 12.23. Unfortunately, the successful projection is not in the diagonal basis. To solve this problem we perform measurement on both outputs, which leads to the projection on states $|{+}{+}\rangle{+}|{-}{-}\rangle$ and $|{+}{+}\rangle{-}|{-}{-}\rangle$. Successful application of the gate on a single photon in each pair of logical qubits results in projection on the

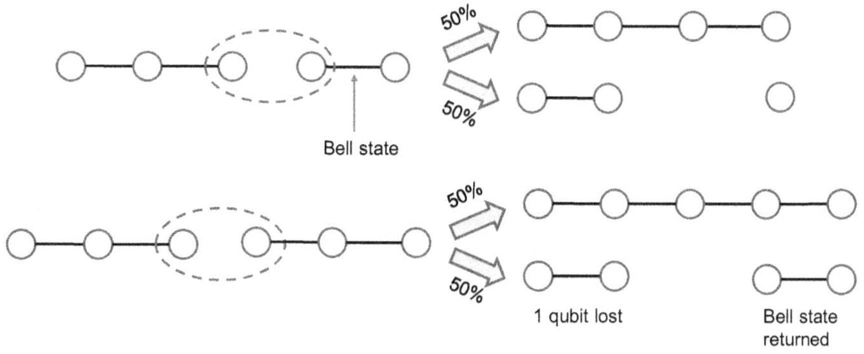

FIGURE 12.21

Two illustrative examples of the fusion process.

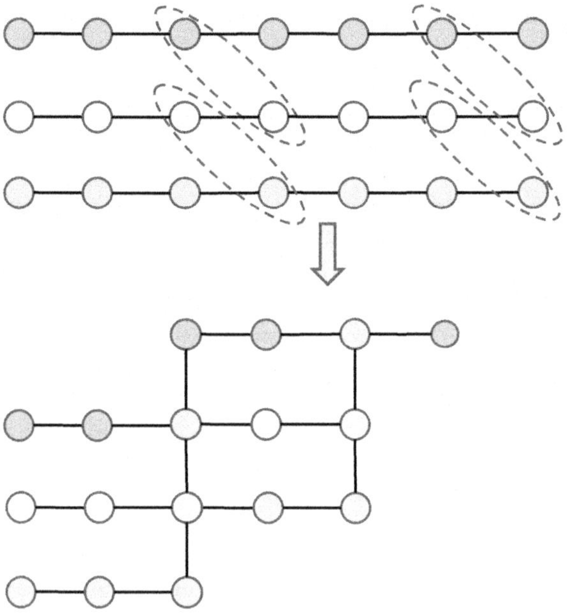

FIGURE 12.22

Employment of linear clusters to create the desired 2D cluster.

maximally entangled state. The failure effectively performs the X-measurement. This type of fusion is called type II by the authors in Ref. [14]. As an illustrative example, see Fig. 12.24. We start with two linear clusters and apply the X-measurement in one of the qubits of the top cluster, which results in removal of that qubit and creation of a logical qubit from two closest neighbors. Type II fusion is then applied between the logic qubit and one of the qubits from the lower cluster. With probability 0.5 the fusion will be successful, and we will obtain the desired shape. On the other hand, with the same probability the fusion will fail. Nevertheless, the

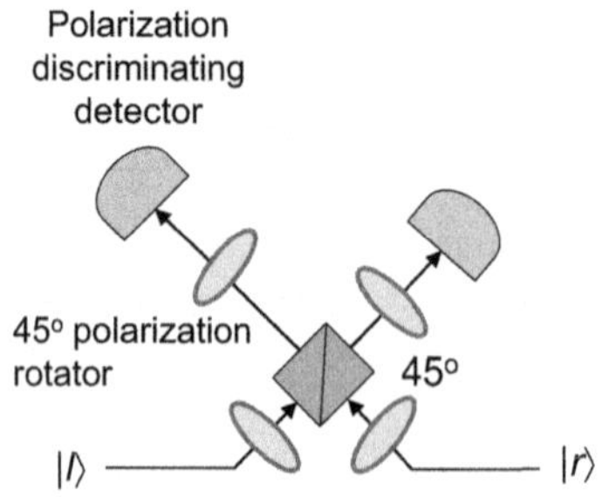

FIGURE 12.23

Type II fusion.

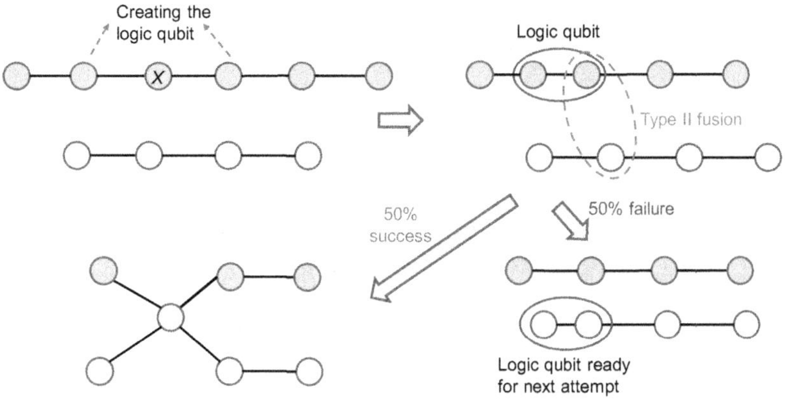

FIGURE 12.24

Type II fusion.

X-measurement on the logic qubit removes the redundancy, but at the same time creates a new logic qubit in the bottom cluster, which can then be used for another attempt at fusion. Clearly, type II fusion is more resilient to fusion failure.

Gilbert et al. proposed in Ref. [12] to use local unitaries and type I fusion only to create the 2D cluster state, with the key idea being summarized in Fig. 12.25. To create the box cluster state, as shown in Fig. 12.25A, we start with a four-qubit linear cluster state, relabel the qubits 2 and 3, and apply the Hadamard gates to qubits 2 and 3, which effectively establish the bond between qubits 1 and 4. Namely, relabeling the qubits is equivalent to the SWAP gate action. To create the box-on-chain cluster state, we start with a longer linear chain of qubits and apply the same approach as in a box-state creation, which is illustrated in Fig. 12.25B. By further performing the *Z*-measurement on qubit 3, we obtain the T-shape cluster state, which is suitable for use to create various 2D cluster states. Two T-shape cluster states can be fused together to get the H-shape cluster state, as illustrated in Fig. 12.26. Unfortunately, e type I fusion can fail with probability 0.5, and in that case we can recover two linear chains to be reused in another attempt.

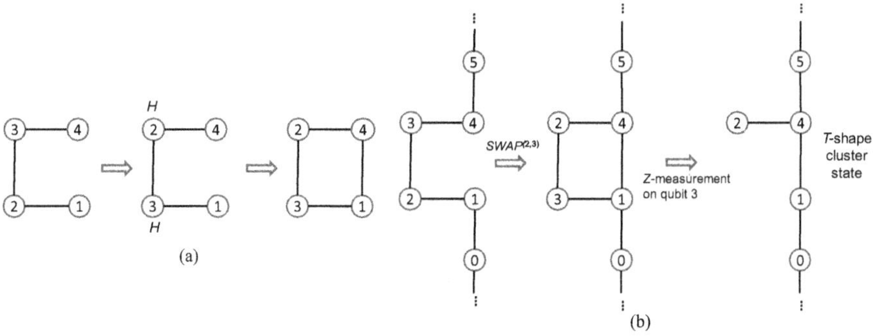

FIGURE 12.25

Gilbert's approach to the creation of the T-shape cluster state.

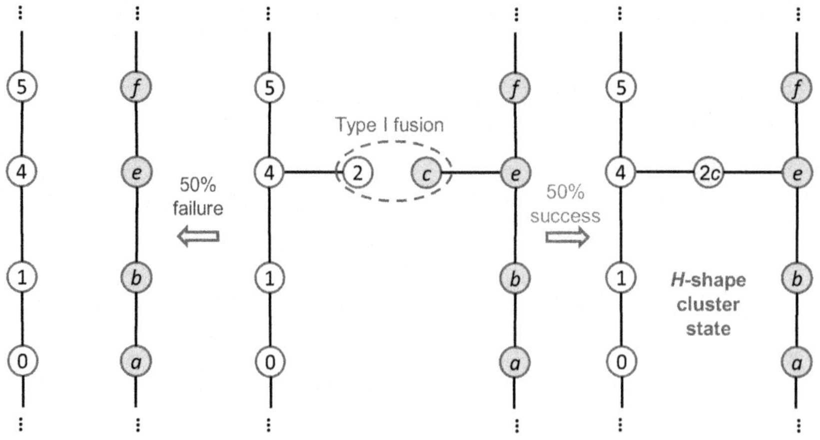

FIGURE 12.26

Gilbert's approach to the creation of the H-shape cluster state.

One of the first photonic cluster-state demonstrations was conducted by Zhang et al. in Ref. [44]. In this experiment, two pairs of entangled photons were created via spontaneous parametric down-conversion (SPDC) with the help of β-barium borate (BBO) crystal. Opposite photon pairs were used as inputs to the PBS, serving the role of type I fusion operator, as already described in Fig. 12.20. Another early experimental demonstration was conducted by Walther et al. [10], where a four-qubit linear chain cluster state (Eq. 12.26) was demonstrated, which can be represented in the computation basis as follows:

$$
\begin{aligned}
|\phi\rangle_{LC,4} &= \frac{1}{2}(|+\rangle|0\rangle|+\rangle|0\rangle + |+\rangle|0\rangle|-\rangle|1\rangle + |-\rangle|1\rangle|-\rangle|0\rangle + |-\rangle|1\rangle|+\rangle|1\rangle) \\
&= \frac{1}{2}(|0\rangle|0\rangle|0\rangle|0\rangle + |0\rangle|0\rangle|1\rangle|1\rangle + |1\rangle|1\rangle|0\rangle|0\rangle - |1\rangle|1\rangle|1\rangle|1\rangle),
\end{aligned}
\tag{12.53}
$$

with $|0\rangle$ representing the horizontally polarized photon, while $|1\rangle$ represents the vertically polarized photon. The entangled photons are created by employing type II SPDC with the help of BBO crystal. The box cluster state can be obtained by relabeling qubits 2 and 3 (or equivalently applying the SWAP operation) and applying the local unitary $H_1H_2H_3H_4$. Quantum state tomography was used to verify the cluster state. The key idea of quantum tomography is to perform discrete measurements on cluster state and extract the density operator. The authors in Ref. [10] applied four-photon projection measurements and maximum-likelihood reconstruction [45]. As expected, they found the following four dominant terms $|0\rangle|0\rangle|0\rangle|0\rangle$, $|0\rangle|0\rangle|1\rangle|1\rangle$, $|1\rangle|1\rangle|0\rangle|0\rangle$, and $|1\rangle|1\rangle|1\rangle|1\rangle$. They further found that the fidelity of the cluster state was $\langle\phi|\rho|\phi\rangle = 0.63 \pm 0.02$. They further studied the effect of loss of the first qubit by applying the *entanglement witness* approach. In this approach, we determine the expected value of the entanglement witness for arbitrary state $|\psi\rangle$ as follows:

$$\langle W\rangle = \langle\psi|W|\psi\rangle = \begin{cases} <0, & \text{for entangled state} \\ \geq 0, & \text{for separable state} \end{cases}. \tag{12.54}$$

As an illustration, the entanglement witness for $|\mathrm{GHZ}\rangle = (|000\rangle + |111\rangle)/\sqrt{2}$ state can be defined by:

$$\begin{aligned} W &= \frac{1}{2}I - |\mathrm{GHZ}\rangle\langle\mathrm{GHZ}| \\ &= \frac{1}{2}I - \frac{1}{2}(|000\rangle\langle 000| + |000\rangle\langle 111| + |111\rangle\langle 000| + |111\rangle\langle 111|). \end{aligned} \tag{12.55}$$

For $|\psi\rangle = |\mathrm{GHZ}\rangle$, the expected value of entanglement witness W is:

$$\begin{aligned} \langle W\rangle &= \langle\mathrm{GHZ}|\left(\frac{1}{2}I - |\mathrm{GHZ}\rangle\langle\mathrm{GHZ}|\right)|\mathrm{GHZ}\rangle \\ &= \langle\mathrm{GHZ}|\left(\frac{1}{2}|\mathrm{GHZ}\rangle - |\mathrm{GHZ}\rangle\underbrace{\langle\mathrm{GHZ}|\mathrm{GHZ}\rangle}_{=1}\right) = -\frac{1}{2}\langle\mathrm{GHZ}|\mathrm{GHZ}\rangle = -\frac{1}{2}. \end{aligned} \tag{12.56}$$

By tracing out the first qubit from the density operator of the cluster state (Eq. 12.53), we can determine the expected value of the reduced density state ρ by $\langle W\rangle = \mathrm{Tr}(\rho W)$, where the entanglement witness was defined as [10]:

$$\begin{aligned} W &= \frac{1}{4}I - \frac{1}{2}\left(|0\rangle_2\langle 0|_2 \otimes |\Phi^+\rangle_{34} + |1\rangle_2\langle 1|_2 \otimes |\Phi^-\rangle_{34}\right), \\ |\Phi^\pm\rangle_{34} &= \frac{1}{\sqrt{2}}\left(|0\rangle_3|0\rangle_4 \pm |1\rangle_3|1\rangle_4\right). \end{aligned} \tag{12.57}$$

They have shown that $\langle W\rangle = -0.115 \pm 0.007$, indicating that the entanglement persists in qubits 2, 3, and 4 even after the loss of qubit 1. In the same paper the authors demonstrated C_Z gate operation.

Another relevant recent research topic is continuous variable (CV) cluster state-based quantum computing [46–49], in which the squeezed coherent states [40,50] are employed. A squeezed state is a Gaussian state that has unequal fluctuations in quadratures. Namely, when we pump a nonlinear crystal in the nondegenerate regime, the crystal will generate pairs of photons in two different modes, commonly referred to as the signal and the idler, and this process is commonly referred to as SPDC. If the photons have the same polarization they belong to type I correlation; otherwise, they have perpendicular polarizations and belong to type II. As the SPDC is stimulated by random vacuum fluctuations, the photon pairs are created at random time instances. The output of a type I down-converter is known as a squeezed vacuum and it contains only even numbers of photon. The output of the type II down-converter is known as a two-mode squeezed vacuum state. When this process occurs inside an optical cavity, the corresponding source is known as the optical parametric oscillator (OPO). Yokoyama et al. used in Ref. [46] two OPOs as sources of squeezed pulses to mix them in an unbalanced Mach–Zehnder interferometer to generate a dual-rail squeezed state. The corresponding setup to generate extended Einstein–Podolsky–Rosen (XEPR) states was composed of four steps. In step (1), the squeezed light beams are divided into time bins of duration T, with rate $1/T$ being sufficiently lower than the bandwidth of identical OPOs. Such obtained series of EPR states are combined by a 50:50 BS, representing step (2). The bottom rail of EPR states is delayed for T seconds by an optical delay line, which represents step (3). In step (4), two staggered EPR states are combined by the second 50:50 BS to obtain the desired XEPR state. Such obtained XEPR state is equivalent, up to local phase shifts, to topological 1D CV cluster states as defined in Ref. [51] and can therefore be used as a source for 1WQC.

12.5 SUMMARY

This chapter was devoted to cluster state-based quantum computing. In Section 12.1, the definition of cluster states was introduced, followed by a description of their relationship to the graph states and stabilizer formalism. In the same section, the relationship of cluster states and qubit teleportation circuits was discussed. The conditions to be met for a corresponding set of unitaries to be able to generate the cluster state was formulated. The problem of generating the cluster state was related to the Ising model. In Section 12.2, the focus moved to the universality of cluster state-based quantum computing. We proved that the 1D cluster state is sufficient to implement arbitrary single-qubit operation, while the 2D cluster state is needed for arbitrary two-qubit operation by applying the proper sequence of measurements. We demonstrated that a five-qubit linear chain cluster is enough to implement arbitrary single-qubit rotation by employing the Euler decomposition theorem. We also demonstrated that T-shape and H-shape cluster states can be used to implement the CNOT gate. In Section 12.3, the cluster state processing was discussed by first describing the roles of X-, Z-, and Y-measurements. The generic 2D cluster state to be used in one-way quantum computation (1WQC) was then introduced and details of the corresponding quantum computational model were provided. During one-way quantum computation, the random Pauli by-product operator naturally arises, and we showed that the performance of 1WQC was not affected by this. Various physical implementations suitable for 1WQC were discussed in Section 12.4. Special focus was devoted to photonic 1WQC implementations, including resource-efficient linear optics implementation. We described how the Bell states can be used to form linear chain, T-shape, H-shape, and arbitrary 2D cluster states through type I and type II fusion processes. We described how the implementation of cluster states and 1WQC can be experimentally demonstrated. In the same section, we briefly described the basic concepts of continuous variable cluster state-based quantum computing. In the following section, we provide a set of problems for self-study.

12.6 PROBLEMS

1. Prove that the controlled-phase gate is symmetric, that is, that the following is valid:

$$C_Z|+\rangle|+\rangle = 2^{-1/2}(|0\rangle|+\rangle + |1\rangle|-\rangle) = 2^{-1/2}(|+\rangle|0\rangle + |-\rangle|1\rangle).$$

2. Let us consider the cluster state for a linear chain of three qubits described by Eq. (12.25). Show that this state is identical to the output of the quantum circuit shown in Fig. 12.2. Prove Eq. (12.26), describing the four-state linear chain cluster state.
3. Let us consider the cluster state described by the star graph shown in Fig. 12.1C. By using Eq. (12.27) can you relate it to the GHZ state?
4. Prove that arbitrary single-qubit gate U can be decomposed in terms of Euler angles α, β, γ, up to the global phase, as follows:

$$U(\alpha, \beta, \gamma) = R_X(\gamma)R_Z(\beta)R_X(\alpha).$$

5. Prove the Pauli propagation relations:

$$R_Z(\vartheta)Z = ZR_Z(\vartheta),\ \ R_Z(\vartheta)X = XR_Z(-\vartheta),\ \ R_X(\vartheta)X = XR_X(\vartheta),$$
$$R_X(\vartheta)Z = ZR_X(-\vartheta).$$

6. Prove the equivalency provided in Fig. 12.8.
7. Derive Eq. (12.35) starting from Eq. (12.33).
8. The cluster state on four qubits, shown in Fig. 12.P8, can be used to implement the CNOT gate. Once the cluster state is created, in which basis should qubits 1 and 2 be measured? Provide the detailed procedure to implement the CNOT gate.

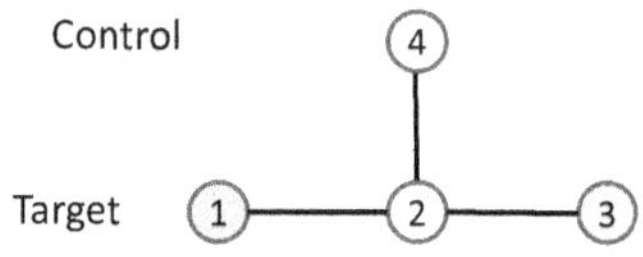

FIGURE 12.P8

Cluster state on four qubits to be used in implementing the CNOT gate.

9. Prove that the cluster state on 15 qubits, shown in Fig. 12.10, implements the CNOT gate by applying the measurement pattern provided in the same figure.
10. Let us observe the cluster state shown in Fig. 12.P10 with the measurement pattern being provided in the same figure. Demonstrate that this measurement pattern, in one-way quantum computing, simulates the controlled-phase gate.

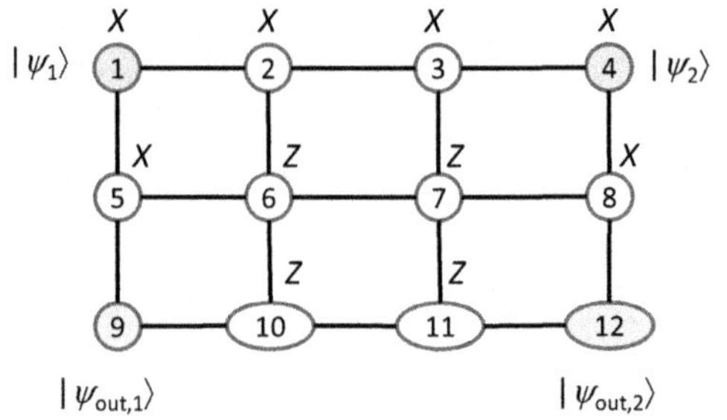

FIGURE 12.P10

A cluster state suitable to implement a controlled-phase gate in one-way quantum computing.

11. The Pauli-X gate does not commute with the controlled-phase gate. Prove that the following identity is valid:

$$C_Z^{(i,j)} X_i = X_i Z_j C_Z^{(i,j)}.$$

12. Prove the action of the Y-measurement on a qubit in the linear chain cluster state illustrated in Fig. 12.14.

13. Prove Eq. (12.38).

14. Prove the identities from Fig. 12.16.

15. Prove the following CNOT gate propagation rules:

$$C_X^{(c,t)} X_t = X_t C_X^{(c,t)}, \quad C_X^{(c,t)} X_c = X_c X_t C_X^{(c,t)}, \quad C_X^{(c,t)} Z_c = Z_c C_X^{(c,t)},$$
$$C_X^{(c,t)} Z_t = Z_c Z_t C_X^{(c,t)}.$$

16. Prove the following identities:

$$U(\alpha, \beta, \gamma) X = X U(\alpha, -\beta, \gamma), \quad U(\alpha, \beta, \gamma) Z = Z U(-\alpha, \beta, -\gamma).$$

17. By using the quantum circuit model, prove the validity of Gilbert's approach to the creation of the box cluster state (see Fig. 12.25A).

18. For entanglement whiteness defined by Eq. (12.57), determine the expected value $\langle W \rangle = \mathrm{Tr}(\rho W)$, where ρ is the reduced density operator of $|\phi\rangle_{LC,4}\langle\phi|_{LC,4}$ obtained by averaging out the first qubit. Compare this expected value against the experimental result in Ref. [10].

References

[1] R. Raussendorf, H.J. Briegel, A one-way quantum computer, Phys. Rev. Lett. 86 (2001) 5188–5191.

[2] H.J. Briegel, R. Raussendorf, Persistent entanglement in arrays of interacting particles, Phys. Rev. Lett. 86 (2001) 910–913.

[3] H.J. Briegel, Cluster States, in: D. Greenberger, K. Hentschel, F. Weinert (Eds.), Compendium of Quantum Physics, Springer, Berlin-Heidelberg, 2009, pp. 96–105.

[4] F. Verstraete, J.I. Cirac, Valence bond solids for quantum computation, Phys. Rev. A 70 (2004) 060302(R).

[5] P. Jorrand, S. Perdrix, Unifying quantum computation with projective measurements only and one-way quantum computation, in: Proc. SPIE 5833, Quantum Informatics 2004, 2005, pp. 44–51.

[6] O. Mandel, M. Greiner, A. Widera, T. Rom, T.W. Hänsch, I. Bloch, Coherent transport of neutral atoms in spin-dependent optical lattice potentials, Phys. Rev. Lett. 91 (2003) 010407.

[7] W.K. Hensinger, S. Olmschenk, D. Stick, D. Hucul, M. Yeo, M. Acton, L. Deslauriers, C. Monroe, J. Rabchuk, T-junction ion trap array for two-dimensional ion shuttling, storage, and manipulation, Appl. Phys. Lett. 88 (3) (2006) 034101.

[8] M.A. Neilsen, Cluster-state quantum computation, Rep. Math. Phys. 57 (1) (2006) 147–161.

[9] D. McMahon, Quantum Computing Explained, John Wiley & Sons, Inc., Hoboken, NJ, 2008.

[10] P. Walther, K.J. Resch, T. Rudolph, E. Schenck, H. Weinfurter, V. Vedral, M. Aspelmeyer, A. Zeilinger, Experimental one-way quantum computing, Nature 434 (2005) 169–176.

[11] N. Kiesel, C. Schmid, U. Weber, G. Tóth, O. Gühne, R. Ursin, H. Weinfurter, Experimental analysis of a 4-qubit cluster state, Phys. Rev. Lett. 95 (21) (2005) 210502.

[12] G. Gilbert, M. Hamrick, Y.S. Weinstein, Efficient construction of photonic quantum-computational clusters, Phys. Rev. A 73 (2006) 064303.

[13] V. Scarani, A. Acín, E. Schenck, M. Aspelmeyer, Nonlocality of cluster states of qubits, Phys. Rev. 71 (4) (2005) 042325.

[14] D.E. Browne, T. Rudolph, Resource-efficient linear optical quantum computation, Phys. Rev. Lett. 95 (2005) 010501.

[15] O. Gühne, G. Tóth, P. Hyllus, H.J. Briegel, Bell inequalities for graph states, Phys. Rev. Lett. 95 (12) (2005) 120405.

[16] A. Mantri, T. Demarie, J. Fitzsimons, Universality of quantum computation with cluster states and (X, Y)-plane measurements, Sci. Rep. 7 (2017) 42861.

[17] C. Reimer, M. Kues, S. Sciara, P. Roztocki, M. Islam, L.R. Cortés, Y. Zhang, B. Fischer, S. Loranger, R. Kashyap, A. Cino, S.T. Chu, B.E. Little, D.J. Moss, L. Caspani, W.J. Munro, J. Azaña, R. Morandotti, High-dimensional one-way quantum computation operations with on-chip optical d-level cluster states, in: Conference on Lasers and Electro-Optics, OSA Technical Digest (Optical Society of America), 2019 paper FTh1A.4.

[18] C. Reimer, S. Sciara, P. Roztocki, et al., High-dimensional one-way quantum processing implemented on *d*-level cluster states, Nat. Phys. 15 (2019) 148–153.

[19] M. Hein, W. Dür, J. Eisert, R. Raussendorf, M. van den Nest, H.J. Briegel, Entanglement in graph states and its applications,, in: G. Casati, et al. (Eds.), International School of Physics Enrico Fermi (Varenna, Italy), Course CLXII, Quantum Computers, Algorithms and Chaos, 2006, pp. 115–218.
[20] G.H. Wannier, Statistical Physics, Dover Publications, Inc., New York, 1987.
[21] I.B. Djordjevic, Quantum Information Processing and Quantum Error Correction: An Engineering Approach, Elsevier/Academic Press, 2012.
[22] R. Raussendorf, D.E. Browne, H.J. Briegel, Measurement-based quantum computation with cluster states, Phys. Rev. A 68 (2003) 022312.
[23] R. Jozsa, An introduction to measurement based quantum computation, in: NATO science series, III: Computer and systems sciences, Quantum Information Processing-From Theory to Experiment, vol. 199, 2006, pp. 137–158.
[24] R. Raussendorf, H.J. Briegel, Computational Model for the One-Way Quantum Computer: Concepts and Summary, Available at: https://arxiv.org/abs/quant-ph/0207183.
[25] D. Gottesman, Theory of fault-tolerant quantum computation, Phys. Rev. 57 (1) (1998) 127–137.
[26] G. Burkard, D. Loss, D.P. DiVincenzo, Coupled quantum dots as quantum gates, Phys. Rev. B 59 (3) (1999) 2070–2078.
[27] M.L. Katz, J. Wang, Cluster state computation with quantum-dot charge qubits, in: Advances in Mathematical Physics, special issue on Quantum Information and Entanglement, vol. 2010, 2010, p. 482598.
[28] T.-C. Wei, Quantum spin models for measurement-based quantum computation, Adv. Phys. 3 (1) (2018) 1461026.
[29] M. Borhani, D. Loss, Cluster states from Heisenberg interactions, Phys. Rev. 71 (3) (2005) 034308.
[30] Y.S. Weinstein, C.S. Hellberg, J. Levy, Quantum-dot cluster-state computing with encoded qubits, Phys. Rev. 72 (2) (2005) 020304.
[31] G.-P. Guo, H. Zhang, T. Tu, G.-C. Guo, One-step preparation of cluster states in quantum-dot molecules, Phys. Rev. 75 (5) (2007) 050301(R).
[32] T.P. Friesen, D.L. Feder, One-way quantum computing in optical lattices with many-atom addressing, Phys. Rev. 78 (3) (2008) 032312.
[33] P.A. Ivanov, N.V. Vitanov, M.B. Plenio, Creation of cluster states of trapped ions by collective addressing, Phys. Rev. 78 (1) (2008) 012323.
[34] J.I. Cirac, P. Zoller, Quantum computations with cold trapped ions, Phys. Rev. Lett. 74 (20) (1995) 4091–4094.
[35] J.-Q. Li, G. Chen, J.-Q. Liang, One-step generation of cluster states in microwave cavity QED, Phys. Rev. 77 (1) (2008) 014304.
[36] J.Q. You, X.-B. Wang, T. Tanamoto, F. Nori, Efficient one-step generation of large cluster states with solid-state circuits, Phys. Rev. 75 (5) (2007) 052319.
[37] D. Jaksch, J.I. Cirac, P. Zoller, S.L. Rolston, R. Côté, M.D. Lukin, Fast quantum gates for neutral atoms, Phys. Rev. Lett. 85 (10) (2000) 2208–2211.
[38] E. Knill, R. Laflamme, G.J. Milburn, A scheme for efficient quantum computation with linear optics, Nature 409 (2000) 46–52.
[39] E. Knill, Quantum gates using linear optics and postselection, Phys. Rev. 66 (5) (2002) 052306.
[40] I.B. Djordjevic, Physical-Layer Security and Quantum Key Distribution, Springer International Publishing, Switzerland, 2019.

[41] D. Gottesman, I.L. Chuang, Demonstrating the viability of universal quantum computation using teleportation and single-qubit operations, Nature 402 (6760) (1999) 390–393.

[42] A.J.F. Hayes, A. Gilchrist, C.R. Myers, T.C. Ralph, Utilizing encoding in scalable linear optics quantum computing, J. Opt. B Quantum Semiclassical Opt. 6 (12) (2004) 533–541.

[43] T.B. Pittman, B.C. Jacobs, J.D. Franson, Demonstration of quantum error correction using linear optics, Phys. Rev. 71 (5) (2005) 052332.

[44] A.-N. Zhang, C.-Y. Lu, X.-Q. Zhou, Y.-A. Chen, Z. Zhao, T. Yang, J.-W. Pan, Experimental construction of optical multiqubit cluster states from Bell states, Phys. Rev. 73 (2) (2006) 022330 (received 9 January 2005).

[45] K. Banaszek, G.M. D'Ariano, M.G.A. Paris, M.F. Sacchi, Maximum-likelihood estimation of the density matrix, Phys. Rev. A 61 (1999) 010304(R).

[46] S. Yokoyama, R. Ukai, S.C. Armstrong, C. Sornphiphatphong, T. Kaji, S. Suzuki, J. Yoshikawa, H. Yonezawa, N.C. Menicucci, A. Furusawa, Ultra-large-scale continuous-variable cluster states multiplexed in the time domain, Nat. Photon. 7 (2013) 982–986.

[47] M. Chen, N.C. Menicucci, O. Pfister, Experimental realization of multipartite entanglement of 60 modes of a quantum optical frequency comb, Phys. Rev. Lett. 112 (12) (2014) 120505.

[48] S.D. Bartlett, B.C. Sanders, S.L. Braunstein, K. Nemoto, Efficient classical simulation of continuous variable quantum information processes, Phys. Rev. Lett. 88 (2002) 097904.

[49] D. Su, K.K. Sabapathy, C.R. Myers, H. Qi, C. Weedbrook, K. Brádler, Implementing quantum algorithms on temporal photonic cluster states, Phys. Rev. A 98 (2018) 032316.

[50] A.I. Lvovsky, Squeezed Light, Available at: https://arxiv.org/abs/1401.4118.

[51] N.C. Menicucci, Temporal-mode continuous-variable cluster states using linear optics, Phys. Rev. A 83 (6) (2011) 062314.

CHAPTER

Physical Implementations of Quantum Information Processing

13

CHAPTER OUTLINE

This chapter is devoted to several promising physical implementations of quantum information processing. The chapter starts with a description of physical implementation basics, *di Vincenzo criteria*, and an overview of physical implementation concepts (Section 13.1). The next section (Section 13.2) is related to nuclear magnetic resonance (NMR) implementation, whose basic concepts are used in various implementations. In Section 13.3, the use of ion traps in quantum computing is described. Next, various photonic implementations (Section 13.4) are described, followed by quantum relay implementation (Section 13.5) and the implementation of quantum encoders and decoders (Section 13.6). Implementation based on optical cavity electrodynamics is further described in Section 13.7. Finally, the use of quantum dots in quantum computing is discussed in Section 13.8. After a short summary (Section 13.9), a set of problems (Section 13.10) is provided for self-study.

Quantum Information Processing, Quantum Computing, and Quantum Error Correction
https://doi.org/10.1016/B978-0-12-821982-9.00001-0

13.1 PHYSICAL IMPLEMENTATION BASICS

We are still in the early stages of the implementation of quantum information processing devices and quantum computers. In the following sections, several successful implementations are described. However, from this point, it is difficult to predict which of them will actually be used in the foreseeable future. The main requirements, also known as *di Vincenzo criteria*, to be fulfilled before a particular physical system can be considered a viable candidate for quantum information processing can be summarized as follows:

1. *Initialization capability.* Initialization capability is related to the ability of preparing the physical system for the desired initial state, say $|00 \ldots 0\rangle$.
2. *Quantum register.* The quantum register represents a physical system with well-defined states—qubits—composing the storage device.
3. *Universal set of gates.* This requirement is related to the ability to perform a universal family of unitary transformations. The most popular sets of universal quantum gates, as discussed in Chapter 3, are: {Hadamard (H), phase (S), CNOT, Toffoli (U_T)} gates, {H, S, $\pi/8$ (T), CNOT} gates, Barenco gate [1], and Deutsch gate [2].
4. *Low error and decoherence rate criterion.* This requirement is related to high-fidelity quantum gates, which operate with reasonably low probability of error P_e, say $P_e < 10^{-3}$. In addition, the quantum decoherence time must be much longer than gate operation duration so that a meaningful quantum computation is possible. Decoherence is the number one enemy of quantum information processing. Because of decoherence, computations must be performed in a time interval smaller than decoherence time $\tau_{\text{decoherence}}$. Let $\tau_{\text{operation}}$ denote the duration of elementary operation performed by a given quantum gate. The maximum number of operations $N_{\text{operations}}$ that a quantum computer can perform is given by $N_{\text{operations}} = \tau_{\text{decoherence}}/\tau_{\text{operation}}$.
5. *Read-out capability.* This requirement refers to the ability to perform measurements on qubits in computational basis to determine the state of qubits at the end of the computation procedure.

In addition to these requirements, in quantum teleportation, quantum superdense coding, quantum key distribution (QKD), and distributing quantum computation applications, one additional requirement is needed. This requirement is related to communication capability.

6. *Communication capability.* Communication capability is the ability to teleport qubits between distant locations.

In the following sections of this chapter we describe several possible technologies, including: (1) NMR, which is provided from a conceptual point of view; (2) trapped ion-based implementations; (3) photonic quantum implementations; (4) cavity quantum electrodynamics (CQED)-based quantum computation; and (5) quantum dot-based implementations. We also discuss the physical implementation of quantum relay and quantum encoders and decoders.

In many of these implementations, the theory of harmonic oscillators, described in Chapter 2 (see Section 2.9), is used. With this approach, the energy levels $|0\rangle$, $|1\rangle,\ldots,|2^N\rangle$ of a single quantum harmonic oscillator can be used to represent N qubits. An arbitrary unitary operation U can be implemented by matching the eigenvalues spectrum of U to that of harmonic oscillator Hamiltonian $H=\hbar\omega a^{\dagger}a$, where $a^{\dagger}$ and a are creation and annihilation operators, respectively (see Section 2.9 for additional details). The *annihilation* a and *creation* $a^{\dagger}$ operators are related to the momentum p and position z operators as follows:

$$a=\sqrt{\frac{M\omega}{2\hbar}}\left(z+\frac{jp}{M\omega}\right)\quad a^{\dagger}=\sqrt{\frac{M\omega}{2\hbar}}\left(z-\frac{jp}{M\omega}\right). \tag{13.1}$$

The Hamiltonian of a particle in a 1D parabolic potential well $V(z)=M\omega^2z^2/2$ is given by:

$$H=\frac{p^2}{2M}+\frac{M\omega^2z^2}{2}. \tag{13.2}$$

The energy eigenvalues E_m can be determined from:

$$H|m\rangle=\hbar\omega\left(a^{\dagger}a+\frac{1}{2}\right)|m\rangle=\hbar\omega\left(m+\frac{1}{2}\right)|m\rangle=E_m|m\rangle. \tag{13.3}$$

The time evolution of an arbitrary state is given by:

$$|\psi(t)\rangle=\sum_m c_m(0)e^{-jm\omega t}|m\rangle, \tag{13.4}$$

where $c_n(0)$ is the expansion coefficient for $|\psi(0)\rangle$. Let the logical states be defined as:

$$|00\rangle_L=|0\rangle,\ |01\rangle_L=|2\rangle,\ |10\rangle_L=(|4\rangle+|1\rangle)/\sqrt{2},\ |11\rangle_L=(|4\rangle-|1\rangle)/\sqrt{2}. \tag{13.5}$$

By setting $t=\pi/(\hbar\omega)$, the basis states evolve as follows:

$$|m\rangle\rightarrow e^{-\frac{j}{\hbar}Ht}|m\rangle=\exp\left(-j\pi a^{\dagger}a\right)|m\rangle=(-1)^m|m\rangle. \tag{13.6}$$

The logic states, on the other hand, evolve as:

$$\begin{aligned}&|00\rangle_L\rightarrow(-1)^0|00\rangle_L=|00\rangle_L,\quad |01\rangle_L\rightarrow(-1)^2|01\rangle_L=|01\rangle_L\\&|10\rangle_L\rightarrow\left((-1)^4|4\rangle+(-1)^1|1\rangle\right)/\sqrt{2}=(|4\rangle-|1\rangle)/\sqrt{2}=|11\rangle_L,\\&|11\rangle_L\rightarrow\left((-1)^4|4\rangle-(-1)^1|1\rangle\right)/\sqrt{2}=(|4\rangle+|1\rangle)/\sqrt{2}=|10\rangle_L.\end{aligned} \tag{13.7}$$

By interpreting the first qubit as the control qubit and the second qubit as the target qubit, from Eq. (13.7) we conclude that when the control qubit is in the $|1\rangle$ state, the target qubit is flipped, which corresponds to CNOT gate operation.

13.2 NUCLEAR MAGNETIC RESONANCE IN QUANTUM COMPUTING

In this section, we describe the use of nuclear magnetic resonance (NMR) in quantum information processing (QIP) [15,50]. Even though that QIP based on NMR is not practical, the basic concepts from NMR are used in ion traps and CQED applications. Let us assume that spin-1/2 systems are located in a strong time-invariant magnetic field B_0, in the order of a few teslas. The time-independent Hamiltonian is given by [50]:

$$\widehat{H} = \begin{bmatrix} \hbar\omega' & 0 \\ 0 & \hbar\omega'' \end{bmatrix}, \tag{13.8}$$

where $\hbar\omega'$ and $\hbar\omega''$ represent the energy levels of the spin-1/2 systems. From the time-evolution equation:

$$j\hbar\frac{dU}{dt} = \widehat{H}U, \tag{13.9}$$

Since the Hamiltonian is time invariant, by separation of variables we obtain the following expression for time-evolution operator $U(t,t_0)$:

$$U(t,\ t_0) = \exp\left[-j\widehat{H}(t-t_0)/\hbar\right] = \begin{bmatrix} e^{-j\omega'(t-t_0)} & 0 \\ 0 & e^{-j\omega''(t-t_0)} \end{bmatrix}. \tag{13.10}$$

The time evolution of arbitrary state $|\psi(t=0)\rangle = a|0\rangle + b|1\rangle$ is then given by:

$$|\psi(t)\rangle = e^{-j\omega' t}a|0\rangle + e^{-j\omega'' t}b|0\rangle = a(t)|0\rangle + b(t)|0\rangle;$$

$$a(t) = ae^{-j\omega' t},\ \ b(t) = be^{-j\omega'' t}. \tag{13.11}$$

Given the arbitrariness of the absolute phase, what really matters is the phase difference, or equivalently the frequency difference $\omega' - \omega'' = \omega_0$, so that the Hamiltonian can be rewritten as:

$$\widehat{H} = \begin{bmatrix} \hbar\omega_0 & 0 \\ 0 & -\hbar\omega_0 \end{bmatrix}. \tag{13.12}$$

The energy $\hbar\omega_0$ is often referred to as the resonant energy and the frequency ω_0 as the *resonant frequency* (*Larmor frequency*). This Hamiltonian is also applicable to two-level atoms, where ω' represents the ground state and level ω'' ($\omega'' > \omega'$) represents the excited state. If the two-level atom is raised to the excited state, it returns spontaneously to the ground state by emitting a photon of energy $\hbar(\omega'' - \omega') = \hbar\omega_0$. On the other hand, if the two-level atom is enlightened by a laser beam of frequency $\omega \cong \omega_0$, a *resonance phenomenon* will be observed. Namely, the closer ω is to ω_0, the stronger the absorption will be.

Let now the proton be placed in the magnetic field containing both constant and periodic components as follows:

$$\boldsymbol{B} = B_0\widehat{z} + B_1(\widehat{x}\cos\omega t - \widehat{y}\sin\omega t), \tag{13.13}$$

which is the same as that in NMR or (nuclear) *magnetic resonance imaging* ([N] MRI). Its spin is associated with the following operator $\hbar\boldsymbol{\sigma}/2$, where $\boldsymbol{\sigma} = [X\,Y\,Z]^{\mathrm{T}}$ (X, Y, and Z are Pauli operators). The corresponding *magnetic moment* $\boldsymbol{\mu}$ is given by: $\boldsymbol{\mu} = \gamma_p\boldsymbol{\sigma}/2$, where γ_p is the *gyromagnetic ratio* to be determined experimentally ($\gamma_p \cong 5.59 q_p\hbar/(2m_p)$, where q_p is the proton charge and m_p is the proton mass). The proton Hamiltonian in this magnetic field can be written as:

$$\hat{H} = -\boldsymbol{\mu}\cdot\boldsymbol{B} = -\frac{1}{2}\gamma_p\boldsymbol{\sigma}\cdot\boldsymbol{B} = -\frac{1}{2}\gamma_p[B_0 Z + B_1(X\cos\omega t - Y\sin\omega t)]. \tag{13.14}$$

By substituting: $\gamma_p B_0 = \hbar\omega_0$ and $\gamma_p B_1 = \hbar\omega_1$ we obtain:

$$\hat{H} = -\frac{1}{2}\hbar\omega_0 Z - \frac{1}{2}\hbar\omega_1(X\cos\omega t - Y\sin\omega t). \tag{13.15}$$

Clearly, the frequency ω_1, known as the *Rabi frequency*, is proportional to B_1. The corresponding matrix representation of the Hamiltonian, based on Eq. (13.15), is given by:

$$\hat{H} = -\frac{\hbar}{2}\begin{bmatrix} \omega_0 & \omega_1 e^{j\omega t} \\ \omega_1 e^{-j\omega t} & -\omega_0 \end{bmatrix}. \tag{13.16}$$

For an arbitrary state: $|\psi(t=0)\rangle = a|0\rangle + b|1\rangle$ we can use the Hamiltonian (13.16) to solve the evolution equation:

$$j\hbar\frac{d|\psi(t)\rangle}{dt} = \hat{H}|\psi(t)\rangle, \quad |\psi(t)\rangle = \begin{bmatrix} a(t) \\ b(t) \end{bmatrix}. \tag{13.17}$$

After the substitution [50] $a(t) = \tilde{a}(t)e^{j\omega_0 t/2}$, $b(t) = \tilde{b}(t)e^{-j\omega_0 t/2}$,the solution of evolution Eq. (13.17) is obtained as:

$$\tilde{a}(t) = A\cos\left(\frac{\omega_1 t}{2}\right) + B\sin\left(\frac{\omega_1 t}{2}\right), \quad \tilde{b}(t) = jA\sin\left(\frac{\omega_1 t}{2}\right) - jB\cos\left(\frac{\omega_1 t}{2}\right), \tag{13.18}$$

where the constants A and B are dependent on the initial state. For example, if the initial state at $t = 0$ was $|0\rangle$, the constants are obtained as $A = 1$, $B = 0$. At time instance $t = \pi/(2\omega_1)$, the $|0\rangle$ state evolves to the superposition state:

$$|\psi(t)\rangle|_{t=\pi/(2\omega_1)} = \frac{1}{2}\left(e^{j\omega_0 t/2}|0\rangle + e^{-j\omega_0 t/2}|1\rangle\right). \tag{13.19}$$

By redefining the $|0\rangle$and $|1\rangle$ states we can obtain the conventional superposition state: $(|0\rangle + |1\rangle)/\sqrt{2}$. This transformation is often called $\pi/2$-pulse transformation. On the other hand, if we allow t to be π/ω_1, the initial state $|0\rangle$ is transformed to the $|1\rangle$ state, and this transformation is known as π-pulse (because $\omega_1 t = \pi$). In general, the probability of finding the initial state $|0\rangle$ into the $|1\rangle$ state, denoted as $p_{01}(t)$, is given by:

$$p_{01}(t) = \left(\frac{\omega_1}{\Omega}\right)^2\sin^2\left(\frac{\Omega t}{2}\right), \quad \Omega = \left[(\omega - \omega_0)^2 + \omega_1^2\right]^{1/2}, \quad \delta = \omega - \omega_0. \tag{13.20}$$

These oscillations between states $|0\rangle$ and $|1\rangle$ are known as *Rabi oscillations*. The oscillations have maximum amplitude when detuning δ is zero.

A similar manipulation, which is fundamental in quantum computing applications, can be performed on two-level atoms. In this case, $\hbar\omega_a$ represents the energy difference between excited and ground states, while Rabi frequency ω_1 is proportional to the electric dipole moment of the atom $\boldsymbol{d}$ and the electric field $\boldsymbol{E}$ of the laser beam, namely $\omega_1 \approx \boldsymbol{d}\cdot\boldsymbol{E}$.

The Hamiltonian given by Eq. (13.15) can be rewritten as:

$$\hat{H} = \underbrace{-\frac{1}{2}\hbar\omega_0 Z}_{\hat{H}_0} \underbrace{-\frac{1}{2}\hbar\omega_1\left(\sigma_+ e^{j\omega t} + \sigma_- e^{-j\omega t}\right)}_{\hat{H}_1(t)} = \hat{H}_0 + \hat{H}_1(t),\ \ \sigma_\pm = (X \pm jY)/2, \tag{13.21}$$

where σ_+ and σ_- denote the *lowering* and *raising* operators, respectively (see Chapter 2), defined as $\sigma_\pm^{(i)} = (X_i \pm jY_i)/2;\ i = 1, 2.$To facilitate the evolution study of state $|\psi(t)\rangle$, we redefine the state in the *rotating reference frame* as follows:

$$|\tilde{\psi}(t)\rangle = e^{-j\omega Zt/2}|\psi(t)\rangle. \tag{13.22}$$

Since:

$$|\psi(t)\rangle = e^{-j\hat{H}_0 t/\hbar}|\psi(0)\rangle = e^{-j\omega_0 Zt/2}|\psi(0)\rangle, \tag{13.23}$$

when $\omega = \omega_0$ and $\omega_1 = 0$ we obtain:

$$|\tilde{\psi}(t)\rangle\Big|_{\omega=\omega_0,\ \omega_1=0} = e^{-j\omega Zt/2}|\psi(t)\rangle = e^{-j\omega Zt/2}e^{j\omega_0 Zt/2}|\psi(0)\rangle = |\psi(0)\rangle, \tag{13.24}$$

indicating that in this reference frame the state is time invariant as long as $\omega = \omega_0$ and $\omega_1 = 0$. Because in the referent frame the raising and lower operators are given by:

$$\tilde{\sigma}_\pm = e^{-j\omega Zt/2}\sigma e^{j\omega Zt/2} = e^{\mp j\omega t}\sigma_\pm, \tag{13.25}$$

the Hamiltonian in the referent frame can be related to the Hamiltonian given by Eq. (13.21) as:

$$\tilde{H}(t) = \frac{1}{2}\hbar\omega_0 Z + e^{-j\omega Zt/2}\hat{H}(t)e^{j\omega Zt/2} = \frac{\hbar}{2}\delta Z - \frac{\hbar}{2}\omega_1 X,\ \ \delta = \omega - \omega_0, \tag{13.26}$$

and it is time invariant. Notice that for $\omega = \omega_0$, the evolution operator becomes:

$$e^{-j\tilde{H}t/\hbar} = e^{-j\omega_1 Xt/2}, \tag{13.27}$$

and represents the rotation by an angle $-\omega_1 t$ around the x-axis. Therefore to perform the rotation around the x-axis for a given angle, the duration of radiofrequency (RF) pulse needs to be properly adjusted.

We are now concerned with the study of the interaction of two spins within the same molecule. As an example, the first spin can be carried by a proton and the second spin can be carried by a ^{13}C nucleus. As these two spins have different magnetic moments, their resonant $\omega_0^{(i)}(i = 1, 2)$ and Rabi frequencies $\omega_1^{(i)}(i = 1, 2)$will be different. This type of interaction can be described by the following operator: $\hbar J_{12}Z_1Z_2$, where $|J_{12}| \ll \omega_1^{(i)}$ $(i = 1, 2)$.The Hamiltonian can be obtained by generalizing Eq. (13.21) as follows:

$$\widehat{H}_{12} = -\frac{1}{2}\hbar\omega_0^{(1)}Z_1 - \frac{1}{2}\hbar\omega_0^{(2)}Z_2 - \frac{1}{2}\hbar\omega_1^{(1)}\left(\sigma_+^{(1)}e^{j\omega^{(1)}t} + \sigma_-^{(1)}e^{-j\omega^{(1)}t}\right)$$
$$-\frac{1}{2}\hbar\omega_1^{(2)}\left(\sigma_+^{(2)}e^{j\omega^{(2)}t} + \sigma_-^{(2)}e^{-j\omega^{(2)}t}\right) + \hbar J_{12}Z_1Z_2. \tag{13.28}$$

Given the fact that resonant and Rabi frequencies for two different spins are different, the fields applied to different spins should be properly adjusted as:

$$\left|\omega^{(i)} - \omega_0^{(i)}\right| = \left|\delta^{(i)}\right| \ll \omega_1^{(i)};\ i = 1, 2. \tag{13.29}$$

The state of this two-spin system can be described in the rotating referent frame by generalizing Eq. (13.22) in the following way:

$$|\widetilde{\psi}_1(t)\rangle \otimes |\widetilde{\psi}_2(t)\rangle = e^{-j\omega^{(1)}Z_1t/2}e^{-j\omega^{(2)}Z_2t/2}|\psi_1(t)\rangle \otimes |\psi_2(t)\rangle. \tag{13.30}$$

In this rotating referent frame, the Hamiltonian is again time invariant:

$$\widetilde{H} = \frac{\hbar}{2}\delta^{(1)}Z_1 + \frac{\hbar}{2}\delta^{(2)}Z_2 - \frac{\hbar}{2}\omega_1^{(1)}X_1 - \frac{\hbar}{2}\omega_1^{(2)}X_2 + \hbar J_{12}Z_1Z_2. \tag{13.31}$$

Attention is moved now to *quantum logic gates*. The qubits are represented as 1/2-spin systems. To manipulate qubits, it is sufficient to apply an RF field for a suitable time interval, with the frequency of the field being in close proximity of the resonant frequency $\omega_0^{(i)}$,which corresponds to the ith qubit to be manipulated. The CNOT gate can be implemented by using Z_1Z_2 interaction between two qubits (1/2-spin systems). This interaction is internal to the system, while in many other implementations this kind of interaction is introduced externally. Since this interaction is internal to the system, it is always present, and in certain situations we need to suppress its presence. By using the fact that the Z_1Z_2-operator is the second-order Hermitian, namely $(Z_1Z_2)^2 = (Z_1Z_2)(Z_1Z_2) = Z_1^2 \otimes Z_2^2 = I$,the evolution operator can be represented as:

$$e^{-jJZ_1Z_2t} = \cos(J_{12}t)I - j\sin(J_{12}t)Z_1Z_2. \tag{13.32}$$

Eq. (13.32) suggests that the action of the evolution operator is in fact the rotation operation, the time needed to perform interaction is typically in the order of milliseconds, about two orders of magnitude of the time interval needed to perform rotation on a single-qubit (~ 10 μs), which causes an interoperability problem.

To explain how the CNOT gate can be implemented based on NMR, let us introduce the following operator:

$$O_{12}(t) = e^{jJZ_1Z_2t}R_z^{(1)}(\pi/2)R_z^{(2)}(\pi/2), \tag{13.33}$$

where $R_z(\theta)$ is the rotation operator for angle θ around the z-axis, introduced in Chapter 3 as follows:

$$R_z(\theta) = e^{-j\theta Z/2} = \cos\left(\frac{\theta}{2}\right)I - j\sin\left(\frac{\theta}{2}\right)Z, \tag{13.34}$$

which for $\theta = \pi/2$ becomes:

$$R_z(\pi/2) = \cos\left(\frac{\pi}{4}\right)I - j\sin\left(\frac{\pi}{4}\right)Z = \frac{1}{\sqrt{2}}(I - jZ).$$

For $t = \pi/(4J)$ (or equivalently $Jt = \pi/4$), the operator given by Eq. (13.33) becomes:

$$O_{12}(\pi/(4J_{12})) = e^{j\pi Z_1Z_2/4}R_z^{(1)}(\pi/2)R_z^{(2)}(\pi/2) = \underbrace{[\cos(\pi/4)I + j\sin(\pi/4)Z_1Z_2]}_{\frac{1}{\sqrt{2}}(I + jZ_1Z_2)}\frac{1}{2}(I - jZ_1)(I - jZ_2)$$

$$= \frac{1-j}{2^{3/2}}(I + Z_1I_2 + I_1Z_2 - Z_1Z_2)$$

$$= \frac{1-j}{2^{3/2}}\left\{\begin{bmatrix}1 & 0\\0 & 1\end{bmatrix}\otimes\begin{bmatrix}1 & 0\\0 & 1\end{bmatrix} + \begin{bmatrix}1 & 0\\0 & -1\end{bmatrix}\otimes\begin{bmatrix}1 & 0\\0 & 1\end{bmatrix}\right.$$

$$\left.+\begin{bmatrix}1 & 0\\0 & 1\end{bmatrix}\otimes\begin{bmatrix}1 & 0\\0 & -1\end{bmatrix} - \begin{bmatrix}1 & 0\\0 & -1\end{bmatrix}\otimes\begin{bmatrix}1 & 0\\0 & -1\end{bmatrix}\right\}$$

$$= \frac{1-j}{\sqrt{2}}\begin{bmatrix}1 & 0 & 0 & 0\\0 & 1 & 0 & 0\\0 & 0 & 1 & 0\\0 & 0 & 0 & -1\end{bmatrix} = \frac{1-j}{\sqrt{2}}\begin{bmatrix}I & 0\\0 & Z\end{bmatrix}$$

$$= \frac{1-j}{\sqrt{2}}C(Z),\ C(Z) = \begin{bmatrix}I & 0\\0 & Z\end{bmatrix}, \tag{13.35}$$

which is equivalent to controlled-Z ($C(Z)$). The Hadamard gate can be implemented by a rotation for π rad around the axis $(1/\sqrt{2},0,1/\sqrt{2})$ as follows:

$$R_{\hat{n}=(1/\sqrt{2},\,0,\,1/\sqrt{2})}(\pi) = \exp(-j\pi\hat{\boldsymbol{n}}\cdot\boldsymbol{\sigma}/2) = \cos\left(\frac{\pi}{2}\right)I - j\sin\left(\frac{\pi}{2}\right)(n_xX + n_yY + n_zZ)$$

$$= -\frac{j}{\sqrt{2}}(X + Z) = -\frac{j}{\sqrt{2}}\begin{bmatrix}1 & 1\\1 & -1\end{bmatrix} = -jH,\ H = \frac{1}{\sqrt{2}}\begin{bmatrix}1 & 1\\1 & -1\end{bmatrix}. \tag{13.36}$$

Furthermore, the S gate can be obtained (up to the phase constant) by rotation for $\pi/2$ rad around the z-axis:

$$R_z(\pi/2) = \begin{bmatrix} e^{-j\pi/4} & 0 \\ 0 & e^{j\pi/4} \end{bmatrix} = e^{-j\pi/4}\begin{bmatrix} 1 & 0 \\ 0 & e^{j\pi/2} \end{bmatrix} = e^{-j\pi/4}\begin{bmatrix} 1 & 0 \\ 0 & j \end{bmatrix} = e^{-j\pi/4}S, \quad S = \begin{bmatrix} 1 & 0 \\ 0 & j \end{bmatrix}. \tag{13.37}$$

Finally, by rotating for $\pi/4$ rad around the z-axis, the $\pi/8$ gate is obtained:

$$R_z(\pi/2) = \begin{bmatrix} e^{-j\pi/8} & 0 \\ 0 & e^{j\pi/8} \end{bmatrix} = T, \tag{13.38}$$

which completes the implementation of the universal set of gates based on NMR. If the CNOT gate is used instead of the $C(Z)$ gate, two Hadamard gates must be applied on the second qubit before and after the $C(Z)$ gate, as shown in Fig. 3.6 of Chapter 3. In other words, we perform the following transformation of the $C(Z)$ gate:

$$(I \otimes H_2)C(Z)(I \otimes H_2) = \begin{bmatrix} H & 0 \\ 0 & H \end{bmatrix}\begin{bmatrix} I & 0 \\ 0 & Z \end{bmatrix}\begin{bmatrix} H & 0 \\ 0 & H \end{bmatrix} = \begin{bmatrix} H^2 & 0 \\ 0 & HZH \end{bmatrix}$$
$$= \begin{bmatrix} I & 0 \\ 0 & X \end{bmatrix} = \text{CNOT}. \tag{13.39}$$

Based on this discussion, the overall implementation of the CNOT gate based on NMR is shown in Fig. 13.1. Two molecules, chloroform for two-qubit operations and perfluorobutadienyl iron complex for seven-qubit operations [3], are given in Fig. 13.2. The atoms used in quantum computing are encircled. For additional examples, interested readers are referred to [4–6] (and references therein). The authors in [3] used the perfluorobutadienyl iron complex (Fig. 13.2B) to implement Shor's factorization algorithm in factorizing integer 15. Since b in the b^x mod N function can take any value from $\{2,4,7,8,11,13,14\}$, the largest period is $r = 4$ for $b = 2,7,8$, and 13. To see two periods, the x should take $0,1,\ldots,2^3 - 1$ values, while the corresponding $f(x)$ values are $0,1,\ldots,2^4 - 1$. Therefore a three-qubit input register and four-qubit output quantum register are needed, indicating that seven-bit

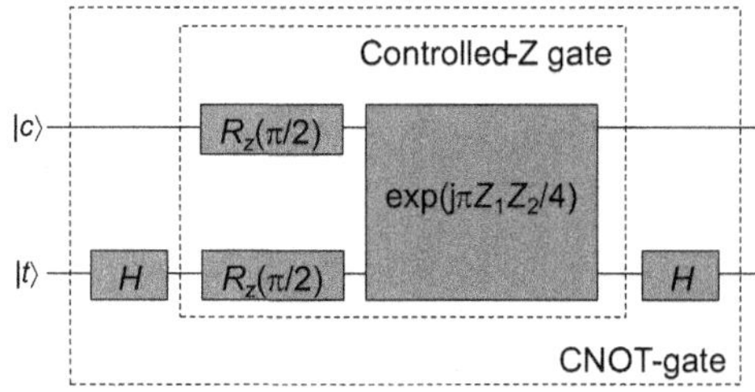

FIGURE 13.1

Nuclear magnetic resonance-based controlled-Z gate and CNOT gate implementations.

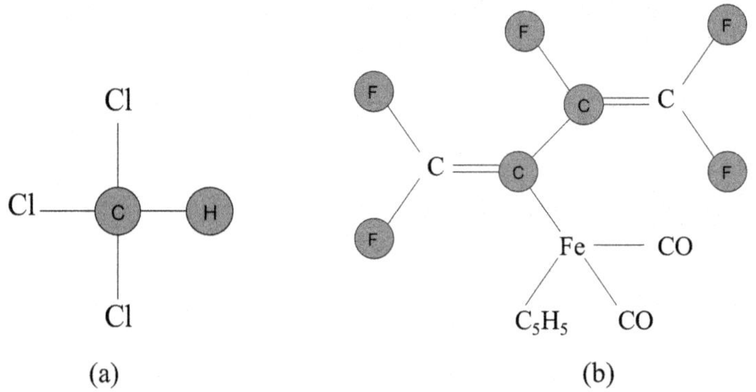

FIGURE 13.2

Nuclear magnetic resonance-based quantum commuting examples: (a) chloroform-based two-qubit example and (b) perfluorobutadienyl iron complex for seven-qubit operations.

NMR-based molecules are sufficient. Although recently more than 10-qubit operations based on NMR have been demonstrated, there are many challenges to be solved. For example, the synthesis of a molecule with as many distinguishable sites as the number of qubits and ability to select the appropriate frequencies to act on different qubits, for any meaningful computation (in the order of thousands of qubits) would be quite challenging. Furthermore, NMR does not employ individual quantum systems (objects), but a collection of $>10^{18}$ active molecules diluted in a solvent. Therefore the resulting signal is a result of collective action. Very complicated initialization, which is out of the scope of this book, is needed. Moreover, the signal decreases as the number of qubits increases. Another interesting problem, as already mentioned, is related to different evolution times for rotation and Hadamard gates (~ 10 μs) and the interaction term $J_{12}Z_1Z_2$ (several milliseconds). This problem can be solved by the *refocusing technique*, used in NMR and MRI. Let us place the evolution term due to $J_{12}Z_1Z_2$ between two rotations of spin 1 $R_x^{(1)}(-\pi) = jX$and $R_x^{(1)}(\pi) = -jX$as follows:

$$\begin{aligned} R_x^{(1)}(-\pi)\exp(-jJ_{12}tZ_1Z_2)R_x^{(1)}(\pi) &= (jX_1)(I\cos J_{12}t - jZ_1Z_2\sin J_{12}t)(-jX_1) \\ &= I\cos J_{12}t + jZ_1Z_2\sin J_{12}t = \exp(jJ_{12}tZ_1Z_2). \end{aligned} \tag{13.40}$$

If two spins have evolved during a time interval t as $\exp(-jJ_{12}tZ_1Z_2)$, the overall result will be:

$$\left[R_x^{(1)}(-\pi)\exp(-jJ_{12}tZ_1Z_2)R_x^{(1)}(\pi)\right]\exp(-jJ_{12}tZ_1Z_2) = \exp(jJ_{12}tZ_1Z_2)\exp(-jJ_{12}tZ_1Z_2) = I, \tag{13.41}$$

indicating that it is possible to cancel out the evolution of the qubits different from those involved in the CNOT gate. Therefore the purpose of refocusing methods is to effectively turn off undesired couplings for a desired interval of time. For additional examples, interested readers are referred to [4].

Before concluding this section, it is interesting to describe how to determine the coupling coefficients in the Hamiltonian. The simplest way would be to observe the NMR spectrum and determine the coefficients by reading off the positions of peaks (in frequency scale). For example, the three ^{13}C nuclei of alanine in 9.4 T are governed by the following Hamiltonian [4]:

$$\widetilde{H}_{\text{alanine}} = \pi\left[10^8(Z_1 + Z_2 + Z_3) - 12580Z_2 + 3440Z_3\right] + \frac{\pi}{2}(53Z_1Z_2 + 38Z_1Z_3 + 1.2Z_2Z_3), \tag{13.42}$$

where the frequencies are given in Hz.

13.3 TRAPPED IONS IN QUANTUM COMPUTING

In the ion trap-like quantum computer [7–9, 50], the qubits are represented by the internal electronic states of single alkali-like ions such as ions from group II of the system of periodical elements ($^{40}Ca^+$ or $^{9}Be^+$). Two states of a qubit are represented by the ground state of an ion $|g\rangle \equiv |0\rangle$ and an excited state $|e\rangle \equiv |1\rangle$ of a very long lifetime (it could be $\sim$1 s). The excited state is either a metastable state or a hyperfine state. For example, $^{40}Ca^+$ ion with $S_{1/2}$ state as the ground state and $D_{5/2}$ metastable state as the excited state (with lifetime in the order of 1 s) can be used as a qubit in quantum computing applications. The individual qubits are manipulated by a laser beam at a given frequency. A similar model to the previous section can be used, where now $\hbar\omega_0$ represents the energy difference between excited and ground states, while Rabi frequency ω_1 is proportional to the electric dipole moment of the ion $\boldsymbol{d}$ and the electric field $\boldsymbol{E}$ of the laser beam; in other words, $\omega_1 \approx \boldsymbol{d}\cdot\boldsymbol{E}$. The transition between excited and ground state in the $^{40}Ca^+$ ion corresponds to a wavelength of 729 nm. The ions (representing the qubits) are placed in a linear Paul trap (named after their inventor) as shown in Fig. 13.3. The ion trap is formed by using four parallel electrodes of radius comparable to the distance from electrodes to the common axis. The sinusoidal RF voltage signal is applied to two opposite electrodes, while other two electrodes are grounded. To confine ions in the

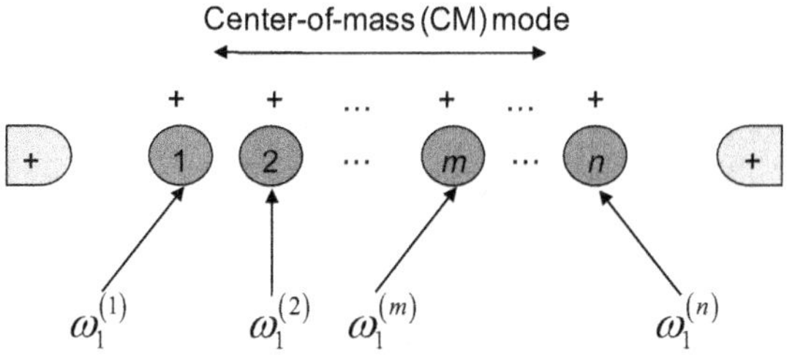

FIGURE 13.3

Operation principle of an ion trap-based quantum register with N qubits.

axial direction, two positive static potentials are applied from opposite sides of the ion chain as illustrated in Fig. 13.3. The interaction among ions is performed by *vibration motion* of the ions, which represents the *external degree of freedom* of the ions. The chain of ions is kept at sufficiently low temperature so that ions can be trapped in the transversal direction. The overall result is the location of ions in a harmonic potential:

$$V(x, y, z) = \frac{1}{2} M_I \left(\omega_x^2 + \omega_y^2 + \omega_z^2 \right), \tag{13.43}$$

where M_I is the ion mass and (x,y,z) denotes the position of the ion in the ion trap. In practice, the trap frequencies (in the order of MHz) satisfy the following conditions:

$$\omega_x^2 \approx \omega_y^2 \gg \omega_z^2, \tag{13.44}$$

so that the ion moves in the direction of the z-axis in a potential and we can write:

$$V(z) \cong \frac{1}{2} M_I \omega_z^2. \tag{13.45}$$

The corresponding Hamiltonian is given by:

$$\hat{H} = \frac{p_z^2}{2M_I} + \frac{1}{2} M_I \omega_z^2, \tag{13.46}$$

where $p_z^2/(2M_I)$is the kinetic energy term, with coordinate z and momentum p_z components being Hermitian and satisfying the commutation relation $[z, p_z] = j\hbar I$. By using Eq. (13.1), the Hamiltonian can be expressed in terms of creation $a^\dagger$ and annihilation a operators as follows:

$$\hat{H} = h\omega_z (a^\dagger a + 1/2). \tag{13.47}$$

The eigenvalues of the Hamiltonian are $\hbar\omega(m + 1/2) = E_m (m = 0, 1, 2, \ldots)$ $\hbar\omega(m + 1/2) = E_m \ (m = 0, 1, 2, \ldots)$, as given by Eq. (13.3), and corresponding eigenkets are denoted by $|m\rangle$. The action of the creation (annihilation) operator is to increase (decrease) the energy level (see Chapter 2 for more details):

$$a^\dagger |m\rangle = \sqrt{m+1} |m+1\rangle, \ \ a|m\rangle = \sqrt{m} |m-1\rangle; \ \ a|0\rangle = 0. \tag{13.48}$$

The energy corresponding to the $|0\rangle$ state is nonzero, namely $E_0 = \hbar\omega_z/2$, and the integer m can be called the vibrational quantum number of the ion trap. If we use the Heisenberg inequality $\Delta z \Delta p_z \sim \hbar/2$(see Chapter 2) and by replacing the coordinate z and momentum p_z operators by corresponding standard deviations Δz and Δp_z we can provide a heuristic interpretation as follows:

$$E \sim \frac{(\Delta p_z)^2}{2M_I} + \frac{M_I \omega_z^2}{2} (\Delta z)^2 \overset{\Delta p_z \sim \hbar/(2\Delta z)}{\sim} \frac{\hbar^2}{8M_I (\Delta z)^2} + \frac{M_I \omega_z^2}{2} (\Delta z)^2. \tag{13.49}$$

By differentiating E with respect to Δz and setting the derivative to zero we obtain the minimum energy as:

$$E_{\mathrm{opt}} = \hbar\omega_z/2, \quad (\Delta z)^2 = \hbar/(2M_I\omega_z). \tag{13.50}$$

Therefore the minimum energy is the best compromise between kinetic and potential energies, while the region where the ion is most probably going to be located is determined by $\Delta z = \sqrt{\hbar/(2M_I\omega_z)}$. This mode, known as normal mode, corresponds to the case when all ions oscillate together as a rigid body with frequency ω_z, that is, it is a center-of-mass (CM) mode.

Let us now model the ion as a two-level system trapped in the potential $V(z) \cong M_I\omega_z^2/2$,and affected by an oscillating electric field of the form:

$$\mathbf{E} = 2E_1\widehat{\mathbf{x}}\cos(\omega t - kz - \phi) = E_1\widehat{\mathbf{x}}\left[e^{j(\omega t-\phi)}e^{-jkz} + e^{-j(\omega t-\phi)}e^{jkz}\right]. \tag{13.51}$$

The corresponding Hamiltonian will have three terms: the first term $\widehat{H}_0$ corresponding to the absence of the oscillating field ($E_1 = 0$), and the other two terms originating from the following expansion: $\exp(\pm jkz) \approx I \pm jkz$, which is valid when:

$$\eta = k\Delta z_0 = k\sqrt{\hbar/(2M_I\omega_z)} \ll 1,$$

where η is the so-called *Lamb–Dicke parameter.* It is a common practice to redefine the zero of the vibrational energy such that the ground state has zero energy. The Hamiltonian operator in the absence of an electric field can be written as:

$$\widehat{H}_0 = -\frac{\hbar\omega_0}{2}Z + \hbar\omega_z a^\dagger a. \tag{13.52}$$

The internal ion states are now $|0\rangle$ state of energy $-\hbar\omega_0/2$ and $|1\rangle$ state of energy $\hbar\omega_0/2$. Similarly to the NMR-based quantum computation we can use the rotating reference frame approach to facilitate the study. Given an operator A, the corresponding operator frame in the rotating reference frame will be:

$$A(t) = e^{j\hat{H}_0 t/\hbar}Ae^{-j\hat{H}_0 t/\hbar}, \tag{13.53}$$

which is essentially the Heisenberg picture representation (see Chapter 2 for more details). Following a similar procedure to that in the previous section we can relate the annihilation, creation, lowering, and raising operators in original and rotating frames as follows:

$$\widetilde{a}(t) = ae^{-j\omega_z t}, \; \widetilde{a}^\dagger(t) = a^\dagger e^{j\omega_z t}, \; \widetilde{\sigma}_-(t) = \sigma_- e^{j\omega_0 t}, \; \widetilde{\sigma}_+(t) = \sigma_+ e^{-j\omega_0 t}. \tag{13.54}$$

Based on Eq. (13.51) and the previous section, the interaction Hamiltonian can be represented as:

$$\widehat{H}_{\mathrm{interaction}} = -\frac{\hbar\omega_1}{2}\underbrace{(\sigma_- + \sigma_+)}_{X}\underbrace{\left[e^{j(\omega t-\phi)}e^{-jkz} + e^{-j(\omega t-\phi)}e^{jkz}\right]}_{2\cos(\omega t - kz - \phi)}, \tag{13.55}$$

where ω_1 is the Rabi frequency, as indicated earlier. The approximation

$\exp(\pm jkz) \approx I \pm jkz$ is now employed. The first term of expansion (the identity operator term) yields to the following contribution in the interaction Hamiltonian:

$$\hat{H}_1 = -\frac{\hbar\omega_1}{2}(\sigma_- + \sigma_+)\left[e^{j(\omega t-\phi)} + e^{-j(\omega t-\phi)}\right], \tag{13.56}$$

which in the rotating reference frame becomes:

$$\tilde{H}_1 = -\frac{\hbar\omega_1}{2}\left(\sigma_- e^{j\omega_0 t} + \sigma_+ e^{j\omega_0 t}\right)\left[e^{j(\omega t-\phi)} + e^{-j(\omega t-\phi)}\right]. \tag{13.57}$$

By employing now the *rotating-wave approximation*, in which the terms of the form $\exp[\pm j(\omega + \omega_0)t]$ can be neglected as they average to zero because of their rapid oscillation in comparison with other terms, we can approximate Eq. (13.57) as:

$$\begin{aligned}\tilde{H}_1 &\simeq -\frac{\hbar\omega_1}{2}\left[\sigma_- e^{-j(\underbrace{(\omega-\omega_0)}_{\delta}t-\phi)} + \sigma_+ e^{j(\underbrace{(\omega-\omega_0)}_{\delta}t-\phi)}\right] \\ &= -\frac{\hbar\omega_1}{2}\left[\sigma_- e^{-j(\delta t-\phi))} + \sigma_+ e^{j(\delta t-\phi)}\right],\ \delta = \omega - \omega_0.\end{aligned} \tag{13.58}$$

By comparing to Eq. (13.26) we conclude that the form is similar, and therefore similar methods can be used to manipulate the qubits. At the resonance ($\delta = 0$), the time-evolution operator is the rotation operator:

$$e^{-j\theta(\sigma_- e^{j\phi}+\sigma_+ e^{-j\phi})} = e^{-j\theta(X\cos\phi + Y\sin\phi)/2},\quad \theta = -\omega_1 t, \tag{13.59}$$

which represents the rotation around the axis: $\hat{\boldsymbol{n}} = (\cos\phi,\ \sin\phi,\ 0)$

The second term in expansion ($\pm jkz$) yields to the following interaction Hamiltonian (based on Eq. (13.55)):

$$\begin{aligned}\hat{H}_2 &\simeq -\frac{\hbar\omega_1}{2}(\sigma_- + \sigma_+)\left[-je^{j(\omega t-\phi)}kz + je^{-j(\omega t-\phi)}kz\right] \overset{z=\sqrt{\hbar/(2M_I\omega_z)}(a^\dagger+a)}{\simeq} \\ &\frac{j\eta\hbar\omega_1}{2}(\sigma_- + \sigma_+)\left(a^\dagger + a\right)\left[e^{j(\omega t-\phi)} - e^{-j(\omega t-\phi)}\right].\end{aligned} \tag{13.60}$$

In the rotating reference frame, the corresponding Hamiltonian term becomes:

$$\begin{aligned}\tilde{H}_2 \simeq &\frac{j\eta\hbar\omega_1}{2}\left(\sigma_- e^{j\omega_0 t}a^\dagger e^{j\omega_z t} + \sigma_- e^{j\omega_0 t}ae^{-j\omega_z t} + \sigma_+ e^{-j\omega_0 t}a^\dagger e^{j\omega_z t} + \sigma_+ e^{-j\omega_0 t}ae^{-j\omega_z t}\right) \\ &\left[e^{j(\omega t-\phi)} - e^{-j(\omega t-\phi)}\right] = \frac{j\eta\hbar\omega_1}{2}\left(\sigma_- a^\dagger e^{j(\omega_0+\omega_z)t} + \sigma_- ae^{j(\omega_0-\omega_z)t}\right. \\ &\left.+\sigma_+ a^\dagger e^{-j(\omega_0-\omega_z)t} + \sigma_+ ae^{-j(\omega_0+\omega_z)t}\right)\left[e^{j(\omega t-\phi)} - e^{-j(\omega t-\phi)}\right].\end{aligned} \tag{13.61}$$

By choosing the laser operating frequency to be $\omega = \omega_0 + \omega_z$ we select the upper ("blue") side-band frequency, while the lower side band can be neglected according to the aforementioned rotating-wave argument, so that the interaction Hamiltonian becomes:

$$\widetilde{H}_2 \simeq \frac{j\eta\hbar\omega_1}{2}\left(-e^{j\phi}\sigma_- a^\dagger + e^{-j\phi}\sigma_+ a\right). \tag{13.62}$$

The state of the single ion in the trap can be represented as $|n,m\rangle$, where $n = 0,1$ is related to the spin (internal) state and $m = 0,1$ is related to the vibrational state of the harmonic oscillator. The action of Hamiltonian terms in Eq. (13.62) is to introduce transitions from $|00\rangle$ to $|11\rangle$ and vice versa (Fig. 13.4), as follows:

$$\sigma'_- a^\dagger|00\rangle = |11\rangle, \quad \sigma'_+ a|11\rangle = |00\rangle. \tag{13.63}$$

On the other hand, by choosing the laser operating frequency to be $\omega = \omega_0 - \omega_z$ we select the lower ("red") side-band frequency, so that the interaction Hamiltonian (13.61) becomes:

$$\widetilde{H}_2 \simeq \frac{j\eta\hbar\omega_1}{2}\left(-e^{j\phi}\sigma'_- a + e^{-j\phi}\sigma'_+ a^\dagger\right). \tag{13.64}$$

The action of operator terms in Eq. (13.64) is to introduce the transformation $|01\rangle \leftrightarrow |10\rangle$ (Fig. 13.4):

$$\sigma_- a|01\rangle = |10\rangle, \quad \sigma_+ a^\dagger|10\rangle = |01\rangle. \tag{13.65}$$

The four basis states for quantum computation are $|00\rangle$, $|01\rangle$, $|10\rangle$, and $|11\rangle$. To facilitate the construction of two-qubit gates, for spin we are going to introduce an auxiliary state, say $|2\rangle$, which is illustrated in Fig. 13.4. The laser tuned to the frequency $\omega = \omega_a + \omega_z$ stimulates the transitions $|20\rangle \leftrightarrow |11\rangle$:

$$\sigma'_-\, a^\dagger|20\rangle = |11\rangle, \quad \sigma'_+\, a|11\rangle = |20\rangle, \tag{13.66}$$

and the corresponding Hamiltonian is given by:

$$\widetilde{H}_{\text{auxilary}} \simeq \frac{j\eta\hbar\omega'_1}{2}\left(-e^{-j\phi}\sigma'_-\, a^\dagger + e^{j\phi}\sigma'_+\, a\right). \tag{13.67}$$

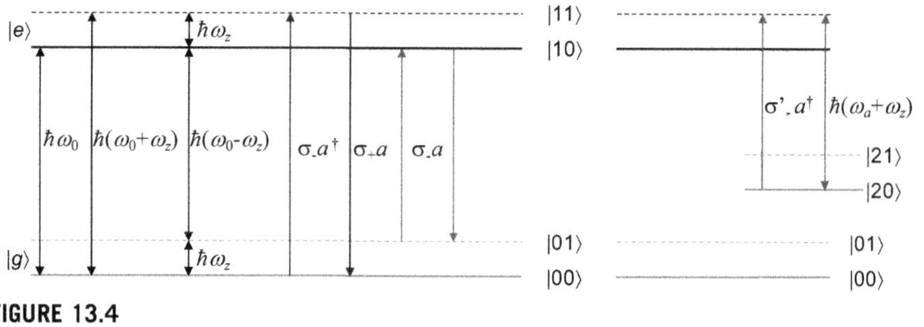

FIGURE 13.4

Energy levels and transitions describing coupling of spin and vibrational degrees of freedom.

The laser beam is applied for a time duration to introduce the $R_x(2\pi)$ rotation, which corresponds to the mapping $|11\rangle \rightarrow -|11\rangle$ as $R_x(2\pi) = \cos\pi I - j\sin\pi X = -I$ Since the other states are not affected, the matrix representation of this operation is:

$$\begin{bmatrix} I & 0 \\ 0 & Z \end{bmatrix} = C(Z), \tag{13.68}$$

which implements the controlled-*Z* gate. However, given the fact that information is imposed on spin states not on vibrational states, this type of gate does not seem to be useful. On the other hand, if this gate is combined with a SWAP gate, the vibrational states can be used as intermediate states, and the controlled-*Z* gate on two ion qubits can be implemented. Let ion 1 serve as a control qubit and ion 2 as a target qubit, while the vibrational state can be used as an external state.

The key idea is to apply first the SWAP gate on the target qubit and vibrational state, followed by the controlled-*Z* gate on the control and vibrational states given by Eq. (13.68). In the final stage, the SWAP gate on the vibrational state and target qubit is applied such that the overall action is equivalent to the controlled-*Z* gate on ions 1 and 2. The initial state of this system can be represented by:

$$\underbrace{(|c\rangle \otimes |t\rangle)}_{\text{ion qubits}} \otimes \underbrace{|0\rangle}_{\text{vibrational state}} = \underbrace{(c_0|0\rangle + c_1|1\rangle)((t_0|0\rangle + t_1|1\rangle))}_{a|00\rangle + b|01\rangle + c|10\rangle + d|11\rangle} \otimes$$

$$|0\rangle = a|000\rangle + b|010\rangle + c|100\rangle + d|110\rangle;\ a = c_0t_0,\ b = c_0t_1,\ c = c_1t_0,\ d = c_1t_1. \tag{13.69}$$

By applying the SWAP gate on the target and vibrational states we obtain:

$$\text{SWAP}^{(23)}(a|000\rangle + b|010\rangle + c|100\rangle + d|110\rangle) = a|000\rangle + b|001\rangle + c|100\rangle + d|101\rangle, \tag{13.70}$$

where the superscript is used to denote the swapping of qubits 2 and 3. In the second stage we apply the *C*(*Z*) gate, with qubit 1 serving as control and vibrational qubit 3 serving as target and obtain:

$$\text{C}^{(13)}(\text{Z})(a|000\rangle + b|001\rangle + c|100\rangle + d|101\rangle) = a|000\rangle + b|001\rangle + c|100\rangle - d|101\rangle. \tag{13.71}$$

In the third stage, the SWAP gate on the second and third qubits is applied leading to:

$$\text{SWAP}^{(23)}(a|000\rangle + b|001\rangle + c|100\rangle - d|101\rangle)$$
$$= a|000\rangle + b|010\rangle + c|100\rangle - d|110\rangle = (a|00\rangle + b|01\rangle + c|10\rangle - d|11\rangle) \otimes |0\rangle, \tag{13.72}$$

and the overall action of ion qubits 1 and 2 is that of the controlled-Z gate. The CNOT gate can be implemented based on Eq. (13.39), while the Hadamard gate can be implemented by a rotation for π rad around the axis $(1/\sqrt{2},0,1/\sqrt{2})$ as given by Eq. (13.36). What remains is to describe how to implement the SWAP gate on two ions in the trap. By tuning the laser to the lower (red) side-band frequency ($\omega=\omega_0-\omega_z$), allowing the duration of the laser beam to introduce the rotation for π rad, and by setting ϕ in Eq. (13.51) to $-\pi/2$, the overall action can be represented in matrix form as follows:

$$\begin{bmatrix} 1 & 0 & 0 & 0 \\ 0 & 0 & -1 & 0 \\ 0 & 1 & 0 & 0 \\ 0 & 0 & 0 & 1 \end{bmatrix} = U'_{\mathrm{SWAP}}, \tag{13.73}$$

which is the same as that of the SWAP gate, except for the sign. By redefining the state of the target qubit by $|1\rangle \to -|1\rangle$, the conventional SWAP gate is obtained.

The ion trap-based N-qubit register has already been shown in Fig. 13.3. The potential energy of this register is given by:

$$V(z_1,\cdots,z_n)=\frac{M_I\omega_z^2}{2}\sum_{n=1}^{N} z_n^2+\frac{q^2}{4\pi\varepsilon_0}\sum_{m\neq n}\frac{1}{|z_n-z_m|}, \tag{13.74}$$

under the assumption that the ion trap is linear. To ensure that the ion trap is confined in the transversal direction, the temperature must be kept sufficiently low so that ions are in the vibrational state with $m=0$. If $k_BT\geq\hbar\omega_z$ (k_B is the Boltzmann constant), $m\neq 0$ and the ions must be cooled down, which can be achieved by *Doppler cooling*. Namely, the ion is placed between two laser beams directed in opposite directions. When the ion moves in the direction opposite to the direction of one of the beams, it experiences more energetic photons due to the Doppler effect, the transition is coming closer to the resonance, and absorption becomes more pronounced. On the other hand, when it moves in the opposite direction, it experiences the same effect, which leads to the slowing down of ions until the temperature $T\simeq\hbar 2\pi\Delta\nu/k_B$, where $\Delta\nu$ is the laser linewidth. The lowest mode, CM mode, corresponding to ω_z, refers to the motion of ensembles of ions. The first excited mode, known as the *breathing* mode, corresponds to ions oscillating at frequency $\sqrt{3}\omega_z$with amplitude proportional to the distance from the axial axis. Namely, by setting $z_i=z_0+u_i$ and by using Taylor expansion up to two terms, Eq. (13.74) for $N=2$ can be written as:

$$\begin{aligned} V(z_1,\, z_2) &\simeq \frac{M_I\omega_z^2}{2}\left(2z_0^2+2z_0(u_1-u_2)+u_1^2+u_2^2\right) \\ &+\frac{q^2}{4\pi\varepsilon_0 z_0}\left[1-\frac{u_1-u_2}{2z_0}+\frac{(u_1-u_2)^2}{4z_0^2}\right]. \end{aligned} \tag{13.75}$$

The equilibrium is obtained by setting the linear terms to zero:

$$M_I\omega_z^2 z_0 - \frac{q^2}{4\pi\varepsilon_0 z_0}\frac{1}{2z_0} = 0, \tag{13.76}$$

which yields to:

$$z_0 = 2^{-1/3}l, \; l = \left(\frac{q^2}{4\pi\varepsilon_0 M_I\omega_z^2}\right)^{1/3}. \tag{13.77}$$

The normal modes are obtained by studying the quadratic terms, which leads to the potential:

$$V_2 \simeq \frac{M_I\omega_z^2}{2}\left(u_1^2 + u_2^2\right) + \frac{q^2}{4\pi\varepsilon_0 z_0}\frac{(u_1 - u_2)^2}{4z_0^2}. \tag{13.78}$$

The corresponding equations of motion are given by:

$$M_I\ddot{u}_1 = M_I\omega_z^2 u_1 + \frac{q^2}{4\pi\varepsilon_0 z_0}\frac{u_1 - u_2}{2z_0^2}, \quad M_I\ddot{u}_2 = M_I\omega_z^2 u_2 - \frac{q^2}{4\pi\varepsilon_0 z_0}\frac{u_1 - u_2}{2z_0^2}. \tag{13.79}$$

The CM mode, $(u_1 + u_2)/2$, oscillates at frequency ω_z as:

$$\ddot{u}_1 + \ddot{u}_2 = -\omega_z^2(u_1 + u_2). \tag{13.80}$$

On the other hand, the breathing mode, $u_1 - u_2$, oscillates at frequency $\sqrt{3}\omega_z$ as:

$$\ddot{u}_1 - \ddot{u}_2 = -3\omega_z^2(u_1 - u_2). \tag{13.81}$$

One very important problem is how to accurately illuminate a single ion by a laser beam. Currently, acousto-optic modulators (AOMs) [10,11] are used, driven by appropriate RF signals, to perform controlled-Z, SWAP, and Hadamard gate operations. The AOMs are unfortunately slow devices and highly temperature sensitive. Moreover, the extinction ratio of AOMs is low and power consumption is high. A possible solution is to use an electro-optical switch such as an optical cross-point switch implemented by means of active vertical coupler (AVC) structures [12–14]. The use of this kind of device will provide ample bandwidth (not the main problem here) with high reliability. Moreover, this approach is able to significantly reduce the overall power consumption. This is possible as power is consumed only when the photonic switching cell is ON. The AVC-based switching technique is a low loss technique and, more importantly, insensitive to temperature when compared to similar devices such AOM-based devices. Finally, the extinction ratio is significantly better than that of AOMs.

13.4 PHOTONIC QUANTUM IMPLEMENTATIONS

Photonic technologies can be used in many different ways to perform quantum computation, including [15–17]: (1) the cavity superposition of zero or one photon $|\psi\rangle = c_0|0\rangle + c_1|1\rangle$; (2) the photon can be located in the first cavity $|01\rangle$ or in the second cavity $|10\rangle$ (dual-rail representation), while the qubit representation is given by: $|\psi\rangle = c_0|01\rangle + c_1|10\rangle$; and (3) spin-angular momentum can be used to carry quantum information, while the qubit is represented as $|\psi\rangle = c_0|0\rangle + c_1|1\rangle$, where the logical "0" is represented by a horizontal (H) photon $|\mathrm{H}\rangle \equiv |0\rangle = (1\ 0)^{\mathrm{T}}$ and the logical "1" is represented by a vertical (V) photon $|\mathrm{V}\rangle \equiv |1\rangle = (0\ 1)^{\mathrm{T}}$. Bulky optics implementation [15] is described first, not because of its practicality, but because of the simplicity of its description.

13.4.1 Bulky Optics Implementation

The laser output state is known as the *coherent state* and represents the right eigenket of the *annihilation operator a* (the operator that decreases the number of photons by one):

$$a|\alpha\rangle = \alpha|\alpha\rangle, \tag{13.82}$$

where α is the complex eigenvalue. The coherent state ket $|\alpha\rangle$ can be represented in terms of orthonormal eigenkets $|n\rangle$ (the number or Fock state) of the number operator $a^{\dagger}a$ as follows (see Chapter 2):

$$|\alpha\rangle = \exp\left[-|\alpha|^2/2\right]\sum_{n=0}^{+\infty}(n!)^{-1/2}\alpha^n|n\rangle, \tag{13.83}$$

and the mean energy $\langle\alpha|n|\alpha\rangle = |\alpha|^2$ has the Poisson distribution. When the laser beam is sufficiently attenuated, the coherent state becomes the weak coherent state. By properly choosing the eigenvalue α, the weak coherent state behaves as a single-photon state with a high probability. For example, by setting $\alpha = 0.1$, the weak coherent state $|0.1\rangle = \sqrt{0.90}|0\rangle + \sqrt{0.09}|1\rangle + \sqrt{0.002}|2\rangle + \ldots$ behaves as the single-photon state (with high probability). Moreover, the implementation of single-photon sources is an active research topic, see [18–21] (and references therein).

Spontaneous parametric down-conversion (SPDC) and Kerr effect are now briefly described as they are key effects in the implementation of optical quantum computers. By sending the photons at frequency ω_p (pump signal) into a nonlinear optical medium, say KH_2PO_4 or beta-barium borate crystal, we generate the photon pairs at frequencies ω_s (signal) and ω_i (idler) satisfying the energy conservation principle $\hbar(\omega_s + \omega_i) = \hbar\omega_p$ and momentum conservation principle $\boldsymbol{k}_s + \boldsymbol{k}_i = \boldsymbol{k}_p$, which is illustrated in Fig. 13.5. Namely, a nonlinear crystal is used to split photons into pairs of photons that satisfy the law of conservation of energy, have combined energies and

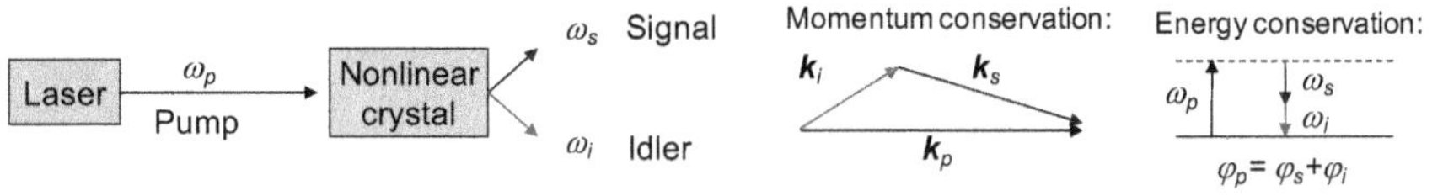

FIGURE 13.5

Spontaneous parametric down-conversion process.

momenta equal to the energy and momentum of the original photon, are phase matched in the frequency domain, and have correlated polarizations. If the photons have the same polarization, they belong to Type I correlation; otherwise, they have perpendicular polarizations and belong to Type II. As the SPDC is stimulated by random vacuum fluctuations, the photon pairs are created at random time instances. The output of a Type I down-converter is known as a squeezed vacuum and it contains only even numbers of photons. The output of the Type II down-converter is a two-mode squeezed vacuum. Modern methods are based on periodically poled $LiNbO_3$ (PPLN), and highly nonlinear fiber (HNLF) [22,23]. PPLN and HNLF are also used in optical phase conjugation and parametric amplifiers. PPLN employs parametric difference frequency generation, while HLNF employs the four-wave mixing (FWM) effect [24]. FWM occurs in HNLFs during the propagation of a composite optical signal, such as the wavelength-division multiplexing signal. The three optical signals with different carrier frequencies f_i, f_j, and f_k $(i,j,k = 1,\ldots,M)$ interact and generate the new optical signal at frequency $f_{ijk} = f_i + f_j - f_k$, providing that the phase-matching condition is satisfied: $\beta_{ijk} = \beta_i + \beta_j - \beta_k$, where β_m is the propagation constant. The phase-matching condition is in fact the requirement for momentum conservation. The FWM process can be considered as the annihilation of two photons with energies $\hbar\omega_i$ and $\hbar\omega_j$, and generation of two new photons with energies $\hbar\omega_k$ and $\hbar\omega_{ijk}$. In the FWM process, the indices i and j are not necessarily distinct, which indicates that only two modes can interact to create the new one; this case is known as a *degenerate*.

For a sufficiently high intensity of light, the index of refection is not only a function of frequency (responsible for dispersion), but also a function of the intensity of light traveling through the nonlinear media: $n(\omega,I) = n(\omega) + n_2 I$, where n_2 is the Kerr coefficient. The phase shift by the nonlinear waveguide of length L is determined by complex factor $\exp(jn_2 IL\omega/c_0)$ (c_0 is the speed of the light in a vacuum).

Regarding quantum–level photodetectors, avalanche photodiodes (APDs) and photomultiplier tubes can be used. Moreover, single-photodetector development is an active research topic, see [25–28] (and references therein).

Regarding the photonic quantum gate's implementation, the most commonly used devices to manipulate the photon states are: (1) mirrors, which are used to change the direction of propagation; (2) phase shifters, which are used to introduce a given phase shift; and (3) beam splitters, which are used to implement various quantum gates. The *phase shifter* is a slab of transparent medium, say borosilicate glass, with index of refraction n being higher than that of air and it is used to perform the following operation:

$$|\psi_{out}\rangle = \begin{bmatrix} e^{j\phi} & 0 \\ 0 & 1 \end{bmatrix} |\psi_{in}\rangle; \quad \phi = kL, \ k = n\omega/c. \tag{13.84}$$

The corresponding equivalent scheme is shown in Fig. 13.6. The *beam splitter* is a partially silvered piece of glass with reflection coefficient parameterized as follows: $R = \cos\theta$, so that its action is to perform the following operation:

$$|\psi_{out}\rangle = \begin{bmatrix} \cos\theta & \sin\theta \\ -\sin\theta & \cos\theta \end{bmatrix} |\psi_{in}\rangle. \tag{13.85}$$

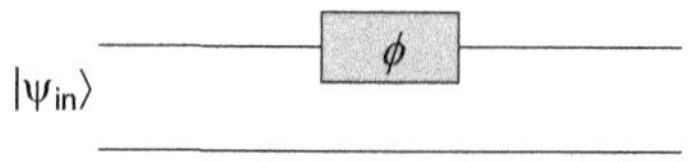

FIGURE 13.6

Equivalent scheme of the phase shifter.

The 50:50 beam splitter is obtained for $\theta = \pi/4$, and its implementation is shown in Fig. 13.7.

By using these basic building blocks (the phase shifter and beam splitter) we can implement various single-qubit gates. As an illustration, in Fig. 13.8 we provide the Hadamard gate implementation by concatenating a 50:50 beam splitter and π rad phase shifter. The operation principle of the gate from Fig. 13.8 is as follows:

$$|\psi_{out}\rangle = \begin{bmatrix} e^{j\pi} & 0 \\ 0 & 1 \end{bmatrix} \begin{bmatrix} \cos(\pi/4) & \sin(\pi/4) \\ -\sin(\pi/4) & \cos(\pi/4) \end{bmatrix} |\psi_{in}\rangle = -\frac{1}{\sqrt{2}} \begin{bmatrix} 1 & 1 \\ 1 & -1 \end{bmatrix}$$
$$= -H, \quad H = \frac{1}{\sqrt{2}} \begin{bmatrix} 1 & 1 \\ 1 & -1 \end{bmatrix}. \tag{13.86}$$

Let us observe the concatenation of two phases and one beam splitter as shown in Fig. 13.9. The overall action of this gate is given by:

$$U_3U_2U_1 = \begin{bmatrix} e^{-j\beta} & 0 \\ 0 & 1 \end{bmatrix} \begin{bmatrix} \cos(\gamma/2) & -\sin(\gamma/2) \\ \sin(\gamma/2) & \cos(\gamma/2) \end{bmatrix} \begin{bmatrix} e^{j[\alpha-\delta/2-(\alpha+\delta/2)]} & 0 \\ 0 & 1 \end{bmatrix}$$
$$= e^{-j(\alpha+\beta/2+\delta/2)} \begin{bmatrix} e^{-j\beta}/2 & 0 \\ 0 & e^{j\beta/2} \end{bmatrix} \begin{bmatrix} \cos(\gamma/2) & -\sin(\gamma/2) \\ \sin(\gamma/2) & \cos(\gamma/2) \end{bmatrix}$$
$$\times \begin{bmatrix} e^{j(\alpha-\delta/2)} & 0 \\ 0 & e^{j(\alpha+\delta/2)} \end{bmatrix}$$
$$= e^{-j(\alpha+\beta/2+\delta/2)} \begin{bmatrix} e^{j(\alpha-\beta/2-\delta/2)} \cos\frac{\gamma}{2} & -e^{j(\alpha-\beta/2+\delta/2)} \sin\frac{\gamma}{2} \\ e^{j(\alpha+\beta/2-\delta/2)} \cos\frac{\gamma}{2} & e^{j(\alpha+\beta/2+\delta/2)} \cos\frac{\gamma}{2} \end{bmatrix}$$
$$= e^{-j(\alpha+\beta/2+\delta/2)} e^{j\alpha} R_z(\beta) R_y(\gamma) R_z(\delta), \tag{13.87}$$

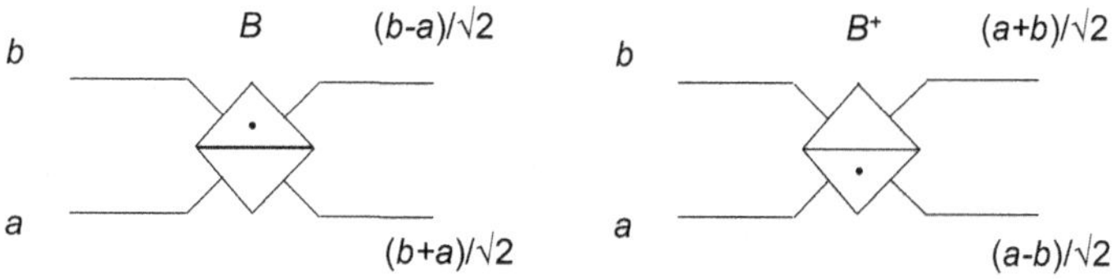

FIGURE 13.7

Operation principle of a 50/50 beam splitter (B) and its Hermitian conjugate ($B^{\dagger}$).

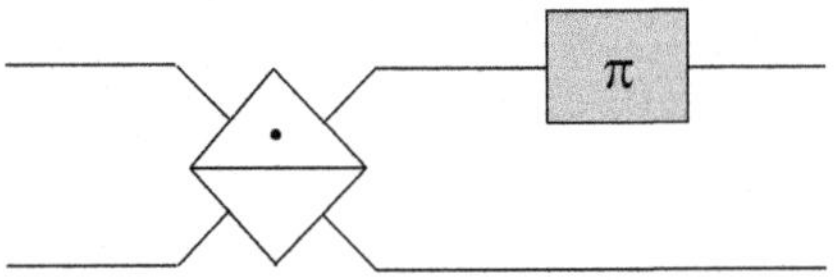

FIGURE 13.8

Hadamard gate implementation in bulky optics based on a 50:50 beam splitter and π rad phase shifter.

which is the same as that of Y–Z decomposition theorem from Chapter 3, except from the global phase shift. Therefore by using the circuit from Fig. 13.9, an arbitrary single-qubit gate can be implemented. To complete the set of universal quantum gates, implementation of a controlled-Z gate or CNOT gate is needed.

Let the logical "0" be represented by a horizontal (H) photon $|\mathrm{H}\rangle \equiv |0\rangle = (1\ 0)^{\mathrm{T}}$ and the logical "1" be represented by a vertical (V) photon $|\mathrm{V}\rangle \equiv |1\rangle = (0\ 1)^{\mathrm{T}}$. Implementation of the CNOT gate by bulky optics is shown in Fig. 13.10. One polarization beam splitter (PBS) is used per each input port, while one polarization beam combiner (PBC) is used per each output port. The output control $|C_{\mathrm{o}}\rangle = [c_{\mathrm{H,o}}\ c_{\mathrm{V,o}}]^{\mathrm{T}}$ and target qubits $|T_{\mathrm{o}}\rangle = [t_{\mathrm{H},o}\ t_{\mathrm{V},o}]^{\mathrm{T}}$ are related to the corresponding input qubits by:

$$\begin{bmatrix} c_{\mathrm{H},o} \\ c_{\mathrm{V},o} \\ t_{\mathrm{H},o} \\ t_{\mathrm{V},o} \end{bmatrix} = \underbrace{\frac{1}{\sqrt{2}}\begin{bmatrix} 1 & 1 & 0 & 0 \\ 1 & -1 & 0 & 0 \\ 0 & 0 & 1 & 1 \\ 0 & 0 & 1 & -1 \end{bmatrix}}_{I\otimes H} \underbrace{\begin{bmatrix} 1 & 0 & 0 & 0 \\ 0 & 1 & 0 & 0 \\ 0 & 0 & 1 & 0 \\ 0 & 0 & 0 & -1 \end{bmatrix}}_{K} \underbrace{\frac{1}{\sqrt{2}}\begin{bmatrix} 1 & 1 & 0 & 0 \\ 1 & -1 & 0 & 0 \\ 0 & 0 & 1 & 1 \\ 0 & 0 & 1 & -1 \end{bmatrix}}_{I\otimes H} \begin{bmatrix} c_{\mathrm{H}} \\ c_{\mathrm{V}} \\ t_{\mathrm{H}} \\ t_{\mathrm{V}} \end{bmatrix}$$

$$= U_{\mathrm{CNOT}} \begin{bmatrix} c_{\mathrm{H}} \\ c_{\mathrm{V}} \\ t_{\mathrm{H}} \\ t_{\mathrm{V}} \end{bmatrix}; \quad U_{\mathrm{CNOT}} = \begin{bmatrix} 1 & 0 & 0 & 0 \\ 0 & 1 & 0 & 0 \\ 0 & 0 & 0 & 1 \\ 0 & 0 & 1 & 0 \end{bmatrix}, \tag{13.88}$$

where with K we denote the Kerr nonlinearity-based gate. The Kerr nonlinearity device in Fig. 13.10 performs the controlled-Z operation. In the absence of the control c_{V} photon, the target qubit is unaffected because $H^2 = I$ (identity operator). In the presence of the control c_{V} photon, thanks to the cross-phase modulation in HNLF, the target vertical photon experiences the phase shift χL, corresponding to the complex phase term $\exp(j\chi L)$, where χ is the third-order nonlinearity susceptibility coefficient and L is the nonlinear crystal length. By selecting appropriately

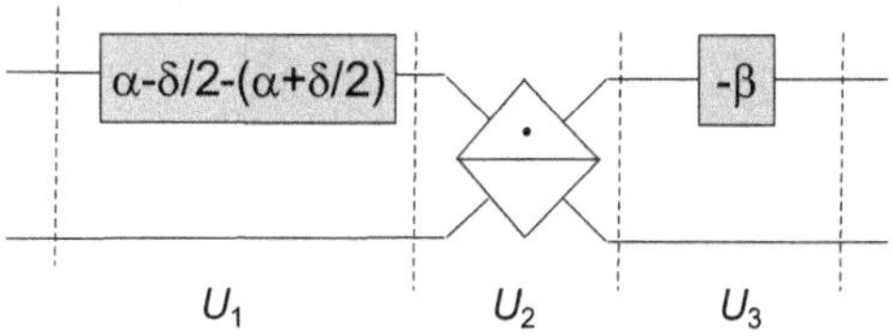

FIGURE 13.9

Y–Z decomposition theorem by bulky optics devices.

the crystal length we obtain $\chi L = \pi$ and the overall action on the target qubit is $HZH = X$, which corresponds to the CNOT gate action. Therefore when only both control and qubit photons are vertical, the phase shift of π rad is introduced. By omitting the Hadamard gates from Fig. 13.10, the corresponding circuit operates as a controlled-Z gate.

In the previous discussion, it was assumed that matrix representation given by Eqs. (13.84) and (13.85) and represe;ntation of the K operator in Eq. (13.88) are correct, without providing quantum mechanical justification. The rest of this section is related to the quantum mechanical interpretation of various bulky optics devices used earlier. The phase shifter P of length L introduces delay in light propagation by an amount of $\Delta = (n - n_0)L/c$. The action of the phase shift circuit P on the vacuum state and single-photon state is as follows:

$$P|0\rangle = |0\rangle, \quad P|1\rangle = e^{j\Delta}|1\rangle. \tag{13.89}$$

On the other hand, the action of P on the dual-rail state is:

$$P|\psi\rangle = P(c_{01}|01\rangle + c_{10}|10\rangle) = c_{01}e^{-j\Delta/2}|01\rangle + c_{10}e^{j\Delta/2}|10\rangle. \tag{13.90}$$

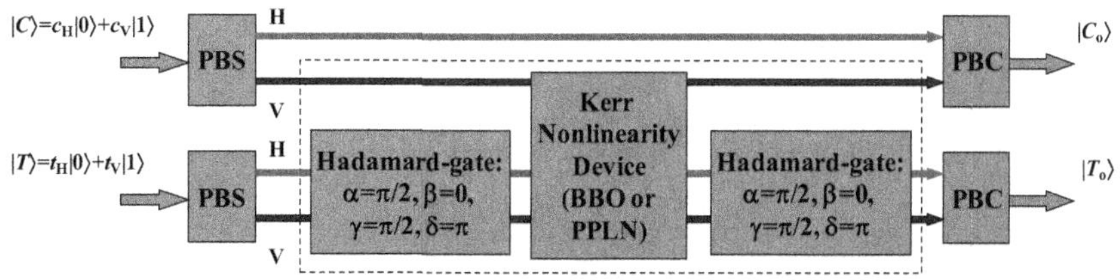

FIGURE 13.10

Implementation of the CNOT gate by using bulky optics when photon polarization is used to carry the quantum information. *BBO*, Beta-barium borate; *PBC*, polarization beam combiner; *PBS*, polarization beam splitter; *PPLN*, periodically poled $LiNbO_3$.

The action of P can be described in terms of the rotation operator around the z-axis by:

$$R_z(\Delta)=e^{-jZ\Delta/2}. \tag{13.91}$$

The corresponding Hamiltonian is given by:

$$\hat{H}=\hbar(n_0-n)Z, \tag{13.92}$$

So that the action of P can be interpreted as the evolution described by the Hamiltonian:

$$P=e^{-j\hat{H}L/\hbar c_0} \tag{13.93}$$

The beam splitter acts on two modes, which can be described by the creation (annihilation) operators a ($a^\dagger$) and b ($b^\dagger$). The corresponding Hamiltonian is given by:

$$\hat{H}_{bs}=j\hbar\omega\left(ab^\dagger-a^\dagger b\right). \tag{13.94}$$

The action of the beam splitter can be represented by the evolution operation as:

$$B=e^{-j\hat{H}t/\hbar}=e^{\theta(ab^\dagger-a^\dagger b)}=e^{\theta G},\ G=ab^\dagger-a^\dagger b,\ \theta=\omega t. \tag{13.95}$$

The Baker–Campbell–Hausdorf formula is further applied:

$$e^{\lambda G}Ae^{-\lambda G}=\sum_{n=0}^{\infty}\frac{\lambda^n}{n!}C_n;\ C_0=A,\ C_n=[G,\ C_{n-1}],\ n=1,\ 2,\ \cdots. \tag{13.96}$$

By noticing that:

$$\left[a,\ a^\dagger\right]=\left[b,\ b^\dagger\right]=I,\ [G,\ a]=-b,\ [G,\ b]=a, \tag{13.97}$$

it can be concluded that:

$$C_0=a,\ C_1=[G,\ a]=-b,\ C_2=[G,\ C_1]=-a,\ G_3=[G,\ C_2]=-[G,\ C_0]=b, \tag{13.98}$$

and consequently:

$$C_n=\begin{cases} j^n a, & n-\text{even} \\ j^{n+1}b, & n-\text{odd} \end{cases}. \tag{13.99}$$

By using the Baker–Campbell–Hausdorf formula, the following is obtained:

$$BaB^\dagger=e^{\theta G}ae^{-\theta G}=\sum_{n=0}^{\infty}\frac{\theta^n}{n!}C_n=\underbrace{\sum_{n-\text{even}}\frac{(j\theta)^n}{n!}}_{\cos\theta}a+j\underbrace{\sum_{n-\text{odd}}\frac{(j\theta)^n}{n!}}_{\sin\theta}b=a\cos\theta-b\sin\theta. \tag{13.100}$$

In similar fashion, it can be shown that:

$$BbB^{\dagger} = a \sin\theta + b \cos\theta, \tag{13.101}$$

so that the matrix representation of B is given by:

$$B = \begin{bmatrix} \cos\theta & -\sin\theta \\ \sin\theta & \cos\theta \end{bmatrix} = e^{j\theta Y}, \tag{13.102}$$

which is the same as the rotation operator around the y-axis. The action of operator B on dual-rail states is given by:

$$B|01\rangle = Ba^{\dagger}|00\rangle = Ba^{\dagger}\underbrace{B^{\dagger}B}_{I}|00\rangle \overset{B|00\rangle=|00\rangle}{=} Ba^{\dagger}B^{\dagger}|00\rangle = \left(a^{\dagger}\cos\theta - b^{\dagger}\sin\theta\right)|00\rangle$$

$$= \cos\theta|01\rangle - \sin\theta|10\rangle, \quad B|10\rangle = \cos\theta|10\rangle + \sin\theta|01\rangle, \tag{13.103}$$

which leads to the same matrix representation as given by Eq. (13.102).

Cross-phase modulation (XPM) is another effect caused by the intensity dependence of the refractive index, and occurs during propagation of a composite signal. The nonlinear phase shift of a specific optical mode is affected not only by the intensity of the observed mode but also by the intensity of the other optical modes. Quantum mechanically, the XPM between two modes a and b is described by the following Hamiltonian:

$$\hat{H}_{\text{XPM}} = -\hbar\chi a^{\dagger}ab^{\dagger}b. \tag{13.104}$$

The propagation of these two modes over nonlinear media of length L is governed by the following evolution operator:

$$K = e^{-j\hat{H}L/\hbar} = e^{j\chi La^{\dagger}ab^{\dagger}b}. \tag{13.105}$$

Therefore the operator K is a unitary transform of nonlinear media of length L. The action of K on single-photon states is as follows:

$$K|00\rangle = |00\rangle, \; K|01\rangle = |01\rangle, \; K|10\rangle = |10\rangle, \; K|11\rangle = e^{j\chi L}|11\rangle,$$

$$\chi L = \pi: \; K|11\rangle = -|11\rangle, \tag{13.106}$$

and the corresponding matrix representation is given by:

$$K = \begin{bmatrix} 1 & 0 & 0 & 0 \\ 0 & 1 & 0 & 0 \\ 0 & 0 & 1 & 0 \\ 0 & 0 & 0 & -1 \end{bmatrix} = \begin{bmatrix} I & 0 \\ 0 & Z \end{bmatrix} = C(Z), \tag{13.107}$$

which is the same as that of the controlled-Z gate ($C(Z)$). Since the Pauli X-operator can be implemented as $X = HZH$ (Fig. 13.10), this implementation of controlled-Z

gates completes the implementation of the universal set of gates $\{H, S, \pi/8\ (T), C(Z)\}$ or equivalently $\{H, S, T, \text{CNOT}\}$. On the other hand, for dual-rail representation, if the base kets are selected as follows [15]:

$$|e_{00}\rangle = |1001\rangle,\ \ |e_{01}\rangle = |1010\rangle,\ \ |e_{10}\rangle = |0101\rangle,\ \ |e_{11}\rangle = |0110\rangle, \tag{13.108}$$

then the action of the K operator on base kets is given by:

$$K|e_i\rangle = |e_i\rangle,\ \ i \neq 11;\ \ K|e_{11}\rangle = -|e_{11}\rangle, \tag{13.109}$$

and the corresponding matrix representation is given by Eq. (13.107). Before concluding this section we describe the implementation of the controlled-SWAP gate also known as Fredkin gate. Fredkin gate implementation for single-photon states by using bulky optics is shown in Fig. 13.11. The operation principle of this circuit is given by: $U = B^{\dagger}KB$, where B is the 50:50 beam splitter $B = B(\pi/2)$. Based on Eqs. (13.95) and (13.105), the operator U can be represented as:

$$U = \exp\left(j\chi L c^{\dagger}c \frac{b^{\dagger} - a^{\dagger}}{2} \frac{b - a}{2}\right) = e^{j\frac{\pi}{2}b^{\dagger}b} \underbrace{e^{\frac{\chi L}{2}c^{\dagger}c(ab^{\dagger} - b^{\dagger}a)}}_{U_{\text{Fredkin}}(\chi L)} e^{-j\frac{\pi}{2}b^{\dagger}b} e^{j\frac{\chi L}{2}a^{\dagger}ac^{\dagger}c} e^{j\frac{\chi L}{2}b^{\dagger}bc^{\dagger}c}. \tag{13.110}$$

The second term in Eq. (13.110):

$$U_{\text{Fredkin}}(\chi L) = e^{\frac{\chi L}{2}c^{\dagger}c(ab^{\dagger} - b^{\dagger}a)}, \tag{13.111}$$

operates as the Fredkin gate for $\chi L = \pi$ on single-photon states. In the absence of a photon in input port c, the output ports are related to the input ports as $a_{\text{out}} = a$, $b_{\text{out}} = b$, since $B^{\dagger}B = I$. On the other hand, in the presence of a photon in input port c, $a_{\text{out}} = b$, $b_{\text{out}} = a$, which is the same as the action of a SWAP gate. Therefore this

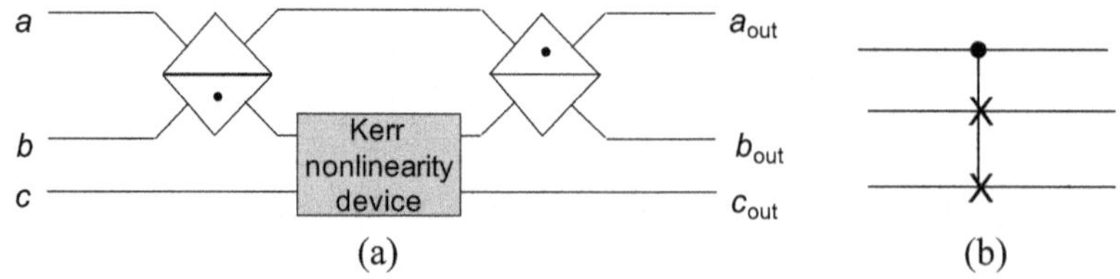

FIGURE 13.11

Implementation of the Fredkin gate by using bulky optics: (a) implementation architecture and (b) quantum circuit representation.

circuit behaves as a controlled-SWAP (Fredkin) gate with input port c serving as the control qubit. The corresponding matrix representation is given by:

$$U_{\text{Fredkin}} = \begin{bmatrix} 1 & 0 & & \cdots & & & & & 0 \\ 0 & 1 & 0 & & & & & & \\ & 0 & 1 & 0 & & & & & \vdots \\ & & 0 & 1 & 0 & & & & \\ \vdots & & & 0 & 1 & 0 & 0 & 0 & \\ & & & & 0 & 0 & 1 & 0 & \\ & & & & 0 & 1 & 0 & 0 & \\ 0 & & \cdots & & 0 & 0 & 0 & 1 & \end{bmatrix}. \quad (13.112)$$

13.4.2 Integrated Optics Implementation

This section is devoted to the integrated optics and all-fiber implementations of universal quantum gates [29–31]. If the waveguides supporting two modes are used, then a Hadamard gate can be implemented based on a Y-junction as shown in Fig. 13.12.

However, since dual-mode waveguides are required for this implementation we restrict our attention to the single-mode solution, as the integrated optics for single-mode devices is more mature. Further study of dual-mode universal quantum gates is postponed until the problem section. In what follows, the logical "0" is represented by a horizontal (H) photon $|H\rangle \equiv |0\rangle = (1\ 0)^T$ and the logical "1" is represented by a vertical (V) photon $|V\rangle \equiv |1\rangle = (0\ 1)^T$. An arbitrary single-qubit gate can be implemented based on a *directional coupler* as shown in Fig. 13.13. We use a PBS at the input of the quantum gate and a PBC at the output of the gate. The horizontal output (input) of the PBS (PBC) is denoted by H, while the vertical output (input) of the PBS (PBC) is denoted by V. The input qubit is denoted by $|\psi\rangle = \psi_H|0\rangle +$

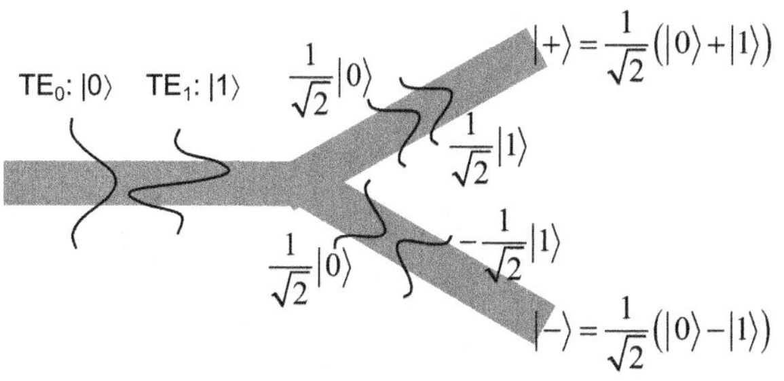

FIGURE 13.12

Hadamard gate implementation in integrated optics based on a Y-junction.

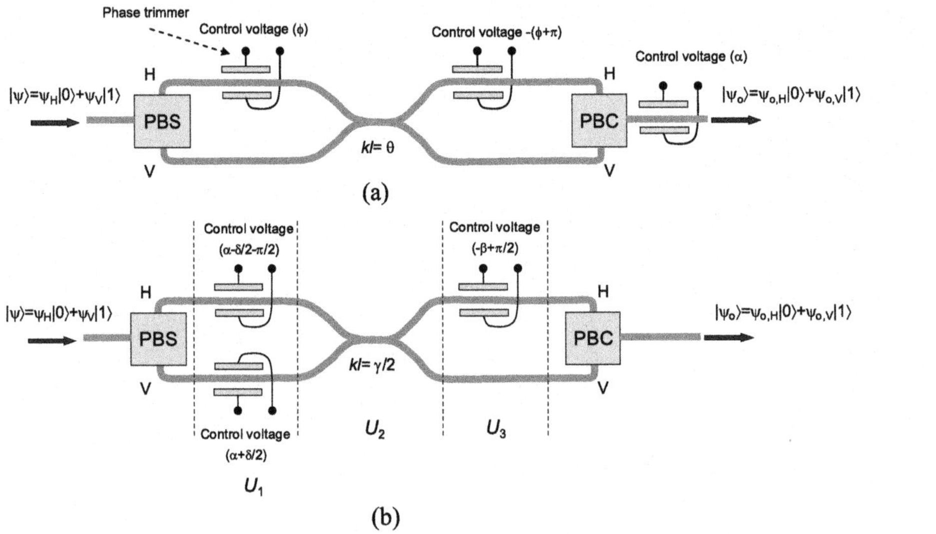

FIGURE 13.13

Integrated optics implementation of arbitrary single-qubit gate based on a single directional coupler: (a) Barenco-type gate and (b) *Z*–*Y* decomposition theorem-based type. *PBS/C*, Polarization beam splitter/combiner.

$\psi_V|1\rangle = [\psi_H\,\psi_V]^{\mathrm{T}}$, while the output qubit is denoted by $|\psi_o\rangle = \psi_{o,H}|0\rangle + \psi_{o,V}|1\rangle = [\psi_{o,H}\,\psi_{o,V}]^{\mathrm{T}}$. In Fig. 13.13A we show an implementation based on a Barenco-type proposal [1]. It can be shown that the output qubit is related to the input qubit by:

$$\begin{bmatrix}\psi_{H,\,o}\\ \psi_{V,\,o}\end{bmatrix} = U_B\begin{bmatrix}\psi_H\\ \psi_V\end{bmatrix},\quad U_B = \begin{bmatrix} e^{j\alpha}\cos(\theta) & -je^{j(\alpha+\phi)}\sin(\theta)\\ -je^{j(\alpha-\phi)}\sin(\theta) & e^{j\alpha}\cos(\theta)\end{bmatrix}. \tag{13.113}$$

In Eq. (13.113), $\theta = kl$, where k is the coupling coefficient and l is the coupling region length. By setting appropriately phase trimmer voltages we can perform the arbitrary single-qubit operation. For example, by setting $\alpha = \phi = 0$ rad and $\theta = 2\pi$ we obtain an identity gate, while by setting $\alpha = 0$ rad, $\phi = \pi$, and $\theta = \pi/2$ we obtain the X gate. In Fig. 13.13B we show an implementation based on Y–Z decomposition theorem. The output qubit is related to the input qubit by:

$$\begin{bmatrix}\psi_{H,\,o}\\ \psi_{V,\,o}\end{bmatrix} = U\begin{bmatrix}\psi_H\\ \psi_V\end{bmatrix},\quad U = \begin{bmatrix} \cos\left(\frac{\Upsilon}{2}\right)e^{j(\alpha-\beta/2-\delta/2)} & -\sin\left(\frac{\Upsilon}{2}\right)e^{j(\alpha-\beta/2+\delta/2)}\\ \sin\left(\frac{\Upsilon}{2}\right)e^{j(\alpha+\beta/2-\delta/2)} & \cos\left(\frac{\Upsilon}{2}\right)e^{j(\alpha+\beta/2+\delta/2)}\end{bmatrix}. \tag{13.114}$$

In Eq. (13.114), $\gamma = 2kl$. By setting $\alpha = \pi/4$, $\beta = \pi/2$, $\gamma = 2\pi$, and $\delta = 0$ rad, the U gate described by Eq. (13.114) operates as the phase gate:

$$S = \begin{bmatrix} 1 & 0 \\ 0 & j \end{bmatrix}. \tag{13.115}$$

By setting $\alpha = \pi/8$, $\beta = \pi/4$ rad, $\gamma = 2\pi$, and $\delta = 0$ rad, the U gate operates as a $\pi/8$ gate:

$$T = \begin{bmatrix} 1 & 0 \\ 0 & e^{j\pi/4} \end{bmatrix}. \tag{13.116}$$

Finally, by setting $\alpha = \pi/2$, $\beta = 0$ rad, $\gamma = \pi/2$, and $\delta = \pi$, the U gate given by Eq. (13.114) operates as a Hadamard gate:

$$H = \frac{1}{\sqrt{2}} \begin{bmatrix} 1 & 1 \\ 1 & -1 \end{bmatrix}. \tag{13.117}$$

To complete the implementation of the set of universal quantum gates, implementation of the CNOT gate is needed. The authors in [32,51] proposed the use of directional couplers to implement the CNOT gate. For completeness of presentation, in Fig. 13.14A, a simplified version of the CNOT gate proposed in [32,51] is provided. We see that the control output qubit $[c_{H,o}, c_{V,o}]^T$ is related to the input control qubit $[c_H, c_V]^T$ and input target qubit $(t_H, t_V)^T$ by [51]:

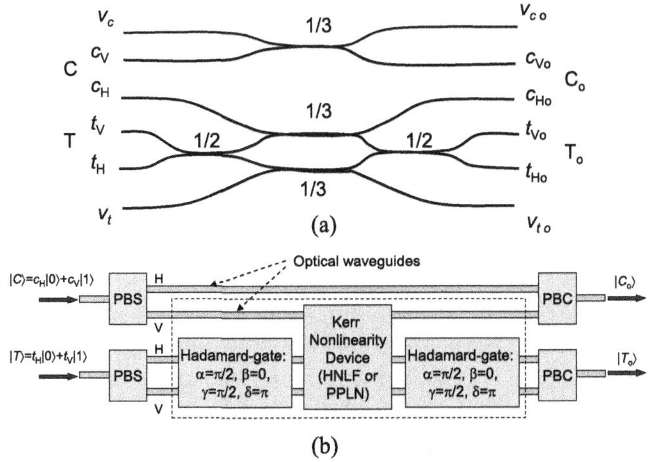

FIGURE 13.14

Integrated optics (or all-fiber) implementation of a CNOT gate: (a) probabilistic gate proposed in [25] and (b) deterministic gate. *HNLF*, Highly nonlinear fiber; *PBC*, polarization beam combiner; *PBS*, polarization beam splitter; *PPLN*, periodically poled $LiNbO_3$.

$$\left[c_{\mathrm{H,o}}=\left(1/\sqrt{3}\right)\left(\sqrt{2}\,v_{\mathrm{c}}+c_{\mathrm{H}}\right),\ c_{\mathrm{V,o}}=\left(1/\sqrt{3}\right)\left(-c_{\mathrm{V}}+t_{\mathrm{H}}+t_{\mathrm{V}}\right)\right]^{\mathrm{T}}. \tag{13.118}$$

Because the output control qubit is affected by the input target qubit, definition of CNOT gate operation (control qubit must be unaffected by target qubit) is violated. This gate operates correctly only with a probability of 1/9, and is essentially a probabilistic gate. In Fig. 13.14B we show the deterministic implementation of the CNOT gate based on the single-qubit gate shown in Fig. 13.13 and HNLF. Notice that different implementations of the CNOT gate in [32,33] employ multirail/dual-rail representation, while the implementation form in Fig. 13.13 is based on single-photon polarization (spin angular momentum). Because different quantum gate implementations in [32,33] require either two modes for dual-rail representation [25,32] or $2n$ modes for multirail representation [33], they cannot be implemented at all using conventional single-mode fibers or single-mode devices. This is a key disadvantage of multirail representation compared to Fig. 13.14B. It can be shown that output control $|C_{\mathrm{o}}\rangle=[c_{\mathrm{H,o}}\ c_{\mathrm{V,o}}]^{\mathrm{T}}$ and target qubits $|T_{\mathrm{o}}\rangle=[t_{\mathrm{H,o}}\ t_{\mathrm{V,o}}]^{\mathrm{T}}$ are related to the corresponding input qubits by:

$$\begin{bmatrix} c_{\mathrm{H,o}} \\ c_{\mathrm{V,o}} \\ t_{\mathrm{H,o}} \\ t_{\mathrm{V,o}} \end{bmatrix} = \underbrace{\frac{1}{\sqrt{2}}\begin{bmatrix} 1 & 1 & 0 & 0 \\ 1 & -1 & 0 & 0 \\ 0 & 0 & 1 & 1 \\ 0 & 0 & 1 & -1 \end{bmatrix}}_{I\otimes H} \underbrace{\begin{bmatrix} 1 & 0 & 0 & 0 \\ 0 & 1 & 0 & 0 \\ 0 & 0 & 1 & 0 \\ 0 & 0 & 0 & -1 \end{bmatrix}}_{K} \underbrace{\frac{1}{\sqrt{2}}\begin{bmatrix} 1 & 1 & 0 & 0 \\ 1 & -1 & 0 & 0 \\ 0 & 0 & 1 & 1 \\ 0 & 0 & 1 & -1 \end{bmatrix}}_{I\otimes H} \begin{bmatrix} c_{\mathrm{H}} \\ c_{\mathrm{V}} \\ t_{\mathrm{H}} \\ t_{\mathrm{V}} \end{bmatrix}$$

$$= U_{\mathrm{CNOT}} \begin{bmatrix} c_{\mathrm{H}} \\ c_{\mathrm{V}} \\ t_{\mathrm{H}} \\ t_{\mathrm{V}} \end{bmatrix}, \quad U_{\mathrm{CNOT}} = \begin{bmatrix} 1 & 0 & 0 & 0 \\ 0 & 1 & 0 & 0 \\ 0 & 0 & 0 & 1 \\ 0 & 0 & 1 & 0 \end{bmatrix}. \tag{13.119}$$

The Kerr nonlinearity device in Fig. 13.14B performs the controlled-Z operation. In the absence of a control c_{V} photon, the target qubit is unaffected because $H^2=I$ (identity operator). In the presence of a control c_{V} photon, thanks to the cross-phase modulation in HNLF, the target vertical photon experiences the phase shift χL, where χ is the third-order nonlinearity susceptibility coefficient and L is the HNLF length. By selecting appropriately the fiber length we obtain $\chi L=\pi$ and the overall action on the target qubit is $HZH=X$, which corresponds to the CNOT gate action. The Toffoli gate can simply be obtained as a generalization of the aforementioned CNOT gate by adding an additional control qubit, while a Deutsch gate can be obtained by employing three control qubits, instead of the one used here.

The implementation of Pauli gates X, Y, and Z in integrated optics based on a single-qubit gate shown in Fig. 13.13B is described now. By appropriately setting the phase shifts of the U gate α, β, γ, and δ we can obtain the corresponding Pauli gates. The Y gate is obtained by setting $\alpha = \pi/2$, $\beta = \delta = 0$ rad, and $\gamma = \pi$:

$$Y = \begin{bmatrix} 0 & -j \\ j & 0 \end{bmatrix}. \tag{13.120}$$

The Z gate is obtained by setting $\alpha = \pi/2$, $\beta = \pi$, $\gamma = 2\pi$, and $\delta = 0$ rad:

$$Z = \begin{bmatrix} 1 & 0 \\ 0 & -1 \end{bmatrix}. \tag{13.121}$$

Finally, the X gate is obtained by setting $\alpha = \pi/2$, $\beta = -\pi$, $\gamma = \pi$, and $\delta = 0$ rad:

$$X = \begin{bmatrix} 0 & 1 \\ 1 & 0 \end{bmatrix}. \tag{13.122}$$

Another device that can be used for single-qubit manipulation, when information is imposed on the photon by using spin angular momentum, is the *optical hybrid.* The optical hybrid is shown in Fig. 13.15A. Let the electrical fields at the input ports be denoted as E_{i1} and E_{i2}, respectively. The output electrical fields, E_{o1} and E_{o2}, are related to the input electrical fields by:

$$\begin{aligned} E_{o,1} &= \left(E_{i,1} + E_{i,2}\right)\sqrt{1-k} \\ E_{o,2} &= \left(E_{i,1} + E_{i,2}\exp(-j\phi)\right)\sqrt{k}, \end{aligned} \tag{13.123}$$

where k is the power-splitting ratio $(0 \le k \le 1)$ between the Y-junction output ports of the optical hybrid, and ϕ is the phase shift introduced by the phase trimmer (Fig. 13.15). Eq. (13.123) can be rewritten in matrix form as follows:

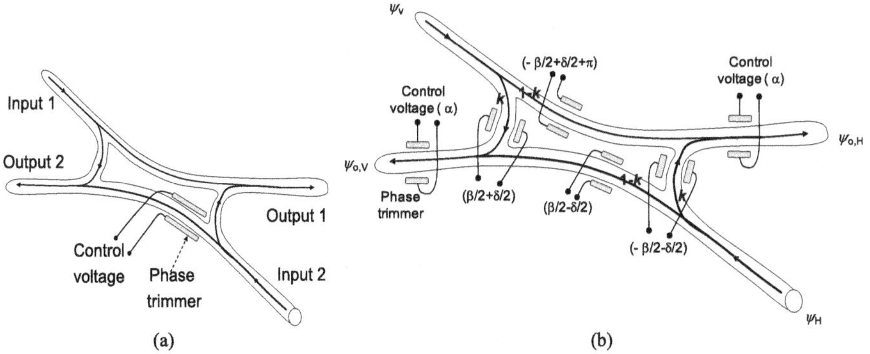

FIGURE 13.15

Optical hybrid as a basic building block to implement various single-qubit gates: (a) implementation of Hadamard gate ($k = 1/2$, $\phi = \pi$) and (b) implementation of arbitrary single-qubit quantum gate in integrated optics based on a single optical hybrid. (α, β, and δ are the phase shifts introduced by the corresponding phase trimmers). *PBS/C*, Polarization beam splitter/combiner.

$$\begin{bmatrix} E_{o,1} \\ E_{o,2} \end{bmatrix} = \begin{bmatrix} \sqrt{1-k} & \sqrt{1-k} \\ \sqrt{k} & e^{-j\phi}\sqrt{k} \end{bmatrix} \begin{bmatrix} E_{i,1} \\ E_{i,2} \end{bmatrix} = U \begin{bmatrix} E_{i,1} \\ E_{i,2} \end{bmatrix}, \quad U = \begin{bmatrix} \sqrt{1-k} & \sqrt{1-k} \\ \sqrt{k} & e^{-j\phi}\sqrt{k} \end{bmatrix}. \tag{13.124}$$

By setting the power-splitting ratio to $k = 1/2$ and phase shift to $\phi = \pi$ rad, the matrix U becomes:

$$U(k = 1/2,\ \phi = \pi) = \frac{1}{\sqrt{2}} \begin{bmatrix} 1 & 1 \\ 1 & -1 \end{bmatrix} = H, \tag{13.125}$$

which is the same as that of the Hadamard gate matrix representation, providing the PBS/PBC is used at the input and output ports, respectively. The optical hybrid can be generalized to the circuit shown in Fig. 13.15B, which can be used to implement an arbitrary single-qubit gate. In this scheme, the output qubit $[\psi_{o,H}\ \psi_{o,V}]^T$ is related to the input qubit $[\psi_H\ \psi_V]^T$ by:

$$\begin{bmatrix} \psi_{o,H} \\ \psi_{o,V} \end{bmatrix} = U \begin{bmatrix} \psi_H \\ \psi_V \end{bmatrix}, \quad U = \begin{bmatrix} \cos\left(\frac{\gamma}{2}\right) e^{j(\alpha-\beta/2-\delta/2)} & -\sin\left(\frac{\gamma}{2}\right) e^{j(\alpha-\beta/2+\delta/2)} \\ \sin\left(\frac{\gamma}{2}\right) e^{j(\alpha+\beta/2-\delta/2)} & \cos\left(\frac{\gamma}{2}\right) e^{j(\alpha+\beta/2+\delta/2)} \end{bmatrix}. \tag{13.126}$$

The U matrix in Eq. (13.126) represents the matrix representation of an arbitrary single-qubit quantum gate according to the $Z-Y$ decomposition theorem. For OH, the corresponding phase shifts α, β, δ can be introduced by the phase trimmer either thermally or electro-optically, while the proper power-splitting ratio $k = \cos^2(\gamma/2)$ should be set in the fabrication phase. By setting $\gamma = \delta = 0$ rad, $\alpha = \pi/4$, and $\beta = \pi/2$ rad, the U gate described by Eq. (13.126) operates as the phase gate; by setting $\gamma = \delta = 0$ rad, $\alpha = \pi/8$, and $\beta = \pi/4$ rad, the U gate operates as a $\pi/8$ gate; while by setting $\gamma = \pi/2$, $\alpha = \pi/2$, $\beta = 0$ rad, and $\delta = \pi$, the U gate given by Eq. (13.126) operates as a Hadamard gate. To complete the set of universal quantum gates, controlled-Z or CNOT gates are needed, which can be implemented as shown in Fig. 13.14B, but now with Hadamard gates implemented by using the optical hybrid shown in Fig. 13.15B.

The single-qubit gates can also be implemented based on a Mach–Zehnder interferometer (MZI) as shown in Fig. 13.16. By using a similar approach as before, the output and input ports can be related by Eq. (13.126), proving that an arbitrary single-qubit gate can indeed be based on a single MZI.

In the previous discussion we assumed that directional coupler theory and the S-matrix could also be applied on a quantum level, without formal proof. As an illustration, in the rest of this section the derivation of Eq. (13.126) is provided by using quantum-mechanical concepts for a directional coupler-based gate from

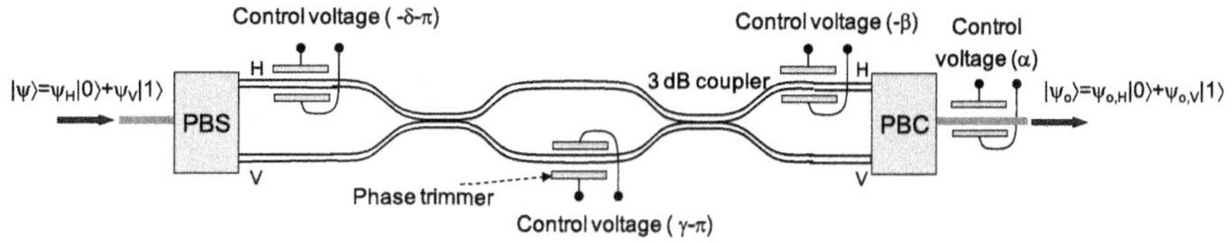

FIGURE 13.16

Photonic implementation of an arbitrary single-qubit gate based on a Mach–Zehnder interferometer. *PBS/C*, Polarization beam splitter/combiner.

Fig. 13.13B. The unitary operator U given by Eq. (13.126) can be written as the product of three unitary operators: (1) U_1 corresponding to the first phase section, (2) U_2 corresponding to the directional coupler section, and (3) U_3 corresponding to the second phase section. When the photon is present in the upper branch of the first phase section, it will experience the phase shift $\exp[i(-\alpha - \delta/2 - \pi/2)]$, and the phase shift $\exp[i(\alpha + \delta/2)]$ when in the lower branch, meaning that the matrix representation of this section is:

$$U_1 = \begin{bmatrix} e^{i(\alpha-\delta/2-\pi/2)} & 0 \\ 0 & e^{i(\alpha+\delta/2)} \end{bmatrix}. \tag{13.127}$$

In similar fashion, when the photon is present in the upper branch of the second phase section, it will experience the phase shift $\exp[i(-\beta + \pi/2)]$, and no phase shift when in the lower branch, indicating that the matrix representation of this section is:

$$U_3 = \begin{bmatrix} e^{i(-\beta+\pi/2)} & 0 \\ 0 & 1 \end{bmatrix}. \tag{13.128}$$

The directional coupler action on H and V photons can be described by the creation (annihilation) operators a ($a^\dagger$) and b ($b^\dagger$). The action of the directional coupler is given by:

$$U_2 = e^{-i(\phi/2)(a^\dagger b + ab^\dagger)} = e^{-i(\phi/2)G}, \quad G = a^\dagger b + ab^\dagger. \tag{13.129}$$

By using the Baker–Campbell–Hausdorf formula:

$$e^{\lambda G} A e^{-\lambda G} = \sum_{n=0}^{\infty} \frac{\lambda^n}{n!} C_n; \quad C_0 = A_0, \quad C_n = [G, C_{n-1}], \quad n = 1, 2, \cdots; \quad C_n$$
$$= \begin{cases} a, & n - \text{even} \\ -b, & n - \text{odd} \end{cases}, \tag{13.130}$$

we can show that:

$$U_2 a U_2^\dagger = e^{-i(\phi/2)G} a e^{+i(\phi/2)G} = \sum_{n=0}^{\infty} \frac{(-i\phi/2)^n}{n!}$$

$$C_n = \sum_{n-\text{even}} \frac{(-i\phi/2)^n}{n!} a - \sum_{n-\text{odd}} \frac{(-i\phi/2)^n}{n!}$$

$$b = a\cos(\phi/2) + ib\sin(\phi/2),$$

$$U_2 b U_2^\dagger = e^{-i(\phi/2)G} b e^{+i(\phi/2)G} = ia\sin(\phi/2) + b\cos(\phi/2). \tag{13.131}$$

The corresponding matrix representation is given by:

$$U_2 = \begin{bmatrix} \cos(\phi/2) & i\sin(\phi/2) \\ i\sin(\phi/2) & \cos(\phi/2) \end{bmatrix}, \tag{13.132}$$

which is the same as the corresponding expression derived from directional coupler theory. The overall operation of the gate from Fig. 13.13B is then:

$$\begin{bmatrix} \psi_{o,H} \\ \psi_{o,V} \end{bmatrix} = U_3 U_2 U_1 \begin{bmatrix} \psi_H \\ \psi_V \end{bmatrix} = \begin{bmatrix} e^{i(-\beta+\pi/2)} & 0 \\ 0 & 1 \end{bmatrix} \begin{bmatrix} \cos(\phi/2) & i\sin(\phi/2) \\ i\sin(\phi/2) & \cos(\phi/2) \end{bmatrix} \times \begin{bmatrix} e^{i(\alpha-\delta/2-\pi/2)} & 0 \\ 0 & e^{i(\alpha+\delta/2)} \end{bmatrix} \begin{bmatrix} \psi_H \\ \psi_V \end{bmatrix}, \tag{13.133}$$

proving therefore Eq. (13.126).

13.5 PHOTONIC IMPLEMENTATION OF QUANTUM RELAY

In this section we first describe the implementation of a Bell-state preparation circuit in integrated optics, required in quantum teleportation systems, which is shown in Fig. 13.17. Among many possible versions of Hadamard and CNOT gates we have chosen two with similar propagation times. The upper circuit operates as a Hadamard gate, while the rest of the circuit operates as a CNOT gate as explained in the previous section. It can be shown that output quantum state $|\beta_{ij}\rangle$ is related to the input state $|\psi_{\text{in}}\rangle$ by:

$$|\beta_{ij}\rangle = \frac{1}{\sqrt{2}} \begin{bmatrix} 1 & 0 & 0 & 0 \\ 0 & 1 & 0 & 0 \\ 0 & 0 & 0 & 1 \\ 0 & 0 & 1 & 0 \end{bmatrix} \begin{bmatrix} 1 & 0 & 1 & 0 \\ 0 & 1 & 0 & 1 \\ 1 & 0 & -1 & 0 \\ 0 & 1 & 0 & -1 \end{bmatrix} |\psi_{\text{in}}\rangle = \frac{1}{\sqrt{2}} \begin{bmatrix} c_H t_H + c_V t_V \\ c_H t_V + c_V t_V \\ c_V t_H - c_V t_V \\ c_H t_H - c_V t_H \end{bmatrix}. \tag{13.134}$$

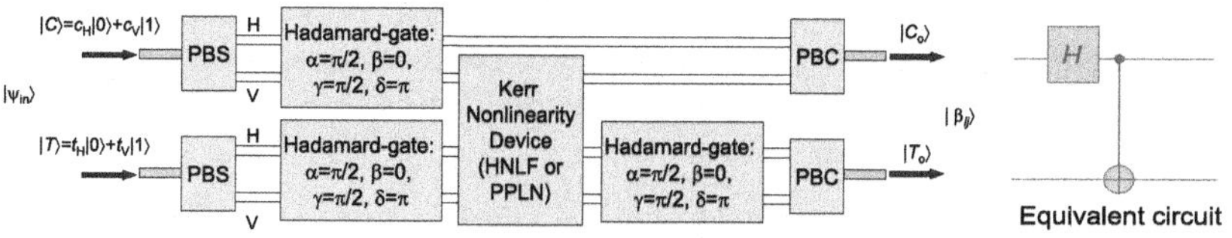

FIGURE 13.17

Photonic implementation of a Bell-state preparation circuit. *HNLF*, Highly nonlinear fiber; *PBC*, polarization beam combiner; *PBS*, polarization beam splitter; *PPLN*, periodically poled $LiNbO_3$.

For example, by setting $c_H = t_H = 1$ and $c_V = t_V = 0$ we obtain the Bell state $|\beta_{00}\rangle = [1 \quad 0 \quad 0 \quad 1]^T/\sqrt{2} = (|00\rangle + |11\rangle)/\sqrt{2}$. In Fig. 13.18 we describe how to implement the quantum relay based on the Bell-state preparation circuit (shown in Fig. 13.2B), Hadamard, controlled-*X*, and controlled-*Z* gates described earlier. We employ the principle of differed measurement and perform corresponding measurements only in the last intermediate node. The measurement circuits in

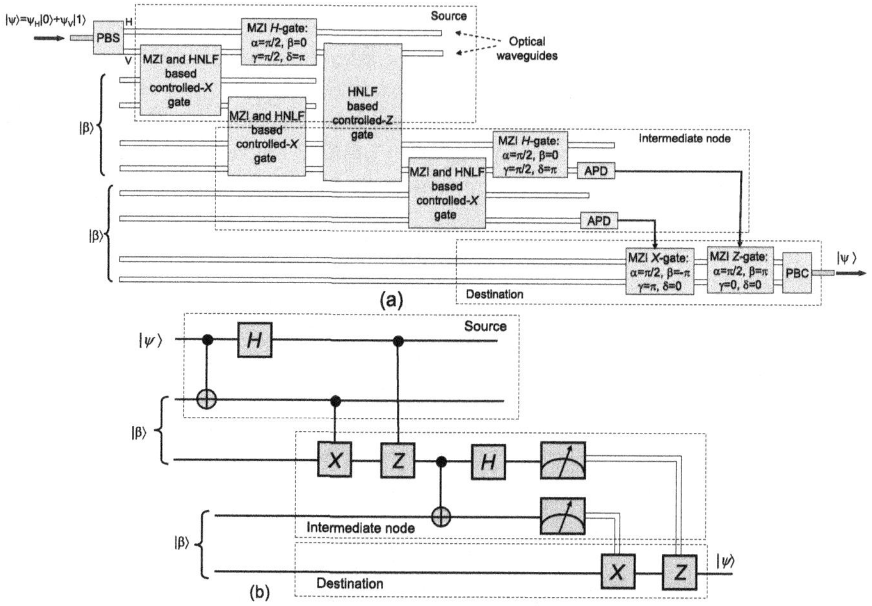

FIGURE 13.18

Photonic implementation of a quantum relay: (a) integrated optics implementation and (b) equivalent circuit. *APD*, Avalanche photodiode; *HNLF*, highly nonlinear fiber; *MZI*, Mach–Zehnder interferometer; *PBC*, polarization beam combiner; *PBS*, polarization beam splitter;

Fig. 13.3 represent the APDs, which are used to detect the presence of c_V photons in corresponding control qubits. The detection of c_V photons triggers the application of required control voltages on phase trimmers to perform controlled-X and controlled-Z operation at the destination node.

13.6 IMPLEMENTATION OF QUANTUM ENCODERS AND DECODERS

In this section we describe two classes of sparse-graph quantum codes: (1) quantum dual-containing low-density parity-check (LDPC) codes and (2) entanglement-assisted LDPC codes, and show that corresponding encoders and decoders can be implemented in using various technologies. For more details of various classes of quantum LDPC codes, interested readers are referred to Chapter 10.

The block scheme of entanglement-assisted quantum code, which requires a certain number of entangled qubits to be shared between the source and destination, is shown in Fig. 13.19. The number of needed preexisting entanglement qubits (also known as ebits [34]) can be determined by $e = \text{rank}(\boldsymbol{H}\boldsymbol{H}^{\text{T}})$, where $\boldsymbol{H}$ is the parity-check matrix of a classical code (and rank(.) is the rank of a given matrix). The source encodes quantum information in state $|\psi\rangle$ with the help of local ancilla qubits $|0\rangle$ and the source-half of shared ebits, and then sends the encoded qubits over a noisy quantum channel (e.g., free-space or fiber-optic channel). The receiver performs decoding on all qubits to diagnose the channel error and performs a recovery unitary operation to reverse the action of the channel. Notice that the channel does not affect at all the receiver's half of the shared ebits. By omitting the ebits, the conventional quantum coding scheme is obtained.

Most practical quantum codes belong to the class of Calderbank, Shor, and Steane (CSS) codes [34–37], and can be designed using a pair of conventional linear codes satisfying the twisted property (one code includes the dual of another code). Their quantum-check matrix has the form:

$$A = \begin{bmatrix} \boldsymbol{H} & | & \boldsymbol{0} \\ \boldsymbol{0} & | & \boldsymbol{G} \end{bmatrix}, \quad \boldsymbol{H}\boldsymbol{G}^{T} = \boldsymbol{0}, \tag{13.135}$$

where $\boldsymbol{H}$ and $\boldsymbol{G}$ are $M \times N$ matrices. The condition $\boldsymbol{H}\boldsymbol{G}^{\text{T}} = \boldsymbol{0}$ ensures that the twisted product condition is satisfied. Each row in Eq. (13.135) represents a stabilizer, with

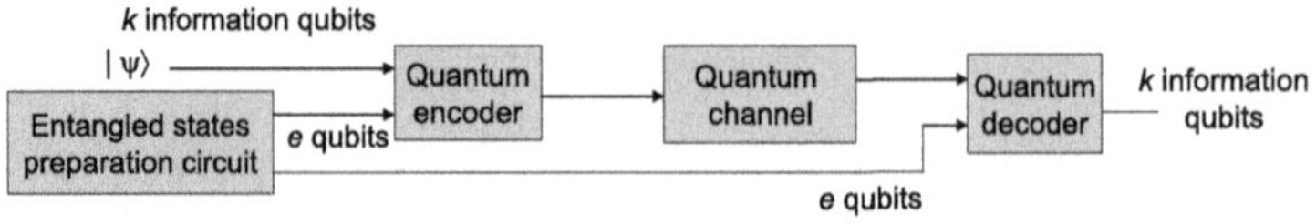

FIGURE 13.19

Entanglement-assisted quantum code.

ones in the left half of $\boldsymbol{A}$ corresponding to the positions of X-operators, and ones in the right half ($\boldsymbol{G}$) corresponding to the positions of Z-operators. As there are $2M$ stabilizer conditions applying to N-qubit states, $N-2M$ qubits are encoded in N qubits. The commutativity of stabilizers now appears as an *orthogonality of rows* with respect to a *twisted (sympletic) product*, formulated as follows: if the kth row in $\boldsymbol{A}$ is $r_k=(x_k;\,z_k)$, where x_k is the X binary string and z_k is the Z binary string, then the twisted product of rows k and l is defined by [35]:

$$r_k \odot r_l = x_k \cdot z_l + x_l \cdot z_k \ \mathrm{mod}2, \tag{13.136}$$

where $x_k \cdot z_l$ is the dot (scalar) product defined by $x_k \cdot z_l = \sum_j x_{kj} z_{lj}$. The twisted product is zero if and only if there is an even number of places where the operators corresponding to rows k and l differ (and neither is the identity), i.e., if the operators commute. The CSS codes based on dual-containing codes are the simplest to implement. Their (quantum)-check matrix can be represented by [34–37]:

$$\boldsymbol{A} = \left[\begin{array}{c|c} \boldsymbol{H} & \boldsymbol{0} \\ \boldsymbol{0} & \boldsymbol{H} \end{array}\right], \tag{13.137}$$

where $\boldsymbol{H}\boldsymbol{H}^{\mathrm{T}}=0$, which is equivalent to $C^{\perp}(\boldsymbol{H}) \subset C(\boldsymbol{H})$, where $C(\boldsymbol{H})$ is the code having $\boldsymbol{H}$ as the parity-check matrix, and $C^{\perp}(\boldsymbol{H})$ is its corresponding dual code. The quantum LDPC codes have many advantages over other classes of quantum codes, thanks to the sparseness of their parity-check matrices [17,35,36]. From Eq. (13.137), it follows that by providing that the $\boldsymbol{H}$-matrix of a dual-containing code is sparse, the corresponding $\boldsymbol{A}$-matrix will be sparse as well, while corresponding stabilizers will be of low weight. For example, the following $\boldsymbol{H}$-matrix satisfies the condition $\boldsymbol{H}\boldsymbol{H}^{\mathrm{T}}=0$, and can be used in quantum-check matrix (13.135) as a dual-containing code:

$$\boldsymbol{H} = \begin{bmatrix} 0 & 0 & 1 & 0 & 1 & 0 & 1 & 1 \\ 1 & 0 & 0 & 1 & 1 & 0 & 0 & 1 \\ 0 & 1 & 0 & 1 & 0 & 0 & 1 & 1 \\ 1 & 0 & 0 & 0 & 0 & 1 & 1 & 1 \\ 0 & 0 & 1 & 1 & 0 & 1 & 0 & 1 \\ 1 & 1 & 1 & 0 & 0 & 0 & 0 & 1 \\ 0 & 1 & 0 & 0 & 1 & 1 & 0 & 1 \end{bmatrix}.$$

The main drawback of dual-containing LDPC codes is the fact that they are essentially girth-4 codes,[1] which do not perform well under a sum-product algorithm (commonly used in the decoding of LDPC codes). On the other hand, it was shown in [34] that through the use of entanglement, arbitrary classical codes can be used in the correction of quantum errors, not only girth-4 codes. Because QKD and quantum

[1]Girth represents the shortest cycle in corresponding bipartite graph representation of a parity-check matrix of classical code.

teleportation systems assume the use of entanglement, this approach does not increase the complexity of the system at all.

The number of ebits needed in entanglement-assisted LDPC codes is $e = \text{rank}(\boldsymbol{H}\boldsymbol{H}^{\mathrm{T}})$, as indicated earlier, so that the minimum number of required EPR pairs (Bell states) is one, exactly the same as already in use in certain QKD schemes. For example, the following LDPC code has $\text{rank}(\boldsymbol{H}_1\boldsymbol{H}_1^{\mathrm{T}}) = 1$ and girth 6:

$$\boldsymbol{H}_1 = \begin{bmatrix} 0 & 0 & 1 & 0 & 1 & 0 & 1 \\ 1 & 0 & 0 & 1 & 1 & 0 & 0 \\ 0 & 1 & 0 & 1 & 0 & 0 & 1 \\ 1 & 0 & 0 & 0 & 0 & 1 & 1 \\ 0 & 0 & 1 & 1 & 0 & 1 & 0 \\ 1 & 1 & 1 & 0 & 0 & 0 & 0 \\ 0 & 1 & 0 & 0 & 1 & 1 & 0 \end{bmatrix},$$

and requires only one ebit to be shared between source and destination. Since arbitrary classical codes can be used with this approach, including LDPC code of girth $g \geq 6$, the performance of quantum LDPC codes can significantly be improved. Notice that this $\boldsymbol{H}$-matrix is obtained from the $\boldsymbol{H}_1$-matrix by adding an all-ones column.

Because two Pauli operators on N qubits commute if and only if there is an even number of places in which they differ (neither of which is the identity I-operator) we can extend the generators in $\boldsymbol{A}$ (for $\boldsymbol{H}_1$) by adding an $e = 1$ column so that they can be embedded into a larger Abelian group; the procedure is known as Abelianization in abstract algebra. Some may use the stabilizer version of the Gram–Schmidt orthogonalization algorithm to simplify this procedure, as indicated in [34].

For example, by performing Gauss–Jordan elimination, the quantum-check matrix (13.137) can be put in standard form [37] (see also Chapter 8):

$$\boldsymbol{A} = \begin{bmatrix} 1 & 0 & 0 & 0 & 0 & 1 & 1 & 1 & 0 & 0 & 0 & 0 & 0 & 0 & 0 & 0 \\ 0 & 1 & 0 & 0 & 1 & 1 & 0 & 1 & 0 & 0 & 0 & 0 & 0 & 0 & 0 & 0 \\ 0 & 0 & 1 & 0 & 1 & 0 & 1 & 1 & 0 & 0 & 0 & 0 & 0 & 0 & 0 & 0 \\ 0 & 0 & 0 & 1 & 1 & 1 & 1 & 0 & 0 & 0 & 0 & 0 & 0 & 0 & 0 & 0 \\ 0 & 0 & 0 & 0 & 0 & 0 & 0 & 0 & 0 & 0 & 1 & 0 & 1 & 0 & 1 & 1 \\ 0 & 0 & 0 & 0 & 0 & 0 & 0 & 0 & 1 & 0 & 0 & 0 & 0 & 1 & 1 & 1 \end{bmatrix}.$$

The corresponding generators in standard form are:

$$g_1 = X_1X_6X_7X_8 \quad g_2 = X_2X_5X_6X_8 \quad g_3 = X_3X_5X_7X_8$$
$$g_4 = X_4X_5X_6X_7 \quad g_5 = Z_3Z_5Z_7Z_8 \quad g_6 = Z_1Z_6Z_7Z_8,$$

where the subscripts are used to denote the positions of corresponding X- and Z-operators. The encoding circuit is shown in Fig. 13.20. We use the efficient implementation of encoders introduced by Gottesman [37]. It is clear from

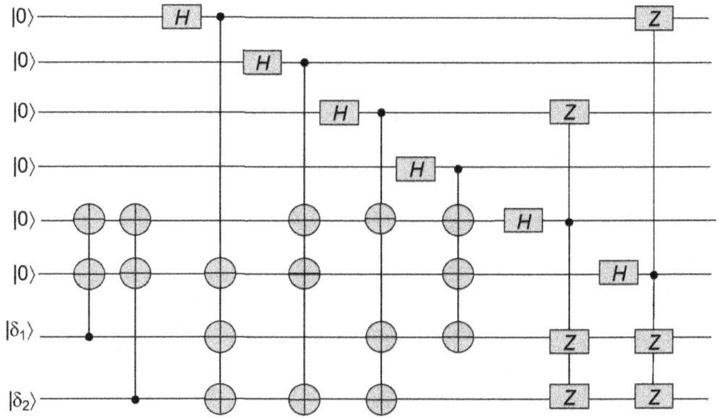

FIGURE 13.20

Encoding circuit for quantum (8,2) low-density parity-check code. $|\delta_1\delta_2\rangle$ are information qubits.

Fig. 13.20 that for encoder implementation of quantum LDPC codes, Hadamard (H), CNOT ($\oplus$), and controlled-Z gates are sufficient, whose implementation in integrated optics, NMR, and ion traps was already discussed in previous sections. In the following section, other implementation-based CQED will be described as well. For example, from Fig. 13.14B it is clear that the controlled-X gate in CQED technology is typically implemented based on one controlled-Z and two H gates. Therefore by using Eq. (13.39), the quantum LDPC encoder can be implemented based only on Hadamard and controlled-Z gates, and their implementation is therefore compatible with various technologies described in previous sections. Because $H^2 = I$, the corresponding circuit based on H gates and controlled-Z gates can be simplified. It can be shown in similar fashion that a corresponding decoder can be implemented based only on H gates and controlled-Z gates or CNOT gates.

As another illustrative example, the following $\boldsymbol{H}$-matrix satisfies the condition $\boldsymbol{HH}^{\mathrm{T}} = 0$, and can be used in quantum-check matrix (13.137) as dual-containing code:

$$\boldsymbol{H} = \begin{bmatrix} 1 & 0 & 0 & 1 & 1 & 1 \\ 1 & 1 & 1 & 0 & 0 & 1 \\ 0 & 1 & 1 & 1 & 1 & 0 \end{bmatrix}.$$

By performing Gauss–Jordan elimination, the quantum-check matrix (13.137) can be put in standard form as follows:

$$\boldsymbol{A} = \begin{bmatrix} 1 & 0 & 0 & 1 & 1 & 1 & 0 & 0 & 0 & 0 & 0 & 0 \\ 0 & 1 & 1 & 1 & 1 & 0 & 0 & 0 & 0 & 0 & 0 & 0 \\ 0 & 0 & 0 & 0 & 0 & 0 & 1 & 1 & 1 & 0 & 0 & 1 \\ 0 & 0 & 0 & 0 & 0 & 0 & 1 & 0 & 0 & 1 & 1 & 1 \end{bmatrix}.$$

The corresponding generators in standard form are:

$$g_1 = X_1X_4X_5X_6 \quad g_2 = X_2X_3X_4X_5 \quad g_3 = Z_1Z_2Z_3Z_6 \quad g_4 = Z_1Z_4Z_5Z_6.$$

The encoding circuit is shown in Fig. 13.21A, while the decoding circuit is shown in Fig. 13.21B. We use the efficient implementation of encoders and decoders introduced by Gottesman [37] (see also Chapter 8). Again, it is clear from Fig. 13.21 that for encoder and decoder implementation of quantum LDPC codes, Hadamard (H) and CNOT ($\oplus$) gates are sufficient.

By closer inspection of Eq. (13.137) we can conclude that generators for CSS design employ exclusively either X or Z gates but not both. This is also evident from the foregoing example (generators g_1, g_2, g_3, and g_4 contain only X-operators, while generators g_5 and g_6 contain only Z-operators). Therefore quantum LDPC decoders of CSS type can be implemented based on controlled-Z gates only, as shown in Fig. 13.22. For entanglement-assisted LDPC decoder implementation, we need to follow the procedure in the text related to Fig. 13.20.

We turn our attention now to the design of quantum LDPC codes based on Steiner triple systems (STSs) [38]. The STS represents a particular instance of a balanced incomplete block design (BIBD) [36,38]. The BIBD(v,k,λ) is defined as a collection of blocks of length k for the set $V$of integers of size v, such that each pair of elements of V occur together in exactly λ of the blocks. STS(v) is defined as a BIBD(v,3,1). It has been shown in [36] that a BIBD of even λ can be used to design quantum LDPC codes belonging to CSS codes by using dual-containing classical codes. It has also been shown in [36] that BIBDs of unitary index ($\lambda = 1$) can be used to design entanglement-assisted LDPC codes that require only one ebit to be shared between source and destination, since rank($\boldsymbol{HH}^{\mathrm{T}}$) = 1. For example, STS(7) is given by the following collection of blocks of length $k = 3$: {2, 4, 6}, {6, 3, 7}, {5, 6, 1}, {3, 2, 5}, {1, 7, 2}, {7, 5, 4}, {4, 1, 3}. By identifying these blocks with nonzero

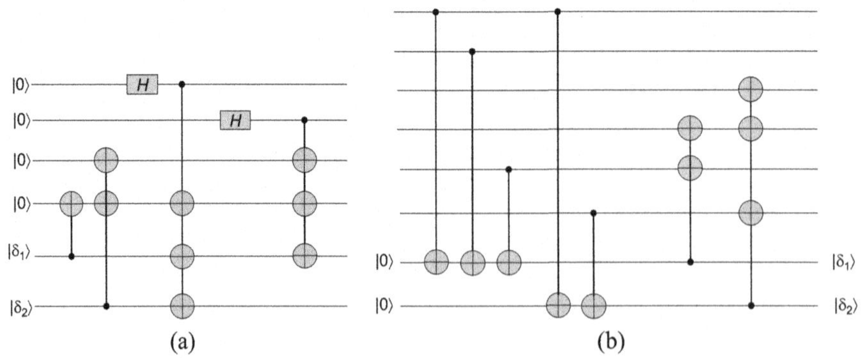

FIGURE 13.21

Encoding and decoding circuits for quantum (6,2) low-density parity-check code: (a) encoder configuration and (b) decoder configuration.

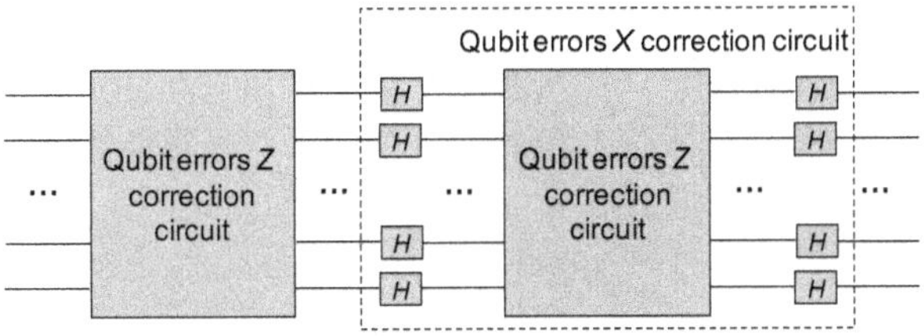

FIGURE 13.22

Quantum low-density parity-check decoder of Calderbank, Shor, and Steane-type implementation based only on controlled-Z gates.

positions of the corresponding parity-check matrix we obtain an LDPC code of girth 6 satisfying the property rank($\boldsymbol{HH}^{\mathrm{T}}$) $= 1$. For example, the preceding $\boldsymbol{H}_1$-matrix is obtained from STS(7). By adding an all-ones column to such matrix we obtain a dual-containing code since rank($\boldsymbol{HH}^{\mathrm{T}}$) $= 0$. The preceding $\boldsymbol{H}$-matrix is obtained using this simple approach. Therefore both entanglement-assisted codes and dual-containing quantum codes can be obtained by using STSs. By selectively removing the blocks from STSs we can increase the girth of corresponding LDPC code, and therefore improve bit error rate (BER) performance at the expense of increasing the complexity of an equivalent entanglement-assisted LDPC code since now rank($\boldsymbol{HH}^{\mathrm{T}}$) > 1. For more details of various STSs, interested readers are referred to [38]. Notice that the codes from STSs are easy to implement because the column weight of corresponding classical parity-check matrices is only 3, while the parity-check matrices' column weight of projective geometry (PG)-based codes [38–40], which also satisfy the rank($\boldsymbol{HH}^{\mathrm{T}}$) $= 1$ property, is huge (Fig. 13.23).

In Fig. 13.23 we provide a comparison of entanglement-assisted LDPC codes of girth $g = 6$, 8, 10, and 12 against dual-containing LDPC code ($g = 4$). With parameter c we denote the column weight of a corresponding parity-check matrix. The dual-containing code of girth 4 is designed based on BIBDs [36]. The entanglement-assisted codes from PGs are designed as described in [30,39]. These codes are of girth 6 but require only one ebit to be shared between source and destination. The entanglement-assisted codes of girth 10 and 12 are obtained in a similar fashion to STS described earlier by selectively removing the blocks from the design.

From Fig. 13.23A it is clear that entanglement-assisted LDPC codes outperform for more than order in magnitude, in terms of cross-over probability, the corresponding dual-containing LDPC code. It is also evident that as we increase the girth we obtain better performance at the expense of increased entanglement-complexity since rank($\boldsymbol{HH}^{\mathrm{T}}$) > 1 for $g > 6$ codes and increases as girth increases. Notice, however, that finite geometry codes [39,40] typically have large column weight (Fig. 13.23), so that although they require a small number of ebits, their decoding complexity is high because of huge column weight. In practice, we will need to make a compromise between decoding complexity, number of required ebits, and BER

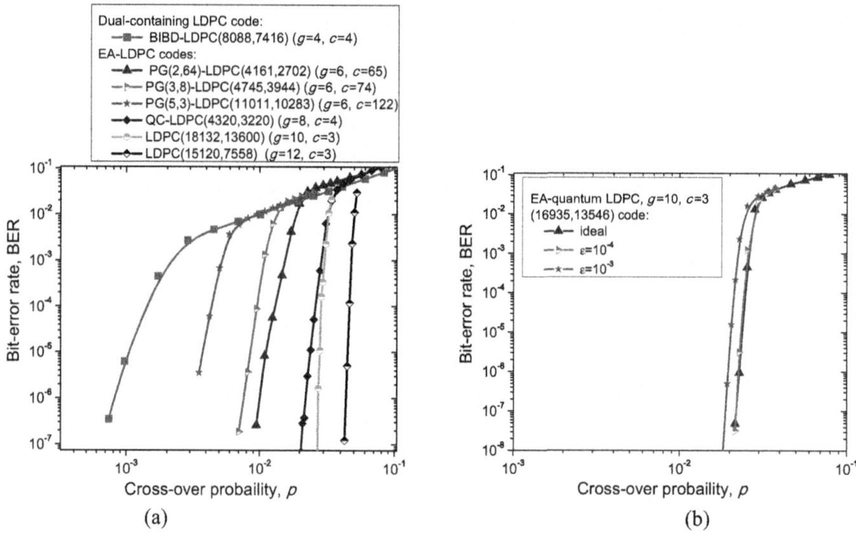

FIGURE 13.23

Bit error rate performance of various quantum codes: (a) assuming that gates are perfect and (b) assuming that gates are imperfect. The parameters *g* and *c* denote girth and column weight of the corresponding parity-check matrix, respectively. *BIBD*, Balanced incomplete block design; *EA*, entanglement-assisted codes; *PG*, projective geometry; *QC*, quasi-cyclic code.

performance. The results shown in Fig. 13.23A are obtained by assuming that quantum gates are perfect. In Fig. 13.23B we study the influence of imperfect quantum gates on BER performance for girth-8 entanglement-assisted LDPC (16935,13546) code. Namely, some practical problems such as decoherence will affect the operation of gates, causing the controlled-Z gate (or equivalently CNOT gate) to fail with certain probability. When gates fail with probability $\varepsilon = 10^{-4}$, the BER performance loss is negligible. On the other hand, when the gates fail with probability $\varepsilon = 10^{-3}$, the BER performance loss is small but noticeable.

13.7 CAVITY QUANTUM ELECTRODYNAMICS-BASED QUANTUM INFORMATION PROCESSING

CQED techniques can be used in many different ways to perform quantum computation, including [41–46]: (1) quantum information can be represented by photon states wherein the cavities with atoms are used to provide the nonlinear interaction between photons [42,43]; (2) quantum information can be represented using atoms wherein the photons can be used to communicate between atoms [44]; and (3) quantum information can be represented using the quantum interface between a single photon and the spin state of an electron trapped in a quantum dot [45].

Before describing one particular example of CQED-based quantum computing, a simple model for atom–field interaction based on the Jaynes–Cummings model (proposed by Edwin Jaynes and Fred Cummings in 1963) is introduced (see [46] for more details). This model describes the interaction between a two-level atom (only two-relevant levels in an atom are observed) and a quantized field. Let the ground state and excited state of the atom be denoted by $|g\rangle$ and $|e\rangle$, respectively, and let the photon number states $|0\rangle$ and $|1\rangle$ represent logic 0 and 1, respectively. The atom energy levels are separated by $\hbar\omega_a$, the atomic transition energy. The photon energy levels are separated by $\hbar\omega_0$. Two-atom states are of opposite parity and mutually orthogonal, and can be used from the following basis: $\{|g\rangle\langle g|,\ |g\rangle\langle e|,\ |e\rangle\langle g|,\ |e\rangle\langle e|\}$An arbitrary operator O can be expanded in this basis as follows:

$$O = O_{gg}\sigma_{gg} + O_{ge}\sigma_{ge} + O_{eg}\sigma_{eg} + O_{ee}\sigma_{ee};\ O_{mn} = \langle m|O|n\rangle,\ \sigma_{mn} = |m\rangle\langle n|,\ m,\ n\in\{e,\ g\}. \tag{13.138}$$

Because two-atom states have opposite parity, the dipole operator can be represented by:

$$\hat{\boldsymbol{d}} = \hat{\boldsymbol{d}}_{ge}\sigma_{ge} + \hat{\boldsymbol{d}}_{eg}\sigma_{eg};\ \left\langle g\middle|\hat{\boldsymbol{d}}\middle|g\right\rangle = \left\langle e\middle|\hat{\boldsymbol{d}}\middle|e\right\rangle = 0. \tag{13.139}$$

Initially, the atom and photon do not interact, and the atom–field state can be represented as the tensor product atom and photon states. The corresponding Hamiltonian is given by:

$$\hat{H} = \underbrace{\hbar\omega_a|e,\ 0\rangle\langle e,\ 0|}_{\hat{H}_{\text{atom}}} + \underbrace{\hbar\omega_0|g,\ 1\rangle\langle g,\ 1|}_{\hat{H}_{\text{photon}}} = \hbar\omega_a\sigma_{ee} + \hbar\omega_0 a^\dagger a = \hat{H}_{\text{atom}} + \hat{H}_{\text{photon}}, \tag{13.140}$$

or in matrix form as:

$$\hat{H} \doteq \begin{bmatrix} \hbar\omega_0 & 0 \\ 0 & \hbar\omega_a \end{bmatrix}, \tag{13.141a}$$

which is the same as that given by Eq. (13.8). In Eq. (13.140), we use a $(a^\dagger)$ to denote the photon annihilation (creation) operator, as before. The energy levels of the atom–photon system before the interaction are shown in Fig. 13.24A. The interaction Hamiltonian is a function of dipole moment $\boldsymbol{d}$ and electrical field $\boldsymbol{E}$ as follows:

$$\hat{H}_{\text{int}} = -\hat{\boldsymbol{d}}\cdot\boldsymbol{E} = -\left(\hat{\boldsymbol{d}}_{ge}\sigma_{ge} + \hat{\boldsymbol{d}}_{eg}\sigma_{eg}\right)\sqrt{\frac{\hbar\omega_0}{2\varepsilon_0 V}}[u(\boldsymbol{r})a + h.c.] = -\hbar\Omega_0(\sigma_{ge} + \sigma_{eg})\left(a + a^\dagger\right), \tag{13.141b}$$

where V is the "photon mode volume" (the volume of space occupied by the photon)

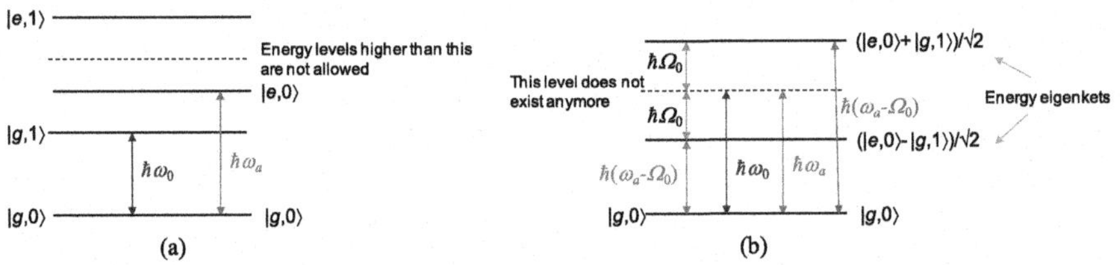

FIGURE 13.24

Atom–photon energy levels: (a) before interaction and (b) after interaction.

and $u(\boldsymbol{r})$ is the mode field at the CM $\boldsymbol{r}$. The interaction Hamiltonian (13.141) can be simplified as follows:

$$\hat{H}_{\text{int}} = -\hbar\Omega_0(\sigma_{ge}+\sigma_{eg})(a+a^\dagger) = -\hbar\Omega_0(\sigma_{ge}a+\sigma_{ge}a^\dagger+\sigma_{eg}a+\sigma_{eg}a^\dagger) \simeq -\hbar\Omega_0(\sigma_{ge}a+\sigma_{eg}a^\dagger), \tag{13.142}$$

as the terms $\sigma_{ge}a^\dagger$, $\sigma_{eg}a$ can be neglected by using the rotating-wave approximation. The operator $\sigma_{ge}a$ raises the atom to the excited state, while annihilating a photon. On the other hand, the operator $\sigma_{eg}a^\dagger$ brings the atom to the ground state, while creating a photon. The overall Hamiltonian can now be represented by:

$$\hat{H} = \hbar\omega_a\sigma_{ee} + \hbar\omega_0 a^\dagger a - \hbar\Omega_0(\sigma_{ge}a+\sigma_{eg}a^\dagger), \tag{13.143}$$

with corresponding matrix representation given by:

$$\hat{H} \doteq \begin{bmatrix} \hbar\omega_0 & \hbar\Omega_0 \\ \hbar\Omega_0 & \hbar\omega_a \end{bmatrix}. \tag{13.144}$$

To make this interaction stronger, we have to tune the light frequency to the atomic transition frequency. In addition, since $\Omega_0 \sim d\sqrt{\omega_0/V}$ we can make Ω_0 larger by increasing the dipole moment (the proper choice of the atom and corresponding states) and by decreasing the volume of the cavity. The energy eigenkets of Hamiltonian (13.144) are given by:

$$\begin{aligned} |E_+ = \hbar(\omega_a+\Omega_0)\rangle &= \frac{(|e,0\rangle + |g,1\rangle)}{\sqrt{2}}, \\ |E_- = \hbar(\omega_a-\Omega_0)\rangle &= \frac{(|e,0\rangle - |g,1\rangle)}{\sqrt{2}}, \end{aligned} \tag{13.145}$$

and the corresponding energy levels are illustrated in Fig. 13.24B. These energy eigenkets can be interpreted as superposition states of having and not having a photon.

We turn our attention now to the CQED implementation of the following set $\{H, P, T, U_{\text{CNOT}}$, or controlled-$Z\}$ of universal quantum gates using CQED technology by employing option (1) from the first paragraph of this section, namely by representing the quantum information by photon states and by using the cavities with atoms to provide the nonlinear interaction between photons [42,43]. For option (2), interested readers are referred to [44], and for option (3) to [45]. The H, P, and T gates are single-qubit gates and can be implemented based on one mode of radiation field inside the cavity by passing a two-level atom through the cavity. In the middle of the passage of the atom through the cavity, a short classical pulse of amplitude A_p is to be applied. Let the ground state and excited state of the atom also be denoted by $|g\rangle$ and $|e\rangle$, respectively, and let the photon states $|0\rangle$ and $|1\rangle$ represent logic 0 and 1, respectively. The interaction Hamiltonian can be represented by [46]:

$$H_{\text{int}} = \hbar\Omega\left(a|e\rangle\langle g| + a^{\dagger}|g\rangle\langle e|\right), \tag{13.146}$$

where $a$and $a^{\dagger}$denote the photon annihilation and creation operators and Ω is the corresponding vacuum Rabi frequency associated with the interaction of the cavity mode with atom states. Based on Eq. (13.146), the time-evolution operator can be derived [46]:

$$U(t) = \cos\left(\Omega t\sqrt{a^{\dagger}a+1}\right)|g\rangle\langle g| + \cos\left(\Omega t\sqrt{a^{\dagger}a+1}\right)|e\rangle\langle e| - i\frac{\sin\left(\Omega t\sqrt{a^{\dagger}a+1}\right)}{\sqrt{a^{\dagger}a+1}}a|e\rangle\langle g|$$
$$-ia^{\dagger}\frac{\sin\left(\Omega t\sqrt{a^{\dagger}a+1}\right)}{\sqrt{a^{\dagger}a+1}}|g\rangle\langle e|, \tag{13.147}$$

and the time evolution of the initial state $|\psi(0)\rangle$ can be described by $|\psi(t)\rangle = U(t)|\psi(0)\rangle$. The atom–field state $|\psi(0)\rangle = |e,0\rangle$ is unaffected after $\Delta t = \pi/2\Omega$, while $|\psi(0)\rangle = |e,1\rangle$ moves to $-i|g,0\rangle$. After the initial time $\Delta t = \pi/2\Omega$, the pulse of amplitude A_p is applied, which prepares the atom in the superposition state [43]:

$$|g\rangle \to \cos\theta|g\rangle + ie^{-i\phi}\sin\theta|e\rangle, \quad |e\rangle \to ie^{i\phi}\sin\theta|g\rangle + \cos\theta|e\rangle; \quad \theta = \omega t/2,$$
$$\omega = |d|A_p/\hbar, \tag{13.148}$$

where $d = |d|e^{i\phi}$ is the dipole moment. The atom again interacts with the cavity field for the same duration $\Delta t = \pi/2\Omega$ so that the initial cavity modes are transformed to [43]:

$$|0\rangle \to \cos\theta|0\rangle + e^{i\phi}\sin\theta|1\rangle, \quad |1\rangle \to e^{-i\phi}\sin\theta|0\rangle - \cos\theta|1\rangle, \tag{13.149}$$

which is equivalent to the one-qubit unitary operator $U(\theta,\phi)$:

$$U(\theta,\ \phi) = \begin{bmatrix} \cos\theta & e^{i\phi}\sin\theta \\ e^{-i\phi}\sin\theta & -\cos\theta \end{bmatrix}. \tag{13.150}$$

For example, by setting $\theta = \pi/4$ and $\phi = 0$, the unitary gate $U(\pi/4,0)$ becomes the Hadamard gate H:

$$U(\pi/4, 0) = \frac{1}{\sqrt{2}} \begin{bmatrix} 1 & 1 \\ 1 & -1 \end{bmatrix} = H. \tag{13.151}$$

The Z gate is obtained by setting $\theta = \phi = 0$:

$$U(0, 0) = \begin{bmatrix} 1 & 0 \\ 0 & -1 \end{bmatrix} = Z. \tag{13.152}$$

The P and T gates and other Pauli gates can be obtained by properly selecting θ and ϕ and/or by concatenation of two U gates with properly chosen parameters. The quantum phase-shift gate based on CQED, in which two qubits are represented as two radiation modes inside a cavity in combination with a three-level atom that provides the desired control interaction, is described in [43]. Namely, the quantum phase-shift gate operation can be described by:

$$C(U_\alpha) = |0_1 0_2\rangle\langle 0_1 0_2| + |0_1 1_2\rangle\langle 0_1 1_2| + |1_1 0_2\rangle\langle 1_1 0_2| + e^{i\alpha}|1_1 1_2\rangle\langle 1_1 1_2|. \tag{13.153}$$

By setting $\alpha = \pi$, the controlled-Z, $C(Z)$, gate is obtained.

To complete the implementation of the set $\{H, P, T, U_{\mathrm{CNOT}}$, or controlled-$Z\}$ of universal quantum gates, implementation of the CNOT gate/controlled-Z gate is needed. One possible implementation based on CQED, compatible with photon polarization states, is shown in Fig. 13.25. To enable the interaction of vertical photons we use an optical cavity with a single trapped three-level atom as illustrated in Fig. 13.25. The atom has three relevant levels: the ground $|g\rangle$, the intermediate $|i\rangle$, and the excited $|e\rangle$ states. The ground and intermediate states are close to each other and can be the hyperfine states. The atom has initially been prepared in superposition state $|\psi_A\rangle = (|g\rangle + |i\rangle)/\sqrt{2}$. The transition $|i\rangle \rightarrow |e\rangle$ is coupled to a cavity mode in vertical polarization and it is resonantly driven by the vertical photon from the input.

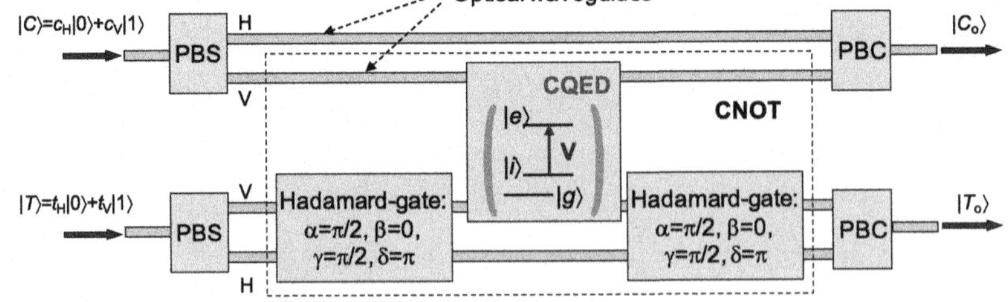

FIGURE 13.25

Deterministic CNOT gate implementation based on cavity quantum electrodynamics (CQED). *PBC*, Polarization beam combiner; *PBS*, polarization beam splitter.

When the incoming photon is in vertical polarization and the atom is in the ground state, the incoming photon is resonant with the cavity mode; it interacts with the atom and after interaction the atom returns to the initial state while the V-photon acquires the phase shift of π rad. If, on the other hand, the atom was in the intermediate state, the frequency corresponding to the entangled mode is significantly detuned from the frequency of the input photon, and the photon leaves the cavity without any phase change. Operation of the gate shown in Fig. 13.25, by ignoring the Hadamard gates, can be described by:

$$U_{AC}R_A(-\theta)U_{AT}R_A(\theta)U_{AC}|CT\rangle|\psi_A\rangle, \quad (13.154)$$

where $|\psi_A\rangle$ denotes the initial atom state $(|g\rangle + |i\rangle)/\sqrt{2}$, $|CT\rangle$ is the input two-qubit state, U_{AC} (U_{AT}) denotes the operator describing atom-control photon (atom-target photon) interaction, and $U_A(\theta)$ is the atom-rotation operator performed by applying the θ pulse on the atom. In the absence of the vertical-control photon, the action of the foregoing operator is simply an identity operator since $R_A(-\theta)R_A(\theta) = I$and $U_{AC}^2 = I$. In the presence of the vertical-control photon, the sequence of operators is as follows: (1) the vertical-control photon interacts with the atom, (2) the rotation operator is applied on the atom, (3) the vertical-target photon interacts with the atom, (4) the derotation operator is applied on the atom, and (5) the vertical-target photon interacts with the atom. After this sequence of operators, the control photon and the atom return to initial states, while the target photon achieves π phase shift. Therefore the overall action is controlled-Z operation.

The additional two Hadamard gates are used to perform the following transformation: $HZH = X$, resulting in CNOT gate operation. Some may follow a more rigorous derivation by applying a similar procedure to that provided in [43]. From Fig. 13.24, it is evident that the use of the controlled-Z gate instead of the CNOT gate for quantum computing applications and quantum teleportation is more appropriate in CQED technology, since controlled-Z gate implementation is simpler (the two Hadamard gates from Fig. 13.25 are not needed for controlled-Z gate implementation). The quantum gate shown in Fig. 13.25 is suitable for implementation in photon crystal technology [47,48]. For proper operation, on-chip integration is required. The first step toward this implementation would be achieving the coherent control of quantum $|CT\rangle$ states from Eq. (13.154). Quantum error correction-based fault-tolerant concepts should be used to facilitate this implementation.

An important practical problem to be addressed in future is related to the trapping of the atom within cavity that requires ultra-cold conditions [49]. There are several research groups performing research in this area: R. Schoelkopf and S. Girvin at Yale, J. Kimble at Caltech, S. Haroche at ENS, D. Stamper-Kurn at Berkley, and P. S. Jessen at the University of Arizona, just to mention a few. As an alternative solution, the quantum-dot approach can be used as indicated in [47].

13.8 QUANTUM DOTS IN QUANTUM INFORMATION PROCESSING

A quantum dot is the structure on a semiconductor that is able to confine electrons in three dimensions such that discrete energy levels are obtained. The quantum dot behaves as an artificial atom, whose properties can be controlled. Quantum dots can be formed spontaneously by depositing a semiconductor material on a substrate with different lattice spacing (this configuration is known as a heterostructure). The quantum dots have a bowl-like shape with a diameter ~100 nm and height ~30 nm, as illustrated in Fig. 13.26. The layer of AlGaAs of ~5 nm thickness is placed between two thicker layers of GaAs. The lower GaAs layer is *n* doped to provide free electrons at the upper GaAs—AlGaAs interface, forming a so-called2D) "electron gas". This process of formation of 2D electron gas is equivalent to total internal reflection, as two interface layers serve as boundaries of layers with different refractive indices. Furthermore, an array of metal contacts (gates) is placed on the top of the upper GaAs layer (50—10 nm above the 2D electron gas). External voltages are applied on these gates to restrict movement of the electrons in the horizontal plane. 3D potential wells are induced in the region among metal electrodes.

There are many different schemes that can be used by employing this concept. One approach is based on *electron—hole pairs*, also known as *excitons*, whose energy is given by $E_{excitons} = E_{bg} - E_{be}$, where E_{bg} is the band-gap energy and E_{be} is the binding energy of electron—hole pairs. Now by using two quantum dots, the entangled qubits can be formed. The electron and hole can be in the first dot, representing the state $|0\rangle$, or in the second dot, representing the state $|1\rangle$. Let the first

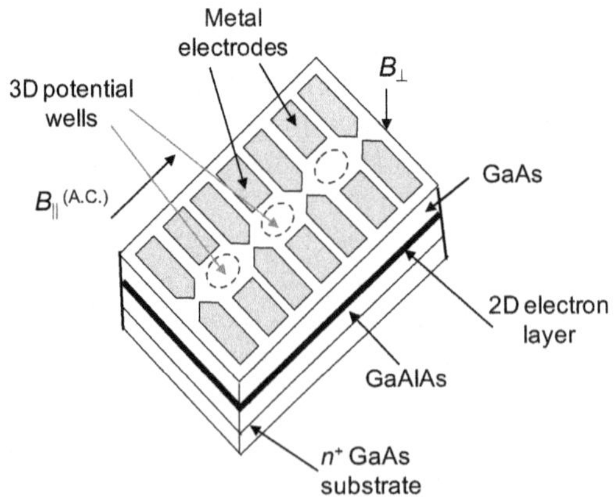

FIGURE 13.26

Semiconductor heterostructure for implementation of quantum computation by using spins of electrons, confined in individual quantum dots, as qubits.

qubit correspond to the electron and the second qubit correspond to the hole. The state $|00\rangle$ denotes that the electron—hole pair is located in the first dot, while the $|11\rangle$ state denotes that the electron—hole pair is located in the second dot. The state $|01\rangle$ denotes that the electron is in the first dot and the hole is in the second dot. Finally, the state $|10\rangle$ denotes that the hole is in the first dot and the electron is in the second dot. By providing that the distance between two interacting dots is ~5 nm, the electron and holes can tunnel from one dot to the other dot. The electron—hole recombines at a time interval ~1 ns, and the wavelength is dependent on the state occupied by particles before recombination.

Another approach is to use the *spin of an electron* as the qubit. There are two key advantages of this approach: (1) the Hilbert space is 2D and (2) the decoherence time is in the order of milliseconds. The Hamiltonian for a set of electron spins localized in a coupler array of quantum dots can be represented by:

$$\hat{H} = \sum_{\langle m,n\rangle} J_{mn}(t)\boldsymbol{\sigma}_m \cdot \boldsymbol{\sigma}_n + \frac{1}{2}\mu_B \sum_m g_m(t)\boldsymbol{B}_m(t)\cdot\boldsymbol{\sigma}_m, \quad \mu_B = e\hbar/2m_e, \qquad (13.155)$$

where $\boldsymbol{\sigma}_m$ denotes the Pauli spin matrix associated with the mth electron, $\boldsymbol{\sigma}_m = (X_m, Y_m, Z_m)$, and J_{mn} denotes the coupling interaction between the mth and nth spins, which can be used to build two-qubit gates. The$\boldsymbol{\sigma}_m \cdot \boldsymbol{\sigma}_n$ stands for simultaneous scalar and tensor product, and $\sum_{<m,n>}$ denotes the summation over closest neighbors. The second term corresponds to the coupling related to the external magnetic field $\boldsymbol{B}_m$. The coupling interaction $J_{mn}(v)$ (v is the parameter used to control the coupling interaction) can be switched on and off adiabatically, providing that $|v^{-1}\mathrm{d}v/\mathrm{d}t| << \Delta E/\hbar$, where ΔE is energy separation. The single-qubit gates can be implemented by properly varying the magnetic field $\boldsymbol{B}_m$ or Landé factor g_m so that the resonant frequency is dependent on qubit position. The two-qubit Hamiltonian, based on Eq. (13.155), can be written as:

$$\hat{H}_{12}(t) = J_{12}(\tau)\boldsymbol{\sigma}_1 \cdot \boldsymbol{\sigma}_2. \qquad (13.156)$$

In Problem 9 in Chapter 3, it was shown that the action of the operator $\boldsymbol{\sigma}_1 \cdot \boldsymbol{\sigma}_2$is to swap the qubits:

$$\boldsymbol{\sigma}_1 \cdot \boldsymbol{\sigma}_2|00\rangle = |00\rangle, \quad \boldsymbol{\sigma}_1 \cdot \boldsymbol{\sigma}_2|11\rangle = |11\rangle, \quad \boldsymbol{\sigma}_1 \cdot \boldsymbol{\sigma}_2|01\rangle = |10\rangle, \quad \boldsymbol{\sigma}_1 \cdot \boldsymbol{\sigma}_2|10\rangle = |01\rangle. \qquad (13.157)$$

By allowing the evolution to last such that:

$$\int_0^{T_s} J(t)dt = J_0 T_s = \pi \bmod 2\pi, \qquad (13.158)$$

the SWAP gate operation is performed. The controlled-Z gate, $C(Z)$, can be implemented by proper combination of rotation and SWAP gates as follows:

$$C(Z) = e^{j\pi Z_1/4}e^{-j\pi Z_2/4}U_{\mathrm{SWAP}}^{1/2}e^{j\pi Z_1/2}U_{\mathrm{SWAP}}^{1/2},$$

$$U_{\mathrm{SWAP}} = \frac{1}{1+j}\begin{bmatrix} 1+j & 0 & 0 & 0 \\ 0 & 1 & j & 0 \\ 0 & j & 1 & 0 \\ 0 & 0 & 0 & 1+j \end{bmatrix}. \tag{13.159}$$

If the CNOT gate is used instead of the controlled-Z gate, the transformation given by Eq. (13.39) should be used.

13.9 SUMMARY

This chapter was devoted to several promising physical implementations of quantum information processing. The introductory section provided di Vincenzo criteria and an overview of physical implementation concepts. Section 13.2 was related to NMR implementation. In Section 13.3, the use of ion traps in quantum computing was described. Section 13.4 was devoted to various photonic implementations, including bulky optics implementation (Subsection 13.4.1), and integrated optics implementation (Subsection 13.4.2). Section 13.5 was related to quantum relay implementation. The implementation of quantum encoders and decoders was discussed in Section 13.6. Furthermore, the implementation of quantum computing based on optical cavity electrodynamics was discussed in Section 13.7. The use of quantum dots in quantum computing was discussed in Section 13.8. In the following section, several interesting problems are provided for self-study.

13.10 PROBLEMS

1. Prove that the circuit shown in Fig. 13.P1 for dual-rail representation operates as a CNOT gate.

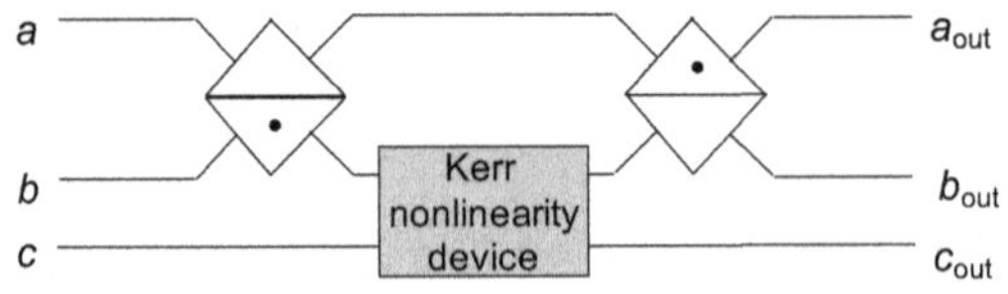

FIGURE 13.P1

2. Using quantum-mechanical arguments, prove that the circuit shown in Fig. 13.P2 for photon polarization states behaves as the CNOT gate.

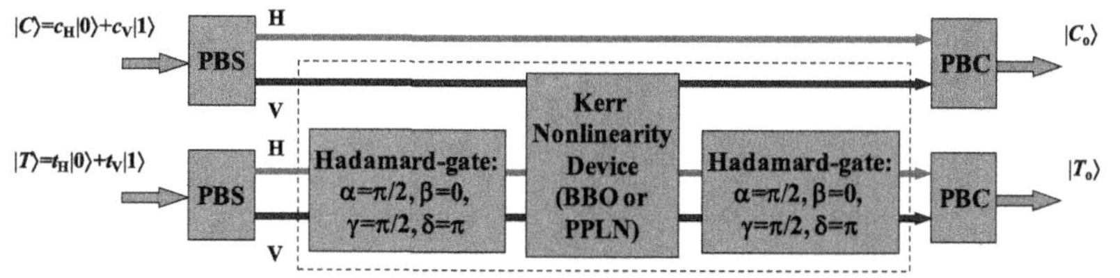

FIGURE 13.P2

BBO, Beta-barium borate; *PBC*, polarization beam combiner; *PBS*, polarization beam splitter; *PPLN*, periodically poled $LiNbO_3$

3. The Hadamard gate can be implemented in integrated optics with two-mode waveguides as shown in Fig. 13.P3 By using quantum-mechanical interpretation, prove this claim. By using this two-mode technology, describe how the following universal set of quantum gates can be implemented: $\{H,\ S,\ \pi/8,$ controlled-$Z\}$.

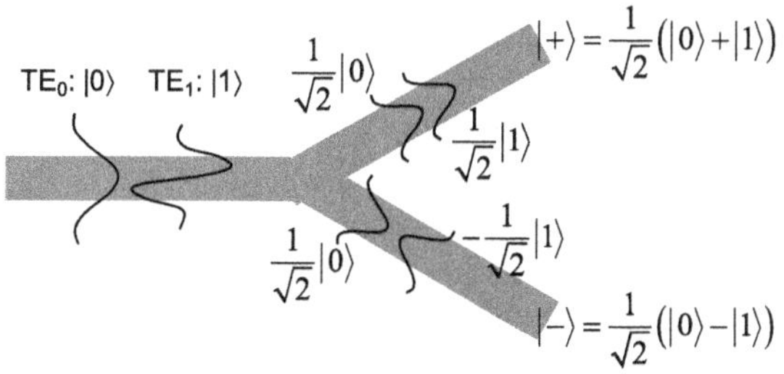

FIGURE 13.P3

4. Determine the energy eigenkets of the Hamiltonian given by:

$$\hat{H} \doteq \begin{bmatrix} \hbar\omega_0 & \hbar\Omega_0 \\ \hbar\Omega_0 & \hbar\omega_a \end{bmatrix}.$$

5. The Hamiltonian used in CQED can be represented by:

$$\hat{H} = \underbrace{\hbar\omega_a\sigma_{ee} + \hbar\omega_0 a^\dagger a}_{\hat{H}_0} \underbrace{-\hbar\Omega_0\left(\sigma_{ge}a + \sigma_{ge}a^\dagger + \sigma_{eg}a + \sigma_{eg}a^\dagger\right)}_{\hat{H}_{\text{int}}},$$

where corresponding operators are defined in Section 13.7. In the interaction picture, the Hamiltonian can be represented as:

$$\hat{H}_{\text{int}}^{(I)} = e^{j\hat{H}_0 t/\hbar}\hat{H}_{\text{int}}e^{-j\hat{H}_0 t/\hbar}.$$

By using the rotating-wave approximation, prove that the interaction Hamiltonian can be approximated as:

$$\hat{H}_{\text{int}} \simeq -\hbar\Omega_0\left(\sigma_{ge}a + \sigma_{eg}a^\dagger\right).$$

6. Let the Hamiltonian be given by:

$$\widetilde{H} = \frac{\hbar}{2}\delta Z - \frac{\hbar}{2}\omega_1 X, \ \ \delta = \omega - \omega_0.$$

Show that $\exp\left(-j\widetilde{H}t/\hbar\right)$ can be written as:

$$\exp\left(-j\widetilde{H}t/\hbar\right) = \exp[-j\Omega t(\delta Z/\Omega - \omega_1 X/\Omega)/2], \ \ \Omega = \sqrt{\delta^2 + \omega_1^2}.$$

Show further that $\exp\left(-j\widetilde{H}t/\hbar\right)$ can be rewritten as:

$$\exp\left(-j\widetilde{H}t/\hbar\right) = [\cos(\Omega t/2) - j(\delta/\Omega)\sin(\Omega t/2)]|0\rangle\langle 0|$$

$$+ j(\omega_1/\Omega)\sin(\Omega t/2)(|0\rangle\langle 1| + |1\rangle\langle 0|) + [\cos(\Omega t/2) + j(\delta/\Omega)\sin(\Omega t/2)]|1\rangle\langle 1|.$$

7. The Hamiltonian for a CQED system can be written as:

$$\widehat{H} = \hbar\omega_a\sigma_{ee} + \hbar\omega_0 a^\dagger a - \hbar\Omega_0\left(\sigma_{eg}a + \sigma_{ge}a^\dagger\right),$$

which possesses the constant of motion:

$$N = \langle\sigma_{ee}\rangle + \langle a^\dagger a\rangle.$$

The state with $N = n$ contains only two states $|n-1, e\rangle$ and $|n, g\rangle$, which are the eigenkets of the uncoupled system described by Hamiltonian $\widehat{H}_0 = \hbar\omega_a\sigma_{ee} + \hbar\omega_0 a^\dagger a$.For strong atom–field coupling, these states no longer represent a proper description of an interacting atom–field system. We can decompose the Hamiltonian matrix into 2 × 2 submatrices. The nth submatrix is spanned by $|n-1, e\rangle$ and $|n, g\rangle$. Provide its matrix representation. Determine the energy eigenvalues and corresponding eigenkets. Determine also how the initial state $|e, 0\rangle$ evolves in time. What would be the probability that a photon will be emitted in an empty cavity?

8. Provide the sequence of operations to be performed using ion trap technology to implement the Toffoli gate and controlled-Z gate, which is controlled with two qubits.
9. Provide the sequence of operation to be performed using NMR to implement four-qubit GHZ states.
10. Let us observe the quantum computation based on single photons and assume the following representation of states: $|0\rangle = |0\ldots 0\rangle$, $|1\rangle = |0\ldots 1\rangle$, $|2\rangle = |0\ldots 10\rangle$, …, $|2^n - 1\rangle = |11\ldots 1\rangle$. Show that arbitrary uniform transformation can be implemented by using only beam splitters and phase shifters without any Kerr nonlinearity medium.

References

[1] A. Barenco, A universal two-bit quantum computation, Proc. Roy. Soc. Lond. A 449 (1937) (1995) 679–683.
[2] D. Deutsch, Quantum computational networks, Proc. Roy. Soc. Lond. A 425 (1868) (1989) 73–90.
[3] L.M.K. Vandersypen, M. Steffen, G. Breyta, C.S. Yannoni, M.H. Sherwood, I.L. Chuang, Experimental realization of Shor's quantum factoring algorithm using nuclear magnetic resonance, Nature 414 (2001) 883–887.
[4] D.G. Cory, N. Boulant, G. Boutis, E. Fortunato, M. Pravia, Y. Sharf, R. Laflamme, E. Knill, R. Martinez, C. Negrevergne, W.H. Zurek, T.F. Havel, L. Viola, S. Lloyd, Y.S. Weinstein, G. Teklemariam, NMR based quantum information processing: achievements and prospects, in: S.L. Braunstein, H.-K. Lo, P. Kok (Eds.), Scalable Quantum Computers: Paving the Way to Realization, Wiley-VCH Verlag GmbH & Co. KGaA, Weinheim, FRG, 2005, https://doi.org/10.1002/3527603182.ch8.
[5] S.J. Glaser, NMR quantum computing, Angew. Chem. Int. Ed. 40 (1) (2001) 147–149.
[6] L.M.K. Vandersypen, I. Chuang, NMR techniques for quantum control and computation, Rev. Mod. Phys. 76 (4) (2004) 1037–1069.
[7] M.G. Raizen, J.M. Gilligan, J.C. Bergquist, W.M. Itano, D.J. Wineland, Ionic crystals in a linear Paul trap, Phys. Rev. A 45 (1992) 6493.
[8] J.I. Cirac, P. Zoller, Quantum computations with cold trapped ions, Phys. Rev. Lett. 74 (1995) 4091.
[9] A. Steane, The ion trap quantum information processor, Appl. Phys. B 64 (1997) 623.
[10] G.P. Agrawal, Lightwave Technology: Components and Devices, John Wiley & Sons, Inc., 2004.
[11] Acousto-Optic Components, Brimrose Co. Available at: http://www.brimrose.com/
[12] I.B. Djordjevic, R. Varrazza, M. Hill, S. Yu, Packet switching performance at 10 Gb/s across a 4x4 optical crosspoint switch matrix, IEEE Photon. Technol. Lett. 16 (2004) 102–104.
[13] R. Varrazza, I.B. Djordjevic, S. Yu, Active vertical-coupler based optical crosspoint switch matrix for optical packet-switching applications, IEEE/OSA J. Lightwave Technol. 22 (2004) 2034–2042.
[14] R. Varrazza, I.B. Djordjevic, M. Hill, S. Yu, 4x4 optical crosspoint packet switch matrix with minimized path-dependent optical gain, Opt. Lett. 28 (22) (2003) 2252–2254 (umina).
[15] M.A. Neilsen, I.L. Chuang, Quantum Computation and Quantum Information, Cambridge University Press, Cambridge, 2000.
[16] J.L. O'Brien, Optical quantum computing, Science 318 (2007) 1567–1570.
[17] I.B. Djordjevic, Photonic quantum dual-containing LDPC encoders and decoders, IEEE Photon. Technol. Lett. 21 (13) (2009) 842–844.
[18] T.M. Babinec, B.J.M. Hausmann, M. Khan, Y. Zhang, J.R. Maze, P.R. Hemmer, M. Loncar, A diamond nanowire single-photon source, Nat. Nanotechnol. 5 (2010) 195–199.
[19] I. Friedler, C. Sauvan, J.P. Hugonin, P. Lalanne, J. Claudon, J.M. Gérard, Solid-state single photon sources: the nanowire antenna, Optic Express 17 (2009) 2095–2110.
[20] B. Lounis, W.E. Moerner, Single photons on demand from a single molecule at room temperature, Nature 407 (2000) 491–493.

[21] A. Beveratos, R. Brouri, T. Gacoin, A. Villing, J.-P. Poizat, P. Grangier, Single photon quantum cryptography, Phys. Rev. Lett. 89 (2002) 187901.
[22] S. Radic, C.J. McKinistrie, Optical amplification and signal processing in highly nonlinear optical fiber, IEICE Trans. Electron. E88-C (2005) 859–869.
[23] S.L. Jansen, D. van den Borne, P.M. Krummrich, S. Spälter, G.-D. Khoe, H. de Waardt, Long-haul DWDM transmission systems employing optical phase conjugation, IEEE J. Sel. Top. Quant. Electron. 12 (2006) 505–520.
[24] M. Cvijetic, Optical Transmission Systems Engineering, Artech House, Inc., 2004.
[25] R.H. Hadfield, Single-photon detectors for optical quantum information applications, Nat. Photon. 3 (2009) 696–705.
[26] G.N. Gol'tsman, O. Okunev, G. Chulkova, A. Lipatov, A. Semenov, K. Smirnov, B. Voronov, A. Dzardanov, C. Williams, R. Sobolewski, Picosecond superconducting single-photon optical detector, Appl. Phys. Lett. 79 (6) (2001) 705.
[27] E.J. Gansen, M.A. Rowe, D. Rosenberg, M. Greene, T.E. Harvey, M.Y. Su, R.H. Hadfield, S.W. Nam, R.P. Mirin, Single-photon detection using a semiconductor quantum dot, optically gated, field-effect transistor, in: Conference on Lasers and Electro-Optics/Quantum Electronics and Laser Science Conference and Photonic Applications Systems Technologies, Technical Digest (CD), Optical Society of America, 2006 paper JTuF4.
[28] S. Komiyama, Single-photon detectors in the terahertz range, IEEE Sel. Top. Quant. Electron. 17 (1) (2011) 54–66.
[29] I.B. Djordjevic, On the photonic implementation of universal quantum gates, Bell states preparation circuit, quantum relay and quantum LDPC encoders and decoders, IEEE Photon. J. 2 (1) (2010) 81–91.
[30] I.B. Djordjevic, Photonic entanglement-assisted quantum low-density parity-check encoders and decoders, Opt. Lett. 35 (9) (2010) 1464–1466.
[31] I.B. Djordjevic, Photonic implementation of quantum relay and encoders/decoders for sparse-graph quantum codes based on optical hybrid, IEEE Photon. Technol. Lett. 22 (19) (2010) 1449–1451.
[32] A. Politi, M. Cryan, J. Rarity, S. Yu, J.L. O'Brien, Silica-on-silicon waveguide quantum circuits, Science 320 (2008) 646.
[33] G.J. Milburn, Quantum optical Fredking gate, Phys. Rev. Lett. 62 (1988) 2124.
[34] T. Brun, I. Devetak, M.-H. Hsieh, Correcting quantum errors with entanglement, Science 314 (2006) 436–439.
[35] D.J.C. MacKay, G. Mitchison, P.L. McFadden, Sparse-graph codes for quantum error correction, IEEE Trans. Inf. Theor. 50 (2004) 2315–2330.
[36] I.B. Djordjevic, Quantum LDPC codes from balanced incomplete block designs, IEEE Commun. Lett. 12 (2008) 389–391.
[37] D. Gottesman, Stabilizer Codes and Quantum Error Correction (Ph.D. Dissertation), California Institute of Technology, Pasadena, CA, 1997.
[38] I. Anderson, Combinatorial Designs and Tournaments, Oxford University Press, 1997.
[39] I.B. Djordjevic, S. Sankaranarayanan, B. Vasic, Projective plane iteratively decodable block codes for WDM high-speed long-haul transmission systems, IEEE/OSA J. Lightwave Technol. 22 (2004) 695–702.
[40] S. Sankaranarayanan, I.B. Djordjevic, B. Vasic, Iteratively decodable codes on *m*-flats for WDM high-speed long-haul transmission, IEEE/OSA J. Lightwave Technol. 23 (2005) 3696–3701.

[41] I. B. Djordjevic, Cavity quantum electrodynamics (CQED) based quantum LDPC encoders and decoders, IEEE Photon. J., (accepted for publication).
[42] Q.A. Turchette, C.J. Hood, W. Lange, H. Mabuchi, H.J. Kimble, Measurement of conditional phase shifts for quantum logic, Phys. Rev. Lett. 75 (1995) 4710–4713.
[43] M.S. Zubairy, M. Kim, M.O. Scully, Cavity-QED-based quantum phase gate, Phys. Rev. A 68 (2003) 033820.
[44] C.-H. Su, A.D. Greentree, W.J. Munro, K. Nemoto, L.C.L. Hollenberg, High-speed quantum gates with cavity quantum electrodynamics, Phys. Rev. A 78 (2008) 062336.
[45] C. Bonato, F. Haupt, S.S.R. Oemrawsingh, J. Gudat, D. Ding, M.P. van Exter, D. Bouwmeester, CNOT and Bell-state analysis in the weak-coupling cavity QED regime, Phys. Rev. Lett. 104 (2010) 160503.
[46] M.O. Scully, M.S. Zubairy, Quantum Optics, Cambridge University Press, 1997.
[47] I. Fushman, D. Englund, A. Faraon, N. Stoltz, P. Petroff, J. Vuckovic, Controlled phase shifts with a single quantum dot, Science 320 (5877) (2008) 769–772.
[48] A. Faraon, A. Majumdar, D. Englund, E. Kim, M. Bajcsy, J. Vuckovic, Integrated quantum optical networks based on quantum dots and photonic crystals, New J. Phys. 13 (2011) paper no. 055025.
[49] K.L. Moore, Ultracold Atoms, Circular Waveguides, and Cavity QED with Millimeter-Scale Electromagnetic Traps (Ph.D. Dissertation), University of California, Berkley, 2007.
[50] M. Le Bellac, An Introduction to Quantum Information and Quantum Computation, Cambridge University Press, 2006.
[51] T.C. Ralph, N.K. Langford, T.B. Bell, A.G. White, Linear optical controlled-NOT gate in the coincidence basis, Phys. Rev. A 65 (2002) 062324.

Further reading

[1] D.P. DiVincenzo, The physical implementation of quantum computation, Fortschr. Phys. 48 (2000) 771.
[2] G. Jaeger, Quantum Information: An Overview, Springer, 2007.
[3] D. Petz, Quantum Information Theory and Quantum Statistics, Theoretical and Mathematical Physics, Springer, Berlin, 2008.
[4] P. Lambropoulos, D. Petrosyan, Fundamentals of Quantum Optics and Quantum Information, Springer-Verlag, Berlin, 2007.
[5] G. Johnson, A Shortcut Thought Time: The Path to the Quantum Computer, Knopf, N. York, 2003.
[6] J. Preskill, Quantum Computing, 1999. Available at: http://www.theory.caltech.edu/~preskill/.
[7] J. Stolze, D. Suter, Quantum Computing, Wiley, N. York, 2004.
[8] R. Landauer, Information is physical, Phys. Today 44 (5) (1991) 23–29.
[9] R. Landauer, The physical nature of information, Phys. Lett. 217 (1991) 188–193.
[10] D.P. DiVincenzo, Two-bit gates are universal for quantum computation, Phys. Rev. A 51 (2) (1995) 1015–1022.
[11] A. Barenco, C.H. Bennett, R. Cleve, D.P. DiVincenzo, N. Margolus, P. Shor, T. Sleator, J.A. Smolin, H. Weinfurter, Elementary gates for quantum computation, Phys. Rev. A 52 (5) (1995) 3457–3467.

CHAPTER

Quantum Machine Learning

14

CHAPTER OUTLINE

Quantum Information Processing, Quantum Computing, and Quantum Error Correction
https://doi.org/10.1016/B978-0-12-821982-9.00007-1

14.1 MACHINE LEARNING FUNDAMENTALS

Machine learning (ML) is a multidisciplinary science (employing artificial intelligence, computer science, and statistics) that studies different algorithms to extract meaningful information from available data and provide automated solutions to complex computational problems [1–3]. The ML algorithms' strength lies in their ability to learn from available data and as such ML is data driven rather than model driven. Various learning models can be grouped into three generic categories:

- *Supervised learning*, in which the ML algorithm employs labeled data samples. In other words, we provide the machine with a dataset together with the correct answer (label) to each data point and expect the ML algorithm to determine the key characteristics, so that the next time the unknown data point is provided, the algorithm will be able to predict correctly the corresponding outcome. Classification and regression problems belong to this category. In the *classification* problem, the machine should be able to classify the data sequence into several distinct categories (classes). In the *regression* problem, the machine should be able to correctly predict the values of a continuous variable. The most famous algorithms under this category include *K*-nearest neighbors (*K*-NNs), support vector machines (SVMs), artificial neural networks (ANNs), decision threes, random forest, linear regression, and logistic regression, to mention a few.
- *Unsupervised learning*, in which a sequence of data is given to the machine and it is expected to determine the structure in the data. For instance, in clustering, the algorithm must find the data points of similar type and group them into different clusters. The clustering algorithm can be used to implement the spam filter, which will be able to identify which e-mail represents junk e-mail. Well-known algorithms in this category include principal component analysis, *K*-means clustering, and hierarchical clustering, to mention a few.
- *Reinforcement learning* (RL), in which the machine needs to sense the environment and based on its current state and environment choose the appropriate action. RL is applicable in time-varying conditions, when the external situation changes all the time. It is also applicable in solving problems for which the state space is huge. RL is popular in data processing, industrial automation, and developing training systems.

14.1.1 Machine Learning Basics

The key idea behind *supervised learning* is to use the label training data to determine the predictive model, which will allow us to make accurate predictions about future data, as illustrated in Fig. 14.1. As an illustration, the filtering spam problem can also be solved by supervised learning. In this example, we will be sending a series of e-mails to the ML algorithm that are properly labeled as spam or not spam. The ML algorithm will build a predictor model, which will be able properly to classify an unknown e-mail as spam or not spam.

A supervised classification problem with discrete labels is commonly referred to as the *classification task*. In the *regression task*, on the other hand, the predictor is able to generate the continuous outcome signal. In the classification task we are concerned with the determination of a category to which an input belongs, among possible alternatives. The *binary classifier* is concerned with determination of the decision boundary to partition the corresponding vector space into two classes. In a *multiclass classifier* problem we are concerned with the partition of the input vector space into multiple sets of vectors, with each set representing a different class. In the binary classification example provided in Fig. 14.2, seven training examples are labeled with the negative sign, while the other seven are labeled with the positive sign. This binary classification problem is clearly 2D. We can use the supervised ML algorithm to determine the decision boundary (the classification rule), represented by the line with the positive slope. By using two independent binary classifiers we can create the multiclass classifier with four labels. In certain situations, the decision boundary might be *nonlinear*, and in this case SVMs can be used, which belong to the class of supervised learning algorithms. The key idea behind SVMs is to construct the hyperplane in a larger, high-dimensional (HD) hyperspace such that the margin separation between two classes, in a binary classification problem, is maximized, as illustrated in Fig. 14.3. In this HD hyperspace, each datum is represented as a point, and the purpose of the SVM is to construct the hyperplane maximizing the separation between the points from two classes, wherein the separation is defined in terms of the distance between two margins. Therefore we have avoided the nonlinear boundary problem by mapping it into HD hyperspace, where there is more space and it is easier to determine the hyperplane (the dimensionality

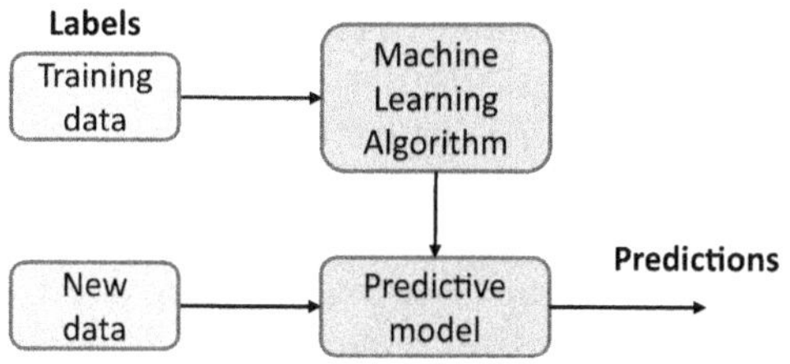

FIGURE 14.1

Supervised learning.

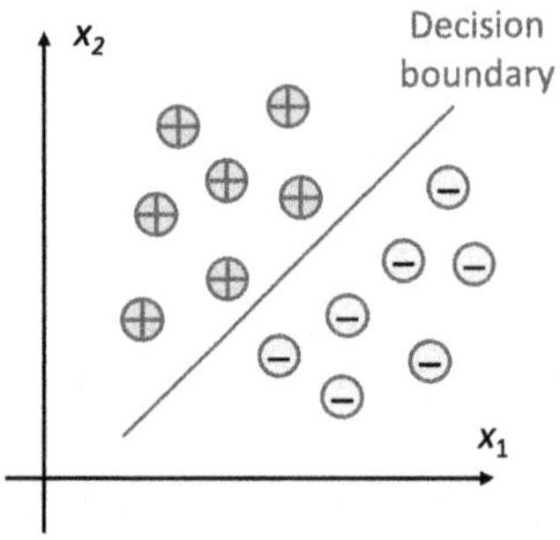

FIGURE 14.2

Binary classification problem.

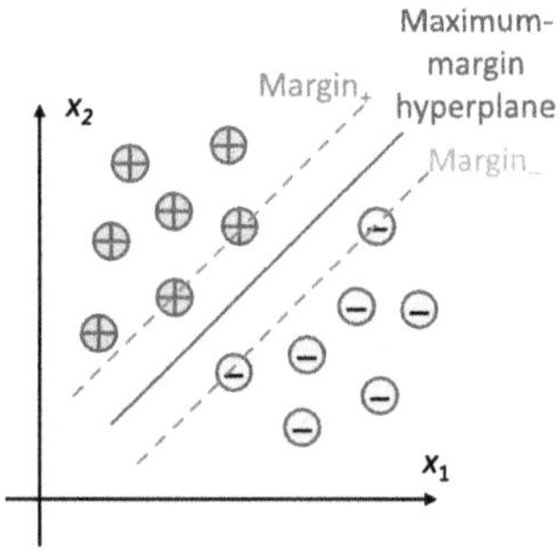

FIGURE 14.3

Optimum margin classifier maximizing the separation between two different classes, wherein the separation is defined as the distance between two margins.

of the hyperplane is one dimension smaller than that of the hyperspace). The SVM is applicable to *linear separable problems* as well, for which there exists the hyperplane that clearly separates the points belonging to two different classes, and in this case the transformation to HD space is not needed.

In the *regression* problem we are concerned with determining the approximate function, given the finite number of training examples, that can accurately predict an outcome. Contrary to the multitask classification problem, where the outcome is discrete, the outcome in regression is continuous. In ML, the predictor variables are commonly referred to as *features*, while the response variables are referred to as the *target* variables, and in this chapter we will adopt the same terminology. In a *linear regression* problem, given a feature variable x and a target variable y, we are interested in drawing a straight line, minimizing the average squared distance between the data points and fitted line, as illustrated in Fig. 14.4.

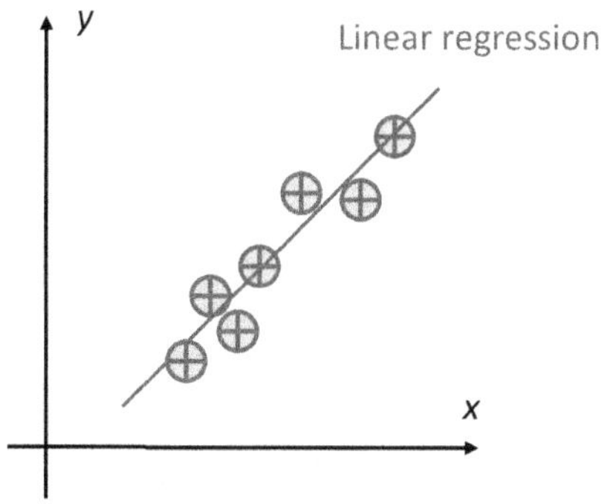

FIGURE 14.4

Linear regression problem.

In *reinforcement learning* we are interested in developing an *agent* (system) that improves its performance and accuracy through interactions with the environment, as illustrated in Fig. 14.5. Compared to supervised learning, the feedback is not the label or value, but how well the action was measured by the reward function. By interacting with the environment, the agent can learn different strategies to maximize the reward. The state of the environment is related to the positive or negative reward, and the reward could be the overall goal. For instance, in chess the reward could be the winning of the game, while the state can be the outcome of each move. The agent in this example is the chess engine, which decides the series of moves based on the state of the board (the environment).

In unsupervised learning we are dealing with the unknown structure or unlabeled data sequence, and we are trying to explore the data structure to extract the meaningful information. In *clustering*, also known as unsupervised classification, we perform exploratory data analysis so that we can group data points into meaningful clusters without having any prior knowledge of the membership to the cluster. The cluster represents a group of data points that share a certain degree of similarity, but at the same time are dissimilar from the members in other classes. The popular unsupervised learning algorithm from this class is the *K-means* algorithm, in which we partition the data points into K classes based on the pairwise distances, with one particular example provided in Fig. 14.6.

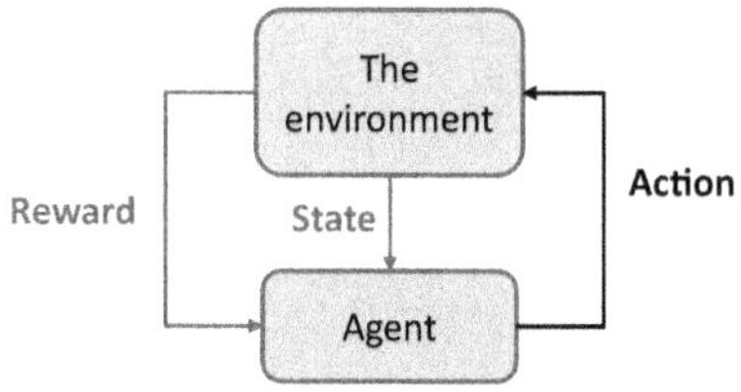

FIGURE 14.5

Reinforcement learning problem.

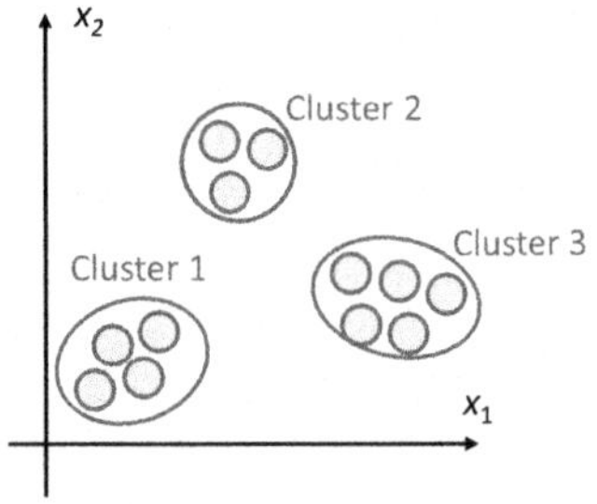

FIGURE 14.6

Clustering problem.

Very often we are faced with data of large dimensionality, or in other words, each observation comes with a large number of dimensions. In the preprocessing stage we often try to reduce the noise from the data and thus improve the predictive performance of the corresponding algorithm, while at the same time compressing the data onto a smaller dimensional subspace. The corresponding unsupervised algorithm is known as the *dimensionality reduction* algorithm. Dimensionality reduction is also very useful in better visualization by projecting to the lower dimensionality space. A very popular algorithm from this class is *principal component analysis* (PCA), which is able to reveal the geometrical structure that is not that obvious. When the ML algorithm does not have any idea of the data structure, it must figure out the structure on its own, and the intuitive way will be to determine the eigenvectors of the available data to get a geometric insight into prevailing directions in the predictive space. PCA builds on this idea by transforming the original ("natural") basis into a lower-dimensionality basis, better aligned with the available data, which is illustrated in Fig. 14.7. Directions in the new basis are related to the actual data structure, while the length of the basis vector is dictated by the variance in that direction. The basis vectors with small variance can be discarded, and thus we can

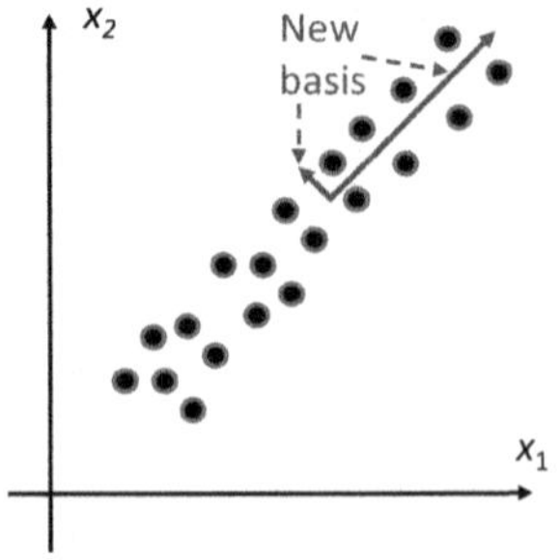

FIGURE 14.7

Principal component analysis concept.

reduce the dimensionality. In this particular example we reduced 2D data to 1D data. In PCA we organize data in terms of matrix $\boldsymbol{X} = (x_{ij})_{m \times n}$, where n is the number of experiments and m denotes the number of measurement values for each experiment. Therefore each row represents measurements of the same type. By removing the mean value for each row we can determine the *covariance matrix* by $\boldsymbol{C} = (1/n)\boldsymbol{X}\boldsymbol{X}^{\mathrm{T}}$. To determine the principal component vectors we need to diagonalize the covariance matrix. PCA employs the following steps during the diagonalization procedure [1]: (1) choose the direction of the maximum variance, (2) out of available directions orthogonal to the previous basis vectors, select one with the largest variance, and (3) repeat step (2) until we run out of dimensions. The resulting basis vectors will be the *principal components*. Additional details on PCA will be provided in the next subsection.

The ML predictive modeling workflow is composed of the following stages [2]: preprocessing, learning, evaluation, and prediction, as summarized in Fig. 14.8. The raw data rarely come in the form suitable for use in ML, and preprocessing represents a crucial step in ML. We can interpret the raw data as a series of objects from which we want to extract relevant features. The *feature* can be any measurable property of the phenomenon under study. Properly selected features are discriminating, and they help the ML algorithm identify different patterns and distinguish different data instances. For optimum performance of the underlying ML algorithms, the features must be on the same scale. We often use a geometric approach, and group different features into a *feature vector*. The feature vectors leave in HD *feature space*. Some of the features might be highly correlated, and we need to employ dimensionality reduction techniques. To evaluate the performance of the ML algorithm we randomly split the available dataset into training dataset and test

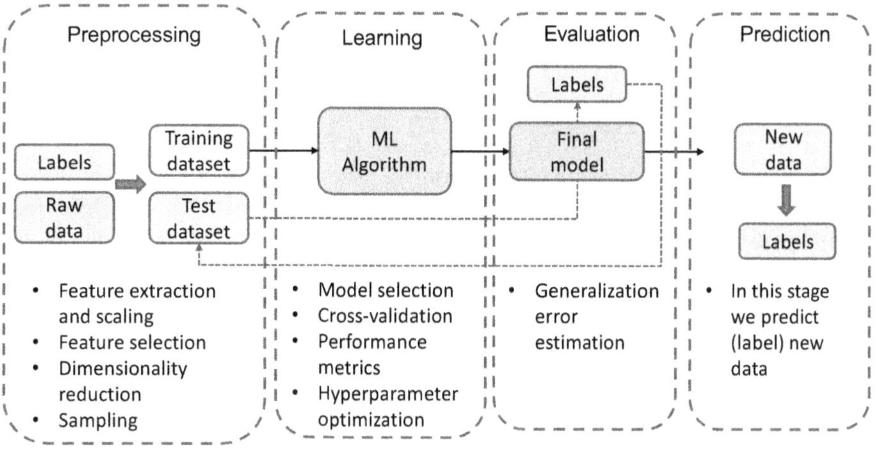

FIGURE 14.8

Roadmap for machine learning (ML).

dataset. The *training dataset* is used to train and optimize the ML model, while the *test dataset* is used to evaluate and finalize the ML model. To estimate how well the predictor model works on unseen data we evaluate a so-called generalization (test) error, and if we are happy with the performance of the predictor we use it to predict new/future data.

The feature vector with d features can be represented as the d-dimensional column vector:

$$\boldsymbol{x}_j = \sum_{i=1}^{d} x_{ij}\boldsymbol{e}_i = \begin{bmatrix} x_{1j} \\ x_{2j} \\ \vdots \\ x_{dj} \end{bmatrix}, \tag{14.1}$$

where the ith basis vector $\boldsymbol{e}_i$ corresponds to the ith feature. The index j corresponds to the jth data instance. In the design of ML algorithms, the distance functions, norms, and angles between the feature vectors are commonly used. As an illustration, the *Euclidian distance* between the jth and kth data instances is defined by:

$$d(\boldsymbol{x}_j, \boldsymbol{x}_k) = \sqrt{\sum_{i=1}^{d} (x_{ij} - x_{ik})^2}. \tag{14.2}$$

The cosine angle between two vectors is known as *cosine similarity*, defined as:

$$\cos(\boldsymbol{x}_j, \boldsymbol{x}_k) = \frac{\boldsymbol{x}_j^{\mathrm{T}}\boldsymbol{x}_k}{\|\boldsymbol{x}_j\|\|\boldsymbol{x}_k\|}, \tag{14.3}$$

where $\boldsymbol{x}_j^{\mathrm{T}}\boldsymbol{x}_k$ denotes the dot product $\boldsymbol{x}_j^{\mathrm{T}}\boldsymbol{x}_k = \sum_{i=1}^{d} x_{ij}x_{ik}$. The *Hamming distance* between two vectors is defined as the number of positions in which they differ, that is:

$$d_H(\boldsymbol{x}_j, \boldsymbol{x}_k) = \sum_{i=1}^{d} x_{ij} \oplus x_{ik}, \tag{14.4}$$

where we use $\oplus$ to denote the mod-2 addition.

To characterize the learner's prediction efficiency on unknown data, generalization performance is used. If we denote the label by y, the training set can be represented as the collection of pairs of data points with corresponding labels; in other words, we can write $\{(\boldsymbol{x}_1, y_1), (\boldsymbol{x}_2, y_2), \ldots, (\boldsymbol{x}_n, y_n)\}$. We are interested in the family of functions, denoted as f, which approximates well the function $g(\boldsymbol{x}) = y$ that generates the data based on a sample, which suffers from the random noise of zero mean and the variance of σ^2. We can define different loss functions $L(y, f(\boldsymbol{x}))$ depending on whether the value of y is discrete or continuous. If the response y is continuous, such as in a regression model, we can use the *squared error* as the loss function:

$$L(y_j,f(x_j)) = [y_j - f(x_j)]^2. \tag{14.5}$$

The *absolute error* can also be used instead of the squared error, defined as:

$$L(y_j,f(\boldsymbol{x}_j)) = |y_j - f(\boldsymbol{x}_j)|. \tag{14.6}$$

In binary classification problems we can use the *0 − 1 loss function*, defined as:

$$L(y_j, f(\boldsymbol{x}_j)) = \mathscr{I}(y_j \neq f(\boldsymbol{x}_j)), \ \mathscr{I}(e) = \begin{cases} 1, \text{ if } e \text{ is true} \\ 0, \text{ if } e \text{ is true} \end{cases}, \tag{14.7}$$

where we use $\mathscr{I}$ to denote the indicator function. Clearly, y_j is the true label, while $f(\boldsymbol{x}_j)$ is the predicted label. Unfortunately, as shown by Feldman et al. [4,5], the classification problem optimization based on the 0 − 1 loss function is NP-hard, and very often we use the convex approximation instead, such as the *hinge loss*, defined as:

$$L(y_j,f(\boldsymbol{x}_j)) = \max(0, 1 - f(\boldsymbol{x}_j)y), \tag{14.8}$$

whose range is the set of real numbers $\mathscr{R}$. Once the loss function is defined we can introduce the *training error*, also known as the *empirical risk*, as follows:

$$E = \frac{1}{n}\sum_{j=1}^{n} L(y_j,f(\boldsymbol{x}_j)). \tag{14.9}$$

To measure how accurately an ML algorithm is able to predict outcomes of predictor for previously unseen data we use the *generalization error* (GE). The GE can be defined as:

$$E_{GE} = \int_{\boldsymbol{x} \in \mathbf{X}, y \in \mathrm{Y}} L(y,f(\boldsymbol{x}))p(y,\boldsymbol{x})d\boldsymbol{x}dy, \tag{14.10}$$

where $\boldsymbol{x}$ is the feature vector from the feature space $\boldsymbol{X}$, and $p(y,\boldsymbol{x})$ is the joint distribution of $\boldsymbol{x}$ and y. Typically, the training error is lower than the GE. Unfortunately, this distribution is not known, and we cannot compute the GE. However, we can estimate it by calculating the *test error* (TE) by using the testing dataset we left aside, as follows:

$$E_{TE} = \frac{1}{n_{TE}}\sum_{j=1}^{n_{TE}} L(y_j,f(\boldsymbol{x}_j)), \tag{14.11}$$

where averaging is performed over the testing set with n_{TE} instances. The TE gives an estimate of the GE, and when the testing set represents the unbiased sample from the joint distribution $p(y,\boldsymbol{x})$, in limit it will approach the GE, that is, $\lim_{n_{TE} \to \infty} E_{TE} = E_{GE}$. It might be a good idea to bound the GE in probability sense as follows:

$$\Pr(E_{GE} - E_{TE} \leq \varepsilon) \geq 1 - \delta, \tag{14.12}$$

where the goal is to characterize the probability $1 - \delta$ that the GE is smaller than some error bound ε, which is typically dependent on δ and the number of instances in the test dataset.

14.1.2 Principal Component Analysis

In PCA we organize data in terms of matrix $\boldsymbol{X}' = \left(x'_{ij}\right)_{m \times n}$, as follows:

$$\boldsymbol{X}' = \begin{bmatrix} x'_{11} & x'_{12} & & \cdots & x'_{1n} \\ x'_{21} & x'_{22} & x'_{23} & \cdots & x'_{2n} \\ \vdots & & & & \\ & & & \ddots & \\ x'_{m1} & x'_{m2} & x'_{m3} & \cdots & x'_{mn} \end{bmatrix}, \tag{14.13}$$

where n is the number of experiments and m denotes the number of measurement values for each experiment. Therefore each row represents the measurements of the same type, and we can determine the *mean value* of each row by:

$$\mu_i = \frac{1}{n} \sum_{j=1}^{n} x_{ij};\ i = 1, 2, \ldots, m. \tag{14.14}$$

Now we create new matrix $\boldsymbol{X} = (x_{ij})_{m \times n}$ by removing the mean value for each row, that is, $x_{ij} = x'_{ij} - \mu_i$. Clearly, the mean values of the rows in $\boldsymbol{X}$ are now equal to 0. We further calculate the *covariance matrix* by $\boldsymbol{C} = (1/n)\boldsymbol{X}\boldsymbol{X}^{\mathrm{T}}$. To determine the principal component vectors we need to diagonalize the covariance matrix. Let us look at the example shown in Fig. 14.9.

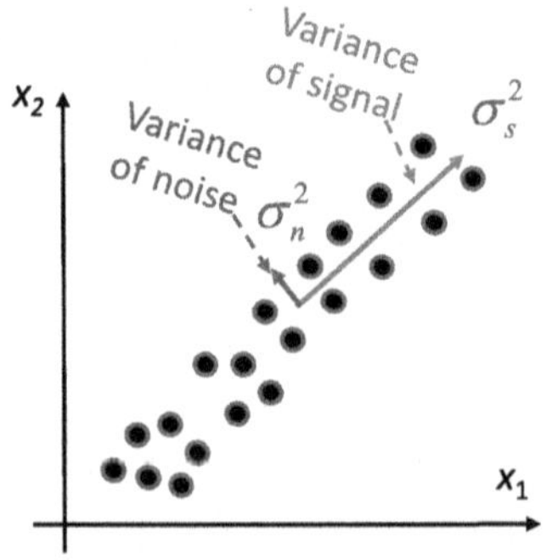

FIGURE 14.9

Principal component analysis. The direction of the largest variance does not coincide with the original ("native") basis.

The coordinates x_1 and x_2 could be interpreted as the height and weight of individuals in a studied population of people. We can clearly identify two dominant (principal) directions, one represented with the green (long arrow) vector of length σ_s^2 (the signal variance) and the second one with the red (short arrow) vector of length σ_n^2 (the noise variance), which is orthogonal to the first one. Evidently, $\sigma_s^2 \gg \sigma_n^2$, or in other words the signal-to-noise ratio (SNR), defined as SNR $= \sigma_s^2/\sigma_n^2$, is much larger than one. We can use these two vectors, well aligned with the data, as the new basis. The direction denoted with the green (long arrow) vector is the dominant one and carries the most significant information, and we can discard the other vector direction, thus reducing the complexity of the problem. What we have essentially done is to reduce the dimensionality of the system while minimizing the information loss. Interpreted differently we translated and rotated the original basis to get the new basis better aligned with the actual data. The length of the vector in a given direction is determined by estimating the spread in the same direction. As we have shown in Chapter 2, this change of the basis can be represented by the unitary matrix U, satisfying the condition $U^{\dagger}U=I$, with I being the identity matrix. (For real-valued problems, the Hermitian transposition $\dagger$ can be replaced by just transposition T.) This change of the base corresponds to the diagonalization of the correlation matrix, which is the essence of the PCA. Intuitively, we can summarize the PCA procedure as follows [1,6]: (1) choose the direction of the maximum variance, (2) out of available directions orthogonal to the previous basis vectors, select one with the largest variance, and (3) repeat step (2) until we run out of dimensions. The resulting basis vectors, obtained after completing the PCA procedure, will be the *principal components*. PCA relies on the following three assumptions:

- *Linearity assumption.* The change of basis is a linear operation, but some problems are inherently nonlinear.
- *Assumption that directions with large variances are the most interesting.* The directions with large variances are considered to belong to the signal vector, while directions with low variances belong to the noise vector. Unfortunately, this is not true for all problems.
- *Orthogonality assumption.* To make the problem easy to solve we assume that the principal components are orthogonal, which is not true for all datasets. As an illustrative example, to explore the city by using the PCA approach, we can make only 90-degree angled turns. This situation is suitable for Phoenix and Tucson, but not for cities with angled (curved) streets such as Washington, DC.

Let us provide two examples when PCA is not beneficial. In Fig. 14.10 we provided two illustrative examples where PCA is not useful. In Fig. 14.10 (left), the dataset is nonorthogonal, and we do not expect PCA to be useful, given that one of the three assumptions discussed earlier is not satisfied. On the other hand, in Fig. 14.10 (right), known as the Ferris wheel dataset, there is no dominant direction. This problem can be solved by moving to the polar coordinate system by the following transformation $x = r\cos\theta$, $y = r\sin\theta$; unfortunately, this is a nonlinear transformation.

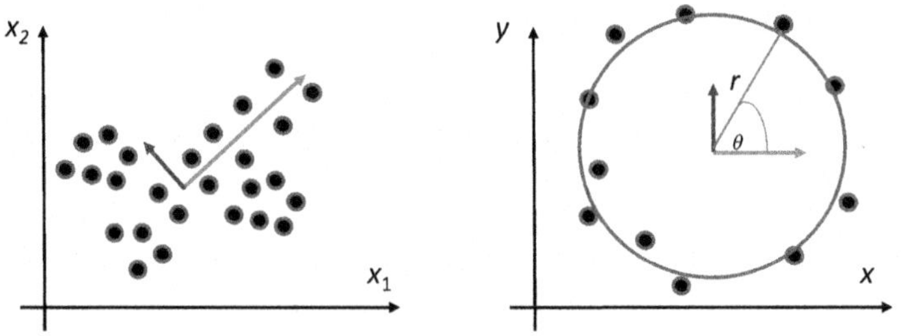

FIGURE 14.10

Datasets when principal component analysis is not beneficial.

Problems in lower-dimensional space, in particular 2D, are easy to solve, and we can apply the best fitting approach. Unfortunately, many problems in ML can have hundreds or even thousands of components. In HD space, PCA can be used to reveal the dataset structure, and reduce the dimensionality of the problem, given that low-variance dimensions can be discarded. By studying the eigenvector structure of the correlation matrix we can make PCA efficient and practical. Given that the correlation matrix $\boldsymbol{C}$ is the square matrix of size $m \times m$ we can perform *spectral (eigenvalue) decomposition* [7–10]:

$$\boldsymbol{C} = \boldsymbol{U}\boldsymbol{D}\boldsymbol{U}^{\dagger}, \tag{14.15}$$

where $\boldsymbol{D}$ is the diagonal matrix of eigenvalues of the correlation matrix $\boldsymbol{C}$, that is, $\boldsymbol{D} = \text{diag}(\lambda_1, \ldots, \lambda_m)$, while $\boldsymbol{U}$ is the unitary matrix with columns corresponding to the orthonormalized eigenvectors of $\boldsymbol{C}$. The eigenvalues of $\boldsymbol{C}$, denoted as λ_i, are solved by solving the *characteristic polynomial*:

$$\det(\boldsymbol{A} - \lambda\boldsymbol{I}) = 0. \tag{14.16}$$

The eigenvectors are now used as the principal components. The eigenvectors with small eigenvalues are discarded from the study, thus effectively performing *dimensionality reduction*. Unfortunately, there are ML problems in which the number of measurements m is too large, in particular in image signal processing problems, when m can represent the number of pixels (which can be in the order of thousands). It can happen that $m >> n$, then the correlation matrix will be huge, and the spectral decomposition procedure is of too high complexity. To avoid this problem, instead of defining the correlation matrix as $\boldsymbol{C} = (1/n)\boldsymbol{X}\boldsymbol{X}^{\mathrm{T}}$, we can redefine it as $\boldsymbol{R} = (1/n)\boldsymbol{X}^{\mathrm{T}}\boldsymbol{X}$, which has the size $n \times n$ (and $n << m$). The eigenvalue λ, eigenvector $\boldsymbol{u}$, and correlation matrix $\boldsymbol{R}$ satisfy the characteristic equation $\boldsymbol{R}\boldsymbol{u} = \lambda\boldsymbol{u}$. If we multiply this equation by matrix $\boldsymbol{X}$ from the left we obtain:

$$\lambda\boldsymbol{X}\boldsymbol{u} = \boldsymbol{X}\underbrace{\boldsymbol{R}}_{(1/n)\boldsymbol{X}^{\mathrm{T}}\boldsymbol{X}}\boldsymbol{u} = \underbrace{\frac{1}{n}\boldsymbol{X}\boldsymbol{X}^{\mathrm{T}}}_{\boldsymbol{C}}\boldsymbol{X}\boldsymbol{u} = \boldsymbol{C}\boldsymbol{X}\boldsymbol{u}, \tag{14.17}$$

which indicates that the λ is also the eigenvalue of $\boldsymbol{C}$, with the corresponding eigenvector being $\boldsymbol{X}\boldsymbol{u}$. Therefore to avoid high complexity of the spectral decomposition of $\boldsymbol{C}$ we perform the spectral decomposition of $\boldsymbol{R}$ instead, and transform eigenvectors $\boldsymbol{u}$ by multiplying them with matrix $\boldsymbol{X}$, from the left, and these new eigenvectors represent the principal directions.

A more general solution to the determination of the principal components can be obtained by *singular value decomposition* (SVD) of the $\boldsymbol{A}$-matrix. In SVD we decompose the $m \times n$ matrix $\boldsymbol{A}$ as follows:

$$\boldsymbol{A} = \boldsymbol{U}\boldsymbol{\Sigma}\boldsymbol{V}^{\dagger}, \tag{14.18}$$

where the $m \times n$ diagonal $\sum$-matrix corresponds to the scaling (stretching) operation and is given by $\sum = \mathrm{diag}(\sigma_1,\sigma_2,\ldots)$, with $\sigma_1 \geq \sigma_2 \geq \ldots \geq 0$ being the *singular values* of $\boldsymbol{A}$. The singular values are in fact the square roots of eigenvalues of $\boldsymbol{A}\boldsymbol{A}^{\dagger}$. The $m \times m$ matrix $\boldsymbol{U}$ and $n \times n$ matrix $\boldsymbol{V}$ are unitary matrices, which describe the rotation operations, as illustrated in Fig. 14.11. The left singular vectors, the columns of unitary matrix $\boldsymbol{U}$, are obtained as the eigenvectors of the Wishart matrix $\boldsymbol{A}\boldsymbol{A}^{\dagger}$. The other rotation matrix $\boldsymbol{V}$ has for columns the eigenvectors of $\boldsymbol{A}^{\dagger}\boldsymbol{A}$ and thus represents the right singular vectors.

PCA by SVD can be summarized as follows:

- Organize the dataset into a real-valued matrix $\boldsymbol{X}$ of size $m \times n$ in which the mean per measurement type (row) has been subtracted.
- Scale this matrix by $\sqrt{n}$ as follows: $\boldsymbol{X}/\sqrt{n} = \boldsymbol{Y}$. Clearly, $\boldsymbol{Y}\boldsymbol{Y}^{\mathrm{T}}$ correspond to the covariance matrix since $\boldsymbol{Y}\boldsymbol{Y}^{\mathrm{T}} = (1/n)\boldsymbol{X}\boldsymbol{X}^{\mathrm{T}} = \boldsymbol{C}$.
- Apply the SVD procedure to $\boldsymbol{Y} = \boldsymbol{X}/\sqrt{n}$ to obtain $\boldsymbol{Y} = \boldsymbol{U}\boldsymbol{\Sigma}\boldsymbol{V}^{\mathrm{T}}$.
- The columns in $\boldsymbol{U}$, denoted by $\boldsymbol{u}_i$ ($i = 1,2,\ldots,m$), are eigenvectors of the covariance matrix $\boldsymbol{C}$ and thus represent the principal components.

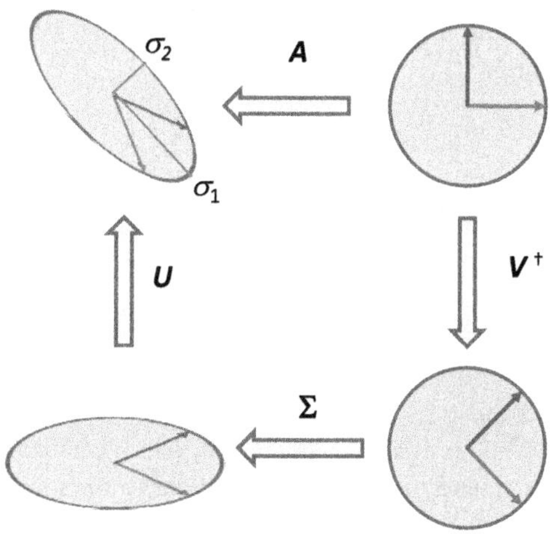

FIGURE 14.11

Singular value decomposition procedure.

- Determine the *scoring matrix* $\boldsymbol{S}$ as follows:

$$\boldsymbol{S} = \boldsymbol{U}^{\mathrm{T}}\boldsymbol{X}. \tag{14.19}$$

The jth column of the scoring matrix is clearly:

$$S_j = \left(\boldsymbol{U}^{\mathrm{T}}\boldsymbol{X}\right)_j = \begin{bmatrix} \boldsymbol{u}_1^{\mathrm{T}} \cdot \boldsymbol{X}_j \\ \boldsymbol{u}_2^{\mathrm{T}} \cdot \boldsymbol{X}_j \\ \vdots \\ \boldsymbol{u}_m^{\mathrm{T}} \cdot \boldsymbol{X}_j \end{bmatrix}, \tag{14.20}$$

with components representing the projections along the principal components and $\boldsymbol{X}_j$ denoting the jth column of matrix-$\boldsymbol{X}$. The rows in $\boldsymbol{S}$ corresponding to small eigenvalues are discarded, and the system dimensionality is reduced to $D < m$.

These steps complete the *training phase*. In the *scoring phase*, for each new data vector $\boldsymbol{y}' = [y'_1\, y'_2 \ldots y'_m]^{\mathrm{T}}$ we subtract the mean value $\mu = (1/n)\Sigma_j y'_j$, to obtain $\boldsymbol{y} = [y_1\, y_2 \ldots y_m]^{\mathrm{T}} = [y'_1 - \mu\; y'_2 - \mu \ldots y'_m - \mu]^{\mathrm{T}}$. The scoring phase has the following steps:

- Calculate the weight (scoring) vector for $\boldsymbol{y}$ vector by:

$$\boldsymbol{S}(\boldsymbol{y}) = \begin{bmatrix} \boldsymbol{u}_1^{\mathrm{T}} \cdot \boldsymbol{y} \\ \boldsymbol{u}_2^{\mathrm{T}} \cdot \boldsymbol{y} \\ \vdots \\ \boldsymbol{u}_D^{\mathrm{T}} \cdot \boldsymbol{y} \end{bmatrix}. \tag{14.21}$$

- Calculate the Euclidean distances of this weight vector from all columns of the scoring matrix, that is:

$$d_j = d(\boldsymbol{S}(\boldsymbol{y}), \boldsymbol{S}_j);\ j = 1, 2, \ldots, n. \tag{14.22}$$

- The score is defined as:

$$\mathrm{score}(y) = \min_j d_j \geq 0. \tag{14.23}$$

Clearly, when we score a vector from the training dataset we will be getting the minimum possible score.

PCA has applications in image compression, data visualization, facial recognition, malware detection, and image spam detection, to mention a few.

14.1.3 Support Vector Machines

The ML algorithms based on SVMs, proposed by Vapnik, Lerner, and Chervonenkis in 1963–64 [11,12], belong to the class of supervised learning algorithms. The key idea behind SVM is to construct a so-called *maximum-margin hyperplane* such that the margin separation between two classes, in a binary classification problem, is maximized, as illustrated in Fig. 14.12 for a 2D case. Let us assume that we are given the dataset of n points as follows:

$$\{(\boldsymbol{x}_1, y_1), (\boldsymbol{x}_2, y_2), \ldots, (\boldsymbol{x}_n, y_n)\}; y_j \in \{-1, +1\}, \boldsymbol{x}_j = \begin{bmatrix} x_{j1} & x_{j2} & \ldots & x_{jd} \end{bmatrix}^{\mathrm{T}}, \tag{14.24}$$

where y_j are labels corresponding to d-dimensional real-valued data points $\boldsymbol{x}_j$. Our goal is to derive the hyperplane that can separate the data points for which the label is $+1$ and from the data points with the label -1. The equation of the hyperplane in d-dimensional hyperspace R^d is given by:

$$\boldsymbol{w}^{\mathrm{T}} \cdot \boldsymbol{x} - b = 0, \ \boldsymbol{w} = \begin{bmatrix} w_1 & w_2 & \ldots & w_d \end{bmatrix}^{\mathrm{T}}, \tag{14.25}$$

where $\boldsymbol{w}$ is the weight vector (not necessarily normalized) orthogonal to the hyperplane, while b is the bias describing the offset from the origin. As shown in Fig. 14.12, the offset from the origin is given by $b/\|\boldsymbol{w}\|$, where $\|\boldsymbol{w}\|$ is the length of the normal vector $\boldsymbol{w}$. When the training dataset is linearly separable we can select two parallel hyperplanes completely separating two classes of data points so that the distance is as large as possible, such as illustrated in Fig. 14.12. The region between these two hyperplanes is known as the *margin*, and the hyperplane intersecting the margin region exactly in the middle is commonly referred to as the maximum-margin hyperplane. The parallel hyperplanes are chosen in such a way that there is no point between them, and we maximize the separation between them. These hyperplanes are described by the following equations in the hyperspace:

$$\begin{aligned} \boldsymbol{w}^{\mathrm{T}} \cdot \boldsymbol{x} - b &= +1 \ (\text{margin}_+) \\ \boldsymbol{w}^{\mathrm{T}} \cdot \boldsymbol{x} - b &= -1 \ (\text{margin}_-). \end{aligned} \tag{14.26}$$

All the data points lying on or above the boundary hyperplane denoted as margin$_+$ belong to the class with the label $y = +1$. On the other hand, all the data points lying on or below the boundary hyperplane denoted as margin$_-$ belong to the class with the label $y = -1$. The data points lying on either of the boundary hyperplanes are known as the *support vectors*. Therefore the linear SVM is the hyperplane that separates data points with positive labels from the data points with negative labels with maximum margin. The separation between boundary hyperplanes is $2/\|\boldsymbol{w}\|$, as illustrated in Fig. 14.12, and by minimizing the length of the normal vector $\|\boldsymbol{w}\|$ we will at the same time maximize the separation between boundary hyperplanes $2/\|\boldsymbol{w}\|$. (If we set $b = 0$ and align the maximum-margin hyperplane with the x_1-axis we conclude that the distance between boundary

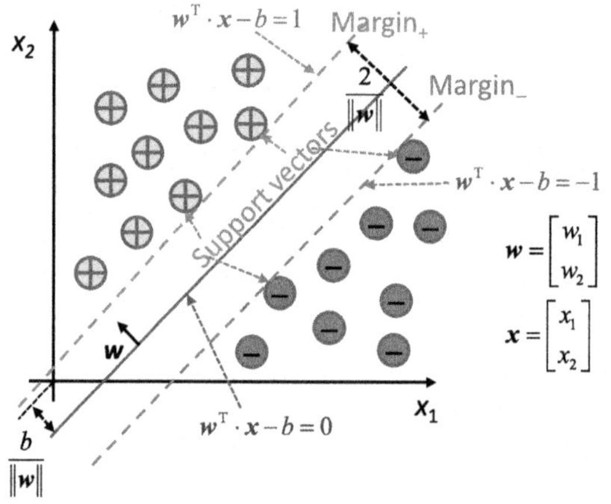

FIGURE 14.12

Defining the optimum margin classifier problem, maximizing the separation between two different classes. The samples located on the margins are known as the support vectors.

hyperplanes is 2, when the normal vector is of unit length.) For the jth training data point $\boldsymbol{x}_j$, the output of linear SVM is determined by:

$$y_j = \text{sign}\left(\boldsymbol{w}^T \cdot \boldsymbol{x}_j - b\right), \ \text{sign}(z) = \begin{cases} +1, & z > 0 \\ -1, & z < 0 \end{cases}. \tag{14.27}$$

To prevent data points from falling within the margin region we impose the following two constraints:

$$\begin{aligned} \boldsymbol{w}^T \cdot \boldsymbol{x}_j - b &\geq +1, \ \text{for } y_j = +1 \\ \boldsymbol{w}^T \cdot \boldsymbol{x}_j - b &\leq -1, \ \text{for } y_j = -1. \end{aligned} \tag{14.28}$$

These two constraints can be combined as follows:

$$y_j\left(\boldsymbol{w}^T \cdot \boldsymbol{x}_j - b\right) \geq 1, \ \text{for } j = 1, 2, \ldots, n. \tag{14.29}$$

The *optimization problem* can now be formulated as:

$$\left(\widehat{\boldsymbol{w}}, \widehat{b}\right) = \underset{\boldsymbol{w}, b}{\arg\min} \|\boldsymbol{w}\|^2, \ \text{subject to: } y_j\left(\boldsymbol{w}^T \cdot \boldsymbol{x}_j - b\right) \geq 1, \ \forall j = 1, 2, \cdots, n. \tag{14.30}$$

This optimization problem has linear complexity in time. When unknown data point $\boldsymbol{x}$ arrives it will be classified by the *linear SVM classifier* as follows:

$$\boldsymbol{x} \rightarrow y = \text{sign}\left(\widehat{\boldsymbol{w}}^T \cdot \boldsymbol{x} - \widehat{b}\right). \tag{14.31}$$

We can also formulate the optimization problem by employing the *Lagrangian method*, which is also known as the *dual* problem, as follows:

$$\left(\widehat{\boldsymbol{w}},\ \widehat{b}\right) = \arg\min_{\boldsymbol{w},\ b}\max_{\lambda_j} \overbrace{\left\{\frac{\|\boldsymbol{w}\|^2}{2} - \sum_{j=1}^{n}\lambda_j\left[y_j\left(\boldsymbol{w}^{\mathrm{T}}\cdot\boldsymbol{x}_j - b\right) - 1\right]\right\}}^{L(\boldsymbol{w},\ b,\ \boldsymbol{\lambda})}$$

$$= \arg\min_{\boldsymbol{w},\ b}\max_{\lambda_j} L(\boldsymbol{w},\ b,\ \boldsymbol{\lambda}), \text{subject to: } y_j\left(\boldsymbol{w}^{\mathrm{T}}\cdot\boldsymbol{x}_j - b\right) \geq 1,\ \forall\, j = 1,\ 2,\ldots,\ n \tag{14.32}$$

where λ_j $(j = 1,\ldots,n)$ are Lagrangian multipliers, which are set to zero for all nonsupport vectors. By taking the derivatives of the $L(\boldsymbol{w},b,\boldsymbol{\lambda})$ with respect to w_i and setting them to zero we obtain the following set of equations:

$$\frac{\partial L(\boldsymbol{w},\ b,\ \boldsymbol{\lambda})}{\partial w_i} = w_i - \sum_{j=1}^{n}\lambda_j y_j x_{j,\ i} = 0. \tag{14.33}$$

By solving for w_i we obtain $w_i = \sum_{j=1}^{n}\lambda_j y_j x_{j,\ i} (i = 1,\ldots,\ d)$, which can be written in compact vector form as:

$$\boldsymbol{w} = \sum_{j=1}^{n}\lambda_j y_j \boldsymbol{x}_j. \tag{14.34}$$

On the other hand, by taking the derivatives of the $L(\boldsymbol{w},b,\boldsymbol{\lambda})$ with respect to b and setting it to zero we obtain the following constraint:

$$\frac{\partial L(\boldsymbol{w},\ b,\ \boldsymbol{\lambda})}{\partial b} = \sum_{j=1}^{n}\lambda_j y_j = 0. \tag{14.35}$$

Now by taking the derivative of L with respect to λ_j and setting to zero we obtain:

$$\frac{\partial L(\boldsymbol{w},\ b,\ \boldsymbol{\lambda})}{\partial \lambda_j} = 1 - y_j\left(\boldsymbol{w}^{\mathrm{T}}\cdot\boldsymbol{x}_j - b\right) = 0, \tag{14.36}$$

from which we obtain:

$$\boldsymbol{w}^{\mathrm{T}}\cdot\boldsymbol{x}_j - b = 1/y_j. \tag{14.37}$$

By substituting Eqs. (14.34) and (14.37) into Eq. (14.32), the optimization problems are simplified to:

$$\widehat{\boldsymbol{\lambda}} = \max_{\lambda_1,\ldots,\lambda_n}\left(\sum_{j=1}^{n}\lambda_j - \frac{1}{2}\sum_{j=1}^{n}\sum_{k=1}^{n}\lambda_j\lambda_k y_j y_k \boldsymbol{x}_j^{\mathrm{T}}\cdot\boldsymbol{x}_k\right)$$

$$\text{subject to: } \sum_{j=1}^{n}\lambda_j y_j = 0,\ \lambda_j \geq 0. \tag{14.38}$$

Typically, only a few λ_js will be nonzero. The optimum $\boldsymbol{w}$ is now determined after substituting $\widehat{\boldsymbol{\lambda}}$ into Eq. (14.34) to obtain:

$$\widehat{\boldsymbol{w}} = \sum_{j=1}^{n} \widehat{\lambda}_j y_j \boldsymbol{x}_j. \tag{14.39}$$

The offset b is finally determined after substituting Eq. (14.39) into Eq. (14.37) to obtain:

$$\widehat{b} = \widehat{\boldsymbol{w}}^{\mathrm{T}} \cdot \boldsymbol{x}_j - \underbrace{1/y_j}_{y_j} = \widehat{\boldsymbol{w}}^{\mathrm{T}} \cdot \boldsymbol{x}_j - y_j. \tag{14.40}$$

To obtain a better estimate of the offset b we should average out over all support vectors by:

$$\widehat{b} = \frac{1}{|\{j|\lambda_j \neq 0\}|} \sum_{j:\ \lambda_j \neq 0} \left(\widehat{\boldsymbol{w}}^{\mathrm{T}} \cdot \boldsymbol{x}_j - y_j\right), \tag{14.41}$$

where the averaging is performed over all js for which $\lambda_j \neq 0$, while the $|\{\ldots\}|$ denotes the cardinality of the corresponding set.

When unknown data point $\boldsymbol{x}$ arrives, it will be classified by the *dual SVM classifier* as follows:

$$\boldsymbol{x} \to y = \operatorname{sign}\left(\widehat{\boldsymbol{w}}^{\mathrm{T}} \cdot \boldsymbol{x} - \widehat{b}\right). \tag{14.42}$$

14.1.3.1 Soft Margin

In certain classification problems, it might happen that several instances of a given class can be located with the points from the other class. To deal with such problems we can introduce the nonnegative *slack variables*, which specify how much a particular point deviates from the margin region. After introducing the slack variables $\zeta_j \geq 0$ we can redefine the constraints as:

$$y_j\left(\boldsymbol{w}^{\mathrm{T}} \cdot \boldsymbol{x}_j - b\right) \geq 1 - \zeta_j, \quad \text{for } j = 1, 2, \ldots, n, \tag{14.43}$$

which essentially are enlarging the margin, commonly referred to as the *soft margin*. The primal optimization problem can be now formulated as:

$$\left(\widehat{\boldsymbol{w}}, \widehat{b}\right) = \underset{\boldsymbol{w},\, b}{\arg\min} \left(\frac{1}{2}\|\boldsymbol{w}\|^2 + C\sum_j \zeta_j\right), \text{subject to: } y_j\left(\boldsymbol{w}^{\mathrm{T}} \cdot \boldsymbol{x}_j - b\right) \geq 1 - \zeta_j, \quad \forall j = 1, 2, \ldots, n. \tag{14.44}$$

By using the Lagrangian approach we can formulate the optimization problem by:

$$\left(\widehat{\boldsymbol{w}},\ \widehat{b}\right) = \underset{\boldsymbol{w},\ b\quad \lambda_j}{\text{arg minmax}}\left\{\frac{\|\boldsymbol{w}\|^2}{2} + C\sum_{j=1}^{n}\zeta_j - \sum_{j=1}^{n}\lambda_j\left[y_j\left(\boldsymbol{w}^{\mathrm{T}}\cdot\boldsymbol{x}_j - b\right) - 1 + \zeta_j\right] - \sum_{j=1}^{n}\eta_j\zeta_j\right\},$$

$$\text{subject to: } y_j\left(\boldsymbol{w}^{\mathrm{T}}\cdot\boldsymbol{x}_j - b\right) \geq 1 - \zeta_j,\ \forall\, j = 1,\ 2, \ldots,\ n. \tag{14.45}$$

By repeating the foregoing procedure, with an additional equation obtained from the derivative with respect to multiplier η_j, that is, $\partial L/\partial\zeta_j = C - \lambda_j - \eta_j = 0$, we conclude that $\lambda_i \leq C$ and the corresponding dual optimization problem become:

$$\widehat{\boldsymbol{\lambda}} = \max_{\lambda_1,\ldots,\lambda_n}\left(\sum_{j=1}^{n}\lambda_j - \frac{1}{2}\sum_{j=1}^{n}\sum_{k=1}^{n}\lambda_j\lambda_k y_j y_k \boldsymbol{x}_j^{\mathrm{T}}\cdot\boldsymbol{x}_k^{\mathrm{T}}\right),\ \text{subject to: } \sum_{j=1}^{n}\lambda_j y_j = 0,\ 0 \leq \lambda_j \leq C. \tag{14.46}$$

14.1.3.2 The Kernel Method

When the decision boundaries are nonlinear we can apply the *kernel trick (method)*, applied to the maximum-margin hyperplane. The key idea is to use a similar algorithm in which the dot product is replaced by the *nonlinear kernel function*. This approach allows us to fit the maximum-margin hyperplane in transformed feature space. Very often the transformation is nonlinear and corresponding transformed space is HD. In nonlinear SVM we are interested in determining the nonlinear classification rule that corresponds to a linear classification in the transformed space, while the corresponding nonlinear mapping can be represented by $\boldsymbol{x}_j \rightarrow \varphi(\boldsymbol{x}_j)$. The corresponding kernel function k maps the dot product of $\boldsymbol{x}_j$ and $\boldsymbol{x}_k$ as follows: $k(\boldsymbol{x}_j,\ \boldsymbol{x}_k) = \varphi(\boldsymbol{x}_j)^{\mathrm{T}}\cdot\varphi(\boldsymbol{x}_k)$. Based on the foregoing, the classification vector can be represented by:

$$\boldsymbol{w} = \sum_{j=1}^{n}\lambda_j y_j \varphi(\boldsymbol{x}_j), \tag{14.47}$$

wherein the λ_j multipliers are obtained by the following optimization problem:

$$\begin{aligned}\widehat{\boldsymbol{\lambda}} &= \max_{\lambda_1,\ldots,\lambda_n}\left(\sum_{j=1}^{n}\lambda_j - \frac{1}{2}\sum_{j=1}^{n}\sum_{k=1}^{n}\lambda_j\lambda_k y_j y_k \underbrace{\varphi\left(\boldsymbol{x}_j^{\mathrm{T}}\right)\cdot\varphi\left(\boldsymbol{x}_k^{\mathrm{T}}\right)}_{k(\boldsymbol{x}_j,\ \boldsymbol{x}_k)}\right)\\ &= \max_{\lambda_1,\ldots,\lambda_n}\left(\sum_{j=1}^{n}\lambda_j - \frac{1}{2}\sum_{j=1}^{n}\sum_{k=1}^{n}\lambda_j\lambda_k y_j y_k k(\boldsymbol{x}_j,\ \boldsymbol{x}_k)\right),\end{aligned}$$

$$\text{subject to: } \sum_{j=1}^{n}\lambda_j y_j = 0,\ 0 \leq \lambda_j \leq C. \tag{14.48}$$

The corresponding algorithm is known as the *coordinate descent*, and it is suitable when dealing with large and sparse datasets.

The offset b for nonlinear classification can be determined by using a similar equation as follows:

$$\hat{b} = \hat{w}^{\mathrm{T}} \cdot \varphi(\boldsymbol{x}_k) - y_k = \sum_{j=1}^{n} \lambda_j y_j \underbrace{\varphi(\boldsymbol{x}_j)^T \cdot \varphi(\boldsymbol{x}_j)}_{k(\boldsymbol{x}_j,\, \boldsymbol{x}_k)} - y_k = \sum_{j=1}^{n} \lambda_j y_j k(\boldsymbol{x}_j, \boldsymbol{x}_k) - y_k. \tag{14.49}$$

When unknown data point $\boldsymbol{x}$ arrives, it will be classified by the *dual nonlinear SVM classifier* as follows:

$$\boldsymbol{x} \to y = \mathrm{sign}\left(\hat{\boldsymbol{w}}^{\mathrm{T}} \cdot \varphi(\boldsymbol{x}) - \hat{b}\right) = \mathrm{sign}\left(\sum_{j=1}^{n} \lambda_j y_j k(\boldsymbol{x}_j, \boldsymbol{x}) - \hat{b}\right). \tag{14.50}$$

The popular *nonlinear kernels* include:

- *Polynomial*: $k(\boldsymbol{x}_j, \boldsymbol{x}) = \left(\mathrm{x}_j^{\mathrm{T}} \cdot \boldsymbol{x} + c\right)^p$, $c \in \{0, 1\}$, wherein $c = 0$ corresponds to the homogeneous case, while $c = 1$ corresponds to the inhomogeneous case.
- *Gaussian radial basis function*: $k(\boldsymbol{x}_j, \boldsymbol{x}) = \exp\left(-\gamma \|\boldsymbol{x}_j, \boldsymbol{x}\|^2\right)$, $\gamma > 0$. Sometimes the parameter γ is selected by $1/(2\sigma^2)$.
- *Hyperbolic tangent*: $k(\boldsymbol{x}_j, \boldsymbol{x}) = \tanh(c_1 \boldsymbol{x}_j \cdot \boldsymbol{x} + c_2)$, for some (but not every) $c_1 > 0$ and $c_2 < 0$.

14.1.4 Clustering

Clustering refers to the grouping of objects in a meaningful way. Typically, we group together the points in datasets that are in close proximity to the feature space. There exist numerous clustering algorithms that can be classified as either extrinsic or intrinsic, hierarchical or partitional, agglomerative or divisive, hard or soft. Extrinsic clustering depends on category labels and requires the preprocessing of data. Intrinsic clustering, on the other hand, does not require training labels on objects. Hierarchical clustering has the parent–child structure and can be represented by dendrograms. In partition clustering we partition data points into disjoint clusters, and there is no hierarchical relationship among them. In agglomerative clustering we apply the bottom-up approach, as we interpret each object as a cluster and then merge the existing clusters. In divisive clustering we apply the top-down approach, wherein initially all objects belong to the same cluster, and we split the initial cluster into different clusters according to certain criterion. In hard clustering, the clusters do not overlap; in other words, a particular object from the dataset belongs to only one cluster. In soft clustering, on the other hand, the clusters may overlap, and we associate with each data point the probabilities of being in different clusters. In this section we will describe the *K*-means and expectation maximization (EM) belonging to the class of unsupervised learning algorithms, as well as the *K*-NN algorithm, belonging to the class of supervised learning algorithms.

14.1.4.1 K-Means Clustering

In the *K-means clustering* algorithm, which is a hard-clustering algorithm, we partition the dataset points into *K* clusters based on their pairwise distances. We typically use the Euclidean distance, defined by Eq. (14.2), that is, for two data points $\boldsymbol{x}_i = (x_{i1} \ldots x_{id})$ and $\boldsymbol{x}_j = (x_{j1} \ldots x_{jd})$, the Euclidian distance is defined by $d(\boldsymbol{x}_i, \boldsymbol{x}_j) = \left[\sum_{k=1}^{d} (\boldsymbol{x}_{ik} - \boldsymbol{x}_{jk})^2\right]^{1/2}$. Other distance definitions can be used, including the Manhattan (taxicab) distance $d_M(\boldsymbol{x}_i, \boldsymbol{x}_j) = \sum_{k=1}^{d} |\boldsymbol{x}_{ik} - \boldsymbol{x}_{jk}|$. The starting point is the dataset of points $\{\boldsymbol{x}_1, \boldsymbol{x}_2, \ldots, \boldsymbol{x}_n\}$ that we want to partition into *K* clusters. We then portion them into *K* initial clusters either at random or by applying certain heuristics. For each cluster *k* we specify the *centroid* $\boldsymbol{c}_k$, calculated as follows:

$$\boldsymbol{c}_k = \frac{1}{N_{\boldsymbol{c}_k}} \sum_{j=1}^{N_{\boldsymbol{c}_k}} \boldsymbol{x}_{\boldsymbol{c}_k}^{(j)}, \tag{14.51}$$

where $N_{\boldsymbol{c}_k}$ denotes the number of points in the *k*th cluster, while $\boldsymbol{x}_{\boldsymbol{c}_k}^{(j)}$ denotes the points in that cluster. We then perform a new partition by assigning each data point to the closest centroid. The centroids are then recalculated for each new cluster. This procedure is repeated until the convergence is achieved, when the data points no longer switch the clusters. The *K-means clustering algorithm* can be summarized as follows (inspired by [1]):

- *Given*: the number of clusters *K* and data point set $\{\boldsymbol{x}_1, \boldsymbol{x}_2, \ldots, \boldsymbol{x}_n\}$.
- *Initialization*: Partition the dataset into *K* initial clusters selected at random or in a heuristic way.
- *Centroids determination step*: For each cluster determine the centroid as a center of mass by using Eq. (14.51).
- *Reassigning points step*: Assign data point $\boldsymbol{x}_i$ $(i = 1, \ldots, n)$ to the cluster $\boldsymbol{c}_j$ so that $d(\boldsymbol{x}_i, \boldsymbol{c}_j) \leq d(\boldsymbol{x}_i, \boldsymbol{c}_k)$, for all $k \in \{1, \ldots, K\}$.
- Repeat the previous two steps until the stopping criterion is met (the data points no longer switch the clusters).

Clearly, this algorithm belongs to the class of "hill-climbing" algorithms, which converge to a local minimum. To avoid this problem we could restart the algorithm multiple times with different initial centroids and select one that minimizes the overall intracluster *distortion* (or compactness) *D*, which is defined as:

$$D = \sum_{i=1}^{n} d(\boldsymbol{x}_i, \boldsymbol{c}(\boldsymbol{x}_i)), \tag{14.52}$$

where we use $\boldsymbol{c}(\boldsymbol{x}_i)$ to denote the centroid to which the data point $\boldsymbol{x}_i$ belongs.

One of the key questions is: how can we select the optimum K? The choice of optimum K is a tricky one. If K is too large it can overfit the dataset. We have to try with different K_s, and for each K rerun the algorithm for different initial centroids, while selecting the best result. (This is not a very scientific approach.) Ideally, we could select the K that minimizes the distortion function (14.52); unfortunately, this problem is NP-complete. Alternatively, we can use different variants such as K-mediods (or K-medians) and fuzzy K-means. In K-mediods, the centroids are placed at an actual median point. In fuzzy K-means, the point can belong to multiple clusters.

14.1.4.2 Cluster Quality

Another relevant question is: how can we measure the *cluster quality*? For this purpose we can use internal (topological) or external validation. In internal validation we specify the cluster quality based on clusters themselves, such as spacing within or between clusters. In external validation we determine the quality of clusters based on the knowledge of labels. Popular topological (internal) measurements include [1]: cluster correlation, similarity matrix, sum of squares error, cohesion and separation, and silhouette coefficient.

Given the dataset and the clusters we define and determine two matrices:

- *Distance matrix* $\mathbf{D} = (d_{ij})_{n\times n}$, where d_{ij} is the distance between data points $\boldsymbol{x}_i$ and $\boldsymbol{x}_j$ and
- *Adjacency matrix* $\mathbf{A} = (a_{ij})_{n\times n}$, where $a_{ij} = 1$ when the data points $\boldsymbol{x}_i$ and $\boldsymbol{x}_j$ belong to the same cluster, and $a_{ij} = 0$ otherwise.

We calculate the *correlation* between these two matrices as follows:

$$\rho_{\mathbf{AD}} = \frac{\sum_{i=1}^{n}\sum_{j=1}^{n}(d_{ij}-\mu_{\mathbf{D}})(a_{ij}-\mu_{\mathbf{A}})}{\sqrt{\sum_{i=1}^{n}\sum_{j=1}^{n}(d_{ij}-\mu_{\mathbf{D}})^2\sum_{i=1}^{n}\sum_{j=1}^{n}(a_{ij}-\mu_{\mathbf{A}})^2}}, \tag{14.53}$$

where $\mu_{\mathbf{D}}$ ($\mu_{\mathbf{A}}$) denotes the mean value of elements of matrix- $\mathbf{D}$ ($\mathbf{A}$). The large negative correlation indicates that clusters are well separated.

The *similarity matrix* provides the visual representation of similarity within the clusters, and can be obtained from the distance matrix by grouping rows and columns by the clusters.

The *cluster cohesion* tells us how the cluster is tightly packed, while the *cluster separation* describes the average distance between the clusters. The distortion parameter, defined by Eq. (14.52), can be used as the measure of cohesion. Alternatively, we could use the average distance between all pairs of points as the measure of cohesion. To characterize the distance separation we have more options, including the average distance among centroids, the average distance among all points in the clusters, and the average distance from centroids to a midpoint, to mention a few.

The *silhouette coefficient* combines cluster cohesion and cluster separation in a single parameter, as illustrated in Fig. 14.13. Let C_i denote the cluster to which the data point $\boldsymbol{x}_i$ belongs. Cluster *cohesion* $c(\boldsymbol{x}_i)$ is defined as the average (intracluster) distance between the data point x_i and all other points in the same cluster C_i:

$$c(\boldsymbol{x}_i) = \frac{1}{|\{\boldsymbol{y}|\boldsymbol{y} \in C_i,\ \boldsymbol{y} \neq \boldsymbol{x}_i\}|} \sum_{\substack{x_i \in C_i \\ y \neq x_i}}^{|C_i|} d(\boldsymbol{x}_i,\ \boldsymbol{y}). \tag{14.54}$$

Cluster separation d_j represents the average intercluster distance between the data point $\boldsymbol{x}_i$ and all points in cluster C_j that is different from C_i, that is:

$$d_j(\boldsymbol{x}_i) = \frac{1}{|\{\boldsymbol{y}|\boldsymbol{y} \in C_j\}|} \sum_{y \in C_j}^{|C_j|} d(\boldsymbol{x}_i,\ \boldsymbol{y}). \tag{14.55}$$

Let the minimum of all d_j's be dented by d, which can be calculated by:

$$d(\boldsymbol{x}_i) = \min_{j:\ C_j \neq C_i} d_j(\boldsymbol{x}_i). \tag{14.56}$$

The *silhouette coefficient* for data point $\boldsymbol{x}_i$ is now defined as:

$$s(\boldsymbol{x}_i) = \frac{d(\boldsymbol{x}_i) - c(\boldsymbol{x}_i)}{\max(c(\boldsymbol{x}_i),\ d(\boldsymbol{x}_i))} = 1 - \frac{c(\boldsymbol{x}_i)}{d(\boldsymbol{x}_i)}, \tag{14.57}$$

where we assume that $d > c$; otherwise, the clustering algorithm would be badly designed. When the silhouette coefficient is close to 1, the points in the cluster are

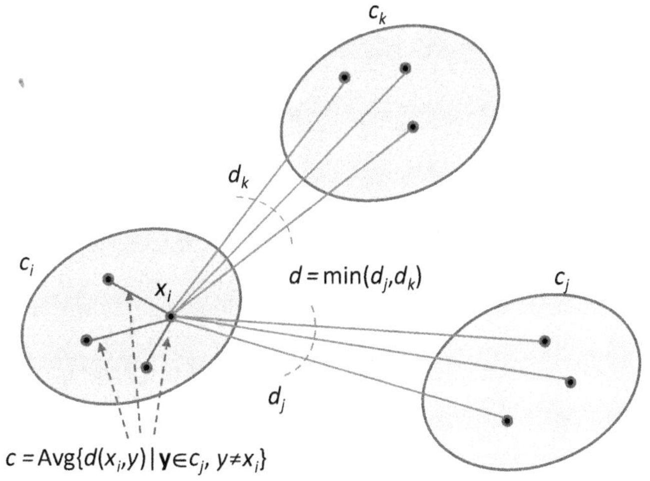

FIGURE 14.13

Defining the silhouette coefficient.

much closer together than the points from any other neighboring cluster. The average *silhouette coefficient* of the whole cluster is defined by:

$$s = \frac{1}{n}\sum_{i=1}^{n} s(\boldsymbol{x}_i), \tag{14.58}$$

and it is clear that this computation is extensive for large datasets.

The popular external measures of clustering quality are entropy and purity. *Entropy* is the measure of uncertainty (randomness), and low intracluster entropy and high intercluster entropy define a good clustering algorithm. External clustering validation requires the knowledge of labels, as mentioned earlier. Let n_{ik} denote the number of labels of type i in cluster C_k, and n_k the number of objects in cluster C_k. Then, the probability (more precisely the frequency) of occurrence of object i in cluster C_k is determined by $p_{ik} = n_{ik}/n_k$. The entropy of the kth cluster is defined now by:

$$H_k = -\sum_{i=1}^{L} p_{ik} \log_2 p_{ik}, \tag{14.59}$$

where L is the number of different object types (labels). The intracluster entropy is obtained by averaging the entropies of different clusters:

$$H = \frac{1}{K}\sum_{k=1}^{K} n_k H_k. \tag{14.60}$$

The smaller intracluster entropy is the better, because this parameter indicates better uniformity of labels within the clusters.

Purity is another useful external validation parameter [1]. The *purity of the cluster k*, denoted as P_k, is defined as $P_k = \max_i p_{ik}$. When $P_k = 1$, there is no ambiguity about the object type, all objects will have the same label. The overall (intracluster) purity is obtained by averaging the purity of different clusters, that is:

$$P = \frac{1}{K}\sum_{k=1}^{K} n_k P_k. \tag{14.61}$$

14.1.4.3 Expectation Maximization Clustering

EM clustering employs probability distributions in the clustering algorithm, instead of Euclidean distance that is used in K-means, and thus represents a soft-clustering algorithm. The Euclidean distance might result in poor clustering results when the clusters are not circular symmetric, and zero-mean Gaussian distribution does not accurately describe the dataset points. EM clustering is composed of two steps: (1) an expectation (E) step and (2) a maximization (M) step, and can be summarized as follows:

- *E-step*: Recalculate the probabilities needed in the M-step.
- *M-step*: Recalculate the maximum likelihood estimates for parameters of the corresponding distributions.

Clearly, there is a certain similarity compared to K-means that can also be summarized as a two-step procedure: (1) redesign the cluster based on the current centroids and (2) recompute the new centroids based on current clusters. When the corresponding distributions are circular Gaussian and clustering is hard, EM clustering essentially reduces to K-means. Namely, the maximum likelihood detector for additive white Gaussian noise channels has the same performance as the Euclidean distance-based detector as shown in [7,10]. This indicates that K-means is a special case of EM clustering, when the hard-clustering approach is applied during EM. In clustering we use EM to determine the hidden parameters of the corresponding distributions in K clusters. In EM clustering, each cluster is described by a generative model, such as Gaussian or multivariate. Each cluster corresponds to the probability density function, and we need to discover through EM the parameters of the distribution, such as mean values and covariance matrix.

In EM clustering, we are given a dataset $D = \{\boldsymbol{x}_1, \boldsymbol{x}_2, \ldots, \boldsymbol{x}_n\}$, and we are interested in clustering this dataset in K clusters $\{C_1, \ldots, C_K\}$ that partition D. EM clustering is a soft-clustering algorithm, so we need to associate with the ith data point the probability of being in the kth cluster ($k = 1, \ldots, K$), denoted with $p_{k,i}$, that is, $p_{k,i} = P(\boldsymbol{x}_i \in C_k)$. Clearly, $\sum_k p_{k,i} = 1$. For each cluster k we associate the density function over R^p, where p is the number of unknown parameters, denoted as $f_{\theta_k(\boldsymbol{x})}$. We assume that the density for $(x_i)_{1 \le i \le n}$ can be described by:

$$f(\boldsymbol{x}) = \sum_{k=1}^{K} p_k f_{\theta_k}(\boldsymbol{x}), \ \forall \ \boldsymbol{x} \in R^p, \tag{14.62}$$

where $p_k \geq 0$ ($k = 1, \ldots, K$) denotes the fraction of the kth distribution from which the sample $\boldsymbol{x}$ is drawn, and $\boldsymbol{\theta}_k$ is used to denote the unknown parameters of the distribution corresponding to the kth cluster. Because the sample must be drawn from one of the K distributions we have that $\sum_k p_k = 1$. Eq. (14.62) describes the mixture model. A very popular model of this type is the *Gaussian mixture model*, for which the parameters are the mean values $\boldsymbol{\mu}_k$ and covariance matrices $\boldsymbol{\Sigma}_k$, that is, we can specify the parameters as $\boldsymbol{\theta}_k = (\boldsymbol{\mu}_k, \boldsymbol{\Sigma}_k)$. The corresponding multivariate Gaussian distribution is given by:

$$f_{\theta_k}(\boldsymbol{x}) = \frac{1}{\sqrt{(2\pi)^n \det(\boldsymbol{\Sigma}_k)}} \exp\left[-\frac{1}{2}(\boldsymbol{x} - \boldsymbol{\mu}_k)^T \boldsymbol{\Sigma}_k^{-1} (\boldsymbol{x} - \boldsymbol{\mu}_k)\right]. \tag{14.63}$$

We can now apply Bayes rule to determine the degree of membership of point $\boldsymbol{x}$ as follows:

$$P(C_k|\boldsymbol{x}) = \frac{P(C_k) P(\boldsymbol{x}|C_k)}{P(\boldsymbol{x})} = \frac{p_k f_{\theta_k}(\boldsymbol{x})}{\sum_{k'=1}^{K} p_{k'} f_{\theta_{k'}}(\boldsymbol{x})}. \tag{14.64}$$

Partition of the dataset is now performed by:

$$C_k = \left\{ x | p_k f_{\theta_k}(\boldsymbol{x}) > \max_{k' \neq k} p_{k'} f_{\theta_{k'}}(\boldsymbol{x}) \right\};\ k = 1,\ 2, \ldots,\ K. \tag{14.65}$$

Let us define the *likelihood function* as follows:

$$\Lambda(\boldsymbol{\theta}|\mathscr{D}) = P_{\boldsymbol{\theta}}(\mathscr{D}) = \prod_{n=1}^{n} P_{\boldsymbol{\theta}}(\boldsymbol{x}_i). \tag{14.66}$$

We can formulate the *maximum likelihood estimation* (MLE) as the optimization problem:

$$\widehat{\boldsymbol{\theta}} = \underset{\boldsymbol{\theta}}{\operatorname{argmax}} \Lambda(\boldsymbol{\theta}|\mathscr{D}). \tag{14.67}$$

To avoid the resolution problem when multiplying many probabilities we use the log-likelihood function instead:

$$L(\boldsymbol{\theta}|\mathscr{D}) = \log\ \Lambda(\boldsymbol{\theta}|\mathscr{D}) = \log \prod_{i=1}^{n} P_{\boldsymbol{\theta}}(\boldsymbol{x}_i) = \sum_{i=1}^{n} \underbrace{\log\ P_{\boldsymbol{\theta}}(\boldsymbol{x}_i)}_{L_{\boldsymbol{\theta}}(\boldsymbol{x}_i)} = \sum_{i=1}^{n} L_{\boldsymbol{\theta}}(\boldsymbol{x}_i). \tag{14.68}$$

The MLE can be now reformulated as:

$$\widehat{\boldsymbol{\theta}} = \underset{\boldsymbol{\theta}}{\operatorname{argmax}} L(\boldsymbol{\theta}|\mathscr{D}). \tag{14.69}$$

For the mixture model, given by Eq. (14.62), the MLE problem can be represented as:

$$\left(\widehat{\boldsymbol{p}},\ \widehat{\boldsymbol{\theta}}\right) = \underset{(\boldsymbol{p},\ \boldsymbol{\theta})}{\operatorname{argmax}} L((\boldsymbol{p},\ \boldsymbol{\theta})|\mathscr{D}),\ L((\boldsymbol{p},\ \boldsymbol{\theta})|\mathscr{D}) = \sum_{i=1}^{n} \log\left(\sum_{k=1}^{K} p_k f_{\theta_k}(\boldsymbol{x}) \right), \quad \boldsymbol{p} = (p_1, \ldots, p_K). \tag{14.70}$$

This complicated optimization problem can be solved by the gradient ascent in $(\boldsymbol{p},\boldsymbol{\theta})$ providing that the log-likelihood function is convex, which is not necessarily true. The alternative solution will be the EM algorithm [13].

By skipping some derivations we can formulate the *log-domain EM algorithm* as follows:

- *Given*: Dataset $D = \{\boldsymbol{x}_1, \boldsymbol{x}_2, \ldots, \boldsymbol{x}_n\}$ and parametric model $f_{\boldsymbol{\theta}}$.
- *Initialization*: Select an initial distribution $\boldsymbol{p}^{(0)}$ and parameter vector $\boldsymbol{\theta}^{(0)}$ randomly in the corresponding state space.
- *E-step*: Recalculate the probabilities that the ith point ($i = 1, \ldots, n$) belongs to the kth cluster ($k = 1, \ldots, K$) as follows:

$$p_{k,i}^{(t)} = \frac{p_k^{(t-1)} f_{\theta_k^{(t-1)}}(\boldsymbol{x}_i)}{\sum_{k'=1}^{K} p_{k'}^{(t-1)} f_{\theta_{k'}^{(t-1)}}(\boldsymbol{x}_i)},$$

where (t) denotes the index of the current iteration.

- M-step: Recalculate the parameters $p^{(t)}$ and $\boldsymbol{\theta}^{(t)}$ by the following maximizations:

$$\widehat{\boldsymbol{p}} = \arg \max_{\boldsymbol{p}:\, \sum p_k = 1} \sum_{k=1}^{K} \log(p_k) \sum_{i=1}^{n} p_{k,i}^{(t)} \tag{14.71}$$

$$\widehat{\boldsymbol{\theta}}_k^{(t)} = \operatorname*{argmax}_{\boldsymbol{\theta}} \sum_{i=1}^{n} p_{k,i}^{(t)} \log f_{\boldsymbol{\theta}_k}(\boldsymbol{x}_i);\ \forall\, k = 1, \ldots, K. \tag{14.72}$$

- Repeat the E- and M-steps until the convergence is achieved.

Eq. (14.71) can be simplified by applying the Lagrangian method:

$$\frac{\partial}{\partial p_k}\left[\sum_{k=1}^{K} \log(p_k) \sum_{i=1}^{n} p_{k,i} + \lambda\left(1 - \sum_{k=1}^{n} p_k\right)\right] = 0, \tag{14.73}$$

to obtain:

$$p_k = \frac{1}{n} \sum_{i=1}^{n} p_{k,i}. \tag{14.74}$$

For a Gaussian mixture for unknown means and covariance matrix, Eq. (14.72) in the M-step simplifies to:

$$\boldsymbol{\mu}_k^{(t)} = \sum_{i=1}^{n} p_{k,i}^{(t)} \boldsymbol{x}_i,\quad \Sigma_k^{(t)} = \sum_{i=1}^{n} p_{k,i}^{(t)} \left(\boldsymbol{x}_i - \boldsymbol{\mu}_k^{(t)}\right)^{\mathrm{T}} \left(\boldsymbol{x}_i - \boldsymbol{\mu}_k^{(t)}\right). \tag{14.75}$$

14.1.4.4 K-Nearest Neighbor Algorithm

The K-NN algorithm represents the simplest learning algorithm, which does not require any calculating in training, and all calculation is postponed for the scoring phase. The existing labeled data are used for training. Once a new, unlabeled, object is presented to the learner, it searches for the K closest neighbors and labels the new incoming object by applying the majority rule. As an illustration, in Fig. 14.14 the object denoted in the circle (in red color) is presented to the learner. For $K = 1$ (on the left), the learner searches for the closest neighbor and labels the unknown object as a "triangle" (in blue color). On the other hand, for $K = 3$ (on the right), the learner searches for three nearest neighbors and applies the majority rule, thus classifying the unknown object as the "square" (in orange color). Given that there is no training needed, just keeping the track of the labeled data, this algorithm if often called the *lazy algorithm.*

We can also associate the weight for contributions from different neighbors, which will give importance to closeness of neighbors. Thus it makes sense to use the reciprocal of the distance for this weight. Let us again observe the three nearest neighbors example in Fig. 14.14 by taking the distance weight into account. Given that $1/d(x,t_1) > 1/d(x,s_1) + 1/d(x,s_2)$, the unknown object will be classified as the triangle (blue). Alternatively, we can use the frequency weight, with number of squares being $S = 9$, while number of triangles is $T = 3$. We can weight each square as 1, while the triangle as $R/T = 3$. By going back to the three nearest neighbors example, the "square score" will be 2 and the triangle score 3, and the unknown object will be classified as the triangle.

Given its simplicity, K-NN can be combined with other ML algorithms. For instance, PCA can be used to reduce dimensionality first, and then we can apply K-NN for the classification in lower-dimensional space.

K-NN also has some limitations, and it is not effective for data with skewed classes. K-NN is also sensitive to all features, even though some may not be relevant. Finally, finding the K closest neighbors can be computationally extensive when the Euclidian distance is used. For $K = 1$, this problem is known as the nearest post office problem, where for a given address we need to identify the closest post office. For d-dimensional data the complexity of finding the K-NN is $O(nd)$. We can reduce complexity by either dimensionality reduction or not comparing to all training examples. By employing the K-D tree construction and searching (described later), the number of operations can be reduced to $O(d \log_2 n)$. Unfortunately, the K-D tree can miss some neighbors, and it is applicable when $d << n$. For HD, but discrete and sparse datasets, we should use inverted lists, wherein the complexity is in the order $O(n'd')$, $n' << n$, $d' << d$. Finally, for HD, but real-valued or discrete problems that are not sparse, we should use locality-sensitive hashing whose complexity is in the

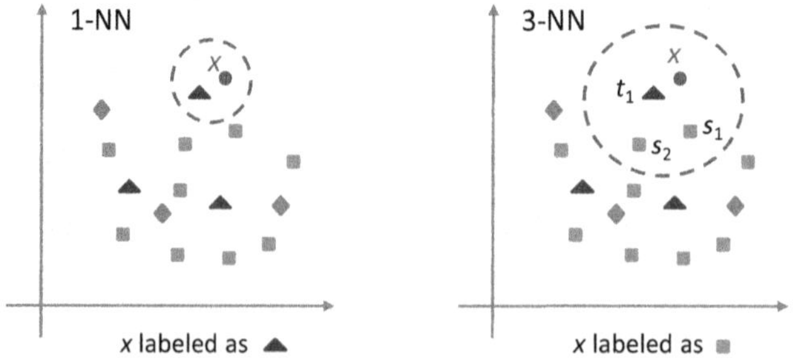

FIGURE 14.14

K-nearest neighbor algorithm for: (left) $K = 1$ and (right) $K = 3$.

order $O(n'd)$, $n' << n$. In the K-D tree we select an attribute at random and then find the median (for that dimension) among all instances in the dataset and split it into two equal parts. In the next iteration we select a different feature at random and repeat the procedure. This procedure is repeated until we reach a predetermined number of points left in each branch. This tree is then used to search for the nearest neighbors. The depth of the tree cannot be larger than $\log_2 n$.

The K-NN algorithm can also be used in *regression problems*, where the input is the same as in the classification problem, while the output is a real-valued target. As the new data point $\boldsymbol{x}$ arrives we run the same algorithm by calculating the distance to every training point $d(\boldsymbol{x},\boldsymbol{x}_i)$, and finding the K-NN, say $\boldsymbol{x}_{i1},\ldots,\boldsymbol{x}_{iK}$. Now, instead of applying the majority vote rule to determine the label of the incoming object we output the means of the labels, that is, $\hat{y} = (1/K)\sum_{k=1}^{k} y_{ik}$.

14.1.5 Boosting

Different learning algorithms have their own weaknesses and strengths. To combine multiple models into a model with better generalization performance, *ensemble learning* is used. The constituent models are known as *weak* (base) *learners*, while the composite model is commonly referred to as a *strong learner.* The generic procedure in ensemble learning can be described as a two-step procedure. In the first step we design K weak learners. In the second step we combine appropriately weak learners to obtain a composite, strong learner. Namely, we train K weak classifiers using the (same as before) dataset $\{(\boldsymbol{x}_1,y_1),\ldots,(\boldsymbol{x}_n,y_n)\}$. The subsets used for the training of weak classifiers may overlap. The weak classifiers should be stronger (have better accuracy) than random guessing. Popular ensemble learning methods include bagging, random forests, and boosting, to mention a few. In bagging, which is short for bootstrap aggregating, the week learners simply vote, and a decision is made by the majority rule. To improve the performance, the training of weak leaners is performed by random training subsets.

Boosting represents an ensemble learning method that builds a stronger classifier from weak classifiers complementing each other. To achieve this goal, the weak classifiers are trained iteratively and sequentially so that the next weak classifier learns what the previous weak classifier cannot do. Thus the next iteration weak classifier concentrates on a data instance being misclassified in the previous classifiers. The strong classifier is created by properly selecting and weighting the weak classifiers. Boosting algorithms are different in the way they weight classifiers and training data objects.

AdaBoost, short for adaptive boosting, does not need the knowledge of accuracy of previous weak learners, but adapts to the accuracies at hand. At each iteration, AdaBoost identifies the key weakness in the composite classifier that has been built in the previous iteration and selects one out of several unused classifiers that can fix that weakness. Once selected, it is properly weighted before being incorporated into the composite classifier. Therefore AdaBoost is an iterative algorithm in which we generate a series of composite classifiers C_t ($t = 1,2,\ldots,T$), using the K weak classifiers c_k ($k = 1,\ldots,K$). For iteration index t the composite classifier is created by:

$$C_t(\boldsymbol{x}_i) = C_{t-1}(\boldsymbol{x}_i) + \alpha_t k_t(\boldsymbol{x}_i),\ \alpha_t \geq 0, \tag{14.76}$$

where k_t is a weak classifier selected from the list of unused weak classifiers and α_t denotes the corresponding weight. To determine the "exact role" of the k_t classifier, that is, the corresponding weight, we us the *exponential loss function* [1]:

$$E_t = \sum_{i=1}^{n} \exp\{-y_i[C_{t-1}(\boldsymbol{x}_i) + \alpha_t k_t(\boldsymbol{x}_i)]\}, \tag{14.77}$$

where y_i is the label that corresponds to $\boldsymbol{x}_i$. Clearly, any error in the composite classifier is penalized by increasing the loss function significantly. To select the weak classifier to be used in iteration t we need to look to the terms in Eq. (14.77) increasing the loss function and selecting the k_t that minimizes the loss function. To simplify the selection of the weak classifier we can split the exponential loss function into two terms:

$$\begin{aligned} E_t &= e^{-\alpha_t} \underbrace{\sum_{i:\, y_i = k_t(\boldsymbol{x}_i)} \exp[-y_i C_{t-1}(\boldsymbol{x}_i)]}_{W_1^{(t)}} + e^{\alpha_t} \underbrace{\sum_{i:\, y_i \neq k_t(\boldsymbol{x}_i)} \exp[-y_i C_{t-1}(\boldsymbol{x}_i)]}_{W_2^{(t)}} \\ &= e^{-\alpha_t} W_1^{(t)} + e^{\alpha_t} W_2^{(t)}. \end{aligned} \tag{14.78}$$

Given that the second term is related to the number of misses, out of unused weak classifiers, we select one that minimizes the W_2 term, and we denote it as k_t. Once the weak classifier k_t is selected, W_1 and W_2 are fixed, and the optimum weight can be determined by determining the derivative $dE_t/d\alpha_t$ as follows:

$$\frac{dE_t}{d\alpha_t} = -e^{-\alpha_t} W_1^{(t)} + e^{\alpha_t} W_2^{(t)}, \tag{14.79}$$

setting it to zero and solving for α_t to obtain:

$$\alpha_t = \frac{1}{2} \log \left[\frac{W_1^{(t)}}{W_2^{(t)}} \right]. \tag{14.80}$$

Therefore the AdaBoost algorithm iteration t can be summarized as follows:

- *Step 1*: Out of unused weak classifiers, select one, denoted as k_t, which minimizes the number of misses, specified by W_2.
- *Step 2*: Once the weak classifier k_t is selected, calculate W_1 and W_2 using the foregoing definitions, followed by calculation of the weight factor α_t.
- Repeat steps 1 and 2 until all weak classifiers are employed.

14.1.6 Regression Analysis

Given the finite training dataset $\{(\boldsymbol{x}_1,y_1),\ldots,(\boldsymbol{x}_n,y_n)\}$, in regression analysis we are concerned with finding an approximate function $f(\boldsymbol{x}_i)$ that will accurately represent the training instances. It resembles the supervised classification; however, the range of y_i can take any value in the set of real numbers R, while in supervised classification the labels y_i are discrete. To evaluate the performance of an approximating function we can use the residual sum of squares (RSS) [14—17]:

$$E=\sum_{i=1}^{n}(y_i-f(\boldsymbol{x}_i))^2. \tag{14.81}$$

Minimization of the error function over arbitrary families of functions yields a large number of solutions, and we need to restrict the problem to a smaller set of functions. Moreover, the zero error does not necessarily mean that the generalization performance will be good; instead we might be running to the overfitting problem. Very often, a good solution could be to search in the family of functions characterized by certain parameters, which is known as parametric function optimization.

The method of *linear least squares* is a parametric model, for which the approximation function can be represented as:

$$f(\boldsymbol{x},\ \boldsymbol{\beta})=\boldsymbol{\beta}^T\boldsymbol{x},\ \boldsymbol{x}=[x_1\ldots x_d]^T,\ x_i\in\mathscr{R}, \tag{14.82}$$

where $\boldsymbol{\beta}$ is an unknown parametric vector. The RSS for this problem becomes:

$$E(\boldsymbol{\beta})=\sum_{i=1}^{n}(y_i-f(\boldsymbol{x}_i,\ \boldsymbol{\beta}))^2. \tag{14.83}$$

Let us now arrange the input data vectors (known as regressors) into a matrix $\boldsymbol{X}$, for which the ith input vector $\boldsymbol{x}_i$ is placed in the ith column. We can represent the model output $\boldsymbol{y}$ as follows:

$$\boldsymbol{y}=\boldsymbol{X}\boldsymbol{\beta}+\boldsymbol{z}, \tag{14.84}$$

where $\boldsymbol{z}$ is the noise vector whose components are samples from the zero mean Gaussian distribution with variance σ^2. In the least-squares estimation we minimize the squared error between the actual output $\boldsymbol{y}$ and the estimated output $\boldsymbol{X}\widehat{\beta}$:

$$\widehat{\boldsymbol{\beta}}=\arg\min_{\boldsymbol{\beta}}E(\boldsymbol{\beta})=\arg\min_{\boldsymbol{\beta}}\|\boldsymbol{y}-\boldsymbol{X}\boldsymbol{\beta}\|_2. \tag{14.85}$$

We can determine the optimum parameter vector $\boldsymbol{\beta}$ by solving the equation $\partial E(\boldsymbol{\beta})/\partial\boldsymbol{\beta}=0$. For reasonably high SNR, the model output is approximately $\boldsymbol{y}\cong\boldsymbol{X}\boldsymbol{\beta}$; however, the matrix $\boldsymbol{X}$ of size $d\times n$ is not a square matrix and thus cannot be inverted. If we premultiply this equation with $\boldsymbol{X}^{\mathrm{T}}$ from the left we obtain $\boldsymbol{X}^{\mathrm{T}}\boldsymbol{y}\cong\boldsymbol{X}^{\mathrm{T}}\boldsymbol{X}\boldsymbol{\beta}$, the matrix $\boldsymbol{X}^{\mathrm{T}}\boldsymbol{X}$ is of size $d\times d$ and thus invertible. By multiplying now the modified model by $(\boldsymbol{X}^{\mathrm{T}}\boldsymbol{X})^{-1}$ from the left we obtain:

$$(X^T X)^{-1} X^T y \cong \underbrace{(X^T X)^{-1} X^T X}_{I}\, \beta, \tag{14.86}$$

from which we determine the optimum parameter vector by:

$$\widehat{\beta} = (X^T X)^{-1} X^T y. \tag{14.87}$$

An alternative way is to use the *pseudoinverse* of X, denoted as X^+. For this purpose we need to perform the SVD of the real-valued X matrix first, that is, $X = U\Sigma V^T$, where U and V are orthogonal matrices, and Σ is the diagonal matrix $\Sigma = \text{diag}(\sigma_1, \sigma_2, \ldots)$, with $\sigma_1 \geq \sigma_2 \geq \ldots \geq 0$ being the *singular values* of X. Given that the pseudoinverse of diagonal matrix Σ is simply $\Sigma^+ = \text{diag}\left(\sigma_1^{-1}, \ldots, \sigma_r^{-1}, 0, \ldots, 0\right)$, where r is the number of nonzero singular values, we can define the pseudoinverse of X as follows:

$$X^+ = \begin{cases} V\begin{bmatrix} \Sigma^+ \\ 0 \end{bmatrix} U^T, & d \leq n \\ V[\Sigma^+ \quad 0]U^T, & d > n \end{cases}. \tag{14.88}$$

Pseudoinverse is a unique solution of the minimization problem:

$$X^+ = \arg\min_X \|X\beta - I_d\|_2, \tag{14.89}$$

where I is the identity matrix of size $d \times d$. When the $\mathbf{0}$-matrix is not required in Eq. (14.88), then the pseudoinverse of X will simply be:

$$X^+ = (X^T X)^{-1} X^T, \tag{14.90}$$

which is easy to verify by using the SVD of X, namely $X = U\Sigma V^T$. To summarize, the optimum parameter vector can also be determined by:

$$\widehat{\beta} = X^+ y. \tag{14.91}$$

The least-squares estimation is also applicable for higher-order polynomial approximation, which for $x \in \mathscr{R}$ has the form:

$$f(x, \beta) = \beta_0 + \beta_1 x + \cdots + \beta_p x^p, \; \beta = [\beta_0 \ldots \beta_p]^T. \tag{14.92}$$

However, for a higher-order polynomial, the orthogonal polynomial might provide better accuracy results.

To ensure either sparsity or smoothness of the solution we can regularize the optimization problem as follows:

$$\widehat{\beta} = \arg\min_\beta \left[\|y - X\beta\|^2 + \|\Gamma X\|^2\right], \tag{14.93}$$

where Γ is a properly chosen Tikhonov matrix. Very often we can choose the

Tikhonov matrix as $\delta \boldsymbol{I}$ (δ is a positive constant) and in this case the minimization problem reduces to:

$$\widehat{\beta} = \arg\min_{\beta}\left[\|\boldsymbol{y} - \boldsymbol{X}\boldsymbol{\beta}\|^2 + \delta^2\|\boldsymbol{X}\|^2\right],\ \delta > 0. \tag{14.94}$$

The optimization problem given by the previous equation is also known as the *ridge regression* problem, which is used to shrink the regression coefficients by imposing a penalty on their size [16]. When we replace the ridge penalty term, defined using L_2-norm, by a corresponding term using L_1-norm, the optimization problem is known as the *LASSO regression* problem:

$$\widehat{\beta} = \arg\min_{\beta}\left[\|\boldsymbol{y} - \boldsymbol{X}\boldsymbol{\beta}\|_2^2 + \delta\|\boldsymbol{X}\|_1\right],\ \delta > 0, \tag{14.95}$$

(LASSO is an abbreviation for the least absolute shrinkage and selection operator.) The second constraint now makes the solutions nonlinear in the y_i.

14.1.7 Neural Networks

A neural network is a network of nodes called *neurons*. To every input data vector $\boldsymbol{x} = [x_1\ \dots\ x_d]^{\mathrm{T}}$, the neural network produces an output vector $\boldsymbol{y} = f(\boldsymbol{x},\boldsymbol{\theta})$, which is nonlinear with respect to corresponding parameters $\boldsymbol{\theta}$. The neurons are connected by the weighted edges called the *synapses*. A neuron receives the weighted inputs and generates a response that can be multivariate or univariate, and we refer to this process as the activation of the neuron, while the corresponding function describing the output/input relationship is called the *activation function*. ANNs are inspired by the central nervous system of mammals; however, the update rule is different from that of natural neural networks. A common characteristic of various ANNs is that the activation function is always nonlinear.

14.1.7.1 Perceptron and Activation Functions

The neural system composed of only a single neuron is called the *perceptron*, and its operation is illustrated in Fig. 14.15. (Strictly speaking, it is not a network.) Mathematically, perceptron operation can be described as:

$$y = f(\boldsymbol{x}) = \phi\left(\boldsymbol{w}^{\mathrm{T}}\boldsymbol{x} + b\right),\ \boldsymbol{w}^{\mathrm{T}}\boldsymbol{x} = \sum_{i=1}^{d} w_i x_i, \tag{14.96}$$

where $\phi(u)$ is the activation function, $\boldsymbol{w} = [w_1\ \dots\ w_d]^{\mathrm{T}}$ is the weight vector describing the connections strength, and b is the bias. $\sum$ is the schematic representation of the neuron, with operation being $\Sigma = \boldsymbol{w}^{\mathrm{T}}\boldsymbol{x} + b$. The possible *activation functions* include:

- The identity function:

$$\phi(u) = u.$$

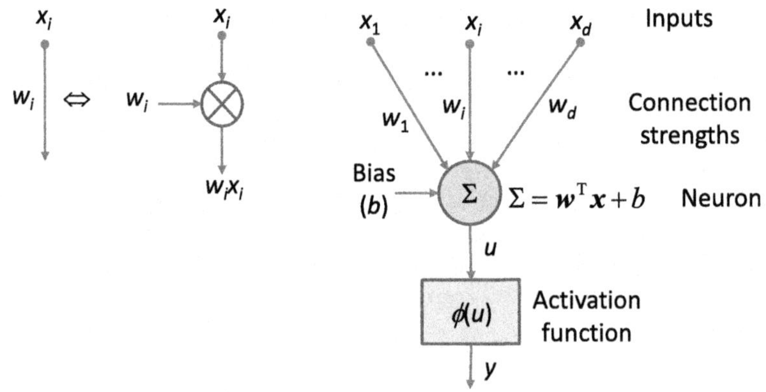

FIGURE 14.15

Operation principle of the perceptron.

- The sigmoid (logistic) function:

$$\phi(u) = \frac{1}{1+e^{-u}}.$$

- The hyperbolic tangent function:

$$\phi(u) = \tanh(u) = \frac{e^u - e^{-u}}{e^u + e^{-u}} = \frac{e^{2u} - 1}{e^{2u} + 1}.$$

- The hard threshold function:

$$\phi_{u_{tsh}}(u) = \mathcal{I}(u \geq u_{tsh}).$$

- The rectified linear unit (ReLU) function:

$$\phi(u) = \max(0,\ x).$$

The aforementioned activation functions are provided in Fig. 14.16. The sigmoid function, denoted in Fig. 14.17 as $s(u)$, has been historically most frequently used, since it is differentiable and has the range [0,1]. The first derivative of the sigmoid function is simple, $s'(u) = s(u)[1 - s(u)]$. However, its derivative is close to zero when $|u|$ is larger than 4, as shown in Fig. 14.17, and it causes numerical problems for large numbers of hidden layers for multilayer feedforward ANNs. To avoid this problem, the ReLU function can be used instead. However, its derivative is zero for $u < 0$, and to ensure that the gradient is nonzero for small values of u, the following modification can be used:

$$\phi(u) = \max(0,\ x) + \delta\min(0,\ x),\ \delta > 0,$$

where δ is a small positive constant.

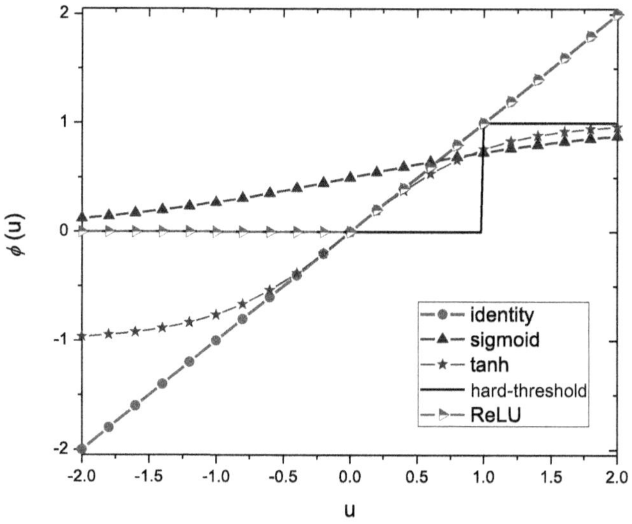

FIGURE 14.16

Possible activation functions. *ReLU*, Rectified linear unit.

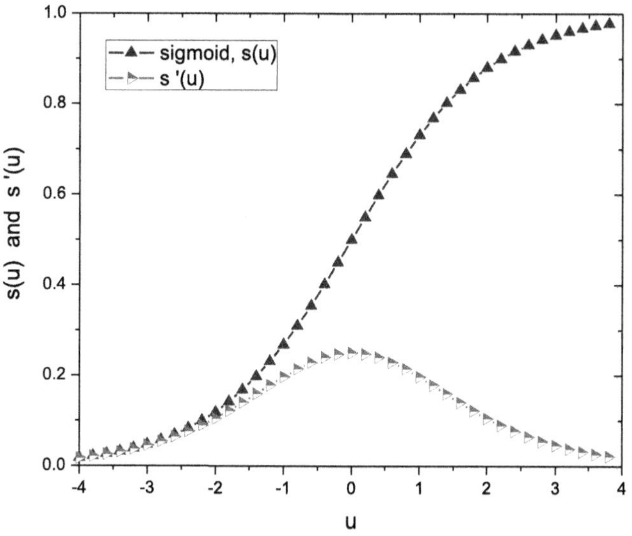

FIGURE 14.17

Sigmoid function and its first derivative.

The perceptron for the threshold (in the hard-threshold function) being set to 0 has the following output/input function:

$$y=f(\boldsymbol{x})=\phi_0\left(\boldsymbol{w}^{\mathrm{T}}\boldsymbol{x}+b\right)=\begin{cases}1, & \boldsymbol{w}^{\mathrm{T}}\boldsymbol{x}+b>0\\ 0, & \text{otherwise}\end{cases}, \tag{14.97}$$

and can be used as a binary classifier. The purpose of the bias here is to change the decision boundary. Training of the perceptron, to determine the weight and bias, will converge only if the problem is linearly separable, and in that case we can use the so-called *delta rule* to update the weights, which is just a particular instance of the gradient descent. To determine the weight vector we can minimize the error function, defined as:

$$E=\frac{1}{2}\sum_{i=1}^{n}(y_i-f(\boldsymbol{x}_i))^2, \tag{14.98}$$

by taking the first derivative of E to obtain:

$$\frac{\partial E}{\partial w_j}=-(y_i-f(\boldsymbol{x}_i))\underbrace{\frac{\partial(\boldsymbol{w}^{\mathrm{T}}\boldsymbol{x}+b)}{\partial w_j}}_{x_j}=-(y_i-f(\boldsymbol{x}_i))x_j. \tag{14.99}$$

We need to update the weights by:

$$\Delta w_j=-\mu(y_i-f(\boldsymbol{x}_i))x_j;\ j=1,\ldots,d, \tag{14.100}$$

where μ is the learning rate. The weight vector update rule is then:

$$\boldsymbol{w}^{(i)}=\boldsymbol{w}^{(i-1)}-\mu(y_i-f(\boldsymbol{x}_i))\boldsymbol{x}_i;\ i=1,\ldots,n. \tag{14.101}$$

14.1.7.2 Feedforward Networks

In the *feedforward network*, the neurons do not form any loop; the information flows in one direction, from input layer, through one or more intermediate levels, known as hidden levels, to the output layer, as illustrated in Fig. 14.18. In other words, each neuron at a given layer is connected to all neurons on the next layer, but there is no connection among neurons lying in the same layer. The ANNs with more than one hidden layer are called "deep" ANNs, and the corresponding learning algorithm is commonly referred to as *deep learning*. The feedforward ANNs are suitable for nonlinear classification and regression problems.

In principle, we can apply different activation functions at different levels. In practice, typically, all hidden layers use the same activation function, very often the sigmoidal activation function. In regression problems, the activation function in the output layer is not needed; in other words, we can use the identity activation function. In binary classification problems, only one output neuron is needed, and the output gives a prediction of $P(y=1|x)$, and since the range for the sigmoidal activation function is [0,1], it is suitable for use in the output layer. On the other hand, in multiclass classification problems we need one output neuron per class c, and the corresponding output gives the prediction $P_c = P(y=c|x)$. These output probabilities must sum to 1, that is, $\sum_c P_c = 1$, indicating that the sigmoidal activation function is not suitable for this problem. The multidimensional *softmax function* should be used instead, which is defined by:

$$\text{soft max}(z)_i = \frac{e^{z_i}}{\sum_j e^{z_j}}. \tag{14.102}$$

For feedforward ANNs there exist numerous training methods, with the most popular one being *backpropagation*, which represents the generalization of the delta rule for the single perceptron [14–18]. Let us go back to Fig. 14.18, which has only one hidden layer. We adopted notation similar to that used in [18], which simplifies the explanations. The indices i, j, and k are used to denote the particular neuron at output, hidden, and input layers, respectively. The weight coefficient related to the edge between the jth node in the hidden layer and the kth node in the input layer is denoted by v_{jk}. In similar fashion, the weight coefficient related to the edge between the ith output node and the jth node in the hidden layer is denoted by w_{ij}. Based on Fig. 14.18 we conclude that the input to the jth node of the hidden layer, denoted by h_j, is related to the inputs x_k as follows:

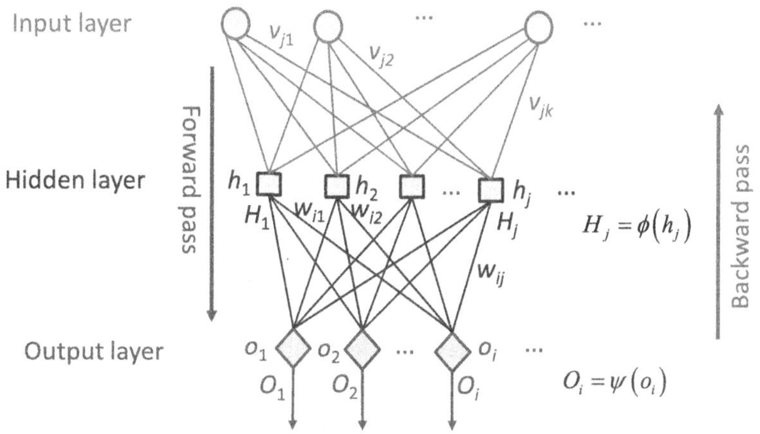

FIGURE 14.18

Feedforward neural network with one hidden layer.

$$h_j = \sum_k v_{jk} x_k, \tag{14.103}$$

where to facilitate explanations we set the bias to zero. The output from the hidden neuron j, denoted as H_j, is given by:

$$H_j = \phi(h_j) = \phi\left(\sum_k v_{jk} x_k\right), \tag{14.104}$$

where $\phi(u)$ is the activation function of the hidden neurons. From Fig. 14.18 we conclude that the input to the ith neuron of the output layer, denoted by o_i, is related to the hidden layer outputs H_j as follows:

$$o_i = \sum_j w_{ij} H_j. \tag{14.105}$$

The output from the neuron i at the output layer, denoted as O_i, is given by:

$$O_i = \psi(o_i), \tag{14.106}$$

where $\psi(u)$ is the activation function of the output neurons.

We can use the sum of squares to evaluate the error performance, defined as:

$$E = \frac{1}{2}\sum_i (O_i - y_i)^2, \tag{14.107}$$

where y_i is the desired output that corresponds to the input data vector $\boldsymbol{x}$, obtained from the training dataset. In the backpropagation method, representing the steepest descent algorithm, each weight is changed in the direction opposite to the gradient of the error, that is, we can write:

$$\Delta w_{ij} = -\mu \frac{\partial E}{\partial w_{ij}}, \quad \Delta v_{jk} = -\mu \frac{\partial E}{\partial v_{jk}}, \tag{14.108}$$

where μ is the learning rate of the ANN. The change in the weights between the hidden and output layers can be calculated by applying the chain rule:

$$\frac{\partial E}{\partial w_{ij}} = \underbrace{\frac{\partial E}{\partial O_i}}_{O_i - y_i} \underbrace{\frac{\partial O_i}{\partial o_i}}_{\psi'(o_i)} \underbrace{\frac{\partial o_i}{\partial w_{ij}}}_{H_j} = (O_i - y_i)\psi'(O_i)H_j. \tag{14.109}$$

The change in weights between the input and hidden layer can also be calculated by applying again the chain rule as follows:

$$\frac{\partial E}{\partial v_{jk}} = \frac{\partial E}{\partial H_j}\frac{\partial H_j}{\partial v_{jk}} = \frac{\partial E}{\partial H_j} \underbrace{\overbrace{\frac{\partial H_j}{\partial h_j}}^{\phi'(h_j)} \overbrace{\frac{\partial h_j}{\partial v_{jk}}}^{x_k}}_{(\partial/\partial v_{jk})H_j} = \phi'(h_j) x_k \frac{\partial E}{\partial H_j}. \tag{14.110}$$

To calculate $\partial E/\partial H_j$ we also apply the chain rule:

$$\frac{\partial E}{\partial H_j}=\sum_i \frac{\partial E}{\partial o_i}\frac{\partial o_i}{\partial H_j}=\sum_i \underbrace{\frac{\partial E}{\partial o_i}}_{\frac{\partial E}{\partial O_i}\frac{\partial O_i}{\partial o_i}}\underbrace{\frac{\partial o_i}{\partial H_j}}_{w_{ij}}=\sum_i \underbrace{\frac{\partial E}{\partial O_i}}_{O_i-y_i}\underbrace{\frac{\partial O_i}{\partial o_i}}_{\psi'(o_i)} w_{ij}=\sum_i (O_i-y_i)\psi'(o_i)w_{ij}. \tag{14.111}$$

Finally, after substituting Eq. (14.111) into Eq. (14.110) we obtain:

$$\frac{\partial E}{\partial v_{jk}}=\phi'(h_j)x_k\sum_i (O_i-y_i)\psi'(o_i)w_{ij}. \tag{14.112}$$

In the learning stage we calculate the outputs O_i as well as all intermediate values, and update the weights, based on partial derivatives calculated in the previous iteration, as follows:

$$w_{ij}^{(\text{new})}=w_{ij}^{(\text{old})}-\mu\frac{\partial E}{\partial w_{ij}},\quad v_{jk}^{(\text{new})}=v_{jk}^{(\text{old})}-\mu\frac{\partial E}{\partial v_{jk}}. \tag{14.113}$$

We refer to this stage as the *forward pass*. In the *backward pass* we determine error terms (O_i-y_i) and then update the partial derivatives, based on Eqs. (14.109) and (14.112). These partial derivatives are propagated to the weights for the next iteration. The error is backpropagated as soon as each new training instance is received.

By close inspection of Eqs. (14.109) and (14.112) we conclude that in the backward pass, the partial derivatives calculated at the deeper level are needed in the shallower level and should be reused for efficient implementation. As an illustration, for example, provided in Fig. 14.18, we provide the sequence of partial derivatives calculation (in the backward direction):

$$\frac{\partial E}{\partial o_i}=\underbrace{\frac{\partial E}{\partial O_i}}_{O_i-y_i}\underbrace{\frac{\partial O_i}{\partial o_i}}_{\psi'(o_i)}=(O_i-y_i)\psi'(o_i) \tag{14.114.1}$$

$$\frac{\partial E}{\partial w_{ij}}=\frac{\partial E}{\partial o_i}\underbrace{\frac{\partial o_i}{\partial w_{ij}}}_{H_j}=H_j\frac{\partial E}{\partial w_{ij}} \tag{14.114.2}$$

$$\frac{\partial E}{\partial h_j}=\sum_i \frac{\partial E}{\partial o_i}\underbrace{\frac{\partial o_i}{\partial h_j}}_{\frac{\partial o_i}{\partial H_j}\frac{\partial H_j}{\partial h_j}}=\sum_i \frac{\partial E}{\partial o_i}\phi'(h_j)w_{ij} \tag{14.114.3}$$

$$\frac{\partial E}{\partial v_{jk}} = \frac{\partial E}{\partial h_j} \underbrace{\frac{\partial h_j}{\partial v_{jk}}}_{x_k} = x_k \frac{\partial E}{\partial h_j}. \tag{14.114.4}$$

In the forward step we use the partial derivatives given by Eqs. (14.114.2) and (14.114.4) to update the weights in Eq. (14.113).

In deep learning, the number of hidden layers can be large, so it is important to study the backward pass for feedforward ANNs containing $L > 1$ hidden layers, which is illustrated in Fig. 14.19, wherein we distinguish the neurons by the corresponding outputs (rather than inputs used in Fig. 14.18). The input layer can be interpreted as the 0th hidden level, thus setting $\boldsymbol{H}^{(0)} = \boldsymbol{x}$ and $w_{jk}^{(0)} = v_{jk}$. Based on the previous example and Eq. (14.104), the output of the ith neuron at layer $(l+1)$ is given by:

$$H_j^{(l+1)} = \phi\left(h_j^{(l+1)}\right) = \phi\left(\sum_i w_{ji}^{(l)} H_i^{(l)}\right). \tag{14.115}$$

The partial derivative of E with respect to $H_j^{(l)}$ is given by:

$$\frac{\partial E}{\partial H_j^{(l)}} = \sum_i \frac{\partial E}{\partial H_i^{(l+1)}} \underbrace{\frac{\partial H_i^{(l+1)}}{\partial H_j^{(l)}}}_{\phi'\left(h_i^{(l+1)}\right) w_{ij}^{(l)}} = \sum_i \phi'\left(h_i^{(l+1)}\right) w_{ij}^{(l)} \frac{\partial E}{\partial H_i^{(l+1)}}. \tag{14.116}$$

On the other hand, the partial derivative of E with respect to $w_{jk}^{(l-1)}$ is given by:

$$\frac{\partial E}{\partial w_{jk}^{(l-1)}} = \frac{\partial E}{\partial H_j^{(l)}} \underbrace{\frac{\partial H_j^{(l)}}{\partial w_{jk}^{(l-1)}}}_{\phi'\left(h_j^{(l)}\right) H_k^{(l-1)}} = \frac{\partial E}{\partial H_j^{(l)}} \phi'\left(h_j^{(l)}\right) H_k^{(l-1)}. \tag{14.117}$$

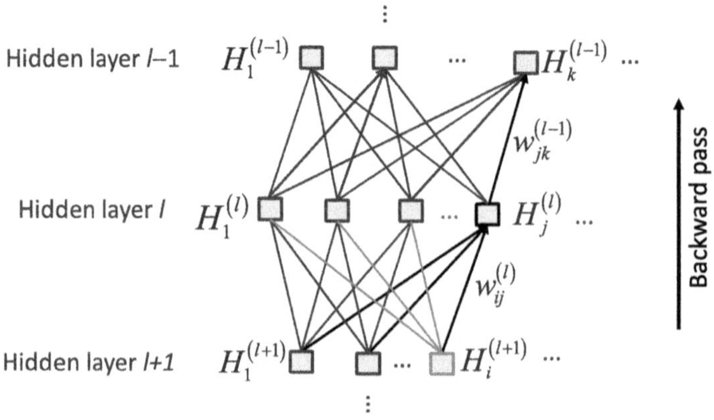

FIGURE 14.19

Backward pass for deep feedforward neural networks.

The weight coefficients can be now updated by:

$$w_{jk}^{(l-1,\,\text{new})} = w_{jk}^{(l-1,\,\text{old})} - \mu \frac{\partial E}{\partial w_{jk}^{(l-1)}}. \tag{14.118}$$

For the output layer, the weights can be updated according to Eq. (14.109) as follows:

$$w_{ij}^{(\text{new})} = w_{ij}^{(\text{old})} - \mu \frac{\partial E}{\partial w_{ij}}, \quad \frac{\partial E}{\partial w_{ij}} = \underbrace{\frac{\partial E}{\partial O_i}}_{O_i - y_i} \underbrace{\frac{\partial O_i}{\partial o_i}}_{\psi'(o_i)} \underbrace{\frac{\partial o_i}{\partial w_{ij}}}_{H_j^{(L)}} = (O_i - y_i)\psi'(o_i) H_j^{(L)}. \tag{14.119}$$

14.2 THE ISING MODEL, ADIABATIC QUANTUM COMPUTING, AND QUANTUM ANNEALING

Many problems in ML can be mapped to the Ising problem formalism; this is why we first describe the *Ising model* [19].

14.2.1 The Ising Model

To study the magnetic dipole moments of atoms in statistical physics, the *Ising model* is introduced, named after German physicist Ernst Ising. The magnetic dipole moments of the atomic spins can be in one of two states, either $+1$ or -1. The spins, represented as the discrete variables, are arranged in a graph, typically the *lattice* graph, and this model allows each spin to interact with the closest neighbors. The spins in the neighborhood having the same value tend to have lower energy; however, the heat disturbs this tendency so that there exist different phase transitions. Let the individual spin on the lattice be denoted by $\sigma_i \in \{-1,1\}$; $i = 1, 2,\ldots,L$ with L being the lattice size. To represent the energy of the Ising model the following Hamiltonian is typically used:

$$H(\boldsymbol{\sigma}) = -\sum_{\langle i,j \rangle} J_{ij}\sigma_i\sigma_j - \mu \sum_i h_i \sigma_i, \tag{14.120}$$

where J_{ij} denotes the interaction strength between the ith and jth sites, while h_i donates the strength of the external magnetic field interacting with the ith spin site in the lattice, and μ is the magnetic moment. We use the notation $\langle i,j \rangle$ to denote the direct nearest neighbors. When $J_{ij} > 0$ the interaction is *ferromagnetic*, when $J_{ij} < 0$ it is *antiferromagnetic*, and when $J_{ij} = 0$ there is no interaction between spins i and j. On the other hand, when $h_i > 0$, the ith spin tends to line up with the external field in the positive direction, while for $h_i < 0$, it tends to line up in the negative direction. The *configuration probability* is described by the *Boltzmann distribution*:

$$P_\beta(\boldsymbol{\sigma}) = \frac{1}{Z_\beta} e^{-\beta H(\boldsymbol{\sigma})}, \quad \beta = 1/(k_B T), \tag{14.121}$$

where $\beta = 1/(k_B T)$ (k_B is the Boltzmann constant and T is the absolute temperature) is the *inverse temperature*, while Z_β denotes the *partition function*:

$$Z_\beta = \sum_{\boldsymbol{\sigma}} e^{-\beta H(\boldsymbol{\sigma})}. \tag{14.122}$$

By denoting the interaction matrix by $\boldsymbol{J} = (J_{ij})$ we can determine the *optimum configuration* as follows:

$$\hat{\boldsymbol{\sigma}} = \arg\min_{\boldsymbol{\sigma}} \left(\boldsymbol{\sigma}^{\mathrm{T}} \boldsymbol{J} \boldsymbol{\sigma} + \boldsymbol{h}^{\mathrm{T}} \boldsymbol{\sigma}\right), \tag{14.123}$$

and this problem is similar to the *quadratic unconstrained binary optimization* (QUBO) problem, which is known to be NP-hard, and can be formulated by:

$$\hat{\boldsymbol{x}} = \arg\min_{\boldsymbol{x}} \left(\sum_{i,j=1}^{N} w_{ij} x_i x_j + \sum_{i=1}^{N} b_i x_i \right), \quad x_i \in \{0, 1\}, \tag{14.124}$$

wherein $\sigma_i = 2x_i - 1$, $\mu = 1$, and $b_i = h_i$ is the bias. By introducing the matrix $\boldsymbol{W} = (w_{ij})$ we can reformulate the QUBO problem in a fashion similar to Eq. (14.123) by:

$$\hat{\boldsymbol{x}} = \arg\min_{\boldsymbol{x}} \left(\boldsymbol{x}^{\mathrm{T}} \boldsymbol{W} \boldsymbol{x} + \boldsymbol{b}^{\mathrm{T}} \boldsymbol{x}\right), \ \boldsymbol{x} \in \{0, 1\}^N. \tag{14.125}$$

By replacing the spin variables with corresponding Pauli operators (matrices) we can express the Ising module using quantum mechanical interpretation:

$$\hat{H} = -\sum_{\langle i,j \rangle} J_{ij} \sigma_i^Z \sigma_j^Z - \sum_i h_i \sigma_i^Z, \tag{14.126}$$

where the Pauli $\sigma^Z = Z$ matrix is defined by $\sigma^Z = Z = \begin{bmatrix} 1 & 0 \\ 0 & -1 \end{bmatrix}$, and the action on the computation basis vector will be $\sigma^Z|0\rangle = |0\rangle$, $\sigma^Z|1\rangle = -|1\rangle$. The expected value of the Ising Hamiltonian will be $\langle \hat{H} \rangle = \langle \psi | \hat{H} | \psi \rangle$, for arbitrary state $|\psi\rangle$. The evolution is described by the unitary operator U to obtain $U|\psi\rangle$. The corresponding Schrödinger equation is given by:

$$j\hbar \frac{d}{dt} U = HU. \tag{14.127}$$

The solution for the time-invariant Hamiltonian is simply $U = e^{-jHt/\hbar}$, which is clearly unitary.

When the external field is transversal, the *transverse-field Hamiltonian* will be:

$$\widehat{H} = -\sum_{\langle i,j \rangle} J_{ij}\sigma_i^Z\sigma_j^Z - \sum_i h_i\sigma_i^X, \qquad (14.128)$$

where the Pauli X-operator is defined by $\sigma^X = X = \begin{bmatrix} 0 & 1 \\ 1 & 0 \end{bmatrix}$. The Pauli X and Z operators do not commute since $XZ = -ZX$.

14.2.2 Adiabatic Quantum Computing

An *adiabatic process* is a process that changes slowly (gradually) so that the system can adapt its configuration accordingly. If the system was initially in the ground state of an initial Hamiltonian H_0, it will, after the adiabatic change, end up in the ground state of the final Hamiltonian H_1. This claim is commonly referred to as the *adiabatic theorem*, proved by Born and Fock in 1928 [20]. For initial Hamiltonian H_0 we typically choose the Hamiltonian whose ground state is easy to obtain, for Hamiltonian H_1 one for which ground state we are interested in. To determine its ground state we consider the following Hamiltonian:

$$H(t) = (1-t)H_0 + tH_1, \quad t \in [0, 1], \qquad (14.129)$$

and start with the ground state of $H(0) = H_0$, and then gradually increase the parameter t, representing the time normalized with the duration of the whole process. According to the quantum adiabatic theorem, the system will gradually evolve to the ground state of H_1, providing that there is no degeneracy for the ground state energy. This transition depends on the time Δt during which the adiabatic change takes place and depends on the minimum energy gap between ground state and the first excited state of the adiabatic process. When applied to the Ising model we select the initial Hamiltonian as $H_0 = \sum_i X_i$, describing the interaction of the transverse external field with the spin sites. On the other hand, the final Hamiltonian is given by $H_1 = -\sum_{\langle i,j \rangle} J_{ij}Z_iZ_j - \sum_i h_iX_i$, representing the classical Ising model. The time-dependent Hamiltonian combines two models by Eq. (14.129). We gradually change the parameter t from 0 to 1, so that the adiabatic pathway is chosen, and the final state will be the ground state of Hamiltonian H_1, encoding the solution of the problem we are trying to solve. The gap ΔE between the ground state and the first excited state is parameter t dependent. The timescale (the runtime of the algorithm) is reversely proportional to $\min[\Delta E(t)]^2$ [21]. So if the minimum gap during the adiabatic pathway is too small, the runtime of the entire algorithm could be large.

In *adiabatic quantum computing*, the Hamiltonian is defined as follows:

$$H = \underbrace{-\sum_{\langle i,j \rangle} J_{ij}Z_iZ_j - \sum_i h_iZ_i}_{\text{classical Ising model}} - \underbrace{\sum_{\langle i,j \rangle} g_{ij}X_iX_j}_{\text{interaction between transverse fields}}, \qquad (14.130)$$

and it can be shown that this model can simulate *universal* quantum computation.

14.2.3 Quantum Annealing

In metallurgy, *annealing* represents the process of creating crystals without defects. To achieve this, the crystal is locally heated, and the cooling is done in a controllable manner so that the internal stress is relieved and imperfectly placed atoms can return to their minimum energy positions. The temperature dictates the rate of diffusion. In *simulated annealing* we map the optimization problem to the atomic configuration, with random dislocation representing the change in the solution. To avoid being trapped into a local minimum, simulating annealing accepts the intermediate solution with a certain probability. So the probability here has the same role as a temperature in metallurgy; the higher it is, the more random dislocations are allowed. Therefore simulated annealing represents a metaheuristic approach to the approximate global optimization problem in a large search space. It is suitable when the search space is discrete.

Because thermal annealing has a higher barrier than quantum tunneling, *quantum annealing* applies quantum tunneling to control the acceptance probabilities [22,23]. The quantum tunneling field strength therefore has the same role as the temperature in metallurgy. Quantum annealing is a metaheuristic method of finding a global minimum of an objective function over a given set of candidate solutions (states) by employing quantum fluctuation-based computation. In adiabatic quantum computing, determination of the minimum energy gap can be hard to calculate. To solve this problem we can apply a similar approach to adiabatic quantum computing, but annealing repeatedly, while probably violating the adiabatic condition, and selecting the result with the lowest energy. By employing quantum annealing we can possibly bypass the higher-level models for quantum computing.

Following the problem definition we need to map the problem to the Ising model, which essentially is the QUBO problem. The next step is to perform *minor embedding* [24], which is similar to quantum compilation. In quantum hardware not all pairs of qubits are entangled, and connectivity is typically sparse. Very often we can use the arrangement of nodes known as the *Chimera graph* [25,26]. In the Chimera graph we create an eight-node cluster, representing the bipartite full graph, as illustrated in Fig. 14.20. The nodes can be further connected to the north and south clusters in the grid. Due to fabrication imperfections, some of the qubits might not be operational. Let us now consider the minimization problem $\min(\sigma_1\sigma_2 + \sigma_2\sigma_3 + \sigma_1\sigma_3)$, which is the triangle graph of three nodes. However, there are no three nodes in the graph on the right creating a triangle. We can group the physical qubits 1 and 6 into a logical qubit σ_3. We can interpret the qubits 1 and 6 as being strongly coupled to create the logical qubit. After that we execute the quantum annealing algorithm on the quantum processing unit (QPU).

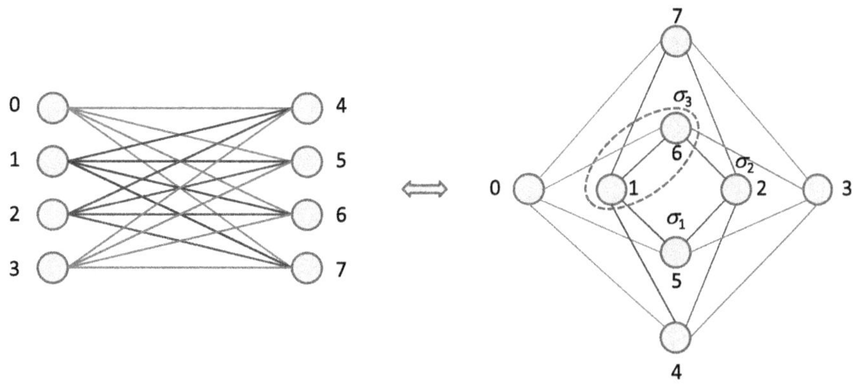

FIGURE 14.20

Eight-node cluster in the Chimera graph.

14.3 QUANTUM APPROXIMATE OPTIMIZATION ALGORITHM AND VARIATIONAL QUANTUM EIGENSOLVER

Quantum circuits are imperfect, which prevents us from running well-known quantum algorithms using the gates-based quantum computing approach. To overcome this problem, a new breed of quantum algorithms has been introduced [27–30], employing the parametrized shallow quantum circuits, which can be called *variational (quantum) circuits*. The variational circuits are suitable for noisy and imperfect quantum computers, which can also be called noisy intermediate-scale quantum devices [31]. They employ the *variational method* to find the approximation to the lowest energy eigenket or ground state, which is illustrated in Fig. 14.21. The QPU is embedded in the environment (bath), and there is an interaction between the QPU and the environment causing decoherence. The decoherence problem limits the ability to control the circuit and therefore the quantum circuit depth. To avoid this problem we can run a short burst calculation on the QPU, extract the results for the classical processor unit, which will optimize the parameters of the QPU to reduce the energy of the system, and feed back a new set of parameters to the QPU for the next iteration. We iterate this hybrid quantum classical algorithm until global minimum energy of the system is achieved. Any Hamiltonian H can be decomposed in terms of Hamiltonians of corresponding subsystems, that is, $H=\sum_i H_i$, where H_i is the Hamiltonian of the ith subsystem. The expected value of the Hamiltonian can then be determined as:

$$\langle H\rangle = \langle\psi|H|\psi\rangle = \langle\psi|\sum_i H_i|\psi\rangle = \sum_i\langle\psi|H_i|\psi\rangle = \sum_i\langle H_i\rangle, \tag{14.131}$$

where we exploit the linearity of quantum observables and $|\psi\rangle$ denotes the arbitrary input state. By applying this approach we effectively replace the long coherent

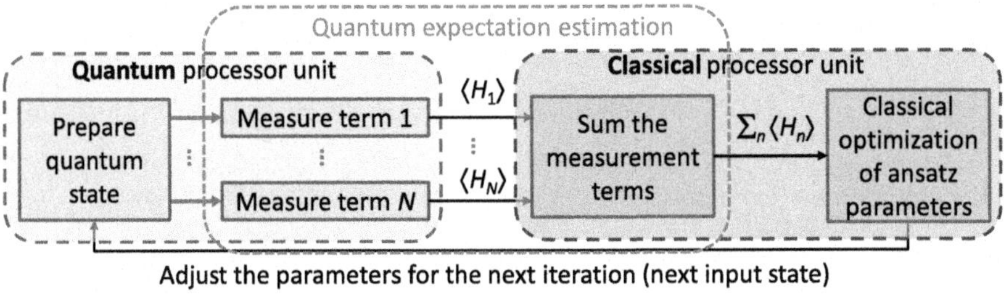

FIGURE 14.21

Variational quantum eigensolver principle.

evolution of QPU with multiple short coherent evolutions of the subsystems. The expected value of the Hamiltonian can also be decomposed in terms of corresponding eigenkets of the Hamiltonian $|\lambda_i\rangle$ as follows:

$$\langle H\rangle = \langle\psi|H|\psi\rangle = \sum_{\lambda_i,\,\lambda_j \in \text{Spectrum}(H)} \langle\psi|(|\lambda_i\rangle\langle\lambda_i|)H(|\lambda_j\rangle\langle\lambda_j|)|\psi\rangle = \sum_{\lambda_i,\,\lambda_j}\langle\psi|\lambda_i\rangle\langle\lambda_i|H|\lambda_j\rangle\langle\lambda_j|\psi\rangle$$

$$= \sum_{\lambda_i}|\langle\psi|\lambda_i\rangle|^2 \underbrace{\langle\lambda_i|H|\lambda_i\rangle}_{\lambda_i} \geq \sum_{\lambda_i}|\langle\psi|\lambda_i\rangle|^2 E_0 = E_0, \tag{14.132}$$

where E_0 is the lowest energy eigenket of the Hamiltonian. Varying over the entire Hilbert space is too complicated; we can choose a subspace of the entire Hilbert space, parametrized by some real-valued parameters p_i $(i = 1,2,\ldots,N)$, which is known as the ansatz. The choice of ansatz is relevant, as some choices of ansatzes can lead to better approximations.

The key idea behind the *quantum approximate optimization algorithm* (QAOA) is to approximate the quantum adiabatic pathway [27,32]. This algorithm is suitable for maximum satisfiability (MAX-SAT), MAX-CUT, and number partitioning problems; however, it can potentially be used to solve various NP-complete and NP-hard problems. As before, the optimization problem should be mapped to the Ising model with the desired (final) Hamiltonian being $H_f = -\sum_{\langle i,j\rangle} J_{ij}Z_iZ_j - \sum_i h_iX_i$, while the initial Hamiltonian is $H_0 = \sum_i X_i$. As before we apply the adiabatic quantum computing approach and introduce the combined Hamiltonian by $H(t) = (1 - t/T)H_0 + (t/T)H_f$, where T is the duration of the computation and $t \in [0,T]$. The corresponding unitary evolution operator is denoted by $U(t)$. The authors in [21] proved the lemma that the unitary evolution operator at the end of the process $U(T)$ can be approximated by $U'(T)$ if the distance between Hamiltonians is limited by $||H(t) - H'(t)|| \leq \delta$ for every t. The same authors proved the theorem claiming that the unitary transformation $U(T)$ can be approximated by p consecutive unitary transformations $U'_1,\ldots,U'_p$, wherein U'_i $(i = 1,\ldots,p)$ is defined by [21]:

$$U'_i = e^{-j\frac{T}{p}\overbrace{H\left(i\frac{T}{p}\right)}^{\left(1-\frac{i}{p}\right)H_0+\frac{i}{p}H_f}} = e^{-j\frac{T}{p}\frac{i}{p}H_f - j\frac{T}{p}\left(1-\frac{i}{p}\right)H_0}$$
$$\cong \underbrace{e^{-j\frac{T}{p}\frac{i}{p}H_f}}_{U\left(H_f,\ \frac{i}{p}\frac{T}{p}\right)} \underbrace{e^{-j\frac{T}{p}\left(1-\frac{i}{p}\right)H_0}}_{U\left(H_0,\ \left(1-\frac{i}{p}\right)\frac{T}{p}\right)} = U\left(H_f,\ \frac{i}{p}\frac{T}{p}\right)U\left(H_0,\ \left(1-\frac{i}{p}\right)\frac{T}{p}\right), \tag{14.133}$$

where in the second line we applied the Campbell–Baker–Hausdorff theorem claiming that we can approximate the parallel Hamiltonians with consecutive ones; in other words, $\|e^{A+B} - e^A e^B\|_2 \cong O(\|AB\|_2)$ [33]. So the key idea is to discretize the computation time T into p intervals of duration T/p, and then apply H_0 and H_f of duration $(1 - i/p)T/p$ and $(i/p)T/p$, respectively, p times. The discretized time steps do not need to be uniformly spaced as illustrated in Fig. 14.22. In other words, the ith step can be represented as:

$$U'_i = U(H_f,\ \gamma_i)U(H_0,\ \beta_i),\ \ \beta_i + \gamma_i = \Delta T_i,\ \ \sum_i \Delta T_i = T;\ i = 1, \ldots, p. \tag{14.134}$$

Therefore we apply H_0 and H_f Hamiltonians of duration β_i and γ_i as horizontal and vertical approximations of the adiabatic pathway in an alternative fashion as follows:

$$U(T) \cong U(H_f,\ \gamma_p)U(H_0,\ \beta_p)\ldots U(H_f,\ \gamma_1)U(H_0,\ \beta_1). \tag{14.135}$$

Overall, the final ground state in both adiabatic and approximating pathways would be the same. This approximation is known as the trotterization

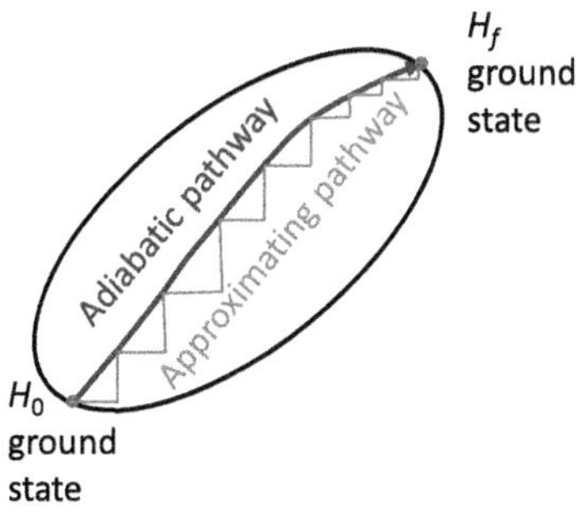

FIGURE 14.22

Approximating the adiabatic pathway.

(Trotter–Suzuki) decomposition. The accuracy can be improved by increasing the number of steps p. The Hamiltonian H_0 has for the ground state the tensor product of diagonal states $|+\rangle = (|0\rangle + |1\rangle)/\sqrt{2}$ (lowest energy state), which is one of two eigenkets of the Pauli X-operator, that is:

$$|\psi_0\rangle = |+\rangle^{\otimes N} = |+\rangle_{N-1} \otimes \ldots \otimes |+\rangle_0 = \frac{1}{\sqrt{2^n}} \sum_z |z\rangle, \ z \in \{0, 1\}^N. \quad (14.136)$$

The final state in the p step QAOA will then be:

$$|\psi_f(\beta, \gamma)\rangle = |\beta, \gamma\rangle = U(H_f, \gamma_p) U(H_0, \beta_p) \ldots U(H_f, \gamma_1) U(H_0, \beta_1) |\psi_0\rangle. \quad (14.137)$$

As an illustration, QAOA can simply be used to solve the following optimization problem:

$$C_{\max} = \max_z C(z) = \max_z \sum_{k=1}^{m} C_k(z), \ z \in \{0, 1\}^n, \quad (14.138)$$

where $C_k : \{0, 1\}^N \to \{0, 1\}$ for Boolean constraints (with 1 meaning that the constraint is satisfied) and $C_k : \{0, 1\}^N \to \mathscr{R}$ ($\mathscr{R}$ is the set of real numbers) for weighted constraints. The objective function can be represented as an operator:

$$C|z\rangle = \sum_{k=1}^{m} C_k(z)|z\rangle = f(z)|z\rangle. \quad (14.139)$$

Based on the previous equation we conclude that the operator C has the eigenvalues $f(z) = \sum_z C_k(z)$. The largest eigenvalue of $f(z)$ can be denoted $C_{\max} = f(z_{\max})$. The expectation of operator C for an arbitrary state $|\psi\rangle = \sum_z \alpha_z |z\rangle$ is given by:

$$\langle\psi|C|\psi\rangle = \sum_{z \in \{0,1\}^N} f(z)|\alpha_z|^2 \le \sum_{z \in \{0,1\}^N} f(z_{\max})|\alpha_z|^2 = f(z_{\max}) \underbrace{\sum_{z \in \{0,1\}^N} |\alpha_z|^2}_{1} = f(z_{\max}). \quad (14.140)$$

For the superposition state given by Eq. (14.136), the expectation of C becomes:

$$\langle\psi_0|C|\psi_0\rangle = \sum_{z \in \{0,1\}^N} f(z)\frac{1}{2N} = \frac{1}{2N} \sum_{z \in \{0,1\}^N} f(z). \quad (14.141)$$

The probability of obtaining $|z_{\max}\rangle$ from each measurement is 2^{-N}. The corresponding unitary evolution operator for C is given by:

$$U(C, \gamma) = e^{-j\gamma C} = e^{-j\gamma \sum_{k=1}^{m} C_k} = \prod_{k=1}^{m} e^{-j\gamma C_k}, \quad (14.142)$$

where the second line comes from the Baker–Campbell–Hausdorff theorem since $[C_i, C_j] = 0$. Let us now observe the action of $U(C,\gamma)$ on arbitrary state $|\psi\rangle$:

$$\underbrace{U(C,\gamma)}_{\prod_{k=1}^{m} e^{-j\gamma C_k}} \underbrace{|\psi\rangle}_{\sum_z \alpha_z |z\rangle} = \sum_z \alpha_z \prod_{k=1}^{m} e^{-j\gamma C_k} |z\rangle, \quad e^{-j\gamma C_k} = \begin{cases} I, & C_k = 0 \\ e^{-j\gamma} I, & C_k = 1 \end{cases}, \tag{14.143}$$

indicating that for any $|z\rangle$, the phase term $\exp(-j\gamma)$ is added when the condition C_k is satisfied.

The evolution of initial Hamiltonian $B = H_0 = \sum_i X_i$ is given by:

$$\begin{aligned} U(H_0, \beta) &= e^{-j\beta H_0} = e^{-j\beta \sum_{i=1}^{N} X_i} \\ &= \prod_{i=1}^{N} e^{-j\beta X_i} = \prod_{i=1}^{N} U(B_i, \beta), \quad U(B_i, \beta) = e^{-j\beta X_i}. \end{aligned} \tag{14.144}$$

where the second line is obtained from the commutativity of X_i and X_j, that is, $[X_i, X_j] = 0$. Clearly, $U(B_i, \beta)$ is just the rotation operator for 2β with respect to the x-axis [34], that is, $R_x(2\beta)$. Now by applying Eq.(14.137), the final state is given by:

$$|\boldsymbol{\beta}, \boldsymbol{\gamma}\rangle = U(C, \gamma_p) U(B, \beta_p) \ldots U(C, \gamma_1) U(B, \beta_1) |\psi_0\rangle. \tag{14.145}$$

Now we have to choose the parameters $\boldsymbol{\beta}$ and $\boldsymbol{\gamma}$ to maximize the expectation of C with respect to state $|\boldsymbol{\beta}, \boldsymbol{\gamma}\rangle$; in other words:

$$C_p = \max_{(\boldsymbol{\beta}, \boldsymbol{\gamma})} \langle \boldsymbol{\beta}, \boldsymbol{\gamma} | C | \boldsymbol{\beta}, \boldsymbol{\gamma} \rangle. \tag{14.146}$$

As we mentioned earlier, the parameters must be properly chosen such that accuracy increases as we increase p, that is, $C_p \geq C_{p-1}$. As the number of steps tends to infinity, we have that:

$$\lim_{p \to \infty} C_p = \max_z C(z) = C_{\max}. \tag{14.147}$$

The QAOA can be summarized as illustrated in Fig. 14.23.

Let us now apply the QAOA to the MAX-CUT problem, which represents the problem of partitioning the nodes (vertices) of a given undirected graph $G = \{V, E\}$ (V is the set of vertices and E is the set of edges connecting them) into two sets, say S and S, such that the number of edges connecting the nodes in opposite sets is maximized. The *cut* δ is defined as the set of the edges in the graph separating S and S. As an illustration, let us consider the square ring graph in Fig. 14.24 (left). Since

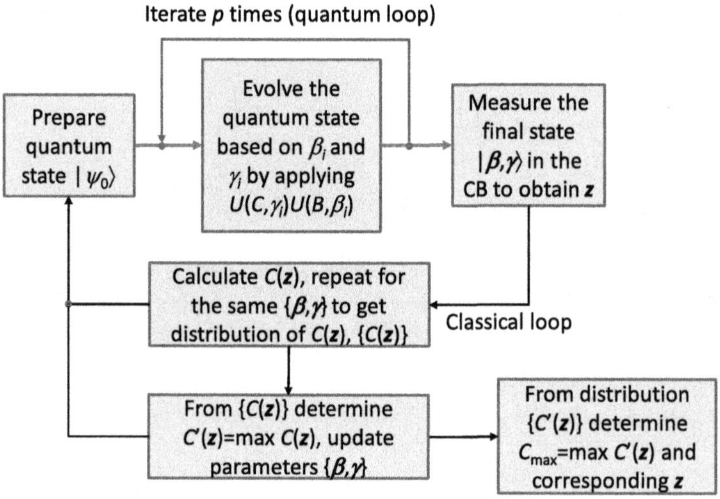

FIGURE 14.23

Summary of the quantum approximate optimization algorithm. *CB*, Computational basis.

there are $N = 4$ nodes, there exist 2^N representations of the graph partitioning, with two being illustrated in Fig. 14.24. We can use 4-bit strings to represent different partitions. If a particular bit position belongs to the set S we assign the corresponding bit 1, otherwise 0.

Any undirected graph is described by N vertices (nodes) and the edge set $\{\langle j,k\rangle\}$. Let the $C_{\langle j,k\rangle}$ function return 1 when the edge spans two nodes j and k in different sets, and 0 otherwise. The corresponding cost function will then be:

$$C = \sum_{\langle j,\, k\rangle} C_{\langle j,\, k\rangle},\ C_{\langle j,\, k\rangle} = \frac{1}{2}(1 - z_j z_k), \tag{14.148}$$

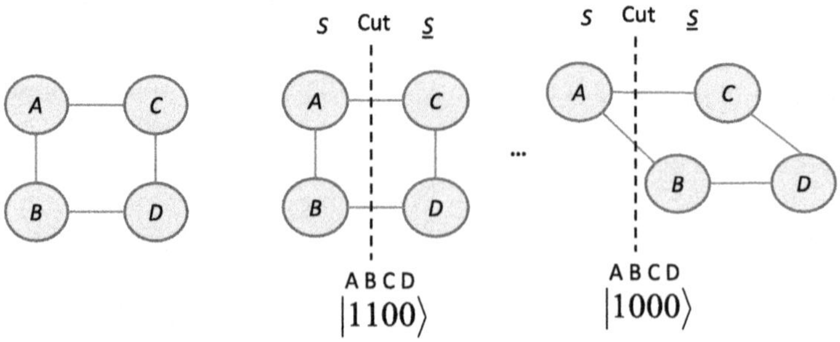

FIGURE 14.24

MAX-CUT problem.

where $z_i = +1$ if the ith ($i \in \{j,k\}$) node is located in the set S, otherwise $z_i = -1$. In other words, if the edge $\langle j,k \rangle$ is cut we get $z_j z_k = -1$; if the same edge is not cut we have that $z_j z_k = 1$. The objective function can be represented as an operator:

$$C|z\rangle = \sum_{\langle i,\, k\rangle} C_{\langle i,\, k\rangle}(z)|z\rangle = C(z)|z\rangle, \tag{14.149}$$

where the action of operator $C_{\langle j,k\rangle}$ on state $|z\rangle$ is given by:

$$C_{\langle i,\, k\rangle}|z\rangle = \frac{1}{2}(I - Z_i Z_k)|z\rangle = \begin{cases} |z\rangle, \text{ if the edge}\langle i,\, k\rangle \text{is cut, that is } Z_i Z_k|z\rangle = -|z\rangle \\ 0, \text{ if the edge}\langle i,\, k\rangle \text{is not cut, that is } Z_i Z_k|z\rangle = |z\rangle \end{cases}. \tag{14.150}$$

The corresponding unitary evolution operator for C is given by:

$$U(C,\, \gamma) = e^{-j\gamma C} = e^{-j\gamma \sum_{\langle i,\, k\rangle} C_{\langle i,\, k\rangle}} = \prod_{\langle i,\, k\rangle} \underbrace{e^{-j\gamma C_{\langle i,\, k\rangle}}}_{U\left(C_{\langle i,\, k\rangle},\, \lambda\right)} = \prod_{\langle i,\, k\rangle} U\left(C_{\langle i,\, k\rangle},\, \lambda\right), \tag{14.151}$$

where:

$$U\left(C_{\langle i,\, k\rangle},\, \lambda\right) = e^{-j\gamma C_{\langle i,\, k\rangle}} = e^{-j\frac{\gamma}{2}(I - Z_i Z_k)} = e^{-j\frac{-\gamma}{2} Z_i Z_k} e^{-j\frac{\gamma}{2} I}. \tag{14.152}$$

On the other hand, the evolution of initial Hamiltonian $B = H_0 = \sum_i X_i$ is given by Eq. (14.144). The workflow for the MAX-CUT problem is similar to Eq. (14.145). To simplify the problem we can exploit some properties of the graphs as discussed in [27].

14.4 QUANTUM BOOSTING

In quantum boosting (QBoost) we use a different loss function compared to the exponential loss function used in AdaBoost, described in Section 14.1.1. Similarly as before, for a given dataset $\{(\boldsymbol{x}_1,y_1),\ldots,(\boldsymbol{x}_n,y_n)\}$ we are also given K weak learners, say $\{h_k(\boldsymbol{x})\colon \mathscr{R}^d \to \{-1,1\};\ k = 1,\ldots,K\}$. Now we need to combine appropriately those weak learners to get a composite, strong learner C_m ($m \le K$) as follows:

$$C_m(\boldsymbol{w},\, \boldsymbol{x}_i) = C_{m-1}(\boldsymbol{w},\, \boldsymbol{x}_i) + w_m h_m(\boldsymbol{x}_i), \tag{14.153}$$

where w_m is the weight derived by minimizing the corresponding cost function. The boosting problem can be interpreted as the following optimization problem [26,35]:

$$\widehat{\boldsymbol{w}} = \arg\min_{\boldsymbol{w}} \left[\frac{1}{N} \sum_{n=1}^{N} \left(\sum_{k=1}^{K} w_k h_k(\boldsymbol{x}_n) - y_n \right)^2 + \lambda \|\boldsymbol{w}\|_0 \right], \tag{14.154}$$

where the second term is the regularization term with the zero-norm counting the number of nonzero entries. The zero weight means that a particular weak learner is not included in the model. The smaller hyperparameter λ indicates that the penalty of including more weak learners is lower. This model is continuous, since the weights are real valued; however, what matters is whether a particular weak learner is included or not. Therefore we can discretize the model by representing the weight with a certain number of bits. We need to do some modifications to the minimization problem to make it more suitable for Hamiltonian encoding—the Ising model representation. Therefore we can take the square of terms in brackets to get:

$$\hat{\boldsymbol{w}} = \arg\min_{\boldsymbol{w}} \left\{ \frac{1}{N} \sum_{n=1}^{N} \left[\left(\sum_{k=1}^{K} w_k h_k(\boldsymbol{x}_n) \right)^2 - 2y_n \sum_{k=1}^{K} w_k h_k(\boldsymbol{x}_n) + y_n^2 \right] + \lambda \|\boldsymbol{w}\|_0 \right\}. \tag{14.155}$$

Given that y_n^2 is just the offset, not w dependent, we can omit it, and after squaring the summation term and rearranging the summation operations we obtain:

$$\hat{\boldsymbol{w}} = \arg\min_{\boldsymbol{w}} \left\{ \frac{1}{N} \sum_{k=1}^{K} \sum_{l=1}^{K} w_k w_l \left(\sum_{n=1}^{N} h_k(\boldsymbol{x}_n) h_l(\boldsymbol{x}_n) \right) - \frac{2}{N} \sum_{k=1}^{K} w_k \sum_{n=1}^{N} h_k(\boldsymbol{x}_n) y_n + \lambda \|\boldsymbol{w}\|_0 \right\}. \tag{14.156}$$

The second term is the bias term, which corresponds to the $\sum_k b_k \sigma_k$ term (the bias term) in the Ising model, with b_k given by $b_k = \sum_n h_k(\boldsymbol{x}_n) y_n$. The b_k is related to the correlation between h_k and $\boldsymbol{y}$. The first term can be related to the Ising model interaction term $\sum \sigma_k \sigma_l J_{kl}$, wherein the coupling strength is given by $J_{kl} = \sum_n h_k(\boldsymbol{x}_n) h_l(\boldsymbol{x}_n)$, representing therefore the correlation between the kth and lth weak learners. Upon the quantization of weights, the boosting problem becomes the QUBO problem. If we use the superscript (Q) to denote the quantization terms we can rewrite the previous optimization problem as follows:

$$\begin{aligned} \hat{\boldsymbol{w}} = \arg\min_{\boldsymbol{w}} \Bigg\{ & \frac{s}{N} \sum_{k=1}^{K} \sum_{l=1}^{K} w_k^{(Q)} w_l^{(Q)} \left(\sum_{n=1}^{N} h_k(\boldsymbol{x}_n) h_l(\boldsymbol{x}_n) \right) \\ & - \frac{2s}{N} \sum_{k=1}^{K} w_k^{(Q)} \sum_{n=1}^{N} h_k(\boldsymbol{x}_n) y_n + \lambda \sum_{k=1}^{K} w_k^{(Q)} \Bigg\}, \end{aligned} \tag{14.157}$$

where we introduced the scaling factor s to maintain the desired margins, that is, to ensure the mappings h_k: $\boldsymbol{x}_n \rightarrow \{-1/N, 1/N\}$. For 1-bit quantization the minimization problem becomes the QUBO problem. By setting $\sigma_k = 2w_k^{(Q)} - 1$ and using the definitions for b_k and J_{kl} from earlier, we converted this QUBO problem into the Ising model. Now we can employ either quantum annealing or QAOA to solve this problem efficiently. For more than one quantization bit we need to introduce an axillary variable $w_{k,\text{aux}}$ for each weight w_k to enforce the regularization condition by [26,35]:

$$\lambda\left\|w^{(Q)}\right\| = \sum_{k=1}^{K} \kappa w_k^{(Q)}\left(1 - w_{k,\text{ aux}}\right) + \lambda w_{k,\text{ aux}}, \tag{14.158}$$

where $w_{k,\text{aux}}$ serves as the indicator having value 1 when $w_k > 0$, and 0 otherwise. The normalization term κ needs to be sufficiently large for proper regularization.

14.5 QUANTUM RANDOM ACCESS MEMORY

Classical random access memory (RAM) allows us to access a memory unit with a unique address. With an n-bit address we can access 2^n memory units. On the other hand, quantum RAM (QRAM) allows us to address a superposition of memory cells with the help of a quantum index (address) register [36]. In other words, the address register is a superposition of addresses $\sum_i p_i|i\rangle_a$, while the output register is a superposition of information in data register d, entangled to the address register as follows [36]:

$$\sum_i p_i|i\rangle_a \overset{\text{QRAM}}{\rightarrow} \sum_i p_i|i\rangle_a|m_i\rangle_d, \tag{14.159}$$

where m_i is the content of the ith memory cell. By employing bucket-brigade architecture, the authors in [36] have shown that to retrieve an item in QRAM with the address register composed of n qubits we will need $O(\log 2^n)$ switching cells instead of 2^n needed in conventional (classical or quantum) RAM. The operation principle of conventional RAM is illustrated by the bifurcation graph in Fig. 14.25. We assume that RAM is composed of 2^n memory cells, placed at the bottom of the bifurcation graph. The address register is composed of n flip-flops as shown in Fig. 14.25, for $n = 4$. The content of the ith flip-flop corresponds to the ith level of the graph. If the ith component of the address is 0 we move to the next level using the left edge; otherwise the right edge. The binary representation of the third memory location (with 4 bits) is 0011, and from the root node we follow the following sequence of edges: left, left, right, and right to reach the correct memory unit. The same architecture can be applied to the quantum level, as suggested in [37]; however, the requirement for coherence is too stringent. In the bucket-brigade, the authors in [36] proposed to place the "trit" memory elements at the tree nodes, which can be in one of three states: *waiting*, *left*, and *right*. Initially, they are all initialized to the low-energy waiting state as illustrated in Fig. 14.26. The address bits are sent to the graph in a serial fashion. When an incoming bit encounters a waiting state, 0-bit transforms it to the left state and 1-bit to the right state. If a trit memory unit on a given level is not in a waiting state, it just routs the current bit. When all address bits are sent, the corresponding route is carved into the tree as illustrated in Fig. 14.27 for address 011.

In QRAM, the trits are replaced by qutrits, being in one of three possible quantum states: $|\text{waiting}\rangle$, $|\text{left}\rangle$, and $|\text{right}\rangle$. The address qubits are sent in similar

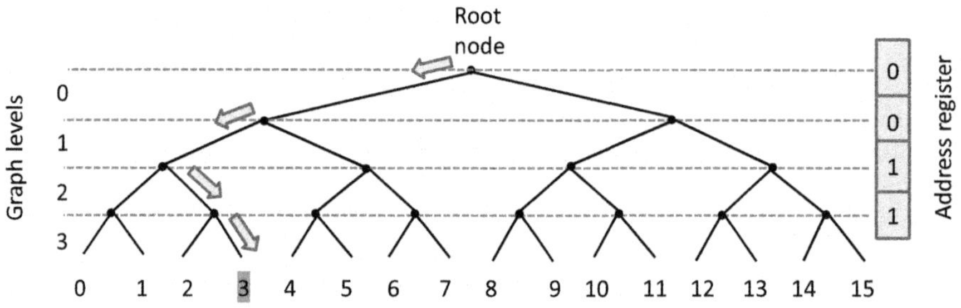

FIGURE 14.25

Operation principle of conventional random access memory.

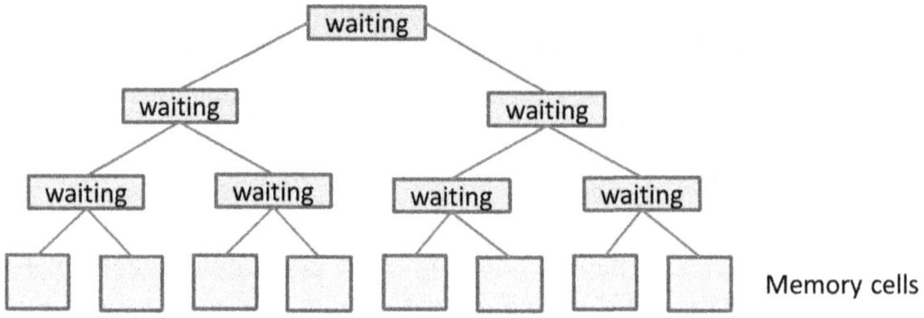

FIGURE 14.26

Initialization of the bifurcation graph for the bucket-brigade architecture.

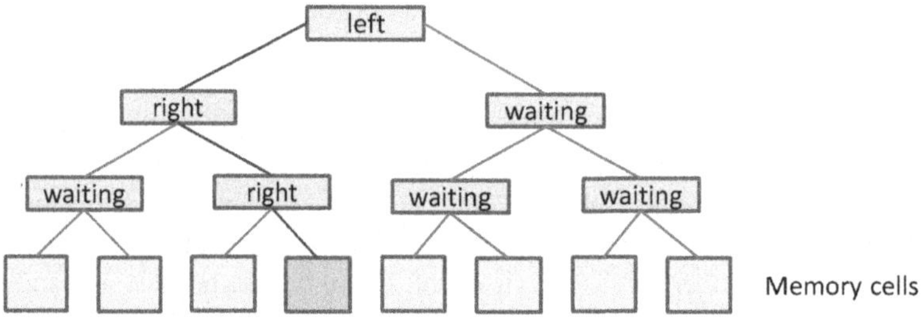

FIGURE 14.27

Addressing of the bifurcation graph for the bucket-brigade architecture for address 011.

fashion as described earlier. However, the key difference is that when the route is carved through the tree, the quantum superposition of addresses will carve the superpositions of routes through the graph. The bus qubit is then injected to interact with the memory cells along the routes being activated. The result of interaction is then sent back through the graph to write it into the output register through the root node. Finally, the reverse evolution of states on the qutrits is performed to reset all the states to the $|\text{waiting}\rangle$ state.

The key advantage of the bucket-brigade procedure is the small number of qutrits needed in the retrieval of information, which is $\log 2^n$. One possible realization, based on ion traps, is proposed in [36].

14.6 QUANTUM MATRIX INVERSION

The quantum matrix inversion algorithm, also known as the *Harrow–Hassidim–Lloyd (HHL) algorithm* [38], has often been used as an ingredient of other quantum ML algorithms, including quantum support vector machine (QSVM) [39] and quantum principal component analysis (QPCA) [40–42], and as such it shall be briefly described here. We are interested in solving the following linear system of equations $\boldsymbol{Ax} = \boldsymbol{b}$, wherein $\boldsymbol{A}$ is an $N \times N$ Hermitian matrix, and $\boldsymbol{b}$ is a column vector $\boldsymbol{b} = [b_1 \ldots b_N]^{\mathrm{T}}$. In other words we are interested in finding $\boldsymbol{x} = \boldsymbol{A}^{-1}\boldsymbol{b}$. If matrix-$\boldsymbol{A}$ is not Hermitian or a square matrix we can do matrix inversion of the $\boldsymbol{A}'$-matrix instead:

$$\boldsymbol{A}' = \begin{bmatrix} \boldsymbol{0} & \boldsymbol{A}^\dagger \\ \boldsymbol{A} & \boldsymbol{0} \end{bmatrix}, \tag{14.160}$$

which is clearly Hermitian and satisfies the following linear system of equations:

$$\begin{bmatrix} \boldsymbol{0} & \boldsymbol{A}^\dagger \\ \boldsymbol{A} & \boldsymbol{0} \end{bmatrix} \begin{bmatrix} \boldsymbol{x} \\ \boldsymbol{0} \end{bmatrix} = \begin{bmatrix} \boldsymbol{0} \\ \boldsymbol{b} \end{bmatrix}. \tag{14.161}$$

We map this problem to the corresponding quantum problem as follows:

$$A|x\rangle = |b\rangle, \ |b\rangle = \begin{bmatrix} b_1 \\ \vdots \\ b_N \end{bmatrix} = \sum_{i=1}^{N} b_i |i\rangle, \tag{14.162}$$

where the matrix representation of Hermitian operator A is given by matrix $\boldsymbol{A}$. To do so, we need to amplitude encode $\boldsymbol{b}$ by $|b\rangle = \sum_i b_i |i\rangle$ and apply Hermitian encoding for $\boldsymbol{A}$. The corresponding evolution operator will be defined by $U = \exp(\mathrm{j}At)$, which is a unitary operator with eigenvalues $\exp(\mathrm{j}\lambda_i t)$, where λ_i are eigenvalues of matrix $\boldsymbol{A}$ with corresponding eigenkets being $|u_i\rangle$. The spectral decomposition of A will be $A = \sum_i \lambda_i |u_i\rangle\langle u_i|$, so the goal is to find $|x\rangle = A^{-1}|b\rangle$. If the matrix $\boldsymbol{A}$ is either sparse or well conditioned the runtime of the quantum matrix inversion algorithm will be $O(\log N\,\kappa^2)$ (κ is the condition number) [38], which represents the exponential speed-up over the corresponding classical algorithm whose runtime is $O(N\kappa)$. In quantum machine learning (QML), very often we are not interested in finding the values of $\boldsymbol{x}$ themselves, but rather the result of applying the operator M to $|x\rangle$, or expectation of

M, $\langle M \rangle$. The condition number is defined as the ratio of the maximum and minimum eigenvalues of $\boldsymbol{A}$.

When the $\boldsymbol{A}$ matrix is diagonal, that is, $\boldsymbol{A} = \text{diag}(\lambda_1,\ldots,\lambda_N)$, we need to perform the following transformation $|i\rangle \rightarrow \lambda_i^{-1}|i\rangle$, which can be done probabilistically assuming that the normalization constant c is chosen such that $c/|\lambda| \leq 1$. Now by performing the mapping:

$$|i\rangle \rightarrow \left(\sqrt{1-\frac{c^2}{|\lambda_i|^2}}|0\rangle + \frac{c}{\lambda_i}|1\rangle\right) \otimes |i\rangle, \tag{14.163}$$

and measuring the first qubit until we get 1, the resulting state will be $A^{-1}|b\rangle$. By applying the unitary operator $U = \exp(\text{j}At)$ to $|b\rangle$ for a superposition of different time instances we can decompose $|b\rangle$ into eigenkets of $\boldsymbol{A}$, denoted by $|u_i\rangle$, while at the same time we can determine the corresponding eigenvalue λ_i with the help of the quantum phase estimation (QPE) algorithm as illustrated in Fig. 14.28. (Readers not familiar with quantum Fourier transform and QPE should consult first Chapter 5.) After that decomposition, the system output can be represented as follows:

$$|b\rangle|\mathbf{0}\rangle \rightarrow \sum_i \beta_i |u_i\rangle|\lambda_i\rangle. \tag{14.164}$$

We then perform the conditional rotation on ancilla $|0\rangle$:

$$\sum_i \beta_i |u_i\rangle|\lambda_i\rangle \otimes |0\rangle \rightarrow \sum_i \beta_i |u_i\rangle|\lambda_i\rangle \left(\sqrt{1-\frac{c^2}{|\lambda_i|^2}}|0\rangle + \frac{c}{\lambda_i}|1\rangle\right), \tag{14.165}$$

which is clearly not unitary. Now we need to uncompute the eigenvalue register (by applying QPE^{-1}) to uncorrelate it from the data register ($|b\rangle$ register). Given that the

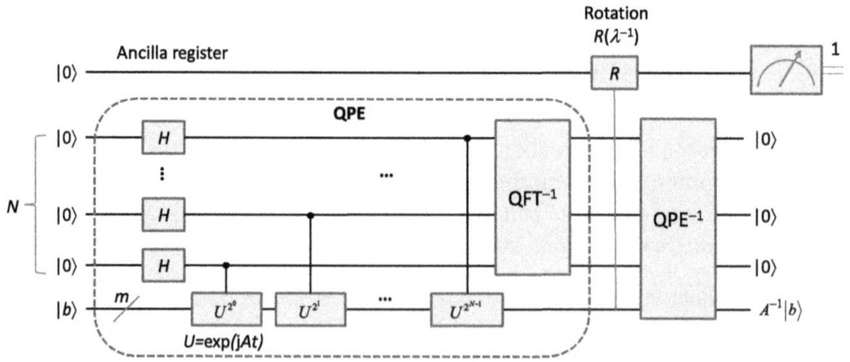

FIGURE 14.28

Quantum matrix inversion. *QPE*, Quantum phase estimation; *QFT*, quantum Fourier transform.

foregoing mapping is not unitary we need to apply rejection sampling on the ancilla (by repeating) until we get 1, which effectively performs the following mapping on the data register:

$$\sum_i \beta_i |u_i\rangle \to \sum_i \beta_i \lambda_i^{-1} |u_i\rangle = A^{-1}|b\rangle. \tag{14.166}$$

14.7 QUANTUM PRINCIPAL COMPONENT ANALYSIS

QPCA is based on the spectral decomposition of a matrix, which is equivalent to simulating the Hamiltonian [26,40]. The quantum simulator approximates the evolution operator exp(jHt), where the Hamiltonian H can be expressed in terms of simpler Hamiltonians $H = \sum_i H_i$. The goal is to simulate the evolution operator exp(jHt) by a sequence of exponentials exp($-jH_i t$) such that the error of the final state does not exceed certain $\varepsilon > 0$. Approximation of the evolution operator by a sequence of exponentials is based on the Baker–Campbell–Hausdorff formula [33]:

$$e^X e^Y = e^{X+Y+\frac{1}{2}[X,\ Y]+\frac{1}{12}[X,\ [X,\ Y]]-\frac{1}{12}[Y,\ [X,\ Y]]+\cdots} \tag{14.167}$$

When X and Y are sufficiently small in Lie algebra, the series on the right side of the exponential is convergent, and we can apply the Lie product formula:

$$e^X e^Y = \lim_{n\to\infty}\left(e^{X/n}e^{Y/n}\right)^n. \tag{14.168}$$

To simulate the evolution operator we can now use the following Trotter decomposition [43]:

$$e^{-jHt} = \left(e^{-jH_1\Delta t}e^{-jH_2\Delta t}\cdots e^{-jH_l\Delta t}\right)^n + \frac{t^2}{2n}\sum_{i>k}[H_i,\ H_k] + \sum_{p=3}^{\infty}\varepsilon(p),\quad \Delta t = t/n, \tag{14.169}$$

wherein we have split the total simulation time t into n short time intervals $\Delta t = t/n$. When the dimensionality of Hilbert space H_i is d_i, the number of operations needed to simulate exp(j$H_i t$) is $\sim d_i^2$, according to Lloyd [44]. When the simpler Hamiltonians H_i act on local qubits only, the evaluation of exp(j$H_i t$) is more efficient. Based on Lloyd's interpretation, the number of operations needed to simulate exp($-$jHt) is:

$$n\left(\sum_{i=1}^{l} d_i^2\right) \le nld^2,\ d = \max d_i. \tag{14.170}$$

The higher-order terms in Eq. (14.169) are bounded by [43]:

$$\|\varepsilon(p)\|_{\sup} \le \frac{n}{p!}\|H\Delta t\|_{\sup}^p, \tag{14.171}$$

where we use $||\bullet||_{\sup}$ to denote the supremum (maximum expectation of a given operator over all possible states). The total error for approximating exp(−j*Ht*) with the first term in Eq. (14.169) is bounded by: $\left\|n\left(e^{-jH\Delta t}-1-jH\Delta t\right)\right\|_{\sup}$. For a given threshold error ε, from the second term in Eq. (14.169) we conclude that $\varepsilon \sim t^2/n$. Therefore the total number of operations required to simulate the evolution operator is $O(t^2ld^2/\varepsilon)$. If every Hamiltonian acts on no more than k local qubits, the parameter l is bounded by $l_{\max} = \binom{n}{k} < n^k/k!$, indicating that the number of operations is polynomial in the problem size.

By applying the density operator ρ as a Hamiltonian on another density operator σ, the complexity needed for component Hamiltonian can be reduced to $O(\log d)$ [42]. By taking the partial trace of the first variable and the SWAP operator S we obtain the following [42]:

$$\begin{aligned}\mathrm{Tr}_P\left(e^{-jS\Delta t}\rho\otimes\sigma e^{-jS\Delta t}\right) &= \left(\cos^2\Delta t\right)\sigma+\left(\sin^2\Delta t\right)\rho - j\sin\Delta t\cos\Delta t[\rho,\sigma]\\ &\cong \sigma - j\Delta t[\rho,\sigma]+O\left(\Delta t^2\right).\end{aligned} \tag{14.172}$$

Given that the SWAP matrix S is sparse, $\exp(-jS\Delta t)$ can be computed efficiently. Assuming that n copies of ρ are available we repeat the SWAP operations on $\rho\otimes\sigma$ to construct the unitary operators $\exp(-jn\rho\Delta t)$ and thus simulate unitary time evolution $\exp(-jn\rho\Delta t)\sigma\exp(jn\rho\Delta t)$. Based on Suzuki–Trotter theory, to simulate $\exp(-j\rho t)$ to accuracy ε, we need $O(t^2\varepsilon^{-1}||\rho-\sigma||^2)\le O(t^2\varepsilon^{-1})$ steps, where $t=n\Delta t$. To retrieve ρ using QRAM we need $O(\log d)$ operations. To implement $\exp(-j\rho t)$ we need $O(t^2\varepsilon^{-1})$ copies of ρ. Therefore to implement $\exp(-j\rho t)$ with accuracy ε we will need $O(t^2\varepsilon^{-1}\log d)$ operations.

Let us now apply the *quantum phase estimation algorithm* for $U=\exp(-j\rho t)$ to determine the eigenkets and eigenvalues of the unknown density matrix. QPE conditionally executes U for different time intervals and maps the initial state $|\psi\rangle|0\rangle$ to $\sum_i \psi_i|\lambda_i\rangle|\tilde{r}_i\rangle$, where $|\lambda_i\rangle$ are eigenkets of ρ, $\psi_i=\langle\lambda_i|\psi\rangle$, and $\tilde{r}_i$ are estimates of corresponding eigenvalues. By applying QPE on ρ itself we essentially perform the following mapping [42]:

$$\rho|0\rangle\langle 0|\rightarrow\sum_i r_i|\lambda_i\rangle\langle\lambda_i|\otimes|\tilde{r}_i\rangle\langle\tilde{r}_i|. \tag{14.173}$$

Now by sampling from this state we can reveal the features of the eigenkets and eigenvalues of ρ. This usage of multiple copies of ρ to construct eigenvalues and eigenkets is commonly referred to as QPCA.

QCPA is suitable for state discrimination and assignment problems, relevant in both supervised and unsupervised learning problems, in particular classification and clustering. As an illustration, let as assume that two sets of data vectors are available represented by $\{|\psi_i\rangle\}$ $(i=1,\ldots,m)$ and $\{|\phi_i\rangle\}$ $(i=1,\ldots,n)$, which can also be represented using density-matrix formalism as follows $\rho=(1/m)\sum_i|\psi_i\rangle\langle\psi_i|$ and

$\sigma = (1/n)\sum_i |\phi_i\rangle\langle\phi_i|$. To classify the incoming state $|\zeta\rangle$ we can represent it in the eigenbasis of $\rho - \sigma$. For positive eigenvalues we assign it to the first set; otherwise to the second set.

14.8 QUANTUM OPTIMIZATION-BASED CLUSTERING

In clustering, for a given dataset $D = \{\boldsymbol{x}_i\}$ $(i = 1,\ldots,N)$, $\boldsymbol{x}_i \in \mathscr{R}^d$ we need to assign the data points to clusters; in other words, to assign labels based on their similarity. We can use the distance measure between two points $\boldsymbol{x}_i$ and $\boldsymbol{x}_j$ as a measure of similarity/dissimilarity, and we can create the Gram matrix $\boldsymbol{K} = (K_{ij})$, $K_{ij} = d(\boldsymbol{x}_i, \boldsymbol{x}_j)$. The Euclidean distance is traditionally used for this purpose; however, other similarity measures can be used such as the scalar product $\boldsymbol{x}_i^{\mathrm{T}}\boldsymbol{x}_j$. Now we can introduce the weighted undirected graph $G = (V,E)$, where V is the set of vertices and E is the set of the edges connecting the vertices. The set of the vertices is represented by the points from the dataset. We can simply create the fully connected graph, such as the one provided in Fig. 14.29 for $N = 4$. In clustering, the distant data points are assigned to different clusters, therefore maximizing the sum of distances of points with different labels can be used as a clustering approach. Clearly, this problem is equivalent to the MAX-CUT problem discussed in Section 14.3. The weight of the edge between points i and j, denoted as w_{ij}, corresponds to the distance $d(\boldsymbol{x}_i, \boldsymbol{x}_j)$. In the MAX-CUT problem the goal is to partition the vertices of the graph G into two sets S and S, such that the number of edges connecting the nodes in opposite sets is maximized. The *cut* δ is defined as the set of the edges in the graph separating S and S, which is illustrated in Fig. 14.29. The *cost of the cut* can be defined as the sum of weight of edges connecting the vertices in S with vertices in S, that is, we can write:

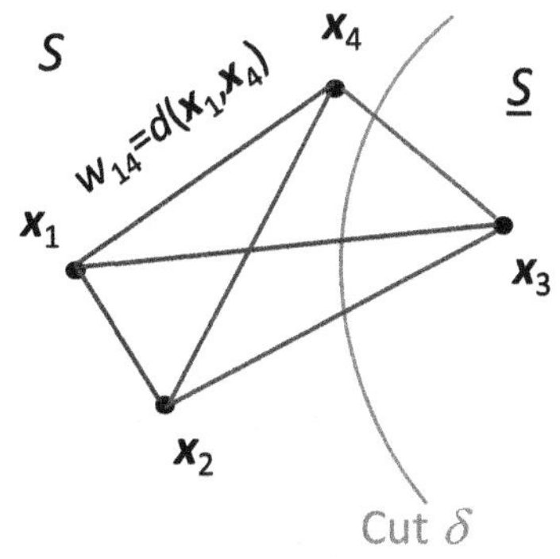

FIGURE 14.29

Relating clustering to the MAX-CUT problem.

$$w(\delta(S)) = \sum_{i\in S,\, j\in \underline{S}} w_{ij} = \sum_{i\in S,\, j\in \underline{S}} d(\boldsymbol{x}_i,\, \boldsymbol{x}_j). \tag{14.174}$$

The MAX-CUT optimization problem can now be defined as:

$$\max_{S\subset V} w(\delta(S)) = \sum_{i\in S,\, j\in \underline{S}} d(\boldsymbol{x}_i,\, \boldsymbol{x}_j). \tag{14.175}$$

Let us now map this optimization problem to the Ising model by introducing the Ising spin variables $\sigma_i \in \{-1,1\}$, wherein $\sigma_i = +1$ when $\boldsymbol{x}_i \in S$; otherwise $\sigma_i = -1$ (when $\boldsymbol{x}_i \in \underline{S}$). Given the symmetry of the distances, $w_{ij} = w_{ji}$, we can redefine the cost of the cut as follows:

$$w(\delta(S)) = \sum_{i\in S,\, j\in \underline{S}} w_{ij} = \frac{1}{4}\sum_{i,\, j\in V} w_{ij} - \frac{1}{4}\sum_{i,\, j\in V} w_{ij}\sigma_i\sigma_j = \frac{1}{4}\sum_{i,\, j\in V} w_{ij}(1-\sigma_i\sigma_j). \tag{14.176}$$

Clearly, when both data points belong to the same cluster (be it S or $\underline{S}$) their contribution to the cut cost is 0. By replacing the spin variables with spin operators we obtain the following Hamiltonian of the corresponding Ising model:

$$H_S = -\frac{1}{4}\sum_{i,\, j\in V} w_{ij}\left(1-\sigma_i^z\sigma_j^z\right), \tag{14.177}$$

wherein we introduce a minus sign to ensure that the optimal solution is the minimum energy state. Regarding the summation term, in the Ising model the nearest neighbors are considered in summation, denoted as $\langle i,j\rangle$. Given that the graph G here is fully connected, $\sum_{\langle i,j\rangle}$ is the same as $\sum_{i,j}$. Now we can use the QAOA to determine the minimum energy state as was done in Section 14.3. Alternatively, quantum annealing can be used instead.

14.9 GROVER ALGORITHM-BASED GLOBAL QUANTUM OPTIMIZATION

The Grover search algorithm, as already explained in Chapter 5, is used to search for an entry in an unstructured database, and provides quadratic speed-up compared to the classical counterpart [45]. The Grover search operator is given by:

$$G = H^{\otimes n}\left(2|0\rangle^{\otimes n}\langle 0|^{\otimes n} - I\right)H^{\otimes n}O, \tag{14.178}$$

where O is the quantum *oracle* operator whose action is given by:

$$O_a|x\rangle = (-1)^{f(x)}\left|x\right\rangle,\ f(x) = \begin{cases} 1, & x = a \\ 0, & x \neq a \end{cases}, \tag{14.179}$$

with $f(x)$ being the *search function* that generates 1 when the desired item a is found. The action of $H^{\otimes n}$ on $|0\rangle^{\otimes n}$ states is the superposition state:

$$H^{\otimes n}|0\rangle^{\otimes n} = H|0\rangle \otimes \ldots \otimes \underbrace{H|0\rangle}_{(|0\rangle+|0\rangle)/\sqrt{2}} = \frac{1}{\sqrt{2^n}} \sum_{x=0}^{2^n} |x\rangle = |s\rangle, \tag{14.180}$$

so that the Grover search operator can also be represented by:

$$G = \underbrace{(2|s\rangle\langle s| - I)}_{G_d} O_a = G_d O_a, \; G_d = 2|s\rangle\langle s| - I, \tag{14.181}$$

where G_d is known as the *Grover diffusion operator.* An alternative representation of the oracle operator is to consider it a conditional operator with an ancillary qubit being the target qubit, and the corresponding action of the oracle will then be:

$$O_a|x\rangle|y\rangle = |x\rangle|y \oplus f(x)\rangle, \tag{14.182}$$

and for $|y\rangle = |-\rangle = H|1\rangle = (|0\rangle - |1\rangle)/\sqrt{2}$ we obtain:

$$O_a|x\rangle|-\rangle = \frac{1}{\sqrt{2}}|x\rangle|(|0\rangle - |1\rangle) \oplus f(x)\rangle = \begin{cases} \frac{1}{\sqrt{2}}|x\rangle|(|1\rangle - |0\rangle)\rangle = -|x\rangle|-\rangle, \; f(x) = 1 \\ \frac{1}{\sqrt{2}}|x\rangle|(|0\rangle - |1\rangle)\rangle = |x\rangle|-\rangle, \; f(x) = 0 \end{cases}. \tag{14.183}$$

The simplified version of the Grover search algorithm is provided in Fig. 14.30. To get the desired entry we need to apply the Grover search operator $\sim(\pi/4)2^{n/2}$ times, as we have already shown in Chapter 5. The measurement at the end of algorithm will reveal the searched item a with a very high probability.

By redefining the oracle operator, the Grover search algorithm can also be used in the optimization of the objective function $f(x)$: $\{0,1\}^n \to \mathscr{R}$, as proposed in [46,47], and the corresponding algorithm can be called *Grover algorithm-based global optimization.* The corresponding *minimization oracle operator* is given by:

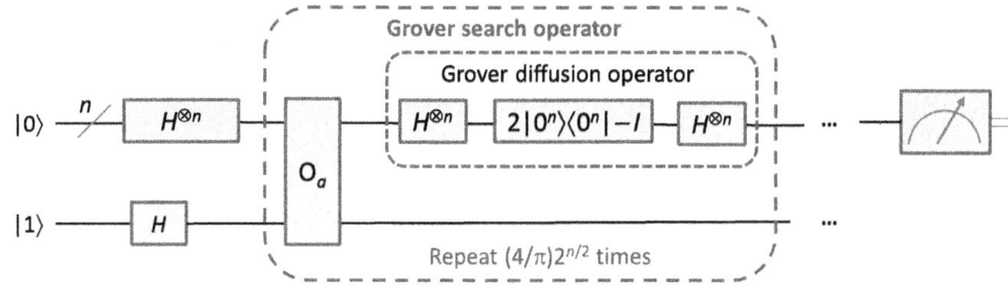

FIGURE 14.30

Quantum circuit to perform the Grover search algorithm.

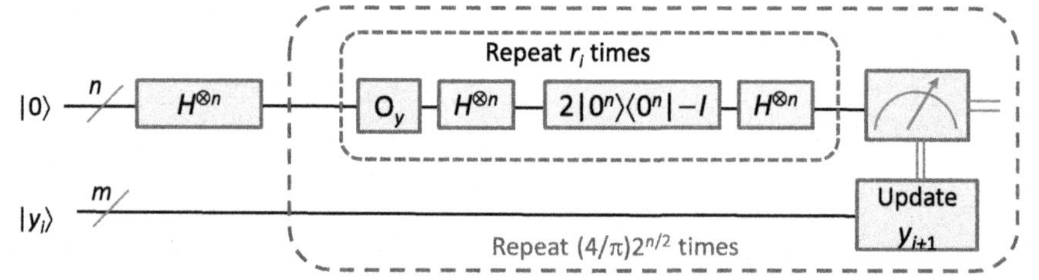

FIGURE 14.31

Quantum circuit to perform the Grover search-based minimization algorithm.

$$O_y|x\rangle = (-1)^{h(x)}|x\rangle,\ h(x) = \begin{cases} 1,\ f(x) < y \\ 0,\ f(x) \geq y \end{cases}. \tag{14.184}$$

To initialize the oracle operator we randomly select input x_0 and set $y_0 = f(x_0)$ to be a starting point in the algorithm. We also create the superposition state as described earlier. In the ith iteration we apply the Grover search operator G r_i times. The output of the Grover algorithm will be state $|x\rangle$, and we then calculate $y = f(x)$ and compare to y_i. If $y < y_i$ we set $x_{i+1} = x$ and $y_{i+1} = y$, which represents the classical step. We then update the threshold in the oracle function, and move to the next iteration, as illustrated in Fig. 14.31. After approximately $O(2^{n/2})$ Grover algorithm iterations we measure the upper register to determine the $x_{\min}$. The number of inner loops r_i is significantly lower than the number of outer loops 2^n. The inner loops are needed to avoid the system being trapped in a local minimum.

14.10 QUANTUM *K*-MEANS

Many quantum clustering algorithms employ the Grover search algorithm [48]; however, they do not provide exponential speed-up over the corresponding classical clustering algorithm. In particular, finding the distance of vectors to the centroids in large dimensional space is computationally extensive on a classical computer, and the corresponding quantum algorithm for distance calculation can provide an exponential speed-up. To describe the quantum *K*-means algorithm we need first to explain how to calculate the dot product efficiently and how to calculate the distance between vectors using the quantum procedure [41].

14.10.1 Scalar Product Calculation

Scalar product calculation is based on the quantum SWAP circuit, introduced in Chapter 3, which is provided in Fig. 14.32. After the first Hadamard gate, we can represent the state of the quantum circuit by:

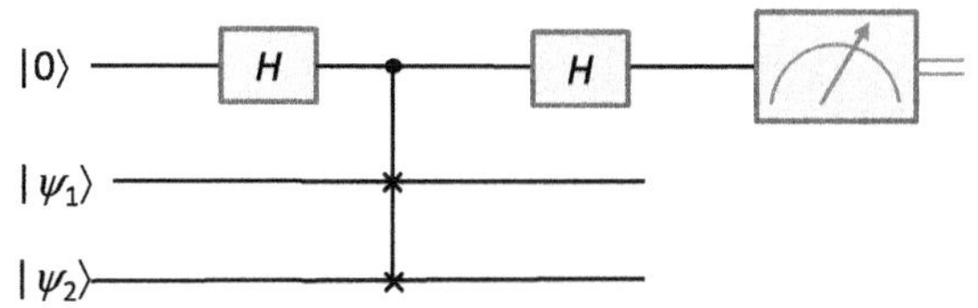

FIGURE 14.32

Quantum SWAP testing circuit.

$$|\Psi\rangle = \frac{1}{\sqrt{2}}(|0\rangle|\psi_1\rangle|\psi_2\rangle + |1\rangle|\psi_1\rangle|\psi_2\rangle). \tag{14.185}$$

Upon applying the SWAP gate we obtain:

$$|\Psi'\rangle = \frac{1}{\sqrt{2}}(|0\rangle|\psi_1\rangle|\psi_2\rangle + |1\rangle|\psi_2\rangle|\psi_1\rangle). \tag{14.186}$$

After the second Hadamard gate, the overall state becomes:

$$\begin{aligned} |\Psi''\rangle &= \frac{1}{2}[(|0\rangle + |1\rangle)|\psi_1\rangle|\psi_2\rangle + (|0\rangle - |1\rangle)|\psi_2\rangle|\psi_1\rangle] \\ &= \frac{1}{2}|0\rangle(|\psi_1\rangle|\psi_2\rangle + |\psi_2\rangle|\psi_1\rangle) + \frac{1}{2}|1\rangle(|\psi_1\rangle|\psi_2\rangle - |\psi_2\rangle|\psi_1\rangle). \end{aligned} \tag{14.187}$$

The probability of the control qubit being in the $|0\rangle$ state is given by:

$$\begin{aligned} P(|0\rangle) = |\langle 0|\Psi''\rangle|^2 &= \left| \frac{1}{2}\underbrace{\langle 0|0\rangle}_{1}(|\psi_1\rangle|\psi_2\rangle + |\psi_2\rangle|\psi_1\rangle) \right. \\ &\quad \left. + \frac{1}{2}\underbrace{\langle 0|1\rangle}_{0}|1\rangle(|\psi_1\rangle|\psi_2\rangle - |\psi_2\rangle|\psi_1\rangle) \right|^2 \\ &= \frac{1}{4}(\langle\psi_1|\langle\psi_2| + \langle\psi_2|\langle\psi_1|)(|\psi_1\rangle|\psi_2\rangle + |\psi_2\rangle|\psi_1\rangle) \\ &= \frac{1}{4}[2\langle\psi_1|\psi_1\rangle\langle\psi_2|\psi_2\rangle + 2\langle\psi_1|\psi_2\rangle\langle\psi_2|\psi_1\rangle] \\ &= \frac{1}{2} + \frac{1}{2}|\langle\psi_1|\psi_2\rangle|^2. \end{aligned} \tag{14.188}$$

$P(|0\rangle) = 0.5$ means that corresponding states are orthogonal, while $P(|0\rangle) = 1$ means that the states are identical. By repeating the measurement multiple times we can obtain an accurate estimate of $P(|0\rangle)$ and we can determine the magnitude squared of the scalar product by:

$$|\langle\psi_1|\psi_2\rangle|^2 = 2P(|0\rangle) - 1. \tag{14.189}$$

14.10.2 Quantum Distance Calculation

The classical information contained in the vector $\boldsymbol{x}$ can be represented as a quantum state by employing amplitude encoding, as introduced earlier:

$$\|\boldsymbol{x}\|^{-1}\boldsymbol{x} \to |x\rangle = \sum_{i=0}^{d-1} \|\boldsymbol{x}\|^{-1} x_i |x_i\rangle, \tag{14.190}$$

and to reconstruct the state we need to query the QRAM $\log_2 d$ times. To determine the distance between two data points $\boldsymbol{x}_i$ and $\boldsymbol{x}_j$ we initialize the following two states [41]:

$$\begin{aligned} |\psi\rangle &= \frac{1}{\sqrt{2}}(|0\rangle|x_i\rangle + |1\rangle|x_j\rangle), \\ |\phi\rangle &= Z^{-1/2}(\|\boldsymbol{x}_i\||x_i\rangle + \|\boldsymbol{x}_i\||x_j\rangle), \ Z = \|\boldsymbol{x}_i\|^2 + \|\boldsymbol{x}_j\|^2. \end{aligned} \tag{14.191}$$

To create the first state we use the QRAM, while for the second state we create the state $H|1\rangle|0\rangle$ and let it evolve by applying the following Hamiltonian:

$$\hat{H} = (\|\boldsymbol{x}_i\||0\rangle\langle 0| + \|\boldsymbol{x}_j\||1\rangle\langle 1|) \otimes \sigma_X, \tag{14.192}$$

to get:

$$\begin{aligned} e^{j\hat{H}t} \underbrace{H|1\rangle}_{(|0\rangle-|1\rangle)/\sqrt{2}} |0\rangle &= 2^{-1/2}[\cos(\|\boldsymbol{x}_i\|t)|0\rangle - \cos(\|\boldsymbol{x}_j\|t)|1\rangle]|0\rangle \\ &- j2^{-1/2}[\sin(\|\boldsymbol{x}_i\|t)|0\rangle - \sin(\|\boldsymbol{x}_j\|t)|1\rangle]|1\rangle. \end{aligned} \tag{14.193}$$

For sufficiently small t, by measuring the ancilla qubit we get the desired state $|\phi\rangle$ with probability $0.5Z^2t^2$. To determine the Euclidian distance squared between $\boldsymbol{x}_i$ and $\boldsymbol{x}_j$ we need first to run the quantum SWAP test from Fig. 14.32 by setting $|\psi_1\rangle = |\psi\rangle$ and $|\psi_2\rangle = |\phi\rangle$ to determine the $|\langle\phi|\psi\rangle|^2 = 2P(|0\rangle) - 1$. The Euclidean distance squared can now be determined by:

$$\|\boldsymbol{x}_i - \boldsymbol{x}_j\|^2 = 2Z|\langle\phi|\psi\rangle|^2. \tag{14.194}$$

The dot product between $\boldsymbol{x}_i$ and $\boldsymbol{x}_j$ can now be calculated simply by $\left(Z - \|\boldsymbol{x}_i - \boldsymbol{x}_j\|^2\right)/2 = \boldsymbol{x}_i^{\mathrm{T}}\boldsymbol{x}_j$.

14.10.3 Grover Algorithm-Based *K*-Means

We are now in a position to formulate the Grover algorithm-based quantum K-means algorithm. We randomly initialize the cluster centroids to $\boldsymbol{c}_k$ $(k = 1,\ldots,K)$ such as in the K-means++ algorithm [49]. We iterate the following two inner loops until convergence is achieved:

- *Determine the closest cluster centroid.* Loop over training samples and for each training sample $\boldsymbol{x}_i$ ($i = 1,\ldots,N$) use the *quantum distance calculation* procedure, described in Subection 14.10.2, to calculate the distances to each cluster centroids $\|\boldsymbol{x}_i - \boldsymbol{c}_k\|$ ($k = 1,\ldots,K$). Use the *Grover algorithm-based optimization* procedure, described in Section 14.9 (Fig. 14.31), to determine the index of cluster centroid $c^{(k)}$ that minimizes the distance between training sample and cluster centroid:

$$c^{(\hat{k})} = \arg\min_k \|\boldsymbol{x}_i - \boldsymbol{c}_k\|^2. \tag{14.195}$$

- *Calculate new cluster centroids.* Loop over clusters C_k ($k = 1,\ldots,K$) and calculate the means of the data points assigned to each cluster as follows:

$$\boldsymbol{c}_k = \frac{1}{|C_k|}\sum_{j \in C_k} \boldsymbol{x}_j, \tag{14.196}$$

where $|C_k|$ denotes the number of vectors assigned to the cluster C_k. Clearly, this step is purely classical.

Convergence is achieved when the distances between cluster centroids at current and previous iterations are lower than a given threshold ζ:

$$\left\|\boldsymbol{c}_k^{(\text{current})} - \boldsymbol{c}_k^{(\text{previous})}\right\|^2 < \zeta. \tag{14.197}$$

The complexity of classical Lloyd's K-means algorithm is $O(N\log(Nd))$ [41], with factor N coming from the fact that every vector is tested for assignment. To reduce the complexity further we can formulate the K-means problem as the quadratic unconstrained optimization problem, and apply quantum annealing or QAOA in similar fashion as we have done in Section 14.8. The complexity can be reduced to $O(K\log(Nd))$ or even possibly $O(\log(KNd))$ according to [41]. Namely, to reduce the complexity from $O(N\log N)$ to $O(\log N)$, the output of quantum K-means should not be in the list of N vectors and their assignments to clusters, but instead the following superposition clustering state [41]:

$$|\psi_c\rangle = N^{-1/2}\sum_i |c_i\rangle|i\rangle = N^{-1/2}\sum_c\sum_{i \in c_i}|c\rangle|i\rangle, \tag{14.198}$$

wherein the state denotes the superposition of clusters with each cluster being the superposition of vectors assigned to it. To obtain the statistical picture of the clustering algorithm we need to sample from this state. The procedure to construct this clustering state by adiabatic algorithm is provided in the supplementary material of [41], and constructing the clustering state with accuracy ε has $O(\varepsilon^{-1}K\log(KNd))$ complexity. When the clusters are well separated so that the adiabatic gap is constant during adiabatic computing, the complexity can be further reduced to $O(\varepsilon^{-1}\log(KNd))$.

14.11 QUANTUM SUPPORT VECTOR MACHINE

The simplest form of QSVM discretizes the cost function and performs Grover algorithm-based optimization [50] in a similar fashion as described in Section 14.9. The real strength of such an algorithm is the ability to optimize the nonconvex objective function; however, the speed-up over corresponding classical soft-margin SVM is not exponential.

For exponential speed-up we should use the QSVM with the least-squares formulation as proposed in [51] and experimentally studied in [39]. The key idea behind classical least-squares SVM is to modify the soft-margin optimization problem (Eq. 14.44) as follows [52]:

$$\left(\widehat{\boldsymbol{w}}, \widehat{b}\right) = \underset{\boldsymbol{w}, b}{\arg\min}\left(\frac{1}{2}\|\boldsymbol{w}\|^2 + \frac{\gamma}{2}\sum_{j=1}^{N}\zeta_j^2\right), \tag{14.199}$$
$$\text{subject to: } y_j\left(\boldsymbol{w}^{\mathrm{T}}\cdot\varphi(\boldsymbol{x}_j) - b\right) = 1 - \zeta_j,\ \zeta_j \geq 0;\ \forall j = 1, 2, \ldots, n$$

where $\{\boldsymbol{x}_j, y_j\}$ ($\boldsymbol{x}_j \in \mathscr{R}^d$, $y_j \in \{-1,1\}$) represents the dataset, while the nonlinear mapping $\varphi(\boldsymbol{x}_j)$ is used in the kernel function $k(\boldsymbol{x}_j, \boldsymbol{x}_k) = \varphi(\boldsymbol{x}_j)^{\mathrm{T}}\cdot\varphi(\boldsymbol{x}_k)$. We can reformulate the optimization problem using the Lagrangian method as follows:

$$L(\boldsymbol{w}, b, \boldsymbol{\alpha}, \zeta) = \frac{1}{2}\boldsymbol{w}^{\mathrm{T}}\boldsymbol{w} + \frac{\gamma}{2}\sum_{j=1}^{N}\zeta_j^2 - \sum_{j=1}^{N}\alpha_i\left\{\left[\boldsymbol{w}^{\mathrm{T}}\cdot\varphi(\boldsymbol{x}_j) - b\right] - y_j + \zeta_j\right\},$$
$$\boldsymbol{\alpha} = \left[\alpha_1 \ldots \alpha_N\right]^{\mathrm{T}}. \tag{14.200}$$

Now by taking the first derivatives of L with respect to $\boldsymbol{w}$, b, α_i, and ζ_i and setting them to zero we obtain the following optimality conditions:

$$\begin{aligned}
&\frac{\partial L}{\partial \boldsymbol{w}} = 0 \Rightarrow \boldsymbol{w} = \sum_{j=1}^{N}\alpha_j\varphi(\boldsymbol{x}_j),\\
&\frac{\partial L}{\partial b} = 0 \Rightarrow \sum_{j=1}^{N}\alpha_j = 0,\\
&\frac{\partial L}{\partial \alpha_j} = 0 \Rightarrow y_j = \boldsymbol{w}^{\mathrm{T}}\cdot\varphi(\boldsymbol{x}_j) + b + \zeta_j,\\
&\frac{\partial L}{\partial \zeta_j} = 0 \Rightarrow \alpha_j = \gamma\zeta_j;\ j = 1, \ldots, N.
\end{aligned} \tag{14.201}$$

After eliminating $\boldsymbol{w}$ and ζ_j we obtain the following system of linear equations to solve instead of the QUBO problem as follows:

$$\begin{bmatrix} 0 & \mathbf{1}^T \\ 1 & \boldsymbol{K}+\gamma^{-1}\boldsymbol{I} \end{bmatrix}\begin{bmatrix} b \\ \boldsymbol{\alpha} \end{bmatrix}=\begin{bmatrix} 0 \\ \boldsymbol{y} \end{bmatrix},\ \boldsymbol{y}=[y_1...y_N]^T,\ \boldsymbol{K}=[K_{ij}],\ K_{ij}=\varphi(\boldsymbol{x}_i)^T\varphi(\boldsymbol{x}_j), \tag{14.202}$$

where $\boldsymbol{K}$ is the kernel matrix of size $N \times N$ and $\boldsymbol{I}$ is the identity matrix (of the same size).

To solve for $[b\boldsymbol{\alpha}]^{\mathrm{T}}$ we need to invert the matrix:

$$\boldsymbol{F}=\begin{bmatrix} 0 & \mathbf{1}^T \\ 1 & \boldsymbol{K}+\gamma^{-1}\boldsymbol{I} \end{bmatrix}=\underbrace{\begin{bmatrix} 0 & \mathbf{1}^T \\ 1 & 0 \end{bmatrix}}_{\boldsymbol{J}}+\underbrace{\begin{bmatrix} 0 & 0 \\ 1 & \boldsymbol{K}+\gamma^{-1}\boldsymbol{I} \end{bmatrix}}_{\boldsymbol{K}_\gamma}=\boldsymbol{J}+\boldsymbol{K}_\gamma, \tag{14.203}$$

whose complexity on the classical computer is $O[(N+1)^3]$. On the other hand, by using quantum matrix inversion (HHL algorithm), the complexity can be significantly reduced. We map this linear system equation-solving problem to the corresponding quantum problem as follows:

$$\widehat{F}|b,\,\alpha\rangle=|y\rangle,\quad \widehat{F}=\boldsymbol{F}/\mathrm{Tr}(\boldsymbol{F}),\quad \left\|\widehat{F}\right\|<1. \tag{14.204}$$

To perform the quantum matrix inversion we first apply the Lie product by:

$$e^{-j\widehat{F}\Delta t}=e^{-j\Delta t\boldsymbol{J}/\mathrm{Tr}(\boldsymbol{K}_\gamma)}e^{-j\Delta t\gamma^{-1}\boldsymbol{I}/\mathrm{Tr}(\boldsymbol{K}_\gamma)}e^{-j\Delta t\boldsymbol{K}/\mathrm{Tr}(\boldsymbol{K}_\gamma)}+O\left(\Delta t^2\right). \tag{14.205}$$

Given that the matrix $\boldsymbol{J}$ is sparse we can apply the method due to Berry et al. [40]. However, if the matrix $\boldsymbol{K}$ is not sparse we can apply QPCA by creating several copies of density matrix ρ and performing the $\exp(-j\rho t)$ as explained in Section 14.7. To create the normalized kernel matrix we encode the training points by employing Eq. (14.190) and then determine the dot products between $\boldsymbol{x}_i$ and $\boldsymbol{x}_j$ as explained in Section 14.10. The training data oracle performs the following mapping [51]:

$$N^{-1/2}\sum_{i=1}^{N}|i\rangle \rightarrow |\chi\rangle = C_\chi^{-1/2}\sum_{i=1}^{N}\|\boldsymbol{x}_i\||i\rangle|\boldsymbol{x}_i\rangle,\quad C_\chi=\sum_{i=1}^{N}\|\boldsymbol{x}_i\|^2, \tag{14.206}$$

with complexity $O(\log(Nd))$. By disregarding the dataset register we obtain the desired kernel matrix:

$$\mathrm{Tr}_2(|\chi\rangle\langle\chi|)=\sum_{j=1}^{N}\sum_{i=1}^{N}\langle\boldsymbol{x}_j|\boldsymbol{x}_i\rangle\|\boldsymbol{x}_i\|\|\boldsymbol{x}_j\||i\rangle\langle j|=\frac{\boldsymbol{K}}{\mathrm{Tr}(\boldsymbol{K})}=\widehat{K}. \tag{14.207}$$

Now we perform the exponentiation in $O(\log N)$ steps by [51]:

$$\begin{aligned} e^{-jL_{\widehat{K}}\Delta t}(\rho)&=e^{-j\widehat{K}\Delta t}\rho e^{j\widehat{K}\Delta t}\cong\mathrm{Tr}_1\left(e^{-jS\Delta t}\widehat{K}\rho e^{jS\Delta t}\right)\\ &=\rho-j\Delta t\left[\widehat{K},\,\rho\right]+\boldsymbol{O}\left(\Delta t^2\right),\ S=\sum_{m=1}^{N}\sum_{n=1}^{N}|m\rangle\langle n|\otimes|n\rangle\langle m|, \end{aligned} \tag{14.208}$$

with S being the SWAP matrix. Regarding the vector $|\boldsymbol{y}\rangle$, it can be expanded into eigenkets $|u_k\rangle$ of F, with corresponding eigenvalues being λ_k, with the help of PCA, described in Section 14.8, as follows: $|\widetilde{y}\rangle = \sum_{k=1}^{N+1} \langle u_k|\widetilde{y}\rangle |u_k\rangle$. Now we follow the steps from Section 14.8, including controlled rotation of the ancilla qubit and eigenvalue register uncomputation, effectively performing the following mappings [51]:

$$|\widetilde{y}\rangle|0\rangle \rightarrow \sum_{k=1}^{N+1} \langle u_k|\widetilde{y}\rangle |u_k\rangle|\lambda_k\rangle \rightarrow \sum_{k=1}^{N+1} \frac{\langle u_k|\widetilde{y}\rangle}{\lambda_k} |u_k\rangle. \tag{14.209}$$

The $|b, \boldsymbol{\alpha}\rangle$ state can be expressed in the training set labels basis by [51]:

$$|b, \boldsymbol{\alpha}\rangle = C^{-1/2}\left(b|0\rangle + \sum_{k=1}^{N} \alpha_k|k\rangle\right), \; C = b^2 + \sum_{k=1}^{N} \alpha_k^2. \tag{14.210}$$

Based on the parameters of the $|b, \boldsymbol{\alpha}\rangle$ state we construct the state:

$$|\widetilde{u}\rangle = C_{\widetilde{u}}^{-1/2}\left(b|0\rangle|0\rangle + \sum_{i=1}^{N} \alpha_i \|\boldsymbol{x}_i\| |i\rangle|\boldsymbol{x}_i\rangle\right), \; C_{\widetilde{u}} = b^2 + \sum_{i=1}^{N} \alpha_i^2 \|\boldsymbol{x}_i\|^2, \tag{14.211}$$

which will be used in *classifying the query state* $|\boldsymbol{x}\rangle$, which can be constructed in QRAM as:

$$|\widetilde{x}\rangle = C_{\widetilde{x}}^{-1/2}\left(|0\rangle|0\rangle + \sum_{i=1}^{N} \|\boldsymbol{x}\| |i\rangle|\boldsymbol{x}\rangle\right), \; C_{\widetilde{u}} = N\|\boldsymbol{x}\|^2 + 1. \tag{14.212}$$

For classification purposes we perform the SWAP text as explained in Section 14.10. The key idea is to construct the state $|\psi\rangle = 2^{-1/2}\left(|0\rangle|\widetilde{u}\rangle + |1\rangle\left|\widehat{x}\right\rangle\right)$ and measure the overlap with $|\phi\rangle = 2^{-1/2}(|0\rangle + |1\rangle)$. The success probability will be:

$$P = |\langle\phi|\psi\rangle|^2 = \frac{1}{2}(1 - \langle\widetilde{u}|\widetilde{x}\rangle), \; \langle\widetilde{u}|\widetilde{x}\rangle = C_{\widetilde{x}}^{-1/2} C_{\widetilde{u}}^{-1/2}\left(b + \sum_{i=1}^{N} \alpha_i \|\boldsymbol{x}_i\| \|\boldsymbol{x}\| \langle \boldsymbol{x}_i|\boldsymbol{x}\rangle\right). \tag{14.213}$$

When $P < 1/2$ we classify $|\boldsymbol{x}\rangle$ as $+1$; otherwise we classify it as -1.

14.12 QUANTUM NEURAL NETWORKS

In this section we describe several representative examples of quantum neural networks (QNNs), including feedforward QNNs, the perceptron, and quantum convolutional neural networks (QCNNs).

14.12.1 Feedforward QNNs

Feedforward QNNs represent straightforward generalization of the classical ANNs [53,54]. However, there are two key differences with respect to classical neural networks: (1) quantum operations are reversible and unitary, while the classical activation function is not reversible, and (2) weighting of the quantum states does not have any practical importance given that quantum states are of unit length. However, if we interpret the neurons as qubits, we can use the Hamiltonian to describe the QNN, with weights now representing the interaction strength. To solve the reversibility problem we can add extra inputs denoted as 0 inputs and propagate the data points to the output layer as illustrated in Fig. 14.33.

In this particular example we assume that the numbers of neurons (of the irreversible neural network) at input, hidden, and output layers are I, J, and K, respectively. For a neural network to be reversible we need to propagate the inputs to the output layer.

To represent the classical data input x_i as a quantum state we can use the binary representation $x_i = [x_{i,0},\ldots, x_{i,L-1}]^{\mathrm{T}}$ and encode it to the quantum state as follows $|x_i\rangle = |x_{i,0}\rangle \otimes \ldots \otimes |x_{i,L-1}\rangle$. In similar fashion, the kth output quantum state will be $|y_k\rangle = |y_{k,0}\rangle \otimes \ldots \otimes |y_{k,L-1}\rangle$. The quantum feedforward neural network can now be obtained by replacing the nodes in the hidden layer by corresponding unitary quantum gates U_j ($j = 1,\ldots,J$) as illustrated in Fig. 14.34. The action of the jth unitary operator on $M = I + K$ qubits can be represented by:

$$U_i = \exp\left[j \sum_{m_1,\ldots,\, m_M=0\ldots0}^{3\ldots3} \alpha^{(i)}_{m_1,\ldots,m_N} \sigma^{(i)}_{m_1} \otimes \ldots \otimes \sigma^{(i)}_{m_N}\right],\ \sigma_{m_l} \in \{I,\ X,\ Y,\ Z\}, \quad (14.214)$$

where σ_{m_l} ($l = 1,\ldots,M$) represent the Pauli matrices. The cost function can be defined as follows:

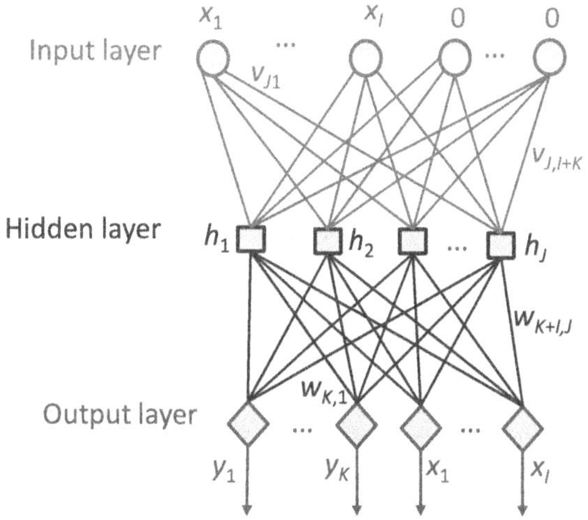

FIGURE 14.33

Reversible feedforward neural network example.

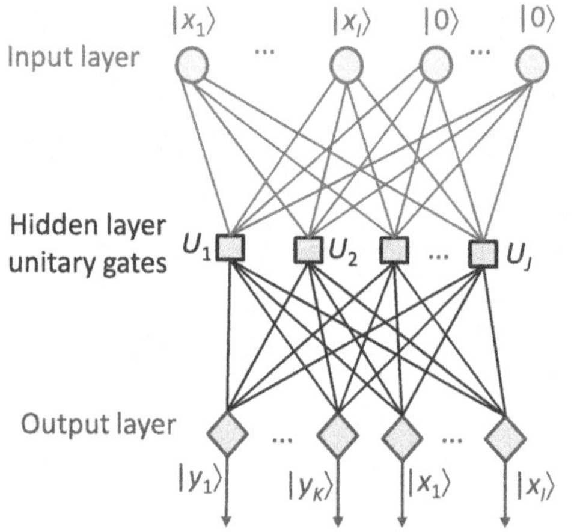

FIGURE 14.34

Quantum feedforward neural network example.

$$C = -\sum_{n=1}^{N}\left\langle \mathbf{y}_U^{(n)} \middle| \mathbf{y}^{(n)} \right\rangle, \tag{14.215}$$

where $\mathbf{y}_U^{(n)}$ is dependent on the unitary operator parameters $\alpha_{m_1,\ldots,m_N}^{(i)}$ and $\mathbf{y}^{(n)}$ represents the labels at the output neurons for the nth training instance ($n = 1,\ldots,N$). We can use the gradient descent to update the α parameters in the ith unitary matrix as follows:

$$\Delta\alpha_{m_1,\ldots,m_N}^{(i)} = -\eta\frac{\partial C}{\partial\alpha_{m_1,\ldots,m_N}^{(i)}}, \tag{14.216}$$

where η dictates the rate of the update. The quantum gradient descent algorithms would be more adequate for this purpose [55,56]. Another version of the cost function was proposed in [53]:

$$C = \sum_{n=1}^{N}\sum_{i=1}^{3} f_{i,\,n}\left[\langle\sigma_i\rangle_{\text{actual}} - \langle\sigma_i\rangle\right],\ f_{i,\,n} \geq 0, \tag{14.217}$$

where the $\langle\sigma_i\rangle$ are expected values of Pauli operators at individual outputs $y^{(n)}$ (assuming that there exists a single output node $|y\rangle$), while $\langle\sigma_i\rangle_{\text{actual}}$ represent actual expected values of Pauli operators (obtained by the measurements on QNN outputs).

The state of the QNN composed of N quantum neurons can be represented as the multiparticle state living in 2^N-dimensional Hilbert space [57]:

$$|\Psi\rangle = \sum_{i=1}^{2^N}\alpha_i|i_1 i_2 \ldots i_N\rangle, \tag{14.218}$$

where α_i represents the complex amplitude of the ith QNN basis state. Zak and Williams introduced quantum formalism that was able to capture the nonlinearity and dissipation properties of the QNN by replacing the activation function with a quantum measurement [58]. They were concerned with the unitary quantum walk between computational basis states rather than evolution of quantum neurons. The evolution, in their formalism, maps the amplitude vector $\boldsymbol{\alpha} = (\alpha_1 \ldots \alpha_i \ldots \alpha_{2^N})$ to $\boldsymbol{\alpha}'$ by applying the unitary operator U, that is, we can write:

$$\boldsymbol{\alpha}' = U\boldsymbol{\alpha}. \tag{14.219}$$

The unitary matrix U is of size $2^N \times 2^N$ and describes the transitions among QNN basis states $|i_1 \ \ldots \ i_N\rangle$. The projective measurement σ collapses the QNN's superposition state (Eq. 14.218) into one of the QNN basis states, that is:

$$\boldsymbol{\alpha} = (\alpha_1 \ldots \alpha_i \ldots \alpha_{2^N}) \rightarrow (0_1 \ldots 1_i 0_{i+1} \ldots 0_{2^N}), \tag{14.220}$$

which happens with the probability $|\alpha_i|^2$ and represents the nonlinear, dissipative, and irreversible mapping, thus playing the role of quantum activation function. The QNN dynamics is represented by the QNN amplitude vector update from t to $(t + 1)$ time instances as follows:

$$\boldsymbol{\alpha}^{(t+1)} = \sigma\left(U\boldsymbol{\alpha}^{(t)}\right). \tag{14.221}$$

14.12.2 Quantum Perceptron

The concept of the *quantum perceptron* was introduced by Altaisky in 2001 [59], which was modeled by the following quantum updating function:

$$\left|y^{(t)}\right\rangle = \widehat{F} \sum_{i=1}^{m} \widehat{w}_{iy}^{(t)} |x_i\rangle, \tag{14.222}$$

where $\widehat{F}$ is an arbitrary quantum operator, while $\widehat{w}_{iy}$ operators represent synaptic weight operators acting on m input qubits $|x_i\rangle$. The quantum perceptron training rule is a generalization of the corresponding classical learning rule:

$$\widehat{w}_{jy}^{(t+1)} = \widehat{w}_{jy}^{(t)} + \eta\left(|d\rangle - \left|y^{(t)}\right\rangle\right)\langle x_i|, \tag{14.223}$$

where $|d\rangle$ is a desired target state, while $|y^{(t)}\rangle$ is the state of the quantum neuron at discrete time instance t (and η dictates the learning rate). Unfortunately, the foregoing learning rule is not a unitary operator. To solve this problem, the author in [60] proposed to use the projector instead of the QNN updated function, defined as $P = |\psi\rangle\,\langle\psi|$, wherein $\langle\psi|x_1 \ldots, x_N\rangle = |d|$, with d being the desired target output. As an illustration, according to the author, the XOR operation can be performed with the help of the two-input quantum perceptron employing the projector

$P = \underbrace{|00\rangle\langle 00| + |11\rangle\langle 11|}_{P_-} + \underbrace{|01\rangle\langle 01| + |10\rangle\langle 10|}_{P_+} = P_- + P_+$. Given that P_- and P_+ are complete and orthogonal, all input states can be unambiguously classified.

14.12.3 Quantum Convolutional Neural Networks

In the rest of this section, we describe recently proposed QCNN [61], which can be used in quantum phase recognition and quantum error correction. The classical CNNs are often used in image recognition problems (see, for example, [62]). The CNN is composed of an input layer, multiple hidden layers, and an output layer. The hidden layers further consist of a sequence of interleaved layers known as convolutional layers, pooling layers, and fully connected layers, which are illustrated in Fig. 14.35. The mapping between different layers is known as the feature maps. Convolutional layers "convolve" the input and forward the corresponding results to the next layer. The operation that convolution layers apply is not convolution at all, but rather the sliding scalar product calculation. In the lth convolution layer, the new pixel values $x_{i,j}^{(l)}$ are not calculated from the corresponding ones in the neighborhood but from the previous layer, so that we can write $x_{i,j}^{(l)} = \sum_{m,n}^{w} w_{m,n} x_{i+m,j+n}^{(l-1)}$, wherein $w_{m,n}$ are corresponding weights from a window of $w \times w$ pixels. The purpose of the pooling layers is to reduce the dimensions of the hidden layer by combining the outputs of neuron clusters at the previous layer into a single neuron in the next layer. Local pooling combines small clusters of typical sizes 2×2. On the other hand, global pooling acts on all the neurons of the previous hidden layer. The pooling operation may compute either a max or an average operation of small neuron clusters in the previous layer. Max pooling calculates the maximum value from each of a cluster of neurons at the prior layer. On the other hand, average pooling calculates the average value from each of a cluster of neurons at the prior layer. Fully

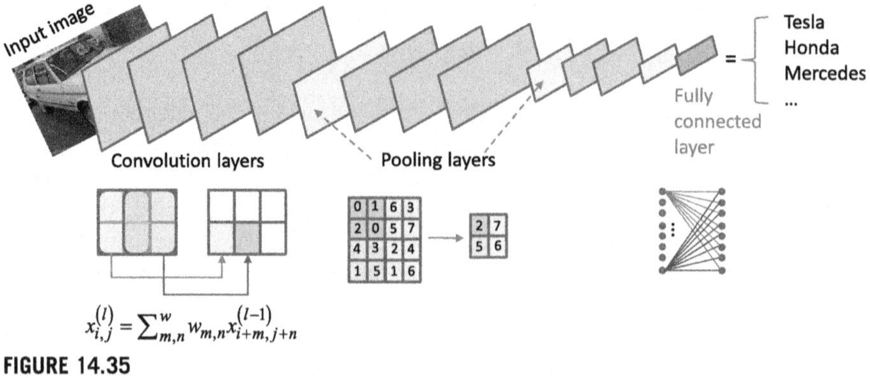

FIGURE 14.35

Illustrative (simplified) classical convolutional neural network example.

connected layers connect every neuron in the previous layer to every neuron in the next layer. Namely, once the size of the feature map becomes reasonable small, the final output is computed based on all remaining neurons in the previous layer. The weights of the fully connected map are optimized based on a training by a large dataset, while the weights on convolutional and pooling layers are typically fixed.

The QCNN introduced in [61] applies the CNN concept but on quantum states rather than images, which is illustrated in Fig. 14.36. The input to the QCNN is a quantum state $|\psi_{\text{in}}\rangle$. The convolution layer is composed of quasi-local unitary operators U_i, which are translationally invariant. In the pooling level, some of the qubits undergo quantum measurement and their classical output is used to control unitary rotation operators V_i. Quantum measurements introduce nonlinearity to the system and at the same time allow us to reduce the number of qubits. The convolution and pooling layers are applied in an interleaved fashion until the system size is sufficiently small so that a fully connected level is reached in which a unitary, nonlocal operator F is applied on all remaining qubits, followed by the quantum measurement on selected qubits. The learning procedure has a certain similarity with classical CNNs. We need to parametrize the unitaries in convolutional, pooling, and fully connected layers and apply a gradient-like approach to estimate these parameters. As an illustration, given the training dataset $\{(|\psi_i\rangle, y_i) \,|i = 1,\ldots,N\}$, wherein $|\psi_i\rangle$ are the input states and $y_i \in \{-1,1\}$ are corresponding labels for the binary classification problem, we can calculate the mean square error by [61]:

$$\text{MSE} = \frac{1}{N}\sum_{i=1}^{N}\left[y_i - f_{(U_j,\, V_j,\, F)}(|\psi_i\rangle)\right], \tag{14.225}$$

where $f_{(Uj,Vj,F)}\,(|\psi_i\rangle)$ is the expected QCNN output for input state $|\psi_i\rangle$. The QCNN has certain similarities with the multiscale entanglement renormalization ansatz (MERA) [63], which can be used to transform the product state $|0\rangle^{\otimes N}$ into a desired state $|\psi\rangle$, as illustrated in Fig. 14.37. 1D MERA requires the use of $2N - 1$ gates that

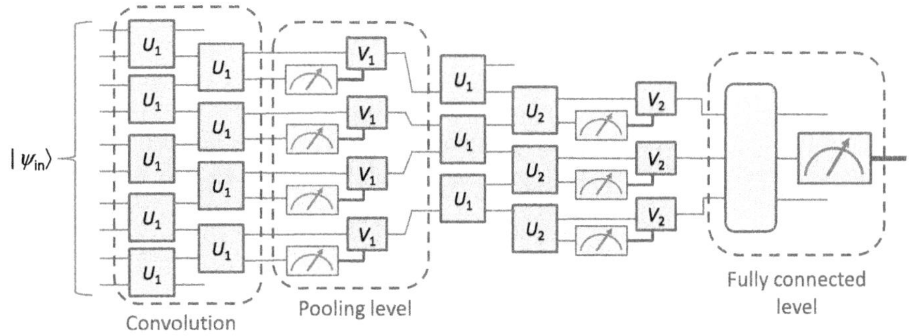

FIGURE 14.36

An illustrative quantum convolutional neural network example.

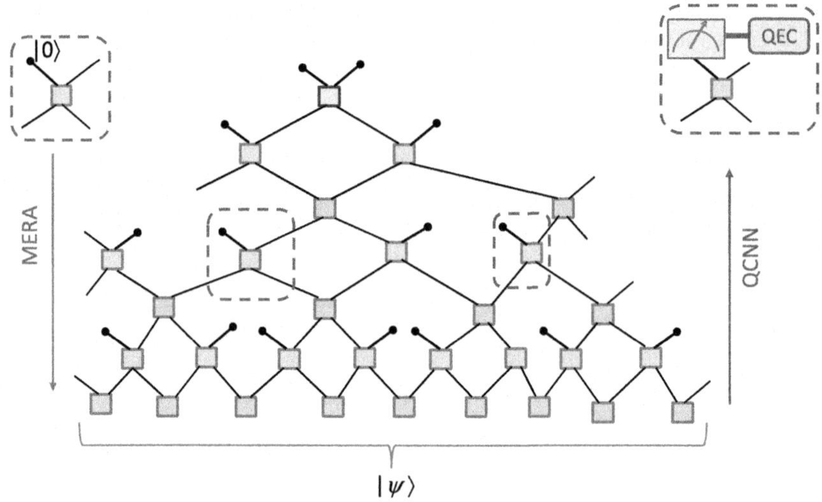

FIGURE 14.37

Relationship between quantum convolutional neural network (QCNN) and multiscale entanglement renormalization ansatz (MERA). *QEC*, Quantum error correction.

are organized in $O(\log N)$ layers. The MERA is composed of three types of tensors [63]: (1) top tensor $t = u|0\rangle|0\rangle$, (2) isometric tensors $w = u|0\rangle$, and (3) unitary gates. The MERA can also be used to transform arbitrary product state $\otimes_{n=1}^{N}|\phi_n^{[n]}\rangle$ into a different entangled state (compared to the transformation of initial state $|0\rangle^{\otimes N}$), denoted as $|\psi_{\{\phi_n\}}\rangle$. Therefore the QCNN can be interpreted as a particular realization of MERA that is run in the opposite direction. Some of the measurements in QCNN can be related to quantum syndrome measurements, relevant in quantum error correction, thus improving the reliability of the QCNN.

The authors have shown in [61] that QCNN can be successfully used to solve the quantum phase recognition problem, that is, to verify whether a given quantum many-body system in ground state $|\psi_g\rangle$ belongs to the quantum phase P. If yes, the output of the final measurement, after the fully connected layer, will give the result 1.

The QCNN can also be used in quantum error correction, as suggested by the authors in [61]. Given the quantum error channel model N, the goal is determining the encoding E_q and decoding D_q quantum circuits such that recovery fidelity is maximized:

$$F_q = \sum_{|\psi\rangle} \langle\psi|\mathscr{D}_q\{\mathscr{N}[\mathbf{E}_q(|\psi\rangle\langle\psi|)]\}|\psi\rangle. \tag{14.225}$$

To demonstrate the efficiency of QCNN in quantum error correction they have designed [1,9] quantum error correction with configuration provided in Fig. 14.38,

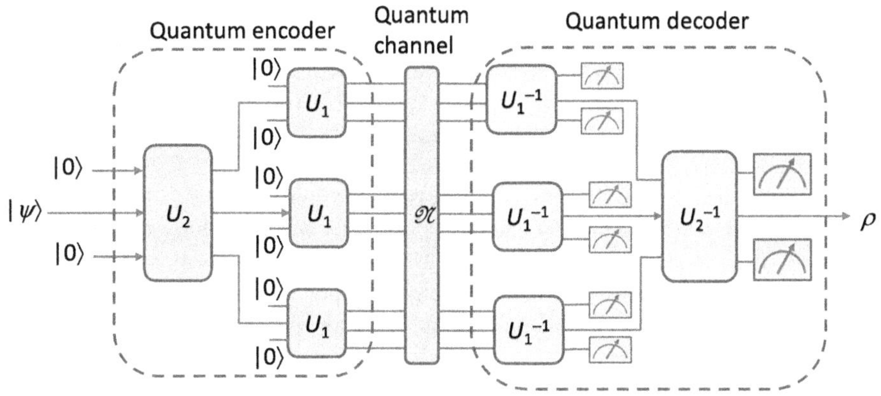

FIGURE 14.38

The quantum convolutional neural network (QCNN)-based [1,9] quantum error correction scheme that encodes the single information qubit into a nine-qubit codeword. The goal is to design QCNN to maximize $\langle\psi|\rho|\psi\rangle$.

significantly outperforming the corresponding Shor's code in a correlated quantum channel model.

Another interesting QNN example, suitable for classification problems, is provided in [64].

14.13 SUMMARY

This chapter was devoted to both classical and QML algorithms. In Subsection 14.1.1, the relevant classical ML algorithms, categorized as supervised, unsupervised, and reinforcement learning algorithms, were briefly introduced. In the same section the concept of feature space was introduced as well as the corresponding generalization performance of ML algorithms. The following topics from classical ML were described: PCA, SVMs, clustering, boosting, regression analysis, and neural networks. In the PCA section (Subsection 14.1.2), it was described how to determine the principal components from the correlation matrix, followed by a description of SVD-based PCA as well as the PCA scoring phase. Subsection 14.1.3 was related to SVM-based algorithms, which were described from both geometric and Lagrangian method-based points of view. For classification problems that are not linearly separable, the concept of soft margins was introduced. For classification problems with nonlinear decision boundaries, the kernel method was described. Subsection 14.1.4 was devoted to clustering problems. The following clustering algorithms were described: *K*-means (Subsection 14.1.4.1), expectation maximization (Subsection 14.1.4.3), and *K*-nearest neighbors (Section 14.1.4.4). In the same section, clustering quality evaluation (Subsection 14.1.4.2) was discussed. In the boosting section (Subsection 14.1.5), the procedure on how to build strong learners from weak learners was described, with special attention being devoted to the AdaBoost algorithm. In the regression analysis section (Subsection 14.1.6), the least-squares estimation, pseudoinverse approach, and ridge regression method were described. Subsection 14.1.7 was related to ANNs. The following topics were described in detail: the perceptron (Subsection 14.1.7.1), activation functions (Subsection 14.1.7.1), and feedforward networks (Subsection 14.1.7.2).

The focus then moved to QML algorithms. In Section 14.2, the Ising model was introduced, followed by adiabatic computing and quantum annealing. In Subsection 14.2.1, the Ising model was related to the QUBO problem. The focus moved then to solving the QUBO problem by adiabatic quantum computing and quantum annealing in corresponding Subsections 14.2.2 and 14.2.3, respectively. The variational quantum eigensolver and QAOA were discussed in Section 14.3 as approaches to perform QML using imperfect and noisy quantum circuits. To illustrate the impact of QAOA, in the same section, it was applied to the combinatorial optimization and MAX-CUT problems. Quantum boosting was discussed in Section 14.4 and was related to the QUBO problem. Section 14.5 was devoted to QRAM, allowing us to address the superposition of memory cells with the help of a quantum register. Quantum matrix inversion, also known as the HHL algorithm, was described in Section 14.6, which can be used as a basic ingredient for other QML algorithms such as the QPCA, which was described in Section 14.7. In the quantum optimization-based clustering Section (14.8) we described how the MAX-CUT problem can be related to clustering, and thus be solved by adiabatic computing, quantum annealing, and QAOA. In Section 14.9, Grover algorithm-based quantum optimization was discussed. In the quantum K-means section (Section 14.10) we described how to calculate the dot product and quantum distance, and how to use them in the Grover search-based K-means algorithm. In the QSVM Section (14.11), we formulated the SVM problem using least squares and described how to solve it with the help of the HHL algorithm. In the QNNs Section (14.12) we described feedforward QNNs, quantum perceptron, and quantum convolutional networks. For a deeper understanding of underlying concepts related to ML and QML, a set of problems is provided in the following section for self-study.

14.14 PROBLEMS

1. The Jacobi eigenvalue method is an iterative method to calculate eigenvalues and eigenvectors of a real-valued symmetric matrix through diagonalization [65,66]. Let the training dataset be given by $\{\boldsymbol{x}_1 = [1, -3, -5, 4, -4, 0]^{\mathrm{T}}, \boldsymbol{x}_2 = [1, 5, -1, -2, 4, 4]^{\mathrm{T}}, \boldsymbol{x}_3 = [1, 5, -1, 0, 4, -4]^{\mathrm{T}}, \boldsymbol{x}_4 = [-3, -7, 7, 4, -4, 0]^{\mathrm{T}}\}$.
   - **(a)** Determine the eigenvalues and eigenvectors of the corresponding correlation matrix for PCA by SVD.
   - **(b)** Determine the eigenvalues and eigenvectors of the corresponding correlation matrix for PCA by the Jacobi eigenvalue method. Discuss the complexity and efficiency of the Jacobi method against that of conventional SVD.

(c) Perform the PCA training phase for both methods.
(d) For data vector $\boldsymbol{x} = [1\ 1\ 0\ 2\ 1\ 1]^{\mathrm{T}}$, determine the scorings for both methods and discuss the results.

2. Let us assume that the dataset $\{\boldsymbol{x}_i | i = 1,2,\ldots,n\}$ is given where $\boldsymbol{x}_i$ are feature vectors of length m. Let us create the $\boldsymbol{X}$-matrix with columns being the feature vectors. By subtracting the mean per measurement type we obtain the matrix $\boldsymbol{X}'$. Let us now perform PCA by SVD as described in Subsection 14.1.2. In this procedure, the columns of matrix $\boldsymbol{U}$ represent the eigenvectors of the covariance matrix. Let us now truncate $\boldsymbol{U}$ by preserving the d dominant principal components to obtain $\boldsymbol{U}'$. Now we can compress the data by computing $\boldsymbol{X}_{\mathrm{c}} = \boldsymbol{U}'^{\mathrm{T}}\boldsymbol{X}'$, and decompress by computing $\boldsymbol{U}'\boldsymbol{X}_{\mathrm{c}}$ and adding the corresponding means. Can this approach be used as an efficient compression method? Discuss it and determine the compression rate.

3. Let us revisit the SVM optimization problem given by Eq. (14.38), but now with the kernel function $K(\boldsymbol{x}_i,\boldsymbol{x}_j)$ taken into account:

$$\widehat{\boldsymbol{\lambda}} = \arg\max_{\boldsymbol{\lambda}} L(\boldsymbol{\lambda}) = \arg\max_{\boldsymbol{\lambda}} \left\{ \sum_{j=1}^{n} \lambda_j - \frac{1}{2} \sum_{i=1}^{n} \sum_{j=1}^{n} \lambda_i \lambda_j y_i y_j K(\boldsymbol{x}_i, \boldsymbol{x}_j) \right\}$$

$$\text{subject to:} \sum_{j=1}^{n} \lambda_j y_j = 0,\ C \geq \lambda_j \geq 0,$$

where $C > 0$ is the regularization parameter related to the softness of the margin. This problem can also be formulated in matrix form as follows:

$$\widehat{\boldsymbol{\lambda}} = \arg\max_{\boldsymbol{\lambda}} \left(\mathbf{1}^{\mathrm{T}}\boldsymbol{\lambda} - \frac{1}{2}\boldsymbol{\lambda}^{\mathrm{T}}\boldsymbol{P}\boldsymbol{\lambda} \right),$$

$$\text{subject to:}\ \boldsymbol{y}^{\mathrm{T}}\boldsymbol{\lambda} = 0 \sum_{j=1}^{n} \lambda_j y_j = 0,\ \boldsymbol{C} \geq \boldsymbol{\lambda} \geq \mathbf{0},$$

where:

$$\boldsymbol{P} = \begin{bmatrix} y_1 y_1 K(\boldsymbol{x}_1, \boldsymbol{x}_1) & y_1 y_2 K(\boldsymbol{x}_1, \boldsymbol{x}_2) & \ldots & y_1 y_n K(\boldsymbol{x}_1, \boldsymbol{x}_n) \\ y_2 y_1 K(\boldsymbol{x}_2, \boldsymbol{x}_1) & y_2 y_2 K(\boldsymbol{x}_2, \boldsymbol{x}_2) & \ldots & y_2 y_n K(\boldsymbol{x}_2, \boldsymbol{x}_n) \\ \vdots & \cdots & \ddots & \vdots \\ y_n y_1 K(\boldsymbol{x}_n, \boldsymbol{x}_1) & y_n y_2 K(\boldsymbol{x}_n, \boldsymbol{x}_2) & \ldots & y_n y_n K(\boldsymbol{x}_n, \boldsymbol{x}_n) \end{bmatrix}, \quad \boldsymbol{C} = \begin{bmatrix} C \\ C \\ \vdots \\ C \end{bmatrix},$$

$$\mathbf{1} = \begin{bmatrix} 1 \\ 1 \\ \vdots \\ 1 \end{bmatrix}, \quad \mathbf{0} = \begin{bmatrix} 0 \\ 0 \\ \vdots \\ 0 \end{bmatrix}, \quad \boldsymbol{y} = \begin{bmatrix} y_1 \\ y_2 \\ \vdots \\ y_n \end{bmatrix}.$$

Verify the matrix form. Discuss the advantages and disadvantages of the matrix form compared to the scalar form.

4. Let us assume that the dataset $\{(\boldsymbol{x}_i,\ y_i)|y_i \in \{-1,\ 1\};\ i = 1,\ \dots,\ n\}$ is provided to us and we use it to train the SVM. For a linear SVM, the scoring function for new incoming vector $\boldsymbol{x}$ is given by:

$$s(\boldsymbol{x}) = \sum_{j=1}^{n} \lambda_j y_j \boldsymbol{x}_j^{\mathrm{T}} \boldsymbol{x} - b.$$

Determine the weight associated with each component of a scored vector. How can these weights be used to reduce the dimensionality of training data? To reduce the system dimensionality, discuss the advantages and disadvantages of the SVM as compared to PCA.

5. Let as assume that the clusters C_k; $k = 1,\dots,K$ have been specified with corresponding centroids being $\boldsymbol{c}_k$. For a given data point $\boldsymbol{x}$, let us now determine the following metrics:

$$p_k = \frac{1}{K-1}\left[1 - \frac{d(\boldsymbol{x},\ \boldsymbol{c}_k)}{\sum_i d(\boldsymbol{x},\ \boldsymbol{c}_i)}\right],$$

where $d(\boldsymbol{x},\boldsymbol{c}_k)$ is the Euclidean distance. Clearly, as Euclidian distance $d(\boldsymbol{x},\boldsymbol{c}_k)$ increases, the parameter p_k decreases. Show that the p_k $(k = 1,\dots,K)$ satisfy the requirements for a discrete probability distribution. Let us now define the reassigning rule as follows: assign the $\boldsymbol{x}$ to the cluster with the largest probability p_k. Discuss the effectiveness of such clustering method compared to the conventional K-means algorithm.

6. In this problem we are concerned with the K-means++ algorithm introduced in [49]. Let $D(\boldsymbol{x})$ denote the shortest distance of data point $\boldsymbol{x}$ from the centroids already selected. We can summarize the K-means++ algorithm as follows:
   - Take one centroid $\boldsymbol{c}_1$ selected uniformly at random from the set of data points X.
   - Take a new centroid $\boldsymbol{c}_k$ by selecting $\boldsymbol{x} \in X$ with probability $D(x)^2 / \Sigma_{x\in X} D(x)^2$.
   - Repeat the previous step until all centroids have been selected.
   - Continue as in the standard K-means algorithm.

Discuss the advantages of the K-means++ algorithm with respect to the standard K-means algorithm.

7. Describe how the Hadamard gate can be implemented using adiabatic quantum computing.

8. Describe how the CNOT gate can be implemented using adiabatic quantum computing.
9. Describe how the Toffoli gate can be implemented using adiabatic quantum computing.
10. Describe how the Fredkin gate can be implemented using adiabatic quantum computing.
11. Describe how quantum optimization can be used in multiclass classifying problems.
12. Describe how QPCA can be used in multiclass classifying problems.
13. To reduce the complexity of the Grover search-based K-means algorithm from $O(N\log N)$ to $O(\log N)$, the output of quantum K-means should not be the list of N vectors and their assignments to clusters, but instead in the following superposition clustering state:

$$|\psi_c\rangle = N^{-1/2}\sum_i |c_i\rangle|i\rangle = N^{-1/2}\sum_c\sum_{i\in c_i} |c\rangle|i\rangle,$$

wherein the state denotes the superposition of clusters with each cluster being the superposition of vectors assigned to it. To obtain the statistical picture of the clustering algorithm we need to sample from this state. Describe the procedure to construct this clustering state by the adiabatic algorithm.

14. Let us observe an undirected graph $G(V, E)$, where V is the set of vertices and E is the set of edges. We are interested in partitioning the set of vertices V into two subsets S_1 and S_2 such that the number of edges between the two sets is as large as possible. To represent this problem, let us introduce the random variable z_i, which takes the value 1 when the vertex i is in S_1, and 0 otherwise. Formulate the optimization problem and apply it to Fig. 14.P14. Describe how the problem from Fig. 14.P14 can be solved using QAOA.

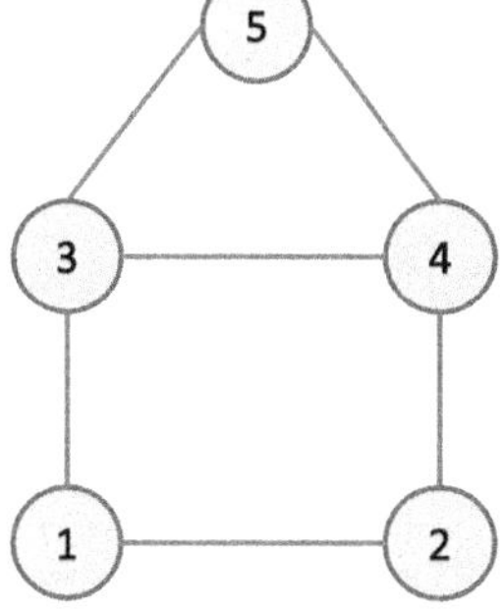

FIGURE 14.P14

Undirected graph under study with five vertices and six edges.

15. The HLL algorithm is used to solve the system of linear equations given by $\boldsymbol{Ax} = \boldsymbol{b}$, where $\boldsymbol{A}$ is an $n \times n$ real-valued matrix, providing that the matrix is well conditioned, that is, when the condition number κ, defined as the ratio between maximum and minimum eigenvalues of $\boldsymbol{A}$, is reasonably low. Discuss the complexity of the HLL algorithm when the condition number is of the form n^p, where p is a positive integer.

References

[1] M. Stamp, Introduction to Machine Learning with Applications in Information Security, CRC Press, Boca Raton, 2018.

[2] S. Raschka, V. Mirjalili, Python Machine Learning, third ed., Packt Publishing, Birmingham-Mumbai, 2019.

[3] J. Unpingco, Python for Probability, Statistics, and Machine Learning, second ed., Springer Nature Switzerland AG, Cham, Switzerland, 2019.

[4] V. Feldman, V. Guruswami, P. Raghavendra, Y. Wu, Agnostic learning of monomials by halfspaces is hard, in: Proc. 2009 50th Annual IEEE Symposium on Foundations of Computer Science, Atlanta, GA, 2009, pp. 385–394.

[5] V. Feldman, V. Guruswami, P. Raghavendra, Y. Wu, Agnostic learning of monomials by halfspaces is hard, SIAM J. Comput. 41 (6) (2012) 1558–1590.

[6] J. Shlens, A Tutorial on Principal Component Analysis, April 2014. Available at: https://arxiv.org/abs/1404.1100.

[7] I.B. Djordjevic, Advanced Optical and Wireless Communications Systems, Springer International Publishing, Cham: Switzerland, December 2017.

[8] E. Biglieri, Coding for Wireless Channels, Springer, 2005.

[9] A. Goldsmith, Wireless Communications, Cambridge University Press, Cambridge, 2005.

[10] M. Cvijetic, I.B. Djordjevic, Advanced Optical Communications and Networks, Artech House, January 2013.

[11] V. Vapnik, A. Lerner, Pattern recognition using generalized portrait method, Autom. Remote Control 24 (1963) 774–780.

[12] V. Vapnik, A.Y. Chervonenkis, A class of algorithms for pattern recognition learning, Avtom. i Telemekh 25 (6) (1964) 937–945.

[13] A.P. Dempster, N.M. Laird, D.B. Rubin, Maximum likelihood from incomplete data via the EM algorithm, J. R. Stat. Soc., Series B (Methodol.) 39 (1) (1977) 1–38.

[14] D. Forsyth, Applied Machine Learning, Springer Nature Switzerland, Cham, Switzerland, 2019.

[15] P. Wittek, Quantum Machine Learning, Elsevier/Academic, Amsterdam-Boston, 2014.

[16] T. Hastie, R. Tibshirani, J. Friedman, The Elements of Statistical Learning: Data Mining, Inference, and Prediction, second ed., Springer Science+Business Media, New York, NY, 2009.

[17] C. Giraud, Introduction to High-Dimensional Statistics, CRC Press, Boca Raton, FL, 2015.

[18] O. Lund, M. Neilsen, C. Lundergaard, C. Kesmir, S. Brunak, Biological Bioinformatics, MIT Press, Cambridge, MA, 2005.

[19] G.H. Wannier, Statistical Physics, Dover Publications, Inc., New York, 1987.

[20] M. Born, V. Fock, Beweis des adiabatensatzes, Z. Phys. 51 (1-3) (1928) 165–180.

[21] W. van Dam, M. Mosca, U. Vazirani, How powerful is adiabatic quantum computation?, in: Proc. the 42nd Annual Symposium on Foundations of Computer Science, 2001, pp. 279–287. Newport Beach, CA, USA.

[22] T. Kadowaki, H. Nishimori, Quantum annealing in the transverse Ising model, Phys. Rev. E 58 (5) (1998) 5355–5363.

[23] A.B. Finnila, et al., Quantum annealing: a new method for minimizing multidimensional functions, Chem. Phys. Lett. 219 (5–6) (1994) 343–348.

[24] W. Vinci, T. Albash, G. Paz-Silva, I. Hen, D.A. Lidar, Quantum Annealing Correction with Minor Embedding, Phys. Rev. A 92 (4) (2015) 042310.

[25] C.C. McGeoch, C. Wang, Experimental evaluation of an adiabatic quantum system for combinatorial optimization, in: Proc. ACM International Conference on Computing Frontiers (CF '13), 2013, pp. 1–23, 11.

[26] P. Wittek, Quantum Machine Learning: What Quantum Computing Means to Data Mining, Elsevier/Academic Press, Amsterdam-Boston, 2014.

[27] E. Farhi, J. Goldstone, S. Gutmann, A Quantum Approximate Optimization Algorithm, 2014, pp. 1–16. Available at: https://arxiv.org/abs/1411.4028.

[28] A. Peruzzo, et al., A variational eigenvalue solver on a photonic quantum processor, Nat. Comm. 5 (July 23, 2014) 4213.

[29] J. R McClean, J. Romero, R. Babbush, A. Aspuru-Guzik, The theory of variational hybrid quantum-classical algorithms, New J. Phys. 18 (February 2016) 023023.

[30] J.S. Otterbach, et al., Unsupervised Machine Learning on a Hybrid Quantum Computer, December 15, 2017. Available at: https://arxiv.org/abs/1712.05771.

[31] J. Preskill, Quantum computing in the NISQ era and beyond, Quantum 2 (July 30, 2018) 79.

[32] J. Choi, J. Kim, A Tutorial on Quantum Approximate Optimization Algorithm (QAOA): fundamentals and applications, in: Proc. 2019 International Conference on Information and Communication Technology Convergence (ICTC), 2019, pp. 138–142. Jeju Island, S. Korea.

[33] R. Bhatia, Matrix Analysis, in: Graduate Texts in Mathematics, vol. 169, Springer-Verlag, New York, 1997.

[34] I.B. Djordjevic, Quantum Information Processing: An Engineering Approach, Elsevier/Academic Press, Amsterdam-Boston, 2012.

[35] H. Neven, V.S. Denchev, G. Rose, W.G. Macready, Training a Binary Classifier With the Quantum Adiabatic Algorithm, November 4, 2008. Available at: https://arxiv.org/abs/0811.0416.

[36] V. Giovannetti, S. Lloyd, L. Maccone, Quantum random access memory, Phys. Rev. Lett. 100 (16) (April 25, 2008) 160501.

[37] M.A. Nielsen, I.L. Chuang, Quantum Computation and Quantum Information, Cambridge University Press, Cambridge, 2000.

[38] A.W. Harrow, A. Hassidim, S. Lloyd, Quantum algorithm for solving linear systems of equations, Phys. Rev. Lett. 103 (15) (October 7, 2009) 150502.

[39] Z. Li, X. Liu, N. Xu, J. Du, Experimental realization of a quantum support vector machine, Phys. Rev. Lett. 114 (April 8, 2015) 140504.

[40] D.W. Berry, G. Ahokas, R. Cleve, B.C. Sanders, Efficient quantum algorithms for simulating sparse hamiltonians, Commun. Math. Phys. 270 (2) (March 2007) 359–371.

[41] S. Lloyd, M. Mohseni, P. Rebentrost, Quantum Algorithms for Supervised and Unsupervised Machine Learning, November 4, 2013. Available at: https://arxiv.org/abs/1307.0411.
[42] S. Lloyd, M. Mohseni, P. Rebentrost, Quantum principal component analysis, Nat. Phys. 10 (July 27, 2014) 631−633.
[43] G. Benenti, G. Casati, D. Rossini, G. Strini, Principles of Quantum Computation and Information: A Comprehensive Textbook, World Scientific, Singapore, 2019.
[44] S. Lloyd, Universal quantum simulators, Science 273 (5278) (1996) 1073−1078.
[45] L.K. Grover, A Fast Quantum Mechanical Algorithm for Database Search, November 19, 1996. Available at: https://arxiv.org/abs/quant-ph/9605043.
[46] C. Dürr, P. Høyer, A Quantum Algorithm for Finding the Minimum, July 18, 1996. Available at: https://arxiv.org/abs/quant-ph/9607014.
[47] A. Ahuja, S. Kapoor, A Quantum Algorithm for Finding the Maximum, November 18, 1999. Available at: https://arxiv.org/abs/quant-ph/9911082.
[48] E. Aïmeur, G. Brassard, S. Gambs, Quantum speed-up for unsupervised learning, Mach. Learn. 90 (2) (2013) 261−287.
[49] D. Arthur, S. Vassilvitskii, k-means++: the advantages of careful seeding, in: Proc. 18th Annual ACM-SIAM Symposium on Discrete Algorithms, Society for Industrial and Applied Mathematics, Philadelphia, PA, USA, 2007, pp. 1027−1035.
[50] D. Anguita, S. Ridella, F. Rivieccio, R. Zunino, Quantum optimization for training support vector machines, Neural Netw. 16 (5-6) (2003) 763−770.
[51] P. Rebentrost, M. Mohseni, S. Lloyd, Quantum Support Vector Machine for Big Data Classification, July 10, 2014. Available at: https://arxiv.org/abs/1307.0471v3.
[52] J.A. Suykens, J. Vandewalle, Least squares support vector machine classifiers, Neural Process. Lett. 9 (3) (1999) 293−300.
[53] K.H. Wan, O. Dahlsten, H. Kristjnsson, R. Gardner, M.S. Kim, Quantum generalisation of feedforward neural networks, npj Quantum Inf. 3 (36) (September 14, 2017).
[54] G. Verdon, M. Broughton, J. Biamonte, A Quantum Algorithm to Train Neural Networks Using Low-Depth Circuits, 2017. Available at: https://arxiv.org/abs/1712.05304.
[55] I. Kerenidis, A. Prakash, Quantum Gradient Descent for Linear Systems and Least Squares, April 17, 2017. Available at: https://arxiv.org/abs/1704.04992.
[56] P. Rebentrost, M. Schuld, L. Wossnig, F. Petruccione, S. Lloyd, Quantum Gradient Descent and Newton's Method for Constrained Polynomial Optimization, December 6, 2016. Available at: https://arxiv.org/abs/1612.01789.
[57] M. Schuld, I. Sinayskiy, F. Petruccione, The quest for a quantum neural network, Quantum Inf. Process. 13 (11) (2014) 2567−2586.
[58] M. Zak, C.P. Williams, Quantum neural nets, Int. J. Theor. Phys. 3 (2) (February 1998) 651−684.
[59] M.V. Altaisky, Quantum Neural Network, July 5, 2001. Available at: https://arxiv.org/abs/quant-ph/0107012.
[60] M. Siomau, A quantum model for autonomous learning automata, Quantum Inf. Process. 13 (January 9, 2014) 1211−1221.
[61] I. Cong, S. Choi, M.D. Lukin, Quantum convolutional neural networks, Nat. Phys. 15 (2019) 1273−1278.
[62] A. Krizhevsky, I. Sutskever, G.E. Hinton, ImageNet classification with deep convolutional neural networks, Commun. ACM 60 (6) (June 2017) 84−90.

[63] G. Vidal, Class of quantum many-body states that can be efficiently simulated, Phys. Rev. Lett. 101 (September 12, 2008) 110501.

[64] E. Farhi, H. Neven, Classification With Quantum Neural Networks on Near Term Processors, February 16, 2018. Available at: https://arxiv.org/abs/1802.06002.

[65] C.G.J. Jacobi, Über ein leichtes Verfahren, die in der Theorie der Säkularstörungen vorkommenden Gleichungen numerisch aufzulösen, Crelle's J. 30 (1846) 51−94 (in German).

[66] G.H. Golub, H.A. van der Vorst, Eigenvalue computation in the 20th century, J. Comput. Appl. Math. 123 (1−2) (2000) 35−65.

CHAPTER

Quantum Key Distribution 15

CHAPTER OUTLINE

Quantum Information Processing, Quantum Computing, and Quantum Error Correction
https://doi.org/10.1016/B978-0-12-821982-9.00002-2

15.1 CRYPTOGRAPHY BASICS

The basic key-based cryptographic system is provided in Fig. 15.1. The source emits the message (plaintext) $\boldsymbol{M}$ toward the encryption block, which with the help of key $\boldsymbol{K}$, obtained from key source, generates the cryptogram. On the receiver side, the cryptogram transmitted over an insecure channel is processed by the decryption algorithm together with the key $\boldsymbol{K}$ obtained through the secure channel, which reconstructs the original plaintext to be delivered to the authenticated user. The encryption process can be mathematically described as $\boldsymbol{E_K}(\boldsymbol{M}) = \boldsymbol{C}$, while the decryption process is $\boldsymbol{D_K}(\boldsymbol{C}) = \boldsymbol{M}$. The composition of decryption and encryption functions yields to identity mapping $\boldsymbol{D_K}(\boldsymbol{E_K}(\boldsymbol{M})) = \boldsymbol{M}$. The key source typically generates the key randomly from the *keyspace* (the range of possible key values).

The key-based algorithms can be categorized into two broad categories:

- *Symmetric algorithms*, in which the decryption key can be derived from the encryption key and vice versa. Alternatively, the same key can be used for both encryption and decryption stages. Symmetric algorithms are also known as one-key (single-key) or secret-key algorithms. The well-known system employing these types of algorithms is the digital encryption standard [1–4].
- *Asymmetric algorithms*, in which the encryption and decryption keys are different. Moreover, the decryption key cannot be determined from the encryption

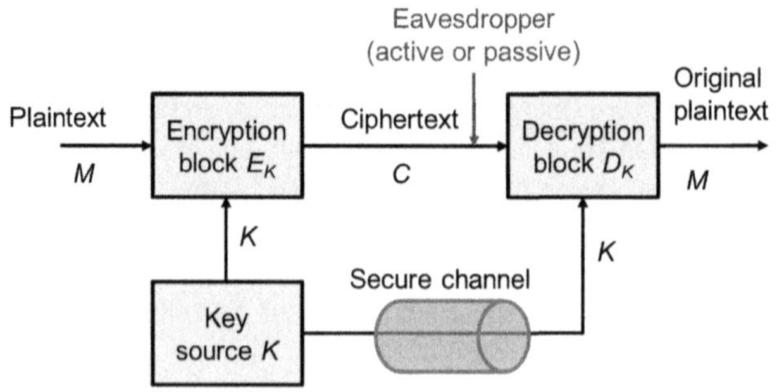

FIGURE 15.1

Basic key-based cryptographic scheme.

key, at least in any reasonable amount of time. Because of this fact, the encryption keys can even be made public, wherein the eavesdropper will not be able to determine the decryption key. The *public-key systems* [5] are based on this concept. In public-key systems, the encrypted keys have been made public, while the decryption key is known only to the intended user. The encryption key is then called the *public* key, while the decryption key is called the *secret (private)* key. The keys can be applied in arbitrary order to create the cryptogram from plaintext, and to reconstruct the plaintext from the cryptogram.

The simplest private-key cryptosystem is the *Vernam cipher* (*one-time pad*). In the *one-time pad* [6], a completely random sequence of characters, with the sequence length being equal to the message sequence length, is used as a key. When for each new message another random sequence is used as a key, the one-time pad scheme provides perfect security. Namely, the brute-force search approach would require verifying m^n possible keys, where m is the employed alphabet size and n is the length of the intercepted cryptogram. In practice, in digital and computer communications we typically operate on binary alphabet GF(2) = {0,1}. To obtain the key we need a special random generator and to encrypt using a one-time pad scheme we simply perform addition mod-2, i.e., XOR operation, as illustrated in Fig. 15.2. Even though the one-time pad scheme offers perfect security, it has several drawbacks [7–9]: it requires secure distribution of the key, the length of the key must be at least as long as the message, the key bits cannot be reused, and the keys must be delivered in advance, securely stored until used, and destroyed after use.

According to Shannon [10], *perfect security*, also known as *unconditional security*, has been achieved when the messages $\boldsymbol{M}$ and cryptograms $\boldsymbol{C}$ are statistically independent so that the corresponding mutual information is equal to zero:

$$I(\boldsymbol{M},\boldsymbol{C}) = H(\boldsymbol{M}) - H(\boldsymbol{M}|\boldsymbol{C}) = 0 \quad \Leftrightarrow \quad H(\boldsymbol{M}|\boldsymbol{C}) = H(\boldsymbol{M}). \tag{15.1}$$

By employing the chain rule [2,11], given by $H(\boldsymbol{X},\boldsymbol{Y}) = H(\boldsymbol{X}) + H(\boldsymbol{Y}|\boldsymbol{X})$, we can write:

$$\begin{aligned} H(\boldsymbol{M}) = H(\boldsymbol{M}|\boldsymbol{C}) \le H(\boldsymbol{M},\boldsymbol{K}|\boldsymbol{C}) = H(\boldsymbol{K}|\boldsymbol{C}) + \underbrace{H(\boldsymbol{M}|\boldsymbol{C},\boldsymbol{K})}_{=0} \\ = H(\boldsymbol{K}|\boldsymbol{C}) = H(\boldsymbol{K}), \end{aligned} \tag{15.2}$$

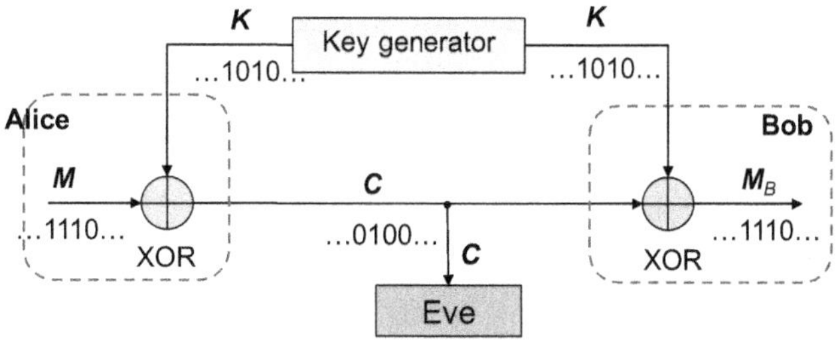

FIGURE 15.2

One-time pad encryption scheme.

where we use the fact that the message is completely specified when both the cryptogram and the key are known, that is, $H(\boldsymbol{M}|\boldsymbol{K},\boldsymbol{C}) = 0$. The perfect secrecy condition can therefore be summarized as:

$$H(\boldsymbol{M}) \leq H(\boldsymbol{K}). \tag{15.3}$$

In other words, the entropy (uncertainty) of the key cannot be lower than the entropy of the message for an encryption scheme to be perfectly secure. Given that in the Vernam cipher the length of the key is at least equal to the message length, it appears that the one-time pad scheme is perfectly secure.

However, given that this condition is difficult to satisfy, in conventional cryptography instead of information-theoretic security, *computational security* is used [1,4,8,12–14]. Computational security introduces two relaxations with respect to information-theoretic security [4]:

- Security is guaranteed against an efficient eavesdropper running the cryptanalytic attacks for a certain limited amount of time. Of course, when the eavesdropper has sufficient computational resources and/or sufficient time he/she will be able to break the security of the encryption scheme.
- Eavesdroppers can be successful in breaking the security protocols, but with small success probability.

Readers interested in learning more about computational security are referred to an excellent book by Katz and Lindell [4]. However, by using quantum computing, any conventional cryptographic scheme can be broken in a reasonable amount of time by employing Shor's factorization algorithm [15–19].

15.2 QKD BASICS

Significant achievements have been recently made in quantum computing [7–9]. There are many companies currently working on development of medium-scale quantum computers. Given that most cryptosystems depend on the computational hardness assumption, quantum computing represents a serious challenge to modern cybersecurity systems. As an illustration, to break the Rivest–Shamir–Adleman protocol [20], one needs to determine the period r of the function $f(x) = m^x \bmod n = f(x + r)$ ($r = 0,1,\ldots,2^l - 1$; m is an integer smaller than $n - 1$). This period is determined in one of the steps of Shor's factorization algorithm [7–9,17–19].

QKD with symmetric encryption can be interpreted as one of the physical-layer security schemes that can provide provable security against quantum computer-initiated attack [21]. The first QKD scheme was introduced by Bennett and Brassard, who proposed it in 1984 [15,16], and is now known as the BB84 protocol. The security of QKD is guaranteed by quantum mechanics laws. Different photon degrees of freedom (DOF), such as polarization, time, frequency, phase, and orbital angular momentum, can be employed to implement various QKD protocols. Generally speaking, there are two generic QKD schemes, discrete variable (DV)-QKD and continuous variable (CV)-QKD, depending on the strategy applied on Bob's side. In DV-QKD schemes, a single-photon detector (SPD) is applied on Bob's side, while in CV-QKD, the field quadratures are measured with the help of homodyne/heterodyne detection. The DV-QKD scheme achieves unconditional security by employing no-cloning theorem and theorem on indistinguishability of arbitrary quantum states. The no-cloning theorem claims that arbitrary quantum states cannot be cloned, indicating that Eve cannot duplicate nonorthogonal quantum states even with the help of a quantum computer. On the other hand, the second theorem claims that non-orthogonal states cannot be unambiguously distinguished. Namely, when Eve interacts with the transmitted quantum states, trying to get information on transmitted bits, she will inadvertently disturb the fidelity of the quantum states that will be detected by Bob. On the other hand, CV-QKD employs the uncertainty principle claiming that both in-phase and quadrature components of coherent states cannot be simultaneously measured with complete precision. We can also classify different QKD schemes as either entanglement-assisted or prepare-and-measure (PM) types.

The research in QKD is gaining momentum, in particular after the first satellite-to-ground QKD demonstration [22]. Recently, QKD over 404 km of ultralow-loss optical fiber has been demonstrated, but only with ultralow secure-key rate ($3.2 \cdot 10^{-4}$ b/s). Given that quantum states cannot be amplified, fiber attenuation limits the distance. On the other hand, the deadtime (the time over which an SPD remains unresponsive to incoming photons due to long recovery time) of the SPDs, typically in the 10–100 ns range, limits the baud rate and therefore the secure-key rate. CV-QKD schemes, since they employ homodyne/heterodyne detection, do not have deadtime limitation; however, typical distances are shorter.

By transmitting nonorthogonal qubit states between Alice and Bob, and by checking for disturbance in the transmitted state, caused by the channel or Eve's activity, they can establish an upper bound on noise/eavesdropping in their quantum communication channel [7]. The *threshold for maximum tolerable error rate* is dictated by the efficiency of the best information reconciliation and privacy amplification steps [7].

QKD protocols can be categorized into several general categories:

- *Device-dependent QKD*, in which, typically, the quantum source is placed on Alice's side and the quantum detector on Bob's side. Popular classes include: DV-QKD, CV-QKD, entanglement-assisted QKD (EA-QKD), distributed phase reference, etc. For EA-QKD, the entangled source can be placed in the middle of the channel to extend the transmission distance.
- *Source device-independent QKD*, in which the quantum source is placed at Charlie's (Eve's) side, while the quantum detectors are placed at both Alice's and Bob's sides.

- *Measurement device-independent QKD (MDI-QKD)*, in which the quantum detectors are placed at Charlie's (Eve's) side, while the quantum sources are placed at both Alice's and Bob's sides. The quantum states are prepared at both Alice's and Bob's sides and are transmitted toward Charlie's detectors. Charlie performs partial Bell-state measurements and announces when the desired partial Bell state is detected, with details to be provided later.

The *classical postprocessing steps* are summarized in Fig. 15.3 [2,7–9].

The raw key is imperfect, and we need to perform *information reconciliation* and *privacy amplification* to increase the correlation between sequence $\boldsymbol{X}$ (generated by Alice) and sequence $\boldsymbol{Y}$ (received by Bob) strings, while reducing eavesdropper Eve's mutual information about the result to a desired level of security.

Information reconciliation is nothing more than error correction performed over a public channel, which reconciles errors between $\boldsymbol{X}$ and $\boldsymbol{Y}$ to obtain a shared bit string $\boldsymbol{K}$ while divulging information as little as possible to Eve.

Privacy amplification [23,24] is used between Alice and Bob to distil from K a smaller set of bits S whose correlation with Eve's string $\boldsymbol{Z}$ is below a desired threshold. One way to accomplish privacy amplification is through use of the *universal hash functions* G, which map the set of n-bit strings A to the set of m-bit strings B such that for any distinct $a_1,a_2 \in A$, when g is chosen uniformly at random from G, the probability of having $g(a_1) = g(a_2)$ is at most $1/|B|$.

Additional details on information reconciliation and privacy amplification will be provided in Section 15.9 of this chapter.

15.3 NO-CLONING THEOREM AND DISTINGUISHING OF QUANTUM STATES

No-Cloning Theorem: No quantum copier exists that can clone an arbitrary quantum state.

Proof: If the input state is $|\psi\rangle$, then the output of the copier will be $|\psi\rangle|\psi\rangle$. For an arbitrary quantum state, such a copier raises a fundamental contradiction. Consider two arbitrary states $|\psi\rangle$ and $|\chi\rangle$ that are input to the copier. When they are inputted

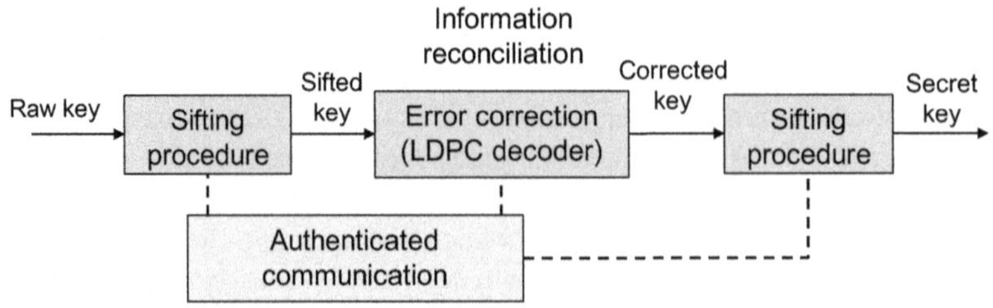

FIGURE 15.3

Classical postprocessing steps. *LDPC*, Low-density parity-check.

individually we expect to get $|\psi\rangle|\psi\rangle$ and $|\chi\rangle|\chi\rangle$. Now consider a superposition of these two states given by:

$$|\varphi\rangle = \alpha|\psi\rangle + \beta|\chi\rangle \Rightarrow |\varphi\rangle|\varphi\rangle = (\alpha|\psi\rangle + \beta|\chi\rangle)(\alpha|\psi\rangle + \beta|\chi\rangle)$$

$$= \alpha^2|\psi\rangle|\psi\rangle + \alpha\beta|\psi\rangle|\chi\rangle + \alpha\beta|\chi\rangle|\psi\rangle + \beta^2|\chi\rangle|\chi\rangle. \tag{15.4}$$

On the other hand, linearity of quantum mechanics tells us that the quantum copier can be represented by a unitary operator that performs the cloning. If such a unitary operator were to act on the superposition state $|\varphi\rangle$, the output would be a superposition of $|\psi\rangle|\psi\rangle$ and $|\chi\rangle|\chi\rangle$, that is:

$$|\varphi'\rangle = \alpha|\psi\rangle|\psi\rangle + \beta|\chi\rangle|\chi\rangle. \tag{15.5}$$

The difference between the previous two equations leads to the contradiction mentioned earlier. As a consequence, there is no unitary operator that can clone $|\varphi\rangle$.

This result raises a related question: Do there exist some specific states for which cloning is possible? The answer to this question is (surprisingly) yes. The cloning is possible only for mutually orthogonal states.

Theorem: It is impossible to unambiguously distinguish nonorthogonal quantum states.

In other words, there is no measurement device we can create that can reliably distinguish nonorthogonal states. This fundamental result plays an important role in quantum cryptography.

Proof: Its proof is based on contradiction. Let us assume that the measurement operator M is a Hermitian operator, with corresponding eigenvalues m_i and corresponding projection operators P_i, which allows us to unambiguously distinguish between two nonorthogonal states $|\psi_1\rangle$ and $|\psi_2\rangle$. The eigenvalue m_1 (m_2) unambiguously identifies the state $|\psi_1\rangle$ ($|\psi_2\rangle$). We know that for projection operators, the following properties are valid:

$$\langle\psi_1|P_1|\psi_1\rangle = 1, \langle\psi_2|P_2|\psi_2\rangle = 1, \langle\psi_1|P_2|\psi_1\rangle = 0, \langle\psi_2|P_1|\psi_2\rangle = 0. \tag{15.6}$$

Given that $|\psi_2\rangle$ can be represented in terms of $|\psi_1\rangle$ and another state $|\chi\rangle$ that is orthogonal to $|\psi_1\rangle$ as follows:

$$|\psi_2\rangle = \alpha|\psi_1\rangle + \beta|\chi\rangle, \tag{15.7}$$

in order satisfy Eq. (15.6) $|\psi_2\rangle$ must be equal to $|\chi\rangle$ representing the contradiction.

15.4 DISCRETE VARIABLE QKD PROTOCOLS

In this section, we describe some relevant DV-QKD protocols, including BB84, B92, time-phase encoding, E91, and EPR protocols.

15.4.1 BB84 Protocols

The BB84 protocol was named after Bennett and Brassard, who proposed it in 1984 [15,16]. Three key principles being employed are: no-cloning theorem, state collapse during measurement, and irreversibility of the measurements [7]. The BB84 protocol can be implemented using different DOF, including polarization DOF [25] or the phase of the photons [22]. Experimentally, the BB84 protocol has been demonstrated over both fiber-optics and free-space optical (FSO) channels [25,26]. The polarization-based BB84 protocol over fiber-optics channel is affected by polarization mode dispersion, polarization-dependent loss, and fiber loss, which affect the transmission distance. To extend the transmission distance, phase encoding is employed in [26]. Unfortunately, the secure-key rate over 405 km of ultralow-loss fiber is extremely low, only 6.5 b/s. In an FSO channel, the polarization effects are minimized; however, atmospheric turbulence can introduce wavefront distortion and random phase fluctuations. Previous QKD demonstrations include a satellite-to-ground FSO link demonstration and a demonstration over an FSO link between two locations in the Canary Islands [25].

Two bases that are used in the BB84 protocol are computational basis (CB) = $\{|0\rangle, |1\rangle\}$ and diagonal basis (DB):

$$DB=\left\{|+\rangle=(|0\rangle+|1\rangle)/\sqrt{2},\ |-\rangle=(|0\rangle-|1\rangle)/\sqrt{2}\right\}. \tag{15.8}$$

These bases belong to the class of *mutually unbiased bases* (MUBs) [27–31]. Two orthonormal bases $\{|e_1\rangle, \ldots, |e_N\rangle\}$ and $\{|f_1\rangle, \ldots, |f_N\rangle\}$ in Hilbert space C^N are MUBs when the square of magnitude of the inner product between any two basis states $\{|e_m\rangle\}$ and $\{|f_n\rangle\}$ is equal to the inverse of the dimension:

$$|\langle e_m|f_n\rangle|^2=\frac{1}{N}\ \forall m,n\in\{1,\cdots,N\}. \tag{15.9}$$

The keyword unbiased means that if a system is prepared in a state belonging to one of the bases, all outcomes of the measurement with respect to the other basis are equally likely.

Alice randomly selects the MUB, followed by random selection of the basis state. The logical 0 is represented by $|0\rangle$, $|+\rangle$, while logical 1 is represented by $|1\rangle$, $|-\rangle$. Bob measures each qubit by randomly selecting the basis, computational or diagonal. In the sifting procedure, Alice and Bob announce the bases being used for each qubit and keep only instances when they use the same basis.

We now describe the *polarization-based BB84 QKD protocol*, with corresponding four states being employed provided in Fig. 15.4. The bulky optics implementation of the BB84 protocol is illustrated in Fig. 15.5.

Laser output state is known as the *coherent state* [31–36], representing the right eigenket of the annihilation operator a; namely $a\,|\alpha\rangle=\alpha\,|\alpha\rangle$, where α is the complex eigenvalue. The coherent state vector $|\alpha\rangle$ can be represented in terms of orthonormal eigenkets $|n\rangle$ (the number or Fock state) of the number operator $N=a^\dagger a$ as follows:

$$|\alpha\rangle=\exp\left[-|\alpha|^2/2\right]\sum_{n=0}^{+\infty}(n!)^{-1/2}\alpha^n|n\rangle. \tag{15.10}$$

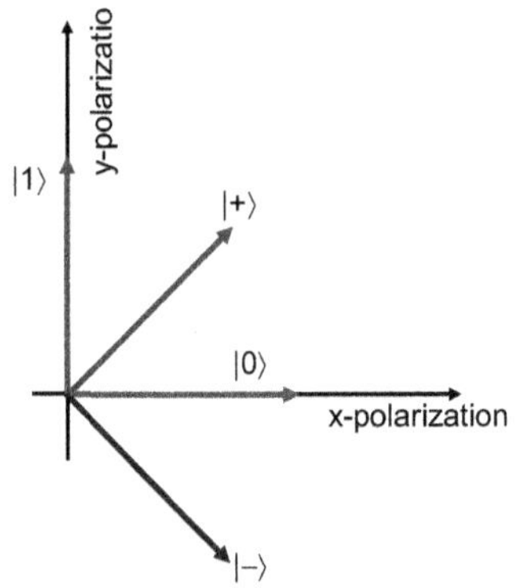

FIGURE 15.4

Four states being employed in the BB84 protocol.

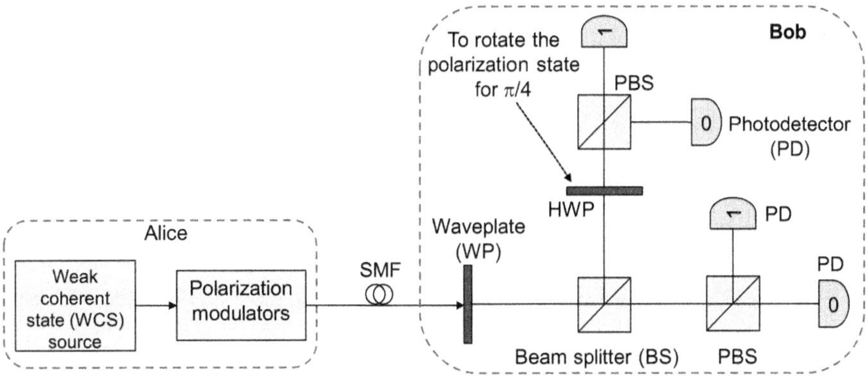

FIGURE 15.5

Polarization-based BB84 protocol. *HWP*, half-waveplate; *PBS*, polarizing beam splitter; *SMF*, single-mode fiber.

The coherent states are not orthogonal since the following is valid:

$$
\begin{aligned}
|\langle\alpha|\beta\rangle|^2 &= \left| e^{-\left(|\alpha|^2+|\beta|^2\right)/2} \sum_{n=0}^{+\infty}\sum_{m=0}^{+\infty} (n!)^{-1/2}(m!)^{-1/2}\alpha^n(\beta^*)^m\langle n|m\rangle \right|^2 \\
&= \left| e^{-\left(|\alpha|^2+|\beta|^2\right)/2} \sum_{n=0}^{+\infty} (n!)^{-1}\alpha^n(\beta^*)^n \right|^2 = \left| e^{-\left(|\alpha|^2+|\beta|^2-2\alpha\beta^*\right)/2} \right|^2. \qquad (15.11) \\
&= e^{-|\alpha-\beta|^2}.
\end{aligned}
$$

By sufficiently attenuating the coherent state we will obtain the weak coherent state, for which the probability of generating more than one photon is very low. For example, for $\alpha = 0.1$ we obtain $|0.1\rangle = \sqrt{0.90}\ |0\rangle + \sqrt{0.09}\ |1\rangle + \sqrt{0.002}\ |2\rangle + \ldots$. The number state $|n\rangle$ can be expressed in terms of the ground state ($|0\rangle$) with the help of the creation operator by:

$$|n\rangle = \frac{(a^{\dagger})^{n}|0\rangle}{(n!)^{1/2}}. \tag{15.12}$$

The density operator of a coherent state is given by:

$$\rho = |\alpha\rangle\langle\alpha| = e^{-|\alpha|^2}\sum_{n=0}^{+\infty}\sum_{m=0}^{+\infty}(n!)^{-1/2}(m!)^{-1/2}\alpha^{n}(\alpha^{*})^{m}|n\rangle\langle m|. \tag{15.13}$$

The diagonal terms of ρ are determined by:

$$\langle n|\rho|n\rangle = e^{-|\alpha|^2}\frac{|\alpha|^{2n}}{n!}, \tag{15.14}$$

which is clearly a Poisson distribution for the average number of photons $|\alpha|^2$. Therefore the probability that n photons can be found in coherent state $|\alpha\rangle$ follows the Poisson distribution:

$$P(n) = |\langle n|\alpha\rangle|^2 = e^{-\mu}\frac{\mu^{n}}{n!}, \tag{15.15}$$

where we use $\mu = |\alpha|^2$ to denote the average number of photons. Given that the probability that multiple photons are transmitted is now nonzero, Eve can exploit this fact to gain information from the channel known as the photon number splitting (PNS) attack. Assuming that Eve is capable of detecting the number of photons that reach her without performing a measurement on the system, when multiple photons are detected, Eve will take a single photon and place it into quantum memory until Alice and Bob perform time sifting and basis reconciliation. By learning Alice's and Bob's measurement basis from the discussion over the authenticated public channel, the photon stored in quantum memory will be measured in the correct basis and provide Eve with all the information contained in the photon.

To overcome the PNS attack, the use of decoy-state-based quantum key distribution was proposed in [37]. In decoy-state QKD, the average number of photons transmitted is increased during random time slots, allowing Alice and Bob to detect if Eve is stealing photons when multiple photons are transmitted.

15.4.2 B92 Protocol

The B92 protocol, introduced by Bennet in 1992 [38], employs only two nonorthogonal states, as illustrated in Fig. 15.6. Alice randomly generates a classical bit d, and depending on its value (0 or 1), she sends the following state to Bob [7]:

$$|\psi\rangle = \begin{cases} |0\rangle, d = 0 \\ |+\rangle = (|0\rangle + |1\rangle)/\sqrt{2}, d = 1 \end{cases}. \tag{15.16a}$$

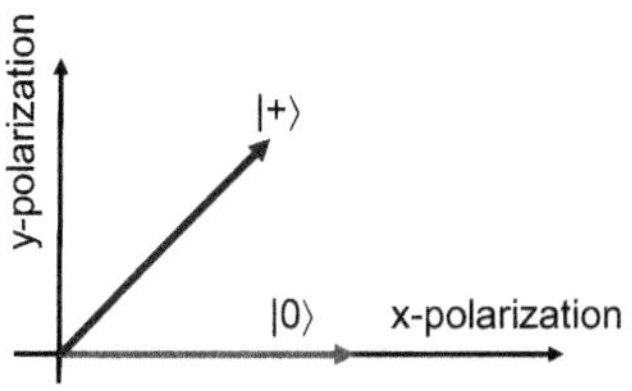

FIGURE 15.6

Two states being employed in the B92 protocol.

Bob randomly generates a classical bit d' and subsequently measures the received qubits in the CB = $\{|0\rangle, |1\rangle\}$ if $d' = 0$ or DB = $\{|+\rangle, |-\rangle\}$ if $d' = 1$, with the result of the measurement being $r = 0$ or 1, corresponding to -1 and $+1$ eigenstates of Z and X observables, which is illustrated in Table 15.1. Clearly, when the result of a measurement is $r = 0$, the possible states sent by Alice are $|0\rangle$ and $|+\rangle$ and Bob does not know which state Alice has sent. On the other hand, when the result of the measurement is $r = 1$, the possible states detected by Bob are $|1\rangle$ and $|-\rangle$, and Bob's detected bit is a complement of Alice's bit. Bob publicly announces r (and of course keeps d' secret). Alice and Bob keep only those pairs $\{d,d'\}$ for which the result of the measurement was $r = 1$. The final key bit is d for Alice and $1 - d'$ for Bob. After that, the information reconciliation and privacy amplification stages take place.

15.4.3 Ekert (E91) and EPR Protocols

The Ekert (E91) protocol was introduced by Ekert in 1991 [39] (see also [7]). Suppose Alice and Bob share n entangled pairs of qubits in the Bell state (Einstein–Podolski–Rosen [EPR] pair) $|B_{00}\rangle$:

$$|B_{00}\rangle = \frac{|00\rangle + |11\rangle}{\sqrt{2}}. \qquad (15.16b)$$

Table 15.1 Explanation of the B92 protocol.

Bits generated by Alice	***d* = 0**		***d* = 1**	
States sent by Alice	$|0\rangle$		$|+\rangle$	
Bits generated by Bob	$d' = 0$	$d' = 1$	$d' = 0$	$d' = 1$
Bob's measurement base	CB: $\{|0\rangle,|1\rangle\}$	DB: $\{|+\rangle,|-\rangle\}$	CB: $\{|0\rangle,|1\rangle\}$	DB: $\{|+\rangle,|-\rangle\}$
The possible resulting states to be detected by Bob	$|0\rangle$ $|1\rangle$	$|+\rangle$ $|-\rangle$	$|0\rangle$ $|1\rangle$	$|+\rangle$ $|-\rangle$
The probability of Bob measuring a given state	1 0	1/2 1/2	1/2 1/2	1 0
Result of Bob's measurement r	0 –	0 1	0 1	0 –

Alice generates a random classical bit b to determine the base, and measures her half on the Bell state in either CB $= \{|0\rangle, |1\rangle\}$ basis for $b = 0$ or DB $= \{|+\rangle, |-\rangle\}$ basis for $b = 1$ to obtain the data bit d. On the other hand, Bob in similar fashion randomly generates the basis bit b', and performs the measurement in either CB or DB to obtain d'. Alice and Bob compare the bases being used, namely b and b', and keep for their raw key only those instances of $\{d,d'\}$ for which the basis was the same $b = b'$. Clearly, the raw key so generated is truly random; it is undetermined until Alice and Bob perform measurements on their Bell state half. For this reason, QKD is essentially a secret-key generation method, as it is dictated by the entangled source. If instead the Bell state $|B_{01}\rangle = (|01\rangle + |10\rangle)/\sqrt{2}$ was used, Bob's raw key sequence will be complementary to Alice's raw key sequence.

The EPR protocol is a three-state protocol that employs Bell's inequality to detect the presence of Eve as a hidden random variable, as described in [40]. Let $|\theta\rangle$ denote the polarization state of a photon that is linearly polarized at an angle θ. Three possible polarization states for the EPR pair include [40]:

$$\begin{aligned} |B_0\rangle &= \frac{|0\rangle|3\pi/6\rangle + |3\pi/6\rangle|0\rangle}{\sqrt{2}}, \\ |B_1\rangle &= \frac{|\pi/6\rangle|4\pi/6\rangle + |4\pi/6\rangle|\pi/6\rangle}{\sqrt{2}}, \\ |B_2\rangle &= \frac{|2\pi/6\rangle|5\pi/6\rangle + |5\pi/6\rangle|2\pi/6\rangle}{\sqrt{2}}. \end{aligned} \tag{15.17}$$

For each of these EPR pairs we apply the following encoding rules [40]:

$$\begin{aligned} |\psi\rangle_1 &= \begin{cases} |0\rangle, d = 0 \\ |3\pi/6\rangle, d = 1 \end{cases} \\ |\psi\rangle_2 &= \begin{cases} |\pi/6\rangle, d = 0 \\ |4\pi/6\rangle, d = 1 \end{cases} \\ |\psi\rangle_3 &= \begin{cases} |2\pi/6\rangle, d = 0 \\ |5\pi/6\rangle, d = 1 \end{cases} \end{aligned} \tag{15.18}$$

The corresponding measurement operators will be [40]:

$$M_0 = |0\rangle\langle 0|, M_1 = |\pi/6\rangle\langle \pi/6|, M_2 = |2\pi/6\rangle\langle 2\pi/6|. \tag{15.19}$$

The EPR protocol can now be described as follows. The EPR state $|B_i\rangle$ ($i = 0,1,2$) is randomly selected. The first photon in the EPR pair is sent to Alice, the second to Bob. Alice and Bob select at random, independently and with equal probability, one of the measurement operators M_0, M_1, and M_2. They subsequently measure their respective photon (qubit), and the results of the measurement determine the corresponding bit for the sifted key. Clearly, the selection of EPR pairs as in Eq. (15.17) results in complementary raw keys. Alice and Bob then communicate over an authenticated public channel to determine instances when they used the same measurement operator and keep those instances. The corresponding shared key represents the sifted key. The instances when they used different measurement operators represent the rejected key, and these are used to detect Eve's presence by employing Bell's inequality on the rejected key. Alice and Bob then perform information reconciliation and privacy amplification steps.

Let us now describe how the rejected key can be used to detect Eve's presence [40]. Let $P(\neq|i,j)$ denote the probability that Alice's and Bob's bits in the rejected key are different given that Alice's and Bob's measurement operators are either M_i and M_j or M_j and M_i, respectively. Let further $P(=|i,j) = 1 - P(\neq|i,j)$. Let us denote the difference of these two probabilities by $\Delta P(i,j) = P(\neq|i,j) - P(=|i,j)$. Bell's parameter can be defined by [40]:

$$\beta = 1 + \Delta P(1,2) - |\Delta P(0,1) - \Delta P(0,2)|, \tag{15.20}$$

which for the foregoing measurement operators reduces to $\beta \geq 0$. However, the quantum mechanics prediction gives $\beta = -1/2$, which is clearly in violation of Bell's inequality.

15.4.4 Time-Phase Encoding

The BB84 protocol can also be implemented using different DOF, in addition to polarization states, such as time-phase encoding [41–43] and orbital angular momentum [44,45]. The time-phase encoding states for the BB84 protocol are provided in Fig. 15.7. The time basis corresponds to the CB, while the phase basis corresponds to the DB. The pulse is localized within the time bin of duration $\tau = T/2$. The time basis is like pulse-position modulation (PPM). The state in which the photon is placed in the first time bin is denoted by $|t_0\rangle$, while the state in which the photon is placed in the second time bin is denoted by $|t_1\rangle$. The phase MUB states are defined by:

$$|f_0\rangle = \frac{1}{\sqrt{2}}(|t_0\rangle + |t_1\rangle), \quad |f_1\rangle = \frac{1}{\sqrt{2}}(|t_0\rangle - |t_1\rangle). \tag{15.21}$$

Alice randomly selects either the time basis or the phase basis, followed by random selection of the basis state. The logical 0 is represented by $|t_0\rangle$, $|f_0\rangle$, while the logical one is represented by $|t_1\rangle$, $|f_1\rangle$. Bob measures each qubit by randomly

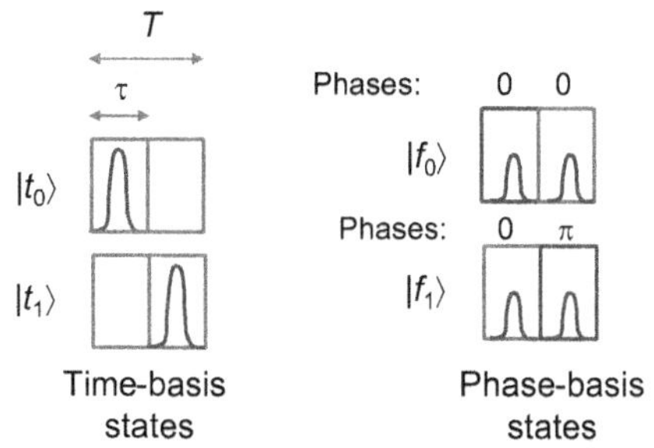

FIGURE 15.7

Time-phase encoding states to be used in the BB84 protocol.

selecting the basis, the time basis, or the phase basis. In the sifting procedure, Alice and Bob announce the bases being used for each qubit and keep only instances when they use the same basis. To implement the time-phase encoder either the electro-optical polar modulator, composed of a concatenation of an amplitude modulator and phase modulator, or the *I/Q* modulator can be used [46]. The amplitude modulator (Mach–Zehnder modulator) is used for pulse shaping and the phase modulator is used to introduce the desired phase shift, as illustrated in Fig. 15.8. The arbitrary waveform generator (AWG) is used to generate the corresponding radio-frequency (RF) waveforms as needed. The variable optical attenuator (VOA) is used to attenuate the modulated beam down to the single-photon level.

On the receiver side, the time basis states can be detected with the help of a single-photon counter and properly designed electronics. On the other hand, to detect the phase basis states, a time-delay interferometer can be used as described in [47] (Fig. 15.9). The difference in the path between the two arms is [47] $\Delta L = \Delta L_0 + \delta L$, where ΔL_0 is the path difference equal to $c\tau$ (c is the speed of light), and $\delta L << \Delta L_0$ is the small path difference used to adjust the phase since $\phi = k\delta L$ (k is the wave number). When the phase-state $|f_0\rangle$ is incident to the time-phase decoder, the outputs

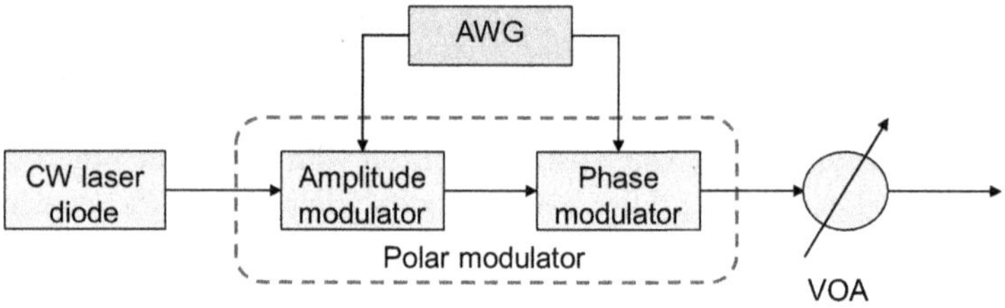

FIGURE 15.8

Time-phase encoder for BB84 quantum key distribution protocol. *AWG*, Arbitrary waveform generator; *VOA*, variable optical attenuator.

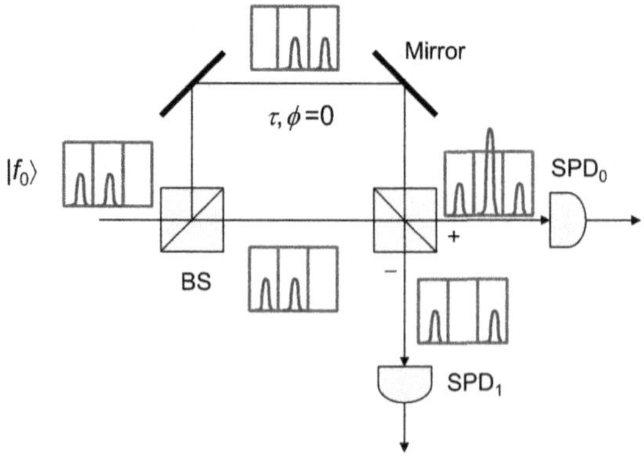

FIGURE 15.9

Time-phase decoder for BB84 quantum key distribution protocol. *BS*, 50:50 beam splitter; *CW*, continuous wave; *SPD*, single-photon detector.

of a second 50:50 beam splitter (BS) occupy three time slots, and the $+$ output denotes the case when the interferometer outputs interfere constructively, while the $-$ output denotes that the corresponding interferometer outputs interfere destructively. For constructive interference, the middle pulse is doubled, while for destructive interference, middle pulses cancel each other. Therefore the click of the SPD in the middle slot at the $+$ output identifies the $|f_0\rangle$ state, while the corresponding click at the $-$ output identifies the $|f_1\rangle$ state.

15.5 QKD SECURITY

As discussed in the BB84 protocol section, after the raw key transmission phase is completed, Alice and Bob perform a sifting procedure and parameter estimation, so that they are left with n symbols out of N transmitted, and the corresponding key is known as the sifted key. After that, they perform the classical postprocessing of information reconciliation and privacy amplification. In the information reconciliation stage, they employ error correction to correct for the error introduced by both the quantum channel and Eve, and the corresponding key after this stage is completed is commonly referred to as the corrected key. During privacy amplification, they remove the correlation remaining with Eve, and they are left with m symbols out of n, and the corresponding key is the secure key. Clearly, the fractional rate of the secret key r can be determined as $r = m/n$ (as $N \rightarrow \infty$), and it is often referred as the secret fraction [48]. The *secret-key rate* (SKR) is then determined as the product of the *secret fraction* r and the raw key rate R_{raw}, and we can write [48]:

$$SKR = r \cdot R_{\text{raw}}. \tag{15.22}$$

Interestingly enough, in many papers the term SKR refers to the secret fraction r. Given that the raw key rate is dictated by devices being employed, in particular the SP***D*** and the quantum channel over which transmission takes place, and not by quantum physics laws, the fraction rate can also be called the normalized SKR, given that $r = \text{SKR}/R_{\text{raw}}$.

The raw key rate can be determined as the product of signaling rate R_s and the probability of Bob accepting the transmitted symbol on average, denoted as Pr(Bob accepting), so that we can write:

$$R_{\text{raw}} = R_s \cdot \Pr(\text{Bob accepting}). \tag{15.23}$$

On the other hand, the signaling rate is determined by [48]:

$$R_s = \min\left(R_{s,\max}, 1/T_{dc}, \frac{\mu T T_B \eta}{\tau_{\text{dead}}}\right), \tag{15.24}$$

where $R_{s,\max}$ is the maximum signaling rate allowed by the source, T_{dc} is the duty cycle, and τ_{dead} is the dead time of the SPD (the time required for SPD to reset after a photon is detected). The term $\mu T T_B \eta$ corresponds to Bob's probability of detection, with μ being the average number of photons generated by the source, η being the

SPD efficiency, T being the transmissivity of the quantum channel, and T_B being the losses in Bob's receiver. The quantum channel is typically either a fiber-optics channel or an FSO channel. Attenuation in the fiber-optics channel is determined by $T = 10^{-\alpha L/10}$, where α is the attenuation coefficient in dB/km, while L is the transmission distance in km. The typical attenuation coefficient in standard single-mode fiber (SSMF) is 0.2 dB/km at 1550 nm. Regarding the FSO link, if we ignore the atmospheric turbulence effects, for the line-of-sight link of transmission distance L, we can estimate the FSO link loss by [49] $T = [d_{Rx}/(d_{Tx} + DL)]^2 \cdot 10^{-\alpha L/10}$, where d_{Tx} and d_{Rx} represent the diameters of transmit and receive telescope apertures, while D is the divergence. Therefore the first term represents the geometric losses, while the second term is the average attenuation due to scattering, with scattering attenuation coefficient being <0.1 dB/km in clear sky condition at 1520−1600 nm.

In the BB84 protocol, the probability that Bob accepts the transmitted symbol is related to the probability that he used the same basis, denoted as p_{sift}, so that the signaling rate becomes $R_s p_{\text{sift}}$, and the raw key rate when Alice sends n photons to Bob will be [48]:

$$R_{\text{raw},n} = R_s p_{\text{sift}} p_{A,n} p_n, \tag{15.25}$$

where $p_{A,n}$ is the probability that the pulse sent by Alice contains n photons (see the following section) and p_n is the probability that Eve sends Bob a pulse with n photons. The overall total raw key rate will be $R_{\text{raw}} = \sum_n R_{\text{raw},n}$.

What remains to be determined is the fraction rate. QKD can achieve unconditional security, which means that its security can be verified without imposing any restrictions on either Eve's computational power or eavesdropping strategy. The bounds on the fraction rate are dependent on the classical postprocessing steps. The most common is one-way postprocessing, in which either Alice or Bob holds the reference key and sends the classical information to the other party through the public channel, while the other party performs a certain procedure on data without providing the feedback. The most common one-way processing consists of two steps: information reconciliation and privacy amplification. The expression for secret fraction, obtained by one-way postprocessing, is given by [49]:

$$r = I(A;B) - \min_{\text{Eve's strategies}} (I_{EA}, I_{EB}), \tag{15.26}$$

where $I(A;B)$ is the mutual information between Alice and Bob, while the second term corresponds to Eve's information I_E about Alice's or Bob's raw key, where minimization is performed over all possible eavesdropping strategies. Alice and Bob will decide to employ either direct or reverse reconciliation so that they can minimize Eve's information.

We now describe different eavesdropping strategies that Eve may employ, which determine Eve's information I_E.

15.5.1 Independent (Individual) or Incoherent Attacks

This is the most constrained family of attacks, in which Eve attacks each qubit independently, and interacts with each qubit by applying the same strategy. Moreover, she measures the quantum states before the classical postprocessing takes place. The security bound for incoherent attacks is the same as that for classical physical-layer security [2], and is determined by the Csiszár–Körner bound [50], given by Eq. (15.26), wherein the mutual information between Alice and Eve is given by:

$$I_{EA} = \max_{\text{Eve's strategies}} I(A;E), \tag{15.27}$$

where the maximization is performed over all possible incoherent eavesdropping strategies. A similar definition holds for I_{BE}.

An important family of incoherent attacks are the *intercept-resend* (IR) attacks in which Eve intercepts the quantum signal sent by Alice, performs the measurement on it, and based on the measurement result prepares a new quantum signal (in the same MUB as the measured quantum state) and sends such prepared quantum signal to Bob. In the BB84 protocol, Eve's basis choice will match Alice's basis 50% of the time. When she uses the wrong basis, there is still a 50% chance of guessing the correct bit. When Bob does the measurement, there is a 50% chance that Bob will use the same basis as Eve. However, these bits will correlate with Eve's sequence not Alice's, and overall probability of error will be 1/4, while the mutual information between Alice and Eve will be $I(A;E) = 1/2$. Given the high probability of error introduced by the IR attack, Alice and Bob can easily identify Eve's activity. However, Eve might choose to apply the IR attack with probability of p_{IR}. In that case, the probability of error will be $q = p_{IR}/4$. The mutual information between Alice and Eve will then be $I(A;E) = h(p_{IR}/4)$, where $h(\bullet)$ is the binary entropy function $h(q) = -q \log_2 q - (1-q)\log_2(1-q)$. If all errors are contributed to Eve, the secret fraction can be estimated as:

$$r = \underbrace{1 - h(q)}_{I(A;B)} - h(q) = 1 - 2h(q). \tag{15.28}$$

Another important family of incoherent attacks are the *photon number splitting* (PNS) attacks [51], in which Eve acts on multiphoton quantum states. In a weak coherent state-based QKD system, the quantum states are generated by modulating the beam from the coherent light source, which is then attenuated so that on average one photon per pulse is transmitted by Alice. However, as discussed earlier, the coherent source emits the photons based on Poisson distribution so that the probability of emitting n photons in a state generated with a mean photon number μ is determined by the following equation:

$$p_{A,n} = p(n|\mu) = e^{-\mu}\frac{\mu^n}{n!}. \tag{15.29}$$

As the parameter μ increases, the probability of getting more than one photon increases as well. To perform the PNS attack, Eve can employ the BS to take one of the photons from the multiphoton state, and pass the rest to Bob, as illustrated in

Fig. 15.10. She can measure the photon by randomly selecting the basis. Eve can even replace the quantum link with ultralow-loss fiber so that Alice and Bob cannot figure that the transmitted signal is attenuated. To solve the PNS attack, Alice and Bob can employ a decoy-state-based protocol [37], in which Alice transmits the quantum states with different mean photon numbers, representing signal and decoy states. Eve cannot distinguish between decoy state and signal state, and given that Alice and Bob know the decoy signal level, they can identify the PNS attack.

15.5.2 Collective Attacks

Collective attacks represent generalization of the incoherent attacks given that Eve's interaction with each qubit is also independent and identically distributed (i.i.d.). However, in these attacks, Eve can store her ancilla qubits in a quantum memory until the end of classical postprocessing steps. For instance, given that information reconciliation requires the exchange of parity bits over an authenticated classical channel, Eve can apply the best known classical attacks to learn the content of the parity bits. Based on all information available to her, Eve can perform the best measurement strategy on her ancilla qubits (stored in the quantum memory).

The PNS attack is stronger when Eve applies quantum memory. Namely, from the sifting procedure, Eve can learn which basis was used by Alice and can apply the correct basis on her photon stored in the quantum memory and thus double the number of identified bits compared to the case when quantum memory is not used.

The security bound for collective attacks, assuming one-way postprocessing, is given by Eq. (15.26), wherein Eve's information about Alice's sequence is determined from Holevo information as follows [52] (see also [48]):

$$I_{EA} = \max_{\text{Eve's strategies}} \chi(A;E), \tag{15.30}$$

where maximization is performed over all possible collective eavesdropping strategies. A similar definition holds for I_{BE}. This bound is also known as the Devetak–Winter bound. The Holevo information, introduced in [53], is defined here as:

$$\chi(A;E) = S(\rho_E) - \sum_a S\left(\rho_{E|a}\right)p(a), \tag{15.31}$$

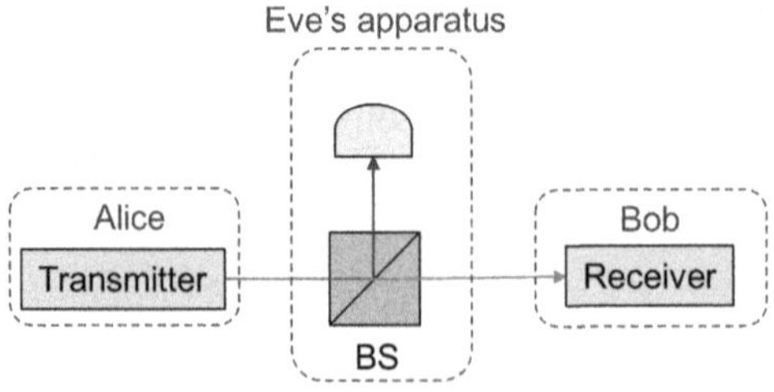

FIGURE 15.10

Eve's realization of the photon number splitting attack. *BS*, Beam splitter.

where $S(\rho)$ is the von Neumann entropy defined as $S(\rho) = -\text{Tr}(\log(\rho)) = -\sum_i \lambda_i \log \lambda_i$, with λ_i being the eigenvalues of the density operator (state) ρ. In Eq. (15.31), $p(a)$ represents the probability of occurrence of symbol a from Alice's classical alphabet, while $\rho_{E|a}$ is the corresponding density operator of Eve's ancilla. Finally, ρ_E is Eve's partial density state defined by $\rho_E = \sum_a p(a)\rho_{E|a}$. In other words, the Holevo information corresponds to the average reduction in von Neumann entropy given that we know how ρ_E is prepared.

15.5.3 Quantum Hacking Attacks and/or Side-Channel Attacks

Quantum hacking attacks exploit the weakness of the particular QKD system implementation. Many hacking attacks are feasible with current technology. In a *Trojan horse attack* [54], Eve sends the bright laser beam toward Alice's encoder and measures the reflected photons to gain the information about the secret. This attack can be avoided by using an optical isolator. It was also noticed that silicon-based photon counters employing avalanche photodiodes (APDs) emit some light at different wavelengths when they detect a single photon, which can be exploited by Eve [55]. In a *time-shifting attack* [56], Eve exploits the efficiency mismatch of Bob's single-photon detectors to estimate Bob's basis selection. In a *blinding attack* [57], Eve exploits the APD-based single-photon counter properties to force Bob to pick up the same basis as Eve does. The APD, for photon counting, often operate in gated mode so that the APD is active only when photon arrival is expected. When in off-mode, the APD operates in a linear regime, meaning that the photocurrent is proportional to the received optical power, and therefore it behaves as the classical photodetector. To exploit this, Eve can send a weak classical light beam, sufficiently strong to generate the click when Bob employs the same basis, and not to register the click otherwise. In such a scenario, 50% of events will be recorded as inconclusive. Fortunately, the common property of quantum hacking attacks is that they are preventable once the attack is known. However, the threat from an unknown quantum hacking attack is still present.

Many *side-channel attacks* are at the same time zero-error attacks [48]. Namely, Eve can exploit the quantum channel loss to hide her side-channel attack. The beam-splitting attack exploits the losses of the quantum channel, and Eve can place the BS just after Alice's transmitter, take the portion of the beam, and pass the rest to Bob. Since Alice does not modify the optical mode, this attack will not be detected by the Bob. Interestingly enough, the Poisson distribution is preserved during the PNS attack [58]. In realistic implementations, Eve can perform unambiguous state discrimination followed by signal resending, when Alice employs the linearly independent signal states, and achieve better than beam-slitting efficiency as shown in [59]. Fortunately, the existence of side-channel attacks does not compromise security as long as the corresponding side-channel attacks are taken into account during the privacy amplification stage [48].

15.5.4 Security of the BB84 Protocol

The *secret fraction* that can be achieved with the BB84 protocol is lower bounded by:

$$r = q_1^{(Z)}\left[1 - h_2\left(e_1^{(X)}\right)\right] - q^{(Z)} f_e h_2\left(e^{(Z)}\right), \tag{15.32}$$

where $q_1^{(Z)}$ denotes the probability of declaring a successful result ("the gain") when Alice sends a single photon and Bob detects it in the Z basis, f_e denotes the error correction inefficiency ($f_e \geq 1$), $e^{(X)}$ [$e^{(Z)}$] denotes the quantum bit error rate (QBER) in the X basis (Z basis), and $h_2(x)$ is the binary entropy function defined as:

$$h_2(x) = -x\log_2(x) - (1-x)\log_2(1-x). \tag{15.33}$$

The second term $q^{(Z)}h_2[e^{(X)}]$ denotes the amount of information Eve was able to learn during the raw key transmission, and this information is typically removed from the final key during the privacy amplification stage. The last term $q^{(Z)}f_e h_2[e^{(Z)}]$ denotes the amount of information revealed during the information reconciliation (error correction) stage, typically related to the parity bits exchanged over the noiseless public channel (when systematic error correction is used).

15.6 THE DECOY-STATE PROTOCOLS

When a weak coherent state (WCS) source is used instead of a single-photon source, this scheme is sensitive to PNS attack. Given that the probability that multiple photons are transmitted from WCS is now nonzero, Eve can exploit this fact to gain information from the channel. To overcome the PNS attack, decoy-state QKD systems can be used [37,60–64,66–70]. In decoy-state QKD, the average number of photons transmitted is increased during random timeslots, allowing Alice and Bob to detect if Eve is stealing photons when multiple photons are transmitted. There exist different versions of decoy-state-based protocols. In one-decoy-state-based protocol, in addition to the signal state with mean photon number μ, the decoy state with mean photon number $\nu < \mu$ is employed. Alice first decides on signal and decoy mean photon number levels, and then determines the optimal probabilities for these two levels to be used, based on corresponding SKR expression. In a weak-plus-vacuum decoy-state-based protocol, in addition to μ and ν levels, the vacuum state is also used as the decoy state. Both protocols have been evaluated in [67], both theoretically and experimentally. It was concluded that the use of one signal state and two decoy states is enough, which agrees with findings in [68,69].

The probability that Alice can successfully detect the photon when Alice employs the WCS with mean photon number μ, denoted as q_μ, also known as the gain, can be determined by:

$$q_\mu = \sum_{n=0}^{\infty} y_n e^{-\mu}\frac{\mu^n}{n!}, \tag{15.34}$$

where y_n, also known as the yield, is the probability of Bob's successful detection when Alice has sent n photons. A similar expression holds for decoy states with mean photon number σ_i, that is:

$$q_{\sigma_i} = \sum_{n=0}^{\infty} y_n e^{-\sigma_i} \frac{\sigma_i^n}{n!}. \tag{15.35}$$

Let e_n denote the QBER corresponding to the case when Alice has sent n photons, the average QBER for the coherent state $|\mu \exp(j\phi)\rangle$ (where ϕ is the arbitrary phase) can then be estimated by:

$$\overline{e}_\mu = \frac{1}{q_\mu} \sum_{n=0}^{\infty} y_n e_n e^{-\mu} \frac{\mu^n}{n!}. \tag{15.36}$$

If T is the overall transmission efficiency, composed of the channel transmissivity T_{ch}, photodetector efficiency η, and Bob's optics transmissivity T_{Bob}, the transmission efficiency of an n-photon pulse will be given by:

$$T_n = 1 - (1 - T)^n. \tag{15.37}$$

In the absence of Eve, for $n = 0$, the yield y_0 is affected by the background detection rate, denoted as p_d, so that we can write $y_0 = p_d$. On the other hand, the yield for $n \geq 1$ is contributed by both signal photons and background rate, as follows:

$$y_n = T_n + y_0 - T_n y_0 = T_n + y_0(1 - T_n) = 1 - (1 - y_0)(1 - T)^n. \tag{15.38}$$

The error rate of n-photon states e_n can be estimated by [37,70]:

$$e_n = \frac{1}{y_n}\left(e_0 y_0 + e_d \underbrace{T_n}_{\cong y_n - y_0} \right) \cong e_d + \frac{y_0}{y_n}(e_0 - e_d), \tag{15.39}$$

where e_d is the probability of a photon hitting an erroneous detector. For time-phase encoding, this represents the intrinsic misalignment error rate. In the absence of signal photons, assuming that two detectors have equal background rates, we can write $e_0 = 1/2$. In similar fashion, the average QBER can be estimated by:

$$\overline{e}_\mu = e_d + \frac{y_0}{q_\mu}(e_0 - e_d). \tag{15.40}$$

The secrecy fraction can now be lower bounded by modifying the corresponding expression for BB84 protocols as follows:

$$r = q_1^{(Z)}\left[1 - h_2\left(e_1^{(X)}\right)\right] - q_\mu^{(Z)} f_e h_2\left(\overline{e}_\mu^{(Z)}\right), \tag{15.41}$$

where we used the subscript 1 to denote the single-photon pulses and μ to denote the pulse with the mean photon number μ.

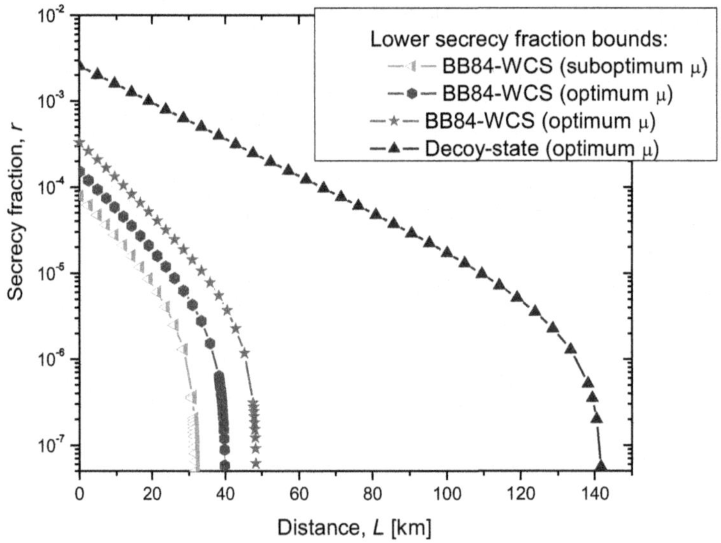

FIGURE 15.11

Secrecy fraction versus transmission distance for decoy-state-based protocol and weak coherent-state (WCS)-based BB84 protocol.

As an illustration, in Fig. 15.11 we provide secrecy fraction results for decoy-state protocol and the WCS-based, PM, BB84 protocol, assuming that the SSMF with attenuation coefficient 0.21 dB/km is used.

Other parameters used in simulations are set as follows [37,71–73]: the detector efficiency $\eta = 0.045$, the dark count rate $p_d = 1.7 \cdot 10^{-6}$, the probability of a photon hitting the erroneous detector $e_d = 0.033$, and the reconciliation inefficiency $f_e = 1.22$. For a decoy-state-based BB84 protocol to calculate the secrecy fraction we use Eq. (15.41) for the optimum value of mean photon number per pulse μ, wherein q_μ is determined by $q_\mu = 1 - (1 - y_0)\exp(-\mu T)$, $\bar{e}_\mu$ is determined by Eq. (15.40), e_1 is determined by Eq. (15.39), and q_1 is the argument of Eq. (15.34) for $n = 1$, that is, by $q_1 = y_1 \mu \exp(-\mu)$. For the WCS-based BB84 protocol, without the decoy state, we use Eq. (15.41), assuming that Eve controls the channel, performs IR attacks, and is transparent to multiphoton states. In this lower-bound scenario, the error from single-photon events dominates. Now by calculating $q^{(Z)}$ by $q_\mu = 1 - (1 - y_0)\exp(-\mu T)$, $e^{(Z)}$ by $E_\mu = e_d + (e_0 - e_d)y_0/q_\mu$, and assuming suboptimum μ is proportional to T, we obtain the yellow (triangles pointed left) curve in Fig. 15.11, which is consistent with [37]. When using the optimum μ instead (maximizing the secrecy fraction) we obtain the violet (circle) curve, which is consistent with [71]. For these two curves we estimate the $q_1^{(Z)}$ from Eq. (15.34) by setting $y_n = 1$ for $n \geq 2$ (see [70]) as follows:

$$q_1^{(X)} \geq q_\mu - \sum_{n=2}^{\infty} e^{-\mu} \frac{\mu^n}{n!}. \tag{15.42}$$

Given that multiphoton states do not introduce errors, that is, $e_n = 0$ for $n \geq 2$, we can bound the e_1 as follows:

$$e_1^{(X)} \leq q_\mu E_\mu / q_1^{(X)}. \tag{15.43}$$

A more accurate expression for q_μ, for WCS-based BB84 protocol, is the following:

$$q_\mu = y_0 e^{-\mu} + y_1 e^{-\mu}\mu + \sum_{n=2}^{\infty} e^{-\mu}\frac{\mu^n}{n!}, \tag{15.44}$$

which is obtained from Eq. (15.34) by setting $y_n = 1$ for $n \geq 2$. The corresponding curve is denoted in pink (stars), which indicates that the distance of $L = 48.62$ km is achievable when the WCS-based BB84 protocol is used. In contrast, for the decoy-state-based BB84 protocol and optimum μ, it is possible to achieve a distance close to 141.8 km (see blue curve [triangles pointed up in printed version]), which agrees with [71].

15.7 MEASUREMENT DEVICE-INDEPENDENT QKD PROTOCOLS

Any discrepancy in parameters of devices from assumptions made in QKD protocols can be exploited by adversaries through side-channel and quantum hacking attacks [54–58] to compromise security of the corresponding protocols. One way to solve this problem is to invent different ways to overcome the known side-channel/quantum hacking attacks. Unfortunately, the eavesdropper can always come up with a more sophisticated attack. Another, more effective strategy is to invent new protocols in which minimal assumptions have been made on devices being employed in the protocol. One such protocol is known as the device-independent QKD (DI-QKD) protocol [74–77] in which the statistics of detection has been determined to secure the key without any assumption being made on the operation of devices.

15.7.1 Photonic Bell-State Measurements

The Bell-state measurements (BSMs) play a key role in photonic quantum computation and communication, including quantum teleportation [78], superdense coding [79], quantum swapping [80], quantum repeaters [81], and QKD [82–84], to mention a few. A *complete* BSM represents a projection of any two-photon state to maximally entangled Bell states, defined by:

$$|\psi^\pm\rangle = \frac{1}{\sqrt{2}}(|01\rangle \pm |10\rangle), \quad |\phi^\pm\rangle = \frac{1}{\sqrt{2}}(|00\rangle \pm |11\rangle). \tag{15.45}$$

Unfortunately, a complete BSM is impossible to achieve using only linear optics without auxiliary photons. It has been shown in [85] that the probability of a successful BSM for two photons being in complete mixed input states is limited to 50%. The conventional approach to Bell-state analysis is to employ a 50:50 BS, followed by single-photon detectors that allow one to discriminate between orthogonal states. As an illustration, the setup to perform BSM on polarization states is provided in Fig. 15.12. The states emitted by Alice and Bob are denoted by density operators ρ_A and ρ_B, respectively.

When photons at BS inputs ***1*** and ***2*** are of the same polarization, that is, $|HH\rangle = |00\rangle$ or $|VV\rangle = |11\rangle$, according to the Hong–Ou–Mandel (HOM) effect [2,86,87], they will appear simultaneously at either output port ***3*** or port 4, and the Bell states $|\phi^{\pm}\rangle$ are transformed to:

$$|\phi^{\pm}\rangle = \frac{1}{\sqrt{2}}\left(|00\rangle_{12} \pm |11\rangle_{12}\right) \overset{BS}{\rightarrow} |\phi^{\pm}_{\text{out}}\rangle$$

$$= \frac{j}{2\sqrt{2}}\left(|00\rangle_{33} + |00\rangle_{44} \pm |11\rangle_{33} \pm |11\rangle_{44}\right), \tag{15.46}$$

and clearly cannot be distinguished by corresponding polarizing beam splitters (PBS). On the other hand, when photons at the BS input ports ***1*** and ***2*** are of different polarizations, they can exit either the same output port or different ports. When both photons exit the output port 3, after the PBS, SPDs D_{10} and D_{11} will simultaneously click. When both photons exit the output port 4, after the PBS, SPDs D_{20} and D_{21} will simultaneously click. When input photons of different polarizations exit different ports, there exist two options. When a horizontal photon exits port ***3*** and vertical photon port 4, SPDs D_{11} and D_{20} will simultaneously click. Finally, when a horizontal photon exits port ***4*** and vertical photon port 3, SPDs D_{10} and D_{21} will simultaneously click. Given that the Bell states $|\psi^{\pm}\rangle$ are transformed to (by the BS):

$$|\psi^{+}\rangle = \frac{1}{\sqrt{2}}\left(|01\rangle_{12} + |10\rangle_{12}\right) \overset{BS}{\rightarrow} |\psi^{+}_{\text{out}}\rangle = j\left(|01\rangle_{33} + |10\rangle_{44}\right),$$

$$|\psi^{-}\rangle = \frac{1}{\sqrt{2}}\left(|01\rangle_{12} - |10\rangle_{12}\right) \overset{BS}{\rightarrow} |\psi^{-}_{\text{out}}\rangle = |01\rangle_{34} - |10\rangle_{34}, \tag{15.47}$$

they can be distinguished by PBSs and SPDs. Clearly, based on the foregoing discussion we conclude that the projection on $|\psi^{+}\rangle$ occurs when both SPDs after any

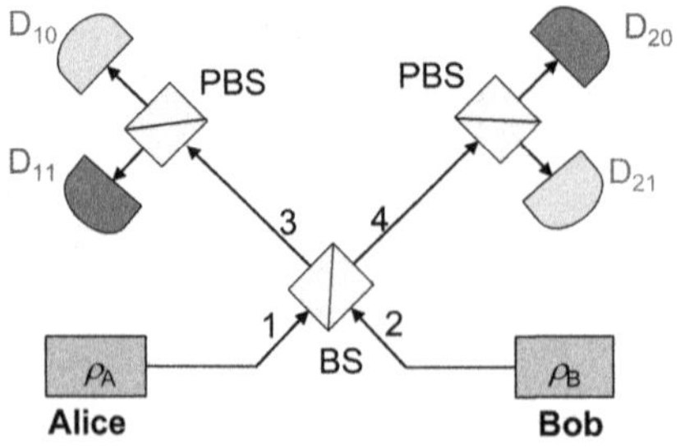

FIGURE 15.12

Setup to perform Bell-state measurements on polarization states. *BS*, Beam splitter; *PBS*, polarizing beam splitter.

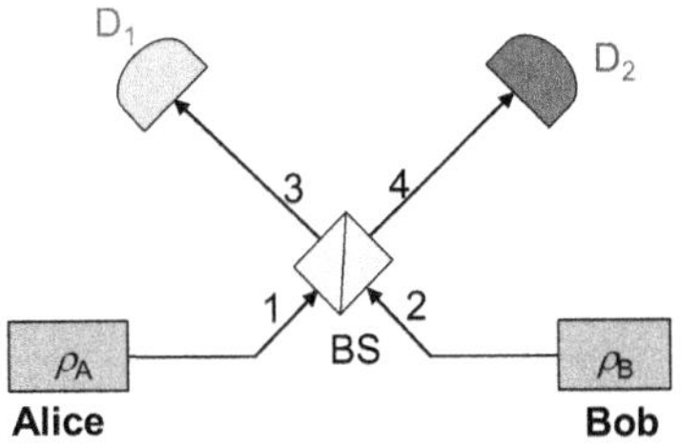

FIGURE 15.13

Setup to perform Bell-state measurements on time-bin states. *BS*, Beam splitter.

of two PBSs simultaneously click. In other words, when either D_{10} and D_{11} or D_{20} and D_{21} simultaneously click, the projection on $|\psi^+\rangle$ is identified. Similarly, when either D_{10} and D_{21} or D_{20} and D_{21} simultaneously click, the projection on the $|\psi^-\rangle$ state is identified. In conclusion, when ideal SPDs are used, the BSM efficiency is 1/2. If the SPDs have nonideal detection efficiency η, the BSM efficiency will be $0.5\eta^2$.

Another widely used DO*F* to encode quantum states is time-bin DOF. For 2D quantum communications, the photon can be encoded as the superposition of the early state, denoted as $|0\rangle$, and late state, denoted as $|1\rangle$, as follows $|\xi\rangle = \alpha|0\rangle + \beta|1\rangle$, $|\alpha|^2 + |\beta|^2 = 1$. To perform the BSM we can use the setup provided in Fig. 15.13, which requires only two SPDs in addition to the BS. Similarly to polarization, the Bell states are defined by Eq. (15.45), but with different basis kets. Again, when two photons arrive at the same time at BS input ports, they will exit the same output port according to the HOM effect, and we cannot distinguish the $|\phi^\pm\rangle$ states. On the other hand, the projection on $|\psi^+\rangle$ is identified when either detector registers the click in early and late time bins. Similarly, the projection on $|\psi^-\rangle$ occurs when either detectors D_1 and D_2 or D_2 and D_1 click in consecutive (early/late) time bins.

15.7.2 Description of MDI-QKD Protocol

In MDI-QKD protocols [82,88–91], Alice and Bob are connected to a third party—Charlie—through a quantum channel, such as an FSO or a fiber-optics link. The SPDs are located at Charlie's side, as illustrated in Fig. 15.14, for a polarization-based MDI-QKD protocol.

Alice and Bob can employ either single-photon sources or WCS sources, and randomly select one of four states $|0\rangle$, $|1\rangle$, $|+\rangle$, $|-\rangle$ to be sent to Charlie, with the help of a polarization modulator (PolM), which is very similar to that used in the BB84 protocol. The intensity modulator is employed to impose different decoy states. Charlie can essentially be Eve as well. Charlie performs the partial BSM with the help of one BS, two PBSs, and four SPDs as described in Fig. 15.14, and announces that the events for each measurement result in either the $|\psi^+\rangle = (|01\rangle + |10\rangle)/\sqrt{2}$ state or the $|\psi^-\rangle = (|01\rangle - |10\rangle)/\sqrt{2}$ state. Charlie also discloses the result of the BSM. Given that Charlie's BSMs are used only to check the parity of Alice's and Bob's bits, these measurements will not disclose any information related to the individual bits themselves [88].

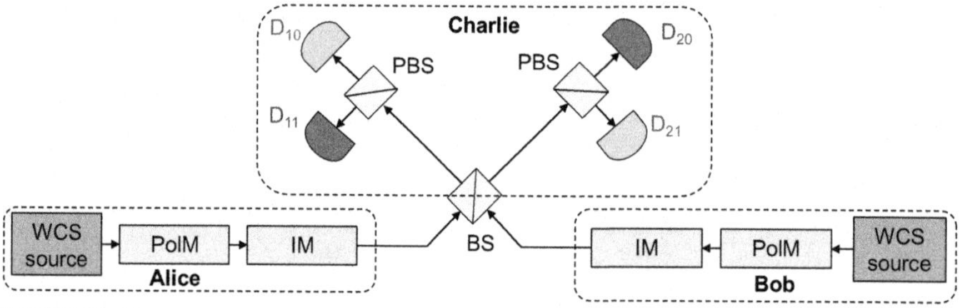

FIGURE 15.14

Polarization states-based measurement device-independent quantum key distribution system configuration. *BS*, Beam splitter; *PBS*, polarizing beam splitter; *IM*, intensity modulator; *PolM*, polarization modulator; *WCS*, weak coherent state.

A successful BSM corresponds to the observation when precisely two detectors (with orthogonal polarizations) are triggered as follows:

- Detectors D_{10} and D_{21} clicks identify the Bell state $|\psi^-\rangle$.
- On the other hand, detectors D_{11} and D_{20} clicks identify the Bell state $|\psi^+\rangle$.

Other detection patterns are related to $|\phi^\pm\rangle$ states, which cannot be distinguished with SPDs, and therefore cannot be considered as successful and as such are discarded from consideration. Clearly, based on the foregoing description we conclude that the MDI-QKD protocol represents generalization of both the EPR-based QKD protocol converted into a PM scheme and the decoy-state-based protocol. By comparing the portion of their sequences, Alice and Bob can verify Charlie's honesty.

Let us now describe one possible implementation of the PolM, that is, decoy-state-based BB84 encoder configuration, which is provided in Fig. 15.15. The laser beam signal is split by a 1:4 coupler into four branches, each containing the polarization controller to impose the following polarization states $|0\rangle, |1\rangle, |+\rangle, |-\rangle$. With the help of a 4:1 optical space switch we randomly select the polarization state to be sent to Charlie. For this purpose, the use of switches with electro-optic switching is desirable, such as those reported in [92,93], since in these switches the switching state can be changed within a few ns. The phase modulator at the output is used to perform the phase randomization of the coherent state $|\alpha\rangle = \exp(-|\alpha|^2/2)\sum_n \alpha^n |n\rangle/\sqrt{n!}$ as follows:

$$\frac{1}{2\pi}\int_0^{2\pi} \left||\alpha|e^{j\vartheta}\right\rangle\left\langle|\alpha|e^{-j\vartheta}\right| d\vartheta = \frac{1}{2\pi}\sum_{n=0}^{\infty}\sum_{m=0}^{\infty}\frac{|\alpha|^{m+n}}{\sqrt{m!n!}}e^{-|\alpha|^2}|n\rangle\langle m|\underbrace{\int_0^{2\pi} e^{j(n-m)\vartheta}d\vartheta}_{\delta_{nm}}$$

$$= \sum_{n=0}^{\infty}\frac{\left(|\alpha|^2\right)^n}{n!}e^{-|\alpha|^2}|n\rangle\langle n| = \sum_{n=0}^{\infty}\frac{\mu^n}{n!}e^{-\mu}|n\rangle\langle n|, \mu = |\alpha|^2. \tag{15.48}$$

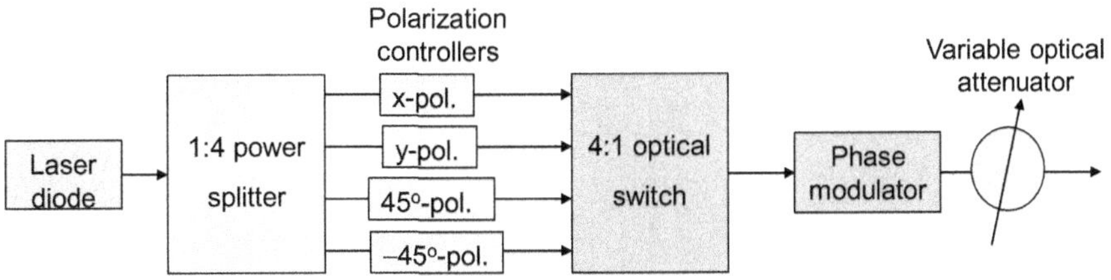

FIGURE 15.15

Decoy-state-based BB84 encoder (polarization modulator) configuration.

where $\mu = |\alpha|^2$ is the mean photon number per pulse. A VOA is then used to reduce the mean photon number to the quantum level. Other PolM configurations can be used, such as the one described in [88], employing four laser diodes. Alice and Bob then exchange the information over an authenticated classical, noiseless, channel about the bases being used: Z basis spanned by $|0\rangle$ and $|1\rangle$(CB) states, or X basis spanned by $|+\rangle$ and $|-\rangle$ (DB) states. Alice and Bob then keep only events in which they use the same basis, while at the same time Charlie announces the BSM states $|\psi^-\rangle$ or $|\psi^+\rangle$. One of them (say Bob) flips the sent bits except when both use the DB and Charlie announces the state $|\psi^+\rangle$. The remaining events are discarded.

Furthermore, the X key is formed out of those bits when Alice and Bob prepare their photons in the X basis (DB). The error rate on these bits is used to bound the information obtained by Eve. Next, Alice and Bob form the Z key out of those bits for which they both use the Z basis (CB). Finally, they perform information reconciliation and privacy amplification on the Z key to get the secret key.

In addition to being insensitive to the SPDs used by Charlie, the MDI-QKD protocol allows one to connect multiple participants in a star-like QKD network, in which Charlie is located in the center (serving as a hub) [94,95]. In this QKD network, illustrated in Fig. 15.16, the most expensive equipment (SPDs) is shared among multiple users, and the QKD network can easily be upgraded without disruption to other users. With the help of an N:2 optical switch, any two users can participate in the MDI-QKD protocol to exchange the secure keys.

15.7.3 Time-Phase-Encoding-Based MDI-QKD Protocol

The time-phase-encoding basis states for the BB84 protocol ($N = 2$) are provided in Fig. 15.17.

The time basis corresponds to the CB $\{|0\rangle,|1\rangle\}$, while the phase basis corresponds to the DB $\{|+\rangle,|-\rangle\}$. The pulse is localized within the time bin of duration $\tau = T/2$. The time basis is similar to PPM. The state in which the photon is placed in the first time bin (early state) is denoted by $|0\rangle = |e\rangle$, while the state in which the photon is placed in the second time bin (late state) is denoted by $|1\rangle = |l\rangle$. The time-phase states

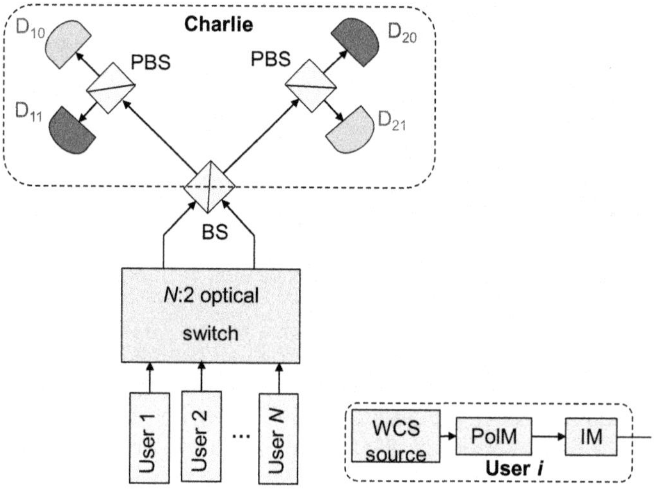

FIGURE 15.16

Measurement device-independent quantum key distribution-based network. *BS*, Beam splitter; *PBS*, polarizing beam splitter; *IM*, intensity modulator; *PolM*, polarization modulator; *WCS*, weak coherent state.

are defined by $|+\rangle = (|0\rangle + |1\rangle)/\sqrt{2}$, $|-\rangle = (|0\rangle - |1\rangle)/\sqrt{2}$. Alice randomly selects either the time basis or the phase basis, followed by random selection of the basis state. Logical 0 is represented by $|0\rangle$, $|+\rangle$, while logical 1 is represented by $|1\rangle$, $|-\rangle$. Alice and Bob randomly select one of four states $|0\rangle$, $|1\rangle$, $|+\rangle$, $|-\rangle$ to be sent to Charlie. Charlie performs the partial BSM with the help of one 50:50 BS and two SPDs as shown in Fig. 5.18, and announces that the events for each measurement result in either the $|\psi^+\rangle = (|01\rangle + |10\rangle)/\sqrt{2}$ state or the $|\psi^-\rangle = (|01\rangle - |10\rangle)/\sqrt{2}$ state. Charlie also discloses the result of the BSM. A successful BSM corresponds to the observation when the SPDs are triggered as follows:

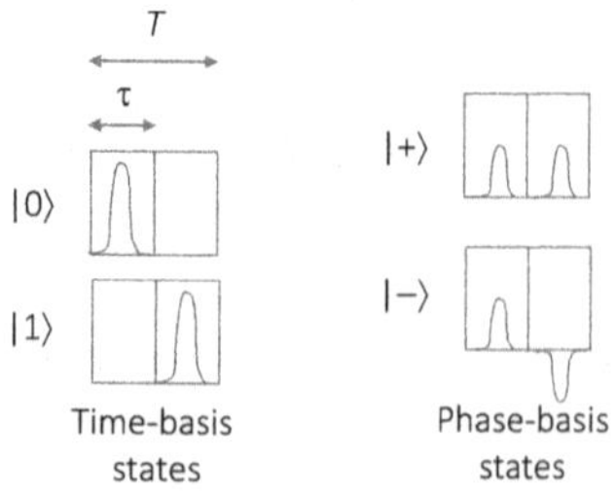

FIGURE 15.17

Time-basis states and phase-basis states used in time-phase encoding for the BB84 protocol.

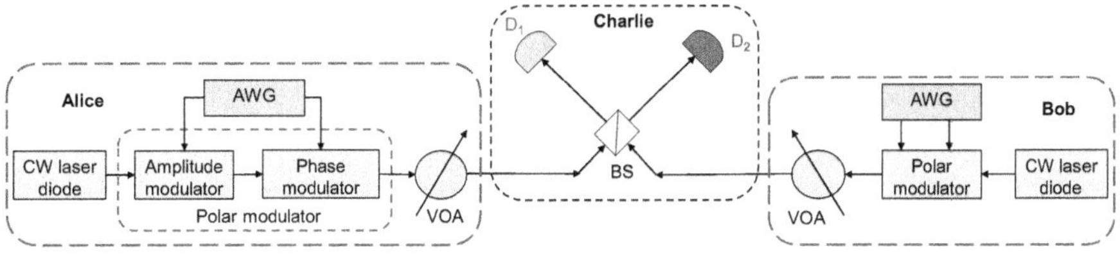

FIGURE 15.18

Time-phase-encoding-based measurement device-independent quantum key distribution system configuration. *AWG*, Arbitrary waveform generator; *BS*, beam splitter; *CW*, continuous wave; *VOA*, variable optical attenuator.

- Two consecutive (both early and late time bins) clicks on one of the SPDs identifies the Bell state $|\psi^-\rangle$.
- On the other hand, when one SPD detects the presence of a photon in the early time bin and the second one in the late time bin we identify the Bell state $|\psi^+\rangle$.
- Other detection patterns are related to $|\phi^\pm\rangle$ states and are discarded from consideration.

The rest of the protocol is similar to the polarization states-based MDI-QKD protocol. To implement the time-phase encoder, either Alice or Bob can employ either the electro-optical polar modulator, composed of a concatenation of an amplitude modulator and phase modulator, which is illustrated in Fig. 15.18, or alternatively an electro-optical *I/Q* modulator can be used [46,96]. The amplitude modulator (Mach–Zehnder modulator) is used for pulse shaping and the phase modulator is used to introduce the desired phase shift, according to phase-basis states (Fig. 15.17). The AWGs are used to generate the corresponding RF waveforms as needed. The VOAs are used to attenuate the modulated beam down to single-photon level. For proper operation, the AWGs must be synchronized. Furthermore, continuous wave (CW) laser diodes, used by Alice and Bob, should be frequency locked. Finally, incoming photons (from Alice and Bob) to the BSM must be of the same polarization.

15.7.4 The Secrecy Fraction of MDI-QKD Protocols

The secret fraction of the MDI-QKD protocol, when both Alice and Bob employ single-photon sources, is similar to that for the BB84 protocol:

$$r = q^{(Z)}\left[1 - h_2\left(e^{(X)}\right)\right] - q^{(Z)} f_e h_2\left(e^{(Z)}\right), \tag{15.49}$$

where $q^{(Z)}$ now denotes the probability of a successful BSM result ("the gain"). On the other hand, when both Alice and Bob employ WCS sources, the decoy-state method must be employed, so that the secret fraction can be estimated by [89]:

$$r = q_{11}^{(Z)}\left[1 - h_2\left(e_{11}^{(X)}\right)\right] - q_{\mu\sigma}^{(Z)} f_e h_2\left(e_{\mu\sigma}^{(Z)}\right), \tag{15.50}$$

where we introduce the following notation:

- $q_{11}^{(Z)}$ denotes the probability that Charlie declares a successful result ("the gain") when both Alice and Bob sent a single photon each in the Z basis,
- $e_{11}^{(X)}$ denotes the phase error rate of the single-photon signals, both sent in the Z basis,
- $q_{\mu\sigma}^{(Z)}$ denotes the gain in the Z basis when Alice sent the WCS of intensity μ to Charlie, while Bob at the same time sent the WCS of intensity σ.

In Eq. (15.50), the information removed from the final key in the privacy amplification step is described by the following term $q_{11}^{(Z)} h_2\left(e_{11}^{(X)}\right)$. The information revealed by Alice in the information reconciliation step is described by the following term:

$$q_{\mu\sigma}^{(Z)} f_e h_2\left(e_{\mu\sigma}^{(Z)}\right),$$

with f_e being the error correction inefficiency ($f_e \geq 1$), as before. The error rates $q_{11}^{(Z)}$ and $e_{11}^{(X)}$ cannot be measured directly, but instead are bounded by employing the decoy-state approach in a similar fashion to that described in the previous section.

15.8 TWIN-FIELD QKD PROTOCOLS

The quantum channel can be characterized by the transmissivity T, which can also be interpreted as the probability of a photon being successfully transmitted by the channel. The secret-key capacity of the quantum channel can be used as an upper bound on the maximum possible SKR. Very often, the Pirandola–Laurenza–Ottaviani–Banchi (PLOB) bound on a linear key rate is used given by [97] $r_{\text{PLOB}} = -\log_2(1 - T)$. To overcome the rate-distance limit of DV-QKD protocols, two approaches have been pursued until recently: (1) development of quantum relays [98] and (2) employment of trusted relays [99]. Unfortunately, quantum relays require the use of long-time quantum memories and high-fidelity entanglement distillation, which are still out of reach with current technology. On the other hand, trusted-relay methodology assumes that the relay between two users can be trusted, and this assumption is difficult to verify in practice. The MDI-QKD approach, described in the previous section, was able to close the detection loopholes and extend the transmission distance; however, its SKR is still bounded by $O(T)$ dependence of the upper limit.

Recently, twin-field (TF) QKD has been proposed to overcome the rate-distance limit [100]. The authors in [100] showed that TF-QKD upper limit scales with the square root of transmittance, that is, $r \sim O(\sqrt{T})$. In TF-QKD, Alice and Bob generate a pair of optical fields with random phases and phase encoding in X and Y bases at remote destinations and send them to untrusted Charlie (or Eve), who then performs single-photon detection. This scheme retains the advantages of the MDI-QKD scheme such as immunity to detector attacks, multiplexing of SPDs, and star network architecture, but at the same time overcomes the PLOB bound. The unconditional security proof was missing in [100], and in a series of papers it was proved that this scheme was unconditionally secure [101–104]. The TF-QKD scheme can be interpreted as the generalization of both (1) the time-phase-encoding-based MDI-QKD scheme shown in Fig. 15.18 and (2) the phase-encoding-based BB84 (see Subsection 15.4.4).

The authors have shown in [104] that the TF-QKD scheme can be interpreted as a particular MDI-QKD scheme, implemented in a 2D subspace of vacuum state $|0\rangle$ and one-photon state $|1\rangle$, as illustrated in Fig. 15.19, representing an equivalent model for TF-QKD with single-photon sources and laser diodes. Given that the vacuum state is insensitive to the channel loss, this state can always be detected, namely no click means successful detection of the vacuum state. This improvement of detection probability of the vacuum state actually corresponding improvements is the reason for higher SKR compared to the traditional MDI-QKD and performance better than the linear key rate bound. The CB or Z basis is defined by $\{|0\rangle, |1\rangle\}$ (with $|0\rangle$ and $|1\rangle$ representing the vacuum and one-photon state, respectively, as mentioned earlier). The X basis (DB) is defined by $\{|+\rangle = (|0\rangle + |1\rangle)/\sqrt{2}, |-\rangle = (|0\rangle - |1\rangle)/\sqrt{2}\}$, while the Y basis is defined by $\{|+j\rangle = (|0\rangle + \mathrm{j}|1\rangle)/\sqrt{2}, |-j\rangle = (|0\rangle - j|1\rangle)/\sqrt{2}\}$. The Bell states are defined in a similar fashion as before, that is, $|\psi^{\pm}\rangle = (|01\rangle \pm |10\rangle)/\sqrt{2}$.

Alice (Bob) with the help of a second optical switch randomly selects whether to use Z basis or X/Y basis. When Z basis is selected, Alice (Bob) employs the first optical switch to randomly select either the vacuum state $|0\rangle$ or single-photon state, and the selected state is sent over an optical channel (SMF or FSO link) toward Charlie. When the X/Y basis branch is selected, the CW laser generates the coherent state $|\alpha\rangle$,which is attenuated by VOA, so that the superposition state of the vacuum and single-photon state is obtained, that is $|+\rangle = (|0\rangle + |1\rangle)/\sqrt{2}$. Alice (Bob) with the help of a phase modulator randomly selects one of four phases from the set $\{0,\pi,\pi/2,3\pi/2\}$. In such a way, both states for the X basis and Y basis are selected at random. (Alternatively, instead of two 2:1 optical switches, one 4:1 optical switch can be used.) Charlie performs the BSM measurements. The coincidence detection with SPD R click and no click on SPD L indicates the Bell state $|\psi^{-}\rangle$. On the other

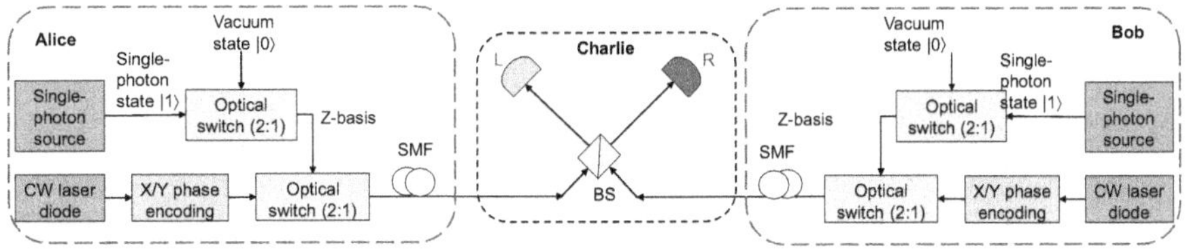

FIGURE 15.19

Conceptual twin-field quantum key distribution system with single photon and continuous wave laser sources. *BS*, Beam splitter; *CW*, continuous wave; *SMF*, single-mode fiber.

hand, the coincidence detection with SPD L click and no click on SPD R indicates the Bell state $|\psi^+\rangle$. Alice and Bob then exchange the basis being used through an authenticated public channel. Bob will flip his sequence when the Z basis is used and Charlie identifies the Bell states $|\psi^{\pm}\rangle$. Bob will also flip his sequence when either the X or Y basis is used and Charlie identifies the Bell state $|\psi^-\rangle$. After the sifting procedure is completed, Alice and Bob will use the sequences corresponding to the Z basis for the key, while the sequences corresponding to the X basis/Y basis for QBER estimation. Finally, Alice and Bob will perform classical postprocessing steps, information reconciliation, and privacy amplification to get the common secure key.

We move our attention now to a more realistic, *practical TF-QKD* scheme, shown in Fig. 15.20, which does not require the use of single-photon sources. To explain this scheme we use interpretation similar to [104]. Both Alice and Bob employ stabilized CW lasers of low linewidth to generate the global-phase-stabilized optical pulses with the help of an amplitude modulator. With the help of phase modulators, Alice and Bob choose the random phases $\phi_A \in [0,2\pi]$ and $\phi_B \in [0,2\pi]$, respectively. The random phase differences between Alice and Bob are discretized so that:

$$\phi_{A,B} \in \Delta\phi_{k_{A,B}} = \left[\frac{2\pi}{M}k_{A,B}, \frac{2\pi}{M}(k_{A,B}+1)\right], \tag{15.51}$$

where the phase bin $\Delta\phi_{k_{A,B}}$ is discretized by $k_{A,B} \in \{0,1,\ldots,M-1\}$. Alice (Bob) then randomly selects whether to use the Z basis or X/Y basis. When the Z basis is selected, the phase-randomized coherent state is sent with intensity either μ or 0, representing logic bits 1 and 0, with probability t and $1-t$, respectively. Given that for lossy channel the TF state $|1\rangle_A|1\rangle_B$ can create Bell-state detection with errors, logic 0 and ***1*** are not sent with the same probability, but with optimized probabilities instead. When $X(Y)$ basis encoding is selected, Alice and Bob employ corresponding phase and amplitude modulators to randomly select 0 ($\pi/2$) and π ($3\pi/2$) representing logic bits 0 and 1, and such phase-encoded pulses are sent with randomly selected intensities $\{v/2,w/2,0\}$. The corresponding quantum states, generated by Alice and Bob, are sent toward Charlie over the optical (SMF or FSO) link. The polarization controllers ensure that Alice's and Bob's pulses have the same polarization. Charlie performs BSM and announces the results in the same fashion as already described. Alice and Bob then disclose their phase information, that is, $k_{A,B}$

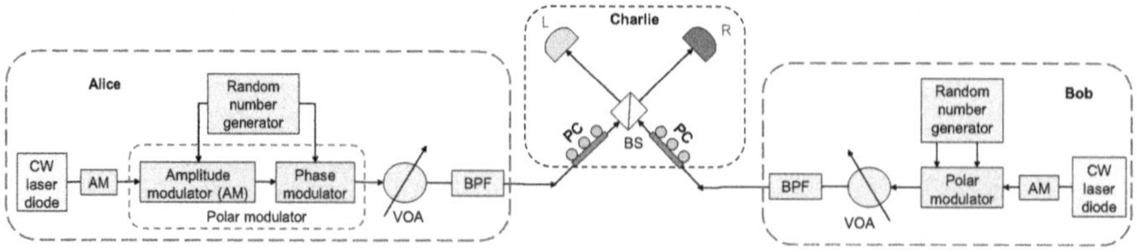

FIGURE 15.20

Practical twin-field quantum key distribution system configuration. *BPF*, Optical bandpass filter; *BS*, Beam splitter; *PC*, polarization controller; *VOA*, variable optical attenuator.

and intensities, and these are used for parameter estimation. They keep the information related to the Z basis confidential to Charlie, and these data are used for the raw key. Alice and Bob then perform classical postprocessing steps, information reconciliation, and privacy amplification to get the common secure key.

For the TF-based BB84 protocol, Alice and Bob keep raw data for phase encoding only if $|k_A - k_B| = 0$ or $M/2$, when both use the X basis, and other raw data results are discarded. For $|k_A - k_B| = 0$, Bob flips his key bit when Charlie detects the Bell state $|\psi^-\rangle$. In similar fashion, for $|k_A - k_B| = M/2$, Bob flips his key bit when Charlie detects the Bell state $|\psi^+\rangle$. The secret fraction for the TF-based BB84 protocol can be estimated by [104]:

$$r_{\text{TF-BB84}} = 2t(1-t)\mu e^{-\mu} y_{1ZZ}^{(TF)} \left[1 - h_2\left(e_{XX}^{(b_1)}\right)\right] - Q_{ZZ} f\, h_2(E_{ZZ}), \tag{15.52}$$

where Q_{ZZ} is the gain when both Alice and Bob use the Z basis and can be determined experimentally. The parameter f denotes error reconciliation inefficiency ($f \geq 1$). ($h_2(\bullet)$ denotes the binary entropy function, defined earlier.) The gain Q_{ZZ} and the QBER for the Z basis can be determined by [104]:

$$\begin{aligned}
Q_{ZZ} &= 2p_d(1-p_d)(1-t)^2 + 4(1-p_d)e^{-\frac{\mu}{2}T_{tot}}\left[1-(1-p_d)e^{-\frac{\mu}{2}T_{tot}}\right]t(1-t) \\
&\quad + 2(1-p_d)e^{-\mu T_{tot}}\left[I_0(\mu T_{tot}) - (1-p_d)e^{-\mu T_{tot}}\right]t^2, \\
E_{ZZ} &= \frac{2p_d(1-p_d)(1-t)^2 + 2(1-p_d)e^{-\mu T_{tot}}\left[I_0(\mu T_{tot}) - (1-p_d)e^{-\mu T_{tot}}\right]t^2}{Q_{ZZ}},
\end{aligned} \tag{15.53}$$

where $T_{\text{tot}} = T\eta_d$, with T being transmissivity and η_d being detector efficiency. Parameter p_d denotes the dark current rate of the threshold detector. Based on Eqs. (15.34) and (15.35), by setting $\mu = v/2$, $\sigma_1 = w/2$, and $\sigma_2 = 0$, the yield $y_{1ZZ}^{(TF)}$ becomes lower bounded by [2]:

$$y_{1ZZ}^{(TF)} \geq \frac{2v}{vw - w^2}\left[q_{w/2}e^{w/2} - \frac{w^2}{v^2}q_{v/2}e^{v/2} - q_0\left(1 - \frac{w^2}{v^2}\right)\right], \tag{15.54}$$

where $q_0 = 2p_d(1 - p_d)$. By employing the decoy-state method [37–64,66–69,104,105], the yield $y_{1XX}^{(TF)}$ and QBER $e_{XX}^{(b_1)}$ are bounded as follows:

$$\begin{aligned}
y_{1XX}^{(TF)} &\geq \frac{v}{vw - w^2}\left[q_w^{(XX)}e^w - \frac{w^2}{v^2}q_v^{(XX)}e^v - q_0\left(1 - \frac{w^2}{v^2}\right)\right] = y_{1XX}^{(TF,LB)}, \\
e_{XX}^{(b_1)} &\leq \frac{q_w^{(XX)}e^w Q_w^{(XX)} - q_0 e^{b_0}}{w y_{1XX}^{(TF,LB)}},
\end{aligned} \tag{15.55}$$

where for the TF vacuum state, $e^{b_0} = 1/2$ and $Q_w^{(XX)}$ should be determined experimentally.

As an illustration, in Fig. 15.21 we provide comparisons of the phase-matching TF-QKD protocol introduced in [101] against the MDI-QKD protocol [83] and decoy-state-based BB84 protocol [37]. The system parameters are selected as follows: detector efficiency $\eta_d = 0.25$, reconciliation inefficiency $f_e = 1.15$, dark count rate $\boldsymbol{pd} = 8 \times 10^{-8}$, misalignment error $e_d = 1.5\%$, and number of phase slices for phase-matching TF-QKD is set to $M = 16$. Regarding the transmission medium, it is assumed that recently reported ultralow-loss fiber of attenuation 0.1419 dB/km (at 1560 nm) is used [106]. In the same figure, the PLOB bound on a linear key rate is provided as well. Clearly, the phase-matching TF-QKD scheme outperforms the decoy-state BB84 protocol for distances larger than 162 km. It outperforms the MDI-QKD protocol for all distances, and exceeds the PLOB bound at distance 322 km. Finally, the phase-matching TF-QKD protocol can achieve a distance of even 623 km. Unfortunately, the normalized SKR value is rather low at this distance. For calculations, the expressions provided in Appendix B of [101] are used, with a few obvious typos being corrected.

15.9 INFORMATION RECONCILIATION AND PRIVACY AMPLIFICATION

As discussed earlier, the raw key is imperfect, and we need perform information reconciliation and privacy amplification to increase the correlation between Alice's and Bob's sequences, while at the same time reduce eavesdropper Eve's mutual information about the result to a desired level of security.

15.9.1 Information Reconciliation

EA-QKD has certain similarities to the source-type secret-key generation (agreement) protocol, relevant in physical-layer security [2]. In this model, Alice, Bob, and Eve have access to the respective output ports X, Y, Z of the source characterized by the joint probability density function (PDF) f_{XYZ}. Each of them sees the i.i.d. realizations of the source denoted, respectively, as X^n, Y^n, and Z^n. Given that Alice's and Bob's realizations are not perfectly correlated, they need to correct the discrepancies' errors. This step in secret-key generation as well as in QKD is commonly referred to as *information reconciliation* (or just reconciliation). The reconciliation process can be considered as a special case of *source coding with side information* [11,107]. To encode the source X we know from Shannon's source-coding theorem that the rate R_x must be as small as possible but not lower than the entropy of the source; in other words, $R_x > H(X)$. To jointly encode the source (XY, f_{XY}) we need the rate $R > H(X,Y)$. If we separately encode the sources, the required rate would be $R > H(X) + H(Y)$. However, Slepian and Wolf have shown in [107] that the rate $R = H(X,Y)$ is still sufficient even for separate encoding of correlated sources. This claim is commonly referred to as the Slepian–Wolf theorem, and can be formulated

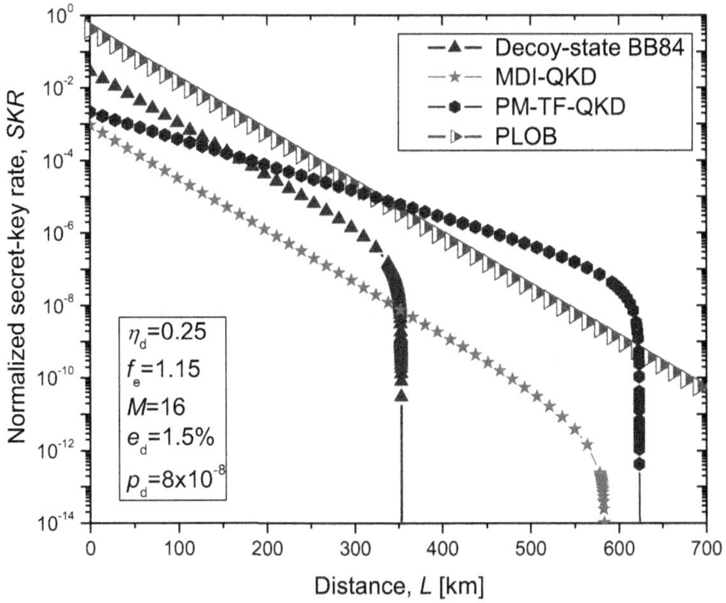

FIGURE 15.21

Phase-matching twin-field quantum key distribution (PM-TF-QKD) protocol against the measurement device-independent quantum key distribution (MDI-QKD) and decoy-state BB84 protocols in terms of normalized secret-key rate versus the transmission distance. *PLOB*, Pirandola–Laurenza–Ottaviani–Banchi.

as follows [11,107]: The *achievable rate* region R (for which the error probability tends to zero) for the separate encoding of the correlated source (XY, f_{XY}) is given by:

$$R \doteq \left\{ (R_x, R_y): \begin{array}{c} R_x \geq H(X|Y) \\ R_y \geq H(Y|X) \\ R_x + R_y \geq H(X,Y) \end{array} \right\}. \tag{15.56}$$

The typical shape of the Slepian–Wolf region is provided in Fig. 15.22. Now if we assume that (X, f_X) should be compressed when (Y, f_Y) is available as the side information, then $H(X|Y)$ is sufficient to describe X.

Going back to our reconciliation problem, Alice compresses her observations X^n and Bob decodes them with the help of correlated side information Y^n, which is illustrated in Fig. 15.23. The source encoder can be derived from capacity-achieving channel code. For an (n,k) low-density parity-check (LDPC) code the parity-check matrix $\boldsymbol{H}$ of size $(n-k) \times n$ can be used to generate the syndrome by $\boldsymbol{s} = \boldsymbol{Hx}$, where $\boldsymbol{x}$ is Alice's DMS observation vector of length n. Since $\boldsymbol{x}$ is not a codeword, the syndrome vector is different from the all-zeros vector. The syndrome vector is of

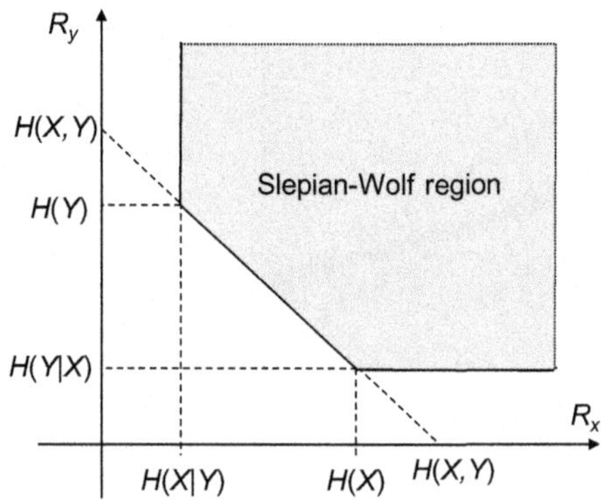

FIGURE 15.22

Typical Slepian–Wolf achievable rate region for a correlated source (XY, f_{XY}).

length $n - k$ and can be represented as $\boldsymbol{s} = [s_1\ s_2\ \dots\ s_{n-k}]^{\mathrm{T}}$. Clearly, the syndrome vector is transmitted toward Bob over the (error-free) authenticated public channel.

Once Bob receives the syndrome vector $\boldsymbol{s}$, his decoder tries to determine the vector $\boldsymbol{x}$ based on $\boldsymbol{s}$ and Bob's observation vector $\boldsymbol{y}$, which maximizes the posterior probability:

$$\widehat{\boldsymbol{x}} = \max P(\boldsymbol{y}|\boldsymbol{x}, \boldsymbol{s}), \tag{15.57}$$

and clearly, the decoding problem is similar to the maximum a posteriori probability (MAP) decoding. The compression rate is $(n - k)/n = 1 - k/n$, while the linear code rate is k/n. The number of syndrome bits required is:

$$n - k \geq H(\boldsymbol{X}^n|\boldsymbol{Y}^n) = nH(X|Y). \tag{15.58}$$

In practice, the practical reconciliation algorithms introduce the overhead $OH > 0$, so that the number of transmitted bits over the public channel is $nH(X|Y)(1 + OH)$. Given that Alice and Bob at the end of the reconciliation step

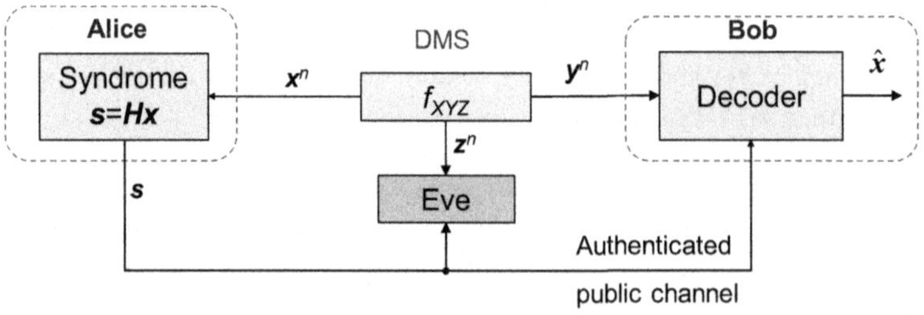

FIGURE 15.23

Reconciliation problem represented as a source coding with syndrome-based side information. *DMS*, discrete memoryless source.

share with high probability the common sequence X^n with entropy $nH(X)$, the reconciliation efficiency β will then be:

$$\beta = \frac{nH(X) - nH(X|Y)(1+OH)}{nI(X,Y)} = 1 - OH\frac{H(X|Y)}{I(X,Y)} \leq 1. \tag{15.59}$$

The MAP decoding complexity is prohibitively high when LDPC coding is used; instead, the low-complexity sum-product algorithm (SPA) (also known as belief-propagation algorithm) should be used as described in [108]. Given that Alice's observation x is not a codeword, the syndrome bits are nonzero. To account for this problem, we need to change the sign of the log-likelihood to be sent from the cth check node to the vth variable node when the corresponding syndrome bit s_c is 1. Based on [109] we can summarize the log-domain SPA next. To improve the clarity of presentation we use the index $v = 1,\ldots,n$ to denote variable nodes and index $c = 1,\ldots,n-k$ to denote check nodes. Additionally, we use the following notation: (1) $N(c)$ to denote $\{v$ nodes connected to c node $c\}$, (2) $N(c)\backslash\{v\}$ denoting $\{v$ nodes connected to c node c except v node $v\}$, (3) $N(v)$ to denote $\{c$ nodes connected to v node $v\}$, and (4) $N(v)\backslash\{c\}$ representing $\{c$ nodes connected to v node v except c node $c\}$.

The formulation of the *log-domain SPA for source coding with the side information* is provided now:

1) Initialization: For the variable nodes $v = 1,2,\ldots,n$, initialize the extrinsic messages L_{vc} to be sent from variable node (v node) v to the check node (c node) c to the channel log-likelihood ratios, denoted as $L_{\text{ch}}(v)$, namely $L_{vc} = L_{\text{ch}}(v)$.
2) The check node (c node) update rule: For the check nodes $c = 1,2,\ldots,n-k$, compute $L_{cv} = (1-2s_c) \underset{N(c)\backslash\{v\}}{\boxplus} L_{vc}$. The box-plus operator is defined by $L_1 \boxplus L_2 = \prod_{k=1}^{2}\text{sign}(L_k)\cdot\phi\left(\sum_{k=1}^{2}\phi(|L_k|)\right)$, where $\phi(x) = -\log\tanh(x/2)$. The box operator for $|N(c)/\{v\}|$ components is obtained by recursively applying the two-component version defined previously.
3) The variable node (v node) update rule: For $v = 1,2,\ldots,n$, set $L_{vc} = L_{\text{ch}}(v) + \sum_{N(v)/\{c\}} L_{cv}$ for all c nodes for which the cth row/vth column element of the parity-check matrix H is one, that is, $h_{cv} = 1$.
4) Bit decisions: Update $L(v)$ ($v = 1,\ldots,n$) by $L(v) = L_{\text{ch}}(v) + \sum_{N(v)} L_{cv}$ and make decisions as follows: $\hat{x}_v = 1$ when $L(v) < 0$ (otherwise set $\hat{x}_v = 0$).

So, the only difference with respect to conventional log-domain SPA is the introduction of the multiplication factor $(1 - 2s_c)$ in the c node update rule. Moreover, the channel likelihoods $L_{\text{ch}}(v)$ should be calculated from the joint distribution f_{XY} by $L_{\text{ch}}(v) = \log[f_{XY}(x_v = 0|y_v)/f_{XY}(x_v = 1|y_v)]$ Because the c node update rule involves log and tanh functions, it can be computationally intensive, and to lower the complexity the reduced complexity approximations should be used.

15.9.2 Privacy Amplification

The purpose of privacy amplification is to extract the secret key from the reconciled (corrected) sequence. The key idea is to apply a well-chosen compression function, which Alice and Bob have agreed on, to the reconciled binary sequence of length n_r as follows g: $(0,1)^n \rightarrow (0,1)^k$, where $k < n$, so that Eve gets negligible information about the secret key [8,23,110–112]. Typically, g is selected at random from the set of compression functions G so that Eve does not know which g has been used. In practice, the set of functions G is based on the family of universal hash functions, introduced by Carter and Wegman [113,114]. We say that the family of functions G: $(0,1)^n \rightarrow (0,1)^k$ is a universal$_2$ family of hash functions if for any two distinct x_1 and x_2 from $(0,1)^n$, the probability that $g(x_1) = g(x_2)$ when g is chosen uniformly at random from G is smaller than 2^{-k}. As an illustrative example, let us consider the multiplication in GF(2^n) [23]. Let a and x be two elements from GF(2^n). Let the g function be defined as the mapping $(0,1)^n \rightarrow (0,1)^k$ that assigns to argument x the first k bits of the product $ax \in$ GF(2^n). It can be shown that the set of such functions is a universal class of functions for $k \in [1,n]$.

The analysis of privacy amplification does not rely on Shannon entropy, but Rényi entropy [115–117]. For discrete random variable X, taking the values from the sample space $\{x_1,\ldots,x_n\}$ and distributed according to $P_X = \{p_1,\ldots,p_n\}$, the *Rényi entropy* of X of order α is defined by [116]:

$$R_\alpha(X) = \frac{1}{1-\alpha}\log_2\left\{\sum_x [P_X(x)]^\alpha\right\}. \tag{15.60}$$

The Rényi entropy of order two is also known as the collision entropy. From Jensen's inequality [11] we conclude that the *collision entropy* is upper limited by the Shannon entropy, that is, $R_2(X) \leq H(X)$. In limit when $\alpha \rightarrow \infty$, the Rényi entropy becomes the *min entropy*:

$$\lim_{\alpha\rightarrow\infty} R_\alpha(X) = -\log_2 \max_{x\in X} P_X(x) = H_\infty(X). \tag{15.61}$$

Finally, by applying l'Hôpital's rule as $\alpha \rightarrow 1$, the Rényi entropy becomes the *Shannon entropy*:

$$\lim_{\alpha\rightarrow 1} R_\alpha(X) = -\sum_x P_X(x)\log_2 P_X(x) = H(X). \tag{15.62}$$

The properties of the Rényi entropy are listed in [116]. Regarding the conditional Rényi entropy, there exist different definitions; three of them have been discussed in detail in [117]. We adopt the first one. Let the joint distribution for (X, Y) be denoted as $P_{X,Y}$, and the distribution for X be denoted as P_X. Furthermore, let the Rényi entropy of the conditional random variables $X|Y = y$, distributed according to $P_{X|Y=y}$, be denoted as $R_\alpha(X|Y = y)$. Then, the conditional Rényi entropy can be defined by:

$$R_\alpha(X|Y) = \sum_y P_Y(y)R_\alpha(X|Y=y) = \frac{1}{1-\alpha}\sum_y P_Y(y)\log_2\sum_x P_{X|Y}(x|y)^\alpha. \tag{15.63}$$

The mutual Rényi information can be defined then as:

$$I_\alpha(X,Y) = R_\alpha(X) - R_\alpha(X|Y), \tag{15.64}$$

and it is not symmetric to $R_\alpha(Y) - R_\alpha(Y|X)$. Moreover, it can even be negative as discussed in [116,117].

Bennett et al. proved the following *privacy amplification theorem* in [23] (see Theorem 3): Let X be a random variable with distribution P_X and Rényi entropy of order two $R_2(X)$. Furthermore, let G be the random variable representing uniform selection at random of a member of universal hash functions $(0,1)^n \rightarrow (0,1)^k$. Then, the following inequalities are satisfied [23]:

$$H(G(X)|G) \geq R_2(G(X)|G) \geq k - \log_2\left[1 + 2^{k-R_2(X)}\right]$$

$$\geq k - \frac{2^{k-R_2(X)}}{\ln 2}. \tag{15.65}$$

Bennett et al. also proved the following *corollary of the privacy amplification theorem* [23] (see also [111]): Let $S \in \{0,1\}^n$ be a random variable representing the sequence shared between Alice and Bob, and let E be a random variable representing the knowledge Eve was able to get about the S, with a particular realization of E denoted by e. If the conditional Rényi entropy of order two, $R_2(S|E=e)$ is known to be at least r_2, and Alice and Bob choose the secret key by $K = G(S)$, where G is a hash function chosen uniformly at random from the universal$_2$ family of hash functions $(0,1)^n \rightarrow (0,1)^k$, then the following is valid:

$$H(K|G, E=e) \geq k - \frac{2^{k-r_2}}{\ln 2}. \tag{15.66}$$

Cachin proved the following Rényi entropy lemma in [118] (see also [119]): Let X and Q be two random variables and $s > 0$. Then with probability of at least $1 - 2^{-s}$, the following is valid [118,119]:

$$R_2(X) - R_2(X|Q=q) \leq \log|Q| + 2s + 2. \tag{15.67}$$

The total information available to Eve is composed of her observations $\mathbf{Z}^n$ from the source of common randomness and additional bits exchanged in the information reconciliation phase over the authenticated public channel, represented by random variable Q. Based on Cachin's lemma we can write:

$$R_2(S|\mathbf{Z}^n = z^n) - R_2(S|\mathbf{Z}^n = z^n, Q=q) \leq \log|Q| + 2s + 2 \text{ with probability } 1 - 2^{-s}. \tag{15.68}$$

We know that $R_2(X) \leq H(X)$, so we conclude that $R_2(S|\mathbf{Z}^n = z^n) \leq H(S|\mathbf{Z}^n = z^n)$ and therefore we can upper bound $R_2(S|\mathbf{Z}^n = z^n)$ as follows:

$$R_2(S|\mathbf{Z}^n = z^n) \leq nH(X|Z). \tag{15.69}$$

Based on Eq. (15.68) we can lower bound $R_2(S|\mathbf{Z^n} = \mathbf{z^n}, Q = q)$ by:

$$R_2(S|Z^n = z^n, Q = q) \geq nH(X|Z) - \log|Q| - 2s - 2 \text{ with probability } 1 - 2^{-s}. \tag{15.70}$$

Since the number of bits exchanged over the public channel $\log|Q|$ is approximately $nH(X|Y)$ $(1 + OH)$, for sufficiently large n the previous inequality becomes:

$$R_2(S|Z^n = z^n, Q = q) \geq \underbrace{nH(X|Z) - nH(X|Y)(1 + OH) - 2s - 2}_{r_2} \tag{15.71}$$

with probability $1 - 2^{-s}$.

From Eq. (15.66) we conclude that $k - r_2 = -k_2 < 0$, which guarantees that for $k_2 > 0$, Eve's uncertainty of the key is lower bounded by:

$$H(K|E) \geq k - 2^{-k_2}/\ln 2 \text{ with probability } 1 - 2^{-s} \tag{15.72}$$

15.10 QUANTUM OPTICS AND GAUSSIAN QUANTUM INFORMATION THEORY

To facilitate the description of CV-QKD schemes, in this section we describe the basics of quantum optics and Gaussian quantum information theory [2,8,32,33,120–122].

15.10.1 Quadrature Operators, Gaussian States, and Squeezed States

The continuous variable quantum system lives in an infinite-dimensional Hilbert space, characterized by the observables with continuous spectra [122]. The CV quantum system can be represented in terms of M quantized radiation modes of the electromagnetic field (M bosonic modes). The quantum theory of radiation treats each radiation (bosonic) mode as a harmonic oscillator [32,33]. The bosonic modes live in a tensor-product Hilbert space $\mathcal{H}^{\otimes M} = \bigotimes_{m=1}^{M} \mathcal{H}_m$. The *radiation (bosonic)* modes are associated with M pairs of bosonic field operators $\left\{a_m^\dagger, a_m\right\}_{m=1}^{M}$, which are known as the creation and annihilation operators, respectively. Clearly, the Hilbert space is infinitely dimensional, but separable. Namely, a single-mode Hilbert space is spanned by a countable basis of number (Fock) states $\{|n\rangle\}_{n=1}^{\infty}$, wherein the Fock state $|n\rangle$ is the eigenstate of the number operator $\hat{n} = a^\dagger a$ satisfying the eigenvalue equation $\hat{n}|n\rangle = n|n\rangle$. Now the creation and annihilation operators can be respectively defined as follows:

$$\begin{aligned} a^\dagger|n\rangle &= (n+1)^{1/2}\left|n+1\right\rangle, \; n \geq 0 \\ a|n\rangle &= n^{1/2}|n-1\rangle, \; n \geq 1, \; a|0\rangle = 0. \end{aligned} \tag{15.73}$$

Clearly, the creation operator raises the number of photons by one, while the annihilation operator reduces the number of photons by one. The right eigenvector of the annihilation operator is commonly referred to as the *coherent state*, and is defined as:

$$a|\alpha\rangle = \alpha|\alpha\rangle,\ \alpha = \alpha_I + j\alpha_Q, \tag{15.74}$$

where α is a complex number. The energy of the mth mode is given by $H_m = \hbar\omega_m|\alpha_m|^2$. The coherent state vector can be expressed in terms of number states as follows [2]:

$$|\alpha\rangle = e^{-|\alpha|^2/2}\sum_n \frac{\alpha^n}{\sqrt{n!}}|n\rangle,\ \langle\alpha|\alpha\rangle = 1. \tag{15.75}$$

The coherent states are complete since the following is valid:

$$\frac{1}{\pi}\int |\alpha\rangle\langle\alpha| d^2\alpha = \mathbf{1},\ d^2\alpha = d\alpha_I d\alpha_Q, \tag{15.76}$$

where we use **1** to denote the identity operator. The coherent states $|\alpha\rangle$ and $|\beta\rangle$ are not orthogonal since their inner product is nonzero:

$$\langle\beta|\alpha\rangle = \exp\left(\alpha\beta^* - |\alpha|^2/2 - |\beta|^2/2\right). \tag{15.77}$$

The creation and annihilation operators can also be expressed in terms of *quadrature operators* $\hat{I}$ and $\hat{Q}$ as follows:

$$a^\dagger = \frac{\hat{I} - j\hat{Q}}{2\sqrt{N_0}},\ a = \frac{\hat{I} + j\hat{Q}}{2\sqrt{N_0}}, \tag{15.78}$$

where N_0 represents the variance of vacuum fluctuation (shot noise). Very often the quadrature operators are expressed in terms of the shot noise unit (SNU), where N_0 is normalized to 1, that is, $a^\dagger = \left(\hat{I} - j\hat{Q}\right)/2,\ a = \left(\hat{I} + j\hat{Q}\right)/2$. In other words, the quadrature operators, related to the in-phase and quadrature components of the electromagnetic field, are related to the creation and annihilation operators (in SNU) by $\hat{I} = a^\dagger + a,\ \hat{Q} = j(a^\dagger - a)$. The quadrature operators satisfy the commutation relation $\left[\hat{I}, \hat{Q}\right] = 2j\mathbf{1}$. For the coherent state, both quadratures have the same uncertainty/variance:

$$V(\hat{I}) = \mathrm{Var}(\hat{I}) = V\left(\hat{Q}\right) = \mathrm{Var}\left(\hat{Q}\right) = \left\langle \hat{I}^2 \right\rangle_\alpha - \langle \hat{I} \rangle_\alpha^2 = 4\alpha_I^2 + 1 - 4\alpha_I^2 = 1, \tag{15.79}$$

which represents the minimum uncertainty expressed in SNU, commonly referred to as the shot noise. Therefore the uncertainty relation in SNU can be represented as:

$$V(\hat{I})V\left(\hat{Q}\right) \geq 1. \tag{15.80}$$

Given that the number state $|n\rangle$ can be represented in terms of ground state by [2,9], $|n\rangle = (a^\dagger)^n|0\rangle \big/ \sqrt{n!}$, another convenient form of the coherent state will be:

$$|\alpha\rangle = e^{-|\alpha|^2/2}\sum_n \frac{\alpha^n}{\sqrt{n!}}(a^\dagger)^n|0\rangle \,/\, \sqrt{n!} = e^{-|\alpha|^2/2}\underbrace{\sum_n \frac{(\alpha a^\dagger)^n}{n!}}_{\exp(\alpha a^\dagger)}|0\rangle$$

$$= \underbrace{\exp(\alpha a^\dagger - \alpha^* a)}_{D(\alpha)}\,|0\rangle = D(\alpha)|0\rangle,\ D(\alpha) = \exp(\alpha a^\dagger - \alpha^* a), \quad (15.81)$$

where $D(\alpha)$ is the *displacement operator.* The displacement operator is used to describe one coherent state as the displacement of another coherent state in the *I-Q* diagram.

The CV system composed of M bosonic modes can also be described by the quadrature operators $\{\hat{I}_m, \hat{Q}_m\}_{m=1}^{M}$, which can be arranged in the vector form:

$$\hat{\boldsymbol{x}} \doteq [\hat{I}_1\hat{Q}_1\cdots\hat{I}_M\hat{Q}_M]^T. \quad (15.82)$$

As a generalization of commutation relation $\left[\hat{I}, \hat{Q}\right] = 2j\mathbf{1}$, these quadrature operators satisfy the commutation relation:

$$[\hat{x}_m, \hat{x}_n] = 2j\Omega_{mn}, \quad (15.83)$$

where Ω_{mn} is the mth row, nth column element of a $2M \times 2M$ matrix:

$$\Omega \doteq \bigoplus_{m=1}^{M} \omega = \begin{bmatrix} \omega & & \\ & \ddots & \\ & & \omega \end{bmatrix}, \omega = \begin{bmatrix} 0 & 1 \\ -1 & 0 \end{bmatrix}, \quad (15.84)$$

commonly referred to as the *symplectic form.*

For the quadrature operators, the first moment of $\hat{\boldsymbol{x}}$ represents the displacement and can be determined as:

$$\bar{\boldsymbol{x}} = \langle\hat{\boldsymbol{x}}\rangle = \mathrm{Tr}(\hat{\boldsymbol{x}}\hat{\rho}), \quad (15.85)$$

where the density operator $\hat{\rho}$ is the trace-one positive operator, which maps $\mathcal{H}^{\otimes M}$ to $\mathcal{H}^{\otimes M}$. The space of all density operators is known as the state space. When the density operator is the projector operator, that is, $\hat{\rho}^2 = \hat{\rho}$, we say that the density operator is pure and can be represented as $\hat{\rho} = |\phi\rangle\langle\phi|,\ |\phi\rangle \in \mathcal{H}^{\otimes M}$. On the other hand we can determine the elements of *covariance matrix* $\boldsymbol{\Sigma} = (\Sigma_{mn})_{2M\times 2M}$ as follows [122]:

$$\Sigma_{mn} = \frac{1}{2}\langle\{\Delta\hat{x}_m, \Delta\hat{x}_n\}\rangle = \frac{1}{2}(\langle\hat{x}_m, \hat{x}_n\rangle + \langle\hat{x}_n, \hat{x}_m\rangle) - \langle\hat{x}_m\rangle\langle\hat{x}_n\rangle,$$

$$\Delta\hat{x}_m = \hat{x}_m - \langle\hat{x}_m\rangle, \quad (15.86)$$

where $\{\bullet,\bullet\}$ denotes the anticommutator. The diagonal elements of the covariance matrix are clearly the variances of the quadratures:

$$V(\widehat{x}_i) = \Sigma_{ii} = \left\langle (\Delta\widehat{x}_i)^2 \right\rangle = \left\langle (\widehat{x}_i)^2 \right\rangle - \langle \widehat{x}_i \rangle^2. \tag{15.87}$$

The covariance matrix is a real, symmetric, and positive definite ($\boldsymbol{\Sigma} > 0$) matrix that satisfies the uncertainty principle, expressed as $\boldsymbol{\Sigma} + j\boldsymbol{\Omega} \geq 0$. When we observe only diagonal terms we arrive at the usual uncertainty principle of quadratures, that is, $V(\widehat{I}_m)V(\widehat{Q}_m) \geq 1$.

When the first two moments are enough to completely characterize the density operator, that is, $\widehat{\rho} = \widehat{\rho}(\overline{\boldsymbol{x}}, \boldsymbol{V})$, we say that they represent the *Gaussian states*. The Gaussian states are bosonic states for which the quasi-probability distribution, commonly referred to as the *Wigner function*, is multivariate Gaussian distribution:

$$W(\boldsymbol{x}) = \frac{1}{(2\pi)^M (\det\boldsymbol{\Sigma})^{1/2}} \exp\left[- (\boldsymbol{x} - \overline{\boldsymbol{x}})^T \boldsymbol{\Sigma}^{-1} (\boldsymbol{x} - \overline{\boldsymbol{x}}) / 2 \right], \tag{15.88}$$

where $\boldsymbol{x} \in R^{2M}$ (R is the set of real numbers) are eigenvalues of the quadrature operators $\widehat{\boldsymbol{x}}$. The Wigner function of the thermal state is zero-mean Gaussian with covariance matrix $\boldsymbol{\Sigma} = (2N+1)\mathbf{1}$, where N is the average number of photons. The Gaussian CVs span a real symplectic space (R^{2M},$\boldsymbol{\Omega}$) commonly referred to as the phase space. The corresponding characteristic function is given by:

$$\chi(\zeta) = \exp\left[- j(\boldsymbol{\Omega}\,\overline{\boldsymbol{x}})^T \zeta - \frac{1}{2}\zeta^T (\Omega\Sigma\Omega^T)\zeta \right], \zeta \in R^{2M}. \tag{15.89}$$

Not surprisingly, the Wigner function and the characteristic function are related by the Fourier transform:

$$W(\boldsymbol{x}) = \frac{1}{(2\pi)^{2M}} \int_{R^{2M}} \chi(\zeta) e^{-j\boldsymbol{x}^T \Omega \zeta} d^{2M}\zeta. \tag{15.90}$$

Let us introduce the *Weyl operator* as follows:

$$D(\zeta) = e^{j\widehat{\boldsymbol{x}}^T \Omega \zeta}. \tag{15.91}$$

Clearly, the displacement operator $D(\alpha)$ is just the complex version of the Weyl operator. The *characteristic function* is related to the Weyl operator by:

$$\chi(\zeta) = \mathrm{Tr}[\widehat{\rho} D(\zeta)], \tag{15.92}$$

and represents the expected value of the Weyl operator. It can be applied to both Gaussian and non-Gaussian states. Given that the Wigner function can be determined as the Fourier transform of the characteristic function, it can be used as an equivalent representation of an arbitrary quantum state, defined over $2M$-dimensional phase space. When the Wigner function is nonnegative, the pure state will be Gaussian.

A *squeezed state* is a Gaussian state that has unequal fluctuations in quadratures:

$$V(\widehat{I}) = z, \; V(\widehat{Q}) = z^{-1}, \tag{15.93}$$

where z is the squeezing parameter. (The foregoing fluctuations are expressed in SNU.)

For a single mode, the PDF of the in-phase component, denoted as $f(I)$, can be determined by averaging the Wigner function over the quadrature component, that is, $f(I) = \int_Q W(I,Q)dQ$. On the other hand, if someone measures the quadrature component, the result will follow the following PDF $f(Q) = \int_I W(I,Q)dI$. Therefore we can interpret the Wigner function of the coherent state $|\alpha\rangle$ as the 2D Gaussian distribution centered at $(\mathrm{Re}\{\alpha\},\mathrm{Im}\{\alpha\})=(\alpha_I,\alpha_Q)$ with variance being 1 in SNU (and of zero covariance). Therefore in the (I,Q)-plane the coherent state is represented by the circle, with radius related to the standard deviation (uncertainty), which is illustrated in Fig. 15.24. For the squeezed states, the I-coordinate is scaled by factor $z^{1/2}$, while the Q-coordinate is scaled by factor $z^{-1/2}$.

The squeezed coherent state $|\alpha,s\rangle$ can also be represented in terms of the ground state $|0\rangle$ as follows:

$$|\alpha, s\rangle = D(\alpha)S(s)|0\rangle,\ S(s) = e^{-\frac{1}{2}(sa^{\dagger 2} - s^* a^2)}, \tag{15.94}$$

where $S(s)$ is the squeezing operator. The squeezing factor s in quantum optics is typically a real number, so that the squeezing operator is simplified to $S(s) = \exp\left[-s\left(a^{\dagger 2} - a^2\right)/2\right]$. The parameter z is related to the parameter s by $z = \exp(s)$. By applying the squeezing operator on the vacuum state we obtain the squeezed vacuum state [2,122]:

$$|0, s\rangle = D(0)S(s)|0\rangle = (\cosh s)^{-1/2} \sum_{n=0}^{\infty} \frac{\sqrt{(2n)!}}{2^n n!} \tanh s^n |2n\rangle. \tag{15.95}$$

15.10.2 Gaussian Transformations and Generation of Quantum States

A given quantum state undergoes the transformation described by the quantum operation (superoperator) [9]. Let the composite system C be composed of quantum register Q and environment E. This kind of system can be modeled as a closed quantum system. When the composite system is closed, its dynamic is unitary, and

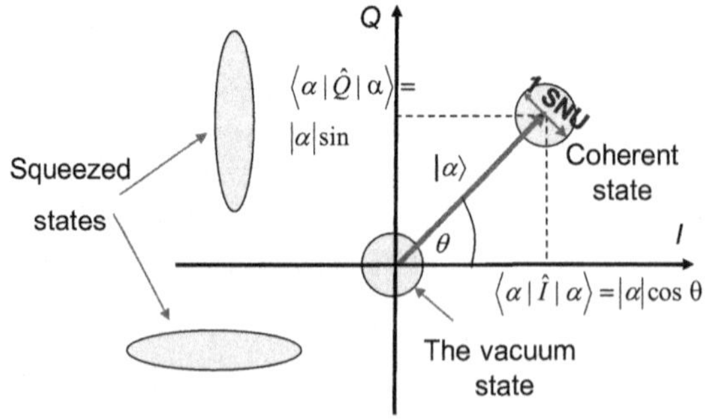

FIGURE 15.24

Coherent and squeezed states. *SNU*, Shot noise unit.

the final state is specified by a unitary operator U as follows: $U(\widehat{\rho} \otimes \varepsilon_0)U^{\dagger}$, where ε_0 is the initial density operator of the environment E. The reduced density operator of Q upon interaction ρ_f can be obtained by tracing out the environment:

$$\widehat{\rho}_f = \mathrm{Tr}_E\left[U(\widehat{\rho} \otimes \varepsilon_0)U^{\dagger}\right] \doteq \xi(\widehat{\rho}). \tag{15.96}$$

The *superoperator* (quantum operation) maps the initial density operator ρ to the final density operator ρ_f, denoted as $\xi : \widehat{\rho} \to \widehat{\rho}_f$. We say that the superoperator is *Gaussian* when it maps Gaussian states into Gaussian states as well. The Gaussian channels preserve the Gaussian nature of the quantum state. The unitary transformations for Gaussian channels are like the time-evolution operator, that is, $U = \exp(-j\widehat{H})$ with $\widehat{H}$ being the Hamiltonian. The Hamiltonian is the second-order polynomial of the creation and annihilation operators' vectors, defined as $\boldsymbol{a}^{\dagger} = \left[a_1^{\dagger} \cdots a_M^{\dagger}\right], \boldsymbol{a} = [a_1 \cdots a_M]$,and can be represented as follows [122]:

$$\widehat{H} = j\left(\boldsymbol{a}^{\dagger}\boldsymbol{\alpha} + \boldsymbol{a}^{\dagger}\boldsymbol{F}_1\boldsymbol{a} + \boldsymbol{a}^{\dagger}\boldsymbol{F}_2\boldsymbol{a}\right) + H.c., \tag{15.97}$$

where $\boldsymbol{\alpha} \in C^M$ (C is the set of complex numbers), while $\boldsymbol{F}_1$ and $\boldsymbol{F}_2$ are $M \times M$ complex matrices (and H.c. stands for the Hermitian conjugate). In the Heisenberg picture, these kinds of transformations correspond to a unitary *Bogoliubov transformation* [122–124]:

$$\boldsymbol{a} \to \boldsymbol{a}_f = U^{\dagger}\boldsymbol{a}U = \boldsymbol{A}\boldsymbol{a} + \boldsymbol{B}\boldsymbol{a}^{\dagger} + \boldsymbol{\alpha}, \tag{15.98}$$

where $\boldsymbol{A}$ and $\boldsymbol{B}$ are $M \times M$ complex matrices satisfying $\boldsymbol{A}\boldsymbol{B}^T = \boldsymbol{B}\boldsymbol{A}^T$, $\boldsymbol{A}\boldsymbol{A}^{\dagger} = \boldsymbol{B}\boldsymbol{B}^{\dagger} + \mathbf{1}$. Bogoliubov transformation induces an autoequivalence on the respective representations. As an illustration, let us consider the creation and annihilation operators, which satisfy the commutation relation $\left[a, a^{\dagger}\right] = \mathbf{1}$. Let us define new pairs of operators $\widehat{b} = ua + va^{\dagger}$, $\widehat{b}^{\dagger} = u^*a^{\dagger} + v^*a$, where u and v are complex constants. Given that the Bogoliubov transformation preserves the commutativity relations, the new commutator $\left[\widehat{b}, \widehat{b}^{\dagger}\right]$ must be $\mathbf{1}$, and consequently $|u|^2 - |v|^2$ must be 1. Since the hyperbolic identity $\cosh^2 x - \sinh^2 x = 1$ has the same form, we can parametrize u and v by $u = e^{j\theta_u}\cosh s$, $v = e^{j\theta_v}\sinh s$. This transformation is therefore a linear symplectic transformation of the phase space. Two angles correspond to the orthogonal symplectic transformations (such as rotations) and the squeezing factor s corresponds to the diagonal transformation. For the quadrature operators, Gaussian transformation is described by an affine map $U_{S,c}$:

$$U_{S,c} : \boldsymbol{x} \to \boldsymbol{x}_f = \widehat{\boldsymbol{S}}\boldsymbol{x} + \boldsymbol{c}, \tag{15.99}$$

where $\boldsymbol{S}$ is a $2M \times 2M$ real matrix and $\boldsymbol{c}$ is a real vector of length $2M$. This transformation preserves the commutation relations, given by Eq. (15.83), when the matrix $\boldsymbol{S}$ is a symplectic matrix, that is, when it satisfies the following property:

$$\boldsymbol{S}\boldsymbol{\Omega}\boldsymbol{S}^T = \boldsymbol{\Omega}. \tag{15.100}$$

The affine mapping can be decomposed as:

$$U_{S,c} = D(c)U_S,\ D(c): x \rightarrow x + c,\ U_S: x \rightarrow Sx, \tag{15.101}$$

where $D(c)$ is the Weyl operator representing the translation in the phase space, while U_S is a linear symplectic mapping. The action of Gaussian transformation on the first two moments of the quadrature field operators will be simply:

$$\bar{x} \rightarrow S\bar{x} + c,\ \Sigma \rightarrow S\Sigma S^T. \tag{15.102}$$

Clearly, the action on the Gaussian state is completely characterized by Eq. (15.102).

As an illustration, let us consider *the beam-splitter (BS) transformation*. The BS, illustrated in Fig. 15.25, is a partially silvered piece of glass with reflection coefficient parameterized by $\cos^2\theta$. The transmittivity of the BS is given by $T = \cos^2\theta \in [0,1]$. The BS acts on two modes, which can be described by the creation (annihilation) operators a ($a^\dagger$) and b ($b^\dagger$). In the Heisenberg picture, the annihilation operators of the modes are transformed by the following Bogoliubov transformation:

$$\begin{bmatrix} a \\ b \end{bmatrix} \rightarrow \begin{bmatrix} a_f \\ b_f \end{bmatrix} = \begin{bmatrix} \sqrt{T} & \sqrt{1-T} \\ -\sqrt{1-T} & \sqrt{T} \end{bmatrix} \begin{bmatrix} a \\ b \end{bmatrix} = \begin{bmatrix} \cos\theta & \sin\theta \\ -\sin\theta & \cos\theta \end{bmatrix} \begin{bmatrix} a \\ b \end{bmatrix}. \tag{15.103}$$

The action of the BS can be described by the following transformation:

$$BS(\theta) = e^{-j\hat{H}} = e^{\theta(a^\dagger b - ab^\dagger)}. \tag{15.104}$$

The quadrature operators of the modes $\hat{x} = (\hat{x}_1\hat{x}_2\hat{x}_3\hat{x}_4)^T = (\hat{I}_a\hat{Q}_a\hat{I}_b\hat{Q}_b)^T$ are transformed by the following symplectic transformation:

$$\hat{x} \rightarrow \hat{x}_f = \begin{bmatrix} \sqrt{T}\mathbf{1} & \sqrt{1-T}\mathbf{1} \\ -\sqrt{1-T}\mathbf{1} & \sqrt{T}\mathbf{1} \end{bmatrix} \hat{x}, \tag{15.105}$$

where now $\mathbf{1}$ is a 2×2 identity matrix.

As another illustrative example, let us consider the *phase rotation* operator. Phase rotation can be described by the free-propagation Hamiltonian $\hat{H} = \theta a^\dagger a$ as follows: $\hat{R}(\theta) = e^{-j\hat{H}} = e^{-j\theta a^\dagger a}$. For the annihilation operator, the corresponding

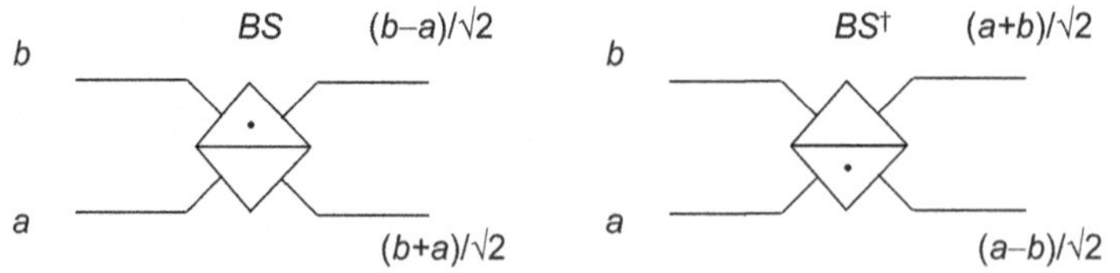

FIGURE 15.25

Operation principle of a 50/50 beam splitter (BS) and its Hermitian conjugate (BS†).

transformation is simply $a \rightarrow a_f = ae^{-j\theta}$. For the quadrature operators, the corresponding transformation is symplectic mapping:

$$\widehat{x} \rightarrow \widehat{x}_f = \widehat{R}(\theta)\widehat{x},\ \widehat{R}(\theta) = \begin{bmatrix} \cos\theta & \sin\theta \\ -\sin\theta & \cos\theta \end{bmatrix}. \tag{15.106}$$

When we pump a nonlinear crystal in the nondegenerate regime, the crystal will generate pairs of photons in two different modes, commonly referred to as the signal and the idler. This process is known as spontaneous parametric down-conversion (SPDC) [9,120–122]. By sending the photons at frequency ω_p (pump signal) into a nonlinear (NL) crystal, such as KH_2PO_4, beta-barium borate, or periodically poled LiNbO3 crystal, we generate the photon pairs at frequencies ω_s (signal) and ω_i (idler) satisfying the energy conservation principle $\hbar(\omega_s + \omega_i) = \hbar\omega_p$, and momentum conservation principle: $\boldsymbol{k}_s + \boldsymbol{k}_i = \boldsymbol{k}_p$, which is illustrated in Fig. 15.26. The nonlinear crystal is used to split photons into pairs of photons that satisfy the law of conservation of energy, have combined energies and momenta equal to the energy and momentum of the original photon, are phase matched in the frequency domain, and have correlated polarizations. This interaction is commonly referred to as *three-wave mixing* (TWM), and can be described by the following interaction Hamiltonian:

$$\widehat{H}_{\mathrm{TWM}} = j\eta a_s^\dagger a_i^\dagger a_p + H.c., \tag{15.107}$$

where we use a_p to denote the pump photon annihilation operators, while a_s^+ and a_i^+ denote the signal and idle photon creation operators. The parameter η is related to the conversion efficiency. The corresponding evolution operator is given by $\widehat{U} = \exp\left(-j\int \widehat{H}_{\mathrm{TWM}}dt\right)$. If the photons have the same polarization, they belong to Type I correlation; otherwise, they have perpendicular polarizations and belong to Type II. As the SPDC is stimulated by random vacuum fluctuations, the photon pairs are created at random time instances. The output of a Type I down-converter is known as a squeezed vacuum and it contains only an even number of photons. The output of the Type II down-converter is known as a two-mode squeezed vacuum state. Alternatively, the highly nonlinear fiber (HNLF) can also be used as an SPDC source [2,9]. HNLF employs the *four-wave mixing* (FWM) effect [2,9,46], in which two bright pump photons interact and generate the signal and idler photons, and can be described by the following interaction Hamiltonian:

$$\widehat{H}_{FWM} = j\eta a_s^\dagger a_i^\dagger a_{p_1} a_{p_2} + H.c. \tag{15.108}$$

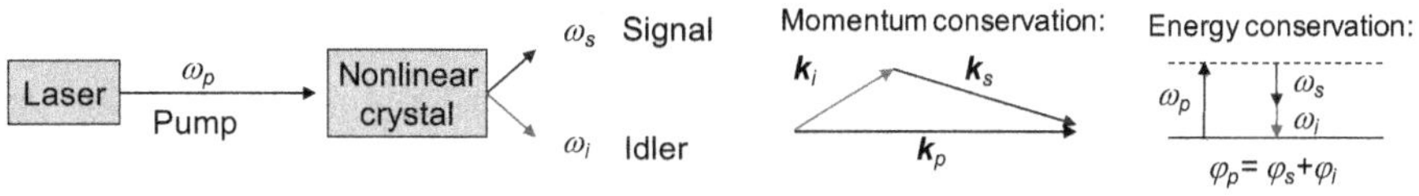

FIGURE 15.26

Spontaneous parametric down-conversion process.

The second- and third-order nonlinear effects are typically weak, and we need to use the strong pump beams to generate any observable quantum effect. When the pump beams are very strong they become classical and can be represented by the corresponding complex variables; for instance, we can use the complex variable A_p instead of a_p. The corresponding two-photon processes are known as the parametric process. By substituting ηA_p in the TWM Hamiltonian with ζ, or $\eta A_{p1} A_{p2}$ in the FWM Hamiltonian with ksi as well, the parametric processes Hamiltonian can be written as:

$$\widehat{H}_P = j\zeta a_s^\dagger a_i^\dagger + H.c. \tag{15.109}$$

The corresponding evolution operator becomes:

$$\widehat{U}_P = \exp\left(-j\int \widehat{H}_P dt\right) = \exp\left(\underbrace{\zeta t}_{-\xi^*}\; a_s^\dagger a_i^\dagger - \zeta^* t a_s a_i\right)$$

$$= \exp\left(-\xi^* a_s^\dagger a_i^\dagger + \xi a_s a_i\right), \xi = -\zeta^* t. \tag{15.110}$$

When both signal and idler photons are in vacuum states, the state at the output of the NL device is the *two-mode squeezed* (TMS) state:

$$|\Psi\rangle = \exp\left(-\xi^* a_s^\dagger a_i^\dagger + \xi a_s a_i\right)|0\rangle|0\rangle. \tag{15.111}$$

In this case, the parameter ξ is a real number and can be denoted by z to specify the level of squeezing. The corresponding Hamiltonian in this case is commonly referred to as the *two-mode squeezing operator*:

$$S_2(s) = \exp\left(-s\left(a_s^\dagger a_i^\dagger - a_s a_i\right)\right). \tag{15.112}$$

In the Heisenberg picture, the quadrature operators $\widehat{\boldsymbol{x}} = (\widehat{x}_1 \widehat{x}_2 \widehat{x}_3 \widehat{x}_4)^T = (\widehat{I}_a \widehat{Q}_a \widehat{I}_b \widehat{Q}_b)^T$ are transformed to [122]:

$$\widehat{\boldsymbol{x}} \rightarrow \widehat{\boldsymbol{x}}_f = \widehat{\mathbf{S}}_2(s)\widehat{\boldsymbol{x}},\ \widehat{\mathbf{S}}_2(s) = \begin{bmatrix} \cosh s\,\mathbf{1} & \sinh s\,\mathbf{Z} \\ \sinh s\,\mathbf{Z} & \cosh s\,\mathbf{1} \end{bmatrix}, \tag{15.113}$$

where $\mathbf{Z} = \text{diag}(1,-1)$ is the Pauli $\mathbf{Z}$-matrix. Now by applying $S_2(s)$ on two vacuum states, we can represent the TMS state, also known as the EPR state, as follows [122]:

$$|\Psi\rangle = \sqrt{1-\lambda^2}\sum_{n=0}^{\infty}\lambda^n |n\rangle_a |n\rangle_b = \sqrt{\frac{2}{v+1}}\sum_{n=0}^{\infty}\left(\frac{v-1}{v+1}\right)^{n/2}|n\rangle_a|n\rangle_b,$$

$$\lambda^2 = \frac{v-1}{v+1}, \tag{15.114}$$

with $\lambda = \tanh(s)$ and $v = \cosh(2s)$ representing the variances of the quadratures. The corresponding density operator is given by $\rho_{EPR} = |\Psi\rangle\langle\Psi|$. The EPR state represents the Gaussian state with zero mean and covariance matrix given by:

$$\Sigma_{EPR} = \begin{bmatrix} v\mathbf{1} & \sqrt{v^2-1}\mathbf{Z} \\ \sqrt{v^2-1}\mathbf{Z} & v\mathbf{1} \end{bmatrix}, v = \cosh 2s, \tag{15.115}$$

wherein the elements of the covariance matrix are determined by applying the definition Eqs. (15.86) and (15.87) to the EPR state. The EPR state, given by Eq. (15.114), is directly applicable to the CV-QKD systems.

15.10.3 Thermal Decomposition of Gaussian States

The thermal state is a bosonic state that maximizes the von Neumann entropy:

$$S(\hat{\rho}) = -\operatorname{Tr}(\hat{\rho}\log\hat{\rho}) = -\sum_i \eta_i \log \eta_i, \tag{15.116}$$

where η_i are eigenvalues of the density operator. The thermal state can be represented in terms of the number states as follows [2,125]

$$\hat{\rho}_{\text{thermal}} = \sum_{n=0}^{\infty}(1-v)v^n|n\rangle\langle n| = \frac{1}{\mathcal{N}+1}\sum_{n=0}^{\infty}\left(\frac{\mathcal{N}}{\mathcal{N}+1}\right)^n|n\rangle\langle n|,\ v = \frac{\mathcal{N}}{\mathcal{N}+1}, \tag{15.117}$$

where $\mathcal{N}$ is the average number of noise photons. So, the entropy of the thermal state is given by [125]:

$$S(\hat{\rho}_{\text{thermal}}) = -\sum_i \eta_i \log \eta_i = \mathcal{N}\log\left(\frac{\mathcal{N}+1}{\mathcal{N}}\right) - \log(\mathcal{N}+1) = g(\mathcal{N}),$$

$$g(x) = (x+1)\log(x+1) - x\log x,\ x > 0, \tag{15.118}$$

where the g function is a monotonically increasing concave function. The density operator of the whole field, composed of M thermal states, can be represented as the direct product of modes, which can also be written as:

$$\hat{\rho}_{\text{total}} = \bigotimes_{m=1}^{M} \hat{\rho}_{\text{thermal}}(\mathcal{N}_m). \tag{15.119}$$

Given that thermal states are independent, the von Neumann entropy is additive, and we can write:

$$S(\hat{\rho}_{\text{total}}) = S\left[\bigotimes_{m=1}^{M}\hat{\rho}_{\text{thermal}}(\mathcal{N}_m)\right] = \sum_{m=1}^{M} S(\hat{\rho}_{\text{thermal}}(\mathcal{N}_m)) = \sum_{m=1}^{M} g(\mathcal{N}_m), \tag{15.120}$$

where $\mathcal{N}_m$ is the average number of photons in the mth thermal state.

By applying the Williamson's theorem [126] on the M-mode covariance matrix $\mathbf{\Sigma}$, claiming that every positive-definite real matrix can be placed in a diagonal form by a suitable symplectic transformation, we can represent the covariance matrix $\sum$, with the help of a symplectic matrix $\boldsymbol{S}$, as follows [122,125]:

$$\mathbf{\Sigma} = \boldsymbol{S}\mathbf{\Sigma}^{\oplus}\boldsymbol{S}^T,\ \mathbf{\Sigma}^{\oplus} = \bigoplus_{m=1}^{M} \nu_m \mathbf{1}, \tag{15.121}$$

where $\mathbf{\Sigma}^{\oplus}$ is the diagonal matrix known as Williamson's form, and ν_m $(m = 1,\ldots,M)$ are symplectic eigenvalues of the covariance matrix, determined as the modulus of ordinary eigenvalues of the matrix $\boldsymbol{j}\mathbf{\Omega\Sigma}$. Given that only the modulus of the eigenvalues a are relevant, the $2M \times 2M$ covariance matrix $\sum$ will have exactly M symplectic eigenvalues. The representation (15.121) is in fact the thermal decomposition of a zero-mean Gaussian state, described by the covariance matrix $\mathbf{\Sigma}$. The uncertainty principle can now be expressed as $\mathbf{\Sigma}^{\oplus} \geq 1$, $\mathbf{\Sigma} > 0$. Given that the covariance matrix is positive definite, its symplectic eigenvalues are larger or equal to 1, and the entropy can be calculated by using Eq. (15.118) as follows:

$$S(\widehat{\rho}(\mathbf{0}, \mathbf{\Sigma}^{\oplus})) = S\left[\bigotimes_{m=1}^{M} \widehat{\rho}_{\text{thermal}}\left(\frac{\nu_m - 1}{2}\right)\right] = \sum_{m=1}^{M} S\left(\widehat{\rho}_{\text{thermal}}\left(\frac{\nu_m - 1}{2}\right)\right)$$
$$= \sum_{m=1}^{M} g\left(\frac{\nu_m - 1}{2}\right). \tag{15.122}$$

The thermal decomposition of an arbitrary Gaussian state can be represented as:

$$\widehat{\rho}(\overline{\boldsymbol{x}}, \mathbf{\Sigma}^{\oplus}) = D(\overline{\boldsymbol{x}})U_S\left[\widehat{\rho}(\mathbf{0}, \mathbf{\Sigma}^{\oplus})\right]U_S^{\dagger}D(\overline{\boldsymbol{x}})^{\dagger}, \tag{15.123}$$

where $D(\overline{\boldsymbol{x}})$ represents the phase-space translation, while U_S corresponds to the symplectic mapping described by Eq. (15.121).

15.10.4 Two-Mode Gaussian States and Measurements

Two-mode Gaussian states $\widehat{\rho} = \widehat{\rho}(\overline{\boldsymbol{x}}, \Sigma)$ have been intensively studied [2,122,127–131], and simple analytical formulas have been derived. The correlation matrix for the two-mode Gaussian states can be represented as [122,129,131]:

$$\mathbf{\Sigma} = \begin{bmatrix} \boldsymbol{A} & \boldsymbol{C} \\ \boldsymbol{C}^T & \boldsymbol{B} \end{bmatrix}, \tag{15.124}$$

where $\boldsymbol{A}$, $\boldsymbol{B}$, and $\boldsymbol{C}$ are 2×2 real matrices, while $\boldsymbol{A}$ and $\boldsymbol{B}$ are also orthogonal matrices. The symplectic eigenvalues' spectrum $\{\lambda_1,\ \lambda_2\} = \{\lambda_-, \lambda_+\}$ can be computed by [122,129,131]:

$$\lambda_{1,2} = \sqrt{\frac{\Delta(\boldsymbol{\Sigma}) \mp \sqrt{\Delta^2(\boldsymbol{\Sigma}) - 4\det\boldsymbol{\Sigma}}}{2}}, \tag{15.125}$$

$$\det\boldsymbol{\Sigma} = \lambda_1^2\lambda_2^2, \Delta(\boldsymbol{\Sigma}) = \det\boldsymbol{A} + \det\boldsymbol{B} + 2\det\boldsymbol{C} = \lambda_1^2 + \lambda_2^2.$$

The uncertainty principle is equivalent to the following conditions [131]:

$$\boldsymbol{\Sigma} > 0,\ \det\boldsymbol{\Sigma} \geq 1,\ \Delta(\boldsymbol{\Sigma}) \leq 1 + \det\boldsymbol{\Sigma}. \tag{15.126}$$

An important class of two-mode Gaussian states are those whose covariance matrix can be represented in the standard form, such as the EPR state, as follows [127,128]:

$$\boldsymbol{\Sigma} = \begin{bmatrix} a\mathbf{1} & \boldsymbol{C} \\ \boldsymbol{C} & b\mathbf{1} \end{bmatrix}, \boldsymbol{C} = \begin{bmatrix} c_1 & 0 \\ 0 & c_2 \end{bmatrix}; a, b, c \in R. \tag{15.127}$$

In particular, when $c_1 = -c_2 = c \geq 0$, the symplectic eigenvalues are simplified to:

$$\lambda_{1,2} = \left[\sqrt{(a+b)^2 - 4c^2} \mp (b-a)\right]/2. \tag{15.128}$$

The corresponding symplectic matrix $\boldsymbol{S}$ performing the symplectic decomposition $\boldsymbol{\Sigma} = \boldsymbol{S}\boldsymbol{\Sigma}^{\oplus}\boldsymbol{S}^T$ is given by [122]:

$$\boldsymbol{S} = \begin{bmatrix} d_+\mathbf{1} & d_-\boldsymbol{Z} \\ d_-\boldsymbol{Z} & d_+\mathbf{1} \end{bmatrix}, d_\pm = \sqrt{\frac{a+b \pm \sqrt{(a+b)^2 - 4c^2}}{2\sqrt{(a+b)^2 - 4c^2}}}. \tag{15.129}$$

A *quantum measurement* is specified by the set of measurement operators given by $\{M_m\}$, where index m stands for possible measurement result, satisfying the completeness relation $\sum_m M_m^\dagger M_m = \mathbf{1}$. The probability of finding the measurement result m, given the state $|\psi\rangle$, is given by [2,9,132,133]:

$$p_m = \Pr(m) = \mathrm{Tr}\left(\widehat{\rho} M_m^\dagger M_m\right). \tag{15.130}$$

After the measurement, the system will be left in the following state:

$$\widehat{\rho}_f = M_m \widehat{\rho} M_m^\dagger / p_m. \tag{15.131}$$

If we are only interested in the outcome of the measurement we can define a set of positive operators $\{\Pi_m = M_m^\dagger M_m\}$ and describe the measurement as the *positive operator-valued measure* (POVM). The probability of a getting the mth measurement result will then be [132,133]:

$$p_m = \Pr(m) = \mathrm{Tr}(\widehat{\rho}\Pi_m). \tag{15.132}$$

In addition to being positive, the Π_m-operators satisfy the completeness relation $\sum_m \Pi_m = \mathbf{1}$.

In the case of CV systems, the measurement results are real, that is, $m \in R$, so that the probability of outcome becomes the PDF. When measuring the Gaussian states, the measurement distribution will be Gaussian. If we perform the Gaussian measurements on M modes out of $M+N$ available, the distribution of measurements will be Gaussian, while the remaining unmeasured N modes will be in a Gaussian state [2,122,134].

Let B be the mode on which measurement is performed, and A denote the remaining $M-1$ modes of the Gaussian state described by $2M \times 2M$ covariance matrix $\boldsymbol{\Sigma}$. Similar to Eq. (15.124) we can decompose the covariance matrix as follows [122,134]:

$$\boldsymbol{\Sigma} = \begin{bmatrix} \boldsymbol{A} & \boldsymbol{C} \\ \boldsymbol{C}^T & \boldsymbol{B} \end{bmatrix}, \tag{15.133}$$

where $\boldsymbol{B}$ is the real matrix of size 2×2, $\boldsymbol{A}$ is the real matrix of size $2(M-1) \times 2(M-1)$, and $\boldsymbol{C}$ is the real matrix of size $2(M-1) \times 2$. For homodyne detection of in-phase I or quadrature Q, the partial measurement of $\boldsymbol{B}$ transforms the covariance matrix of $\boldsymbol{A}$ as follows [122,134]:

$$\boldsymbol{\Sigma}_{A|B}^{(\hat{I},\hat{Q})} = \boldsymbol{A} - \boldsymbol{C}\left(\boldsymbol{\Pi}_{I,Q}\boldsymbol{B}\boldsymbol{\Pi}_{I,Q}\right)^{-1}\boldsymbol{C}^T,\ \boldsymbol{\Pi}_I = \mathrm{diag}(1,0),\ \boldsymbol{\Pi}_Q = \mathrm{diag}(0,1), \tag{15.134}$$

where we use $(\bullet)^{-1}$ to denote the pseudoinverse operation. Given that the $\boldsymbol{\Pi}_I\boldsymbol{B}\boldsymbol{\Pi}_I = \begin{bmatrix} B_{11} & 0 \\ 0 & 0 \end{bmatrix} = \begin{bmatrix} V(\hat{I}) & 0 \\ 0 & 0 \end{bmatrix}$ and $\boldsymbol{\Pi}_Q\boldsymbol{B}\boldsymbol{\Pi}_Q = \begin{bmatrix} 0 & 0 \\ 0 & B_{22} \end{bmatrix} = \begin{bmatrix} 0 & 0 \\ 0 & V(\hat{Q}) \end{bmatrix}$ matrices are diagonal, the previous equation simplifies to:

$$\boldsymbol{\Sigma}_{A|B}^{(\hat{I},\hat{Q})} = \boldsymbol{A} - \frac{1}{V\left(\hat{I},\hat{Q}\right)}\boldsymbol{C}\boldsymbol{\Pi}_{I,Q}\boldsymbol{C}^T. \tag{15.135}$$

In heterodyne detection we measure both quadratures, and given that an additional balanced BS is used, with one of the inputs being in the vacuum state, we need to account for fluctuations due to the vacuum state and modify the partial correlation matrix as follows [122,134]:

$$\boldsymbol{\Sigma}_{A|B} = \boldsymbol{A} - \boldsymbol{C}(\boldsymbol{B}+\boldsymbol{1})^{-1}\boldsymbol{C}^T. \tag{15.136}$$

15.11 CONTINUOUS VARIABLE QKD

CV-QKD can be implemented by employing either homodyne detection, where only one quadrature component is measured at a time (because of the uncertainty principle) as illustrated in Fig. 15.27, or with heterodyne detection, where one BS and two balanced photodetectors are used to measure both quadrature components

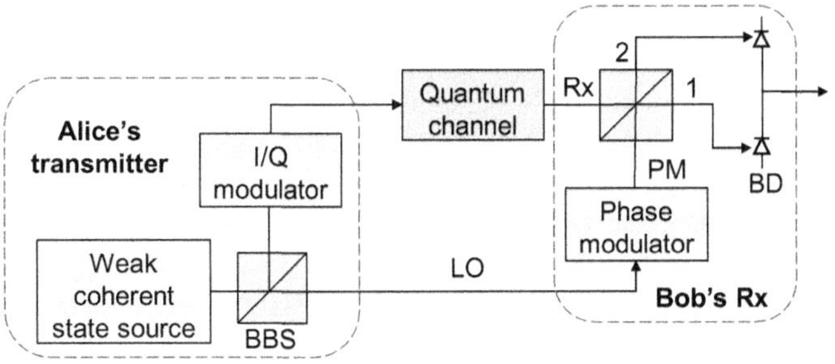

FIGURE 15.27

Gaussian modulation-based continuous variable quantum key distribution protocols with homodyne detection. *BBS*, Balanced beam splitter; *BD*, balanced detector.

simultaneously as illustrated in Fig. 15.28. Heterodyne detection can double the mutual information between Alice and Bob compared to the homodyne detection scheme at the expense of an additional 3 dB loss of the BS. To reduce the phase noise, the quantum signals are typically copropagated together with the time-domain multiplexed high-power pilot tone to align Alice's and Bob's measurement bases. CV-QKD studies are gaining momentum judging by the increasing number of papers related to CV-QKD [8,135–161], thanks to their compatibility with state-of-the-art optical communication technologies. The CV-QKD system experiences the 3 dB loss limitation in transmittance when direct reconciliation is used. To avoid this problem either the reverse reconciliation [162] or postelection [143] methods are used. Different eavesdropping strategies discussed earlier are applicable to CV-QKD systems as well. It has been shown that for Gaussian modulation (GM), Gaussian attack is the optimum attack for both individual attacks [149] and collective attacks [150,151]. In this section of the chapter we focus our attention on CV-QKD with GM. To implement CV-QKD, both squeezed states and coherent states can be employed.

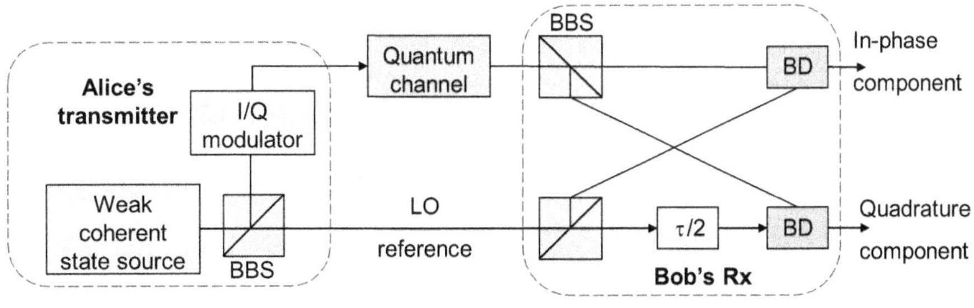

FIGURE 15.28

Gaussian modulation-based continuous variable quantum key distribution protocols with heterodyne detection. *BBS*, Balanced beam splitter; *BD*, balanced detector, shown in Fig. 15.27.

15.11.1 Homodyne and Heterodyne Detection Schemes

In the homodyne detection scheme, provided in Fig. 15.27, which represents the most common measurement in CV systems, we measure one of the two quadratures, since we cannot measure both quadratures simultaneously, according to the uncertainty principle. The measurement operators are projectors over the quadrature basis $|I\rangle\langle I|$ or $|Q\rangle\langle Q|$, and corresponding outcomes have PDF $f(I)$ or $f(Q)$, obtained by the marginal integration of the Wigner function, introduced by Eq. (5.88), which for a single mode becomes:

$$f(I) = \int_Q W(I,Q)dQ,\ f(Q) = \int_I W(I,Q)dI.$$

When more than one bosonic mode is in the optical beam we have to apply partial homodyning and also perform integration over both quadratures of all other modes.

In GM-based CV-QKD with homodyne detection [2,8], shown in Fig. 15.27, only one quadrature component is measured at a time. This protocol is commonly referred to as the GG02 CV-QKD protocol, named after the authors Grosshans and Grangier who proposed it in 2002 [162]. Alice sufficiently attenuates the laser beam, which serves as a WCS source, and employs the BBS to split the signal into two parts. The upper output of the BBS is used as an input to the I/Q modulator to place the Gaussian state to the desired location in the I-Q plane to get $|\alpha\rangle = |I + jQ\rangle$ as well as to control Alice's variance. The second output of BBS is used to send the local oscillator (LO) reference signal for Bob. Since the action of the BS is described by $BBS = \frac{1}{\sqrt{2}}\begin{bmatrix} 1 & 1 \\ -1 & 1 \end{bmatrix}$, the annihilation operators out of Bob's BBS will be:

$$\begin{bmatrix} a_1 \\ a_2 \end{bmatrix} = \frac{1}{\sqrt{2}}\begin{bmatrix} 1 & 1 \\ -1 & 1 \end{bmatrix}\begin{bmatrix} a_{Rx} \\ a_{PM} \end{bmatrix} = \frac{1}{\sqrt{2}}\begin{bmatrix} a_{Rx} + a_{PM} \\ -a_{Rx} + a_{PM} \end{bmatrix}, \tag{15.137}$$

where the creation (annihilation) operators at the receive BBS inputs Rx and PM are denoted by $a_{Rx}^{\dagger}(a_{Rx})$ and$a_{PM}^{\dagger}(a_{PM})$, respectively. The number of operators at the receive BBS outputs 1 and 2 can be expressed now as follows:

$$\begin{aligned} \widehat{n}_1 &= a_1^{\dagger}a_1 = \frac{1}{2}\left(a_{PM}^{\dagger} + a_{Rx}^{\dagger}\right)(a_{PM} + a_{Rx}), \\ \widehat{n}_2 &= a_2^{\dagger}a_2 = \frac{1}{2}\left(a_{PM}^{\dagger} - a_{Rx}^{\dagger}\right)(a_{PM} - a_{Rx}). \end{aligned} \tag{15.138}$$

The output of the balanced detector (BD), assuming the photodiode responsivity of $P_{\text{ph}} = 1$ A/W, is given by (in SNU):

$$\widehat{i}_{BD} = 4(\widehat{n}_1 - \widehat{n}_2) = 4\left(a_{PM}^{\dagger}a_{Rx} + a_{Rx}^{\dagger}a_{PM}\right). \tag{15.139}$$

The phase modulation output can be represented in SNU by $a_{PM} = |\alpha|\exp(j\theta)\mathbf{1}/2$, and after substitution into Eq. (15.139) we obtain:

$$\hat{i}_{BD} = 4\left(\frac{|\alpha|e^{-j\theta}}{2}a_{Rx} + a_{Rx}^{\dagger}\frac{|\alpha|e^{j\theta}}{2}\right) = 2|\alpha|\left(e^{j\theta}a_{Rx}^{\dagger} + e^{-j\theta}a_{Rx}\right). \tag{15.140}$$

In the absence of Eve, the receive signal constellation will be $a_{Rx} = (I,Q)/2$, which after substitution into the previous equation we obtain:

$$\hat{i}_{BD} = |\alpha|\left[(\cos\theta, \sin\theta)\mathbf{1}\cdot\left(I\hat{I}, -Q\hat{Q}\right) + (\cos\theta, \sin\theta)\mathbf{1}\cdot\left(I\hat{I}, Q\hat{Q}\right)\right]$$

$$= |\alpha|\left(\cos\theta\cdot I\hat{I} + \sin\theta\cdot Q\hat{Q}\right). \tag{15.141}$$

Bob measures the in-phase component by selecting $\theta = 0$ to get $|\alpha|I$ and the quadrature component by setting $\theta = \pi/2$ to get $|\alpha|Q$.

The theory of optical heterodyne detection has been developed in [163]. Heterodyne detection corresponds to projection onto coherent states, that is, $M(\alpha) \sim |\alpha\rangle\langle\alpha|$. This theory has been generalized to POVM based on projections over pure Gaussian states [122,134]. For heterodyne detection-based GM CV-QKD, the transmitter side is identical to the coherent detection version. In the heterodyne detection receiver, shown in Fig. 15.28, the quantum signal is split using the BBS; one output is used to measure the in-phase component (upper branch), while the other output (lower branch) is used as input to the second BD to detect the quadrature component. The LO classical reference signal is also split by the BBS, with the upper output being used as the input to the in-phase BD, and the other output, after phase modulation to introduce the $\pi/2$-phase shift, is used as the LO input to the quadrature BD to detect the quadrature component. In such a way, both quadratures can be measured simultaneously, and mutual information can be doubled. On the other hand, the BBS introduces 3 dB attenuation of the quantum signal.

15.11.2 Squeezed States-Based Protocols

Squeezed states-based protocols employ the Heisenberg uncertainty principle, claiming that it is impossible to measure both quadratures with complete precision. To impose the information, Alice randomly selects to use either in-phase or quadrature DOF, as illustrated in Fig. 15.29A. In Alice's encoding rule I (when in-phase DOF is used), the squeezed state is imposed on the in-phase component (with squeezed parameter $s_I < 1$) by (in SNU):

$$X_{A,I} \to |X_{A,I}, z_I\rangle,\ z_I < 1,\ X_{A,I} \sim \mathcal{N}\left(0, \sqrt{v_{A,I}}\right). \tag{15.142}$$

The amplitude is selected from a zero-mean Gaussian source of variance $v_{A,I}$. On the other hand, in Alice's encoding rule Q (when the quadrature is used), the

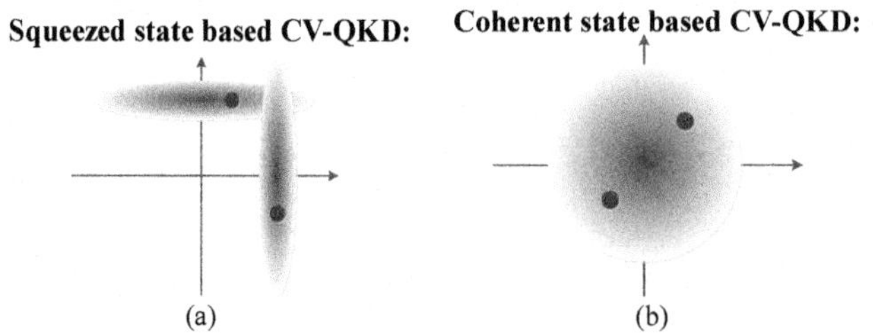

FIGURE 15.29

Operation principles of squeezed (a) and coherent (b) states-based protocols. *CV-QKD,* Continuous variable quantum key distribution.

squeezed state is imposed on the quadrature (with squeezed parameter $s_Q > 1$) and we can write (in SNU):

$$X_{A,Q} \to |jX_{A,Q}, z_Q\rangle,\ z_Q > 1,\ X_{A,Q} \sim \mathcal{N}\left(0, \sqrt{v_{A,Q}}\right). \tag{15.143}$$

The variances of squeezed states I and Q will be $V\left(\widehat{I}\right) = \sigma_I^2 = z_I$ and $V\left(\widehat{Q}\right) = \sigma_Q^2 = 1/z_Q$, respectively (as described earlier). On the receiver side, Bob randomly selects whether to measure either the in-phase or quadrature component. Alice and Bob exchange the encoding rules being used by them to measure the quadrature for every squeezed state and keep only instances when they measure the same quadrature in the sifting procedure. Therefore this protocol is very similar to the BB84 protocol. After that, information reconciliation takes place, followed by privacy amplification.

From Eve's point of view, the corresponding mixed states when Alice employs either I or Q DOF will be [8]:

$$\begin{aligned}\rho_I &= \int f_N\left(0, \sqrt{v_{A,I}}\right)|x, z_I\rangle\langle x, z_I|dx,\\ \rho_Q &= \int f_N\left(0, \sqrt{v_{A,Q}}\right)|jQ, z_Q\rangle\langle jQ, z_Q|dx\end{aligned}, \tag{15.144}$$

where $f_N(0,\sigma)$ is zero-mean Gaussian distribution with variance σ^2. When the condition $\rho_I = \rho_Q$ is satisfied, Eve cannot distinguish whether Alice was imposing the information on the squeezed state I or Q. Given that squeezed states are much more difficult to generate than coherent states we focus our attention on coherent states-based CV-QKD protocols.

15.11.3 Coherent States-Based CV-QKD Protocols

In coherent states-based protocols, with the operation principle provided in Fig. 15.29B, there is only one encoding rule for Alice:

$$(X_{A,I}, X_{A,Q}) \to |X_{A,I} + jX_{A,Q}\rangle; X_{A,I} \sim \mathcal{N}\left(0, \sqrt{v_A}\right),\ X_{A,Q} \sim \mathcal{N}\left(0, \sqrt{v_A}\right). \tag{15.145}$$

In other words, Alice randomly selects a point in 2D (I,Q) space from a zero-mean circular symmetric Gaussian distribution with the help of apparatus provided in Fig. 15.28. Clearly, both quadratures have the same uncertainty. Here we again employ the Heisenberg uncertainty, claiming that it is impossible to measure both quadratures with complete precision. For the homodyne receiver, Bob performs the random measurement on either $\hat{I}$ or $\hat{Q}$, and lets the result of his measurement be denoted by Y_B. When Bob measures $\hat{I}$, his measurement result Y_B is correlated $X_{A,I}$. On the other hand, when Bob measures $\hat{Q}$, his measurement result Y_B is correlated $X_{A,Q}$. Clearly, Bob is able to measure a single coordinate of the signal constellation point sent by Alice using the Gaussian coherent state. Bob then announces which quadrature he measured in each signaling interval, and Alice selects the coordinate that agrees with Bob's measurement quadrature. The rest of the protocol is the same as for the squeezed states-based protocol.

From Eve's point of view, the mixed state will be:

$$\rho = \int f_N(0, \sqrt{v_A}) f_N(0, \sqrt{v_A}) |I + jQ\rangle\langle I + jQ| dI dQ, \tag{15.146}$$

where v_A is Alice's modulation variance in SNU. For lossless transmission, the total variance of Bob's random variable in SNU will be:

$$V(Y_B) = \mathrm{Var}(Y_B) = v_A + 1, \tag{15.147}$$

which is composed of Alice's modulation term v_A and the variance of intrinsic coherent state fluctuations is 1.

For lossy transmission, the channel can be modeled as a BS with attenuation $0 \leq T \leq 1$, as illustrated in Fig. 15.30. From this figure we conclude that the output state can be represented in terms of input state x_A and ground state x_0 as follows:

$$\boldsymbol{x}_B = \sqrt{T}\boldsymbol{x}_A + \sqrt{1-T}\boldsymbol{x}_0 = \sqrt{T}\left(\boldsymbol{x}_A + \sqrt{\chi_{\mathrm{line}}}\,\boldsymbol{x}_0\right), \ \chi_{\mathrm{line}} = (1-T)/T = 1/T - 1. \tag{15.148}$$

For lossy transmission when the GM of coherent states is employed instead of vacuum states, the previous expression needs to be modified as follows:

$$\begin{aligned} \boldsymbol{x}_B &= \sqrt{T}\left(\boldsymbol{x}_A + \sqrt{\chi_{\mathrm{line}}}\,\boldsymbol{x}_0\right), \ \chi_{\mathrm{line}} = (1-T)/T + \varepsilon \\ &= 1/T - 1 + \varepsilon, \varepsilon \geq 0. \end{aligned} \tag{15.149}$$

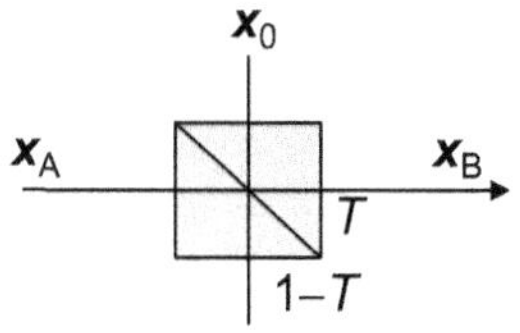

FIGURE 15.30

Lossy transmission channel.

The first term in χ_{line} is contributed to the channel attenuation, while the second term ε, commonly referred to as the excess noise, is due to other noisy factors such as the laser phase noise, imperfect orthogonality of quadratures, and so on. The total variance of Bob's random variable will be now composed of three terms:

$$V(Y_B) = \text{Var}(Y_B) = T\left(\underbrace{v_A + 1}_{v} + \chi_{\text{line}}\right) = T(v + \chi_{\text{line}}), \tag{15.150}$$

where $v = v_A + 1$.

Mutual information between Alice and Bob, based on Shannon's formula, can be calculated by:

$$\begin{aligned} I(A;B) \doteq I(X_A;Y_B) &= \frac{1}{2}\log\left(1 + \frac{T v_A}{T(1 + \chi_{\text{line}})}\right) \\ &= \frac{1}{2}\log\left(1 + \frac{v_A}{1 + \chi_{\text{line}}}\right). \end{aligned} \tag{15.151}$$

Eve's information, in CV-QKD schemes, is determined by reconciliation strategy.

Assuming that all sources to excess noise are statistically independent and additive, the total variance (in SNU) can be represented as a summation of individual contributions:

$$\varepsilon = \varepsilon_{\text{modulator}} + \varepsilon_{PN} + \varepsilon_{RIN} + \cdots. \tag{15.152}$$

where the first term denotes the modulation imperfections, the second term corresponds the phase noise (PN), while the third term corresponds to the relative intensity noise (RIN) of the LO reference signal.

Referring now to Bob's input (or equivalently to the channel output) we can represent Bob's equivalent detector noise variance by:

$$\chi_{\text{det}} = \frac{d(1 + v_{\text{el}}) - \eta}{\eta}, \tag{15.153}$$

where $d = 1$ for homodyne detection (either in-phase or quadrature can be measured at a given signaling interval in a homodyne detection scheme), and $d = 2$ for heterodyne detection (both quadratures can be measured simultaneously). The parameter η denotes the detection efficiency, while v_{el} denotes the variance of equivalent electrical noise. The variance of the total noise can be determined by summing up the variance of the channel noise and detection noise, which can be expressed by referring to the channel input as follows:

$$\chi_{\text{total}} = \chi_{\text{line}} + \frac{\chi_{\text{det}}}{T} = \underbrace{\frac{1-T}{T} + \varepsilon}_{\chi_{\text{line}}} + \frac{d(1+v_{el}) - \eta}{T\eta}. \tag{15.154}$$

When referring to the channel output, variance of the total noise at Bob's side can be determined by:

$$\begin{aligned}\chi_{\text{total, Bob}} &= T\eta\left(\chi_{\text{line}} + \frac{\chi_{\text{det}}}{T}\right) = T\eta\frac{1-T}{T} + T\eta\varepsilon + T\eta\frac{d(1+v_{el}) - \eta}{T\eta} \\ &= \eta(1-T) + T\eta\varepsilon + d(1+v_{el}) - \eta.\end{aligned} \tag{15.155}$$

For the bipartite system composed of two separable subsystems A and B, each described by corresponding density matrices $\boldsymbol{\rho}_A$ and $\boldsymbol{\rho}_B$, the density operator of the bipartite system can be represented by $\boldsymbol{\rho} = \boldsymbol{\rho}_A \otimes \boldsymbol{\rho}_B$. For a Gaussian bipartite system, composed of two separable subsystems A and B, with the A subsystem being a Gaussian state with M modes represented by quadrature operators $\hat{\boldsymbol{x}}_A$, while the B subsystem being a Gaussian state with N modes represented by quadrature operators $\hat{\boldsymbol{x}}_B\hat{\boldsymbol{x}}_B$, the Gaussian state of the bipartite system will contain $M+N$ modes and can be described by the quadrature operators $\hat{\boldsymbol{x}} = [\hat{\boldsymbol{x}}_A\hat{\boldsymbol{x}}_B]^T$, with a corresponding covariance matrix of size $2(M+N) \times 2(M+N)$ being represented as:

$$\boldsymbol{\Sigma} = \boldsymbol{\Sigma}_A \oplus \boldsymbol{\Sigma}_B = \begin{bmatrix} \boldsymbol{\Sigma}_A & \mathbf{0} \\ \mathbf{0} & \boldsymbol{\Sigma}_B \end{bmatrix}, \tag{15.156}$$

where 0 denotes the all-zeros matrix.

Let us assume that the EPR state, defined by Eq. (15.114), is used in the CV-QKD system, with the covariance matrix defined by Eq. (15.115). The identity matrix can be used to describe the thermal state at the input of the BS (Fig. 15.30). Based on the previous equation, the corresponding bipartite system will have the following correlation matrix:

$$\boldsymbol{\Sigma} = \begin{bmatrix} \boldsymbol{\Sigma}_{\text{EPR}} & \mathbf{0} \\ \mathbf{0} & \mathbf{1} \end{bmatrix} = \begin{bmatrix} v\mathbf{1} & \sqrt{v^2-1}\mathbf{Z} & \mathbf{0} \\ \sqrt{v^2-1}\mathbf{Z} & v\mathbf{1} & \mathbf{0} \\ \mathbf{0} & \mathbf{0} & \mathbf{1} \end{bmatrix}. \tag{15.157}$$

The BS acts on Bob's qubit and thermal state, while it leaves Alice's qubit unaffected; therefore its action can be described as:

$$\begin{aligned} BS_{\text{combined}}(T) &= \mathbf{1} \oplus \begin{bmatrix} \sqrt{T}\mathbf{1} & \sqrt{1-T}\mathbf{1} \\ -\sqrt{1-T}\mathbf{1} & \sqrt{T}\mathbf{1} \end{bmatrix} \\ &= \begin{bmatrix} \mathbf{1} & \mathbf{0} & \mathbf{0} \\ \mathbf{0} & \sqrt{T}\mathbf{1} & \sqrt{1-T}\mathbf{1} \\ \mathbf{0} & -\sqrt{1-T}\mathbf{1} & \sqrt{T}\mathbf{1} \end{bmatrix}. \end{aligned} \tag{15.158}$$

To determine the covariance matrix after the BS we apply the symplectic operation by BS_{combined} on the input covariance matrix to obtain:

$$\boldsymbol{\Sigma}' = BS_{\text{combined}}(T)\boldsymbol{\Sigma}[BS_{\text{combined}}(T)]^T$$

$$= \begin{bmatrix} \mathbf{1} & \mathbf{0} & \mathbf{0} \\ \mathbf{0} & \sqrt{T}\mathbf{1} & \sqrt{1-T}\mathbf{1} \\ \mathbf{0} & -\sqrt{1-T}\mathbf{1} & \sqrt{T}\mathbf{1} \end{bmatrix}$$

$$\times \begin{bmatrix} v\mathbf{1} & \sqrt{v^2-1}\mathbf{Z} & \mathbf{0} \\ \sqrt{v^2-1}\mathbf{Z} & v\mathbf{1} & \mathbf{0} \\ \mathbf{0} & \mathbf{0} & \mathbf{1} \end{bmatrix} \begin{bmatrix} \mathbf{1} & \mathbf{0} & \mathbf{0} \\ \mathbf{0} & \sqrt{T}\mathbf{1} & \sqrt{1-T}\mathbf{1} \\ \mathbf{0} & -\sqrt{1-T}\mathbf{1} & \sqrt{T}\mathbf{1} \end{bmatrix}^T. \quad (15.159)$$

Now by keeping only Alice's and Bob's submatrices we obtain:

$$\boldsymbol{\Sigma}'_{AB} = \begin{bmatrix} v\mathbf{1} & \sqrt{T(v^2-1)}\mathbf{Z} \\ \sqrt{T(v^2-1)}\mathbf{Z} & (vT+1-T)\mathbf{1} \end{bmatrix}. \quad (15.160)$$

In the presence of excess noise we need to replace the $(\Sigma_{AB})_{22}$ element by $vT + 1 + \varepsilon' - T$ to get the following covariance matrix for *homodyne detection*:

$$\boldsymbol{\Sigma}_{AB}^{(\text{homodyne})} = \begin{bmatrix} v\mathbf{1} & \sqrt{T(v^2-1)}\mathbf{Z} \\ \sqrt{T(v^2-1)}\mathbf{Z} & (vT+1+\varepsilon'-T)\mathbf{1} \end{bmatrix}. \quad (15.161)$$

In *heterodyne detection*, Bob employs the BBS to split the quantum signal, while the second input to the BBS is the vacuum state, and the operation of the BBS can be represented by Eq. (15.105) by setting $T = 1/2$. The corresponding covariance matrix at the input of the BBS will be:

$$\boldsymbol{\Sigma}_{\text{in}}^{(\text{BBS})} = \begin{bmatrix} \boldsymbol{\Sigma}_{\text{AB}}^{(\text{homodyne})} & \mathbf{0} \\ \mathbf{0} & \mathbf{1} \end{bmatrix} = \begin{bmatrix} v\mathbf{1} & \sqrt{T(v^2-1)}\mathbf{Z} & \mathbf{0} \\ \sqrt{T(v^2-1)}\mathbf{Z} & (vT+1+\varepsilon'-T)\mathbf{1} & \mathbf{0} \\ \mathbf{0} & \mathbf{0} & \mathbf{1} \end{bmatrix}. \quad (15.162)$$

The covariance matrix and the BBS output can be determined by applying the symplectic operation by $BS_{\text{combined}}(1/2)$ on the input covariance matrix to obtain:

$$\Sigma_{AB}^{(het)} = BS_{combined}(1/2)\Sigma_{in}^{(BBS)}[BS_{combined}(1/2)]^T = \begin{bmatrix} \mathbf{1} & \mathbf{0} & \mathbf{0} \\ \mathbf{0} & \frac{1}{\sqrt{2}}\mathbf{1} & \frac{1}{\sqrt{2}}\mathbf{1} \\ \mathbf{0} & -\frac{1}{\sqrt{2}}\mathbf{1} & \frac{1}{\sqrt{2}}\mathbf{1} \end{bmatrix}$$

$$\begin{bmatrix} v\mathbf{1} & \sqrt{T(v^2-1)}\mathbf{Z} & \mathbf{0} \\ \sqrt{T(v^2-1)}\mathbf{Z} & (vT+1+\varepsilon'-T)\mathbf{1} & \mathbf{0} \\ \mathbf{0} & \mathbf{0} & \mathbf{1} \end{bmatrix} \begin{bmatrix} \mathbf{1} & \mathbf{0} & \mathbf{0} \\ \mathbf{0} & \frac{1}{\sqrt{2}}\mathbf{1} & \frac{1}{\sqrt{2}}\mathbf{1} \\ \mathbf{0} & -\frac{1}{\sqrt{2}}\mathbf{1} & \frac{1}{\sqrt{2}}\mathbf{1} \end{bmatrix}^T$$

$$= \begin{bmatrix} v\mathbf{1} & \sqrt{\frac{T}{2}(v^2-1)}\mathbf{Z} & -\sqrt{\frac{T}{2}(v^2-1)}\mathbf{Z} \\ \sqrt{\frac{T}{2}(v^2-1)}\mathbf{Z} & \frac{vT+2+\varepsilon'-T}{2}\mathbf{1} & -\frac{vT+\varepsilon'-T}{2}\mathbf{1} \\ -\sqrt{\frac{T}{2}(v^2-1)}\mathbf{Z} & -\frac{vT+\varepsilon'-T}{2}\mathbf{1} & \frac{vT+2+\varepsilon'-T}{2}\mathbf{1} \end{bmatrix}. \quad (15.163)$$

As discussed in the previous sections, instead of employing the EPR state, Alice employs the PM scheme, in which she employs a WCS source, and with the help of an *I*/*Q* modulator, generates the Gaussian state with variance v_A. Given that the minimum uncertainty of the coherent source is 1 SNU, Bob's variance will be $v_A + 1$, and we can represent Bob's in-phase operator by:

$$\hat{I}_B \sim \hat{I}_A + \mathcal{N}(0,1), \quad (15.164)$$

where $N(0,1)$ is a zero-mean Gaussian source of variance ***1*** SNR. Clearly, the variance of Bob's in-phase operator is $V(\widehat{I}_B) = V(\widehat{I}_A) + 1 = v_A + 1$, as expected. On the other hand, the covariance of Alice's and Bob's in-phase operators will be:

$$\mathrm{Cov}(\hat{I}_A,\hat{I}_B) = \left\langle \left(\hat{I}_A - \underbrace{\langle \hat{I}_A \rangle}_{=0}\right)\left(\hat{I}_B - \underbrace{\langle \hat{I}_B \rangle}_{0}\right)\right\rangle = \left\langle \hat{I}_A \underbrace{\hat{I}_B}_{\hat{I}_A+\mathcal{N}(0,1)} \right\rangle = \langle \hat{I}_A^2 \rangle + \underbrace{\langle \hat{I}_A \mathcal{N}(0,1) \rangle}_{=0}$$

$$= v_A = \mathrm{Cov}(\hat{I}_B, \hat{I}_A). \quad (15.165)$$

Therefore the *covariance matrix* for the PM protocol is given as follows:

$$\boldsymbol{\Sigma}_{\mathrm{AB}}^{(\mathrm{PM\ protocol})}=\begin{bmatrix}\underbrace{V(\hat{I}_A)}_{v_A}\mathbf{1} & \underbrace{\mathrm{Cov}(\hat{I}_A,\hat{I}_B)}_{v_A}\mathbf{1}\\ \underbrace{\mathrm{Cov}(\hat{I}_B,\hat{I}_A)\mathbf{1}}_{v_A} & \underbrace{V(\hat{I}_B)\mathbf{1}}_{v_B=v_A+1}\end{bmatrix}=\begin{bmatrix}v_A\mathbf{1} & v_A\mathbf{1}\\ v_A\mathbf{1} & \underbrace{(v_A+1)}_{v}\mathbf{1}\end{bmatrix}=\begin{bmatrix}(v-1)\mathbf{1} & (v-1)\mathbf{1}\\ (v-1)\mathbf{1} & v\mathbf{1}\end{bmatrix}. \tag{15.166}$$

By using Eq. (15.163) we can demonstrate that the covariance matrix for the corresponding entanglement-assisted protocol has the same form as that for the PM protocol.

15.11.4 Secret-Key Rate of CV-QKD with Gaussian Modulation Under Collective Attacks

In this subsection we describe how to calculate the secret fraction for a coherent-state homodyne detection-based CV-QKD scheme when Eve employs the Gaussian collective attack [147–151], which is known to be optimum for GM. This attack is completely characterized by estimating the covariance matrix by Alice and Bob Σ_{AB}, which is based on Eq. (15.161) for homodyne detection and Eq. (15.163) for heterodyne detection. Let us rewrite the covariance matrix for homodyne detection as follows:

$$\boldsymbol{\Sigma}_{\mathrm{AB}}^{(\mathrm{homodyne})}=\begin{bmatrix}v\mathbf{1} & \sqrt{T(v^2-1)}\mathbf{Z}\\ \sqrt{T(v^2-1)}\mathbf{Z} & T\left(v+1/T-1+\underbrace{\varepsilon'/T}_{\varepsilon}\right)\mathbf{1}\end{bmatrix}=\begin{bmatrix}v\mathbf{1} & \sqrt{T(v^2-1)}\mathbf{Z}\\ \sqrt{T(v^2-1)}\mathbf{Z} & T\left(v+\underbrace{1/T-1+\varepsilon}_{\chi_{\mathrm{line}}}\right)\mathbf{1}\end{bmatrix}$$
$$=\begin{bmatrix}v\mathbf{1} & \sqrt{T(v^2-1)}\mathbf{Z}\\ \sqrt{T(v^2-1)}\mathbf{Z} & T(v+\chi_{\mathrm{line}})\mathbf{1}\end{bmatrix}, \tag{15.167}$$

where $\varepsilon=\varepsilon'/T$ is the excess noise and the $1/T-1$ is the channel loss term, both observed from Alice's side (at the channel input). The total variance of both terms is denoted by $\chi_{\mathrm{line}}=1/T-\varepsilon$. Now by adding the detector noise term, the covariance matrix for homodyne detection becomes:

$$\boldsymbol{\Sigma}_{AB}^{(\mathrm{hom.,\ complete})}=\begin{bmatrix}v\mathbf{1} & \sqrt{T\eta(v^2-1)}\mathbf{Z}\\ \sqrt{T\eta(v^2-1)}\mathbf{Z} & T\eta(v+\chi_{\mathrm{total}})\mathbf{1}\end{bmatrix},\ v=v_A+1, \tag{15.168}$$

where η is detection efficiency and χ_{total} was defined earlier by Eq. (15.154), that is, $\chi_{\mathrm{total}}=\chi_{\mathrm{line}}+\chi_{\mathrm{det}}/T=\chi_{\mathrm{line}}+[d(1+v_{el})-\eta]/(T\eta)$, where v_{el} is the variance of the equivalent electrical noise.

In CV-QKD for direct reconciliation with nonzero SKR, the channel loss is lower limited by $T\geq 1/2$ [2,8], and for higher channel losses we need to use reverse reconciliation. In reverse reconciliation, the secret fraction is given by:

$$r=\beta I(A;B)-\chi(B;E), \tag{15.169}$$

where β is the reconciliation efficiency and $I(A;B)$ is the mutual information between Alice and Bob that is identical for both individual and collective attacks [2,8]:

$$I(A;B) = \frac{d}{2}\log_2\left(\frac{v+\chi_{\text{total}}}{1+\chi_{\text{total}}}\right),\ d = \begin{cases} 1, \text{homodyne detection} \\ 2, \text{heterodyne detection} \end{cases}. \tag{15.170}$$

which is nothing else but the definition formula for channel capacity of the Gaussian channel. What remains to be determined is the Holevo information between Bob and Eve, denoted as $\chi(B;E)$, which is much more challenging to derive, and will be described next. After that, the SKR can be determined as the product of the secret fraction and signaling rate. Given that homodyne/heterodyne detection is not limited by the dead time, typical SKRs for CV-QKD schemes are significantly higher than that of DV-QKD schemes.

The Holevo information between Eve and Bob can be calculated by:

$$\chi(E;B) = S_E - S_{E|B}, \tag{15.171}$$

where S_E is the von Neumann entropy of Eve's density operator $\widehat{\rho}_E$, defined by $S(\widehat{\rho}_E) = -\text{Tr}(\widehat{\rho}_E \log\widehat{\rho}_E)$, while $S_{E|\text{B}}$ is the von Neumann entropy of Eve's state when Bob performs the projective measurement, be it homodyne or heterodyne. In principle, the symplectic eigenvalues of corresponding correlation matrices can be determined first to calculate the von Neumann entropies, as described in Section 15.10.3. To simplify the analysis, typically we assume that Eve possesses the purification [7,9] of Alice's and Bob's joint state $\widehat{\rho}_{AB}$, so that the resulting state is a pure state $|\psi\rangle$ and the corresponding density operator will be:

$$\widehat{\rho}_{ABE} = |\psi\rangle\langle\psi|. \tag{15.172}$$

By tracing out Eve's subspace we obtain the Alice–Bob mixed state, that is, $\widehat{\rho}_{AB} = \text{Tr}_E\widehat{\rho}_{ABE}$. In a similar fashion, by tracing out the Alice–Bob subspace we get Eve's mixed state $\widehat{\rho}_E = \text{Tr}_{AB}\widehat{\rho}_{ABE}$. By applying the Schmidt decomposition theorem [2,7,9] we can represent the bipartite state $|\psi\rangle \in H_{AB} \otimes H_E$ in terms of Schmidt bases for subsystem AB, denoted as $\{|i_{AB}\rangle\}$, and subsystem E, denoted as $\{|i_E\rangle\}$, as follows:

$$|\psi\rangle = \sum_i \sqrt{p_i}|i_{AB}\rangle|i_E\rangle, \tag{15.173}$$

where $\sqrt{p_i}$ are Schmidt coefficients satisfying relationship $\sum_i p_i = 1$. By tracing out Eve's subsystem we obtain the following mixed density operator for the Alice–Bob subsystem:

$$\widehat{\rho}_{AB} = \text{Tr}_E(|\psi\rangle\langle\psi|) = \sum_i p_i|i_{AB}\rangle\langle i_{AB}|. \tag{15.174}$$

On the other hand, by tracing out the Alice–Bob subsystem we obtain the following mixed state for Eve:

$$\widehat{\rho}_E = \mathrm{Tr}_{AB}(|\psi\rangle\langle\psi|) = \sum\nolimits_i p_i |i_E\rangle\langle i_E|. \tag{15.175}$$

Clearly, both Alice–Bob's and Eve's density operators have the same spectra (of eigenvalues), and therefore the same von Neumann entropy:

$$S_{AB} = S_E = -\sum\nolimits_i p_i \log p_i. \tag{15.176}$$

Calculation of the Holevo information is now greatly simplified since the following is valid:

$$\chi(E;B) = S_E - S_{E|B} = S_{AB} - S_{A|B}. \tag{15.177}$$

To determine the von Neuman entropy for the Alice–Bob subsystem, which is independent on Bob's measurement, we use the following covariance matrix from Eq. (15.167), observed at the channel input:

$$\mathbf{\Sigma}_{AB} = \begin{bmatrix} v\mathbf{1} & \sqrt{T(v^2-1)}\mathbf{Z} \\ \sqrt{T(v^2-1)}\mathbf{Z} & T(v+\chi_{\mathrm{line}})\mathbf{1} \end{bmatrix}. \tag{15.178}$$

To determine the symplectic eigenvalues we employ the theory from Section 15.10.4. Clearly, the covariance matrix has the same form as Eq. (15.127), with $a = v$, $b = T(v+\chi_{\mathrm{line}})$, and $c = \sqrt{T(v^2-1)}$. We can directly apply Eq. (15.128) to determine the symplectic eigenvalues. Alternatively, given that symplectic eigenvalues satisfy the following:

$$\lambda_1^2 + \lambda_2^2 = a^2 + b^2 - 2c^2,\ \lambda_1^2\lambda_2^2 = \left(ab - c^2\right)^2, \tag{15.179}$$

by introducing:

$$\begin{aligned} A &= a^2 + b^2 - 2c^2 = v^2(1-2T) + 2T + T^2(v+\chi_{\mathrm{line}}), \\ B &= \left(ab - c^2\right)^2 = T^2(1 + v\chi_{\mathrm{line}}), \end{aligned} \tag{15.180}$$

the corresponding symplectic eigenvalues for the Alice–Bob covariance matrix can be expressed as:

$$\lambda_{1,2} = \sqrt{\frac{1}{2}\left(A \pm \sqrt{A^2 - 4B}\right)}, \tag{15.181}$$

which represents a form very often found in open literature [66,145–147,164,165]. The von Neumann entropy, based on thermal state decomposition theorem and Eq. (15.122), is given by:

$$S_{AB} = S(\widehat{\rho}_{AB}) = g\left(\frac{\lambda_1 - 1}{2}\right) + g\left(\frac{\lambda_2 - 1}{2}\right). \tag{15.182}$$

The calculation of $S_{A|B}$ is dependent on Bob's measurement strategy, homodyne or heterodyne, and we can apply the theory of partial measurements from Section 15.10.4. For homodyne detection, based on Eq. (15.168), we conclude that $\boldsymbol{A} = v\mathbf{1}$, $\boldsymbol{B} = T(v + \chi_{\text{line}})\mathbf{1}$, and $\boldsymbol{C} = \sqrt{T(v^2 - 1)}\boldsymbol{Z}$ so that Bob's homodyne measurement transforms the foregoing covariance matrix to:

$$\boldsymbol{\Sigma}_{A|B}^{(\hat{I})} = \boldsymbol{A} - \frac{1}{V(\hat{I})}\boldsymbol{C\Pi}_I\boldsymbol{C}^T = \begin{bmatrix} v - \dfrac{v^2-1}{v+\chi_{\text{line}}} & 0 \\ 0 & v \end{bmatrix}. \tag{15.183}$$

The symplectic eigenvalue is determined now by solving for:

$$\det\left(j\Omega\boldsymbol{\Sigma}_{\text{A|B}} - \lambda\mathbf{1}\right) = \lambda^2 + v\left(\frac{v^2-1}{v+\chi_{\text{line}}} - v\right) = 0, \tag{15.184}$$

and by solving for λ we obtain:

$$\lambda_3 = \sqrt{v\left(v - \frac{v^2-1}{v+\chi_{\text{line}}}\right)}, \tag{15.185}$$

so that $S_{A|B}$ becomes $g(\lambda_3)$. This expression is consistent with that provided in [164], except that a slightly different correlation matrix was used. Finally, for homodyne detection, the Holevo information is given by:

$$\chi(E;B) = S_{AB} - S_{A|B} = g\left(\frac{\lambda_1 - 1}{2}\right) + g\left(\frac{\lambda_2 - 1}{2}\right) - g\left(\frac{\lambda_3 - 1}{2}\right). \tag{15.186}$$

In heterodyne detection, Bob measures both quadratures, and given that an additional BBS is used, with one of the inputs being in the vacuum state, we need to account for fluctuations due to the vacuum state and modify the partial correlation matrix as follows [2,163]:

$$\sum\nolimits_{A|B} = \boldsymbol{A} - \boldsymbol{C}(\boldsymbol{B}+\mathbf{1})^{-1}\boldsymbol{C}^T = K\mathbf{1},\ K = v - \frac{T(v^2-1)}{T(v+\chi_{\text{line}})+1}. \tag{15.187}$$

The symplectic eigenvalue can be determined from the following characteristic equation:

$$\det\left(j\Omega\boldsymbol{\Sigma}_{\text{A|B}}^{(\text{heterodyne})} - \lambda\mathbf{1}\right) = \lambda^2 - K^2 = 0, \tag{15.188}$$

to get:

$$\lambda_3^{(\text{heterodyne})} = K = v - \frac{T(v^2-1)}{T(v+\chi_{\text{line}})+1}. \tag{15.189}$$

Finally, for heterodyne detection, the Holevo information is given by:

$$\chi(E;B)^{(\text{heterodyne})} = g\left(\frac{\lambda_1 - 1}{2}\right) + g\left(\frac{\lambda_2 - 1}{2}\right) - g\left(\frac{\lambda_3^{(\text{heterodyne})} - 1}{2}\right). \quad (15.190)$$

15.11.5 Reverse Reconciliation Results for Gaussian Modulation-Based CV-QKD

In this section we provide illustrative secrecy fractions (normalized SKRs) for different scenarios by employing the derivations from the previous subsection. We consider the reverse reconciliation case only. In Fig. 15.31 we provide normalized SKRs versus channel loss when electrical noise variance in SNU is used as a parameter. The excess noise variance is set to $\varepsilon = 10^{-3}$, detector efficiency is $\eta = 0.85$, and reconciliation efficiency is set to $\beta = 0.85$. Interestingly enough, when electrical noise variance $v_{el} \leq 0.1$, there is not much degradation in SKRs,

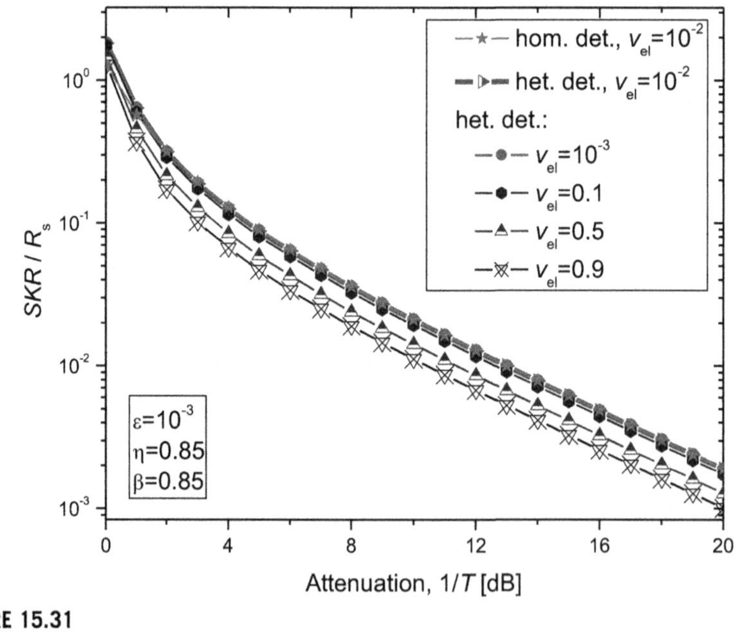

FIGURE 15.31

The secrecy fraction for CV-QKD scheme with GM vs. channel loss when electrical noise variance is used as a parameter.

which indicates that commercially available p-i-n photodetectors can be used in CV-QKD.

In Fig. 15.32 we show normalized SKRs versus channel loss when excess noise variance (expressed in SNU) is used as a parameter. The electrical noise variance is set to $v_{el} = 10^{-2}$, detector efficiency is $\eta = 0.85$, and reconciliation efficiency is set to $\beta = 0.85$. Evidently, the excess noise variance has a high impact on SKR performance.

In Fig. 15.33 we provide normalized SKRs versus channel loss when detector efficiency η is used as a parameter. The excess noise variance is set to $\varepsilon = 10^{-3}$, the electrical noise variance is set to $v_{el} = 10^{-2}$, and reconciliation efficiency is set to $\beta = 0.85$. Clearly, the detection efficiency does not have a high impact on SKRs.

In Fig. 15.34 we show normalized SKRs versus channel loss when reconciliation efficiency β is used as a parameter. The excess noise variance is set to $\varepsilon = 10^{-3}$, the electrical noise variance is set to $v_{el} = 10^{-2}$, and detector efficiency is set to $\eta = 0.85$. Evidently, the reconciliation efficiency does have a relevant impact on SKRs.

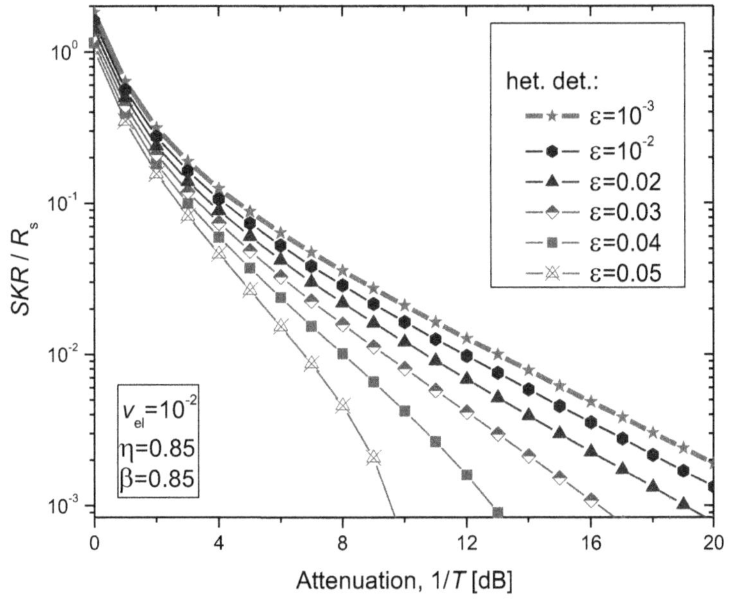

FIGURE 15.32

The secrecy fraction for CV-QKD scheme with GM vs. channel loss when electrical noise variance is used as a parameter.

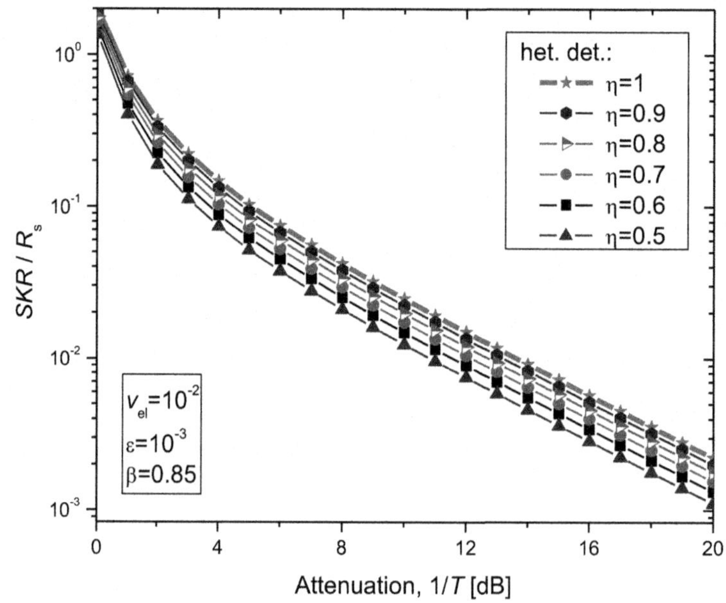

FIGURE 15.33

The secrecy fraction for CV-QKD scheme with GM vs. channel loss when detector efficiency is used as a parameter.

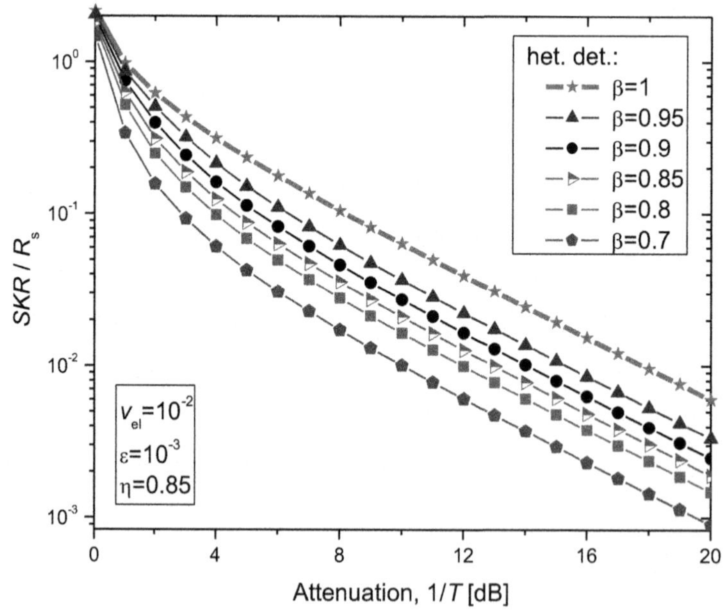

FIGURE 15.34

The secrecy fraction for CV-QKD scheme with GM vs. channel loss when reconciliation efficiency is used as a parameter.

15.12 SUMMARY

This chapter was devoted to the fundamental concepts of QKD. The main differences between conventional cryptography and QKD were described in Section. 15.1. In the section on QKD basics (Section 15.2), after a historical overview, different QKD types were briefly introduced. Two fundamental theorems on which QKD relies, namely the no-cloning theorem and the theorem of the inability to unambiguously distinguish nonorthogonal quantum states, were described in Section 15.3. In Section 15.4, related to DV-QKD systems, the BB84, B92, Ekert (E91), EPR, and time-phase encoding protocols were described in corresponding Subsections 15.4.1−15.4.4, respectively. Section 15.5 was devoted to QKD security, wherein the SKR was represented as the product of raw key rate and fractional rate. Moreover, the generic expression for the fractional rate was provided, followed by a description of different eavesdropping strategies, including individual attacks (Section 15.5.1), collective attacks (Section 15.5.2), and quantum hacking/side-channel attacks (Section 15.5.3). For individual and coherent attacks, the corresponding secrete fraction expressions were provided, including the secret fraction for the BB84 protocol in Section 15.5.4. After that, in Section 15.6, the decoy-state protocol was described together with the corresponding SKR calculation.

Next, the key concepts for MDI-QKD protocols were introduced in Section 15.7, including photonic Bell-state measurements (Section 15.7.1), polarization-based MDI-QKD protocol (Section 15.7.2), and time-phase encoding-based MDI-QKD protocol (Section 15.7.3), while secrecy fraction calculation was provided in Section 15.7.4. Twin-field QKD protocols were further described in Section 15.8, whose performance was evaluated against decoy-state and MDI-QKD protocols. The focus then moved in Section 15.9 to the classical processing steps, including information reconciliation (Section 15.9.1) and privacy amplification (Section 15.9.2) steps.

To facilitate the description of CV-QKD protocols, the fundamentals of quantum optics and Gaussian information theory were introduced in Section 15.10. Among different topics, quadrature operators and Gaussian and squeezed states were described in Section 15.10.1, Gaussian transformation and the generation of quantum states were introduced in Section 15.10.2, while the thermal decomposition of Gaussian states was introduced in Section 15.10.3. Finally, two-mode Gaussian states and measurements on Gaussian states were discussed in Section 15.10.4. Section 15.11 was devoted to CV-QKD protocols, wherein the homodyne and heterodyne detection schemes were described in Section 15.11.1, following a brief description of squeezed states-based protocols in Section 15.11.2. Given that the coherent states are much easier to generate and manipulate, the coherent states-based protocols were described in detail in Section 15.11.3. For the lossy transmission channel, the corresponding covariance matrices were derived for both homodyne and coherent detection schemes, followed by SKR derivation for a PM GM-based CV-QKD in Section 15.11.4. Some illustrative reconciliation results were provided in Section 15.11.5 related to the GM-based CV-QKD schemes.

15.13 PROBLEMS

1. Derive the SKR expression for the B92 protocol.
2. Derive the SKR expression for the Ekert protocol.
3. Assuming that the detector efficiency is 0.2, the dark count rate is 10^{-6}, the reconciliation inefficiency is 1.2, and that the probability of a photon hitting an erroneous detector is 0.04, plot secrecy fraction versus attenuation for optimized BB84 and decoy-state protocols.
4. Assuming the same system parameters as in Problem 3, plot the secrecy fraction versus attenuation for an MDI-QKD protocol.
5. For a number of phase slices $M = 16$ in a phase-matching TF-QKD protocol, assuming that the misalignment error is 4%, and for the remaining parameters being the same as in Problem 3, plot secrecy fraction versus attenuation for a PM TF-QKD protocol. Discuss the results.
6. For a (3,15) quasi-cyclic LDPC code of girth 10 and codeword length ~100,000, determine the reconciliation efficiency β, given by Eq. (15.59) for a BB84 protocol.
7. We say that the key is ε-secure with respect to an eavesdropper E if the trace distance between the joint state of key K and Eve E, denoted as $\rho_{K,E}$, and the product state of a completely mixed state (CMS) with respect to the set K of all possible secure keys, denoted as ρ_{CMS}, and Eve's state, denoted as ρ_E, is smaller than or equal to ε. In other words, we can write:

$$D(\rho_{K,E}, \rho_{CMS} \otimes \rho_E) = \frac{1}{2}\|\rho_{K,E} - \rho_{CMS} \otimes \rho_E\|_1 = \frac{1}{2}\mathrm{Tr}|\rho_{K,E} - \rho_{CMS} \otimes \rho_E| \le \varepsilon,$$

where $|\rho| = \sqrt{\rho^\dagger \rho}$. With this definition, the parameter ε has quite a clear interpretation; it represents the maximum tolerable failure probability for the key extraction process. Given that the composability of this definition is satisfied we can decompose the security of the final key ε in terms of securities for error correction (EC) ε_{EC}, privacy amplification (PA) ε_{PA}, and parameter estimation (PE) ε_{PE} step, as well as the Rényi entropies estimates failure probability, denoted as ε_R. In other words we can write $\varepsilon = \varepsilon_{EC} + \varepsilon_{PA} + \varepsilon_{PE} + \varepsilon_R$. Prove the following inequality representing the generalization of the Chernoff bound:

$$\Pr(D(\rho_{K,E}, \rho_{CMS} \otimes \rho_E) > \varepsilon) \le e^{m - f(\rho_{K,E}, \varepsilon)},$$

where m is the length of the secret key being extracted wherein the constant factors are omitted.

8. Fidelity-based ε-security is defined by:

$$1 - F\left(\rho_{AB}^m, |B_{00}\rangle^{\otimes m}\right) \le \varepsilon', \; F^2\left(\rho_{AB}^n, |B_{00}\rangle^{\otimes n}\right) = {}^{\otimes n}\langle B_{00}|\rho_{AB}^n|B_{00}\rangle^{\otimes n},$$

$$|B_{00}\rangle = \frac{|00\rangle + |11\rangle}{\sqrt{2}},$$

where ρ_{AB}^n is the shared density state between Alice and Bob. How is this definition of security related to trace-distance-based security? Discuss the relationship.

9. Let us consider the CB given by:

$$CB = \{|0\rangle, |1\rangle, \cdots, |D-1\rangle\},$$

and the corresponding DB given by:

$$DB = \left\{|\hat{0}\rangle, |\hat{1}\rangle, \cdots, |\widehat{D-1}\rangle\right\},$$

$$\left|\hat{i}\right\rangle = \frac{1}{\sqrt{D}}\sum_{d=0}^{D-1} e^{-j\frac{2\pi}{D}di}|i\rangle = \frac{1}{\sqrt{D}}\sum_{d=0}^{D-1} W_D^{-di}|i\rangle,\ W_D = e^{j\frac{2\pi}{D}}.$$

Demonstrate that CB and DB can be used as MUBs.

10. Let us define the cyclic X- and Z-operators as follows:

$$Z|k\rangle = W_D^k|k\rangle,\ X|\hat{i}\rangle = W_D^i|\hat{i}\rangle,\ X^D = Z^D = \mathbf{1}.$$

Prove that these operators satisfy the following commutation rule:

$$X^l Z^k = W_D^{-lk} Z^k X^l.$$

11. For all kets $|\psi\rangle$ and bras $\langle\phi|$, we introduce the conjugate $|\psi^*\rangle$ and bras $\langle\phi^*|$, through the dot product:

$$\langle\phi|\psi\rangle = \langle\psi|\phi\rangle^* = \langle\psi^*|\phi^*\rangle.$$

Unfortunately, this introduction is not a unique one; nevertheless, different versions for mapping $|\psi\rangle \rightarrow |\psi^*\rangle$ are related to each by a unitary transformation. Given that conjugate kets live in the same space as the bras, the following one-to-one correspondence is useful:

$$|\psi\rangle\langle\phi| \leftrightarrow |\phi^*\rangle|\psi\rangle = |\phi^*, \psi\rangle. \qquad (*)$$

As a consequence, the operators $A = |\psi_1\rangle\langle\phi_1|$ and $B = |\psi_2\rangle\langle\phi_2|$ are transformed to:

$$A = |\psi_1\rangle\langle\phi_1| \rightarrow |\phi_1^*\rangle|\psi_1\rangle = |\phi_1^*, \psi_1\rangle = |a\rangle,\ B = |\psi_2\rangle\langle\phi_2| \rightarrow |\phi_2^*, \psi_2\rangle = |b\rangle.$$

Prove that the following is valid:

$$\begin{aligned} A \leftrightarrow |a\rangle &\Rightarrow BA \leftrightarrow (\mathbf{1} \otimes B)|a\rangle, \\ &\Rightarrow AB^\dagger \leftrightarrow (B^* \otimes \mathbf{1})|a\rangle. \end{aligned}$$

12. This problem is a continuation of Problem 11. By using the mapping given by Eq. (*) in Problem 11, the identity operator ***1*** will be mapped to the generalized

Bell state $|B_{00}\rangle$. By applying the mapping rule given by Eq. (*) in Problem 11 on completeness relation, prove that the following mapping is valid:

$$\mathbf{1}=\sum_d |d\rangle\langle d| = \sum_d \left|e_d^{(i)}\right\rangle\left\langle e_d^{(i)}\right| \leftrightarrow \sum_d |d^*\rangle|d\rangle = \sum_d \left|e_d^{(i)*}, e_d^{(i)}\right\rangle = D^{1/2}|B_{00}\rangle,$$

where $D^{-1/2}$ is the normalization factor that normalizes the generalized Bell state $|B_{00}\rangle$ to the unit length. (Namely, $\mathrm{Tr}(\mathbf{1}) = D$, so we need to normalize.)

13. Let the operator V_i^k be defined by:

$$V_i^k = \sum_{d=0}^{D-1} \omega^{(d\oplus i)\odot k}|d\oplus i\rangle\langle d|,$$

where $\omega = \exp(j2\pi/p)$ (p is the prime), while the addition and multiplication operations in the corresponding GF(p^m) are denoted by $\oplus$ and $\odot$, respectively. Prove that Eq. (*) in Problem 11 introduces the following mapping:

$$D^{-1/2}V_m^n = D^{-1/2}\sum_d \omega^{(d\oplus m)\odot n}|d\oplus m\rangle\langle d| \leftrightarrow D^{-1/2}\sum_d \omega^{(d\oplus m)\odot n}|d^*\rangle|d\oplus m\rangle.$$

The state on the right side of the foregoing equivalence is known as the generalized Bell state $|B_{mn}\rangle$. Prove that the generalized Bell states are orthonormal.

14. Let the W_{mn} operator be defined by:

$$W_{mn} = V_m^n = \sum_{d=0}^{D-1} \omega^{d\oplus n}|d\oplus m\rangle\langle d|.$$

The eigenkets of W_{mn} can be used to create the MUBs. For a two-basis protocol we can select the following set $\{W_{01}, W_{10}\}$, corresponding to CB and DB as introduced earlier. On the other hand, for $(D+1)$ basis protocols the following set $\{W_{01}; W_{10}, \ldots, W_{1,D-1}\}$ can be used. Prove that the Alice–Bob density operator ρ_{AB} is diagonal in the generalized Bell basis:

$$\rho_{AB} = \sum_{m,n=0}^{D-1} \lambda_{mn}|B_{mn}\rangle\langle B_{mn}|; \sum_{m,n=0}^{D-1} \lambda_{mn} = 1.$$

15. The parameters to be estimated in entanglement-assisted protocols based on Problem 14 are related to probabilities that Alice's and Bob's outcomes, denoted as a and b, differ by $d \in \{0,1,\ldots,D-1\}$; when both Alice and Bob choose to use randomly the basis of W_{mn}, observed per mod D, denoted as $q_{mn}(d)$, and can be determined as:

$$q_{01}(d) = \sum_{n=0}^{D-1} \lambda_{d,n},\ q_{1n}(d) = \sum_{n=0}^{D-1} \lambda_{m,(mn-d)\mathrm{mod}D}.$$

Eve's accessible information, representing the maximum of mutual information where maximization is performed over all generalized POVM schemes, is upper bounded by the Holevo information:

$$\chi(A:E|\rho_{AB}) = S(\rho_E) - \sum_{a=0}^{D-1} p(a)S\left(\rho_{E|a}\right),\ S(\rho) = -Tr(\rho \log \rho) = -\sum\nolimits_{\lambda_i} \lambda_i \log \lambda_i.$$

Determine the Holevo information when $q_{mn}(d)$ is given by:

$$q_{mn}(d) = \begin{cases} 1-q,\ d=0 \\ q/(D-1),\ d \neq 0 \end{cases}.$$

Determine the SKR of the corresponding D-dimensional entanglement-assisted protocol.

16. Demonstrate that an entanglement-assisted CV-QKD scheme is equivalent to the PM CV-QKD scheme in terms of cross variance matrix.

17. The optimum Gaussian eavesdropping strategy is based on an entangling-cloning machine, which is illustrated in Fig. P1. Determine the Holevo information between Bob and Eve. Determine the corresponding SKR.

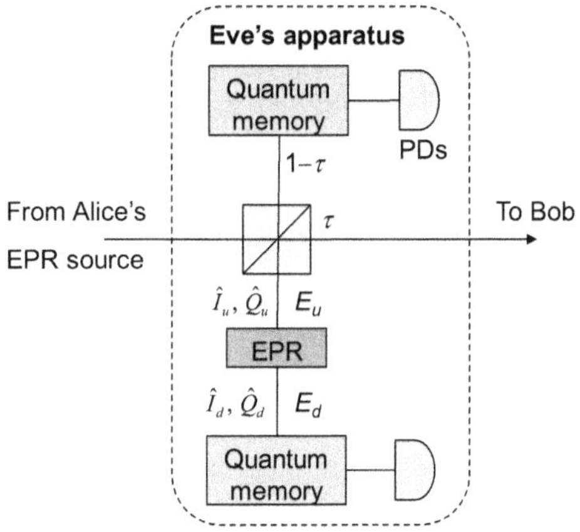

FIGURE P1

Optimum Gaussian eavesdropping strategy based on an entangling–cloning machine. *EPR*, Einstein–Podolski–Rosen; *PDs*, Photodetectors.

18. The entanglement-assisted Gaussian coherent state-based CV-QKD scheme is illustrated in Fig. P2. Determine the SKR for this scheme. Compare it against that of the PM CV-QKD scheme.

19. The four-state discrete modulation protocol can be summarized as follows. Alice sends at random one of four coherent states $|\alpha_k\rangle = |\alpha \exp[j(2k+1)\pi/4]\rangle$ ($k = 0,1,2,3$) to Bob over the quantum channel. The channel is characterized by transmissivity T and excess noise ε so that the total channel added noise, referred at the channel input, can be expressed in SNU as $\chi_{\text{line}} = 1/T - 1 + \varepsilon$. On the receiver side, once the coherent state is received, Bob can perform either homodyne or heterodyne detection, with a detector being characterized by the detector efficiency η and electric noise variance v_{el}. Derive the covariance matrix for this case at Bob's side. Derive the corresponding SKR assuming that the channel is linear.

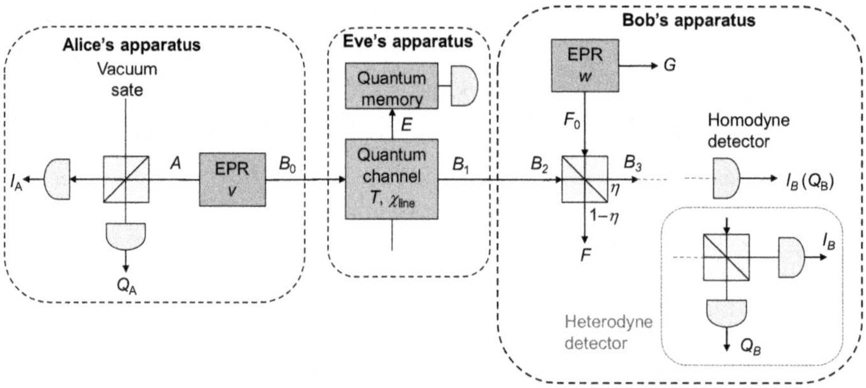

FIGURE P2

Entanglement-assisted Gaussian coherent state-based continuous variable quantum key distribution scheme. *EPR*, Einstein–Podolski–Rosen.

20. For an eight-state protocol, Alice sends at random one of eight coherent states $|\alpha_k\rangle = |\alpha\ \exp(jk\pi/4)\rangle$ $(k = 0,1,\ldots,7)$ to Bob over the quantum channel. Assuming the same channel model as in Problem 19, derive the covariance matrix for this case at Bob's side. Derive the corresponding SKR assuming that the channel is linear. Compare the SKR for the eight-state protocol against that of the four-state protocol from Problem 19.

References

[1] B. Schneier, Applied Cryptography, in: Protocols, Algorithms, and Source Code in C, second ed., John Wiley & Sons, Indianapolis, IN, 2015.

[2] I.B. Djordjevic, Physical-Layer Security and Quantum Key Distribution, Springer Nature Switzerland AG, Cham: Switzerland, 2019.

[3] D. Drajic, P. Ivanis, Introduction to Information Theory and Coding, third ed., Akademska Misao, Belgrade, Serbia, 2009 (in Serbian).

[4] J. Katz, Y. Lindell, Introduction to Modern Cryptography, second ed., CRC Press, Boca Raton, FL, 2015.

[5] W. Diffie, M.E. Hellman, New direction in cryptography, IEEE Trans. Inf. Theor. IT-22 (1976) 644–654.

[6] D. Kahn, The Codebreakers: The Story of Secret Writing, Macmillan Publishing Co., New York, 1967.
[7] M.A. Neilsen, I.L. Chuang, Quantum Computation and Quantum Information, Cambridge University Press, Cambridge, 2000.
[8] G. van Assche, Quantum Cryptography and Secrete-Key Distillation, Cambridge University Press, Cambridge-New York, 2006.
[9] I.B. Djordjevic, Quantum Information Processing and Quantum Error Correction: An Engineering Approach, Elsevier/Academic Press, Amsterdam-Boston, 2012.
[10] C.E. Shannon, Communication theory of secrecy systems, Bell Syst. Tech. J. 28 (1949) 656–715.
[11] T.M. Cover, J.A. Thomas, Elements of Information Theory, John Wiley and Sons, New York, 1991.
[12] J.-P. Aumasson, Serious Cryptography: A Practical Introduction to Modern Encryption, No Starch Press, San Francisco, CA, 2018.
[13] J. Sebbery, J. Pieprzyk, Cryptography: An Introduction to Computer Security, Prentice Hall, New York, 1989.
[14] H. Delfs, H. Knebl, Introduction to Cryptography: Principles and Applications (Information Security and Cryptography), third ed., Springer, Heidelberg •New York, 2015.
[15] C.H. Bennet, G. Brassard, Quantum Cryptography: public key distribution and coin tossing, in: Proc. IEEE International Conference on Computers, Systems, and Signal Processing, Bangalore, India, 1984, pp. 175–179.
[16] C.H. Bennett, Quantum cryptography: uncertainty in the service of privacy, Science 257 (1992) 752–753.
[17] M. Le Bellac, An Introduction to Quantum Information and Quantum Computation, Cambridge University Press, 2006.
[18] P.W. Shor, Polynomial-time algorithms for prime number factorization and discrete logarithms on a quantum computer, SIAM J. Comput. 26 (5) (1997) 1484–1509.
[19] A. Ekert, R. Josza, Quantum computation and Shor's factoring algorithm, Rev. Mod. Phys. 68 (3) (1996) 733–753.
[20] R.L. Rivest, A. Shamir, L. Adleman, A method for obtaining digital signatures and public-key cryptosystems, Commun. ACM 21 (2) (1978) 120–126.
[21] S.M. Barnett, Quantum Information, Oxford University Press, Oxford, 2009.
[22] S.-K. Liao, et al., Satellite-to-ground quantum key distribution, Nature 549 (2017) 43–47.
[23] C.H. Bennett, G. Brassard, C. Crepeau, U. Maurer, Generalized privacy amplification, IEEE Inf. Theor. 41 (6) (1995) 1915–1923.
[24] A. Chorti, et al., Physical layer security: a paradigm shift in data confidentiality, in: Physical and Data-Link Security Techniques for Future Communications Systems, Springer, 2016, pp. 1–15. Ser. Lecture Notes in Electrical Engineering 358.
[25] R. Ursin, et al., Entanglement-based quantum communication over 144 km, Nat. Phys. 3 (7) (2007) 481–486.
[26] A. Boaron, et al., Secure quantum key distribution over 421 km of optical fiber, Phys. Rev. Lett. 121 (2018) 190502.
[27] J. Schwinger, Unitary operator bases, Proc. Natl. Acad. Sci. U.S.A. 46 (1960) 570–579.
[28] W.K. Wootters, B.D. Fields, Optimal state-determination by mutually unbiased measurements, Ann. Phys. 191 (1989) 363–381.

[29] I.D. Ivanovic, Geometrical description of quantal state determination, J. Phys. A. 14 (1981) 3241–3245.

[30] I. Bengtsson, Three ways to look at mutually unbiased bases, AIP Conference Proceedings 889 (1) (2007) 40.

[31] I.B. Djordjevic, FBG-based weak coherent state and entanglement assisted multidimensional QKD, IEEE Photon. J. 10 (4) (2018) 7600512.

[32] C.W. Helstrom, Quantum Detection and Estimation Theory, Academic Press, New York, 1976.

[33] C.W. Helstrom, J.W.S. Liu, J.P. Gordon, Quantum-mechanical communication theory, Proc. IEEE 58 (10) (1970) 1578–1598. Oct. 1970.

[34] C. Vilnrotter, C.-W. Lau, Quantum Detection Theory for the Free-Space Channel, 2001, pp. 1–34. IPN Progress Report 42–146.

[35] I.B. Djordjevic, LDPC-coded optical coherent state quantum communications, IEEE Photon. Technol. Lett. 19 (24) (2007) 2006–2008.

[36] I.B. Djordjevic, LDPC-coded M-ary PSK optical coherent state quantum communication, IEEE/OSA J. Lightwave Technol. 27 (5) (2009) 494–499.

[37] H.-K. Lo, X. Ma, K. Chen, Decoy state quantum key distribution, Phys. Rev. Lett. 94 (2005) 230504.

[38] C.H. Bennett, Quantum cryptography using any two nonorthogonal states, Phys. Rev. Lett. 68 (21) (1992) 3121–3124.

[39] A.K. Ekert, Quantum cryptography based on Bell's theorem, Phys. Rev. Lett. 67 (6) (1991) 661–663.

[40] S.J. Lomonaco, A quick glance at quantum cryptography, Cryptologia 23 (1) (1999) 1–41.

[41] P.D. Townsend, J.G. Rarity, P.R. Tapster, Single photon interference in 10 km long optical fibre interferometer, Electron. Lett. 29 (7) (1993) 634–635.

[42] C. Marand, P.D. Townsend, Quantum key distribution over distances as long as 30 km, Opt. Lett. 20 (16) (1995) 1695–1697.

[43] N.T. Islam, C.C.W. Lim, C. Cahall, J. Kim, D.J. Gauthier, Provably secure and high-rate quantum key distribution with time-bin qudits, Sci. Adv. 3 (11) (2017) e1701491.

[44] F.M. Spedalieri, Quantum key distribution without reference frame alignment: exploiting photon orbital angular momentum, Optic Commun. 260 (1) (2006) 340–346.

[45] I.B. Djordjevic, Multidimensional QKD based on combined orbital and spin angular momenta of photon, IEEE Photon. J. 5 (6) (2013) 7600112.

[46] I.B. Djordjevic, Advanced Optical and Wireless Communications Systems, Springer International Publishing, Switzerland, 2017.

[47] N.T. Islam, High-Rate, High-Dimensional Quantum Key Distribution Systems (Ph.D. Dissertation), Duke University, 2018.

[48] V. Scarani, H. Bechmann-Pasquinucci, N.J. Cerf, M. Dušek, N. Lütkenhaus, M. Peev, The security of practical quantum key distribution, Rev. Mod. Phys. 81 (2009) 1301.

[49] S. Bloom, E. Korevaar, J. Schuster, H. Willebrand, Understanding the performance of free-space optics [Invited], J. Opt. Netw. 2 (6) (2003) 178–200.

[50] I. Csiszár, J. Körner, Broadcast channels with confidential messages, IEEE Trans. Inf. Theor. IT-24 (3) (1978) 339–348.

[51] G. Brassard, N. Lütkenhaus, T. Mor, B.C. Sanders, Limitations on practical quantum cryptography, Phys. Rev. Lett. 85 (2000) 1330.

[52] I. Devetak, A. Winter, Distillation of secret key and entanglement from quantum states, Proc. R. Soc. London Ser. A 461 (2053) (2005) 207−235.

[53] A.S. Holevo, Bounds for the quantity of information transmitted by a quantum communication channel, Probl. Inf. Transm. 9 (3) (1973) 177−183.

[54] M. Lucamarini, I. Choi, M.B. Ward, J.F. Dynes, Z.L. Yuan, A.J. Shields, Practical security bounds against the Trojan-horse attack in quantum key distribution, Phys. Rev. X 5 (2015) 031030.

[55] C. Kurtsiefer, P. Zarda, S. Mayer, H. Weinfurter, The breakdown flash of silicon avalanche photodiodes - back door for eavesdropper attacks? J. Mod. Optic. 48 (13) (2001) 2039−2047.

[56] B. Qi, C.-H.F. Fung, H.-K. Lo, X. Ma, Time-shift attack in practical quantum cryptosystems, Quant. Inf. Comput. 7 (1) (2006) 73−82. Available at: https://arxiv.org/abs/quant-ph/0512080.

[57] L. Lydersen, C. Wiechers, S. Wittmann, D. Elser, J. Skaar, V. Makarov, Hacking commercial quantum cryptography systems by tailored bright illumination, Nat. Photon. 4 (10) (2010) 686.

[58] N. Lütkenhaus, M. Jahma, Quantum key distribution with realistic states: photon-number statistics in the photon-number splitting attack, New J. Phys. 4 (2002) 44.

[59] M. Dušek, M. Jahma, N. Lütkenhaus, Unambiguous state discrimination in quantum cryptography with weak coherent states, Phys. Rev. A 62 (2) (2000) 022306.

[60] X. Ma, C.-H.F. Fung, F. Dupuis, K. Chen, K. Tamaki, H.-K. Lo, Decoy-state quantum key distribution with two-way classical postprocessing, Phys. Rev. A 74 (2006) 032330.

[61] Y. Zhao, B. Qi, X. Ma, H.-K. Lo, L. Qian, Experimental quantum key distribution with decoy states, Phys. Rev. Lett. 96 (2006) 070502.

[62] D. Rosenberg, J.W. Harrington, P.R. Rice, P.A. Hiskett, C.G. Peterson, R.J. Hughes, A.E. Lita, S.W. Nam, J.E. Nordholt, Long-distance decoy-state quantum key distribution in optical fiber, Phys. Rev. Lett. 98 (1) (2007) 010503.

[63] Z.L. Yuan, A.W. Sharpe, A.J. Shields, Unconditionally secure one-way quantum key distribution using decoy pulses, Appl. Phys. Lett. 90 (2007) 011118.

[64] J. Hasegawa, M. Hayashi, T. Hiroshima, A. Tanaka, A. Tomita, Experimental Decoy State Quantum Key Distribution With Unconditional Security Incorporating Finite Statistics, 2007. Available at: arXiv:0705.3081.

[65] Z. Qu, I.B. Djordjevic, RF-assisted coherent detection based continuous variable (CV) QKD with high secure key rates over atmospheric turbulence channels, in: Proc. ICTON 2017, paper Tu.D2.2, Girona, Spain, July 2−6, 2017, 2017 (Invited Paper.).

[66] T. Tsurumaru, A. Soujaeff, S. Takeuchi, Exact minimum and maximum of yield with a finite number of decoy light intensities, Phys. Rev. A 77 (2008) 022319.

[67] M. Hayashi, General theory for decoy-state quantum key distribution with an arbitrary number of intensities, New J. Phys. 9 (2007) 284.

[68] Y. Zhao, B. Qi, X. Ma, H. Lo, L. Qian, Simulation and implementation of decoy state quantum key distribution over 60km telecom fiber, in: Proc. 2006 IEEE International Symposium on Information Theory, Seattle, WA, 2006, pp. 2094−2098.

[69] X.-B. Wang, Three-intensity decoy-state method for device-independent quantum key distribution with basis-dependent errors, Phys. Rev. 87 (1) (2013) 012320.

[70] X. Sun, I.B. Djordjevic, M.A. Neifeld, Secret key rates and optimization of BB84 and decoy state protocols over time-varying free-space optical channels, IEEE Photon. J. 8 (3) (2016) 7904713.

[71] X. Ma, Quantum Cryptography: From Theory to Practice (Ph.D. Dissertation), University of Toronto, 2008.

[72] C.-H.F. Fung, K. Tamaki, H.-K. Lo, Performance of two quantum-key-distribution protocols, Phys. Rev. A 73 (2006) 012337.

[73] M. Ben-Or, M. Horodecki, D.W. Leung, D. Mayers, J. Oppenheim, The universal composable security of quantum key distribution, in: Theory of Cryptography: Second Theory of Cryptography Conference, TCC 2005, Lecture Notes in Computer Science vol. 3378, Springer Verlag, Berlin, 2005, pp. 386–406.

[74] L. Masanes, S. Pironio, A. Acín, Secure device-independent quantum key distribution with causally independent measurement devices, Nat. Commun. 2 (2011) 238.

[75] U. Vazirani, T. Vidick, Fully device-independent quantum key distribution, Phys. Rev. Lett. 113 (14) (2014) 140501.

[76] S. Pironio, A. Acin, N. Brunner, N. Nicolas Gisin, S. Massar, V. Scarani, Device-independent quantum key distribution secure against collective attacks, New J. Phys. 11 (4) (2009) 045021.

[77] C.C.W. Lim, C. Portmann, M. Tomamichel, R. Renner, N. Gisin, Device-independent quantum key distribution with local Bell test, Phys. Rev. X 3 (3) (2013) 031006.

[78] C.H. Bennett, G. Brassard, C. Crépeau, R. Jozsa, A. Peres, W.K. Wootters, Teleporting an unknown quantum state via dual classical and Einstein-Podolsky-Rosen channels, Phys. Rev. Lett. 70 (13) (1993) 1895.

[79] K. Mattle, H. Weinfurter, P.G. Kwiat, A. Zeilinger, Dense coding in experimental quantum communication, Phys. Rev. Lett. 76 (25) (1996) 4656.

[80] M. Źukowski, A. Zeilinger, M.A. Horne, A.K. Ekert, "Event-ready-detectors" Bell experiment via entanglement swapping, Phys. Rev. Lett. 71 (26) (1993) 4287.

[81] N. Sangouard, C. Simon, H. De Riedmatten, N. Gisin, Quantum repeaters based on atomic ensembles and linear optics, Rev. Mod. Phys. 83 (1) (2011) 33.

[82] H.-L. Yin, et al., Measurement-device-independent quantum key distribution over a 404 km optical fiber, Phys. Rev. Lett. 117 (2016) 190501.

[83] H.-K. Lo, M. Curty, B. Qi, Measurement-device-independent quantum key distribution, Phys. Rev. Lett. 108 (13) (2012) 130503.

[84] R. Valivarthi, I. Lucio-Martinez, A. Rubenok, P. Chan, F. Marsili, V.B. Verma, M.D. Shaw, J.A. Stern, J.A. Slater, D. Oblak, S.W. Nam, W. Tittel, Efficient Bell state analyzer for time-bin qubits with fast-recovery WSi superconducting single photon detectors, Optic Express 22 (2014) 24497–24506.

[85] N. Lütkenhaus, J. Calsamiglia, K.-A. Suominen, Bell measurements for teleportation, Phys. Rev. A 59 (1999) 3295–3300.

[86] C.K. Hong, Z.Y. Ou, L. Mandel, Measurement of subpicosecond time intervals between two photons by interference, Phys. Rev. Lett. 59 (1987) 2044.

[87] Z.-Y.J. Ou, Quantum Optics for Experimentalists, World Scientific, New Jersey-London-Singapore, 2017.

[88] F. Xu, M. Curty, B. Qi, H.-K. Lo, Measurement-device-independent quantum cryptography, IEEE J. Sel. Top. Quant. Electron. 21 (3) (2015) 148–158.

[89] P. Chan, J.A. Slater, I. Lucio-Martinez, A. Rubenok, W. Tittel, Modeling a measurement-device-independent quantum key distribution system, Optic Express 22 (11) (2014) 12716–12736.

[90] M. Curty, F. Xu, C.C.W. Lim, K. Tamaki, H.-K. Lo, Finite-key analysis for measurement-device-independent quantum key distribution, Nat. Commun. 5 (2014) 3732.
[91] R. Valivarthi, et al., Measurement-device-independent quantum key distribution: from idea towards application, J. Mod. Optic. 62 (14) (2015) 1141–1150.
[92] I.B. Djordjevic, R. Varrazza, M. Hill, S. Yu, Packet switching performance at 10Gb/s across a 4x4 optical crosspoint switch matrix, IEEE Photon. Technol. Lett. 16 (2004) 102–104.
[93] R. Varrazza, I.B. Djordjevic, S. Yu, Active vertical-coupler based optical crosspoint switch matrix for optical packet-switching applications, IEEE/OSA J. Lightwave Technol. 22 (2004) 2034–2042.
[94] Y.-L. Tang, et al., Measurement-device independent quantum key distribution over untrustful metropolitan network, Phys. Rev. X 6 (1) (2016) 011024.
[95] R. Valivarthi, et al., Quantum teleportation across a metropolitan fibre network, Nat. Photon. 10 (2016) 676–680.
[96] M. Cvijetic, I.B. Djordjevic, Advanced Optical Communications and Networks, Artech House, 2013.
[97] S. Pirandola, R. Laurenza, C. Ottaviani, L. Banchi, Fundamental limits of repeaterless quantum communications, Nat. Commun. 8 (2017) 15043.
[98] L.-M. Duan, M. Lukin, J.I. Cirac, P. Zoller, Long-distance quantum communication with atomic ensembles and linear optics, Nature 414 (2001) 413–418.
[99] J. Qiu, et al., Quantum communications leap out of the lab, Nature 508 (2014) 441–442.
[100] M. Lucamarini, Z.L. Yuan, J.F. Dynes, A.J. Shields, Overcoming the rate–distance limit of quantum key distribution without quantum repeaters, Nature 557 (2018) 400–403.
[101] X. Ma, P. Zeng, H. Zhou, Phase-matching quantum key distribution, Phys. Rev. X 8 (2018) 031043.
[102] K. Tamaki, H.-K. Lo, W. Wang, M. Lucamarini, Information Theoretic Security of Quantum Key Distribution Overcoming the Repeaterless Secret Key Capacity Bound, 2018. Avalable at: arXiv:1805.05511 [quant-ph].
[103] J. Lin, N. Lütkenhaus, Simple security analysis of phase-matching measurement-device-independent quantum key distribution, Phys. Rev. A 98 (2018) 042332.
[104] H.-L. Yin, Y. Fu, Measurement-device-independent twin-field quantum key distribution, Sci. Rep. 9 (2019) 3045.
[105] W.-Y. Hwang, Quantum key distribution with high loss: toward global secure communication, Phys. Rev. Lett. 91 (2003) 057901.
[106] Y. Tamura, et al., The first 0.14-dB/km loss optical fiber and its impact on submarine transmission, J. Lightwave Technol. 36 (2018) 44–49.
[107] D. Slepian, J.K. Wolf, Noiseless coding of correlated information sources, IEEE Trans. Inf. Theor. 19 (4) (1973) 471–480.
[108] A.D. Liveris, Z. Xiong, C.N. Georghiades, Compression of binary sources with side information using low-density parity-check codes, in: Proc. IEEE Global Telecommunications Conference 2002 vol. 2, GLOBECOM '02, Taipei, Taiwan, 2002, pp. 1300–1304.
[109] I.B. Djordjevic, On advanced FEC and coded modulation for ultra-high-speed optical transmission, IEEE Commun. Surv. Tutorials 18 (3) (2016) 1920–1951, https://doi.org/10.1109/COMST.2016.2536726.

[110] M. Bloch, Physical-layer Security (Ph.D. Dissertation), School of Electrical and Computer Engineering, Georgia Institute of Technology, 2008.
[111] M. Bloch, J. Barros, Physical-layer Security: From Information Theory to Security Engineering, Cambridge University Press, Cambridge, 2011.
[112] C. Cachin, U.M. Maurer, Linking information reconciliation and privacy amplification, J. Cryptol. 10 (1997) 97–110.
[113] J.L. Carter, M.N. Wegman, Universal classes of hash functions, J. Comput. Syst. Sci. 18 (2) (1979) 143–154.
[114] M.N. Wegman, J. Carter, New hash functions and their use in authentication and set equality, J. Comput. Sci. Syst. 22 (1981) 265–279.
[115] A. Rényi, On measures of entropy and information, in: Proc. 4th Berkeley Symp. On Mathematical Statistics and Probability, vol. 1, Univ. Claifornia Press, 1961, pp. 547–561.
[116] I. Ilic, I.B. Djordjevic, M. Stankovic, On a general definition of conditional Rényi entropies, in: Proc. 4th International Electronic Conference on Entropy and its Applications, 21 November–1 December 2017; Sciforum Electronic Conference Series, vol. 4, 2017, https://doi.org/10.3390/ecea-4-05030.
[117] A. Teixeira, A. Matos, L. Antunes, Conditional Rényi entropies, IEEE Trans. Inf. Theor. 58 (7) (2012) 4273–4277.
[118] C. Cachin, Entropy Measures and Unconditional Security in Cryptography (Ph.D. Dissertation), ETH Zurich, Hartung-Gorre Verlag, Konstanz, 1997.
[119] U. Maurer, S. Wolf, Information-theoretic key agreement: from weak to strong secrecy for free, in: B. Preneel (Ed.), Advances in Cryptology — EUROCRYPT 2000, Lecture Notes in Computer Science, vol. 1807, Springer, Berlin, Heidelberg, 2000.
[120] L. Mandel, E. Wolf, Optical Coherence and Quantum Optics, Cambridge University Press, Cambridge-New York-Melbourne, 1995.
[121] M.O. Scully, M.S. Zubairy, Quantum Optics, Cambridge University Press, Cambridge-New York-Melbourne, 1997.
[122] C. Weedbrook, S. Pirandola, R. García-Patrón, N.J. Cerf, T.C. Ralph, J.H. Shapiro, L. Seth Lloyd, Gaussian quantum information, Rev. Mod. Phys. 84 (2012) 621.
[123] N.N. Bogoliubov, On the theory of superfluidity, J. Phys. 11 (1947) 77 (USSR) 11: 23, (Izv. Akad. Nauk Ser. Fiz.).
[124] N.N. Bogoljubov, On a new method in the theory of superconductivity, Il Nuovo Cimento 7 (6) (1958) 794–805.
[125] A.S. Holevo, M. Sohma, O. Hirota, Capacity of quantum Gaussian channels, Phys. Rev. A 59 (3) (1999) 1820–1828.
[126] J. Williamson, On the algebraic problem concerning the normal forms of linear dynamical systems, Am. J. Math. 58 (1) (1936) 141–163.
[127] R. Simon, Peres-Horodecki separability criterion for continuous variable systems, Phys. Rev. Lett. 84 (2000) 2726.
[128] L.-M. Duan, G. Giedke, J.I. Cirac, P. Zoller, Inseparability criterion for continuous variable systems, Phys. Rev. Lett. 84 (2000) 2722.
[129] A. Serafini, F. Illuminati, S. De Siena, Symplectic invariants, entropic measures and correlations of Gaussian states, J. Phys. B 37 (2004) L21–L28.
[130] A. Serafini, Multimode uncertainty relations and separability of continuous variable states, Phys. Rev. Lett. 96 (2006) 110402.

[131] S. Pirandola, A. Serafini, S. Lloyd, Correlation matrices of two-mode bosonic systems, Phys. Rev. A 79 (2009) 052327.

[132] I.B. Djordjevic, Quantum Biological Information Theory, Springer International Publishing, Switzerland, Cham-Heidelberg-New York, 2016.

[133] D. McMahon, Quantum Computing Explained, John Wiley & Sons, Hoboken, NJ, 2008.

[134] G. Giedke, J.L. Cirac, Characterization of Gaussian operations and distillation of Gaussian states, Phys. Rev. A 66 (2002) 032316.

[135] T.C. Ralph, Continuous variable quantum cryptography, Phys. Rev. A 61 (1999) 010303.

[136] F. Grosshans, P. Grangier, Continuous variable quantum cryptography using coherent states, Phys. Rev. Lett. 88 (2002) 057902.

[137] F. Grosshans, Collective attacks and unconditional security in continuous variable quantum key distribution, Phys. Rev. Lett. 94 (2005) 020504.

[138] A. Leverrier, P. Grangier, Unconditional security proof of long-distance continuous-variable quantum key distribution with discrete modulation, Phys. Rev. Lett. 102 (2009) 180504.

[139] A. Becir, F.A.A. El-Orany, M.R.B. Wahiddin, Continuous-variable quantum key distribution protocols with eight-state discrete modulation, Int. J. Quant. Inf. 10 (2012) 1250004.

[140] Q. Xuan, Z. Zhang, P.L. Voss, A 24 km fiber-based discretely signaled continuous variable quantum key distribution system, Optic Express 17 (26) (2009) 24244–24249.

[141] D. Huang, P. Huang, D. Lin, G. Zeng, Long-distance continuous-variable quantum key distribution by controlling excess noise, Sci. Rep. 6 (2016) 19201.

[142] Z. Qu, I.B. Djordjevic, High-speed free-space optical continuous-variable quantum key distribution enabled by three-dimensional multiplexing, Optic Express 25 (7) (2017) 7919–7928.

[143] C. Silberhorn, T.C. Ralph, N. Lütkenhaus, G. Leuchs, Continuous variable quantum cryptography: beating the 3 dB loss limit, Phys. Rev. A 89 (2002) 167901.

[144] K.A. Patel, J.F. Dynes, I. Choi, A.W. Sharpe, A.R. Dixon, Z.L. Yuan, R.V. Penty, J. Shields, Coexistence of high-bit-rate quantum key distribution and data on optical fiber, Phys. Rev. X 2 (2012) 041010.

[145] Z. Qu, I.B. Djordjevic, Four-dimensionally multiplexed eight-state continuous-variable quantum key distribution over turbulent channels, IEEE Photon. J. 9 (6) (2017) 7600408.

[146] Z. Qu, I.B. Djordjevic, Approaching Gb/s secret key rates in a free-space optical CV-QKD system Affected by atmospheric turbulence, in: Proc. ECOC 2017, P2.SC6.32, Gothenburg, Sweden, 2017.

[147] Z. Qu, I.B. Djordjevic, High-speed free-space optical continuous variable-quantum key distribution based on Kramers-Kronig scheme, IEEE Photon. J. 10 (6) (2018) 7600807.

[148] M. Heid, N. Lütkenhaus, Efficiency of coherent-state quantum cryptography in the presence of loss: influence of realistic error correction, Phys. Rev. A 73 (2006) 052316.

[149] F. Grosshans, N.J. Cerf, Continuous-variable quantum cryptography is secure against non-Gaussian attacks, Phys. Rev. Lett. 92 (2004) 047905.

[150] R. García-Patrón, N.J. Cerf, Unconditional optimality of Gaussian attacks against continuous-variable quantum key distribution, Phys. Rev. Lett. 97 (2006) 190503.

[151] M. Navascués, F. Grosshans, A. Acín, Optimality of Gaussian attacks in continuous-variable quantum cryptography, Phys. Rev. Lett. 97 (2006) 190502.

[152] F. Grosshans, G. Van Assche, J. Wenger, R. Tualle-Brouri, N.J. Cerf, P. Grangier, Quantum key distribution using Gaussian-modulated coherent states, Nature 421 (2003) 238.

[153] F. Grosshans, Communication et Cryptographie Quantiques avec des Variables Continues (Ph.D. Dissertation), Université Paris XI, 2002.

[154] J. Wegner, Dispositifs Imulsionnels pour la Communication Quantique à Variables Contines (Ph.D. Dissertation), Université Paris XI, 2004.

[155] Z. Qu, Secure High-Speed Optical Communication Systems (Ph.D. Dissertation), University of Arizona, 2018.

[156] I.B. Djordjevic, Hybrid QKD protocol outperforming both DV- and CV-QKD protocols, IEEE Photon. J. 12 (1) (2019) 7600108.

[157] I.B. Djordjevic, On the photon subtraction-based measurement-device-independent CV-QKD protocols, IEEE Access 7 (2019) 147399–147405.

[158] T.-L. Wang, I.B. Djordjevic, J. Nagel, Laser beam propagation effects on secure key rates for satellite-to-ground discrete modulation CV-QKD, Appl. Optic. 58 (29) (2019) 8061–8068.

[159] I.B. Djordjevic, Proposal for slepian-states-based DV- and CV-QKD schemes suitable for implementation in integrated photonics platforms, IEEE Photon. J. 11 (4) (2019) 7600312.

[160] I.B. Djordjevic, Optimized-eight-state CV-QKD protocol outperforming Gaussian modulation based protocols, IEEE Photon. J. 11 (4) (2019) 4500610.

[161] I.B. Djordjevic, On the discretized Gaussian modulation (DGM)-based continuous variable-QKD, IEEE Access 7 (2019) 65342–65346.

[162] F. Grosshans, P. Grangier, Reverse Reconciliation Protocols for Quantum Cryptography With Continuous Variables, 2002. Available at: arXiv:quant-ph/0204127.

[163] H.P. Yuen, J.H. Shapiro, Optical communication with two-photon coherent states–part III: quantum measurements realizable with photoemissive detectors, IEEE Trans. Inf. Theor. 26 (1) (1980) 78–92.

[164] S. Fossier, E. Diamanti, T. Debuisschert, R. Tualle-Brouri, P. Grangier, Improvement of continuous-variable quantum key distribution systems by using optical preamplifiers, J. Phys. B 42 (2009) 114014.

[165] R. Garcia-Patron, Quantum Information with Optical Continuous Variables: From Bell Tests to Key Distribution (Ph.D. thesis), Université Libre de Bruxelles, 2007.

Abstract Algebra Fundamentals

This appendix is devoted to abstract algebra fundamentals [1–12]. Only the most important topics needed for better understanding of quantum information processing and quantum error correction are covered. The appendix is organized as follows. In Section A.1 we introduce the concept of groups and provide their basic properties. Group action on the set is discussed in Section A.2. Group mapping, in particular homomorphism, is described in Section A.3. In Section A.4 we introduce the concept of fields, while in Section A.5 we introduce the concept of vector spaces. Furthermore, in Section A.6 we provide basic facts on character theory, which is important in nonbinary quantum error correction. Section A.7 is devoted to algebra of finite fields, which has high importance in both classical and quantum error correction. In Section A.8 we introduce the concept of metric spaces, while in Section A.9 the concept of Hilbert space is introduced.

A.1 GROUPS

Definition D1: A group is the set G that together with an operation, denoted by "+", satisfies the following axioms:

1. Closure: $\forall\ a, b \in G \Rightarrow a + b \in G$
2. Associative law: $\forall\ a, b, c \in \mathrm{G} \Rightarrow a + (b + c) = (a + b) + c$
3. Identity element: $\exists\ e \in G$ such that $a + e = e + a = a\ \forall\ a \in G$
4. Inverse element: $\exists\ a^{-1} \in G\ \forall\ a \in G$, such that $a + a^{-1} = a^{-1} + a = e$

We call a group *Abelian* or *commutative* if the operation "+" is also commutative: $\forall\ a, b \in G \Rightarrow a + b = b + a$.

Theorem T1: The identity element in a group is a unique one, and each element in the group has a unique inverse element.

Proof. If the identity element e is not a unique one, then there exists another one e': $e' = e' + e = e$, implying that e' and e are identical. To show that a group element has a unique inverse a^{-1}, let us assume that it has another inverse element a_1^{-1}: $a_1^{-1} = a_1^{-1} + e = a_1^{-1} + (a + a^{-1}) = (a_1^{-1} + a) + a^{-1} = e + a^{-1} = a^{-1}$, thus implying that the inverse element in the group is unique.

Examples:

- $F_2 = \{0,1\}$ and "+" operation is in fact the modulo-2 addition, defined by: $0+0=0$, $0+1=1$, $1+0=1$, $1+1=0$. The closure property is satisfied, 0 is the identity element, 0 and 1 are their own inverses, and operation "+" is associative. The set F_2 with modulo-2 addition forms therefore a group.
- The set of integers forms a group under the usual addition operation on which 0 is the identity element, and for any integer n, $-n$ is its inverse.
- Consider the set of codewords of a *binary linear (N,K) block code*. Any two codewords added per modulo-2 form another codeword (according to the definition of the linear block code). All-zeros codeword is an identity element, and each codeword is its own inverse. The associative property also holds. Therefore the all-group properties are satisfied, and the set of codewords forms a group, and this is the reason why this it is called the *group code*.

The number of elements in the group is typically called the *order* of the group. If the group has a finite number of elements, it is called the *finite* group.

Definition D2: Let H be a subset of elements of group G. We call H subgroup of G if the elements of H themselves form the group under the same operation as that defined on elements from G.

To verify that the subset S is the subgroup, it is sufficient to check the closure property and that an inverse element exists for every element of the subgroup.

Theorem T2 (Lagrange theorem): The order of a finite group is an integer multiple of any of its subgroup order.

Example: Consider the set $Y$of N-tuples (received words on a receiver side of a communication system), with each element of the N-tuple being 0 or 1. It can easily be verified that elements of Y form a group under modulo-2 addition. Consider a subset C of Y with elements being the codewords of a binary (N,K) block code. Then, C forms a subgroup of Y, and the order of group Y is divisible by the order of subgroup C.

There exist groups, whose elements can be obtained as a power of some element, say a. Such a group G is called the *cyclic group*, and the corresponding element a is called the *generator* of the group. The cyclic group can be denoted by $G = <a>$.

Example: Consider an element α of a finite group G. Let S be the set of elements $S = \{\alpha,\alpha^2,\alpha^3,\ldots,\alpha^i,\ldots\}$. Because G is finite, S must be finite as well, and therefore not all powers of α are distinct. There must be some l,m $(m > l)$ such that $\alpha^m = \alpha^l$ so that $\alpha^m\,\alpha^{-1} = \alpha^1\,\alpha^{-1} = 1$. Let k be the smallest such power of α for which $\alpha^k = 1$, meaning that $\alpha,\alpha^2,\ldots,\alpha^k$ are all distinct. We can verify now that set $S = \{\alpha,\alpha^2,\ldots,\alpha^k = 1\}$ is a subgroup of G. S contains the identity element, and for any element α^i, the element α^{k-i} is its inverse. Given that any two elements α^i, $\alpha^j \in S$, the corresponding element obtained as their product, $\alpha^{i+j} \in S$ if $i+j \leq k$. If $i+j > k$, then $\alpha^{i+j}\cdot 1 = \alpha^{i+j}\cdot\alpha^{-k}$, and because $i+j-k \leq k$; the closure property is clearly satisfied. S is therefore the subgroup of group G. Given the definition of cyclic group, the subgroup S is also cyclic. The set of codewords of a cyclic (N,K) block code can be obtained as a cyclic subgroup of the group of all N-tuples.

Theorem T3: Let G be a group and $\{H_i \mid i \in I\}$ be a nonempty collection of subgroups with index set I. The intersection $\bigcap_{i \in I} H_i$ is a subgroup.

Definition D3: Let G be a group and X be a subset of G. Let $\{H_i \mid i \in I\}$ be the collection of subgroups of G that contain X. Then, the intersection $\bigcap_{i \in I} H_i$ is called the subgroup of G generated by X and is denoted by $\langle X \rangle$.

Theorem T4: Let G be a group and X a nonempty subset of G with elements $\{x_i | i \in I\}$ (I is the index set). The subgroup of G generated by X, $\langle X \rangle$, consists of all finite products of the x_i. The x_i elements are known as generators.

Example: A *quantum stabilizer code* with stabilizer group S and generators $g_1, g_2, \ldots, g_{N-K}$ is a good illustration of this theorem. Since S is an Abelian group, and the generators have order 2 (meaning that $g^2_i = I$, where I is the identity operator) any element $s \in S$ can be represented by $s = g_1^{c_1} \ldots g_{N-K}^{c_{N-K}}$, $c_i \in \{0, 1\}$. For $c_i = 1$ ($c_i = 0$), the corresponding generator is included (excluded).

Definition D4: Let G be a group and H be a subgroup of G. For any element $a \in G$, the set $aH = \{ah \mid h \in H\}$ is called the left coset of H in G. Similarly, the set $Ha = \{ha \mid h \in H\}$ is called the right coset of H in G.

Theorem T5: Let G be a group and H be a subgroup of G. The collection of right cosets of H, $Ha = \{ha \mid h \in H\}$ forms a partition of G.

Instead of formal proof we provide the following justification. Let us create the following table. In the first row we list all elements of subgroup H, beginning with the identity element e. The second row is obtained by selecting an arbitrary element from G, not used in the first row, as the leading element of the second row. We then complete the second row by "multiplying" this element with all elements of the first row from the right. We then arbitrarily select an element from G, not previously used, as the leading element of the third row. We then complete the third row by multiplying this element with all elements of the first row from the right. We continue this procedure until we exploit all elements from G. The resulting table is as follows.

$h_1 = e$	h_2	...	h_{m-1}	h_m
g_2	h_2g_2	...	$h_{m-1}g_2$	$h_m\, g_2$
g_3	h_2g_3	...	$h_{m-1}g_3$	$h_m\, g_3$
...	...	...	...	...
g_n	h_2g_n	...	$h_{m-1}g_n$	$h_m\, g_n$

Each row in this table represents a *coset*, and the first element in each row is a coset leader. The number of cosets of H in G is in fact the number of rows, and it is called the *index* of H in G, typically denoted by $[G{:}H]$. It follows from this table that $|H| = m$, $[G{:}H] = n$ and $|G| = nm = [G{:}H]|H|$, which can be used as the proof of Lagrange's theorem. In other words, $[G{:}H] = |G|/|H|$.

Definition D5: Let G be a group and H be a subgroup of G. H is a normal subgroup of G if it is invariant under conjugation, that is, $\forall\ h \in H$ and $g \in G$, the element $ghg^{-1} \in H$.

In other words, H is fixed under conjugation by the elements from G, namely $gHg^{-1} = H$ for any $g \in G$. Therefore the left and right cosets of H in G coincide: $\forall\ g \in G$, $gH = Hg$.

Theorem T6: Let G be a group and H a normal subgroup of G. If $G \mid H$ denotes the set of cosets of H in G, then the set $G \mid H$ with coset multiplication forms a group, known as the quotient group of G by H.

The coset multiplication of aH and bH is defined as: $aH * bH = abH$. It follows from Lagrange's theorem that: $|G/H| = [G{:}H]$. This theorem can simply be proved by using the foregoing table.

A.2 GROUP ACTING ON THE SET

We first study the action of a group G on an arbitrary set S, then consider the action of G on itself. A group G is said to act on a set S if:

1. Each element $g \in G$ implements a map $s \rightarrow g(s)$, where $s, g(s) \in S$;
2. The identity element e in G produces the identity map: $e(s) = s$;
3. The map produced by g_1g_2 is the composition of the maps produced by g_1 and g_2, namely $g_1g_2(s) = g_1(g_2(s))$.

The set of elements from S that are images of s under the action of G is called the *orbit* of $s \in S$, and often denoted as orb(s). The orbit is formally defined by orb(s) $= \{g(s) \mid g \in G\}$. Another important concept is that of the stabilizer. The *stabilizer* of $s \in S$, denoted as S_s, is the set of elements from G that fixes s: $S_s = \{g \in G \mid g(s) = s\}$. It can be shown that (1) each orbit defines an equivalence class on S so that the collection of orbits partitions S, and (2) the stabilizer S_s is a subgroup of G.

Let us consider the Pauli group G_N on N-qubits (consisting of all Pauli matrices together with factors ± 1, $\pm j$) and the set S with the subspace spanned by the basis codewords. Each basis codeword $|i\rangle$ defines its own stabilizer S_i, and the code stabilizer is obtained as an intersection of all S_i.

If $S = G$, in quantum mechanical applications, the action is usually a conjugation action, namely $g \rightarrow xgx^{-1}$. The corresponding orbit of g is called the *conjugacy class* of g, and the corresponding stabilizer is called the *centralizer* of g in G, denoted as $C_G(g)$. The centralizer of g in G is formally defined by $C_G(g) = \{x \in G \mid xg = gx\} = \{x \in G \mid xgx^{-1} = g\}$. Therefore the centralizer of g in G, $C_G(g)$, contains all elements in G that commute with g.

If $S \subseteq G$, the centralizer of S in G, denoted as $C_G(S)$, is defined by: $C_G(S) = \{x \in G \mid xS = Sx\}$.

The center of G, denoted as $Z(G)$, is the set of all elements from G that commutes with all elements from G, that is, $Z(G) = \{x \in G \mid xg = gx, \forall\ g \in G\}$.

The normalizer of S in G, denoted as $N_G(S)$, is defined by: $N_G(S) = \{x \in G \mid xSx^{-1} = S\}$. Let H be a subgroup of G. The *self-normalizing* subgroup of G is one that satisfies: $N_G(H) = H$.

A.3 GROUP MAPPING

Suppose A and B are two sets, and f is a function or mapping (map) from $A \rightarrow B$ that assigns to each element from A a corresponding element from B. The set A is called the domain, and the set of images B is called the range. If each element of B is the image of no more than one element from A, then f is said to be injective or one-to-one. In other words, the mapping is injective if distinct elements from A have distinct images, that is, the function f: $A \rightarrow B$ is *injective* if and only if $f(x_1) = f(x_2)$ implies $x_1 = x_2$. If each element of B is the image of at least one element from A, then f is said to be *surjective* or *onto*. In other words, the range is the whole set B. If f is both injective and surjective, the f is said to be *bijective* (one-to-one correspondence). The function f: $A \rightarrow B$ has an inverse only if it is bijective. The inverse function f^{-1} is a bijective function from B to A. The bijective mapping from A to A is typically called the *permutation*. After this short reminder, in the rest of this section we study the mapping from group G to group H.

Homomorphism

Definition D6: Let G and H be groups and f be a function that maps $G \rightarrow H$. The function f is a homomorphism from G to H if $\forall\ a,b \in G$: $f(ab) = f(a)f(b)$.

If f maps G onto H, H is said to be a homomorphic image of G. In other words, f is said to be a homomorphism if the image of a product ab is the product of images of a and b. If f is a homomorphism and it is bijective, then f is said to be an isomorphism from G to H, and is commonly denoted by $G \cong H$. Finally, if $G = H$ we call the isomorphism an automorphism.

Let f be a homomorphism from $G \rightarrow H$. The kernel of f, denoted Ker(f), is the set of elements from G that are mapped to the identity element e_H of H:

$$\mathrm{Ker}(f) = \{x \in G | f(x) = e_H\}.$$

The image of f, denoted Im(f), is the set of elements in H that are the image of some $a \in G$:

$$\mathrm{Im}(f) = \{b \in H | b = f(a) \text{for some } a \in G\}.$$

Theorem T7: Let f: $G \rightarrow H$ be a homomorphism with kernel K, then $f(a) = f(b)$ if and only if $Ka = Kb$.

Instead of formal proof of this theorem we provide the following justification. Clearly, $f(a) = f(b)$ iff $f(a)[f(b)]^{-1} = e$, which is equivalent to $f(ab^{-1}) = e$. Based on the definition of the kernel we conclude that $ab^{-1} \in K$, which is equivalent to $Ka = Kb$. The foregoing theorem can be reformulated as follows: any two elements a and b from G have the same image if and only if they belong to the same kernel. This theorem will be used to prove the following important theorem.

Theorem T8 (fundamental homomorphism theorem): Let G and H be groups. If f: $G \rightarrow H$ is a homomorphism, then f induces an isomorphism between H and the quotient group $G/\mathrm{Ker}(f)$:

$$H \cong G/\mathrm{Ker}(f).$$

Proof: Let the function that matches $\mathrm{Ker}(f)x$ with $f(x)$ be called ϕ, namely we can write $\phi(\mathrm{Ker}(f)x) = f(x)$. This function must ensure that if $\mathrm{Ker}(f)a$ is the same coset as $\mathrm{Ker}(f)b$, then $\phi(\mathrm{Ker}(f)a) = \phi(\mathrm{Ker}(f)b)$, which is correct because of T7. If $\phi(\mathrm{Ker}(f)a) = \phi(\mathrm{Ker}(f)b)$, then $f(a) = f(b)$, and based on T7 we conclude that $\mathrm{Ker}(f)a = \mathrm{Ker}(f)b$, which indicates that the function ϕ from $G/\mathrm{Ker}(f)$ to H is injective. Because every element from H has the form $f(x) = \phi(\mathrm{Ker}(f)x)$, the function ϕ is clearly surjective. Finally, from the coset multiplication definition we obtain: $\phi(\mathrm{Ker}(f)a*\mathrm{Ker}(f)b) = \phi(\mathrm{Ker}(f)ab) = f(ab) = f(a)f(b) = \phi(\mathrm{Ker}(f)a)$ $\phi(\mathrm{Ker}(f)b)$, which proves that function ϕ is the isomorphism from $G/\mathrm{Ker}(f)$ to H.

A.4 FIELDS

Definition D7: A *field* is a set of elements F with two operations, addition "+" and multiplication "·", such that:

1. F is an Abelian group under addition operation, with 0 being the identity element.
2. The nonzero elements of F form an Abelian group under the multiplication operation, with 1 being the identity element.
3. The *multiplication operation is distributive over the addition operation*:

$$\forall a, b, c \in F \Rightarrow a \cdot (b + c) = a \cdot b + a \cdot c$$

Examples:

- The set of real numbers, with operation + as ordinary addition, and operation · as ordinary multiplication, satisfy the foregoing three properties and it is therefore a field.
- The set consisting of two elements {0,1}, with modulo-2 multiplication and addition, given in the following table, constitutes a field known as the *Galois field*, and is denoted by GF(2).

+	0	1
0	0	1
1	1	0

·	0	1
0	0	0
1	0	1

- The set of integers modulo-p, with modulo-p addition and multiplication, forms a field with p elements, denoted by GF(p), providing that p is a prime.
- For any q that is an integer power of prime number p ($q = p^m$, m is an integer), there exists a field with q elements, denoted as GF(q). (The arithmetic is not modulo-q arithmetic, except when $m = 1$.) GF(p^m) contains GF(p) as a subfield.

Addition and multiplication in GF(3) is defined as follows:

+	0	1	2
0	0	1	2
1	1	2	0
2	2	0	1

·	0	1	2
0	0	0	0
1	0	1	2
2	0	2	1

Addition and multiplication in GF(2^2) is defined as follows:

+	0	1	2	3
0	0	1	2	3
1	1	0	3	2
2	2	3	0	1
3	3	2	1	0

·	0	1	2	3
0	0	0	0	0
1	0	1	2	3
2	0	2	3	1
3	0	3	1	2

Theorem T9: Let Z_p denote the set of integers $\{0,1,\ldots,p-1\}$, with addition and multiplication defined as ordinary addition and multiplication modulo-p. Then, Z_p is a field if and only if p is a prime.

A.5 VECTOR SPACES

Definition D8: Let V be a set of elements with a binary operation "+" and let F be a field. Furthermore, let an operation "·" be defined between the elements of V and the elements of F. Then, V is said to be a *vector space* over F if the following conditions are satisfied: $\forall\ a, b \in F$ and $\forall\ \boldsymbol{x}, \boldsymbol{y} \in V$:

1. V is an Abelian group under the addition operation.
2. $\forall\ a \in F$ and $\forall\ \boldsymbol{x} \in V$, then $a \cdot \boldsymbol{x} \in V$.

3. Distributive law:

$$a\cdot(\boldsymbol{x}+\boldsymbol{y}) = a\cdot\boldsymbol{x} + a\cdot\boldsymbol{y}$$
$$(a+b)\cdot\boldsymbol{x} = a\cdot\boldsymbol{x} + b\cdot\boldsymbol{x}.$$

4. Associative law:

$$(a\cdot b)\cdot\boldsymbol{x} = a\cdot(b\cdot\boldsymbol{x}).$$

If 1 denotes the identity element of F, then $1\cdot\boldsymbol{x}=\boldsymbol{x}$. Let 0 denote the zero element (identity element under $+$) in F, and $\mathbf{0}$ the additive element in V, -1 the additive inverse of 1, and the multiplicative identity in F. It can be easily shown that the following two properties hold:

$$0\cdot\boldsymbol{x} = \mathbf{0}$$
$$\boldsymbol{x} + (-1)\cdot\boldsymbol{x} = \mathbf{0}.$$

Examples:

- Consider the set V, whose elements are n-tuples of the form $\boldsymbol{v}=(v_0,v_1,\ldots,v_{n-1})$, $v_i \in F$. Let us define the addition of any two n-tuples as another n-tuple obtained by component-wise addition and multiplication of an n-tuple by an element from F. Then, V forms vector space over F, and it is commonly denoted by F^n. If $F=R$ (R is the field of real numbers), then R^n is called the Euclidean n-dimensional space.
- The set of n-tuples whose elements are from GF(2), again with component-wise addition and multiplication by an element from GF(2), forms a vector space over GF(2).
- Consider the set V of polynomials whose coefficients are from GF(q). Addition of two polynomials is the usual polynomial addition being performed in GF(q). Let the field F be GF(q). Scalar multiplication of a polynomial by a field element from GF(q) corresponds to the multiplication of each polynomial coefficient by the field element, carried out in GF(q). V is then a vector space over GF(q).

Consider a set V that forms a vector space over a field F. Let $\boldsymbol{v}_1,\boldsymbol{v}_2,\ldots,\boldsymbol{v}_k$ be vectors from V, and $a_1,a_2,\ldots,a_k$ be field elements from F. The *linear combination* of the vectors $v_1,v_2,\ldots,v_k$ is defined by:

$$a_1\boldsymbol{v}_1 + a_2\boldsymbol{v}_2 + \cdots + a_k\boldsymbol{v}_k.$$

The set of vectors $\{v_1,v_2,\ldots,v_k\}$ is said to be *linearly independent* if there does not exist a set of field elements $a_1,a_2,\ldots,a_k$, not all $a_i=0$, such that:

$$a_1\boldsymbol{v}_1 + a_2\boldsymbol{v}_2 + \cdots + a_k\boldsymbol{v}_k = \mathbf{0}.$$

Example: The vectors (0 0 1), (0 1 0), and (1 0 0) (from F^3) are linearly independent. However, the vectors (0 0 2), (1 1 0), and (2 2 1) are linearly dependent over GF(3) because they sum to zero vector.

Let V be a vector space and S be subset of the vectors in V. If S is itself a vector space over F under the same vector addition and scalar multiplication operations applicable to V and F, then S is said to *a subspace* of V.

Theorem T10: Let $\{v_1, v_2, \ldots, v_k\}$ be a set of vectors from a vector space V over a field F. Then the set consisting of all linear combinations of $\{v_1, v_2, \ldots, v_k\}$ forms a vector space over F, and is therefore a subspace of V.

Example: Consider the vector space V over GF(2) given by the set {(0 0 0), (0 0 1), (0 1 0), (0 1 1), (1 0 0), (1 0 1), (1 1 0), (1 1 1)}. The subset S = {(0 0 0), (1 0 0), (0 1 0), (1 1 0)} is a subspace of V over GF(2).

Example: Consider the vector space V over GF(2) given by the set {(0 0 0), (0 0 1), (0 1 0), (0 1 1), (1 0 0), (1 0 1), (1 1 0), (1 1 1)}. For the subset B = {(0 1 0), (1 0 0)}, the set of all linear combinations is given by S = {(0 0 0), (1 0 0), (0 1 0), (1 1 0)}, and forms a subspace of V over F. The set of vectors B is said to *span* S.

Definition D9: A *basis* of a vector space V is a set of linearly independent vectors that spans the space. The number of vectors in a basis is called the *dimension* of the vector space.

In the foregoing example, the set {(0 0 1), (0 1 0), (1 0 0)} is the basis of vector space V, and has dimension 3.

A.6 CHARACTER THEORY

Character theory is often used to facilitate the explanation of nonbinary quantum codes, Clifford's codes, and subsystems codes.

The general linear group of vector space V (over a field F), denoted as GL(V), is the group of all automorphisms of V. In other words, GL(V) is the set of all bijective linear transformations $V \rightarrow V$ together with functional composition as the group operation.

A representation of a group G on a vector space V (over a field F) is a group homomorphism from G to GL(V). In other words, the representation is a mapping:

$$\rho: G \rightarrow GL(V) \text{ such that } \rho(g_1 g_2) = \rho(g_1)\rho(g_2)\, \forall g_1,\, g_2 \in G.$$

The vector space V is called the representation space and the dimension of V is called the dimension of the representation.

Definition D10: Let V be a finite-dimensional vector space over a field F and let ρ: $G \rightarrow$ GL(V) be a representation of a group G on V. The character of ρ is the function χ_ρ: $G \rightarrow F$ given by $\chi_\rho(g) = \mathrm{Tr}(\rho(g))$.

A character χ_ρ is called irreducible if ρ is an irreducible representation. The character χ_ρ is called linear if dim(V) = 1.

The kernel of χ_ρ is defined by: Ker $\chi_\rho = \{g \in G \mid \chi_\rho(g) = \chi_\rho(1)\}$, where $\chi_\rho(1)$ is the value of $\chi_\rho(1)$ on group identity. If ρ is the representation of G of dimension k and 1 is the identity of G, then:

$$\chi_\rho(1) = \mathrm{Tr}(\rho(1)) = \mathrm{Tr}I = k = \dim\rho.$$

Definition D11: Let N be a group, $Z(N)$ be the center of N, and Irr(N) be the set of irreducible characters. The scalar product of two characters $\chi, \psi \in N$ is defined by:

$$(\chi, \psi)_N = \frac{1}{|N|}\sum_{n\in N}\chi(n)\psi(n^{-1}).$$

It can be shown that Irr(N) forms an orthonormal basis in this space.

Group representation can also be related to the *matrix group*, that is, the set of complex square $n \times n$ matrices, denoted by M_n, that satisfies the properties of a group under matrix multiplication. In this formalism, the representation ρ of a group G is the mapping ρ: $G \to M_n$, which preserves the matrix multiplication. The character is introduced as the function defined by $\chi_\rho(g) = \mathrm{Tr}(\rho(g))$.

A.7 ALGEBRA OF FINITE FIELDS

A *ring* is a set of elements R with two operations, addition "+" and multiplication "·", such that: (1) R is an Abelian group under addition operation, (2) multiplication operation is associative, and (3) multiplication is associative over addition.

The quantity a is said to be congruent to b to modulus n, denoted as $a \equiv b$ (mod n), if $a - b$ is divisible by n. If $x \equiv a$ (mod n), then a is called a residue to x to modulus n. A class of residues to modulus n is the class of all integers congruent to a given residue (mod n), and every member of the class is called a representative of the class. There are n classes, represented by (0), (1), (2), ..., ($n - 1$), and the representative of these classes is called a complete system of incongruent residues to modulus n. If i and j are two members of a complete system of incongruent residues to modulus n, then addition and multiplication between i and j are defined by:

$$i + j = (i + j)(\mathrm{mod}\ n) \quad \text{and} \quad i \cdot j = (i \cdot j)(\mathrm{mod}\ n).$$

A complete system of residues (mod n) forms a commutative ring with a unity element. Let s be a nonzero element of these residues. Then, s is inverse if and only if n is a prime, p. Therefore if p is a prime, a complete system of residues (mod p) forms a Galois (or finite) field, and is denoted by GF(p).

Let $P(x)$ be any given polynomial in x of degree m with coefficients belonging to GF(p), and let $F(x)$ be any polynomial in x with integral coefficients. Then, $F(x)$ may be expressed as:

$$F(x) = f(x) + p \cdot q(X) + P(x) \times q(x), \text{ where}$$

$$f(x) = a_0 + a_1x + a_2x^2 + \cdots + a_{m-1}x^{m-1}, a_i \in GF(p).$$

This relationship may be written as $F(x) \equiv f(x) \bmod \{p,P(x)\}$, and we say that $f(x)$ is the residue of $F(x)$ modulus p and $P(x)$. If p and $P(x)$ are kept fixed but $f(x)$ varies, p^m classes may be formed, because each coefficient of $f(x)$ may take p values of GF(p). The classes defined by $f(x)$ form a commutative ring, which will be a field if and only if $P(x)$ is irreducible over GF(p) (not divisible with any other polynomial of degree $m - 1$ or less).

The finite field formed by p^m classes of residues is called a Galois field of order p^m and is denoted by GF(p^m). The function $P(x)$ is said to be *minimum polynomial* for generating the elements of GF(p^m) (the smallest degree polynomial over GF(p) having a field element $\beta \in$ GF(p^m) as a root). The nonzero elements of GF(p^m) can be represented as polynomials of degree at most $m - 1$ or as powers of a primitive root α such that:

$$\alpha^{p^m-1} = 1, \quad \alpha^d \neq 1 \quad (\text{for } d \text{ dividing } p^m - 1).$$

A primitive element is a field element that generates all nonzero field elements as its successive powers. A primitive polynomial is an irreducible polynomial that has a primitive element as its root.

Theorem T11: Two important properties of GF(q), $q = p^m$ are:

1. The roots of polynomial $x^{q-1} - 1$ are all nonzero elements of GF(q).
2. Let $P(x)$ be an irreducible polynomial of degree m with coefficients from GF(p) and β be a root from the extended field GF($q = p^m$). Then, all the m roots of $P(x)$ are:

$$\beta, \beta^p, \beta^{p^2}, \ldots, \beta^{p^{m-1}}.$$

To obtain a minimum polynomial we divide $x^q - 1$ ($q = p^m$) by the least common multiple of all factors like $x^d - 1$, where d is a divisor of $p^m - 1$. Then we get the *cyclotomic equation*—the equation having for its roots all primitive roots of equation $x^{q-1} - 1 = 0$. The order of this equation is $\phi(p^m - 1)$, where $\phi(k)$ is the number of positive integers less than k and relatively prime to it. By replacing each coefficient in this equation by least nonzero residue to modulus p we get the cyclotomic polynomial of order $\phi(p^m - 1)$. Let $P(x)$ be an irreducible factor of this polynomial, then $P(x)$ is a minimum polynomial, which is in general not the unique one.

Example: Let us determine the minimum polynomial for generating the elements of GF(2^3). The cyclotomic polynomial is:

$$\left(x^7 - 1\right)/(x - 1) = x^6 + x^5 + x^4 + x^3 + x^2 + x + 1 = \left(x^3 + x^2 + 1\right)\left(x^3 + x + 1\right).$$

Hence, $P(x)$ can be either $x^3 + x^2 + 1$ or $x^3 + x + 1$. Let us choose $P(x) = x^3 + x^2 + 1$.

$$\phi(7) = 6, \ \deg[P(x)] = 3.$$

Three different representations of GF(2^3) are given in the following table.

Power of α	Polynomial	3-tuple
0	0	000
α^0	1	001
α^1	α	010
α^2	α^2	100
α^3	$\alpha^2 + 1$	101
α^4	$\alpha^2 + \alpha + 1$	111
α^5	$\alpha + 1$	011
α^6	$\alpha^2 + \alpha$	110
α^7	1	001

A.8 METRIC SPACES

Definition D12: A linear space X, over a field K, is a *metric* space if there exists a nonnegative function $u \rightarrow \|u\|$ called the *norm* such that ($\forall\ u,v \in X,\ c \in K$):

1. $\|u\| = 0 \Leftrightarrow u = 0$ (0 is the fixed point)
2. $\|cu\| = |c| \cdot \|u\|$ (homogeneity property)
3. $\|u + v\| \leq \|u\| + \|v\|$ (triangle inequality)

Definition D13: The metric is defined as $\rho(u,v) = \|u - v\|$
The metric has the following important properties:

1. $\rho(u,v) = 0 \Leftrightarrow u = v$ (reflectivity)
2. $\rho(u,v) = \rho(v,u)$ (symmetry)
3. $\rho(u,v) + \rho(v,w) \geq \rho(u,w)$ (triangle inequality)

Examples: $K = R$ (the field of real numbers), $X = R^n$ (Euclidean space). The various norms can be defined as follows:

$$\|\boldsymbol{x}\|_p = \left(\sum_{k=1}^{n} |x_k|^p\right)^{1/p} \quad 1 \leq p \leq \infty\ (p - \text{norm})$$

$$\|\boldsymbol{x}\|_\infty = \max_k |x_k| (\text{absolute value norm})$$

$$\|\boldsymbol{x}\|_2 = \left(\sum_{k=1}^{n} |x_k|^2\right)^{1/2} \quad (\text{Euclidean norm}).$$

Example: For $X = C[a,b]$ (complex signals on interval $[a,b]$), the norm can be defined by either of the following ways:

$$\|u\| = \max_{a \le t \le b} |u(t)|$$

$$\|u\| = \int_a^b |u(t)| dt.$$

Example: For $X = L^r[a,b]$, the norm can be defined by:

$$\|u\| = \left(\int_a^b |u(t)|^r dt \right)^{1/r}.$$

Definition D14: Let $\{u_n\}_{n \in N}$ be a sequence of points in the metric space X and let $u \in X$ be such that

$\lim_{n \to \infty} \|u_n - u\| = 0$; we say that u_n converges to u.

Definition D15: A series $\{u_n\}_{n \in N}$ satisfying:

$$\lim_{n,m \to \infty} \|u_n - u_m\| = 0$$

is called the *Cauchy series.*

Definition D16: Metric space is *complete* if every Cauchy series converges.

Definition D17: A complete metric space is called a *Banach space.*

A.9 HILBERT SPACE

Definition D18: A vector space X over the complex field C is called unitary (or space with a dot [scalar] product) if there exists a function $(u,v) \to C$ satisfying:

1. $(u,u) \ge 0$
2. $(u,u) = 0,\ u = 0$
3. $(u + v,w) = (u,w) + (v,w)$
4. $(cu,v) = c(u,v)$
5. $(u,v) = (v,u)^*$

$\forall\ u,v,w \in X$ and $c \in C$. The function (u,v) is a dot (scalar) product.

This function has the following properties:

$$(u, cv) = c^*(u, v)$$

$$(u, v_1 + v_2) = (u, v_1) + (u, v_2)$$

$$|(u, v)|^2 \le (u, u)(v, v).$$

Definition D19: Define a norm:

$$\|u\| = \sqrt{(u,u)}.$$

A unitary space with this norm is called a pre-Hilbert space. If it is complete, then it is called a *Hilbert space*.

Example: R^n is a Hilbert space with the dot product introduced by:

$$(\boldsymbol{x},\boldsymbol{y}) = \sum_{k=1}^{n} x_k y_k = \left(\boldsymbol{y}^{\mathrm{T}},\boldsymbol{x}^{\mathrm{T}}\right).$$

Example: C^n is a Hilbert space with the dot product introduced by:

$$(\boldsymbol{x},\boldsymbol{y}) = \sum_{k=1}^{n} x_k y_k^* = \left(\boldsymbol{y}^{\dagger},\boldsymbol{x}^{\dagger}\right)^*.$$

In this space, the following important inequality, known as Cauchy–Schwarz(–Buniakowsky) inequality, is valid:

$$\left|\sum_{k=1}^{n} x_k y_k^*\right|^2 \le \left(\sum_{k=1}^{n} |x_k|^2\right)\left(\sum_{k=1}^{n} |y_k|^2\right).$$

Example: $L^2(a,b)$ is a Hilbert space with the dot product introduced by:

$$(u,v) = \int_a^b u(t)v^*(t)dt.$$

Orthogonal Systems in Hilbert Space

Definition D20: A set of vectors $\{u_k\}_{k\in I}$ in the Hilbert space forms an orthogonal system if:

$$(u_n,u_k) = \delta_{n,k}\|u_k\|^2 \quad \forall n,k\in J$$

$$\delta_{n,k} = \begin{cases} 1, & n=k \\ 0, & n\neq k \end{cases} \qquad \|u_k\| = \sqrt{(u_k,u_k)}.$$

The set of indices J can be finite, countable, or uncountable. If $\|u_k\| = 1$, the system is orthonormal.

Gram–Schmidt Orthogonalization Procedure

The Gram–Schmidt procedure assigns an orthogonal system of vectors $\{u_0,u_1,\ldots\}$ starting from countable many linearly independent vectors $\{v_0,v_1,\ldots\}$. For the first basis vector we select $u_0 = v_0$. We express the next basis vector as v_1 plus projection along basis u_0:

$$u_1 = v_1 + \lambda_{10}u_0.$$

The projection λ_{10} can be obtained by multiplying the previous equation with u_0 and by determination of the dot product as follows:

$$(u_1, u_0) = (v_1, u_0) + \lambda_{10}(u_0, u_0) = 0.$$

By solving this equation per λ_{10} we obtain:

$$\lambda_{10} = -(v_1, u_0)/(u_0, u_0).$$

Suppose that we have already determined the basis vectors $u_0, u_1, \ldots, u_{k-1}$. We can express the next basis vector u_k in terms of vector v_k and projections along the already determined basis vectors as follows:

$$u_k = v_k + \lambda_{k0}u_0 + \lambda_{k1}u_1 + \cdots + \lambda_{k,k-1}u_{k-1}.$$

By multiplying the previous equation with u_i from the right side and by evaluating the dot product we obtain:

$$(u_k, u_i) = (v_k, u_i) + \sum_{j=0}^{k-1} \lambda_{k,j}(u_j, u_i) = 0; \quad i = 0, 1, \ldots, k-1.$$

By invoking the principle of orthogonality, the summation in the foregoing equation is nonzero only for $j = i$ so that the projection along the ith basis function can be obtained by:

$$\lambda_{k,i} = -\frac{(v_k, u_i)}{(u_i, u_i)}.$$

Finally, the kth basis vector can be obtained by:

$$u_k = v_k - \sum_{j=0}^{k-1} \frac{(v_k, u_j)}{(u_j, u_j)} u_j; \quad k = 1, 2, \ldots.$$

The orthonormal basis is obtained by simply normalizing u_k by $\|u_k\| = \sqrt{(u_k, u_k)}$:

$$\phi_k = \frac{u_k}{\|u_k\|}.$$

Example: Let us observe the following vectors:

$$v_1 = (1, 1, 1, 1) \quad v_2 = (1, 2, 4, 5) \quad v_3 = (1, -3, -4, -2)$$

By applying the Gram–Schmidt procedure we obtain the following basis:

$$u_1 = v_1 = (1, 1, 1, 1)$$

$$u_2 = v_2 - \frac{(v_2, u_1)}{(u_1, u_1)} u_1 = (-2, -1, 1, 2)$$

$$u_3 = v_3 - \frac{(v_3, u_1)}{(u_1, u_1)} u_1 - \frac{(v_3, u_2)}{(u_2, u_2)} u_2 = \left(\frac{8}{5}, -\frac{17}{10}, -\frac{13}{10}, \frac{7}{5}\right).$$

The orthonormal basis can be obtained by normalization as follows:

$$\phi_1 = \frac{u_1}{\|u_1\|} = \frac{1}{2}(1,1,1,1)$$

$$\phi_2 = \frac{u_2}{\|u_2\|} = \frac{1}{\sqrt{10}}(-2,-1,1,2)$$

$$\phi_3 = \frac{u_3}{\|u_3\|} = \frac{1}{\sqrt{910}}(16,-17,-13,14).$$

The following table is an illustration of equivalence of conventional vector space with the signal space, often used in communication theory, and Hilbert space, often used in quantum mechanics.

	Vector space	**Signal space**	**Hilbert space**
Basis	$\vec{u} = \sum_{k=1}^{n} u_k \vec{b}_k$	$u(t) = \sum_{k=1}^{n} u_k \phi_k(t)$	$u = \sum_{k=1}^{n} u_k b_k$
	$u_k = \vec{u} \cdot \vec{b}_k$	$u_k = (u(t),\, \phi_k(t))$	$u_k = (u, b_k)$
Scalar product	$\vec{u} \cdot \vec{v} = \sum_{k=1}^{n} u_k v_k$	$(u(t),v(t)) = \int_a^b u(t)v^*(t)dt$	(u,v) – defined
Norm	$\|\vec{u}\| = \sqrt{\sum_{k=1}^{n} u_k^2}$	$\|\vec{u}\| = \sqrt{\int_a^b |u(t)|^2 dt}$	$\|u\| = \sqrt{(u,u)}$
Metric distance	$d(\vec{u}\,\vec{v})$	$d(u(t),v(t))$	$\|u - v\| = \rho(u,v)$
	$= \|\vec{u} - \vec{v}\|$	$= \|u(t) - v(t)\|$	
	$= \sqrt{\sum_{k=1}^{n}(u_k - v_k)^2}$	$= \sqrt{\int_a^b |u(t) - v(t)|^2 dt}$	
Triangle inequality	$\|\vec{u} + \vec{v}\| \leq \|\vec{v}\| + \|\vec{v}\|$	$\|u(t) + v(t)\| \leq \|u(t)\| + \|v(t)\|$	$\|u + v\| \leq \|u\| + \|v\|$
Schwartz inequality	$\|\vec{u} \cdot \vec{v}\| \leq \|\vec{v}\| \cdot \|\vec{v}\|$	$|(u(t),v(t))| \leq \|u(t)\|\|v(t)\|$	$|(u,v)| \leq \|u\|\|v\|$

References

[1] C.C. Pinter, A Book of Abstract Algebra, Dover Publications, 2010 (reprint).

[2] J.B. Anderson, S. Mohan, Source and Channel Coding: An Algorithmic Approach, Kluwer Academic Publishers, 1991.

[3] S. Lin, D.J. Costello, Error Control Coding: Fundamentals and Applications, Prentice Hall, 2004.

[4] F. Gaitan, Quantum Error Correction and Fault Tolerant Quantum Computing, CRC Press, 2008.
[5] P.A. Grillet, Abstract Algebra. Springer, 2007.
[6] A. Chambert-Loir, A Field Guide to Algebra. Springer, 2005.
[7] B. Vasic, Digital Communications I, Lecture Notes, University of Arizona, 2005.
[8] M.A. Neilsen, I.L. Chuang, Quantum Computation and Quantum Information, Cambridge University Press, Cambridge, 2000.
[9] D. Raghavarao, Constructions and Combinatorial Problems in Design of Experiments, Dover Publications, Inc., New York, 1988 (reprint).
[10] I.B. Djordjevic, W. Ryan, B. Vasic, Coding for Optical Channels. Springer, 2010.
[11] D.B. Drajic, P.N. Ivanis, Introduction to Information Theory and Coding, Academic Mind, 2009 (in Serbian).
[12] S. Lang, Algebra, Reading: Addison-Wesley Publishing Company, 1993.

Index

'*Note*: Page numbers followed by "f" indicate figures and "t" indicate tables.'

F

Q

R

Printed in the United States
By Bookmasters